W0193142

Potassium Channels in Cardiovascular Biology

Potassium Channels in Cardiovascular Biology

Edited by

Stephen L. Archer

Walter McKenzie Health Science Centre
University of Alberta Hospital
Edmonton, Alberta, Canada

and

Nancy J. Rusch

Medical College of Wisconsin
Milwaukee, Wisconsin

Kluwer Academic/Plenum Publishers
New York, Boston, Dordrecht, London, Moscow

Library of Congress Cataloging-in-Publication Data

Archer, Stephen L. and Rusch, Nancy J.
    Potassium channels in cardiovascular biology/Stephen L. Archer and Nancy J. Rusch.
      p. cm.
    Includes bibliographical references and index.
    ISBN 0-306-46402-0
     1.   2.

ISBN 0-306-46402-0

©2001 Kluwer Academic / Plenum Publishers, New York
233 Spring Street, New York, N.Y. 10013

http://www.wkap.nl

10 9 8 7 6 5 4 3 2 1

A C.I.P. record for this book is available from the Library of Congress.

All rights reserved

No part of this book may be reproduced, stored in a retrieval system, or transmitted in
any form or by any means, electronic, mechanical, photocopying, microfilming, recording,
or otherwise, without written permission from the Publisher

Printed in the United States of America

To our families, who have supported and encouraged us through the "Year(s) of the Book"

The family of Stephen L. Archer: wife Kathie Doliszny, Ph.D., children Elizabeth, Ben and Anya Archer and parents Lawrence and Barbara Archer

The family of Nancy J. Rusch: husband Steven C. Matson, M.D., and children Daniel and Sandra Matson

We hope you enjoy this book and that it assists you in your research. Albert Einstein distilled the essence of the quest for understanding which inspired this book in the following quote:

> "In the light of knowledge attained, the happy achievement seems almost a matter of course, and any intelligent student can grasp it without too much trouble. But the years of anxious searching in the dark, with their intense longing, their alterations, of confidence and exhaustion, and the final emergence into the light—only those who have themselves experienced it can understand that."

We look forward to your comments and suggestions and hope this will be a useful addition to your library.

Contributors

Stephen Archer • Cardiology Division, Department of Medicine, University of Alberta, Edmonton, Alberta Canada, T6G 2B7 Canada

Peter H. Backx • Department of Medicine, The Toronto Hospital, Centre for Cardiovascular Research, and Department of Physiology and Institute of Medical Sciences, University of Toronto, Toronto, Ontario, Canada M5G 2C4

D. J. Beech • School of Biomedical Sciences, University of Leeds, Leeds LS2 9JT, United Kingdom

Marcie G. Berger • Department of Cardiology, Sinai Samaritan Medical Center, Milwaukee, Wisconsin, 53233

Michael A. Blanar • Department of Metabolic and Cardiovascular Drug Discovery, Bristol-Myers Squibb Pharmaceutical Research Institute, Princeton, New Jersey 08543-4000

Eric Blanc • Architecture et Fonction des Macromolécules Biologiques, CNRS UPR 9039, 1FR1, 13402 Marseille Cedex 20, France

Mulugu V. Brahmajothi • Department of Pharmacology, Duke University Medical Center, Durham, North Carolina 27710

Joseph E. Brayden • Department of Pharmacology, The University of Vermont, Burlington, Vermont 05405-9998,

Donald L. Campbell • Department of Physiology and Biophysics, University of Buffalo — SUNY, Buffalo, New York 14212

Fuhua Chen • Department of Pediatrics, Division of Cardiology, University of California at Los Angeles School of Medicine, Los Angeles, California 90095

A. Cheong • School of Biomedical Sciences, University of Leeds, Leeds LS2 9JT, United Kingdom

Hee-Cheol Cho • Department of Medicine, The Toronto Hospital, Centre for Cardiovascular Research, and Department of Physiology and Institute of Medical Sciences, University of Toronto, Toronto, Ontario, Canada M5G 2C4

W. C. Cole • Smooth Muscle Research Group, Department of Pharmacology and Therapeutics, Faculty of Medicine, University of Calgary, Calgary, Alberta, Canada T2N 4N1

Laura Conforti • Department of Internal Medicine, Division of Nephrology and Hypertension, University of Cincinnati, Cincinnati, Ohio 45267-0585

David N. Cornfield • Department of Pediatrics, University of Minnesota, Minneapolis, Minnesota 55455

Albert J. D'Alonzo • Department of Metabolic and Cardiovascular Drug Discovery, Bristol-Myers Squibb Pharmaceutical Research Institute, Princeton, New Jersey 08543-4000

Hervé Darbon • Architecture et Fonction des Macromolécules Biologiques, CNRS UPR 9039, 1FR1, 13402 Marseille Cedex 20, France

Guy Droogmans • Laboratorium voor Fysiologie, KU Leuven, Campus Gasthuisberg, B-3000 Leuven, Belgium

David Fedida • Department of Physiology, University of British Columbia, Vancouver, British Columbia, Canada V6T 1Z3

Michel Félétou • Département de Diabétologie, Institut de Recherches Servier, 92150 Suresnes, France

R. Flemming • School of Biomedical Sciences, University of Leeds, Leeds LS2 9JT United Kingdom

Akikazu Fujita • Department of Pharmacology II, Faculty of Medicine and Graduate School of Medicine, Osaka University, Suita, Osaka 565-0871, Japan

Warren J. Gallin • Department of Biological Sciences, University of Alberta, Edmonton, Alberta, Canada T6G 2E9 *and* Bamfield Marine Station, Bamfield, British Columbia, Canada V0R 1B0

Maria L. Garcia • Membrane Biochemistry and Biophysics, Merck Research Laboratories, Rahway, New Jersey 07065

Kathryn M. Gauthier • Department of Pharmacology and Toxicology, The Medical College of Wisconsin, Milwaukee, Wisconsin 53226

Craig H. Gelband • Department of Physiology, College of Medicine, University of Florida, Gainesville, Florida 32610

Maik Gollasch • Franz Volhard Clinic at the Max Delbrück Center for Molecular Medicine, Charité University Hospitals, Humboldt University Berlin, D-13125 Berlin, Germany

Garrett J. Gross • Department of Pharmacology & Toxicology, The Medical College of Wisconsin, Milwaukee, Wisconsin 53226

Gary J. Grover • Division of Metabolic Diseases, Bristol-Myers Squibb Pharmaceutical Research Institute, Princeton, New Jersey 08543-4000

Antonio Guia • Department of Physiology, University of Montreal, Montreal, Quebec, Canada H3C 3J7 *and* Research Centre, Montreal Heart Institute, Montreal, Quebec Canada H1T 1C8

C. Guibert • School of Biomedical Sciences, University of Leeds, Leeds LS2 9JT, United Kingdom

B. M. Heath • Department of Pharmacology, College of Physicians and Surgeons of Columbia University, New York, New York 10032

Joanne T. Hulme • Department of Physiology, Colorado State University, Fort Collins, Colorado 80523

William F. Jackson • Department of Biological Sciences, Western Michigan University, Kalamazoo, Michigan 49008

Gregory J. Kaczorowski • Membrane Biochemistry and Biophysics, Merck Research Laboratories, Rahway, New Jersey 07065

R. S. Kass • Department of Pharmacology, College of Physicians and Surgeons of Columbia University, New York, New York 10032

Thomas S. Klitzner • Department of Pediatrics, Division of Cardiology, University of California at Los Angeles School of Medicine, Los Angeles, California 90095

Hans-Guenther Knaus • Institute for Biochemical Pharmacology, The University of Innsbruck, A-6020 Innsbruck, Austria

Robert O. Koch • Institute for Biochemical Pharmacology, The University of Innsbruck, A-6020 Innsbruck, Austria

Alexandra Koschak • Institute for Biochemical Pharmacology, The University of Innsbruck, A-6020 Innsbruck, Austria

Yoshihisa Kurachi • Department of Pharmacology II, Faculty of Medicine and Graduate School of Medicine, Osaka University, Suita, Osaka 565-0871, Japan

Normand Leblanc • Department of Physiology, University of Montreal, Montreal, Quebec, Canada H3C 3J7 *and* Research Centre, Montreal Heart Institute Montreal, Quebec Canada H1T 1C8

Paul C. Levesque • Department of Metabolic and Cardiovascular Drug Discovery, Bristol-Myers Squibb Pharmaceutical Research Institute, Princeton, New Jersey 08543-4000

Edwin S. Levitan • Department of Pharmacology, University of Pittsburgh, Pittsburgh, Pennsylvania 15261

Barry London • Cardiovascular Institute, University of Pittsburgh Medical Center, Pittsburgh, Pennsylvania 15213

A. N. Lopatin • Department of Cell Biology and Physiology, Washington University School of Medicine St. Louis, Missouri 63110

Sergey M. Marchenko • The Bogomoletz Institute of Physiology, Ukraninian Academy of Sciences, Kiev 24, 252601 GSP, Ukraine

Jeffrey A. Martens • Department of Physiology, Colorado State University, Fort Collins, Colorado 80523

Pratap Meera • Department of Anesthesiology, University of California at Los Angeles, Los Angeles, California 90095-1778

Evangelos D. Michelakis • Department of Medicine, University of Alberta, Edmonton, Alberta, Canada T6G 2B7

Michael J. Morales • Department of Physiology and Biophysics, University at Buffalo — SUNY, Buffalo, New York 14212

Ricardo A. Navarro-Polanco • Department of Physiology, Colorado State University, Fort Collins, Colorado 80523

Mark T. Nelson • Department of Pharmacology, University of Vermont, Burlington, Vermont 05405

Jeanne M. Nerbonne • Department of Molecular Biology and Pharmacology, Washington University Medical School, St. Louis, Missouri 63110

C. G. Nichols • Department of Cell Biology and Physiology Washington University School of Medicine St. Louis, Missouri 63110

Bernd Nilius • Laboratorium voor Fysiologie, KU Leuven, Campus Gasthuisberg, B-3000 Leuven, Belgium

Atsushi Nishiyama • Department of Physiology, Colorado State University, Fort Collins, Colorado 80523

Christopher Parker • Department of Physiology, Queen's University, Kingston, Ontario, Canada K7L 3N6

Olaf Pongs • Institut für Neurale Signalverarbeitung, Zentrum für Molekulare Neurobiologie, 20251 Hamburg, Germany

Helen L. Reeve • Departments of Medicine and Physiology, University of Minnesota, Minneapolis, Minnesota 55455

Carmelle V. Remillard • Department of Physiology, University of Montreal, Montreal, Quebec, Canada H3C 3J7 *and* Research Centre, Montreal Heart Institute, Montreal, Quebec, Canada H1T 1C8

Kenneth J. Rhodes • CNS Disorders, Wyeth-Ayerst Research, Princeton, New Jersey 08543

Brian Robertson • Electrophysiology Group, Department of Biochemistry, Imperial College of Science, Medicine and Technology, London SW7 2AY, United Kingdom

Lewis J. Rubin • Division of Pulmonary and Critical Care Medicine, Department of Medicine, University of California — San Diego, San Diego, California 92103-8382

Nancy J. Rusch • Department of Pharmacology, Cardiovascular Research Center, The Medical College of Wisconsin, Milwaukee, Wisconsin 53226

Stewart O. Sage • Department of Physiology, University of Cambridge, Cambridge CB2 3EG, United Kingdom

Michael C. Sanguinetti • Department of Pediatrics and Medicine, Eccles Program in Human Molecular Biology and Genetics, University of Utah and Primary Children's Medical Center, Salt Lake City, Utah 84113

Steven C. Sansom • Department of Physiology and Biophysics, University of Nebraska Medical Center Omaha, Nebraska 68198-4575

S.-L. Shyng • Department of Cell Biology and Physiology, Washington University School of Medicine, St. Louis, Missouri 63110 *Permanent address:* Center for Research on Occupational and Environmental Toxicology Portland, Oregon 97201

Andrew N. Spencer • Department of Biological Sciences, University of Alberta, Edmonton, Alberta, Canada T6G 2E9 *and* Bamfield Marine Station, Bamfield, British Columbia, Canada V0R 1B0

Nicholas Sperelakis • Department of Molecular and Cellular Physiology, University of Cincinnati, Cincinnati, Ohio 45267-0576

James D. Stockand • The Center for Cellular and Molecular Signaling, Department of Physiology, Emory University School of Medicine Atlanta, Georgia 30322

Harold C. Strauss • Department of Physiology and Biophysics, University of Buffalo — SUNY, Buffalo, New York 14212

Koichi Takimoto • Department of Pharmacology, University of Pittsburgh, Pittsburgh, Pennsylvania 15261

Michael M. Tamkun • Departments of Physiology and Biochemistry and Molecular Biology, Colorado State University, Fort Collins, Colorado 80523

Andre Terzic • Division of Cardiovascular Diseases, Department of Medicine and Pharmacology, Mayo Clinic, Mayo Foundation, Rochester, Minnesota 55905

Ligia Toro • Department of Anesthesiology, Department of Molecular and Medical Pharmacology, and Brain Research Institute, University of California at Los Angeles, Los Angeles, California 90095-1778

Maria Trieb • Institute for Biochemical Pharmacology, The University of Innsbruck, A-6020 Innsbruck, Austria

Christopher R. Triggle • Department of Pharmacology and Therapeutics, University of Calgary, Calgary, Alberta, Canada T2N 4N1

James S. Trimmer • Department of Biochemstry and Cell Biology and Institute for Cell and Developmental Biology, State University of New York, Stony Brook, New York 11794

Martin Tristani-Firouzi • Department of Pediatrics and Medicine, Eccles Program in Human Molecular Biology and Genetics, University of Utah and Primary Children's Medical Center, Salt Lake City, Utah 84113

Paul M. Vanhoute • Institut de Recherches Internationales Servier, 92410 Courbevoie, France

Michel Vivaudou • Laboratory of Molecular and Cellular Biophysics, CEA-DBMS/BMC, URA CNRS 520, 38054 Grenoble, France

Martin Wallner • Department of Anesthesiology, University of California at Los Angeles, Los Angeles, California 90095-1778

M. P. Walsh • Smooth Muscle Research Group, Department of Biochemistry and Molecular Biology, Faculty of Medicine, University of Calgary, Calgary, Alberta, Canada T2N 4N1

Siegmund G. Wanner • Institute for Biochemical Pharmacology, The University of Innsbruck, A-6020 Innsbruck, Austria

X. Wehrens • Department of Pharmacology, College of Physicians and Surgeons of Columbia University, New York, New York 10032

Kenneth E. Weir • Department of Medicine, VA Medical Center, Minneapolis, Minnesota 55417

George C. Wellman • Department of Pharmacology, University of Vermont, Burlington, Vermont 05405

S. Z. Xu • School of Biomedical Sciences, University of Leeds, Leeds LS2 9JT, United Kingdom

Jason Xiao-Jian Yuan • Division of Pulmonary and Critical Care Medicine, Department of Medicine, University of California—San Diego, San Diego, California 92103-8382

Preface

Potassium channels (K^+) conduct the electricity of life, and many of the important functions of the cardiovascular system are orchestrated by the opening and closing of these channels. Although the evolution of K^+ channels is subject to conjecture, one might conclude in the beginning, there were K^+ channels. Indeed, cells cannot exist without the constant exchange of ions with their environment. Diverse gene families have evolved in eukaryotic and prokaryotic cells to provide multimeric K^+ channel structures equipped with a highly conserved K^+ selectivity filter, that allows selective, transmembrane flux of K^+ ions, through a pore that is only 3 at its narrowest point. Other parts of the channel confer sensitivity to voltage, redox state, pH, polyamines, and G-proteins. Modular molecular additions of regulatory proteins may confer sensitivity to calcium, voltage and second messengers to the primary, pore-forming α-subunit and alter its expression level in the plasma membrane. Relying on their elegant selectively for K^+ ions, K^+ channels mediate the transmembrane efflux of K^+ to establish the resting cell membrane potential, and thereby regulate voltage-gated Ca^{2+} influx and modulate the electrical gradient that alters ion fluxes critical to the function of cardiovascular cells. These functions may include, among others, pacemaker activity and contraction in myocardial cells, excitation-contraction coupling in vascular smooth muscle cells, and the release of vasoactive factors from endothelial cells.

The Earth's venomous creatures including scorpions, honeybees and several arachnids, have exploited the critical role of K^+ channels by targeting them with highly specific toxins that bind and inhibit K^+ channels. The ubiquitous distribution of K^+ channels in cardiovascular tissues attests to their importance, as does the high degree of conservation of channel structure and function across phylo. Humans and fruit flies not only share many of the same K^+ channels but both have diseases due to mutations in these proteins. Indeed, mutations in the *Drosophila* K^+ channel genes *S. taker*, *H. perkinetic* and *Ether-a-go-go*, which causes the flies to exhibit leg shaking under ether anesthesia, or to have heightened metabolic rates and shortened life spans, that is associated with lethal cardiac arrhythmias in humans (e.g., Long QT Syndrome). An emerging concept is that many cardiovascular diseases may, at least in part, be channelopathies. In addition to arrhythmic disorders, more common cardiovascular diseases including congestive heart failure, primary pulmonary hypertension, and essential hypertension may have an element of K^+ channelopathy in their pathogenesis, either representing primary events or perhaps as a consequence of more fundamental aspects of the disease. Nonetheless, the concept of channelopathy, together with the ability to transfer normal K^+ channel genes to cells to restore normal excitable properties or to modulate channel expression, is regarded as a promising direction for future therapeutic interventions. The recent crystallization of a K^+ channel has provided new information about the anatomical structure that will accelerate the design of drugs to modulate K^+ channel function, another promising avenue for the treatment of cardiovascular pathologies.

If the molecular view of K^+ channels shows us their conserved structure, the physiologist's view emphasizes their diversity. Not only are there many types and subtypes of K^+ channels, but there is geographical diversity in the cardiovascular system generated by cell and site-specific expression profiles. This heterogeneity of K^+ channel expression confers unique electrical profiles to the different cells of the heart and the vascular network. Diversity is generated not only by the existence of diverse pore forming α-subunits, but also provided by smaller, regulatory β subunits that modify K^+ channel function and expression. Fundamental differences between the electrical profiles of cardiovascular cells reflect the complement of K^+ channels expressed in their plasma membranes. The identity and function of these channels, and the methodologies available for studying their role in cardiovascular physiology and pathologies are emphasized in this text.

Why would a cardiovascular physiologist, pharmacologist, geneticist or physician care to learn about K^+ channels? Cardiovascular disease is the leading cause of death for men and women in North America, and K^+ channel abnormalities are implicated in the pathogenesis of many cardiovascular diseases. This text provides the first comprehensive review of K^+ channels in the cardiovascular system, drawing on the expertise of an international authorship of contributors who have been pioneers in developing what is now one of the fastest growing areas of ion channel research. This book has relevance for students and scientists working in basic science laboratories as well as physicians and translational researchers interested in human cardiovascular diseases, including heart failure, myocardial preconditioning and ischemia, arrhythmia-induced sudden death, neuropreservation-stroke, hypertension (pulmonary and systemic), developmental cardiovascular biology, and conditions of chronic hypoxia. To provide background for readers at all career levels, the introductory chapters include discussions of K^+ channel evolution and diversity, followed by sections focused on the molecular biology and pharmacology of K^+ channels. For those readers new to the study of K^+ channels, key sets of chapters includes tutorials on using the patch-clamp technique to measure K^+ channel currents in mammalian cells and expression systems, the application of techniques for evaluating K^+ channel mRNA and protein expression levels and distribution, and the use of transgenic and knockout approaches for evaluating K^+ channel function. Subsequent sections provide comprehensive and detailed discussions on K^+ channels in the heart, the vascular smooth muscle and endothelium, and the role of K^+ channels in cardiac and vascular diseases.

We thank the talented authors for their excellent contributions to this book. To the reader, we hope this book becomes a useful part of your clinical or research efforts. Please contact us with your comments on the text, and your suggestions for future editions.

ACKNOWLEDGEMENTS

The editors are indebted to their contributors and colleagues for their diligence in preparing the chapters, and for their dedication to sharing their knowledge with others. It also is a pleasure to thank Michael F. Hennelly, our senior editor at Plenum Publishing Corporation, Mr. Kevin Sequeira, for his guidance in preparing and editing the text, and Ms. Jennifer Stevens for her production editing assistance with the text.

Contents

Chapter 3

Molecular Biology of Voltage-Gated K$^+$ Channels

Olaf Pongs

Chapter 4

Molecular Biology of High-Conductance, Ca^{2+}-Activated Potassium Channels

Pratap Meera, Martin Wallner, and Ligia Toro

Chapter 5

Molecular Biology of Inward Rectifier and ATP-Sensitive Potassium Channels

S. L. Shyng, A. N. Lopatin, and C. G. Nichols

Part II.   Potassium Channel Expression and Function

Chapter 6

Design and Use of Antibodies for Mapping K^+ Channel Expression in the Cardiovascular System

*Robert O. Koch, Maria Trieb, Alexandra Koschak, Siegmund G. Wanner,
Kathryn M. Gauthier, Nancy J. Rusch, and Hans-Guenther Knaus*

Chapter 7

Molecular Methods for Evaluation of K^+ Channel Expression and Distribution in the Heart

*Michael J. Morales, Mulugu V. Brahmajothi, Donald L. Campbell,
and Harold C. Strauss*

Chapter 8

**Concepts for Patch-Clamp Recording of Whole-Cell and
Single-Channel K$^+$ Currents in Cardiac and Vascular Myocytes**

Antonio Guia, Carmelle V. Remillard, and Normand Leblanc

Chapter 9

**The Patch-Clamp Technique for Measurement of K$^+$ Channels in
Xenopus Oocytes and Mammalian Expression Systems**

Laura Conforti and Nicholas Sperelakis

Chapter 10

Heteromultimer Formation in Native K^+ Channels

James S. Trimmer and Kenneth J. Rhodes

Chapter 11

Use of Transgenic and Gene-Targeted Mice to Study K^+ Channel Function in the Cardiovascular System

Barry London

Part III.   Pharmacology of Potassium Channels

Chapter 15

Molecular Pharmacology of ATP-Sensitive K$^+$ Channels: How and Why?

Andre Terzic and Michel Vivaudou

Part IV.   Potassium Channels in the Heart

Chapter 16

Overview: Molecular Physiology of Cardiac Potassium Channels

B. M. Heath, X. Wehrens, and R. S. Kass

Chapter 17

Molecular Mechanisms Controlling Functional Voltage-Gated K^+ Channel Diversity and Expression in the Mammalian Heart

Jeanne M. Nerbonne

Chapter 18

Voltage-Gated Potassium Channels in the Myocardium

*Joanne T. Hulme, Jeffrey R. Martens, Ricardo A. Navarro-Polanco,
Atsushi Nishiyama, and Michael M. Tamkun*

Chapter 19

Inward Rectifying and ATP-Sensitive K$^+$ Channels in the Ventricular Myocardium

Akikazu Fujita and Yoshichisa Kurachi

Chapter 20

Cholinergic and Adrenergic Modulation of Cardiac K$^+$ Channels

Christopher Parker and David Fedida

Chapter 21

Cardiac K$^+$ Channel Expression and Function at Birth and in the Neonate

Fuhua Chen and Thomas S. Klitzner

Part V. Potassium Channels in Vascular Smooth Muscle

Chapter 22
Overview: Physiological Role of K^+ Channels in the Regulation of Vascular Tone

Joseph E. Brayden

Chapter 23

Modulation of Vascular K^+ Channels by Extracellular Messengers

D. J. Beech, A. Cheong, R. Flemming, C. Guibert, and S. Z. Xu

Chapter 24

Delayed Rectifier K$^+$ Channels of Vascular Smooth Muscle: Characterization, Function, and Regulation by Phosphorylation

W. C. Cole and M. P. Walsh

Chapter 25

Potassium Channels in the Circulation of Skeletal Muscle

William F. Jackson

Chapter 26

Regulation of Cerebral Artery Diameter by Potassium Channels

George C. Wellman and Mark T. Nelson

Chapter 27

The Role of Potassium Channels in the Control of the Pulmonary Circulation

Stephen Archer

Chapter 28

Potassium Channels in the Renal Circulation

James D. Stockand and Steven C. Sansom

Chapter 29

Potassium Channels in the Coronary Circulation

Maik Gollasch

Chapter 30

Vascular K^+ Channel Expression and Function at Birth and in the Neonate

Helen L. Reeve and David N. Cornfield

Part VI.   Potassium Channels in the Endothelium

Chapter 33

Endothelial Cell K$^+$ Channels, Membrane Potential, and the Release of Vasoactive Factors from the Vascular Endothelium

Christopher R. Triggle

Chapter 34

Activation of Vascular Smooth Muscle K$^+$ Channels by Endothelium-Derived Factors

Michel Félétou and Paul M. Vanhoutte

Part VII.   Potassium Channels in Cardiac Disease

Chapter 35

Overview: Role of Potassium Channels in Cardiac Arrthythmias

Albert J. D'Alonzo, Paul C. Levesque, and Michael A. Blanar

Chapter 36

The Molecular Basis of the Long QT Syndrome

Martin Tristani-Firouzi and Michael C. Sanguinetti

Chapter 37

Altered K^+ Channel Expression in the Hypertrophied and Failing Heart

Koichi Takimoto and Edwin S. Levitan

Chapter 38

Role of ATP-Sensitive K^+ Channels in Cardiac Preconditioning

Garrett J. Gross

Chapter 39

Therapeutic Potential of ATP-Sensitive K^+ Channel Openers in Cardiac Ischemia

Gary J. Grover

Part VIII.   Potassium Channels in Vascular Disease

Chapter 40

Altered Expression and Function of Kv Channels in Primary Pulmonary Hypertension

Jason Xiao-Jian Yuan and Lewis J. Rubin

Chapter 41

Anorectic Drugs and the Vasculature

Evangelos D. Michelakis and E. Kenneth Weir

Chapter 42

Induction of Ca^{2+}-Activated K^+ Channel Expression during Systemic Hypertension: Protection against Pathological Vasoconstriction

Marcie G. Berger and Nancy J. Rusch

Chapter 43

Antisense Approaches and the Modulation of Potassium Channel Function in the Cardiovascular System

Craig H. Gelband

Abbreviations

α_{1A} receptors	a subtype of α_1-adrenoceptors, alternatively called $\alpha_{1A/c}$
$\alpha_{1a/d}$, α_{1b}, and α_{1c}	subtypes of α_1-adrenoceptors
β-subunit	auxiliary subunits which do not form Kv channels by themselves in heterologous *in vitro* expression systems but specifically associate with Kv channel α-subunits
τ_{rec}	time constant for channel inactivation
$[Ca^{2+}]_i$	cytosolic Ca^{2+} concentration
$[K^+]_o$ or K_o	extracellular K^+ concentration
2-TM	two-transmembrane-domain
4-AP	4-aminopyridine
4-TM	four-transmembrane-domain
5-HD	hydroxydecanoic acid or sodium 5-hydroxydecanoate
6-TM	six-transmembrane-domain
8-SPT	8-sulfophenyl theophylline
11,12-EET	11,12-epoxyeicosatrienoic acid
20-HETE	20-hydroxyeicosatetraenoic acid
AA	arachidonic acid
ABC	ATP-binding cassette
AC	adenylate cyclase
ACE	angiotensin-converting enzyme
ACh	acetylcholine
ADP	adenosine diphosphate
ANF	atrial naturietic factor
Ang II	angiotensin II
ANP	atrial natriuretic peptide
APD	action potential duration
ARs	adrenoceptors
AsODNs	antisense oligodeoxynucleotides
AT-1	an atrial tumor cell line
AT_1 receptor	angiotensin 1 receptor
ATP	adenosine triphosphate
ATPγS	adenosine S'-*O*-(3-thiotriphosphate), a nonhydrolyzable ATP analog
ATP_i	intracellular ATP levels
AV node	atrioventricular node
Ba^{2+}	barium

BK_{Ca} channels	large-conductance Ca^{2+}-activated K^+ channels
C-region	cytoplasmic region
C10-TEA	decyltriethylammonium
CaMKII	calmodulin-dependent protein kinase type II
cAMP	cyclic adenosine monophosphate
CCPA	2-chloro-N^6-cyclopentyladenosine
cDNA	complementary deoxynucleic acid
cfu	colony-forming units
cGMP	cyclic guanosine monophosphate
CGRP	calcitonin gene-related peptide
CHO cells	Chinese hamster ovary cells
cIsK	cat IsK channel
CK	creatine kinase
CO	carbon monoxide
COS cells	African green monkey kidney cell line
CP	immunogenic competing peptide
CPC	calcium-induced preconditioning
CPP	coronary perfusion pressure
Cs^+	cesium
CTX or ChTx	charybdotoxin
DA	ductus arteriosus
DAG	diacylglycerol
di-4ANEPPS	a voltage-sensitive dye
DIDS	4,4′-diisothiocyanostilbene-2,2′-disulfonic acid, a chloride channel blocker
DOCA rats	deoxycorticosterone acetate-treated rat models
DSS	disuccinimidyl suberate
DTX	dendrotoxins, selective Kv1.1-Kv,1.2- and Kv1.6 blocking toxins derived from black and green mamba snakes
Eag	*ether-a-go-go* gene
EC_{50}	effective drug concentrations causing half-maximal contraction
EDHF	endothelium-derived hyperpolarizing factor
EDRF	endothelium-derived relaxing factor
EETs	epoxyeicosatrienoic acids
E_K	equilibrium potential for potassium
EKG	electrocardiogram
E_m	membrane potential
E_{rev}	membrane potential at which direction of ion flow reverses; reversal potential
ERG	*ether-a-go-go*-related gene
ERG channels	*ether-a-go-go*-related K^+ channels
ERG-1A	*ether-a-go-go*-related channel 1A
ERG2 and ERG3	human *ether-a-go-go*-related genes which have been cloned from brain
ES cells	embryonic stem cells

ET-1	endothelin-1
f	fraction of wild-type subunits
FIAU	an analog of gancyclovir
FiO_2	inspired oxygen tension
GDPβS	guanosine 5'-O-(2-thiodiphosphate), an inhibitor of G-protein activation
GFP	green fluorescent protein
GFR	glomerular filtration rate
G_i	inhibitory G-proteins
GIRK1	G-protein-sensitive, inward rectifier K^+ channel, Kir 3.1
GIRK4	G-protein-sensitive, inward rectifier K^+ channel, Kir 3.4/CIR
Glib	glibenclamide
gpIsK	guinea pig IsK current
G-protein	GTP-binding protein
GRK	G-protein-coupled receptor kinase
G_s	stimulatory G-protein
GSH/GSSG	glutathione (reduced/oxidized)
GTP	guanosine triphosphate
GTPγS	guanosine 5'-O-(3-thiotriphosphate), a nonhydrolyzable GTP analog
GYG	amino acid "signature" sequence that is critical for K^+ channel pore selectivity
H	hypoxia
HCN channels	hyperpolarization-activated, cyclic-nucleotide-gated ion channels
HEK293 cells	human embryonic kidney cell line
HERG	human *ether-a-go-go*-related gene
HERG channels	molecular correlate to ventricular I_{Kr}
HETEs	hydroxyeicosatetraenoic acids
hIsK	human IsK
Hk gene	*Drosophila Hyperkinetic* gene; mutations in this *Drosophila* Kvβ subunit alter synaptic transmission through effects on *Shaker* channels
HP	holding potential
HPTX	heteropodatoxin
HpTX1–3	heteropodatoxins isolated from the venom of Malaysian *Heteropoda venatoria* spiders, which prolong ventricular action potentials by blocking I_{to}
HSP	heat shock protein
IBMX	isobutyl-1-methylxanthine
IBTX	iberiotoxin
IC_{50}	concentration of a drug required to achieve 50% inhibition
I_{Ca}	calcium current
I_D	a rapidly activating, slowly inactivating K^+ current in mammalian central neurons

I_f	pacemaker current
I_h (also known as I_f, for "funny")	a slow inward current in the sinoatrial node, activated following the hyperpolarization phase of an action potential. It is a mixed Na^+ and K^+ conductance current
I_K	cardiac delayed rectifier current [a composite of I_{Kur}, I_{Kr}, and I_{Ks} (ultrarapidly, rapidly, and slowly activating components)]
I_K	whole cell K^+ current
I_{k1} or I_{K1} channels	strongly inward rectifying cardiac K^+ channels
$I_{K(Ach)}$ or IK_{Ach}	muscarinic K^+ current
I_{KATP}	ATP-sensitive K^+ current
IK_{Ca} channels	intermediate-conductance, Ca^{2+}-activated K^+ channels
I_{KDTX}	slowly inactivating current in rat atrial myocytes
I_{Kir}	inward rectifier K^+ current
I_{Kp}	novel guinea pig myocyte K^+ current that activates very rapidly on depolarization, does not inactivate, and is sensitive to block by barium
I_{Kr} or I_{kr} or $I_{I,rapid}$	fast component of the cardiac delayed rectifier K^+ current
I_{Ks} or $I_{K,slow}$ or I_{Kslow}	slow component of the cardiac delayed rectifier K^+ current; I_{Kslow} is a rapidly activating and slowly inactivating K^+ current with kinetic and pharmacological properties distinct from those of $I_{to,f}$ and $I_{to,s}$
I_{Kss}	a noninactivating, steady-state, outward cardiac K^+ current
I_{Kur}(ultrarapid), I_{ss} (steady-state) or I_{sus} (sustained guinea pig I_{Kp})	various names for similar rapidly activating, noninactivating components of the cardiac delayed rectifier K^+ currents
I_{max}	maximal predicted current
IP_3	inositol 1,4,5-trisphosphate
IP_4	inositol 1,3,4,5-tetrakisphosphate
IPC	ischemic preconditioning
I_{sK} or minK	a gene which encodes a protein of 130 amino acids with a single membrane-spanning domain that interacts with KvLQT1 to produce a heteromultimeric cardiac K^+ channel with slowly activating K^+ currents
I_{slow}	rapidly activating, slowly-inactivating, 4-aminopyridine-sensitive cardiac K^+ current
I_{ss}	delayed rectifier-like current from isolated rat ventricular myocytes
I_{to}	transient outward cardiac K^+ current
$I_{to,fast}$, ($I_{to,f}$) and $I_{to,slow}$ ($I_{to,s}$)	fast- and slow-inactivating components of the cardiac transient outward K^+ current
I_{to2}	a cardiac chloride current

k	the slope of a current–voltage plot
$K_{(Ach)}$ or K_{Ach} channel	G-protein-activated muscarinic K^+ channels
K_{ATP} channels	adenosine triphosphate-sensitive K^+ channels
K_{Ca} channels	calcium-activated potassium channels
KChAP	voltage-gated K^+ channel accessory protein
KCNA1B, KCNA2B, KCNA3B	genes which encode $Kv\beta$ subunits
KCNE1	minK
K_{CNG}	cyclic-nucleotide-gated K^+ channels
KCNQ1/KCNE1 channels	molecular correlate to cardiac I_{Ks}
KCNQ2 and KCNQ3	members of the KvLQT (KCNQ) subfamily that have been cloned from brain
KCO	K^+ channel opener
KcsA channel	a recently crystallized bacterial K^+ channel from *Streptomyces lividans*
K_d	dissociation constant
K_{DR} channel	delayed rectifier K^+ channel
Kir or K_{IR} channel	inward rectifier K^+ channel
K_{Na} channels	Na^+-activated K^+ channels
$Kv\beta1.1$, $Kv\beta1.2$, and $Kv\beta1.3$	splice variants encoded by the gene KCNA1B
Kv channel	voltage-gated K^+ channel
KvLQT1 or KCNQ1	mutations in this K^+ channel gene underlie an inherited form of long QT syndrome (LQT1)
L-NNA	N^G nitro-L-arginine, a nitric oxide synthase inhibitor
LNSV	a retroviral vector
LQT	long QT interval
LQT2	mutations of HERG leading to one form of familial long QT syndrome
LQTS	long QT syndrome
LTX	Leiurotoxin I
L-type Ca^{2+} channels	"long-lasting", dihydropyridine-sensitive, voltage-dependent Ca^{2+} channels
M_2-AChRs	M_2-type muscarinic cholinergic receptors
MAPS	Multiple Antigenic Peptide System
MERG1	the mouse homolog of HERG1
α-MHC promoter	myosin heavy-chain promoter
minK (or IsK)	accessory membrane protein that associates with the KvLQT1 channel α-subunits to produce a Kv current similar to cardiac myocyte I_{KS}
MiRP1	minimal K^+ channel related protein (minK-related protein)
mito K_{ATP}	mitochondrial K_{ATP} channel
MLA	monophosphoryl lipid A
mRNA	messenger ribonucleic acid
MS222	tricaine methanesulfonate
MTX	maurotoxin
NADH/NAD^+	nicotinamide adenine dinucleotide (reduce/oxidized

	forms)
NADPH	nicotinamide adenine dinucleotide phosphate
NBD	nucleotide-binding domains
NBFs	nucleotide-binding folds
Neo_R, gene	the neomycin resistance gene, an antibiotic resistance cassette
NO	nitric oxide
NS1619	a BK_{Ca} channel opener
N-terminus	amino terminus
N-type inactivation	Mammalian inactivating domains contain a conserved cysteine residue. The presence of this residue in the amino terminus confers sensitivity to oxidation of the inactivating domains.
P	pore
PA	pulmonary artery
PASMC	pulmonary artery smooth muscle cells
PDD	phorbol 12,13-didecanoate
PGI_2	prostaglandin I_2
PHT	pulmonary hypertension
PIP	phosphatidylinositol-4-phosphate
PIP_2	phosphatidylinositol-4,5-bisphosphate
PKA	cyclic adenosine monophosphate-dependent protein kinase
PKC	phospholipid-dependent protein kinase C
PKG	cyclic guanosine monophosphate-dependent protein kinase
PLA_2	phospholipase A_2
PLC	phospholipase C
PLD	phospholipase D
P-loop	pore-forming loop of an ion channel
PMA	phorbol 12-myristate 13-acetate
P_{Na}/P_K ratio	the ratio of sodium to potassium permeability
Po	open-state probability
PO_2	partial pressure of oxygen
PP1	protein phosphatase 1
PP2A	phosphatase 2A
pS	picosiemens (a unit of channel conductance)
PTX	pertussis toxin
P-type proteins	a superfamily of pore-forming membrane proteins which includes K^+, Na^+, and Ca^{2+} channels.
R_m	membrane resistance
R_s	resistance in series
RT-PCR	reverse-transcription polymerase chain reaction
RyR	ryanodine receptor
S1 segment	first transmembrane segment of a K^+ channel
SA node	sinoatrial node
sAHP	slow afterhyperpolarization

SAR	structure–activity relationship
ScTX	scyllatoxin
SDS	sodium dodecyl sulfate
SH	sulfhydryl
Sh genes	*Shaker* genes from *Drosophila* sp. that correspond to the Kv1 channel family in mammals
SHR	spontaneously hypertensive rats
SK_{Ca} channels	small conductance, Ca^{2+}-activated K^+ channels
SM	smooth muscle
SNP	sodium nitroprusside
STOCs	spontaneous transient outward currents
STX	saxitoxin; a selective blocker of Na^+ channels
SU	sulfonylurea derivatives
SUR	sulfonylurea receptor
SWOP	window of protection
T1 domain	a tetramerization domain in K^+ channels which determines the specificity of subunit assembly and participates in subunit assembly
TBA	tetrabutylammonium
TEA	tetraethylammonium
TK	tyrosine kinase
TK gene	thymidine kinase gene
TM	transmembrane
TMD	transmembrane domain
TOK	tandem pore outwardly rectifying K^+ channel
TPA	12-*O*-tetradecanoylphorbol 13-acetate, a protein kinase C activator
TpeA	tetrapentylammonium ion
TTX	tetrodotoxin, a selective blocker of Na^+ channels
TWIK-1	human tandem pore weak inward rectifier K^+ channel
UDP	uridine diphosphate
$V_{1/2}$	membrane potential at which half-maximal activation of a channel occurs
VIP	vasoactive intestinal polypeptide
V_m	the test potential
WKY rats	Wistar Kyoto rats

Potassium Channels in
Cardiovascular Biology

Part I

Molecular Biology of Potassium Channels

Chapter 1

Evolution of Potassium Channel Proteins

Warren J. Gallin and Andrew N. Spencer

1. INTRODUCTION

The activities of a variety of K^+ channel proteins combine to control electrical excitability, the shape of action potentials, and the firing frequency of cardiac cells. Because the multichambered heart of vertebrates evolved relatively recently in the history of metazoans, it is probable that the suite of K^+ channels required to meet the specific functional requirements of a complex fluid pump were "recruited" from other excitable tissues during the course of evolution. Thus, it is not surprising to find that all of the K^+ channel proteins that have been isolated from vertebrate heart are expressed in other tissues and have homologs in other phyla. During the course of evolution, a series of natural experiments occurred to select the various adaptive functional characteristics of these channel proteins. By judicious study of the sequences and physiological properties of homologous channels in a variety of organisms, it is possible to deduce some elements of how structure determines the function of these channels. A detailed comparative and phylogenetic analysis of related K^+ channels has the potential to provide us with insights into K^+ channel function in the heart that are not apparent from detailed structural and functional studies focused solely on the mammalian myocardium.

Membrane excitability is a characteristic of many cell types in all animals. K^+ channels are the primary determinants of the resting membrane potential of the cell and the major modulators of the shape, duration, and frequency of action potentials in excitable cells (Hille, 1991). Recent results from studies of K^+ currents in a wide spectrum of metazoan phyla indicate that the physiological currents that are common to most animals are based on the activities of a similar set of K^+ channels that have arisen during evolution by a combination of gene duplication and divergence.

Warren J. Gallin and Andrew N. Spencer • Department of Biological Sciences, University of Alberta, Edmonton, Alberta, Canada T6G 2E9 and Bamfield Marine Station, Bamfield, British Columbia, Canada V0R 1B0.

Potassium Channels in Cardiovascular Biology, edited by Archer and Rusch. Kluwer Academic/Plenum Publishers, New York, 2001.

The physiological functions of excitable cells and the molecular mechanisms that underlie these functions have arisen during evolution, by accretion of heritable changes in the structure and activity of the relevant proteins. As a result, these proteins reflect the history of their evolution. To understand why an extant organism manifests some suite of physiological functions, it is helpful to understand how they evolved and the constraints that are placed upon them by their evolutionary heritage. Thus, structural and functional analysis of K^+ channels from a variety of vertebrates and from invertebrate phyla can provide us with a clearer understanding of the electrophysiology of the vertebrate heart.

In addition, comparative analyses of the range of K^+ channels present in the animal kingdom provide more examples for establishing the relationship between structure and function of these channels. Because of evolutionary constraints on the function of K^+ channels, the range of observable variation and the absence of many possible variants are strong indicators of the structure–function relationships in this gene family. Examination of K^+ channels from diverse phyla can also provide a broader range of information on interactions between pharmacological agents and their target sites on channels (MacKinnon *et al.*, 1988, 1998), complementing the more commonly used *in vitro* mutagenesis approaches. Not only does this examination provide more information about structure–function relationships in K^+ channels, but it can also contribute to the design of new drugs that are more specific to a set of target channels.

In this chapter, we will discuss the evolution of several families of K^+ channels that are expressed in cardiac cells: inward rectifiers of the Kir family; voltage-gated (Vg) channels of the Kv family; ether-a-go-go (EAG) family channels that also act as inward rectifiers but whose activity is modulated by intracellular cAMP; and hyperpolariz-ation-activated (Ih) channels that act as pacemakers. All of these channel types are also expressed in excitable cells that are noncardiac, and all have homologs in organisms that may not have multichambered hearts.

From the data that are presently available, it appears that a complete suite of K^+ channel families evolved in the last common ancestor of all multicellular animals. After the divergence of the various metazoan phyla, a variety of phylum-specific evolutionary changes occurred within the different gene families, including loss of some genes in some lineages, duplication and divergence to produce multigene families in some lineages, and evolution of alternative splicing of a single gene in yet other lineages.

2. ORIGIN OF MAJOR K^+ CHANNEL FAMILIES

Figure 1 shows a summary scheme illustrating a general hypothesis of how the various families of K^+ channels have evolved. The evolution of the various families of K^+ channels appears to have started with a channel protein with two transmembrane (TM) helices (M1 and M2) and an extracellular connecting loop that forms the ion pore by passing part way into the membrane in association with the two transmembrane helices; both the C-terminus and the N-terminus of these proteins are cytoplasmic.

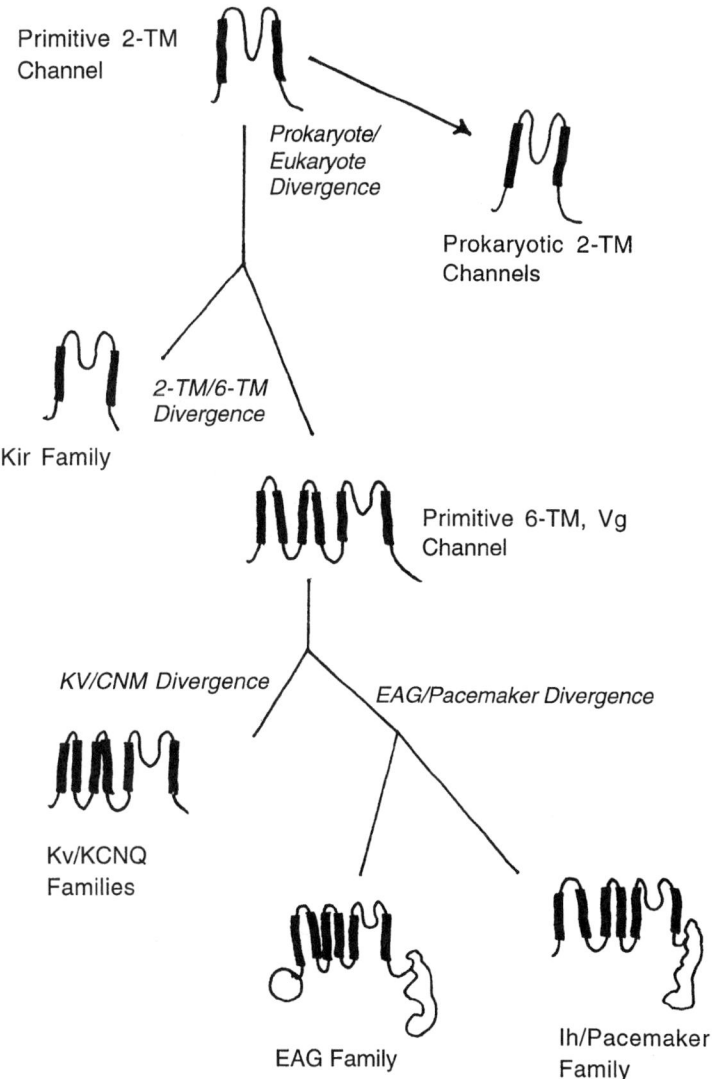

Figure 1. Hypothetical scenario for major structural changes during evolution of K$^+$ channels. The first recognizable K$^+$ channels are 2-TM forms that are present in both prokaryotes and eukaryotes. The prokaryotes and eukaryotes appear to have diverged prior to the evolution of 4-TM or 6-TM variants in eukaryotic ancestors. The family of 6-TM channels arose by addition of four N-terminal transmembrane domains to a 2-TM ancestral channel. This addition provided the structural basis for a voltage-dependent gating mechanism intrinsic to the channel protein. Subsequently, an additional C-terminal segment that confers cyclic-nucleotide sensitivity to the channels was added to a 6-TM voltage-gated (Vg) backbone ancestor in one lineage, followed by the addition of an N-terminal PAS domain in the lineage leading to the EAG channel family. Abbreviations: TM, Transmembrane; Kir, inward rectifier; Vg, voltage-gated; KV, voltage-gated K$^+$; CNM, cyclic nucleotide modulated; EAG, *ether-a-go-go*; KCNQ, potassium, cyclic nucleotide modulated, long QT; Ih, hyperpolarization activated.

Channels of this type comprise the Kir inward rectifier family in animals and plants and are also found in prokaryotes. The three-dimensional structure of such a channel has been recently described (Doyle *et al.*, 1998), confirming the models that had been formulated based on more indirect evidence. The wide distribution of the 2-TM K^+ channel, in both prokaryotes and eukaryotes, suggests that this is the most primitive of the K^+ channel types. Thus, in Fig. 1 a hypothetical common ancestor is shown as diverging into a prokaryotic and a eukaryotic subdivision; this chapter is concerned exclusively with the eukaryotic branch.

Within the eukaryotic lineage, and particularly within the vertebrate lineage of metazoans, the 2-TM K^+ channels have undergone extensive divergence. The members of the Kir inward rectifier family of channels are 2-TM proteins. The wide variation of function of the Kir channels is due to the different sequences that have evolved in this family, leading to differences in strength of rectification, single-channel conductance, and response to a variety of modulating factors.

The ancestor of most other families of K^+ channel proteins appears to have arisen by the addition of four additional transmembrane segments to the N-terminus of a 2-TM precursor, yielding a channel with six transmembrane helices (S1–S6). The 6-TM channels have two general features that distinguish them from the 2-TM channels. First, they have intrinsic gating structures as opposed to the Kir inward rectifiers, which are gated by extrinsic cytoplasmic factors, Mg^{2+} ions, and/or polyamines (Lopatin *et al.*, 1994). Second, the fourth transmembrane helix (S4) has positively charged residues (either lysine or arginine) at every third position in this S4 segment, producing a cylinder with a single row of positive charge extending through the membrane. This charged cylinder confers intrinsic voltage sensitivity to the channels by providing a charged element in the protein that will move in response to changes in membrane potential. This combination of intrinsic gating and voltage sensitivity within a single channel protein confers voltage dependence of channel opening that is independent of soluble cytoplasmic factors. The S2 and S3 segments also contribute to voltage sensitivity through charged amino acid residues and hydrophobic residues that interact with the S4 segment and stabilize the open or closed conformation of the channel protein ((Papazian *et al.*, 1995; also see below). A 6-TM K^+ channel was probably ancestral to the voltage-gated Ca^{2+} and Na^{2+} channels, which consist of four 6-TM modules (Strong *et al.*, 1993).

After the origination of the 6-TM structure, which includes the S4 voltage sensor, the ancestor of the EAG/pacemaker family diverged first, by the addition of a C-terminal cytoplasmic domain that binds cyclic nucleotides and modulates channel function. This cyclic-nucleotide-modulated channel family then split into the EAG family, with the addition of an N-terminal PAS domain (an interaction domain found in the *Per*, *Arnt*, and *Sim* genes and in a variety of prokaryotic proteins) that confers a hyperpolarizing current phenotype on a channel that is intrinsically activated by depolarizing voltage (Morais *et al.*, 1998), and the Ih pacemaker family of channels, which also pass a K^+ current under a hyperpolarizing potential, although the gating mechanism is not defined as yet. The KCNQ (*K*$^+$ channel, *c*yclic *n*ucleotide modulated, mutated in long *Q*T syndrome) family of delayed rectifiers then diverged from the lineage that gave rise to the various members of the Kv/*Shaker*-like family. This is not to say that fully functional cyclic-nucleotide-modulated, voltage-gated, 6-TM channels evolved before the structurally more simple Kv channels but, more likely, that the extant family of Kv channels is a descendant of a single precursor that underwent substantial evolutionary change after the origin of the first Vg 6-TM channel prior to the first duplications and divergences that gave rise to the Kv1, Kv2, Kv3, and Kv4 families.

The other striking result is that the KCNQ1/KVLQT1 channel that has been designated as KV1.9 is not a close relative of the other Kv1/*Shaker* types of channels. In vertebrate heart, mutations in the KVLQT1 channel (also designated as KCNQ1 and Kv1.9) are responsible for one form of long QT syndrome (Wang *et al.*, 1996). Although this channel has been designated as a member of the Kv1 family, it has clearly evolved separately from other members of this channel family, instead being strongly grouped with the KCNQ2/3 family of delayed rectifiers. This whole family of delayed rectifiers clearly has an ancient origin, because it robustly groups outside the Kv1 through Kv4 families which originated in a common ancestor of all metazoans. Thus, this channel should probably be considered the exemplar of a distinctive family of voltage-gated K^+ channels.

Figure 2 shows an evolutionary tree, generated using an extensive set of K^+ channel proteins isolated from humans, that confirms the scheme shown in Fig. 1. This tree is based on a comparison of the only homologous structures that are shared between all the channel families in the tree, namely, the M2/S6 transmembrane helix and the H5 pore region. The M1/S5 transmembrane helix and the extracellular loop that connects this helix to the pore region are so poorly conserved between the 2-TM and 6-TM channel proteins that a reliable alignment of sequences cannot be obtained. Because of the small size of the data set and the extremely long time period that this tree represents (assuming that divergence of 2-TM and 6-TM channels predates the evolution of metazoans), the fine structure of this tree is ambiguous. The shapes of subsets of the tree can be resolved with larger alignments of sequences between the more closely related members of the subfamilies, as discussed in detail below.

Separation of the 2-TM and 6-TM subgroups is extremely robust, indicating that divergence within these two groups probably occurred after a single duplication event. With that assumption, the whole tree can be rooted and the relative time of origin of the various subfamilies can be deduced. The rooting of the Kir channel segment of the tree shown in Fig. 2 is suspect, however. The branching pattern seen in this tree is different from the branching pattern seen in the more extensive analysis of solely Kir channels (see below, Fig. 3). Because the latter analysis is based on more channels and more sequence data from each channel, it must be viewed as the more acceptable result. The rooting and branching of the 6-TM channel proteins, on the other hand, is similar in Fig. 2 and in later analyses (see Figs. 4 and 5).

Given this general outline of how the different channel protein types have arisen, we will now discuss in more detail how evolution has shaped the ensemble of channel proteins within each group.

3. EVOLUTION OF THE KIR FAMILY OF K^+ CHANNELS

The 2-TM inward rectifier channels (Kir family) have evolved a significantly different pore sequence from the 6-TM K^+ channels and are susceptible to a number of modulatory factors, including phosphorylation by cytoplasmic kinases (Fakler *et al.*,

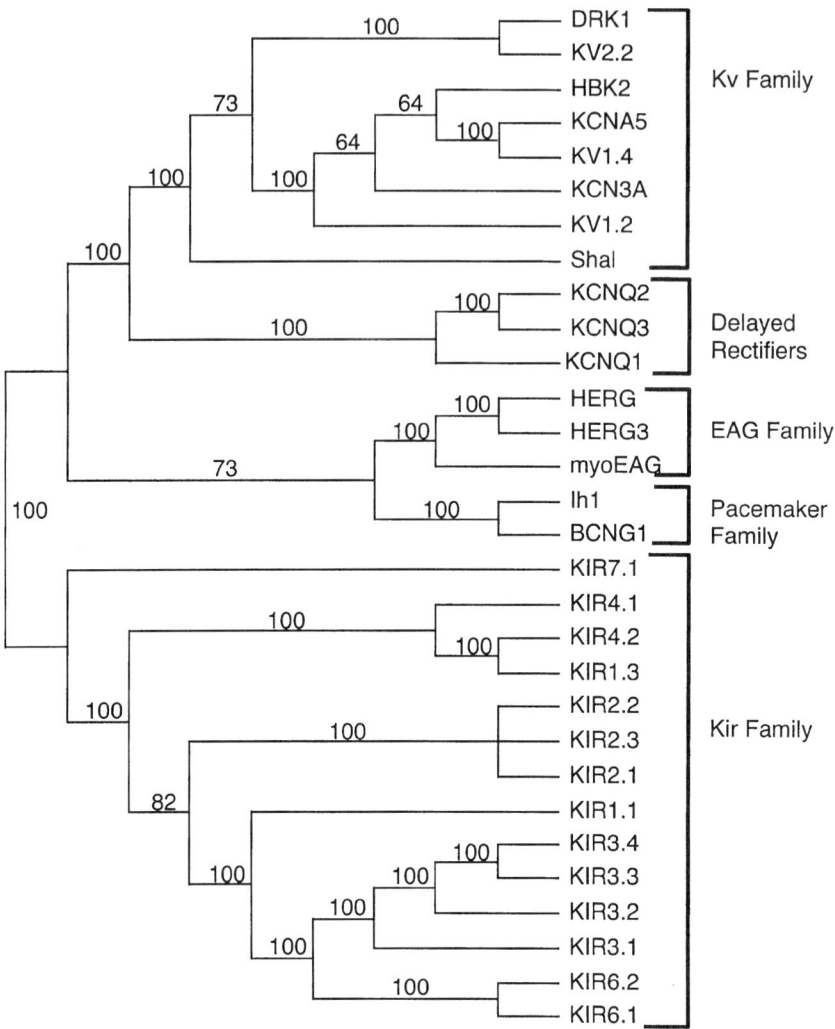

Figure 2. Phylogenetic tree of human K⁺ channel proteins. This is a consensus summary of the shortest 11 trees generated by a maximum parsimony analysis of the aligned H5 and S6/M2 regions of the channel proteins. This analysis involves finding which of the possible phylogenetic trees requires the smallest number of nucleotide changes to account for the sequence differences between the various proteins. The numbers on the branches indicate the percentage of these 11 trees that contain the grouping of channels shown above the labeled branch. *Homo sapiens* possesses all of the channel types that are discussed in this chapter. Preliminary analysis indicates that these various channel types originated before the origin of multicellular animals; therefore, the analysis of human homologs will yield a tree that is representative of the evolution of the complete gene family. The tree has been rooted between the 2-TM and 6-TM families on the assumption that the differences between these two families only arose once during evolution. The fact that the different families are robustly separated is consistent with this assumption. The 2-TM inward rectifier (Kir) families all arose from a single ancestral sequence by duplication and divergence. Similarly, the 6-TM channel type arose once, and then the various families of voltage-gated channels arose through a process of duplication, divergence, and possibly fusion with domains having different modulator activities.

1994), cytoplasmic ATP concentration (Takumi *et al.*, 1995), and G-proteins (Sanchez *et al.*, 1998). The M1 transmembrane segment of Kir channels has considerable sequence differences from the homologous region (S5) of 6-TM channels; however, within the Kir group of channels, the M1 segment is sufficiently similar between channels to yield a robust alignment and thus a more extensive data set for phylogenetic analysis than was possible for the analysis shown in Fig. 2. Because parts of the cytoplasmic segments of the channel proteins are involved in interactions with intracellular Mg^{2+} ions or polyamines, there is considerable sequence similarity of these segments among the various channels in this family.

Figure 3 represents the shortest tree derived from 37 Kir channel sequences from a variety of vertebrate species and *Caenorhabditis elegans*. It has been rooted with Kir7.1 as an outgroup, as indicated in Fig. 2; however, since the positions of the

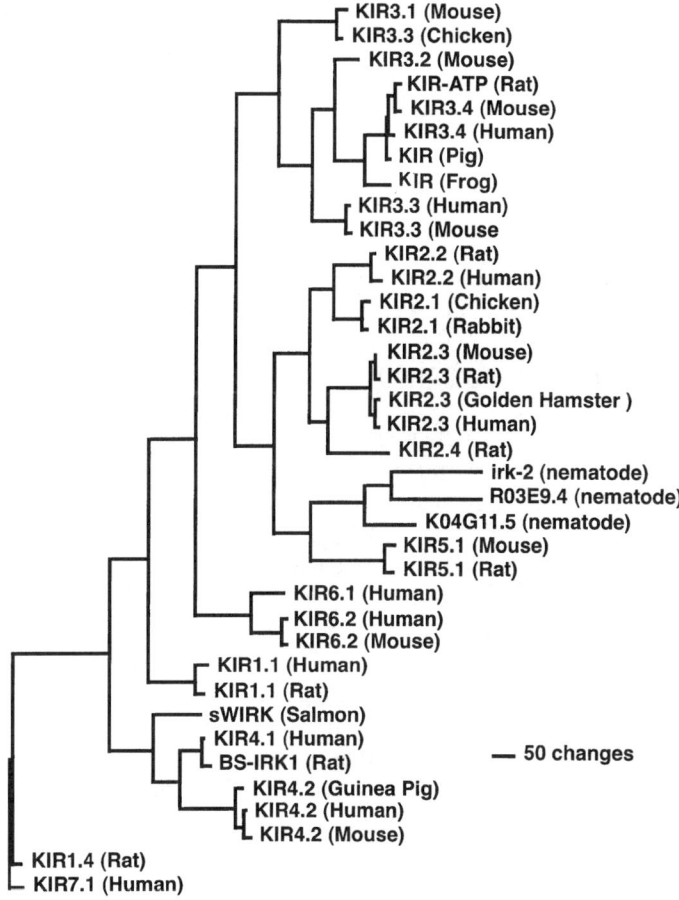

Figure 3. Phylogenetic tree of Kir 2-TM K$^+$ channels. Rooting is provisional, based on the position of the root of this family in Fig. 2. This is a phylogram, where the lengths of the branches are proportional to the number of nucleotide changes that are inferred to have occurred along that branch. The scale bar is proportional to 50 changes in underlying nucleotide sequence.

different Kir families are different in this more extensive tree than in the summary tree shown in Fig. 2, this rooting must be viewed skeptically. One striking feature of this tree is that channels of the same type from different species group more closely to other channels of the same kind than do different types of channels from the same species. This implies that most of the Kir channels are paralogs, arising by duplication and divergence prior to the speciation events that led to the evolution of different organisms. Therefore, with the data available, it is likely that the common ancestor of vertebrates had a full suite of 2-TM inward rectifier channel genes.

Variation in the sequence in the pore region has altered the ion-transfer capabilities (Krapivinsky *et al.*, 1998), and variation in the M2 domains of the proteins produced altered degrees of inward rectification by altering the affinity of the cytoplasmic domains for Mg^{2+} ions and intracellular polyamines, which control the gating capabilities of the Kir channel family. For example, mutation of a single residue in the M2 segment of ROMK1 (originally isolated as a *r*at kidney *o*uter *m*edullary K^+ channel), from asparagine to aspartate, is sufficient to convert this channel from a weakly rectifying to a strongly rectifying phenotype (Lu and MacKinnon, 1994; Wible *et al.*, 1994) and to alter the ion selectivity of the channel (Reuveny *et al.*, 1996). However, although a single amino acid alteration will cause this physiological change, that is only true within the context of a specific channel; at least one other Kir channel has glutamate at this position and yet is still a weak rectifier (Kubo *et al.*, 1996).

Similarly, alterations in the M2 and C-terminal cytoplasmic domain of Kir channels can alter their capacity for forming heteromultimers with other Kir family members (Tinker *et al.*, 1996), thus extensively altering the suite of inward rectifying currents that can be expressed by the organism. Changes in the C-terminal cytoplasmic domain can also affect the kinetics of opening and closing of Kir channels (Trapp *et al.*, 1998).

Recent analysis of the complete genome of *C. elegans* indicates that there are only three 2-TM Kir channel proteins in the whole organism (Bargmann, 1998). All three of these predicted channel proteins group with the Kir5 family, as would be expected of a paralog family that had originated prior to the divergence of the ancestral organisms of nematodes and vertebrates; if this is the case, then many of the Kir family channels must have been lost during *C. elegans* evolution; this would be consistent with the finding that *C. elegans* has lost all genes for voltage-gated Na^+ channels (Bargmann, 1998). It is also possible that there has been a strongly convergent evolution of this group of channels in nematodes and vertebrates, implying that the function of the Kir5 family in vertebrates is the only Kir function that is strongly selected for in *C. elegans*. A third possibility is that the Kir gene tree should be rooted between the vertebrate Kir5 family and the *C. elegans* 2-TM channels, indicating that the divergence of a large number of 2-TM inward rectifier channels is a relatively late evolutionary event and unique to the vertebrate lineage of organisms. The significance of this result can only be resolved with a wider search for 2-TM channel proteins in other invertebrate organisms and a more extensive evaluation of the phylogeny of the Kir channels from a variety of vertebrates and invertebrates. These channel proteins must also be functionally characterized.

4. EVOLUTION OF THE KV FAMILIES OF K⁺ CHANNELS

Figure 4A shows a simplified summary of an original, extensive phylogenetic analysis of the voltage-gated K^+ channels of the *Shaker* superfamily and the KCNQ family of delayed rectifier channels. All of the K^+ channel types in these families have been found in all phyla for which they have been studied. The subfamilies of channel types within different phyla, however, appear to have evolved independently, indicating that within the various phylogenetic lineages, different evolutionary pathways have been followed by the various channel families. For example, the *Shaker* subtype, or Kv1 family, of voltage-gated K^+ channels has been found in diverse nonvertebrate multicellular animals, including the Cnidaria, Platyhelminthes, Nematoda, Mollusca, Arthropoda, and Chordata. Within the vertebrate lineage, there are at least seven subfamilies of the Kv1 family of channels, each encoded by a distinct gene, and all of these subfamilies have been found in the vertebrates studied to date. However, although there are a number of functional Kv1-type channels in nonvertebrate taxa, these subfamilies appear to have arisen independently, after the divergence of the various phyla. In some cases, for example, *Drosophila melanogaster*, there is a single Kv1 type gene that yields a variety of channel proteins by alternative splicing. In *Polyorchis penicillatus*, a cnidarian (jellyfish), there are at least two different *Shaker*-type channel genes, but these appeared to have diverged from an ancestral channel gene after the separation of the cnidarian lineage from the other metazoans (Jegla *et al.*, 1995). Thus,

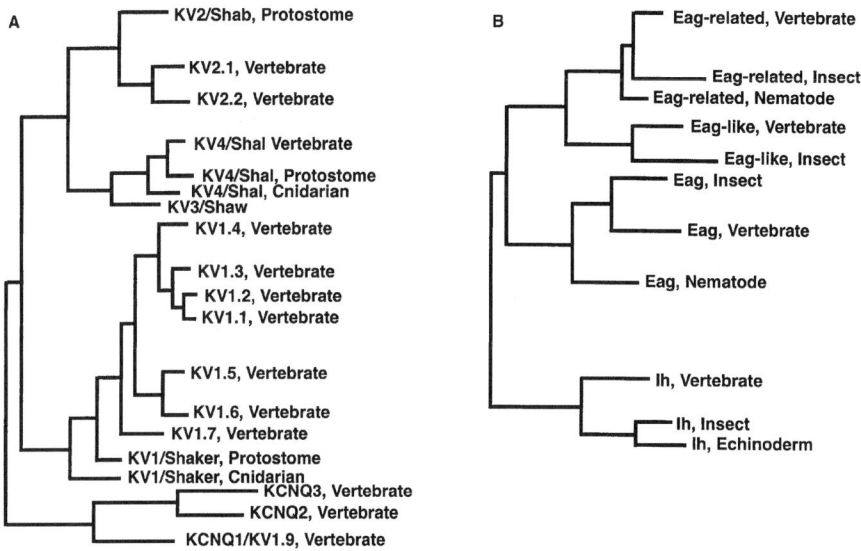

Figure 4. (A) Summary phylogenetic tree of KV family of 6-TM channels, based on an extensive analysis of 51 protein sequences. This family is rooted based on the more comprehensive tree shown in Fig. 2. (B) Summary phylogenetic tree of cyclic-nucleotide-modulated, voltage-gated 6-TM, K^+ channels, based on an extensive analysis of 28 protein sequences. Rooting is based on the position of the root of this family in Fig. 2.

it appears that ancestral members of the Kv1, Kv2, Kv3, and Kv4 families of K$^+$ channels arose in the common ancestor of all multicellular animals and then these various families evolved independently in different phyla.

5. EVOLUTION OF EAG-RELATED AND PACEMAKER K$^+$ CHANNELS

The Ih/EAG family of channels are 6-TM voltage-sensitive K$^+$ channels whose activities are modified by cyclic nucleotides binding to cytoplasmic portions of the proteins (Bruggemann *et al.*, 1993; Gauss *et al.*, 1998; Ludwig *et al.*, 1998). They have different functions in the heart and nervous system, with the Ih family functioning as pacemakers and the EAG family providing rectification as a complex function of holding potential and second-messenger levels. Figure 4B shows a simplified summary of a phylogram originally generated using 28 members of these two related families of channels. The rooting of this tree is predicated on the presence of a root between the EAG superfamily and the pacemaker family as seen in Fig. 2.

Both of these families of channels have positively charged S4 segments, yet both pass current when the membrane is hyperpolarized. This activity is not apparently due to a major difference in the way that the voltage-sensing mechanism is coupled to the gating mechanism, but rather to a unique interplay between the kinetics of activation, inactivation, and deactivation. In the HERG (*human EAG-related gene*) channel, a member of the EAG superfamily, the N-terminus of the protein has a characteristic domain that has recently been shown to have a three-dimensional structure similar to that of the PAS domain (Morais *et al.*, 1998). This domain has a high affinity for the open state of the channel, so channels that are opened by depolarization do not pass an outward current because they are rapidly blocked. However, when the membrane is hyperpolarized, the PAS domain releases from the channel, but the deactivation of the core structure is slow, allowing a transient inward K$^+$ current. From this structure–function analysis, we can see that a change in the direction that a channel passes current can be caused, not by a wholesale reorganization of the primitive gating mechanism, but rather by addition of new structures that modify the more primitive mechanism. Most other members of the EAG family of channels, however, act as delayed rectifiers, with outward currents on depolarization.

As yet there is not a good explanation for the activation of K$^+$ currents in the pacemaker channel family by hyperpolarization, since they do not have an N-terminal PAS domain. However, it seems likely that this family has acquired a different structural modification that is modulating the primitive depolarization gating of the 6-TM channel core. It is possible to cause a shift from depolarization activation to hyperpolarization activation with a small number of mutations that affect gating mechanism but not movement of the charge sensor (Miller and Aldrich, 1996). Further analysis will be required to determine how the hyperpolarization activation of these channels might have evolved from the more typical voltage response in 6-TM Vg K$^+$ channels.

6. FUTURE DIRECTIONS FOR EVOLUTIONARY STUDIES

Evolutionary studies of K^+ channels require comprehensive sequence and physiological data from a range of widely diverged organisms. Degenerate polymerase chain reaction (PCR) amplification has proven to be a sufficiently sensitive method for amplification of a variety of K^+ channels from the most diverse extant animal phyla (Jegla *et al.*, 1995; Jegla and Salkoff, 1995). A second approach is to use computational methods to discover K^+ genes from sequences obtained during genome sequencing projects. The complement of K^+ channel proteins has been most extensively catalogued in *C. elegans*. An extensive structural comparison of K^+ channels from part of the *C. elegans* genome indicates that examples of the channel families found in vertebrates and insects are also present in the nematode (Wei *et al.*, 1996). The complete genomic sequence of *C. elegans* has been recently reported, and an intensive search revealed a total of 80 different K^+ channel genes (Bargmann, 1998). The channels from *C. elegans* have not yet been expressed and characterized electrophysiologically. Although the full sequence of a single metazoan genome is valuable for intensive study of a single organism, it is also important to have data from a diverse set of organisms. *C. elegans* illustrates this point well because it appears to lack voltage-gated Na^+ channels (Bargmann, 1998), and it has only three members of the Kir family, all closely related to the Kir5 family of vertebrates. This raises the question of whether sodium selectivity was lost or never arose by modification of Ca^{2+} channels in the nematode lineage and of whether the paucity of Kir channels is due to loss of genes from a larger ancestral population or to a failure to evolve in the first place within the lineage leading to *C. elegans*. Thus, although the complete sequence of the genome of any metazoan is invaluable for understanding many physiological processes, any single organism can have unique physiological adaptations that are terminally derived and can thus be misleading if they are generalized.

Given the diversity of K^+ channels in each channel family and the capacity to express these channel proteins in heterologous systems, it will be possible to use the structural and functional diversity of these channels to evaluate the structural basis for a variety of biophysical characteristics of the channels. Application of this approach has already proved a useful adjunct for studying voltage-sensing mechanisms of the Kv1 family of channels. An example of this kind of analysis, applied to defining which parts of the channel are responsible for setting the voltage of half-activation ($V_{1/2}$) is shown in Fig. 5. The sequences of 13 members of the Kv1 (*Shaker*) family of genes, with $V_{1/2}$ ranging from -34 to $+22$ mV, were aligned. The correlation between $V_{1/2}$ and the identity of the amino acid residue at each position was then evaluated by computing the variance of the $V_{1/2}$ values of all channels that have identical residues at that position and computing a pooled variance for each position. In Fig. 5, the positions that correspond to the lowest 20 variances are mapped on a schematic of the channel backbone. In principle, the marked positions should be involved in the molecular changes that occur during the channel protein's response to a change in membrane potential. However, because only 13 channels are used and the selection of organisms is limited, some of the results may be confounded by common heritage unrelated to functional conservation.

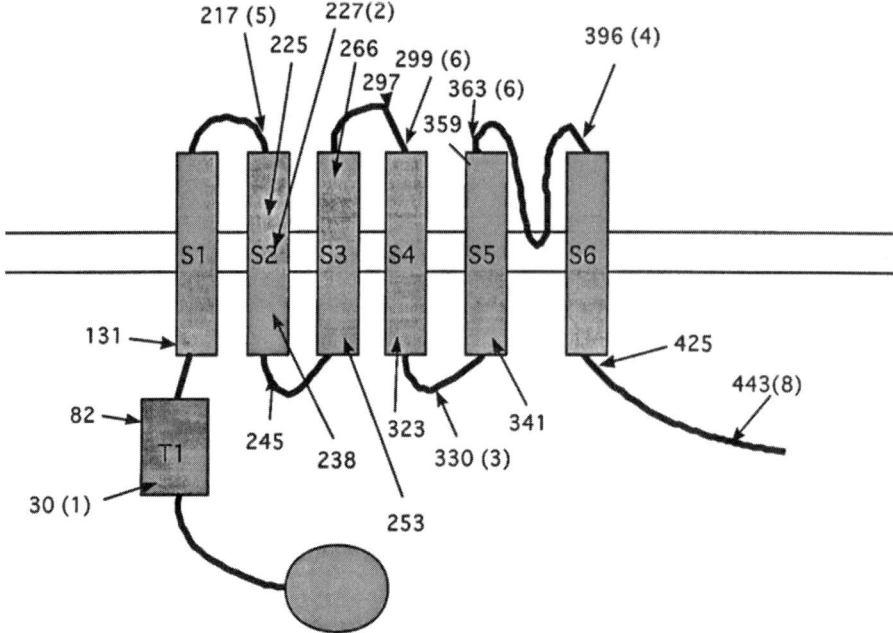

Figure 5. Locations of the amino acid positions that co-vary most closely with half-activation voltage ($V_{1/2}$) in 13 *Shaker*-type channels (mammalian and cnidarian). The numbers represent the position in the multiple alignment of the 13 channels, not in any single channel; the arrows represent the approximate location of the residue in the generally accepted topology of these channels. Numbers in parentheses are the rank order of variance, with 1 having the lowest variance; residues without a parenthetical number had equal values, thus ranking tied for ninth. For example, the position labeled 227(2) within the S2 transmembrane domain is position 227 in the multiple alignment of channels and has the second lowest covariance of voltage sensitivity with residue identity at that position. This means that the residues at this position in the channel protein are likely to be involved in setting the voltage sensitivity of the channel. Note that the S2 segment is the only one of the six transmembrane segments with low-scoring residues in the interior of the segment and that 25% of the low-variance residues are within or immediately adjacent to S2.

    Interestingly, three residues within the S2 domain are among the positions with lowest variance. One of the residues (labeled 225) is a glutamate residue in most channels that has been proposed to act as a countercharge to positively charged S4 residues during channel opening (Papazian *et al.*, 1995). The other two positions have not been previously proposed to be involved in the voltage response. This preliminary analysis suggests that the S2 segment interacts directly with the S4 segment in the membrane, through electrostatic and hydrophobic interactions. The individual identities of the various residues in these three positions in the various channels provide us with a well-defined set of possible changes that can be used in *in vitro* mutagenesis experiments to test this hypothetical model.

    With the wider characterization of cloned K^+ channels, as molecular studies of cellular excitability are extended to a wider range of tissues and organisms, and with the development of more sophisticated comparative methods, we will be able to formulate more precise methods for using the natural experiments of evolution to extend our understanding of the molecular basis of K^+ channel function, in conjunction with the more traditional biophysical and pharmacological approaches.

REFERENCES

Bargmann, C. I., 1998, Neurobiology of the *Caenorhabditis elegans* genome, *Science* **282:**2028–2033.

Bruggemann, A., Pardo, L. A., Stuhmer, W., and Pongs, O., 1993, Ether-a-go-go encodes a voltage-gated channel permeable to K^+ and Ca^{2+} and modulated by cAMP, *Nature* **365:**445–448.

Doyle, D. A., Cabral, J. M., Pfuetzner, R. A., Kuo, A., Gulbis, J. M., Cohen, S. L., Chait, B. T., and MacKinnon, R., 1998, The structure of the potassium channel:Molecular basis of K^+ conduction and selectivity, *Science* **280:**69 77.

Fakler, B., Brandle, U., Glowatzki, E., Zenner, H. P., and Ruppersberg, J. P., 1994, Kir2.1 inward rectifier K^+ channels are regulated independently by protein kinases and ATP hydrolysis, *Neuron* **13:**1413–1420.

Gauss, R., Seifert, R., and Kaupp, U. B., 1998, Molecular identification of a hyperpolarization-activated channel in sea urchin sperm, *Nature* **393:**583–587.

Hille, B., 1991, *Ionic Channels of Excitable Membranes*, 2nd ed., Sinauer Associates Inc., Sunderland, Massachusetts.

Jegla, T., and Salkoff, L., 1995, A multigene family of novel K^+ channels from *Paramecium tetraurelia*, *Recept. Channels* **3:**51–60.

Jegla, T., Grigoriev, N., Gallin, W. J., Salkoff, L., and Spencer, A. N., 1995, Multiple *Shaker* potassium channels in a primitive metazoan, *J. Neurosci.* **15:**7989–7999.

Krapivinsky, G., Medina, I., Eng, L., Krapivinsky, L., Yang, Y., and Clapham, D. E., 1998, A novel inward rectifier K^+ channel with unique pore properties, *Neuron* **20:**995–1005.

Kubo, Y., Miyashita, T., and Kubokawa, K., 1996, A weakly inward rectifying potassium channel of the salmon brain:Glutamate 179 in the second transmembrane domain is insufficient for strong rectification, *J. Biol. Chem.* **271:**15729–15735.

Lopatin, A. N., Makhina, E. N., and Nichols, C. G., 1994, Potassium channel block by cytoplasmic polyamines as the mechanism of intrinsic rectification, *Nature* **372:**366–369.

Lu, Z., and MacKinnon, R., 1994, Electrostatic tuning of Mg^{2+} affinity in an inward-rectifier K^+ channel, *Nature* **371:**243–246.

Ludwig, A., Zong, X., Jeglitsch, M., Hofmann, F., and Biel, M., 1998, A family of hyperpolarization-activated mammalian cation channels, *Nature* **393:**587–591.

MacKinnon, R., Reinhart, P. H., and White, M. M., 1988, Charybdotoxin block of *Shaker* K^+ channels suggests that different types of K^+ channels share common structural features, *Neuron* **1:**997–1001.

MacKinnon, R., Cohen, S. L., Kuo, A., Lee, A., and Chait, B. T., 1998, Structural conservation in prokaryotic and eukaryotic potassium channels, *Science* **280:**106–109.

Miller, A. G., and Aldrich, R. W., 1996, Conversion of a delayed rectifier K^+ channel to a voltage-gated inward rectifier K^+ channel by three amino acid substitutions, *Neuron* **16:**853–858.

Morais, J. H., Lee, A., Cohen, S. L., Chait, B. T., Li, M., and MacKinnon, R., 1998, Crystal structure and functional analysis of the HERG potassium channel N terminus:A eukaryotic PAS domain, *Cell* **95:**649–655.

Papazian, D. M., Shao, X. M., Seoh, S. A., Mock, A. F., Huang, Y., and Wainstock, D. H., 1995, Electrostatic interactions of S4 voltage sensor in Shaker K^+ channel, *Neuron* **14:**1293–1301.

Reuveny, E., Jan, Y. N., and Jan, L. Y., 1996, Contributions of a negatively charged residue in the hydrophobic domain of the IRK1 inwardly rectifying K^+ channel to K^+-selective permeation, *Biophys. J.* **70:**754–761.

Sanchez, J. A., Gonoi, T., Inagaki, N., Katada, T., and Seino, S., 1998, Modulation of reconstituted ATP-sensitive K^+-channels by GTP-binding proteins in a mammalian cell line, *J. Physiol. (London)* **507:**315–324.

Strong, M., Chandy, K. G., and Gutman, G. A., 1993, Molecular evolution of voltage-sensitive ion channel genes—on the origins of electrical excitability, *Mol. Biol. Evol.* **10:**221–242.

Takumi, T., Ishii, T., Horio, Y., Morishige, K., Takahashi, N., Yamada, M., Yamashita, T., Kiyama, H., Sohmiya, K., Nakanishi, S., *et al.*, 1995, A novel ATP-dependent inward rectifier potassium channel expressed predominantly in glial cells, *J. Biol. Chem.* **270:**16339–16346.

Tinker, A., Jan, Y. N., and Jan, L. Y., 1996, Regions responsible for the assembly of inwardly rectifying potassium channels, *Cell* **87:**857–868.

Trapp, S., Proks, P., Tucker, S. J., and Ashcroft, F. M., 1998, Molecular analysis of ATP-sensitive K channel gating and implications for channel inhibition by ATP, *J. Gen. Physiol.* **112:**333–349.

Wang, Q., Curran, M. E., Splawski, I., Burn, T. C., Millholland, J. M., VanRaay, T. J., Shen, J., Timothy, K. W., Vincent, G. M., de Jager, T., Schwartz, P. J., Toubin, J. A., Moss, A. J., Atkinson, D. L., Landes, G. M., Connors, T. D., and Keating, M. T., 1996, Positional cloning of a novel potassium channel gene:KVLQT1 mutations cause cardiac arrhythmias, *Nat. Genet.* **12:**17–23.

Wei, A., Jegla, T., and Salkoff, L., 1996, Eight potassium channel families revealed by the *C. elegans* genome project, *Neuropharmacology* **35:**805–829.

Wible, B. A., Taglialatela, M., Ficker, E., and Brown, A. M., 1994, Gating of inwardly rectifying K$^+$ channels localized to a single negatively charged residue, *Nature* **371:**246–249.

Chapter 2

Three-Dimensional Structure of the K^+ Channel Pore: Basis for Ion Selectivity and Permeability

Hee-Cheol Cho and Peter H. Backx

1. INTRODUCTION

Physiologists have long known that ions play a central role in the excitability of nerve and muscle. In a series of papers between 1881 and 1887, Sidney Ringer showed that the solution perfusing a frog heart must contain salts of sodium, potassium, and calcium mixed in a definite proportion if the heart is to continue beating (Hille, 1992). Later, Hodgkin and Huxley, in their classical studies on squid giant axons (Hodgkin and Huxley, 1952;a,b), found that the conductance changes of membranes could be separated into two distinct "permeation" pathways, one for sodium and another for potassium. This early recognition that distinct proteins conferred selective passageways for different ions was fully established with the discovery and use of pharmacological agents and toxins. For example, the toxins tetrodotoxin (TTX) and saxitoxin (STX) selectively and reversibly block Na^+ current without affecting K^+ conductance (Hille, 1966; Hille, 1968; Nakamura *et al.*, 1965). These early studies laid the foundation for the concept of channel "gating," wherein the availability of the channel's active site or pore is allosterically regulated in a time- and voltage-dependent manner (Hodgkin and Huxley, 1952c).

The development of the patch-clamp technique provided the opportunity to dissect the permeation and gating properties of ion channels at the single-channel level. These studies revealed that K^+ channels support transport rates of between 10^6 and 10^8 K^+ ions per second across the membrane with a driving force of about 100 mV, while exquisitely discriminating against other biologically relevant cations like Na^+ and Ca^{2+}. Indeed, under normal physiological conditions, only a single Na^+ ion passes through a K^+ channel pore for every 10,000 K^+ ions that are conducted. The

Hee-Cheol Cho and Peter H. Backx ● Department of Medicine, The Toronto Hospital, Centre for Cardiovascular Research, and Department of Physiology and Institute of Medical Sciences, University of Toronto, Toronto, Ontario, Canada M5G 2C4.

Potassium Channels in Cardiovascular Biology, edited by Archer and Rusch. Kluwer Academic/Plenum Publishers, New York, 2001.

remarkable selectivity of these pores suggests intimate contact between the K^+ ion and the channel pore. On the other hand, electrostatic calculations predict that ion movement through very narrow membrane pores encounters an enormous energy barrier of about $200kT$, where k is the Boltzmann constant and I is absolute temperature, compared with the average kinetic energy available to an ion of $\frac{3}{2}kT$. This energy barrier originates from differences in polarization between the aqueous and lipid bilayers (Parsegian, 1975). The ability of K^+ channels to support K^+ ion fluxes at near "diffusion-limited" rates, while virtually excluding Na^+ ions, remains poorly understood despite decades of research. The recent report of the crystal structure of the KcsA channel, a bacterial K^+ channel (Doyle *et al.*, 1998), offers new opportunities to understand the fundamental basis of K^+ channel permeation and selectivity. However, the precise connection between the structure and function of ion channels cannot be gleaned directly from their static structure. This knowledge will require the development and implementation of appropriate dynamic computer modeling of channel proteins embedded in a lipid bilayer and surrounded by a sea of water and ions.

K^+ channels are the most widespread and diverse class of membrane channel proteins, and they are found in most eukaryotic and prokaryotic cells. Currently, there are at least 35 distinguishable K^+ channel clones documented as belonging to the superfamily of pore-forming membrane proteins referred to as "P-type" (Jan and Jan, 1997; Nichols and Lopatin, 1997). Other classes of ion channels like Na^+, Ca^{2+}, and various nonselective channels also belong to this superfamily of membrane proteins, suggesting a highly conserved, unique, and efficient molecular structure for producing the selective passage of ions across lipid bilayers. One common feature of P-type channels is a pseudosymmetric arrangement of four protein subunits or equivalent motifs in the plasma membrane, each having two or six transmembrane segments. These transmembrane segments anchor a pore-forming loop, or "P-loop," that forms the selectivity region of the channel's active site (Fig. 1). These fascinating proteins play a unique and crucial role in the cardiovascular system by mediating cellular excitability and signal transduction in a controlled and tissue-specific manner. Alterations in the expression and function of these proteins occur in a number of cardiovascular pathologies including hypertension and heart disease (Billman, 1994; Kowey *et al.*, 1997; Silvestry and Kimmel, 1996; Singh, 1999), whereas mutations of ion channels are linked to life-threatening disorders such as the long QT syndromes (Keating and Sanguinetti, 1996). Accordingly, these proteins are the primary or indirect targets of many therapeutic agents used in the treatment of cardiovascular disease. The availability of a reliable structural model for ion channel proteins will unquestionably facilitate the development of new and better agents for the modulation of ion channel function.

This chapter focuses on the molecular mechanisms of K^+ channel permeability and selectivity and other aspects of pore function and will relate these mechanisms to the recently solved X-ray crystallographic structure of the bacterial K^+ channel, KcsA, which is expressed in *Streptomyces lividans* (Doyle *et al.*, 1998).

2. MOLECULAR DIVERSITY OF MAMMALIAN K^+ CHANNELS

It is evident that K^+ currents are rather diverse and heterogeneous. They show a wide range of voltage-dependent, biophysical, and kinetic properties, suggesting genetic and

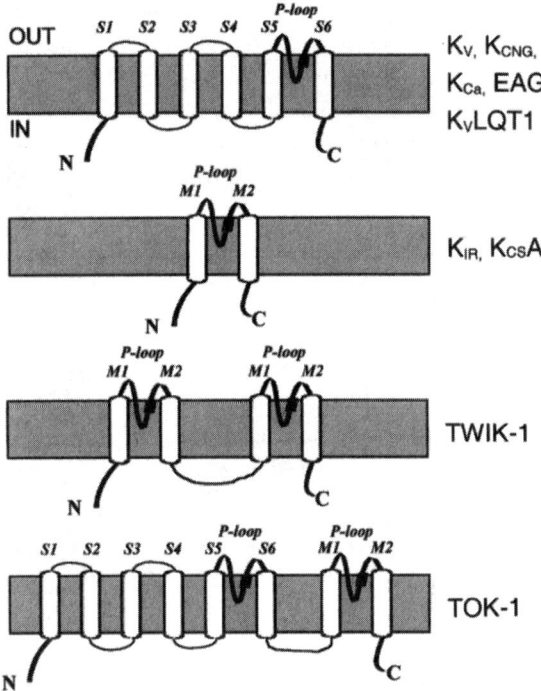

Figure 1. Putative transmembrane topologies of various K$^+$ channel subunits viewed from the plane of the cell membrane. *Top*: K$^+$ channel subunits containing six transmembrane (TM) domains include the voltage-gated K$^+$ (Kv) channels, cyclic nucleotide-gated K$^+$ (K$_{CNG}$) channels, Ca^{2+}-activated K$^+$ (K$_{Ca}$) channels, and *ether-a-go-go*-related (ERG) K$^+$ channels. Each of the four domains of voltage-gated Na$^+$ and Ca^{2+} channels has a similar basic membrane topology as Kv channels, but their four domains are linked in tandem as one long open reading frame. *Second from the top*: K$^+$ channels composed of two transmembrane domains include the inward rectifier (Kir) channel and the bacterial K$^+$channel from *Streptomyces lividans*, KcsA. *Bottom*: TWIK-1 and TOK-1 are K$^+$ channel subunits containing two pores consisting of two 2-TM domain structures (TWIK-1) or one 6-TM domain plus one 2-TM domain (TOK-1). The extracellular loop (P-loop) between S5 (M1) and S6 (M2) provides the pore-lining residues with the selectivity filter (box).

molecular diversity. The first K$^+$ channel to be cloned was a voltage-gated K$^+$ potassium (Kv) channel in *Drosophila* called the *Shaker* channel (Papazian *et al.*, 1987), which corresponds to the mammalian Kv1 channel subfamily (Chandy, 1991). This channel has six putative α-helical transmembrane segments (S1–S6) and resembles the structure of a single internal repeat of the four homologous domains seen in voltage-gated Na$^+$ and Ca^{2+} channels (Figs. 1 and 2). As expected from this similarity with Na$^+$ and Ca^{2+} channels, functional K$^+$ channels require the assembly of four K$^+$ channel subunit proteins to form the channel pore (Papazian, 1999). Subsequently, three other members of the Kv channel gene family were cloned from *Drosophila* and classified into the subfamilies called *Shab*, *Shaw*, and *Shal*. The mammalian counterparts of these subfamilies are designated Kv2, Kv3, and Kv4, respectively (Chandy and Gutman, 1995; Salkoff *et al.*, 1992). In general, each subfamily has multiple members. For example, the Kv1 subfamily has many members, labeled Kv1.1, Kv1.2, and so on. At present, about 20 distinct mammalian Kv channel genes have been identified and

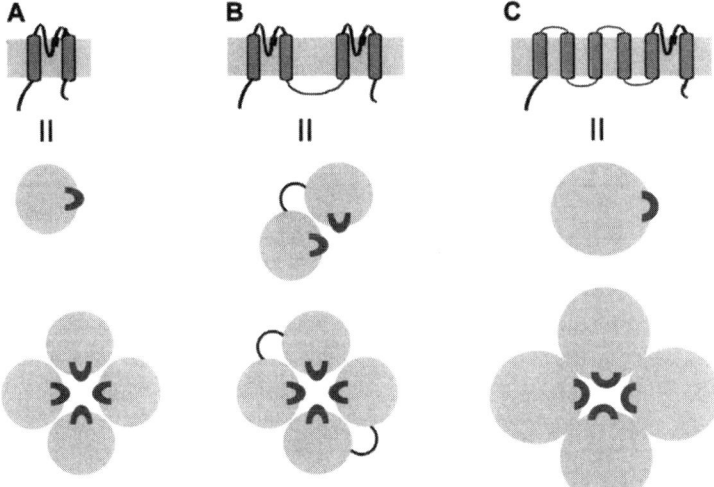

Figure 2. Functional K^+ channels are formed by pseudosymmetric assembly of four monomeric subunits. Tetrameric assembly of two-transmembrane (A), two-pore (B), and six-transmembrane (C) K^+ channel proteins is deproted (*bottom row*). A monomeric subunit is viewed from the side (*top row*) and from the top of the channel (*middle row*), normal to the cell membrane.

classified into nine subfamilies (Salinas *et al.*, 1997). Using the scorpion toxin charybdotoxin, the pore-forming region of these channels has been identified as an external loop (i.e., pore-forming loop or P-loop) between transmembrane segments S5 and S6 (MacKinnon and Miller, 1989). This P-loop is present in all K^+ channels and voltage-gated Na^+ and Ca^{2+} channels. This superfamily of structurally related channels has recently been coined "P-type" channels (Moczydlowski, 1998).

As another major family of K^+ channels is encoded by the inward rectifier K^+ potassium channel (Kir) genes, which have a simple architecture consisting of only two transmembrane domains (called M1 and M2) and a pore-forming region (P-region). Their structure is analogous to the S5–P–S6 region of Kv channel genes. Similarly to Kv channels, the Kir channels also have a tetrameric structure. This family shows no voltage-dependent gating, but the activity of members of this family often is regulated directly by extracellular and intracellular signals. For example, the Kir subfamily is activated directly by $G_{\beta\gamma}$ subunits of G-proteins (Reuveny *et al.*, 1994; Wickman *et al.*, 1994). Kir6.2 channels are coupled to the sulfonylurea receptor and are closed by adenosine triphosphate (ATP) (Aguilar-Bryan *et al.*, 1995; Inagaki *et al.*, 1995), whereas Kir2.1 channels are blocked by Mg^{2+} and polyamines at voltages above the equilibrium potential for K^+ (Kubo *et al.*, 1993). To date, ~ 15 different Kir channels have been cloned and are classified into seven subfamilies (Krapivinsky *et al.*, 1998; Nichols and Lopatin, 1997).

As discussed later, additional types of K^+ channel genes have been cloned that appear to have two pore domains. For example, the outwardly rectifying TOK1 clone of yeast is a tandem combination of S1–S6 motif and a Kir-like motif (Ketchum *et al.*, 1995). Also, the human weak inward rectifier TWIK-1 K^+ channel clone is a tandem combination of two Kir-like motifs (Lesage et al, 1996).

3. THE CRYSTAL STRUCTURE OF THE KcsA PROTEIN

The recent publication of the X-ray crystallographic structure of the KcsA K$^+$ channel to a resolution of ~ 3.2 Å provides the first glimpse of the structure of a P-type ion channel. As expected, the KcsA channel, which is a member of the two-membrane-spanning family of P-type channels, is a tetrameric complex with fourfold symmetry around a central pore. The KcsA monomer contains two transmembrane helices. The outer transmembrane helix (M1, residues 28–50) faces the surrounding lipids, and the inner transmembrane helix (M2, residues 87–114) lines large sections of the pore. These helices are connected by a ~ 30-residue P-loop that is comprised of three distinct structures. Starting from the carboxyl-terminal end of M1, the first ~ 12 residues of the P-region emerge from the membrane to form an extended loop or "turret." This region is followed by a short pore helix (residues 63–73) that dips down toward the center of the pore, while interdigitating laterally with the external portion of M2 and thereby contributing to a large portion of the outer vestibule. The next five to six residues of the P-loop form an extended loop that leads back toward the extracellular face of the channels and forms the selectivity filter. Another short loop leads to the amino-terminal end of M2. The M2 segments are tilted by about 25° with respect to the plane of the lipid bilayer and pack together to form an "inverted teepee" with the base opening toward the extracellular face. The expansion of the M2 helices on the extracellular face of the channel allows interdigitation of the pore helices between the M2 segments, thereby forming a scaffold to support the loop segments (Fig. 3).

The permeation pathway of KcsA channels coincides astonishingly well with the models constructed during past decades of the functional, biophysical, and molecular properties of K$^+$ channels. The KcsA channel has an hourglass appearance with long internal and shallow external vestibules connected by a short narrow selectivity region. The overall length of the pore is 45 nm and it shows a varying diameter along its distance. The pore begins on the inner face as an 18-nm tunnel with hydrophobic walls that are primarily lined by the M2 inner helix. The tunnel opens into a wide cavity with a diameter of about 10 Å, which is centered at the middle of the lipid bilayer. The significance of this cavity will be discussed more fully later. The cavity is separated from the extracellular solution by a selectivity filter with a diameter such that only K$^+$ ions stripped of water can pass. The selectivity filter is lined by the polar oxygens of the carbonyl backbone of the signature sequence residues and opens on the extracellular side into a shallow outer vestibule.

As pointed out by Doyle et al. (1998), ion permeation is expected to be optimized in these proteins for a number of reasons. First, the high-resistance selectivity region is kept very short. Second, rapid access to and from the selectivity filter by permeant ions on the extracellular face is ensured by the shallow nonrestrictive contour of the external vestibule and by the presence of negatively charged and polar residues on the surface that should enhance the local K$^+$ ion concentration. On the intracellular face, low-resistance access to the selectivity filter is aided by the presence of inert hydrophobic walls and the presence of negative residues at the entrance of the narrow tunnel. Third, the channel pore can accommodate three K$^+$ ions; one in the wide cavity and two at either end of the selectivity filter. This will enhance ion movement along the axis of the pore by electrostatic repulsion (Hess and Tsien, 1984; Neyton and Miller, 1988) and will promote the highly coordinated and unidirectional movement of ions.

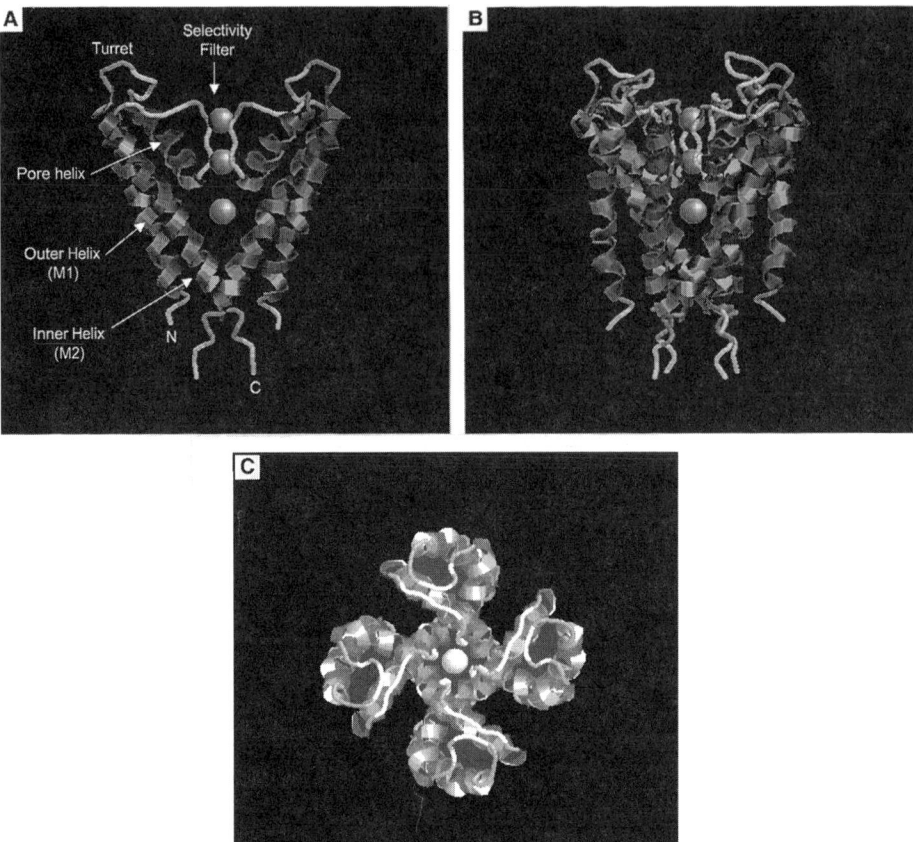

Figure 3. Structure of the KcsA channel from *Streptomyces lividans*. (A) Ribbon representation of the channel subunits viewed from the side with the extracellular side up. The two subunits at the front and back of the view are omitted for clarity. Various structural regions are labeled for the subunit on the left. The secondary structures represent α-helices β-turns, random coils. The three K^+ ions are shown as balls. (B) The "inverted teepee" architecture of the tetramer, seen from the side. (C) Channel viewed from the extracellular side with the ion-conducting pore at the center of the channel protein. The structures were generated using RasMol 2.7.1 with the coordinates acquired from Protein Data Bank (PDB # 1BL8).

The general structural features of the KcsA channel could be universally applicable to other P-type K^+ channels, as well as possibly P-type Na^+ and Ca^{2+} channels. This assertion is based on the high level of sequence conservation between different K^+ channel clones in the core region pictured in the crystal structure of the KcsA channels. In addition, replacement of selected residues of the charybdotoxin-insensitive KcsA channel with those present in *Shaker* channels confers sensitivity to block by charybdotoxin (MacKinnon *et al.*, 1998). There are, however, certain to be some differences between KcsA structure and that of other K^+ channels, as discussed more fully below. Regardless of these inherent limitations of using crystal structures, many of the structural features of the KcsA channel pore appear to relate remarkably well to the known properties of P-type K^+ channels.

4. ROLE OF THE PORE IN SUBUNIT ASSEMBLY

Analogous to voltage-gated Na$^+$ and K$^+$ channels, functional K$^+$ channels require the assembly of four subunits arranged symmetrically around the central pore (Heginbotham *et al.*, 1997; Liman *et al.*, 1992; MacKinnon, 1991; Yang *et al.*, 1995). In voltage-gated channels, previous studies have established that structural elements in the N-terminus just prior to S1 are critical for homotetrameric and heterotetrameric assembly of K$^+$ channel α-subunits (Li *et al.*, 1992; Shen *et al.*, 1993; Yu *et al.*, 1996). These elements are referred to as NAB1 or T1 and NAB2. They control the assembly between α-subunits from the same gene family (i.e., between Kv1.4 and Kv1.5 from the Kv1 gene family) (Li *et al.*, 1992; Shen and Pfaffinger, 1995; Yu *et al.*, 1996), between members of different gene families (i.e., between Kv2.1 and Kv9) (Stocker *et al.*, 1999), and between α- and β-subunits (Yu *et al.*, 1996). T1 elements are capable of forming tetrameric complexes when expressed as polypeptides. Their self-assembly depends primarily on electrostatic interactions for the *Shaker* channels (Kreusch *et al.*, 1998) and on an interpeptide Zn^{2+}-finger binding motif in the *Drosophila* equivalents of Kv2, Kv3, and Kv4 channels (Bixby *et al.*, 1999). However, it is clear that other channel regions also contribute to tetrameric assembly of some voltage-gated K$^+$ channels. For example, portions of the hydrophobic core of these proteins (i.e., S1 and S6 including the pore) (Hopkins *et al.*, 1994; Tu *et al.*, 1996; Ludwig *et al.*, 1997; Sheng *et al.*, 1997; Stocker *et al.*, 1999; Mathur *et al.*, 1999), as well as the C-terminal tails, may directly influence channel assembly.

In the inward rectifier K$^+$ channel Kir2.1, the control of homotetrameric and heterotetrameric assembly is linked to the M2 segment and a conserved region in the C-terminal domain (amino acids 220–300) (Tinker *et al.*, 1996). The intrasubfamily incompatibility for coassembly between Kir2.1 and Kir 2.2 is overcome by replacing either the M2 or the C-terminal hydrophilic domain of Kir2.1 with the corresponding region of Kir2.2, whereas intersubfamily incompatibility between IRK1 and ROMK1 (Kir1.1) is effectively overcome by substituting the C-terminal hydrophilic domain of IRK1 with that of ROMK1 (Tinker *et al.*, 1996). By contrast, the self-association of TWIK-1 channels depends on the presence of an extracellular disulfide bond between dimers in the P-loop region (Lesage *et al.*, 1996).

The process of assembling K$^+$ channels is clearly complex with a number of different motifs and elements being involved. This should not be too surprising, because the subunit assembly and folding of the *Shaker* channels are interspersed during biogenesis (Nagaya and Papazian, 1997). Therefore, interference with the assembly of functional channels may arise at many levels. The crystal structure of the KcsA channel reveals that the pore helices (see below) interdigitate with the M2 helices, resulting in multiple intersubunit interactions. This observation suggests that proper pore structure is necessary for appropriate subunit assembly, as previously observed.

5. ROLE OF K$^+$ CHANNEL PORES IN GATING AND BLOCK

In single-channel patch-clamp recordings (Hamill *et al.*, 1981), ion channels typically fluctuate stochastically between conducting or open state(s) and nonconducting state(s). These nonconducting states are classified as either closed (deactivated) or inactivated states. For voltage-gated K^+ channels, the probability of being in the open state (Po) is steeply voltage-dependent and is controlled by "voltage sensors" (Hodgkin and Huxley, 1952c) in the channel protein. In voltage-gated P-type channels, S4 contains 4–6 basic residues (Arg or Lys) at every third position and is essential for channel opening from the closed state in response to membrane depolarization (Noda et al., 1984; Catterall, 1986; Guy and Seetharamulu, 1986). Indeed, basic residues in S4 are known to change from intracellular to extracellular positions as a function of membrane potential (Liman et al., 1991; Yang and Horn, 1995; Larsson et al., 1996; Seoh et al., 1996; Yang et al., 1996) by moving within a "gating pore" (Yang et al., 1996). In the Shaker channel, activation appears to be directly coupled to channel opening by an opening of a "trap door" at the intracellular entrance to the channel pore (Holmgren et al., 1997; Liu et al., 1997). These observations are consistent with previous studies in K^+ channels showing that access to and from the binding site for tetraethylammonium (TEA), and its analogs was controlled by the activation gate (Armstrong, 1971). Recent results demonstrate that channel gating in the KcsA channel is associated with outward and inward movements of the M2 helix (Perozo et al., 1998, 1999), consistent with M2 forming not only part of the permeation pathway, but also the activation gate.

As part of their depiction of the gating process, Hodgkin and Huxley called the decrease of the Na^+ current during maintained depolarizations the "inactivation" process (Hodgkin and Huxley, 1952a). Subsequent studies established that inactivation could be eliminated by intracellular application of proteases, suggesting that a peptide ball stopped current flow by plugging the pore (i.e., the ball-and-chain mechanism) (Armstrong et al., 1973). Inactivation gating in K^+ channels is shown to depend critically on the presence of permeant ions and blockers within the pore as well as on pore structure. In voltage-dependent K^+ channels, at least two forms of inactivation have been identified: N-type and C-type. N-type inactivation depends on the first 20 amino acids at the N-terminal end of many voltage-gated K^+ channels (Hoshi et al., 1990). This N-terminal region has been shown to occlude the channel pore by interaction with the intracellular S4-S5 linker region (Isacoff et al., 1991; Holmgren et al., 1996; Murrell-Lagnado and Aldrich, 1993). Pore occlusion is impeded by binding of TEA^+ to the inner vestibule of the pore (Choi et al., 1991) and destabilized by external K^+ binding within the pore (Demo and Yellen, 1991), consistent with an electrostatic "knockoff" of the inactivation particle by charges within the pore (Armstrong, 1971).

C-type inactivation also depends critically on a number of pore properties (Yellen et al., 1994; Liu et al., 1996). Taken together, previous studies suggest that when ions and blockers occupy selected extracellular sites within the channel pore, they can prevent the conformational changes associated with C-type inactivation by a "foot-in-the-door" mechanism (MacKinnon and Yellen, 1990; Choi et al., 1991). Consistent with this idea, the open-state binding of TEA^+ binding at an extracellular position (i.e., T449) just beyond the selectivity filter in the Shaker channel and elevated extracellular K^+ both impede entry into this inactivated state from the open conformation. On the other hand, introducing a cysteine residue at the T449 site makes channels exquisitely sensitive to extracellular Cd^{2+} (i.e., $IC_{50} \approx 0.2 \ \mu M$) as a result of stabilization of the inactivated state by coordinated metal binding while having a much lower affinity in the open state. Also consistent with the foot-in-the-door mechanism is the observation that C-type inactivation is accelerated when TEA$^+$ binds to its internal pore binding site and cuts off access of internal K$^+$ to the external K$^+$ binding site, thereby preventing the conformational changes in the pore associated with C-type inactivation (Baukrowitz and Yellen, 1995, 1996). By contrast, reducing occupancy of an internal K$^+$ binding site by introducing a point mutation (A463C) in the *Shaker* channel, which lowers the internal K$^+$ binding affinity by 1000-fold, severely impairs C-type inactivation (Hurst *et al.*, 1995; Ogielska and Aldrich, 1998, 1999). This observation can be readily explained by lowered occupancy of the external K$^+$ binding site, which is expected to result from the reduced ion–ion interaction within the pore of multi-ion K$^+$ channels (see below). Interestingly, during C-type inactivation, the channel is still capable of conducting Na$^+$ but not K$^+$ (Starkus *et al.*, 1997), suggesting that there is localized conformational change rather than a general collapse of the pore structure.

6. MOLECULAR BASIS OF PERMEATION AND SELECTIVITY IN K$^+$ CHANNELS

6.1. Selectivity in K$^+$ Channels

Early channel pore models postulated that a "bracelet" of negatively charged or dipolar oxygen atoms provided the necessary electrostatic energy to replace the polarization energy in the aqueous phase (Benzanilla and Armstrong, 1972; Hille, 1973). In this model, the channel selects for ions on the basis of size and hydration energy. Large ions cannot enter the pore owing to steric factors, whereas the ions that are too small cannot optimally interact with appropriate numbers of surrogate pore oxygens to overcome the energy of interaction with dipolar water. Another feature of these models is their "hourglass" appearance, wherein the selectivity bracelet is much shorter than the width of the lipid bilayer and is connected to two large, water-filled vestibules located on either face of the channel pore (Hille, 1970). With this morphology, ions only need to diffuse a very short distance through a narrow constricted region of the channel pore, thereby minimizing resistance to ion flow. The hourglass appearance of K$^+$ channels is supported by subsequent mutagenesis and TEA$^+$ blocking studies of the *Shaker* K$^+$ channel (MacKinnon and Yellen, 1990; Yellen *et al.*, 1991), showing that more than 80% of the voltage drop across the pore occurs between the intracellular threonine 441 (internal TEA$^+$ binding site) and the extracellular threonine 449 (external TEA$^+$ binding site). These studies also were the first to demonstrate that this highly conserved eight amino acid segment (T^{441}TVGYGDMT^{449}) of the P-loop in *Shaker* K$^+$ channels is involved in forming the pore (Yellen *et al.*, 1991). Depending on the secondary structure of this region, these findings also suggest that the selectivity region may be as short as 1.2 nm in length for an α-helical structure or as long as 2.7 nm for a β-strand (Tomaselli *et al.*, 1993). Subsequent studies confirmed that P-loops play an essential role in pore formation (Hartmann *et al.*, 1991; Yool and Schwarz, 1991) and further showed that the "signature" sequence (GYG) is absolutely critical for K$^+$ selectivity (Heginbotham *et al.*, 1992, 1994). The crystal structure of the KcsA K$^+$channel (Doyle *et al.*, 1998) confirmed the critical role of this signature sequence in forming the channel pore, as predicted earlier from molecular modeling studies (Guy and Durell, 1995).

Indeed, the narrow, rigid selectivity filter of the channel is formed by the carbonyl oxygens in the peptide backbone of the GYG signature sequence, having a length of only about 1.2 nm. Difference electron density maps using crystals exchanged with the electron-dense alkali cations Rb^+ and Cs^+ showed that the selectivity filter contains two K^+ ions located 0.75 nm apart at either end of the narrow selectivity region and separated by a single water molecule. As predicted from earlier models, the diameter of the selectivity filter allows optimal coordination of water-stripped K^+ ions by the carbonyl oxygen atoms, thus compensating for the energy required for dehydration. A sheet of aromatic residues (Trp and Tyr) appears to create a rigid, open pore structure by stacking with the tyrosine in the GYG signature and forming a network of hydrophobic interactions. The rigidity of this selectivity filter ensures that smaller monovalent ions, like Li^+ and Na^+, are relatively impermeable by preventing the carbonyl oxygen atoms from collapsing onto the ion and thereby compensating for dehydration. Rigidity also prevents large ions like Cs^+ from traversing the pore owing to steric factors.

A tantalizing feature of K^+ channels is that ions do not move independently through the pore (Hodgkin and Keynes, 1955), as assessed using the measured K^+ influx: K^+ efflux ratio as a function of the electrochemical driving forces (Ussing, 1949). These studies suggest that, at any given moment, the K^+ channel pore accommodates about three ions that traverse the channel in single file. Direct evidence for the presence of three ions within the channel was observed in the KcsA channel, wherein two K^+ ions are located at either end of the selectivity filter and one is localized within the large aqueous cavity in the inner vestibule. This multi-ion feature readily accounts for many, often unique, properties of K^+ channels, such as intense rectification (Hille and Schwarz, 1978), complex voltage-dependent blocking properties due to interactions of blockers with permeant ions (Hille and Schwarz, 1978), and discrepancies in selectivity estimates based on flux versus reversal potential measurements (Hagiwara and Takahashi, 1974; Hagiwara et al., 1977).

6.2. The Inner Vestibule of the K^+ Channel Pore

Block from the cytoplasmic (internal) side of K^+ channels by various alkyl derivatives of TEA^+ reveals that potency increases strongly with hydrophobicity (Armstrong, 1971). For example, decyltriethylammonium ($C10$-TEA^+) blocks Shaker channels with a dissociation constant (K_d) of 0.75 μM compared with 390 μM for TEA^+ from the cytoplasmic side (Choi et al., 1993). Moreover, open-channel block by these positively charged molecules is very voltage-dependent, with block increasing strongly in response to membrane depolarization. Taken together, these results suggest that the internal vestibules are hydrophobic and penetrate deeply into the plane of the membrane (Armstrong, 1971; Armstrong and Hille, 1972; Hille, 1992). The crystal structure of the KcsA channels reveals that entry into the internal vestibule from the inner face begins with a relatively narrow passageway, 1.8 nm in length (the internal pore), lined by the hydrophobic M2 helix. This passage opens into a wide cavity (1.0 nm in diameter) strategically centered near the middle of the membrane, which is lined by a helical portion of the P-loop (the pore helix), the amino end of the selectivity filter, and the central portion of the hydrophobic M2 transmembrane helix. A K^+ ion is located at the center of this large cavity and is surrounded by 60 to 100 water molecules, thereby ensuring maximal polarizability at the bilayer center, where the cation's energy would otherwise be the highest. In addition, the dipolar axes of the pore helices are oriented toward the center of this cavity and contribute sufficient electrostatic energy to create a stable monovalent binding site (Roux and MacKinnon, 1999). Therefore, the structure of the inner vestibule is optimized not only to lower the natural electrostatic barrier to ion movement across the lipid bilayer but also to act as a reservoir for K$^+$ ion entry into the selectivity filter. The passage of K$^+$ ions into the central reservoir is thought to be greatly facilitated by the mainly hydrophobic lining of the narrow passageway, which has an inert binding-free surface, ensuring low resistance and rapid diffusion. The entrance of the narrow passage also is lined by negative charges, which will enhance the local permeant ion concentration.

An intriguing aspect of internal block by TEA$^+$ and its analogs in some voltage-gated K$^+$ channels is the observation that block requires the channel to open and that the blocking particles can become trapped following channel closure (Armstrong, 1971; Armstrong and Hille, 1972). Release of the trapped blocker requires reopening of the channel. This suggests that the binding site in these channels is only accessible when the channel opens and that the docking site is sufficiently deep within the channel to allow the "activation gate" to close around the blocker. Control of access by S6 residues lining the inner vestibule is further supported by eloquent mutagenesis studies in the *Shaker* channel, which also show that activation is largely unimpeded by the presence of blockers within the inner vestibule. The nature of the activation gating is unclear but may be related to the observation in KcsA channels that opening and closing involves the outward and inward movements of M2 (Perozo *et al.*, 1998, 1999). Based on the analogy between S6 and M2, these results suggest that the activation gating might involve similar movements of S6 which are likely to be coupled to voltage-dependent movements of S4 (Holmgren *et al.*, 1997; Liu *et al.*, 1997).

It seems likely that many aspects of inner vestibule structure in KcsA channels are conserved in other types of K$^+$ channels. Indeed, a number of residues in the transmembrane α-helix S6 of the *Shaker* channel, analogous to M2 of KcsA, also appear to line the inner vestibule of Kv channels (Liu *et al.*, 1997; Taglialatela *et al.*, 1994; Lopez *et al.*, 1994). The structure of the KcsA channel also suggests that the site of binding of many internal blockers of K$^+$ channels occurs within the large cavity of the inner vestibule. For example, most K$^+$ channel blockers are monovalent particles, and electrostatic stabilization in the large cavity is optimal for such ions (Roux and MacKinnon, 1999). Consistent with this assertion, quinidine block of K$^+$ channels is sensitive to residue replacements in S6 which are predicted, from the KcsA channel structure, to line the inner vestibule cavity (Yeola *et al.*, 1996). On the other hand, the rectification properties of some mammalian inward rectifier channels (e.g., Kir1.1 and Kir2.1) involve the block of polyamines that depends critically on a charged residue in a region of M2 (Wible *et al.*, 1994; Yang *et al.*, 1995) predicted to line the large cavity in the inner vestibule.

6.3. The Outer Vestibule of K$^+$ Channels

Block of various K$^+$ channels by TEA$^+$ and selected scorpion venom toxins has been used to define the surface topology of the K$^+$ channel's outer vestibule. Blockade by externally applied TEA$^+$ is voltage-independent and, unlike internal block, is not responsive to increased hydrophobicity of the blocking molecule, suggesting a more polar environment (Armstrong and Hille, 1972). Later "molecular caliper" studies involving simultaneous pairwise mutagenesis of residues in the channel and in rigid scorpion toxins, such as charybdotoxin and agitoxin-2, allowed complementary pairs of interacting residues to be identified using thermodynamic mutant cycle analysis (Hidalgo and MacKinnon, 1995; MacKinnon and Miller, 1989). From these studies, the relationships of various residues on the extracellular face of the channel to one another were determined. The external vestibule appears to form a shallow funnel with four-fold symmetry and possesses a hydrophilic, negatively charged surface. An identical molecular picture is seen in the KcsA crystallographic structure, where the vestibule is made up of contributions from the extracellular portion of M2, the pore helix, the P-loop turret, and the outer portion of the selectivity filter. This structure is likely applicable to other voltage-gated K^+ channels, because KcsA channels are made sensitive to block by the scorpion toxin agitoxin-2 by mutation of several residues to analogous residues found in toxin-sensitive *Shaker* channels (MacKinnon *et al.*, 1998). The shape of the external vestibule probably ensures easy escape of permeant ions from the narrow selectivity region to the extracellular solution (Hille, 1992). On the other hand, the charged groups in the outer vestibule likely contribute to the surface charge effects commonly observed in Na^+ (Hille *et al.*, 1975; Green *et al.*, 1987; Ravindran and Moczydlowski, 1989), Ca^{2+} (Cota and Stefani, 1984; Zhou and Jones, 1995), and K^+ (Elinder *et al.*, 1996) channels. For example, voltage-dependent gating and channel conductance are influenced by extracellular cation concentrations as a result of specific and nonspecific shielding of electrical fields originating from negatively charged surface residues. These charged residues within the vestibule also are expected to enhance the local permeant ion levels, thereby increasing channel conductance as well as possibly facilitating directional ion movement by electrostatic steering of ions, as seen in enzymes (Wade *et al.*, 1998).

7. LIMITATIONS OF THE KcsA STRUCTURE AND SOME UNANSWERED QUESTIONS

Based on the degree of P-loop sequence conservation and the striking similarities in the permeation properties among different K^+ channels, it seems likely that the pore structure determined from KcsA K^+ channels is representative of most, if not all, other K^+-selective channels (Doyle *et al.*, 1998; MacKinnon *et al.*, 1998). However, the known diversity of K^+ channel pore properties suggests that important fundamental differences in pore structure also must exist between different types of K^+ channels. For example, the unitary conductance of K^+ channels ranges between 10 and 300 pS, and K^+ channels display varying degrees of selectivity and sensitivity to K^+ channel drugs and blockers. For example, previous studies demonstrate that Kv2.1 has a 3-fold higher conductance than Kv1.2 channels but has a 20-fold higher affinity for TEA^+ block (Hartmann *et al.*, 1991), and these differences are linked to highly conservative residue replacements in the P-loop. In addition, unlike other K^+ channels, Kv2.1 channels become permeable to Na^+ in the absence of K^+ while becoming insensitive to TEA^+ block, suggesting large conformational changes in the pore (Kiss *et al.*, 1998; Immke *et* al., 1999). To complicate matters further, it now appears that channel gating can modulate the ion selectivity of Kv2.1 (Kiss et al., 1998) and *Shaker* channels (Starkus et al., 1997).

Crystallization of proteins is often performed under very unphysiological conditions, which may result in distortions of the protein. Indeed, close inspection of the KcsA crystal structure reveals the presence of hydrocarbon side chains in the inner tunnel, which is likely to preclude significant conductance. This observation may be related to the fact that the KcsA channel crystals were generated after removal of the N- and C-termini of the channel. In fact, the C-terminus, in KcsA channels, is known to influence the widening and narrowing of the inner tunnel as a result of M2 movements associated with channel opening and closing (Perozo et al., 1998, 1999). In other inward rectifier channels, these termini influence channel opening and block (Koster et al., 1999; Lee et al., 1999), suggesting that the published KcsA crystal structure corresponds to a closed conformation or state. In this regard, a recent study found that M1 helices of Kir2.1 channels are situated between two M2 helices, where they may participate in intersubunit interactions (Minor et al., 1999). By contrast, the M1 helices in KcsA channels are situated away from the pore and only contact the M2 within the same subunit. These results can be reconciled if the open state of the KcsA channel has a structure like that suggested for Kir2.1 channels whereas during the closed state, the transmembrane domain arrangements are as shown for the KcsA crystal structure.

Another limitation of X-ray crystal structures of proteins is the loss of dynamic information that is clearly important for understanding fundamental dynamic functional properties, like the permeation, selectivity, and gating of ion channels. In order to make the connection between the static structure and the functional properties, sophisticated dynamic computer modeling of these proteins is required (Dorman et al., 1996; Jakobsson, 1998; Jordan, 1990). These calculations are theoretically and computationally demanding endeavors, but, fortunately, the tools for performing these investigations have been developed (Jakobsson, 1998).

8. CONCLUSIONS

The recent crystallization of the KcsA channel provides the first definitive structure of a P-type ion channel. This structure readily explains a number of important functional properties of K$^+$ channels. For example, selectivity and multi-ion properties appear to originate from a narrow, rigid selectivity filter and the presence of three K$^+$ ion binding sites within the permeation pathway. This crystal structure will certainly be used in the design of future studies into ion channel structure and function, including mechanisms of ion channel block and gating. A reliable model of ion channel pore structure also has a number of important practical implications. For example, a number of aromatic pharmacological agents, used clinically, including local anesthetics, Ca^{2+} channel blockers, and quinidine, are known to bind to the inner face of channels in a voltage-dependent manner and to become trapped by channel closure (Ragsdale et al., 1991; Johnson et al., 1996; Yeola et al., 1996; Hockerman et al., 1997), thereby having effects similar to those of TEA$^+$ and its analogs on K$^+$ channels. Subsequent mutagenesis studies show that the binding of these drugs depends critically on aromatic hydrophobic residues located on the face of the S6 helix facing the pore (Yeola *et al.*, 1996; Peterson *et al.*, 1996, 1997; Zhang *et al.*, 1998). By analogy with the M2 of the KcsA channel, the residues involved in drug block are at positions predicted to line the large cavity of the internal vestibule, which coincide with the residue in Kir channels that is critical for Mg^{2+} and polyamine block (Roux and MacKinnon, 1999). The availability of a molecular structure of the KcsA channel pore will undoubtedly, if applicable to other ion channels, lead to novel insights and approaches for the development of new therapeutic agents targeting ion channels.

REFERENCES

Aguilar-Bryan, L., Nichols, C. G., Wechsler, S. W., Clement, J. P. T., Boyd, A. E., Gonzalez, G., Herrera-Sosa, H., Nguy, K., Bryan, J., and Nelson, D. A., 1995, Cloning of the beta cell high affinity sulfonylurea receptor: A regulator of insulin secretion, *Science* **268**:423–426.

Armstrong, C. M., 1971, Interaction of tetraethylammonium ion derivatives with the potassium channels of giant axons, *J. Gen. Physiol.* **58**:413–437.

Armstrong, C. M., and Hille, B., 1972, The inner quaternary ammonium ion receptor in potassium channels of the node of Ranvier, *J. Gen. Physiol.* **59**:388–400.

Baukrowitz, T., and Yellen, G., 1995, Modulation of K^+ current by frequency and external $[K^+]$: A tale of two inactivation mechanisms, *Neuron* **15**:951–960.

Baukrowitz, T., and Yellen, G., 1996, Use-dependent blockers and exit rate of the last ion from the multi-ion pore of a K^+ channel, *Science* **271**:653–656.

Bezanilla, F., and Armstrong, C. M., 1972, Negative conductance caused by entry of sodium and cesium ions into the potassium channels of squid axons, *J. Gen. Physiol.* **60**:588–608.

Billman, G. E., 1994, Role of ATP sensitive potassium channel in extracellular potassium accumulation and cardiac arrhythmias during myocardial ischemia, *Cardiovasc. Res.* **28**:762–769.

Bixby, K. A., Nanao, M. H., Shen, N. V., Kreusch, A., Bellamy, H., Pfaffinger, P. J., and Choe, S., 1999, Zn^{2+} binding and molecular determinants of tetramerization in voltage-gated K^+ channels, *Nat. Struct. Biol.* **6**:38–43.

Catterall, W. A., 1986, Molecular properties of voltage-sensitive sodium channels, *Annu. Rev. Biochem.* **55**:953–985.

Chandy, K. G., 1991, Simplified gene nomenclature [letter], *Nature* **352**:26.

Chandy, K. G., and Gutman, G. A., 1995, Voltage-gated K^+ channel genes, in: *Handbook of Receptors and Channels: Ligand- and Voltage-Gated Ion Channels* (R. A. North, ed.), CRC Press, Boca Raton, Florida, pp. 1–77.

Choi, K. L., Aldrich, R. W., and Yellen, G., 1991, Tetraethylammonium blockade distinguishes two inactivation mechanisms in voltage-activated K^+ channels, *Proc. Natl. Acad. Sci. U.S.A.* **88**:5092–5095.

Choi, K. L., Mossman, C., Aube, J., and Yellen, G., 1993, The internal quaternary ammonium receptor site of Shaker potassium channels, *Neuron* **10**:533–541.

Cota, G., and Stefani, E., 1984, Saturation of calcium channels and surface charge effects in skeletal muscle fibres of the frog, *J. Physiol. (London)* **351**:135–154.

Dorman, V., Partenskii, M. B., and Jordan, P. C., 1996, A semi-microscopic Monte Carlo study of permeation energetics in a gramicidin-like channel: The origin of cation selectivity, *Biophys. J.* **70**:121–134.

Doyle, D. A., Morais Cabral, J., Pfuetzner, R. A., Kuo, A., Gulbis, J. M., Cohen, S. L., Chait, B. T., and MacKinnon, R., 1998, The structure of the potassium channel: Molecular basis of K^+ conduction and selectivity [see comments], *Science* **280**:69–77.

Elinder, F., Madeja, M., and Arhem, P., 1996, Surface charges of K channels: Effects of strontium on five cloned channels expressed in *Xenopus* oocytes, *J. Gen. Physiol.* **108**:325–332.

Green, W. N., Weiss, L. B., and Andersen, O. S., 1987, Batrachotoxin-modified sodium channels in planar lipid bilayers: Ion permeation and block, *J. Gen. Physiol.* **89**:841–872.

Guy, H. R., and Durell, S. R., 1995, Structural models of Na$^+$, Ca^{2+}, and K$^+$ channels, *Soc. Gen. Physiol. Ser.* **50:**1–16.

Guy, H. R., and Seetharamulu, P., 1986, Molecular model of the action potential sodium channel, *Proc. Natl. Acad. Sci. U.S.A.* **83:**508–512.

Hagiwara, S., and Takahashi, K., 1974, The anomalous rectification and cation selectivity of the membrane of a starfish egg cell, *J. Membr. Biol.* **18:**61–80.

Hagiwara, S., Miyazaki, S., Krasne, S., and Ciani, S., 1977, Anomalous permeability of the egg cell membrane of a starfish in K$^+$–T1$^+$ mixtures, *J. Gen. Physiol.* **70:**269–281.

Hamill, O. P., Marty, A., Neher, E., Sakmann, B., and Sigworth, F. J., 1981, Improved patch-clamp techniques for high-resolution current recording from cells and cell-free membrane patches, *Pflügers Arch.* **391:**85–100.

Hartmann, H. A., Kirsch, G. E., Drewe, J. A., Taglialatela, M., Joho, R. H., and Brown, A. M., 1991, Exchange of conduction pathways between two related K$^+$ channels, *Science* **251:**942–944.

Heginbotham, L., Abramson, T., and MacKinnon, R., 1992, A functional connection between the pores of distantly related ion channels as revealed by mutant K$^+$ channels, *Science* **258:**1152–1155.

Heginbotham, L., Lu, A., Abramson, T., and MacKinnon, R., 1994, Mutations in the K$^+$ channel signature sequence, *Biophys. J.* **66:**1061–1067.

Heginbotham, L., Odessey, E., and Miller, C., 1997, Tetrameric stoichiometry of a prokaryotic K$^+$ channel, *Biochemistry.* **36:**10335–10342.

Hess, P., and Tsien, R. W., 1984, Mechanism of ion permeation through calcium channels, *Nature* **309:**453–456.

Hidalgo, P., and MacKinnon, R., 1995, Revealing the architecture of a K$^+$ channel pore through mutant cycles with a peptide inhibitor, *Science* **268:**307–310.

Hille, B., 1966, Common mode of three agents that decrease the transient change in sodium permeability in nerves, *Nature* **210:**1220–1222.

Hille, B., 1968, Pharmacological modifications of the sodium channels of frog nerve, *J. Gen. Physiol.* **51:**199–219.

Hille, B., 1970, Ionic channels in nerve membranes, *Prog. Biophys. Mol. Biol.* **21:**1–32.

Hille, B., 1973, Potassium channels in myelinated nerve: Selective permeability to small cations, *J. Gen. Physiol.* **61:**669–686.

Hille, B., 1992, *Ionic Channels of Excitable Membranes*, Sinauer Associates, Inc., Sunderland, Massachusetts.

Hille, B., and Schwarz, W., 1978, Potassium channels as multi-ion single-file pores, *J. Gen. Physiol.* **72:**409–442.

Hille, B., Woodhull, A. M., and Shapiro, B. I., 1975, Negative surface charge near sodium channels of nerve: Divalent ions, monovalent ions, and pH, *Trans. R. Soc. London* **270:**301–318.

Hockerman, G. H., Johnson, B. D., Abbott, M. R., Scheuer, T., and Catterall, W. A., 1997, Molecular determinants of high affinity phenylalkylamine block of L-type calcium channels in transmembrane segment IIIS6 and the pore region of the α1 subunit, *J. Biol. Chem.* **272:**18759–18765.

Hodgkin, A. L., and Huxley, A. F., 1952a, Currents carried by sodium and potassium ions through the membrane of the giant axon of *Loligo, J. Physiol. (London)* **116:**449–472.

Hodgkin, A. L., and Huxley, A. F., 1952b, Measurement of current–voltage relations in the membrane of the giant axon of *Loligo, J. Physiol. (London)* **116:**424–448.

Hodgkin, A. L. and Huxley, A. F., 1952c, A quantitative description of membrane current and its application to conduction and excitation in nerve, *J. Physiol. (London)* **117:**500–544.

Hodgkin, A. L., and Keynes, R. D., 1955, The potassium permeability of a giant nerve fibre, *J. Physiol. (London)* **128:**61–88.

Holmgren, M., Jurman, and M. E., Yellen, G. 1996, N-type inactivation and the S4–S5 region of the Shaker K$^+$ channel, *J. Gen. Physiol,* **108:**195–206.

Holmgren, M., Smith, P. L., and Yellen, G., 1997, Trapping of organic blockers by closing of voltage-dependent K$^+$ channels, *J. Neurosci.* **14:**1385–1393.

Hopkins, W. F., Demas, V., and Tempel, B. L., 1994, Both N- and C-terminal regions contribute to the assembly and functional expression of homo- and heteromultimeric voltage-gated K$^+$ channels, *J. Neurosci.* **14:**1385–1393.

Hoshi, T., Zagotta, W. N., Aldrich, R. W. 1990, Biophysical and molecular mechanism of *Shaker* potassium channel inactivation, *Science* **250:** 533–538.

Hurst, R. S., Latorre, R., Toro, L., and Stefani, E., 1995, External barium block of Shaker potassium channels: Evidence for two binding sites, *J. Gen. Physiol.* **106:**1069–1087.

Immke, D., Wood, M., Kiss, L., and Korn, S. J., 1999, Potassium-dependent changes in the conformation of the Kv2.1 potassium channel pore, *J. Gen. Physiol.* **113:**819–836.

Inagaki, N., Gonoi, T., Clement, J. P. T., Namba, N., Inazawa, J., Gonzalez, G., Aguilar-Bryan, L., Seino, S., and Bryan, J., 1995, Reconstitution of IK_{ATP}: An inward rectifier subunit plus the sulfonylurea receptor [see comments], *Science* 270:1166–1170.

Isacoff, E. Y., and Jan, Y. N., 1991, Putative receptor for the cytoplamic inactivation gate in the Shaker K^+ channel *Nature*, **353:**86–90.

Jakobsson, E., 1998, Using theory and simulation to understand permeation and selectivity in ion channels, *Methods* **14:**342–351.

Jan, L. Y., and Jan, Y. N., 1997, Cloned potassium channels from eukaryotes and prokaryotes, *Annu. Rev. Neurosci.* **20:**91–123.

Johnson, B. D., Hockerman, G. H., Scheuer, T., and Catterall, W. A., 1996, Distinct effects of mutations in transmembrane segment IVS6 on block of L-type calcium channels by structurally similar phenylalkylamines, *Mol. Pharmacol.* **50:**1388–1400.

Jordan, P. C., 1990, Ion–water and ion–polypeptide correlations in a gramicidin-like channel: A molecular dynamics study, *Biophys. J.* **58:**1133–1156.

Keating, M. T., and Sanguinetti, M. C., 1996, Molecular genetic insights into cardiovascular disease, *Science* **272:**681–685.

Ketchum, K. A., Joiner, W. J., Sellers, A. J., Kaczmarek, L. K., and Goldstein, S. A., 1995, A new family of outwardly rectifying potassium channel proteins with two pore domains in tandem, *Nature* **376:**690–695.

Koster, J. C., Sha, Q., Shyng, S., and Nichols, C. G., 1999, ATP inhibition of K_{ATP} channels: Control of nucleotide sensitivity by the N-terminal domain of the Kir6.2 subunit, *J. Physiol. (London)* **515:**19–30.

Kowey, P. R., Marinchak, R. A., Rials, S. J., and Bharucha, D., 1997, Pharmacologic and pharmacokinetic profile of class III antiarrhythmic drugs, *Am. J. Cardiol.* **80:**16–23.

Krapivinsky, G., Medina, I., Eng, L., Krapivinsky, L., Yang, Y., and Clapham, D. E., 1998, A novel inward rectifier K^+ channel with unique pore properties, *Neuron* **20:**995–1005.

Kreusch, A., Pfaffinger, P. J., Stevens, C. F., and Choe, S., 1998, Crystal structure of the tetramerization domain of the Shaker potassium channel, *Nature* **392:**945–948.

Kubo, Y., Baldwin, T. J., Jan, Y. N., and Jan, L. Y., 1993, Primary structure and functional expression of a mouse inward rectifier potassium channel, *Nature* **362:**127–133.

Larsson, H. P., Baker, O. S., Dhillon, D. S., and Isacoff, E. Y., 1996, Transmembrane movement of the Shaker K^+ channel S4, *Neuron* **16:**387–397.

Lee, J. K., John, S. A., and Weiss, J. N., 1999, Novel gating mechanism of polyamine block in the strong inward rectifier K channel Kir2.1, *J. Gen. Physiol.* **113:**555–564.

Lesage, F., Reyes, R., Fink, M., Duprat, F., Guillemare, E., and Lazdunski, M., 1996, Dimerization of TWIK-1 K^+ channel subunits via a disulfide bridge, *EMBO J.* **15:**6400–6407.

Li, M., Jan, Y. N., and Jan. L. Y., 1992, Specification of subunit assembly by the hydrophilic amino-terminal domain of the Shaker potassium channel, *Science* **257:**1225–1230.

Liman, E. R., Hess, P., Weaver, F., and Koren, G., 1991, Voltage-sensing residues in the S4 region of a mammalian K^+ channel, *Nature* **353:**752–756.

Liman, E. R., Tytgat, J., and Hess, P., 1992, Subunit stoichiometry of a mammalian K^+ channel determined by construction of multimeric cDNAs, *Neuron* **9:**861–871.

Lipkind, G. M., Hanck, D. A., and Fozzard, H. A., 1995, A structural motif for the voltage-gated potassium channel pore, *Proc. Natl. Acad. Sci. U.S.A.,* **92:**9215–9219.

Liu, Y., Jurman, M. E., and Yellen, G., 1996, Dynamic rearrangement of the outer mouth of a K^+ channel during gating. *Neuron* **16:**859–867.

Liu, Y., Holmgren, M., Jurman, M. E. and Yellen, G., 1997, Gated access to the pore of the voltage-dependent K^+ channel, *Neuron* **19:**175–184.

Lopez, G. A., Jan, Y. N., and Jan, L. Y., 1994, Evidence that the S6 segment of the Shaker voltage-gated K^+ channel comprises part of the pore, *Nature* **367:**179–182.

Ludwig, J., Owen, D., and Pongs, O., 1997, Carboxy-terminal domain mediates assembly of the voltage-gated rat ether-a-go-go potassium channel, *EMBO J.* **16:**6337–6345.

MacKinnon, R., 1991, Determination of the subunit stoichiometry of a voltage-activated potassium channel, *Nature* **350:**232–235.

MacKinnon, R., and Miller, C., 1989, Mutant potassium channels with altered binding of charybdotoxin, a pore-blocking peptide inhibitor, *Science* 245:1382–1385.

MacKinnon, R., and Yellen, G., 1990, Mutations affecting TEA blockade and ion permeation in voltage-activated K$^+$ channels, *Science* **250**:276–279.

MacKinnon, R., Cohen, S. L., Kuo, A., Lee, A., and Chait, B. T., 1998, Structural conservation in prokaryotic and eukaryotic potassium channels, *Science* **280**:106–109.

Mathur, R., Zhou, J., Babble, T., and Koren, G., 1999, Ile-177 and Ser-180 in the S1 segment are critically important in Kv1.1 channel function, *J. Biol. Chem.* **274**:11487–11493.

Minor, D. L., Jr., Masseling, S. J., and Jan, Y. N., and Jan, L. Y., 1999, Transmembrane structure of an inwardly rectifying potassium channel, *Cell* **96**:879–891.

Moczydlowski, E., 1998, Chemical basis for alkali cation selectivity in potassium-channel proteins, *Chem. Biol.* **5**:R291-R301.

Murrell-Lagnado, R. D., Aldrich, R. W. 1993, Interactions of amino terminal domains of Shaker K channels with a pore blocking site studied with synthetic peptides, *J. Gen. Physiol.* **102**:949–975.

Nagaya, N., and Papazian, D. M., 1997, Potassium channel alpha and beta subunits assemble in the endoplasmic reticulum, *J. Biol. Chem.* **272**:3022–3027.

Nakamura, Y., Nakajima, S., and Grundfest, H., 1965, The action of tetrodotoxin on electrogenic components of squid giant axons, *J. Gen. Physiol.* **48**:985–996.

Neyton, J., and Miller, C., 1988, Discrete Ba^{2+} as a probe of ion occupancy and pore structure in the high-conductance Ca^{2+}-activated K$^+$ channel, *J. Gen. Physiol.* **92**:569–586.

Nichols, C. G., and Lopatin, A. N., 1997, Inward rectifier potassium channels, *Annu. Rev. Physiol.* **59**:171–191.

Noda, M., Shimizu, S., Tanabe, T., Takai, T., Kayano, T., Ikeda, T., Takahashi, H., Nakayama, H., Kanaoka, Y., Minamino, N., *et al.*, 1984, Primary structure of *Electrophorus electricus* sodium channel deduced from cDNA sequence, *Nature* **312**:121–127.

Ogielska, E. M., and Aldrich, R. W., 1998, A mutation in S6 of Shaker potassium channels decreases the K$^+$ affinity of an ion binding site revealing ion–ion interactions in the pore, *J. Gen. Physiol.* **112**:243–257.

Ogielska, E. M., and Aldrich, R. W., 1999, Functional consequences of a decreased potassium affinity in a potassium channel pore: Ion interactions and C-type inactivation, *J. Gen. Physiol.* **113**:347–358.

Papazian, D. M., 1999, Potassium channels: Some assembly required, *Neuron* **23**:7–10.

Papazian, D. M., Schwarz, T. L., Tempel, B. L., Jan. Y. N., and Jan, L. Y., 1987, Cloning of genomic and complementary DNA from *Shaker*, a putative potassium channel gene from *Drosophila*, *Science* **237**:749–753.

Parsegian, V. A., 1975, Ion–membrane interactions as structural forces, *Ann. N.Y. Acad. Sci.* **264**:161–171.

Perozo, E., Cortes, D. M., and Cuello, L. G., 1998, Three-dimensional architecture and gating mechanism of a K$^+$ channel studied by EPR spectroscopy, *Nat. Struct. Biol.* **5**:459–469.

Perozo, E., Cortes, D. M., and Cuello, L. G., 1999, Structural rearrangements underlying K$^+$ channel activation gating, *Science* **285**:73–78.

Peterson, B. Z., Tanada, T. N., and Catterall, W. A., 1996, Molecular determinants of high affinity dihydropyridine binding in L- type calcium channels, *J. Biol. Chem.* **271**:5293–5296.

Peterson, B. Z., Johnson, B. D., Hockerman, G. H., Acheson, M., Scheuer, T., and Catterall, W., 1997, Analysis of the dihydropyridine receptor site of L-type calcium channels by alanine-scanning mutagenesis, *J. Biol. Chem.* **272**:18752–18758.

Ragsdale, D. S., Scheuer, T., and Catterall, W. A., 1991, Frequency and voltage-dependent inhibition of type IIA Na$^+$ channels, expressed in a mammalian cell line, by local anesthetic, antiarrhythmic, and anticonvulsant drugs, *Mol. Pharmacol.* **40**:756–765.

Ravindran, A., and Moczydlowski, E., 1989, Influence of negative surface charge on toxin binding to canine heart Na channels in planar bilayers, *Biophys. J.* **55**:359–365.

Reuveny, E., Slesinger, P. A., Inglese, J., Morales, J. M., Iniguez-Lluhi, J. A., Lefkowitz, R. J., Bourne, H. R., Jan, Y. N., and Jan, L. Y., 1994, Activation of the cloned muscarinic potassium channel by G protein $\beta\gamma$ subunits, *Nature* **370**:143–146.

Roux, B., and MacKinnon, R., 1999, The cavity and pore helices in the KcsA K$^+$ channel: Electrostatic stabilization of monovalent cations, *Science* **285**:100–102.

Salinas, M., Duprat, F., Heurteaux, C., Hugnot, J. P., and Lazdunski, M., 1997, New modulatory α subunits for mammalian Shab K$^+$ channels, *J. Biol. Chem.* **272**:24371–24379.

Salkoff, L., Baker, K., Butler, A., Covarrubias, M., Pak, M. D., and Wei, A., 1992, An essential 'set' of K$^+$ channels conserved in flies, mice and humans, *Trends Neurosci.* **15**:161–166.

Seoh, S. A., Sigg, D., Papazian, D. M., and Bezanilla, F., 1996, Voltage-sensing residues in the S2 and S4 segments of the Shaker K$^+$ channel, *Neuron* **16**:1159–1167.

Shen, N. V., and Pfaffinger, P. J., 1995, Molecular recognition and assembly sequences involved in the subfamily- specific assembly of voltage-gated K$^+$ channel subunit proteins, *Neuron* **14**:625–633.

Shen, N. V., Chen, X., Boyer, M. M., and Pfaffinger, P. J., 1993, Deletion analysis of K$^+$ channel assembly, *Neuron* **11**:67–76.

Sheng, Z., Skach, W., Santarelli, V., and Deutsch, C., 1997, Evidence for interaction between transmembrane segments in assembly of Kv1.3, *Biochemistry* **36**:15501–15513.

Silvestry, F. E., and Kimmel, S. E., 1996, Calcium-channel blockers in ischemic heart disease, *Curr. Opin. Cardiol.* **11**:434–439.

Singh, B. N., 1999, Current antiarrhythmic drugs: An overview of mechanisms of action and potential clinical utility, *J. Cardiovasc. Electrophysiol.* **10**:283–301.

Starkus, J. G., Kuschel, L., Rayner, M. D., Heinemann, and S. H., 1997, Ion conduction through C-type inactivated Shaker channels, *J. Gen. Physiol.* **110**:539–550.

Stocker, M., Hellwig, M., and Kerschensteiner, D., 1999, Subunit assembly and domain analysis of electrically silent K$^+$ channel α-subunits of the rat Kv9 subfamily, *J. Neurochem.* **72**:1725–1734.

Taglialatela, M., Champagne, M. S., Drewe, J. A., and Brown, A. M., 1994, Comparison of H5, S6 and H5–S6 exchanges on pore properties of voltage-dependent K$^+$ channels, *J. Biol. Chem.* **269**:13867–13873.

Tinker, A., Jan. Y. N., and Jan, L. Y., 1996, Regions responsible for the assembly of inwardly rectifying potassium channels, *Cell* **87**:857–868.

Tomaselli, G. F., Backx, P. H. and Marban, E., 1993, Molecular basis of permeation in voltage-gated ion channels, *Circ. Res.* **72**:491–496.

Tu, L., Santarelli, V., Sheng, Z., Skach, W., Pain, D., and Deutsch, C., 1996, Voltage-gated K$^+$ channels contain multiple intersubunit association sites, *J. Biol. Chem.* **271**:18904–18911.

Ussing, H. H., 1949, The distinction by means of tracers between active transport and diffusion: The transfer of iodide across the isolated frog skin, *Acta Physiol. Scand.* **19**:43–56.

Wible, B. A., Taglialatela, M., Ficker, E., and Brown, A. M., 1994, Gating of inwardly rectifying K$^+$ channels localized to a single negatively charged residue, *Nature* **371**:246–249.

Wickman, K. D., Iniguez-Lluhl, J. A., Davenport, P. A., Taussig, R., Krapivinsky, G. B., Linder, M. E., Gilman, A. G., and Clapham, D. E., 1994, Recombinant G-protein βγ-subunits activate the muscarinic-gated atrial potassium channel, *Neuron* **16**:113–122.

Yang, J., Jan, Y. N., and Jan, L. Y., 1995, Control of rectification and permeation by residues in two distinct domains in an inward rectifier K$^+$ channel, *Neuron* **14**:1047–1054.

Yang, N., and Horn, R., 1995, Evidence for voltage-dependent S4 movement in sodium channels, *Neuron* **15**:213–218.

Yang, N., George, A. L., Jr., and Horn, R., 1996, Molecular basis of charge movement in voltage-gated sodium channels, *Neuron* **16**:113–122.

Yellen, G., Jurman, M. E., Abramson, T., and MacKinnon, R., 1991, Mutations affecting internal TEA blockade identify the probable pore-forming region of a K$^+$ channel, *Science* **251**:939–942.

Yellen, G., Sodickson, D., Chen, T. Y., and Jurman, M. E., 1994, An engineered cysteine in the external mouth of a K$^+$ channel allows inactivation to be modulated by metal binding, *Biophys. J.* **66**:1068–1075.

Yeola, S. W., Rich, T. C., Uebele, V. N., Tamkun, M. M., and Snyders, D. J., 1996, Molecular analysis of a binding site for quinidine in a human cardiac delayed rectifier K$^+$ channel: Role of S6 in antiarrhythmic drug binding, *Circ. Res.* **78**:1105–1114.

Yool, A. J., and Schwarz, T. L., 1991, Alteration of ionic selectivity of a K$^+$ channel by mutation of the H5 region, *Nature* **349**:700–704.

Yu, W., Xu, J., and Li, M., 1996, NAB domain is essential for the subunit assembly of both alpha–alpha and alpha–beta complexes of Shaker-like potassium channels, *Neuron* **16**:441–453.

Zhang, H., Zhu, B., Yao, J. A., and Tseng, G. N., 1998, Differential effects of S6 mutations on binding of quinidine and 4-aminopyridine to rat isoform of Kv1.4: Common site but different factors in determining blockers' binding affinity, *J. Pharmacol. Exp. Ther.* **287**:332–343.

Zhou, W., and Jones, S. W., 1995, Surface charge and calcium channel saturation in bullfrog sympathetic neurons, *J. Gen. Physiol.* **105**:441–462.

Chapter 3

Molecular Biology of Voltage-Gated K^+ Channels

Olaf Pongs

1. INTRODUCTION

Voltage-gated K^+ (Kv) channels may be assembled from various subunits as homo- or heteromultimers. The pore-forming α-subunits are integral membrane proteins, which express functional tetrameric Kv channels in heterologous expression systems. Three main families encoding Kv channel α-subunits have been detected related to the *Drosophila* genes *Shaker* and *ether-a-go-go* and the human KvLQT1 (KCNQ1) gene. Members of each family contribute to cardiac Kv channels and to cardiac action potential repolarization. Auxiliary subunits do not express functional Kv channels by themselves. They associate with α-subunits and may modulate Kv channel properties, including voltage dependence of activation and inactivation, deactivation, single-channel conductance, recovery from inactivation, and pharmacology. Auxiliary β-subunits have a structure which suggests that they may function as NADPH-dependent oxidoreductases. Whether this putative enzymatic activity is independent of the association of β-subunits with the pore-forming α-subunits is not known. Auxiliary γ-subunits are similar in sequence and topology to *Shaker*−related α-subunits but yield functional Kv channels only when coexpressed with certain α-subunits. In most cases, however, the exact subunit compositions of native Kv channels have not been elucidated. Therefore, it is still difficult to know which of the cloned Kv channels contribute to the different components of outward K^+ current in cardiac myocytes. In only a few cases has the combination of human genetics, molecular biology, electrophysiology, and pharmacology provided a clear-cut identification of the α and auxiliary subunits that contribute to native K^+ currents.

Olaf Pongs • Institut für Neurale Signalverarbeitung, Zentrum fr Molekulare Neurobiologie, 20251 Hamburg, Germany.

Potassium Channels in Cardiovascular Biology, edited by Archer and Rusch. Kluwer Academic/Plenum Publishers, New York, 2001.

2. BACKGROUND

Kv channels play important roles in many aspects of cardiac action potential generation. For example, Kv channel activities are involved in setting and varying the resting membrane potential, in shaping action potential waveforms, and in determining action potential frequencies (Hille, 1992). Recently, remarkable progress has been made in the molecular physiology of cardiac K^+ channels. Thanks to a happy marriage of human genetic and molecular biology, human Kv channel genes have been identified whose mutations are associated with inherited cardiac arrhythmias. These important findings have made it possible to correlate Kv channel dysfunction with the occurrence of certain heart diseases related to the long QT (LQT) syndrome (Attali, 1996). Furthermore, they have significantly advanced our understanding of the contributions of two Kv channel subtypes, the rapidly inactivating K^+ current (I_{Kr}) and the slowly inactivating K^+ current (I_{Ks}), to cardiac action potential waveforms and frequency regulation.

The Kv channels underlying I_{Kr} and I_{Ks}, respectively, have been cloned and characterized *in vitro* in heterologous expression systems (Barhanin *et al.*, 1996; Sanguinetti *et al.*, 1995, 1996; Abbott *et al.*, 1999). These data, in conjunction with human genetics and cardiac electrophysiology, have made it possible to make clear-cut correlations between Kv channel genotype (i.e., mutations in the KCNQ1/KCNE1 genes encoding KvLQT1 channels or in the KCNH2/KCNE2 genes encoding HERG channels) and the occurrence of LQT syndromes. I_{Kr} and I_{Ks} are not the only Kv channel-related currents that have been recorded from cardiac tissue (Barry and Nerbonne, 1996). Rather, many cloned Kv channel α- and β-subunits, for example, KCNA2, KCNA4, KCNA5, KCNB1, KCND3, KCNG3, Helk2, Helk3, Kvβ1.2, and Kvβ1.3 (England *et al.*, 1995; Dixon *et al.*, 1996; Engeland *et al.*, 1998), have been detected in the human myocardium by molecular biological techniques. However, it has not yet been possible to associate mutations in these Kv channel genes with human diseases, although some correlations are likely. For example, KCND3 expressed *in vitro* mediates a voltage-activated, rapidly inactivating transient outward current reminiscent of I_{to} the transient outward cardiac K^+ current (Kong *et al.*, 1998). I_{to} presents a prominent K^+ outward current that makes an important contribution to the early repolarization phase of the cardiac action potential. Probably, KCND3 channels mediate I_{to} in human cardiac cells (Kong *et al.*, 1998; Zhu *et al.*, 1999a). A direct link between KCND3 channel activity and I_{to} may come from mutational analysis of the KCND3 gene.

Extrapolations from the properties of cloned Kv channels to cardiac outward currents are often difficult and not as straightforward as one would like. Unfortunately, there generally is either insufficient physiological information or insufficient structural knowledge, or both, for an unambiguous correlation between a certain cardiac outward current and a cloned human Kv channel subunit. To understand the precise function of cloned Kv channel subunits in cardiac action potential generation, the exact subunit composition of the Kv channels in the different cardiac cells must be defined. In the absence of this knowledge, we have a few facts, but mainly fiction, about many Kv channel subunits and their role in cardiac physiology. In this chapter, I will try to make some educated guesses regarding the physiological role of certain Kv channel subunits, particularly the auxiliary β- and γ-subunits. For a more detailed discussion of the structure and general function of Kv channels, I refer the reader to the many excellent reviews on this topic (e. g., Chandy and Gutman, 1995; Jan and Jan, 1997; Yellen, 1998).

3. MIX AND MATCH OF CARDIAC VOLTAGE-GATED K$^+$ CHANNEL SUBUNITS

Extensive studies including cDNA cloning, reverse-transcription polymerase chain reaction (RT-PCR), ribonuclease protection assays, *in situ* hybridization, and immunocytochemistry have been employed to study Kv channel mRNA and protein expression in cardiac cells (Deal *et al.*, 1996; Kong *et al.*, 1998; Zhu *et al.*, 1999b; Pongs, 1999; Dixon *et al.*, 1996; Barry *et al.*, 1995). The results of these studies should be regarded with some caution. First, most data have been obtained with RNA preparations from crude atrial and/or ventricular tissue samples, which inadvertently, may not have contained a homogeneous cell population. For example, ventricular cells isolated from the cardiac endocardium and those isolated from the epicardium are significantly different in their electrophysiological properties, including the density of I_{to}-like currents (Nabauer *et al.*, 1996; Wickenden *et al.*, 1999). Also, RNA preparations from dissociated and cultivated myocardial cells are problematic. They most likely differ in their expression profile, which may have been altered by the experimental manipulations. Second, cardiac action potential waveforms differ markedly among animal species. This difference is functionally significant and probably is correlated also with the expression of a different set of Kv channel mRNAs and Kv channel subunits or with the expression of different amounts of the same set of Kv channels (Dixon *et al.*, 1996). It certainly is useful to study cardiac Kv channels in the mouse, but it should be kept in mind that the physiology of human heart differs from that of mouse heart. Third, the hormonal and metabolic status of the host animal may have an important influence on Kv channel expression, both at the level of mRNA synthesis, transport, and translation and at the level of Kv channel activity at the cellular surface membrane. For example, it has been shown that thyroid status and diabetes modulate the expression of Kv currents in rat ventricle (Shimoni *et al.*, 1992).

With these cautionary notes in mind, a list of putative cardiac Kv channel mRNAs is presented in Table 1. It is not clear how accurate or incomplete this list is. Also, it is uncertain how mRNA levels correspond to the levels of Kv channel subunit expression in the plasma membrane. Nevertheless, most Kv channel subunit mRNAs in this table probably contribute to the various outward K$^+$ currents that have been described electrophysiologically in cardiac cells (Table 2). However, it appears that there are more cardiac Kv channel subunits expressed than there are currents. This demonstrates the difficulty in correlating gene expression with a particular K$^+$ current, especially when the underlying Kv channels are heteromultimeric complexes. Note that some of the mRNAs listed in Table 1 may not give rise to atrial or ventricular Kv channels, but may rather have originated from contaminating tissues. In this respect, expression profiles from single electrophysiologically identified cardiac cells may be useful as an alternative technique for Kv channel mRNA analysis. Also, the delivery of dominant-negative constructs to cardiac muscle, and their subsequent expression, may be a

Table 1
Kv Channel Subunit mRNAs Detected in Cardiac Tissue[a]

α-Subunits		Auxiliary subunits
Kv1.2 (KCNA2)		Kv6.2 (KCNG3)
		Kvβ1.2 (KCNA1B)
Kv1.4 (KCNA4)	KvLQT1 (KCNQ1)	Kvβ1.3 (KCNA1B)
Kv1.5 (KCNA5)	HERG1 (KCNH2)	MinK (KCNE1)
Kv2.1 (KCNB1)	Helk2	MiRP1 (KCNE2)
Kv4.1 (KCND1)	Helk3	
Kv4.2 (KCND2)		
Kv4.3 (KCND3)		

[a] For references, see discussion of each Kv-channel subunit in the text. Nomenclature in parentheses refers to the human Kv channel genes.

powerful alternative strategy for determining the molecular nature of native currents in the heart (Johns *et al.*, 1997; London *et al.*, 1998; Barry *et al.*, 1998).

The combination of molecular genetics, electrophysiological phenotypes, and a specific pharmacology may be regarded as the magic triad for correlating Kv channel clones with physiological function (Fig. 1). In this respect, the correlation of KCNQ1/KCNE1 channels with the delayed rectifier currents that generate I_{Ks} may serve as a representative example (Attali, 1996; Barhanin *et al.*, 1996; Sanguinetti *et al.*, 1996). The LQT phenotype is correlated with a delayed repolarization phase of the cardiac action potential (for details, see Chapters 17 and 36). Patients suffering from a LQT1 syndrome may have mutations in their KCNQ1 or KCNE1 genes leading to the generation of dysfunctional Kv channel subunits. KCNQ1 and KCNE1 proteins are

Table 2
Cardiac Kv Channels: Facts and Fiction

Kv channel	Related disease	Native current candidate[a]
KCND1/KCND3[b]	?	$I_{to\,f}$
KCNA2/KCNA5/KCNA1B[b]	?	I_{Kur}, $I_{K,slow}$
KCNB1/KCNG3[b]	?	$I_{K,ss}$; $I_{K,slow}$
KCNH2/KCNE2[c]	Long QT syndrome	I_{Kr}
KCNQ1/KCNE1[c]	Long QT syndrome	I_{Ks}
KCNA4/KCNA1B[b]	?	$I_{to,s}$
HELK2/HELK3[b]	?	I_b

[a] $I_{to,f}$, Fast transient outward current; $I_{to,s}$, slow transient outward current; I_{Kur}, ultrarapid activating, slowly inactivating outward current; I_{Kr}, rapid I_K; I_{Ks}, slow I_K; $I_{K,slow}$, slowly activating and slowly inactivating outward current; I_{Kss}, sustained steady-state outward current; I_b, background current.

[b] Correlations are circumstantial or even speculative. It is assumed that the underlying Kv channels are heteromultimers. The assumption that Kv channel subunits form heteromultimers has been inferred from heterologous expression experiments.

[c] Convincing evidence has been published that these Kv channels are the molecular correlates for I_{Kr} and I_{Ks} (for references, see text).

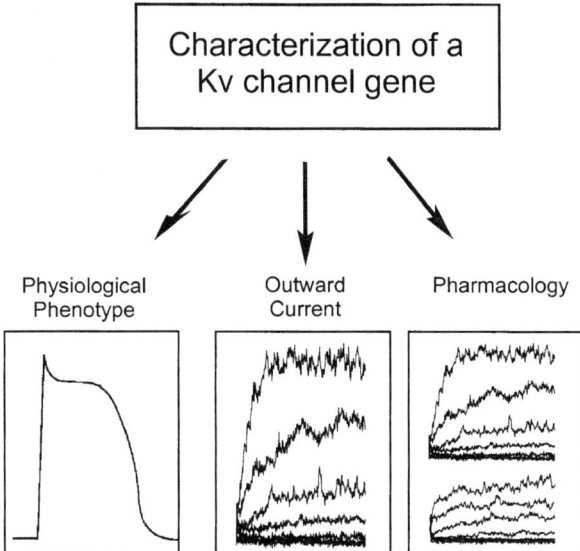

Figure 1. Characterization of a Kv channel gene. Many Kv channel subunit genes have been cloned, but only rarely has it been possible to correlate a physiological phenotype to mutation(s) in a Kv channel gene, the underlying electrophysiologically recorded outward current, and a specific Kv channel pharmacology. In the case of cardiac Kv channels, such an unambiguous situation may apply for Kv channels associated with a long QT syndrome (e.g., HERG/MiRP and KvLQT1/MinK).

subunits of the same Kv channel (Barhanin *et al.*, 1996; Sanguinetti *et al.*, 1996). When expressed *in vitro*, the KCNQ1/KCNE1 channels mediate slowly activating voltage-dependent currents having I_{Ks} (slow I_K)-like properties. Also, both native I_{Ks} and the *in vitro* expressed KCNQ1/KCNE1 channels are blocked by the same drug antagonists. Thus, it is very likely that KCNQ1/KCNE1 channels are the molecular correlate to I_{Ks}. Note, however, that these findings do not exclude the possibility that I_{Ks} channels may contain other auxiliary subunits, such as the Kvβ subunits of *Shaker*-related channels. Similarly, the combination of results obtained from analyzing gene mutations, *in vitro* expression, and pharmacology have demonstrated that HERG/MiRP1 channels mediate ventricular I_{Kr} (rapid I_K) (Sanguinetti *et al.*, 1995; Abbott *et al.*, 1999). Again, I_{Kr} channels mediating I_{Kr} may well contain other subunits than HERG and MiRP1 that have not been identified yet. Note that auxiliary subunits, including the β-subunit, were only found by biochemical purification of Kv channels (Scott *et al.*, 1994). In fact, this method still may be the only approach for learning the genuine subunit compositions of ionic channels.

In addition to I_{Kr} and I_{Ks}, some other outward K$^+$ currents have been recorded from atrial and ventricular myocytes, including an ultrarapid I_{Kr} component (I_{Kur}), other slow I_K components ($I_{K,slow1}$, $I_{K,slow2}$), a sustained steady-state outward current (I_{Kss}), a fast inactivating component ($I_{to,f}$), and a slow inactivating component ($I_{to,}{}^s$) (Hille, 1992; Giles and Imaizumi, 1988; Deal *et al.*, 1996; Nabauer *et al.*, 1996; Kong *et al.*, 1998; Xu *et al.*, 1999). Recently, a transgenic approach has significantly advanced our knowledge about possible molecular Kv channel correlates of the diverse outward

K^+ currents (see Table 2). In this approach, constructs under the control of a heart-specific promoter have been used to express mutated or truncated Kv channel subunits. These mutant α-subunits inhibit the function of native Kv channels in a dominant-negative manner (i.e., the incorporation of even a single mutant α-subunit into a tetramer impairs or destroys channel function) (Johns *et al.*, 1997; London *et al.*, 1998; Barry *et al.*, 1998). The results of these studies may be regarded with some caution. For example, ectopic expression of mutant Kv channel subunits may influence the expression of other unrelated Kv channel subunits. Because the ectopic expression usually generates a disease status, the properties and the densities of voltage-gated K^+ currents may change, as is known for cardiovascular disease states (Boyden and Jeck, 1995; Van Wagoner *et al.*, 1997; Näbauer and Kääb, 1998).

As summarized in Table 2, Kv channels containing Kv1.4 subunits may mediate a slow component of the transient outward current I_{to} (Wang *et al.*, 1999). Considering the observation that low stimulation frequencies can lead to permanent inactivation of homomeric Kv1.4 channels (Roeper *et al.*, 1997), it is very likely that the slow inactivating component ($I_{to,}$) of outward current in the heart may represent subunits other than Kv1.4. Possible[s] candidates are Kv1.2 as well as the auxiliary subunits, Kvβ1.2 or Kvβ1.3. In this regard, assembly with other subunits to form heteromultimeric channels would affect (i) the time constant of inactivation; (ii) the steady-state current component; and (iii) the type and kinetics of inactivation, resulting in reduced cumulative inactivation. Given the discrepancies between Kv1. 4 channel properties and I_{to}s, the channels that contribute to $I_{to,s}$ most likely are heteromultimers that contain additional subunits besides Kv1.4.

The ultrarapidly inactivating and slowly inactivating outward K^+ current components, I_{Kur} and $I_{K,slow}$ respectively, have been associated with the occurrence of Kv1.2, Kv1.5, and Kvβ1. 3 subunits in cardiac myocytes (Deal *et al.*, 1996; London *et al.*, 1998; Feng *et al.*, 1997; Kwak *et al.*, 1999). Attenuation of Kv1.2 and Kv1.5 expression has been correlated with a reduction in I_{Kur} and $I_{K,slow}$. Apparently, the attenuation leads to a prolonged action potential resulting in prolonged QT intervals, increased frequencies of premature ventricular beats, and arrhythmias. Yet these correlations require further experimentation. In particular, the question of which types of heteromultimeric Kv1 channels occur in atrial and ventricular myocytes needs to be explored. Furthermore, investigations should be undertaken to determine how the subunit compositions of Kv1 channels vary between particular cardiac tissues.

The transgenic approach also has been used to study the possible role of Kv2.1 subunits (Xu *et al.*, 1999a). Expressing a truncated, dominant-negative Kv2.1 subunit in mouse heart leads to an attenuation of $I_{K,slow}$ as well as I_{Kss}. The former current is slowly inactivating, whereas the latter is noninactivating. A simple explanation of these observations might be that Kv channels containing Kv2.1 subunits indeed contribute to both $I_{K,slow}$ and $I_{K,ss}$. However, the channels clearly must be different in order to mediate different types of outward K^+ current. Probably, Kv2.1 subunits are associated with other subunits to form heteromultimeric Kv channels with different biophysical properties. Indeed, mouse atrial cells express an auxiliary γ-subunit (Kv6.2, see below) which specifically assembles with Kv2.1 subunits (Zhu *et al.*, 1999). The assembly confers noninactivating properties to the channel. Thus, it might be possible that Kv2.1 subunits participate in the formation of $I_{K,slow}$ as well as of $I_{K,ss}$ channels, depending on their association with other auxiliary subunits.

4. AUXILIARY VOLTAGE-GATED K$^+$ CHANNEL SUBUNITS

Auxiliary Kv channel subunits may be defined as subunits that do not form Kv channels by themselves in heterologous *in vitro* expression systems. Rather, they specifically associate with Kv channel α-subunits. Auxiliary subunits may increase or decrease Kv channel surface expression or influence biophysical properties such as voltage dependence of gating, activation, inactivation, and deactivation kinetics, and conductance, as well as pharmacological properties. To date, three different types of auxiliary Kv channel subunits have been described: Kvβ subunits, Kvγ subunits, and MinK-related subunits. MinK subunits are integral membrane proteins containing one membrane-spanning segment. MinK (KCNE1) associates specifically with KCNQ1 subunits (Barhanin *et al.*, 1996; Sanguinetti *et al.*, 1996), and MiRP1 (KCNE2) with KCNH2 subunits (Abbott *et al.*, 1999). Their influence on Kv channel pore and gating properties is discussed in detail in Chapter 36.

The Kvγ subunits also are membrane-integrated proteins. They exhibit a topology and sequence similar to those of the Kv α-subunits of the *Shaker* superfamily (Robertson, 1997). Thus, Kvγ subunits appear to have six hydrophobic, most likely membrane-spanning, segments (S1--S6) and a membrane-inserted domain (called P or H5) between segments S5 and S6, which participates in pore formation. Typically, the P-domain contains a K$^+$ channel "signature" sequence –GYGD–, which plays an important role in the selectivity of K$^+$ channels for the potassium ion relative to other cations (Heginbotham *et al.*, 1994). The membrane-spanning core region is flanked by hydrophobic amino- and carboxyl-terminal sequences facing the cytoplasmic side of the plasma membrane. The amino terminus contains a tetramerization domain (T1), which determines the specificity of subunit assembly and participates in subunit assembly (Papazian, 1999). An important feature of T1-domains is that they can interact with each other and self-tetramerize. These interactions appear to be subfamily-specific (Kreusch *et al.*, 1998; Bixby *et al.*, 1999). Therefore, it has been proposed that the T1-domain structure determines which α-subunits may coassemble and which ones may not. Kv subunits seem to contain T1-domain variants that cannot interact with each other (Zhu *et al.*, 1999; Robertson, 1997). In contrast, Kvγ T1-domains may interact and assemble with the T1-domains of Kv2.1 and Kv2.2 α-subunits. Amino-terminal sequences between the T1-domain and the first transmembrane segment S1 may be responsible for the influence of Kvγ subunits on Kv channel gating behavior. Thus, the resulting Kvα/Kvγ heteromultimeric Kv channels may have, in comparison to Kv2.1 channels, altered activation kinetics, altered deactivation kinetics, and altered pharmacology. As discussed above, the results of *in vitro* expression suggest that a decrease in the expression of Kv2.1 subunits leads to an attenuation of $I_{K,slow}$ as well as of $I_{K,s}{}^s$ (Xu *et al.*, 1999a). The different inactivation behaviors of $I_{K,slow}$ as well as of $I_{K,ss}$ would fit well with the hypothesis that the absence and presence of Kv6.2 γ-subunits (Zhu *et al.*, 1999b), respectively, underlie the different inactivation kinetics of $I_{K,slow}$ as well as of $I_{K,ss}$.

Shaker-related Kv1 subunits (e.g., Kv1.2, Kv1.4, and Kv1.5) are associated with Kvβ subunits. So far, three genes—KCNA1B, KCNA2B, and KCNA3B—that encode different Kvβ subunits are known (Leicher *et al.*, 1996, 1998). The KCNA1B gene gives rise to at least three splice variants: Kvβ1.1, Kvβ1.2 and Kvβ1.3 (Leicher *et al.*, 1996). These three proteins have a variant amino terminus. Kvβ1.2 and Kvβ1.3 have been detected in cardiac tissue (Deal *et al.*, 1996; England *et al.*, 1995; Kwak *et al.*, 1999; De Biasi *et al.*, 1997). Kvβ subunits bind to the amino terminus of Kv1 α-subunits, which faces the cytoplasm (Yu *et al.*, 1996; Sewing *et al.*, 1996). The binding site lies within the amino-terminal T1 domain of the Kv1 channels, and binding of the Kvβ subunits may shift the voltage dependence of activation of Kv1 channels, generally in a more hyperpolarized direction. Association of Kvβ subunits may influence cell-surface expression of Kv1 α-subunits when coexpressed in heterologous cells (Shi *et al.*, 1996; Nagaya and Papazian, 1997). In this context, we have observed a three- to fourfold increase in cell-surface expression of Kv channels when Kv1.2 and Kv1.5 subunits are coexpressed with a Kvβ subunit but a threefold decrease when Kv1.4/Kvβ subunits are coexpressed (R. Bäehring and O. Pongs, unpublished results). It is possible that the cellular level of Kvβ expression *in vivo* influences the cell-surface expression of certain Kv1α subunits in opposite directions. This observation needs to be further explored. Interestingly, it has been reported that interleukin-2 significantly elevates the transcription of the Kvβ2 gene (Cohen *et al.*, 1992). Combined with the observation that Kvβ subunit expression may have opposing effects on the cell-surface expression of Kv α-subunits (Fig. 2), this finding suggests that different physiological situations may exert very complex effects on Kv channel subunit composition and Kv channel surface expression. It is an attractive hypothesis that the control of Kvβ subunit expression is a key player in this scenario.

The presence of N-terminal inactivating domains in Kvβ1 and Kvβ3 subunits may confer inactivation to Kv1 channels that otherwise would not inactivate (Rettig *et al.*, 1994; England *et al.*, 1995; Leicher *et al.*, 1996, 1998; De Biasi *et al.*, 1997). Indeed, an alignment of N-terminal α- and β-inactivating domains shows that they are comparable in structure and function. Remarkably, mammalian inactivating domains contain a conserved cysteine residue. The presence of this residue confers to the inactivating domains a sensitivity to oxidation (Rettig *et al.*, 1994; Ruppersberg *et al.*, 1991b). In the oxidized state, the inactivating domains are inactive. Only in the reduced state do the domains have the ability to bind to a receptor at or near the inner entrance of the Kv

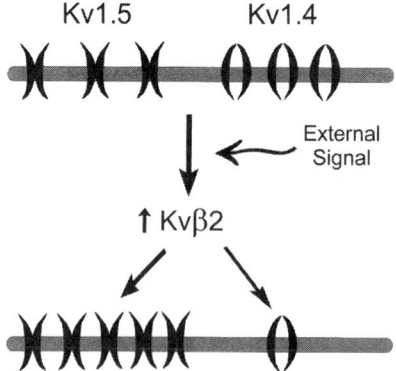

Figure 2. Possible influence of Kvβ2-subunit expression on cell-surface expression of Kvα subunits. External signals like interleukin-2 may increase the production of Kvβ2 subunits and thereby increase the cell-surface expression of some Kvα subunits (e.g., Kv1.5) but decrease the expression of others (e.g., Kv1.4).

channel pore upon depolarization. This binding leads to a rapid closure of the pore, thereby freezing the pore in an open conformation. Because of this, unbinding of the inactivating domain at hyperpolarized membrane potentials may be accompanied by a K$^+$ inward current through the not yet closed Kv channel pore (Ruppersberg et al., 1991a). Whether this behavior contributes to the afterhyperpolarization phenomena in vivo is not known. It is possible, however, that N-type inactivation of mammalian Kv channels is sensitive to the cellular redox status or the concentration of reactive oxygen intermediates, or both. Thus, N-type inactivation of Kv channels may serve as a link between cellular metabolism, free-radical-induced oxidant stress, oxygen concentration, and membrane excitability.

Mutations in the *Drosophila Hyperkinetic* (*Hk*) gene affect synaptic transmission at the larval neuromuscular junction (Stern and Ganetzky, 1989). The similarity in phenotype between *Hk* and *Shaker* (*Sh*) suggested that *Hk* exerts its effect on synaptic transmission by affecting the function of the rapidly inactivating Kv channels encoded by *Sh*. Subsequently, this predicted function of *Hk* was elegantly proven by showing that the *Hk* gene encodes a *Drosophila* Kvβ subunit (Chouinard et al., 1995). Coexpression of *Hk* with *Sh* in *Xenopus* oocytes increased current amplitudes and changed the voltage dependence and kinetics of activation and inactivation of the currents mediated by *Sh* Kv channels. A mouse Kvβ1.1 knockout line has been generated recently (Giese et al., 1998). The most striking phenotype of the mutant mice to date is that they have a largely reduced slow afterhyperpolarization (sAHP) amplitude, which is indicative of an altered Ca^{2+} homeostasis. It is not clear how this may relate to a reduced Kv channel activity. Collectively, the results so far have produced a very heterogeneous picture of Kvβ subunit functions. This impression may have changed dramatically with the advent of a Kvβ crystal structure, which has been published recently (Gulbis et al., 1999).

For some time, the most puzzling observation regarding Kvβ subunits has been that their sequences have significant homology with members of the aldose reductase family (McCormack and McCormack, 1994). The solution of the crystal structure of a Kvβ subunit has provided important insights into this relationship. Kvβ subunits appear to be genuine oxidoreductase enzymes, indeed having a functional NADPH-cofactor binding site and a complete active site that is structurally competent to mediate hydride-transfer chemistry (Gulbis et al., 1999). In all likelihood, the crystallographic data imply that the Kvβ subunits may, in fact, represent voltage-dependent enzymes that couple cellular redox status or metabolism, or both, to changes in membrane potential. The presence of Kvβ subunits may link Kv channel activity to O$_2$ or lipid metabolism or both, in some cases apparently through an identified NADPH oxidase pathway. However, the transduction mechanisms and the specific Kvβ substrates involved remain unclear (McCormack and McCormack, 1994).

Hypoxic conditions have been shown previously to inhibit Kv channels (Lopez-Barneo et al., 1998). Thus, Kv channels have been implicated in oxygen sensing in various cells. For example, it has been observed that an H$_2$O$_2$-generating heme protein may control the open-state probability of Kv channels in the smooth muscle cells of pulmonary vessels (Weir and Archer, 1995). The heme protein involved in H$_2$O$_2$ formation has apparent similarities with components of NADPH oxidase (Acker, 1998). Of special interest in this context is the observation that cysteine oxidation by H$_2$O$_2$ attenuates the activity of the N-terminal inactivating domains found in Kv1.4, Kv3.4,

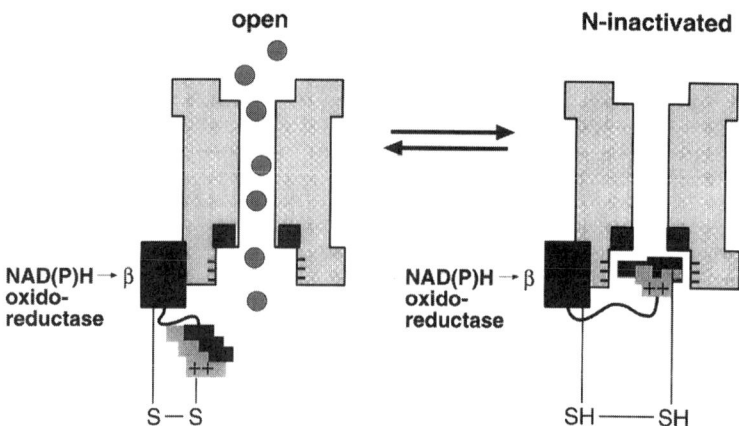

Figure 3. Model of a Kv channel composed of Kvα and Kvβ subunits containing N-terminal inactivation gate(s) with a redox-sensitive cysteine. The model proposes that a Kvβ NADPH oxidoreductase activity renders N-type inactivation of Kv channels redox-sensitive. For a detailed description of this model, see Acker, H. 1998.

Kvβ1.1, and Kvβ1.3 subunits (Ruppersberg *et al.*, 1991b). Thereby, rapid inactivation of Kv channels with N-terminal inactivating domains may be redox-sensitive. It is quite possible that the Kvβ subunits acting as NADPH oxidoreductases might produce high and low levels of reactive oxygen intermediates under conditions of normoxia and hypoxia, respectively. Thus, the amount of cysteine oxidation in the inactivating domain(s) might change, thereby conferring oxygen sensitivity to Kv channel activity. In fact, this type of mechanism, which is illustrated in Fig. 3, has been suggested previously (Acker, 1998). The model describes a redox-dependent activation/inactivation of the N-terminal inactivating domains. A cystine bridge keeps the inactivating domains of Kvα and Kvβ subunits in an inactive conformation. Reduction of the cystine bridge to cysteines converts the domains into their active conformation. If this model is correct, cardiac Kv channels containing Kv1.4 and/or Kvβ1.2 or Kvβ1.3 subunits would behave as oxygen-sensing devices. In normoxic conditions, these Kv channels would rapidly inactivate and presumably would accumulate in an inactivated state at normal heart rate because of their slow recovery from inactivation (Roberds *et al.*, 1993). Under hypoxic conditions, the inactivating domain of the Kv channels would undergo an oxidation reaction leading to quick recovery. Thereby, the Kv channels might contribute to action potential repolarization under hypoxic conditions. According to this hypothesis, Kv1 channels containing Kvβ subunits might function as redox-sensitive Kv channels and may assist in protecting the cell against oxidative stress.

Alternatively, Kvβ subunits may have two independent functions. One function would consist of their role as auxiliary Kv channel subunits, and the other as NADPH oxidoreductases. A similar observation has been made recently for an auxiliary subunit of glycine receptors. Glycine receptors are anchored at inhibitory chemical synapses by a cytoplasmic protein, gephyrin. The protein has a structural similarity to a cofactor required for activity of molybdoenzymes. Indeed, gene targeting in mice shows that gephyrin has a dual activity for clustering glycine receptors in neural tissue and for molybdoenzyme activity in non-neural tissue (Feng *et al.*, 1998). It is quite possible that Kvβ subunits also have a dual activity and function as NADPH oxidoreductases independently of their association with Kvα subunits. To explore this oxidoreductase function further, it will be important to identify the substrates for Kvβ NADPH oxidoreductases. Recent reports that Kvβ subunits affect the activities of Kv channels to which they do not bind indicate the possibility of an independent Kvβ function (Wilson *et al.*, 1998; Pérez-Garciá *et al.*, 1999).

5. IMPLICATIONS FROM CRYSTAL STRUCTURES

Members of the oxidoreductase family typically exhibit a three dimensional structure (TIM)-barrel fold (McCormack and McCormack, 1994). TIM barrels have eight parallel β-strands connected by α-helices encircling the β-barrels perimeter. The NADPH binding site is situated along one face of the TIM barrel, and the catalytic site is located at the C-terminal edge of the β-strands. The Kvβ2 subunit structure is very similar (Gulbis *et al.*, 1999), showing a fourfold-symmetric tetramer of TIM barrels with approximate dimensions of 90 Å × 90 Å × 40 Å. They are arranged end to side, so that one face of a barrel is wedged against the side of an adjacent barrel. The opposite faces are 30 Å–35 Å away from the fourfold axis and provide, together with a helical subdomain, the NADPH binding site. Fourfold molecular symmetry is a shared feature of the recently crystallized KcsA channels (Doyle *et al.*, 1998), the T1-domains (Kreusch et al,. 1998; Bixby *et al.*, 1999), and the Kvβ2 subunit (Gulbis *et al.*, 1999). It is likely that the integral membrane Kvα subunits, the Kvβ2 subunits, and the tetramerization domains share the same molecular symmetry. Possibly, the fourfold axes coincide when Kvβ subunits bind to Kvα subunits, and it has been suggested that the Kvβ subunits dock against the cytoplasmic face of the tetrameric Kvα subunits to maintain the overall molecular symmetry. Using the dimensions of the KcsA tetramer to estimate the space required for the S5–P–S6 pore-forming domain, it was proposed that the β-subunit active sites (i.e., part of the TIM barrels and the C-terminal helical subdomains) lie beneath segments S1 to S4, which include the voltage sensor of Kv channels (Gulbis *et al.*, 1999).

A direct interaction of Kvβ subunits with the voltage sensor of Kvα subunits may underlie the observation that association of Kvβ subunits with Kvα subunits affects the voltage-dependent activation of Kv channels. On the other hand, it has been shown that Kvβ-subunit binding sites are located within the T1-domain (Yu *et al.*, 1996; Sewing *et al.*, 1996). This domain has been crystallized in the form of a tetramer having a fourfold symmetry whose structure is similar to that of KcsA tetramers and Kvβ tetramers (Kreusch *et al.*, 1998; Bixby *et al.*, 1999). The T1-domain also affects the voltage dependence of Kv channel activation with interactions involving Kvβ subunits. It is not clear whether both effects have a similar molecular basis. Without knowing the structure of a T1/Kvβ complex, however, it is difficult at present to correlate the separate crystal structures of the KcsA pore, the T1-domain, and the Kvβ tetramer.

REFERENCES

Abbott, G. W., Sesti, F., Splawsi, I., Buck, M., Lehmann, M. H., Timothy, K. W., Keating, M. H., and Goldstein, S. A. N., 1999, MiRP1 forms I_{Kr} potassium channels with HERG and is associated with cardiac arrhythmias, *Cell* **97**:175–187.

Acker, H., 1998, Reactive oxygen intermediates as mediators for regulating ion channel activity, in: *Oxygen Regulation of Ion Channels and Gene Expression* (J.Lopez-Barneo and E. K. Weir, eds.), Futura, Armonk, N.Y. pp. 9–18.

Attali, B., 1996, Ion channels: A new wave for heart rhythms, *Nature* **384**:24–25.

Barhanin, J., Lesage, F., Guillemare, E., Fink, M., Lazdunski, M., and Romey, G., 1996, Kv LQT1 and IsK (minK) proteins associate to form the I_{Ks} cardiac potassium current, *Nature* **384**:78–80.

Barry, D. M., and Nerbonne, J. M., 1996, Myocardial potassium channels: Electrophysiological and molecular diversity, *Annu. Rev. Physiol.* **58**:363–394.

Barry, D. M., Trimmer, J. S., Merlie, J. P., and Nerbonne, J. M., 1995, Differential expression of voltage-gated K^+ channel subunits in adult rat heart. Relation to functional K^+ channels?, *Circ. Res.* **77**:361–369.

Barry, D. M., Xu, H., Schuessler, R. B., and Nerbonne, J. M., 1998, Functional knockout of the transient outward current, long QT syndrome, and cardiac remodeling in mice expressing a dominant-negative Kv4 α subunit, *Circ. Res.* **83**:560–567.

Bixby, K. A., Nanao, M. H., Shen, N. V., Kreusch, A., Bellamy, H., Pfaffinger, P. J., and Choe, S., 1999, Zn^{2+}-binding and molecular determinants of tetramerization in voltage-gated K^+ channels, *Nat. Struct. Biol.* **6**:38–43.

Boyden, P. A., and Jeck, C. D., 1995, Ion channel function in disease, *Cardiovasc. Res.* **29**:312–318.

Chandy, K. G., and Gutman, G. A., 1995, Voltage-gated potassium channel genes, in: *Handbook of Receptors and Channels: Ligand- and Voltage-Gated Ion Channels* (R. A. North, ed.) CRC Press, Boca Raton, Florida, pp. 1–71.

Chouinard S. W., Wilson, G. F., Schlimgen, A. K., and Ganetzky, B., 1995, A potassium channel β subunit related to the aldo-keto reductase superfamily is encoded by the *Drosophila* Hyperkinetic locus, *Proc. Natl. Acad. Sci. U.S.A.* **92**:6763–6767.

Cohen, J. A., Arai, M., Prak, E. L., Brooks, S. A., Young, L. H., and Prystowsky, M. B., 1992, Characterization of a novel mRNA expressed by neurons in mature brain, *J. Neurosci. Res.* **31**:273–284.

Deal, K. K., England, S. K., and Tamkun, M. M., 1996, Molecular physiology of cardiac potassium channels, *Physiol. Rev.* **76**:49–67.

De Biasi, M., Wang, Z., Accili, E., Wible, B., and Fedida, D., 1997, Open channel block of human heart hKv1.5 by the β-subunit hKvβ1.2, *Am. J. Physiol.* **272**:H2932–H2941.

Dixon, J. E., Shi, W., Wang, H. S., McDonald, C., Yu, H., Wymore, R. S., Cohen, I. S., and McKinnon, D., 1996, Role of the Kv4.3 K^+ channel in ventricular muscle: A molecular correlate for the transient outward current, *Circ. Res.* **79**:659–668.

Doyle, D. A., Morais Cabral, J. H., Pfuetzner, R. A., Kuo, A., Gulbis, J. M., Cohen, S. L., Chait, B. T., and MacKinnon, R., 1998, The structure of the potassium channel: Molecular basis of K^+ conduction and selectivity, *Science* **280**:69–77.

Engeland, B., Neu, A., Ludwig, J., Roeper, J., and Pongs, O., 1998, Cloning and functional expression of rat *ether-à-go-go*-like K^+ channel genes, *J. Physiol. (London)* **513**:647–654.

England, S., Uebele, V., Shear, H., Kodali, J., Bennett, P., and Tamkun, M., 1995, Characterization of voltage-gated K^+ channel β subunit expressed in human heart, *Proc. Natl. Acad. Sci. U.S.A.* **92**:6309–6313.

Feng, J., Wible, B., Li, G. R., Wang, Z., and Nattel, S., 1997, Antisense oligonucleotide directed against Kv1.5 mRNA specifically inhibits ultrarapid delayed rectifier K^+ current in cultured human atrial myocytes, *Circ. Res.* **80**:572–579.

Feng, G., Tintrup, H., Kirsch, J., Nichol, M. C., Kuhse, J., Betz, H., and Sanes, J. R., 1998, Dual requirement for gephyrin in glycine receptor clustering and molybdoenzyme activity, *Science* **282**:1321–1324.

Giese, K. P., Storm, J. F., Reuter, D., Fedorov, N. B., Shao, L-R., Leicher, T., Pongs, O., and Silva, A. J., 1998, Reduced K^+ channel inactivation, spike broadening, and after-hyperpolarization in Kvβ1.1-deficient mice with impaired learning, *Learning Memory* **5**:257–273.

Giles, W. R., and Imaizumi, Y., 1988, Comparison of potassium currents in rabbit atrial and ventricular cells, *J. Physiol. (London)* **405**:123–145.

Gulbis J. N., Mann, S., and MacKinnon, R., 1999, Structure of a voltage-dependent K$^+$ channel β subunit, *Cell* **97**:943–952.

Heginbotham, L., Lu, Z., Abramson, Z., and MacKinnon, R., 1994, Mutations in the K$^+$ channel signature sequence, *Biophys. J.* **66**:1061–1067.

Hille, B., 1992, *Ionic Channels of Excitable Membranes*, 2nd ed., Sinauer Associates, Inc., Sunderland, Massachusetts.

Jan, Y. N., and Jan, L. Y., 1997, Cloned potassium channels from eukaryotes and prokaryotes, *Annu. Rev. Neurosci.* **20**:91–123.

Johns, D. C., Nuss, H. B., and Marban, E., 1997, Suppression of neuronal and cardiac transient outward currents by viral gene transfer of dominant-negative Kv4.2 constructs, *J. Biol. Chem.* **272**:31598–31603.

Kong, W., Po, S., Yamagishi, T., Ashen, M. D., Stetten, G., and Tomaselli, G. F., 1998, Isolation and characterization of the human gene encoding I_{to}: Further diversity by alternative MRNA splicing, *Am. J. Physiol.* **275**:H1963–H1970.

Kreusch, A., Pfaffinger, P. J., Stevens, C. F., and Choe, S., 1998, Crystal structure of the tetramerization domain of the Shaker potassium channel, *Nature* **392**:945–948.

Kwak, Y. G., Hu, N. N., Wie, J., George, A. L., Grobaski, T. D., Tamkun, M. M., and Murray, K. T., 1999, Protein kinase A phosphorylation alters Kvβ1.3 subunit-mediated inactivation of the Kv1.5 potassium channel, *J. Biol. Chem.* **274**:13928–13932.

Leicher, T., Roeper, J., Weber, K., Wang, X., and Pongs, O., 1996, Structural and functional characterization of human potassium channel subunit β1 (KCNA1B), *Neuropharmacology* **35**:787–795.

Leicher, T., Bähring, R., Isbrandt, B., and Pongs, O., 1998, Coexpression of the KCNA3B gene product with Kv1.5 leads to a novel A-type potassium channel, *J. Biol. Chem.* **273**:35095–35101.

London, B., Jeron, A., Zhou, A., Buckett, P., Han, X., Mitchell, G. F., and Koren, G., 1998, Long QT and ventricular arrhythmias in transgenic mice expressing the N terminus and the first transmembrane segment of a voltage-gated potassium channel, *Proc. Natl. Acad. Sci. U.S.A.* **95**:2926–2931.

Lopez-Barneo, J., Montoro, R., Ortega-Saenz, P., and Urena, J., 1998, Oxygen-regulated ion channels, in: *Oxygen Regulation of Ion Channels and Gene Expression* (J. Lopez-Barneo and E. K. Weir, eds.), Futura Press, Armonk, N.Y. pp. 127–144.

McCormack, T., and McCormack, K., 1994, *Shaker* K$^+$ channel β subunits belong to an NAD(P)H-dependent oxidoreductase superfamily, *Cell* **79**:1133–1135.

Näbauer, M., and Käb, M., 1998, Potassium channel down regulation in heart failure, *Cardiovasc. Res.* **37**:324–334.

Nabauer, M., Beuckelmann, D. J., Uberfuhr, P., and Steinbeck, G., 1996, Regional differences in current density and rate-dependent properties of the transient outward current in subepicardial and subendocardial myocytes of human left ventricle, *Circulation* **93**:168–177.

Nagaya, N., and Papazian, D. M., 1997, Potassium channel α and β subunits assemble in the endoplasic reticulum, *J. Biol. Chem.* **272**:3022–3027.

Papazian, D. M., 1999, Potassium channels: Some sssembly required, *Neuron* **23**:7–10.

Pérez-Garciá, M. T., López-López, J. R., and González, C., 1999, Kvβ1.2 subunit coexpression in HEK293 cells confers O_2 sensitivity to Kv4.2 but not to *Shaker* channels, *J. Gen. Physiol.* **113**:897–907.

Pongs, O., 1999, Voltage-gated potassium channels: From hyperexcitability to excitement, *FEBS Lett.* **452**:31–35.

Rettig, J, Heinemann, S. H., Wunder, F., Lorra, C., Parcej, D. N., Dolly, J. O., and Pongs, O., 1994, Inactivation properties of voltage-gated K$^+$ channels altered by presence of β-subunit, *Nature* **369**:289–294.

Roberds, S. L., Knoth, K. M., Po, S., Blair, T. A., Bennett, P. B., Hartshorne, R. P., Snyders, D. J., and Tamkun, M. M., 1993, Molecular biology of the voltage-gated potassium channels of cardiovascular system. *J. Cardiovasc. Physiol.* **4**:68–80.

Robertson, B., 1997, The real life of voltage-gated K$^+$ channels: More than model behaviour, *Trends Pharmacol. Sci.* **18**:474–483.

Roeper, J., Lorra, C., and Pongs, O., 1997, Frequency-dependent inactivation of mammalian A-type K$^+$ channel Kv1.4 regulated by Ca^{2+}/calmodulin-dependent protein kinase, *J. Neurosci.* **17**:3379–3391.

Ruppersberg, J. P., Frank, R., Pongs, O., and Stocker, M., 1991a, Cloned neuronal I_k(A) channels reopen during recovery from inactivation, *Nature* **353**:657–660.

Ruppersberg, J. P., Stocker, M., Pongs, O., Heinemann, S. H., Frank R., and Koenen, 1991b, Regulation of fast inactivation of cloned mammalian I_k(A) channels by cysteine oxidation, *Nature* **352**:711–714.

Sanguinetti, M. C., Jiang, C., Curran, M. E., and Keating, M. T., 1995, A mechanistic link between an inherited and an acquired cardiac arrhythmia: HERG encodes the I_{Ks} potassium channel, *Cell* **81**:299–307.

Sanguinetti, M. C., Curran, M. E., Zou, A., Shen, J., Spector, P. S., Atkinson, D. L., and Keating, M. T., 1996, Coassembly of KvLQT1 and minK (IsK) proteins to form cardiac I_{Ks} potassium channel, *Nature* **384**:80–83.

Scott, V. E., Rettig, J., Parcej, D. N., Keen, J. N., Findlay, F. B. C., Pongs, O., and Dolly, J. O., 1994, Primary structure of a β subunit of α-dendrotoxin-sensitive K^+ channels from bovine brain, *Proc. Natl. Acad. Sci. U.S.A.* **91**:1637–1641.

Sewing, S., Roeper, J., and Pongs, O., 1996, Kvβ1 subunit binding specific for *Shaker*-related potassium channel α subunits, *Neuron* **16**:455–463.

Shi, G., Nakahira, K., Hammond, S., Rhodes, K. J., Schechter, L. E., and Trimmer, J. S., 1996, Beta subunits promote K^+ channel surface expression through effects early in biosynthesis, *Neuron* **16**:843–852.

Shimoni, Y., Severson, D., and Giles, W. R., 1992, Thyroid status and diabetes modulate regional differences in potassium currents in rat ventricle, *J. Physiol. (London)* **488**:673–688.

Stern, M., and Ganetzky, B., 1989, Altered synaptic transmission in *Drosophila Hyperkinetic* mutants, *J. Neurogenet.* **5**:215–228.

Van Wagoner, D. R., Pond, A. L., McCarthy, P. M., Trimmer, J. S., and Nerbonne, J. M., 1997, Outward K^+ current densities and Kv1.5 expression are reduced in chronic human atrial fibrillation, *Circ. Res.* **80**:772–781.

Wang, Z., Feng, J., Pond, A. L., Nerbonne, J. M., and Nattel, S., 1999, The potential molecular basis of different physiological properties of transient outward K^+ current in rabbit and human atrial myocytes, *Circ. Res.* **84**:551–561.

Weir, E. K., and Archer, S. L., 1995, The mechanism of acute hypoxic pulmonary vasoconstriction: The tale of two channels, *FASEB J.* **9**:183–189.

Wickenden, A. D., Jegla, T. J., Kaprielian, R., and Backx, P. H., 1999, Regional contributions of Kv1.4, Kv4.2, and Kv4.3 to transient outward K^+ current in rat ventricle, *Am. J. Physiol.* **276**:H1599–H1607.

Wilson, G. F., Wang, Z., Chouinard, S. W., Griffith, L. C., and Ganetzky, B., 1998, Interaction of the K^+ channel β subunit, Hyperkinetic, with eag family members, *J. Biol. Chem.* **273**:6389–6394.

Xu, H., Barry, D. M., Li, H., Brunet, S., Guo, W., and Nerbonne, J. M., 1999a, Attenuation of the slow component of delayed rectification, action potential prolongation, and triggered activity in mice expressing a dominant negative Kv2 α subunit, *Circ. Res.* **85**:623–633.

Xu, H., Guo, W., and Nerbonne, J. M., 1999b, Four kinetically distinct depolarization-activated outward K^+ currents in adult mouse ventricular myocytes, *J. Gen. Physiol,* **113**:661–678.

Yellen, G., 1998, The moving parts of voltage-gated ion channels, *Rev. Biophys.* **31**:239–295.

Yu, W., Xu, J., and Li, M., 1996, NAB domain is essential for the subunit assembly of both alpha-alpha and alpha-beta complexes of Shaker-like potassium channels, *Neuron* **16**:441–453.

Zhu, S. R., Wulf, A., Schwarz, M., Isbrandt, D., and Pongs, O., 1999a, Characterization of human Kv4.2 mediating a rapidly-inactivating transient voltage-sensitive K^+ current, *Recept. Channels,* **6**:387–400.

Zhu, X.-R., Netzer, R., Blke, K., Liu, A., and Pongs, O., 1999b, Structural and functional characterization of Kv6.2, a new -subunit of voltage-gated potassium channel, *Recept. Channels,* **6**:337–350.

Chapter 4

Molecular Biology of High-Conductance, Ca^{2+}-Activated Potassium Channels

Pratap Meera, Martin Wallner, and Ligia Toro

1. MOLECULAR PROPERTIES OF THE BK_{Ca} CHANNEL PORE-FORMING α-SUBUNIT (SLO1) AND STRUCTURALLY RELATED GENES (SLO2 AND SLO3)

1.1. Slo1, the BK_{Ca} Channel α-Subunit

The pore-forming α-subunit of the large-conductance, voltage- and Ca^{2+}-activated K^+ channel (BK, BK_{Ca}, or Slo1) was first cloned by utilizing the *Drosophila Slowpoke* (Slo) mutant (Atkinson *et al.*, 1991), which carries homozygous mutant *Slowpoke* alleles. Flight muscles from this mutant fly lacked a Ca^{2+}-activated K^+ current (Elkins *et al.*, 1986), whereas functional expression of the *Slowpoke* wild-type cDNA (dSlo1) yielded voltage- and Ca^{2+}-activated K^+ currents in *Xenopus* oocytes (Adelman *et al.*, 1992). A large variety of dSlo1 isoforms are generated by alternative splicing, and some of them show large functional differences, including changes in kinetics, single-channel conductance, and Ca^{2+} and voltage sensitivities (Lagrutta *et al.*, 1994). A series of vertebrate BK_{Ca} channel α-subunit clones including human hSlo1 (Butler *et al.*, 1993; Dworetzky *et al.*, 1994; Pallanck and Ganetzky, 1994; Tseng-Crank *et al.*, 1994; McCobb *et al.*, 1995; Wallner *et al.*, 1995; Vogalis *et al.*, 1996; Jiang *et al.*, 1997; Morita *et al.*, 1997; Jones *et al.*, 1998) were isolated by homology screening, and an ortholog in *Caenorhabditis. elegans*, nSlo1 (Wei *et al.*, 1996), has been identified (Fig. 1A). The deduced protein of these cDNA clones shows similarities to other Kv channels in the voltage sensor and in the "pore" region, which determines ionic conductance and selectivity (Figs. 1B, 1C, and 2A). Functionally expressed vertebrate Slo1 channels have a conductance of ~ 250

Pratap Meera Martin Wallner, and Ligia Toro ● Department of Anesthesiology, Department of Molecular and Medical Pharmacology, and Brain Research Institute, University of California at Los Angeles, Los Angeles, California 90095–1778.

Potassium Channels in Cardiovascular Biology, edited by Archer and Rusch. Kluwer Academic/Plenum Publishers, New York, 2001.

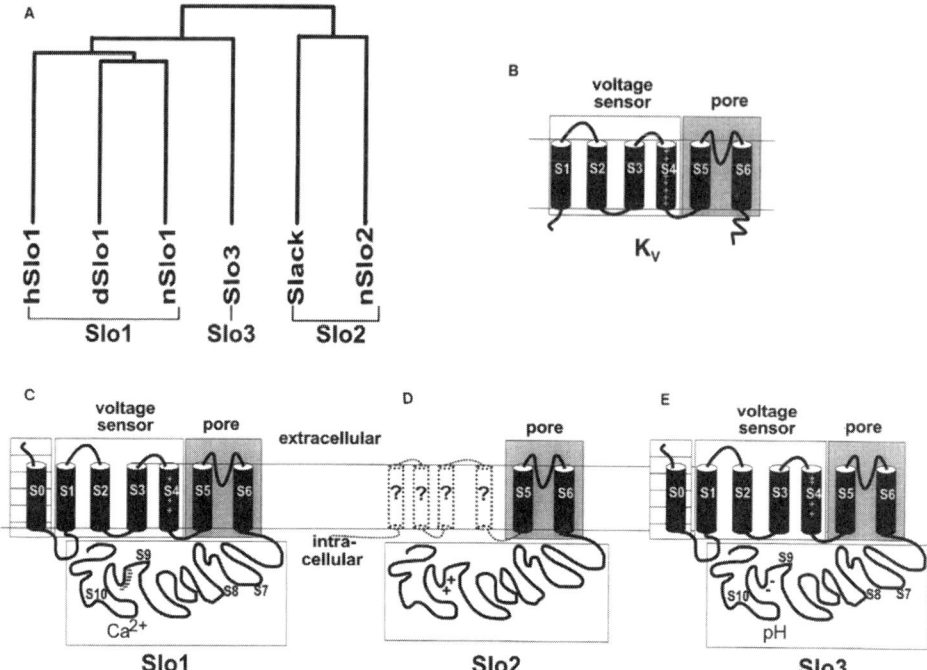

Figure 1. Molecular structure of BK_{Ca} channel α-subunits (Slo1) and evolutionary related sequences. (A) The length of the horizontal bars in the dendrogram represents the evolutionary distance among protein sequences. The dendrogram was generated from a multiple sequence alignment using the GCG program "pileup." GenBank Accession numbers are: hSlo1, U11058; dSlo1, JH0697; nSlo1, as reported by Wei *et al.* (1996); Slo3, AF039213 (mouse); Slack, AF089730 (rat); nSlo2, F08B12. 3A; h (human), d (*Drosophila*), n (nematode, *C. elegans*). (B) Membrane topology of voltage-dependent K^+ (Kv) channels. (C) Suggested membrane topology of Slo1, BK_{Ca} channel α-subunit. (D) Possible structure of Slo2 channels. The dotted boxes and question marks refer to the unknown membrane topology. (E) Proposed topology of Slo3 based on sequence similarity to Slo1. −, Negatively charged residues; +, positively charged residues.

pS (in symmetrical 110 mM K^+), their activity is increased by raising intracellular Ca^{2+} ($[Ca^{2+}]_i$) and depolarizing voltages, and they are identified by their sensitivity to iberiotoxin blockade ($IC_{50} = 1$ nM) and by their insensitivity to block by 4–aminopyridine (Wallner *et al.*, 1995; Meera *et al.*, 1997). They have no absolute $[Ca^{2+}]_i$ dependence for opening, as they can open by depolarization when $[Ca^{2+}]_i$ is practically zero, and their open-state probability is independent of $[Ca^{2+}]_i$ from 0 to 100 nM. However, their opening is facilitated by $[Ca^{2+}]_i$ above 100 nM (Meera *et al.*, 1996). Gating current measurements showed that these channels indeed have an intrinsic voltage sensor, and thus are voltage-gated (Stefani *et al.*, 1997). This finding is consistent with the presence of a conserved sequence identical to the voltage sensor of voltage-dependent channels. However, the pore-forming α-subunit of BK_{Ca} channels (Slo1) is strikingly different from the α-subunit of Kv channels at both the amino and carboxy termini (Figs. 1B and 1C). The carboxy terminus is very long, with additional hydrophobic regions (S7, S8, S9, and S10) (Fig. 1C). Hydrophobic regions S9 and S10, and most likely also S7 and S8, are not membrane-spanning (Meera *et al.*, 1997), and

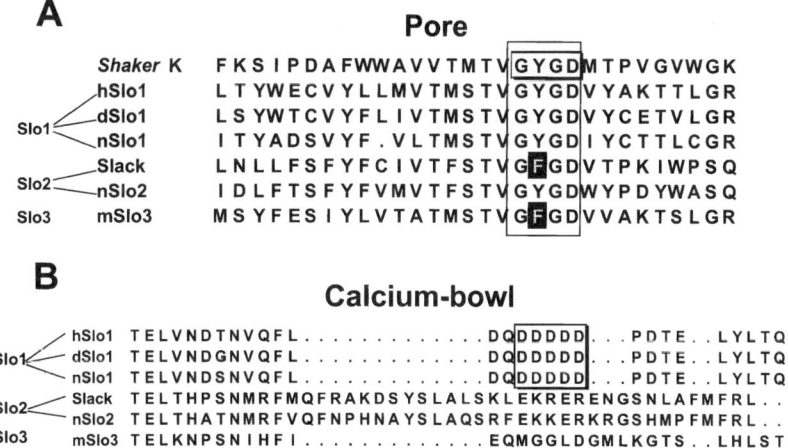

Figure 2. Pore and "Ca^{2+}-bowl" sequence alignments. (A) Sequence comparison of a voltage-dependent K^+ channel pore (*Shaker* K^+) with Slo1, Slo2 (nSlo2 and Slack), and Slo3. The signature sequence for a K^+-selective pore, GYGD, is highlighted. Slack and Slo3 have an imperfect K^+ channel pore sequence, GFGD. (B) Sequence comparison of "Ca^{2+}-bowl" region in Slo1 channels with corresponding sequences in Slo2 and Slo3. String of aspartic acids conserved in Slo1 channels is boxed. . . . , Spaces introduced for a better alignment.

they may form the hydrophobic interior of globular intracellular protein domains. A region between hydrophobic segments S9 and S10 contains a stretch of negatively charged aspartate residues. This region is called the "Ca^{2+} bowl" (Figs. 2B and 3) and has been shown to contribute to the Ca^{2+} sensitivity of Slo1 channels (Wei *et al.*, 1994; Schreiber and Salkoff, 1997). The "Ca^{2+} bowl" is identical in all cloned Slo1 channels (Figs. 2B and 3), consistent with its importance in Ca^{2+} sensing. Furthermore, functional channels can be expressed from two separable domains: the "core" comprising 50–58 and the "tail" that includes the rest of the protein and contains the "Ca^{2+}-bowl" (Fig. 3) (Wei *et al.*, 1994). We have recently shown that Slo1 channels are unique not only at the C-terminus but also at the N-terminus, as they possess an additional N-terminal transmembrane region (S0), which leads to an extracellular N-terminus (Wallner *et al.*, 1996) (Fig. 1C). This membrane topology was confirmed by antibody binding to c-myc epitope-tagged hSlo1 channels (Meera *et al.*, 1997).

1.2. Slo2 and Slack, Novel BK_{Ca} Channel Homologs

Sequences related to Slo1, the BK_{Ca} channel α-subunit, encoded by different mammalian and *C. elegans* genes have been recently identified using information obtained from screening EST databases. These clones can be grouped into two subfamilies, Slo2 and Slo3 (Fig. 1A, 1D, and 1E).

So far, two Slo2-related sequences have been reported: nSlo2, found in *C. elegans* (n for nematode) (Wei *et al.*, 1996; Wallner *et al.*, 1997), and a potential mammalian

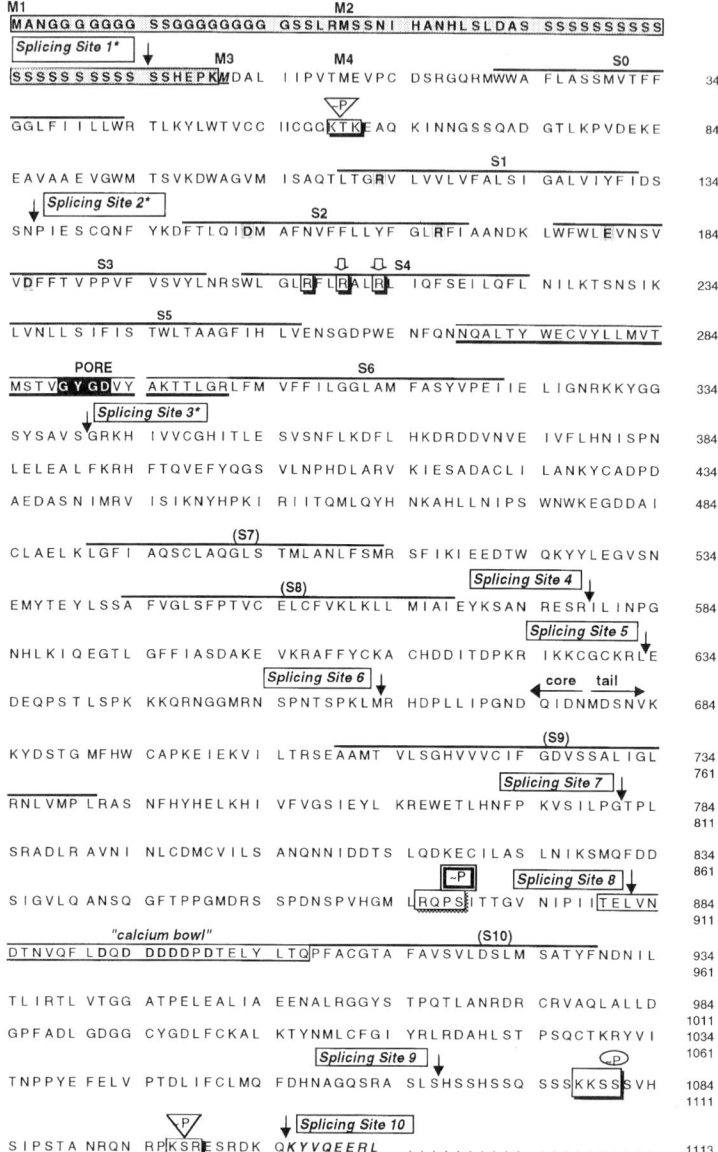

Figure 3. Human Slo1 (hSlo1) amino acid sequence. M1–M4, are possible start sites for translation. Numbering in hSlo1 starts at M3. Alternative splice sites are indicated by arrows. Splice sites 1–3, marked with asterisks, have been reported in chicken Slo1. Splice site 1 is absent in human and mouse Slo1. Transmembrane segments S0–S6 and hydrophobic regions (S7)–(S10) are marked with lines. Conserved charged residues in S1–S3 are shaded. In S4, the boxed arginines contribute to charge movement in *Shaker* K[+] channels; arrows mark the arginines that contribute to the gating valence of Slo1 (Diaz *et al.*, 1998). The "pore" signature sequence GYGD is highlighted. The "core" and "tail" regions are delimited by horizontal arrows. Within the "tail" region, the "calcium bowl" is boxed. ~P designates a potential phosphorylation site: ~P inside a square, a PKA-dependent phosphorylation site (Nara *et al.*, 1998); inside an oval, a PKG-dependent phosphorylation consensus site, and inside a triangle, a PKC-dependent phosphorylation consensus site. Splice inserts at splice site 10 in italics.

ortholog Slack (*sequence like a calcium-activated K^+ channel*) (Joiner *et al.*, 1998). nSlo2 and Slack show sequence homology in the pore region (Fig. 2A) and over most of the long C-terminus to other members of the Slo family (Figs. 1C–1E). However, Slack and nSlo2 are distinguished from other Slo family members by their lack of both the S0 domain and the voltage sensor, as judged by the absence of discernible sequence similarity in these regions (S0–S4) (Fig. 1D). This suggests that Slo2 channels lack the type of voltage sensor characteristic of Kv channels (Fig. 1B). Moreover, the lack of a clear hydrophobicity plot in this region makes the prediction of four membrane-spanning regions highly uncertain (question marks in Fig. 1D). Slack and nSlo2 also differ from Slo1 channels in the corresponding "Ca^{2+}–bowl" region as they carry an insertion of 13 amino acids, and most of the conserved negatively charged residues are replaced by positively charged amino acids (Fig. 2B).

Yuan et al. (1998; 1999) reported that nSlo2 forms K^+-selective channels, consistent with its pore signature sequence, GYGD, of 80-pS conductance. They reported that the channel requires both Cl^- and Ca^{2+} for activation, and they mapped the Cl^- sensing domain to the positively charged region corresponding to the "Ca^{2+}-bowl" in Slo1 channels.

Unlike Slo1, Slack channels have a conductance of 25–65 pS, are inhibited by intracellular Ca^{2+} concentration, and are insensitive to the Slo1 channel opener NS-1619 and the blocker iberiotoxin (Joiner *et al.*, 1998). They have a decreased selectivity to K^+ over Na^+, which is consistent with the lack of a perfect K^+-selective pore sequence (GFGD instead of GYGD) (Fig. 2A). Slack not only forms channels by itself but also forms functionally distinct heteromultimeric channels when coexpressed with the large-conductance Slo1, giving rise to channels with intermediate conductance, of 60–189 pS. These heteromultimeric channels gain some of the properties of Slo1; they are activated by intracellular Ca^{2+} and by NS-1619 and are sensitive to iberiotoxin, but only at the onset of depolarizing command potentials. Therefore, expression of Slack channels may contribute to the functional diversity of BK_{Ca} channels found in native tissues.

1.3. Slo3, a Voltage- and pH-Sensitive BK_{Ca} Channel Homolog

Slo3, a paralog of Slo1 channels, is abundantly expressed in mouse and human spermatocytes (Schreiber *et al.*, 1998). Unlike the Slo2 members, Slo3 shows sequence homology to Slo1, the BK_{Ca} channel α-subunit, over most of its coding region, including S0–S4, and it therefore is likely to share a common membrane topology with Slo1, as shown in Fig. 1E. Functional expression shows that these channels are activated by voltage, have a conductance of 90 pS, and exhibit a decreased selectivity to K^+ over Na^+, likely because they have a GFGD motif instead of the typical GYGD sequence of K^+-selective pores (Fig. 2A). Slo3 channels also differ from Slo1 channels in that they are activated by intracellular pH and are insensitive to intracellular Ca^{2+} (Schreiber *et al.*, 1998). The insensitivity of Slo3 channels to intracellular Ca^{2+} is consistent with the lack of most of the negative charges in the region corresponding to the "Ca^{2+}-bowl" domain in Slo1 (Fig. 2B); this region might have evolved for Slo3 regulation by intracellular pH.

TABLE 1
Properties of SLO1 (BK$_{Ca}$ Channel α-Subunit), SLO2 and SLO3

Property	Slo1[a]	Slo2 Slo2[b]	Slo2 Slack[c]	Slo3[d]
Distribution	Ubiquitous	?[e]	Neurons	Testes
Conductance	250 pS	80 pS	25–65 pS	90 pS
Response to depolarization	Activated	Activated	Activated	Activated
Requirement for/effect of intracellular Ca^{2+}	Stimulated	Required	Inhibited	Insensitive
Requirement for/effect of intracellular Cl$^-$	No effect	Required	?	?
Effect of changes in intracellular pH	?	?	?	Stimulated by Increase in pH
Pore sequence	GYGD	GYGD	GFGD	GFGD
K$^+$ selectivity	High		Decreased relative to Slo1	Decrease relative to Sl01
Ca^{2+}-bowl[f]	Many (−)[g]	(+)[h]	(+)	Few (−)
Effect of iberiotoxin	Blocks	?	No effect	?
Effect of tetraethylammonium	Blocks	?	?	?
Effect of 4-aminopyridine	Insensitive	?	?	?

[a] This chapter and Toro et al., 1998).
[b] Yuan et al., 1999, 1998.
[c] Joiner et al., 1998.
[d] Schreiber et al., 1998.
[e] ?, Not known.
[f] See Figure 2B.
[g]° (−) Negatively charged residues.
[h] (+) Positively charged residues.

In summary, the identification of K$^+$ channels with sequence homologies to Slo1 (Table 1) that are regulated by intracellular factors other than Ca^{2+} shows that there is an evolutionary conserved machinery which serves as the biosensor for intracellular signals. Slo1 channels are gated by voltage, and their opening is facilitated by an increase in intracellular Ca^{2+}. They provide a negative feedback for Ca^{2+} entry into cells by closing Ca^{2+} channels via hyperpolarization. Slo1 channels are expressed in almost all tissues since this negative-feedback mechanism is ubiquitous, whereas Slo3 channels may exist to serve a function distinct from Slo1 channels as they are activated by intracellular pH and depolarizing voltages but are insensitive to Ca^{2+}. It is suggested that Slo3 channels play a role in sperm capacitation and the acrosome reaction. With respect to the Slo2 family, nSlo2 requires intracellular Cl$^-$ for its activation, and both nSlo2 and Slack sense Ca^{2+}. However, nSlo2 is activated whereas Slack is inhibited by intracellular Ca^{2+}; such differences in the functional properties of these channels argue against an orthologous relationship between Slack and nSlo2. The physiological relevance of Slack channels may be the formation of heteromultimers with ubiquitously expressed Slo1 channels, thereby contributing to the large diversity of BK$_{Ca}$ channels found in native cells.

2. DIVERSITY OF SLO1 GENERATED BY ALTERNATIVE SPLICING

Vertebrate as well as the *Drosophila* BK_{Ca} channel α-subunits (Slo1) display numerous isoforms generated by alternative splicing. Alternative splicing in the BK_{Ca} channel α-subunit leads to insertion of short stretches of amino acids (from 3 to ~60) or to replacement of homologous regions in the primary transcript (Fig. 3 and Table 2).

2.1. *Drosophila* Splice Variants

Drosophila Slo1 (dSlo1) has at least six alternative splice sites, one at the N-terminus and five at the C-terminus (Adelman *et al.*, 1992; Brenner *et al.*, 1996). Splice variants at the carboxyl end of the S6 transmembrane region show differences in channel kinetics and conductance (Lagrutta *et al.*, 1994), consistent with the finding that the S6 segment influences the pore properties of Kv channels (Lopez *et al.*, 1994). Splice variants between segments S7 and S8 lead to differences in apparent Ca^{2+} sensitivities (Lagrutta *et al.*, 1994) although they are not within the "Ca^{2+}-bowl" region. Usage of different promoters and alternate splicing leads to N-terminal variants in dSlo1 (Brenner *et al.*, 1996); however, functional analysis of these variants has not been reported.

2.2. Vertebrate Splice Variants

Slo1 seems to be derived from a single gene, and vertebrate orthologs share >96% sequence identity. Isoforms of Slo1 are primarily generated by alternative splicing; in vertebrates, a total of 10 splice sites have been reported (Fig. 3). However, only several of them have been studied. Two splice variants show clear functional changes in apparent Ca^{2+}/voltage sensitivity (Table 2) and channel kinetics.

Splice site 1 has been proposed to exist in the chicken Slo1 (chSlo1) (Rosenblatt *et al.*, 1997). However, in mammals, variants at the N-terminus (site 1) arise from mechanisms other than alternative splicing, since in human and mouse Slo1 genes the N-terminal region does not contain introns (McCobb *et al.*, 1995; Martin Waller, unpublished results).

Splice site 2 described in chSlo1 (Rosenblatt *et al.*, 1997) is in the region homologous to voltage-dependent K^+ channels and lies between transmembrane segments S1 and S2. Insertion of an epitope tag at the equivalent position in the human channel (hSlo1) lowers the apparent voltage/Ca^{2+} sensitivity (Meera *et al.*, 1997). It will be interesting to examine the functional consequences of this variant.

Splice site 4, located between S8 and S9, shows several variants which may arise from the usage of several alternative exons and/or splice sites. Insertion of four amino acids (SRKR) at this splice site resulted in an apparent decrease in voltage/Ca^{2+} sensitivity by 10 mV in a human variant (Tseng-Crank *et al.*, 1994), whereas the same

Table 2
Splice Variants in Vertebrate BK$_{Ca}$ Channel α-Subunit, Slo1

Site	Region[a]	Sequence[b]	h^c	m^d	r^e	c^f	rb^g	t^h	$ch^{i,j}$	$\Delta V_{1/2}$ (mV)[k]
1[f]	NH$_2$	MKPFEVSLPPPPPS (14)							$\equiv^l$	
		MSNNINANNLNTDSSSS (17)							$\equiv$	
2	S1–S2	SRTADSLI (8)							$\equiv$	
3	S6–S7	GRKAMFARYVPEIAALILNRKKYGGTFNSTR (31)							$\equiv$	
4	S8–S9	SRKR (4)	$\equiv$	$\equiv$					$\equiv$	$\pm 10^m$
		SRKRHRAQKKEAMPMSYKRCAAGIHTDPTRC (31)						$\equiv$		
		SRKRYALFVNFSSNLHPLSSLLTTGLTICQEFKKREIIYI (40)							$\approx(20)^n$	
5	S8–S9	IYF (3)	$\equiv$	$\equiv$					$\equiv$	
		KVAARSRYSKDPFEFKKETPNSRLVTEPV (29)	$\equiv$	$\approx(29)$						
		IYFKVAARSRYSKDPFEFKKETPNSRLVTEPV (32)	$\equiv$							
		RPKMSIYKRMRRACCFDCGRSERDCSCMSGRVRG NVDTLERAFPLSSVSVNDCSTSFRAF (60)	$\equiv$		$\approx(58–61)$		$\approx(58)$	$\approx(61)$	$\approx(59–61)$	-20 to -40°
6	S8–S9	RFSCPFLP (8)	$\equiv$						$\equiv$	
7	S9–S10	QGMHLGVTQHQIYAVX (fs/sc)[p]				$\equiv$				
8	S9–S10	AKPAKLPLVSVNQEKNSGTHILMITEL (27)	$\equiv$	$\approx(27)$		$\equiv$			$\approx(28)$	
9	>S10	QKHS (0)								
10	-COOH	KYVQEDRL(8)	$\approx(8–9)$							
		NRKEMVY (7)	$\equiv$	$\approx(61)$						
		(NSTRMNRMGGQEKKWFTDEPDNAYPRNIQIKP MSTHMANQINQYKSTSSLIPPIREVEDEC (60)							$\approx(28)$	

Observed in: (h^c, m^d, r^e, c^f, rb^g, t^h, $ch^{i,j}$)

[a] **-NH₂, Amino terminus; S1–S2**, between transmembrane segments S1 and S2; S6–S7, between transmembrane segment S6 and hydrophobic segment S7; S8–S9, between hydrophobic segments S8 and S9; S9–S10, between hydrophobic segments S9 and S10. -COOH, >S10, downstream S10 carboxyl terminus.

[b] In parentheses are the number of amino acids inserted. Insertless splice variations have been reported in sites 2 and 4–8 but are only shown for site 9.

[c] h, Human; Tseng-Crank et al., 1994; McCobb et al., 1995; Pallanck and Ganetzky, 1994.

[d] m, Mouse; Pallanck and Ganetzky, 1994; Butler et al., 1993; Ferrer et al., 1996.

[e] r, Rat; Saito et al., 1997; Xie and McCobb, 1998.

[f] c, Canine; Vogalis et al., 1996.

[g] rb, Rabbit; Morita et al., 1997.

[h] t, Turtle; Jones et al., 1998.

[i] ch, Chicken; Rosenblatt et al., 1997; Navaratnam et al., 1997.

[j] Spice site 1 proposed for chicken slo (chSlo1) is absent in hSlo1 and mSlo1 (McCobb et al., 1995).

[k] $\Delta V_{1/2}$, Shift in the half-activation potential induced by the insert; (−) leftward shift reflects an increase in voltage/Ca^{2+} sensitivity of the variant.

[l] ≡, Identical.

[m] ±, Tseng-Crank et al. (1994) reported a decrease whereas Rosenblatt et al. (1997) reported an increase in voltage/Ca^{2+} sensitivity by this insert. The opposite results may be due to differences in the background clones.

[n] ≈, Homologous.

[o] Saito et al., 1997; Xie and McCobb, 1998; Ramanathan et al., 1999.

[p] fs/sc, Frame shift induced by 104-bp deletion leads to 15 different amino acids, a premature stop codon, and lack of function.

[q] Unpublished.

insertion increased the apparent voltage/Ca^{2+} sensitivity by 10 mV in a chicken variant (Rosenblatt *et al.*, 1997). These opposite results may be due to the splice variant background used in these studies.

The most common splice variants are found in splice site 5, located in the nonconserved linker region between S8 and S9 (Fig. 3), and arise from the usage of several alternative splice junctions or exons. In human tissues, four splice variants with inserts of between 3 and 60 amino acids have been found. As discussed below, one of these splice variants ($\sim$60-amino-acid insert), first isolated from human pancreatic islets (Ferrer *et al.*, 1996), profoundly changes the functional properties of Slo1, including an increase in voltage/Ca^{2+} sensitivity and kinetics. Besides being found in humans, this variant has also been reported in rat, chicken, rabbit, and turtle (Table 2). In rat Slo1 (rSlo1) channels, this insert increases the voltage/Ca^{2+} sensitivity, as reflected by a leftward shift of $\sim$20–30 mV in the voltage activation curve and a change in the channel activation and deactivation kinetics (Saito *et al.*, 1997). Xie and McCobb (1998) found that the mRNA of this splice variant decreases in hypophysectomized rats and concluded that the alternative splicing machinery producing Slo channel variants in the adrenal gland is controlled by stress hormones. A homologous splice insertion in the chicken Slo1 channels shows similar functional properties: the voltage activation curve is shifted to more negative potentials by 30 mV and deactivation rates are slower (Ramanathan *et al.*, 1999). In rabbit, this splice variant also shows an increase in apparent voltage/Ca^{2+} sensitivity of the channel (Morita *et al.*, 1997).

Splice site 7, between hydrophobic segments S9 and S10, has an insert of 27 amino acids in human, mouse, and chicken. The functional consequences of this insert should be interesting to explore, because it lies in the most conserved region of the Slo1 channels, close to the "Ca^{2+}-bowl." In rabbit, at splice site 8, located at the N-terminus of the "Ca^{2+}-bowl," a 104-base-pair splicing deletion introduces a frame shift and a premature termination codon and leads to loss of channel function (Morita *et al.*, 1997).

2.3. Physiological Impact

Because heterologous expression of splice variants introduces changes in BK_{Ca} channel phenotype (Table 2), investigators have begun to determine if these changes, together with a differential distribution, may explain marked differences in BK_{Ca} channel function found in native cells.

A differential distribution of splice variants was first reported in human brain (Tseng-Crank *et al.*, 1994), but the functional role of splice variants in different brain regions has yet to be explored. In hair cells from the inner ear, this issue has received more attention. Hair cells in the inner ear of turtles and chickens are electrically tuned by a feedback mechanism involving Slo1 channels and voltage-dependent Ca^{2+} channels: whether these cells respond to low or high frequency stimulation is determined by the *density* and *kinetic* properties of BK_{Ca} channels. Cells responding to low frequency have slowly activating Slo1 channels in low density, whereas those responding to high frequency have fast-activating Slo1 channels in high densities and are arranged in a tonotopic manner according to their frequency (Howard and Hudspeth, 1988; Fuchs and Evans, 1990). Several investigators have explored the possibility that splice variants with different kinetics and Ca^{2+}/voltage sensitivities underlie the diverse properties of BK$_{Ca}$ channels along the cochlea tonotopic map. Indeed, mRNAs of Slo1 splice variants are differentially distributed along the tonotopic map, but a close correlation between distribution and function needs to be established (Rosenblatt *et al.*, 1997; Navaratnam *et al.*, 1997; Jones *et al.*, 1998). These findings emphasize that alternative splicing of BK$_{Ca}$ channels may provide a mechanism to produce channel isoforms with the distinct properties demanded by the physiological requirements of specific cells.

3. MODULATORY SUBUNITS OF SLO1

3.1. Molecular Properties of Transmembrane β1- and β2-Subunits

In most tissues, similar to the K$^+$ currents measured in expressed Slo1 channels (Fig. 4A), BK$_{Ca}$ channels produce sustained "noninactivating" currents upon depolarization. However, in native cells they show large variations in their voltage/Ca^{2+} sensitivities. Furthermore, in chromaffin cells and hippocampal neurons, transient "inactivating" BK$_{Ca}$ currents are also present. As discussed in this section, molecular cloning of β-subunits and their heterologous expression with Slo1 has made it evident that association of β-subunits with α-subunits is one of the mechanisms underlying large variations in voltage/Ca^{2+} sensitivities and inactivation of BK$_{Ca}$ channels (Table 3).

The β1-subunit of Slo1 (BK$_{Ca}$ channel α-subunit) was first found by biochemical purification of the channel complex from bovine trachea and aorta (Knaus *et al.*, 1994b,c). Cloning and functional coexpression of this smooth muscle β1-subunit with the pore-forming Slo1 α-subunit yields noninactivating currents (Fig. 4B) that show an enormous increase in the apparent voltage/Ca^{2+} sensitivity (in symmetrical 110 mM K$^+$) (McManus *et al.*, 1995; Wallner *et al.*, 1995). This sensitization by the β-subunit is "switched on" by micromolar [Ca^{2+}]$_i$ (see Fig. 7A) (Meera *et al.*, 1996). β1-Subunit mRNA is abundant in smooth muscles whereas in other tissues, including brain and skeletal muscle, it is low (Jiang *et al.*, 1999). It is also found in heart; however, the functional role of β1-subunits in this tissue is intriguing, as the pore-forming BK$_{Ca}$ channel α-subunit (Slo1) is absent in cardiac myocytes.

We have recently identified a novel human β-subunit (β2) that shows 43% sequence identity to the β1-subunit and is characterized by a longer and unique N-terminus with charged residues (Fig. 4C,D). Both β1- and β2-subunits have two membrane-spanning regions separated by a large extracellular loop containing consensus sequences for N-linked glycosylation (marked with ψ in Fig. 4B–D). Both subunits have four conserved cysteines in the extracellular loop that may form structurally important disulfide bridges (Fig. 4D). The β-subunit models in Fig. 4B,C are based on hydrophobicity analysis (Knaus *et al.*, 1994b), *in vitro* translation (Jiang *et al.*, 1999; Wallner *et al.*, 1999), and cross-linking studies (Knaus *et al.*, 1994a).

3.2. Molecular Determinants of β2-Induced Fast Inactivation of Slo1 Channels

The β2-subunit yields inactivating currents when coexpressed with the human BK$_{Ca}$ channel α-subunit, hSlo1 (Fig. 4C), similar to those found in native tissues like

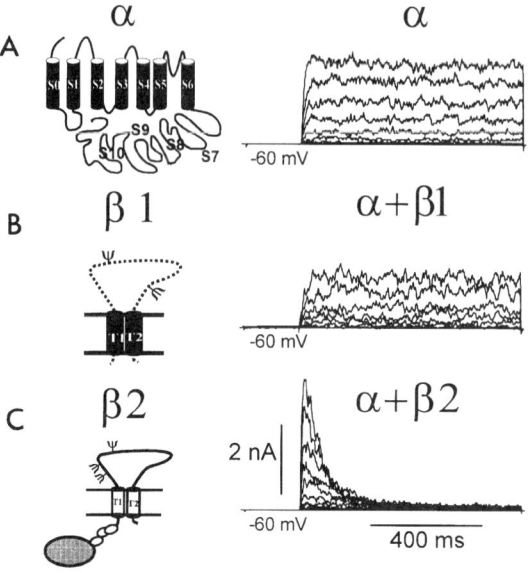

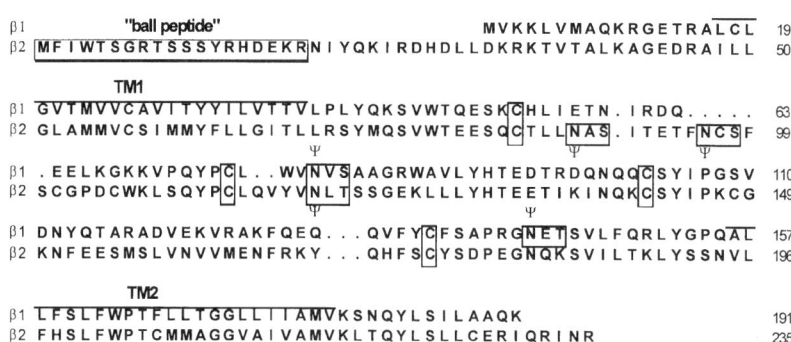

Figure 4. Transmembrane β1- and β2-subunits modulate hSlo1 channel function. (A) Schematic representation of the BK_{Ca} channel α-subunit, Slo1. Noninactivating K^+ currents are from oocytes expressing hSlo1 (α). (B) Schematic representation of smooth muscle β1-subunit. Coexpression of β1 with hSlo1 ($\alpha + \beta$1) produces noninactivating macroscopic currents. (C) Topology of β2 subunit. Inactivating BK_{Ca} currents are obtained after coexpression of hslo1 and β2 subunits ($\alpha + \beta$2). Currents from inside-out patches were elicited after a 200-ms prepulse to -60 mV (to relieve inactivation) followed by test pulses from -20 mV to 110 mV in increments of 10 mV. For comparison purposes, the same protocol was applied in (A) and (B). Holding potential was 0 mV. Currents were measured in symmetrical 110 mM potassium methanesulfonate (Kmes) and 100 nM intracellular Ca^{2+}. (D) Sequence alignment of β1 and β2 subunits. Transmembrane regions, TM1 and TM2, are marked with lines. . . . , Spaces introduced for better alignment. Ψ, Consensus sequences for N-linked glycosylation are boxed. Conserved cysteines in the extracellular loop are boxed. Nineteen N-terminal amino acids that form the "ball peptide" are marked.

Table 3

Slo1(BK$_{Ca}$ α-Subunit) Interacting Subunits

Subunit	Topology[a]	Necessary region in α-subunit[b]	Interacting region in β-subunit[b]	Change in α-subunit function[c]	$\Delta V_{1/2}$ (mV)[d]	Reference
β1	TM	Part of NH$_2$-terminus and transmembrane domain SO	?	Activation kinetics, increase in Ca^{2+}/V, pharmacology (IbTX, CTX, DHS-I)	−100[e]	Toro et al., 1998 (review)
β2	TM	?	?	Inactivation, activation kinetics increase in Ca^{2+}/V, pharmacology (CTX, DHS-I)	−100[e]	Wallner et al., 1999
Slob[f]	CYT	—COOH	?	Increase in Po, cellular redistribution	?	Schopperle et al., 1998
dSLIP1[g]	CYT	—COOH (132), >S10	—COOH (100)	Decrease in current density, cellular redistribution	None	Xia et al., 1998
Slack[h]	TM	?	?	Decrease in channel amplitude, pharmacology	?	Joiner et al., 1998

[a] TM, Transmembrane; CYT, cytosolic.

[b] —COOH, Carboxyl terminus; in parentheses are the number of C-terminal residues necessary for interaction; >S10, downstream hydrophobic segment S10; ?, unknown.

[c] Increase in Ca^{2+}/V, Increase in calcium and voltage sensitivities; IbTX, iberiotoxin; CTX, charybdotoxin; DHS-I, dehydrosoyasaponin I; Increase in Po, Increase in open-state probability (reflects an increase in voltage sensitivity); decrease.

[d] $\Delta V_{1/2}$: Shift in half-activation potential; (−) negative shift reflects an increase in V/Ca^{2+} sensitivity.

[e] Measured in isotonic 110 mM K$^+$.

[f] Slob interacts with *Drosophila* Slo (dSlo1) and EAG K$^+$ channel, but not with mSlo1 or hSlo1.

[g] dSLIP1 (*Drosophila* slo interacting protein 1) interacts with dSlo1 and hSlo1.

[h] Slack (sequence like a calcium-activated K$^+$ channel) forms K$^+$ channels by itself (25–65 pS) that are inhibited by cytosolic Ca^{2+} but may form heteromultimers with hSlo1.

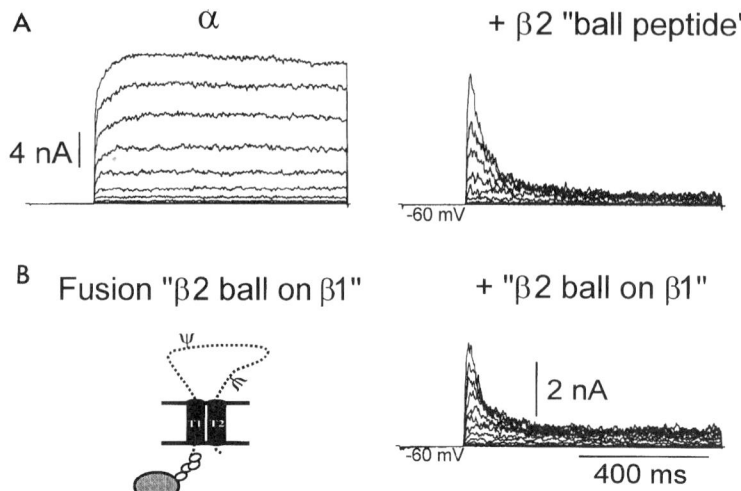

Figure 5. Determinants of $\beta2$-subunit-induced inactivation in Slo1. (A) Currents from an inside-out patch expressing hSlo1 (α-subunit), prior to (*left*) and after (*right*) application of 3 μM synthetic $\beta2$ "ball" peptide (19 N-terminal amino acids of $\beta2$ subunit; see Fig. 4D). (B) *Left*: Model showing a fusion construct consisting of the bb "ball peptide" attached to the noninactivating $\beta1$ subunit. *Right*: Currents from an inside-out patch expressing hSlo1 and the fusion construct. Pulse protocol is the same as in Fig. 4. Currents were measured in symmetrical 110 mM potassium methanesulfonate (Kmes) and 500 nM free Ca^{2+}. (Reproduced with permission from Wallner *et al.*, 1999.

chromaffin and hippocampal neurons (Wallner *et al.*, 1999). This inactivation is removed by the intracellular application of trypsin, as observed in the inactivating *Shaker* B voltage-dependent K^+ channel (Hoshi *et al.*, 1990). This property suggests the involvement of a cytosolic domain of the $\beta2$-subunit and a "ball and chain" type of mechanism for inactivation (Bezanilla and Armstrong, 1977). Consistent with this notion, deletion of 19 N-terminal amino acids of the $\beta2$-subunit [$\beta2$-IR (IR for inactivation removed)] eliminates the inactivation in hSlo1 BK_{Ca} channels (*see* Fig. 6C). Further, addition of a synthetic peptide encompassing 19 N-terminal amino acids of the $\beta2$-subunit (Fig. 4D) induced inactivation of otherwise noninactivating hSlo1 currents (Fig. 5A). Moreover, fusion of these 19 N-terminal amino acids to the smooth muscle noninactivating $\beta1$-subunit caused inactivation in hSlo1 channels (Fig. 5B). Thus, the molecular domain of the $\beta2$-subunit that causes inactivation of hSlo1 is the 19 N-terminal amino acids, which form an inactivation "ball peptide" analogous to "ball peptides" located at the N-terminus of voltage-gated K^+ channel α-subunits or their *cytosolic β-subunits*.

The results summarized above show that the $\beta2$-subunit is the first identified member of a possible new family of transmembrane β-subunits that cause inactivation of BK_{Ca} channels in native tissues.

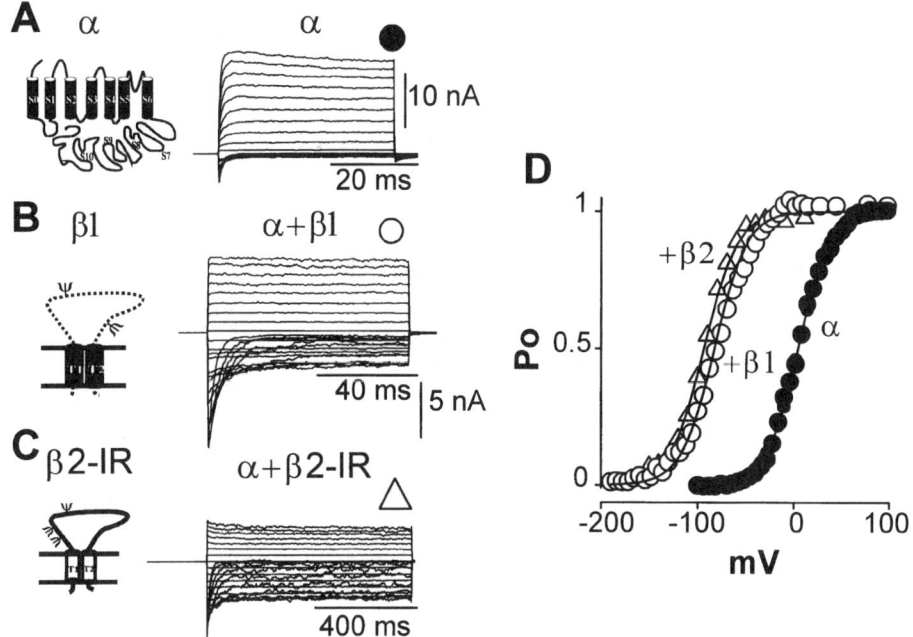

Figure 6. β1- and β2-subunits increase the voltage/Ca^{2+} sensitivities of hSlo1. (A) *Left*: Model of hslo1. *Right*: Currents measured from oocytes expressing hSlo1 (α-subunit). Test pulses are from -100 mV to 116 mV with intervals of 12 mV. (B) *Left*: Model of β1-subunit. *Right*: Currents obtained from oocytes expressing $\alpha + \beta$1 subunits. Test pulses are from -199 mV to 98 mV in steps of 12 mV. (C) *Left*: Model of β2-subunit with inactivation removed (β2-IR). *Right*: Currents obtained from oocytes coexpressing α-subunit and β2-IR. Test pulses are from -150 mV to $+90$ mV in steps of 10 mV. Note that currents from oocytes expressing β1 or β2-IR show large tail currents at very negative potentials (e.g., -200 mV), likely due to the large driving force and high open-state probability of the channel at the holding potential of 0 mV. Therefore, to construct voltage activation curves (open-state probability, Po, vs. voltage), currents were measured at the steady-state value. (D) Po vs. potential plots of hSlo1 alone (α, $\bullet$) or in the presence of β1-subunit (O) or β2-subunit ($\triangle$). Half-activation potentials, $V_{1/2}$ are 4 mV, -82 mV, and -93 mV for α-subunit alone, $+\beta$1-subunit and $+\beta$2-subunit, respectively. All measurements were performed in symmetrical 110 mM K^+, with $[Ca^{2+}]_i = 10\,\mu M$, from inside-out patches of *Xenopus* oocytes.

3.3. Facilitation of Channel Opening and Pharmacological Changes Caused by β-Subunits

The β1- and β2-subunits share some modulatory properties. Both cause a profound increase in the apparent voltage/Ca^{2+} sensitivity of the pore-forming α-subunit of hSlo1 channels (Fig. 6D). This is evident when currents are measured after removal of the inactivating "ball" from the β2-subunit (Fig. 6C; β2-IR). Unlike channels formed by the α-subunit alone ($\bullet$), the channels associated either with the β1-(O) or

$\beta2$-subunit (Δ) open at very negative potentials in 10 μM Ca^{2+}. This is clear from the open-state probability versus potential plots (Fig. 6D), which show that the half-activation potential ($V1/2$) for the α-subunit alone is 4 mV ($\bullet$) whereas for the $\beta1$-subunit it is -82 mV (O) and for the $\beta2$-subunit it is -93 mV (Δ). In addition, both $\beta1$- and $\beta2$-subunits modify activation and deactivation kinetics of the hSlo1 channels (Meera *et al.*, 1996; Dworetzky *et al.*, 1996; Wallner *et al.*, 1999).

Coexpression of β-subunits also changes the pharmacology of the pore-forming α-subunit of the BK$_{Ca}$ channel, Slo1. Previously, it was thought that dehydrosoyasaponin I (ΔHS-I) specifically activated BK$_{Ca}$ channels formed by α- and $\beta1$-subunits (McManus *et al.*, 1995). However, we have recently shown that β-subunits increase the sensitivity of the α-subunit (hSlo1) to DHS-I from the micromolar (EC$_{50}$ of 7. 6 μM) to the nanomolar range (EC$_{50}$ for $\beta1 \sim 129$ nM; EC$_{50}$ for $\beta2 \approx 165$ nM) (Tanaka *et al.*, 1997; Wallner *et al.*, 1999).

The effect of the pore blocker charybdotoxin (CTX) on the α-subunit (Slo1) is also altered by the presence of β-subunits. Electrophysiological experiments show that the $\beta2$-subunit decreases the sensitivity of the α-subunit to CTX from an EC$_{50}$ of 1 nM (α) to 58 nM ($\alpha + \beta2$) (Wallner *et al.*, 1999). However, the $\beta1$-subunit apparently does not alter the sensitivity of the α-subunit to CTX (Dworetzky *et al.*, 1996). The $\beta1$-subunit also lowers the sensitivity of hSlo1 channels ~10-fold to another pore blocker, iberiotoxin (Dworetzky *et al.*, 1996).

3.4. Molecular Determinants for β1-Subunit Modulation in Slo1 Channels

As mentioned before, the β-subunits dramatically increase the apparent voltage and Ca^{2+} sensitivity of mammalian hSlo1 α-subunits. This is illustrated in Figs. 6D and 7A, where the $\beta1$-subunit coexpressed with hSlo1 α-subunits leads to a ~100-mV leftward shift of the voltage activation curve when measured at Ca^{2+} concentrations above 3 μM. However, the mammalian $\beta1$-subunit has no effect when coexpressed with the *Drosophila* Slo1 homolog (Fig. 7B) even when Ca^{2+} concentrations are increased up to 4.3 mM. We utilized this difference and constructed chimeras between hSlo1 and dSlo1 and tested for β-subunit regulation (Fig. 7C). The minimum region required for gain of $\beta1$–subunit regulation in dSlo1 comprises 41 N-terminal amino acids, including the hydrophobic region S0 and the extracellular N-terminus from hSlo1. These results and the fact that both $\beta1$- and $\beta2$-subunits increase the voltage/Ca^{2+} sensitivity of Slo1 channels suggest that the unique N-terminus of Slo1 including S0 may form the functional interaction site with modulatory β-subunits.

3.5. Other Slo-Interacting Subunits

Recently, interacting proteins of Slo1 channels called Slob (for *Slowpoke* binding protein) (Schopperle *et al.*, 1998) and dSLIP1 (for dSlo interacting protein) (Xia *et al.*, 1998) were isolated by two-hybrid screening using carboxyl-terminal domains as bait

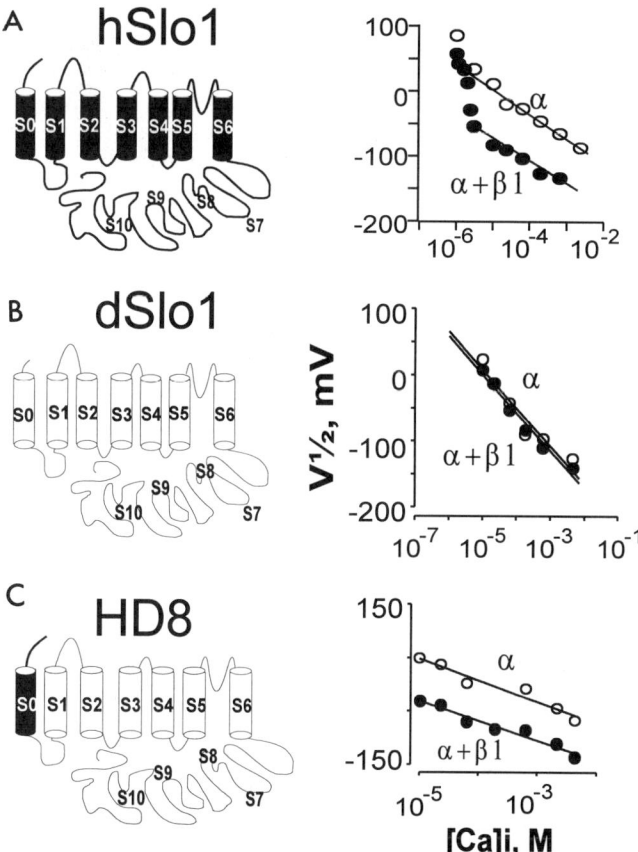

Figure 7. hSlo1 N-terminus along with its transmembrane segment S0 mediates the functional effect of $\beta1$-subunit. (A) *Left*: Topology of hSlo1. *Right*: Plot of $V_{1/2}$ vs. $[Ca^{2+}]_i$. The $\beta1$-subunit causes a dramatic increase of the voltage/Ca^{2+} sensitivity of hSlo1 (α) above 3 M Ca^{2+}. (B) *Left*: Model of dSlo1 channels. *Right*: Plot of $V_{1/2}$ vs. $[Ca^{2+}]_i$. Coexpression of $\beta1$ subunit ($\alpha + \beta1$) has no effect on dSlo1 voltage/Ca^{2+} sensitivity up to 4.3 mM $[Ca^{2+}]_i$. (C) *Left*: Model of a chimeric construct, called HD8, with N-terminus and S0 from hSlo1 and the rest of the protein from dSlo1. *Right*: Plot of $V_{1/2}$ vs. $[Ca^{2+}]_i$. The N-terminus and S0 from hSlo1 made dSlo1 channels responsive to the $\beta1$-subunit, as evident from the large increase in the voltage/Ca^{2+} sensitivity of HD8 when coexpressed with the $\beta1$ subunit. (Reproduced with permission from Wallner *et al.*, 1996).

proteins (Table 3). dSLIP1 and the dSlo1 mRNA are expressed throughout the *Drosophila* nervous system. dSLIP1 seems to reduce the number of dSLo1 or hSlo1 channels when coexpressed in the *Xenopus* oocytes, which is reflected in the reduction of current amplitude when compared to the currents obtained from Slo1 alone, whereas Slob seems to increase dSlo1 channel activity but not that of hSlo1. Both Slob and dSLIP1 have *PSO-9S, Dlg* and *Zo-1* (PDZ) protein domains (Ponting *et al.*, 1997), suggesting a possible role in bringing Slo1 channels in the stream of cellular signaling pathways.

3.6. Physiological Impact

Recent studies have shown that the mRNA of the modulatory β1-subunit is differentially distributed along the tonotopic map of cochlear cells, being more abundant in the low-frequency region of the hair cell. These results are in agreement with the presence of slowly activating BK_{Ca} channels in the low-frequency region (Ramanathan et al., 1999), since the β1-subunit slows down the activation kinetics of Slo1 (Meera et al., 1996). Thus, the function and expression of Slo1 channels can be exquisitely modulated by a differential expression and association of modulatory subunits. A multitude of channel variants can be imagined by the combination of splice variants, modulatory subunits, and heteromultimerization. Our task is now to define the molecular anatomy and function of these proteins in native tissues.

4. PHYSIOLOGICAL FUNCTION OF SLO1 CHANNELS

Molecular biology not only has enabled us to understand the basic mechanisms and molecular determinants of Slo1 function and regulation, but also has opened new avenues to determine their cellular function of the Slo1 channels. In general, the physiological role of Slo1 BK_{Ca} channels is to regulate intracellular Ca^{2+}, which is pivotal for cell function. Activation of BK_{Ca} channels causes hyperpolarization, thereby reducing cellular excitability; conversely, their inhibition leads to depolarization and an increase in cellular activity. Therefore, Slo1 channel distribution, abundance and colocalization with other channels and receptors, as well as the association of Slo1 channels with modulatory subunits, should determine their specific role in each cell type. In particular, Slo1 colocalization with its β1-subunit may profoundly sensitize BK_{Ca} channels to Ca^{2+} and allow their opening at low Ca^{2+} concentrations.

The physiological role of Slo1 channels has been explored in detail in neurons and smooth muscle. In neurons, they are colocalized with voltage-dependent Ca^{2+} channels and serve as Ca^{2+} sensors regulating neuronal firing and neurotransmitter release. In fact, molecular anatomy studies have shown that Slo1 protein is preferentially distributed in major projection tracts and their terminal areas, consistent with a functional role in neural transmission (Knaus et al., 1996). In neurons, the presence of the β1-subunit does not appear to be crucial for Slo1 channel function, because the mRNA levels of this subunit are low or absent. This may explain the necessity for Slo1 to be colocalized with Ca^{2+} channels; Ca^{2+} influx through Ca^{2+} channels would raise $[Ca^{2+}]_i$ to levels on the order of hundreds of micromolar, thereby facilitating Slo1 activity and the participation of Slo1 channels in neuronal function. However, as yet unidentified brain-specific subunits may associate with Slo1 channels in neurons.

In smooth muscles, Slo1 channels are found at high density and are key regulators of vascular tone. Their activation causes vasorelaxation, whereas their inhibition causes vasoconstriction. Thus, BK_{Ca} channels are critical in the maintenance of normal oxygen supply and cardiac function. Biochemical (Knaus et al., 1994b) and electrophysiological studies (Tanaka et al., 1997) in smooth muscles have revealed that Slo1 channels are tightly associated with β1-subunits. This association may be necessary for their efficient opening in smooth muscle, which enables them to sense Ca^{2+} released from intracellular stores (Nelson et al., 1995).

Notable is the *absence* of Slo1 channels in heart myocytes, which is consistent with the prolonged action potentials in these cells. The lack of Slo1 channels in cardiac myocytes may be critical to avoid an early termination of the action potential and of Ca^{2+}-induced Ca^{2+} release from intracellular stores, which is indispensable for heart contraction.

Thus, the abundance of Slo1, the differential distribution of Slo1 versus its splice variants, and the association of Slo1 with modulatory proteins will all determine the level of BK_{Ca} channel activity, as required for the fine-tuning of cellular excitability and function.

SUMMARY

BK_{Ca} channels are widely expressed in tissues and are especially abundant in smooth muscles. Their activation by voltage and calcium provides a negative-feedback mechanism to limit further Ca^{2+} entry and thereby regulate cellular excitability. In vertebrates, BK_{Ca} channels are formed by two subunits, a pore-forming α-subunit and a modulatory transmembrane β-subunit. They also interact with cytosolic subunits and structurally related channels. The BK_{Ca} channel pore-forming α-subunit has seven transmembrane segments (S0–S6). Segments S1–S6 are homologous to voltage-dependent K^+ (Kv) channels with a positively charged S4 region and represent the essential part of the voltage sensor. Segments S5 and S6 flank a pore loop and constitute the K^+-selective pore with a GYGD signature sequence. BK_{Ca} channels deviate from Kv channels in possessing (a) an additional transmembrane region (S0) at the N-terminus and (b) a long C-terminus implicated in Ca^{2+} binding and regulation. BK_{Ca} channels found in native tissues show enormous diversity in pharmacology and biophysical properties, which may arise from alternative splicing and association with modulatory subunits and by formation of heteromultimeric channels with α-subunit homologs. Transmembrane β-subunits do not form channels by themselves but profoundly influence the function of the pore-forming α-subunit, affecting the kinetics, Ca^{2+} and voltage sensitivities, and pharmacology of the channel. Cytosolic subunits primarily alter BK_{Ca} channel cellular distribution. Structurally related K^+ channels can form heteromultimers with the BK_{Ca} channel α-subunit, thereby modifying single-channel conductance and pharmacology. Recently, Molecular cloning and functional expression are providing the tools to investigate and understand the functional role of BK_{Ca} channels in physiological and pathological conditions.

ACKNOWLEDGEMENTS

This work was supported by NIH grant HL54970 (L.T.) and AHA National Center Grant in-Aid 9750745N (P.M.). L.T. is an Established Investigator of the American Heart Association.

REFERENCES

Adelman, J. P., Shen, K. Z., Kavanaugh, M. P., Warren, R. A., Wu, Y. N., Lagrutta, A., Bond, C. T., and North, R. A., 1992, Calcium-activated potassium channels expressed from cloned complementary DNAs, *Neuron* **9**:209–216.

Atkinson, N. S., Robertson, G. A., and Ganetzky, B., 1991, A component of calcium-activated potassium channels encoded by the *Drosophila slo* locus, *Science* **253**:551–555.

Bezanilla, F. and Armstrong, C. M., 1977, Inactivation of the sodium channel. I. Sodium current experiments, *J. Gen. Physiol.* **70**:549–566.

Brenner, R., Thomas, T. O., Becker, M. N., and Atkinson, N. S., 1996, Tissue-specific expression of a Ca^{2+}-activated K^+ channel is controlled by multiple upstream regulatory elements, *J. Neurosci.* **16**:1827–1835.

Butler, A., Tsunoda, S., McCobb, D. P., Wei, A., and Salkoff, L., 1993, *mSlo*, a complex mouse gene encoding "maxi" calcium-activated potassium channels, *Science* **261**:221–224.

Diaz, F., Meera, P., Amigo, J., Stefani, E., Alvarez, O., Toro, L., and Latorre, R., 1998, Role of the S4 segment in a voltage-dependent calcium-sensitive potassium (*hSlo*) channel, *J. Biol. Chem.* **273**:32430–32436.

Dworetzky, S. I., Trojnacki, J. T., and Gribkoff, V. K., 1994, Cloning and expression of a human large-conductance calcium-activated potassium channel, *Brain Res. Mol. Brain Res.* **27**:189–193.

Dworetzky, S. I., Boissard, C. G., Lum-Ragan, J. T., McKay, M. C., Post-Munson, D. J., Trojnacki, J. T., Chang, C. P., and Gribkoff, V. K., 1996, Phenotypic alteration of a human BK (hSlo) channel by hSloβ subunit coexpression: Changes in blocker sensitivity, activation/relaxation and inactivation kinetics, and protein kinase A modulation, *J. Neurosci.* **16**:4543–4550.

Elkins, T., Ganetzky, B., and Wu, C. F., 1986, A *Drosophila* mutation that eliminates a calcium-dependent potassium current, *Proc. Natl. Acad. Sci. U.S.A.* **83**:8415–8419.

Ferrer, J., Wasson, J., Salkoff, L., and Permutt, M. A., 1996, Cloning of human pancreatic islet large conductance Ca^{2+}-activated K^+ channel (hSlo) cDNAs: Evidence for high levels of expression in pancreatic islets and identification of a flanking genetic marker, *Diabetologia* **39**:891–898.

Fuchs, P. A. and Evans, M. G., 1990, Potassium currents in hair cells isolated from the cochlea of the chick, *J. Physiol.* (*London*) **429**:529–551.

Hoshi, T., Zagotta, W. N., and Aldrich, R. W., 1990, Biophysical and molecular mechanisms of *Shaker* potassium channel inactivation, *Science* **250**:533–538.

Howard, J., and Hudspeth, A. J., 1988, Compliance of the hair bundle associated with gating of mechanoelectrical transduction channels in the bullfrog's saccular hair cell, *Neuron* **1**:189–199.

Jiang, G. J., Zidanic, M., Michaels, R. L., Michael, T. H., Griguer, C., and Fuchs, P. A., 1997, CSlo encodes calcium-activated potassium channels in the chick's cochlea, *Proc. R. Soc. London, Ser. B.* **264**:731–737.

Jiang, Z., Wallner, M., Meera, P., and Toro, L., 1999, Cloning and characterization of human and rodent MaxiK channel β subunit genes: Cloning and characterization, *Genomics* **55**:57–67.

Joiner, W. J., Tang, M. D., Wang, L. -Y., Dworetzky, S. I., Boissard, C. G., Gan, L., Gribkoff, V. K., and Kaczmarek, L. K., 1998, Formation of intermediate-conductance calcium-activated potassium channels by interaction of Slack and Slo subunits, *Nat. Neurosci.* **1**:462–469.

Jones, E. M., Laus, C., and Fettiplace, R., 1998, Identification of Ca^{2+}-activated K^+ channel splice variants and their distribution in the turtle cochlea, *Proc. R. Soc. London Ser. B* **265**:685–692.

Knaus, H. G., Eberhart, A., Kaczorowski, G. J., and Garcia, M. L., 1994a, Covalent attachment of charybdotoxin to the β-subunit of the high conductance Ca^{2+}-activated K^+ channel: Identification of the site of incorporation and implications for channel topology, *J. Biol. Chem.* **269**:23336–23341.

Knaus, H. G., Folander, K., Garcia-Calvo, M., Garcia, M. L., Kaczorowski, G. J., Smith, M., and Swanson, R., 1994b, Primary sequence and immunological characterization of β-subunit of high conductance Ca^{2+}-activated K^+ channel from smooth muscle, *J. Biol. Chem.* **269**:17274–17278.

Knaus, H. G., Garcia-Calvo, M., Kaczorowski, G. J., and Garcia, M. L., 1994c, Subunit composition of the high conductance calcium-activated potassium channel from smooth muscle, a representative of the *mSlo* and *slowpoke* family of potassium channels, *J. Biol. Chem.* **269**:3921–3924.

Knaus, H. G., Schwarzer, C., Koch, R. O., Eberhart, A., Kaczorowski, G. J., Glossmann, H., Wunder, F., Pongs, O., Garcia, M. L., and Sperk, G., 1996, Distribution of high-conductance Ca^{2+}-activated K^+ channels in rat brain: Targeting to axons and nerve terminals, *J. Neurosci.* **16**:955–963.

Lagrutta, A., Shen, K. Z., North, R. A., and Adelman, J. P., 1994, Functional differences among alternatively spliced variants of *Slowpoke*, a *Drosophila* calcium-activated potassium channel, *J. Biol. Chem.* **269**:20347–20351.

Lopez, G. A., Jan, Y. N., and Jan, L. Y., 1994, Evidence that the S6 segment of the *Shaker* voltage-gated K$^+$ channel comprises part of the pore, *Nature* **367**:179–182.

McCobb, D. P., Fowler, N. L., Featherstone, T., Lingle, C. J., Saito, M., Krause, J. E., and Salkoff, L., 1995, A human calcium-activated potassium channel gene expressed in vascular smooth muscle, *Am. J. Physiol.* **269**:H767–H777.

McManus, O. B., Helms, L. M., Pallanck, L., Ganetzky, B., Swanson, R., and Leonard, R. J., 1995, Functional role of the beta subunit of high conductance calcium-activated potassium channels, *Neuron* **14**:645–650.

Meera, P., Wallner, M., Jiang, Z., and Toro, L., 1996, A calcium switch for the functional coupling between α (*hslo*) and β subunits (K$_{V,Ca}$β) of maxi K channels, *FEBS Lett.* **382**:84–88.

Meera, P., Wallner, M., Song, M., and Toro, L., 1997, Large conductance voltage- and calcium-dependent K$^+$ channel, a distinct member of voltage-dependent ion channels with seven N-terminal transmembrane segments (S0–S6), an extracellular N terminus, and an intracellular (S9–S10) C terminus, *Proc. Natl. Acad. Sci. U.S.A.* **94**:14066–14071.

Morita, T., Hanaoka, K., Morales, M. M., Montrose-Rafizadeh, C., and Guggino, W. B., 1997, Cloning and characterization of maxi K$^+$ channel alpha-subunit in rabbit kidney, *Am. J. Physiol.* **273**:F615–F624.

Nara, M., Dhulipala, P. D., Wang, Y. X., and Kotlikoff, M. I., 1998, Reconstitution of β-adrenergic modulation of large conductance, calcium-activated potassium (maxi-K) channels in *Xenopus* oocytes: Identification of the cAMP-dependent protein kinase phosphorylation site, *J. Biol. Chem.* **273**:14920–14924.

Navaratnam, D. S., Bell, T. J., Tu, T. D., Cohen, E. L., and Oberholtzer, J. C., 1997, Differential distribution of Ca^{2+}-activated K$^+$ channel splice variants among hair cells along the tonotopic axis of the chick cochlea, *Neuron* **19**:1077–1085.

Nelson, M. T., Cheng, H., Rubart, M., Santana, L. F., Bonev, A. D., Knot, H. J., and Lederer, W. J., 1995, Relaxation of arterial smooth muscle by calcium sparks, *Science* **270**:633–637.

Pallanck, L., and Ganetzky, B., 1994, Cloning and characterization of human and mouse homologs of the *Drosophila* calcium-activated potassium channel gene, *slowpoke*, *Hum. Mol. Genet.* **3**:1239–1243.

Ponting, C. P., Phillips, C., Davies, K. E., and Blake, D. J., 1997, PDZ domains: Targeting signalling molecules to sub-membranous sites, *Bioessays* **19**:469–479.

Ramanathan, K., Michael, T. H., Jiang, G. J., Hiel, H., and Fuchs, P. A., 1999, A molecular mechanism for electrical tuning of cochlear hair cells, *Science* **283**:215–217.

Rosenblatt, K. P., Sun, Z. P., Heller, S., and Hudspeth, A. J., 1997, Distribution of Ca^{2+}-activated K$^+$ channel isoforms along the tonotopic gradient of the chicken's cochlea, *Neuron* **19**:1061–1075.

Saito, M., Nelson, C., Salkoff, L., and Lingle, C. J., 1997, A cysteine-rich domain defined by a novel exon in a slo variant in rat adrenal chromaffin cells and PC12 cells, *J. Biol. Chem.* **272**:11710–11717.

Schopperle, W. M., Holmqvist, M. H., Zhou, Y., Wang, J., Wang, Z., Griffith, L. C., Keselman, I., Kusinitz, F., Dagan, D., and Levitan, I. B., 1998, Slob, a novel protein that interacts with the Slowpoke calcium-dependent potassium channel, *Neuron* **20**:565–573.

Schreiber, M., and Salkoff, L., 1997, A novel calcium-sensing domain in the BK channel, *Biophys. J.* **73**:1355–1363.

Schreiber, M., Wei, A., Yuan, A., Gaut, J., Saito, M., and Salkoff, L., 1998, Slo3, a novel pH-sensitive K$^+$ channel from mammalian spermatocytes, *J. Biol. Chem.* **273**:3509–3516.

Stefani, E., Ottolia, M., Noceti, F., Olcese, R., Wallner, M., Latorre, R., and Toro, L., 1997, Voltage-controlled gating in a large conductance Ca^{2+}-sensitive K$^+$ channel (hslo), *Proc. Natl. Acad. Sci. U.S.A.* **94**:5427–5431.

Tanaka, Y., Meera, P., Song, M., Knaus, H.-G., and Toro, L., 1997, Molecular constituents of maxi K$_{Ca}$ channels in human coronary smooth muscle: Predominant α + β subunit complexes, *J. Physiol. (London)* **502**:545–557.

Toro, L., Meera, P., Wallner, M., and Tanaka, Y., 1998, MaxiK$_{Ca}$, a unique member of the voltage-gated K channel superfamily, *News Physiol. Sc.* **13**:112–117.

Tseng-Crank, J., Foster, C. D., Krause, J. D., Mertz, R., Godinot, N., DiChiara, T. J., and Reinhart, P. H., 1994, Cloning, expression, and distribution of functionally distinct Ca^{2+}-activated K$^+$ channel isoforms from human brain, *Neuron* **13**:1315–1330.

Vogalis, F., Vincent, T., Qureshi, I., Schmalz, F., Ward, M. W., Sanders, K. M., and Horowitz, B., 1996, Cloning and expression of the large-conductance Ca^{2+}-activated K^+ channel from colonic smooth muscle, *Am. J. Physiol.* **271:**G629–G639.

Wallner, M., Meera, P., Ottolia, M., Kaczorowski, G., Latorre, R., Garcia, M. L., Stefani, E., and Toro, L., 1995, Characterization of and modulation by a β-subunit of a human maxi K_{Ca} channel cloned from myometrium, *Recept. Channels* **3:**185–199.

Wallner, M., Meera, P., and Toro, L., 1996, Determinant for β-subunit regulation in high-conductance voltage-activated and Ca^{2+}-sensitive K^+ channels: An additional transmembrane region at the N terminus, *Proc. Natl. Acad. Sci. U.S.A.* **93:**14922–14927.

Wallner, M., Meera, P., and Toro, L., 1997, A new family of potassium channels: Relatives of voltage and calcium activated maxi K channels, *Biophys. J.* **72:**A224

Wallner, M., Pratap, M., and Toro, L. 1999, Molecular basis of fast inactivation in voltage and Ca^{2+}-activated K^+ channels: A transmembrane β subunit homolog, *Proc. Natl. Acad. Sci. U.S.A.* **96:**4137–4142.

Wei, A., Solaro, C., Lingle, C., and Salkoff, L., 1994, Calcium sensitivity of BK-type K_{Ca} channels determined by a separable domain, *Neuron* **13:**671–681.

Wei, A., Jegla, T., and Salkoff, L., 1996, Eight potassium channel families revealed by the *C. elegans* genome project, *Neuropharmacology* **35:**805–829.

Xia, X., Hirschberg, B., Smolik, S., Forte, M., and Adelman, J. P., 1998, dSLo interacting protein 1, a novel protein that interacts with large-conductance calcium-activated potassium channels, *J. Neurosci.* **18:**2360–2369.

Xie, J. and McCobb, D. P., 1998, Control of alternative splicing of potassium channels by stress hormones, *Science* **280:**443–446.

Yuan, A., Dourado, M., Butler, A., and Salkoff, L., 1998, A novel K^+ channel requiring both chloride and calcium for activation, *Biophys. J.* **74:**A206.

Yuan, A., Dourado, M., Butler, A., and Salkoff, L., 1999, A chloride sensing domain in a large conductance K^+ channel, *Biophys. J.* **76:**A329.

Chapter 5

Molecular Biology of Inward Rectifier and ATP-Sensitive Potassium Channels

S. L. Shyng, A. N. Lopatin, and C. G. Nichols

1. FUNCTIONAL DIVERSITY OF CARDIAC K$^+$ CHANNELS: THE ROLE OF INWARD RECTIFICATION

Many different types of K$^+$ channels are involved in cardiac electrical activity (Noble, 1965). Various depolarization-activated K$^+$ channels (Kv channels) determine the shape of the cardiac action potential. Cardiac cells also contain Kir channels (Fig. 1) that are open at very negative potentials but show a reduced conductance at positive membrane potentials, a phenomenon termed inward, or anomalous, rectification (Fig. 2A; Katz, 1949). Work over the past 30 years has described three main types of Kir channel in cardiac tissue. (1) The classical *inward rectifier*, I_{K1}, present in atrial and ventricular myocytes, shows "strong" inward rectification. Essentially no current flows through these channels at potentials positive to -40 mV (Noble, 1965; Vandenberg, 1994). (2) The K$_{ATP}$ *channel* is found in ventricular, atrial, and nodal cells. K$_{ATP}$ channels display "weak" rectification, allowing substantial outward current to flow at positive potentials (Noma, 1983; Nichols and Lederer, 1991). (3) The *muscarinic-receptor-activated K$^+$ channel*, K$_{(ACh)}$, is a strong inward rectifier found predominantly in atrial tissue (Bond *et al.*, 1994). K$_{(ACh)}$ channels open in response to acetylcholine released from the vagus nerve and underlie the resultant slowing of the heart rate.

The degree of rectification exhibited by Kir channels is fundamental to their respective roles. For example, because the K$_{ATP}$ channel rectification is "weak," K$_{ATP}$ channel activation causes marked shortening of the action potential, reducing voltage-dependent Ca^{2+} entry and hence conserving ATP in conditions of metabolic stress (Nichols and Lederer, 1991). The strong inward rectification of I_{K1}, on the other hand,

S. L. Shyng, A. N. Lopatin, and C. G. Nichols • Department of Cell Biology and Physiology, Washington University School of Medicine, St. Louis, Missouri 63110. S. L. Shyng's permanent address is Center for Research on Occupational and Environmental Toxicology, Portland, Oregon.
Potassium Channels in Cardiovascular Biology, edited by Archer and Rusch. Kluwer Academic/Plenum Publishers, New York, 2001.

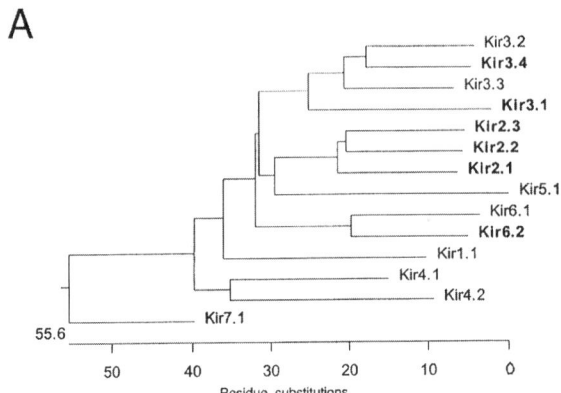

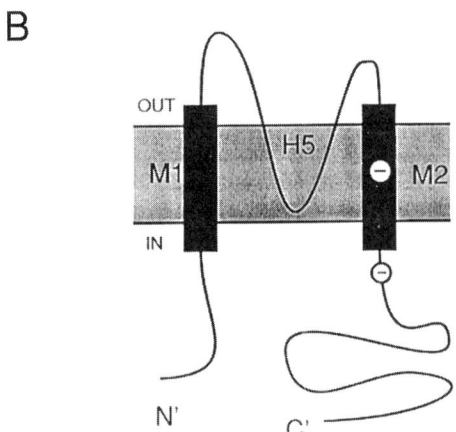

Figure 1. (A) Phylogenetic tree of the Kir channel family. The phylogenetic tree is depicted by an unbalanced branch cladogram generated using Megalign software (DNAStar Inc.). The scale indicates residue substitutions from the ancestral channel. Members of the same subfamily (e.g., Kir3.1 to Kir3.4) are typically 50–70% identical; members of different subfamilies (e.g., Kir2.1 and Kir1.1) are approximately 40% identical. Clones reported from heart are highlighted in bold. (B) Schematic diagram of the proposed topology of Kir channels, showing the approximate positions of negatively charged residues that have been demonstrated to be involved in high-affinity Mg^{2+} and polyamine block.

means that very little current flows through this channel during the action potential, even though it is the dominant conductance at the resting potential (Vandenberg, 1994). Over the past five years, a large number of cDNAs encoding pore-forming subunits of Kir channels have been cloned, and this has greatly advanced our understanding of the mechanisms and structural requirements of channel activity. The purpose of this chapter is to review recent progress in understanding the molecular basis of Kir channel function, particularly mechanisms of inward rectification and of nucleotide regulation of K_{ATP} channels.

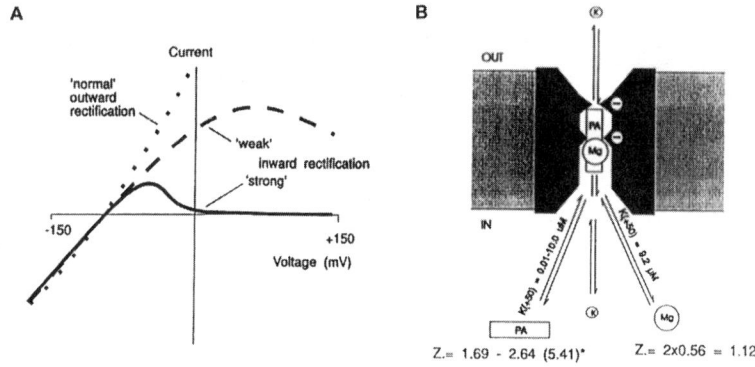

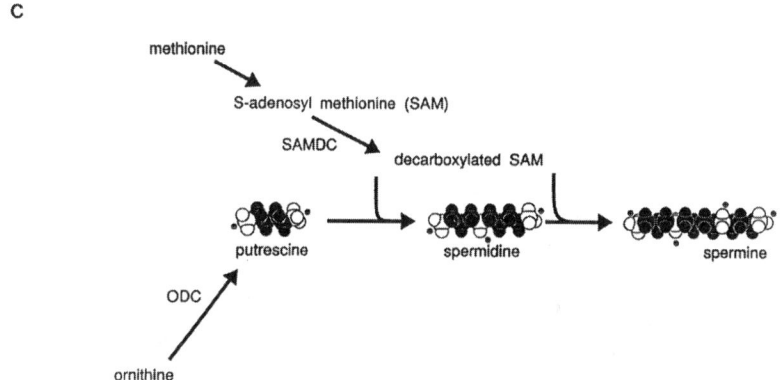

Figure 2. (A). Idealized current-voltage relationship of strong and weak inward rectifier K channels. Both conduct significantly at diastolic potentials, but strong inward rectifiers pass little or no current during the action potential plateau. (B). Schematic diagram of proposed pore blocking mechanisms causing inward rectification. Mg^{2+} and polyamines block strong inward rectifiers very potently (K(+50)-concentration causing 50% inhibition at +50 mV). Blocking ions enter the pore from the inside, and binding is relieved by potassium ions entering the pore from the outside. Although the effective valency (Z) of Mg^{2+} block (Z = 1.2) is consistent with one Mg^{2+} ion permeating about 50% of the voltage field, polyamines (PA) may enter more deeply. Experimental data suggest that more than one PA actually enters the field, or additional ions mobe in the blocking process, since although single Boltmann fits to conductance-voltage relationships are in the range 1.69–2.64 for different PAs, the effective valency (Z) of spermine[4+] block approaches 5.4 (*) as the block approaches saturation. (C). Schematic diagram of polyamine structure and outline of synthetic pathway in animal cells. Amines are shown white, methyl groups are shown in black. All amines are charged at neutral pH. ODC—Ornithine decarboxylase; SAM—S-adenosyl methionine; SAMDC—SAM decarboxylase.

2. THE KIR CHANNEL FAMILY: TWO-TRANSMEMBRANE-DOMAIN K^+ CHANNELS

Cloned Kv channel subunits consist of six transmembrane domains with cytoplasmic amino and carboxyl termini and a conserved hydrophobic P- or H5-loop between the fifth and sixth transmembrane domains. Four such subunits are required to form a functional channel, and activation (channel opening) is steeply dependent on membrane depolarization. In contrast, Kir channel subunits have only two transmembrane domains (Ho et al., 1993; Kubo et al., 1993a; Choe et al., 1995; Nichols, 1993), but they retain the H5-loop that is responsible for K^+ selectivity (Heginbotham et al., 1992). Utilizing mutations that express channels with altered rectification properties, there is evidence that, like Kv channels (MacKinnon, 1991), Kir channel α-subunits form tetramers (Glowatzki et al., 1995; Yang et al., 1995a; Shyng and Nichols, 1997). There are now at least seven Kir channel subfamilies (Nichols and Lopatin, 1997; Krapivinsky et al., 1998), with approximately 40% interfamily amino acid homology. Within each Kir subfamily, there is approximately 60% identity between individual channel types (Fig. 1).

2.1. Kir1 Subfamily

Kir1.1 encodes a "weak" inward rectifier and is expressed predominantly in the kidney, but also in various brain tissues (Ho et al., 1993; Boim et al., 1995). Alternate splicing at the 5′ end generates multiple Kir1.1 splice variants (Shuck et al., 1994; Yano et al., 1994; Zhou et al., 1994).

2.2. Kir2 Subfamily

Three distinct Kir2 subfamily members have been cloned to date, all encoding "strong" inward rectifiers that differ in single-channel conductance. The approximate conductances of Kir2.x channels, measured in symmetrical K^+ (external K^+, 140 mM), are 20 pS for Kir2.12 35 pS for Kir2.32, and 10 pS for Kir2.3. These channels also differ in their sensitivity to phosphorylation and other second messengers (Fakler et al., 1994b; Henry et al., 1996; Makhina et al., 1994). Kir2 subfamily members are expressed in the heart and nervous system (Ishii et al., 1994; Kubo et al., 1993a; Perier et al., 1994; Pessia et al., 1996; Wible et al., 1994). The time- and voltage-dependent rectification of the expressed channels is virtually indistinguishable from that of native I_{K1} channels in the heart (Ishihara et al., 1989; Ishihara and Hiracka, 1994; Kurachi, 1985; Oliva et al., 1990; Stanfield et al., 1994).

2.3. Kir3 Subfamily

Members of the Kir3 family all express G-protein-activated strong inward rectifier K^+ channels (Kubo et al., 1993b; Dascal et al., 1993; Lesage et al., 1994). There is substantial evidence that they are the basis for express G-protein-coupled, receptor-activated currents in heart, brain, and endocrine tissues (Kubo et al., 1993b; Karschin et al., 1994; Ferrer et al., 1995). Krapivinsky et al. (1995) demonstrated that Kir3.4 subunits coassemble with Kir3.1 (also called GIRK1) to form the cardiac muscarinic-receptor-activated $I_{K(ACh)}$ channel. Additional studies have provided evidence for a promiscuous coupling between the various members of the Kir3 subfamily (Duprat *et al.*, 1995; Ferrer *et al.*, 1995; Isomoto *et al.*, 1996; Kofuji *et al.*, 1995; Spauschus *et al.*, 1996) .

2.4. Kir4 and Kir5 Subfamilies

Two additional subfamilies of Kir channels have been discovered in brain and other tissues (Kir4 and Kir5,). Kir4.1 forms weak inward rectifier K^+ channels when expressed alone. Kir5.1 does not form homomeric channels in an oocyte expression system (Bond *et al.*, 1994). However, Kir4.1 and Kir5.1 form novel channels when they are coexpressed. Tandem dimers and tetramers in a specific Kir4.1–Kir5.1–Kir4.1–Kir5.1 arrangement reproduce the characteristics of these channels (Pessia *et al.*, 1996). Intriguingly, a Kir4.1–Kir4.1–Kir5.1–Kir5.1 tetrameric arrangement produces channels with the properties of homomeric Kir4.1 channels. These data are evidence for the importance of subunit position in determining the properties of heterotetrameric Kir channels.

2.5. Kir6 Subfamily

Inagaki *et al.* (1995a) isolated a novel, ubiquitously expressed gene, which they named uK_{ATP1} (Kir6.1 in the unified nomenclature). A pancreatic-specific isoform (Kir6.2) was subsequently found to encode a weak inward rectifier K_{ATP} channel (Inagaki *et al.*, 1995b). Expression of active K_{ATP} channels requires coexpression of Kir6.2 (or Kir6.1) with a sulfonylurea receptor (SUR1 or SUR2).

2.6. Kir7 Subfamily

Krapivinsky *et al.* (1998) recently reported a novel Kir channel that is primarily expressed in brain. This channel has an apparently very low single-channel conductance and is insensitive to intracellular Mg^{2+}.

2.7. Inward Rectification in Other K^+ Channels

Ketchum *et al.* (1995) described a novel yeast K^+ channel subunit (TOK1) which appears to be formed from a Kir subunit in tandem with a six-transmembrane-domain Kv subunit. TOK expresses an outwardly rectifying K^+ current in *Xenopus* oocytes. Lesage *et al.* (1996) reported the cloning and expression of a similarly structured channel (TWIK-1). TWIK-1 consists of two Kir subunits in tandem and expresses a weak inward rectifier K^+ current. It seems likely that these tandems were formed from gene duplications, providing a whole new series of possibilities for the generation of novel Kir channels. It should also be noted that many Kv channels show "weak" inward rectification caused primarily by Mg^{2+} or Na^+ block (French and Wells, 1977; Forsythe et al., 1992; Rettig et al., 1992; Lopatin and Nichols, 1994). A recently cloned Kv channel shows a much higher degree of inward rectification and behaves essentially as an inward rectifier that deactivates rapidly upon hyperpolarization (Sanguinetti et al., 1995; Trudeau et al., 1995). This human ether-a-go-go-related channel (HERG) is likely to be a constituent of the human, cardiac delayed rectifier channel (I_{Kr}). Mutations in this gene are responsible for certain inherited forms of long QT syndrome (Curran et al., 1995). Smith et al. (1996) have examined the rectification properties of expressed HERG and concluded that rectification results from "C-type" voltage-dependent inactivation. C-type inactivation is slow in onset, and the mechanism by which it occurs is incompletely understood, although it is also present in other Kv channels (Hoshi et al., 1991). However, C-type inactivation is distinguishable from the strong inward rectification manifested by many Kir channels.

3. THE NATURE OF INWARD RECTIFICATION: NEW INSIGHTS FROM CLONED CHANNELS

3.1. Polyamines as the Cause of Strong Inward Rectification

Armstrong (1969) suggested that inward rectification of K^+ channels might result from a voltage-dependent block of the channel pore by cytoplasmic cations. In support of this hypothesis, application of tetraethylammonium ions to the cytoplasmic surface of Kv channels induces an inward rectification by blocking the channel pore. Subsequently, Mg^{2+} and Na^+ ions were shown to cause inward rectification of weakly inward rectifying K_{ATP} channels (Ciani and Ribalet, 1988; Horie et al., 1987) and of cardiac I_{K1} channels (Vandenberg, 1987; Matsuda et al., 1987). However, a seemingly intrinsic voltage dependence of the conductance is also clearly a dominant cause of inward rectification in "I_{K1}-like" channels (Kelly et al., 1992; Kurachi, 1985; Matsuda, 1991; Matsuda et al., 1989; Oliva et al., 1990; Silver and DeCoursey, 1990). For both Mg^{2+}-induced, and "intrinsic," rectification, a strong dependence on external K^+ concentration (Ko) was evident such that increasing K_o relieves the rectification. For Mg^{2+}-induced rectification, this effect is explained by the ability of the K^+ ion to bind at external sites and thereby "knock-off" Mg^{2+} from sites deeper inside a multi-ion pore (Armstrong, 1971; Hille and Schwarz, 1978; Yellen, 1984). An intriguing observation made by Matsuda in 1988 is that the "intrinsic" rectification of cardiac Kir channels gradually disappears with time after excision of a membrane patch into the inside-out configuration. Following the cloning of strong inward rectifier K^+ channel genes (Kir2.x gene family members), it was possible to observe high levels of expressed inward rectifier currents. In macropatch experiments on Kir2.3 channels expressed in Xenopus oocytes, we also observed that rectification disappeared when patches were isolated (Lopatin et al., 1994) but was restored when the patch was moved back toward the oocyte. This indicated that rectification disappeared because some factor, or factors, was being lost from the oocyte interior and that these "intrinsic rectifying factors" were actually being released from intact oocytes. Conditioning solutions by exposure to intact oocytes allowed us to make some rudimentary biochemical characterization of "intrinsic rectifying factors," sufficient to indicate that they are actually polyamines (spermine, spermidine, putrescine; Fig. 2), metabolites of amino acids that are found in almost all cells (Tabor and Tabor, 1984). Application of these polyamines to inside-out patches containing Kir2.x channels restores all the essential features of "intrinsic" rectification (Lopatin et al., 1994, 1995; Ficker et al., 1994; Fakler et al., 1995). Less potent than spermine and spermidine, putrescine and cadaverine also cause rectification with similar efficacy to the rectification caused by Mg^{2+}. The voltage dependence of spermine and spermidine block is steeper than that of Mg^{2+} block (Lopatin et al., 1994, 1995; Fakler et al., 1994a; Ficker et al., 1994), explaining why inward rectification in endogenous cells is steeper than that produced by Mg^{2+} ions (Hille, 1992; see Fig. 1).

The voltage dependence of spermine and spermidine unblock rates matches the rate constants of channel activation in cell-attached patches (Lopatin et al., 1995). Weak inward rectifiers such as Kir1.1 and Kir4.1, as well as K_{ATP} (Kir6.2), channels and delayed rectifier Kv2.1 (previously called DRK1) channels all show only "weak" inward rectification. In contrast to Kir2.x channels, they are only blocked by millimolar concentrations of Mg^{2+} and polyamines (Lopatin et al., 1994; Fakler et al., 1994a; Nichols et al., 1994; Shyng et al., 1997b), and the block is only weakly voltage-dependent. The steepness of the voltage dependence of channel block by polyamines increases as the charge on the polyamine increases (Lopatin et al., 1994), and mutations that alter Mg^{2+} block sensitivity also alter polyamine blocking affinity (Fakler et al., 1994a; Yang et al., 1995b). As expected for a channel blocker that interacts with permeant ions inside the pore, external potassium ions substantially relieve rectification (Lopatin and Nichols, 1996).

3.2. The Structure of the Kir Channel Pore: Binding Sites for Polyamines

Mg^{2+} ions are spherical charges, similar in diameter to K^+ ions, and it is reasonable to suggest that these ions block the channel by occupying K^+ ion binding sites within the pore. On the other hand, spermine is a very long (almost 20 Å long) and thin molecule (diameter ≈ 3 Å), with spatially distributed positive charges. It is a possibility that in blocking Kir channels, spermine lies in a long pore, each charge associating with a site that might otherwise be occupied by K^+ ions (Lopatin et al., 1995). Yang et al. (1995b) examined steady-state polyamine block of Kir2.1 channels over a wide concentration range, and their data suggest that at least two polyamines bind in the channel, each with a different affinity. We also initially proposed that two polyamines independently enter the channel pore, partly in order to account for the very large charge movement (more than 5 elementary charges) that accompanies spermine block (Lopatin et al., 1995).

All K^+ channels contain a highly conserved region which includes an extracellular loop (H5- or P-loop) with a —Gly-X-Gly triplet that forms the K^+ selectivity filter within the pore region (Hartmann et al., 1991; MacKinnon and Yellen, 1990; Yool and Schwartz, 1991). Mutagenesis followed by biophysical analysis demonstrates that the transmembrane region that follows the P-loop is also involved in forming the permeation pathway (Aiyar et al., 1994; Liu et al., 1997). Multiple studies have indicated that a specific residue in the second transmembrane domain M2 of Kir2.1 (IRK1) is a major determinant of the potency of Mg^{2+} or polyamine block, and hence of whether a channel will show classical strong inward rectification. When this residue is a negatively charged glutamate or aspartate, high-affinity block is observed, and neutralization of this residue reduces or abolishes both Mg^{2+} and polyamine blocking affinity (Fakler *et al.*, 1994a; Lopatin *et al.*, 1994; Lu and MacKinnon, 1994; Wible *et al.*, 1994; Ficker *et al.*, 1994). A histidine residue at this site also leads to permanent rectification at low internal pH (Lu and MacKinnon, 1995). The rectification is progressively diminished at higher pH, as the histidine residue is neutralized. However, rectification is insensitive to external pH, indicating that internal, but not external, protons have free access to the regulatory site. This is consistent with the idea that a tight selectivity filter, formed by the H5 region, exists at the outer mouth of the channel and blocks access of ions other than K^+ to the long inner vestibule. Studies with chimeras between "weakly" rectifying Kir1.1 (previously called ROMK1) and "strongly" rectifying Kir2.1 (also called iRK1) indicate that the C-terminal region, beyond M2, might contain necessary structural elements for strong inward rectification and high-affinity Mg^{2+} block (Pessia *et al.*, 1995; Taglialatela *et al.*, 1994). Yang *et al.* (1995b) demonstrated that E224 (in the carboxy terminus of Kir2.1) is also a determinant of both Mg^{2+} and polyamine sensitivity. Ruppersberg *et al.* (1996) subsequently demonstrated that both absolute and relative off-rates of different polyamines and Mg^{2+} from the channel also depend critically on the amino acid at residue 84 (in iRK1). This amino acid is positioned at the entrance to the M1 transmembrane domain. These latter results suggest that the region immediately before M1 (containing residue 84) and the region immediately after M2 (containing residue 224) participate in the formation of the internal entrance to the pore.

Very recently, these predictions have been dramatically confirmed by determination of the crystal structure of KcsA, a K^+ channel from *Streptomyces lividans* (Doyle *et al.*, 1998; Fig. 2). Although there is presently little functional characterization of this channel, it is structurally a member of the Kir channel family and contains two transmembrane domains with an H5-loop, or P-loop, containing the K^+ channel signature Gly-X-Gly motif. The crystal structure demonstrates that the P-loop region forms a shallow disc at the outer surface of the membrane with a long inner vestibule that extends at least 20 A into and through the membrane. This vestibule is long enough to accommodate a spermine molecule in extended form (Fig. 2). The width of the inner vestibule is variable, but the maximum diameter is approximately 10 Å. It is a tantalizing possibility that the binding site for the blocking polyamine that causes inward rectification is physically in this vestibule (Ruppersberg *et al.*, 1994; Lopatin *et al.*, 1995; Shyng *et al.*, 1997a).

4. NUCLEOTIDE REGULATION OF K_{ATP} CHANNELS: NEW INSIGHTS FROM CLONED CHANNELS

4.1. The "Classical View" of Nucleotide Regulation

K_{ATP} channels are unique among K^+ channels in being rapidly, and reversibly, inhibited by the nonhydrolytic binding of cytoplasmic adenosine nucleotides. They are also activated in complex ways by nucleotide tri- and diphosphates (Ashcroft, 1988;

Nichols and Lederer, 1991). It is clear that neither nucleotide hydrolysis nor Mg^{2+} ions are required for K_{ATP} channel inhibition to occur. Some confusion has existed as to whether K_{ATP} channels are inhibited by ATP in the presence of Mg^{2+}. Certainly, MgATP can both inhibit and activate these channels, probably through several distinct mechanisms. Thus, one may observe mixed effects of ATP when it is applied to membrane patches in the presence of Mg^{2+}. This may also confound the measurement of inhibitory dose–response relationships. Inhibition of K_{ATP} channels by nucleotides requires the phosphate group of the nucleotide contain at least one phosphate. The nucleotide's inhibitory potency increases significantly from AMP to ADP and is greatest for ATP. The half-maximal inhibitory concentrations, $K_{1/2}$, of AMP, ADP, and ATP in native K_{ATP} channels from heart or pancreas are 10,000, 250, and 10–25 μM, respectively. The adenosine moiety seems to be critical to achieving K_{ATP} inhibition because GTP and pyrimidine triphosphates are essentially ineffective. The "classical" efforts to describe nucleotide activation of the channel were rather less successful at defining the underlying mechanisms. This difficulty no doubt reflects the multiple and overlapping mechanisms involved. Because MgATP could activate channels, ATP hydrolysis and possibly phosphorylation were initially presumed to be necessary. However, MgADP can also activate channels by unknown mechanisms that antagonize the inhibitory effects of ATP (Findlay, 1988; Lederer and Nichols, 1989; Terzic et al., 1995).

4.2. Separable Roles of the Sulfonylurea Receptor (SUR) and Kir6 Subunits: SUR Controls Nucleotide Activation

The realization that K_{ATP} channels are composed of an ATP-binding cassette (ABC) protein (the sulfonylurea receptor, SUR1 or SUR2), in addition to pore-forming subunits Kir6.x (Fig. 3; Aguilar-Bryan et al., 1995; Inagaki et al., 1995, 1996), led to the natural presumption that inhibitory ATP binding occurs at the nucleotide-binding folds (NBFs) of the SUR. However, in other ABC proteins, these NBFs not only bind but also hydrolyze MgATP. In contrast, the inhibitory effect of ATP does not require ATP hydrolysis, since it is Mg^{2+}-independent and can be mimicked by nonhydrolyzable ATP analogs, AMP, and ADP in the absence of Mg^{2+}. Loss of K_{ATP} channel function in pancreatic β-cells is associated with the disease persistent hyperinsulinemic hypoglycemia of infancy (PHHI) (Thomas et al., 1995; Kane et al., 1996). The first insight into the role of the NBFs in regulating channel activity came from examination of a disease causing a mutation, G1479R, in NB (Nichols et al., 1996). When K_{ATP} channels were reconstituted from SUR1[G1479R] + Kir6.2 subunits, active channels were not observed in intact cells, measured using the rubidium efflux assay. However, in inside-out patches, channels were present and displayed conductances, channel density, and ATP sensitivity similar to those of wild-type channels. Stimulation by MgADP and by diazoxide was abolished by this SUR mutation (Nichols et al., 1996). Like diazoxide, MgADP activates K_{ATP} channels in the presence of inhibitory concentrations of ATP, and both processes require Mg^{2+} and hydrolyzable nucleotides (Kakei et al., 1986; Dunne and Petersen, 1986; Misler et al., 1986; Findlay, 1988; Lederer and Nichols, 1989; Larsson et al., 1993). Gribble et al. (1997) then showed that mutation of the conserved

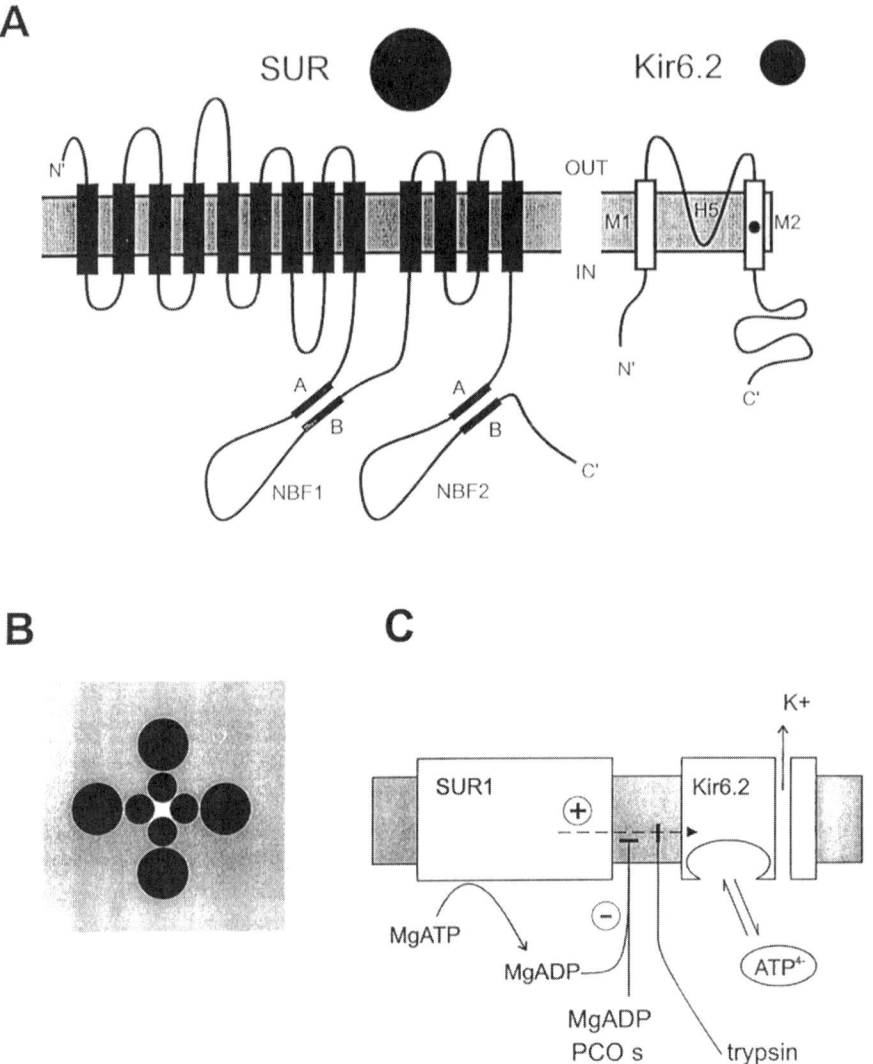

Figure 3. (A) Schematic diagram of the proposed topology of SUR and Kir6 subunits that form K_{ATP} channels. SUR subunits contain multiple membrane-spanning domains with external N-terminus and cytoplasmic C-terminus. Two nucleotide-binding folds (NBF1, NBF2) hydrolyze ATP and are required for MgADP and K^+ channel openers (PCO) stimulation of channel activity. (B) Schematic diagram illustrating proposed octameric arrangement of subunits in the K_{ATP} channel (viewed from above, looking through the channel). A tetramer of Kir6.2 subunits forms the channel pore, each subunit being associated with one SUR subunit. (C) Schematic diagram illustrating subunit involvement in nucleotide regulation of channel activity. The inhibitory ATP binding site is proposed to be on the pore-forming Kir6 subunit. The SUR subunits are proposed to sensitize the channel to ATP, this sensitivity being relieved by MgATP hydrolysis in the SUR NBFs. The ATP-desensitizing effect is then stabilized by MgADP or PCOs.

lysine residues in the Walker A motifs of either NBF1 (mutation K719A) or NBF2 (mutation K1384M) which are predicted to reduce ATP hydrolytic activity, blocks the stimulatory effect of MgADP. In their study, diazoxide stimulated K1384M (NBF2 mutant) channels but failed to stimulate K719A (NBF1 mutant) channels. Consequently, they suggested that hydrolysis at the NBF1 site is most critical for diazoxide stimulation. However, we also demonstrated that whereas mutations in the linker and Walker B motifs of NBF2 abolish channel activation by diazoxide or MgADP, analogous mutations in NBF1 control the kinetics of activation. Based on these results, we suggest that nucleotide hydrolysis at both NBFs is involved. We propose a model for gating of the K_{ATP} channel in which MgADP and diazoxide stimulate channel activity by stabilizing an open state resulting from this hydrolysis (Fig. 3; Shyng et al., 1997b; Gribble et al., 1997).

4.3. Kir6.2 Controls Nucleotide Inhibition

Whereas the SUR subunit seemed the obvious candidate for providing nucleotide sensitivity, the Kir6 subunits, being homologous to other homomeric K^+ channels, seemed obvious candidates for providing the channel pore. Utilizing mutations in a putative pore-lining residue (amino acid 160) in Kir6.2 and SUR1–Kir6.2 dimeric constructs, it is clear that Kir6.2 subunits do form the ion-conducting pathway (Shyng et al., 1997a) and that a tetrameric arrangement of Kir6.2 subunits forms the pore, each one probably associated with one SUR1 subunit (Shyng and Nichols, 1997; Clement et al., 1997; Inagaki et al., 1997). We recently examined the effect of SUR mutations on the intrinsic sensitivity of expressed channels to ATP inhibition. To our initial surprise, we found that mutations of residue 160 not only controlled sensitivity to polyamine-induced rectification but also shifted the $K_{1/2}$ for ATP from ~ 10 μM (wild type) to $\sim 50\,\mu M$ (Kir6.2[N160D]) (Shyng et al., 1997a). Tucker et al. (1997) reported that a C-terminally truncated Kir6.2 (Kir6.2ΔC36) could actually express ATP-sensitive channels in the absence of ATP and also reported a significant reduction of ATP sensitivity for mutation Kir6.2[K185Q]. Taken together, these results demonstrate that Kir6.2 forms the pore of the K_{ATP} channel and further suggests that ATP may inhibit the channel through a direct binding to the Kir6.2 subunit.

Systematic mutagenesis of the Kir6.2 subunit has now been attempted, in order to gain further insight into the mechanism of ATP sensitivity. Truncations of the N-terminus (Kir6.2ΔN2–30; Koster et al., 1999) can reduce apparent ATP sensitivity by 10-fold. Tucker et al. (1998) introduced random mutations throughout the N- and C-terminal regions of Kir6.2ΔC26 and examined ATP sensitivities of channels expressed in the absence of SUR1. They demonstrated that mutations clustered between residues 39 and 51 in the N-terminal region and between residues 166 and 185 at the end of the M2 segment and beginning of the C-terminal region can significantly decrease ATP sensitivity. Of these mutations, R50G and K185Q reduce ATP sensitivity without changing the open-state probability in the absence of nucleotides (Po_{zero}), suggesting that these residues might actually reside within an ATP binding site. Drain et al. (1998) identified two regions in the C-terminus which, when mutated, caused reduction of ATP sensitivity. The first included residues 171–182 at the end of M2, mutations in this region being associated with reduced Po_{zero}. The second region included residues 334–337 in the distal C-terminus. As with R50G and K185Q (Tucker *et al.*, 1998; Koster *et al.*, 1999), mutations in this region are not associated with reduced Po_{zero}, leading to the suggestion that this region may be involved in ATP binding rather than channel gating.

We find that the introduction of a cysteine residue at almost every amino acid position in M2, between positions 157 and 175, results in normal expression of K_{ATP} channels (Loussouarn *et al.*, 2000). Striking variability of ATP sensitivity is apparent in different mutants (ranging from 30 μM for I162C to 100,000 μM for L164C) (Loussouarn *et al.*, 2000). Amazingly, the two positions that show the most extreme disparity of ATP sensitivity are only two residues apart. Trapp *et al.* (1998) also examined the effects on ATP sensitivity and gating behavior of multiple mutations at position 166 within M2. Their study provides further evidence that altered ATP sensitivity of M2 mutants is uniformly associated with a change in Po_{zero}.

4.4. ATP Stabilizes the Closed Channel

For N-terminal deletions, mutations at position 160 and 166, and mutations throughout M2, there is an inverse correlation between $K_{1/2}$ and Po_{zero} (Shnyg *et al.*, 1997a; Trapp *et al.*, 1998; Tucker *et al.*, 1998; Drain *et al.*, 1998; Loussouarn *et al.*, 2000). This correlation can be explained by assuming that ATP serves to stabilize the closed state. Thus, these mutations alter the stability of the closed state relative to the open state. The effects are analogous to those of mutations in cyclic-nucleotide-activated ion channels which alter the activating efficacy, but not the affinity, of cyclic nucleotides for these channels. Gordon and Zagotta (1995) demonstrated that in chimeras between olfactory and rod cyclic-nucleotide-gated channels, cyclic nucleotide activating potency could vary over three orders of magnitude, simply as a consequence of shifts in the allosteric conformational change between closed and open states. In this case, regions throughout the channel protein could affect this transition (Gordon and Zagotta, 1995). A reasonable gating scheme to explain such allosteric effects predicts saturation of Po_{zero} without saturation of $K_{1/2}$ at one extreme and saturation of minimum $K_{1/2}$ once Po_{zero} falls below about 0.8 (for channels coexpressed with SUR1) (Shyng *et al.*, 1997a; Koster *et al.*, 1999; Loussouarn *et al.*, 2000). Thus, although an increase in Po_{zero} will decrease the apparent $K_{1/2}$ for any scheme in which ATP binds to, or stabilizes, a closed state (Shyng *et al.*, 1997a), the relationship between $K_{1/2}$ and Po_{zero} can be extremely nonlinear. This nonlinearity makes it difficult to predict $K_{1/2}$ quantitatively from experimentally measured Po_{zero}.

4.5. Membrane Phospholipid Control of ATP Sensitivity: Getting to the Underlying Mechanism

In the absence of convincing biochemical evidence of ATP binding, a picture is nevertheless emerging whereby ATP binding, probably to multiple residues in the cytoplasmic N- and C-terminal portions of the Kir6.2 subunit, is associated with channel closure, by stabilizing the channel *closed* state. Membrane phospholipids stabilize the *open* state of inward rectifier channels (Fan and Makielski, 1997; Hilgemann and Ball, 1996), and so we suggest that they should antagonize the inhibitory action of ATP on channels. We have now found that phosphatidylinositol 4,5-bisphosphate (PIP_2) profoundly antagonizes ATP inhibition of K_{ATP} channels when applied to inside-out membrane patches (Shyng and Nichols, 1998; Baukrowitz *et al.*, 1998); 5 μM PIP_2 reduced sensitivity of Kir6.2 + SUR1 channels to ATP by 340-fold after approximately 10 minutes of exposure. Similar effects are observed on both native cardiac and pancreatic K_{ATP} channels (Shyng and Nichols, 1998). The effects depend both on the presence of a lipid tail and on the number of phosphate groups on the phospholipid. We propose that membrane PIP_2 binds to positive charges in the cytoplasmic region of the Kir6.2 subunit, stabilizing the open state of the channel and thereby antagonizing the inhibitory effect of ATP. A mutation in the C-terminus of the Kir6.2 subunit, which reduces PIP_2 binding, leads to lower channel activity in intact cells, demonstrating the physiological significance of PIP_2 regulation. A synthetic Kir6.2 C-terminal protein also specifically inhibits channel activity, and this effect is relieved by PIP_2, consistent with the hypothesis that the C-terminal protein competes with the channel itself for membrane PIP_2. Based on these findings, we propose that ATP inhibition of the channel results from an essentially electrostatic "negative heterotropic cooperativity" with membrane phospholipids (Shyng and Nichols, 1998).

5. CONCLUSIONS

The last five years have seen remarkable progress in understanding the molecular basis of inward rectifying K^+ channel activity. The cloning and expression of members of the Kir gene family have demonstrated the conserved tetrameric structure of these channels, each subunit consisting of two transmembrane domains. Expression of cloned Kir2.x channels has permitted determination of the molecular basis of inward rectification. Expression of Kir3.x channels has provided significant insight into the molecular basis of G-protein gating of Kir channels (not covered in this chapter), and expression of Kir6.x subunits with sulfonylurea receptor subunits has led to rapid advance in understanding the structural basis of nucleotide regulation of K_{ATP} channels.

ACKNOWLEDGMENTS

We are grateful for support of our own experimental work by the National Institutes of Health, the Juvenile Diabetes Foundation, the American Diabetes Association, and the American Heart Association.

REFERENCES

Aguilar Bryan, L., Nichols, C. G., Wechsler, S. W., Clement, J. P., IV, Boyd, A. E., III, Gonzalez, G., Herrera Sosa, H., Nguy, K., Bryan J., and Nelson, D. A., 1995, The β Cell high affinity sulfonylurea receptor: A regulator of insulin secretion, *Science* **268**:423–426.

Aiyar, J., Nguyen, A. N., Chandy, K. G., and Grissmer, S., 1994, The P-region and S6 of Kv3.1 contribute to the formation of the ion conduction pathway, *Biophys. J.* **67**:2261–2264.

Armstrong, C. M., 1969, Inactivation of the potassium conductance and related phenomena caused by quaternary ammonium ion injected in squid axons, *J. Gen. Physiol.* **54**:553–575.

Armstrong, C. M., 1971, Interaction of tetraethylammonium ion derivatives with the potassium channel of giant axons. *J. Gen. Physiol.* **58**:413–437.

Ashcroft, F. M, 1988, Adenosine 5'-triphosphate-sensitive potassium channels, *Annu. Rev. Neurosci.* **11**:97–118.

Baukrowitz, T., Schulte, U., Oliver, D., Herlitze, S., Krauter, T., Tucker, S. J., Ruppersberg, J. P., and Fakler, B., 1998, PIP_2 and PIP as determinants for ATP inhibition of K-ATP channels, *Science* **282**:1141–1144.

Boim, M. A., Ho, K, Shuck, M. E., Bienkowski, M. J., Block, J. H., Slightom, J. L., Yang, Y., *et al.*, 1995, ROMK inwardly rectifying ATP-sensitive K^+ channel. II. Cloning and distribution of alternative forms, *Am. J. Physiol.* **268**:F1132–F1140.

Bond, C. T., Pessia, M., Xia, X. M., Lagrutta, A., Kavanaugh, M. P., and Adelman, J. P., 1994, Cloning and expression of a family of inward rectifier potassium channels, *Receptors Channels* **2**:183–191.

Choe, S., Stevens, C. F., Sullivan, J. M. 1995. Three distinct structural environments of a transmembrane domain in the inwardly rectifying potassium channel ROM-1 defined by perturbation. Proceedings of the National Academy of Sciences of the United States of America. **92**(26):12046–12049.

Ciani, S., and Ribalet, B., 1988, Ion permeation and rectification in ATP-sensitive channels from insulin-secreting cells (RINm5F): Effects of K^+, Na^+ and Mg^{2+}. *J. Membra. Biol.* **103**:171–180.

Clement, J. P., IV, Kunjilwar, K., Gonzalez, G., Schwanstecher, M., Panten, U., Aguilar-Bryan, L., and Bryan, J., 1997, Association and stoichiometry of K_{ATP} channel subunits. *Neuron* **18**:827–838.

Curran, M. E., Splawski, I., Timothy, K. W., Vincent, G. M., Green, E. D. and Keating, M. T., 1995, A molecular basis for cardiac arrhythmia: HERG mutations cause long QT syndrome. *Cell* **80**:795–804.

Dascal, N., Schreibmayer, W., Lim, N. F., Wang, W., Chavkin, C., DiMagno, L., *et al.*, 1993, Atrial G protein-activated K^+ channel: Expression cloning and molecular properties. *Proc. Natl. Acad. Sci. U.S.A.* **90**:10235–10239.

Doyle, D. A., Cabral, J. M., Pfuetzner, R. A., Kuo, A., Gulbis, J. M., Cohen, S. L., Chait, B. T., and MacKinnon, R, 1998, The structure of the potassium channel: Molecular basis of K^+ conduction and selectivity, *Science* **280**:69–77.

Drain, P., Li, L. H., and Wang, J., 1998, K-ATP channel inhibition by ATP requires distinct functional domains of the cytoplsmic C-terminus of the pore-forming subunit, *Proc. Natl. Acad. Sci.* **95**:13953–13958.

Dunne, M. J., and Petersen, O. H., 1986, Intracellular ADP activates K^+ channels that are inhibited by ATP in an insulin-secreting cell line, *FEBS Lett.* **208**:59–62.

Duprat, F., Lesage, F., Guillemare, E., Fink, M., Hugnot, J. P., Bigay, J., *et al.*, 1995, Heterologous multimeric assembly is essential for K^+ channel activity of neuronal and cardiac G-protein-activated inward rectifiers, *Biochem Biophys. Res. Commun.* **212**:657–663.

Fakler, B., Brandle, U., Bond, C., Glowatzki, E., Konig, C., Adelman, J. P., *et al.*, 1994a, A structural determinant of differential sensitivity of cloned inward rectifier K^+ channels to intracellular spermine, *FEBS Lett.* **356**:199–203.

Fakler, B., Brandle, U., Glowatzki, E., Zenner, H. P., and Ruppersberg, J. P., 1994b, Kir2.1 inward rectifier K^+ channels are regulated independently by protein kinases and ATP hydrolysis, *Neuron* **13**:1413–1420.

Fakler, B., Brandle, U., Glowatzki, E., Weidemann, S., Zenner, H. P., and Ruppersberg, J. P., 1995, Strong voltage-dependent inward rectification of inward rectifier K^+ channels is caused by intracellular spermine, *Cell* **80**:149–154.

Fan, Z., And J. C. Makielski, 1997, Anionic phospholipids activate ATP-sensitive potassium channels, *J. Biol. Chem.* **272**:5388–5395.

Ferrer, J., Nichols, C. G., Makhina, E. N., Salkoff, L., Bernstein, J., Gerhard, D., *et al.*, 1995, Pancreatic islet cells express a family of inwardly rectifying K^+ channel subunits which interact to form G-protein-activated channels, *J. Biol. Chem.* **270**:26086–26091.

Ficker, E., Taglialatela, M., Wible, B. A., Henley, C. M., Brown, A. M., 1994, Spermine and spermidine as gating molecules for inward rectifier K channels, *Science* **266**:1068–1072.

Findlay, I., 1988, Effects of ADP upon the ATP-sensitive K^+ channel in rat ventricular myocytes, *J. Membr. Biol.* **101**:83–92.

Forsythe, I. D., Linsdell, P., and Stanfield, P. R., 1992, Unitary A-currents of rat locus coeruleus neurones grown in Cell culture: Rectification caused by internal Mg^{2+} and Na^+, *J. Physiol. (London)* **451**:553–583.

French, R. J., and Wells, J. B., 1977, Sodium ions as blocking agents and charge carriers in the potassium channel of the squid giant axon, *J. Gen. Physiol.* **70**:707–724.

Glowatzki, E., Fakler, G., Brandle, U., Rexhausen, U., Zenner, H. P., Ruppersberg, J. P., and Fakler, B., 1995, Subunit-dependent assembly of inward-rectifier K^+ channels, *Proc. R. Soc. London Ser.B* **261**:251–261.

Gordon, S. E. and W. N. Zagotta, 1995, Localization of regions affecting an allosteric transition in cyclic nucleotide-activated channels, *Neuron* **14**:857–864.

Gribble, F. M., Tucker, S. J., and Ashcroft, F. M. 1997, The essential role of the Walker A motifs of SUR1 in K-ATP channel activation by Mg-ADP and diazoxide, *EMBO J.* **16**:1145–1152.

Hartmann, H. A., Kirsch, G. E., Drewe, J. A., Taglialatela, M., Joho, R. H. and Brown, A. M., 1991, Exchange of conduction pathways between two related K^+ channels, *Science* **251**:942–944.

Heginbotham, L. Abramson, T., and MacKinnon, R., 1992, A functional connection between the pores of distantly related ion channels as revealed by mutant K^+ channels, *Science* **258**:1152–1155.

Henry, P., Pearson, W. L., and Nichols, C. G., 1996, Protein kinase C inhibition of cloned inward (HRK1/Kir2.3) K^+ channels *J. Physiol. (London)* **495**:681–688.

Hilgemann, D. W., and Ball, R., 96, Regulation of cardiac Na^+,Ca^{2+} exchange and KATP potassium channels by PIP_2, *Science* **273**:956–959.

Hille, B., 1992, *Ionic Channels of Excitable Membranes*, Sinauer Associates Inc., Sunderland, Massachusetts.

Hille, B.. and Schwarz, W., 1978, Potassium channels as multi-ion single-file pores, *J. Gen. Physiol.* **72**:409–442.

Ho, K., Nichols, C. G., Lederer, W. J., Lytton, J., Vassilev, P. M., Kanazirska, M. V., *et al.*, 1993, Cloning and expression of an inwardly rectifying ATP-regulated potassium channel, *Nature* **362**:31–38.

Horie, M., Irisawa, H., and Noma, A., 1987, Voltage-dependent magnesium block of adenosine-triphosphate-sensitive potassium channel in guinea-pig ventricular cells, *J. Physiol. (London)* **387**:251–272.

Hoshi, T., Zagotta, W. N., and Aldrich, R. W., 1991, Two types of inactivation in Shaker K^+ channels: Effects of alterations in the carboxy-terminal region, *Neuron* **7**:547–556.

Inagaki, N., Tsuura, Y., Namba, N., Masuda, K., Gonoi, T., Horie, M., *et al.*, 1995a, Cloning and functional characterization of a novel ATP-sensitive potassium channel ubiquitously expressed in rat tissues, including pancreatic islets, pituitary, skeletal muscle, and heart. *J. Biol. Chem.* **270**:5691–5694.

Inagaki, N., Gonoi, T., Clement, J. P., Namba, N., Inazawa, J., Gonzalez, G., *et al.*, 1995b, Reconstitution of I_{KATP}: An inward rectifier subunit plus the sulfonylurea receptor, *Science* **270**:1166–1170.

Inagaki, N., Gonoi, T., Clement IV, J. P., Wang, C. Z., Aguilar-Bryan, L., Bryan, J., and Seino, S., 1996, A family of sulfonylurea receptors determines the pharmacological properties of ATP-sensitive K^+ channels, *Neuron* **16**:1011–1017.

Inagaki, N., Gonoi, T., and Seino, S., 1997, Subunit stoichiometry of the pancreatic ?-cell ATP-sensitive K^+ channel, *FEBS Lett.* **409**:232–236.

Ishihara, K., and Hiraoka, M., 1994, Gating mechanisms of the cloned inward rectifier potassium channel from mouse heart, *J. Membr. Biol.* **142**:55–64.

Ishihara, K., Mitsuiye, A., Noma, A., and Takano, M., 1989, The Mg^{2+} block and intrinsic gating underlying inward rectification of the K^+ current in guinea-pig cardiac myocytes, *J. Physiol. (London)* **419**:297–320.

Ishii, K., Yamagashi, T., and Taira, N., 1994, Cloning and functional expression of a cardiac inward rectifier K^+ channel, *FEBS Lett.* **338**:107–111.

Isomoto, S., Kondo, C., Takahashi, N., Matsumoto, S., Yamada, M., Takumi, T., *et al.*, 1996, A novel ubiquitously distributed isoform of GIRK2 (GIRK2B) enhances GIRK1 expression of the G-protein-gated K^+ current in *Xenopus* oocytes, *Biochem. Biophys. Res. Commun.* **218**:286–291.

Kakei, M., Kelly, R. P., Ashcroft, S. J., and Ashcroft, F. M., 1986, The ATP-sensitivity of K^+ channels in rat pancreatic B-cells is modulated by ADP, *FEBS Lett.* **208**:63–66.

Kane, C., Shepherd, R. M., Squires, P. E., Johnson, P. R. V., James, R. F. L., Milla, P. J., Aynsley-Green, A., Lindey, K. J., and Dunne, M. J., 1996, Loss of functional KATP channels in pancreatic β-cells causes persistent hyperinsulinemic hypoglycemia of infancy, *Nat. Medi.* **2**:1344–1347.

Karschin, C., Schreibmayer, W., Dascal, N., Lester, H., Davidson, N., and Karschin A., 1994, Distribution and localization of a G protein-coupled inwardly rectifying K^+ channel in the rat, *FEBS Lett.* **348**:139–144.

Katz, B., 1949, Les constantes électriques de la membrane du muscle, *Arch. Sci Physiol.* **2**:285–299.

Kelly, M. E, Dixon, S, J., and Sims, S., M., 1992, Inwardly rectifying potassium current in rabbit osteoclasts: A whole-cell and single-channel study, *J. Membr. Biol.* **126:**171–181.

Ketchum, K. A., Joiner, W. J., Sellers, A. J., Kaczmarek, L. K., and Goldstein, S. A., 1995, A new family of outwardly rectifying potassium channel proteins with two pore domains in tandem, *Nature* **376:**690–695.

Kofuji, P., Davidson, N., and Lester, H. A., 1995, Evidence that neuronal G-protein-gated inwardly rectifying K$^+$ channels are activated by G $\beta\gamma$ subunits and function as heteromultimers, *Proc. Natl. Acad. Scie. U.S.A.* **92:**6542–6546.

Koster, J. C., Sha, Q., Shyng, S.-L., and Nichols, C. G., 1999, ATP inhibition of K$_{ATP}$ channels: Control of nucleotide sensitivity by the N-terminal domain of the Kir6.2 subunit, *J. Physiol. (London)* **515:**19–30.

Krapivinsky, G., Gordon, E. A., Wickman, K., Velimirovic, B., Krapivinsky, L., and Clapham, D. E., 1995, The G-protein-gated atrial K$^+$ channel IKACh is a heteromultimer of two inwardly rectifying K(+)-channel proteins, *Nature* **374:**135–141.

Krapivinsky G., Medina, I., Eng, L., Krapivinsky, L., Yang, Y., and Clapham, D. E., 1998, A novel inward rectifier K$^+$ channel with unique pore properties, *Neuron* **20:**995–1005.

Kubo, Y., Baldwin, T. J., Jan, Y. N., and Jan, L. Y., 1993a, Primary structure and functional expression of a mouse inward rectifier potassium channel, *Nature* **362:**127–133.

Kubo, Y., Reuveny, E., Slesinger, P. A., Jan, Y. N., and Jan, L. Y., 1993b, Primary structure and functional expression of a rat G- protein coupled muscarinic potassium channel, *Nature* **364:**802–806.

Kurachi, Y., 1985, Voltage-dependent activation of the inward rectifier potassium channel in the ventricular cell membrane of guinea-pig heart, *J. Physiol. (London)* **366:**365–385.

Larsson, O., Ammala, C., Bokvist, K., Fredholm, B., and Rorsman, P., 1993, Stimulation of the K$_{ATP}$ channel by ADP and diazoxide requires nucleotide hydrolysis in mouse pancreatic β-cells, *J. Physiol. (London)* **463:**349–365.

Lederer, W. J., and Nichols, C. G., 1989, Nucleotide modulation of the activity of rat heart K$_{ATP}$ channels in isolated membrane patches, *J. Physiol. (London)* **419:**193–211.

Lesage, F., Duprat, F., Fink, M., Guillemare, E., Coppola, T., Lazdunski, M., and Hugnot, J. P., 1994, Cloning provides evidence for a family of inward rectifier and G-protein coupled K$^+$ channels in the brain. *FEBS Lett.* **353:**37–42.

Lesage F., Guillemare, E., Fink, M., Duprat, F., Lazdunski, M., Romey, G., and Barhanin, J., 1996, TWIK-1, a ubiquitous human weakly inward rectifying K$^+$ channel with a novel structure, *EMBO J.* **15:**1004–1011.

Liu, Y., Holmgren, M., Jurman, M. E., and Yellen, G., 1997, Gated access to the pore of a voltage-dependent K$^+$ channel, *Neuron* **19:**175–184.

Lopatin, A. N., and Nichols, C. G., 1994, Inward rectification of outward rectifying DRK1 (Kv2.1) potassium channels. *J. Gen. Physiol.* **103:**203–216.

Lopatin, A. N., and Nichols, C. G., 1996, [K$^+$]-dependence of polyamine induced rectification in inward rectifier potassium channels (IRK1, Kir2.1). *J. Gen. Physiol.* **108:**105–113.

Lopatin, A. N., Makhina, E. N., and Nichols, C. G., 1994, Potassium channel block by cytoplasmic polyamines as the mechanism of intrinsic rectification, *Nature* **372:**366–369.

Lopatin, A. N., Makhina, E. N., and Nichols, C. G., 1995, The mechanism of inward rectification of potassium channels. *J. Gen. Physiol.* **106:**923-955.

Loussouarn, G., Makhina, E. N., Rose, T., and Nichols, C. G., 2000, Structure and dynamics of the pore of inward rectifier K$_{ATP}$ channels, *Neuron* J. Biol. Chem. **275:**1137–1144.

Lu, Z., and MacKinnon, R., 1994, Electrostatic tuning of Mg^{2+} affinity in an inward rectifier K$^+$ channel, *Nature* **371:**243–246

Lu, Z., and MacKinnon, R., 1995, Probing a potassium channel pore with an engineered protonatable site, *Biochemistry* **34:**13133–13138.

MacKinnon, R., 1991, Determination of the subunit stoichiometry of a voltage-activated potassium channel, *Nature* **350:**232–235.

MacKinnon, R., and Yellen, G., 1990, Mutations affecting TEA blockade and ion permeation in voltage-activated K$^+$ channels, *Science* **250:**276–279.

Makhina, E. N., Kelly, A. J., Lopatin, A. N., Mercer, R. W., and Nichols, C. G., 1994, Cloning and expression of a novel inward rectifier potassium channel from human brain, *J. Biolo. Chem.* **269:**20468–20474.

Matsuda, H., 1988, Open-state substructure of inwardly rectifying potassium channels revealed by magnesium block in guinea-pig heart cells, *J. Physiol. (London)* **397:**237–258.

Matsuda, H., 1991, Magnesium gating of the inwardly rectifying K$^+$ channel, *Annu. Rev. Physiol.* **53:**289–298.

Matsuda, H., Saigusa, A., and Irisawa, H., 1987, Ohmic conductance through the inwardly rectifying K$^+$ channel and blocking by internal Mg^{2+}, *Nature* **325:**156–159.

Matsuda H., Matsuura, H., and Noma, A., 1989, Triple-barrel structure of inwardly rectifying K$^+$ channels revealed by Cs$^+$ and Rb$^+$ block in guinea-pig heart cells, *J. Physiol. (London)* **413:**139–157.

Misler, S., Falke, L. C., Gillis, K., and McDaniel, M. L., 1986, A metabolite-regulated potassium channel in rat pancreatic β cells, *Proc. Natl. Acad. Sci. U.S.A.* **83:**7119–7123.

Nichols, C. G., 1993, The 'inner core' of inward rectifier potassium channels, *Trends Pharmacol. Sci.* **14:**320–323.

Nichols, C. G., and Lederer, W. J., 1991, ATP-sensitive potassium channels in the cardiovascular system, *Am. J. Physiol.* **261:**H1675–H1686.

Nichols, C. G., and Lopatin, A. N., 1997, Inward rectifier potassium channels, *Annu. Rev. Physiol.* **59:**171–191.

Nichols, C. G., Ho, K., and Hebert, S., 1994, Mg^{2+} dependent inward rectification of ROMK1 potassium channels expressed in *Xenopus* oocytes, *J. Physiol.* **476:**399–409.

Nichols, C. G., Shyng, S.-L. Nestorowicz, A., Glaser, B., Clement IV, J., Gonzalez, G., Aguilar-Bryan, L., Permutt, A. M., and Bryan, J. P., 1996, Adenosine diphosphate as an intracellular regulator of insulin secretion, *Science* **272:**1785–1787.

Noble, D., 1965, Electrical properties of cardiac muscle attributable to inward going (anomalous) rectification, *J. Cell. Comp. Physiol.* **66:**127–136.

Noma, A., 1983, ATP-regulated K$^+$ channels in cardiac muscle, *Nature* **305:**147–148.

Oliva, C., Cohen, I. S., and Pennefather, P., 1990, The mechanism of rectification of I$_{K1}$ in canine Purkinje myocytes. *J. Gen. Physiol.* **96:**299–318.

Perier, F., Radeke, C. M., and Vandenberg, C. A., 1994, Primary structure and characterization of a small-conductance inwardly rectifying potassium channel from human hippocampus. *Proc. Natl. Acad. Sci. U.S.A.* **91:**6240–6244.

Pessia, M, Tucker, S. J., Lee, K., Bond, C. T., and Adelman, J. P., 1996, Subunit positional effects revealed by novel heteromeric inwardly rectifying K$^+$ channels. *EMBO J.* **15:**2980–2987.

Rettig, J., Wunder, F., Stocker, M., Lichtinghagen, R., Mastiaux, F., Beckh, S., *et al.*, 1992, Characterization of a Shaw-related potassium channel family in rat brain, *EMBO J.* **11:**2473–2486.

Ruppersberg, J. P., vanKitzing, E., and Schoepfer, R., 1994, The mechanism of magnesium block of NMDA receptors, *Semin. Neurosci.* **6:**87–96.

Ruppersberg, J. P., Fakler, B., Brandle, U., Zenner, H.-P., and Schultz, J. H., 1996, An N-terminal site controls blocker-release in Kir2.1 channels, *Biophys. J.* **70:**A361.

Snguinetti, M. C., Jiang, C., Curran, M. E., Keating, M. T. 1995. A mechanistic link between an inherited and an acquired cardiac arrhythmia: HERG encodes the Ikr potassium channel. *Cell.* **81**(2):299–307.

Shuck, M. E., Bock, J. H., Benjamin, C. W., Tsai, T. D., Lee, K. S., Slightom, J. L., and Bienkowski, M. J., 1994, Cloning and characterization of multiple forms of the human kidney ROM-K potassium channel, *J. Biol. Chem.* **269:**24261–24270.

Shyng, S.-L. and Nichols, C. G., 1997, Octameric stoichiometry of the K$_{ATP}$ channel complex, *J. Gen. Physiol.* **110:**655–664.

Shyng, S.-L. and Nichols, C. G., 1998, Phosphatidyl inositol phosphates control of nucleotide-sensitivity of K$_{ATP}$ channels, *Science* **282:**1138–1141.

Shyng, S.-L., Ferrigni, T. and Nichols, C. G., 1997a, Control of rectification and gating of cloned K$_{ATP}$ channels by the Kir6.2 subunit, *J. Gen. Physiol.* **110:**141–153.

Shyng, S.-L., Ferrigni, T., and Nichols, C. G., 1997b, Regulation of K$_{ATP}$ channel activity by diazoxide and MgADP: Distinct functions of the two nucleotide binding folds of the sulfonylurea receptor. *J. Gen. Physiol.* **110:**643–654.

Silver, M. R., and DeCoursey, T. E., 1990, Intrinsic gating of inward rectifier in bovine pulmonary artery endothelial cells in the presence or absence of internal magnesium, *J. Gen. Physiol.* **96:**109–133.

Smith, P. L., Baukrowitz, T., and Yellen, G., 1996, The inward rectification mechanism of the HERG cardiac potassium channel, *Nature* **379:**833–836.

Spauschus, A., Lentes, K. U., Wischmeyer, E., Dissmann, E., Karschin, C., and Karschin, A., 1996, A G-protein-activated inwardly rectifying K$^+$ channel (GIRK4) from human hippocampus associates with other GIRK channels, *J. Neurosci.* **16:**930–938.

Stanfield, P. R., Davies, N. W., Shelton, P. A., Sutcliffe, M. J., Khan, I. A., Brammar, W.J., *et al.*, 1994, A single aspartate residue is involved in both intrinsic gating and blockage by Mg^{2+} of the inward rectifier, IRK1, *J. Physiol. (London)* **478:**1–6.

Tabor, C. W., and Tabor, H., 1984, Polyamines. *Annu Rev. Biochem.* **53:**749–790.

Taglialatela, M., Wible, B. A., Caporoso, R., and Brown, A. M., 1994, Specification of the pore properties by the carboxyl terminus of inwardly rectifying K$^+$ channels, *Science* **264:**844–847.

Terzic, A., Jahangir, A. and Kurachi, Y., 1995, Cardiac ATP-sensitive K$^+$ channels: Regulation by intracellular nucleotides and K$^+$ channel-opening drugs, *Am. J. Physiol.* **269:**C525–C545.

Thomas, P. M., Cote, G. J., Wohllk, N., Haddad, B., Mathew, P. M., Rabl, W., Aguilar-Bryan, L., Gagel, R. F., and Bryan, J., 1995, Mutations in the sulfonylurea receptor gene in familial persistent hyperinsulinemic hypoglycemia of infancy, *Science* **268:**426–429.

Trapp, S., Proks, P., Tucker S. J., and Ashcroft, F. M., 1998, Molecular analysis of K$_{ATP}$ channel gating and implications for channel inhibition by ATP, *J. Gen. Physiol.* **112:**333–350.

Trueau, M. C., Warmke, J. W., Ganetzky, B., Robertson, G. A. 1995. HERG, a human inward rectifier in the voltage gated potassium channel family. *Science* **269**(5220):92–95.

Tucker, S. J., Gribble, F. M., Zhao, C., Trapp, S., and Ashcroft, F. M., 1997, Truncation of Kir6.2 produces ATP-sensitive K$^+$ channels in the absence of the sulphonylurea receptor, *Nature* **387:**179–183.

Tucker, S. J., Gribble, F. M., Proks, P., Trapp, S., Ryder, T. J., Haug, T., Reimann, F., and Ashcroft, F. M., 1998, Molecular determinants of K-ATP channel inhibition by ATP, *EMBO J.* **17:**3290–3296.

Vandenberg, C. A., 1987, Inward rectification of a potassium channel in cardiac ventricular cells depends on internal magnesium ions, *Proc. Natl. Acad. Scie. U.S.A.* **84:**2560–2566.

Vandenberg, C. A., 1994, Cardiac inward rectifier potassium channel, in: *Ion Channels in the Cardiovascular System.* (P. M. Spooner and A. M. Brown, eds.) Futura Publishing Co. New York, Chapter 8.

Wible B. A., Taglialatela, M., Ficker, E., and Brown, A. M., 1994, Gating of inwardly rectifying K$^+$ channels localized to a single negatively charged residue, *Nature* **371:**246–249.

Yang, J., Jan, Y. N., and Jan, L. Y., 1995a, Determination of the subunit stoichiometry of an inwardly potassium channel, *Neuron* **15:**1441–1447.

Yang, J., Jan, Y. N., and Jan, L. Y., 1995b, Control of rectification and permeation by residues in two distinct domains in an inward rectifier K$^+$ channel, *Neuron* **14:**1047–1054.

Yano, H., Philipson, L. H., Kugler, J. L., Tokuyama, Y., Davis, E. M., Le Beau, M. M., *et al.,* 1994, Alternative splicing of human inwardly rectifying K$^+$ channel ROMK1 mRNA, *Mol. Pharmacol.* **45:**854–860.

Yellen, G., 1984, Relief of Na$^+$ block of Ca^{2+}-activated K$^+$ channels by external cations, *J. Gen. Physiol.* **84:**187–199.

Yool, A. J., and Schwartz, T. L., 1991. Alteration of ionic selectivity of a K$^+$ channel by mutation in the H5 region, *Nature* **349:**700–704.

Zhou, H., Tate, S. S., and Palmer, L. G., 1994, Primary structure and functional properties of an epithelial K channel, *Am. J. Physiol.* **266:**C809–C824.

Part II

Potassium Channel Expression and Function

Chapter 6

Design and Use of Antibodies for Mapping K^+ Channel Expression in the Cardiovascular System

Robert O. Koch, Maria Trieb, Alexandra Koschak,
Siegmund G. Wanner, Kathryn M. Gauthier,
Nancy J. Rusch, and Hans-Guenther Knaus

1. INTRODUCTION

K^+ channels represent a large and diverse group of ion channels that play a fundamental role in controlling cell excitability. These channels regulate hormone release from endocrine cells, modulate the pattern of transmitter release from neurons, and set the level of contraction in arterial smooth muscle cells. At first glance, K^+ channels can be subdivided into either voltage-gated K^+ (Kv) channels or ligand-gated K^+ channels, depending on the stimulus that triggers the conformational change that leads to channel opening. Within the wide family of K^+ channels gated by voltage, the Kv channels are activated solely by membrane depolarization, whereas high-conductance Ca^{2+}-activated K^+ (BK_{Ca}) channels require both membrane depolarization and an increased level of cytosolic free Ca^{2+} ($[Ca^{2+}]_i$) to activate effectively. The use of molecular biology techniques has greatly extended our current understanding regarding the structure and existence of subfamilies of K^+ channels. However, three major questions still are the subject of extensive investigation: (1) What is the molecular composition of the different types of K^+ channels expressed *in vivo*, (2) what is the physiological function of these K^+ channels, and (3) is the molecular composition and functional profile of K^+ channels altered in pathophysiological states?

To investigate these questions in detail, selective high-affinity modulators of channel function as well as specific antibodies for distinct channel subunits must be

Robert O. Koch, Maria Trieb, Alexandra Koschak, Siegmund G. Wanner, and Hans-Guenther Knaus • Institute for Biochemical Pharmacology, The University of Innsbruck, A-6020 Innsbruck, Austria. *Kathryn M. Gauthier, and Nancy J. Rusch* • Department of Pharmacology, The Medical College of Wisconsin, Milwaukee, Wisconsin 53226.

Potassium Channels in Cardiovascular Biology, edited by Archer and Rusch. Kluwer Academic/Plenum Publishers, New York, 2001.

discovered or developed. The use of high-affinity ligands is possible because of the discovery of K^+ channel modulators in the venom of various organisms and plants. For instance, components isolated from the venom of scorpions, snakes, sea anemones, bees, and spiders, as well as natural products derived from plants, have been highly useful as pharmacological tools to purify K^+ channels from native tissues (Rehm and Lazdunski, 1988; Parcej and Dolly, 1989; Garcia *et al.*, 1994) and to study their structure and function (Vazquez, *et al.*, 1989, 1990; Galvez *et al.*, 1990; Crest *et al.*, 1992; Garcia-Calvo *et al.*, 1993; Koschak *et al.*, 1998). Some of these toxins, and information derived from their use, are discussed in Chapter 14. A recent and excellent review is also available (Darbon and Sabatier, 1999).

A second important class of investigative tools is comprised of sequence-specific antibodies directed against various K^+ channel subunits. The use of monoclonal and polyclonal antibodies has been instrumental in clarifying the subunit composition of both BK_{Ca} and Kv channels (Sheng *et al.*, 1993; Koch et al, 1997; Shamotienko et al, 1997), establishing their cellular and subcellular distribution in various tissues (Sheng et al, 1994; Veh *et al.*, 1995), and conclusively proving that numerous native K^+ channels are formed by heterotetrameric subunit assembly (Shamotienko *et al.*, 1997; Koch *et al.*, 1997; Roeper *et al.*, 1998). In this chapter, we will discuss approaches that have been undertaken to raise selective sequence-specific antibodies against BK_{Ca} and Kv channel subunits and demonstrate their use in detecting the expression of these proteins in the vascular smooth muscle membranes of small human coronary arteries. Other chapters in this book demonstrate the usefulness of polyclonal antibodies for detecting changes in K^+ channel expression in cardiovascular diseases (Chapters 27, 40, and 42) or outline the use of antibodies for detecting K^+ channel subunit association using immunoprecipitation techniques (Chapter 10).

2. ANTIPEPTIDE ANTIBODIES SELECTIVE FOR K^+ CHANNELS

The injection of synthetic peptide antigens to elicit immunogenic responses initially was described over 20 years ago. In recent years, this technique has been adapted to produce antibodies against the protein sequences deduced from cDNA clones. Production of antipeptide antibodies is quite straightforward because of the large quantities of synthetic peptide reagents available. In general, antibodies are obtained that show variable affinities for the native antigens, but these antibodies still may be used successfully in Western blots, immunoaffinity chromatography, and immunoprecipitation. In fact, the injection of most foreign peptide sequences will elicit an immune response in the host animal, but the antibodies produced may not always react with the native antigen. Therefore, to ensure the final development of a successful antibody, it is advisable to synthesize and immunize a number of peptide sequences for the protein of interest.

3. SELECTION OF THE PEPTIDE SEQUENCE

In our laboratory, immunogenic peptides consisting of 18–20 residues provide optimal results. Peptides as small as 13 residues or as large as 22 residues also have worked successfully. However, both shorter and longer peptides have disadvantages.

Although smaller peptides are more soluble, antibodies raised against them are not as likely to recognize and cross-react with their intended epitope on the native K$^+$ channel. On the other hand, although longer peptides may be more specific for the channel of interest, they are less soluble and more difficult to prepare synthetically. A large part of the decision about peptide length is determined by the individual peptide sequence itself. In this regard, important considerations include the lack of cysteine residues that otherwise foster secondary structure in the antigenic peptide and the assurance that the sequence of the immunogenic peptide is specific only for the channel of interest.

Choosing an appropriate immunogenic peptide is critical in obtaining a sequence-directed antibody that cross-reacts with its respective K$^+$ channel subunit. In most cases, the sequence for an immunogenic peptide is chosen based on three main criteria. First, the targeted residue region of the channel that includes and flanks the epitope should be surface-exposed and conformationally flexible. The respective antibody recognition sequences are never placed in close proximity to transmembrane segments, and hydrophilic carboxyl or amino-terminal regions are chosen preferentially. We have not succeeded in raising high-titer antibodies against the fairly short loops connecting putative transmembrane segments. Rather, in most cases, recognition sequences located in the carboxyl-terminal tail domain of K$^+$ channels represent suitable epitope choices. Second, the recognition sequence preferably should have an overall negative net charge and have no known homology with other protein sequences. Third, tyrosine- and/or proline-rich regions appear to enhance immunogenicity, although this criterion is given a lower priority. Our selection of immunogenic peptides is not based on the use of algorithms that predict potential antigenic sites or secondary structures.

An example of a successful epitope chosen for the BK$_{Ca}$ channel α-subunit is shown in Fig. 1A. In this case, amino acid residues 913–926, composing part of the large loop representing the "S9/S10 linker", were identified as the targeted residue region. Using the Multiple Antigenic Peptide System (MAPS), antibodies were raised against this epitope in several rabbits (see next section). Antisera containing the resulting polyclonal antibodies were used to identify BK$_{Ca}$ channels and their expression levels in the rat brain, and also in rat and human vascular smooth muscle (Liu *et al.*, 1997, 1998; also see Fig. 3). Similarly, the hydrophilic amino and carboxy termini of the *Shaker* (Kv1) K$^+$ channel family represent successful sites for antibody development. Figure 1B shows the epitopes that represent the targets of successful sequence-directed antibodies developed against the Kv1.2, Kv1.4, and Kv1.5 channel subtypes. Again, targeting tail regions of the K$^+$ channels that are remote from the transmembrane-spanning segments and free of cysteine residues has been a successful approach.

4. PEPTIDE SYNTHESIS

Numerous commercial enterprises provide peptide synthesis services. Alternatively, peptide synthesis can be performed using a continuous-flow peptide synthesizer, normally at a peptide scale of 100 μmoles using a fourfold excess of amino acids over resin. In our laboratory, we have investigated three different approaches for preparing immunogenic peptides.

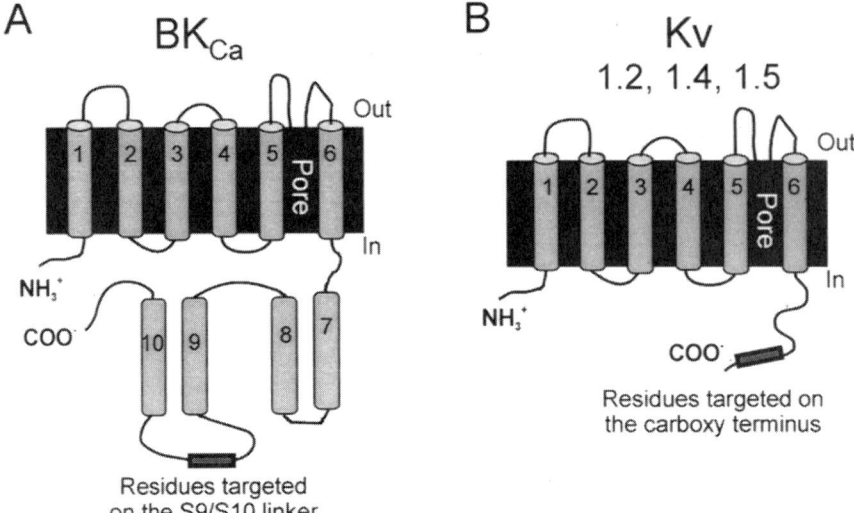

Figure 1. Schematic of amino acid residues that represent the epitopes (antibody-binding regions) in BK$_{Ca}$ channel α-subunits (A) and *Shaker* channel subtypes (B) including the Kv1.2, Kv1.4, and Kv1.5 α-subunits.

The MAPS consists of a small and immunogenic inert core of radially branching lysine derivatives onto which a number of peptide antigens are anchored (Fig. 2). The inert MAP core is attached to a solid-phase, peptide synthesis support, and the desired peptide antigens are synthesized directly on the eight-branched lysine core. After the synthesis is completed, the MAP molecule is cleaved from the support. The result is a large macromolecule with a unique three-dimensional configuration that has a high molar ratio of peptide antigen to core molecule and does not require the use of a carrier protein to elicit an antibody response. We did not further purify the crude cleavage product, as would be done in regular peptide synthesis. In most cases, MAPS induced significantly higher titers than the other two methods exploited (see below). All antibodies discussed in this chapter were generated using this method.

Glutaraldehyde conjugation also is used to cross-link peptides to the carrier protein keyhole limpet hemocyanin. Peptides possessing internal lysine residues are not subjected to this procedure. In our hands, two antibodies raised using this technique did not prove to be of superior quality when compared to immunogenic peptides raised by the MAPS method.

Construction of bacterial fusion proteins is a third method for preparing immunogenic peptides. In this technique, fragments of cDNA encoding 40–110 residues are cloned into the bacterial expression vector pGEX 5X-1 and used to produce a number of channel fusion proteins in *Escherichia coli*. Glutathione S-transferase (GST)-fusion proteins are prepared and purified by standard techniques, including glutathione affinity chromatography or preparative sodium dodecyl sulfate (SDS) gel electrophoresis. This procedure also has yielded good antibodies for BK$_{Ca}$ and Kv channels that are highly suitable for use in Western blots as well as in immunohistochemical distribution studies.

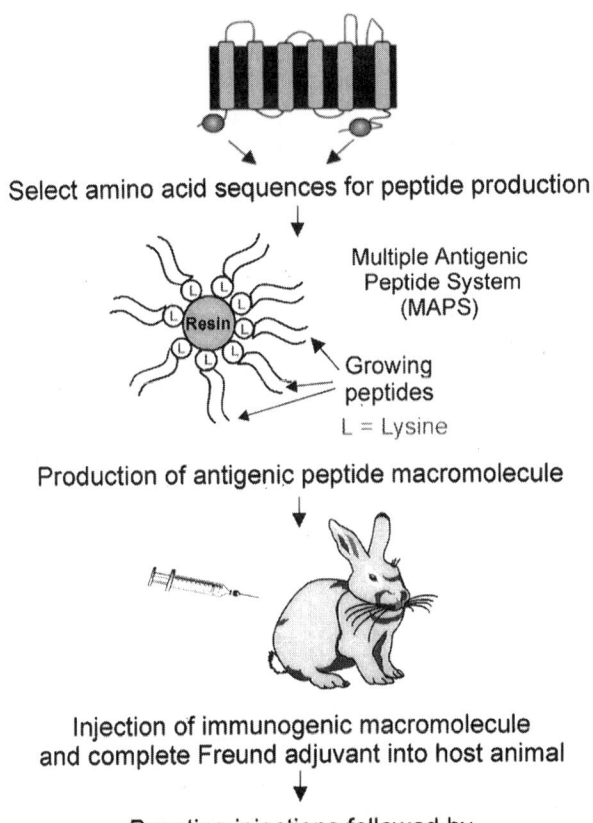

Figure 2. Pathway illustrating the major steps used for antibody production, including the use of the Multiple Antigenic Peptide System (MAPS). ELISA, Enzyme-linked immunosorbent assay.

5. PRODUCTION OF ANTIPEPTIDE ANTIBODIES

All of our antibodies are raised in Chinchilla bastard rabbits. The immunogenic peptides (usually 1–2 mg of peptide for two rabbits) are dissolved in phosphate buffer and, depending on the stage of immunization, mixed with adjuvant and emulsified. For the first immunization, the immunogenic peptide is emulsified with complete Freund adjuvant and subsequently injected in 6–8 sites. Under these conditions, the plasma cells of the host animal will produce a collection of antibodies that recognize different epitopes (antibody binding regions) on the immunogenic peptide. Thus, the rabbit serum will become a source of polyclonal antibodies directed against different sequences of the immunogenic peptide, and an array of channel-directed antibodies will be available to detect the K$^+$ channel protein of interest. For the first boosting injection 14 days later, a 1:1 mixture of complete and incomplete Freund adjuvant is used. This boosting injection enhances the production of antibodies from the sensitized plasma cells, particularly those antibodies that show a high affinity for the immunogenic peptide. All follow-up boosting injections are performed using exclusively incomplete Freund adjuvant. Trial bleeds are drawn 10 days after immunization. The development of specific antibody titers is monitored by enzyme-linked immunosorbent assay (ELISA). Depending on these titers, the rabbits are kept for as long as four months before terminal bleeding.

6. SELECTION OF SUITABLE ANTIBODIES

In a second screening step, high-titer sera are subjected to evaluation in immunoblots. As a general rule, most high-titer antisera will recognize the respective K^+ channel, provided that the K^+ channel of interest is expressed at a density of 100 fmol/mg of membrane protein. The fact that sequence-directed antibodies recognize the corresponding channel after SDS-polyacrylamide gel electrophoresis (SDS-PAGE) and electroblotting is not surprising, because these proteins lack significant secondary structure. The most important requirement for high-quality immunoblots is the purity and density of channels in the membrane preparation. Most preferably, a Ficoll or sucrose density gradient of purified plasma membrane fraction should be used. In such membrane preparations, the channel density is strongly enriched, a fact which will lead to superior noise-to-signal ratios. In plasma membranes derived from most smooth muscle, brain, or neuroendocrine tissues, the immunoblotted K^+ channel proteins are easily visualized by chromogenic substrates with the use of alkaline phosphatase- or peroxidase-labeled secondary antibodies. Alternatively, luminescence-based detection will give comparable results.

Potential problems include strong nonspecific staining patterns or weak cross-reactivity with the channel of interest. Because the antibodies may be directed to relatively small regions of the immunogenic peptide, they may also recognize similar residue sequences in other protein molecules and hence, "cross-react" with these proteins. The identification of the "interfering" protein that contains the common residues that also exist in the channel subunit of interest may not be obvious. Indeed, although the epitope in the interfering protein may represent a continuous residue sequence, it also may consist of a noncontinuous sequence in which the epitope is formed by a three-dimensional folding in the protein that results in the appropriate residues being in close proximity for antibody binding. Thus, amino acid sequences that compose the antibody binding site may originate from one-dimensional or three-dimensional epitopes in a protein and, consequently, may not lend themselves to sequence recognition. The obstacle of cross-reactivity can be overcome by affinity purification of the antisera on the respective peptide columns, as long as certain antibody clones present in crude antiserum unequivocally recognize the K^+ channel of interest. One has to keep in mind that affinity purification will not "generate" better antibodies. Instead, it only will select antibodies that exclusively recognize the immunogenic peptide.

Sequence-specific antibodies that yield specific and strong immunostaining signals in Western blots also may perform well in immunohistochemical or immunocytochemical staining assays, as long as the K$^+$ channel protein is partially denatured by the respective fixation methods to provide a linear epitope. In these latter methods, however, antibody specificity is more difficult to establish because the identity of the antigen may not be as certain as after using size separation on Western immunoblot. To confirm the antigen identity, two antibodies targeted to different epitopes on the antigen can be used to provide independent confirmation of specificity in immunohistochemical or immunocytochemical staining. Generally, however, the potential loss of epitope number or the masking of epitopes by other cellular components has not surfaced as a limiting factor for channel detection. As a general rule, if K$^+$ channel antibodies produce strong and specific signals in immunoblots, they also will succeed in establishing a respective immunohistochemical distribution profile.

In contrast, a very different situation is observed when the individual antibodies are screened in immunoprecipitation assays. For these assays, we prelabel the respective K$^+$ channel population with radioiodinated toxin and solubilize the channels with detergent. For example [^{125}I]hongotoxin is used as a high-affinity ligand for Kv channels, and [^{125}I]iberiotoxin labels the BK$_{Ca}$ channel α-subunits. If the preservation of a native channel conformation is not the direct focus of the study, Triton X-100 is the detergent of choice. This detergent is chemically well defined, inexpensive, and stable in solution. In a concentration of 2%, it partially denatures K$^+$ channels, which often increases the precipitation efficiency of sequence-directed antibodies but does not promote K$^+$ channel subunit dissociation. However, reversible toxin binding studies using a filtration assay cannot be performed in Triton X-100 because this detergent significantly decreases filter retention of the channel complex.

7. USE OF ANTIBODIES TO DETECT K$^+$ CHANNELS OF INTEREST

We have used antibodies developed by the methods described above in immunoprecipitation, immunohistochemical, and Western blot approaches in order to investigate the levels of expression and association of K$^+$ channel subunits in several tissues of interest. The immunoprecipitation and histochemical studies have focused exclusively on the composition of toxin-sensitive Kv1 channels in the cerebellum; the high-affinity K$^+$ ligands [^{125}I]margatoxin or [^{125}I]hongotoxin have been employed in these studies (Koch et al., 1997). The cerebellum offers the definite advantage that its neuroanatomical organization is understood in some detail (Wang et al., 1994; Veh et al., 1995). Importantly, high densities of distinct Kv1 subunits are expressed in this brain region, a fact that greatly facilitates biochemical and immunological studies. The combination of methods mentioned above allowed us to determine the subunit composition of the toxin-sensitive channels in this particular brain region. Kv1.1, Kv1.2, Kv1.3, and Kv1.6 proteins are observed at significant levels in the cerebellum. These studies indicated that all toxin-sensitive channels contain at least a single copy of Kv1.2 protein. This observation, *per se*, is not surprising because homotetrameric Kv1.2 channels form a high-affinity toxin receptor. However, ~80% of toxin receptors are heterotetramers of Kv1.1 and Kv1.2. In addition, some of these Kv1.1/Kv1.2 channels also contain an additional Kv1.3 or Kv1.6 subunit. The fact that individual Kv1 subunits mix and match in native tissue provides nature with the opportunity to generate an enormous variety of biophysically distinct K$^+$ channels while using a limited number of genes.

In other studies using an immunoblot approach, antisera containing polyclonal antibodies directed against the α-subunit of the BK$_{Ca}$ channel were used to document the upregulation of these channels in arterial smooth muscle membranes of hypertensive rats. This overexpression is viewed as a protective mechanism that acts to buffer the abnormal vasoconstriction that develops in hypertensive disease (see Chapter 42). Alternatively, it may be useful to use antibodies in immunoblot studies to assess the K$^+$ channel profile in a given tissue of interest and thereby identify which K$^+$ channels are available to regulate cell excitability. In this regard, immunoblots themselves give a good indication of signal strength and the specificity of the antibody–protein interaction, as long as the protein size is already known. Furthermore, because SDS-PAGE is used to size-separate denatured proteins, the epitope is presented in a linear fashion to the antibody, maximizing the opportunity for suitable binding. It is important to perform appropriate controls to verify that the immunoreactive band observed in the immunoblot represents the intended protein of interest. As a negative control, immunoblots are performed with pre-immune serum, rather than the subsequent antibody-containing sera, which should result in a loss of the band of interest. To further verify the specificity of the antibody for its putative epitope, the antibody also can be incubated with a high concentration (1–3 μM) of its corresponding immunogenic peptide before the immunoblotting step. If the antibody demonstrates specific affinity and binds to the peptide, it will be unavailable for binding to the epitope on the channel protein of interest, and the expected immunoreactive band will be markedly decreased or absent.

As an example, we recently used sequence-specific polyclonal antibodies (those shown in Fig. 1) for the BK$_{Ca}$ channel and *Shaker* K$^+$ (Kv1) channels to identify the specific K$^+$ channels that may be functional in small coronary arteries from the human left ventricle. In these vessels, pharmacological block of BK$_{Ca}$ and Kv channels by iberiotoxin (100 nM) or 4-aminopyridine (4-AP) (1 mM), respectively, triggered a pronounced constriction of cannulated human coronary arteries perfused at physiological blood pressure levels (Fig. 3A). Patch-clamped arterial smooth muscle cells from these arteries also showed two distinct components of K$^+$ current sensitive to pharmacological block by iberiotoxin and 4-AP, suggesting that both BK$_{Ca}$ and Kv channels are actively expressed in these small blood vessels (Fig. 3B).

Subsequent immunoblot screening provided additional insight into the profile of K$^+$ channel expression. A single primary antibody confirmed the expression of the BK$_{Ca}$ channel α-subunit, which originates from a single gene family (McCobb *et al.*, 1995). The first lane of the left immunoblot in Fig. 3C shows that a sequence-specific antibody directed against the S9/S10 linker of the BK$_{Ca}$ channel α-subunit resulted in an immunoreactive band at 125 kDa, which is the estimated molecular size of this pore-forming subunit (Knaus *et al.*, 1995). This immunoreactive band at 125 kDa was abolished in the adjacent lane when the anti-BK$_{Ca}$ antibody was incubated with 1 μM of its immunogenic competing peptide (CP), indicating the specificity of the antibody for its recognition site. For the *Shaker* Kv channel family, which has at least nine subfamilies (Kv1.1–Kv1.9), primary antibodies against the Kv1.2, Kv1.4, and Kv1.5 α-subunits were used, because transcripts corresponding to these channels have been detected in vascular tissues (Wang *et al.*, 1997; Roberds and Tamkun, 1991; Overturf *et al.*, 1994). Incubation of the coronary vascular proteins with anti-Kv1.2 antibody

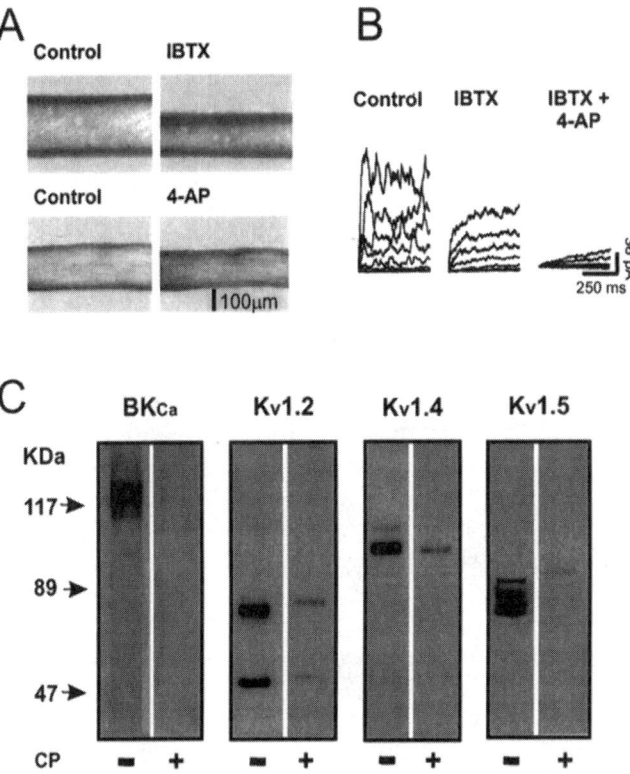

Figure 3. Mapping channels from function to expression in human small coronary arteries. (A) Isolated human coronary arteries contract after the addition of either iberiotoxin (IBTX), a specific blocker of BK$_{Ca}$ channels, or 4-aminopyridine (4-AP), a specific blocker of Kv channels. (B) Whole-cell recordings from patch-clamped human coronary smooth muscle cells show a slowly inactivating K$^+$ current elicited by 10-mV depolarizing pulses from a holding potential of -60 mV. Sequential addition of IBTX and 4-AP additively blocked the K$^+$ current. (C) Western immunoblots of membrane proteins from human coronary arteries show immunoreactive bands characteristic of the BK$_{Ca}$, Kv1.2, Kv1.4 and Kv1.5 α-subunits. Specificity was established by inhibition of band intensity by competition of antibody binding with the corresponding immunogenic competing peptide (CP).

revealed two intense immunoreactive bands at about 84 kDa and 60 kDa, which were markedly diminished by the antigenic peptide. The larger polypeptides recognized by the Kv1.2 antibody approximate the molecular size reported for rat brain Kv1.2 channel α-subunits and likely represent glycosylated products (Sheng *et al.*, 1994). The 60-kDa band may designate a nonglycosylated Kv1.2 species or, more likely, a cross-reacting protein, as supported by the recent finding of a similar band in rat brain membranes that failed to immunoprecipitate with Kv1.2 antibodies (Sheng *et al.*, 1994, 1993). Incubation with anti-Kv1.4 antibody resulted in an intense band at about 100 kDa, whereas incubation with the anti-Kv1.5 antibody directed against the Kv1.5 α-subunit resulted in a diffuse band representing a heterogeneous set of polypeptides between 80 and 90 kDa. These signals also were prevented by incubation of the antibody with the immunogenic CP. A diffuse immunostaining pattern also is observed for Kv1.5 α-subunits heterologously expressed in HEK293 cells (unpublished data). Taken together, these findings are consistent with the expression of the Kv1.2, Kv1.4, and Kv1.5 subtypes of the *Shaker* K^+ channel family in human coronary smooth muscle membranes. Although the estimated molecular mass of the core protein of the α-subunits of these channels is between 55 and 75 kDa, posttranslational modifications of α-subunits, particularly by glycosylation processes, typically enhance the apparent channel sizes on immunoblots (Stuhmer *et al.*, 1989). Further variable posttranslational processing leading to channel microheterogeneity represents one potential explanation for the multiple immunoreactive bands that we observed for Kv1.5.

Although BK_{Ca} channels are known to be highly expressed in coronary arteries from most species, the *Shaker* (Kv1) channels represent a diverse multigene family of channels. Furthermore, the Kv channel subtypes appear to be differentially expressed in arterial smooth muscle cells from different vascular tissues (Berger and Rusch, 1999). Additionally, certain Kv1 channel subtypes share common properties, including similar voltage dependence, kinetic profiles, and pharmacological block by relatively low concentrations of 4-AP ($IC_{50} \leqslant 1$ mM). Thus, standard biophysical or pharmacological analyses of Kv currents may be poor predictors of the Kv channel subtypes expressed in native tissues. For this reason, Western screening may be the preferred approach for providing tissue-specific profiles of Kv channel expression. Indeed, the expression of Kv1.4 α-subunits, which form a rapidly inactivating channel when expressed as a homomultimer structure, would not be expected in human coronary vascular smooth muscle that shows a sustained outward K^+ current in patch-clamp studies (Fig. 3B). However, its detection by Western methods (Fig. 3C) raises the possibility that the Kv1.4 α-subunit may indeed contribute to K^+ current in these small arteries. In this case, its rapidly inactivating phenotype may be masked if it forms heteromultimer channels with other Kv channel α-subunits that show slow inactivation, including the Kv1.2 and Kv1.5 proteins.

8. SUMMARY

Over the past years, antibodies directed against individual ion channel subunits have gained increasing importance as tools to clarify various biochemical, neuroanatomical, or even functional aspects of ion channels. In very many cases, receptor autoradiography and radioligand binding studies *in vitro* established the ion channel's overall distribution profile and density of tissue expression. However, these data were confirmed, or extended, once specific antibodies became available. By using these antibodies, it became feasible to elucidate the respective ion channel distribution down to the subcellular level or to establish the biochemical properties of ion channels through Western blots. Moreover, the fact that Kv channels are capable of forming heterotetrameric channels was originally proposed based on recombinant channel expression in heterologous systems. However, the confirmation that these channels also form heterotetramers in native tissue was finally established by the use of sequence-directed antibodies. As a final example, specific K^+ channel antibodies can be used to compare the regulation of channel expression between physiological and pathophysiological conditions, in order to examine if a certain disease state influences channel density. Conversely, biological systems are capable of modulating ion channel expression in order to compensate for disease-associated malfunctions, an event that also can be investigated with selective antibodies. Thus, selective antibodies are likely to gain increasing importance as molecular biology techniques accelerate the characterization of ion channels. The more we learn about the regulation of gene expression and about the occurrence and distribution of any given mRNA species in a tissue of interest, the more we have to rely on antibody experiments to answer the crucial question: *what happens at the respective protein level?*

REFERENCES

Berger, M. G., Rusch, N. J. 1999, Voltage and calcium-gated potassium channels: Functional expression and therapeutic potential in the vasculature, In: *Perspectives in Drug Discovery and Design*, Y. C. Martin (series ed.), vol. 15/16 Kluwer Academic Plenum Publishers, Dordrecht, The Netherlands, pp. 313–332.

Crest, M., Jacquet, G., Gola, M., Zerrouk, H., Benslimane, A., Rochat, H., Mansuelle P., and Martin-Eauclaire, M. F., 1992, Kaliotoxin, a novel peptidyl inhibitor of neuronal BK-type Ca^{2+}-activated K$^+$ channels characterized from *Androctonus mauretanicus mauretanicus* venom, *J. Biol. Chem.* **267:**1640–1647.

Darbon H., Blanc, E. and Sabatier, J.-M., 1999, Three-dimensional structure of scorpion toyins: Towards a new model of interaction with potassium channels, In: *Perspectives in Drug Discovery and Design* (Y. C. Martin, Series ed.), Vol. 15/16, Kluwer Academic Publishers, Dordrecht, The Netherlands, pp. 41–60.

Galvez, A., Gimenez-Gallego, G., Reuben, J. P., Roy-Contancin, L., Feigenbaum, P., Kaczorowski, G. J., and Garcia, M. L., 1990, Purification and characterization of a unique, potent, peptidyl probe for the high conductance calcium-activated potassium channel from venom of the scorpion *Buthus tamulus*, *J. Biol. Chem.* **265:**11083–11090.

Garcia, M. L., Garcia-Calvo, M., Hidalgo, P., Lee, A., and MacKinnon, R., 1994, Purification and characterization of three inhibitors of voltage-dependent K$^+$ channels from *Leiurus quinquestriatus var. hebraeus* venom, *Biochemistry* **33:**6834–6839.

Garcia-Calvo, M., Leonard, R. J., Novick, J., Stevens, S. P., Schmalhofer, W., Kaczorowski, G. J., and Garcia, M. L., 1993, Purification, characterization, and biosynthesis of margatoxin, a component of *Centruroides margaritatus* venom that selectively inhibits voltage-dependent potassium channels, *J. Biol. Chem.* **268:**18866–18874.

Knaus, H. -G., Eberhart, A., Koch, R. O. A., Munujos, P., Schmalhofer, W. A., Warmke, J. W., Kaczorowski, G. J., and Garcia, M. L., 1995, Characterization of tissue-expressed α subunits of the high conductance Ca^{2+}-activated K$^+$ channel, *J. Biol. Chem.* **270:**22434–22439.

Koch, R. O., Wanner, S. G., Koschak, A., Hanner, M., Schwarzer, C., Kaczorowski, G. J., Slaughter, R. S., Garcia, M. L., and Knaus, H. G., 1997, Complex subunit assembly of neuronal voltage-gated K$^+$ channels: Basis for high-affinity toxin interactions and pharmacology, *J. Biol. Chem.* **272:**27577–27581.

Koschak, A., Bugianesi, R. M., Mitterdorfer, J., Kaczorowski, G. J., Garcia, M. L. and Knaus, H. G., 1998, Subunit composition of brain voltage-gated potassium channels determined by hongotoxin-1, a novel peptide derived from *Centruroides limbatus* venom, *J. Biol. Chem.* **273:**2639–2644.

Liu, Y., Pleyte, K. A., Knaus, H.-G., and Rusch, N. J., 1997, Increased expression of Ca^{2+}-sensitive K$^+$ channels in aorta of hypertensive rats, *Hypertension* **30:**1403–1409.

Liu, Y., Hudetz, A. G., Knaus, H.-G., and Rusch, N. J., 1998, Increased expression of Ca^{2+}-sensitive K$^+$ channels in the cerebral microcirculation of genetically hypertensive rats: Evidence for protection against cerebral vasospasm, *Circ. Res.* **82:**729–737.

McCobb, D. P., Fowler, N. L., Featherstone, T., Lingle, C. J., Saito, M., Krause, J. E., and Salkoff, L., 1995, A human calcium-activated potassium channel gene expressed in vascular smooth muscle, *Am. J. Physiol.* **269:**H767–H777.

Overturf, K. E., Russell, S. N., Carl, A., Vogalis, F., Hart, P. J., Hume, J. R., Sanders, K. M., and Horowitz, B., 1994, Cloning and characterization of a Kv1.5 delayed rectifier K$^+$ channel from vascular and visceral smooth muscles, *Am. J. Physiol.* **267:**C1231–C1238.

Parcej, D. N., and Dolly, J. O., 1989, Dendrotoxin acceptor from bovine synaptic plasma membranes: Binding properties, purification and subunit composition of a putative constituent of certain voltage-activated K$^+$ channels [see comments], *Biochem. J.* **257**:899–903.

Rehm, H., and Lazdunski, M., 1988, Purification and subunit structure of a putative K$^+$-channel protein identified by its binding properties for dendrotoxin I, *Proc. Natl. Acad. Sci. U.S.A.* **85**:4919–4923.

Roberds, S. L., and Tamkun, M. M., 1991, Cloning and tissue-specific expression of five voltage-gated potassium channel cDNAs expressed in the rat heart, *Proc. Natl. Acad. Sci. U.S.A.* **88**:1798–1802.

Roeper, J., Sewing, S., Zhang, Y., Sommer, T., Wanner, S. G., and Pongs, O., 1998, NIP domain prevents N-type inactivation in voltage-gated potassium channels, *Nature* **391**:390–393.

Shamotienko, O. G., Parcej, D. N., and Dolly, J. O.,1997, Subunit combinations defined for K$^+$ channel Kv1 subtypes in synaptic membranes from bovine brain, *Biochemistry* **36**:8195–8201.

Sheng, M., Liao, Y. J., Jan, Y. N., and Jan, L. Y., 1993, Presynaptic A-current based on heteromultimeric K$^+$ channels detected in vivo, *Nature* **365**:72–75.

Sheng, M., Tsaur, M. L., Jan, Y. N., and Jan, L. Y., 1994, Contrasting subcellular localization of the Kv1.2 K$^+$ channel subunit in different neurons of rat brain, J. Neurosci. **14**:2408–2417.

Stuhmer, W., Ruppersberg, J. P., Shroter, K. H., Sakmann, B., Stocker, M., Giese, K. P., Perschke, A., Baumann, A., and Pongs, O., 1989, Molecular basis of functional diversity of voltage-gated potassium channels in mammalian brain, *EMBO J.* **8**:3235–3244.

Vazquez, J., Feigenbaum, P., Katz, G., King, V. F., Reuben, J. P., Roy-Contancin, L., Slaughter, R. S., Kaczorowski, G. J., and Garcia, M. L., 1989, Characterization of high affinity binding sites for charybdotoxin in sarcolemmal membranes from bovine aortic smooth muscle: Evidence for a direct association with the high conductance calcium-activated potassium channel, *J. Biol. Chem.* **64**:20902–20909.

Vazquez, J., Feigenbaum, P., King, V. F., Kaczorowski, G. J., and Garcia, M. L., 1990, Characterization of high affinity binding sites for charybdotoxin in synaptic plasma membranes from rat brain: Evidence for a direct association with an inactivating, voltage-dependent, potassium channel, *J. Biol. Chem.* **265**:15564–15571.

Veh, R. W., Lichtinghagen, R., Sewing, S., Wunder, F., Grumbach, I. M., and Pongs, O., 1995, Immunohistochemical localization of five members of the Kv1 channel subunits: Contrasting subcellular locations and neuron-specific co-localizations in rat brain, *Eur. J. Neurosci.* **7**:2189–2205.

Wang, H., Kunkel, D. D., Schwartzkroin, P. A., and Tempel, B. L, 1994, Localization of Kv1.1 and Kv1.2, two K channel proteins, to synaptic terminals, somata, and dendrites in the mouse brain, *J. Neurosci.* **14**:4588–4599.

Wang, J., Juhaszova, J., Rubin, L. J., and Yuan, X,-J., 1997, Hypoxia inhibits gene expression of voltage-gated K$^+$ channels α-subunits in pulmonary artery smooth muscle cells, *J. Clin. Invest.* **100**:2347–2353.

Chapter 7

Molecular Methods for Evaluation of K⁺ Channel Expression and Distribution in the Heart

Molecular Methods for Evaluation of K^+ Channel Expression and Distribution in the Heart

Michael J. Morales, Mulugu V. Brahmajothi, Donald L. Campbell, and Harold C. Strauss

1. INTRODUCTION

One of the most significant challenges in cardiac physiology is to acquire a detailed understanding of the molecular basis of cardiac repolarization. To approach this problem, it is necessary to identify and fully characterize each of the K^+ currents that are responsible for repolarization. In the last several years, there has been much progress made in phenotypic characterization of these currents (Barry and Nerbonne, 1996; Brown, 1997; Doyle and Stubbs, 1998; Jongsma, 1998; Tseng, 1999). In principle, knowledge of the detailed physiological and pharmacological properties of these currents should have led to ready identification of the corresponding channel molecules. This approach, however, has been complicated by two factors. First, there are many cardiac K^+ currents whose presence, properties, and distribution throughout the heart can vary depending on the species and stage of development. Second, the molecular profile of the K^+ channels responsible for these currents is extremely complex. There have been roughly 30 K^+ channel α-subunit clones reported (Chandy and Gutman, 1995; Jan and Jan, 1997), and many have the ability to "mix and match" to form heteromultimers (Li *et al.*, 1993). Furthermore, many or possibly all of these channel complexes have ancillary subunits which may contribute to their physiological properties (Brown, 1997; Robertson, 1997; Xu and Li, 1998).

Unfortunately, the complexity and low abundance of K^+ channels make it unlikely that they can be purified and characterized directly. Although naturally occurring mutations can help in this respect (Curran *et al.*, 1995; Barhanin *et al.*, 1996; Sanguinetti

Michael J. Morales, Donald L Campbell, and Harold C. Strauss • Department of Physiology and Biophysics, University at Buffalo—SUNY, Buffalo, New York 14212   *Mulugu V. Brahmajothi* • Department of Pharmacology, Duke University Medical Center, Durham, North Carolina 27710

Potassium Channels in Cardiovascular Biology, edited by Archer and Rusch. Kluwer Academic/Plenum Publishers, New York, 2001.

et al., 1996; Wang *et al.*, 1996), it is generally necessary to compile a circumstantial case for correlation of a current with a K^+ channel molecule by comparison of their physiological properties, combined with information about the distribution of the channel's expression in the mammalian heart (Barry and Nerbonne, 1996; Brown, 1997; Doyle and Stubbs, 1998; Jongsma, 1998; Tseng, 1999).

Knowledge of the patterns of K^+ channel gene expression can lead to a more complete understanding of the role of individual K^+ currents. This is because studies of cardiac K^+ currents rarely involve measurements from more than 100 or so cells. In contrast, molecular techniques allow analysis of the entire heart, including regions that may not have been the focus of electrophysiological experiments in single cardiac myocytes. Once correlations have been established, understanding the patterns and mechanisms of K^+ channel gene expression can lead to a more detailed understanding of such issues as the regulation of K^+ channel levels in development and aging and the involvement of K^+ channels in the heart's response to injury and disease.

In general, methods for analysis of gene expression fall into two categories: (1) methods that depend on the biochemical isolation of protein or RNA, which require destruction of the underlying structure of the heart, and (2) techniques that detect RNA or protein in individual isolated myocytes and intact heart tissue. The difference is important. The heart is a complex of many cell types; myocytes, while comprising from 65% to 80% of cardiac cell mass, represent only about 20% of the total number of cells. The rest are neurons, fibroblasts, smooth muscle cells, and endothelial cells (Morkin and Ashford, 1968; Grove *et al.*, 1969; Zak, 1974; Dow *et al.*, 1981; Jacobson and Piper, 1986; Bassingthwaighte, 1991). Of the methods outlined below, Northern hybridization, reverse-transcription polymerase chain reaction (RT-PCR), ribonuclease protection assay, and protein immunoblotting all require solubilization of a large mass of heart tissue. Extracts used for these procedures are a mixture of RNA or protein from different cell types. Therefore, results from these experiments should be interpreted carefully; signals observed do not necessarily come from cardiac myocytes.

Other methods, such as *in situ* hybridization, immunofluorescence, and single-cell PCR, while technically much more challenging, provide a more precise determination of the location of a K^+ channel protein or mRNA. Recently, transgenic mouse technology has made it possible to directly influence K^+ channel gene expression. Although currently applicable only to the mouse, these experiments have provided valuable insight into K^+ channel expression and distribution in the heart.

2. HEART EXTRACT METHODS

2.1. Northern Hybridization

One of the oldest and perhaps most popular methods for analysis of gene expression is Northern hybridization (Alwine *et al.*, 1977, 1979; Farrell, 1998). In this technique, RNA is electrophoretically size separated, usually on a denaturing agarose gel, and transferred by pressure or capillary action to a solid support, such as nitrocellulose. Subsequently, the support with the bound RNA is treated to prevent the probe from binding nonspecifically and then incubated with a radioactively or chemically labeled probe to detect the RNA of interest. In the case of probes that are labeled chemically (through incorporation of biotin or digoxin), fluorescent or chemiluminescent detection is employed (Farrell, 1998).

Northern hybridization has been used extensively for analysis of cardiac K^+ channel transcript expression (Roberds and Tamkun, 1991; Tamkun et al., 1991; Attali et al., 1993; Fedida et al., 1993; Curran et al., 1995; England et al., 1995; Wymore et al., 1996; London et al., 1997; Yang et al., 1997; Kong et al., 1998) and has several important strengths. Its main benefit is that it is the only method that gives the size of the channel mRNA. This information is useful in discovery of multiple forms of an mRNA, generated by either alternative splicing of precursor mRNAs or multiple transcriptional start sites. K^+ channels generally have mRNAs with long untranslated regions, and size differences must be large to be resolved on an agarose gel, as was the case with multiple splice variants of the *ether-a-go-go*-related gene (ERG) (Curran et al., 1995; London et al., 1997).

Another advantage of Northern hybridization is flexibility in the choice of probe, which may be a full-length or partial cDNA, an *in vitro* transcribed antisense RNA, or an oligonucleotide. Also, the probe need not be from the same species as the RNA target, as long as the probe and RNA target have close sequence homology, as is the case with most mammalian K^+ channels (Chandy and Gutman, 1995). However, caution should be exercised when using heterologous probes. If the stringency of the hybridization is too low, transcripts not related to the channel of interest may be detected. However, the selection of conditions that are too stringent or the use of short probes that cover a region less closely related than the entire cDNA carries the risk of missing a closely related transcript entirely. These risks are much higher when an oligonucleotide probe is used.

There are some disadvantages to the Northern hybridization procedure. It is the least sensitive of the RNA detection methods outlined in this chapter, so that large amounts of whole-cell or poly $(A)^+$-enriched RNA may be necessary. This can be a significant burden if small portions of a heart (e.g., the sinoatrial node) are being investigated, or if the RNA comes from a rare sample. Even when large amounts of RNA are available, the amount that can be loaded onto a horizontal agarose gel is limited, depending on the design of the gel apparatus. Under ideal conditions, it is possible to detect RNAs that are 0.01% of total RNA (Ausubel et al., 1987). The RNA also must be in very good condition, because partially degraded RNA will generally give a poor or nonexistent signal.

2.2. Ribonuclease Protection Assay

Another older technology that continues to gain in popularity is the ribonuclease (RNase) protection assay (Dixon and McKinnon, 1994; Dixon et al., 1996; Lees-Miller et al., 1997; Shimoni et al., 1997; Wymore et al., 1997; Demolombe et al., 1998; Hershman and Levitan, 1998; Kääb et al., 1998; Kong et al., 1998; Kupershmidt et al., 1998; Wickenden et al., 1999; Yu et al., 1999). Like the Northern, this is a hybridization assay. Cardiac RNA, either crude or poly $(A)^+$-enriched, is mixed with an antisense radiolabeled RNA probe and allowed to hybridize. The mixture is then incubated with a mixture of RNases specific for single-stranded RNA. The radiolabeled riboprobe that is hybridized to the target RNA will be double-stranded and, therefore, protected from digestion by the RNases. Thus, the protected RNA probe can be recovered by ethanol precipitation and detected by autoradiography or phosphoimaging after electrophoresis (Zinn *et al.*, 1983; Melton *et al.*, 1984).

Because of its many advantages, in most cases the protection assay is the preferred method for detection of K^+ channel transcripts in whole tissue. It is approximately 10-fold more sensitive than Northern hybridization (Farrell, 1998). This sensitivity allows the use of less RNA in the assay or the option of using whole-cell RNA as a target. Furthermore, because the probe covers only a small portion of the target RNA (typically 100–500 bases), the RNA sample does not have to be in pristine condition. This is a potentially important factor when RNA must be isolated from human tissues that are often impossible to procure under ideal laboratory conditions. Because the assay occurs under conditions where all available K^+ channel mRNA targets are available for hybridization, RNase protection lends itself well to quantification of transcript levels (Dixon and McKinnon, 1994; Shimoni *et al.*, 1997; Wymore *et al.*, 1997; Hershman and Levitan, 1998; Kääb *et al.*, 1998; Yu *et al.*, 1999). Designing a probe to span splice junctions allows characterization of alternatively spliced channels (Lees-Miller *et al.*, 1997; Kupershmidt *et al.*, 1998).

Unlike the wide choice of probes in Northern hybridizations, a probe in an RNase protection assay must meet very rigid requirements. It must be a single-stranded antisense RNA probe that comes from the same species as the target. This is because of the specificity of the assay; single-stranded RNAs formed by imperfect duplex formation, as would be the case with a heterologous probe, are recognized by the RNase mixture used in the assay and digested. It is usually not a major burden to isolate homologous partial cDNAs that can be used as probes, and the necessity of having a homologous probe offers the advantage of confidence. There is little chance of an RNase protection assay producing a spurious signal due to cross-hybridization with another mRNA species, as may occur in Northern hybridization.

It is important to consider the possibility of allelic variation when working with outbred species such as human, ferret, or dog. Variation of sequence between the probe and the gene in the individual being studied can give a false-negative result, because the RNases will recognize even one or two base-pair mismatches as single-stranded. This also may be a problem when analyzing gene expression in rats and mice from different strains. In these cases, it may be advisable to check the sequence of the cDNA in that species or strain before embarking on a protection assay.

2.3. Quantitative Polymerase Chain Reaction of Reverse Transcribed mRNA

If little RNA is available, or the quality is low, sometimes even RNase protection assays may be insufficiently sensitive for detection of cardiac K^+ channel mRNA. In this case, the amazing sensitivity of PCR has made reverse transcription of RNA, followed by PCR, a tempting method for detection and quantification of K^+ channel mRNA expression. Unfortunately, there are several factors that make conventional

PCR an inappropriate choice for this type of assay (Farrell, 1998; Freeman *et al.*, 1999). The problem is that the amount of PCR product at the end of the procedure depends on both the efficiency of the reverse transcription of the mRNA, which can have batch-to-batch fluctuations of between 5% and 90% (Ferré *et al.*, 1994), and the reproducibility of the PCR reaction. These parameters are critical, because PCR proceeds in an exponential fashion, thus increasing errors exponentially. Another problem is that the sensitivity of the PCR reaction, combined with the complexity of cardiac tissue, means that transcripts from nonmyocyte cells present at very low abundance might be efficiently amplified. This is possible even in cases in which the most extreme precautions have been taken to prevent contamination.

An attempt to control this sensitivity and the efficiency of the reverse transcription can be made by measuring the amplification efficiency of an internal control, typically a "housekeeping" transcript such as cyclophilin or glyceraldehyde phosphate dehydrogenase (GAPDH) (Matsubara *et al.*, 1993; Ohya *et al.*, 1997; Farrell, 1998; Freeman *et al.*, 1999). However, for this internal control to be useful, it must then be amplified at an identical rate to the gene of interest. Furthermore, the assumption behind the use of such controls is that their expression is invariant within an organism; this assumption has never been thoroughly tested and is almost certainly false.

The best method for measurement of transcript levels by PCR is through the use of an external competitor control (Gilliland *et al.*, 1990; Hengen, 1995; Raeymaekers, 1995; Zimmermann and Mannhalter, 1996; Farrell, 1998; Freeman *et al.*, 1999). This control is usually a very close copy of the channel mRNA under investigation, with only a small deletion, insertion, or restriction site between the primer binding sequences (Gilliland *et al.*, 1990; Matsubara *et al.*, 1993; Comer *et al.*, 1994; Farrell, 1998; Freeman *et al.*, 1999; Ojamaa *et al.*, 1999). It can also be a "mimic," which is a DNA molecule containing the same primer binding sequences as the channel target, but with different intervening DNA (Siebert and Kelloff, 1995; Wang *et al.*, 1999). It is best if an RNA copy of the competitor is transcribed *in vitro* (ideally, it should be approximately the same size as the native transcript) and known amounts are added to the RNA mixture before reverse transcription.

When this technique is used, there are several factors required for validation of the assay. The amplification efficiency of the competitor and target must be identical, and they must enter into the plateau phase of the reaction after the same number of cycles (Raeymaekers, 1993, 1995; Farrell, 1998; Freeman *et al.*, 1999). The PCR reaction must be optimized to minimize heteroduplex formation between the target and the competitor as their concentrations within the reaction increase (Farrell, 1998; Chen *et al.*, 1999; Freeman *et al.*, 1999). Finally, a sensitive method must be developed for detection of the PCR reaction.

Despite the many technical hurdles, quantitative PCR may be the only choice available in cases where small amounts of sample are available or the sample must be isolated from a difficult source. Recently, technology has become available that allows direct fluorescent labeling of accumulating PCR products (Lee *et al.*, 1993; Gibson *et al.*, 1996; Piatek *et al.*, 1998), and instruments have been developed that take advantage of this labeling to measure accumulation of PCR products in real time (Gibson *et al.*, 1996). These developments promise to make quantitative PCR a more facile and common technique for transcript measurement.

2.4. Immunoblotting

All of the above methods analyze K^+ channel gene expression through measurement of mRNA. However, if the question one wishes to address concerns expression of the K^+ channel protein and not its transcript, it is important to consider the possibility that mRNA expression and protein expression do not correlate, as has been shown to be the case for Kv1.4 in mouse (London *et al.*, 1998b), rat (Barry *et al.*, 1994), and ferret heart (Brahmajothi *et al.*, 1999), in which the mRNA has been shown to be abundant, but the protein is nearly absent.

Because of these considerations, it is always advisable to consider detection of the channel proteins directly through immunoblotting (Burnette, 1981). A membrane protein extract is fractionated on a sodium dodecylsulfate (SDS)-polyacrylamide gel, and the proteins are transferred electrophoretically to a solid support, which is then treated to prevent nonspecific binding of antibodies. This support is then incubated with an antibody specific for the K^+ channel of interest. Antibody binding is detected through a label directly attached to the primary antibody or, more commonly, by treatment with a labeled secondary antibody, directed to the first. Common detection methods include labeling the antibody with ^{125}I, a fluorescent compound, alkaline phosphatase, or horseradish peroxidase, with the latter two methods allowing enzymatic detection.

There are many reasons why immunoblotting has not gained wider acceptance. It is time-consuming and expensive, and suitable antibodies are difficult to produce and characterize. Although there are good published protocols for cardiac membrane protein preparation (Barry *et al.*, 1994), these are not as standardized as RNA preparation protocols. Optimization of the first and second antibody binding, as well as the detection method, is also necessary. Despite these difficulties, clear characterization of channel protein expression ultimately is required for positive identification of cardiac K^+ channels (Barry *et al.*, 1994; Xu *et al.*, 1996; Brahmajothi *et al.*, 1997, 1999; Van Wagoner *et al.*, 1997; Guo *et al.*, 1998).

3. METHODS WHICH PRESERVE UNDERLYING STRUCTURE

All of the above methods require total disruption of the cardiac tissue prior to fractionation of RNA or protein. Although it is possible to garner some useful information regarding localization of ion channels by using these methods, detailed analysis is still difficult. In this regard, recent studies suggest that complex localization patterns for cardiac K^+ currents may be the rule. Accordingly, methods that permit the detection of regional patterns of variation in K^+ channel transcript or protein expression may assume increasing importance.

3.1. *In Situ* Hybridization and Immunofluorescence of Individual Myocytes

One of the problems with methods that rely on the lysis of a whole heart is the possibility that the mRNA detected is not present in working myocytes, but rather exists in another cell type found in the heart. Another problem is that channel expression localized to only a small number of cells might be missed entirely, because of overall low levels of expression. One method for avoiding these problems is to perform *in situ* hybridization on isolated myocytes (Wilkinson, 1992; Brahmajothi *et al.*, 1996).

To perform *in situ* hybridization, cells are isolated after perfusion of the heart with a Langendorff apparatus (Campbell *et al.*, 1993). The heart is dissected, and individual myocytes are isolated by enzymatic digestion (Brahmajothi *et al.*, 1997, 1999). Populations of individual cells are fixed in paraformaldehyde, washed, prehybridized, and incubated with an oligonucleotide conjugated to digoxin. After overnight incubation, the probe is washed away and the cells are incubated with fluorescein isothiocyanate-conjugated antidigoxin antibody. The signal is evaluated by direct visualization with a fluorescent microscope (Brahmajothi *et al.*, 1996).

Oligonucleotides make excellent probes for *in situ* hybridization, although it is more common for *in vitro* transcribed RNAs to be used (Wilkinson, 1992). In designing an oligonucleotide probe, it is important to consider factors such as the potential for cross-hybridization and formation of hairpin loops that interfere with hybridization. Oligonucleotides between 45 and 55 nucleotides long that have been gel-purified are optimal (Brahmajothi *et al.*, 1996). In some cases, it has been difficult to design probes that lack potential secondary structure for specific channels. In these cases, up to three inosine bases can be substituted for guanine (G) or cytosine (C) in order to break up these structures.

For each probe designed, it is necessary to design a complementary probe identical to the channel RNA to be tested. Using this sense probe in parallel experiments with the antisense probe ensures that the fluorescent signals are not due to spurious binding to DNA or positively charged molecules within the cell. Furthermore, it always is useful to include a positive control that is expected to be expressed in all myocytes, such as the cardiac form of troponin I (Vallins *et al.*, 1990; Gorza *et al.*, 1993; Brahmajothi *et al.*, 1996). In practice, signals for the positive control will only be detected in 85%–95% of myocytes, probably because some cells are damaged or otherwise rendered incompetent for hybridization during processing. This factor should be considered when interpreting experiment results (Brahmajothi *et al.*, 1996).

A previous study performed in myocytes isolated from the ferret heart hints at an enormous diversity of cardiac K$^+$ channel gene expression (Brahmajothi *et al.*, 1996). In this study, probes for 21 different K$^+$ channel mRNAs were hybridized with cells from different regions of the ferret heart (Fig. 1). All of the 21 different K$^+$ channel mRNAs were expressed in some population of cells within the heart. Channels such as Kv1.4, Kv1.5, and KvLQT1 were expressed in at least half of all myocytes throughout the heart and were observed in more than 85% of myocytes in some compartments. This latter finding was interpreted as a 100% expression incidence, because of possible experimental damage to some myocytes. Other transcripts were found in a much lower number of myocytes. None of the 21 "sense" negative control probes gave a positive signal.

With the caveat that these techniques looked at channel mRNA and not protein expression, this study has important implications. It suggests that many K$^+$ channels previously thought to be absent from the heart may be present in small populations of cells throughout the myocardium. Such a finding, if true for other species, suggests that the range of cardiac K$^+$ current phenotypes may be much broader than previously suspected.

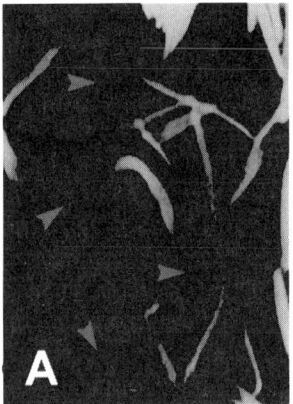 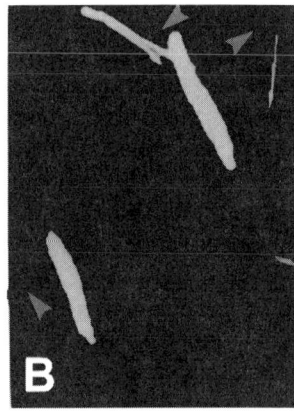 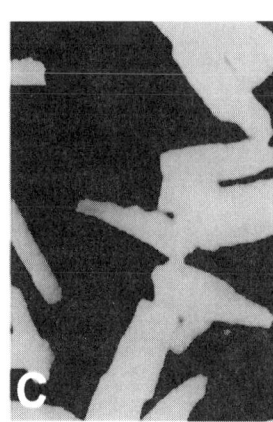

Figure 1. *In situ* hybridization of isolated cardiac myocytes. Cells were isolated, fixed, and hybridized, and imaged as described by Brahmajothi *et al.* (1996). (A) Cells isolated from ferret right atrium and hybridized to an oligonucleotide directed to Kv1.5. (B) Cells isolated from ferret left ventricle and hybridized to an oligonucleotide directed to ERG. (C) Cells isolated from ferret left ventricle and hybridized to an oligonucleotide directed to cardiac troponin I. Red carrots point to cells within the field that show no fluorescent signal.

When antibodies are available, it is possible to perform immunofluorescence on isolated myocytes. The cells are isolated and fixed as above and then treated with a K^+ channel-specific antibody, followed by a fluorescently labeled second antibody. Immunofluorescent studies on single cells have been used to characterize channel proteins that may contribute to the transient outward current (I_{to}) in the ferret heart, including the voltage-gated K^+ (Kv) channels Kv1.4, Kv4.2, and Kv4.3 (Brahmajothi *et al.*, 1999). These studies show that the numbers of ventricular myocytes that express Kv4.2 and Kv4.3 mRNA and protein are roughly equivalent but that Kv1.4 protein is restricted to a much smaller subset of myocytes than is its corresponding mRNA.

This study also produced interesting information concerning the identity of the channels responsible for I_{to}. In this regard, the primary characteristic that distinguishes Kv4.2 and Kv4.3 from Kv1.4 is the rapid recovery rate of Kv4.2 and Kv4.3 from inactivation. Although reports vary widely, time constants for inactivation (τ_{rec}) approximating 250 ms at -80 mV have been reported (Serodio *et al.*, 1994; Tseng *et al.*, 1996; Fiset *et al.*, 1997; Yeola and Snyders, 1997; Kong *et al.*, 1998; Faivre *et al.*, 1999; Franqueza *et al.*, 1999; Guo *et al.*, 1999). In contrast, ferret Kv1.4 shows a τ_{rec} of ~ 4800 ms (Comer *et al.*, 1994)). Furthermore, only the Kv4.2 and Kv4.3 channels are sensitive to block by heteropodatoxin (HPTX) (Sanguinetti *et al.*, 1997). Immunofluorescent measurements of myocytes derived from ferret left ventricular epicardial and endocardial myocytes showed that Kv1.4 was present in $\sim 60\%$ of endocardial myocytes but in only 4% of epicardial myocytes. The data for the Kv4 channels were not as striking, with Kv4.2 being found in 57% of epicardial myocytes and 21% of endocardial myocytes, whereas Kv4.3 was present in 45% of epicardial myocytes and 35% of endocardial myocytes (Brahmajothi *et al.*, 1999). These findings predict that there should be a fast-recovering, HPTX-sensitive I_{to} predominant in the ferret left ventricular epicardium, but a mixture of epicardial I_{to} and slowly recovering HPTX-insensitive I_{to} should be observed in the endocardium. Physiological evidence confirmed this prediction.

3.2. *In situ* Hybridization and Immunofluorescence of Whole Tissue

All of the methodologies discussed rely on dissection of the heart to get information about localization of K⁺ channel expression. However, the resolution of these techniques is limited to what can be resolved through dissection and relies on correct predictions about the localization of interesting patterns. A better approach is to perform *in situ* hybridization or immunofluorescence on sections of the heart (Brahmajothi *et al.*, 1997, 1999).

In this method, preparation of the heart is critical. In the ferret, hearts are perfused with Ringer solution, followed by fixation in paraformaldehyde and overnight incubation in 40% sucrose. The tissue is then incubated in an embedding medium containing polyvinyl alcohol and polyethylene glycol (Tissue-Tek O.C.T. Compound) and frozen in a dry-ice/isobutanol bath. Frozen sections, usually 10 μm in thickness, are produced on a cryostat. The sections are laid out on gelatin-coated slides and processed for either immunofluorescence or *in situ* hybridization (Brahmajothi *et al.*, 1997).

Thus far, studies designed to explore the localization patterns of K⁺ channel subunits have yielded striking patterns of localization (Fig. 2). In the ferret heart, the *ether-a-go-go*-related channel ERG-1A, which is responsible for the generation of the

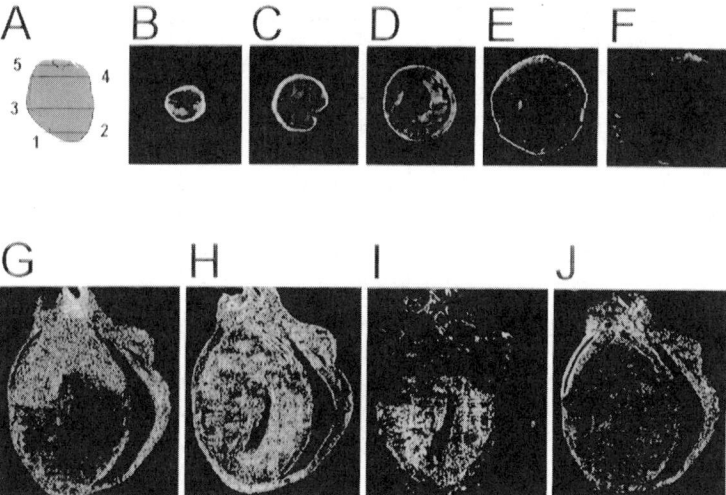

Figure 2. Localization of K⁺ channels in ferret ventricle. (A) Representation of the ferret ventricle showing the locations of the sections in the top row. Lines labeled 1–5 correspond to (B)–(F), respectively. (B)–(F) Cross sections of the ferret ventricle showing immunolocalization of ERG1A (Brahmajothi *et al.*, 1997). (G)–(J) Sagittal sections of the ferret ventricle, showing immunolocalization of Kv4.2 (G), Kv4.3 (H), and Kv1.4 (I), and *in situ* hybridization of Kv1.4 (J). In contrast to ERG, Kv4.2, and Kv4.3, Kv1.4 expression is more widespread than the Kv1.4 protein.

rapid component of the delayed rectifier K$^+$ current (I_{Kr}), is clearly localized to the epicardial layers of the heart. This is very distinct at the ventricular apex; as sections move toward the base of the ventricle, ERG expression is reduced to a very low level. The fast-inactivating channels Kv1.4, Kv4.2, and Kv4.3 also show distinct patterns of localization in the ferret ventricle. Kv1.4 is localized to the left ventricular endocardium and ventricular septum. Kv4.2 is nearly a mirror image, with most of the protein present in the right ventricle and the ventricular base. Kv4.3 is not as intensely localized, yet it is clearly more abundant in the epicardium.

These results, combined with the earlier study showing 21 K$^+$ channel mRNAs in isolated myocytes (Brahmajothi *et al.*, 1996) and the emerging knowledge of K$^+$ channel ancillary subunits in the heart (Robertson, 1997; Xu and Li, 1998), argue for the presence of very complex patterns of cardiac K$^+$ channel expression. When developmental and disease issues are taken into account, the picture seems almost impossibly complex. Yet, despite the significant technical difficulties in performing immunofluorescence on intact hearts, it is clear that such studies will be necessary to completely understand K$^+$ channel gene expression in the mammalian myocardium.

4. *IN VIVO* METHODOLOGIES

The final class of techniques available to study K$^+$ channel expression involves *in vivo* manipulation of the heart or cardiac cells. This may occur either in cultures of heart cells or in transgenic mice. Although these techniques have many limitations, they all have the advantage of yielding direct information on channel biosynthesis.

4.1. Antisense Oligonucleotide Inhibition of Channel Biosynthesis

The antisense approach relies on transfection of cultured heart cells with oligonucleotides complementary to a portion of the K$^+$ channel mRNA, which then inhibit channel synthesis, primarily through direct inhibition of translational initiation (Crooke, 1992). The oligonucleotides are typically short (15–30 bases) and are complementary to the translational start site. To inhibit degradation of the oligonucleotide, it is usually synthesized as a phosphorothioate analog. This approach rarely eliminates channel biosynthesis completely, so it also is very important to measure inhibition in cells treated with an oligonucleotide with a randomized base composition.

Antisense oligonucleotides are being used to characterize cardiac K$^+$ channel gene expression. For example, they were used to inhibit the synthesis of the K$^+$ channel ancillary subunit *minK* in AT-1 cells, a mouse cardiac cell line (Yang *et al.*, 1995). In this case, an antisense oligonucleotide complementary to *minK* mRNA inhibited the I_{Kr} current in these cells. Although recent evidence suggests that a protein related to *minK*, MiRP1, is the ancillary subunit that associates with ERG to form the native I_{Kr} current (Abbott *et al.*, 1999), this study provided important clues to the nature of the native I_{Kr} channel.

In two other studies, antisense oligonucleotides were used to characterize the I_{to} current in rat ventricular myocytes (Fiset *et al.*, 1997) and the ultrarapid component of the delayed rectifier current (I_{Kur}) in human atrial myocytes (Feng *et al.*, 1997). In the case of rat I_{to}, antisense oligonucleotides to Kv4.2 and Kv4.3 were used to inhibit the currents for which these channels are responsible. Similarly, antisense oligonucleotides complementary to Kv1.5 were used to test the role of currents produced by this channel in generation of I_{Kur} in atrial myocytes. In both studies, the currents were reduced by roughly 50%, providing evidence for the participation of these distinct channels in the generation of these important currents.

4.2. Approaches to Manipulation of Mammalian Hearts *In Vivo*

The classic approach to *in vivo* manipulation of gene expression in live animals is through production of transgenic mice. The most commonly used transgenic technology allows overexpression of a foreign or modified protein, whose expression can be directed to the heart with tissue-specific promoters. Several investigators have taken advantage of the capacity of K⁺ channels to heteromultimerize, in order to study the roles of specific K⁺ channels in the mouse heart. One study used a truncated form of the α-subunit of mouse Kv1.1, which included the amino acid sequence responsible for heteromultimerization, but not the sequence for the pore region. Overexpression of this protein in a cultured cell line was shown to inhibit Kv1.4 and Kv1.5 expression through a dominant-negative mechanism, because the overexpressed protein formed complexes with native channels during their biosynthesis. Because the Kv1.1 construct had no pore region, the resulting complexes were incapable of expression of K⁺ currents (Folco et al., 1997). This same effect was seen in transgenic mice expressing this construct, which showed a significant reduction in a rapidly activating, slowly inactivating, 4-aminopyridine-sensitive current in their cardiac myocytes, probably due to a reduction in the Kv1.5 channel (London et al., 1998a; Zhou et al., 1998). A second study used a similar strategy to look at Kv4 expression (Barry et al., 1998). Instead of a truncated channel, a full-length Kv4.2 with a pore mutation that rendered it incapable of conduction was used. Ventricular cardiac myocytes isolated from mice harboring this mutant lacked I_{to}, supporting the idea that Kv4 channels are involved in generation of this current. Interestingly, these mice also showed a K⁺ current not present in nontransgenic mice (Barry et al., 1998). These data showed that introduction of a K⁺ channel transgene can change other aspects of the mouse's developmental program.

These studies used the introduction of dominant-negative mutations to inhibit channel expression. However, a more direct method is to use gene-targeting methodology to specifically remove all or part of a channel gene, thereby ensuring that it is not expressed. This approach has been used to suggest that Kv1.4 plays no part in the generation of cardiac K⁺ currents in the mouse (London et al., 1998b). A variation of this methodology also has been used to map *minK* expression in mice. By replacing the *minK* coding sequence with that of β-galactosidase (and presumably not altering elements responsible for transcriptional regulation), it was possible to establish that *minK* is probably localized to the conduction system in the mouse heart (Kupershmidt et al., 1999).

Aside from the technical sophistication required to produce transgenic mice, the main problem with this technique is that it is generally restricted to mice, although the use of viral systems to introduce foreign genes into hearts may alleviate this limitation

(Kass-Eisler *et al.*, 1993; Donahue *et al.*, 1997; Johns *et al.*, 1997; Morsy *et al.*, 1998; Nuss *et al.*, 1999). The main advantage conferred by transgenics is the ability to manipulate currents *in vivo*, as opposed to relying on circumstantial lines of evidence for matching K^+ currents to specific channels.

5. CONCLUSION

This chapter has outlined several methodologies of varying complexity and utility. Traditional methodologies such as the RNase protection assay and immunoblotting remain a good choice for understanding overall gene expression in large portions of the heart. However, newer techniques such as *in situ* hybridization will assume increasing importance as it becomes necessary to delve more fully into complex issues such as mapping the patterns of K^+ channel expression in the various compartments of the myocardium and elucidating how these patterns evolve during development, aging, and disease states. Finally, as the methodologies for direct genetic manipulation of cardiac myocytes and intact hearts become more available, those investigators interested in understanding K^+ channel gene expression will be able to move beyond simple characterizations of gene expression to a more mechanistic understanding of how specific channels contribute to distinct patterns of repolarizing currents in the heart.

REFERENCES

Abbott, G. W., Sesti, F., Splawski, I., Buck, M. E., Lehmann, W. H., Timothy, K. W., Keating, M. T., and Goldstein, S. A. N., 1999, MiRP1 forms I_{Kr} potassium channels with HERG and is associated with cardiac arrhythmia, *Cell* **97:**175–187.

Alwine, J. C., Kemp, D. J., and Stark, G. R., 1977, Method for detection of specific RNAs in agarose gels by transfer to diazobenzyloxymethyl-paper and hybridization with DNA probes, *Proc. Natl. Acad. Sci. U.S.A.* **74:**5350–5354.

Alwine, J. C., Kemp, D. J., Parker, B. A., Reiser, J., Renart, J., Stark, G. R., and Wahl, G. M., 1979, Detection of specific RNAs or specific fragments of DNA by fractionation in gels and transfer to diazobenzyloxymethyl paper, *Methods Enzymol.* **68:**220–242.

Attali, B., Lesage, F., Ziliani, P., Guillemare, E., Honor, E., Waldmann, R., Hugnot, J.-P., Mattéi, M.-G., Lazdunski, M., and Barhanin, J., 1993, Multiple messenger RNA isoforms encoding the mouse cardiac Kv1-5 delayed rectifier K^+ channel, *J. Biol. Chem.* **268:**24283–24289.

Ausubel, F. M., Brent, R., Kingston, R. E., Moore, D. D., Seidman, J. G., Smith, J. A., and Struhl, K. (eds.), 1987, *Current Protocols in Molecular Biology*, John Wiley & Sons, New York.

Barhanin, J., Lesage, F., Guillemare, E., Fink, M., Lazdunski, M., and Romey, G., 1996, KvLQT1 and IsK (minK) proteins associate for form the I_{Ks} cardiac potassium current, *Nature* **384:**78–80.

Barry, D. M., and Nerbonne, J. M., 1996, Myocardial potassium channels: Electrophysiological and molecular diversity, *Annu. Rev. Physiol.* **58:**363–394.

Barry, D. M., Trimmer, J. S., Merlie, J. P., and Nerbonne, J. M., 1994, Differential expression of voltage-gated K^+ channel subunits in adult rat heart: Relation to functional K^+ channels?, *Circ. Res.* **77:**361–369.

Barry, D. M., Xu, H., Schuessler, R. B., and Nerbonne, J. M., 1998, Functional knockout of the transient outward current, long-QT syndrome, and cardiac remodeling in mice expressing a dominant-negative Kv4 α subunit, *Circ. Res.* **83:**560–567.

Bassingthwaighte, J. B., 1991, The myocardial cell, in: *Cardiology: Fundamentals and Practice* 2nd ed. (E. R. Giuliani, B. J. Gersh, and M. D. McGoon, eds.), Mosby Year Book, St. Louis, pp. 113–149. Mosby Year Book, c1991.

Brahmajothi, M. V., Morales, M. J., Liu, S., Rasmusson, R. L., Campbell, D. L., and Strauss, H. C., 1996, In situ hybridization reveals extensive diversity of K$^+$ channel mRNA in isolated ferret cardiac myocytes, *Circ. Res.* **78:**1083–1089.

Brahmajothi, M. V., Morales, M. J., Reimer, K. A., and Strauss, H. C., 1997, Regional localization of *ERG*, the channel protein responsible for the rapid component of the delayed rectifier K$^+$ current in the ferret heart, *Circ. Res.* **81:**128–135.

Brahmajothi, M. V., Campbell, D. L., Rasmusson, R. L., Morales, M. J., Trimmer, J. S., Nerbonne, J. M., and Strauss, H. C., 1999, Distinct transient outward potassium current (I_{to}) phenotypes and distribution of fast-inactivating potassium channel alpha subunits in ferret left ventricular myocytes, *J. Gen. Physiol.* **113:**581–600.

Brown, A. M., 1997, Cardiac potassium channels in health and disease, *Trends Cardiovasc. Med.* **7:**118–124.

Burnette, W. N., 1981, Western blotting: Electrophoretic transfer of proteins from sodium dodecyl sulfate-polyacrylamide gels to unmodified nitrocellulose and radiographic detection with antibody and radioiodinated protein A, *Anal. Biochem.* **112:**195–203.

Campbell, D. L., Rasmusson, R. L., Qu, Y., and Strauss, H. C., 1993, The calcium-independent transient outward potassium current in isolated ferret right ventricular myocytes. I. Basic characterization and kinetic analysis, *J. Gen. Physiol.* **101:**571–601.

Chandy, K. G., and Gutman, G. A., 1995, Voltage-gated potassium channel genes, in: *Handbook of Receptors and Channels* (R. A. North, ed.), CRC Press, Boca Raton, Florida, pp. 1–71.

Chen, Z. G., Smithberger, J., Sun, B., and Eggerman, T. L., 1999, Prevention of heteroduplex formation in mRNA quantitation by reverse transcription polymerase chain reaction, *Anal. Biochem.* **266:**230–232.

Comer, M. B., Campbell, D. L., Rasmusson, R. L., Lamson, D. R., Morales, M. J., Zhang, Y., and Strauss, H. C., 1994, Cloning and characterization of an I_{to}-like potassium channel from ferret ventricle, *Am. J. Physiol.* **267:**H1383–H1395.

Crooke, S. T., 1992, Therapeutic applications of oligonucleotides, *Annu. Rev. Pharmacol. Toxicol.* **32:**329–376.

Curran, M. E., Splawski, I., Timothy, K. W., Vincent, G. M., Green, E. D., and Keating, M. T., 1995, A molecular basis for cardiac arrhythmia: *HERG* mutations cause long QT syndrome, *Cell* **80:**795–803.

Demolombe, S., Baró, I., Péroén, Y., Bliek, J., Mohammad-Panah, R., Pollard, H., Morid, S., Mannens, M., Wilde, A., Barhanin, J., Charpentier, F., and Escande, D., 1998, A dominant negative isoform of the long QT syndrome 1 gene product, *J. Biol. Chem.* **273:**6837–6843.

Dixon, J. E., and McKinnon, D., 1994, Quantitative analysis of potassium channel mRNA expression in atrial and ventricular muscle of rats, *Circ. Res.* **75:**252–260.

Dixon, J. E., Shi, W. M., Wang, H.-S., McDonald, C., Yu, H., Wymore, R. S., Cohen, I. S., and McKinnon, D., 1996, Role of the Kv4.3 K$^+$ channel in ventricular muscle: A molecular correlate for the transient outward current, *Circ. Res.* **79:**659–668.

Donahue, J. K., Kikkawa, K., Johns, D. C., Marban, E., and Lawrence, J. H., 1997, Ultrarapid, highly efficient viral gene transfer to the heart, *Proc. Natl. Acad. Sci. U.S.A.* **94:**4664–4668.

Dow, J. W., Harding, N. G. L., and Powell, T., 1981, Isolated cardiac myocytes. I. Preparation of adult myocytes and their homology with the intact tissue, *Cardiovasc. Res.* **15:**483–514.

Doyle, J. L., and Stubbs, L., 1998, Ataxia, arrhythmia and ion-channel gene defects, *Trends Genet.* **14:**92–98.

England, S. K., Uebele, V. N., Shear, H., Kodali, J., Bennett, P. B., and Tamkun, M. M., 1995, Characterization of a voltage-gated K$^+$ channel β subunit expressed in human heart, *Proc. Natl. Acad. Sci. U.S.A.* **92:**6309–6313.

Faivre, J.-F., Calmels, T. P. G., Rouanet, S., Javré, J.-L., Cheval, B., and Bril, A., 1999, Characterisation of Kv4.3 in HEK293 cells: Comparison with the rat ventricular transient outward potassium current, *Cardiovasc. Res.* **41:**188–199.

Farrell, R. E., 1998, *RNA Methodologies. A Laboratory Guide for Isolation and Characterization*, 2nd ed., Academic Press, San Diego.

Fedida, D., Wible, B., Wang, Z., Fermini, B., Faust, F., Nattel, S., and Brown, A. M., 1993, Identity of a novel delayed rectifier current from human heart with a cloned K$^+$ channel current, *Circ. Res.* **73:**210–216.

Feng, J. L., Wible, B., Li, G. R., Wang, Z. G., and Nattel, S., 1997, Antisense oligodeoxynucleotides directed against Kv1.5 mRNA specifically inhibit ultrarapid delayed rectifier K$^+$ current in cultured adult human atrial myocytes, *Circ. Res.* **80:**572–579.

Ferré, F., Marchese, A., Pezzoli, P., Griffin, S., Buxton, E., and Boyer, V., 1994, Quantitative PCR. An overview, in: *The Polymerase Chain Reaction* (K. B. Mullis, F. Ferré, and R. A. Gibbs, eds.), Birkhäuser, Boston, pp. 67–88.

Fiset, C., Clark, R. B., Shimoni, Y., and Giles, W. R., 1997, *Shal*-type channels contribute to the Ca^{2+}-independent transient outward K^+ current in rat ventricle, *J. Physiol. (London)* **500**:51–64.

Folco, E., Mathur, R., Mori, Y., Buckett, P., and Koren, G., 1997, A cellular model for long QT syndrome: Trapping of heteromultimeric complexes consisting of truncated Kv1.1 potassium channel polypeptides and native Kv1.4 and Kv1.5 channels in the endoplasmic reticulum, *J. Biol. Chem.* **272**:26505–26510.

Franqueza, L., Valenzuela, C., Eck, J., Tamkun, M. M., Tamargo, J., and Snyders, D. J., 1999, Functional expression of an inactivating potassium channel (Kv4.3) in a mammalian cell line, *Cardiovasc. Res.* **41**:212–219.

Freeman, W. M., Walker, S. J., and Vrana, K. E., 1999, Quantitative RT-PCR: Pitfalls and potential, *Biotechniques* **26**:112–125.

Gibson, U. E. M., Heid, C. A., and Williams, P. M., 1996, A novel method for real time quantitative RT PCR, *Genome Res.* **6**:995–1001.

Gilliland, G., Perrin, S., Blanchard, K., and Bunn, H. F., 1990, Analysis of cytokine mRNA and DNA: Detection and quantitation by competitive polymerase chain reaction, *Proc. Natl. Acad. Sci. U.S.A.* **87**:2725–2729.

Gorza, L., Ausoni, S., Merciai, N., Hastings, K. E. M., and Schiaffino, S., 1993, Regional differences in troponin I isoform switching during rat heart development, *Dev. Biol.* **156**:253–264.

Grove, D., Zak, R., Nair, K. G., and Aschenbrenner, V., 1969, Biochemical correlates of cardiac hypertrophy, *Circ. Res.* **25**:473–485.

Guo, W., Kamiya, K., Hojo, M., Kodama, I., and Toyama, J., 1998, Regulation of Kv4.2 and Kv1.4 K^+ channel expression by myocardial hypertrophic factors in cultured newborn rat ventricular cells, *J. Mol. Cell. Cardiol.* **30**:1449–1455.

Hengen, P. N., 1995, Methods and reagents—quantitative PCR: An accurate measure of mRNA?, *Trends. Biochem. Sci.* **20**:476–477.

Hershman, K. M., and Levitan, E. S., 1998, Cell–cell contact between adult rat cardiac myocytes regulates Kv1.5 and Kv4.2 K^+ channel mRNA expression, *Am. J. Physiol.* **44**:C1473–C1480.

Jacobson, S. L., and Piper, H. M., 1986, Cell cultures of adult cardiomyocytes as models of the myocardium, *J. Mol. Cell. Cardiol.* **18**:661–678.

Jan, L. Y., and Jan, Y. N., 1997, Cloned potassium channels from eukaryotes and prokaryotes, *Annu. Rev. Neurosci.* **20**:91–123.

Johns, D. C., Nuss, H. B., and Marban, E., 1997, Suppression of neuronal and cardiac transient outward currents by viral gene transfer of dominant-negative Kv4.2 constructs, *J. Biol. Chem.* **272**:31598–31603.

Jongsma, H. J., 1998, Sudden cardiac death: A matter of faulty ion channels?, *Curr. Biol.* **8**:R568-R571.

Kääb, S., Dixon, J., Duc, J., Ashen, D., Näbauer, M., Beuckelmann, D. J., Steinbeck, G., McKinnon, D., and Tomaselli, G. F., 1998, Molecular basis of transient outward potassium current downregulation in human heart failure: A decrease in Kv4.3 mRNA correlates with a reduction in current density, *Circulation* **98**:1383–1393.

Kass-Eisler, A., Falck-Pedersen, E., Alvira, M., Rivera, J., Buttrick, P. M., Wittenberg, B. A., Cipriani, L., and Leinwand, L. A., 1993, Quantitative determination of adenovirus-mediated gene delivery to rat cardiac myocytes *in vitro* and *in vivo*, *Proc. Natl. Acad. Sci. U.S.A.* **90**:11498–11502.

Kong, W., Po, S., Yamagishi, T., Ashen, M. D., Stetten, G., and Tomaselli, G. F., 1998, Isolation and characterization of the human gene encoding I_{to}: Further diversity by alternative mRNA splicing, *Am. J. Physiol.* **44**:H1963–H1970.

Kupershmidt, S., Snyders, D. J., Raes, A., and Roden, D. M., 1998, A K^+ channel splice variant common in human heart lacks a C-terminal domain required for expression of rapidly activating delayed rectifier current, *J. Biol. Chem.* **273**:27231–27235.

Kupershmidt, S., Yang, T., Anderson, M. E., Wessels, A., Niswender, K. D., Magnuson, M. A., and Roden, D. M., 1999, Replacement by homologous recombination of the minK gene with lacZ reveals restriction of minK expression to the mouse cardiac conduction system, *Circ. Res.* **84**:146–152.

Lee, L. G., Connell, C. R., and Bloch, W., 1993, Allelic discrimination by nick-translation PCR with fluorogenic probes, *Nucleic Acids Res.* **21**:3761–3766.

Lees-Miller, J. P., Kondo, C., Wang, L., and Duff, H. J., 1997, Electrophysiological characterization of an alternatively processed ERG K^+ channel in mouse and human hearts, *Circ. Res.* **81**:719–726.

Li, M., Isacoff, E., Jan, Y. N., and Jan, L. Y., 1993, Assembly of potassium channels, *Anna. N.Y. Acad. Sci.* **707**:51–59.

London, B., Trudeau, M. C., Newton, K. P., Beyer, A. K., Copeland, N. G., Gilbert, D. J., Jenkins, N. A.,

Satler, C. A., and Robertson, G. A., 1997, Two isoforms of the mouse *ether-a-go-go*-related gene coassemble to form channels with properties similar to the rapidly activating component of the cardiac delayed rectifier K$^+$ current, *Circ. Res.* **81:**870–878.

London, B., Jeron, A., Zhou, J., Buckett, P., Han, X. G., Mitchell, G. F., and Koren, G., 1998a, Long QT and ventricular arrhythmias in transgenic mice expressing the N terminus and first transmembrane segment of a voltage-gated potassium channel, *Proc. Natl. Acad. Sci. U.S.A.* **95:**2926–2931.

London, B., Wang, D. W., Hill, J. A., and Bennett, P. B., 1998b, The transient outward current in mice lacking the potassium channel gene Kv1.4, *J. Physiol. (London)* **509:**171–182.

Matsubara, H., Suzuki, J., and Inada, M., 1993, *Shaker*-related potassium channel, Kv1.4, messenger RNA regulation in cultured rat heart myocytes and differential expression of Kv1.4 and Kv1.5 genes in myocardial development and hypertrophy, *J. Clin. Invest.* **92:**1659–1666.

Melton, D. A., Krieg, P. A., Rebagliati, M. R., Maniatis, T., Zinn, K., and Green, M. R., 1984, Efficient *in vitro* synthesis of biologically active RNA and RNA hybridization probes from plasmids containing a bacteriophage SP6 promoter, *Nucleic Acids Res.* **12:**7035–7056.

Morkin, E., and Ashford, T. P., 1968, Myocardial DNA synthesis in experimental cardiac hypertrophy, *Am. J. Physiol.* **215:**1409–1413.

Morsy, M. A., Gu, M. C., Motzel, S., Zhao, J., Su, Q., Allen, H., Franlin, L., Parks, R. J., Graham, F. L., Kochanek, S., Bett, A. J., and Caskey, C. T., 1998, An adenoviral vector deleted for all viral coding sequences results in enhanced safety and extended expression of a leptin transgene, *Proc. Natl. Acad. Sci. U.S.A.* **95:**7866–7871.

Nuss, H. B., Marban, E., and Johns, D. C., 1999, Overexpression of a human potassium channel suppresses cardiac hyperexcitability in rabbit ventricular myocytes, *J. Clin. Invest.* **103:**889–896.

Ohya, S., Tanaka, M., Oku, T., Asai, Y., Watanabe, M., Giles, W. R., and Imaizumi, Y., 1997, Molecular cloning and tissue distribution of an alternatively spliced variant of an A-type K$^+$ channel α-subunit, Kv4.3 in the rat, *FEBS Lett.* **420:**47–53.

Ojamaa, K., Sabet, A., Kenessey, A., Shenoy, R., and Klein, I., 1999, Regulation of rat cardiac Kv1.5 gene expression by thyroid hormone is rapid and chamber specific, *Endocrinology* **140:**3170–3176.

Piatek, A. S., Tyagi, S., Pol, A. C., Telenti, A., Miller, L. P., Kramer, F. R., and Alland, D., 1998, Molecular beacon sequence analysis for detecting drug resistance in *Mycobacterium tuberculosis*, *Nat. Biotechnol.* **16:**359–363.

Raeymaekers, L., 1993, Quantitative PCR—theoretical considerations with practical implications, *Anal. Biochem.* **214:**582–585.

Raeymaekers, L., 1995, Qantitative PCR, in: *Methods in Molecular Medicine, 16:* Clinical Applications of PCR (Y. M. D. Lo, ed.), Humana Press, Totowa, New Jersey, pp. 27–38.

Roberds, S. L., and Tamkun, M. M., 1991, Cloning and tissue-specific expression of five voltage-gated potassium channel cDNAs expressed in rat heart, *Proc. Natl. Acad. Sci. U.S.A.* **88:**1798–1802.

Robertson, B., 1997, The real life of voltage-gated K$^+$ channels: More than model behaviour, *Trends Pharmacol. Sci.* **18:**474–483.

Sanguinetti, M. C., Curran, M. E., Zou, A., Shen, J., Spector, P. S., Atkinson, D. L., and Keating, M. T., 1996, Coassembly of K$_v$LQT1 and minK (IsK) proteins to form cardiac I_{Ks} potassium channel, *Nature* **384:**80–83.

Sanguinetti, M. C., Johnson, J. H., Hammerland, L. G., Kelbaugh, P. R., Volkmann, R. A., Saccomano, N. A., and Mueller, A. L., 1997, Heteropodatoxins: Peptides isolated from spider venom that block Kv4.2 potassium channels, *Mol. Pharmacol.* **51:**491–498.

Serodio, P., Kentros, C., and Rudy, B., 1994, Identification of molecular components of A-type channels activating at subthreshold potentials, *J. Neurophysiol.* **72:**1516–1529.

Shimoni, Y., Fiset, C., Clark, R. B., Dixon, J. E., McKinnon, D., and Giles, W. R., 1997, Thyroid hormone regulates postnatal expression of transient K$^+$ channel isoforms in rat ventricle, *J. Physiol. (London)* **500:**65–73.

Siebert, P. D., and Kelloff, D. E., 1995, PCR MIMICs: Competitive DNA fragments for use in quantitiative PCR, in: *Polymerase Chain Reaction: A Practical Approach*, 2nd ed. (M. J. McPherson and B. D. Hames, eds.), IRL Press, New York, pp. 135–148.

Tamkun, M. M., Knoth, K. M., Walbridge, J. A., Kroemer, H., Roden, D. M., and Glover, D. M., 1991, Molecular cloning and characterization of two voltage-gated K$^+$ channel cDNAs from human ventricle, *FASEB J.* **5:**331–337.

Tseng, G.-N., 1999, Molecular structure of cardiac I_{to} channels: Kv4.2, Kv4.3, and other possibilities?, *Cardiovasc. Res.* **41:**16–18.

Tseng, G. N., Jiang, M., and Yao, J. A., 1996, Reverse use dependence of Kv4.2 blockade by 4-aminopyridine, *J. Pharmacol. Exp. Ther.* **279**:865–876.

Vallins, W. J., Brand, N. J., Dabhade, N., Butler-Browne, G., Yacoub, M. H., and Barton, P. J., 1990, Molecular cloning of human cardiac troponin I using polymerase chain reaction, *FEBS Lett.* **270**:55–61.

Van Wagoner, D. R., Pond, A. L., McCarthy, P. M., Trimmer, J. S., and Nerbonne, J. M., 1997, Outward K^+ current densities and Kv1.5 expression are reduced in chronic human atrial fibrillation, *Circ. Res.* **80**:772–781.

Wang, Q., Curran, M. E., Splawski, I., Burn, T. C., Millholland, J. M., VanRaay, T. J., Shen, J., Timothy, K. W., Vincent, G. M., de Jager, T., Schwartz, P. J., Towbin, J. A., Moss, A. J., Atkinson, D. L., Landes, G. M., Connors, T. D., and Keating, M. T., 1996, Positional cloning of a novel potassium channel gene: *KvLQT1* mutations cause cardiac arrhythmias, *Nat. Genet.* **12**:17–23.

Wang, Z. G., Feng, J. L., Shi, H., Pond, A., Nerbonne, J. M., and Nattel, S., 1999, Potential molecular basis of different physiological properties of the transient outward K^+ current in rabbit and human atrial myocytes, *Circ. Res.* **84**:551–561.

Wickenden, A. D., Jegla, T. J., Kaprielian, R., and Backx, P. H., 1999, Regional contributions of Kv1.4, Kv4.2, and Kv4.3 to transient outward K^+ current in rat ventricle, *Am. J. Physiol.* **276**:H1599–H1607.

Wilkinson, D. G. (ed.), 1992, *In Situ Hybridization: A Practical Approach*, 2nd ed., IRL Press, Oxford.

Wymore, R. S., Negulescu, D., Kinoshita, K., Kalman, K., Aiyar, J., Gutman, G. A., and Chandy, K. G., 1996, Characterization of the transcription unit of mouse Kv1.4, a voltage-gated potassium channel gene, *J. Biol. Chem.* **271**:15629–15634.

Wymore, R. S., Gintant, G. A., Wymore, R. T., Dixon, J. E., McKinnon, D., and Cohen, I. S., 1997, Tissue and species distribution of mRNA for the I_{Kr}-like K^+ channel, *erg*, *Circ. Res.* **80**:261–268.

Xu, H. D., Dixon, J. E., Barry, D. M., Trimmer, J. S., Merlie, J. P., McKinnon, D., and Nerbonne, J. M., 1996, Developmental analysis reveals mismatches in the expression of K^+ channel α subunits and voltage-gated K^+ channel currents in rat ventricular myocytes, *J. Gen. Physiol.* **108**:405–419.

Xu, J., and Li, M., 1998, Auxiliary subunits of Shaker-type potassium channels, *Trends Cardiovasc. Med.* **8**:229–234.

Yang, T., Kupershmidt, S., and Roden, D. M., 1995, Anti-minK antisense decreases the amplitude of the rapidly activating cardiac delayed rectifier K^+ current, *Circ. Res.* **77**:1246–1253.

Yang, W.-P., Levesque, P. C., Little, W. A., Conder, M. L., Shalaby, F. Y., and Blanar, M. A., 1997, KvLQT1, a voltage-gated potassium channel responsible for human cardiac arrhythmias, *Proc. Natl. Acad. Sci. U.S.A.* **94**:4017–4021.

Yeola, S. W., and Snyders, D. J., 1997, Electrophysiological and pharmacological correspondence between Kv4.2 current and rat cardiac transient outward current, *Cardiovasc. Res.* **33**:540–547.

Yu, H. G., McKinnon, D., Dixon, J. E., Gao, J. Y., Wymore, R., Cohen, I. S., Danilo, P., Shvilkin, A., Anyukhovsky, E. P., Sosunov, E. A., Hara, M., and Rosen, M. R., 1999, Transient outward current, I_{to1}, is altered in cardiac memory, *Circulation* **99**:1898–1905.

Zak, R., 1974, Development and proliferative capacity of cardiac muscle cells, *Circ. Res.* **34/35**(Supple.):II-17–II-26.

Zhou, J., Jeron, A., London, B., Han, X., and Koren, G., 1998, Characterization of a slowly inactivating outward current in adult mouse ventricular myocytes, *Circ. Res.* **83**:806–814.

Zimmermann, K., and Mannhalter, J. W., 1996, Technical aspects of quantitative competitive PCR, *Biotechniques* **21**:268–279.

Zinn, K., DiMaio, D., and Maniatis, T., 1983, Identification of two distinct regulatory regions adjacent to the human β-interferon gene, *Cell* **34**:865–879.

Chapter 8

Concepts for Patch-Clamp Recording of Whole-Cell and Single-Channel K$^+$ Currents in Cardiac and Vascular Myocytes

Antonio Guia, Carmelle V. Remillard, and Normand Leblanc

1. INTRODUCTION

The development of the patch-clamp technique (Hamill *et al.*, 1981; Neher and Sakmann, 1976) in the mid-1970s and early 1980s by Bert Sakmann and Erwin Neher, the Nobel laureates for physiology and medicine of 1991, in parallel with the development of methods to disperse single cells from complex organs, has revolutionized our knowledge about the proteins that regulate transmembrane flux of ions across biological membranes. Its widespread implementation in the 1980s initially served to identify and characterize the various ion channels present in the membrane of various cell types. In the 1990s, molecular identification of several genes encoding for ion channel and transporter proteins provided electrophysiologists with new tools to unravel protein behavior at an unprecedented level of resolution. With the explosion of new information regarding their encoding sequences at the nucleotide and amino acid levels, these proteins can now be expressed and functionally studied in different cell types (*Xenopus* oocytes, mammalian cell lines, etc.), mutated, and even crystallized (Doyle *et al.*, 1998) for investigation of the molecular domains responsible for their behavior in their normal environment.

The main goal of this chapter is to provide an overview of the patch-clamp technique and its application to record K$^+$ channel activity in freshly isolated cardiac and vascular smooth muscle cells. After an initial historical perspective of the development of this technique, a description of methods used in our laboratory to obtain

Antonio Guia, Carmelle V. Remillard, and Normand Leblanc ● Department of Physiology, University of Montreal, Montreal, Quebec, Canada H3C 3J7 and Research Centre, Montreal Heart Institute, Montreal, Quebec, Canada H1T 1C8.

Potassium Channels in Cardiovascular Biology, edited by Archer and Rusch. Kluwer Academic/Plenum Publishers, New York, 2001.

freshly isolated cardiomyocytes and smooth muscle cells is presented. This is followed by a formal description of the patch-clamp technique and its various configurations, with emphasis on advantages and limitations. Using specific examples from actual experiments, the final section of the chapter provides an account of many protocols that can be used to assess the permeation and gating properties of K^+ currents recorded in isolated cardiac and vascular smooth muscle cells.

2. HISTORICAL PERSPECTIVE

Based on findings that certain toxins, enzymes, and agents known to modulate protein function also influenced ionic currents in nerve, it had been suspected in the 1950s and 1960s that ions probably move across biological membranes by passing through hydrophilic pores created by integral transmembrane proteins. As early as 1969, "box-like" electrical events suspected to be due to the behavior of individual channel proteins were recorded from bacterial membrane proteins reincorporated into lipid bilayers (Bean et al., 1969). Although the electronic circuitry available at the time was sufficient to resolve unitary currents with the bilayer method, the resolution afforded by feedback voltage-clamp amplifiers was one to two orders of magnitude less than that required to record such events from cells and tissues.

Neher and Sakmann (1976) recorded the first unitary currents from an intact cell membrane patch of $\sim 1\ \mu m^2$ by applying a large-bore glass micropipette onto denervated frog muscle membranes. In the presence of acetylcholine, nicotinic receptors led to the activation of all-or-none cation currents produced by the stochastic opening and closing of individual channel proteins. However, despite considerable efforts to improve the quality of the electrical junction between the pipette and membrane (enzymatic treatments to clean up the membrane, optimization of micropipette size, geometry, and composition, etc.), seal resistances obtained with this method were at best in the range of 50–100 MΩ, a range that was still too low to resolve single-channel currents in most preparations. In addition to the use of better electronic components and improved amplifier design, a major breakthrough arose from the work of Hamill et al. (1981), who found that the gentle application of negative suction from the back of the micropipette led to the formation of a much tighter seal with resistances in the gigaohm range (so-called gigaseal). This refined the resolution and reproducibility of the measurements by increasing the signal-to-noise ratio by over an order of magnitude, and it allowed the application of local voltage-clamp protocols to investigate the voltage-dependent properties of an ion channel. One unexpected outcome of the latter improvements was that the resulting gigaseal was found to be mechanically very stable, which permitted the development of the excised-patch and whole-cell configurations that will be described later in this chapter. Since the original work of Hamill et al. in 1981, several additional variants (i.e., the perforated patch and giant excised patch methods, among several others) have been developed while electronic design and analytical algorithms have continued to improve. It is now possible to record with reasonable confidence single-channel events in the subpicoampere range and relatively high macroscopic currents from large cells such as cardiac myocytes. For more information on the historical aspects of the development of the patch-clamp technique, the reader is invited to consult the monograph added by Sakmann and Neher (1995) and the review articles by Neher and Sakmann (1992) and Sigworth (1986).

3. SINGLE-CELL ISOLATION PROCEDURES

In the following paragraphs, we briefly describe the techniques that we have employed in our laboratory over the years to isolate cardiac and vascular myocytes with reasonable success (Leblanc et al., 1998; Cole et al., 1997; Remillard and Leblanc, 1996; Leblanc and Leung, 1995; Leblanc et al., 1994). The cell isolation techniques that are typically employed are illustrated in Fig. 1. Both the tissue chunk method (commonly referred to as the "chunk" or "chop" method) and the perfusion ("Langendorff") method are depicted. With every type of cardiovascular tissue, it is possible to use the "chop" or "chunk" method, in which the freshly dissected tissue is sectioned into cubes or squares of just a few hundred cells each (approximately 1 mm thick). This is the usual method of choice to isolate smooth muscle myocytes, but it also has been used with some success to isolate cardiac myocytes, in particular atrial cells because there are relatively few layers of cells in atrial tissue compared to ventricular tissue. Some laboratories successfully use a combination of both methods to isolate cardiac myocytes: the heart is perfused with the enzymes for the first stage of (partial) digestion, and then the area of interest is cut into small chunks that are subsequently immersed in enzymes for the second stage of digestion.

3.1. Isolation of Cardiac Myocytes

The technique for isolating ventricular myocytes is outlined in panels A and C of Fig. 1 and has been successfully used with minor modifications to isolate cells from guinea pig, rabbit, and rat hearts. The heart should be excised immediately after sacrifice of the animal. Because a prolonged dissection time may permit time for coagulation, which will interfere with perfusion, it is helpful to preinject the animal intraperitoneally with heparin about 30 minutes before dissection to reduce the incidence of coagulation in the heart during dissection. The easiest and most secure way of excising the heart is to remove the entire pericardial sac and whatever portions of the lungs come attached to it and then expose the aorta by dissection in a petri dish (cutting at a location adjacent to the first aortic branch), leaving a sufficient length of aorta (3–5 mm) to enable easy and rapid cannulation. If further dissection is necessary before cannulation, then as much blood as possible should be flushed from the heart using a syringe filled with warm (37°C) saline solution. Removal of lungs, pericardium, and excess fatty tissue should be delayed until after the heart is cannulated and perfused. The aorta must be secured to the cannula with a silk thread so as to hold the weight of the heart.

As soon as the aorta is attached to the cannula, the heart is perfused in a retrograde manner (Langendorff configuration) with saline solution under pressure to flush out the remaining blood and other unwanted substances contained in the blood vessels. In our laboratory we use a warm (37°C) HEPES-based solution: 136 mM NaCl; 5 mM KCl; 0.34 mM NaH$_2$PO$_4$; 1 mM MgCl$_2$; 1.8 mM or 0 CaCl$_2$; 5.5 mM dextrose; HEPES-NaOH, pH 7.4) through which 100% O$_2$ is bubbled. The same perfusion solution is used during dissection and for the initial irrigation and flushing. The perfusate is pressurized by gravity using a column height of 65 cm (flow rate of approximately 1–3 ml/min depending on species).

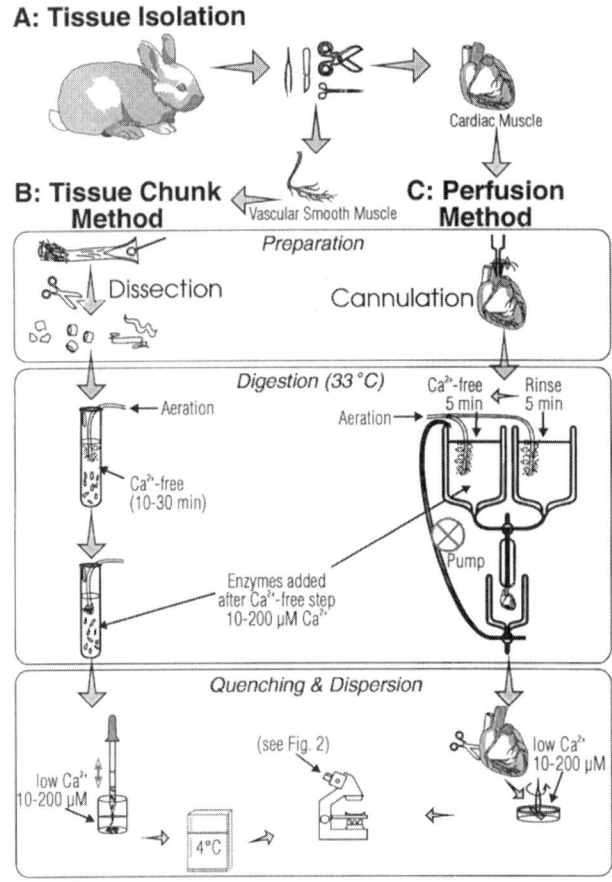

Figure 1. Sketch diagram illustrating general procedures that are commonly used to isolate ventricular and vascular smooth muscle cells. Consult text for a full description of these methods.

After at least 10–15 min of this initial flush, the perfusate is switched to a nominally Ca^{2+}-free perfusing solution for 5 min to reduce the Ca^{2+} concentration in the vessels and interstitial spaces. After 5 min, the apparatus is set to recirculate the perfusate before addition of a mixture of crude collagenase (Worthington Biochemical Corp., type 2, ~0.4–1.0 mg/ml or ~100–400 U/ml depending on species) and protease (Sigma Chemical Co., type XXVII, ~0.2 mg/ml or ~0.2 U/ml). Digestion is allowed to proceed for 15–45 min. Under these conditions, the myocytes are sufficiently dissociated from one another that we can carefully cut out the section that we wish to study and then gently pull the tissue apart and stir it in a petri dish containing an enzyme-free and Ca^{2+}-free perfusing solution. Although it may be more efficient to quench the enzymes by washing out with a volume of Ca^{2+}-free solution, we have found that our method provides cells that appear healthy and respond as expected to pharmacological agents for at least 6 h following dispersion of the cells. Use of the cells more than 6 h after dispersion may produce questionable results owing to dedifferenti-

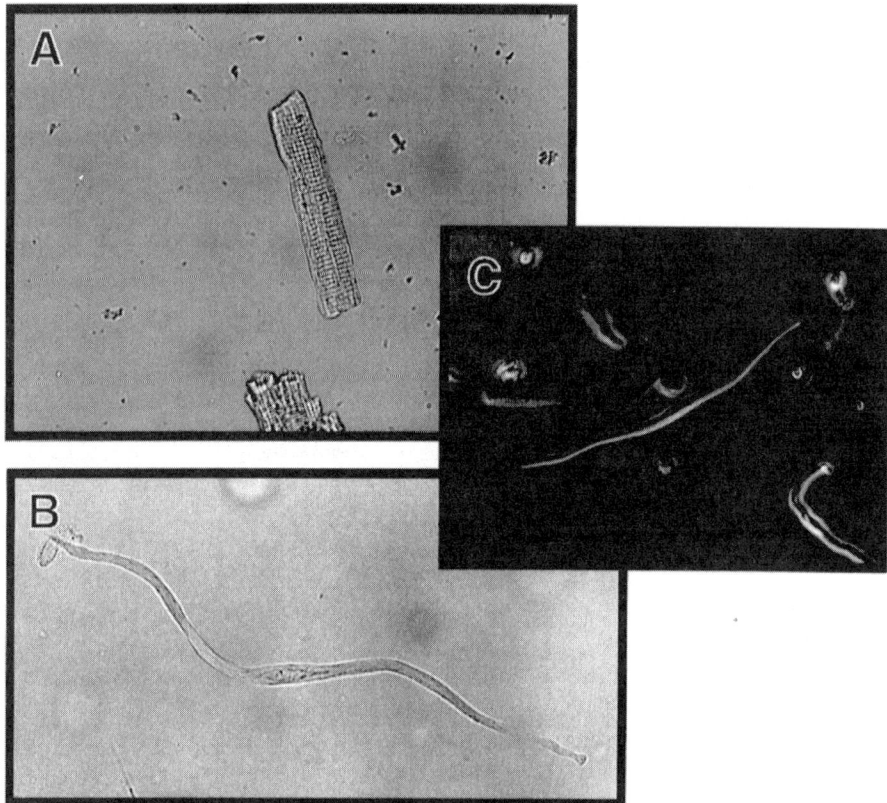

Figure 2. Photomicrographs of single guinea pig ventricular myocyte (A) and rabbit portal vein smooth muscle cells (B and C) freshly dispersed using methods similar to those described in Fig. 1. Photomicrographs in panels A and B were obtained using standard illumination. The picture shown in panel C was taken using a phase-contrast ring.

ation and rundown. Once the cells are dispersed in the petri dish, we slowly elevate the bathing Ca^{2+} concentration to 200 μM and then store the cells in this solution at room temperature until needed. Healthy freshly dispersed cardiac cells, when observed at 400x magnification in regular transmission microscopy, will appear rod-shaped with clear striations and a smooth, well-delineated membrane (Fig. 2A). Contracted ends or blebbing membranes are usually a sign of overdigestion. Underdigestion is usually indicated by tissue that is more difficult to triturate or the persistence of a high number of multicellular segments after trituration. Healthy ventricular myocytes dispersed by the above method usually display the following characteristics: (1) they maintain negative resting membrane potential (E_m) in the range of -70 to -85 mV in normal external saline and do not fire action potentials or do not twitch spontaneously; (2) they contract upon depolarization elicited by an action potential or a voltage-clamp step above -40 mV and relax upon restoration of normal resting E_m; and (3) they respond normally to pharmacological agents and hormones.

3.2. Isolation of Vascular Smooth Muscle Cells

The first step in the isolation of vascular smooth muscle cells is obtaining the tissue and cleaning off adherent fat and connective tissues from the tunica adventitia (see Fig. 1B, Dissection). The cleaned tissue is then cut into small chunks, rings, or helical strips in preparation for the following steps. In a similar manner as in the isolation of cardiac myocytes, the second step involves the incubation of the tissue in a nominally Ca^{2+}-free medium (120 mM NaCl; 25 mM NaHCO$_3$; 4.2 mM KCl; 1.2 mM KH$_2$PO$_4$; 1.2 mM MgCl$_2$; 11 mM dextrose; 25 mM taurine; 0.01 mM adenosine; pH 7.4 when bubbled with 95% O$_2$–5% CO$_2$) at room or physiological temperatures for 10–30 min (Fig. 1B, Digestion—Ca^{2+}-free); this procedure may facilitate subsequent dispersion by separating the outer portion of the basal lamina of the surface membrane and favor disruption of desmosomes connecting adjacent cells. In addition, we have found that including 100 μM EGTA to buffer any Ca^{2+} contaminants present in the solution significantly improves both the yield and the Ca^{2+} tolerance of the myocytes. The third step requires a certain amount of creativity, patience, and good old-fashioned luck. The tissue pieces are transferred to a nominally Ca^{2+}-free or low-Ca^{2+} (10–50 μM) medium containing various enzymes (Fig. 1B, Digestion—enzymes added). Among the wide variety of enzymes that have been tried over the years, collagenase (in our lab, we use type 1A from Sigma Chemical Co., $\sim$0.2–1 mg/ml or $\sim$50–400 U/ml, arteries requiring more enzyme than veins) and papain are still the most popular enzymes for smooth muscle cell isolation. Digestion is usually performed at near-physiological temperatures (33–37°C) for 15–60 min. A low-Ca^{2+} solution is used because the digestive activity of collagenase and other enzymes is enhanced by Ca^{2+}. Because some tissues (e.g., arteries) contain a large amount of connective tissue, additional enzymes such as protease (in our lab, we use type XXVII from Sigma Chemical Co., $\sim$0.05–0.1 mg/ml or 0.25–0.5 U/ml), elastase, and hyaluronidase are sometimes required to separate the cells.

Knowing when to end the digestion requires both a keen eye and a lot of trial and error because even a few minutes' variation can make a big difference in the quality and quantity of cells. We have found that cells tend to start dissociating after approximately 15 min of digestion, at which time we retrieve a few pieces of tissue every 2–5 min, rinse them in fresh low-Ca^{2+} solution, gently triturate the tissue with a Pasteur pipette to release cells, and look for free cells at 400x magnification (Fig. 1B, Quenching & Dispersion). The incubation should be stopped when a few healthy looking cells are liberated after mild trituration ($\sim$1 min). Any pieces remaining at this time are rinsed repeatedly in fresh low-Ca^{2+} medium and triturated to release the cells. In several laboratories, the preferred practice is to remove cellular debris from the medium containing the cells by low-speed centrifugation (and resuspension of the cells) or by filtering through a fine nylon mesh; this minimizes problems associated with debris sticking to the patch pipette during experiments. Finally, storing the cells in a cold (4°C) low-Ca^{2+} (10–200 μM) solution can keep cells viable and usable for 4–8 (and sometimes 24!) h. Vascular myocytes isolated by the above method are Ca^{2+}-tolerant, are spindle-shaped, have clear membranes (Fig. 2B,C), and contract when exposed to high-K$^+$ perfusate or vasoconstrictors such as norepinephrine, endothelin, or angiotensin II. Using the aforementioned technique (with a few minor modifications), we have been able to isolate viable cells from many types of smooth muscle from several species (rabbit coronary artery, portal vein, and mesenteric arteriole, human coronary artery, guinea pig portal vein and ileum, dog airway muscles, and rabbit pulmonary artery).

4. DESCRIPTION OF THE PATCH-CLAMP TECHNIQUE

The patch-clamp technique is not a unique tool but rather a combination of configurations that can be used to record (1) the "normal" electrical behavior of the cell at rest (resting membrane potential or E_m) or during activity (action potential), (2) the activity of all channels present in the cell membrane (often referred to as macroscopic or whole-cell current), (3) the activity of one or more ion channel proteins (microscopic or unitary current), and (4) net current produced by electrogenic carrier systems (Na$^+$/K$^+$ pump, Na$^+$/Ca^{2+} exchange, etc.). In view of the limited space available, the following paragraphs will only introduce the basic elements of the technique. The reader interested in becoming familiar with the building of a patch-clamp station, the fabrication of patch pipettes, electronic design and circuitry, and other practical aspects of the technique, should consult well-documented textbooks and monographs on these subjects (Conn, 1998; Sakmann and Neher, 1995; Rudy and Iverson, 1992).

4.1. Formation of a Gigaseal

Depending on the clamp configuration used, and on the desired results, the pipette solution may be different from the bathing medium. If this is the case, then once the pipette is placed into the bathing medium, there will be a net movement of ions across the tip in both directions, and if the net mobility of ions in the pipette solution is different from that in the bathing medium, there will be a potential at the junction of the two solutions. Because it is necessary to use well-calibrated electrodes, it is imperative that the true junction potential be measured. This is accomplished by using the pipette solution as the bathing medium while offsetting the measured potential to zero and then replacing the bathing medium with the one that will be used during the experiments. The resulting junction potential (averaged over a few trials) is where the offset should be placed each time for that combination of pipette solution and bathing medium as soon as the pipette is placed into the solution and brought toward the cell with the help of a remote-controlled micromanipulator (Fig. 3a). Contact with the cell may be seen as an increased tip resistance (as the cell membrane blocks the tip). The tip resistance is measured by applying a continuous square-wave voltage-clamp pulse, which requires a large current when the tip resistance is low and very little current when the tip resistance is high (following Ohm's law, $R = V/I$; see example in Fig. 3b). The membrane is sealed to the pipette tip when the resistance rises from a few megaohms to gigaohms ('gigaseal'). Unless the cell membranes are overdigested, it is often necessary to press the tip a short distance into the cell membrane and/or apply gentle suction to the back of the pipette (most pipette holders are designed to allow a suction line to be attached) to facilitate formation of the gigaseal (Fig. 3b,c).

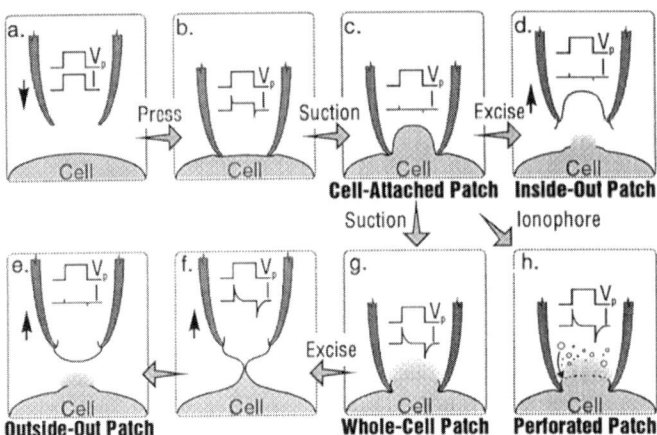

Figure 3. Sketch diagram illustrating the steps involved in the formation of a gigaseal on a cell (panels a–c) with a micropipette and the various configurations of the patch-clamp technique (panels d, e, g, and h) that can be derived from a cell-attached patch (panel c). V_p, is a repetitive command voltage pulse that is applied through to the patch micropipette to monitor pipette current (I) during formation of a gigaohm seal. Notice that the level of the steady-state current decreases when the pipette is pressed against the cell membrane (panel b) and is virtually abolished after formation of the gigaseal (panel c). Also notice the appearance of large and slow transient capacitive currents under whole-cell (panel g) and perforated (panel h) patch modes due to charging and discharging of the membrane capacitance from the entire cell during the onset and relaxation of V_p.

Once a gigaseal is formed, the current records will show a nearly flat line with upward spikes at the rising edge and downward spikes at the falling edge of the square-wave voltage-clamp pulses. These current spikes are due to the amount of charge that must be transferred to the glass surface of the pipette (capacitance) and to the entire volume of the solution (resistance). There are at least two components to this pipette capacitance adjustment on most patch-clamp amplifiers that can almost completely remove these capacitance spikes.

4.2. Patch-Clamp Configurations

Once a gigaseal has been formed (Fig. 3c), it is possible to apply a transmembrane potential to this cell-attached patch of membrane so that individual openings of single channels can be recorded. The resolution of these recordings may be improved by shielding the cell, pipette, and headstage from electrical noise, by reducing the volume of the bathing medium or isolating the tip from the bathing medium with a nonconductive polymer (e.g., Sylgard), by grounding all conductors inside the Faraday cage, by removing or reducing ground loops, and so on. Under ideal conditions, it is possible to resolve unitary channel openings smaller than 1 pA in amplitude with 1–2-kHz multipole filtering. This configuration is useful to study channels under "physiological conditions" because the intracellular environment of the channel remains undisturbed and therefore all cytosolic factors are present at "normal" levels, allowing the study of pharmacological agents that act through cytosolic cofactors or second messengers.

The cell-attached patch configuration has a couple of very important limitations: exact E_m cannot be known or clamped, and cytosolic ionic species are unknown. Specifically, because this method does not allow access to the cytosol, it is impossible to exchange the cytosolic solution with one containing known concentrations of each chemical species, and therefore it is impossible to know the exact electrochemical gradients. Further, because we cannot sample voltage from the cytosolic facet, it is impossible to know the exact resting E_m except by approximation from population averages. In cardiac cells, the resting potential is usually around -70 to -80 mV, and in smooth muscle it is around -30 to -60 mV. This spread of possible resting potentials makes it impossible to identify the true voltage dependence of the measured unitary currents. This point having been made, there are nevertheless several methods of overcoming these limitations. For example, the E_m may be clamped near 0 mV by simply replacing the bathing medium with a depolarizing medium (in the absence of Ca^{2+}), and both E_m and cytosolic ionic composition may be clamped by using a second voltage-clamp electrode in a whole-cell configuration (see below).

From the cell-attached patch as the starting point (Fig. 3c), it is possible, with a reasonable success rate, to pull the tip off the cell while maintaining a gigaseal on the excised membrane patch. This results in an inside-out configuration, where the outer facet of the membrane is sealed within the pipette tip and the inner facet is free in the bathing medium (Fig. 3d). From the cell-attached configuration, it is also possible to apply greater suction so as to stretch and then break the membrane within the pipette tip, giving the pipette access to the entire cell (Fig. 3g). This whole-cell configuration allows measurement of currents from the entire cell membrane. Finally, from the whole-cell configuration the pipette may be slowly drawn away from the cell to stretch the membrane immediately outside the seal until it reseals on the cell and on the tip (Fig. 3e,f). This produces an outside-out configuration, where the outer facet of the membrane patch is in the bathing medium and the inner facet is sealed inside the pipette tip. For each of the excised-patch configurations, it is necessary to use cells that are well attached to the bottom surface of the chamber. If the cells do not themselves attach to the surface, then the chamber surface may be coated with poly(-L-lysine), laminin, or any other compound that helps the cells adhere to the surface.

The excised-patch configurations (inside-out and outside-out) effectively overcome the problems inherent to the cell-attached patch configuration. Specifically, they allow precise control of all ionic species on both sides of the membrane and of the transmembrane potential. It is possible therefore to study with accuracy the permeation process (properties of the pore, such as selectivity and conductance), gating properties (opening and closing behaviors), and regulation by either cytoplasmic constituents (in the inside-out configuration) or extracellular agents (in the outside-out configuration).

Another practical version of the technique allows for the recording of channel activity from the entire membrane of the cell (macroscopic ionic currents) and is referred to as the "whole-cell" configuration (Fig. 3g). There are two variants of this configuration. The "standard" whole-cell mode is the original configuration introduced by Hamill *et al.* (1981), whereby the membrane invagination produced by the tight seal in the cell-attached mode is suddenly ruptured while maintaining the gigaseal by the application of additional negative pressure through the syringe or by supplying a strong voltage surge (1–1.5 V) of variable duration (0.1–10 ms) via the micropipette (a feature called the "ZAP" command is available on many patch-clamp amplifiers). Successful access is manifest as the sudden appearance of capacitive spikes in response to 10–20-mV repetitive voltage-clamp pulses. These spikes are produced by charging of the membrane capacitance (C_m). Rupture of the membrane permits the exchange of cytoplasmic and pipette fluid, or "dialysis." Because the volume of the micropipette is several orders of magnitude larger than the volume of the cell, the micropipette closely approximates an infinite reservoir; after a period of equilibration, pipette solution and cytoplasm have a similar ionic composition.

The possibility of simultaneously recording a great number of channels on the membrane is probably the greatest advantage of this configuration. This allows for measurement of currents carried by channels of very small ionic conductance, channels that would be difficult to resolve using the other configurations. Another advantage of simultaneously measuring an entire population of channels is that the currents resolved are representative of the average channel, hence simplifying the statistics and giving meaningful information in fewer experiments. On the other hand, the large pipette volume also acts as a sink in which metabolic factors and factors important for ion channel function are likely to be washed out of the cell with time. However, attempts can be made to minimize the time-dependent decline of a current due to rundown by providing compounds suspected to be responsible for its regulation (e.g., ATP, or GTP).

To adequately record whole-cell ionic currents, one must satisfy the criterion of "space clamp." Specifically, this is the condition of homogeneous spatial distribution of transmembrane voltage under voltage clamp. This can become a problem when attempting to voltage-clamp long (> 100 μm) and relatively thin (5–20 μm) cardiac and smooth muscle cells, especially at strong depolarizing potentials where large outwardly rectifying K^+ currents may be activated. One way to minimize this problem is to patch-clamp the cell at its center.

A second issue that one must be cautious about during whole-cell voltage-clamp experiments is the presence of a resistance in series (R_s) with membrane resistance (R_m) which results from the sum of two resistive components—the pipette resistance (which is measured in the bath before patch-clamping the cell) and the access resistance (R_{acc}), the latter being the sum of the resistance to dialysis (passage of current at the tip opening) and the internal (cytoplasm) resistivity of the cell. The presence of R_s introduces an error (V_{err}) in the voltage applied to the cell that is proportional to the magnitude of the membrane current (I_m): $V_{err} = R_s I_m$; as an example, if one were to record a cardiac transient outward K^+ current of 4 nA at +40 mV with an uncompensated R_s of 5 MΩ, V_{err} would be as high as 20 mV. In addition, as for simple electrical RC circuits, the introduction of a resistance in series with the parallel arrangement of R_m and C_m will slow down the charging of C_m during a voltage-clamp step (fast current transients observed following step changes in voltage). Compensation of series resistance can reduce the transient charging time during membrane voltage transitions. All modern patch-clamp amplifiers have built-in series resistance compensation circuitry that can compensate for 80–95% of R_s. This compensation procedure minimizes the R_s-induced voltage error and improves C_m charging time and the effective frequency bandwidth of the recorded ionic current.

Another variant method of the whole-cell recording configuration is called the "perforated-patch" method (Fig. 3h). This technique takes advantage of the cell-attached patch configuration to gain whole-cell access through the use of antibiotic ionophores. The pipette tip is filled normally with pipette solution and then back-filled with the same solution to which nystatin or amphotericin B has been added at a concentration of 150–400 μg/ml of pipette solution [prepared from a 60-mg/ml stock solution in dimethyl sulfoxide (DMSO), which should be kept in the dark or under red light] and is maintained in suspension. The pipette solution must be filtered before addition of the ionophores, and the suspension of the ionophores should be ensured by sonicating the mixture for a short time or by vigorous stirring. This technique provides whole-cell access more slowly (over 5–10 min after formation of a gigaseal) and requires that the experimenter form a gigaseal before the ionophores start falling from the tip and onto the cell membrane. As partitioning of the antibiotic occurs, whole-cell access can be visualized by monitoring the development of a slow capacitive current (I_C) produced by the charging of C_m in response to repetitive voltage-clamp pulses applied to the cell-attached patch. As the number of pores formed increases with time, I_C becomes larger and its time course faster, indicating a decrease in R_{acc}. By optimizing the geometry of the pipette tip—both size (the largest pipette tip that allows formation of a gigaseal) and shape (one that favors the typical ohmic shape of the invaginated membrane)—it is possible to routinely obtain values of R_s that are comparable to those obtained with the standard whole-cell technique (<10 MΩ). The perforated-patch technique presents several advantages over the standard method: (1) because enzymes and second-messenger systems are not washed out, it slows down and often eliminates rundown of ion channels that are regulated by cell signaling mechanisms (the whole-cell potassium current I_K in the heart being a good example, see Fig. 5); (2) once partitioning is complete, R_{acc} is more stable so that R_s and C_m compensations are more easily achieved; (3) longer whole-cell recordings are possible (up to 3 h), and this is particularly true for smooth muscle cells; and (4) because divalent cations are not dialyzed, the intracellular homeostasis of at least ions such as Na$^+$, H$^+$, and Ca^{2+} can be studied with the use of fluorescent probes simultaneously with whole-cell currents with little perturbation by divalent cations of the ion transporters that regulate their activity. The choice of one variant over the other will depend on a compromise between the need to control the intracellular milieu (standard method) and the need to preserve the cytoplasm (perforated-patch method).

5. GENERAL PROPERTIES OF ION CHANNELS

There are two processes that control the flow of ions across a single channel pore (Hille, 1984). The first process is the permeation of ions via a conduction pathway. The three-dimensional structure of the channel protein's amino acid chain sequences determine the size, geometry, and net charge within the conduction pathway. These features in turn determine the rate of ion flux when the channel is in the open state and act as a filter to select which ion passes through it. The second process is the opening and closing of the channel, which is commonly referred to as "gating." Because ion channels are proteins that obey the rules of thermal molecular motion, their conformation switches randomly between two discrete levels (all-or-none) of conduction: an open state (O) that allows passage of ions or current flow and a closed state (C) that prevents ion flux.

5.1. Protocols to Analyze the Permeation and Selectivity of K^+ Channels

When an ion channel is in the open state, if there is no driving force or electrochemical energy applied on the ions, no net current flow will ensue. The driving force is defined by $(V_m - E_K)$, where V_m is the transmembrane potential and E_K is the equilibrium potential for potassium as defined by the Nernst equation:

$$E_K = \frac{RT}{zF} \log \frac{[K^+]_o}{[K^+]_i} \tag{1}$$

where R is the gas constant, T is the absolute temperature, z is the valence of the ion, F is the Faraday constant, and $[K^+]_o/[K^+]_i$ is the ratio of extracellular to intracellular concentrations of potassium. At 37°C, the term RT/zF is a constant, approximately equal to 61 mV. Therefore, E_K depends only on the concentration gradient and represents the electrical energy in volts that would be required to maintain the gradient constant. When V_m is forced away from E_K (as is the case under voltage-clamp conditions), net current can be recorded that is produced by ion flux through all the channels in the membrane (whole-cell) or through a single channel in a membrane patch.

In practice, especially exposed to the asymmetric K^+ gradient encountered under physiological conditions, most K^+ channels are not entirely selective for this ion and are permeable to other cations. Selectivity can be studied by determination of the relative permeability (P_X/P_K) of a given channel to various permeating ions. For a typical K^+ channel, the measured reversal potential (E_{rev}) will deviate from E_K, often due to a finite but significant flux of Na^+ driven by the electrochemical driving force $(E_{Na} = +60$ mV$)$. P_X/P_K can be estimated by using the voltage form of the Goldman–Hodgkin–Katz equation:

$$E_{rev} = \frac{RT}{zF} \log \frac{[X]_o + (P_X/P_K)[K^+]_o}{[X]_i + (P_X/P_K)[K^+]_i} \tag{2}$$

Such a determination can be facilitated by taking advantage of the excised and whole-cell variants of the patch-clamp technique to easily fix ionic gradients. The reversal potential of a membrane current should be measured using different sets of ionic conditions to confirm the identity of the permeating ion species. Owing to the fact that the intracellular concentrations of two hypothetical permeating species can be adequately controlled, and provided that ion X and K^+ are the only permeating ionic species on the external and internal sides of the membrane, respectively, a simplified form of Eq. (2) can instead be used:

$$E_{rev} = \frac{RT}{zF} \log \frac{P_X[X]_o}{P_K[K^+]_i} \tag{3}$$

The latter paradigm should, however, be used with caution as external K^+ has been shown to influence the gating of K^+ channels; total removal of this ion from the bathing medium could result in closing of the channels. A good example of this is the well-known sensitivity of the cardiac inward rectifier I_{K1} to changes in $[K^+]_o$ (Sakmann and Trube, 1984).

Estimation of the selectivity and conductance of a K^+ channel at the single-channel level is straightforward provided that the channel is active at, or near, the reversal potential. The selectivity of a K^+ channel can be assessed from excised-patch experiments designed to construct current–voltage ($I-V$) relationships obtained in different concentration gradients. For each concentration gradient, the reversal potential is determined by measuring the voltage at which the unitary current changes direction. These measurements can then be plotted as a function of log $[K^+]_o$ or log ($[K^+]_o/[K^+]_i$). If a channel is perfectly selective for K^+, the slope of this relationship should approximate 61 mV at 37°C. In practice, most cardiac and smooth muscle K^+ channels are imperfectly selective of K^+ and allow other cations to flow through. Under physiological conditions, a finite permeability to Na^+ (in general, less than 10% of total) is responsible for deviation of E_{rev} from E_K in the positive direction.

Under whole-cell conditions, the selectivity properties of a time-independent current can be evaluated by determining the voltage at which the current inverts provided that: (1) the current measured is the product of a single class of K^+ channels (often a limiting condition), (2) contaminating factors such as the junction potential (based on different mobilities of diffusing ions at the interface between the pipette and bath solutions) have been adequately canceled out, and (3) cell dialysis is complete. Voltage step or ramp protocols can be used to assess these properties. Estimation of the selectivity of K^+ currents such as the inward rectifier K^+ current (I_{K1} in cardiac myocytes, I_{IR} in smooth muscle cells) or ATP-sensitive K^+ current (I_{KATP}) can be readily accomplished by measuring the potential at which the current reverses on an $I-V$ plot generated by a step or ramp protocol. Figure 4A shows the results of a typical experiment carried out using a guinea pig ventricular myocyte. On the left, the inset shows a family of inward rectifier currents or I_{K1} elicited by the step protocol shown at the bottom. This current is recognized to play a major role in determining the cardiac resting E_m and to participate in the terminal phase of repolarization of the action potential (Roden and George, 1997; Shimoni et al., 1992). It is evident that I_{K1} channels display strong inward rectification; that is, they allow more current to pass in the inward than in the outward direction. This is well reflected in the current–voltage relationship, where the late current (indicated by filled circles in the inset) is plotted against step potential. Because I_{K1} displays little time dependence above -100 mV, the reversal potential can be estimated by measuring the voltage at which the current crosses the zero-current axis (in this case, -81 mV). From such a plot, one can also derive the conductance (g_{K1}) of this current by measuring the slope (in this case $g_{K1} = 140$ nS) of the linear portion of the $I-V$ curve, as indicated by the dashed line. Because the open-state probability of the underlying channels in this region of the $I-V$ curve is relatively independent of voltage, estimation of g_{ion} can provide useful information on the permeation properties of the pore. A voltage ramp protocol such as shown on the right-hand side of Fig. 4A can also be used to estimate these parameters. The smooth inwardly rectifying current elicited by the protocol shown in the inset was recorded in the same cell. These two protocols yielded similar estimates of E_{rev} and g_{K1}. Consistent with the high selectivity of I_{K1} channels for K^+, estimates of E_{rev} deviated only slightly from the predicted E_K (-86 mV) based on pipette and bath K^+ concentrations.

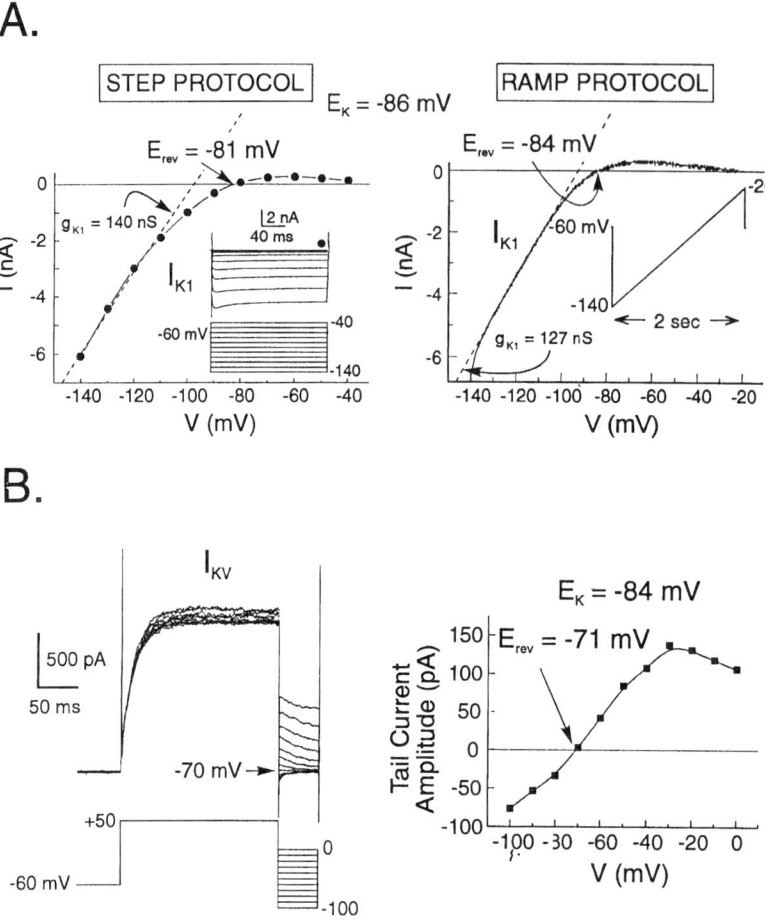

Figure 4. Examples of experimental protocols used to determine the E_{rev} of whole-cell K^+ currents. (A), *Left:* Current–voltage relationship of the cardiac inward rectifier K^+ current I_{K1} recorded from a guinea pig ventricular cell. The dashed line is a linear least-squares fit to the data points between -140 and -110 mV; as indicated, the estimated slope conductance of I_{K1} (g_{K1}) was 140 nS. *Inset:* Family of I_{K1} currents elicited by the step protocol shown below [holding potential (HP) $= -60$ mV]. On the plot, current was measured at the point indicated by the filled circle. *Right:* In the same cell, $I-V$ relationship of I_{K1} evoked by the ramp protocol displayed in the inset. The dashed line is a least-squares fit to the data between -130 and -110 mV; as shown, g_{K1} estimated with this method was similar to that derived with the step protocol. (B) *Left:* Example of the use of a double-pulse protocol (depicted at the bottom) to measure the E_{rev} of the time- and voltage-dependent K^+ current I_{KV} recorded from a rabbit coronary smooth muscle cell. The cell was dialyzed with 5 mM EGTA to buffer intracellular Ca^{2+} and exposed to 1 μM nifedipine. The tail current elicited at -70 mV displayed little time-dependent kinetics, suggesting that I_{KV} reverses near that voltage. *Right:* Plot of tail current amplitude as a function of voltage during the second step for the experiment shown on the left. For both panels, E_{rev} is the reversal potential, and E_K is the equilibrium or Nernst potential for K^+.

For time-dependent currents such as delayed rectifier and transient outward K$^+$ currents, it would be possible to measure the reversal potential (E_{rev}) of these currents by estimating the potential at which the current inverts, provided that the voltage range in which the current exhibits voltage dependence and E_{rev} overlap. However, in general, most investigators study these currents under physiological K$^+$ gradients, giving a predicted E_K value (< -80 mV) that is significantly more negative than the voltages at which their voltage dependence is observed (above -50 mV). If the current under investigation deactivates relatively slowly upon repolarization, a double-pulse protocol can be used to determine its selectivity by tail current analysis. An example of the use of such a protocol is presented in Fig. 4B, which shows the results obtained from experiments with a rabbit coronary myocyte exposed to 1 μM nifedipine to inhibit Ca^{2+} channels and dialyzed with 5 mM EGTA to minimize the activity of large-conductance, Ca^{2+}-dependent K$^+$ channels. The remaining current mainly consists of a time- and voltage-dependent delayed rectifier, or I_{KV}, which may be composed of several molecular gene products. The left-hand side of Fig. 4B shows a family of I_{KV} currents evoked by the double-step protocol shown below them. A constant initial step to $+50$ mV serves to activate K$_V$ channels to their maximum open-state probability. The membrane is subsequently repolarized to various potentials to record deactivating tail currents. The voltage eliciting a tail current that exhibits little or no time dependence during the second pulse (-70 mV in the example) is at or near the E_{rev} of the current. For accuracy, the amplitude of the tail current can be plotted as a function of step potential as shown on the right-hand side of Fig. 4B. As indicated, E_{rev} of I_{KV} for this particular experiment was -71 mV, which indicates that although the underlying channel is quite selective for K$^+$, other cations (most likely Na$^+$) probably permeate it. A similar protocol can be employed to determine the ion selectivity of a fast transient outward current (I_{to}) such as that measured in cardiac (Campbell et $al.$, 1993) and certain types of smooth muscle cells (Imaizumi et $al.$, 1990), by using an initial step of short duration, which interrupts the current before substantial inactivation has taken place.

5.2. Protocols to Investigate Gating Mechanisms

Although channels have, in general, only two measurable conducting states (some channels exhibit subconductance states) — open or closed — a channel protein may, in fact, transit between several more stable conformational states, with each transition involving a jump over an energy barrier. Despite the fact that these conformational states are not directly observable, they will manifest themselves when long periods of single-channel activity are monitored over time at a constant voltage. By analogy to equilibrium chemical reactions, the stochastic behavior of single channel proteins can be analyzed by kinetic modeling. In the simplest case of a channel with single closed (C) and open states (O), the state diagram would be as follows (Hille, 1984):

$$C \underset{\beta}{\overset{\alpha}{\rightleftarrows}} O \qquad\qquad (4)$$

where α and β are the rate constants of the forward and reverse transitions. For the majority of K^+ channels, several more states (open, closed, and inactivated states) are necessary to describe their complex time and voltage dependence, and many of the rate constants accounting for each channel transition are voltage- and/or ligand-dependent. From rate constant determinations, it would be possible to discard many possibilities because predictions for each reaction scheme can be made and tested experimentally. This type of analysis, along with analysis of ion-transfer properties, is extremely useful to identify a particular ion channel and characterize its functional molecular characteristics.

Most drugs and hormones modulate ion channel activity by altering the rate constants of transitions between the different conformational states. These kinetic changes induced by various agents, which are often time- and voltage-dependent (and, in cases of ligand-gated channels, dependent on the concentration of the ligand), can be accurately described by analysis of single-channel data. Of course, ion channel activity is often more complicated than pictured in this chapter (for instance, many channels open in bursts, despite the fact that voltage is maintained constant throughout the experiment), and more sophisticated analytical protocols are required to extract the desired information. For more information on the analysis (determination of open-state probability, lifetime, and first latency distributions, etc.) and modeling of single ion-channel kinetics, the reader is referred to a recent textbook by Sakmann and Neher (1995). In the following paragraphs, voltage-clamp protocols that are commonly utilized to investigate general gating mechanisms of macroscopic K^+ currents will be described.

5.2.1. Current–Voltage Relationships

To gather information on gating mechanisms, the simplest protocol, routinely used by all electrophysiologists, is the one designed to obtain information on the amplitude of the current and its dependence on voltage. This can be accomplished by using a protocol similar to that illustrated in Fig. 4A for the cardiac I_{K1}, where E_m is stepped from a conditioning or holding potential (HP) to a range of potentials imposed in sequence, usually by 10-mV increments, at a given frequency. Maximum or peak current, as well as the current at the end of each step, can then be plotted as a function of step voltage to obtain the $I-V$ curve. In the case of time-dependent currents, the activation and inactivation phases of the current can be fit to single or multiple exponential components to investigate the voltage dependence of the kinetics and to derive some of the rate constants for incorporation into kinetic models.

For voltage-independent channels such as those underlying I_{K1} in the heart or I_{KATP} in cardiac and smooth muscle cells, the value of HP should have little influence on the current measurements. However, for voltage-dependent channels that activate and inactivate with time, careful consideration should be given to the choice of the HP. It is normally set to values more negative than the voltage range of steady-state activation and inactivation. The frequency of stimulation should be slow enough to allow full recovery from inactivation; for example, if steps are applied too quickly, incomplete recovery from inactivation imposed by preceding pulses may cause the true magnitude of the cardiac transient outward K^+ current I_{to} to be underestimated. When working with a new preparation, it is good practice to generate an $I-V$ relationship with a short-step protocol (< 500 ms) and repeat the procedure with a longer step (> 5 s); such an approach minimizes the chances of missing fast and slow kinetic components.

5.2.2. Protocols Designed to Study Activation

Many K$^+$ channels expressed in the cardiovascular system display activation and deactivation kinetics during depolarization and repolarization, respectively. The deactivating current, commonly referred to as the *tail current*, has been widely used as an indicator of not only permeation and selectivity (see Fig. 4 above) but also conductance. One problem associated with the use of the standard $I-V$ protocol to study the gating properties of a K$^+$ current is the fact that both the gating mechanisms and the driving force acting upon K$^+$ vary in concert. Double-pulse protocols are designed to overcome this problem. After an initial voltage step (P1) to activate the channels, E_m is stepped down to a level (P2) that will result in time-dependent closure of the channels. The amplitude of the tail current following the capacitive current during P2 can be measured and used as an index of conductance. Altering the duration or voltage of P1 will result in tail currents whose magnitude will reflect different states of time-dependent or steady-state activation attained at the end of P1. The level of P2 (around -40 mV for the majority of voltage-dependent K$^+$ currents) is normally chosen to provide a reasonable driving force for K$^+$ and to ensure complete deactivation of the underlying channel(s). As opposed to the standard $I-V$ protocol, repolarization to a common P2 level eliminates the driving-force variable because all currents elicited by P1 (displaying distinct parameters such as duration or voltage) can be normalized during P2 against a constant driving force; the conductance reached at the end of P1 ($G_{K(P1)}$)can be calculated using the following equation:

$$(G_{K(P1)}) = \frac{I_{K-\text{tail(P2)}}}{(E_{P2} - E_K)} \tag{5}$$

where $I_{K\text{-tail(P2)}}$ is the magnitude of the tail current (peak — fully relaxed states), and $(E_{P2} - E_K)$ is the K$^+$ driving force during P2. Examples of tail current analysis are described below.

Even though the time-dependent K$^+$ current elicited during a step may appear kinetically relatively straightforward, it may nevertheless be the product of several distinct current (or molecular) components. The *"envelope of tail currents test"* has been useful in determining whether a macroscopic time-dependent K$^+$ current is composed of one or more components. A classical example in the heart was the identification of the rapid component of the delayed rectifier K$^+$ current I_K called I_{Kr}, which is now known to be encoded by the "human *ether-a-gogo*" related gene or HERG (Sanguinetti *et al.*, 1995). Figure 5 shows typical experiments carried out in guinea pig ventricular myocytes. The panel at the upper left of Fig. 5A displays sample traces of I_K currents elicited by the protocol displayed below them. Voltage steps to $+60$ mV of variable duration were applied in sequence to evoke a slowly developing outward I_K. Repolarization to -40 mV after variable intervals at $+60$ mV resulted in a family of tail

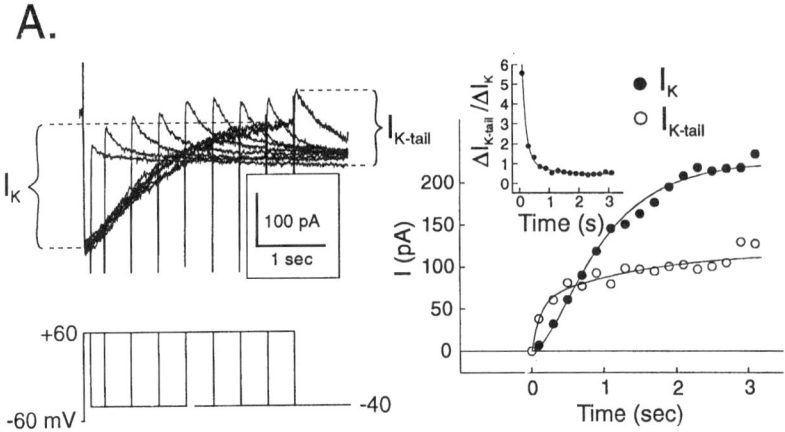

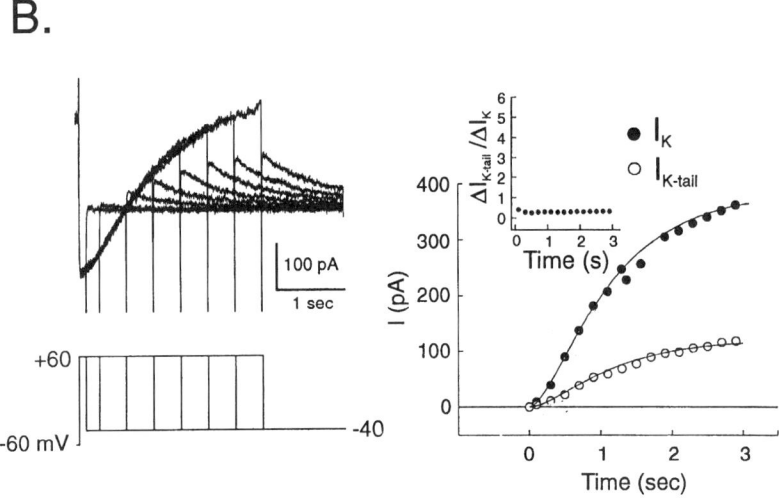

Figure 5. Demonstration of the "envelope of tail current" protocol. (A) *Left*: Family (top traces) of slow delayed rectifier K^+ currents I_K recorded from a guinea pig ventricular myocyte with the nystatin perforated-patch technique and exposed to 100 μM Cd^{2+} to block Ca^{2+} channels; the currents were evoked by the protocol shown below the traces. I_K, Amplitude of I_K current during the step to $+60$ mV; $I_{K\text{-tail}}$, amplitude of the tail current measured after repolarization to -40 mV. *Right*: Graphs showing the time course of development of I_K at $+60$ mV ($\bullet$) and $I_{K\text{-tail}}$ at -40 mV ($\bigcirc$) for the experiment shown on the left. The inset shows a plot of the ratio of $I_{K\text{-tail}}$ to I_K as a function of the duration of the step. Failure of the "envelope of tail current" test is apparent for steps of <1 s as the ratio varies exponentially during this period of time. (B) *Left*: Family of slow delayed rectifier K^+ currents I_{Ks} from a different cell elicited by a protocol identical to that depicted in (A). The myocyte was exposed to 100 μM Cd^{2+} and 500 nM dofetilide, a specific blocker of I_{Kr}. *Right*: Similar plots to those displayed in (A).

currents whose amplitude increased with the duration of the initial step. It is evident from the traces and the plots shown on the right-hand side of Fig. 5A that the tail current initially rose more quickly than the time-dependent current evoked at $+60$ mV. Failure of the "envelope of tail currents test" is demonstrated in the inset, where the ratio of the tail current at -40 mV to the time-dependent current at $+60$ mV ($\Delta I_{\text{K-tail}}/\Delta I_K$) is plotted against the duration of the pulse. This ratio is high early during the onset of I_K and declines exponentially toward a steady state after about 1 s. Sanguinetti and Jurkiewicz (1990) resolved this issue by showing that a rapid component of delayed rectification contaminated the tail current but produced little effect on I_K at positive potentials, a phenomenon ascribed to instantaneous inward rectification (Sanguinetti and Jurkiewicz, 1990), which may be due to fast inactivation (Yang *et al.*, 1997). This important observation was rendered possible by the demonstration that class III antiarrhythmic agents such as D,L-sotalol and its derivative E-4031 were able to selectively suppress I_{Kr}, leaving intact the slow component of I_K, or I_{Ks}. Figure 5B shows the utility of this test in an experiment performed in a different myocyte pretreated with dofetilide (500 nM), another class III antiarrhythmic compound that is more selective and more potent at blocking I_{Kr} (Carmeliet, 1992). As shown on the left-hand side of Fig. 5B, the development of the tail current parallels the time course of I_K during the step. This is illustrated further in the plots shown on the right-hand side of Fig. 5B; in particular, in contrast to control I_K, the ratio $\Delta I_{\text{K-tail}}/\Delta I_K$ exhibits little, if any, time dependence (inset). In summary, analysis of an envelope of tail currents can provide valid information on the time-dependent development of a K$^+$ conductance and can be used, in concert with other tools, to ascertain the existence of one or several components of time- and voltage-dependent current.

Analysis of the tail current can also be used to analyze the steady-state voltage dependence of activation of a K$^+$ current. A variant of the double-pulse protocol described above involves varying the voltage level of P1 while its duration is kept constant; E_m is then returned to a fixed voltage level (P2) that is used to record the tail current. Figure 6 shows a typical analysis of whole-cell I_{KV} currents recorded from a rabbit coronary myocyte. The upper left panel of Fig. 6A shows a family of current traces elicited by the protocol shown below them. The tail currents recorded during P2 are reproduced on the right-hand side of Fig. 6A on an expanded scale. The amplitude of the late current during P1 and that of each tail current elicited during P2, normalized against the highest amplitude $[(I_{\text{KV-tail}}/I_{\text{KV-tail}}^{\max}) \times 100]$, were plotted as a function of P1 voltage to generate the I–V relationship (Fig. 6B, *left*) and steady-state activation curve (Fig. 6B, *right*). As the I–V curve of I_{KV} shows typical outward rectification, which is the product of gating and driving force, the voltage dependence of the tail current is sigmoidal and reflects the intrinsic voltage dependence of gating or open-state probability of the channel(s). This relationship means that the channel under investigation (provided that only one type of channel is contributing to the measured current) is closed at potentials below ~ -40 mV and is fully activated (maximum open-state probability) at potentials beyond $+30$ mV.

5.2.3. Protocols Designed to Study Inactivation

Certain K$^+$ channels in heart and vascular smooth muscle cells tend to shut off after they have been activated by membrane depolarization. Besides analyzing the

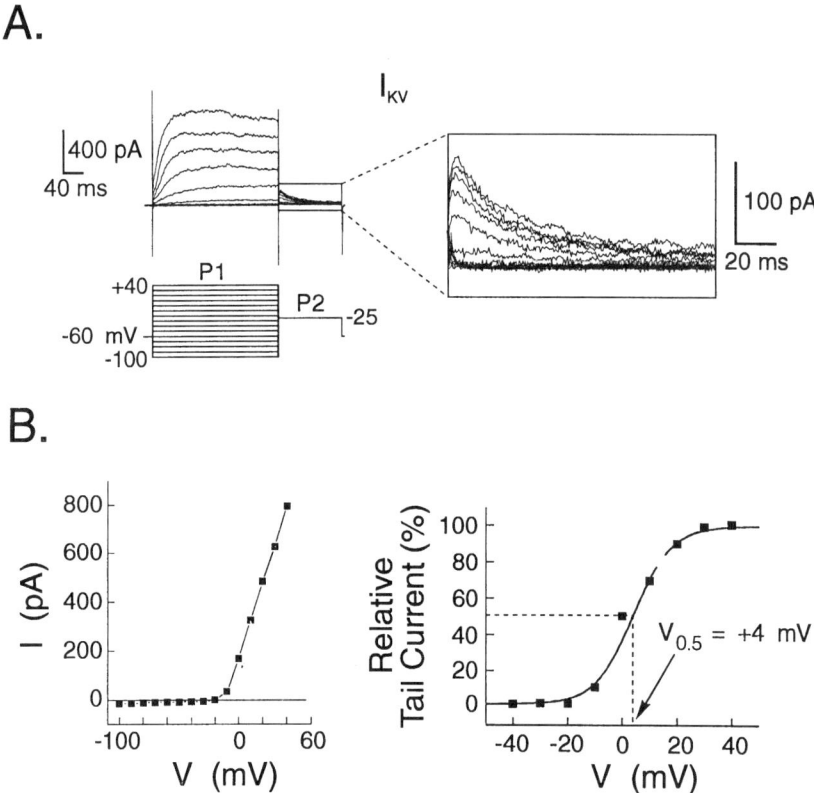

Figure 6. Example of analysis of the voltage dependence of the voltage-dependent K^+ current I_{KV} recorded from a single rabbit coronary artery smooth muscle cell using the standard whole-cell configuration. (A) Family of I_{KV} currents (top traces) elicited by a series of voltage-clamp pulses from a holding potential of -60 mV (bottom traces). To the right, the family of tail currents recorded during P2 are reproduced on an expanded scale for clarity's sake. (B) *Left:* Current–voltage (I–V) relationship obtained by plotting the amplitude of the current measured at the end of P1 as a function of the step voltage during P1. *Right:* Steady-state activation curve of I_{KV} obtained by plotting the amplitude of normalized tail currents (panel A) as a function of P1 voltage; the line passing through the data points is a least-squares fit to a Boltzmann distribution; as indicated, half-maximal activation ($V_{0.5}$) occurred at $+4$ mV.

kinetics of this process, it is also useful to determine the steady-state voltage dependence of inactivation of a K^+ current. In the simplest kinetic scheme, an additional state accounting for inactivation (I) is added, which is, of course, as the closed state, a nonconducting state:

$$C \rightleftharpoons O \atop {\diagdown \; \diagup} \atop I \tag{6}$$

Depending on the voltage dependence of the inactivation gating mechanism under steady-state conditions (such as those imposed by the resting potential of a cell or the holding potential in voltage-clamp experiments), a variable fraction of channels may be unavailable for opening during a depolarizing step. Figure 7A illustrates an example of an experiment designed to investigate this process in rabbit coronary myocytes. Although less evident when studied using short pulses, inactivation of whole-cell I_{KV} current becomes apparent when long depolarizing steps are used. The left-hand side of Fig. 7A shows a family of I_{KV} currents that were evoked by the double-pulse protocol depicted below them. From HP $= -60$ mV, an initial variable step of 10 s in duration (P1) was followed by a constant 250-ms pulse to $+20$ mV (P2), which serves to measure the impact on I_{KV} of the preconditioning P1 step. As more and more I_{KV} channels are inactivated during P1 (arrow pointing up), less channels become available for activation during P2 (arrow pointing down). An important factor in this protocol is the duration and amplitude level of the delay between P1 and P2, which was set to 10 ms at -60 mV in this experiment. The delay is optimally adjusted to allow full or near-complete deactivation of the channels and avoid partial recovery from inactivation. The interval between each series of steps (or frequency of stimulation) should be adjusted to prevent cumulative inactivation, a factor which could also have an influence on the results. The right-hand panel of Fig. 7A shows a plot of outward K$^+$ current measured during P2 as a function of P1 voltage; in the inset, the data were normalized between 0 and 1 and fitted using the following form of the Boltzmann equation:

$$\frac{I_{KV(P2)}}{I_{KV(P2)}^{max}} = \frac{1}{1 + e^{(Va^{P}_1 - V_{0.5})/k}} \tag{7}$$

where $I_{KV(P2)}/I_{KV(P2)}^{max}$ is the ratio of I_{KV} to the maximum I_{KV} during P2, V_{P1} is the voltage during P1, $V_{0.5}$ is the voltage at which 50% of the channels are inactivated, and k is the slope factor (in millivolts) and is an index of the steepness of the steady-state voltage dependence. The inset shows that the K_V channels were fully available at a potential of ~ -80 mV, were half-inactivated at -44 mV, and were apparently fully inactivated at potentials above -20 mV (there are indications that a fraction of I_{KV} may undergo incomplete inactivation).

Another property that is important to investigate for inactivating K$^+$ channels is the process of recovery from inactivation. The standard protocol consists of two pulses whose magnitude and duration are maintained constant while a delay of increasing duration is interspersed between the two. In the example illustrated in Fig. 7B(a), an initial 20-s step to $+20$ mV (P1) was used to activate and inactivate I_{KV} in a rabbit coronary myocyte. E_m was stepped down to -60 mV for a variable delay after which a second step to $+20$ mV of 2-s duration (P2) was applied to evaluate the amount of recovery from inactivation. The rate of recovery can be estimated by measuring the relative magnitude of I_{KV} during P2 against the control current elicited during P1. A graph such as that shown on the right-hand side of Fig. 7B(a) can be generated by plotting the percentage recovery of I_{KV} as a function of the duration of the delay between P1 and P2. The large estimated time constant ($\tau = 10$ s) derived from least-squares monoexponential fitting reveals that I_{KV} exhibits very slow recovery kinetics, requiring over 25 s in this particular cell for complete recuperation from inactivation. Repeating a similar protocol from different holding potentials would be useful in assessing the voltage dependence of the I $\rightarrow$ C step of Eq. (6) [Fig. 7B(b)]; for

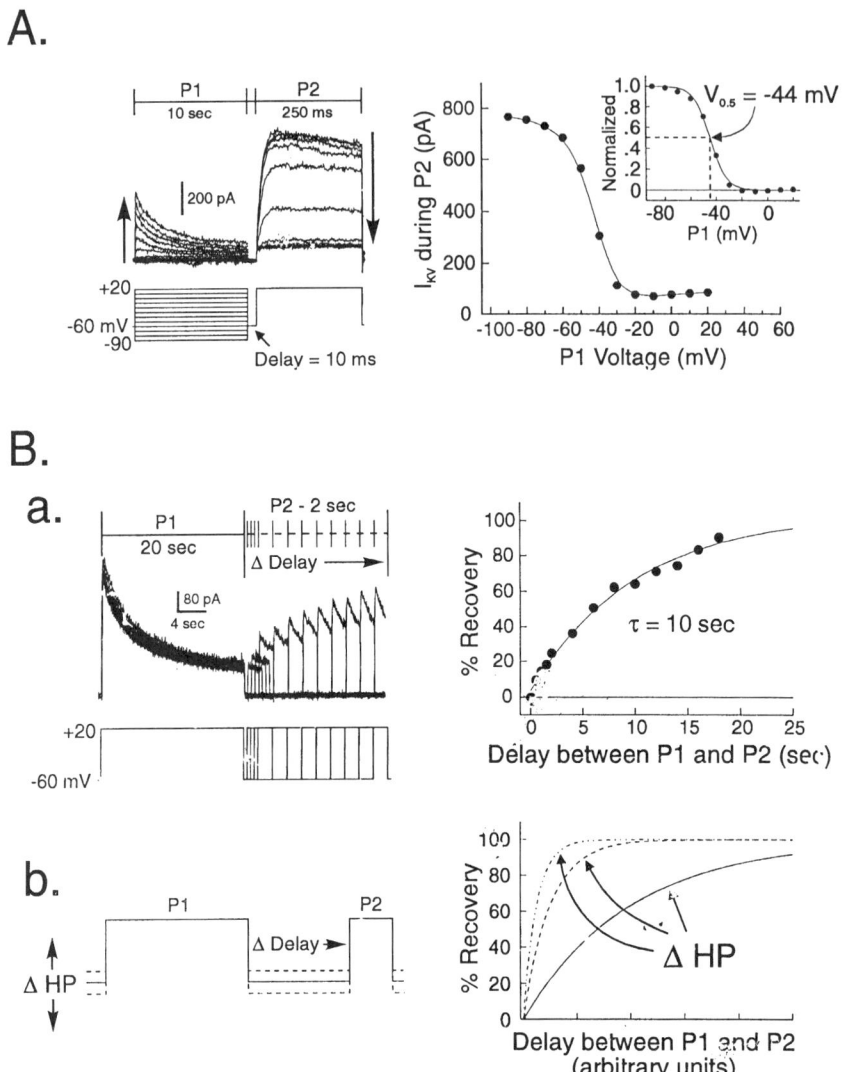

Figure 7. Typical protocols used to investigate the inactivating properties of a voltage-dependent K^+ current displaying inactivation kinetics. (A) Protocol designed to derive information on the steady-state voltage dependence of inactivation of I_{KV} recorded from a rabbit coronary myocyte. The recordings were obtained under conditions identical to those described in Fig. 6. *Left*: Family of I_{KV} currents (top traces) elicited by the double-pulse protocol shown below then. Notice the difference in duration of P1 and P2. Arrows pointing up and down indicate, respectively, the increase of I_{KV} during P1 and the decrease of I_{KV} during P2, as P1 becomes more positive. *Right*: Plot of the voltage dependence of peak I_{KV} during P2 as a function of P1 voltage. *Inset*: Steady-state inactivation curve obtained by plotting normalized I_{KV} during P2 as a function of P1 voltage; the smooth curve is a Boltzmann fit to the data points (see Eq. 7). (B) Protocols designed to study the kinetics of recovery from inactivation. (a) *Left*: Recovery kinetics of rabbit coronary I_{KV} currents (top traces) evoked by the double-pulse protocol depicted below the traces. *Right*: Plot of the percentage recovery of I_{KV} as a function of the duration of the delay between P1 and P2. The curve passing through the data point is a monoexponential fit to the data points yielding a time constant $\tau = 10$ s. (b) This panel shows how a hypothetical alteration of the holding potential (*left*, ΔHP) would help to provide more information about the voltage dependence of the kinetics of recovery from inactivation, which are known to be faster at more negative HP for most K^+ channels (*right*).

most transient K$^+$ channels, recovery kinetics are faster at more negative potentials (from the smooth to the dashed and dotted lines).

6. CONCLUSION

The development of the patch-clamp technique has contributed in a major way in identifying and characterizing the structure and function of K$^+$ channels in the cardiovascular system. The patch-clamp technique is a powerful tool that enables us to study with great accuracy the biophysical properties and modulation of voltage-activated or ligand-gated ion channels present in native membranes of freshly dispersed or cultured smooth muscle cells. It is now clear that the expression of the various K$^+$ channel genes identified so far is under tight genetic control at the posttranslational (e.g., glycosylation) and transcriptional (e.g., alternative splicing) levels. Although we are just beginning to understand the specific gene products that are expressed in cardiac, smooth muscle, and endothelial cells, we still know little about the nature of the molecular arrangement that forms the basis of the native K$^+$ currents recorded in these cells.

The main goal of this chapter was to provide a relatively simple description of the various configurations of the patch-clamp technique and protocols designed to extract information on the permeation and gating properties of macroscopic K$^+$ currents. The technique has served as, and will remain, a useful tool to study the activity of K$^+$ channels in native cells from healthy, diseased, chronically treated (pharmacologically or genetically), and transgenic cardiovascular tissues, as well as heterologous expression cell systems.

ACKNOWLEDGMENTS

The authors wish to thank Marie-Andrée Lupien for her excellent technical skills and Dr. Stanley Nattel for kindly providing dofetilide. This work was supported by grants from the Medical Research Council of Canada, the Heart and Stroke Foundation of Canada, Fonds de la Recherche en Santé du Québec, and the Montréal Heart Institute Fund.

REFERENCES

Bean, R. C., Shepherd, W. C., Chan, H., and Eichner, J., 1969, Discrete conductance fluctuations in lipid bilayer protein membranes, *J. Gen. Physiol.* **53:**741–757.

Campbell, D. L., Rasmusson, R. L., Qu, Y. H., and Strauss, H. C., 1993, The calcium-independent transient outward potassium current in isolated ferret right ventricular myocytes, *J. Gen. Physiol.* **101:**571–601.

Carmeliet, E., 1992, Voltage-dependent and time-dependent block of the delayed K$^+$ current in cardiac myocytes by dofetilide, *J. Pharmacol. Exp. Ther.* **262:**809–817 .

Cole, W. C., Chartier, D., Martin, M., and Leblanc, N., 1997, Ca^{2+} permeation through Na$^+$ channels in guinea pig ventricular myocytes, *Am. J. Physiol.* **273:**H128–H137.

Conn, P. M. (ed.), 1998, Ion Channels Part B, *Methods of Enzymology*, Vol. 293 (J. N. Abelson and M. I. Simon, eds.), Academic Press, San Diego.

Doyle, D. A., Cabral, J. M., Pfuetzner, R. A., Kuo, A. L., Gulbis, J. M., Cohen, S. L., Chait, B. T., and MacKinnon, R., 1998, The structure of the potassium channel: Molecular basis of K^+ conduction and selectivity, *Science*, **280**:69–77.

Hamill, O. P., Marty, A., Neher, E., Sakmann, B., and Sigworth, F. J., 1981, Improved patch-clamp techniques for high-resolution current recording from cells and cell-free membrane patches, *Pflügers Arch.* **391**:85–100.

Hille, 1984, *Ionic Channels of Excitable Membranes*, Sinauer Associates Inc., Sunderland, Massachusetts.

Imaizumi, Y., Muraki, K., and Watanabe, M., 1990, Characteristics of transient outward currents in single smooth muscle cells from the ureter of the guinea-pig, *J. Physiol. (London)* **427**:301–324 .

Leblanc, N., and Leung, P. M., 1995, Indirect stimulation of Ca^{2+}-activated Cl^- current by Na^+/Ca^{2+} exchange in rabbit portal vein smooth muscle, *Am. J. Physiol.* **268**:H1906–H1917.

Leblanc, N., Wan, X. D., and Leung, P. M., 1994, Physiological role of Ca^{2+}-activated and voltage-dependent K^+ currents in rabbit coronary myocytes, *Am. J. Physiol.* **266**:C1523–C1537.

Leblanc, N., Chartier, D., Gosselin, H., and Rouleau, J. L., 1998, Age and gender differences in excitation–contraction coupling of the rat ventricle, *J. Physiol (London).* **511**:533–548.

Neher, E., and Sakmann, B., 1976, Single-channel currents recorded from membrane of denervated frog muscle fibres, *Nature* **260**:799–802.

Neher, E., and Sakmann, B., 1992, The patch-clamp technique, *Sci. Am.* **266**:44–51.

Remillard, C. V., and Leblanc, N., 1996, Mechanism of inhibition of delayed rectifier K^+ current by 4-aminopyridine in rabbit coronary myocytes, *J. Physiol. (London)* **491**:383–400.

Roden, D. M., and George, A. L., 1997, Structure and function of cardiac sodium and potassium channels, *Am. J. Physiol.* **273**:H511–H525.

Rudy, B., and Iverson, L. E. (eds.), 1992, Ion Channels, *Methods of Enzymology*, Vol. 207 (J. N. Abelson and M. I. Simon, eds.), Academic Press, San Diego.

Sakmann, B., and Neher, E. (eds.), 1995, *Single-Channel Recording*, 2nd. ed., Plenum Press, New York.

Sakmann, B., and Trube, G., 1984, Conductance properties of single inwardly rectifying potassium channels in ventricular cells from guinea-pig heart, *J. Physiol. (London)* **347**:641–657.

Sanguinetti, M. C., and Jurkiewicz, N. K., 1990, Two components of cardiac delayed rectifier K^+ current: Differential sensitivity to block by class-III antiarrhythmic agents, *J. Gen. Physiol.* **96**:195–215.

Sanguinetti, M. C., Jiang, C. G., Curran, M. E., and Keating, M. T., 1995, A mechanistic link between an inherited and an acquired cardiac arrhythmia: HERG encodes the $I_K r$ potassium channel, *Cell* **81**:299–307.

Shimoni, Y., Clark, R. B., and Giles, W. R., 1992, Role of an inwardly rectifying potassium current in rabbit ventricular action potential, *J. Physiol. (London)* **448**:709–727.

Sigworth, F. J., 1986, The patch-clamp is more useful than anyone had expected, *Fed. Proc.* **45**:2673–2677.

Yang, T., Snyders, D. J., and Roden, D. M., 1997, Rapid inactivation determines the rectification and $[K^+]_o$ dependence of the rapid component of the delayed rectifier K^+ current in cardiac cells, *Circ. Res.* **80**:782–789.

Chapter 9

The Patch-Clamp Technique for Measurement of K$^+$ Channels in *Xenopus* Oocytes and Mammalian Expression Systems

Laura Conforti and Nicholas Sperelakis

1. INTRODUCTION

Potassium channels are ubiquitous membrane proteins. They are expressed in most cells, both excitable and nonexcitable. In excitable cells, K$^+$ channels preside over a wide range of cell functions such as excitability, propagation of excitation, and neurotransmitter release (Hille, 1992; Rudy, 1988). In nonexcitable cells, they are responsible for regulating cell volume and cell proliferation (Sobko *et al.*, 1998; Hille, 1992). Therefore, most cells express different voltage-dependent K$^+$ (Kv) channels to guarantee certain functions and to optimize the cellular response to different stimuli (Sobko *et al.*, 1998; Conforti and Millhorn, 1997). The heteromultimeric nature of native Kv channels and the possibility that several related types of Kv channels of different compositions are expressed in the same cell complicate the study of an isolated channel in mammalian cells (Shamotienko *et al.*, 1997; Sheng *et al.*, 1993; Wang *et al.*, 1993). Study of the behavior of a particular type of K$^+$ channel in the native cell is further complicated by the limited selectivity of available K$^+$ channel blockers and toxins (Hopkins, 1998; Grissmer *et al.*, 1994). Kv channels of known composition can be expressed in heterologous expression systems, such as the *Xenopus* oocytes and mammalian cell lines. Structural and functional studies of Kv channels in expression systems provide important information for understanding, by extrapolation, the function of similar channels in native cells.

This chapter aims to summarize two different heterologous expression systems used to study Kv channels—*Xenopus* oocytes and mammalian cell lines—and the

Laura Conforti and Nicholas Sperelakis ● Department of Internal Medicine, Division of Nephrology and Hypertension and Department of Molecular and Cellular Physiology, University of Cincinnati, Cincinnati, Ohio 45267-0576.

Potassium Channels in Cardiovascular Biology, edited by Archer and Rusch. Kluwer Academic/Plenum Publishers, New York, 2001.

electrophysiological techniques available for recording from these systems. In particular, we will focus on the applications of these expression systems, the techniques associated with them, and their advantages and limitations.

2. APPLICATIONS OF HETEROLOGOUS EXPRESSION SYSTEMS

Expression systems permit study of the behavior of ion channels whose molecular composition is known *a priori*, and they therefore allow assignment of a particular function to a specific protein. This is possible because the expression systems permit control of the type of coding sequence introduced into the host cell.

Fundamental information has been obtained from studies of the structure–function relationships of ion channels, in particular from studies using oocytes from *Xenopus laevis*. Engineered modifications in the coding sequence of a cloned gene, combined with patch-clamp recordings, allow detailed analysis of the contributions of specific protein regions, or even single amino acids, to the biological activity of the encoded channel protein (Jan and Jan, 1992). Reconstitution of multisubunit channel molecules by the coexpression of the constituent channel subunits also allows the investigation of subunit coassembly processes and interaction sites (Covarrubias *et al.*, 1991; Ruppersberg *et al.*, 1990). The *Xenopus* oocyte expression system has also enabled the optimal expression cloning of unknown ion channels, by screening of DNA libraries or injection of fractionated mRNA (Hart *et al.*, 1993; Frech *et al.*, 1989; McDonald *et al.*, 1989). Heterologous expression systems have also been used to study the regulation of gene expression, posttranslational processing (such as channel glycosylation), and protein sorting and transport (Trimmer, 1998; Nagaya and Papazian, 1997; Thornhill *et al.*, 1996).

2.1. *Xenopus* Oocyte Expression System

The oocytes of the African clawed frog, *Xenopus laevis*, are widely used for studying the structure and function of many channel proteins. These frogs are easily maintained in a laboratory setting (Goldin, 1992). The *Xenopus* oocytes are large cells and have a low expression of endogenous channels; they also have very efficient translational machinery. They are easy to use, and most cells injected with foreign messenger RNAs give expression of the relevant protein within a few days. These proteins are then processed and incorporated into the plasma membrane. Therefore, *Xenopus* oocytes are an excellent tool for the systematic expression of ion channels of known amino acid sequence.

2.1.1. Preparation of Oocytes for Electrophysiological Experiments

Xenopus oocyte size and appearance varies at different developmental stages (Dumont, 1972). Table 1 summarizes the characteristics of *Xenopus* oocytes at different developmental stages and the advantages and disadvantages of using oocytes at

Table 1
Xenopus Oocyte Developmental Stages

Stage	Appearance	Diameter (μm)	Advantages	Disadvantages
I	Transparent	50–100		Too fragile,[a] translation capacity not tested
II	White	300–450 ⎫	Small capacitance, fewer endogenous currents	Technically more difficult to inject (20 nl max.)
III	Pigmented	300–450 ⎭		
IV	Defined V/A poles[b] (pigmented)	600–1000		
V	Defined V/A poles	1000–1200 ⎫	High translation capacity, 100-nl maximum injection volume	Large capacitance
VI	Marginal band between V/A poles	1200–1300 ⎭		

[a] No vitelline membrane.
[b] V/A, Vegetal/animal.

different stages for electrophysiological experiments. From stage IV, these cells are characterized by a dark-brown *animal pole*, with pigmented granules containing a high concentration of melanin, and a white *vegetal pole*. The presence of the two different animal and vegetal poles affects distribution of expressed channels and influences efficiency of channel protein expression (Lupu-Meiri *et al.*, 1988; Oron *et al.*, 1988). Each oocyte is surrounded by several layers of cells as revealed by ultrastructural analysis of oocyte sections (Wischnitzer, 1996). From outside to inside, these layers consist of epithelium, theca, follicle cells, and vitelline membrane. Macrovilli and microvilli traverse the vitelline membrane to establish important gap-junction connections between the oocyte and the follicle cells, providing chemical and electrical communication pathways.

Isolation of oocytes is performed as follows. The frog is anesthetized by immersion in 0.2% tricaine methanesulfonate (MS222) in tap water until immobile. Under hypothermic conditions (on ice), a 1-cm incision is made in the right or left lower abdominal quadrant, and the ovarian lobes are exposed. Two to four lobules containing 300–500 oocytes are excised under sterile conditions. The incision in the body wall and skin are sutured separately with viacryl for the muscle layer and nylon for the skin. Nylon has proven to be softer than other materials, and the frogs rarely develop skin infections. The frog is left partially immersed in shallow water to recover and monitored until fully awake. The same frog can be operated on up to four times at intervals of two months.

The removed oocytes are placed in Ca^{2+}-free OR2 medium.* At this point or after RNA injection, the follicular layer is removed. This procedure allows better access of solutions and drugs to the oocyte itself and facilitates the insertion of low-resistance electrodes for electrophysiological recordings. The follicular layer is not removed when it is necessary to study responses that depend on the presence of follicular cells (e.g., the opening of K^+ channels in response to cyclic nucleotides) (Miledi and Woodward,

---

* The compositions of the solutions mentioned in the text are given in an appendix at the end of this chapter.

1989). Removal of the follicular layer can be done manually by forceps under a dissecting microscope (Miledi and Woodward, 1989) or enzymatically with collagenase in Ca^{2+}-free medium. Collagenase used in the presence of Ca^{2+} might damage the oocytes because Ca^{2+} activates proteases. After collagenase treatment, oocytes are maintained in Barth's medium or ND96, and a subset of oocytes are selected for injection. It is important to carefully select the proper oocytes for injection: oocytes should be as uniform as possible in size and general appearance to ensure a similar stage of development (Table 1). In fact, there are marked differences (at various developmental stages) regarding the presence of important elements necessary for stability and translation efficiency of oocyte mRNAs. Stage V and VI oocytes are most commonly used for expression of ion channels. Stage V oocytes are optimal for injection because of their large size, which allows injection of up to 100 nl of mRNA-containing solution, and also because they possess an efficient translational machinery. The drawback of the large cell size is that these oocytes have a large membrane capacitance, which causes a long capacitive transient during voltage-clamp recordings. Smaller oocytes (stage II and III) have a faster voltage-clamp settling time, but a maximal volume of only about 20 nl can be injected.

2.1.1a. RNA Preparation. The selected oocytes are injected with mRNAs. The injected RNA can be extracted from tissues or synthesized *in vitro* from cloned cDNAs. Injection of total tissue RNA or poly(A)$^+$-mRNA is done to achieve the expression of specific genes not yet cloned or to supply unknown factors that might influence the function of a known channel (Rudy *et al.*, 1988). Total RNA for oocyte microinjection can be prepared using standard procedures (Chomczynski and Sacchi, 1987). Poly(A)$^+$-mRNA can be purified from total RNA by affinity chromatography on oligo(dT)-cellulose columns. Size-fractionated poly(A)$^+$-RNA can be injected to increase the abundance of specific mRNA species and to estimate their size (Chabala *et al.*, 1993). Poly(A)$^+$-mRNA can be fractionated by sucrose density gradient centrifugation or agarose-gel electrophoresis. The sucrose gradient centrifugation has lower resolving power than agarose electrophoresis, but it allows easy recovery of fractionated mRNA without loss of translational activity.

Higher levels of expression and the detailed analysis of specific RNA transcripts and their encoded proteins can be achieved by injecting mRNA obtained by *in vitro* transcription of a cloned cDNA (cRNA). The cDNA is subcloned into a transcription vector from which the RNA polymerase II binding site of phages (such as SP6, T3, or T7) may direct production of synthetic mRNA. Commercially available kits that allow synthesis of capped RNA by including cap analog $m^7G(5')ppp(5')G$ in the *in vitro* transcription reaction are now widely used. Capped RNAs appear to be protected against nucleolytic degradation.

2.1.1b. RNA Injection. The injection of foreign RNA into the oocyte cytoplasm is performed via glass pipettes with a tip of ca.10 μm using a Drummond microdispenser. The injections have to be performed in RNase-free conditions. The use of gloves and RNase-free solutions and glassware is recommended. The amount of RNA required to be injected depends on the source of the RNA and varies from 1 to 10 μg/ml for mRNA from tissue extract and from 0.1 to 1 μg/μl for *in vitro* transcribed cRNA.

Usually, a volume of 50 nl is injected in mature oocytes. It is important to optimize the amount of cRNA injected in oocytes for each particular message. This is because excess quantities of injected synthetic RNA may be inhibitory and may modify ionic current kinetics (Honoré *et al.*, 1992). Also, the injection of high concentrations of mRNA that produce high levels of expression of heterologous membrane proteins increases endogenous ionic currents (Tzounopoulos *et al.*, 1995).

Coinjection of different cRNAs can be performed to study heteromultimeric proteins and subunit interactions (Heinemann *et al.*, 1996). For example, the coinjection of some β-subunits of Kv channels with the corresponding α-subunit increases the level of expression of functional Kv channels, presumably by promoting surface expression (Shi *et al.*, 1996; McCormack *et al.*, 1995). Coinjection is also used to establish the effectiveness and selectivity of a given nucleotide for its target sequence. In this regard, antisense oligonucleotides or dominant negatives (truncated form of the channel with suppression of the functional channel) may be injected to inhibit the expression of the gene product against which it is designed (Li *et al.*, 1992). Additionally, tissue-extracted mRNA may be coinjected with cRNA to identify tissue factors that selectively modulate the channel function (Chabala *et al.*, 1993). Coinjection of membrane-impermeable chemical compounds together with the cRNA of interest also is used to study the regulation of ion channels.

After injection, the oocytes are maintained at 190°C in an isoosmotic solution, such as Barth's solution, OR2 with Ca^{2+}, normal frog Ringer solution, L-15 medium, or ND96 (Goldin, 1992). All these solutions contain sodium pyruvate as a carbon source and gentamicin or penicillin/streptomycin to prevent bacterial contamination of the storage media.

2.1.2. Electrophysiological Recording from Xenopus Oocytes

The function of ion channels expressed in *Xenopus* oocytes is studied electrophysiologically with the two-electrode voltage-clamp technique or patch-clamp techniques. The time course of functional channel expression varies with the type of RNA injected and is monitored experimentally by recording ion currents carried by the protein of interest each day after injection with the use of two-electrode voltage clamping (Fig. 1).

2.1.2a. Two-Electrode Voltage-Clamp Techniques.   Two-electrode voltage clamping allows measurement of the total current carried by all the channels expressed and incorporated in the plasma membrane. This technique and its application using *Xenopus* oocytes have been recently reviewed in detail by Stuhmer (1998). The two-electrode voltage-clamp technique permits the membrane potential of the oocyte to be held constant at some desired level while the current needed to maintain that potential is recorded. The electrodes used are glass electrodes filled with 3 M KCl that show a tip resistance lower than 1 MΩ. These electrodes can be used many times for different oocytes, providing they maintain a low resistance. One electrode is used to record the actual intracellular potential, while the second electrode (using a negative-feedback loop) is used to pass the current necessary to maintain the desired voltage. Basically, a two-electrode voltage-clamp setup includes the voltage-clamp amplifier, a

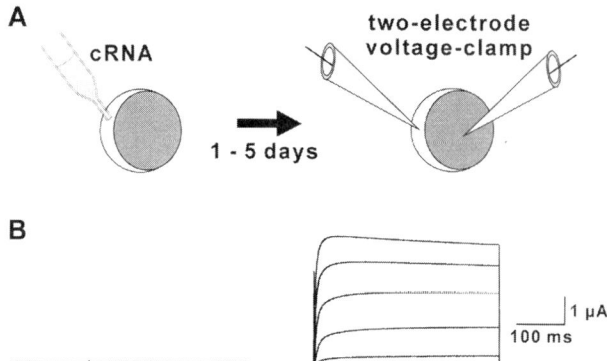

Figure 1. Two-electrode voltage-clamp recording from *Xenopus* oocytes. (A) Expression of ion channels in *Xenopus* oocytes. Oocytes are injected with the cRNA encoding the channel of interest. One to five days after injection, recording of the total current flowing through the expressed channel is performed by two-electrode voltage clamp. (B) Expression of K^+ currents in *Xenopus* oocytes injected with Kv1.2 cRNA. *Left*: Sample recordings from control oocytes (injected with water); *Right*: K^+ currents recorded after injection with Kv1.2 cRNA. A two-electrode voltage-clamped paradigm was used. The K^+ currents were elicited by depolarizing voltage steps from a holding potential (HP) of -80 mV to test potentials between -60 and $+20$ mV in 10-mV increments.

computer-controlled pulse generator, a Bessel low-pass filter, an analog-to-digital interface, and a computer-based data acquisition system. Similar instrumentation is required for patch-clamp recording of whole-cell and single-channel currents (Penner, 1995). In general, the two-electrode voltage clamp is stable for several hours, and extracellular solution changes are well tolerated. The recordings are similar to voltage-clamp experiments performed in smaller cells. The only difference resides in the large capacitive current transient due to the large size of the oocytes. Leak current and capacitive current transients can be partially subtracted using currents elicited by small hyperpolarizing pulses. Figure 1B shows current recordings elicited by step depolarizing pulses from oocytes injected either with vehicle (RNase-free water) or with Kv1.2 cRNA. Outward K^+ currents are elicited only in oocytes injected with the cRNA of interest. Although the two-electrode voltage clamp of oocytes is simple and stable, requires only low channel density, and allows easy manipulation of the external solution, it does not allow study of fast processes because of the large capacitance. For study of fast kinetics, macropatches are a better choice (described below).

2.1.2b. Patch-Clamp Techniques. Patch-clamp techniques are used for single-channel recording. Access to the plasma membrane is obtained by the removal of the vitelline membrane. Removal of this membrane from *Xenopus* oocytes can be performed manually after shrinkage in a hyperosmotic medium (Fig. 2A; Methfessel *et al.*, 1986). After 3–10 min in this solution, the transparent vitelline membrane becomes visible and can be removed with very fine forceps under a dissecting microscope. However, this maneuver can cause damage to the oocyte surface that may prevent the formation of the gigaseal and satisfactory recording of channel activity.

A modified technique has been proposed recently that optimizes the plasma membrane surface for patch recording (Choe and Sackin, 1997). Figure 2A shows the

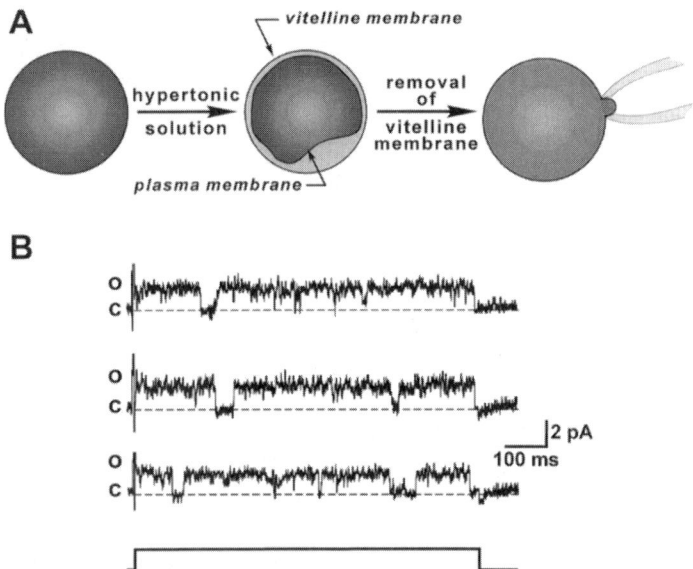

Figure 2. Single-channel recordings from injected *Xenopus* oocytes. (A) Schematic representation of the procedure used to perform single-channel experiments. Injected oocytes are placed in a hypertonic solution. Cell shrinkage results in a clear separation between the vitelline and plasma membranes. After removal of the vitelline membrane, cell-attach configuration can be obtained. (B) Representative traces recorded during step depolarizing pulses (from a holding potential of -60 mV to $+50$ mV) in the cell-attached configuration in an oocyte injected with Kv1.2 cRNA. Experiments were performed in high K$^+$ (140 mM) bath solution and 2.8 mM pipette solution. Leak and capacitive currents were subtracted from the record. The upward current deflections from the zero line (dashed, C) correspond to the opening of the channel (O).

clear separation between the vitelline and plasma membranes in oocytes shrunk in hypertonic solution. After removal of the vitelline membrane, the intact plasma membrane is available for patching. Gigaseals are formed using microelectrodes of 5–10 ΩW, fire-polished and coated with a silicone curing agent to reduce electrode capacitance. Different recording configurations can be used, and current recordings can be obtained by following procedures similar to those used for smaller cells (Hamill *et al.*, 1981). These procedures are described in Section 2.2.2 of this chapter. Figure 2B shows representative single-channel recordings in the cell-attached configuration from an oocyte injected with Kv1.2 cRNA.

Inside-out excised patches are obtained as for other cells (described below). The oocytes offer the unique possibility of reinserting the inside-out patch into the same cell from which the patch was obtained (so-called "patch cramming") (Kramer, 1990). It is also possible to introduce the excised patch into another oocyte that has been exposed to different conditions with a technique called "cross-cramming" (Parekh *et al.*, 1993). These techniques are very useful when studying the regulation of ion channels and when determining the role of cytoplasmic factors. Very stable outside-out patches are also obtainable via standard techniques (described below).

A great advantage of the *Xenopus* oocytes is the possibility of obtaining *macro-patches*. Macropatches can be obtained with the use of pipettes having an opening of

3–8 μm in diameter (Stuhmer *et al.*, 1987). They enable low-noise, fast-clamp patch recordings of many channels and can be used to study channels with fast kinetics. Excised "giant" membrane patches also have been obtained in *Xenopus* oocytes (Hilgemann, 1995). These are patches of 12–40 μm in diameter that have capacitance values of 2–15 pF and seal resistances of 1–10 GΩ. They allow recordings of transporter currents and charge movements and single-channel recordings of low-density channels. Other techniques have been introduced that allow control over the intracellular ionic composition, such as the *cut-open oocyte* (Perozo *et al.*, 1992).

2.1.3. Advantages of the Oocyte Expression System

The *Xenopus* oocyte has a series of advantages over other expression systems. Hundreds of cells can be obtained from one frog, and the same frog can be reused several times. The large size of the oocytes allows direct injection of RNA and membrane-impermeable compounds. The oocyte expression machinery faithfully expresses the injected RNA. The cells are relatively sturdy, and they can survive up to two weeks *in vitro*. Two-electrode voltage-clamp experiments are relatively easy to do and can be used as a tool for fast screening of different wild and mutant cRNAs. The oocytes express only a few endogenous channels that contribute but a small fraction of the total current in injected oocytes. They allow the recording of many channels in *macropatches*.

2.1.4. Limitations of the Oocyte Expression System

The *Xenopus* oocyte expression system is not ideal when large numbers of transfected cells are needed (e.g., for obtaining large amounts of proteins). In fact, each oocyte has to be injected individually. There is a certain degree of variability in the level of expression, even within oocytes simultaneously taken from the same frog and injected with the same amount of RNA. There are also seasonal variations in the expression efficiency. Expressed channels may also display different properties from those of the native channel, owing to possible posttranslational modifications that are different from those in mammalian cells (Shi *et al.*, 1994).

A limitation in recording expressed K^+ channels in oocytes is the fact that these channels are not uniformly distributed in the cell membrane, but tend to cluster in regions. This clustering process complicates single-channel recordings by increasing the chances of empty patches and decreasing the chances of recording from a single channel (Peter *et al.*, 1991). The presence of endogenous channels also may interfere with the recordings of currents, especially small currents. Various endogenous channels have been recorded in *Xenopus* oocytes. Endogenous stretch-activated channels with a conductance of about 35 pS are easily identified by applying negative pressure to the membrane patch, which will cause the channel activity to increase (McBride and Hamill, 1992; Yang and Sachs, 1989; Methfessel *et al.*, 1986). These channels are blocked by 10–100 μM gadolinium (Yang and Sachs, 1989). Ca^{2+}-activated Cl^- channels may be expressed in high density at the animal pole (Lupu-Meiri *et al.*, 1988). These channels may be blocked by replacing Cl^- ions with methanesulfonic acid, by injecting Ca^{2+} chelators (BAPTA or EGTA), or by using Cl^- channel blockers (niflumic acid or flufenamic acid). Endogenous delayed rectifier K^+ currents (30–400 nA at membrane potential of $+30$ mV) that are only partially blocked by tetraethylammonium (TEA), have been recorded (Lu *et al.*, 1990). A depolarization-induced Na^+ current blocked by tetrodotoxin and an endogenous voltage-dependent Ca^{2+} channel blocked by Cd^{2+} (but not by organic Ca^{2+} channel blockers) also have been observed (Lory *et al.*, 1990).

2.2. Mammalian Cell Lines

Mammalian cells with very low expression of endogenous ion channels have been used as heterologous expression systems. They provide similar machinery and pathways as the mammalian cells carrying the native channel. The introduction of foreign DNA into the host cells is called transfection and can be performed by various techniques, such as calcium phosphate precipitation or lipofection. The choice of the host cell to be transfected is critical and depends on the type of endogenous channels present and the type of channel to be expressed. Various mammalian cell lines have been used for transfection and expression of Kv channels. These include fibroblast-type cells (NIH/3T3, COS cells, L cells, and BHK), epithelial cells, and epithelial-like cells [HeLa cells, Chinese hamster ovary (CHO) cells, and human embryonic kidney cells, HEK293]. The rat basophilic leukemia cell line RBL-1 is also used for expression of outward K^+ currents (Ikeda *et al.*, 1992). A list of the most commonly used cell lines for transfection of Kv channels, their origin, and the type of endogenous ion channels and auxiliary Kv channel subunits expressed is given in Table 2. The RBL-1 cell line expresses an endogenous inward-rectifying K^+ channel (Lindau and Fernandez, 1986). HEK293 cells express endogenous voltage-activated Ca^{2+} channels (Berjukow at al., 1996), inward rectifier K^+ channels (Ammala *et al.*, 1996), and an endogenous TEA-sensitive

Table 2
Mammalian Cell Lines for K^+ Channel Expression

Name	Cell type	Origin	Endogenous ion currents and auxiliary channel subunits
L cells	Fibroblast	Mouse	Kvβ2.1 subunit of Kv channels[a]
NIH/3T3	Fibroblast	Mouse	—
COS-7	Fibroblast	Monkey	—
BHK	Fibroblast	Hamster	—
HeLa	Epithelial	Human	—
CHO	Epithelial-like	Hamster	Na^+ current and Ca^{2+}-activated K^+ current[b]
HEK293	Kidney transformed	Human	Inward rectifier and delayed rectifier K^+ currents; Ca^{2+} current[c]
RBL-1	Lymphoblast	Rat	Inward-rectifying K current[d]

[a] Uebele *et al.*, 1996.

[b] Lalik *et al.*, 1993; Skryma *et al.*, 1994.

[c] Berjukow *et al.*, 1996; Ammala *et al.*, 1996; Yu and Kerchner, 1998.

[d] Lindau and Fernandez, 1986.

and 4-aminopyridine (4-AP)-sensitive delayed rectifier K^+ current (Yu and Kerchner, 1998). In contrast, little or no K^+ currents are present in CHO cells (Yu and Kerchner, 1998), but they express endogenous Na^+ channels (Skryma *et al.* 1994; Lalik *et al.*, 1993). It has been reported that CHO cells also express Ca^{2+} channels and a charybdotoxin (CTX)-sensitive and Ca^{2+}-sensitive K^+ conductance (Skryma *et al.*, 1994).

2.2.1. Transfection of Mammalian Cell Lines

The process of transfection consists of introducing a foreign gene or transgene into a host cell using a plasmid vector in which the cDNA of interest has been cloned. Many variations of commercially available plasmids used as expression vectors usually contain early promoter/enhancer sequences for a high level of transcription of the cDNA, a polyadenylation signal, and transcription termination sequences to enhance RNA stability (Fig. 3). Dominant selectable markers, genes encoding resistance to some drugs, such as the Tn5 aminoglycoside phosphotransferase (*neo*^r) gene that confers resistance to neomycin, kanamycin, and similar compounds such as G418, are necessary for stable transfection. Plasmids with multiple expression cassettes have been constructed and used for coexpression of different channel subunits (Ahring *et al.*, 1997).

Cells can be transiently or stably transfected (Fig. 3). In transient transfection, the plasmid is taken up by the nucleus but not incorporated in the chromosome. In this case, only a fraction of the cells express the channel, and this feature is lost over succeeding generations as the plasmids are not replicated when the cells divide. During the process of transfection, a very low proportion of cells incorporate the DNA into their chromosomes. If the vector used contains a selectable marker gene, then these cells can be selected for by using a growth medium containing the appropriate drug. Only the cells with the marker gene that confers resistance to the drug used will survive, thus leading to the establishment of a stable cell line in which all the cells and their descendants express the channel.

The most commonly used types of transfection are all described in Fig. 4 (Claudio, 1992). Many mammalian transfection kits are now commercially available. The

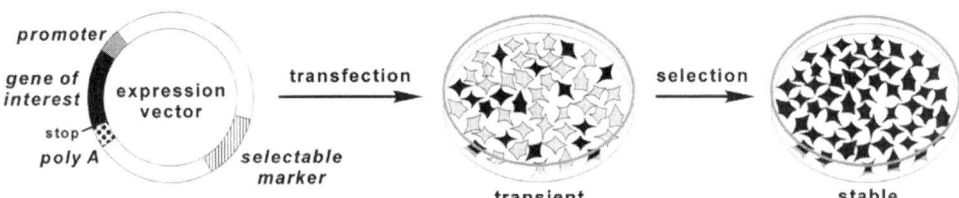

Figure 3. Expression of gene of interest in mammalian cell lines. The gene of interest is subcloned into an expression vector that contains a promoter/enhancer, a stop codon, a polyadenylation site, and selectable markers (necessary for stable transfection). During the first few days after transfection, only a subfraction of the cells carry the gene of interest (black cells; transient transfection). In a few cells, the gene of interest is incorporated in the chromosome. These cells survive the application of the appropriate drug against which the marker gene confers resistance (selection). The cells that do not integrate the transfected DNA are killed by this drug, and only transfected cells propagate in culture (stable transfection).

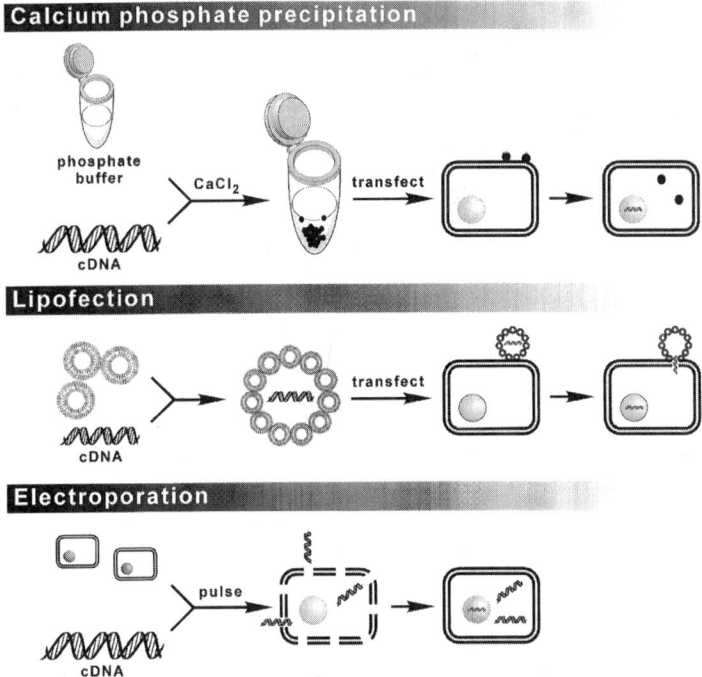

Figure 4. Methods used to introduce a foreign DNA into the host cell. *Calcium phosphate precipitation:* The DNA of interest, subcloned into an expression vector, is mixed with a buffered solution containing phosphate. The addition of calcium chloride forms a fine precipitate of calcium phosphate and DNA. This precipitate attaches to the cell surface and is then endocytosed by an unknown mechanism. *Lipofection:* The positively charged cationic liposomes bind with the negatively charged DNA. This complex binds to the cell surface, and the DNA is introduced into the cytoplasm, presumably via endocytosis. *Electroporation:* The DNA of interest enters the host cell through holes in the cell membrane. These holes are produced by exposing the cells to a high-voltage electric pulse, whose intensity depends on the cell type (usually 1.5 kV at 25 μF).

transfection using the calcium phosphate precipitation technique consists of introducing the foreign cDNA into cells in a monolayer via a precipitate that adheres to the cell surface. The basic protocol uses a HEPES-buffered solution which, in the presence of $CaCl_2$, allows the formation of a cDNA-calcium phosphate precipitate that is directly layered onto the cells and the cDNA is endocytosed into the cells.

An alternative method of transfection, *lipofection*, uses cationic liposomes to introduce the plasmid DNA into the host cell. It is still not clear how this technique works. The negatively charged phosphate groups on the cDNA bind to the positively charged surface of the liposome. The residual positive charge then presumably mediates binding of the liposomes to negatively charged sialic acid residues on the cell membrane.

Another method of delivering the cDNA to the cell, *electroporation*, uses high-voltage electric shocks to open up pores in the cell membrane through which the DNA presumably enters the cell. This technique is mostly used with suspension cultures and not adherent cells (Potter, 1993).

Very efficient transfections have been obtained using viruses such as vaccina virus (Karschin *et al.*, 1992). However, the cloning of the cDNA into the vaccina virus genome and packaging into mature virions is difficult and time-consuming.

The efficiency with which the DNA is introduced into the host cell is an important consideration when choosing the type of transfection protocol. The choice of the type of transfection (transient transfection or stable transfection) depends on the type of experiments planned. Both types of transfection have, in fact, their own advantages and disadvantages. When large quantities of identical cells are necessary for biochemical and immunological studies, stable transfections are optimal. However, the long time (6–8 weeks) required to establish a new cell line has to be considered. Stably transfected cells are an unlimited source of the channel of interest, and they provide a consistent and reproducible expression.

Transient transfections can be obtained in 1–2 days, with the limitation that only a small percentage of cells are actually transfected. This inconvenience is now overcome by the wide use of transfection markers, such as the *green fluorescent protein* (GFP). GFP is a 26.8-kDa protein from *Aequorea victoria* jellyfish that contains a fluorochrome that emits green light when excited with light in the blue range of the spectrum (Chalfie *et al.*, 1994). Formation of the fluorochrome occurs posttranslationally by spontaneous cyclization of a Ser-Tyr-Gly tripeptide, followed by oxidation (Heim *et al.*, 1994). Mutations of (or near) the tripeptide improve fluorescence and may also change the fluorescence properties (Heim and Tsien, 1996). Because no additional cofactors or substrates are required for visualization, GFP is a very suitable marker of transfected cells. Cells can be cotransfected with two vectors, one expressing the ion channel of interest and the other GFP; the intensity of fluorescence may correlate with the level of expression of functional ion channels (Fig. 5; Marshall *et al.*, 1995). The fraction of fluorescent cells that express the ion channel depends on the efficiency of cotransfection. In order to increase the efficiency of identifying the transfected cells, *fusion proteins* between the channel and GFP have been used (Marshall, 1995; John *et al.*, 1998). Fluorescently tagged ion channels also are used to study the mobility of channels in plasma membranes (Levitan, 1999). The limitation in using fusion proteins resides in the possibility that the attached GFP might affect the biosynthesis, stability, or function of the ion channel itself. Recently, a *bicistronic expression system* that allows simultaneous expression of a K^+ channel and red-shifted GFP as separate proteins has been proposed as a more accurate transfection method, having 100% efficiency in selecting cells that express the transfected cDNA (Trouet *et al.*, 1997). New expression vectors that allow the simultaneous and separate expression of the channel protein and GFP are now commercially available (pTracer vectors, Invitrogen, Carlsbad, California).

2.2.2. *Electrophysiological Recordings of Transfected Cells*

Standard patch-clamp recording configurations are used to perform electrophysiological experiments in transfected mammalian cells (Fig. 6; Hamill *et al.*, 1981). The whole-cell configuration system allows for the recording of voltage or current changes from the membrane of the whole cell, with control of the internal composition because the cytosol rapidly equilibrates with the solution in the electrode. This technique also allows for the introduction into the cytosol of a substance of

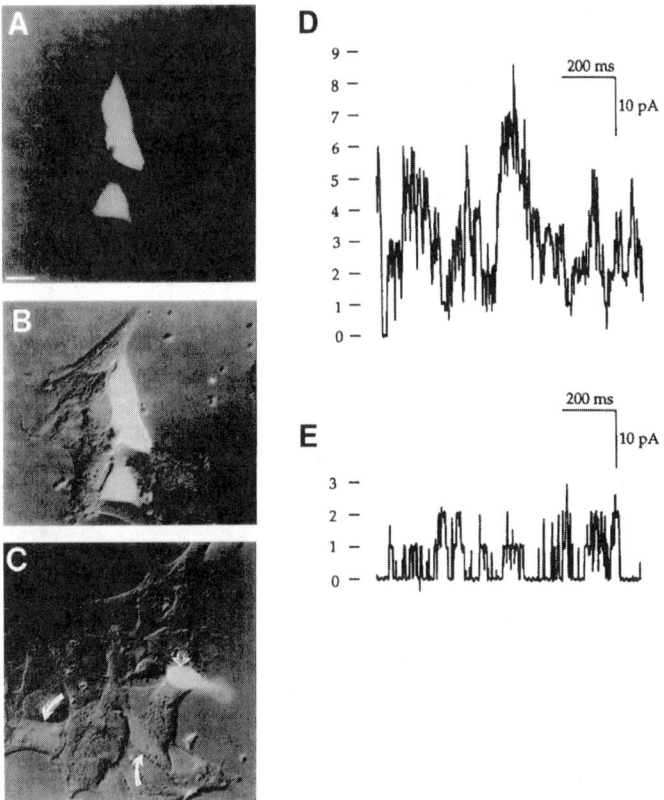

Figure 5. Cotransfection of HEK293 cells with plasmids encoding green fluorescent protein (GFP) and the Ca^{2+}-activated K^+ channel bSlo. Cotransfected cells are characterized by a strong green fluorescence (A and B). Only a fraction of cells are cotransfected (B and C). There are differences in the intensity of the signal: bright fluorescence (C, hollow arrow) and dim fluorescence (C, solid arrows). (D) and (E:) Inside-out patch-clamp recordings from HEK293 cells cotransfected with plasmids encoding GFP and the Ca^{2+}-activated K^+ channel bSlo. The currents shown in (D) were recorded in a patch taken from an intensely fluorescent cell, and the currents in (E) were recorded in a patch taken from a dimly fluorescent cell. Currents were recorded at $+20$ mV in symmetrical KCl solutions. Current levels indicative of the number of channels open simultaneously are shown at the left of each record. (Reprinted with permission, from Marshall *et al.*, 1995. Copyright 1995 Cell Press.)

interest present in the pipette; in a more sophisticated version, a perfusion patch pipette is used to allow recording of voltage or current during a control period before the substance is introduced (Ohya and Sperelakis, 1988). The whole-cell configuration can be used in the *voltage-clamp mode*, in which the membrane potential is held constant at some desired level and the current required to maintain the selected voltage is measured. The same configuration is also used in the *current-clamp mode*. In this mode, the membrane current is controlled (often at zero) while the membrane potential is recorded, so that it is possible to study changes in membrane potential in response to selected stimuli.

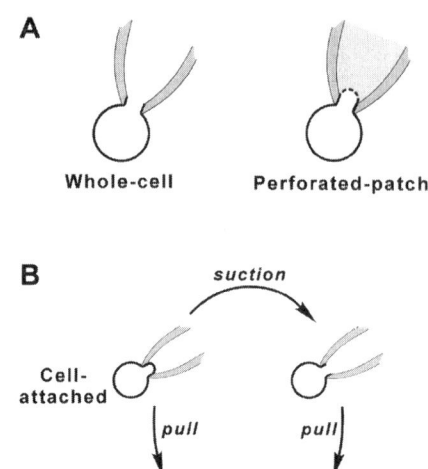

Figure 6. Recording configurations in single cells. (A) Whole-cell and perforated-patch configurations used to record whole-cell current. (B) Single-channel recording configurations. In the cell-attached configuration, the gigaseal allows recording of current flowing through a single or a few channels in the patch. From this configuration, rapid withdrawal of the patch pipette leads to the formation of an inside-out patch. In this configuration, the inside portion of the channel is exposed to the bath solution. To obtain an outside-out patch, the cell-attached patch must be broken by applying negative pressure. After obtaining the whole-cell configuration, the pipette is pulled back until the membrane seals in the outside-out configuration. In this configuration, the outer portion of the channel is exposed to the bath solution.

An alternative method of recording from the whole cell is provided by the perforated-patch configuration (Rae *et al.*, 1991; Wang and Large, 1991). To obtain this configuration, a pore-forming antibiotic (amphotericin or nystatin) is added to the pipette solution. This antibiotic makes the membrane patch permeable to ions but not to large proteins. Therefore, most of the soluble cytoplasmic constituents do not diffuse out of the cell into the patch pipette. Because of the low resistance to current flow across the patch, the whole cell can be voltage-clamped. Alternatively, the whole cell may be studied in the current-clamp mode to monitor changes in membrane potential in cytosol-intact cells.

In both whole-cell voltage-clamp and perforated-patch voltage-clamp configurations, it is possible to record the total current that flows through all the ion channels in the entire cell membrane. Different configurations can be used to record single-channel activity. All these configurations allow voltage clamp of a single channel in the patch and thus permit recording of the unitary current flowing through it. The different patch configurations and the techniques to obtain them are described in Fig. 6B. The cell-attached configuration consists of forming a tight seal (gigaseal) between the plasma membrane and the pipette. This is done by pushing the pipette gently onto the cell membrane and then applying light negative pressure. In the cell-attached patch configuration, the intracellular environment is maintained intact. The inside-out configuration is obtained by abruptly withdrawing the patch pipette from the cell-attached configuration. To obtain an outside-out patch, the cell-attached patch must be broken by applying negative pressure. From this whole-cell configuration, the pipette is pulled back until the membrane seals in the outside-out configuration. In excised patches, such as inside-out and outside-out configurations, the channel is studied independently of many soluble cytoplasmic factors. Isolated patches have the advantage of allowing exposure of the intracellular or extracellular surface of the channel to different conditions. A similar advantage is shared by the open cell-attached patch configuration (Yokoshiki et al., 1997). This is obtained by mechanically disrupting one end of the cell using a glass electrode (containing the bath solution) while in the cell-attached configuration. This technique allows one to minimize the cytoskeletal damage caused by patch excision and to alter the composition of the cytoplasm.

2.2.3. Advantages of Mammalian Cell Lines

Posttranslational modifications occurring in mammalian cells are different than in Xenopus oocytes (Shi et al., 1994). Control of intracellular medium in the whole-cell configuration and single-channel analysis is more conveniently addressed. Furthermore, intracellular transduction cascades have been well characterized in studies of mammalian cells. Finally, stable lines of transfected cells can be obtained, which are useful when a large number of cells is needed (e.g., for biochemical studies).

2.2.4. Limitations of Mammalian Cell Lines

In transient transfection, only a fraction of cells will pick up the expression vector and express the ion channels. The selection of transfected cells is time-consuming, although this is less problematic if one uses GFP as a detection tool. The presence of endogenous ion channels (Table 2) should be taken into account when choosing a mammalian cell line for expression of the desired channel. The possibility that unknown endogenous Kv pore-forming α- and auxiliary β-subunits or posttranslational processing in heterologous expression systems might influence the behavior of the expressed channel should always be taken into account (Petersen and Nerbonne, 1999).

3. SUMMARY

The introduction of two expression systems, Xenopus oocytes and mammalian cell lines, has been of extraordinary importance to the discovery of the function and molecular properties of ion channels. In fact, these systems allow expression of functional channel proteins of known amino acid sequence. The activity of the channels expressed in Xenopus oocytes can be studied with two-electrode voltage-clamp and patch-clamp techniques. The electrophysiological techniques used to record from trasfected host cells include whole-cell voltage clamp and current clamp, perforated-patch whole-cell voltage clamp, cell-attached patch, open cell-attached patch, and isolated inside-out or outside-out patch.

Electrophysiological techniques applied to heterologous expression systems are now commonly used in most laboratories interested in studying the behavior of ion channels. In the last two years, for example, approximately 300 manuscripts that utilize the *Xenopus* oocyte expression system have been published. Because of the complexity of native K^+ channels, their heteromultimeric nature, and the expression of multiple channels within the same cell, the need for a simplified system is still very great. The comparison of information obtained from expressed channels and from native channels will continue to provide new insights into the molecular nature and function of ion channels in their native environments.

APPENDIX

The solutions mentioned in the text have the following compositions:

OR2: 82.5 mM NaCl, 2 mM KCl, 1 mM MgCl$_2$, 5 mM HEPES-Tris; pH 7.5.
Barth's solution: 84 mM NaCl, 1 mM KCl, 2.4 mM NaHCO$_3$, 0.82 mM MgSO$_4$, 0.33 mM Ca(NO$_3$)$_2$, 0.41 mM CaCl$_2$ 7.5 mM Tris-HCl; pH 7.5.
Normal frog Ringer solution: 115 mM NaCl, 2.5 mM KCl, 1.8 mM CaCl$_2$, 10 mM HEPES; pH 7.2.
L-15 medium: 70% Leibovitz's L-15 medium, 10 mM HEPES; pH 7.5.
ND96: 96 mM NaCl, 2 mM KCl, 1.8 mM CaCl$_2$, 1 mM MgCl$_2$, 5 mM HEPES; pH 7.5.
Hyperosmotic medium: 200 mM potassium aspartate, 20 mM KCl, 1.0 mM MgCl$_2$, 5 mM EGTA, 10 mM HEPES; pH 7.3.

REFERENCES

Ahring, P. K., Strobaek, D., Christophersen, P., Olesen, S. P., and Johansen, T. E., 1997, Stable expression of the human large-conductance Ca^{2+}-activated K$^+$ channel alpha- and beta-subunits in HEK293 cells, *FEBS Lett.* **415**:67–70.

Ammala, C., Moorhouse, A., and Ashcroft, F. M., 1996, The sulphonylurea receptor confers diazoxide sensitivity on the inwardly rectifying K$^+$ channel Kir6.1 expressed in human embryonic kidney cells, *J. Physiol. (London)* **494**:709–714.

Berjukow, S., Doring, F., Froschmayr, M., Grabner, M., Glossman, H., and Hering, S., 1996, Endogenous calcium channels in human embryonic kidney (HEK293) cells, *Br. J. Pharmacol.* **118**:748–754.

Chabala, L. D., Bakry, N., and Covarrubias, M., 1993, Low molecular weight poly(A)$^+$ mRNA species encode factors that modulate gating of a non-Shaker A-type K$^+$ channel, *J. Gen. Physiol.* **102**:713–728.

Chalfie, M., Euskirchen, G., Ward, W. W., and Prasher, D. C., 1994, Green fluorescent protein as a marker for gene expression, *Science* **263**:802–805.

Choe, H., and Sackin, H., 1997, Improved preparation of *Xenopus* oocytes for patch-clamp recording, *Pflügers Arch.* **433**:648–652.

Chomczynski, P., and Sacchi, N., 1987, Single step method for RNA isolation by acid guanidinium thiocyanate–phenol–chloroform extraction, *Anal. Biochem.* **162**:156–159.

Claudio, T, 1992, Stable expression of heterologous multisubunit protein complexes established by calcium phosphate- or lipid-mediated co-transfection, *Methods Enzymol.* **207**:391–408.

Conforti, L., and Millhorn, D. E., 1997, Selective inhibition of a slow-inactivating voltage-dependent K$^+$ channel in rat PC12 cells by hypoxia, *J. Physiol. (London)* **502**:293–305.

Covarrubias, M., Wei, A., and Salkoff, L., 1991, *Shaker, Shal, Shab* and *Shaw* express independent K$^+$ current systems, *Neuron* **5**:847–856.

Dumont, J. N., 1972, Oogenesis in *Xenopus laevis* (Daudin). I. Stages of oocyte development in laboratory maintained animals, *J. Morphol.* **136:**153–179.

Frech, G. C., VanDongen, A. M. J., Schuster, G., Brown, A. M., and Joho, R. H., 1989, A novel potassium channel with delayed rectifier properties isolated from rat brain by expression cloning, *Nature* **340:**642–645.

Goldin, A. L., 1992, Maintenance of *Xenopus laevis* and oocyte injection, *Methods Enzymol.* **207:**266–297.

Grissmer, S., Nguyen, A. N., Aiyar J., Hanson, D. C., Mather, R. J., Gutman, G. A., Karmilowicz, M. J., Auperin, D. A., and Chandy, K. G., 1994, Pharmacological characterization of five cloned voltage-gated K^+ channels, type Kv1.1, 1.2, 1.2, 1.5 and 3.1, stably expressed in mammalian cell lines, *Mol. Pharmacol.* **45:**1227–1234.

Hamill, O. P., Marty, A., Neher, E., Sakmann, B., and Sigworth, F. J., 1981, Improved patch-clamp techniques for high-resolution current recording from cells and cell-free membrane patches, *Pflügers Arch.* **391:**85–100.

Hart, P. J., Overturf, K. E., Russel, S. N., Carl, A., Hume, J. R., Sanders, K. M., and Horowitz, B., 1993, Cloning and expression of a Kv1.2 class delayed rectifier K^+ channel from canine smooth muscle, *Proc. Natl. Acad. Sci. U.S.A.* **90:**9659–9663.

Heim, R., and Tsien, R. Y., 1996, Engineering green fluorescent protein for improved brightness, longer wavelengths and fluorescence resonance energy transfer, *Curr. Biol.* **6(2):**178–182.

Heim, R., Prasher, D. C., and Tsien, R. Y., 1994, Wavelength mutations and posttranslational autoxidation of green fluorescent protein, *Proc. Natl. Acad. Sci. U.S.A.* **91:**12501–12504.

Heinemann, S. H., Rettig, J., Graack, H.-R., and Pongs, O., 1996, Functional characterization of Kv channel β-subunits from rat brain, *J. Physiol. (London)* **493:**625–633.

Hilgemann, D. W., 1995, The giant membrane patch, in: *Single-Channel Recording*, 2nd Ed. (B. Sakmann and E. Neher, eds.), Plenum Press, New York, pp. 307–327.

Hille, B., 1992, Potassium channels and chloride channels, in: *Ion Channels of Excitable Membranes*, 2nd ed. Sinauer Associates, Sunderland, Massachusetts, pp. 115–139.

Honore, E., Attali, B., Romey, G., Lesage, F., Barhanin, J., and Lazdunski, M., 1992, Different types of K^+ channel current are generated by different levels of a single mRNA, *EMBO J.* **11:**2465–2471.

Hopkins, W. F., 1998, Toxin and subunit specificity of blocking affinity of three peptide toxins for heteromultimeric, voltage-gated potassium channels expressed in *Xenopus* oocytes, *J. Pharmacol. Exp. Ther.* **285:**1051–1060.

Ikeda, S. R., Soler, F., Zuhlke, R. D., Joho, R. H., and Lewis, D. L., 1992, Heterologous expression of the human potassium channel Kv2.1 in clonal mammalian cells by direct cytoplasmic microinjection of cRNA, *Nature* **422:**201–203.

Jan, L. Y., and Jan, Y. N., 1992, Structural elements involved in specific K^+ channel functions, *Annu. Rev. Physiol.* **54:**537–555.

John, S. A., Monck, J. R., Weiss, J. N., and Ribalet, B., 1998, The sulphonylurea receptor SUR1 regulates ATP-sensitive mouse Kir6.2 K^+ channels linked to the green fluorescent protein in human embryonic kidney cells (HEK 293), *J. Physiol. (London)* **510:**333–345.

Karschin, A., Thorne, B. A., Thomas, G., and Lester, H. A., 1992, Vaccina virus vector to express ion channel genes, *Methods Enzymol.* **207:**408–423.

Kramer, R. H., 1990, Patch cramming: Monitoring intracellular messengers in intact cells with membrane patches containing detector ion channels, *Neuron* **2:**335–341.

Lalik, P. H., Krafte, D. S., Volberg, W. A., and Ciccarelli, R. B., 1993, Characterization of endogenous sodium channels gene expressed in Chinese hamster ovary cells, *Am. J. Physiol.* **264:**C803–C809.

Levitan, B., 1999, Tagging potassium channels with fluorescent protein to study mobility and interactions with other proteins, *Methods Enzymol.* **294:**47–59.

Li, M., Jan Y. N., and Jan, L. Y., 1992, Specification of subunit assembly by the hydrophilic amino-terminal domain of the Shaker potassium channel, *Science* **257:**1225–1230.

Lindau, M., and Fernandez, J. M., 1986, Patch-clamp study of histamine secreting cells, *J. Gen. Physiol.* **88:**349–368.

Lory, P., Rassendren, F. A., Richard, S., Tiaho, F., and Nargeot, J., 1990, Characterization of voltage-dependent calcium channels expressed in *Xenopus* oocytes injected with mRNA from rat heart, *J. Physiol. (London)* **429:**95–112.

Lu, L., Montrose-Rafizadeh, C., Hwang, T. C., and Guggino, W. B., 1990, A delayed rectifier potassium current in *Xenopus* oocytes, *Biophys. J.* **57:**1117–1123.

Lupu-Meiri, M., Shapira, H., and Oron, Y., 1988, Hemispheric asymmetry of rapid chloride responses to inositol triphosphate and calcium in *Xenopus* oocytes, *FEBS Lett.* **240**:8387.

Marshall, J., Molloy, R., Moss, G. W. J., Howe, J. R. and Hughes, T. E., 1995, The jellyfish green fluorescent protein: A new tool for studying ion channel expression and function, *Neuron*, **14**:211–215.

McBride, D. W., and Hamill, O. P., 1992, Pressure-clamp: A method for rapid step perturbation of mechanosensitive channels, *Pflügers Arch.* **412**:606–612.

McCormack, K., McKormac, T., Tanouye, M., Rudy, B., and Stuhmer, W., 1995, Alternative splicing of the human *Shaker* K$^+$ channel β1 gene and functional expression of the β2 gene product, *FEBS Lett.* **370**:32–36.

McDonald, J. C., Adelman, J. P., Douglass J., and North, A. L., 1989, Expression of a cloned rat brain potassium channel in *Xenopus* oocytes, *Science* **24**:221–224.

Methfessel, C., Witzemann, V., Takahashi, T., Mishina, M., Numa, S., and Sakmann, B., 1986, Patch clamp measurements on *Xenopus laevis* oocytes: Currents through endogenous channels and implanted acetylcholine receptor and sodium channels, *Pflügers Arch.* **407**:577–588.

Miledi, R., and Woodward, R. M., 1989, Effects of defolliculation on membrane current responses of *Xenopus* oocytes, *J. Physiol. (London)* **416**:601–621.

Nagaya, N., and Papazian, D. M., 1997, Potassium channel α and β subunits assemble in the endoplasmic reticulum, *J. Biol. Chem.* **272**:3022–3027.

Ohya, Y., and Sperelakis N., 1988, Whole-cell voltage clamp and intracellular perfusion technique on single smooth muscle cells, *Mol. Cell. Biochem.* **80**:79–86.

Oron, Y., Gillo, B., and Gershengorn, M. C.,1988, Differences in receptor-evoked membrane electrical responses in native and mRNA-injected *Xenopus* oocytes, *Proc Natl Acad Sci U.S.A.* **85**:3820–3824.

Parekh, A. B., Terlau, H., and Stuhmer, W., 1993, Depletion of InsP$_3$ stores activates a Ca^{2+} and K$^+$ current by means of a phosphatase and a diffusible messenger, *Nature* **364**:814–818.

Penner, R., 1995, A practical guide to patch clamping, in: *Single Channel Recording*, 2nd ed. (B. Sakmann and E. Neher, eds.). Plenum Press, New York, pp. 3–30.

Perozo, E., Papazian, D. M., Stefani, E., and Bezanilla, F., 1992, Gating currents in *Shaker* K$^+$ channels: Implications for activation and inactivation models, *Biophys. J.* **62**:160–168.

Peter, A. B., Schittny, J. C., Niggli, V., Reuter, H., and Segel, E., 1991, The polarized distribution of poly(A$^+$)-mRNA-induced functional ion channels in the *Xenopus* oocyte plasma membrane is prevented by anticytoskeletal drugs, *J. Cell. Biol.* **114**:455–464.

Petersen, K. R., and Nerbonne, J. M., 1999, Expression environment determines K$^+$ current properties: Kv1 and Kv4 alpha-subunit-induced K$^+$ currents in mammalian cell lines and cardiac myocytes, *Pflügers Arch.* **437**:381–392.

Potter, H., 1993, Application of electroporation in recombinant DNA technology, *Methods Enzymol.* **217**:461–478.

Rae, J., Cooper, K., Gates, P., and Watsky, M., 1991, Low access resistance perforated patch recordings using amphotericin B, *J. Neurosci. Methods* **37(1)**:15–26.

Rudy, B., 1988, Diversity and ubiquity of K channels, *Neuroscience* **25**:729–749.

Rudy, B., Hoger, J. H., Lester, H. A., and Davidson, N., 1988, At least two mRNA species contribute to the properties of rat brain A-type potassium channels expressed in *Xenopus* oocytes, *Neuron* **1**:649–658.

Ruppersberg, J. P., Schroter, K. H., Sakmann, B., Stocker, M., Sewing, S., and Pongs, O., 1990, Heteromultimeric channels formed by rat brain potassium-channel proteins, *Nature* **345**:535–537.

Shamotienko, O. G., Parcej, D. N., and Dolly, O., 1997, Subunit combinations defined for K$^+$ channel Kv1 subtypes in synaptic membranes from bovine brain, *Biochemistry* **36**: 8195–8201.

Sheng, M., Liao, Y. J., Jan, Y. N., and Jan, L. Y., 1993, Presynaptic A-current based on heteromultimeric K$^+$ channels detected *in vivo*, *Nature* **365**:72–75.

Shi, G., Kleinklaus, A. K., Marrion, N. V., and Trimmer, J. S., 1994, Properties of Kv2.1 K$^+$ channels expressed in transfected mammalian cells, *J. Biol. Chem.* **269**:23204–23211.

Shi, G., Nakahira, K., Hammond, S., Rhodes, K. J., Schechter, L. E., and Trimmer, J. S., 1996, β Subunits promote K$^+$ channel surface expression through effects early in biosynthesis, *Neuron* **16**:843–852.

Skryma, R., Prevarskaya, N., Vacher, P., and Dufy, B., 1994, Voltage-dependent ionic conductances in Chinese hamster ovary cells, *Am. J. Physiol.* **267**:C544–C553.

Sobko, A., Peretz, A., Shirihai, O., Etkin, S., Cherepanova, V., Dagan, D., and Attali, B., 1998, Heteromultimeric delayed-rectifier K$^+$ channels in Schwann cells: Developmental expression and role in cell proliferation, *J. Neurosci.* **18**:10398–10408.

Stuhmer, W., 1998, Recordings from *Xenopus* oocytes, *Methods Enzymol.* **293**:280–300.

Stuhmer, W., Methfessel, C., Sakmann, B., Noda, M., and, Numa, S., 1987, Patch clamp characterization of sodium channels expressed from rat brain cDNA, *Eur. Biophys. J.* **14**:131–138.

Thornhill, W. B., Wu, M. B., Jiang, X., Wu, X., and Morgan, P. T., 1996, Expression of Kv1.1 delayed rectifier potassium channels in Lec mutant Chinese hamster ovary cell lines reveals a role for sialidation in channel function, *J. Biol. Chem.* **271**:19093–19098.

Trimmer, J. S., 1998, Analysis of K^+ channel biosynthesis and assembly in transfected mammalian cells, *Methods Enzymol.* **293**:32–49.

Trouet, D., Nilius, B., Voets, G. D., and Eggermont, J., 1997, Use of a bicistronic GFP-expression vector to characterize ion channels after transfection in mammalian cells, *Pflügers Arch.* **434**:632–638.

Tzounopoulos, T., Maylie, J., and Adelman, J.P., 1995, Induction of endogenous channels by high levels of heterologous membrane proteins in *Xenopus* oocytes, *Biophys. J.* **69**:904–908.

Uebele, V. N., England, S. K., Chaudhary, A., Tamkun, M. M., and Snyders, D. J., 1996, Functional differences in Kv1.5 currents expressed in mammalian cell lines are due to the presence of endogenous Kvβ2.1 subunits, *J. Biol. Chem.* **271**:2406–2412.

Wang, Q., and Large, W. A., 1991, Noradrenaline-evoked cation conductance recorded with the nystatin whole-cell method in rabbit portal vein cells, *J. Physiol. (London)* **435**:21–39.

Wang, H., Kunkel, D. D., Martin, T. M., Schwartzkroin, P. A., and Temple, B. L., 1993, Heteromultimeric K^+ channels in terminal and juxtaparanodal regions of neurons, *Nature* **365**:75–79.

Wischnitzer, S., 1966, The ultrastructure of the cytoplasm of the developing amphibian egg, in: *Advances in Morphogenesis*, Vol. 5 (M. Abercrombie and J. Brachet, eds.), Academic Press, New York. pp 131–179.

Yang, X. C., and Sachs, F., 1989, Block of stretch-activated ion channels in *Xenopus* oocytes by gadolinium and calcium ions, *Science* **243**:1068–1071.

Yokoshiki H., Katsube Y., Sunugawa M., Seki T., and Sperelakis N., 1997, Disruption of actin cytoskeleton attenuates sulfonylurea inhibition of cardiac ATP-sensitive K^+ channels, *Pflügers Arch.* **434**:203–205.

Yu, S. P., and Kerchner, G. A., 1998, Endogenous voltage-gated potassium channels in human embryonic kidney (HEK293) cells, *J. Neurosci. Res.* **52**:612–617.

Chapter 10

Heteromultimer Formation in Native K$^+$ Channels

James S. Trimmer and Kenneth J. Rhodes

1. INTRODUCTION

Potassium channels in the cardiovascular system are, in all known cases, multi-subunit membrane protein complexes. Typically, multiple pore-forming and voltage-and/or ligand-sensing polytopic transmembrane α-subunits are associated with nonconducting transmembrane or cytoplasmic auxiliary subunits. Cardiac K$^+$ channel complexes can consist of up to 8–10 such polypeptide components whose presence and position within the channel complex determine the precise functional properties of the resultant channel. Unlike many well-characterized multisubunit membrane protein complexes (e.g., the muscle nicotinic acetylcholine receptor, the T-cell antigen receptor), native K$^+$ channel complexes have variable quaternary structures, such that the differential assembly of component subunits can yield channels of distinct subunit composition and function. This complexity presents an especially interesting yet challenging task to biochemists and cell biologists interested in the quaternary structure of these proteins, and in more general aspects of membrane protein biosynthesis and structure. However, K$^+$ channels offer an attractive experimental system for identifying potential novel mechanisms involved in the biosynthetic assembly and maturation of this family of proteins, which are clearly distinct from those multisubunit membrane proteins of invariant subunit composition previously studied. The analysis of the quaternary structure of native K$^+$ channels is also of obvious importance to our understanding of K$^+$ channel function, and therefore the molecular basis of electrical excitability, and for any drug discovery strategies that plan to utilize expression of cloned K$^+$ channels to reconstitute relevant K$^+$ channel protein complexes in cell lines for high-throughput screening.

James S. Trimmer ● Department of Biochemistry and Cell Biology and Institute for Cell and Developmental Biology, State University of New York, Stony Brook, New York 11794. *Kenneth J. Rhodes* ● CNS Disorders, Wyeth-Ayerst Research, Princeton, New Jersey 08543.

Potassium Channels in Cardiovascular Biology, edited by Archer and Rusch. Kluwer Academic/Plenum Publishers, New York, 2001.

Voltage-gated K^+ (Kv) channels represent one of the extreme examples of a system in which differential combinatorial assembly of component subunits is used to generate an astonishing diversity of channel function. Since the cloning 10 years ago of the *Drosophila Shaker* gene, which encodes a Kv channel α-subunit, a remarkably large number of Kv channel genes have been unearthed in mammals, with undoubtedly more to surface in the course of ongoing genome projects. In addition, an array of cytoplasmic auxiliary Kv channel β-subunits have been found that exert important and diverse effects on channel function. Identification of the combinations of α- and β-subunits used to construct native voltage-gated K^+ channel complexes is critical to both understanding their function in excitable tissue and reliably reassembling them from the recombinant component subunits.

Recent work in several laboratories has shown that K^+ channels are integral membrane, hetero-oligomeric glycoprotein complexes composed of four pore-forming and voltage-sensing α-subunits and four cytoplasmic β-subunits (Dolly and Parcej, 1996). It is clear that the differential biosynthetic association and subsequent assembly of α- and β-subunits into the resultant $\alpha_4\beta_4$ complex can be a key mechanism to generate a tremendous diversity of K^+ channel structures and, consequently, functions. The analysis of the contributions of individual K^+ channel α- and β-subunit polypeptides to Kv channel complexes in different areas of the mammalian cardiovascular system is therefore paramount to our understanding of the functions of these channels, their role in shaping excitability, and their potential for pharmacological modulation.

In order to begin to address the expression pattern and subunit composition of Kv channels in mammalian tissues, it is essential to generate reagents that allow for the selective isolation and localization of channel subunits. We and others have generated subunit-specific polyclonal (Trimmer, 1991; Sheng et al., 1992; Wang *et al.*, 1993; Scott *et al.*, 1994) and monoclonal antibodies (Muniz *et al.*, 1992; Bekele-Arcuri *et al.*, 1996) that allow for the selective isolation of the respective subunits by immunoprecipitation and for their localization using immunohistochemical approaches. Here, as a model for similar studies as yet to be undertaken on cardiac tissue, we describe the biochemical and immunohistochemical analysis of the extent and nature of differential assembly of α- and β-subunit polypeptides in mammalian brain K^+ channel complexes.

2. BIOCHEMICAL CHARACTERIZATION OF VOLTAGE-GATED K^+ CHANNEL COMPLEXES

The initial studies on the biochemistry of Kv channels were based on characterization of the acceptor molecules for the peptide neurotoxins α-dendrotoxin (α-DTX), β-bungarotoxin, and mast cell degranulating peptide (Parcej *et al.*, 1989a). These toxins block a class of fast-activating, 4-aminopyridine-sensitive K^+ channels in a variety of cells (Harvey, 1997). The availability of radiolabeled derivatives of these toxins that could be used as high-affinity probes for the corresponding acceptor complexes led to the chromatographic purification of putative K^+ channel complexes. These protein complexes were shown (Parcej *et al.*, 1989b) to be composed of a diffuse glycosylated toxin-binding component of 75–85 kDa (the α-subunit) and an associated non-toxin-binding subunit of 37–39 kDa (the β-subunit). Accurate determination of the oligomeric size of the α-DTX-binding complex revealed that these subunits were present in an $\alpha_4\beta_4$ stoichiometry (Parcej *et al.*, 1992). The fundamental characterization of the biochemical characteristics of "voltage-gated" K$^+$ channels, as reviewed by Dolly and Parcej (1996), established the groundwork for future studies on the subunit composition of neuronal K$^+$ channel complexes. The cloning of the *Drosophila Shaker* gene, and subsequent isolation of cDNAs representing the large Kv gene family in mammals (Chandy and Gutman, 1995), allowed for the generation of the critical subunit-specific immunological probes necessary for these more detailed analyses.

2.1. Biochemical Characterization of α-Subunit Heteromultimerization

The molecular cloning of cDNAs corresponding to genes homologous to the *Drosophila Shaker* gene unearthed a surprisingly large Kv α-subunit gene family in mammals. The most obvious feature of the encoded polypeptides was that they resembled one of the four internally repeated homologous domains of voltage-gated Na$^+$ and Ca^{2+} channel α- and α_1-subunits, respectively. This led to the proposal that four Kv α-subunits would assemble to form a tetrameric complex whose overall structure resembled that of the other voltage-gated cation channels. Initial heterologous expression studies focused on expressing single cRNAs and showed that homo-oligomeric assemblies of Kv α-subunits could indeed form functional channels. It was only a short time before a number of groups were able to show that expressing multiple or tandem Kv α-subunits in a single cell led to the generation of channels whose properties could only be explained by postulating the formation of heteromultimeric assemblies. Although assembly of α-subunits yielding functional heteromultimeric channels has only been reliably shown to occur between members of the Kv1 subfamily (e.g., between Kv1.1 and Kv1.4), it is assumed that Kv2, Kv3, and Kv4 α-subunits are also capable of forming heterotetramers with other members of their respective subfamilies.

Although it is clear that Kv1 α-subunits form heterotetrameric channels when overexpressed in heterologous expression systems, a critical question remained as to whether such assemblies occur in native tissue. The initial studies demonstrating the existence of such heteromultimeric assemblies of Kv α-subunits were performed in the mouse cerebellum (Wang *et al.*, 1993) and in the rat hippocampus (Sheng *et al.*, 1993) and used a reciprocal immunoprecipitation/immunoblot approach (Fig. 1) as an initial step. These studies showed that multiple Kv1 family members could coassemble into native K$^+$ channel complexes with distinct subunit compositions and distributions. Wang *et al.* (1993) showed, based on co-immunoprecipitation of Kv1.1 and Kv1.2 from detergent extracts of mouse brain membranes, that these two *Shaker*-related Kv α-subunits were α-subunit components of the same mouse brain K$^+$ channel complexes. Quantitation of yields from these co-immunoprecipitation experiments suggested that ∼30% of the total immunoprecipitable Kv1.1 pool in mouse brain was associated with Kv1.2, and vice versa. These subunits were shown to extensively colocalize in cerebellar basket cell terminals and at nodes of Ranvier of myelinated axons (see below).

Coincident with the initial studies in mouse brain, Sheng *et al.* (1993) used a powerful combination of protein biochemistry and immunocytochemistry to show that

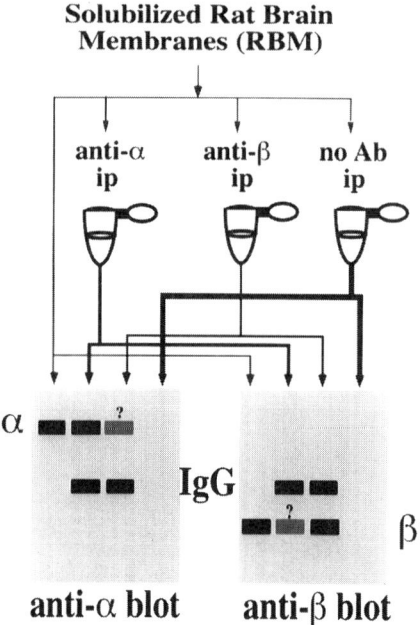

Figure 1. Schematic representation of the reciprocal immunoprecipitation/immunoblot procedure. The cartoon depicts the experimental approach used by us and others to characterize subunit composition of Kv channels in mammalian brain. In this cartoon, the α-subunit-specific immunoprecipitation reaction ("-anti-α ip") should contain α-subunit immunoreactivity on the anti-α blot; whether the same sample will also contain associated β-subunit immunoreactivity is revealed on the anti-β blot (designated by "?" on cartoon of blot). A similar strategy is used to analyze the β-subunit-specific immunoprecipitation reaction ("anti-β ip").

Kv1.2 also associated with Kv1.4 in rat brain K^+ channel complexes. Chromatographic enrichment of Kv channel complexes by diethylaminoethyl (DEAE) ion-exchange chromatography and by lectin affinity chromatography using wheat-germ agglutinin columns resulted in a cofractionation of Kv1.2 and Kv1.4; another Kv α-subunit of the *Shal* subfamily, Kv4.2, did not co-fractionate with Kv1.2 and Kv1.4. Co-immunoprecipitation experiments performed on detergent-solubilized rat brain membranes with Kv1.2, and Kv1.4-specific antibodies revealed association of Kv1.2 and Kv1.4 in hetero-oligomeric Kv channel complexes. The copurification of these two α-subunits through successive chromatographic procedures and the fact that they could be co-immunoprecipitated, taken together with immunohistochemical colocalization data, provided a compelling argument for the existence of Kv1.2 and Kv1.4 in a heteromultimeric complex.

As described briefly above, the characterization of K^+ channel protein complexes originated with studies based on their function as α-DTX-binding proteins (the Kv1.1, Kv1.2, and Kv1.6 α-subunits bind α-DTX; Stuhmer *et al.*, 1988). Such studies converged with molecular cloning studies with the direct protein microsequencing of α-DTX-binding proteins purified from bovine brain. These studies revealed that the Kv1.2 α-subunit was the major component subunit of these K^+ channel complexes (Scott *et al.*, 1990). Using antibodies against the Kv1.1, Kv1.2, Kv1.4, and Kv1.6 α-subunits in combination with α-DTX binding, Scott *et al.* (1994) were then able to show that in bovine brain, the bulk of α-DTX-binding complexes were composed of hetero-oligomeric complexes containing multiple α-subunits. The majority of α-DTX binding sites in brain contained Kv1.2, followed by Kv1.1 and, to a much lesser extent, Kv1.6 and Kv1.4.

More recently, Dolly and his colleagues (Shamotienko *et al.*, 1997) applied a novel and elegant serial multistep immunoprecipitation analysis to completely define the α-subunit composition of at least one heterotetrameric form of an (α-DTX)-binding K$^+$ channel protein complex. This complex was found to consist of one copy each of the Kv1.1, Kv1.2, Kv1.4, and Kv1.6 α-subunits. Complexes containing Kv1.4 but not Kv1.2 were also detected, as were what appeared to be Kv1.2 homotetrameric complexes.

Koch and colleagues (Koch, *et al.* 1997) followed these studies on dendrotoxin-sensitive K$^+$ channels with similar studies on those channels that bind and are blocked by the scorpion neurotoxin margatoxin. Analysis of [^{125}I] margatoxin binding sites in rat cerebellum was combined with biochemical studies employing a panel of subtype-specific antibodies (Koch *et al.*, 1997). Within the cerebellum, high-affinity [^{125}I] margatoxin binding sites were localized by high-resolution receptor autoradiography to basket cell terminals and to sites in the molecular layer. Immunoprecipitation experiments revealed that virtually all of the cerebellar [^{125}I] margatoxin binding sites contained at least one Kv1.2 subunit; the vast majority of binding sites were heteromultimers containing at least one Kv1.1 and one Kv1.2 α-subunit. A portion of the cerebellar [^{125}I] margatoxin binding sites also contained at least one Kv1.3 or one Kv1.6 subunit. Evidence that a small fraction of margatoxin receptors were present as homotetramers of Kv1.2 α-subunits was also presented. These studies showed that, a complex profile of margatoxin-sensitive K$^+$ channels are expressed in the rat cerebellum.

This same group used a similar approach to define the rat brain acceptor complexes for a novel K$^+$ channel-specific scorpion toxin, hongotoxin-1 (Koschak *et al.*, 1998). A major population of hongotoxin binding sites in brain consists of heterotetrameric channels containing only the Kv1.1 and Kv1.2 subunits, whereas another major population is formed by Kv1.1, Kv1.2, and Kv1.4. These studies combining toxin binding with subtype-specific antibodies represent a powerful approach for determining the subunit composition of pharmacologically defined voltage-gated K$^+$ channel complexes.

The studies described above have provided compelling evidence for a diversity of heteromultimeric Kv1 K$^+$ channels in mammalian brain that could contribute to the diverse electrical phenotype of central neurons. Takimoto and Levitan (1996) have provided evidence for the generation of additional diversity through dynamic hormonal regulation of Kv channel subunit composition. They showed that in unstimulated GH$_3$ pituitary cells, the Kv1.4 α-subunit forms heteromeric K$^+$ channels with the Kv1.5 α-subunit. Size fractionation of GH$_3$ K$^+$ channel complexes on Bio-gel A5m columns showed that Kv1.4 and Kv1.5 cofractionated in a single peak between 440 and 669 kDa under mild detergent conditions, in this case after solubilization in the gentle detergent CHAPS. Solubilization in the stronger detergent Zwittergent, however, disrupted the complex such that now Kv1.4 and Kv1.5 eluted from the A5m column at a position near the 158-kDa size marker, indicating that they were probably present as monomers under these conditions of detergent extraction conditions. Co-immunoprecipitation experiments provided further evidence that Kv1.4 and Kv1.5 were present in hetero-meric K^+ channel complexes. Moreover, the majority of the populations of both Kv1.4 and Kv1.5 in GH_3 cells were present in mixed channel complexes, and in association with one another. Steroid hormone treatment of GH_3 cells, which dramatically upregulates Kv1.5 gene and protein expression (Takimoto *et al.*, 1993), was also found to lead to alterations in the subunit composition of GH_3 cell K^+ channels. After hormone treatment, a large pool of K^+ channel complexes containing Kv1.5 but not Kv1.4 was observed. These results showed not only that Kv1.4 and Kv1.5 can coassemble in native K^+ channels, but also that the assembly of K^+ channel subunits, and therefore the composition of Kv channels, is dynamically regulated by external stimuli.

2.2. Biochemical Characterization of β-Subunit Involvement in Kv Channel Complexes

The recent discovery and characterization of a family of auxiliary or β-subunits (Pongs, 1995) of Kv channels indicates that these cytoplasmic polypeptides also make a significant contribution to channel diversity. In particular, the multiple β-subunit isoforms arising from the alternative splicing of Kvβ1 transcripts and the Kvβ3 β-subunit exert dramatic effects on the inactivation properties of *Shaker*-related or Kv1 subfamily K^+ channel α-subunits, whereas the Kvβ2 β-subunit does not (Pongs, 1995). All of these β-subunit isoforms may also play a role in promoting the efficient surface expression of Kv1 subfamily α-subunits (Shi *et al.*, 1996). It is clear from these data that the differential biosynthetic association and subsequent assembly of α- and β-subunits into the resultant $\alpha_4\beta_4$ complex (Shi *et al.*, 1996; Parcej *et al.*, 1992) is a key mechanism used to generate diversity of K^+ channel function and ultimately neuronal excitability.

Although it was clear from these studies that recombinant Kv1 α-subunits could assemble in heterologous expression systems with recombinant β-subunits to generate Kv channel complexes with unique biochemical and biophysical properties, a critical question remained as to the relative contribution of the auxiliary β-subunits to Kv channel complexes in native tissue. To initially address the association of Kv β-subunits with Kv α-subunits, we (Rhodes *et al.*, 1995) raised a rabbit polyclonal antibody against a synthetic peptide corresponding to a carboxyl-terminal region of the bovine Kvβ2 sequence. This antibody would be expected to recognize the Kvβ1.1, Kvβ1.2, Kvβ2, Kvβ3, and Kvβ4 polypeptides. We found that this antibody could specifically recognize four polypeptides in immunoblot analyses of rat brain membranes, a major species of 38 kDa and minor species of 41, 44, and 50 kDa (Nakahira *et al.*, 1996). Subsequent studies (Rhodes *et al.*, 1996, 1997; Shi *et al.*, 1996) with subtype-specific antibodies raised against unique sequences revealed that these polypeptides most probably represent Kvβ2 (38- and 41-kDa species), Kvβ1.1 (44-kDa species), and Kvβ1.2 (50-kDa species).

Using a reciprocal immunoprecipitation/immunoblot approach, we initially found that Kv1.2 and Kv1.4 are extensively associated with Kvβ subunits in rat brain (Rhodes *et al.*, 1995). Immunoprecipitations with the pan-β-subunit antibody yielded im-munopurified K^+ channel complexes that contained a substantial fraction of the immunoprecipitable Kv1.2 and Kv1.4, but not Kv2.1, subunits in brain. Conversely, immunoprecipitations performed with anti-Kv1.2 and anti-Kv1.4, but not anti-Kv2.1, antibodies could effectively immunopurify the bulk of the 38-kDa β-subunit that we tentatively proposed to be Kvβ2, an interpretation which has now been confirmed.

We extended these studies using subtype-specific anti-β-subunit antibodies raised against synthetic peptides corresponding to unique amino-terminal regions of the rat Kvβ1.1 and Kvβ2 primary sequences (Rhodes *et al.*, 1996). These antibodies were then used to begin to characterize the β-subunit pool in rat brain by a reciprocal immunoprecipitation/immunoblot approach (Fig. 2). We found that virtually all of the Kvβ1.1 polypeptide pool in rat brain could be co-immunoprecipitated with the Kvβ2-specific antibody, suggesting that the vast majority of Kvβ1.1-containing $\alpha_4\beta_4$ K$^+$ channel complexes also contained at least one Kvβ2 subunit. However, the reciprocal analysis showed that, by contrast, very few of the Kvβ2-containing K$^+$ channel complexes in rat brain contained Kvβ1.1. We interpreted these data, together with the immunohistochemical data described above, as suggesting that Kvβ2 is a much more abundant and widespread β-subunit in rat brain than is Kvβ1.1. Many K$^+$ channel complexes contain Kvβ2, but very few contain Kvβ1.1, and the bulk of the complexes that contain Kvβ1.1 also contain Kvβ2.

Analyses of the interaction of Kvβ1.1 and Kvβ2 with Kv1 α-subunits provided further evidence supporting this interpretation (Rhodes *et al.*, 1997). Virtually all of the immunoprecipitable pools of Kv1.1, Kv1.2, Kv1.4, and Kv1.6 in rat brain could be immunopurified using anti-Kvβ2 antibodies, showing that the vast majority of the K$^+$ channel complexes composed of these α-subunits also contain at least one Kvβ2 auxiliary subunit (Fig. 2). In contrast, only a small fraction of the Kv1.1, Kv1.2, Kv1.4, and Kv1.6 α-subunits in brain copurified with Kvβ1.1 (Fig. 2), suggesting that quantitatively Kvβ1.1 makes a much smaller contribution to brain K$^+$ channel complexes than does Kvβ2. However, as we pointed out, these data reflect only the overall bulk of all brain Kv1-containing channel complexes and in no way preclude the possibility that important pools of channels in discrete neuronal populations may be under the functional influence of a Kvβ1.1 subunit that modulates channel inactivation.

Shamotienko *et al.* (1997) used a similar approach in their work on bovine brain to show that each of the Kv1 α-subunits examined (Kv1.1–Kv1.6) could be co-immunoprecipitated with a pan-β-subunit antibody; a Kvβ2-specific antibody yielded similar results. As in rat brain, the majority of bovine brain K$^+$ channel protein complexes formed by Kv1 family members were also found to contain Kvβ2.

3. IMMUNOHISTOCHEMICAL LOCALIZATION OF Kv α- AND β-SUBUNIT POLYPEPTIDES IN MAMMALIAN BRAIN

From the above discussion, it is clear that a wealth of information about the coassociation of individual K$^+$ channel subunits in native tissues has been obtained by purification of toxin acceptor sites and by careful application of immunopurification/ co-immunoprecipitation strategies. However, it is also clear that although these approaches define the total range of subunit interactions present in a given tissue, they provide little information about which channel subunits may coassociate to generate a site-specific K$^+$ conductance within a distinct subcellular domain. Because of this, we

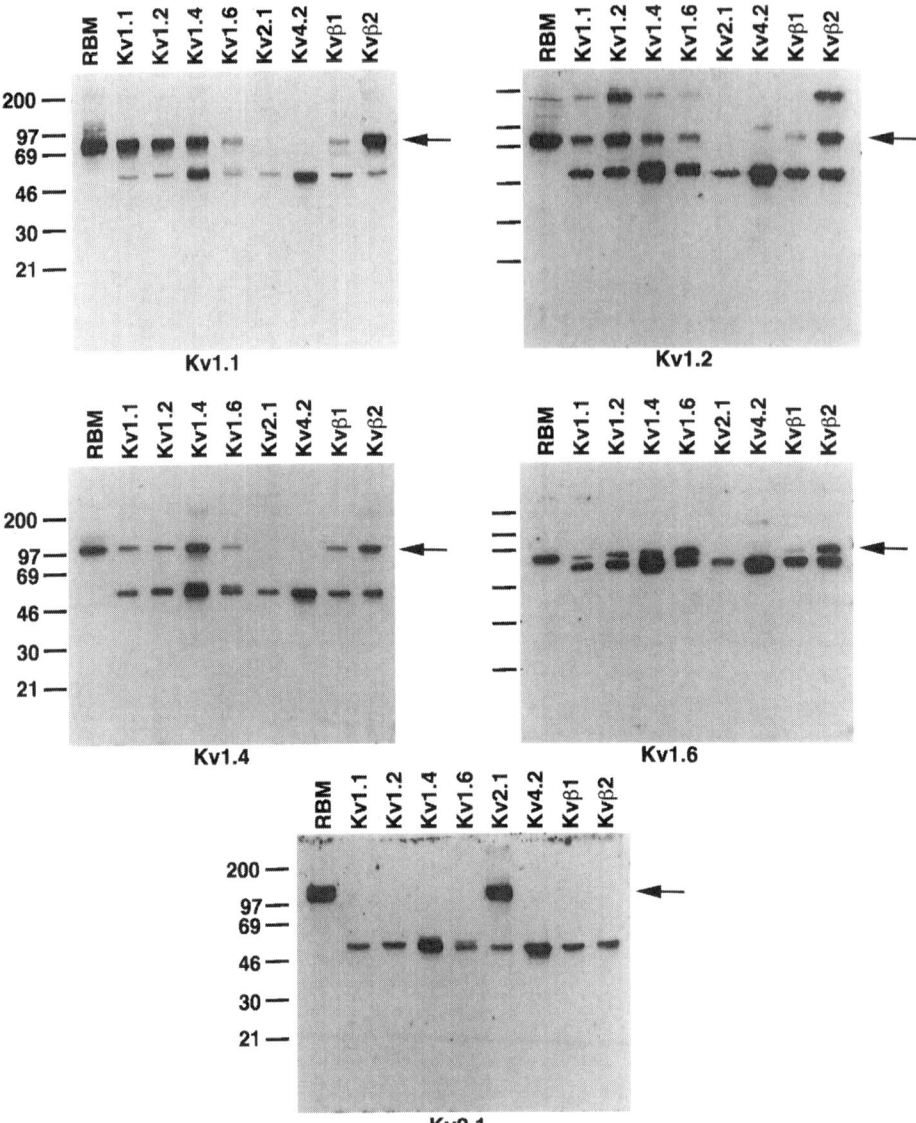

Figure 2. Co-immunoprecipitation of K$^+$ channel α- and β-subunit polypeptides. A detergent lysate of adult rat brain membranes (RBM, 60 μg) and aliquots of products of immunoprecipitation reactions from detergent extracts of 100 μg of RBM performed with polyclonal antibodies specific for the indicated K$^+$ channel α- and β-subunit polypeptides were size-fractionated by 9% (α-subunits) sodium dodecyl sulfate–polyacrylamide gel electrophoresis (SDS-PAGE). Samples were transferred to a nitrocellulose membrane and probed with subunit-specific affinity-purified rabbit antibodies. Bound antibody was detected by enhanced chemiluminescence (ECL)/autoradiography. Arrows point to the respective K$^+$ channel polypeptides; also visible are bands resulting from detection of the rabbit IgG used in the immunoprecipitation reactions by the anti-rabbit secondary antibody used for immunoblotting. Numbers at left of figure denote M_r of prestained molecular-weight standards. Note that extensive association is observed between Kv1 α-subunits and Kv β-subunits, and between different Kv1 α-subunits, whereas Kv2.1 associates with neither Kv1 α-subunits nor Kv β-subunits.

and others have used an immunohistochemical approach, employing subunit-specific antibodies to examine the discrete localization of individual channel subunits in native tissues. Here we focus our immunohistochemical analyses on Kv1 α- and β-subunits, because these subunits have been studied most extensively in terms of α-subunit localization and in terms of α/β-colocalization. Detailed immunohistochemical analyses have been published for other α-subunit families as well and are reported elsewhere for the Kv2 subfamily (Trimmer, 1991; Hwang et al., 1993; Rhodes et al., 1995, 1997; Scannevin et al., 1996; Du et al., 1998), the Kv3 subfamily (Weiser et al., 1995; Moreno et al., 1995; Du et al., 1996; Sekirnjak et al., 1997), and the Kv4 subfamily (Sheng et al., 1992, 1993; Tsaur et al., 1997).

Immunohistochemical studies of mouse (Wang et al., 1993, 1994) and rat brain (Sheng et al., 1992, 1993, 1994; Rhodes et al., 1995, 1997; Veh et al., 1995; Bekele-Arcuri et al., 1996; McNamara et al., 1996; Cooper et al., 1998; Rasband et al., 1998) have revealed that mRNAs encoding Kv1 α- and β-subunits are widely expressed in brain. Although the corresponding polypeptides are found predominantly along axons and at or near presynaptic terminals, it is clear that the subcellular distribution of these subunits varies widely across brain regions and cell types. In myelinated axons, such as those constituting many cortical callosal and commissural pathways, and in peripheral nerve, these subunits are concentrated in the axolemma immediately adjacent to the node of Ranvier (Wang et al., 1993, 1994; Rhodes et al., 1997; Rasband et al., 1998). In some nonmyelinated pathways, such as the hippocampal mossy-fiber pathway, these subunits are distributed fairly evenly along the entire axonal membrane (Sheng et al., 1992; Wang et al., 1994; Rhodes et al., 1995, 1997; Cooper et al., 1998) but, as demonstrated by the elegant electron microscopic studies of Cooper et al. (1998) and Wang et al., (1994), appear to be specifically excluded from the presynaptic bouton. In contrast, in other nonmyelinated pathways, such as the perforant path and cerebellar basket cell terminals, these subunits are found within presynaptic boutons (McNamara et al., 1993; Wang et al., 1994; Cooper et al., 1998). In certain other cell types, including a subpopulation of cortical interneurons, the Kv1.1, Kv1.4, Kv1.6, and Kvβ1 subunits are concentrated in the somata and appear to be present in proximal dendritic membranes (Rhodes et al., 1997). Although the specific subunit combinations that are present in these different pathways and subcellular domains vary, it is clear that Kv1 subunits regulate neuronal excitability in a diverse cell- and pathway-specific manner.

Some Kv β-subunits can dramatically alter the inactivation of coexpressed Kv1 α-subunits, either by further accelerating the inactivation kinetics of rapidly inactivating channels, such as those formed as homotetramers of Kv1.4 α-subunits or by converting channels with little or no inactivatiion, such as those formed from Kv1.1 subunits, into rapidly inactivating channels. Because Kvβ1.1 exerts such profound effects upon inactivation, while Kvβ2 does not, we carefully examined the codistribution and colocalization of these two β-subunits with Kv1 α-subunits using single- and multiple-label immunohistochemistry (Rhodes et al., 1995, 1997). What clearly emerged from these analyses was that Kvβ2 colocalizes with all Kv1 α-subunits studied in virtually all brain regions and subcellular domains, and the distribution of Kvβ2 most closely matches the distribution of Kv1.1 and Kv1.2. This is particularly apparent in the juxtaparanodal membrane of myelinated axons and in cerebellar basket cell terminals, where triple-label immunofluorescence revealed that there is a precise one-to-one overlap in the distribution of Kv1.1, Kv1.2, and Kvβ2 (Rhodes et al., 1997). It was

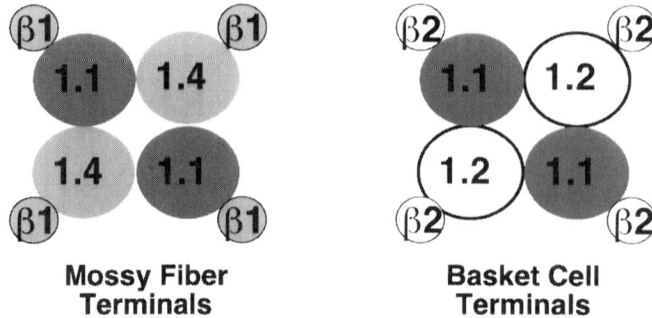

Figure 3. Subunit composition of two mammalian brain Kv channels. Cartoon depicts putative subunit composition of Kv channels revealed by immunoblot/immunoprecipitation assays and by immunohistochemical staining. Precise stoichiometry of α- and β-subunit polypeptides within the complexes is not known.

exceedingly rare to find Kvβ1.1 immunoreactivity in juxtaparanodes, and we never observed Kvβ1.1 staining in basket cell terminals. Since the typical sustained conductance profile of Kv1.1 and Kv1.2, either as homomeric or heteromeric α-subunit complexes, would not be affected by their coassociation with Kvβ2, we would predict that these native channels would form α-DTX-sensitive delayed-rectifier-type K^+ channels (Fig. 3).

Another consistent finding from these studies was that the pattern of Kvβ1.1 immunoreactivity most closely resembled the staining pattern for Kv1.4 and that the majority of structures where Kvβ1.1 and Kv1.4 were colocalized also contained Kv1.1, but not other Kv1 α-subunits. Because the inclusion of Kv1.1 in these complexes would be expected to confer sensitivity to DTX and because the presence of Kv1.4 and Kvβ1.1 would confer the property of rapid inactivation, this specific heteromeric K^+ channel complex (Kv1.1, Kv1.4, Kvβ1.1) may account for the presence of- DTX-sensitive A-type channels in brain (Fig. 3).

Although the biochemical and anatomical approaches described above have clearly helped elucidate critical features of K^+ channel subunit interactions and localization, they are limited by their inability to provide information about the relative stoichiometry of individual subunits in the channel complex. In structures where two or more α- and β-subunits are precisely colocalized, these approaches suffer the additional limitation that they cannot accurately distinguish between subunits that are physically associated and those that are simply colocalized.

4. SUMMARY

A pressing question revolving around the study of any cellular K^+ channel is what component subunits form the channel protein complex. The successful reproduction of physiologically relevant channel complexes from the heterologous expression of recombinant subunits underlies most basic and applied studies of K^+ channels. Recapitulating such channels in transfected cells hinges on expressing and assembling the appropriate subunits into what may be complex arrays consisting of multiple α- and β-subunit polypeptides. In order to rationally design and then generate such a model system, it is imperative to first characterize the subunit composition of the native channel in question. Given the lack of subtype-specific neurotoxins or small-molecule modulatory agents for many K$^+$ channels, these studies must be undertaken using an antibody-based approach. The success of this approach therefore hinges on the design, generation, purification, and characterization of the antibodies themselves and their application in controlled biochemical and immunohistochemical analyses of native channels.

Voltage-gated K$^+$ channels, and perhaps K$^+$ channels in general, represent a new challenge for membrane protein biochemists. Most multisubunit membrane proteins that have been extensively characterized in the past have had fixed subunit compositions, such as the muscle nicotinic acetylcholine receptor, the T-cell receptor antigen, and a host of hormone receptors and model viral glycoproteins. The variety of distinct K$^+$ channel complexes generated from the combinatorial assembly of subunits is theoretically enormous and raises the question of whether it will be possible to obtain a detailed stoichiometric analysis of every type of K$^+$ channel protein complex. However, characterizing the component subunits of specific K$^+$ channels, such as the Kv1.1, Kv1.2, and Kvβ2 complexes present in cerebellar basket cell terminals and at juxtaparanodal regions of myelinated axons, provides critical information necessary for further basic and applied studies of K$^+$ channels. The task is daunting but must continue if we are to truly understand mammalian K$^+$ channel structure–function relationships.

The studies described above have focused primarily on voltage-gated K$^+$ channels in the mammalian central nervous system. However, they serve as a useful template for the characterization of this family of K$^+$ channels in cardiovascular preparations where these channels regulate important physiological processes. In addition, further insights into the subunit composition and expression patterns of the other members of the broader family of K$^+$ channels, such as the KvLQT, HERG, TWIK, and GIRK families, can be revealed using similar approaches.

REFERENCES

Bekele-Arcuri, Z., Matos, M. F., Manganas, L., Strassle, B. W., Monaghan, M. M., Rhodes, K. J., and Trimmer, J. S., 1996, Generation and characterization of subtype-specific monoclonal antibodies to K$^+$ channel α- and β-subunit polypeptides, *Neuropharmacology* **35**:851–865.

Chandy, K. G., and Gutman, G. A., 1995, Voltage-gated potassium channel genes, in: *Handbook of Receptors and Channels: Ligand- and Voltage-Gated Ion Channels* (R. A. North, ed.), CRC Press, Boca Raton, Florida, pp. 1-71.

Cooper, E. C., Milroy, A., Jan, Y. N., Jan, L. Y., and Lowenstein, D. H., 1998, Presynaptic localization of Kv1.4-containing A-type potassium channels near excitatory synapses in the hippocampus, *J. Neurosci.* **18**:965–974.

Dolly, J. O., and Parcej, D. N., 1996, Molecular properties of voltage-gated K$^+$ channels, *J. Bioenerg. Biomembr.* **28**:231–253.

Du, J., Zhang, L., Weiser, M., Rudy, B., and McBain, C. J., 1996, Developmental expression and functional characterization of the potassium-channel subunit Kv3.1b in parvalbumin-containing interneurons of the rat hippocampus, *J. Neurosci.* **16**:506–518.

Du, J., Tao-Cheng, J. H., Zerfas, P., and McBain, C. J., 1998, The K$^+$ channel, Kv2.1, is apposed to astrocytic processes and is associated with inhibitory postsynaptic membranes in hippocampal and cortical principal neurons and inhibitory interneurons, *Neuroscience* **84**:37–48.

Harvey, A. L., 1997, Recent studies on dendrotoxins and potassium ion channels, *Gen. Pharmacol.* **28**:7-12.

Hwang, P. M., Cunningham, A. M., Peng, Y. W., and Snyder, S. H., 1993, CDRK and DRK1 K$^+$ channels have contrasting localizations in sensory systems, *Neuroscience* **55**:613–620.

Koch, R. O., Wanner, S. G., Koschak, A., Hanner, M., Schwarzer, C., Kaczorowski, G. J., and Slaughter, R. S., 1997, Complex subunit assembly of neuronal voltage-gated K$^+$ channels: Basis for high-affinity toxin interactions and pharmacology, *J. Biol. Chem.* **272**:27577–27581.

Koschak, A., Bugianesi, R. M., Mitterdorfer, J., Kaczorowski, G. J., Garcia, M. L., and Knaus, H. G., 1998, Subunit composition of brain voltage-gated potassium channels determined by hongotoxin-1, a novel peptide derived from *Centruroides limbatus* venom, *J. Biol. Chem.* **273**:2639–2644.

McNamara, N. M., Muniz, Z. M., Wilkin, G. P., and Dolly, J. O., 1993, Prominent location of a K$^+$ channel containing the α-subunit Kv1.2 in the basket cell nerve terminals of rat cerebellum, *Neuroscience* **57**:1039–1045.

McNamara, N. M., Averill, S., Wilkin, G. P., Dolly, J. O., and Priestley, J. V., 1996, Ultrastructural bvlocalization of a voltage-gated K$^+$ channel α-subunit (Kv1.2) in the rat cerebellum, *Eur. J. Neurosci.* **8**:688–699.

Moreno, H., Kentros, C., Bueno, E., Weiser, M., Hernandez, A., Vega-Saenz de Miera, E., Ponce, A., Thornhill, W., and Rudy, B., 1995, Thalamocortical projections have a K$^+$ channel that is phosphorylated and modulated by cAMP-dependent protein kinase, *J. Neurosci.* **15**:5486–5501.

Muniz, Z. M., Parcej, D. N., and Dolly, J. O., 1992, Characterization of monoclonal antibodies against voltage-dependent K$^+$ channels raised using α-dendrotoxin acceptors purified from bovine brain, *Biochemistry* **31**:12297–12303.

Nakahira, K., Shi, G., Rhodes, K. J., and Trimmer, J. S., 1996, Selective interaction of voltage-gated K$^+$ channel β-subunits with α-subunits, *J. Biol. Chem.* **271**:7084–7089.

Parcej, D. N., and Dolly, J. O., 1989a, Elegance persists in the purification of K$^+$ channels, *Biochem. J.* **264**:623–624.

Parcej, D. N., and Dolly, J. O., 1989b, Dendrotoxin acceptor from bovine synaptic plasma membranes: Binding properties, purification and subunit composition of a putative constituent of certain voltage-activated K$^+$ channels, *Biochem. J.* **257**:899–903.

Parcej, D. N., Scott, V. E., and Dolly, J. O., 1992, Oligomeric properties of α-dendrotoxin-sensitive potassium ion channels purified from bovine brain, *Biochemistry* **31**:11084–11088.

Pongs, O., 1995, Regulation of the activity of voltage-gated potassium channels by β-subunits, *Semin. Neurosci.* **7**:137–146.

Rasband, M. N., Trimmer, J. S., Schwarz, T. L., Levinson, S. R., Ellisman, M. H., Schachner, M., and Shrager, P., 1998, Potassium channel distribution, clustering, and function in remyelinating rat axons, *J. Neurosci.* **18**:36–47.

Rhodes, K. J., Keilbaugh, S. A., Barrezueta, N. X., Lopez, K. L., and Trimmer, J. S., 1995, Association and colocalization of K$^+$ channel α- and β-subunit polypeptides in rat brain, *J. Neurosci.* **15**:5360–5371.

Rhodes, K. J., Monaghan, M. M., Barrezueta, N. X., Nawoschik, S., Bekele-Arcuri, Z., Matos, M. F., Nakahira, K., Schechter, L. E., and Trimmer, J. S., 1996, Voltage-gated K$^+$ channel β-subunits: Expression and distribution of Kvβ1 and Kvβ2 in adult rat brain, *J. Neurosci.* **16**:4846–4860.

Rhodes, K. J., Strassle, B. W., Monaghan, M. M., Bekele-Arcuri, Z., Matos, M. F., and Trimmer, J. S., 1997, Association and colocalization of the Kvβ1 and Kvβ2 β-subunits with Kv1 α-subunits in mammalian brain K$^+$ channel complexes, *J. Neurosci.* **17**:8246–8258.

Scannevin, R. H., Murakoshi, H., Rhodes, K. J., and Trimmer, J. S., 1996, Identification of a cytoplasmic domain important in the polarized expression and clustering of the Kv2.1 K$^+$ channel, *J. Cell Biol.* **135**:1619–1632.

Scott, V. E., Parcej, D. N., Keen, J. N., Findlay, J. B., and Dolly, J. O., 1990, α-Dendrotoxin acceptor from bovine brain is a K$^+$ channel protein: Evidence from the N-terminal sequence of its larger subunit, *J. Biol. Chem.* **265**:20094–20097.

Scott, V. E., Muniz, Z. M., Sewing, S., Lichtinghagen, R., Parcej, D. N., Pongs, O., and Dolly, J. O., 1994, Antibodies specific for distinct Kv subunits unveil a heterooligomeric basis for subtypes of α-dendrotoxin-sensitive K$^+$ channels in bovine brain, *Biochemistry* **33**:1617–1623.

Sekirnjak, C., Martone, M. E., Weiser, M., Deerinck, T., Bueno, E., Rudy, B., and Ellisman, M., 1997, Subcellular localization of the K$^+$ channel subunit Kv3.1b in selected rat CNS neurons, *Brain. Res.* **766:**173–187.

Shamotienko, O. G., Parcej, D. N., and Dolly, J. O., 1997, Subunit combinations defined for K$^+$ channel Kv1 subtypes in synaptic membranes from bovine brain, *Biochemistry* **36:**8195–8201.

Sheng, M., Tsaur, M. L., Jan, Y. N., and Jan, L. Y., 1992, Subcellular segregation of two A-type K$^+$, channel proteins in rat central neurons, *Neuron* **9:**271–284.

Sheng, M., Liao, Y. J., Jan, Y. N., and Jan, L. Y., 1993, Presynaptic A-current based on heteromultimeric K$^+$ channels detected in vivo, *Nature* **365:**72–75.

Sheng, M., Tsaur, M. L., Jan, Y. N., and Jan, L. Y., 1994, Contrasting subcellular localization of the Kv1.2 K$^+$ channel subunit in different neurons of rat brain, *J. Neurosci.* **14:**2408–2417.

Shi, G., Nakahira, K., Hammond, S., Rhodes, K. J., Schechter, L. E., and Trimmer, J. S., 1996, β-Subunits promote K$^+$ channel surface expression through effects early in biosynthesis, *Neuron* **16:**843–852.

Stuhmer, W., Stocker, M., Sakmann, B., Seeburg, P., Baumann, A., Grupe, A., and Pongs, O., 1988, Potassium channels expressed from rat brain cDNA have delayed rectifier properties, *FEBS Lett.* **242:**199–206.

Takimoto, K., and Levitan, E. S., 1996, Altered K$^+$ channel subunit composition following hormone induction of Kv1.5 gene expression, *Biochemistry* **35:**14149–14156.

Takimoto, K., Fomina, A. F., Gealy, R., Trimmer, J. S., and Levitan, E. S., 1993, Dexamethasone rapidly induces Kv1.5 K$^+$ channel gene transcription and expression in clonal pituitary cells, *Neuron* **11:**359–369.

Trimmer, J. S., 1991, Immunological identification and characterization of a delayed rectifier K$^+$ channel polypeptide in rat brain, *Proc. Natl. Acad. Sci. U.S.A.* **88:**10764–10768.

Tsaur, M. L., Chou, C. C., Shih, Y. H., and Wang, H. L., 1997, Cloning, expression and CNS distribution of Kv4.3, an A-type K$^+$ channel α-subunit, *FEBS Lett.* **400:**215–220.

Veh, R.W., Lichtinghagen, R., Sewing, S., Wunder, F., Grumbach, I. M., and Pongs, O., 1995, Immunohistochemical localization of five members of the Kv1 channel subunits: Contrasting subcellular locations and neuron-specific co-localizations in rat brain, *Eur. J. Neurosci.* **7:**2189–2205.

Wang, H., Kunkel, D. D., Martin, T. M., Schwartzkroin, P. A., and Tempel, B. L., 1993, Heteromultimeric K$^+$ channels in terminal and juxtaparanodal regions of neurons, *Nature* **365:**75–79.

Wang, H., Kunkel, D. D., Schwartzkroin, P. A., and Tempel, B. L., 1994, Localization of Kv1.1 and Kv1.2, two K channel proteins, to synaptic terminals, somata, and dendrites in the mouse brain, *J. Neurosci.* **14:**4588–4599.

Weiser, M., Bueno, E., Sekirnjak, C., Martone, M. E., Baker, H., Hillman, D., Chen, S., Thornhill, W., Ellisman, M., and Rudy, B., 1995, The potassium channel subunit KV3.1b is localized to somatic and axonal membranes of specific populations of CNS neurons, *J. Neurosci.* **15:**4298–4314.

Chapter 11

Use of Transgenic and Gene-Targeted Mice to Study K⁺ Channel Function in the Cardiovascular System

Barry London

1. INTRODUCTION

A large number of K^+ channel genes are expressed in the mammalian heart (for reviews, see Chandy and Gutman, 1995; Deal *et al.*, 1996; Yost, 1999). The diversity of channel expression is enhanced further by coassembly of subunits encoded by different genes to form heteromeric channels, interaction of α-subunits with β-subunits, alternative splicing of K^+ channel genes, posttranslational channel modifications, and factors that control insertion and clustering of channels on the cell membrane (Po *et al.*, 1993; Morales *et al.*, 1995; Kim *et al.*, 1996; London *et al.*, 1997; Zhou *et al.*, 1998). This marked diversity leads to a complex array of K^+ currents within the heart that varies among species and changes during development.

The relationship of individual gene products to cardiac currents, action potential repolarization, and arrhythmias has been difficult to determine. To clarify the role of ion channel genes in physiology and disease *in vitro*, investigators have compared (a) detailed kinetic and pharmacological properties of cardiac currents to those of channel genes expressed in heterologous expression systems such as *Xenopus* oocytes and mammalian cell lines and (b) transmural gradients or developmental changes of the currents to gradients of K^+ channel gene expression (e.g., Antzelevitch *et al.*, 1991; Dixon *et al.*, 1996; Xu *et al.*, 1996). In addition, genetic manipulation of cardiac myocytes has been performed using antisense oligonucleotide and dominant-negative adenoviral strategies (Fiset *et al.*, 1997; Johns *et al.*, 1997).

Mutations of the human K^+ channel genes *KvLQT1*, *IsK*, *HERG*, and *MirP1* have been shown to disrupt the human cardiac currents I_{Kr} and I_{Ks} and cause the inherited

Barry London ● Cardiovascular Institute, University of Pittsburgh Medical Center, Pittsburgh, Pennsylvania 15213.

Potassium Channels in Cardiovascular Biology, edited by Archer and Rusch. Kluwer Academic/Plenum Publishers, New York, 2001.

long QT syndrome (Wang *et al.*, 1996; Curran *et al.*, 1995; Splawski *et al.*, 1997; Schulze-Bahr *et al.*, 1997; Abbott *et al.*, 1999). While invaluable, studies on humans are limited by the small number of affected patients, the small number of mutant genes that have been identified, the difficulty in obtaining cardiac tissue from affected hearts, and the lack of affected cardiac myocytes for *in vitro* electrophysiological analysis.

Genetic manipulation of the mouse has led to great strides toward the understanding of cardiovascular physiology (Izumo and Shioi, 1998). Mice with engineered ion channel mutations provide a powerful mechanism for clarifying the molecular basis of cardiac repolarization *in vivo*. This chapter will review the two major techniques used to alter K^+ channel gene expression in the mouse at the chromosomal level: (1) overexpression of dominant-negative channel subunits in the hearts of transgenic mice and (2) gene targeting of embryonic stem cells to eliminate ("knock out") or mutate individual ion channel genes. A list of published dominant-negative transgenic and gene-targeted mice is presented in Table 1.

2. DOMINANT-NEGATIVE TRANSGENIC MICE

Mice can be engineered to lack one or more K^+ channel gene products by using a dominant-negative transgenic strategy. This is analogous to the techniques used to study other biological systems (Lagna and Jemmati-Brivanlou, 1998). For K^+ channels, we take advantage of the fact that coassembly of four related α-subunits is required to form a functional K^+ channel (MacKinnon, 1991) and that subunits containing certain point mutations or truncations will coassemble with the wild-type subunits. Assuming that a single mutant subunit is sufficient to inhibit functional channel formation and that native and mutant subunits randomly coassemble, then the number of functional channels will be given by f^4, where f is the fraction of wild-type subunits. This means that with equal expression of wild-type and mutant subunits, only 1/16 of the channels will be functional (Fig. 1), and that a heavily expressed transgenic subunit will effectively eliminate wild-type channels.

2.1. Methods

The transgenic construct is engineered using a cardiac-specific promoter, the dominant-negative transgene, and a poly-(A) tail. The most commonly used promoter is from the rat or mouse α-myosin heavy-chain gene, because it directs expression of the transgene specifically to the heart of the adult mouse (Mahdavi *et al.*, 1989). The transgenic construct is injected directly into the pronucleus of fertilized mouse oocytes, and the oocytes are then reimplanted into pseudopregnant females. Founders (F_0 generation) are identified by screening tail DNA from live offspring using the polymerase chain reaction (PCR) or genomic Southern blots and mated with wild-type mice to generate heterozygous F_1 offspring. The F_1 mice are then examined for transgene expression by Northern blot, Western blot, and/or enzyme-linked immunosorbent assay (ELISA). Additional studies are based on the expected phenotype. Thus, only two generations of mice are required, and studies may be completed in as little as one year

Table 1

Dominant Summary of Properties of Negative Transgenic and Gene-Targeted Mice

Mouse[a]	Current	Channel(s)	Long APD[b]	Long QT	Arrhythmia[c]	Reference
Kv4.xDN	$I_{to,fast}$	Kv4.2, Kv4.3	Yes	Yes	No	Barry et al., 1998
Kv1.4KO	$I_{to,slow}$	Kv1.4	No	No	?	London et al., 1998b; Guo et al., 1999
Kv1.xDN	I_{slow}	Kv1.4, Kv1.5	Yes	Yes	PVCs, VT	London et al., 1998a
Kv1.5SWAP	I_{slow}	Kv1.5	No	No	?	London et al., 1995
Kv1.5KO	I_{slow}	Kv1.5	?	No	Drug-induced	Hill et al., 1998
Kv2.xDN	I_{slow}	Kv2.1	Yes	Yes	PVCs	Xu et al., 1999
HERGDN	I_{Kr}	Merg1	Yes	No	No	Babij et al., 1998
Merg1KO	I_{Kr}	Merg1	?	Neonatal	Drug-induced	London et al., 1998c
IsKKO	I_{Ks}	minK	?	Rate-dependent	?	Drici et al., 1998
minKLacZ	I_{Ks}	minK	?	No	?	Kupershmidt et al., 1999
GIRK4KO	I_{KACh}	GIRK4	No	No	Loss of HRV	Wickman et al., 1998

[a] DN, Dominant-negative transgenic; KO, knockout; SWAP, knockin mouse line; LacZ, knockin mouse line.

[b] APD, Action potential duration. ? indicates that the result was not reported.

[c] PVCs, Premature ventricular contractions; VT, ventricular tachycardia; HRV, heart rate variability. ? indicates that the result was not reported. Note that each study used different techniques to document spontaneous arrhythmias and/or promote arrhythmias. As such, comparisons between different lines of mice may not be valid.

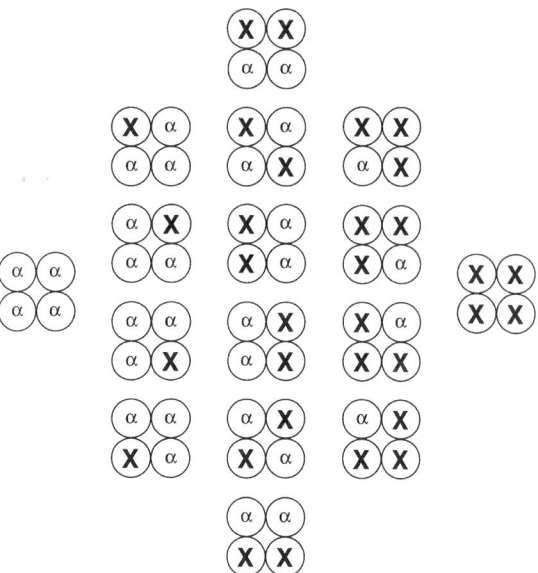

Figure 1. Schematic representation of coassembly of a mutant dominant-negative subunit (X) with a wild-type K$^+$ channel α-subunit (α). Assuming that there are equal numbers of wild-type and mutant subunits, that coassembly is random, and that one mutant subunit prevents channel function, then only one of the 16 possible subunit combinations will form a functional channel.

(Fig. 2, left). In addition, littermate controls are readily available in the F$_1$ generation.

This technique requires expression of a transgene that will function in a dominant-negative manner *in vivo*. The efficacy of the transgene is usually tested first *in vitro* in heterologous expression systems. These include (a) *Xenopus* oocytes coinjected with native and mutant subunit RNA and (b) mammalian cell lines expressing the targeted channel transfected with DNA encoding the mutant subunit driven by a mammalian promoter.

2.2. Advantages and Disadvantages

The promoter determines the expression pattern of the transgene. The α-myosin heavy-chain promoter begins to express near birth and is predominantly expressed in adult ventricle and atrium. This circumvents the problem of embryonic lethality seen in gene-targeted mice. However, this does limit the use of the technique for the study of the roles of channels in early development, pending the development and characterization of other promoters.

Random incorporation of the transgene into a mouse chromosome may disrupt the function of a native gene at the insertion site. Two or more independent lines of mice expressing the transgene are usually studied to ascertain that the observed

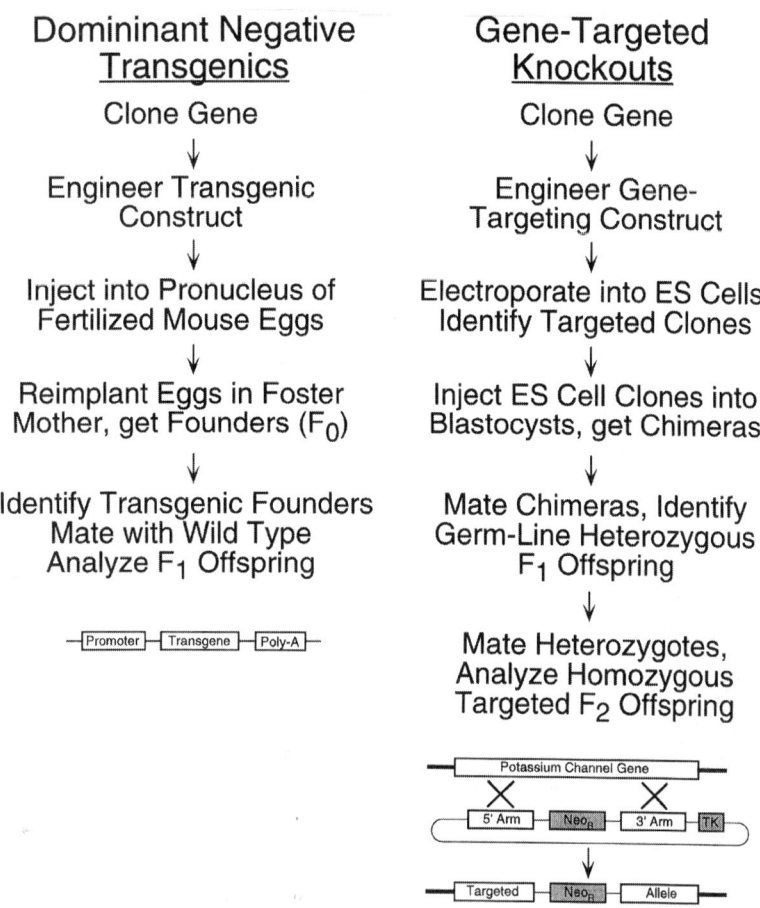

Domininant Negative Transgenics

Clone Gene

↓

Engineer Transgenic Construct

↓

Inject into Pronucleus of Fertilized Mouse Eggs

↓

Reimplant Eggs in Foster Mother, get Founders (F_0)

↓

Identify Transgenic Founders Mate with Wild Type Analyze F_1 Offspring

Gene-Targeted Knockouts

Clone Gene

↓

Engineer Gene-Targeting Construct

↓

Electroporate into ES Cells, Identify Targeted Clones

↓

Inject ES Cell Clones into Blastocysts, get Chimeras

↓

Mate Chimeras, Identify Germ-Line Heterozygous F_1 Offspring

↓

Mate Heterozygotes, Analyze Homozygous Targeted F_2 Offspring

Figure 2. Comparison of the techniques used to engineer dominant-negative transgenic mice (*left*) and gene-targeted knockout mice (*right*). The transgenic construct (*bottom left*) randomly inserts into the chromosome. In contrast, the targeting construct undergoes homologous recombination (a double-crossover event, denoted by X's with the K$^+$ channel gene to form the targeted allele (*bottom right*). Note that gene targeting requires at least three matings to engineer a knockout animal and that each mating requires approximately 3 months (mice become fertile at ~6 weeks of age, and the gestation period is 21 days). Neo$_R$, Neomycin resistance gene; TK, thymidine kinase gene.

phenotype results directly from transgene expression. This also allows examination of lines with varying amounts of mutant subunit protein. Expression levels can also be increased by mating two mice heterozygous for the transgene to obtain homozygotes. Homozygotes are identified by quantitative genomic Southern analysis and confirmed by breeding with wild-type mice. Of note, phenotypic alterations resulting from the insertion site of the transgene are more likely to occur in homozygous lines.

Several channel subunits may be affected by a single transgene. This is advantageous, because it allows for rapid testing of the role of a class of ion channels in the heart. It can, however, be difficult to determine which channels are actually disrupted by the transgene. In addition, different channel subunits may be affected with differing efficiency. Another potential problem arises from the massive overexpression of a mutant channel fragment. The protein may titrate away important factors such as β-subunits or have direct toxic effects on cardiac myocytes.

3. TARGETED DISRUPTION OF K$^+$ CHANNEL GENES IN MICE

Individual genes can be inactivated (gene knockout) or modified (gene knockin, targeted mutagenesis) in the mouse by using homologous recombination in embryonic stem (ES) cells (Bronson and Smithies, 1994; Shastry, 1998; Torres, 1998). This technique has unlocked many secrets regarding the roles of individual gene products in development, physiology, and disease.

3.1. Methods

The technique of homologous recombination in ES cells is summarized on the right-hand side of Fig. 2. ES cells are produced from the inner cell mass of male blastocysts and expanded *in-vitro*. The cells used most frequently were produced from SV129 mice that are homozygous for a dominant gene that codes for the agouti coat color. These cells are unique in that they maintain the ability to differentiate into all cell types, including germ cells. DNA from the K$^+$ channel gene of interest is cloned, and a targeting construct is engineered using a gene fragment of several thousand base pairs 5' to the area to be modified, an antibiotic resistance cassette (usually Neo$_R$, the neomycin resistance gene) for positive selection, a gene fragment of several thousand base pairs 3' to the area of the gene to be modified, and a thymidine kinase (TK) gene cassette for negative selection. The linearized construct is then introduced into ES cells by electroporation (electric shock), and a small fraction of the cells undergo homologous recombination. In those cells, two crossover events replace on one chromosome the part of the K$^+$ channel gene between the homologous arms of the targeting construct with the Neo$_R$ cassette and any other DNA in the targeting construct. Rates of homologous recombination depend on the gene targeted, design of the construct, and experimental conditions but generally range between $1:10^6$ and $1:10^7$. This results in ES cells heterozygous for the targeted allele.

Heterozygous targeted ES cells grow in the presence of the aminoglycoside antibiotic G418 (because they have the Neo$_R$ gene) and gancyclovir (or its analog FIAU, because they do not have the TK gene). Double selection with gancyclovir or FIAU helps to eliminate clones in which the targeting construct is randomly inserted into the chromosome, because these clones will have a functional TK gene that leads to cell death in the presence of these agents. Doubly resistant clones are then screened by PCR and/or genomic Southern blot analysis, expanded, injected into blastocysts harvested from C57BL/6 mice, with black coat color, and reimplanted into pseudo-pregnant females. Chimeric offspring, in part derived from ES cells and in part from the donor blastocysts, are identified by mixed agouti/black coat color. Because the ES cells are male and the male phenotype is dominant, chimeras derived to a large extent from the ES cells are male. The male chimeras are mated with female C57BL/6 mice, and germ line transmission of ES cell-derived sperm is determined by agouti coat color of the offspring. Half of these agouti offspring should carry one copy of the targeted allele. Heterozygous male and female mice are mated to each other to yield homozygous targeted or "knockout" mice.

3.2. Advantages and Disadvantages

This form of K$^+$ channel gene manipulation has several important sequelae. The targeted allele is present in all cell types in which the gene is expressed. Because virtually all cardiac K$^+$ channels are expressed in the brain and other tissues, any phenotype reflects the loss of the channel not only from the heart, but also from the nervous system and potentially other tissues. In addition, the targeted allele is present throughout embryonic development. As a result, the phenotype of the adult mouse may be modified by long-term compensatory changes in other genes, and embryonic lethality may preclude the study of the role of some genes in the adult mouse.

The phenotype produced by the genetic manipulations can depend on the strain of the mouse (Banbury Conference, 1997). When engineered as above, the targeted mice are of mixed background (50% SV129; 50% C57BL/6). Mice of pure SV129 background can be engineered by mating male germ line chimeras with SV129 females. Mice of predominantly C57BL/6 background are produced by backcrossing heterozygous mice of mixed background with wild-type mice of C57BL/6 background. After 5–10 generations, most alleles will match those in the C57BL/6 strain, with the exception of genes linked to the targeted allele (in close proximity on the same chromosome) that cosegregate with that allele.

4. ANALYSIS OF THE PHENOTYPE OF GENETICALLY MODIFIED MICE

Genetically engineered mice are examined for molecular and electrophysiological abnormalities by a variety of techniques (Doevendans et al., 1998). Channel RNA is studied using reverse-transcription PCR (RT-PCR), ribonuclease (RNAse) protection assays, Northern analysis, and in situ hybridization. Channel protein is analyzed using Western blotting and immunohistochemistry. Enzymatically isolated single cardiac myocytes are used for action potential, whole-cell voltage-clamp, and single-ion-channel recordings. Electrocardiograms (EKGs) can be recorded from the mice, and formulas correcting QT interval for heart rate have been developed (Mitchell et al., 1998). Implanted radiotelemetry monitor systems (Data Sciences, MN) are available for long-term EKG monitoring. Cardiac function can be assessed in vivo by echocardiography (Tanaka et al., 1996), gated magnetic resonance imaging (MRI) (Kubota et al., 1997), and invasive catheterization (Zimmer and Millar, 1998). In vitro cardiac function measurements are performed on Langendorff-perfused hearts. Programmed stimulation can be used to assess the conduction system and trigger arrhythmias in open-chested mice, closed-chested mice, or Langendorff-perfused hearts (Berul et al., 1997). Optical mapping using voltage-sensitive or calcium-sensitive dyes can be performed to examine conduction, repolarization, calcium handling, and arrhythmia mechanisms (Kanai and Salama, 1995; Vaidya et al., 1999).

The mouse heart has several inherent disadvantages for the study of repolarization abnormalities. Although its channels are highly homologous to those of the human heart, the basal electrophysiological properties differ substantially. The mouse has a resting heart rate of 600 beats per minute, almost an order of magnitude faster than the human heart rate. In addition, the mouse ventricular action potential and QT intervals are very short, even when corrected for the rapid heart rate. Thus, the sequelae of repolarization abnormalities in the mouse may be markedly different from those in larger mammals. Despite this limitation, genetically altered mice provide a means to determine the role of individual ion channel genes in an *in vivo* system and may unlock the mechanisms by which K^+ channels cause or prevent arrhythmias and sudden death.

5. EXAMPLES

Table 1 outlines the properties of a number of mice with engineered defects in specific K^+ channel gene products. This section will provide examples that illustrate the dominant-negative and gene-targeting techniques aimed at the *Shaker* (Kv1.*x*) family of K^+ channels. In addition, some examples that highlight the problems and pitfalls with the techniques will be presented.

5.1. Dominant-Negative Disruption of Kv1.*x* Channels

We engineered transgenic mice that overexpress a 206-amino-acid N-terminal fragment of the rat brain K^+ channel Kv1.1 under the control of the α-myosin heavy-chain promoter (London *et al.*, 1998a; Fig. 3A). This fragment included the first transmembrane segment (S1) and had previously been shown to inhibit currents from Kv1.1 and Kv1.5 channels in a dominant-negative manner when coexpressed with them in *Xenopus* oocytes (Babila *et al.*, 1994). Because RNA for Kv1.2, Kv1.4, and Kv1.5 is expressed in the heart, we hypothesized that K^+ channels composed of these subunits would be inhibited.

Hearts from the Kv1.*x* dominant-negative transgenic mice expressed high levels of transgene RNA and protein (Fig. 3B, C). Western blot analysis showed that Kv1.5 protein levels in crude cell membrane extracts were decreased, suggesting decreased transport of Kv1.5-containing channels to the membrane or more rapid destruction of heteromeric channels. Action potentials were prolonged owing to the loss of a rapidly activating, slowly inactivating, 4-aminopyridine (4-AP)-sensitive current (I_{slow}) that

---

at the sinus rate of 530 beats/min. (H) Optical map showing the initiation of ventricular tachycardia in a Langendorff perfused heart from a transgenic mouse using the voltage-sensitive dye di-4ANEPPS and programmed ventricular stimulation. The imaged area is 4×4 mm, with the apex of the heart on the left and the base on the right. Isochronal lines of activation are 1 ms apart, and activation spreads from light to dark. A premature stimulus applied at the apex (symbol) is unable to spread directly to the base, which is refractory. It goes around this functional line of block (black arrow) to initiate the arrhythmia. (Reproduced in part from London *et al.*, 1998a, Copyright 1998 National Academy of Sciences, U.S.A).

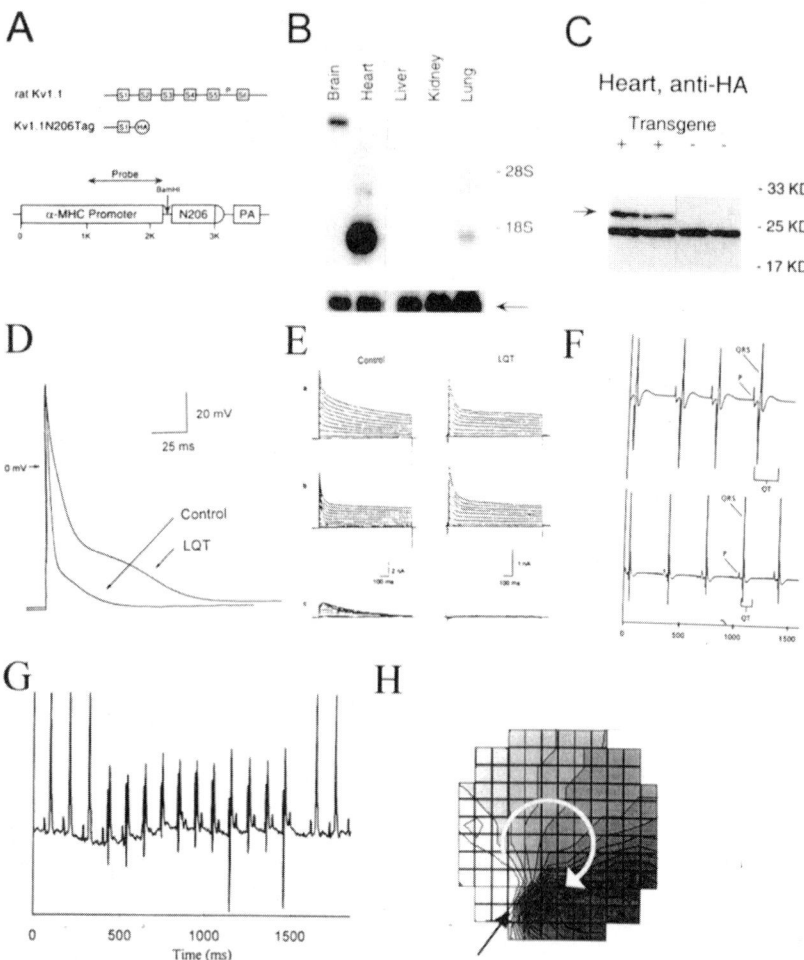

Figure 3. Disruption of the Kv1.x family of cardiac K$^+$ channels using a dominant-negative transgenic fragment of rat Kv1.1. (A) Schematic representation of the rat brain K$^+$ channel rKv1.1 (*top*), the N-terminal truncated fragment of Kv1.1 tagged with the nine-amino-acid hemagglutinin (HA) epitope at its C-terminus (*middle*), and the transgenic construct (*bottom*). The transgenic construct includes the α-myosin heavy-chain promoter (α-MHC), the truncated rKv1.1 fragment (N206), and the SV40 poly-(A) tail (PA). (B) Northern blot analysis of 15 μg of total RNA from the indicated tissues of one transgenic F$_1$ mouse, probed with a fragment from the 5' end of the rKv1.1 coding region. Note the presence of native Kv1.1 only in brain and of the transgene in heart and (to a lesser extent) lung. The blot was reprobed with a 700-base-pair fragment of β-actin (arrow) to confirm equal RNA loading. (C) Western analysis of a crude membrane fraction, incubated with an antibody directed against the HA epitope. Note specific staining of transgene protein in hearts from the transgenic mice (arrow). (D) Action potentials recorded from single cardiac myocytes enzymatically isolated from the hearts of transgenic (LQT) and control mice. (E) Voltage-clamp records of outward K$^+$ currents from isolated control (*left*) and transgenic (*right*) myocytes before (*top*) and following (*middle*) the addition of 50 μM 4-aminopyridine (4-AP). The records at the bottom show the difference currents. Myocytes from transgenic mice lacked the rapidly activating, slowly inactivating, 4-AP-sensitive current designated I_{slow}. (F) Electrocardiograms of anesthetized transgenic (*top*) and wild-type (*bottom*) mice. Note the marked prolongation of the QT interval seen in transgenic mice. (G) An 11-beat run of nonsustained ventricular tachycardia at a rate of 570 beats/min, recorded by an implanted radiotelemetry monitor in a transgenic mouse. Note the atrioventricular dissociation, with P-waves marching through the QRS complexes

resembles currents produced by Kv1.5 channels in heterologous expression systems and the atrial current I_{Kur} (Fig. 3D,E). High-resolution surface EKGs and radiotelemetry implant signals demonstrated QT prolongation and spontaneous runs of nonsustained ventricular tachycardia in transgenic mice (Fig. 3F,G).

Optical mapping studies of Langendorff-perfused transgenic hearts were performed to define the mechanism of arrhythmia initiation in these mice (Baker *et al.*, 1998). This technique, previously applied extensively to the hearts of larger mammals, uses voltage-sensitive dyes to image action potentials from a 4 mm × 4 mm area of the epicardial surface of the left ventricle on a 124-element photodiode array and map the spread of excitation and repolarization (Kanai and Salama, 1995). Programmed stimulation was used to pace the heart at the apex and initiate runs of ventricular tachycardia. Hearts from transgenic mice developed ventricular tachycardia when premature impulses were given at the apex but not at the base (Fig. 3H). Optical maps showed that increased dispersion of refractoriness and repolarization between the base and apex of the hearts from transgenic mice led to functional lines of conduction block within the myocardium and provided the substrate for the initiation of reentrant ventricular tachycardia. Thus, a connection between a defined genetic change, electrophysiological abnormalities, and an arrhythmic mechanism can be determined.

5.2. Targeted Disruption of Kv1.5 and Kv1.4

The above study of dominant-negative transgenic mice suggests that elimination of the current I_{slow} leads to action potential prolongation, QT prolongation, and arrhythmias. To test this hypothesis, we engineered a gene-targeted mouse in which the coding region of Kv1.5 was replaced with the coding region of Kv1.1 followed by the Neo_R cassette (Fig. 4A; London *et al.*, 1995). Unlike Kv1.5, Kv1.1 encodes channels that are insensitive to 4-AP. Homozygous targeted mice appeared normal. The baseline QT interval of these mice was not prolonged (Fig. 4B). However, intraperitoneal injections of 4-AP at doses that prolonged QT interval in the wild-type mice had no effect on the targeted mice. These and additional studies confirmed that the channels responsible for the 4-AP-sensitive component of I_{slow} included Kv1.5 subunits.

We also used gene targeting of ES cells to inactivate (knock out) the K^+ channel Kv1.4 (London *et al.*, 1998b). These mice appeared normal aside from occasional seizure activity and did not have QT prolongation. $I_{to,fast}$, the rapid component of the transient outward cardiac K^+ current I_{to}, was unchanged in these mice. Recent studies do suggest, however, that another current in the mouse ventricle, $I_{to,slow}$, is absent in the ventricular septum of these mice (Guo *et al.*, 1999). Thus, the Kv1.x dominant-negative mice probably lack multiple repolarizing currents, and the relationship of each current to QT prolongation and arrhythmias remains to be determined. Dominant-negative and gene-targeting studies provide complementary data on the roles of individual gene products and their interactions *in vivo*.

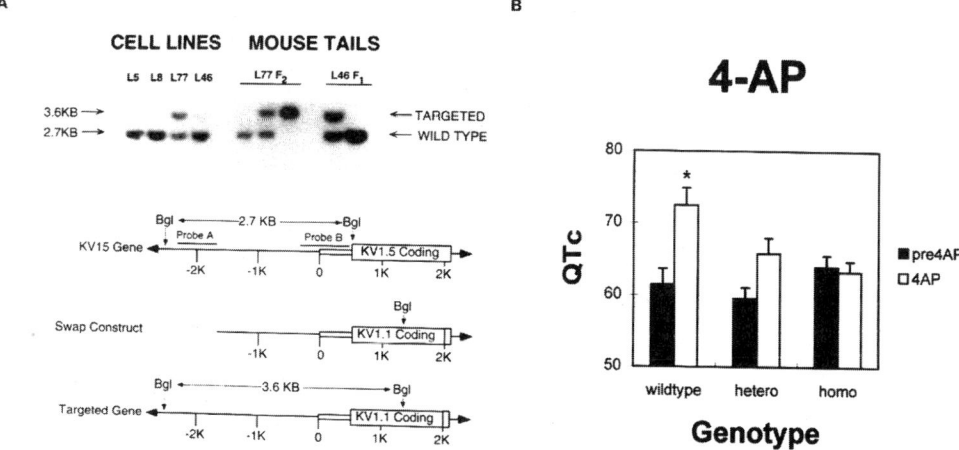

Figure 4. Replacement of the mouse Kv1.5 gene with rat Kv1.1 at the chromosomal level using gene targeting of embryonic stem (ES) cells. (A) *Bottom* Schematic representation of the Kv1.5 gene, the 5′ end of the targeting construct, and the targeted gene that resulted from homologous recombination. The targeting construct consisted of an ~2-kilobase fragment of the Kv1.5 promoter and 5′-UTR, the rat Kv1.1 gene, a neomycin resistance cassette, an ~3-kilobase fragment of the Kv1.5 gene including the 3′-UTR, and the thymidine kinase gene. *Top*: Targeted cell lines heterozygous for the gene replacement were identified by a change in the size of a BglII restriction fragment on genomic Southern analysis. Heterozygous and homozygous mice were identified in a similar manner using DNA isolated from tail biopsies. Note that the targeted band in the L46 line is less intense than the wild-type band. This may indicate the presence of nontargeted ES cells in this clone. (B) QT interval as a function of genotype for wild-type ($n = 9$), heterozygous ($n = 4$), and homozygous ($n = 9$) mice before and after intraperitoneal injections with 4-aminopyridine (4-AP). QT interval was corrected for rate using the formula QTc = QT/(RR/100)$^{1/2}$, as described by Mitchell et al. (1998). Note that QTc is similar in homozygous and wild-type mice at baseline but that homozygous mice fail to develop QT prolongation following exposure to 4-AP. This was statistically significant (*, $p < 0.05$).

5.3. Kv4.*x* Dominant-Negative Mice

Barry et al. (1998) used the α-myosin heavy-chain promoter to overexpress a Kv4.2 channel in the mouse heart that carried a single amino acid mutation in the pore region. They had previously demonstrated dominant-negative suppression of I_{to} in heterologous expression systems using this construct *in vitro*. Myocytes from mice overexpressing the transgene lacked the fast component of I_{to} and had prolonged action potential durations. The mice had QT interval prolongation on EKG. This strongly supports a role for the Kv4.*x* family of channels as the molecular basis of I_{to} in the mouse. The Kv4.*x* dominant-negative transgenic mice did not have spontaneous or inducible ventricular arrhythmias, in contrast to the Kv1.*x* dominant-negative mice, despite a similar degree of action potential and QT prolongation. The Kv4.*x* transgenic mice also showed upregulation of a more slowly inactivating outward current ($I_{to,slow}$). Thus, the study of mice with engineered mutations not only allows assignment of different currents to different channels, but also provides insights into gene regulation and the mechanisms of arrhythmias that are not possible in studies using *in vitro* systems.

Kv4.*x* dominant-negative mice were also engineered using a Kv4.2 transgene that was truncated after the S3 transmembrane domain (Wickenden *et al.*, 1997). As expected, myocytes from these mice lacked I_{to}. Unlike the mice expressing the Kv4.2 construct with a point mutation, however, the mice developed congestive heart failure and died at an early age. The truncated transgene may inhibit other ion channels, titrate away a β-subunit, or have a direct toxic effect on the heart. The phenotype of dominant-negative transgenic mice can depend on subtle experimental details, and interpretations of the etiology of that phenotype must be done with care.

5.4.   HERG Dominant-Negative Mice

Mutations of the human *ether-a-go-go*-related gene (HERG) cause the autosomal dominant long QT syndrome characterized by QT prolongation on the surface EKG, arrhythmias, and sudden cardiac death due to loss of the repolarizing current I_{Kr} (Curran *et al.*, 1995). The HERG point mutant G628S yields nonfunctional channels and inhibits HERG currents in a dominant-negative manner when coexpressed with wild-type subunits *in vitro* (Sanguinetti *et al.*, 1996). Overexpression of this mutant subunit in the heart of transgenic mice prolonged action potential duration and eliminated I_{Kr} in ventricular myocytes isolated from the transgenic mice (Babij *et al.*, 1998). No QT prolongation or arrhythmias were present in the mice, however, despite the similarity to a known mutation that leads to sudden death in humans. This probably reflects differences in the abundance or importance of I_{Kr} in the repolarization of the mouse versus the human heart.

6. FUTURE DIRECTIONS

Until recently, the pattern of gene expression in transgenic mice was limited by the available promoters. Fishman (1998) added a tetracycline-binding element to the α-myosin promoter, making it inactive in the presence of low concentrations of tetracycline. Thus, the initiation and duration of gene expression can be better controlled.

Targeted disruption of genes in ES cells disrupts gene expression throughout development and in all tissues. *Merg1* mutations are embryonic homozygous lethal, making it difficult to fully define the role of the channel in the adult heart (London *et al.*, 1998c). In addition, disruption of channels outside of the cardiovascular system (e.g., in the nervous system) may lead to indirect changes in the heart. The ability to engineer gene knockouts in a time- and tissue-restricted manner has recently been developed using the cre/lox system (Sauer, 1998). In this section, gene targeting is used to surround the desired gene by two 34-base-pair elements. Homozygous targeted mice are then mated with transgenic mice that express the cre recombinase enzyme in a tissue- and time-specific manner. In the presence of the cre recombinase protein, the part of the gene between the elements is removed and the conditional tissue-specific knockout completed.

Mating together of different lines of mice will help to identify the relationships between channels. Additional refinements of the technique and analysis of the mice will occur. Genetically engineered mice will be used to study the roles of K$^+$ channels in other parts of the cardiovascular system, including vascular smooth muscle and endothelium. A major challenge, however, remains in relating the findings in the mouse to larger mammals, including humans. Because of this, the relationship of individual ion channels to the common diseases that cause arrhythmias and sudden death may remain elusive.

REFERENCES

Abbott, G. W., Sesti, F., Splawski, I., Buck, M. E., Lehmann, M. H., Timothy, K. W., Keating, M. T., and Goldstein, S. A. N., 1999, MiRP1 forms I_{Kr} potassium channels with HERG and is associated with cardiac arrhythmia, *Cell* **97**:175–187.

Antzelevitch, C., Sicouri, S., Litovsky, S. H., Lukas, A., Krishnan, S. C., Di Diego, J. M., Gintant, G.A., and Liu, D.-W., 1991, Heterogeneity within the ventricular wall: Electrophysiology and pharmacology of epicardial, endocardial, and M cells, *Circ. Res.* **69**:1427–1450.

Babij, P., Askew, G. R., Nieuwenhuijsen, B., Su, C.-M., Bridal, T. R., Jow, B., Argentieri, T. M., Kulik, J., DeGennaro, L. J., Spinelli, W., and Colatsky, T. J., 1998, Inhibition of cardiac delayed rectifier K$^+$ current by overexpression of the long-QT syndrome HERG G628S mutation in transgenic mice, *Circ. Res.* **83**:668–678.

Babila, T., Moscucci, A., Wang, H., Weaver, F. E., and Koren, G., 1994, Assembly of mammalian voltage-gated potassium channels: Evidence for an important role of the first transmembrane segment, *Neuron* **12**:615–626.

Baker, L. C., London, B., Choi, B-R., Koren, G., and Salama, G., 1998, Optical mapping of reentrant VT in transgenic mice, *Circulation* **98**:-I744 (Abstract).

Banbury Conference on Genetic Background in Mice, 1997, Mutant mice and neuroscience: Recommendations concerning genetic background, *Neuron* **19**:755–759.

Barry, D. M., Xu, H., Schuessler, R. B., and Nerbonne, J. M., 1998, Functional knockout of the transient outward current, long-QT syndrome, and cardiac remodeling in mice expressing a dominant-negative Kv4 α subunit, *Circ. Res.* **83**:560–567.

Berul, C. I., Christe, M. E., Aronovitz, M. J., Seidman, C. E., Seidman, J. G., and Mendelsohn, M. E., 1997, Electrophysiological abnormalities and arrhythmias in alpha MHC mutant familial hypertrophic mice, *J. Clin. Invest.* **99**:570–576.

Bronson, S. K., and Smithies, O., 1994, Altering mice by homologous recombination using embryonic stem cells, *J. Biol. Chem.* **269**:27155–27158.

Chandy, K. G., and Gutman, G. A., 1995, Voltage-gated K$^+$ channel genes, in: *Handbook of Receptors and Channels: Ligand- and Voltage-Gated Ion Channels* (R. A. North, ed.), CRC Press, Boca Raton, Florida, pp. 1–79.

Curran, M. E., Splawski, I., Timothy, K. W., Vincent, G. M., Green, E. D., and Keating, M. T., 1995, A molecular basis for cardiac arrhythmias: *HERG* mutations cause long QT syndrome, *Cell* **80**:795–803.

Deal, K. K., England, S. K., and Tamkun, M. M., 1996, Molecular physiology of cardiac potassium channels, *Physiol. Rev.* **76**:49–67.

Dixon, J. E., Shi, W., Wang. H.-S., McDonald, C., Yu, H., Wymore, R. S., Cohen, I. S., and McKinnon, D., 1996, Role of the Kv4.3 K$^+$ channel in ventricular muscle: A molecular correlate for the transient outward current, *Circ. Res.* **79**:659–668.

Doevendans, P. A., Daemen, M. J., de Muinck, E. D., and Smits, J. F., 1998, Cardiovascular phenotyping in mice, *Cardiovasc. Res.* **39**:34–49.

Drici, M.-D., Arrighi, I., Chouabe, C., Mann, J. R., Lazdunski, M., Romey, G., and Barhanin, J., 1998, Involvement of IsK-associated K$^+$ channel in heart rate control of repolarization in a murine engineered model of Jervell and Lange-Nielsen syndrome, *Circ. Res.* **83**:95–102.

Fiset, C., Clark, R. B., Shimoni, Y., and Giles, W. R., 1997, *Shal*-type channels contribute to the Ca^{2+}-independent transient outward K^+ current in rat ventricle, *J. Physiol. (London)* **500**:51–64.

Fishman, G., 1998, Timing is everything in life: Conditional transgene expression in the cardiovascular system, *Circ. Res.* **82**:837–844.

Guo, W., Xu, H., London, B., and Nerbonne, J. M., 1999, Molecular basis of transient outward K^+ current diversity in mouse ventricular myocytes. *J. Physiol,* **521**: 587–599.

Hill, J. A., Kutschke, W., and London, B., 1998, Differential susceptibilities to class III antiarrhythmic drugs in mice with targeted disruptions in voltage-dependent potassium channel subunits, *Circulation* **98**:I-695 (Abstract).

Izumo, S., and Shioi, T., 1998, Cardiac transgenic and gene-targeted mice as models of cardiac hypertrophy and failure: A problem of (new) riches, *J. Card. Failure* **4**:349–361.

Johns, D. C., Nuss, H. B., and Marban, E., 1997, Suppression of neuronal and cardiac transient outward currents by viral gene transfer of dominant-negative Kv4.2 constructs, *J. Biol. Chem.* **272**:31598–31603.

Kanai, A., and Salama, G., 1995, Optical mapping reveals that repolarization spreads anisotropically and is guided by fiber orientation in guinea pig hearts, *Circ. Res.* **77**:784–802.

Kim, E., Cho, K. O., Rothschild, A., and Sheng, M., 1996, Heteromultimerization and NMDA receptor-clustering activity of Chapsyn-110, a member of the PSD-95 family of proteins, *Neuron* **17**:103–113.

Kubota, T., McTiernan, C. F., Frye, C. S., Slawson, S. E., Koretsky, A. P., Demetris, A. J., and Feldman, A. M., 1997, Dilated cardiomyopathy in transgenic mice with cardiac-specific overexpression of tumor necrosis factor-α, *Circ. Res.* **81**:627–635.

Kupershmidt, S., Yang, T., Anderson, M. E., Wessels, A., Niswender, K. D., Magnuson, M. A., and Roden, D. M., 1999, Replacement by homologous recombination of the *minK* gene with *lacZ* reveals restriction of *minK* expression to the mouse cardiac conduction system, *Circ. Res.* **84**:146–152.

Lagna, G., and Hemmati-Brivanlou, A., 1998, Use of dominant-negative constructs to modulate gene expression, *Curr. Top. Dev. Biol.* **36**:75–98.

London, B., Hill, J. A., Nguyen, H., Schieferl, S., and Nadal-Ginard, B., 1995, *in vivo* mutagenesis of the mouse heart: Targeted replacement of the murine delayed rectifier potassium channel mKv1.5 with the rat brain delayed rectifier channel rKv1.1, *Circulation* **92**:I-155 (Abstract).

London, B., Trudeau, M. C., Newton, K. P., Beyer, A. K., Copeland, N. G., Gilbert, D. J., Jenkins, N. A., Satler, C. A., and Robertson, G. A., 1997, Two isoforms of the *mouse ether-a-go-go-related gene* coassemble to form channels with properties similar to the rapidly activating component of the cardiac delayed rectifier K^+ current, *Circ. Res.* **81**:870–878.

London, B., Jeron, A., Zhou, J., Buckett, P., Han, X., Mitchell, G. F., and Koren, G., 1998a, Long QT and ventricular arrhythmias in transgenic mice expressing the N terminus and first transmembrane segment of a voltage-gated potassium channel, *Proc. Natl. Acad. Sci. U.S.A.* **95**:2926–2931.

London, B., Wang, D. W., Hill, J. A., and Bennett, P. B., 1998b, The transient outward current in mice lacking the potassium channel gene *Kv1.4, J. Physiol. (London)* **509**:171–182.

London, B., Pan, X.-H., Lewarchik, C. M., and Lee, J. S., 1998c, QT interval prolongation and arrhythmias in heterozygous Merg1-targeted mice, *Circulation* **98**:I-56 (Abstract).

MacKinnon, R., 1991, Determination of the subunit stoichiometry of a voltage-activated potassium channel, *Nature* **350**:232–235.

Mahdavi, V., Koren, G., Michaud, S., Pinset, C., and Izumo, S., 1989, Identification of the sequences responsible for the tissue-specific and hormonal regulation of the cardiac myosin heavy chain genes, in: *Cellular and Molecular Biology of Muscle Development* (F. Stockdale and L. Kedes, eds.), Alan R. Liss Inc., New York, pp. 369–379.

Mitchell, G. F., Jeron, A., and Koren, G., 1998, Measurement of heart rate and Q-T interval in the conscious mouse, *Am. J. Physiol.* **274**:H747–H751.

Morales, M. J., Castellino, R. C., Crews, A. L., Rasmusson, R. L., and Strauss, H. C., 1995, A novel β subunit increases rate of inactivation of specific voltage-gated potassium channel α subunits, *J. Biol. Chem.* **270**:6272–6277.

Po, S., Roberds, S., Snyders, D. J., Tamkun, M. M., and Bennett, P. B., 1993, Heteromultimeric assembly of human potassium channels: Molecular basis of a transient outward current?, *Circ. Res.* **72**:1326–1336.

Sanguinetti, M. C., Curran, M. E., Spector, P. S., and Keating, M. T., 1996, Spectrum of HERG K^+-channel dysfunction in an inherited cardiac arrhythmia, *Proc. Natl. Acad. Sci. U.S.A.* **93**: 2208–2212.

Sauer, B., 1998, Inducible gene targeting in mice using the Cre/lox system, *Methods* **14**:381–392.

Schulze-Bahr, E., Wang, Q., Wedekind, H., Haverkanp, W., Chen, Q., and Sun, Y., 1997, KCNE1 mutations cause Jervell and Lange-Nielsen syndrome, *Nat. Genet.* **17**:267–268.

Shastry, B. S., 1998, Gene disruption in mice: Models of development and disease, *Mol. Cell. Biochem.* **181**:163–169.

Splawski, I., Timothy, K. W., Vincent, G. M., Atkinson, D. L., and Keating, M. T., 1997, Molecular basis of the long-QT syndrome associated with deafness, *N. Engl. J. Med.* **336**:1562–1567.

Tanaka, N., Dalton, N., Mao, L., Rockman, H. A., Peterson, K. L., Gottshall, K. R., Hunter, J. J., Chien, K. R., and Ross, J., Jr., 1996, Transthoracic echocardiography in models of cardiac disease in the mouse, *Circulation* **94**:1109–1117.

Torres, M., 1998, The use of embryonic stem cells for the genetic manipulation of the mouse, *Curr. Topi. Dev. Biol.* **36**:99–114.

Vaidya, D., Morley, G. E., Samie, F. H., and Jalife, J., 1999, Reentry and fibrillation in the mouse heart: A challenge to the critical mass hypothesis, *Circ. Res.* **85**:174–181.

Wang, Q., Curran., M. E., Splawski, I., Burn, T. C., Millholland, J. M., VanRaay, T. J., Shen, J., Timothy, K. W., Vincent, G. M., de Jager, T., Schwartz, P. J., Towbin, J. A., Moss, A. J., Atkinson, D. L., Landes, G. M., Connors, T. D., and Keating, M. T., 1996, Positional cloning of a novel potassium channel gene: KVLQT1 mutations cause cardiac arrhythmias *Nat. Genet.* **12**:17–23.

Wickenden, A. D., Huang, Q., Factor, S. M., Backx, P. H., and Fishman, G. I., 1997, Expression of a dominant-negative Kv4.2 channel in the hearts of transgenic mice, *Circulation* **96**:I-422 (Abstract).

Wickman, K., Nemec, J., Gendler, S. J., and Clapham, D. E., 1998, Abnormal heart rate regulation in GIRK4 knockout mice, *Neuron* **20**:103–114.

Xu, H., Dixon, J. E., Barry, D. M., Trimmer, J. S., Merlie, J. P., McKinnon, D., and Nerbonne, J. M., 1996, Developmental analysis reveals mismatches in the expression of K$^+$ channel α subunits and voltage-gated K$^+$ channel currents in rat ventricular myocytes, *J. Gen. Physiol.* **108**:405–419.

Xu, H., Barry, D. M., Li, H., Brunel, S., Guo, W., and Nerbonne, J. M., 1999, Attenuation of the slow component of delayed rectification, action potential prolongation, and triggered activity in mice expressing a dominant-negative Kv2 α subunit, *Circ. Res.* **85**:623–633.

Yost, C. S., 1999, Potassium channels: Basis aspects, functional roles, and medical significance, *Anesthesiology* **90**:1186–1203.

Zhou, Z., Gong, Q., Epstein, M. L., and January, C. T., 1998, HERG channel dysfunction in human long QT syndrome: Intracellular transport and functional defects, *J. Biol. Chem.* **273**:21061–21066.

Zimmer, H. G., and Millar, H. D., 1998, Technology and application of ultraminiature catheter pressure transducers, *Can. J. Cardiol.* **14**:1259–1266.

Part III

Pharmacology of Potassium Channels

Chapter 12

Pharmacology of Voltage-Gated K^+ Channels

Brian Robertson

1. INTRODUCTION

To completely describe the pharmacology of voltage-gated K^+ channels would take an entire volume, and that volume would probably be obsolete before it appeared on bookshelves. This chapter is therefore necessarily a restricted view. Many excellent reviews of K^+ channel pharmacology are available (e.g., Chandy and Gutman, 1995; Fedida, *et al.*, 1998; see elsewhere in this volume). Here, we focus on pharmacology of cloned voltage-gated (Kv) subunits known to be important in cardiac function. Since we now know the molecular targets of many drugs, and in some cases even which regions of the channel protein are necessary for drug action, this chapter deals mainly with results obtained from electrophysiological experiments using expression systems, in which the experimenter can focus on a single, identified channel. Using molecular biological techniques, such as Northern blotting or antibody labeling to determine what channels are present in heart tissues, it is relatively simple to put the cDNA or cRNA for this channel subunit in an expression system and examine its pharmacology. Such approaches to understanding pharmacology have great advantages over approaches that employ isolated myocytes, where one has to use often complex procedures (voltage protocols, blocking agents) to dissect out the K^+ current of interest. The molecular/electrophysiological approach of studying K^+ channels in expression systems has been tremendously successful, as several examples cited here will hopefully show. However, there are potential drawbacks; for instance, *in vivo*, Kv channels probably assemble as heteromultimers, comprised of a mixture of α-subunits and perhaps other subunits, complicating their pharmacology; different expression systems may give different results, for a variety of technical reasons; The IC_{50} values which one has worked so hard to obtain for a particular compound may have little relation to therapeutically relevant concentrations in the working (human) heart at 37°C.

Brian Robertson ● Electrophysiology Group, Department of Biochemistry, Imperial College of Science, Medicine & Technology, London SW7 2AY, United Kingdom.

Potassium Channels in Cardiovascular Biology, edited by Archer and Rusch. Kluwer Academic/Plenum Publishers, New York, 2001.

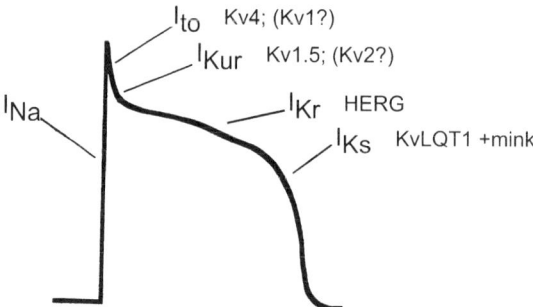

Figure 1. Schematic diagram of a cardiac action potential and genes suggested to be involved at various phases. The rapid depolarization is produced by sodium current. I_{to}, the transient outward K^+ current, is likely to be produced by channels belonging to the Kv4 subfamily, although it has been suggested that some members of the Kv1 subfamily (e.g., Kv1.4) may be involved. I_{Kur}, the ultrarapid component of delayed rectifier K^+ current, is produced by Kv1.5 channels, although it is suggested that some members of the Kv2 subfamily might also play a role. There is also substantial evidence that I_{Kr}, the rapid component of delayed rectifier current, is carried by HERG channels. The last voltage-gated K^+ current involved is I_{Ks}, which is produced by KvLQT1 and the accessory subunit minK. This drawing is based on that of Fedida *et al.*, (1998).

In spite of such difficulties, the last few years have witnessed an unprecedented explosion in our knowledge of cardiac Kv channel pharmacology, and this chapter focuses on some recent, exciting results. We follow pharmacology through different Kv subfamily subunits found in the mammalian heart and discuss how many of these channels relate to different components of the cardiac action potential (see Fig 1).

2. Kv1 CHANNELS

Among all of the mammalian Kv channels, including those in the heart, Kv1 channels are the best characterized, especially pharmacologically. The number of pharmacological tools available for study of these channels is large, but we will focus here on those members of the Kv1 subfamily shown to be important in the heart.

Northern blots and ribonuclease (RNAse) protection assays have shown the presence of RNA for Kv1.2, Kv1.4, and Kv1.5 channels in the heart (Roberds and Tamkun, 1991a,b), and *in situ* hybridization to this RNA has shown signals for Kv1.1 and Kv1.3 (Brahamajothi *et al.*, 1996). However, as pointed out by Barry *et al.*, (1995) and Xu *et al.*, (1996), the channel protein may not always be expressed in the cell membrane, despite the presence of RNA. For example, Kv1.4 protein is undetectable in the heart, whereas its mRNA is clearly present (Barry *et al.*, 1995). Kv1.1 protein also appears not to be transcribed at significant levels. Thus far, there is no substantial evidence for the presence of Kv1.6–Kv1.9 channel subunits, although recent work has shown Kv1.7 to be present in the heart (see below).

The potential presence of Kv1.1 and Kv1.2 protein would lead one to suspect that there would be ionic currents sensitive to dendrotoxins (DTXs) which are selective Kv1.1/1.2 and Kv1.6 blocking toxins derived from black and green mamba snakes (Harvey and Anderson, 1991). DTXs, having strong overall positive charge, bind to key negatively charged amino acids near the ion conduction pathway: in the rat, Kv1.1: AEEAESH, Kv1.2: ADERDSH, Kv1.6: ADDVDSL; however, some species differences do exist (Hurst *et al.*, 1991). Xu *et al.*, (1996) tested one DTX homolog (β-DTX, 200 nM) in rat ventricular myocytes and found no effect on the Kv currents. It is worth pointing out, though, that different DTX homologs can selectively block some, but not other, Kv1 subunits (Robertson *et al.*, 1996). There is only one direct report of a DTX-sensitive current in the heart, from Van Wagoner *et al.*, (1996), although very high concentrations (micromolar levels) of one DTX homolog (DTX-I) change heart rate (John Tippins, personal communication). Further work is necessary to explore whether DTX-sensitive Kv channels represent a significant pharmacological target in the heart.

Kv1.2 channels display an interesting form of block and "enhancement" by quinidine. At voltages just past threshold, quinidine has been reported to act like an "agonist," enhancing Kv1.2 current, whereas at stronger depolarizations (past 0 mV), quinidine blocks the current (Tseng *et al.*, 1996). The authors of this study suggested that the "agonist" action occurs at a site on the outside of the channel, distinct from its normal intracellular blocking site.

Kv1.3 channels also have selective peptide-toxin blockers (e.g., margatoxin), but there are no reports of these being effective in cardiac myocytes.

Kv1.4 subunits produce a channel that is 4-aminopyridine (4-AP) sensitive (IC$_{50}$ $\approx$ 1 mM; Chandy and Gutman, 1995) and shows rapid N-terminal-induced inactivation. These properties made it an early contender for the basis of the transient outward cardiac K$^+$ current, I_{to}, although several studies suggest that Kv4 channels are much more likely to underlie this native current (see Chapter 17). One important functional difference between Kv4.x and Kv1.4 channels is that Kv1.4 channels recover very slowly from inactivation [over a timescale of seconds, at normal physiological extracellular concentrations of K$^+$([K$^+$]$_o$)], whereas I_{to} and authentic Kv4 currents recover much more quickly. Unfortunately, there are no selective, high-affinity pharmacological tools available for Kv1.4 channels to determine finally whether these channels make a significant contribution to cardiac voltage-gated currents; the present view is that they are probably not important in I_{to}. However, as very recent experiments have shown, one must exercise considerable caution in making such blanket statements, as major species differences may exist, as do regional variations. For instance, in a powerful study, Brahmajothi *et al.*, (1999) have shown that I_{to} differs between epicardial and endocardial myocytes from the ferret left ventricle in activation threshold, rate of inactivation, and recovery from inactivation ($\sim$6-fold). Furthermore, whereas I_{to} in epicardial myocytes is blocked by *Heteropoda* toxins (selective inhibitors of Kv4 channels; see below), I_{to} in endocardial myocytes is resistant to these toxins. Based on these differences, with further evidence from immunofluorescence and *in situ* hybridization measurements, Brahmajothi *et al.*, concluded that I_{to} in endocardial myocytes in the ferret ventricle is produced by Kv1.4 channel subunits. Similarly, Wang *et al.*, (1999) have presented compelling evidence to show that in rabbit atrial myocytes, unlike human atrial myocytes, Kv1.4 might be the predominant isoform underlying I_{to}.

2.1. Kv1 Modulation by β-Subunits

The Kv1 α-subunits can have their physiological properties altered through association with accessory Kvβ subunits, which are found in the heart (Castellino et al., 1995; Morales et al., 1995; England et al., 1995; Majumder et al., 1995). The number of β-subunits identified grows quickly. β-Subunits can modulate α-subunits in a number of ways, by increasing inactivation rate or by contributing an N-terminal "inactivation ball" of their own to the oligomeric complex (Rettig et al., 1994; England et al., 1995). β-subunits can also modify channel behavior by altering the activation curve and other elements of channel gating (see, e.g., Uebele et al., 1996; Heinemann et al., 1996) or by "chaperoning" channels to the surface membrane (Shi et al., 1995).

Association with β-subunits can dramatically alter the physiological profile of Kv1 channels, making otherwise noninactivating channels rapidly inactivate and shifting gating. This makes simple extrapolation from cloned channels in expression systems to native currents difficult. Additionally, changes in gating, especially inactivation, can alter the responsiveness of a current to a drug. If the drug shows use dependence, then N-type inactivation may change the "access" to its site of action (see, e.g., Stephens et al., 1994). The cell may equally well change the amount of β-subunits associating with their α-subunit partners under different physiological or pathophysiological states.

Kv1.5 is unquestionably the most important Kv1-subfamily member in the human heart. Its RNA is present in ventricle and in high concentrations in atria (Tamkun et al., 1991; Fedida et al., 1993). Kv1.5 protein is also present in the heart (Barry et al., 1995). The rapid gating of Kv1.5, as well as other properties [see Fedida et al. (1993) and Fedida (1998) for a comprehensive review], strongly suggests that this channel subunit makes the ultrarapidly activating delayed rectifier I_{Kur} in the mammalian atrium. I_{Kur} and Kv1.5 are sensitive to relatively low concentrations of 4-AP (IC$_{50}$ ~ 50 μM; Fedida et al., 1993). The mechanism by which 4-AP inhibits K$^+$ channels has been explored in detail for several Kv channels, including Kv1.5. Binding sites are thought to be intracellular, associated with parts of the S5 and S6 segments (Kirsch et al., 1993). It is the positively charged form of 4-AP which is active (see, e.g., Stephens et al., 1994, and references therein). 4-AP can block open and closed channels and also reduces the gating charge (i.e., the conformational changes preceding final channel opening; Fedida et al., 1996). 4-AP results in a slowing of C-type inactivation in Kv1.5 channels (Fedida et al., 1996).

2.2. Kv1.5 Channels and Channel Antagonists

Kv1.5 channels, like I_{Kur}, are insensitive to external tetraethylammonium (TEA) (only minimal block at tens of millimolar concentrations). This absence of sensitivity to externally applied TEA is explained by lack of the important tyrosine at the latter end of the P-region (GYGDXYP) (MacKinnon and Yellen, 1990; Kavanaugh et al., 1991). In Kv1.5 this residue is an arginine (Fedida et al., 1993). Although there is a direct relationship between the number of tyrosine-containing subunits and the tetrameric channel's affinity for TEA (Heginbotham and MacKinnon, 1992), other residues likely play a role in fine-tuning this affinity. For example, Kv3 channels are slightly more TEA sensitive than Kv1.6 and Kv2.1 channels (Chandy and Gutman, 1995), yet all have the tyrosine residue. Kv1.5 channels are blocked by internal TEA (Fedida et al., 1996) acting at a site on the inner vestibule of the K$^+$ channel pore (e.g., Shieh and Kirsch, 1994). Block by internal TEA is influenced by amino acid residues in S5, S6, and the P-region, as discussed by Gomez-Hernandez et al., (1997).

Verapamil, an inhibitor of L-type Ca^{2+} channels, also inhibits Kv1.5 currents (Rampe et al., 1993b), probably by an open-channel blocking mechanism at the inner pore. Nifedipine, another L-type Ca^{2+} antagonist, also blocks Kv1.5 (Zhang et al., 1997). Threshold effects on the cloned channel were seen at 500 nM, well within the clinical range (Zhang and Fedida, 1998). Nifedipine causes a voltage- and time-dependent block and "crossover" of the deactivating tail currents but has little effect on gating currents. Results from different recording configurations suggest that nifedipine preferentially blocks at the extracellular side of the channel or at some other site that is accessible from the extracellular side.

Quinidine is an antiarrhythmic agent that blocks a variety of K$^+$ currents in the heart, including I_{to} and Kv1.5 (Colatsky, 1982; Snyders et al., 1992; Wang et al., 1995). Fedida (1997) and his colleagues have studied the block of Kv1.5 by quinidine in considerable detail. Quinidine rapidly blocks Kv1.5 channels by binding internally to a site that is exposed only when the channel opens. Once quinidine is bound there, it immobilizes gating charge, which only returns once quinidine dissociates from its binding site. Fedida (1997) suggested that, in contrast to many other blocking drugs (e.g., 4-AP), quinidine is a pure open-channel blocking agent, devoid of actions on closed channels (Chen and Fedida, 1998).

Although Kv1.5 channels are blocked by both flecainide and clofilium, other channels are probably the primary point of action of these compounds.

2.3. Effects of Antihistamines on Kv1.5 Channels

Loratadine is a nonsedating antihistamine that may cause cardiac arrhythmias in some cases. Delpon et al., (1997) and Lacerda et al., (1997) have shown that loratadine blocks Kv1.5, but not human ether-a-go-go-related (HERG), channels at concentrations in the high-nanomolar to micromolar range, reducing the open-state probability of opening of single Kv1.5 channels and accelerating the decay of macroscopic currents. Whether inhibition of Kv1.5 explains the increasing number of reports of arrhythmia associated with loratidine use is unclear, because the concentrations that block the channel are considerably above those seen therapeutically (Delpon et al., 1997). The effects on Kv1.5 of two other nonsedating antihistamines, terfenadine and ebastine, have been examined (Rampe et al., 1993; Crumb et al., 1995; Valenzuela et al., 1997). Terfenadine blocks Kv1.5 in a voltage- and time-dependent manner. However, as is the case with loratadine, the concentrations of terfenadine required to block Kv1.5 (high micromolar) greatly exceed the plasma levels normally achieved with clinical use of this drug (low nanomolar). Ebastine caused almost no block of Kv1.5 at a 3 μM dose (Fig. 2A).

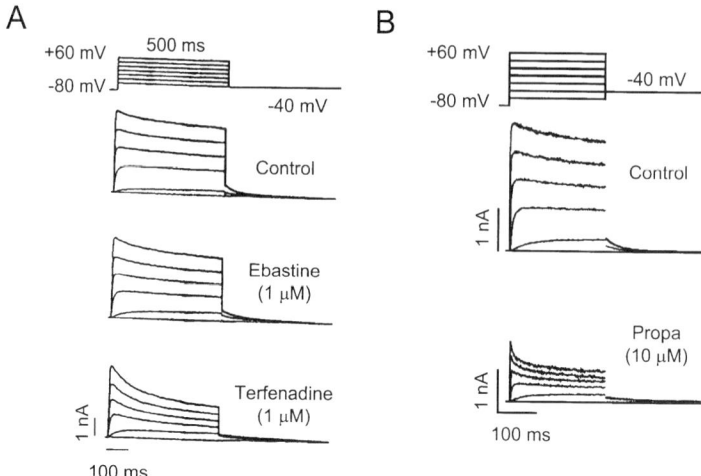

Figure 2. (A) Effects of ebastine and terfenadine, two nonsedating antihistamines, on human Kv1.5 currents in transfected mouse LtK⁻ cells. Ebastine minimally alters Kv1.5 current, whereas terfenadine induces time- and voltage-dependent block of Kv1.5. (Traces reproduced from Valenzuela *et al.*, 1997 with permission from Elsevier Science). (B) Effect of the class Ic antiarrhythmic propafenone (Aopa). Propafenone blocks hKv1.5 currents with an IC_{50} value of ∼4 mM. (Traces reproduced, with permission, from Franqueza *et al.*, 1998.) In all traces voltage was incremented from a holding potential of −80 mV to voltages between −60 and +60 mV in 20-mV steps.

2.4. Miscellaneous Inhibitors of Kv1.5 Channels

Rampe and Murawsky (1997) have suggested that the antibiotic erythromycin, at concentrations achieved following intravenous injection, blocks open Kv1.5 channels from an intracellular site. This may contribute to the prolongation of cardiac repolarization seen with the clinical use of this drug.

Propafenone, a clinically used antiarrhythmic drug, also inhibits Kv1.5. Propafenone and its major metabolite 5-OH-propafenone block Kv1.5, with IC_{50} values of ∼4 and 9 μM respectively, at +60 mV. Propafenone blocks a variety of cardiac voltage-gated channels, including Na^+ and Ca^{2+} types. Block of Kv1.5 is time- and voltage-dependent and appears to occur via an open-channel block mechanism (Franqueza *et al.*, 1998; Fig. 2B). Again, there are questions as to whether the concentrations of propafenone that inhibit Kv1.5 are within the range of drug levels achieved in patients.

2.5. Kv1.7 Channels

Very recently, Kalman *et al.*, (1998) have expressed Kv1.7 channels, which form a fairly rapidly inactivating current. They identified mRNA for Kv1.7 in the heart by Northern blotting. Kv1.7 is blocked by 4-AP ($IC_{50} \approx 250$ μM), tedisamil ($IC_{50} \approx 18$ μM), and nifedipine ($IC_{50} \approx 13$ μM). Kv1.7 is insensitive to external TEA because it has a hydrophobic alanine at the "TEA site." However, Kv1.7 is potently blocked by the *Stichodactyla* sea anemone toxin.

3. Kv2 CHANNELS

Kv2.1 is a delayed rectifier, sensitive to both TEA and 4-AP block, and also blocked by the fairly recently discovered hanatoxins isolated from spider venom. Kv2.1 channels are present at high levels in the rat heart, as shown by Northern blotting and RNase protection assays (Drewe *et al.*, 1992; Xu *et al.*, 1996; Barry and Nerbonne, 1996). Expression of Kv2.1 is developmentally regulated, with levels increasing with postnatal development (Xu *et al.*, 1996). In an elegant study, Swartz and MacKinnon (1995, 1997) showed that hanatoxin shifted the activation curve for Kv2.1 toward positive voltages by a mechanism which appears to involve hanatoxin binding to the S3/S4 linker. Hanatoxin thereby inhibits channel opening until substantially de-polarized potentials are reached. At the time of writing, there have been no studies of hanatoxin in cardiac tissues.

It is worth bearing in mind that there are major species differences in the molecular components of cardiac K$^+$ currents. Kv2.1, but not Kv2.2, is found in rat heart (Dixon and McKinnon, 1994). However, Kääb *et al.*, (1998) could find no Kv2.1 in human ventricle. In dogs, Kv2.1 is expressed at very low levels, whereas Kv2.2 is not expressed at all (Dixon *et al.*, 1996). The picture is different again in ferret, where both Kv2.1 and Kv2.2 have been found in myocytes by *in situ* hybridization (Brahmajothi *et al.*, 1997).

The pharmacology of Kv2 channels is somewhat sparse, but significant detail is available for the site(s) of block of TEA (internal and external) and 4-AP from the elegant work of Kirsch and his colleagues (Kirsch *et al.*, 1993). For instance, Kv2.1 requires about 100 times more 4-AP to achieve block than does Kv3.1. Kirsch *et al.*, (1993) made chimeric channel constructs in which segments of Kv2.1 and Kv3.1 were transferred. They found that much of the 4-AP sensitivity of the chimeric channel was determined by Kv3.1's intracellular S6 segment. This site overlaps the "receptor" for the membrane-permeant long-chain quaternary ammonium derivative tetrapentylam-monium.

3.1. Physiologically "Silent" α-Subunits Modulate Kv2 Channels

Kv2 channels are unusual (so far) in that their behavior is modified by accessory α-subunits (also called γ-subunits), which themselves are physiologically "silent." Association of Kv6.1 with Kv2.1 slows the closure rate of Kv2.1 severalfold and shifts activation ~ -34 mV leftward (Post *et al.*, 1996). Because Kv6.1 has a hydrophobic residue, valine (instead of the tyrosine of Kv2.1), in the external TEA binding site, TEA blocking potency is about 10-fold lower in the heteromultimer. Localization studies indicate that Kv6.1 is present in the heart, suggesting that this subunit might be physiologically relevant in generating more Kv2.1-like phenotypes than previously suspected. For instance, Kv6.1 subunits are quite common in sinoatrial (SA) nodal cells (Brahamajothi *et al.*, 1997). Similarly, Kv8.1 can slow activation and inactivation of

Kv2 channels quite dramatically (Salinas *et al.*, 1997; Castellano *et al.*, 1997). Kramer *et al.*, (1998) also examined the effects of Kv5.1 and Kv6.1 subunits on Kv2.1. Gating and single-channel conductance were changed by both subunits. It appears likely that additional Kv2 channel subunits will be identified with further research. At the time of writing, at least another two have been the subject of preliminary reports.

4. Kv3 CHANNELS

Kv3 channels are common in the central nervous system (CNS), where they are often found in neurons involved in high-frequency firing (e.g., Vega Saenz de Miera *et al.*, 1994; Wang *et al.*, 1998). These channels have high sensitivity to block by 4-AP and TEA (see, e.g., Chandy and Gutman, 1995; Grissmer *et al.*, 1994). Until recently, there were no other pharmacological hallmarks for Kv3 channels, but Diochot *et al.*, (1998) have isolated a novel toxin from a sea anemone that preferentially blocks Kv3.4 channels, and not other Kv3 channels. However, it is unlikely that these channels are important in the heart. Dixon *et al.*, (1996) did not detect mRNA for Kv3.1 and Kv3.2 in canine left ventricle, whereas only very slight levels of Kv3.3 and Kv3.4 mRNA were found. In the rat heart, Kv3 mRNA was present at just above the detection level of the assay, and Dixon and McKinnon (1994) concluded that products of the Kv3 gene family probably do not make an important contribution to the electrophysiological properties of rat myocytes.

5. Kv4 CHANNELS

I_{to} is a rapidly activating and inactivating K^+ current, characterized extensively in a wide variety of cardiac myocytes from an impressive array of species (Nerbonne, 1998). I_{to} is blocked by 4-AP, in a "reverse-use dependent" manner at concentrations above 0.1 mM, with an IC_{50} of approximately 0.2 mM (Campbell *et al.*, 1993). It was previously suggested that Kv1.4 channels (see above), which are also 4-AP sensitive and rapidly activating and inactivating, were a major contender for I_{to} in the heart (especially as Kv1.4 was cloned from human heart). However, Kv1.4 recovers very slowly (seconds) from inactivation in normal $[K]_o$ (e.g., Rasmusson *et al.*, 1995), unlike I_{to} (~ 60 ms; Fiset *et al.*, 1997), and various other doubts began to form. Studies performed in the laboratories of McKinnon and Nerbonne, (Dixon and McKinnon, 1994; Dixon *et al.*, 1996; Barry *et al.*, 1995; Xu *et al.*, 1996) elegantly and powerfully combining electrophysiology, mapping, and molecular biology, have now led to the consensus view that in rat heart, Kv4.2 and Kv4.3 channels contribute to I_{to} in ventricular myocytes. Kv4.2 message levels show a marked expression gradient across the ventricular wall in rat, whereas Kv4.3 levels are more uniform (Dixon and McKinnon, 1994). In contrast, Kv1.4 protein is barely detectable, despite the presence of mRNA (Barry *et al.*, 1995; Xu *et al.*, 1996), whereas Kv4.2 protein is highly expressed. Further supportive evidence for Kv4 channels (perhaps linked as heteromultimers) contributing to I_{to} comes from studies that show parallel changes in the levels of I_{to} and Kv4 channels with antisense oligonucleotides (Fiset *et al.*, 1997). Concordant changes in I_{to} and Kv4 channels also occur in response to changes in thyroid hormone levels in the rat ventricle (Shimoni et al., 1997). Very recently, strong supportive evidence has come from transgenic animal studies. Barry et al., (1998) made a point mutation in Kv4.2 channels (W362F) that functions as a dominant-negative mutant, attenuating wild-type currents. The W-to-F mutant makes channels that gate "normally," but do not pass K$^+$ current. I_{to} was selectively eliminated in this Kv4.2 mutant, with a corresponding increase in ventricular action potential duration and QT interval. There was no change in Kv1.4 expression in these animals. In another study (Kääb et al., 1998), RNase protection assays and electrophysiological techniques were employed to measure changes in K$^+$ channel expression in human ventricular myocytes from normal and "failing" hearts. In congestive heart failure, Kv4.3 RNA was decreased by $\sim 30\%$ compared to control hearts; in contrast, there were no changes in HERG or Kv1.4 channel RNA. Kv4.3 is the most likely basis for I_{to} in the human and dog heart ventricular muscle because there is no detectable Kv4.2 mRNA (Dixon et al., 1996).

The biophysical properties of Kv4 channels and I_{to} are very similar—both activate at negative potentials (above -40 mV) and are rapidly inactivating (with a decay time constant of ~ 50 ms) (Dixon et al., 1996; Fiset et al., 1997). Both are resistant to 5 mM TEA, and the inactivation is resistant to oxidative agents (Dixon et al., 1996), unlike that of Kv1.4 or Kv3.4 channels. Surprisingly, there is not much single-channel data to define the unitary conductances (18 pS for Kv4.2 versus 14 pS for I_{to}; see Fiset et al., 1997). There are reports of discrepancies in the rate of recovery from inactivation of Kv4.2 in Ltk$^-$ cell lines and I_{to} in rat heart, with the cloned channels being about seven times slower to recover than I_{to} (Fiset et al., 1997). It is possible that some of these discrepancies may be explained if additional accessory subunits are involved in changing gating in either the expression system or the native cell (Robertson, 1997).

Sanguinetti et al. (1997) have identified three peptidic toxins (heteropodatoxins; HpTX1–3), isolated from the venom of Malaysian *Heteropoda venatoria* spiders, which prolong ventricular action potentials by blocking I_{to}. HpTX1–3 are about 30 amino acids long and have roughly 40% sequence identity. Six cysteine residues within the sequence suggest tight disulfide bonding in the tertiary structure. HpTX2 (30 nM) lengthens the ventricular action potential and also blocks I_{to} by a voltage-dependent mechanism. The IC$_{50}$ of HpTX2 increases from 35 nM at $+20$ mV to 138 nM at $+50$ mV (see Fig. 3A,B). HpTX3 is quite similar ($\sim 39\%$) to the Chilean tarantula toxin hanatoxin2, which blocks Kv2.1 channels (Swartz and MacKinnon, 1995). In *Xenopus* oocytes, all three HpTXs block Kv4.2 in a voltage-dependent manner and slow its activation and inactivation rate. Sanguinetti et al., showed that HpTXs also shift steady-state inactivation. These authors also reported that Kv4.2 channels are blocked by hanatoxin (73% inhibition by 500 nM at 0 mV), which also blocks Kv2.1 in a voltage-dependent manner. The mechanisms for hanatoxin block have been elucidated in some detail for Kv2.1 channels. Hanatoxin is thought to bind to a site 10–15 Å away from the central pore axis, on the S3–S4 linker. Hanatoxin binding inhibits channel opening, possibly by impeding the outward movement of S4, until strong depolarizing potentials are reached (Swartz and MacKinnon, 1997). Stronger depolarizations also relieved heteropodatoxin block.

Very recently, two new spider toxins have been isolated; these phrixotoxins are also selective for Kv4.2 and Kv4.3 channels. Diochot et al., (1999) have shown that phrixotoxins block Kv4.3 and Kv4.2, at concentrations of 5–70 nM, by shifting the

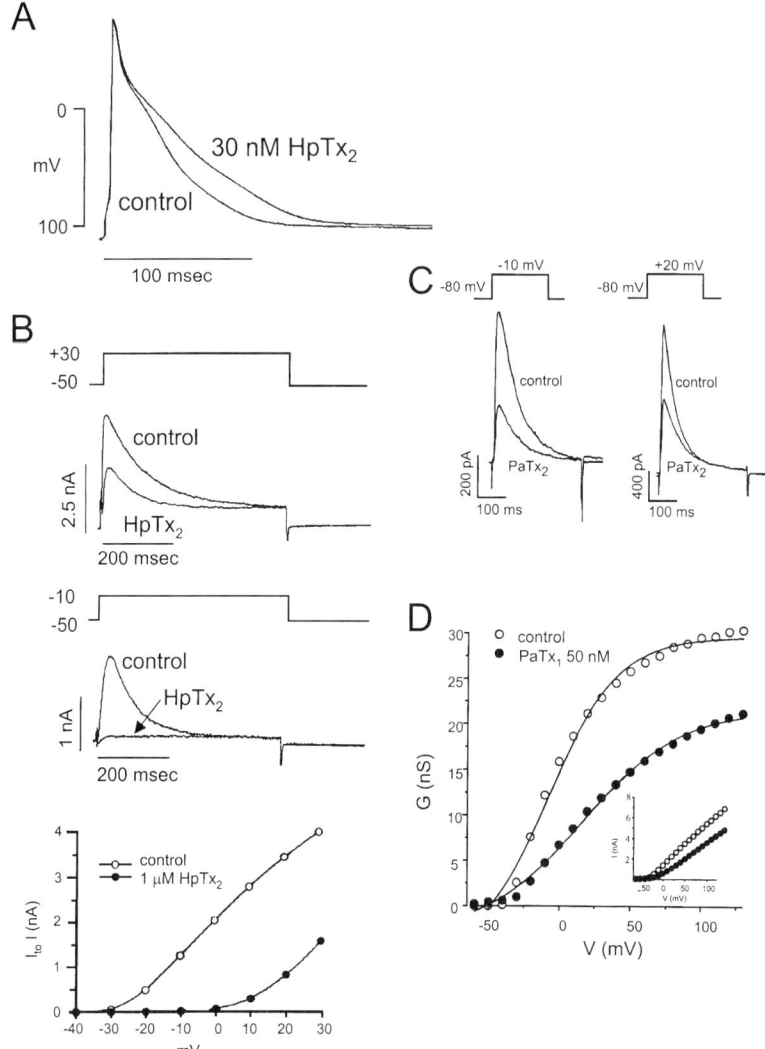

Figure 3. Actions of novel toxins on cardiac voltage-activated conductances and cloned channels. (A) Prolongation of the rat ventricular action potential by 30 nM heteropodatoxin (HpTX2) isolated from spider venom. (Reproduced, with permission, from Sanguinetti *et al.*, 1997.) (B) Voltage-dependent block of I_{to1} in isolated rat ventricular myocytes by HpTX2. Current recorded at -10 mV was completely blocked by 1 μM HpTX2, whereas current at $+30$ mV was only blocked by about half. Lower trace shows the corresponding current–voltage relationship before and after 1 mM HpTX2. (Reproduced, with permission, from Sanguinetti *et al.*, 1997.) (C) Block of Kv4.3 currents transfected into COS cells by phrixotoxins (PaTX1; 50 nM) from the tarantula. Block is more effective at -10 mV compared to $+20$ mV. (Reproduced, with permission, from Diochot *et al.*, 1999.) (D) Conductance–voltage relationship and current–voltage relationship (*inset*) for Kv4.3 current before and after block by 50 nM PaTX1. (Reproduced, with permission, from Diochot *et al.*, 1999.) Note that both toxins shift the current/conductance–voltage curves to the right.

activation curve in the positive direction. Similar to heteropodatoxin block, phrixotoxin block can be relieved by strong depolarizations (Fig. 3C,D). Phrixotoxins are 29–31 amino acids long and are derived from the tarantula *Phrixotrichus auratus*. These toxins have $\sim 50\%$ sequence identity with heteropodatoxins, and 20% with the hanatoxins. More work needs to be done to understand precisely how and where these exciting new tools act on Kv4 channels.

There is a greater amount of information available on the block of Kv4 channels by more conventional blockers. Yeola and Snyders (1997) compared the sensitivities of Kv1.4 and Kv4.2 to two effective blockers of I_{to}, quinidine and flecainide. Flecainide blocked Kv4.2 ($IC_{50} \sim 10 \ \mu M$) but had minimal effects on Kv1.4. Quinidine was about equipotent against each clone ($IC_{50} \approx 10 \ \mu M$).

Wang *et al.*, (1997) made the surprising finding that certain commonly used Cl$^-$ channel blockers previously used to dissect out a Ca^{2+}-activated Cl$^-$ component to I_{to} (e.g., Corabeouf and Carmeliet, 1982), inhibited Kv4.3 and Kv4.2 channels in expression systems. Niflumic acid, at 100 μM, causes a profound reduction of Kv4.3 (expressed in oocytes), accompanied by an ~ 10mV hyperpolarizing shift in steady-state activation and inactivation curves and a slowing of recovery from inactivation. 4,4'-Diisothiocyanatostilbene-2,2'-disulfonic acid (DIDS), at 100 μM, also reduced peak Kv4.3 current and caused a 3.3-fold slowing of inactivation recovery, without major changes in activation and inactivation curves. Kv4.2 channels, despite having a high amino acid identity with Kv4.3 channels ($\sim 75\%$), were much less sensitive to the Cl$^-$ channel blockers. Nonselective effects of Cl$^-$ channel blockers have been reported on other ionic conductances, including I_h a slow inward current in the SA node. Accili and DiFrancesco (1996) suggested that this I_h inhibition might be clinically relevant.

6. HERG CHANNELS

The human *ether-go-go*-related (HERG) channel is widely held to be the molecular correlate of the "rapid delayed rectifier K$^+$ current," I_{Kr}, in the heart (Sanguinetti *et al.*, 1995; Trudeau et al 1995; Curran *et al.*, 1995). Defects in the HERG gene have been shown to underlie LQT2 syndrome (see Chapter 36). HERG has aroused considerable interest as a drug target. HERG currents have a fairly high threshold for activation (above -50 mV, $V_{1/2} \approx -13$ mV), slow activation, and fast inactivation, which produces inward rectification at positive potentials (Sanguinetti *et al.*, 1995)

E-4031 is a methanesulfonamide antiarrhythmic, which has a QT-wave-prolonging effect and is thought to block I_{Kr} selectively. In the original reports, E-4031 was reported to block I_{Kr} in guinea pig myocytes with an IC_{50} of ~ 400 nM (Sanguinetti and Jurkiewicz, 1990), although recent experiments suggest that the IC_{50} value for I_{Kr} is 10 nM (Liu *et al.*, 1996). Similarly, there are large differences in the IC_{50} value of E-4031 for block of HERG in different expression systems [oocytes: 590 nM (Trudeau et al., 1995); HEK cells: ~ 8 nM (Zhou *et al.*, 1998)]. It is not known whether these differences may be explained by variations in experimental technique or preparation. Oocytes may not be ideal for such studies, because the yolk may absorb drug and the complex membranous structures could compromise diffusion.

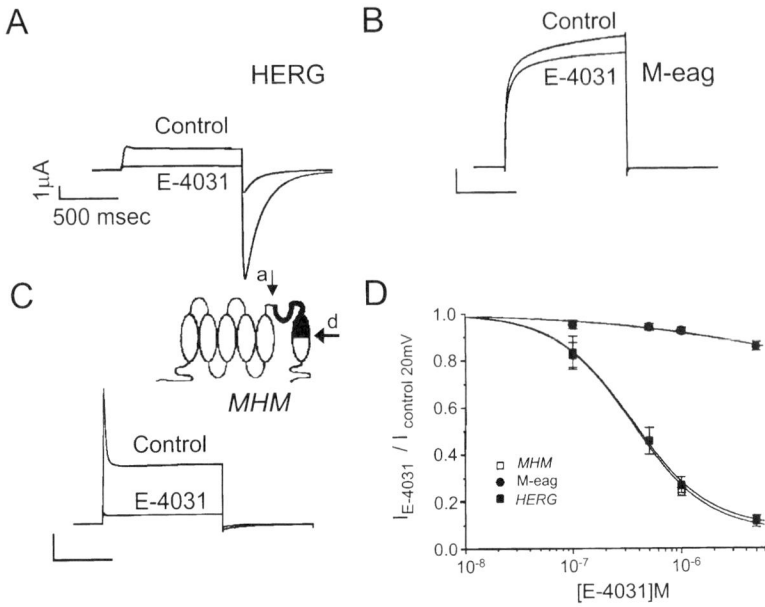

Figure 4. Transfer of the P-region and part of S6 from HERG channels into the M-eag channel background transfers sensitivity block by E-4031. (A)–(C) Steady-state block of HERG (A) and M-eag (B) and MHM chimeras (C) by 5 μM E-4031. All voltage steps are from -80 mV to $+20$ mV. (D) Concentration–response curves for E-4031 block. IC_{50} values are 348 nM for HERG and 346 nM for MHM channels, whereas M-eag channels are effectively insensitive to this drug. MHM contains amino acids 612–650 (the P-domain and half of the S6 domain) of HERG. This differs from the corresponding region in M-eag by only 15 residues. (After Herzberg *et al.*, 1998.)

Herzberg *et al.* (1998) have recently 'transferred' key regions of HERG and another, structurally related member of the *ether-a-go-go* (Eag), family, M-eag, making novel chimeras that have told us much about the regions involved in fast inactivation, as well as sensitivity to E-4031. M-eag channels are substantially insensitive to the antiarrhythmic, but a chimeric channel, having the P-region of HERG and half its S6 segment, in a background of M-eag, had enough of the drug binding site to give the same sensitivity to E-4031 (~ 350 nM each; see Fig. 4). Interestingly, this region is also a major part of the "fast inactivation" domain, in the upstream P-region. These novel chimeric channels, and future developments with these, will be vital in our deciphering of drug action on HERG. The most recent and extensive study of E-4031 on HERG expressed in mammalian cells (Zhou *et al.*, 1998) shows that the drug blocks only open channels.

The class III antiarrhythmic clofilium (a quaternary ammonium derivative) blocks a number of Kv currents in the heart but is "selective" for HERG currents at concentrations several orders of magnitude lower, suggesting that HERG is the main locus of its action in the heart (Suessbrich *et al.*, 1997b). The IC_{50} for clofilium is ~ 150 nM at $+40$ mV, and ~ 250 nM at 0 mV, whereas a tertiary analog of this drug, LY97241, was 10 times more potent still. Curiously, block in oocytes was poorly reversible, even after 2 h.

Dofetilide, another methanesulfonamide, also blocks HERG. Estimates of IC_{50} vary from $\sim 12-15$ nM in mammalian cell lines (Snyders and Chaudhary, 1996; Rampe et al., 1997) to 35 nM in oocyte patches. In "whole" oocytes, the IC_{50} for dofetilide is ~ 600 nM (Kiehn et al., 1996). These findings illustrate that one has to be cautious when comparing results obtained with different expression systems.

Terfenadine (Seldane) is a nonsedating antihistamine that was found to lead to prolonged QT syndrome (see, e.g., Roy et al., 1996). This drug blocks HERG in HEK cells with an IC_{50} of 56 nM (Rampe et al., 1997) and blocks HERG in oocytes less potently (Roy et al., 1996). Interestingly, patients treated with terfenadine occasionally present with long QT symptoms. Similarly, patients receiving the fungicide ketoconazole also occasionally develop long QT symptoms. Both compounds are metabolized by the same cytochrome P-450 pathway. If terfenadine and ketoconazole are coadministered, the competition between the two raises plasma terfenadine levels. Dumaine et al. (1998) recently showed that ketoconazole itself (in the micromolar range) can inhibit Kv1.5 and HERG currents and that coapplication of terfenadine and ketoconazole prolongs the cardiac action potential. These authors also concluded that block of HERG channels occurs via the closed-channel route. This study provides an interesting and important example of how combinations of drugs can lead to more serious effects than one alone.

Cisapride (Propulsid) is a gastrointestinal prokinetic agent commonly-used in patients suffering from reflux. However, there are reports of patients acquiring long QT syndrome at high doses of cisapride (see Rampe et al., 1997). Rampe et al., (1998) showed that these unwanted cardiac side effects might be associated with block of HERG by cisapride. They obtained an IC_{50} for HERG current by cisapride of ~ 45 nM. For the inhibition of HERG, the drug was almost 7 times more potent when it was allowed to equilibrate with the channel during very long depolarizations, again suggesting interaction with open states. In contrast, cisapride only blocked Kv1.5 channels at 1000-fold higher concentrations ($\sim 20~\mu M$).

The class III antiarrythmic agent azimilide blocks a number of voltage-gated channels in the heart, including HERG (see Fig. 5). Busch et al. (1998) have shown that azimilide blocks HERG channels in a reverse-use-dependent and voltage-independent manner, in contrast to all of the other common HERG channel blockers (E-4031, dofetelide, terfenadine, astemizole, clofilium, and haloperidol), which show both positive use dependence and voltage dependence in their blockade. This means that the block and apparent affinity of azimilide decrease with HERG channel activation frequency (e.g., $IC_{50} = 1.4~\mu M$ at 0.1 Hz, 5.2 μM at 1 Hz).

Haloperidol is a butyrophenone antipsychotic drug, used in the treatment of schizophrenia. The use of haloperidol has been complicated by several cases of acquired QT syndrome. Centrally, it acts by blocking dopamine receptors, but Suessbrich et al., (1997a) have shown that it can also effectively block HERG channels expressed in oocytes ($IC_{50} \approx 1~\mu M$). In contrast, it was ineffective against some other cloned Kv channels, including Kv1.5. Block is strongly use-dependent, increasing with activation frequency, and is stronger at more depolarized potentials. Further results suggest that haloperidol preferentially blocks HERG channels in the inactivated state, and this is supported by a fourfold reduction in sensitivity to block in a HERG mutant (S631A), that has markedly reduced inactivation (Suessbrich et al., 1997a).

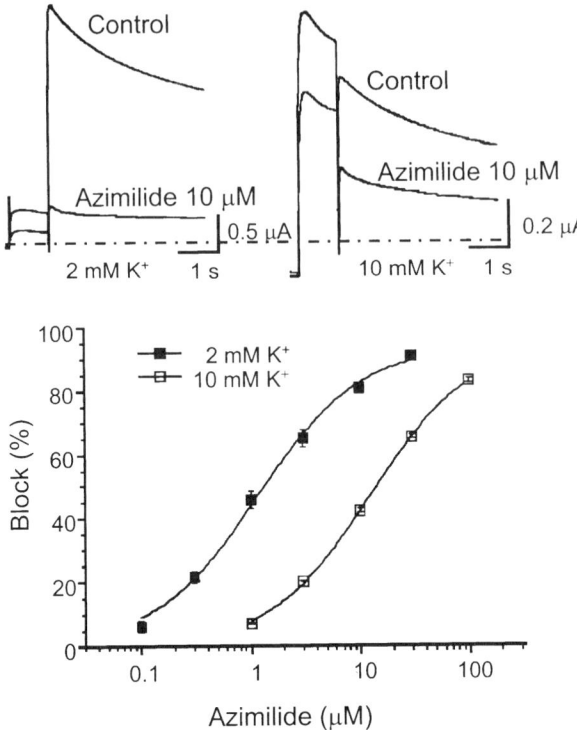

Figure 5. Block of HERG currents expressed in oocytes by azimilide. Block depends on extracellular K^+ concentration, decreasing with modest increases in K^+; contrast currents in 2 mM K^+ with those in 10 mM K^+. Complete concentration–response curve is shown at bottom. HERG channels were activated by a 1-s depolarizing step to +30 mV from −85 mV, and tail currents were measured at −55 mV. (Data from Busch *et al.*, 1998.)

Rampe *et al.*, (1998) have recently shown that the antipsychotic agent sertindole (a new indolylpiperidine which has nanomolar affinities for some dopaminergic, serotonergic, and adrenergic receptors) blocks HERG currents in mammalian cell lines. Sertindole causes HERG inhibition at the nanomolar levels at which it has its effects on adrenergic and serotonergic receptors. This may explain reports of acquired long QT syndrome in patients being treated with sertindole. Sertindole-induced HERG block is poorly reversible, and potency increases as HERG channels are kept open longer. In contrast, 1000-fold higher concentrations of sertindole are required to get significant block of Kv1.5 channels (Rampe *et al.*, 1998).

HERG currents are increased by elevating $[K^+]_o$, despite a decrease in driving force (Sanguinetti *et al.*, 1995; Schonherr and Heinemann, 1996). A similar effect is also seen for I_{Kr} in heart (Sanguinetti and Jurkiewicz, 1990). Increases in $[K^+]_o$ decrease the effectiveness of the HERG blockers clofilium (Suessbrich *et al.*, 1997b) and azimilide (Busch *et al.*, 1998; Fig. 5). Similarly, dofetilide is ∼26 times less effective against I_{Kr} as $[K^+]_o$ increases from 1 to 8 mM (Yang and Roden, 1996). It is likely that the $[K^+]_o$ effect is linked to modulation of C-type inactivation in HERG. Interestingly, in patients suffering from HERG-related long QT syndrome, Compton *et al.*, (1996) found that raised serum K$^+$ concentration reduced the QT interval. Whether these observations are linked is as yet a moot point.

Recently, Wilson *et al.* (1998) suggested that HERG currents may also be modified by the accessory β-subunits. Their *in vitro* study shows not only a tight physical coassociation between the HERG and β-subunits, but also consequent changes in expression and gating of HERG. McDonald *et al.*, (1997) have also demonstrated a tight association between the minK subunit and HERG. It will be important and useful to determine whether such coassembly between accessory subunits occurs in the heart.

Another route to modification of HERG is activation of protein kinase A. This leads to a suppression of HERG currents, due to a rightward shift in the steady-state activation curve, which can be as great as $+35$ mV (Kiehn *et al.*, 1998). It will be of interest to learn under what conditions this kinase effect occurs *in vivo*, since anything which modifies the availability of a channel will also contribute to its drug responsiveness.

7. KvLQT1/IsK OLIGOMERIC CHANNELS

Mutations in the KvLQT1 gene can cause the long QT syndrome. Coexpression of KvLQT1 channel α-subunits in association with the channel "accessory" protein minK (also called IsK) produces a K$^+$ current with characteristics reminiscent of the slow component of the delayed rectifier in cardiac myocytes, I_{Ks} (Barhanin *et al.*, 1996; Sanguinetti *et al.*, 1996). There are few selective blockers of the KvLQT1/IsK hetero-multimeric complex thus far; however, there are some interesting leads.

Chromanol 293B [*trans*-6-cyano-4-(*N*-ethylsulfonyl-*N*-methylamino)-3-hydroxy-1,2-dimethylchromane] is a blocker of I_{Ks} and KvLQT1 (Loussouarn *et al.*, 1997). The block is the same whether or not IsK is present. The IC$_{50}$ for KvLQT1 by chromanol 293B is $\sim 10\ \mu M$ in COS cells, which is similar to its IC$_{50}$ against I$_{Ks}$ in cardiomyocytes (Bosch *et al.*, 1998).

Azimilide blocks I_{Ks} in guinea pig cardiomyocytes, but this agent also blocks HERG currents (see Busch *et al.*, 1998) at marginally lower concentrations; Busch *et al.* (1998) concluded that the relative block of either channel by the drug will vary depending on several factors, such as heart rate and $[K^+]_o$.

Chouabe *et al.*, (1998) have made the significant observation that both KvLQT1/IsK and HERG are targets for certain Ca^{2+}-channel blockers (Fig. 6). HERG is blocked by verapamil, bepridil, and mibefradil at IC$_{50}$ concentrations as low as 1 μM, which are relevant to therapeutic values, whereas KvLQT1/IsK currents are effectively blocked only by bepridil and mibefradil (IC$_{50}$'s of $\sim 10\ \mu M$). With HERG, steady-state activation and inactivation curves were shifted more negative.

8. THE PACEMAKER CURRENT—HYPERPOLARIZATION-ACTIVATED CATION CHANNELS

Lately, we have seen the unraveling of the molecular identity of "pacemaking" currents, including I_h (also known as I_f, for "funny"). In the SA node, pacemaking

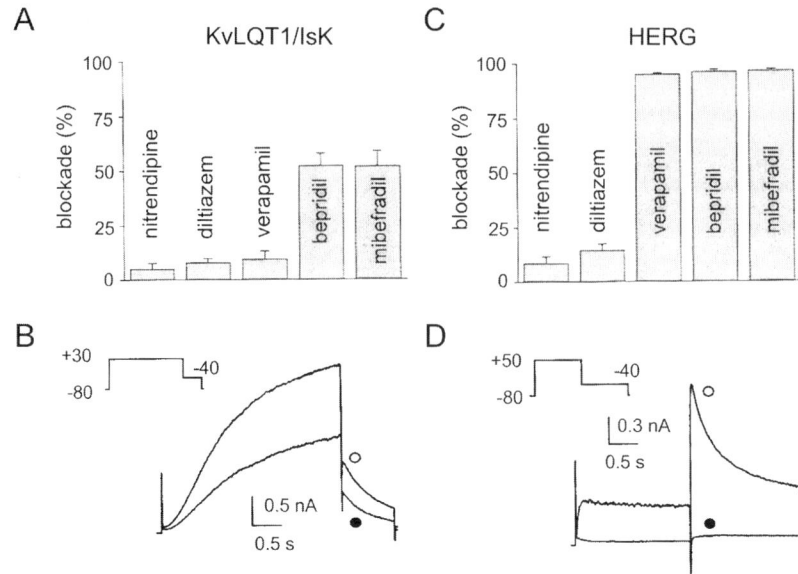

Figure 6. Calcium channel blockers, at clinically relevant concentrations, may block Kv channels expressed in COS cells. (A) Relative blockade of KvLQT1/IsK currents expressed in COS cells by 10 of each calcium channel antagonist. (B) Example traces of KvLQT1/Isk tail current block by 10 μM bepridil (○, control; ●, drug). Voltage step protocol is shown in inset. (C) Relative blockade of HERG currents by antagonists. (D) Representative traces illustrating that HERG current is also, but more dramatically than KvLQT1 current, blocked by 10 μM bepridil (●). Voltage step protocol used to elicit tail current at -40 mV is shown in inset. (All traces reproduced, with permission, from Chouabe *et al.*, 1998.)

activity is partly controlled by a hyperpolarization-activated, mixed Na^+/K^+ current, which in turn is regulated by adrenaline and cAMP (DiFrancesco, 1993). I_h is a slow inward current, activated following the hyperpolarization phase of an action potential. Because it is a mixed Na^+ and K^+ conductance (with an equilibrium potential of -30 mV), at more negative potentials it elicits an inward current, causing a slow depolarization of the cell, leading to another action potential. Figure 7A shows schematically the role played by I_h in SA node cells.

I_h in the heart is finely regulated by cAMP levels, as occurs following adrenergic stimulation. cAMP directly binds to the I_h channel protein, providing a vital clue to the molecular biologists searching for the correlate of I_h. Cyclic-nucleotide-gated channels were cloned in 1989. Three groups cloned the first set of "pacing proteins" in 1998 (Gauss *et al.*, 1998; Ludwig *et al.*, 1998; Santoro *et al.*, 1998), and those identified thus far have been called HCN channels. (hyperpolarization-activated, cyclic-nucleotide-gated ion channels) (Clapham, 1998). An excellent review on the properties of HCN channels has recently been published (Biel *et al.*, 1999).

HCN in heart was found by BLAST (Basic Local Alignment Search Tool) database searching using the cyclic nucleotide binding domain. Three HCN channel clones were found with 779–910 amino acids, a cyclic nucleotide-binding (CNB) domain in the C-terminus, six transmembrane segments (including a voltage-sensing S4 segment), and the classical GYG K^+ channel "signature sequence" in the pore region.

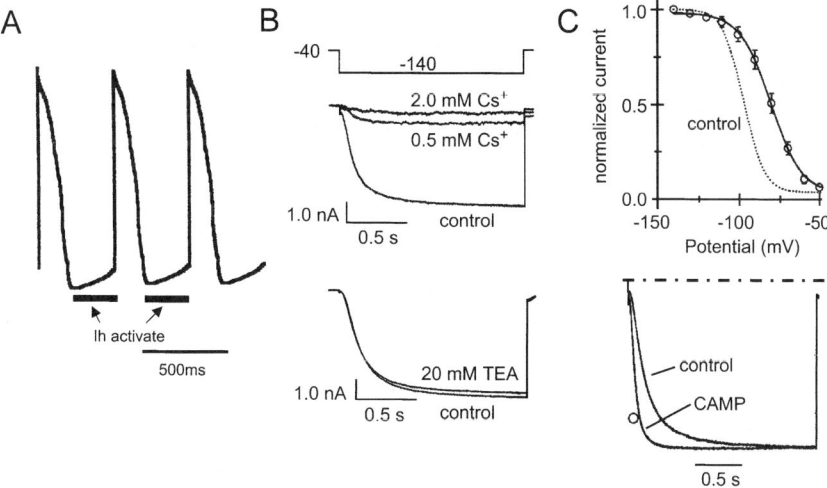

Figure 7. (A) Cartoon illustrating role of I_h in a sinoatrial node cell; it activates on the repolarization phase of the action potential, leading to a slight depolarization, which deactivates the current again. (B) Block of the cloned current (human) HCN2 expressed in HEK293 cells by low concentrations of Cs$^+$ externally (*top*). This current is unaffected by high concentrations of tetraethylammonium (TEA) (*bottom*). (C) Modulation of human HCN2 by cAMP. Top trace shows how the voltage dependence of activation is shifted rightward to more depolarized potentials following intracellular perfusion for 1 min with 1mM cAMP (O). The control activation curve is shown as a dotted line; both fits are Boltzmann curves. Lower trace shows how cAMP also increases the activation rate of this current (O) on a voltage step from -40 to -140 mV. [Traces in (B) and (C) reproduced, with permission, from Ludwig *et al.*, 1999].

Interesting, and potentially exploitable, differences exist between the P-region of the "classical" Kv channels and the HCN channels, especially in the selectivity filter (e.g., HCN: ---CIGYGRQAP---; Kv1.5: ---TVGYGDMRP---; Kv4.2: ---TLGYGDMVP---). The cysteine just before the GYG sequence is unusual, replacing the highly conserved threonine. Another difference in the selectivity filter is the positively charged arginine in HCN channels, rather than the almost ubiquitous aspartate. These substitutions may account for the loss of the otherwise exquisite K$^+$ selectivity of K$^+$ channels, making HCN currents more Na$^+$-permeable.

HCN2 channels were expressed in HEK293 cells (Ludwig *et al.*, 1998), revealing a hyperpolarizing-activated current (Fig. 7B), activating with a time constant of ~ 240 ms at -140 mV (at room temperature). $V_{1/2}$ for activation was ~ -100 mV, E_{rev} was -37mV in 5.4 mM [K$^+$]$_o$, suggesting a mixed Na/K conductance (Ludwig *et al.*, 1998). HCN2 thus is similar to I_h in cardiac cells (DiFrancesco, 1993). Another similarity to I_h is the finding that the HCN current is increased by both cAMP and cGMP (1 mM each), through a $+12$ to 14 mV shift in the steady-state activation curve (Fig. 7C), with an EC$_{50}$ of ~ 0.5 μM (Ludwig *et al.*, 1998; DiFrancesco, 1993, Hoppe and Beuckelmann, 1998).

The preliminary pharmacology of this cloned current is satisfyingly similar to that of I_h (Fig. 7B,C). HCN current is almost entirely abolished by 2 mM [Cs$^+$]$_o$, with EC$_{50}$

values for related HCNs of ~ 200 μM (Santoro *et al.*, 1998; see also Gauss *et al.*, 1998), and 2mM [Cs$^+$]$_o$ also completely eliminates I_h in human atrial myocytes (Hoppe and Beuckelmann, 1998) and reduces the heart rate of the isolated heart by 20% (Leitch *et al.*, 1995). The cloned hyperpolarization-activated current is resistant to external block by 20 mM TEA (owing to the absence of the critical tyrosine in the P-region) and 1 mM 4-AP, whereas 2 mM [Ba^{2+}]$_o$ blocks only about 20% of the current (Ludwig *et al.*, 1998). Thus far, the drug pharmacology of I_h is somewhat limited. *In vitro* studies show that the ZENECA compound, ZD7288 [4-(*N*-ethyl-*N*-phenylamino)-1,2-dimethyl-6-(methylamino)pyrimidinium chloride], a bradycardiac agent that is claimed to affect the SA node without having any other effect in the heart, (0.6 μM) reduces isolated heart rate by ~ 40–50% (Leitch *et al.*, 1995) and is active against I_h in both heart (BoSmith *et al.*, 1993) almost total block at 0.3 μM; and neurons (e.g., Luthi *et al.*, 1998). It is unknown if this compound blocks cloned HCN channels.

Ludwig *et al.* (1999) have recently suggested that the native I_h currents in human heart might be comprised of homomeric assemblies of two gene products (HCN2 and HCN4), which have similar pharmacological responses to cAMP, Cs$^+$, Ba^{2+}, and TEA, but markedly different activation rates.

9. CONCLUDING COMMENTS

We have seen merely a hint of the wealth of pharmacological data relevant to Kv channels in the heart. Advancement in this area has been spectacular since the first cloning of mammalian Kv channels, just over 10 years ago. The heart contains a huge variety of K$^+$ channel subtypes, and many of these are vital therapeutic targets. We hope that by knowing the amino acid sequence of many of these K$^+$ channels, and using many of the approaches employed above, we can dissect out areas which contribute to the channels' physiology and pharmacology and perhaps even selectively design specific agents for different channel subtypes. The next decade of K$^+$ channels in cardiovascular biology will be an even more astonishing one.

ACKNOWLEDGMENTS

Work in my own laboratory is financed by the Medical Research Council (G9802307) and the Wellcome Trust (045812), and I am very grateful for their support.

REFERENCES

Accili, E. A., and DiFrancesco, D. 1996 Inhibition of the hyperpolarization-activated current (I_h) of rabbit SA node myocytes by niflumic acid, *Pflügers Arch.* **431**:757–762.

Barhanin, J., Lesage, F., Guillemare, E., Fink, M., Lazdunski, M., and Romey, G., 1996, KvLQT1 and IsK (minK) proteins associate to form the I_{Ks} cardiac potassium current, *Nature* **384**:78–80.

Barry, D. M., Nerbonne, J. M. 1996, Myocardial potassium channels: electrophysiological and molecular diversity. [Review] [211 refs] *Annual Review of Physiology.* **58**:363–394.

Barry, D. M., Trimmer, J. S., Merlie, J. P. and Nerbonne, J. M., 1995 Differential expression of voltage-gated K$^+$ channel subunits in adult rat heart: Relation to functional K$^+$ channels?, *Circ. Res.* **77:**361–369.

Barry, D. M., Xu, H. D., Schuessler, R. B., and Nerbonne, J. M., 1998, Functional knockout of the transient outward current, long-QT syndrome, and cardiac remodelling in mice expressing a dominant-negative Kv4 α subunit, *Circ. Res.* **83:**560–567.

Biel, M., Ludwig, A., Zong, X., and Hofmann, F., 1999, Hyperpolarization-activated cation channels: A multi-gene family, *Rev. Physiol. Biochem. Pharmacol.* **136:**165–181.

Bosch, R. F., Gaspo, R., Busch, A. E., Lang, H. J., Li, G. R., and Nattel, S., 1998, Effects of the chromanol 293B, a selective blocker of the slow component of the delayed rectifier K$^+$ current, on repolarization in human and guinea pig ventricular myocytes, *Cardiovasc. Res.* **38:**441–450.

BoSmith, R. E., Briggs, I., and Sturgess, N. C., 1993, Inhibitory actions of ZENECA ZD7288 on whole-cell hyperpolarization activated inward current (I_f) in guinea pig dissociated sinoatrial node cells, *Br. J. Pharmacol.* **110:**343–349.

Brahmajothi, M. V., Morales, M. J., Liu, S., Rasmusson, R. L., and Strauss, H. C., 1996, *In situ* hybridization reveals extensive diversity of K$^+$ channel mRNA in isolated ferret cardiac myocytes, *Circ. Res.* **78:**1083–1089.

Brahmajothi, M. V., Morales, M. J., Rasmusson, R. L., Campbell, D. L., and Strauss, H. C., 1997, Heterogeneity in K$^+$ channel transcript expression detected in isolated ferret cardiac myocytes, *Pacing Clin. Electrophysiocol* **20:**388–396.

Brahamajothi, M. V., Campbell, D. L., Rasmusson, R. L., Morales, M. J., Trimmer, J. S., Nerbonne, J. M., and Strauss, H. C., 1999, Direct transient outward potassium current(I_{to}) phenotypes and distribution of fast-inactivating potassium channel α subunits in ferret left ventricular myocytes, *J. Gen. Physiol.* **113:**581–600.

Busch, A. E., Eigenberger, B., Jurkiewicz, N. K., Salata, J. J., Pica, A., Suessbrich, H., and Lang, F., 1998, Blockade of HERG channels by the class III antiarrhythmic azimilide: Mode of action, *Br. J. Pharmacol.* **123:**23–30.

Campbell, D. L., Yu, Y., Rasmusson, R. L., and Strauss, H. C., 1993, The calcium-independent transient outward potassium current in isolated ferret right ventricular myocytes. II. Closed state reverse use-dependent block by 4-aminopyridine, *J. Gen. Physiol.* **101:**603–626.

Castellano, A., Chiara, M. D., Mellstrom, B., Molina, A., Monje, F., Naranjo, J. R., and LopezBarneo, J., 1997, Identification and functional characterization of a K$^+$ channel α-subunit with regulatory properties specific to brain, *J. Neurosci.* **17:**4652–4661.

Castellino, R. C., Morales, M. J., Strauss, H. C., and Rasmusson, R. R., 1995, Time- and voltage-dependent modulation of a Kv1.4 channel by a β subunit (Kvβ3) cloned from ferret ventricle, *Am. J. Physiol.* **269:**H385–391.

Chandy, K. G., and Gutman, G. A., 1995, Voltage-gated potassium channel genes, in: *Handbook of Receptors and Channels: Ligand- and Voltage-Gated Ion Channels* (R. A. North, ed.) CRC Press, Boca Raton, Florida, 1–71.

Chen, F. S. P. and Fedida, D., 1998, On the mechanism by which 4-aminopyridine occludes quinidine block of the cardiac K$^+$ channel, hKv1.5, *J. Gen. Physiol.* **111:**539–554.

Chouabe, C., Drici, M.-D., Romey, G., Barhanin, J., and Lazdunski, M., 1998, HERG and KvLQT1/IsK, the cardiac K$^+$ channels involved in long QT syndromes, are targets for calcium channel blockers, *Mol. Pharmacol.* **54:**695–703.

Clapham, D. E., 1998 Not so funny anymore: Pacing channels are cloned, *Neuron* **21:**5–7.

Colatsky, T., 1982, Mechanisms of action of lidocaine and quinidine on action potential duration in rabbit cardiac Purkinje fibers, *Circ. Res.* **50:**17–27.

Compton, S. J., Lux, R. L., Ramsey, M. R., Strelich, K. R., Sanguinetti, M. C., Green, L. S., Keating, M. T., and Mason, J. W., 1996, Genetically defined therapy of inherited long-QT syndrome — Correction of abnormal repolarization by potassium, *Circulation* **94:**1018–1022.

Coraboeuf, E., and Carmeliet, E., 1982, Existence of two transient outward currents in sheep cardiac Purkinje fibres, *Pflügers Arch.* **392:**353–359.

Crumb, W., Wible, B., Arnold, A., Payne, J., and Brown, A., 1995, Blockade of multiple human cardiac potassium currents by the antihistamine terfenadine: Possible mechanisms for the terfenadine-associated cardiotoxicity, *Mol. Pharmacol.* **47:**181–190.

Curran, M. E., Splawski, I., Tomithy, K. W., Vincent, G. M., Green, E. D., and Keating, M. T., 1995, A molecular basis for cardiac arrhythmia: HERG mutations cause long QT syndrome. *Cell* **80:**795–803.

Delpon, E., Valenzuela, C., Gay, P., Franqueza, L., Snyders, D. J., and Tamargo, J., 1997, Block of human cardiac Kv1.5 channels by loratadine: Voltage-, time- and use-dependent block at concentrations above therapeutic levels, *Cardiovasc. Res.* **35**:341–350.

DiFrancesco, D., 1993, Pacemaker mechanisms in cardiac tissue, *Annu. Rev. Physiol.* **55**:455–472.

Diochot, S., Schweitz, H., Beress, L., and Lazdunski, M., 1998, Sea anemone peptides with a specific blocking activity against the fast inactivating potassium channel Kv3.4., *J. Biol. Chem.* **273**:6744–6749.

Diochot, S., Drici, M.-D., Moinier, D., Fink, M., and Lazdunski, M., 1999, Effects of phrixotoxins on the Kv4 family of potassium channels and implications for the role of I_{to1} in cardiac electrogenesis, *Br. J. Pharmacol.* **126**:251–263.

Dixon, J.E., and McKinnon, D., 1994, Quantitative analysis of potassium channel mRNA expression in atrial and ventricular muscle of rats, *Circ. Res.* **75**:252–260.

Dixon, J. E., Shi, W., Wang, H.-S., McDonald, C., Yu, H., Wymore, R. S., Cohen, I. S., and McKinnon, D., 1996, Role of the Kv4.3 K^+ channel in ventricular muscle, *Circ. Res.* **79**:659–668.

Drewe, J. A., Verma, S., Frech, G., and Joho, R. L., 1992, Distinct spatial and temporal expression patterns of K^+ channel mRNA's from different subfamilies, *J. Neurosci.* **12**:538–548.

Dumaine, R., Roy, M. L., and Brown, A. M., 1998, Blockade of HERG and Kv1.5 by ketoconazole, *J. Pharmacol. Expt. Ther.* **286**:727–735.

England, S. K., Uebele, V. N., Shear, H., Kodali, J., Bennett, P. B., and Tamkun, M. M., 1995, Characterization of a voltage-gated K^+ channel β-subunit expressed in human heart, *Proc. Natl. Acad. Sci. U.S.A.* **92**:6309–6313.

Fedida, D., 1997, Gating charge and ionic currents associated with quinidine block of human Kv1.5 delayed rectifier channels, *J. Physiol. (London)* **499**:661–675.

Fedida, D., Chen, F. S., Zhang, X. 1998, Cardiac K^+ channel gating: cloned delayed rectifier mechanisms and drug modulation, *Can. J. Physiol. Pharmacol.* **76**:77–89.

Fedida, D., Wible, B., Wang, Z., Ffermini, B., Faust, F., Nattel, S., and Brown, A. M., 1993, Identity of a novel delayed rectifier current from human heart with a cloned K^+ channel current, *Circ. Res.*, **73**:210–216.

Fedida, D., Chen, S. P., and Zhang, X., 1998, Cardiac K channel gating: Cloned delayed rectifier mechanisms and drug modulation, *Can. J. Physiol. Pharmacol.* **76**:77–89.

Fedida, D., Bouchard, R., and Chen, F. S. P., 1996, Slow gating charge immobilization in the human potassium channel Kv1.5 and its prevention by 4-aminopyridine, *J. Physiol. (London)* **494**:377–387.

Fiset, C., Clark, R. B., Shimoni, Y., and Giles, W. R., 1997, *Shal*-type channels contribute to the Ca^{2+}-independent transient outward K^+ current in rat ventricle, *J. Physiol. (London)* **500**:51–64.

Franqueza, L., Valenzuela, C., Delpon, E., Logobardo, M., Caballero, R., and Tamargo, J., 1998, Effects of propafenone and 5-hydroxypropafenone on hKv1.5 channels, *Br. J. Pharmacol.* **125**:969–978.

Gauss, R., Seifert, R., and Knapp, U. B., 1998, Molecular identification of a hyperpolarization-activated channel in sea urchin sperm, *Nature* **393**:583–587.

Gomez-Hernandez, J. M., Lorra, C., Pardo, L. A., Stuhmer, W., Pongs, O., Heinemann, S. H., and Elliott, A. A., 1997, Molecular basis for different pore properties of potassium channels from the rat brain Kv1 gene family, *Pflügers Arch.* **434**:661–668.

Grissmer, S., Nguyen, A. N., Aiyar, J., Hanson, D. C., Mather, R. J., Gutman, G. A., Karmilowicz, M. J., Auperin, D. D., and Chandy, K. G., 1994, Pharmacological characterization of five cloned K^+ channels, types Kv1.1, 1.2, 1.3, 1.5 and 3.1, stably expressed in mammalian cell lines, *Mol. Pharmacol.* **45**:1227–1234.

Harvey, A. L., and Anderson, A. J., 1991, Dendrotoxins: Snake toxins that block potassium channels and facilitate neurotransmitter release, in: *Snake Toxins* (A. L. Harvey, ed.), Pergamon Press, New York pp. 79–92.

Heginbotham, L., and MacKinnon, R., 1992, The aromatic binding site for tetraethylammonium ion on potassium channels, *Neuron* **8**:483–491.

Heinemann, S. H., Rettig, J., Graack, H.-R., and Pongs, O., 1996, Functional characterization of Kv channel β-subunits from rat brain, *J. Physiol. (London)* **493**:625–633.

Herzberg, I. M., Trudeau, M. C., and Robertson, G. A., 1998, Transfer of rapid inactivation and sensitivity to the class III antiarrhythmic drug E-4031 from HERG to M-eag channels, *J Physiol. (London)* **511**:3–14.

Hoppe, U. C., and Beuckelmann, D. J., 1998, Characterization of the hyperpolarization-activated inward current in isolated human atrial myocytes, *Cardiovasc. Res.* **38**:788–801.

Hurst, R. S., Busch, A. E., Kavanaugh, M. P., Osborne, P. B., North, R. A., and Adelman, J. P., 1991, Identification of amino acid residues involved in dendrotoxin block of rat voltage dependent potassium channels, *Mol. Pharmacol.* **40**:572–576.

Kääb, S., Dixon, J., Duc, J., Ashen, D., Nabauer, M., Beuckelmann, D. J., Steinbeck, G., McKinnon, D., and Tomaselli, G. F., 1998, Molecular basis of transient outward potassium current downregulation in human heart failure: A decrease in Kv4.3 mRNA correlates with a reduction in current density, *Circulation* **89**:1383–1393.

Kalman, K., Nguyen, A., TsengCrank, J., Dukes, I. D., Chandy, G., Hustad, C. M., Copeland, N. G., Jenkins, N. A., Mohrenweiser, H., Brandriff, B., Cahalan, M., Gutman, G. A., and Chandy, K. G., 1998, Genomic organization, chromosomal localization, tissue distribution, and biophysical characterization of a novel mammalian Shaker-related voltage-gated potassium channel, Kv1.7, *J. Biol. Chem.* **273**:5851–5857.

Kavanaugh, M. P., Varnum, M. D., Osborne, P. B., Christie, M. J., Busch, A. E., Adelman, J. P., and North, R. A., 1991, Interaction between tetraethylammonium and amino acids in the pore of cloned voltage-dependent potassium channels, *J. Biol. Chem.* **266**:7583–7587.

Kiehn, J., Lacerda, A. E., Wible, B., and Brown, A. M., 1996, Molecular physiology and pharmacology of HERG: Single channel currents and block by dofetilide, *Circulation* **94**:2572–2579.

Kiehn, J., Karle, C., Thomas, D., Yao, X., Brachmann, J., and Kubler, W., 1998, HERG potassium channel activation is shifted by phorbol esters via protein kinase A-dependent pathways, *J. Biol. Chem.* **273**:25285–25291.

Kirsch, G. E., Shieh, C. C., Drewe, J. A., Vener, D. F., and Brown, A. M., 1993, Segmental exchanges define 4-aminopyridine binding and the inner mouth of K$^+$ pores, *Neuron* **11**:503–512.

Kramer, J. W., Post, M. A., Brown, A. M., and Kirsch, G. E., 1998, Modulation of potassium channel gating by coexpression of Kv2.1 with regulatory Kv5.1 or Kv6.1 α-subunits, *Am. J. Physiol.* **43**:C1501–C1510.

Lacerda, A. E., Roy, M. L., Lewis, E. W., and Rampe, D., 1997, Interactions of the nonsedating antihistamine with a Kv1.5 type potassium channel cloned from human heart, *Mol. Pharmacol.* **52**:314–322.

Leitch, S. P., Sears, C. E., Brown, H. F., and Paterson, D. J., 1995, Effects of high potassium and the bradycardiac agents ZD7288 and cesium on heart-rate of rabbits and guinea-pigs, *J. Cardiovasc. Pharmacol.* **25**:300–306.

Liu, S., Rasmusson, D. L., Campbell, S., Wang, S., and Strauss, H. C., 1996, Activation and inactivation kinetics of an E-4031-sensitive current from single ferret atrial myocytes, *Biophys. J.* **70**:2704–2715.

Loussouarn, G., Charpentier, F., MohammadPanah, R., Kunzelmann, K., Baro, I., and Escande, D., 1997, KvLQT1 potassium channel but not IsK is the molecular target for *trans*-6-cyano-4-(*N*-ethylsulfenyl-N-methylamino)-3-hydroxy-2,2-dimethylchromane, *Mol. Pharmacol.* **52**:1131–1136.

Ludwig, A., Zong, X., Jeglitsch, M., Hofmann, F., and Biel, M., 1998, A family of hyperpolarization-activated cation channels, *Nature* **393**:587–591.

Ludwig, A., Zong, X., Stieber, J., Hullin, R., Hofmann, F., and Biel, M., 1999, Two pacemaker channels from human heart with profoundly different activation kinetics, *EMBO J.* **18**:2323–2329.

Luthi, A., Bal, T., and McCormick, D. A., 1998, Periodicity of thalamic spindle waves is abolished by ZD7288, a blocker of I_h, *J. Neurophysiol.* **79**:3284–3289.

MacKinnon, R., and Yellen, G., 1990, Mutations affecting TEA blockage and ion permeation in voltage-activated K$^+$ channels, *Science* **250**:276–279.

Majumder, K., DeBiasi, M., Wang, Z. G., and Wible, B. A., 1995, Molecular-cloning and functional expression of a novel potassium channel β-subunit from human atrium, *FEBS Lett.* **361**:13–16.

McDonald, T. V., Yu, Z. H., Ming, Z., Palma, E., Meyers, M. B., Wang, K. W., Goldstein, S. A. N., and Fishman, G. I., 1997, A minK–HERG complex regulates the cardiac potassium current I_{Kr} *Nature*, **388**:289–292.

Morales, M. J., Castellino, R. C., Crews, A. L., Rasmusson, R. L., and Strauss, H. C., 1995, A novel β subunit increases rate of inactivation of specific voltage-gated potassium channel α subunits, *J. Biol. Chem.* **270**:6272–6277.

Nerbonne, J. M., 1998, Regulation of voltage-gated K$^+$ channel expression in the developing mammalian myocardium, *J. Neurobiol.* **37**:37–59.

Post, M. A., Kirsch, G. E., and Brown, A. M., 1996, Kv2.1 and electrically silent Kv6.1 potassium channel subunits combine and express a novel current, *FEBS Lett.* **399**:177–182.

Rampe, D., and Murawsky, M. K., 1997, Blockade of the human cardiac K$^+$ channel Kv1.5 by the antibiotic erythromycin, *Naunyn-Schmeideberg's Arch. Pharmacol.* **355**:743–750.

Rampe, D., Wible, B., Brown, A., and Dage, R., 1993a, Effects of terfenadine and its metabolites on a delayed rectifier cloned from human heart, *Mol. Pharmacol.* **44**:1240–1245.

Rampe, D., Wible, B., Fedida, D., Dage, R. C., and Brown, A. M., 1993b, Verapamil blocks a rapidly activating delayed rectifier K$^+$ channel cloned from human heart, *Mol. Pharmacol.* **44**:642–648.

Rampe, D., Roy, M.-L., Dennis, A., and Brown, A. M., 1997, A mechanism for the proarrhythmic effects of cisapride (Propulsid): High affinity blockade of the human cardiac potassium channel HERG, *FEBS Lett.* **417**:28–32.

Rampe, D., Murawsky, M. K., Grau, J., and Lewis, E. W., 1998, The antipsychotic agent sertindole is a high affinity antagonist of the human cardiac potassium channel HERG, *J. Pharmacol. Exp. Ther.* **286**:788–793.

Rasmusson, R. L., Morales, M. J., Castellino, R. C., Zhang, Y., Campbell, D. L., and Strauss, H. C., 1995, C-type inactivation controls recovery in fast inactivating cardiac K$^+$ channel (Kv1.4) expressed in *Xenopus* oocytes, *J. Physiol. (London)* **489**:709–721.

Rettig, J., Heinemann, S. H., Wunder, F., Lorra, C., Parcej, D. N., Dolly, J. O., and Pongs, O., 1994, Inactivation properties of voltage gated K$^+$ channels altered by presence of β subunit, *Nature* **369**:289–294.

Roberds, S. L., and Tamkun, M. M., 1991a, Cloning and tissue specific expression of five voltage gated potassium channel cDNAs expressed in rat heart, *Proc. Natl. Acad. Sci. U.S.A.* **88**:1798–1802.

Roberds, S. L., and Tamkun, S. L., 1991b, Developmental expression of cloned cardiac potassium channels,, *FEBS Lett.* **284**:152–154.

Robertson, B., 1997, The real life of voltage-gated K$^+$ channels: More than model behaviour, *Trends Pharmacol. Sci.* **18**:474–483.

Robertson, B., Owen, D. G., Stow, J., Butler, C., and Newland, C., 1996, Novel effects of dendrotoxin homologues on subtypes of mammalian Kv1 potassium channels expressed in *Xenopus* oocytes, *FEBS Lett.* **383**:26–30.

Roy, M.-L., Dumaine, R., and Brown, A. M., 1996, HERG, a primary human ventricular target of the nonsedating antihistamine terfenadine, *Circulation* **94**:817–823.

Salinas, M., Duprat, F., Heurteaux, C., Hugnot, J. P., and Lazdunski, M., 1997, New modulatory α subunits for mammalian *Shab* K$^+$ channels, *J. Biol. Chem.* **272**:24371–24379.

Sanguinetti, M. C., and Jurkiewicz, N. K., 1990, Two components of cardiac delayed rectifier current, *J. Gen. Physiol.* **96**:195–215.

Sanguinetti, M. C., Jiang, C., Curran, M. E., and Keating, M. T., 1995, A mechanistic link between inherited and an acquired cardiac arrhythmia: HERG encode the I_{Kr} potassium channel, *Cell* **81**:299–307.

Sanguinetti, M. C., Curran, M. E., Zou, A., Shen, J., Spector, P. S., Atkinson, D. L., and Keating, M. T., 1996, Coassembly of KvLQT1 and minK (IsK) proteins to form the cardiac I_{Ks} potassium channel, *Nature* **384**:80–83.

Sanguinetti, M. C., Johnson, J. H., Hammerland, L. G., Kelbaugh, P. R., Volkmann, R. A., Saccomano, N. A., and Mueller, A. L., 1997, Heteropodatoxins: Peptides isolated from spider venom that block Kv4.2 potassium channels, *Mol. Pharmacol.* **51**:491–498.

Santoro, B., Liu, D. T., Yao, H., Bartsch, D., Kandel, E. R., Siegelbaum, S. A., and Tibbs, G. R., 1998, Identification of a gene encoding a hyperpolarization-activated pacemaker channel of brain, *Cell* **93**:717–729.

Schonherr, R., and Heinemann, S. H., 1996, Molecular determinants for activation and inactivation of HERG, a human inward rectifier potassium channel, *J. Physiol. (London)* **493**:635–642.

Shi, G., Nakahira, K., Hammond, S., Rhodes, K. J., Schechter, L. E., and Trimmer, J. S., 1995, β Subunits promote K$^+$ channel surface expression through effects early in biosynthesis, *Neuron* **16**:843–852.

Shieh, C.-C., and Kirsch, G. E., 1994, Mutational analysis of ion conduction and drug binding sites in the inner mouth of voltage-gated K$^+$ channels, *Biophys. J.* **67**:2316–2325.

Shimoni, Y., Fiset, C., Clark, R. B., Dixon, J. E., McKinnon, D., and Giles, W. R., 1997, Thyroid hormone regulates postnatal expression of transient K$^+$ channel isoforms in rat ventricle, *J. Physiol. (London)* **500**:65–73.

Snyders, D. J., and Chaudhary, A., 1996, High affinity open channel block by dofetilide of HERG expressed in a human cell line, *Mol. Pharmacol.* **49**:949–955.

Snyders, D. J., Knoth, K. M., Roberds, S. L., and Tamkun, M. M., 1992, Time-, voltage- and state-dependent block by quinidine of a cloned human cardiac potassium channel, *Mol. Pharmacol.* **41**:322–330.

Stephens, G. J., Garratt, J. C., Robertson, B., and Owen, D. G., 1994, On the mechanism of 4-aminopyridine action on the cloned mouse brain potassium channel, mKv1.1, *J. Physiol. (London)* **477**:187–196.

Suessbrich, H., Schonherr, R., Heinemann, S. H., Attali, B., Lang, F., and Busch, A. E., 1997a, The inhibitory effect of the antipsychotic drug haloperidol on the HERG potassium channels expressed in *Xenopus* oocytes, *Br. J. Pharmacol.* **120:**968–974.

Suessbrich, H., Schonherr, R., Heinemann, S. H., Lang, F., and Busch, A. E., 1997b, Specific block of cloned HERG channels by clofilium and its tertiary analog LY97241, *FEBS Lett.* **414:**435–438.

Swartz, K. J., and MacKinnon, R., 1995, An inhibitor of the Kv2.1 potassium channel isolated from the venom of a Chilean tarantula, *Neuron* **15:**941–949.

Swartz, K. J., and MacKinnon, R., 1997, Mapping the receptor site for hanatoxin, a gating modifier of voltage-dependent K$^+$ channels, *Neuron* **18:**675–682.

Tamkun, M. M., Knoth, K. M., Walbridge, J. A., Kroemer, H., Roden, D. M., and Glover, D. M., 1991, Molecular-cloning and characterization of 2 voltage-gated K$^+$ channel cDNA's from human ventricle, *FASEB J.* **5:**331–337.

Trudeau, M. C., Warmke, J., Ganetzky, B., and Robertson, G., 1995, HERG, a human inward rectifier in the voltage-gated potassium channel family, *Science* **269:**92–95.

Tseng, G. N., Zhu, B., Ling, S., and Yao, J. A., 1996, Quinidine enhances and suppresses Kv1.2 from outside and inside the cell, respectively *J. Pharmacol. Exp. Ther.* **279:**844–855.

Uebele, V. N., England, S. K., Chaudhary, A., Tamkun, M. M., and Snyders, D. J., 1996, Functional differences in Kv1.5 currents expressed in mammalian cell lines are due to the presence of endogenous Kvβ2.1 subunits, *J. Biol. Chem.* **271:**2406–2412.

Valenzuela, C., Delpon, E., Franqueza, L., Gay, P., Vicente, J., and Tamargo, J., 1997, Comparative effects of nonsedating histamine H1 receptor antagonists, ebastine and terfenadine, on human Kv1.5 channels, *Eur. J. Pharmacol.* **326:**257–263.

Van Wagoner, D. R., Kirian, M., and Lamorgese, M., 1996, Phenylephrine suppresses outward K$^+$ current in rat atrial myocytes, *Am. J. Physiol.* **217:**H772–H781.

Vega-Saenz de Miera, E., Weiser, M., Kentros, C., Lau, D., Moreno, H., Serodio, P., and Rudy, B., 1994, *Shaw*-related K$^+$ channels in mammals, in: *Handbook of Membrane Channels: molecular and cellular physiology*, Academic Press, New York, pp. 41–78.

Wang, H.-S., Dixon, J. E., and McKinnon, D., 1997, Unexpected and differential effects of Cl$^-$ channel blockers on the Kv4.3 and Kv4.2 K$^+$ channels, *Circ. Res.* **81:**711–718.

Wang, L.-Y., Gan, L., Forsythe, I. D., and Kaczmarek, L. K., 1998, Contribution of the Kv3.1 potassium channel to high-frequency firing in mouse auditory neurones, *J. Physiol. (London)* **509:**183–194.

Wang, Z., Fermini, B., and Nattel, S., 1995, Effects of flecainide, quinidine, and 4-aminopyridine on transient outward and ultrarapid delayed rectifier currents in human atrial myocytes, *J. Pharmacol. Exp. Ther.* **272:**184–196.

Wang, Z., Feng, J., Shi, H., Pond, A., Nerbonne, J. M., and Nattel, S., 1999, Potential molecular basis of different physiological properties of the transient outward K$^+$ current in rabbit and human atrial myocytes, *Circ. Res.* **84:**551–561.

Wilson, G. F., Wang, Z., Chouinard, S. W., Griffith, L. C., and Ganetzky, B., 1998, Interaction of the K channel β subunit, Hyperkinetic, with eag family members, *J. Biol. Chem.* **273:**6389–6394.

Xu, H., Dixon, J. E., Barry, D. M., Trimmer, J. S., Merlie, J. P., McKinnon, D., and Nerbonne, J. M., 1996, Developmental analysis reveals mismatches in the expression of K$^+$ channel α-subunits and voltage-gated K$^+$ channel currents in rat ventricular myocytes, *J. Gen. Physiol.* **108:**405–419.

Yang, T., and Roden, D. M., 1996, Extracellular potassium modulation of drug block of I_{Kr}. *Circulation* **93:**407–411.

Yeola, S. W., and Snyders, D. J., 1997, Electrophysiological and pharmacological correspondence between Kv4.2 current and rat cardiac transient outward current, *Cardiovasc. Res.* **33:**540–547.

Zhang, X., and Fedida, D., 1998, Potassium channel-blocking actions of nifedipine: A cause for morbidity at high doses?, *Circulation* **97:**2098.

Zhang, X., Anderson, J. W., and Fedida, D., 1997, Characterization of nifedipine block of the human heart delayed rectifier, hKv1.5, *J. Pharmacol. Exp. Ther.* **281:**1247–1256.

Zhou, Z., Gong, Q., Ye, B., Fan, Z., Makielski, J. C., Robertson, G. A., and January, C. T., 1998, Properties of HERG channels stably expressed in HEK 293 cells studied at physiological temperature, *Biophys. J.* **74:**230–241.

Chapter 13

Pharmacology of High-Conductance, Ca^{2+}-Activated Potassium Channels

Maria L. Garcia and Gregory J. Kaczorowski

1. INTRODUCTION

Potassium channels represent a diverse family of ion channels whose members display the common property of being highly selectivity for potassium as the conducting ion. Support for the idea of the large diversity of potassium channels is gained from sequencing the *Caenorhabditis elegans* genome. It is predicted that about 100 K^+ channel subunits exist in this organism. These K^+ channel subunits belong to eight conserved K^+ channel families (Wei *et al.*, 1996), and the high-conductance, Ca^{2+}-activated K^+ (B_{KCa}) channel is a representative member of one of these families. B_{KCa} channels are widely distributed in both electrically excitable and nonexcitable cells (Latorre *et al.*, 1989; McManus, 1991), are activated by both voltage and intracellular Ca^{2+}, and display a high conductance, as well as a high selectivity for K^+. They regulate excitation–contraction coupling processes in vascular, airway, bladder, and other types of smooth muscle and also control transmitter release from neuroendocrine tissue. In nonexcitable cells, B_{KCa} channels regulate fluid secretion and cell volume. In various tissues, B_{KCa} channels are modulated by exogenous ligands signaling through their respective membrane receptors. Regulatory mechanisms such as phosphorylation, interaction with GTP-binding proteins, or direct modulation by intracellular second messengers have been identified and charcterized.

BK_{Ca} channels have been purified to homogeneity from bovine tracheal and aortic smooth muscle, and their subunit composition has been determined (Garcia-Calvo *et al.*, 1994; Giangiacomo *et al.*, 1995). In these smooth muscle tissues, the molecular components of these channels consist of two noncovalently linked subunits, α and β. The pore-forming a α-subunit is a member of the Slo1 family of K^+ channels (Knaus et al., 1994c), whereas the β-subunit is a unique protein (Knaus *et al.*, 1994b) that alters

Maria L Garcia and Gregory J. Kaczorowski ● Membrane Biochemistry and Biophysics, Merck Research Laboratories, Rahway, New Jersey 07065.

Potassium Channels in Cardiovascular Biology, edited by Archer and Rusch. Kluwer Academic/Plenum Publishers, New York, 2001.

the biophysical, pharmacological, and biochemical properties of the pore. Slo1 channels were initially identified in mammals based on sequence homology with the *Drosophila Slowpoke* gene product (Adelman *et al.*, 1992; Butler *et al.*, 1993). A single Slo1 gene gives rise to multiple functional B_{KCa} channel isoforms owing to the existence of splice variants (Dworetzky *et al.*, 1994; McCobb *et al.*, 1995; Pallanck and Ganetzky, 1994; Tseng-Crank *et al.*, 1994; Vogalis *et al.*, 1996; Wallner *et al.*, 1995). Some splice variants of Slo1 express channels with different Ca^{2+}-sensitivities (Tseng-Crank *et al.*, 1994). It could be argued that differences in Ca^{2+} sensitivity between native B_{KCa} channels could be due to specific tissue expression of a particular splice variant. However, it appears that α- and β-subunits are not always coexpressed together, and this could also account, at least in part, for differences in the properties of native channels (McCobb *et al.*, 1995). Recently, a new potassium channel gene, *Slack*, has been isolated (Joiner *et al.*, 1998). Slack channels have a lower unitary conductance than Slo1 channels, are inhibited by intracellular Ca^{2+}, and are abundantly expressed in the nervous system. Coexpression of Slack and Slo1 α-subunits yields channels with pharmacological and biophysical properties that are not representative of either channel. The Slack/Slo channels have intermediate conductance and are activated by cytoplasmic calcium. This channel phenotype may correspond to some intermediate-conductance, Ca^{2+}-activated K^+ channels found in the nervous system.

With the use of epitope tags, it has been possible to determine the topology of Slo1 channels. These channels appear to consist of seven α-helical transmembrane domains (S0–S6), with an extracellular N-terminus (Fig. 1) (Meera *et al.*, 1997). This topological

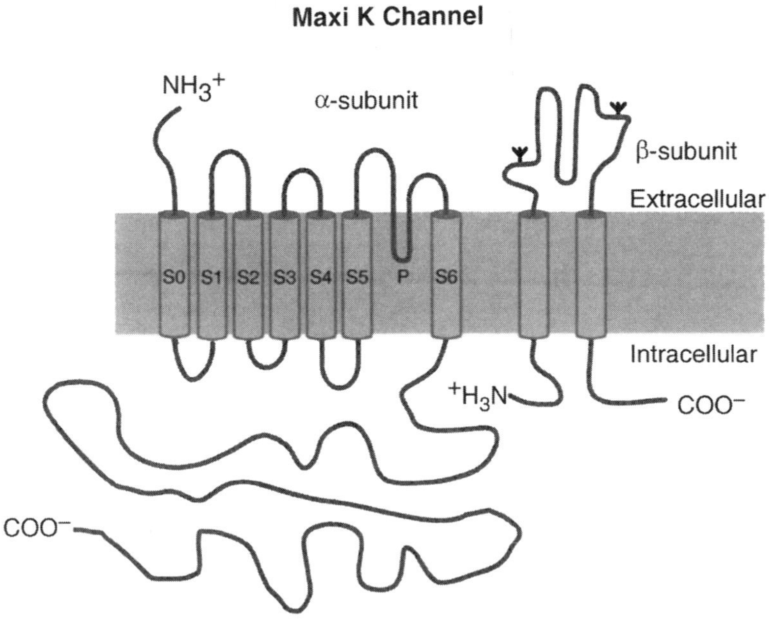

Maxi K Channel

Figure 1. Structure of BK_{Ca} channel α- and β-subunits. The predicted transmembrane topology of the α- and α-subunits of the BK_{Ca} channel is illustrated. Four such complexes are predicted to associate and form a functional BK_{Ca} channel.

arrangement differentiates Slo1 channels from other voltage-gated K^+ channels that are believed to contain six α-helical transmembrane domains, with an intracellular N-terminus. In addition, Slo1 channels have a large C-terminus, with four hydrophobic regions (S7–S10). There is good evidence that S9–S10 are not membrane associated, but rather are cytoplasmic (Meera et al., 1997). Whether S7 and S8 cross the membrane as α-helical segments or are cytoplasmic as well has not been established. A region of negatively charged aspartic acid residues present between S9 and S10 has been identified as one important region for conferring Ca^{2+} sensitivity to the channel and has been termed "the Ca^{2+} bowl" (Schreiber and Salkoff, 1997). However, natural splice variants, outside this aspartic acid-rich region, have different Ca^{2+} sensitivities, suggesting that various regions of Slo1 may contribute to the Ca^{2+} binding site. It is important to note that BK_{Ca} channels can be opened in the absence of Ca^{2+} by large membrane depolarizations (Cui et al., 1997; Meera et al., 1996) because they contain an intrinsic voltage sensor whose rearrangement triggered by voltage induces measurable gating currents. The S4 region in hSlo is part of the voltage sensor, and two charged amino acid residues in this region, Arg_{210} and Arg_{213} appear to contribute to the gating valence of the channel (Diaz et al., 1998). The role of Ca^{2+} is to shift the voltage dependence for channel opening to more hyperpolarized potentials.

The topology of the β-subunit of the smooth muscle BK_{Ca} channel is well established; it contains two α-helical transmembrane domains connected by a large extracellular loop, with both the N- and C-termini being cytoplasmic (Fig. 1) (Knaus et al., 1994a). The β-subunit is glycosylated at two positions, with sugars representing about one-third of the mass of the protein. Recent studies have determined that the four Cys residues located within the extracellular loop of the β-subunit are paired in the form of two disulfide bridges; this arrangement should lead to a more restricted conformation of that loop (Hanner et al., 1998). Slo1 channels are sensitive to a number of peptidyl inhibitors purified from scorpion venoms, such as charybdotoxin (ChTX), iberiotoxin (IbTX), and limbatustoxin (LbTX) (Garcia et al., 1997). These agents bind through a simple bimolecular reaction to a receptor site present in the outer vestibule of the α-subunit and inhibit ion conduction by physical occlusion of the pore. When either $[^{125}I]ChTX$ or $[^{125}I]IbTX$ is bound in the vestibule of the α-subunit, addition of a bifunctional cross-linking reagent causes covalent coupling of Lys_{32} of each peptide with Lys_{69} of the smooth muscle β-subunit (Knaus et al., 1994a; Koschak et al., 1997; Munujos et al., 1995). These data suggest that some residues of the β-subunit are in close proximity to the vestibule of the α-subunit.

Heterologous expression of the α-subunit leads to the appearance of BK_{Ca} channels with defined Ca^{2+} sensitivities, depending on the splice variant used. However, coexpression of both α- and β-subunits causes marked changes in the properties of the α-subunit. Although single-channel conductance is not altered in the presence of the β-subunit, there is a shift of approximately 80 mV in the hyperpolarizing direction for the midpoint of channel activation (Dworetzky et al., 1996; McCobb et al., 1995; McManus et al., 1995; Wallner et al., 1995). This shift is equivalent to that produced by a 10-fold rise in intracellular Ca^{2+} concentration. With the use of chimeric molecules constructed between dSlo, the Drosophila analogue of Slo1, which does not couple functionally to the β-subunit, and hSlo1, which does show coupling, it has been determined that S0 and the N-terminus of hSlo1 are required for the association with the β-subunit that leads to the profound effects on channel activity observed for hSlo1

(Wallner *et al.*, 1996). Under low-ionic-strength conditions, the properties of the interaction of $[^{125}I]$-ChTX with membranes derived from cells transiently transfected with either α or $\alpha + \beta$ subunits of the BK_{Ca} channel are markedly different. Although in each type of membrane $[^{125}I]$ChTX binds to a single class of sites that displays nearly identical density in both types of membranes, the presence of the β-subunit induces a 50-fold increase in $[^{125}I]$-ChTX affinity, due to both a decrease in toxin dissociation kinetics and an increase in the kinetics of ligand association (Hanner *et al.*, 1997, 1998). However, other pharmacological properties of the BK_{Ca} channel are not altered by coexpression with the β-subunit. Alanine scanning mutagenesis has identified four residues in the extracellular loop of the β-subunit, Leu_{90}, Tyr_{91}, Thr_{93}, and Glu_{94}, that are critical for conferring high-affinity $[^{125}I]$ChTX binding (Hanner *et al.*, 1998). Mutations at these positions cause large effects on the kinetics of ligand association and dissociation, but they do not alter the physical interaction of the β-subunit with the α-subunit. It is possible that S0 of Slo1 is involved in α-helical interactions with one of the α-helical transmembrane domains of the β-subunit and that this interaction could account for the large conformational change in the α-subunit that favors transition to the channel open state and alters the biochemical and pharmacological properties of the α-subunit upon binding of the β-subunit.

At the mRNA level, both α- and β-subunits coexist in a number of tissues (Chang *et al.*, 1997; Tseng-Crank *et al.*, 1996; Vogalis *et al.*, 1996). Coexpression of the α- and β-subunits of the BK_{Ca} channel may be a way in which channel activity is regulated *in vivo*. Smooth muscle tissues such as aorta, colon, coronary artery, and trachea express both subunits of the BK_{Ca} channel (Garcia-Calvo *et al.*, 1994; Giangiacomo *et al.*, 1995; Tanaka *et al.*, 1997; Vogalis *et al.*, 1996). In these tissues, it is expected that the channel could respond to subtle changes in Ca^{2+} concentration and play a predominant role in controlling cellular excitability. However, in brain the level of β-subunit mRNA is much lower than the level of α-subunit mRNA. In those tissues in which α-subunits do not appear to be associated with a smooth muscle-like α-subunit, such as in the central nervous system (CNS); (Chang *et al.*, 1997; Tseng-Crank *et al.*, 1996), larger changes in Ca^{2+} concentration may be needed for the channel to open at relevant membrane potentials. It is possible that these CNS α-subunits possess a different subunit that functions equivalently to the smooth muscle β-subunit. Evidence for the existence of a brain-specific BK_{Ca} channel β-subunit has been obtained after covalent incorporation of $[^{125}I]$IbTX into a protein that appears, based on pharmacological data, to be associated with the brain BK_{Ca} channel (Koch *et al.*, 1996). Its identity and functional role(s) remain to be determined. Recently, Wallner *et al.* (1999) identified a novel β-subunit ($\beta2$) after probing databases with the smooth muscle β-subunit sequence. The deduced amino acid sequence of the $\beta2$-subunit shows $\sim 53\%$ sequence similarity and $\sim 43\%$ sequence identity to the smooth muscle β-subunit. Substantial levels of the $\beta2$-subunit mRNA are found in human fetal kidney, but very low levels are present in several other human tissues. Functional expression with the α-subunit in oocytes leads to fast-inactivating currents that resemble the inactivating BK_{Ca} currents recorded from rat chromaffin cells. Inactivation is eliminated by application of trypsin to the intracellular surface of the patch (Meera *et al.*, 1999), a finding similar to that observed with the native chromaffin current (Solaro and Lingle, 1992). Similarly to the smooth muscle β-subunit, the $\beta2$-subunit shifts the voltage activation of the channel to more hyperpolarized potentials. Moreover, $\beta2$-subunit coexpression significantly reduces the channel sensitivity to externally applied ChTX, and this sensitivity resembles that of native BK$_{Ca}$ currents from chromaffin cells (Ding *et al.*, 1998).

2. PHARMACOLOGY OF BK$_{Ca}$ CHANNELS

2.1. Peptidyl Inhibitors

Development of the pharmacology of K$^+$ channels, and in particular the BK$_{Ca}$ channel, has been initiated within the last few years and is still an active area of research. The first potent modulators of the BK$_{Ca}$ channel to be identified were the peptidyl inhibitors ChTX, IbTX and limbatustoxin LbTX, purified from venom of the scorpions *Leiurus quinquestriatus* var. *hebraeus*, *Buthus tamulus*, and *Centruroides limbatus*, respectively. These peptides are 37 amino acids in length, display high homology in their primary amino acid sequences, and contain six conserved cysteine residues that form three disulfide bridges (Fig. 2). The three-dimensional structures of ChTX and IbTX have been determined in solution by nuclear magnetic resonance (NMR) techniques (Bontems *et al.*, 1992, *b*; Johnson and Sugg, 1992). The backbone structure of the two peptides is identical and consists of three anti-parallel β-strands that are connected by disulfide bridges to a helix region composed of residues 10–18 lying behind the plane of the β-sheet surface. All side chains of the amino acids, except for the cysteine residues, are exposed to solvent. ChTX is not a selective probe for BK$_{Ca}$ channels, since it blocks other small-conductance Ca^{2+}-activated K$^+$ channels and also some voltage-dependent K$^+$ channels, such as KV1.2 and KV1.3 (Garcia *et al.*, 1997). However, IbTX and LbTX are highly selective blockers of the BK$_{Ca}$ channel and, under physiologic ionic-strength conditions, display K$_d$ values of 1–5 n*M* (Garcia *et al.*, 1997). These peptides bind with high affinity to residues in the outer vestibule of the BK$_{Ca}$ channel and inhibit channel activity by blocking the ion conduction pathway. In single-channel recordings of BK$_{Ca}$ channels reconstituted into planar lipid bilayers, the peptides added at the external face of the channel cause the appearance of long silent periods interspersed between bursts of normal channel activity (Anderson *et al.*, 1988; Candia *et al.*, 1992; Giangiacomo *et al.*, 1992). These silent periods represent the times at which a single toxin molecule is bound in the mouth of the channel to block ion conduction. Because toxin binding is a freely reversible process, toxin dissociation leads to normal channel behavior because channel gating kinetics are faster than those of toxin binding. Block occurs again when another toxin molecule binds to the channel. Binding of these peptides is driven by electrostatic interactions between positively

ChTX	Z F T N V S C T T S K E C W S V C Q R L H N T S R G K C M N K K C R C Y S
IbTX	Z F T D V D C S V S K E C W S V C K D L F G V D R G K C M G K K C R C Y Q
LbTX	V F I D V S C S V S K E C W A P C K A A V G T D R G K C M G K K C K C Y ...

Figure 2. Comparison of the amino acid sequences of charybdotoxin (ChTX), iberiotoxin (IbTX), and limbatustoxin (LbTX).

charged residues on the toxin molecule and negatively charged residues located in the external vestibule of the channel (Anderson *et al.*, 1988; Giangiacomo *et al.*, 1992; MacKinnon *et al.*, 1989).

Mutagenesis studies have identified residues that are important for ChTX and IbTX interactions with the BK_{Ca} channel (Mullmann *et al.*, 1999; Stampe *et al.*, 1994). These residues are located on the three antiparallel β-strand surface and are conserved between ChTX and IbTX. Lys_{27} in the peptides is an important residue because it is responsible for the interaction between the peptide and K^+ entering the ion conduction pathway from the inside (MacKinnon and Miller, 1988). If position 27 of the peptides carries a positively charged residue, internal K^+ accelerates the dissociation rate of the peptide in a voltage-dependent manner. However, if a neutral asparagine or glutamine is substituted at this position, the dissociation rate is completely insensitive to either internal K^+ or applied voltage (Park and Miller, 1992a, b). These studies suggest that when the peptides are bound to the channel, Lys_{27} lies in close physical proximity to a K^+ specific binding site located in the outer part of the conduction pathway. When K^+ occupies this site, it causes a destabilization of the peptide–receptor interaction through a direct electrostatic repulsion of the ε-amino group of Lys_{27}.

Radiolabeling of ChTX at the Tyr_{36} residue with $[^{125}I]NaI$ leads to a biologically active peptide with high specific activity (Vazquez *et al.*, 1989). This reagent has been used for identifying BK_{Ca} channels in different smooth muscle tissues, for purifying BK_{Ca} channels from native tissues, and for developing the molecular pharmacology of the channel (Garcia *et al.*, 1997). A ChTX mutant containing a cysteine residue at position 19 has been produced and shown to have the same biological activity as the native peptide. This ChTX mutant can be derivatized with thiol-alkylating reagents such as $[^3H]N$-ethylmaleimide $[^3H]NEM$ or fluorescent derivatives of maleimide to yield novel ChTX adducts that retain full biological activity (Shimony *et al.*, 1994). Initial attempts to radiolabel IbTX with $[^{125}I]NaI$ led to a derivative with no biological activity. A biologically active radiolabeled derivative of IbTX has been prepared by substituting cysteine at position 19 in this peptide and reacting that residue with $[^3H]NEM$. $[^3H]NEM$-IbTX has been useful in identifying BK_{Ca} channel receptor sites in rat brain membranes, a tissue in which use of $[^{125}I]ChTX$ to study BK_{Ca} channels is hampered by the presence of relatively high densities of ChTX-sensitive, voltage-gated K^+ channels (Knaus *et al.*, 1996). Recently, it has been possible to obtain a derivative of IbTX with high specific activity by incorporation of iodine into the mutant IbTX[D19Y/Y36F] (Koschak *et al.*, 1997). This IbTX double mutant displays the same potency and selectivity as IbTX, but this mutation allows for the radiolabel to be incorporated at a position within the peptide (i.e., Tyr19) that is not critical for its interaction with the channel. $[^{125}I]IbTX$-[D19Y/Y36F] has been useful for autoradiographic localization of BK_{Ca} channels in brain tissue (Koch *et al.*, 1996).

Peptide inhibitors of BK_{Ca} channels have been used to investigate the functional role of this channel in smooth muscle physiology (Jones *et al.*, 1990; Suarez-Kurtz *et al.*, 1991; Winquist *et al.*, 1989). In some guinea pig smooth muscle tissues such as bladder and ileum, myogenic activity is stimulated by both ChTX and IbTX, suggesting that BK_{Ca} channels provide a major repolarization pathway in such electrically active tissues. In other quiescent tissues, such as guinea pig aorta and trachea, ChTX and IbTX cause contraction, demonstrating that BK_{Ca} channels can also control the resting membrane potential of smooth muscle cells. Interestingly, the peptides do not modify the activity of several other guinea pig smooth muscles that are known to possess BK$_{Ca}$ channels (e.g., portal vein, uterus), indicating that in these cases other ionic conductances are responsible for repolarizing the membrane during action potential generation. In addition, the role of the BK$_{Ca}$ channel in smooth muscle appears to be species-dependent; BK$_{Ca}$ channel inhibitors increase the myogenic activity of rat portal vein but have no effect on the same tissue in guinea pig.

The role of the BK$_{Ca}$ channel in guinea pig airway smooth muscle has been investigated in some detail (Jones et al., 1990, 1993). Relaxation of carbachol-contracted tissue can be accomplished by a variety of agents (e.g., β-adrenergic agonists, phosphodiesterase inhibitors, elevators of cGMP), and both ChTX and IbTX interfere with this process by shifting the dose–response curves for these agents to the right or, in some cases, abolishing the relaxing effect. Similar data have been obtained with human tissue (Miura et al., 1992). These data suggested initially that part or all of the relaxation mechanism of these agents involved activation of BK$_{Ca}$ channels, and it was therefore predicted that BK$_{Ca}$ channel activation would be a novel mechanism by which to cause airway smooth muscle relaxation. Indeed, independent studies (Kume et al., 1992, 1989) have shown that BK$_{Ca}$ channels in trachea can be activated by β-adrenergic receptor agonists, either through protein kinase A-dependent phosphorylation or by direct G-protein modulation. However, part of the effects elicited by the BK$_{Ca}$ channel blocking peptides is due to functional antagonism of the smooth muscle relaxants (Huang et al., 1993). Therefore, it is presently unclear what role BK$_{Ca}$ channels play in the mechanism of various smooth muscle relaxants and whether activation of the BK$_{Ca}$ channel by itself would cause smooth muscle relaxation, as is clearly produced by ATP-dependent K$^+$ channel agonists. Taken together, these studies highlight some of the difficulties associated with using functional studies to address the physiologic role of BK$_{Ca}$ channels in intact tissues.

2.2. Small-Molecule Modulators of BK$_{Ca}$ Channels

The search for BK$_{Ca}$ channel modulators has yielded both small-molecule channel agonists and antagonists using [^{125}I]ChTX binding screens with smooth muscle sarcolemmal membrane vesicles (Fig. 3). Alcoholic extracts of the plant *Desmodium adscendens* inhibit [^{125}I]ChTX binding in tracheal sarcolemmal membranes. It is interesting to note that these extracts are used in Ghana for treatment of asthma, dysmenorrhea, and other conditions associated with smooth muscle dysfunction. Three active components of the plant extract were purified and their structures determined (McManus et al., 1993). The three compounds are known glycosylated triterpenes:dehydrosoyasaponin I (DHS-I), soyasaponin I, and soyasaponin III. The most potent of these agents in blocking [^{125}I]ChTX binding is DHS-I (K_i = 120 nM, 60% maximal inhibition). The other two components display K_I values of 6 and 1 μM, respectively, whereas the aglycone triterpene is inactive. These data suggest a defined structure–activity relationship where both sugar and triterpene moieties are important for interaction with the BK$_{Ca}$ channel. These agents modulate [^{125}I]ChTX binding through an allosteric mechanism and, consistent with this idea, increase the rate of toxin dissociation from its receptor.

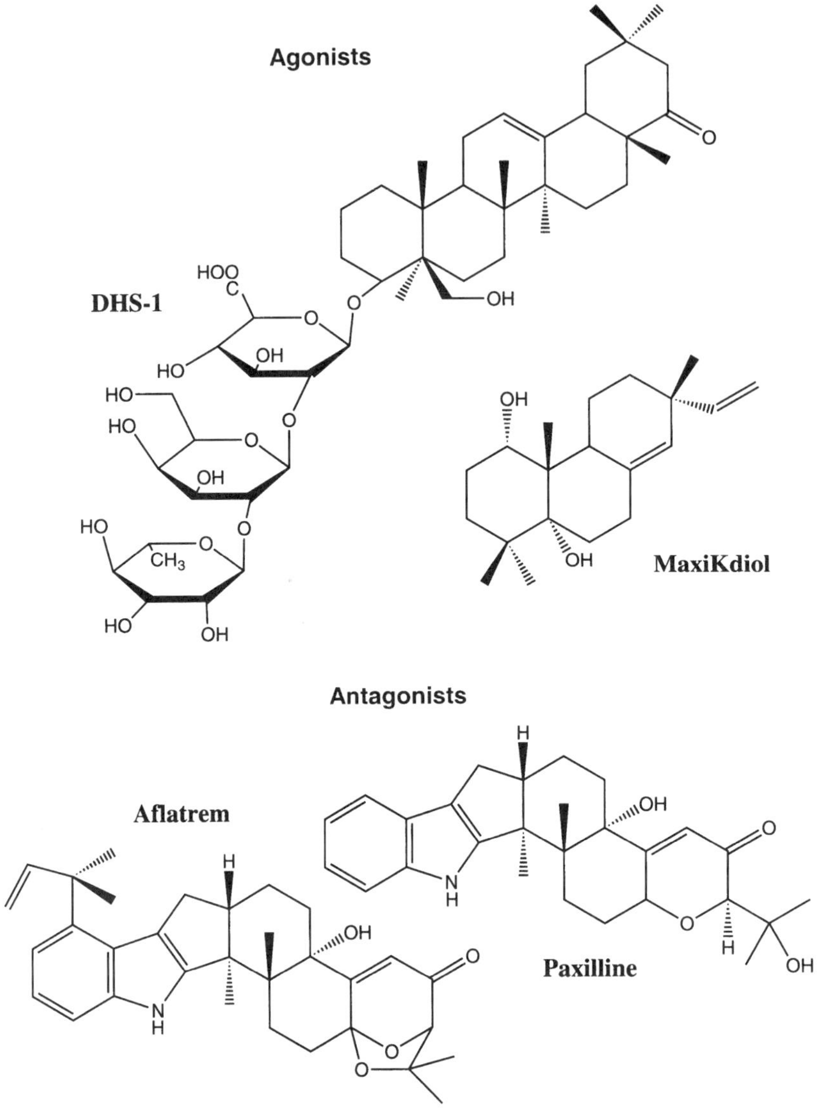

Figure 3. Modulators of BK_{Ca} channels isolated from natural sources.

In single-channel recordings of BK_{Ca} channels incorporated into planar lipid bilayers, addition of DHS-I (10 nM) causes a reversible increase in the open-state probability of the channel. This effect is only observed when DHS-I is added at the intracellular face of the channel. DHS-I added at the cytoplasmic face of the channel causes a 3- to 8-fold increase in the rate of ChTX dissociation from the channel's outer vestibule, similar to results obtained in [125I]ChTX dissociation experiments. These data further illustrate the allosteric coupling between peptide and drug receptors. DHS-I activates BK_{Ca} channels through a high-order reaction; three to four DHS-I molecules bind in order to maximally activate the channel (Giangiacomo et al., 1998). DHS-I modifies the calcium- and voltage-dependent gating of the channel. At 100 nM DHS-I, there is a threefold decrease in the calcium concentration that is required for half-maximum activation of the channel, and a hyperpolarizing shift of 80 mV in the midpoint of channel activation. Changes in the calcium dependence and the voltage dependence of channel activation induced by 100 nM DHS-I correspond to a 10- to 30 -fold increase and a ~ 1000-fold increase in steady-state channel opening, respectively. These findings suggest that DHS-I differentially modulates calcium- versus voltage-dependent gating. Changes in voltage-dependent gating can be explained by a model that considers the existence of four identical, noninteracting binding sites for DHS-I in the BK_{Ca} channel. DHS-I would bind to the open conformation of the channel with 10- to 20-fold higher affinity than for closed conformations (Giangiacomo et al., 1998).

When the BK_{Ca} channel α-subunit alone is expressed in Xenopus oocytes, the resulting channels are insensitive to DHS-I. However, after coexpression of both α- and β-subunits, BK_{Ca} channels are activated by this compound (McManus et al., 1995). These data demonstrate a clear requirement for the β-subunit in conferring channel sensitivity to DHS-I. Because the intracellular domains of the β-subunit are small and DHS-I only acts when added at the intracellular face of the channel, it is most likely that the conformational changes induced by the β-subunit during channel gating result in the appearance of a high-affinity site for DHS-I on the α-subunit. DHS-I does not affect other voltage-gated K^+ channels, ATP-dependent K^+ channels, Na^+ and Ca^{2+} channels, or various membrane transporters, suggesting that DHS-I is a potent and selective BK_{Ca} channel agonist. Activation of BK_{Ca} channels in certain smooth muscle tissues should lead to membrane hyperpolarization, which would diminish electrical excitability and cause smooth muscle relaxation. Unfortunately, the intracellular site of action of DHS-I, together with its poor membrane permeability, have precluded the pharmacological evaluation of this BK_{Ca} channel agonist. It is interesting to note that alcoholic extracts of Desmodium adscendens must be given chronically to elicit therapeutic benefit. Perhaps these agents are metabolized in vivo to other active species that can more easily penetrate cells.

From a fermentation broth of an unidentified coelomycite, a novel dihydroxyisoprimane diterpene termed "maxikdiol" was isolated based on its ability to inhibit $[^{125}I]$ChTX binding to smooth muscle membranes (Singh et al., 1994). Maxikdiol inhibits $[^{125}I]$ChTX binding with a K_i of 1 μM and markedly enhances the rate of dissociation of ChTX from its receptor, suggesting an allosteric interaction between two distinct receptor sites. In functional experiments, maxikdiol (3–10 μM) causes a reversible increase in BK_{Ca} channel activity. Unlike DHS-I, the effects of maxikdiol can be observed when the α-subunit of the channel is expressed alone in Xenopus oocytes (Kaczorowski et al., 1996). Although maxikdiol does not affect other types of K^+ channels investigated, it does display significant Ca^{2+} entry blocker activity at 10 μM. This property complicates interpretation of those in vitro pharmacological experiments in which maxikdiol has been shown to elicit relaxation of several smooth muscle preparations.

The first structural series of potent nonpeptidyl inhibitors of the BK_{Ca} channel was also discovered using $[^{125}I]$ChTX binding to smooth muscle membranes. These agents belong to a family of tremorgenic indole alkaloids (Knaus et al., 1994d). Some members of this family, such as paspalitrem A and C, aflatrem, penitrem A, and paspalinine, inhibit [125I]ChTX binding with a defined rank order of potency and either increase or have no effect on toxin dissociation kinetics. In contrast, paxilline, verruculogen, and paspalicine enhance [125I]ChTX binding in a concentration-dependent fashion and slow the dissociation of [125I]ChTX from its receptor. The indole diterpene alkaloids either abolish or enhance the covalent incorporation of [125I]ChTX and [125I]IbTX[D19Y/Y36F] into the β-subunit of the BK_{Ca} channel, consistent with their effects on binding of these ligands (Knaus et al., 1994d; Koschak et al.,] 1997). These data suggest a specific interaction of indole diterpenes with a unique site on the BK_{Ca} channel that can modulate ChTX binding to its receptor via either positive or negative allosteric interactions. The modulatory effect of indole diterpenes on [125I]ChTX binding is identical whether or not the β-subunit is coexpressed with the α-subunit (Hanner et al., 1997), suggesting that the receptor site resides solely within the α-subunit of the BK_{Ca} channel. Recent data indicate that the site of action of indole diterpenes is located within S0–S6 of the α-subunit (unpublished observations).

Although indole diterpenes display a range of effects as modulators of both [125I]ChTX and [125I]IbTX-[D19Y/Y36F] binding, all potently (0.1–10 nM) block BK_{Ca} channels in functional experiments, and block appears to be state dependent (Knaus et al., 1994d). Thus, when Ca^{2+} concentration is elevated and channel open-state probability is high, these agents are weaker inhibitors than when the channel open-state probability is low. The indole alkaloids are highly selective BK_{Ca} channel blockers; they have no effect on other K^+, Ca^{2+}, or Na^+ channels at micromolar concentrations. It is interesting to note that paspalicine, which lacks a C-19 hydroxyl group, does not have tremorgenic activity in vivo, but it is a potent BK_{Ca} channel blocker. This suggests that the tremorgenic activity of indole diterpenes may not be related to BK_{Ca} channel blockade but could be a consequence of effects of indole alkaloids on GABA-gated Cl- channels (Yao et al., 1989). All these data, taken together, indicate that indole diterpenes are the most potent and selective nonpeptidyl inhibitors of BK_{Ca} channels identified to date.

The pharmacology of indole diterpene BK_{Ca} channel blockers has been characterized in smooth muscle (DeFarias et al., 1996). Compounds such as paxilline and paspalitrem C increase the spontaneous contractility of guinea pig and rat urinary bladder, whereas they cause contraction of guinea pig trachea. The effects of indole diterpenes on smooth muscle tissues are similar to those elicited by peptidyl BK_{Ca} channel blockers. Interestingly, paxilline is able to potentiate the ability of ChTX to increase the myogenic activity of guinea pig detrusor muscle; this finding is consistent with binding data which demonstrate a positive allosteric coupling between indole diterpenes and toxin receptor sites on the channel.

A series of indole carboxylate compounds, CGS 7181 (Fig. 4), CGS 7184, CGS 7590, and CGS 7725, have recently been characterized for their ability to activate BK_{Ca} channels present in smooth muscle cells from vascular (artery) and nonvascular (bladder) tissues (Hu et al., 1997). Under whole-cell recording conditions, an increase in channel activity is observed after addition of the compounds to the bath solution. The minimal effective concentrations of these compounds ranged between 0.1 and 0.5 μM At a concentration of 5 μM, CGS 7181 and CGS 7184 increase the BK_{Ca} channel current of porcine coronary artery cells by 5- and 14-fold, respectively. However, application of the compounds to the intracellular side of the membrane during

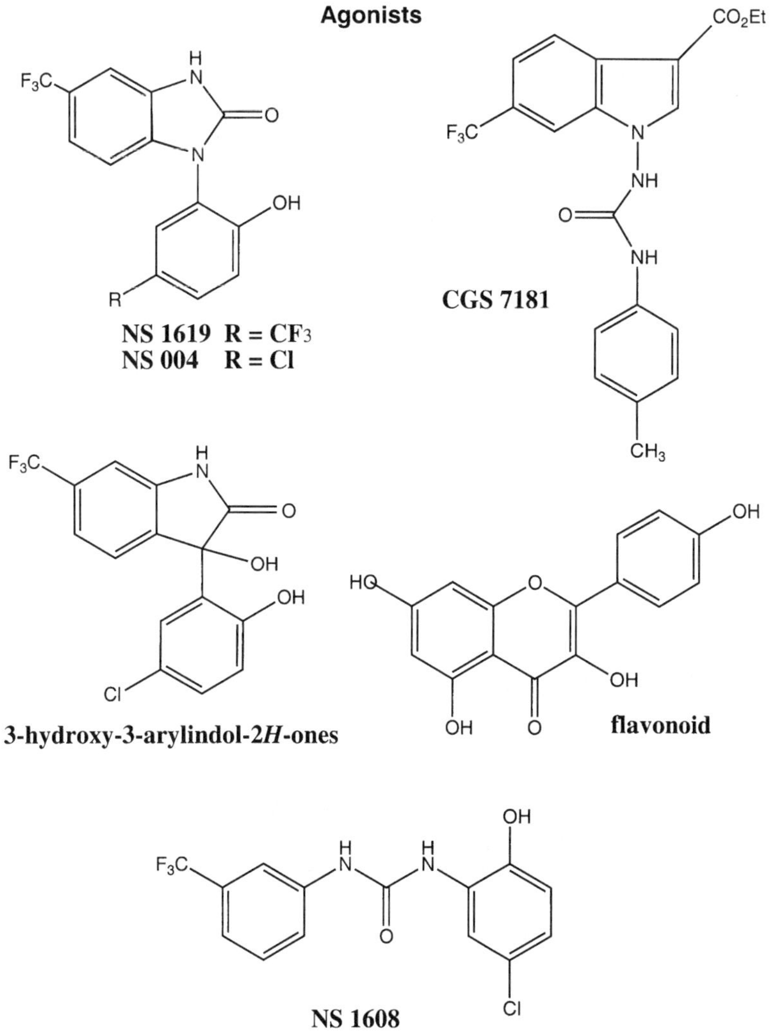

Agonists

NS 1619 R = CF$_3$
NS 004   R = Cl

CGS 7181

3-hydroxy-3-arylindol-2*H***-ones**

flavonoid

NS 1608

Figure 4. Activators of BK$_{Ca}$ channels of synthetic origin. The presence of an electron-withdrawing group attached to the heterocyclic nucleus in NS 004 and NS 1619 is critical for conferring BK$_{Ca}$ channel agonist activity. The 3-hydroxy-3-arylindol-2*H*-ones series also require the presence of an electron-withdrawing group as well as a phenolic hydroxyl for BK$_{Ca}$ channel-opening properties. The structure–activity relationships for the other illustrated classes of compounds are too restricted for defining the parts of the molecule involved in BK$_{Ca}$ channel interaction.

single-channel recordings of inside-out patches decreases the threshold for activation by CGS 7181 and CG5 7184 to 0.01 and 0.1 μM, respectively. These values are comparable to those required for channel activation by DHS-I. It is likely that the site of action of these compounds is at the intracellular face of the channel because the time period between addition of the drugs and development of stimulatory activity is longer in whole-cell recordings than in excised inside-out patch measurements. The recovery of activity after removal of the drug also takes longer in whole-cell as compared to excised-patch recordings. It is not known whether the β-subunit of the BK_{Ca} channel is required for the stimulation of channel activity by these compounds. Although the specificity of these agents with respect to other ion channels has not been established, they appear to have no effect against a panel of membrane-bound receptors and enzymes. The membrane-permeant property of the CGS compounds appears to make them attractive tools with which to evaluate the pharmacological consequences of activating BK_{Ca} channels *in vivo* and for determining whether these channels represent a novel target for therapeutic intervention.

A number of substituted benzimidazolones, typified by NS 004 and NS 1619 (Fig. 4) are also reversible activators of BK_{Ca} channels (Olesen, 1994). The most potent of these agents, NS 1619, causes a dose-dependent shift in the activation curve of aortic smooth muscle BK_{Ca} channels of up to 50 mV in the hyperpolarizing direction, and this effect is fully reversible and antagonized by the BK_{Ca} channel blockers ChTX and tetraethylammonium (Olesen *et al.*, 1994). The site of action of these agents appears to reside on the α-subunit of the channel (Gribkoff *et al.*, 1996). Consistent with its effects on BK_{Ca} channels, NS 1619 causes hyperpolarization of smooth muscle cells. However, the pharmacological evaluation of these compounds is compromised by the fact that they display significant Ca^{2+}-entry blocker activity and that they also affect other ion conduction pathways (Kaczorowski *et al.*, 1996). A series of 3-substituted 2H-indol-2-one derivatives, in which the phenol-bearing nitrogen atom of the heterocycle of NS 004, has been substituted with a carbon atom (Fig. 4), have been prepared (Hewawasam *et al.*, 1997). This synthetic strategy leads to the introduction of an asymmetric center, and it is useful for determination of the absolute stereochemistry associated with BK_{Ca} channel opening. Preliminary structure–activity relationships for this series indicate the importance of both an electron-withdrawing substituent on the oxindole nucleus and the presence of a phenolic hydroxyl for confering BK_{Ca} channel activation properties. A three-dimensional database search based on the pharmacophore of NS 004 has led to the discovery of certain flavonoids (Fig. 4) that act as openers of BK_{Ca} channels (Li *et al.*, 1997). The most potent agents in this series display a slightly higher efficacy at increasing mSlo1 currents expressed in *Xenopus* oocytes than does NS 004.

The substituted diphenylurea NS 1608 (Fig. 4) (Strobaek *et al.*, 1996) has been shown to enhance BK_{Ca} channel activity by shifting the midpoint of channel activation to more negative potentials. The maximum shift produced by NS 1608 is -74 mV with an EC_{50} of 2.1 μM. The stimulatory effect of NS 1608 on BK_{Ca} channels appears to occur through an interaction with the α-subunit, since it does not require the presence of the β-subunit. The effect of NS 1608 is independent of the presence of internal Ca^{2+}, suggesting that this agent does not increase the affinity of Ca^{2+} for the channel.

3. CONCLUSIONS

The search for selective modulators of BK_{Ca} channels continues. The possibility of stably expressing BK_{Ca} channel subunits in cell lines allows for the development of high-capacity functional assays that should lead to the discovery of novel small-molecule templates useful for medicinal chemists to establish defined structure–activity relationships for eliciting channel modulation. Potent and selective BK_{Ca} channel inhibitors have been identified, and these agents have been useful for defining the physiologic role that the channel plays in some target tissues. However, it is not known whether selective activation of BK_{Ca} channels will produce the therapeutic benefits that are predicted based on theoretical arguments. Unfortunately, the BK_{Ca} channel agonists that have been identified so far do not display the appropiate properties to allow this important question to be answered. The development of agonists that have useful mechanisms of action and, very importantly, that are free of Ca^{2+}-entry blocker activity is necessary to better define the pharmacological consequences of stimulating BK_{Ca} channel activity *in vitro*. Likewise, BK_{Ca} channel blockers may yield interesting biological profiles that could be exploited for therapeutic benefit.

REFERENCES

Adelman, J. P., Shen, K.-Z., Kavanaugh, M. P., Warren, R. A., Wu, Y.-N., Lagrutta, A., Bond, C. T., and North, R. A., 1992, Calcium-activated potassium channels expressed from cloned complementary DNAs, *Neuron* **9:**209–216.

Anderson, C. S., MacKinnon, R., Smith, C., and Miller, C., 1988, Charybdotoxin block of single Ca^{2+}-activated K^+ channels. Effects of channel gating, voltage, and ionic strength, *J. Gen. Physiol.* **91:**317–333.

Bontems, F., Roumestand, C., Boyot, P., Gilquin, B., Doljansky, Y., Menez, A., and Toma, F., 1991a, Three-dimensional structure of natural charybdotoxin in aqueous solution by 1H-NMR: Charybdotoxin possesses a structural motif found in other scorpion toxins, *Eur. J. Biochem.* **196:**19–28.

Bontems, F., Roumestand, C., Gilquin, B., Menez, A., and Toma, F., 1991b, Refined structure of charybdotoxin: Common motifs in scorpion toxins and insect defensins, *Science* **254:**1521–1523.

Bontems, F., Gilquin, B., Roumestand, C., Menez, A., and Toma, F., 1992, Analysis of side chain organization on a refined model of charybdotoxin; structural and functional implications, *Biochemistry* **31:**7756–7764.

Butler, A., Tsunoda, S., McCobb, D. P., Wei, A., and Salkoff, L., 1993, mSlo, a complex mouse gene encoding "maxi" calcium-activated potassium channels, *Science* **261:**221–224.

Candia, S., Garcia, M. L., and Latorre, R., 1992, Mode of action of iberiotoxin, a potent blocker of the large conductance Ca^{2+}-activated K^+ channel, *Biophys. J.* **63:**583–590.

Chang, C.-P., Dworetzky, S. I., Wang, J., and Goldstein, M. E., 1997, Differential expression of the α and β subunits of the large-conductance calcium-activated potassium channel: Implications for channel diversity, *Mol. Brain Res.* **45:**33–40.

Cui, J., Cox, D. H., and Aldrich, R. W., 1997, Intrinsic voltage dependence and Ca^{2+} regulation of *mslo* large conductance Ca-activated K^+ channels, *J. Gen. Physiol.* **5:**647–673.

DeFarias, F. P., Carvalho, M. F., Lee, S. H., Kaczorowski, G. J., and Suarez-Kurtz, G., 1996, Effects of the K^+ channel blockers paspalitrem C and paxilline on mammalian smooth muscle, *Eur. J. Pharmacol.* **314:**123–128.

Diaz, L., Meera, P., Amigo, J., Stefani, E., Alvarez, O., Toro, L., and Latorre, R., 1998, Role of the S4 segment in a voltage-dependent calcium-sensitive potassium (hSlo) channel, *J. Biol. Chem.* **273:**32430–32436.

Ding, J. P., Li, Z. W., and Lingle, C. J., 1998, Inactivating BK channels in rat chromaffin cells may arise from heteromultimeric assembly of distinct inactivation-competent and noninactivating subunits, *Biophys. J.* **74:**268–289.

Dworetzky, S. I., Trojnacki, J. T., and Gribkoff, V. K., 1994, Cloning and expression of a human large-conductance calcium-activated potassium channel, *Mol. Brain Res.* **27:**189–193.

Dworetzky, S. I., Boissard, C. G., Lum-Ragan, J. T., McKay, M. C., Post-Munson, D. J., Trojnacki, J. T., Chang, C.-P., and Gribkoff, V. K., 1996, Phenotypic alteration of a human BK (*hSlo*) channel by $hSlo\beta$ subunit coexpression: Changes in blocker sensitivity, activation/relaxation and inactivation kinetics, and protein kinase A modulation, *J. Neurosci.* **16** 4543–4550.

Garcia, M. L., Hanner, M., Knaus, H.-G., Koch, R., Schmalhofer, W., Slaughter, R. S., and Kaczorowski, G. J., 1997, Pharmacology of potassium channels, *Adv. Pharmacol.* **39:**425–471.

Garcia-Calvo, M., Knaus, H.-G., McManus, O. B., Giangiacomo, K. M., Kaczorowski, G. J., and Garcia, M. L., 1994, Purification and reconstitution of the high-conductance calcium-activated potassium channel from tracheal smooth muscle, *J. Biol. Chem.* **269**:676–682.

Giangiacomo, K. M., Garcia, M. L., and McManus, O. B., 1992, Mechanism of iberiotoxin block of the large-conductance calcium-activated potassium channel from bovine aortic smooth muscle, *Biochemistry* **31**:6719–6727.

Giangiacomo, K. M., Garcia-Calvo, M., Knaus, H.-G., Mullmann, T. J., Garcia, M. L., and McManus, O., 1995, Functional reconstitution of the large-conductance, calcium-activated potassium channel purified from bovine aortic smooth muscle, *Biochemistry* **34**:15849–15862.

Giangiacomo, K. M., Kamassah, A., Harris, G., and McManus, O. B., 1998, Mechanism of maxi-K channel activation by dehydrosoyasaponin-I, *J. Gen. Physiol.* **112**:485–501.

Gribkoff, V. K., Lum-Ragan, J. T., Boissard, C. G., Post-Munson, D. J., Meanwell, N. A., Starrett, J. E., Kozlowski, E. S., Romine, J. L., Trojnacki, J. T., McKay, M. C., Zhong, J., and Dworetzky, S. I., 1996, Effects of channel modulators on cloned large-conductance calcium-activated potassium channels, *Mol. Pharmacol.* **50**:206–217.

Hanner, M., Schmalhofer, W. A., Munujos, P., Knaus, H.-G., Kaczorowski, G. J., and Garcia, M. L., 1997, The β subunit of the high conductance calcium-activated potassium channel contributes to the high affinity receptor for charybdotoxin, *Proc. Natl. Acad. Sci. U.S.A.* **94**:2853–2858.

Hanner, M., Vianna-Jorge, R., Kamassah, A., Schmalhofer, W. A., Knaus, H.-G., Kaczorowski, G. J., and Garcia, M. L., 1998, The β subunit of the high conductance calcium-activated potassium channel; identification of residues involved in charybdotoxin binding, *J. Biol. Chem.* **273**:16289–16296.

Hewawasam, P., Meanwell, N. A., Gribkoff, V. K., Dworetzky, S. I., and Boissard, C. G., 1997, Discovery of a novel class of BK channel openers: Enantiospecific synthesis and BK channel opening activity of 3-(5-chloro-2-hydroxyphenyl)-1,3-dihydro-3-hydroxy-6-(trifluoromethyl)-2*H*-indol-2-one, *Bioorga. Medicinal Chem. Lett.* **7**:1255–1260.

Hu, S., Fink, C. A., Kim, H. S., and Lappe, R. W., 1997, Novel and potent BK channel openers: CGS 7181 and its analogs, *Drug Dev. Res.* **41**:10–21.

Huang, J.-C., Garcia, M. L., Reuben, J. P., and Kaczorowski, G. J., 1993, Inhibition of β-adrenoceptor agonist relaxation of airway smooth muscle by Ca^{2+}-activated K^+ channel blockers, *Eur. J. Pharmacol.* **235**:37–43.

Johnson, B. A., and Sugg, E. E., 1992, Determination of the three-dimensional structure of iberiotoxin in solution by ^1H nuclear magnetic resonance spectroscopy, *Biochemistry* **31**:8151–8159.

Joiner, W. J., Tang, M. D., Wang, L.-Y., Dworetzky, S. I., Boissard, C. G., Gan, L., Gribkoff, V. K., and Kaczmarek, L. K., 1998, Formation of intermediate-conductance calcium-activated potassium channels by interaction of Slack and Slo subunits, *Nat. Neurosci.* **1**:462–469.

Jones, T. R., Charette, L., Garcia, M. L., and Kaczorowski, G. J., 1990, Selective inhibition of relaxation of guinea-pig trachea by charybdotoxin, a potent Ca^{2+}-activated K^+ channel inhibitor, *J. Pharmacol. Exp. Ther.* **255**:697–705.

Jones, T. R., Charette, M. L., Garcia, M. L., and Kaczorowski, G. J., 1993, Interaction of iberiotoxin with beta adrenoceptor agonists and sodium nitroprusside on guinea pig trachea, *J. Appli. Physiol.* **74**:1879–1884.

Kaczorowski, G. J., Knaus, H.-G., Leonard, R. J., McManus, O. B., and Garcia, M. L., 1996, High conductance calcium-activated potassium channels; structure, pharmacology and function, *J. Biomembr. Bioenerg.* **28**:255–267.

Knaus, H.-G., Eberhart, A., Kaczorowski, G. J., and Garcia, M. L., 1994a, Covalent attachment of charybdotoxin to the β-subunit of the high-conductance Ca^{2+}-activated K^+ channel, *J. Biol. Chem.* **269**:23336–23341.

Knaus, H.-G., Folander, K., Garcia-Calvo, M., Garcia, M. L., Kaczorowski, G. J., Smith, M., and Swanson, R., 1994b, Primary sequence and immunological characterization of the b-subunit of the high-conductance Ca^{2+}-activated K^+ channel from smooth muscle, *J. Biol. Chem.* **269**:17274–17278.

Knaus, H.-G., Garcia-Calvo, M., Kaczorowski, G. J., and Garcia, M. L., 1994c, Subunit composition of the high conductance calcium-activated potassium channel from smooth muscle, a representative of the *mSlo* and *slowpoke* family of potassium channels, *J. Biol. Chem.* **269**:3921–3924.

Knaus, H.-G., McManus, O. B., Lee, S. H., Schmalhofer, W. A., Garcia-Calvo, M., Helms, L. M. H., Sanchez, M., Giangiacomo, K., Reuben, J. P., Smith A. B., III Kaczorowski, G. J., and Garcia, M. L., 1994d, Tremorgenic indole alkaloids potently inhibit smooth muscle high-conductance Ca^{2+}-activated K^+ channels, *Biochemistry* **33**:5819–5828.

Knaus, H.-G., Schwarzer, C., Koch, R. O. A., Eberhart, A., Kaczorowski, G. J., Glossmann, H., Wunder, F., Pongs, O., Garcia, M. L., and Sperk, G., 1996, Distribution of high-conductance Ca^{2+}-activated K$^+$ channels in rat brain: Targeting to axons and nerve terminals, *J. Neurosci.* **16**:955–963.

Koch, R. O. A., Koschak, A., Wanner, S. G., Kaczorowski, G. J., Wittka, R., Garcia, M. L., and Knaus, H.-G., 1996, High-conductance calcium-activated potassium channels in rat brain: Pharmacological profile, quantification of expression, subunit composition and functional implications, *Soc. Neurosci. Abstr.* **22**:1754.

Koschak, A., Koch, R. O., Liu, J., Kaczorowski, G. J., Reinhart, P. H., Garcia, M. L., and Knaus, H.-G., 1997, [^{125}I]Iberiotoxin-D19Y/Y36F, the first selective, high specific activity radioligand for high-conductance calcium-activated potassium channels, *Biochemistry* **36**:1943–1952.

Kume, H., Tokuno, H., and Tomita, T., 1989, Regulation of Ca^{2+}-dependent K$^+$-channels in trachael myocytes by phosphorylation, *Nature* **341**:152–154.

Kume, H., Graziano, M. P., and Kotlikoff, M. I., 1992, Stimulatory and inhibitory regulation of calcium-activated potassium channels by guanine nucleotide-binding proteins, *Proc. Natl. Acad. Sci. U.S.A.* **89**:11051–11055.

Latorre, R., Oberhauser, A., Labarca, P., and Alvarez, O., 1989, Varieties of calcium-activated potassium channels, *Annu. Rev. Physiol.* **51**:385–399.

Li, Y., Starrett, J. E., Meanwell, N. A., Johnson, G., Harte, W. E., Dworetzky, S. I., Boissard, C. G., and Gribkoff, V. K., 1997, The discovery of novel openers of Ca^{2+}-dependent large-conductance potassium channels: Pharmacophore search and physiological evaluation of flavonoids, *Bioorg, Medicinal Chem. Lett.* **7**:759–762.

MacKinnon, R., and Miller, C., 1988, Mechanism of charybdotoxin block of the high-conductance, Ca^{2+}-activated K$^+$ channel, *J. Gen. Physiol.* **91**:335–349.

MacKinnon, R., Latorre, R., and Miller, C., 1989, Role of surface electrostatics in the operation of a high-conductance Ca^{2+}-activated K$^+$ channel, *Biochemistry,* **28**:8092–8099.

McCobb, D. P., Fowler, N. L., Featherstone, T., Lingle, C. J., Saito, M., Krause, J. E., and Salkoff, L., 1995, A human calcium-activated potassium channel gene expressed in vascular smooth muscle, *Am. J. Physiol.* **269**:H767–H777.

McManus, O. B., 1991, Calcium-activated potassium channels: regulation by calcium, *J. Bioenerg. Biomembr.* **23**:537–560.

McManus, O. B., Harris, G. H., Giangiacomo, K. M., Feigenbaum, P., Reuben, J. P., Addy, M. E., Burka, J. F., Kaczorowski, G. J., and Garcia, M. L., 1993, An activator of calcium-dependent potassium channels isolated from a medicinal herb, *Biochemistry* **32**:6128–6133.

McManus, O. B., Helms, L. M. H., Pallanck, L., Ganetzky, B., Swanson, R., and Leonard, R. J., 1995, Functional role of the β subunit of high-conductance calcium-activated potassium channels, *Neuron* **14**:1–20.

Meera, P., Wallner, M., Jiang, Z., and Toro, L., 1996, A calcium switch for the functional coupling between α (*hslo*) and β subunits (K$_{v + Ca}\beta$) of maxi K channels, *FEBS Lett.* **382**:84–88.

Meera, P., Wallner, M., Song, M., and Toro, L., 1997, Large conductance voltage- and calcium-dependent K$^+$ channel, a distinct member of voltage-dependent ion channels with seven N-terminal transmembrane segments (S0–S6), an extracellular N terminus, and an intracellular (S9–S10) C terminus, *Proc. Natl. Acad. Sci. U.S.A.* **94**:14066–14071.

Meera, P., Wallner, M., and Toro, L., 1999, Molecular determinant of maxi-K channel inactivation, *Biophys. J.* **76**:A267.

Miura, M., Belvesi, M. G., Stretton, C. D., Yacoub, M. H., and Barnes, P. J., 1992, Role of potassium channels in bronchodilator responses in human airways, *Am. Rev. Respir. Dis.* **146**:132–136.

Mullmann, T. J., Munujos, P., Garcia, M. L., and Giangiacomo, K. M., 1999, Electrostatic mutations in iberiotoxin as a unique tool for probing the electrostatic structure of the maxi-K channel outer vestibule, *Biochemistry* , **38**:2395–2402

Munujos, P., Knaus, H.-G., Kaczorowski, G. J., and Garcia, M. L., 1995, Crosslinking of charybdotoxin to high-conductance calcium-activated potassium channels: Identification of the covalently modified toxin residue, *Biochemistry* **34**:10771–10776.

Olesen, S.-P., 1994, Activators of large-conductance Ca^{2+}-dependent K$^+$ channels, *Exp. Opin. Invest. Drugs* **3**:1181–1188.

Olesen, S.-P., Munch, E., Moldt, P., and Drejer, J., 1994, Selective activation of Ca^{2+}-dependent K$^+$ channels by novel benzimidazolone, *Eur. J. Pharmacol.* **251**:53–59.

Pallanck, L., and Ganetzky, B., 1994, Cloning and characterization of human and mouse homologs of the *Drosophila* calcium-activated potassium channel gene, *slowpoke, Hum. Mol. Genet.* **3:**1239–1243.

Park, C.-S., and Miller, C., 1992a, Interaction of charybdotoxin with permeant ions inside the pore of a K$^+$ channel, *Neuron* **9:**307–313.

Park, C.-S., and Miller, C., 1992b, Mapping function to structure in a channel-blocking peptide: Electrostatic mutants of charybdotoxin, *Biochemistry* **31:**7749–7755.

Schreiber, M., and Salkoff, L., 1997, A novel calcium-sensing domain in the BK channel, *Biophys. J.* **73:**1355–1363.

Shimony, E., Sun, T., Kolmakova-Partensky, L., and Miller, C., 1994, Engineering a uniquely reactive thiol into a cysteine-rich peptide, *Protein Eng.* **7:**503–507.

Singh, S. B., Goetz, M. A., Zink, D. L., Dombrowski, A. W., Polishook, J. D., Garcia, M. L., Schmalhofer, W., McManus, O. B., and Kaczorowski, G. J., 1994, Maxikdiol: A novel dihydroxyisoprimane as an agonist of maxi-K channels, *J. Chem. Soc., Perkin Trans.* 1 **1994:**3349–3352.

Solaro, C. R., and Lingle, C. J., 1992, Trypsin-sensitive, rapid inactivation of a calcium-activated potassium channel,*Science* **257:**1694–1698.

Stampe, P., Kolmakova-Partensky, L., and Miller, C., 1994, Intimations of K$^+$ channel structure from a complete functional map of the molecular surface of charybdotoxin, *Biochemistry* **33:**443–450.

Strobaek, D., Christophersen, P., Holm, N. R., Moldt, P., Ahring, P. K., Johansen, T. E., and Olesen, S.-P., 1996, Modulation of the Ca^{2+}-dependent K$^+$ channel, *hslo*, by the substituted diphenylurea NS 1608, paxilline and internal Ca^{2+}, *Neuropharmacology* **35:** 903–914.

Suarez-Kurtz, G., Garcia, M. L., and Kaczorowski, G. J., 1991, Effects of charybdotoxin and iberiotoxin on the spontaneous motility and tonus of different guinea pig smooth muscle tissues, *J. Pharmacol. Exp. Ther.* **259:**439–443.

Tanaka, Y., Meera, P., Song, M., Knaus, H.-G., and Toro, L., 1997, Molecular constituents of maxi K$_{Ca}$ channels in human coronary smooth muscle: Predominant $\alpha + \beta$ subunit complexes, *J. Physiol. (London)* **502:**545–557.

Tseng-Crank, J., Foster, C. D., Krause, J. D., Mertz, R., Godinot, N., DiChiara, T. J., and Reinhart, P. H., 1994, Cloning, expression, and distribution of functionally distinct Ca^{2+}-activated K$^+$ channel isoforms from human brain, *Neuron* **13:**1315–1330.

Tseng-Crank, J., Godinot, N., Johansen, T. E., Ahring, P. K., Strobaek, D., Mertz, R., Foster, C. D., Olesen, S.-P., and Reinhart, P. H., 1996, Cloning, expression, and distribution of a Ca^{2+}-activated K$^+$ channel β-subunit from human brain, *Proc. Natl. Acad. Sci. U.S.A.* **93:**9200–9205.

Vazquez, J., Feigenbaum, P., Katz, G., King, V. F., Reuben, J. P., Roy-Contancin, L., Slaughter, R. S., Kaczorowski, G. J., and Garcia, M. L., 1989, Characterization of high affinity binding sites for charybdotoxin in sarcolemmal membranes from bovine aortic smooth muscle: Evidence for a direct association with the high conductance calcium-activated potassium channel, *J. Biol. Chem.* **264:**20902–20909.

Vogalis, F., Vincent, T., Qureshi, I., Schmalz, F., Ward, M. W., Sanders, K. M., and Horowitz, B., 1996, Cloning and expression of the large-conductance Ca^{2+}-activated K$^+$ channel from colonic smooth muscle, *Am. J. Physiol.* **271:**G629--G639.

Wallner, M., Meera, P., Ottolia, M., Kaczorowski, G. J., Latorre, R., Garcia, M. L., Stefani, E., and Toro, L., 1995, Characterization of and modulation by a β-subunit of a human maxi K$_{Ca}$ channel cloned from myometrium, *Recept. Channels* **3:**185–199.

Wallner, M., Meera, P., and Toro, L., 1996, Determinant for β-subunit regulation in high-conductance voltage-activated and Ca^{2+}-sensitive K$^+$ channels: An additional transmembrane region at the N terminus, *Proc. Natl. Acad. Sci. U.S.A.* **93:**14922–14927.

Wallner, M., Meera, P., and Toro, L., 1999, A novel β subunit leads to inactivating maxiK currents, *Biophys. J.* **76:**A267.

Wei, A., Jegla, T., and Salkoff, L., 1996, Eight potassium channel families revealed by the *C. elegans* genome project, *Neuropharmacology* **35:**805–829.

Winquist, R. J., Heany, L. A., Wallace, A. A., Baskin, E. P., Stein, R. B., Garcia, M. L., and Kaczorowski, G., 1989, Glyburide blocks the relaxation response to BRL 34915 (cromakalim), minoxidil sulfate and diazoxide in vascular smooth muscle, *J. Pharmacol. Exp. Ther.* **248:**149–156.

Yao, Y., Peter, A. B., Baur, R., and Sigel, E., 1989, The tremorigen aflatrem is a positive allosteric modulator of the γ-amonibutyric acid$_A$ receptor channel expressed in *Xenopus* oocytes, *Mol. Pharmacol.* **35:**319–323.

Chapter 14

Pharmacology of Small-Conductance, Calcium-Activated K$^+$ Channels

Eric Blanc and Hervé Darbon

1. INTRODUCTION

1.1. Potassium (K$^+$) Channels

K$^+$ channels are transmembrane proteins dedicated to allowing K$^+$ fluxes through physiological membranes. Topologically, they can be described as transmembrane segments surrounding a pore-forming region directly involved in K$^+$ selectivity and transfer. As originally depicted by Hodgkin and Huxley (1952a–d), K$^+$ channels are involved in the propagation of the action potential. Their opening is regulated by the level of membrane depolarization, and their role is to return the membrane to its resting potential. However, far beyond this unique role, K$^+$ channels form the most diverse ion channel family described so far. They are present in nearly all cell types, and their biophysical as well as their pharmacological profiles are among the most complex ever seen.

The recent *Caenorhabditis elegans* genome sequencing project pointed out the complexity of the K$^+$ channel gene family. *C. elegans* contains about 80 predicted K$^+$ channels in its genome. These 80 genes are members of the three major structural classes of K$^+$ channels that are characterized by the number of transmembrane (TM) domains in the α-subunit: (i) the inward rectifier channels possessing two transmembrane domains (2-TM), (ii) the two pore channels, with four transmembrane domains (4-TM), and (iii) the six-transmembrane-domain family (6-TM). This last family can be further divided into five subclasses: the voltage-gated K$^+$ channels (Kv channels), the *ether-a-go-go* (EAG) and *ether-a-go-go*-related (ERG) channels with cyclic-nucleotide-binding domains, the KQT channels, and the Ca^{2+}-activated K$^+$ channels (Bargmann, 1998; Wei *et al.*, 1996). A fourth structural class exists, containing only the large-

Eric Blanc and Hervé Darbon ● Architecture et Fonction des Macromolécules Biologiques, CNRS UPR 9039, IFRI, 13402 Marseille Cedex 20, France.

Potassium Channels in Cardiovascular Biology, edited by Archer and Rusch. Kluwer Academic/Plenum Publishers, New York, 2001.

conductance, Ca^{2+}-activated K^+ channels (BK_{Ca} channels). The BK_{Ca} channels were previously described as being members of the 6-TM class, but it appears that the topology of the BK_{Ca} channel α-subunit is formed by seven, instead of six, transmembrane domains (Atkinson *et al.*, 1991; Butler *et al.*, 1993; Adelman *et al.*, 1992; Wallner *et al.*, 1996). Because a functional K^+ channel is the result of the assembly of four α-subunits (two in the case of 4-TM channels), another level of complexity is generated by the possibility that distinct subunits coassemble to form heteromultimeric channels with new biophysical and pharmacological properties. Finally, some K^+ channels are composed of the four α-subunits mentioned above but also contain additional β-subunits. These β-subunits are intracellular (in the case of Kv channels) or transmembrane in the case of (BK_{Ca} and SK_{Ca} channels) proteins that modify the gating and the pharmacology of the resulting channel.

All these K^+ channels differ from one another in their respective sensitivity to membrane depolarization, increases of intracellular Ca^{2+} concentration, or responses to other stimuli, such as second messengers or cyclic nucleotides. They also differ in the amount of K^+ released while they are in the open state, the duration of this opening, and even the direction of the K^+ flux. These diverse properties allow K^+ channels to play a key role in numerous biological processes by fine-tuning the K^+ permeability of the cell membrane. Indeed, K^+ channels must be seen as a true linkage between electrical phenomena at the level of the membrane and the chemical messages inside or between cells.

1.2. Overall Topology of K^+ Channels

The functional diversity of K^+ channels principally affects the gating mechanisms, opening, and inactivation of the channels. However, the main feature of all K^+ channels is the selective transfer of K^+ ions. Indeed, all of these channels express a selectivity sequence of $K^+ \approx Rb^+ > Cs^+ \ggg Na^+$ ions. This ability to discriminate between K^+ and the smaller alkali ions (Li^+ or Na^+) is conferred by the hyperconserved sequence GYGD of the so-called selectivity filter. Thus, even if K^+ channels can be classified into different structural classes, depending on the number of transmembrane segments, all of them must have the same pore constitution. Thus, K^+ channels are tetramers with a fourfold symmetry of α-subunits assembled around a central ion pore.

1.3. Interaction of K^+ Channels with Peptidic Toxins

The pore structure of K^+ channels has been investigated for years, through the use of scorpion toxins. Scorpion toxins directed against K^+ channels form a homogeneous family of small basic proteins, 29–39 residues in length, reticulated by three or four disulfide bridges (Fig. 1).Because they bind directly to the pore-forming region, it has been possible to analyze the effect of local mutations of either the toxin or the channel on the potency of their interaction. This use of directed mutagenesis and peptidic synthesis has defined the structural requirements for the interaction between toxins and K^+ channels. Charybdotoxin (CTX), a scorpion toxin purified from the venom of the Israeli scorpion *Leiurus quinquestriatus* var. *hebraeus*, (Miller et al, 1985) was the first

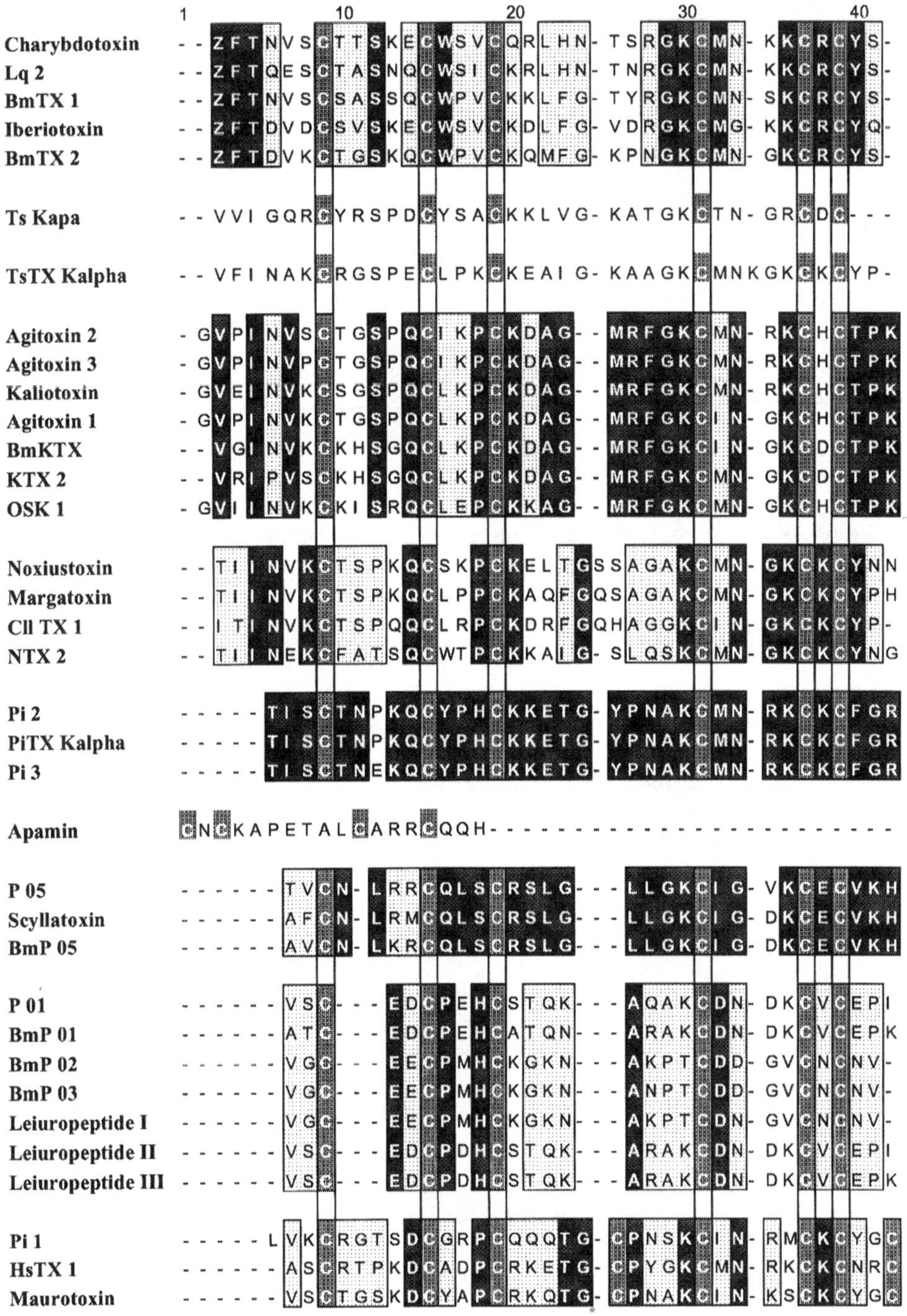

Figure 1. Amino acid sequence alignment of scorpion toxins acting on K$^+$ channels. The residues highlighted by the light gray boxes are closely related in nature and volume. The residues highlighted by the black boxes are conserved for all the sequences. Cystine residues are highlighted in dark gray.

scorpion toxin to be extensively studied for its binding potency toward Kv and BK_{Ca} channels. CTX directly plugs into the pore-forming region, with one of its basic residues, the lysine K27, situated in the vicinity of the selectivity filter. Other residues also mediate the interaction of the toxin with the channel. Structure–activity relationships of scorpion toxins demonstrate that all the critical residues are located on the same face of the molecule to define the functional map of the toxin. CTX acts on both the BK_{Ca} and Kv channels by this same mechanism. The functional map of CTX is basically composed of five residues: the central lysine K27, the methionine M29, the asparagine N30, the arginine R34 and the tyrosine Y36. Three other residues, namely, the serine S10, the tryptophan W14 and the arginine R25, also are critical for the binding of the toxin to BK_{Ca} channels. The mapping of CTX was aided by identification, purification, and studies of the structure–activity relationships of numerous other scorpion toxins belonging to the noxiustoxin or kaliotoxin (KTX) families; these toxins are closely related to CTX but possess slightly different specificities. All this work led to the construction of three dimensional models of the Kv channels (Durell and Guy, 1996; Durell *et al.*, 1998; Lipkind and Fozzard, 1997; Aiyar *et al.*, 1995, 1996; Kerr and Sansom, 1997; Yang *et al.*, 1997). Models do not exist yet for other channel subtypes, largely because high-affinity ligands have yet to be identified for these other channel types. Furthermore, some channel types have higher complexity; for example, in the case of the BK_{Ca} channel, a β-subunit modulates the high-affinity binding site of CTX by modifying the biophysical and pharmacological properties of the α-subunit. Finally, progress has been delayed in the case of some channel types, such as the SK_{Ca} channels, because their sequence has not yet been determined.

There is no doubt that research on the structure–activity relationships of all K^+ channels will grow rapidly, owing in part to the recent determination of the first structure of a K^+ channel (Doyle *et al.*, 1998). The X-ray structure of this K^+ channel, KcsA from *Streptomyces lividans*, shows that it is composed of only the helices S5 and S6 (following the naming convention of the 6-TM channels) surrounding a pore region Fig. 2). This pore region is formed by a turret, a coiled region composing the

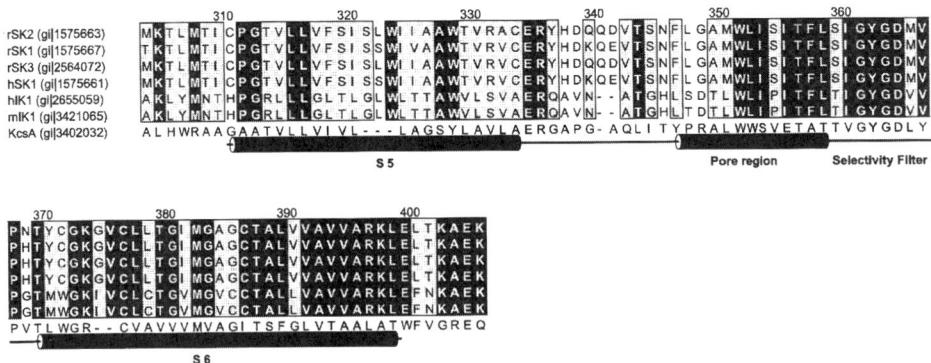

Figure 2. Partial amino acid sequence alignment (region surrounding the pore) and localization of transmembrane segments of SK_{Ca} and IK_{Ca} channels (KcsA sequence is taken as reference). The residues highlighted by the light gray boxes are closely related in nature and volume for at least four of the six sequences. The residues highlighted in the black boxes are conserved for all the six sequences.

extracellular mouth of the pore, followed by a helix (the pore helix) and another coiled region containing the selectivity filter. The definition of this pore structure has permitted the elucidation of the structural basis of numerous processes mediated by K$^+$ channels. Studies of the structure–activity relationships of the inhibition of K$^+$ channels inhibition by scorpion toxins will take advantage of these new data, allowing us to better understand the mechanism of toxin specificity.

Although scorpions provide a whole set of toxins directed against K$^+$ channels, other animal venoms also are able to block K$^+$ efflux in cells, including toxins from the venom of snakes (Harvey *et al.*, 1998; Harvey, 1997), spiders (Swartz and MacKinnon, 1995; Sanguinetti *et al.*, 1997), sea anemones (Schweitz *et al.*, 1995; Pennington *et al.*, 1995; Cotton *et al.*, 1997), and cone snails (Shon *et al.*, 1998). Although CTX remains the reference ligand for BK$_{Ca}$ and certain Kv channels, the bee venom toxin apamin is the reference ligand for the SK$_{Ca}$ channel family.

2. SK$_{Ca}$ CHANNELS

The small-conductance, Ca^{2+}-activated K$^+$ channels were first identified on the basis of their unique pharmacology. They are insensitive to CTX-related scorpion toxins but are highly sensitive to apamin, a toxin present in the venom of the European honeybee *Apis mellifera*, and to a plant alkaloid, *d*-tubocurarine. These two blockers, particularly apamin, provided a useful tool for the study of this class of channels. Thus, the biophysical characterization of the SK$_{Ca}$ channels occurred after their pharmacological definition. SK$_{Ca}$ channels are sensitive to submicromolar concentrations of Ca^{2+} but are only weakly sensitive to voltage, in contrast to the high sensitivity of BK$_{Ca}$ or IK$_{Ca}$ channels to voltage activation.

Using a sequence based on conserved residues within the pore region of previously cloned K$^+$ channels, Köhler *et al.* (1996) identified one partial and three full-length coding sequences in cDNA libraries from human and rat brain. The four identified proteins code for four SK$_{Ca}$ channels, hSK1 (561 residues), rSK2 (580 residues), rSK3 (553 residues), and the partial clone rSK1, homologous to hSK1 (Fig. 2).

2.1. Homologies with the K$^+$ Channel Superfamily

The four putative proteins show high levels of homology with the K$^+$ channel superfamily. They all possess six predicted transmembrane segments, as is the case in Kv-related channels, and a GYGD sequence between the fifth and sixth segments in the pore-forming region. This short sequence is known to be the selectivity filter in other K$^+$ channels (Yool and Schwartz, 1991; Heginbotham and MacKinnon, 1992; Heginbotham *et al.*, 1994; Ranganathan *et al.*, 1996). Moreover, the fourth transmembrane segment (S4) contains three basic residues separated by six and seven residues, respectively. In voltage-gated K$^+$ channels, several basic residues separated by three hydrophobic residues constitute the voltage sensor (for reviews see Guy and Conti, 1990; Jan and Jan, 1992; Goldstein, 1996). The low sensitivity of SK$_{Ca}$ channels to voltage may be the result of the low content of basic residues in the S4 domain (Köhler *et al.*, 1996). Despite these topological similarities, SK_{Ca} channels lie on a distinct evolutionary branch within the K^+ channel superfamily. Indeed, except for a 12-residue stretch within the pore domain, there is no significant amino acid homology between SK_{Ca} channels and other types of cloned K^+ channels.

2.2. Subunits

The transmembrane segments of the SK_{Ca} channels cloned thus far show a high level of homology to one another, whereas the extra- and intracellular loops, as well as the amino and carboxy termini, are more variable. However, these sequences alone do not account for all the observed properties of the channel. The channel function reflects the three-dimensional features of the entire protein. Taking advantage of the high affinity and specificity of apamin for this class of ion channels, some research groups have performed numerous cross-linking experiments aimed at defining the polypeptide seen by the toxin when bound to its receptor. Early experiments employing radiation inactivation, demonstrated that the functional size of the apamin-binding component in SK_{Ca} channels is derived from an oligomeric structure of 250 kDa (Schmid-Antomarchi *et al.*, 1984). Within this oligomer, apamin can be chemically cross-linked to a 33- kDa subunit. Further studies on this system revealed an unexpected complexity. Depending upon the cross-linker used, both high-molecular-weight (86 and 59 kDa) and low-molecular-weight (33 and 30 kDa) subunits have been implicated as being intrinsically involved with SK_{Ca} channels. Thus, using the bifunctional reagent disuccinimidyl suberate (DSS), Schmid-Antomarchi *et al.* (1984) covalently bound [^{125}I]-apamin to a 30-kDa component. Later, Seagar *et al.* (1986) used photolabile monoazidoaryl derivatives of apamin to covalently link SK_{Ca} channels in different preparations. They linked the toxin modified at the αNH_2-Cys_1 position to an 86-kDa subunit in cultured neurons and to 86- and 59- kDa subunits in rat brain synaptic membranes. The toxin modified at the ϵNH_2-Lys_4 position linked the smaller 33- and 22-kDa subunits in both preparations (Seagar *et al.*, 1986). In photoaffinity labeling studies using two types of membrane preparations (rat brain and PC12 cells) and six different ways of grafting the toxin to its receptor, Auguste *et al.* (1989) demonstrated that the-30 kDa subunit is an important element of the SK_{Ca} channels. This 30-kDa subunit was the unique constituent that was always labeled in both preparations. Other subunits of 86 or 59 kDa were also clearly associated with SK_{Ca} channels, whereas the 45-kDa component was only labeled in rat brain membranes. Additional differences between results obtained with brain membranes and with PC12 cell membranes depended on the chemical structure of the photoprobe. Taken together, these experiments indicated that the apamin receptor is similar but not identical in the two types of membranes. It was later proposed that only the 86- and 33-kDa subunits constructed in neuronal SK_{Ca} channels: the other identified subunit species (59, 30, or 27 kDa) were thought to be stable proteolytic fragments derived from the 86- and 33-kDa structures (Levèque *et al.*, 1990; Wadsworth *et al.*, 1994). New photolabile azidoaryl derivatives of apamin helped to clarify the importance of the cross-linker length in photolabeling different apamin-binding polypeptides. The ability to label the 86-, 59-, or 33-kDa SK_{Ca} subunits in rat brain from the α-Cys_1 position depends upon the length of the spacer arm incorporated in the photoprobe. A spacer arm longer than 5.7 Å is necessary to label the 33-kDa subunit. In contrast, all the ε-azido derivatives of ^{125}I[α-formyl-Cys$_1$]apamin were bound to the 33-kDa apamin-binding polypeptide (Wadsworth et al., 1996).

All these photoaffinity labeling studies have established that SK channels are hetero-oligomeric assemblies composed of both high-molecular-weight (86 and 59 kDa) and low-molecular-weight (30 or 33 kDa) polypeptide subunits. These assemblies are tissue-specific, as can be judged from the different labeling patterns obtained in various cell types (Seagar et al., 1986; Wadsworth et al., 1996, 1997). The cloning work of Köhler et al. (1996) recently identified an α-subunit in SK$_{Ca}$ channels. Interestingly, the 59-kDa subunit previously described is similar to the 64-kDa protein product encoded by the longest open reading frame of rSK2. The structural relationship between the 86-kDa and 59-kDa subunits remains unclear. Both subunits can be colabeled in a given cell type. This colabeling could result from the expression of alternative α-subunits leading to the generation of heteromultimeric assemblies, thus contributing to the increase in diversity among SK$_{Ca}$ channel subtypes. The smaller 30-kDa subunit in this assembly could represent an auxiliary β-subunit, analogous to BK$_{Ca}$ channel β-subunits (Wadsworth et al., 1997).

2.3. SK$_{Ca}$ Channel Blockers

2.3.1. Apamin

Generally speaking, the polypeptide animal toxins are useful pharmacological probes for the study of a given receptor. In the case of apamin, this statement has proven to be true. Indeed, this basic octadecapeptide purified from the venom of the European honeybee, *Apis mellifera*, is a powerful and specific blocker of SK$_{Ca}$ channels. Thus, this toxin has been used for years as a tool to characterize and identify SK$_{Ca}$ channel subtypes and as a reference toxin in the screening of new venoms. Owing to its small size, apamin could be easily synthesized, thereby allowing the critical requirements for its toxin activity to be defined. Series of analogs have been used to define the residues responsible for the toxin activity and the binding of apamin to SK$_{Ca}$ channels. Apamin is reticulated by two disulfide bridges: C1—C11 and C3—C15. The overall structure of apamin, as determined by nuclear magnetic resonance (NMR) techniques, is globular. The secondary structure is composed of a type I β-turn from residue N2 to A5 and an α-helix from residue A11 to Q17 (Bystrov et al., .1980; Pease and Wemmer, 1988).

Active Site of Apamin. The free amino groups αNH$_2$-Cys$_1$ and εNH$_2$-Lys$_4$ of the apamin peptide can be modified without loss of activity. Two other charged side chains (the carboxyl group of E7 and the imidazole ring of H18) do not influence the binding of the toxin, as they can be neutralized without modification of the dose that is lethal to 50% of animals exposed (LD$_{50}$). In the contrast, any modification of the two basic residues R13 and R14 induces a drop in activity of the apamin. These residues are mainly responsible for the toxicity and the binding efficiency of apamin for SK$_{Ca}$ channels (Vincent et al., 1975; Habermann and Fischer, 1979).

Further studies of the respective influence of R13 and R14 revealed the importance of both residues. The results obtained with chemically modified synthetic analogs apamin [R13K] and apamin [R14K] and their acetylated or guanidylated derivatives demonstrated that the presence of two adjacent positive charges is needed, that at least one guanidinium group is required, and that the length of the two side chains handling the positive charges is important for the biological activity of the toxin (Vincent *et al.* 1975; Granier *et al.*, 1978; Sandberg, 1979). The determination of the structure of apamin by NMR (Pease and Wemmer, 1988) showed that the three critical residues R13, R14, and H18, although mobile in solution, are located on the same face of the toxin (Fig. 3).

Two other functional groups interact with the apamin receptor. Studies of truncated forms of apamin indicated the importance of the C-terminal residues Q17 and H18 (Labbé-Julié *et al.*, 1991; Devaux *et al.*, 1995). Indeed, apamin (1–17) and apamin (1–16) show relative binding efficiencies of 1.3% and 0.01%, respectively (half-effect displacement, $K_{0.5}$, of 12 pM for natural apamin versus 890 pM and 120 nM for the two truncated analogs) (Labbé-Julié *et al.*, 1991). These results argued for an influence of the Q17 side chain and, to a lesser extent, of H18 amidation on the binding potency of apamin. Despite only a weak activity, truncated forms of apamin are still able to completely inhibit ^{125}I-labeled apamin, suggesting that, although important, Q17 is not essential for the expression of the specific activity of the toxin. It further appears that the important parameter for the full expression of the biological activity of apamin is the C-terminal amidation. Both recombinant and synthetic apamin with a free carboxylic C-terminal extremity were tested and displayed only 0.06% relative binding potency when compared to natural C-terminal amidated apamin (Devaux *et al.*, 1995). The finding in the previous study that a truncated form of apamin [apamin (1-17)] displayed 1.3% of the activity of the native apamin could be due to a slightly different orientation of the Q17 side chain in the truncated apamin(1–17), playing the role of the amidated extremity of the natural toxin.

2.3.2. Scorpion Toxins

2.3.2a. Charybdis and Scylla. The analysis of the effect of the crude venom of *Leiurus quinquestriatus* on guinea pig hepatocytes led to the isolation of a component able to bind specifically to apamin-sensitive K^+ channels (Abia *et al.*, 1986). The fraction of the venom responsible for the activity toward SK_{Ca} channels was later identified as the leiurotoxin I (LTX) (Chicchi *et al.*, 1988), also called scyllatoxin (ScTX) to distinguish this toxin from the previously characterized charybdotoxin (Auguste et al., 1990). It is a 31-residue-long peptide reticulated by three disulfide bridges that inhibits [125I]apamin binding to rat brain membrane with a K_i of 75 pM (Chicchi *et al.*, 1988). [125I]-Scyllatoxin binds to the same membranes with a dissociation constant (K_d) of 130 pM (Auguste *et al.*,1990).

Later, P05, a toxin purified from the venom of the scorpion *Androctonus mauretanicus mauretanicus*, was found to specifically block SK_{Ca} channels (Zerrouk *et al.*, 1993). P05 possesses 31 residues and like LTX is reticulated by three disulfide bridges. In fact, these two toxins are closely related in terms of primary structure and specificity but exhibit different affinities for SK_{Ca} channels. P05, tested in competition experiments with [125I]-apamin, had a $K_{0.5}$ value of 20 pM for the apamin binding site, and thus its binding is 20-fold more efficient than that of LTX (Zerrouk *et al.*,

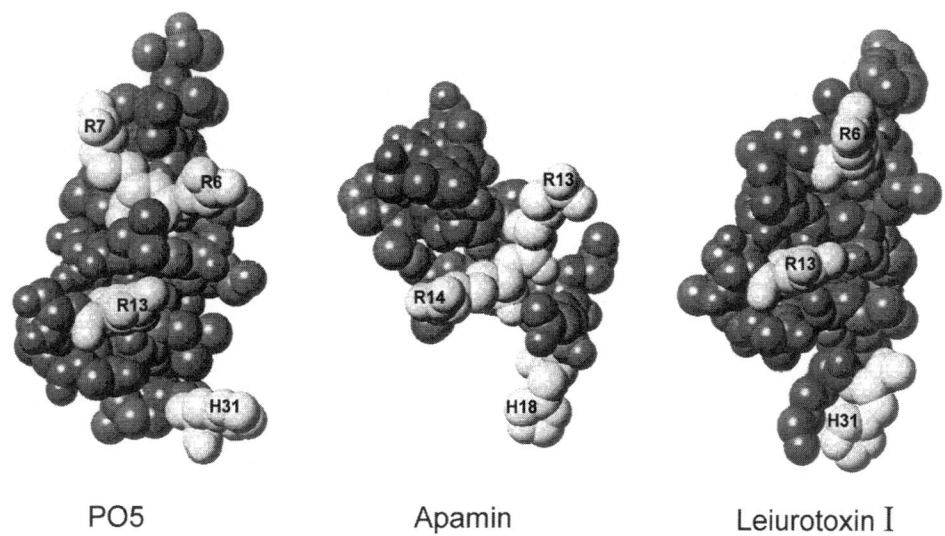

| PO5 | Apamin | Leiurotoxin I |

Figure 3. Functional maps of P05 (*left*), apamin (*center*), and leiurotoxin I (*right*). Residues in light gray are involved in the binding to the pore region.

1993). Structure–activity relationships defined the structural criteria for the activity of these two toxins.

2.3.2b. Leiurotoxin I. The overall tertiary structure of the LTX molecule is described as the CSαβ motif (cysteine-stabilized α–β motif) (Cornet *et al.*, 1995). More precisely, the structure of LTX revealed an α-helix spanning residues L5 to S14 and connected to a two-stranded β-sheet encompassing residues L18 to V29 with an unclassified tight turn centered on residues G23 and D24. The helix is affixed to the β-sheet by two disulfide bridges (Martins *et al.*, 1990, 1995). Its structure is very different from that of apamin. In particular, LTX does not possess the sequential positive charges found to be critical for the binding of apamin to SK$_{Ca}$ channels. Instead, LTX possesses two arginines, R6 and R13, separated by two turns of the α-helix and thus located on the same face of the molecule (Fig. 3).

The structure–activity relationship of LTX defines the surface directly implicated in the binding with the receptor. LTX competitively inhibits the binding of [^{125}I] apamin to its receptor. This suggests that, despite the absence of sequence and even of three-dimensional structure homologies between LTX and apamin (Fig. 1), they interact with the channel by the same mechanism (Auguste *et al.*, 1990). Thus, the influence of the charged residues of LTX on its binding capacity was investigated. The negative charge imparted by E27 to the β-sheet face of LTX can be modified without disturbing the activity of LTX toward SK$_{Ca}$ channels (K_d of 160 pM for the analog and 100 pM for the native toxin).

The three lysines K20, K25 and K30 located in the β-sheet region of the toxin can be chemically modified into homoarginines with only a minor effect on the toxin activity (Auguste *et al.*, 1992). On the other hand any modification of the positive charges on the α-helix face drastically alters the activity of the toxin. Indeed, taking into account the local differences between P05 and LTX, mainly at position 7 in the polypeptide chain (a methionine in LTX and an arginine in P05), different analogs were synthesized or chemical modifications performed. The analog [R6L]-leiurotoxin is 50 times less potent than native LTX, and the double neutralization of the arginines R6 and R13 in P05 induces about a 5000-fold drop in activity (Sabatier *et al.*, 1994). Moreover the C-terminal residue seems to play an important role in the binding process, as it has been demonstrated that the iodination of H31 induces about a 40-fold drop in activity and the LTX with a free carboxylic C-terminal group is 4-fold less potent than the naturally amidated form.

2.3.2c. P05. P05 also adopts the CSαβ scaffold (Cornet *et al.*, 1995) that has been found in all the scorpion toxins directed against K^+ and Na^+ channels. P05 possesses the same secondary structure as LTX, with an α-helix encompassing 10 residues from L5 to S14 connected to a two-stranded β-sheet composed of residues L17 to V29, with a type II' β-turn centered around G23 and V24 (Meunier *et al.*, 1993). In spite of the high level of conserved residues between LTX and P05, one difference is noticeable. The seventh residue is an arginine in P05 but a methionine in LTX (Fig. 1). This difference makes P05 more similar to apamin than LTX is. Indeed, P05 possesses a sequence of two arginines, which was previously shown to be the active site in apamin (Labbé-Julié *et al.*, 1991). After recognizing that a sequence of two arginines was a common feature of P05 and apamin binding, a series of analogs were synthesized to identify residues involved in the binding of P05 to SK_{Ca} channels. The analog [R6L, R7L]-P05, although possessing the same structure as the native toxin (Inisan et al., 1995), displays only a residual activity ($< 0.2\%$) when compared to native P05 (Sabatier *et al.* 1993). Moreover, as was previously demonstrated for apamin, the replacement of the two arginines by lysine residues leads to a toxin with a weaker activity. Thus, the functional map initially described for P05 includes the arginines R6 and R7 and the glutamine Q9 (Sabatier *et al.*, 1993). The structure determination of P05 allowed an investigation of the relative orientation of the side chains and of the overall electrostatic characteristics of the toxin. Interestingly, the charge distribution of P05 is highly asymmetric, with a positively charged surface including the arginines R6, R7 and R13 and the glutamine Q9. The carboxylamidation of the C-terminal residue H31 induces a considerable strengthening of the toxin–receptor interaction, such that to the amidated toxin had the properties of an irreversible SK_{Ca} channel blocker. This histidine H31, which is thought to play a role in the interaction with the receptor (Sabatier*et al.*, 1993) is remote from the positively charged surface. The functional map of P05, composed of these five residues, suggests a multipoint interaction of the toxin with its receptor (Meunier*et al.*, 1993).

2.3.3. The Dipole Moment: A Key to Locating Interacting Surfaces

Results obtained with toxins such as maurotoxin (MTX) provide new insight into the mechanism of the coupling of scorpion toxins to SK_{Ca} channels. MTX is a 34-residue- long toxin extracted from the venom of the Tunisian scorpion *Scorpio maurus*. This toxin is a blocker of Kv1.1, Kv1.2, and Kv1.3 channels and competes with apamin for binding to SK$_{Ca}$ channels (Kharrat *et al.*, 1997). The structure determination of this scorpion toxin demonstrates that the toxin scaffold is compatible with the unique disulfide pairing of this toxin (Kharrat *et al.*, 1997; Blanc *et al.*, 1997; Rochat *et al.*, 1998). Structure–activity relationships of MTX, when compared to other Kv channel blockers, show that neither the structural characteristics of this toxin nor the local differences from the other toxins can explain its specificity. One has to consider the whole distribution of charged residues on the toxin to elucidate the measured basis of the affinities. Thus, it has been shown that this toxin possesses a charge anisotropy that creates a dipole moment directed toward the side chain of a lysine residue. This lysine (K23) is located on the β-sheet face, and its equivalent in other toxins, such as KTX or CTX, is the central residue involved in the interaction of these toxins with Kv channels (Aiyar *et al.*, 1995, 1996; Darbon *et al.*, 1999; Durell and Guy, 1996). Analysis of the charge anisotropy of KTX leads to a similar result. The dipole moment of KTX is directed toward the crucial lysine. Thus, the hypothesis was advanced that this charge anisotropy serves as a guide for a correct orientation of the scorpion toxin in the vicinity of the K$^+$ channel (Blanc *et al.*, 1997).

Applying these calculations to specific SK$_{Ca}$ channel blockers, it has been noted that the dipole moment of LTX presents the face of the toxin molecule containing arginines R6 and R13 and histidine H31 to its receptor (Fremont *et al.*, 1997). A similar result was found with apamin, the dipole moment issuing from the toxin face containing the two critical arginines R13 and R14. Moreover, the dipole moment of that toxin emerges through the α-helix, a result consistent with the original calculation of the charge anisotropy of P05. A more detailed analysis of the orientation of the dipole moment of apamin shows that it presents a surface containing R6, R13, and H31 to SK$_{Ca}$ channels (Fig. 4), the same residues as in the case of LTX. The arginine R7 is located lateral to the new functional map. The difference observed in the binding potencies of the two toxins (P05 and LTX) could result from a better orientation, or a better charge distribution, in P05 than in LTX. This method applied to newly characterized toxins could help us to predict the surface directly involved in the interaction with the K$^+$ channels.

2.3.4. Prediction of Toxic Surfaces

2.3.4a. P01. P01 was identified in the same venom as P05. Although 31 residues are thought to constitute the critical length for the CSαβ motif formation (Martins *et al.*, 1995), the structures of P01 and the leiuropeptide II demonstrate that the critical length can be reduced to 29 residues. The structures of P01 (Blanc *et al.*, 1996) and leiuropeptide II (Buisine *et al.*, 1997) are highly homologous to those of P05 and LTX. Buisine *et al.* (1997) found leiuropeptide II to be devoid of activity toward mammals or insects, which can be explained in terms of an inappropriate electrostatic profile. The α-helix face of the two peptides is globally acidic instead of presenting the two positive charges described for other toxins.

Three different activities have been characterized for P01, depending on the origin of the molecule. The Ki values for the binding of ^{125}I-labeled apamin to its receptor on rat brain synaptosomes are 0.1 µM for natural P01 purified from the venom of *Androctonus mauretanicus mauretanicus*, 2 µM for natural P01 purified from *Androc-*

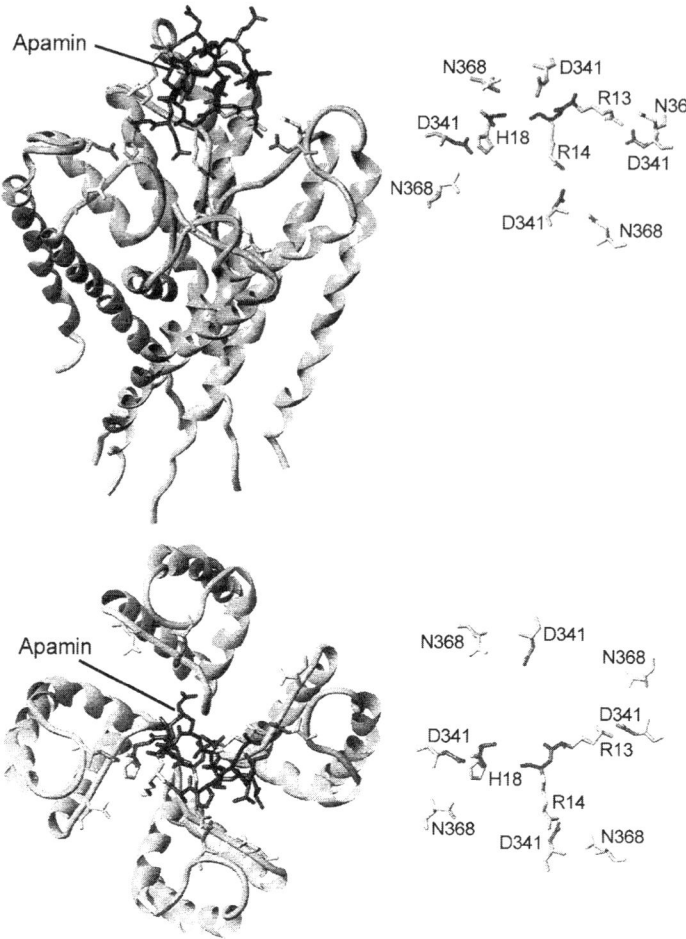

Figure 4. The docking of apamin on the pore region. *Top:* Front view; *Bottom:* view. The pore region backbone is in yellow with the filter of selectivity in green. Apamin is in dark gray On the right of the figure the residues of apamin that interact with the pore region and their channel counterparts are shown in blue.

tonus australis hector and 3 μM for the synthetic P01. The higher affinity of P01 from *Androctonus australis hector* compared to PO1 from other origins could be the result of a contamination by another toxin highly active against SK$_{Ca}$ channels (Zerrouk *et al.*, 1996). In fact, even this residual activity, when compared to that of P05 or apamin, is hard to explain, taking into account the absence of basic residues on the α-helix face. In the first analysis, we suggested that the weak binding potency of P01 could be a result of the presence of a histidine ring, partially charged at physiological pH, in the helical segment, (Blanc *et al.*, 1996). Subsequently, electrostatic analysis of P01 confirmed the acidic character of the α-helix face. According to its molecular orientation resulting from dipole moment orientation, P01 could better interact with its receptor via the two lysines $K14$ and $K18$. The weak affinity of this toxin could be the result of an incorrect orientation of the two basic residues with respect to their acidic counterparts on the receptor or to the absence of the guanidinium group. Further experiments are required to assess this hypothesis.

2.3.4b. TsKapa. TsKapa purified from the venom of the scorpion *Tityus serrulatus* competes with apamin for binding to SK$_{Ca}$ channels ($K_{0.5} = 0.3$ nM). Its primary structure shows only weak homologies with previously described SK$_{Ca}$-directed toxins, but the three-dimensional structure demonstrates that TsKapa possesses two basic residue couples, *R*6, *R*9 and *K*18, *K*19, that could account for its potency (Legros *et al.*, 1997; Blanc *et al.*, 1997). The dipole moment of TsKapa presents these basic residues toward the receptor. Synthetic analogs of TsKapa, in which an alanine replaces the arginine R6 or R9 or the lysine K18 or K19, called [*R6A*]-TsKapa, [*R9A*]-TsKapa, [*K18A*]-TsKapa, and [*K19A*]-TsKapa, were tested for their ability to displace ^{125}I apamin. The analog [*R6A*]-TsKapa is N130-fold less efficient than synthetic TsKapa ($K_{0.5}$ values of 400 nM and 3 nM, respectively), and [R$_9$A]-TsKapa is $\sim$13-fold less potent ($K_{0.5} = 31$ nM). There is little or no influence of the neutralization of the two lysines K18 and K19 (K0.5 values of 5 nM and 6 nM respectively), suggesting that the basic residue couple interacting with the receptor is composed of the two arginines R6 and R9 (Lecomte C. personal communication).

2.3.4c. Other Scorpion Toxins Acting on SK$_{Ca}$ Channels. Other scorpion venoms have been investigated to determine whether they contain potential new SK$_{Ca}$ channel blockers. The venom of the scorpion *Buthus martensi* Karsch contains the toxins BmP01, BmP02, BmP03, and BmP05 (Romi-Lebrun *et al.*, 1997), and that of *Leiurus quinquestriatus hebraeus* contains leiuropeptides I–III (Buisine *et al.*, 1997) (Fig. 1). In addition, toxins belonging to other families also block SK$_{Ca}$ channels. CTX is a blocker of SK$_{Ca}$ channels in aplysia neurons (Hermann and Erxleben, 1987), but these voltage-sensitive channels are insensitive to apamin and tetraethylammonium (TEA). The scorpion *Orthochirus scrobiculosus* produces a toxin named OSK1 that clearly belongs to the kaliotoxin family (Jaravine *et al.*, 1997). This toxin has been shown to specifically block apamin-insensitive, small-conductance K$^+$ channels in neuroblastoma–glioma NG 108-15 hybrid cells and further studies are required to examine its capacity to bind to other K$^+$ channels. As a general rule, toxins are often tested on a limited set of channels. Some toxins are able to block more than one K$^+$ channel subtype, such as CTX, which blocks both Kv and BK$_{Ca}$ channels, and MTX, which inhibits both Kv and SK$_{Ca}$ channels.

2.4. Receptor Counterparts of Apamin Critical Residues

The SK$_{Ca}$ channels have now been cloned. The first four sequences thus identified, namely, hSK1 and rSK1, rSK2, and rSK3 allowed the structural requirements for toxin activity to be investigated. Among the cloned SK$_{Ca}$ channels, rSK1 is resistant to apamin at concentrations up to 100 nM, whereas rSK2 is sensitive to 60 pM apamin. Analysis of the sensitivity of chimera rSK–rSK2 defined the pore-forming region as representing the binding site of apamin. An alignment of the two sequences showed that only three residues differ from one channel to the other. The glutamine *Q*339 aspartic acid $D341$, and asparagine $N368$ in rSK2 are replaced respectively by a lysine, a glutamic acid, and a histidine in rSK1 (Fig. 2). The replacement of the lysine rSK1 by its counterpart in rSK2 ($Q339$) has little effect on the sensitivity to apamin, whereas the two other mutations, E3Y1D and H368N, restore a partial apamin sensitivity. Moreover, the double mutant endows rSK1 with the sensitivity of rSK2. These results, together with the sensitivity of heterotetramers rSK1–rSK2, suggest that the apamin interacts with at least two distinct subunits of the channel. Thus, the arginines $R13$ and $R14$ of the apamin would face two aspartic residues. Glutamine $Q17$ would face one asparagine, $N368$ (Ishii *et al* 1997).

2.4.1. A Lack of Experimental Data

In spite of abundant data on the coupling of toxins and K^+ channels, there remain unresolved questions. The first question concerns the description of the functional map of apamin. It has been suggested that the glutamine residue $Q17$ plays a role in the interaction of apamin with SK_{Ca} channels, but the experimental data do not exclude a role for amidation of the C-terminal extremity in the interaction, instead of an interaction of the side chain itself. To date, no coupling of interacting toxin/channel residues has been identified as in the case of Kv channels. Thus, numerous models have been built to account for the activity and specificity of toxins directed against SK_{Ca} channels. In the most recently proposed model, the arginine $R13$ of apamin faces an aspartate residue of the channel, and the glutamic residue $E7$ of the toxin faces an asparagine on the channel; no mention was made of the importance of glutamine $Q17$ (Vergara *et al.*, 1998).

2.4.2. A K⁺ Channel Structure Available

The interaction of toxins with the SK_{Ca} channel has to be revisited on the basis of new information regarding the channel structure. The recent resolution of the first K^+ channel structure will certainly help to define the structural determinants of this interaction. Based on the conservation of the selectivity filter, centered around a constant GYGD motif, it can be postulated that the overall architecture of the pore-forming region is conserved between the K^+ channel subfamilies. Evidence in support of this hypothesis was obtained by MacKinnon *et al* (1998) who mutated the KcsA K^+ channel in order to render it sensitive to agitoxin-2. Only three point mutations in the sequence of KcsA are necessary to mimic the binding site of agitoxin-2 on the *Shaker* K^+ channel. The three mutations—an alanine in place of the glutamine $Q58$ and the replaement of KcsA residues 61 and 64 their Shaker counterparts were chosen because of the implication of thcsc residues in the interaction between the *Shaker* channel and agitoxin-2. The effects of toxin mutations $K27A$ and $N30A$ showed that the toxin interacts in the same manner with both the *Shaker* and KcsA mutated channels. Moreover, Adelman and co-workers demonstrated that sensitivity to block by TEA can be conferred to SK_{Ca} channels by introducing a tyrosine at a position that mediates external TEA sensitivity in Kv1.1 (Ishii *et al.*, 1998). Thus, KcsA and rSK2 most probably possess the same tertiary structure as Kv-related channels.

2.4.3. Docking Model

Taking the above hypothesis as a basis and starting from the KcsA channel coordinates, we modeled the pore-forming region of the rSK2 channel. As previously described, the local differences between rSK1 and rSK2 are located on the edges of the P-region. What is more surprising is that the two residues responsible for the activity of apamin, namely, D341 and N368 (rSK2 numbering), are within 7 Å of each other. Moreover, the channel has the appearance of an inverted teepee, and the mouth of the pore is large enough to allow deep penetration of toxins like apamin.

We modeled the interaction between apamin and the rSK2 channel (Fig. 4). Different alternatives exist for the binding of apamin. The best result is obtained with the two arginines pointing toward two aspartic residues (D341) two adjacent subunits. In this case a third interaction can exist between the C-terminal histidine side chain and another D341 residue or an asparagine (N368). This last interaction could account for the weaker affinity of the free-carboxylic form of apamin. Indeed, a repulsive interaction between the C-terminal carboxylic group and the D341 channel side chain would destabilize the toxin–receptor complex. Moreover, this docking takes into account the greater sensitivity of rSK2 to apamin, when compared to that of rSK1. An interaction between the asparagine N368 of rSK2 and the C-terminal histidine of apamin cannot exist in the apamin–rSK1 complex, the asparagine N368 in rSK2 being replaced by a histidine in rSK1. Another argument favors this orientation. It has been demonstrated that the truncated form of apamin, apamin (1–17), still possesses 1.3% of the affinity of the native toxin, whereas the carboxylic form of the toxin retains only 0.06% of its native activity. The absence of the histidine H18 shortens the polypeptide chain and thus pulls the C-terminal extremity of the toxin away from the channel residues D341 and N368. This shortening allows the toxin to penetrate deeper into the pore region and the side chain of glutamine Q17 to settle in the vicinity of the selectivity filter, near the acidic ring formed by the four aspartic D364 residues. An interaction between Q17 (toxin) and D364 (channel) could account for part of the affinity of the truncated apamin. Last, this model also takes into account the results of cross-linking experiments. All cross-linker agents are able to link the εNH2-Lys$_4$ to the β-subunit, which is rendered possible because of the lateral position of the K4 side chain. Furthermore, a spacer arm of at least 5.7 Å is necessary to link the αNH2-Cys$_1$ to that β-subunit. In the presently proposed orientation, the position of αNH$_2$-Cys$_1$ is on the fourfold axis of the channel, thus far away from the β-subunit.

Results obtained with scorpion toxins such as LTX and P05 are less clear. As can be seen from structure–activity relationship studies, P05 and LTX share possible pharmacophores — arginines R6 and R13 in LTX and R6 and R7 in P05. In both toxins, the C-terminal amidation plays an important role in the activity of the toxin. On the basis of the high level of homology that exists between P05 and LTX, it can be supposed that these two toxins block the channel by means of the same interactions. However, this hypothesis must be further validated because, whereas it is established that R13 is critical for the binding efficiency of LTX, there is no information regarding the importance of the arginine R13 in P05.

We also modeled a possible interaction between LTX or P05 and rSK2. Different positions for the toxin can be easily imagined. We started by directing the arginine R6 of leiurotoxin I toward an aspartic residue D341. Such an orientation places the arginine R13 in the vicinity of D341 of the adjacent SK_{Ca} channel subunit and H31 in the vicinity of N368 in the opposite subunit. The same network remains valid if we replace the LTX by P05.

Another alternative to this network is to adopt the results previously obtained for the interaction between CTX-like scorpion toxins and Kv channels or BK_{Ca} channels. The toxin plugs directly into the pore, with its critical lysine (K27 in CTX) located in the vicinity of the selectivity filter (Darbon et al., 1999). It has been suggested that the ε-NH2 of the lysine K27 competes with the potassium ion for the last K^+ binding site of the selectivity filter (Park and Miller, 1992a,b). Following such a hypothesis, we guided the arginine R13 of LTX to the level of the acid ring formed by the four aspartic acids of the GYGD sequence. This orientation allows an interaction between R13, Q9, and H31 of the toxin and residue D341 of three different SK_{Ca} channel types. A better orientation can be found for P05, owing to the presence of three arginines R6, R7, and R13. Orienting R6 in the acid ring permits R7 and R13 to come into the vicinity of D341 residues, and the histidine H31 to come close to N368.

This would mean that two highly homologous toxins (i.e., LTX and P05) can interact in different ways with the channel. If P05 is oriented by guiding R13 into the mouth of the pore in the vicinity of the acidic ring, then its arginine R7 is unable to interact with a channel residue. The effect of the C-terminal amidation on the blocking activity of these two toxins could be indicative of this different behavior. Whereas amidation has a small effect on the binding potency of LTX, (Lei-NH2 is 4-fold more potent than Lei-OH), the amidation of H31 in P05 increases its affinity such that it behaves like an irreversible ligand. Interestingly, although the sequences are nearly identical, the overall charge distribution differs from one toxin to the other. The dipole moment of P05 emerges near R6, whereas it emerges near R13 in LTX. These dipole moments present the same surface of the toxin to its receptor, containing the residues R6, R13, and H31, but centered around R6 in the case of P05 and R13 in the case of LTX.

2.5. Organic Compounds

Studies of the structure–activity relationships of peptidic blockers, mainly apamin, revealed the importance of two positive charges in the expression of their activity. Although not restricted to these charges, the presumed pharmacophore of apamin is based upon arginines R13 and R14. This consideration opened the way for research on organic blockers. Series of bisquaternary ammonium compounds were tested for their activity against SK_{Ca} channels. The most potent bisquaternary compound is dequalinium ($IC_{50} = 1.5$ μM), but other compounds, such as atracurium, pancuronium, and tubocurarine, are also effective blockers of SK_{Ca} channels (Fig. 5) (Castle et al., 1993; Galanakis et al., 1995). A series of dequalinium analogs were used to investigate the influence of the charged heterocycles on the blocking potency. The two quinolinium groups are required, as the elimination of one of them leads to a compound expressing impure SK_{Ca} channel-blocking activity. Moreover, although both charges are important for blocking activity, the level of charge delocalization also influences the blocking activity, with the most potent compound being the one having the highest degree of

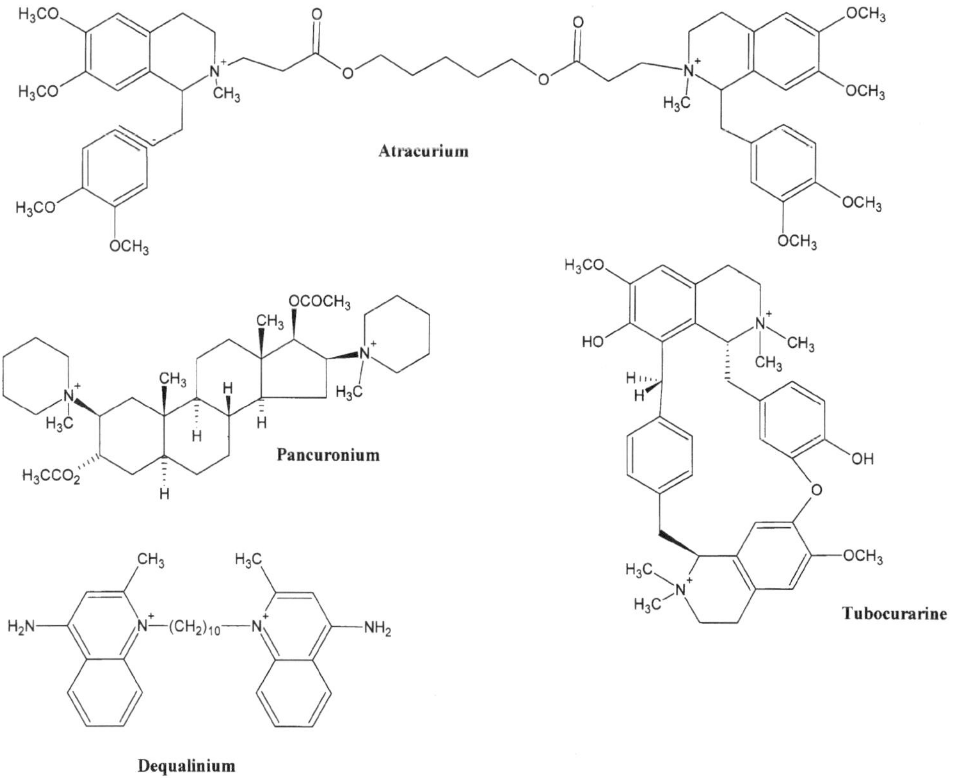

Figure 5. Structures of the nonpeptidic blockers of SK$_{Ca}$ channels.

charge delocalization (Galanakis *et al.*, 1995). As the concentration of quinolinium groups seems to be an important factor governing the potency, a trisquinolinium compound was synthesized, which exhibited blocking activity at nanomolar concentrations. This blocker is one order of magnitude more potent than dequalinium, but the question of whether the third ring directly binds to the channel or only increases the local charge concentration of quinolinium groups is unresolved (Galanakis *et al.*, 1996). Higher-affinity compounds are available, the most potent of which, UCL 1684, displays activity at nanomolar concentrations (Rosa *et al.*, 1998). All these compounds are based upon the basic diad found in apamin. Recently, we analyzed the differential sensitivity of SK$_{Ca}$ channels rSK1 (IC$_{50}$ = 354.3 μM) and rSK2 (IC$_{50}$ = 5.4 μM) to *d*-tubocurarine (dTC). The structural determinants for dTC sensitivity are the same as for apamin sensitivity. Mutating rSK1 at position 341 or 368 (mutants E3141D or H368N; rSK2 numbering) restores a partial sensitivity, whereas the double mutant exhibits the same dTC sensitivity as rSK2 (Ishii *et al.*, 1997). Thus, the positive charges of these organic compounds would play the role of the two arginines of apamin. As confirmation, the distance separating the two guanidinium groups in apamin (12 Å) is comparable to the distance separating the ammonium groups in tubocurarine (11 Å) or pancuronium (12 Å) (Fig. 6).

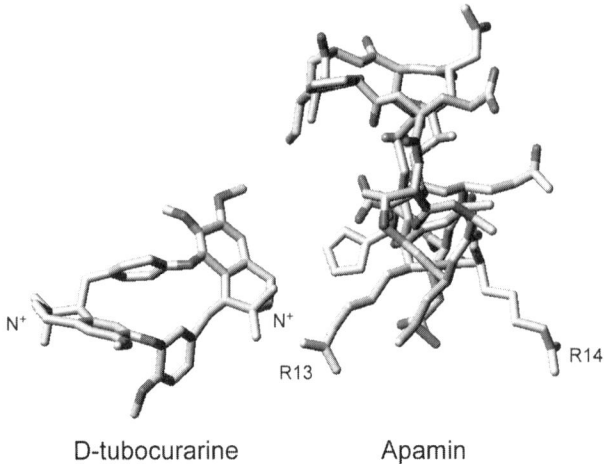

D-tubocurarine   Apamin

Figure 6. Structure of *d*-tubocurarine (*left*) compared with that of apamin (*right*).

2.6. An auxiliary β-Subunit

The studies described above consider the direct binding of a given toxin to the rSK2 pore region. However, as is the case for other channels, SK_{Ca} channels are the result of a heteromultimeric assembly of different subunits. In the case of BK_{Ca} channels, the β-subunit has been implicated as being directly involved in the high-affinity complex formed between CTX and the channel, based upon evidence that the β-subunit modifies the biophysical and pharmacological properties of the α-subunits (Hanner *et al.*, 1997, 1998). Thus, a complete picture of the binding characteristics of scorpion/bee toxins to the SK_{Ca} channel heteromultimer must integrate the potential influence of the β-subunit.

3. CONCLUSION

This chapter has attempted to summarize the current status of the pharmacology of SK_{Ca} channels and their specific ligands. This overview clearly shows that research on these channels lags far behind that on Kv channels. Despite this, current data suggest that SK_{Ca} channels may be involved in diverse diseases. There is considerable evidence pointing to the importance of these channels in vascular smooth muscle function. These channels are also present in rabbit cardiac muscles. Therefore, knowledge of the pharmacology of specific ligands is of primary importance for the design of potent therapeutic drugs. The main challenge in designing such artificial molecules will be to pinpoint the structural basis of the specificity and high binding activity of the natural ligands. From the receptor point of view, a more precise understanding of the modulatory role of the β-subunit on α-subunit function is required.

REFERENCES

Abia, A., Lobaton, C. D., Moreno, A., and Garcia-Sancho, J., 1986, *Leiurus quinquestriatus* venom inhibits different kinds of Ca^{2+}-dependent K$^+$ channels, *Biochim. Biophys. Acta* **856**:403–407.

Adelman, J. P., Shen, Z. K., Kavanaugh, M. P., Warren, R. A., Wu, Y. N., Lagrutta, A., Bond, C. T., and North, R. A., 1992, Calcium-activated K$^+$ channels expressed from cloned complementary DNAs, *Neuron* **9**:209–216.

Aiyar, J., Withka, J. M., Rizzi, J. P., Singleton, D. H., Andrews, G. C., Lin, W., Boyd, J., Hanson, D. G., Simon, M., Dethlefs, B., Lee, C. L., Hall, J. F., Gutman, G. A., and Chandy, K. G., 1995, Topology of the pore-region of a K$^+$ channel revealed by the NMR-derived structures of scorpion toxins, *Neuron* **15**:1169–1181.

Aiyar, J., Rizzi, J. P., Gutman, G. A., and Chandy, K. G., 1996, The signature sequence of voltage-gated K$^+$ channels projects into the external vestibule, *J. Biol. Chem.* **271**:31013–31016

Atkinson, N. S., Robertson, G. A., and Ganetzky, B., 1991, A component of calcium-activated K$^+$ channels encoded by the *Drosophila slo* locus, *Science* **253**:551–555.

Auguste, P., Hugues, M., and Lazdunski, M., 1989, Polypeptide constitution of receptors for apamin, a neurotoxin which blocks a class of Ca^{2+}-activated K$^+$ channels, *FEBS Lett.* **248**:150–154.

Auguste, P., Hugues, M., Gravé, B., Gesquire, J. C., Maes, P., Tartar, A., Romey, G., Schweitz, H., and Lazdunski, M., 1990. Leiurotoxin I (scyllatoxin), a peptide ligand for Ca^{2+}-activated K$^+$ channels, *J. Biol. Chem.* **265**:4753–4759.

Auguste, P., Hugues, M., Mourre, C., Moinier, D., Tartar, A., and Lazdunski, M., 1992, Scyllatoxin, a blocker of Ca^{2+}-activated K$^+$ channels: Structure–function relationships and brain localization of the binding sites, *Biochemistry* **31**:648–654.

Bargmann, C I., 1998, Neurobiology of the *Caenorhabditis elegans* genome, *Science* **282**:2028–2033.

Blanc, E., Fremont, V., Sizun, P., Meunie, S, Van Rietschoten, J., Thevand A., Bernassau, J. M., and Darbon, H., 1996, Solution structure of P01, a natural scorpion peptide structurally analogous to scorpion toxins specific for apamin-sensitive K$^+$ channel, *Proteins* **24**:359–369.

Blanc, E., Sabatier, J. M., Kharrat, R., Meunier, S., el Ayeb, M., Van Rietschoten, J., and Darbon H., 1997, Solution structure of maurotoxin, a scorpion toxin from *Scorpio maurus,* with high affinity for voltage-gated K$^+$ channels, *Proteins* **29**:321–233.

Buisine, E., Wieruszeski, J. M., Lippens, G., Wouters, D., Tartar, A., and Sautiere, P., 1997, Characterization of a new family of toxin-like peptides from the venom of the scorpion *Leiurus quinquestriatus hebraeus.* ^{1}H-NMR structure of leiuropeptide II, *J. Pept. Res.* **49**:545–555.

Butler, A. Tsunoda, S., McCobb, D. P., Wei., A,. and Salkoff, L., 1993, mSlo, a complex mouse gene encoding "maxi" calcium-activated K$^+$ channels, *Science* **261**:221–224.

Bystrov, V. F., Okhanov, V. V., Miroshnikov, A. I., and Ovchinnikov. Y. A., 1980, Solution spatial structure of apamin as derived from NMR study, *FEBS Lett* **119**:113–117.

Castle, N. A., Haylett, D. G., Morgan, J. M., and Jenkinson, D. H., 1993, Dequalinium: A potent inhibitor of apamin-sensitive K$^+$ channels in hepatocytes and of nicotinic responses in skeletal muscle, *Eur. J. Pharmacol.* **236**:201–207.

Chicchi, G. G., Gimenez-Gallego, G., Ber, E., Garcia, M. L., Winquist, R., and Cascieri, M., 1988, Purification and characterization of a unique potent inhibitor of apamin binding from *Leiurus quinquestriatus hebraeus* venom, *J. Biol. Chem.* **263**:10192–10197.

Cornet, B., Bonmatin, J. M., Hetru, C., Hoffmann, J. A., Ptak, M., and Vovelle, F., 1995, Refined three dimensional structure of insect defensin A, *Structure* **3**:435–448.

Cotton, J., Crest, M., Bouet, F., Alessandri, N., Gola, M., Forest, E., Karlsson, E., Castaeda, O., Harvey, A. L., Vita, C., and Menez, A., 1997, A K$^+$ channel toxin from the sea anemone *Bunodosoma granulifera,* an inhibitor for Kv1 channels, *Eur. J. Biochem.* **244**:192–202.

Darbon, H., Blanc, E., and Sabatier, J. M., 1999, Three dimensional structure of scorpion toxins: Towards a new model of interaction with K$^+$ channels, in: *Perspectives in Drug Discovery and Design* Vol. 15/16 (H. Darbon and J. M. Sabatier, eds.), Kluwer Academic Publishers, Dordrecht, The Netherlands, pp. 41–60.

Devaux, C., Knibiehler, M., Defendini, M. L., Mabrouk, K., Rochat, H., Van Rietschoten, J., Baty, D., and Granier, C., 1995, Recombinant and chemical derivatives of apamin: Implication of post-transcriptional C-terminal amidation of apamin in biological activity, *Eur. J. Biochem.* **231**:544–550.

Doyle, D. A., Cabral, J. M., Pluetzner, R. A., Kuo A., Gulbis, J. M., Cohen, S. L., Chait, B. T., and MacKinnon, R., 1998, The structure of the K^+ channel: Molecular basis of K^+ conduction and selectivity, *Science* **280:**69–77.

Durell, S. R., and Guy, H. R., 1996, Structural model of the outer vestibule and selectivity filter of the Shaker voltage-gated K^+ channel, *Neuropharmacolog* **35:**761–773.

Durell, S. R., Hao, Y., and Guy, H. R., 1998, Structural models of the transmembrane region of voltage-gated and other K^+ channels in open, closed, and inactivated conformations, *J. Struct. Biol.* **121:**263–284.

Fremont, V., Blanc, E., Crest, M., Martin-Eauclaire, M.-F., Gola, M., Darbon, H., and Van Rietschoten, J., 1997, Dipole moments of scorpion toxins direct the interaction towards small- or large-conductance Ca^{2+}-activated K^+ channels, *Lett. Pept. Sci.* **4:**305–312.

Galanakis, D., Davis C. A., Del Rey Herrero, B., Ganellin, C. R., Dunn, P. M., and Jenkinson D. H., 1995, Synthesis and structure-activity relationships of dequalinium analogues as K^+ channel blockers: Investigations on the role of the charged heterocycle, *J. Med. Chem.* **38:**595–606.

Galanakis, D., Ganellin, R. C., Dunn, P. M., and Jenkinson, D. H., 1996, On the concept of a bivalent pharmacophore for SK_{Ca} channel blockers: Synthesis, pharmacological testing, and radioligand binding studies on mono-, bis-, and trisquinolinium compounds, *Arch. Pharm.* **329:**524–528.

Goldstein, S. A. N., 1996, A structural vignette common to voltage sensors and conduction pores: Canaliculi, *Neuron* **16:**717–722.

Granier, C., Pedroso Muller, E., and Van Rietschoten, J., 1978, Use of synthetic analogs for a study on the structure–activity relationship of apamin, *Eur. J. Biochem.* **82:**293–299.

Guy, H. R., and Conti, F., 1990, Pursuing the structure and function of voltage-gated channels, *Trends Neurosci.* **13:**201–206.

Habermann, E., and Fischer, K., 1979, Apamin, a centrally acting neurotoxic peptide: Binding and actions, *Adv. Cytopharmacol.* **3:**387–394.

Hanner, M., Schmalhofer, W. A., Munujos, P., Knaus, H. G., Kaczorowski, G. J., and Garcia, M. L., 1997, The β subunit of the high-conductance calcium-activated K^+ channel contributes to the high-affinity receptor for charybdotoxin, *Proc. Natl. Acad. Sci, U.S.A.* **94:**2853–2858.

Hanner, M., Vianna-Jorge, R., Kamassah, A., Schmalhofer, W. A., Knaus, H. G., Kaczorowski, G. J., and Garcia, M. L., 1998, The beta subunit of the high conductance calcium-activated K^+ channel: Identification of residues involved in charybdotoxin binding, *J. Biol. Chem.* **273:**16289–16296.

Harvey, A. L., 1997, Recent studies on dendrotoxins and K^+ ion channels, *Gen Pharmacol.* **28:**7–12.

Harvey, A. L., Bradley, K. N., Cochran, S. A., Rowan, E. G., Pratt, J. A., Quillfeldt, J. A., and Jerusalinsky D. A., 1998, What can toxins tell us for drug discovery?, *Toxicon* **36:**1635–1640.

Heginbotham, L. and MacKinnon, R., 1992, The aromatic binding site for tetraethylammonium ion on K^+ channels, *Neuron* **8:**483–491.

Heginbotham, L., Lu, Z., Abramson, T., and MacKinnon, R., 1994, Mutations in the K^+ channel signature sequence, *Biophys. J.* **66:**1061–1067.

Hodgkin, A. L., and Huxley, A. F., 1952a, Currents carried by sodium and K^+ ions through the membrane of the giant axon of *Loligo, J. Physiol. (London)* **116:**449–472.

Hodgkin, A. L., and Huxley, A.F., 1952b, The components of membrane conductance in the giant axon of *Loligo, J. Physiol. (London)* **116:**473–496.

Hodgkin, A. L., and Huxley, A. F., 1952c, The dual effect of membrane potential on sodium conductance in the giant axon of the squid, *J. Physiol. (London)* **116:**497–506.

Hermann, A. and Erxleben, C. 1987, Charbdotoxin selectively blocks small Ca-activated K channels in Aplysia neurons. *J. Gen. Physiol.* **90**(1):27–47.

Hodgkin, A. L., and Huxley, A. F., 1952d, A quantitative description of membrane current and its application to conduction and excitation in nerve, *J. Physiol. (London)* **117:**500–544.

Inisan, A. G., Meunier, S., Fedelli, O., Altbach, M., Fremont, V., Sabatier, J. M., Thevan, A., Bernassau, J.M., Cambillau, C., and Darbon, H., 1995, Structure–activity relationship study of a scorpion toxin with high affinity for apamin-sensitive K^+ channels by means of the solution structure of analogues, *Int. J. Pept. Protein Res.* **45:**441–450.

Ishii T. M., Maylie J., and Adelman J. P., 1997, Determinants of apamin and *d*-tubocurarine block in SK K^+ channels, *J. Biol. Chem.* **272:**23195–23200.

Jan, L.Y., and Jan, Y. N., 1992, Structural elements involved in specific K^+ channel functions, *Annu. Rev. Physiol.* **54:**537–555.

Jaravine, V. A., Nolde, D. E., Reibarkh, M. J., Korolkova, Y. V., Kozlov, S. A., Pluzhnikov, K. A., Grishin, E. V., and Arseniev, A. S., 1997, Three-dimensional structure of OSK1 from *Orthochirus scrobiculosus* scorpion venom, *Biochemistry* **36:**1223–1232.

Kerr, I. D., and Sansom, M. S., 1997, The pore-lining region of Shaker voltage-gated K$^+$ channels: Comparison of beta-barrel and alpha-helix bundle models, *Biophys. J.* **73:**581–602.

Kharrat, R., Mansuelle, P., Sampieri, F., Crest, M., Martin-Eauclaire, M. F., Rochat, H., and El Ayeb, M., 1997, Maurotoxin, a new four disulfide bridges toxin from *Scorpio maurus* venom: Purification, structure and pharmacology on K$^+$ channels, *FEBS Lett.* **406:**284–290.

Köhler, M., Hirschberg, B., Bond, C. T., Kinzie, J. M., Marrion, N. V., Maylie, J., and Adelman, J. P., 1996, Small-conductance, calcium-activated K$^+$ channels from mammalian brain, *Science* **273:**1709-1714.

Labbé-Jullié, C., Granier, C., Albericio, F., Defendini, M. L., Ceard, B., Rochat, H, and Van Rietschoten, J., 1991, Binding and toxicity of apamin: Characterization of the active site, *Eur. J. Biochem.* **196:**639–645.

Legros, C., Oughuideni, R., Darbon, H., Rochat, H., Bougis, P. E., and Martin-Eauclaire, M. F., 1996, Characterization of a new peptide from *Tityus serrulatus* scorpion venom which is a ligand of the apamin-binding site, *FEBS Lett.* **390:**81–84.

Levèque, C., Marqueze, B., Couraud, F., and Seagar, M., 1990, Polypeptide components of the apamin receptor associated with a calcium activated K$^+$ channel, *FEBS Lett.* **275:**185–189.

Lipkind, G. M., and Fozzard, H. A., 1997, A model of scorpion toxin binding to voltage-gated K$^+$ channels, *J. Membr. Biol.* **158:**187–196.

MacKinnon, R., Cohen, S. E., Kuo, A., Lee, A., and Chait, B. T., 1998, Structural conservation in prokaryotic and eukaryotic K$^+$ channels,*Science* **280:**106–109.

Martins, J. C., Zhang, W., Tartar, A., Lazdunski, M., and Borremans, F., 1990, Solution conformation of leiurotoxin I (scyllatoxin) by ^{1}H nuclear magnetic resonance, *FEBS Lett.* **260:**249–253.

Martins, J. C., Van de Ven, F. J. M., and Borremans, F. A. M., 1995, Determination of the three-dimensional solution structure of scyllatoxin by ^{1}H nuclear magnetic resonance, *J. Mol. Biol.* **253:**590–603.

Meunier, S., Bernassau, J. M., Martin-Eauclaire, M. F., Van Rietschoten, J., Cambillau, C., and Darbon, H., 1993, Solution structure of P05-NH2, a scorpion toxin analog with high affinity for the apamin-sensitive K$^+$ channel, *Biochemistry* **32:**11969–11976.

Miller, C., Moczydlowski, E., Latorre, R., and Philips, M., 1985, Charybdotoxin, a protein inhibitor of single Ca^{2+}-activated K$^+$ channels from mammalian skeletal muscle, *Nature* **313:**316–318.

Park, C. S., and Miller, C., 1992a, Mapping function to structure in a channel-blocking peptide: Electrostatic mutants of charybdotoxin, *Biochemistry* **31:**7749–7755.

Park, C. S., and Miller, C., 1992b, Interaction of charybdotoxin with permeant ions inside the pore of a K$^+$ channel, *Neuron* **9:**307–313.

Pease, J. H., and Wemmer, D. E., 1988, Solution structure of apamin determined by nuclear magnetic resonance and distance geometry, *Biochemistry* **27:**8491–8498.

Pennington, M. W., Byrnes, M. E., Zaydenberg, I., Khaytin, I., De Chastonay, J., Krafte, D. S., Hill, R., Mahnir, V. M., Volberg, W. A., Gorczyca, W., and Kem, W. R., 1995, Chemical synthesis and characterization of ShK toxin: A potent K$^+$ channel inhibitor from a sea anemone, *Int. J. Pept. Protein Res.* **46:**354–358.

Ranganathan, R., Lewis, J. H., and MacKinnon, R., 1996, Spatial localization of the K$^+$ channel selectivity filter by mutant cycle-based structure analysis, *Neuron* **16:**131–139.

Rochat, H., Kharrat, R., Sabatier, J. M., Mansuelle, P., Cres, M., Martin-Eauclaire, M. F., Sampieri, F., Oughideni, R., Mabrou, K., Jacquet, G., Van Rietschoten, J., and El Ayeb, M., 1998, Maurotoxin, a four disulfide bridges scorpion toxin acting on K$^+$ channels, *Toxicon* **36:**1609–1611.

Romi-Lebrun, R., Martin-Eauclaire, M. F., Escoubas, P., Wu, F. Q., Lebrun, B., Hisada, M., Nakajima, T. 1997, Characterization of four toxins from Buthus martens, scorpion venom, which act on apamin-sensitive Ca^{2+} activated K$^+$ channels. *Eur. J. Biochem.* **245**(2):457–464.

Rosa, J. C., Galanaki, D., Ganellin, C. R., Dunn, P. M., and Jenkinson, D. H., 1998, Bis-quinolinium cyclophanes: 6,10-diaza-3(1,3),8(1,4)-dibenzena-1,5(1,4)-diquinolinacyclodecaphane (UCL 1684), the first nanomolar, non-peptidic blocker of the apamin-sensitive Ca^{2+}-activated K$^+$ channel, *J. Med. Chem.* **41:**2–5.

Sabatier, J. M., Zerrouck, H., Darbon, H., Mabrouk, K., Benslimane, A., Rochat, H., Martin-Eauclaire, M. F., and Van Rietschoten, J., 1993, P05, a new leiurotoxin I-like scorpion toxin: Synthesis and structure–activity relationships of the α-amidated analog, a ligand of Ca^{2+}-activated K$^+$ channels with increased affinity, *Biochemistry* **32:**2763–2770.

Sabatier, J. M., Frmont, V., Mabrouk, K., Crest, M., Darbon, H., Rochat, H., Van Rietschoten, J., and Martin-Eauclaire, M. F., 1994, Leiurotoxin I, a scorpion toxin specific for Ca^{2+}-activated K^+ channels, *Int. J. Pept. Protein Res.* **43**:486–495.

Sandberg, B. E., 1979, Solid phase synthesis of 13-lysine-apamin, 14-apamin, and the corresponding guanidinated derivatives, *Int. J. Pept. Protein Res.* **13**:327–333.

Sanguinetti, M. C., Johnson, J. H., Hammerland, L G., Kelbaugh, P. R., Volkmann, R. A., Saccomano, N.A., and Mueller A. L., 1997, Heteropodatoxins: Peptides isolated from spider venom that block Kv4.2 K^+ channels, *Mol Pharmacol.* **51**:491–498.

Schmid-Antomarchi, H., Hugues, M., Norman, R., Ellory, C., Borsotto, M., and Lazdunski, M., 1984, Molecular properties of the apamin-binding component of the Ca^{2+}-dependent K^+ channel, radiation-inactivation, affinity labeling and solubilization. *Eur. J.Biochem.* **142**:1–6.

Schweitz, H., Bruhn, T., Guillemare, E., Moinier, D., Lancelin, J. M., Beress, L., and Lazdunski, M., 1995, Kalicludines and kaliseptine, *J. Biol. Chem.* **270**:25121–25126.

Seagar, M., Labbé-Jullié, C., Granier, C., Goll, A., Glossmann, A., Van Rietschoten, J. and Couraud, F., 1986, Molecular structure of rat brain apamin receptor: differential photoaffinity labeling of putative K^+ channel subunits and target size analysis, *Biochemistry* **25**:4051–4057.

Shon, K. J., Stocker, M., Terlau H., Stuhmer W., Jacobsen R., Walker, C., Grilley M., Watkins M., Hillyard, D. R., Gray, W. R., and Olivera, B. M., 1998, K-Conotoxin PVIIA is a peptide inhibiting the Shaker K^+ channel, *J. Biol. Chem.* **273**:33–38.

Swartz, K. J., and MacKinnon, R., 1995, An inhibitor of the Kv2.1 K^+ channel isolated from the venom of a Chilean tarantula, *Neuron* **15**:941–949.

Vergara, C., Latorre, R., Marrion, N. V., and Adelman, J. P., 1998, Calcium-activated K^+ channels, *Curr. Opin. Neurobiol.* **8**:321–329.

Vincent, J. P., Schweitz, H., and Lazdunski, M., 1975, Structure–function relationships and site of action of apamin, a neurotoxic polypeptide of bee venom with an action on the central nervous system, *Biochemistry* **14**:2521–2525.

Wadsworth, J. D. F., Doorty, K. B., and Strong, P. N., 1994, Comparable 30-kDa apamin binding polypeptides may fulfill equivalent roles within putative subtypes of small conductance Ca^{2+}-activated K^+ channels, *J. Biol. Chem.* **269**:18053–18061.

Wadsworth, J. D. F., Doorty, K. B., Ganellin, C. R., and Strong, P.N., 1996, Photolabile derivatives of ^{125}I-apamin: Defining the structural criteria required for labeling high and low molecular mass polypeptides associated with small conductance Ca^{2+}-activated K^+ channels, *Biochemistry* **35**:7917–7927.

Wadsworth, J. D., Torelli, S., Doorty, K. B., and Strong P. N., 1997, Structural diversity among subtypes of small-conductance Ca^{2+}-activated K^+ channels, *Arch. Biochem. Biophys.* **346**:151–160.

Wallner, M., Meera, P., and Toro, L., 1996, Determinant for β-subunit regulation in high-conductance voltage-activated and Ca^{2+}-sensitive K^+ channels: An additional transmembrane region at the N-terminus, *Proc. Natl Acad. Sci., USA* **93**:14922–14927.

Wei, A., Jegla,, T., and Salkoff, L., 1996, Eight K^+ families revealed by the *C. elegans* project, *Neuropharmacol.* **35**:805–829.

Yang, P. K., Lee, C. Y., and Hwang, M. J., 1997, Shaker pore structure as predicted by annealed atomic simulation using symmetry and novel geometric restraints, *Biophys. J.* **72**:2479–2489.

Yool, A., and Schwartz, T. L., 1991, Alteration of ionic selectivity of a K^+ channel by mutation of the H5 region. *Nature* **349**:700–704.

Zerrouk, H., Mansuelle, P., Benslimane, A., Rochat, H., and Martin-Eauclaire, M. F., 1993, Characterization of a new leiurotoxin I-like scorpion toxin P05 from *Androctonus mauretanicus mauretanicus*, *FEBS Lett.* **320**:389–392.

Zerrouk, H., Laraba-Djebari, F., Fremont, V., Meki, A., Darbon, H., Mansuelle, P., Oughuideni, R., Van Rietschoten, J., Rochat, H, and Martin-Eauclaire, M. F., 1996, Characterization of PO1, a new peptide ligand of the apamin-sensitive Ca^{2+}-activated K^+ channel, *Int. J. Pept. Protein Res.* **48**:514–521.

Chapter 15

Molecular Pharmacology of ATP-Sensitive K^+ Channels: How and Why?

Andre Terzic and Michel Vivaudou

1. K_{ATP} Channels: From Discovery to Structure

ATP-sensitive K^+ (K_{ATP}) channels are recognized by their biophysical fingerprint, unique heteromultimeric structure, and distinct nucleotide-dependent regulation (Noma, 1983; Aguilar-Bryan and Bryan, 1999; Seino, 1999). These weakly inwardly rectifying, high-conductance, potassium-selective channels are kept closed by intracellular ATP and activated by intracellular ADP. Thereby, K_{ATP} channels set the membrane potential according to changes in the cellular metabolic state (Weiss and Venkatesh, 1993; O'Rourke *et al.*, 1994; Dzeja and Terzic, 1998). K_{ATP} channels are distributed in the plasmalemma of various metabolically active tissues, including the heart (Noma, 1983), pancreatic β-cells (Ashcroft, 1996), skeletal (Vivaudou *et al.*, 1991) and smooth (Quayle et al., 1997) muscle, and the brain (Spanswick *et al.*, 1997). A related channel has been recognized in the inner membrane of mitochondria (Inoue *et al.*, 1991; Paucek *et al.*, 1992), underscoring the role of K_{ATP} channels in signaling networks that transduce intracellular metabolic events.

The inwardly rectifying K^+ channel Kir6.2 usually serves as a pore-forming core of K_{ATP} channels (Inagaki *et al.*, 1995; Ashcroft and Gribble, 1998; Seino, 1999). Kir6.2 contains two transmembrane domains (M1 and M2) which flank the pore (P) region (Fig. 1). Although deletion of the very carboxy terminus of Kir6.2 (Tucker *et al.*, 1997; Lorenz *et al.*, 1998) or its overexpression (John *et al.*, 1998) may give rise to measurable K_{ATP} channel current, it is through assembly with regulatory, sulfonylurea receptor (SUR) subunits (Fig. 1), which belong to the family of ATP-binding cassette (ABC) proteins, that the full properties of tissue-specific channel phenotypes are generated (Aguilar-Bryan *et al.*, 1995, 1998; Inagaki *et al.*, 1995, 1996). It is in this way that Kir6.2

Andre Terzic • Division of Cardiovascular Diseases, Department of Medicine and Pharmacology, Mayo Clinic, Mayo Foundation, Rochester, Minnesota 55905. *Michel Vivaudou* • Laboratory of Molecular and Cellular Biophysics, CEA-DBMS/BMC, URA CNRS 520, 38054 Grenoble, France.

Potassium Channels in Cardiovascular Biology, edited by Archer and Rusch. Kluwer Academic/Plenum Publishers, New York, 2001.

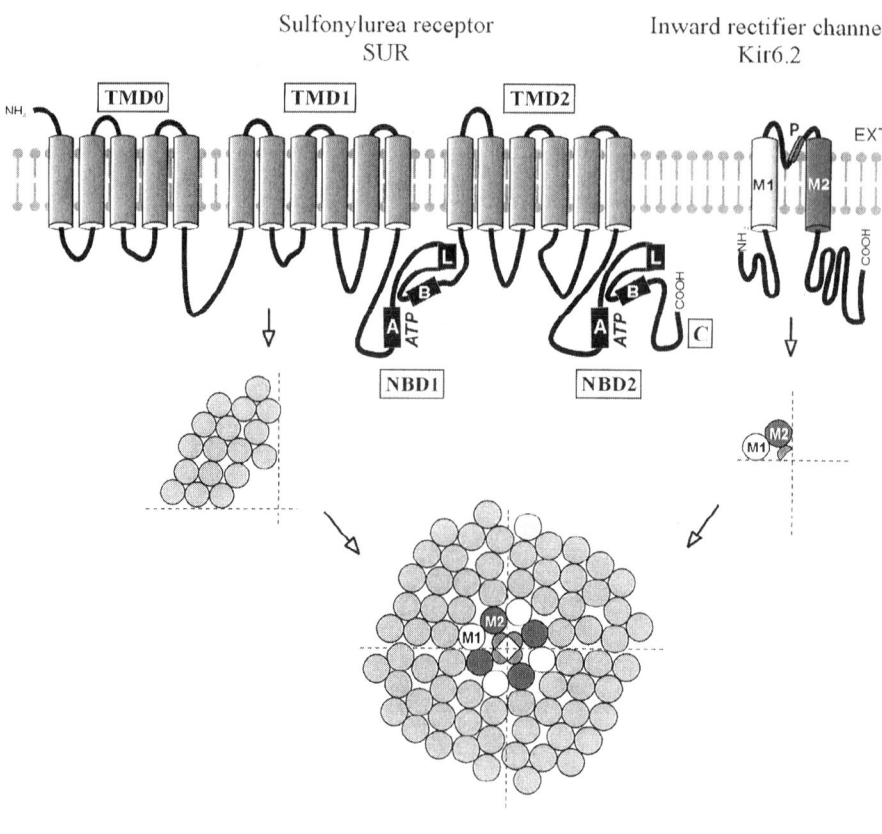

Octameric K_{ATP} channel complex

Figure 1. Kir6.2 and the sulfanylurea receptor (SUR) are the constitutive subunits of the K_{ATP} channel. Four subunits of the inwardly rectifying K^+ channel, Kir6.2, associate with four ATP-binding cassette (ABC) proteins, SUR, to form a functional KATP channel octamer. In Kir6.2, M1 and M2 represent the transmembrane domains, with P identifying the pore region. Like other ABC transporters, SUR is built as a modular protein with clearly defined transmembrane and cytoplasmic domains. TMD0, TMD1, and TMD2, indicate transmembrane domains of the protein, with NBD1 and NBD2 designating nucleotide-binding domains. The NBDs are characterized by conserved sequence motifs involved in the binding and hydrolysis of ATP (Walker motifs: A and B), and in the transmission of this signal to transmembrane domains (Linker: L). The region termed C has been implicated in the action of potassium channel openers.

and SUR1 genes, which are clustered on chromosome 11 (11p15.1), could be cotran-scribed and cotranslated to produce pancreatic K_{ATP} channels (Aguilar-Bryan *et al.*, 1995; Inagaki *et al.*, 1995; Bryan and Aguilar-Bryan, 1997). Translated Kir6.2 and SUR1 form a physical complex (Lorenz et al., 1998) with an octameric stoichiometry (Clement *et al.*, 1997; Shyng and Nichols, 1997; (Fig. 1). Similarly, cardiac K_{ATP} channels can be generated by coexpression of Kir6.2 with SUR2A, an isoform that has 68% identity with SUR1 despite differences in domains outside the highly conserved nucleotide-binding-fold regions (Inagaki *et al.*, 1996; Babenko *et al.*, 1998; Okuyama *et al.*, 1998). Although the SUR2 gene has been mapped to a chromosome (12p11.12) distinct from that of Kir6.2 (Chutkow *et al.*, 1996), Kir6.2 can also physically associate with SUR2A to form a channel complex, provided that the proximal carboxy terminus together with the M2 domain of the Kir6.2 protein is available (Lorenz and Terzic, 1999). Vascular K_{ATP} channels are likely to be composed of the SUR2B isoform in conjunction with either Kir6.2 itself or Kir6.1, a related inwardly rectifying channel (Isomoto *et al.*, 1996; Yamada *et al.*, 1997; Repunte *et al.*, 1999). At present, the structures of skeletal and neuronal, as well as mitochondrial, K_{ATP} channels are less definite, although various combinations of Kir6.2/6.1 subunits with SUR isoforms are considered as the likely constitutive components (Suzuki *et al.*, 1997).

By virtue of their metabolism-sensing property, K_{ATP} channels regulate vital cellular functions as diverse as hormone secretion (Ashcroft, 1996), muscle excitability (Davies *et al.*, 1991) and neurotransmitter release (Amoroso *et al.*, 1990). In particular, their role in mediating glucose-induced insulin release in pancreatic β-cells is best understood. Disruption of the Kir6.2 subunit leads to impaired hormone secretion (Miki *et al.*, 1997; 1998), whereas mutations in Kir6.2 and/or SUR1 genes have been linked to persistent hyperinsulinemia of infancy, a life-threatening channelopathy (Thomas *et al.*, 1995; Dunne *et al.*, 1997; Nestorowicz *et al.*, 1998; Abraham *et al.*, 1999). In the cardiovascular system, K_{ATP} channels contribute to the cardioprotective mechanism of ischemic preconditioning and to regulation of vascular tone (Daut *et al.*, 1990; Nichols and Lederer, 1991; Terzic *et al.*, 1995; Grover, 1997; Gross and Fryer, 1999). Delivery of K_{ATP} channel genes confers resistance to cells otherwise vulnerable to metabolic injury (Jovanovic *et al.*, 1998a). Thus, targeting K_{ATP} channels provides a powerful means to regulate numerous metabolism-dependent cellular functions and to interfere with disease conditions associated with metabolic insults (Terzic, 1999).

2. TARGETS FOR K_{ATP} CHANNELS: FROM ENDOGENOUS LIGANDS TO SYNTHETIC AGENTS

2.1. Endogenous Ligands

2.1.1. Adenine Mononucleotides

The primary endogenous inhibitor of K_{ATP} channels is intracellular ATP, whereas ADP serves as the channel activator (Noma, 1983). The pore-forming Kir6.2 subunit, in conjunction with the regulatory SUR subunit, is involved in the ATP-induced inhibition of channel opening, whereas SUR, which possesses two cytosolic nucleotide-binding folds, is also responsible for the ADP-induced activation of K_{ATP} channels (Fig. 1; Nichols *et al.*, 1996; Bryan and Aguilar-Bryan, 1997; Gribble *et al.*, 1997a; Ueda *et al.*, 1997; Tucker *et al.*, 1998; D'hahan *et al.*, 1999a, b).

Although the defining property of K_{ATP} channels is their adenine nucleotide-dependent gating, it is still unresolved how transition occurs from the ATP-liganded to the ADP-liganded channel state in an environment of millimolar concentrations of intracellular ATP. Such concentrations of ATP would saturate the ATP-binding site(s) on the K_{ATP} channel complex, keeping the channel closed. In fact, the cytosolic concentrations of ATP are one to two orders of magnitude higher than the experimental IC_{50} value required for ATP-induced channel inhibition. Some reports indicate that changes in the cellular ATP/ADP ratio, a more sensitive index of fluctuations in the concentration of adenine nucleotides, are sufficient to cause changes in K_{ATP} channel activity, in particular in cells previously deprived of nutrients (Deutsch et al., 1991). However, under less drastic conditions, such changes are not readily detectable nor correlated with changes in K_{ATP} channel activity (Olson et al., 1996; Decking et al., 1997). Recently, the existence of intracellular phosphotransfer networks capable of transferring phosphoryls between cellular compartments, in the absence of major changes in cytosolic levels of adenine nucleotides or changes in the ATP/ADP ratio, has been established. In light of these findings, transitions between the ATP-liganded (closed) and ADP-liganded (open) K_{ATP} channel states could be accomplished through reversible phosphotransfer between ATP and ADP (Dzeja and Terzic, 1998). Available data suggest that the transition from the ATP- to the ADP-liganded state of the K_{ATP} channel and channel opening could be mediated, in part, through phosphotransfer catalyzed by adenylate kinase (Elvir-Mairena et al., 1996). In intact cells, net adenylate kinase-catalyzed phosphotransfer flux closely correlates with K_{ATP} channel-associated functions (Olson et al., 1996). Transitions from the ATP- to the ADP-liganded state of the K_{ATP} channel could also be catalyzed by an ATPase activity inherent to the channel. Such activity has been anticipated but not yet demonstrated within the K_{ATP} channel complex.

In addition to ADP, intracellular H^+ (Vivaudou and Forestier, 1995), lactate (Keung and Li, 1991), and phospholipids (Baukrowitz et al., 1998; Shyng and Nichols, 1998) activate K_{ATP} channels by decreasing the channel sensitivity to ATP. Such regulation may be of importance during acidotic challenges and lactate accumulation as the result of increased glycolysis during ischemia or muscle fatigue (Davies et al., 1991; Keung and Li, 1991). Phosphatidylinositol 4,5-bisphosphate (PIP_2) and phosphatidylinositol 4-phosphate (PIP), which act through the Kir6.2 subunit, can reduce the ATP sensitivity of K_{ATP} channels by several orders of magnitude, implicating membrane phospholipids in the regulation of membrane excitability (Baukrowitz et al., 1998; Shyng and Nichols, 1998).

2.1.2. Dinucleotide Polyphosphates

The intracellular diadenosine polyphosphates, diadenosine tetra-, penta-, and hexaphosphate, block cardiac K_{ATP} channels (Jovanovic and Terzic, 1995, 1996; Jovanovic et al., 1996, 1997). Diadenosine polyphosphates also inhibit pancreatic K_{ATP} channels (Ripoll et al., 1996), acting through the Kir6.2 subunit (Tucker et al., 1998). Diadenosine polyphosphates may act as "alarmones," alerting the cell to the onset of metabolic stress. Identified within the heart muscle, diadenosine polyphosphates drop in concentration in response to ischemia, which is associated with channel opening (Jovanovic et al., 1998c). Conversely, in pancreatic β-cells, a glucose challenge augments the concentration of diadenosine polyphosphates, leading to K_{ATP} channel closure and insulin release (Ripoll et al., 1996). Thus, diadenosine polyphosphates may serve a signaling role in regulating channel activity in response to stress in conjunction with or separately from adenine mononucleotide-mediated gating (Jovanovic et al., 1998c).

2.1.3. Endosulfine

An endogenous peptide, named endosulfine, has also been identified as a ligand of the K_{ATP} channel (Virsolvy-Vergine *et al.*, 1992). The human cDNA for α-endosulfine encodes a 13-kDa peptide, expressed in muscle, brain, and pancreas. The recombinant protein displaces binding of labeled sulfonylurea, the synthetic channel blocker, and inhibits K_{ATP} channel current. Endosulfine has been found to stimulate insulin secretion in pancreatic β-cells, as expected for a K_{ATP} channel blocker (Heron *et al.*, 1998). In view of this, endosulfine has been proposed to serve as an important regulator of K_{ATP} channels, and is likely to act through the SUR channel subunit (Heron *et al.*, 1998).

2.1.4. Neurohormones

A number of neurohormones regulate K_{ATP} channels. These include galanin (Dunne *et al.*, 1989), somatostain (de Weille *et al.*, 1989), and vasopressin (Martin *et al.*, 1989) in pancreatic β-cells and adenosine and acetylcholine in cardiomyocytes (Terzic *et al.*, 1994). In the vasculature, adenosine (Daut *et al.*, 1990), along with the calcitonin gene-related peptide (Nelson *et al.*, 1990) and possibly an endothelium-dependent hyperpolarizing factor (Standen *et al.*, 1989), may also regulate K_{ATP} or related channels. Pertussis-sensitive GTP-binding proteins (G-proteins) may link respective neurohormone receptors with channel gating (Dunne *et al.*, 1989). This may occur via a membrane-delineated pathway involving a G_α subunit that would modulate the ATP-mediated inhibition of K_{ATP} channels (Terzic *et al.*, 1994; Sanchez *et al.*, 1998).

2.2. Synthetic Agents

2.2.1. Sulfonylureas and Other Blockers

Sulfonylurea drugs are the most recognized K_{ATP} channel blockers (Ashcroft and Ashcroft, 1992). The first-generation sulfonylureas, such as tolbutamide, contain the sulfonylurea moiety (SO_2NHCO); a benzamide (CONH) or related group is added in the second-generation sulfonylureas, exemplified by glyburide (Fig. 2). A number of nonsulfonylurea drugs, including meglitinide, whose structure corresponds to the benzamido half of glyburide (Fig. 2; Zunkler *et al.*, 1988), a lipid component of human milk, 5-hydroxydecanoate (Notsu *et al.*, 1992), and the imidazoline phentolamine (Proks and Ashcroft, 1997), have also been identified as K_{ATP} channel blockers.

2.2.2. Potassium Channel Openers

Potassium channel openers are a chemically diverse group of agents (Fig. 3), which share the property of activating K_{ATP} channels. There has been no evidence for a common pharmacophore principle among such distinct compounds. Based on their structure, potassium channel openers are classified into six subgroups (Fig. 3): benzopyrans (e.g., cromakalim), cyanoguanidines (e.g., pinacidil), nitronicotinamides (e.g., nicorandil), thioformamides (e.g., aprikalim), pyrimidines (e.g., minoxidil), and

K$_{ATP}$ channel blockers

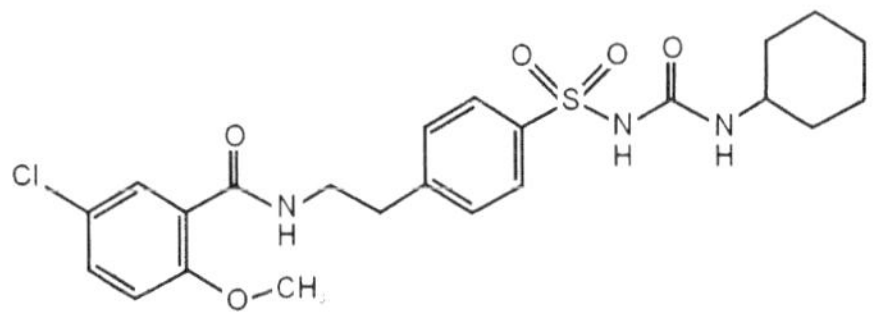

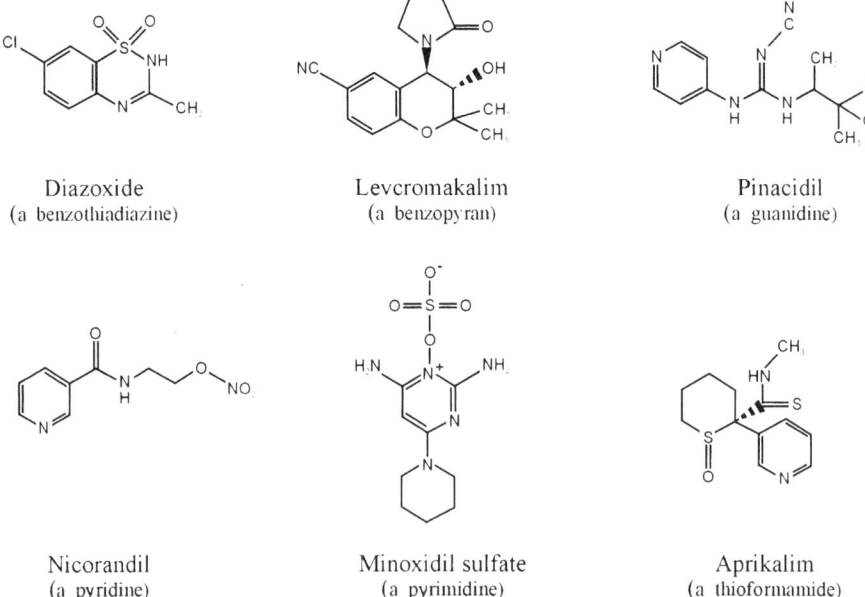

Glyburide
(a sulfonylurea)

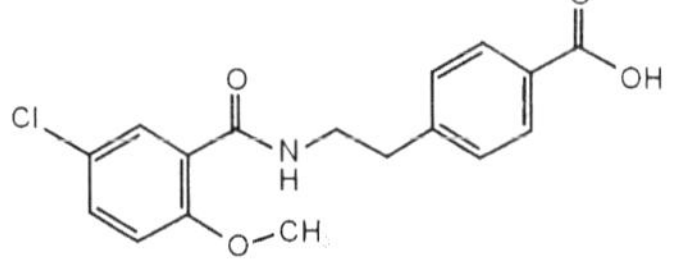

Meglitinide
(a benzoic acid)

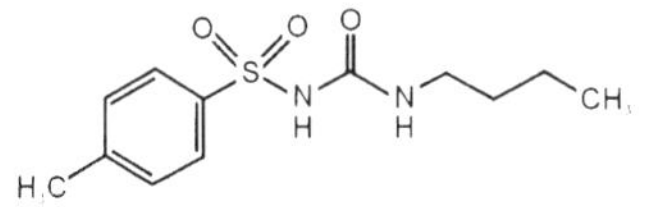

Tolbutamide
(a sulfonylurea)

Figure 2. Chemical structures of selected K$_{ATP}$ channel blockers.

benzothiadiazines (e.g., diazoxide). The benzopyran derivative cromakalim is a trans-racemate containing two enantiomers resulting from the chiral carbon number 3 on the benzopyran moiety. It is the ($-$)-(3S,4R)-enantiomer, known as levcromakalim, that possesses biological activity. The pyridylcyanoguanidine derivative pinacidil contains a chiral carbon in its trimethylpropyl moiety. The biological activity of this racemate is

K$_{ATP}$ channel openers

Diazoxide
(a benzothiadiazine)

Levcromakalim
(a benzopyran)

Pinacidil
(a guanidine)

Nicorandil
(a pyridine)

Minoxidil sulfate
(a pyrimidine)

Aprikalim
(a thioformamide)

Figure 3. Chemical structures of representative potassium-channel-opening drugs.

associated with the $(-)$-(R)-enantiomer. P1075 is a more potent analog of pinacidil and is used as a radioligand for labeling of K_{ATP} channel subunits. The nicotinamide derivative nicorandil was the first among the potassium channel openers to have been shown to promote K^+ channel activity. It is commonly referred to as a hybrid molecule possessing direct potassium channel opening properties as well as the ability to activate soluble guanylate cyclase owing to its nitrate moiety. Aprikalim is the enantiomeric active form of the RP49356 molecule, which is a trans-diastereoisomer, carbothiamide derivative. The benzothiadiazine diazoxide is chemically related to the thiazide diuretic chlorothiazide but does not share its pharmacological properties. It is recognized by a pronounced tropism for pancreatic β-cells, in addition to the vasorelaxant action common to the majority of potassium channel openers. More recently, it has been identified as a potent activator of mitochondrial K_{ATP} channels. The pyrimidine minoxidil is a prodrug, metabolized by sulfotransferases to minoxidil sulfate, the active compound. This opener has a distinctive efficacy in targeting K_{ATP} channels associated with regulation of hair growth. The preclinical pharmacology of potassium channel openers has been previously reviewed (Edwards and Weston, 1993). The purpose of the following section is to summarize advances in the molecular pharmacology of K_{ATP} channels.

3. MOLECULAR PHARMACOLOGY OF K_{ATP} CHANNELS: FROM BINDING TO GATING

Recent advances in the molecular pharmacology of K_{ATP} channels include the demonstration that sulfonylureas and potassium channel openers both act by binding to the SUR subunit of the K_{ATP} channel complex. It is also now established that binding is tightly regulated by nucleotides, yet it remains to be elucidated how such binding translates into gating of channel activity.

3.1. Characterization of Sulfonylurea Binding Sites

Identification of high-affinity binding sites for sulfonylurea drugs (Kaubisch et al., 1982) shortly preceded the discovery of the K_{ATP} channel (Noma, 1983). With the observation that sulfonylureas block K_{ATP} channels in pancreas (Sturgess et al., 1985; Trube et al., 1986) and heart (Belles et al., 1987), with a potency similar to receptor binding (Schmid-Antomarchi et al., 1987; Fosset et al., 1988), a close link between sulfonylurea receptors and K_{ATP} channels was proposed. More recently with the cloning of K_{ATP} channel subunits, high-affinity sulfonylurea receptors have been recognized as the essential regulatory subunit of K_{ATP} channels (Aguilar-Bryan et al., 1995; Inagaki et al., 1995).

Sites with both high and low affinity for the sulfonylurea prototype glyburide exist in the plasma membrane of excitable cells (Ashcroft and Ashcroft, 1992). The high-affinity site has an affinity for glyburide ranging from 0.05 to 10 nM and would represent the SUR subunit (Aguilar-Bryan et al., 1995). Less is known with regard to low-affinity sites, which have an affinity that is at least two orders of magnitude lower than that reported for high-affinity sites. Low-affinity sites may mediate the effects that sulfonylureas have been observed to have on membrane proteins distinct from SUR itself, such as the other K_{ATP} channel subunit Kir6.2 (Gribble et al., 1997b), the cystic fibrosis transmembrane regulator associated with a chloride conductance (Sheppard and and Walsh, 1992), the Na–K-ATPase (Ribalet et al., 1996), or voltage-dependent K^+ channels (Yao et al., 1996). In mitochondrial membranes, sites also exist which have been shown to bind glyburide with rather low affinity (Szewczyk, 1997).

Sulfonylurea binding to high-affinity sites is generally downregulated by intracellular nucleotide di- and triphosphates (Ashcroft and Ashcroft, 1992) through a process which involves ATP hydrolysis and phosphorylation (Schwanstecher et al., 1991, 1992) and possibly, in neuronal cells, a Mg^{2+}-dependent interaction of ADP with a distinct nucleotidic site (Bernardi et al., 1992). Nucleotide diphosphates also reduce the efficacy of glyburide-induced inhibition of cardiac K_{ATP} channels (Alekseev et al., 1998; Brady et al., 1998). In contrast, inhibition of pancreatic K_{ATP} channels by the sulfonylurea tolbutamide is more, rather than less, pronounced in the presence of cytosolic nucleotide diphosphates (Zunkler et al., 1988; Schwanstecher et al., 1992). This observation, reproduced with cloned SUR1/Kir6.2 channels, is due to a reduction by tolbutamide of the stimulatory effect induced by the interaction of MgADP with the nucleotide-binding domains of SUR (Gribble et al., 1997b).

In patch-clamp experiments, sulfonylureas block K_{ATP} channels from either side of the membrane (Trube et al., 1989) with greater efficacy at acidic pH (Zunkler et al., 1989; Findlay, 1992). Correlating the effects of pH on block and the effects of pH on the weakly acidic sulfonylureas, Zunkler et al. (1987) showed that the nonionized forms of sulfonylureas are the active molecules, reaching their receptor from the lipid phase of the membrane. Schwanstecher et al. (1994) further concluded, after investigating chemical derivatives of varying lipid solubility, that the receptor site is on the cytoplasmic face of the membrane.

Evidence based on the differential response of recombinant SUR1 and SUR2A to tolbutamide, glyburide, and the non-sulfonylurea meglitinide has given weight to the hypothesis that SUR could possess separate neighboring sites for sulfonylurea groups (accessible to tolbutamide and the second half of glyburide) and benzamido groups (accessible to meglitinide and the first half of glyburide; Gribble et al., 1998). Recently, the difference in tolbutamide sensitivity of SUR1 and SUR2A has been exploited to pursue a chimeric approach and identify part of the transmembrane domain TMD2 as essential to the interaction between SUR and tolbutamide (Fig. 1; Gribble et al., 1999).

3.2. Closing in on Binding Sites for Potassium Channel Openers

The observation that potassium channel openers activate K_{ATP} channels was first demonstrated for diazoxide, which, when applied to patches from pancreatic β-cells, increase the activity of single K_{ATP} channels (Trube et al., 1986). Single-channel recordings from cardiac myocytes later showed that newer molecules, like cromakalim (Escande et al, 1988) or pinacidil (Arena and Kass, 1989), also increase K_{ATP} channel activity. Although we still do not have a precise understanding of the molecular mechanism underlying channel activation, details are starting to emerge on how openers bind to K_{ATP} channels and how such binding leads to channel opening.

Using radiolabeled pinacidil analogs, P1075 or BayX 9228, specific binding sites for potassium channel openers were identified in vascular smooth muscle (Bray and Quast, 1992; Quast et al., 1993), insulin-secreting cell lines (Hoffman et al., 1993), and skeletal (Dickinson et al., 1997) and cardiac muscle (Atwal et al., 1998; Löffler-Walz and Quast, 1998). The dissociation constant of the radioligand was found to be in the 6–40 nM range, with a Hill constant close to unity. The radiolabeled pinacidil analogs were competitively displaced from the binding sites by other potassium channel openers, including pinacidil, cromakalim, nicorandil, and diazoxide, with an order of potency that correlated with their efficacies as vasorelaxants. Openers did not change the dissociation rate of the radioligand (Dickinson et al., 1997), suggesting that they all compete for the same binding sites. In contrast, the sulfonylurea glyburide, which also inhibits binding but accelerates radioligand dissociation in a dose-dependent manner, is likely to occupy a different site allosterically coupled to the opener site (Bray and Quast, 1992; Dickinson et al., 1997).

Initially, it was necessary to perform binding assays in intact cells because specific binding would disappear as soon as cells were disrupted. The lost cytosolic cofactors have now been identified, as binding of an opener is critically dependent on Mg^{2+} and hydrolyzable ATP. When the activity of endogenous nucleotidases is counteracted by a creatine-phosphokinase ATP-regenerating system which can maintain stable MgATP levels, specific opener binding to skeletal and cardiac muscle membranes increases in a dose-dependent manner with the concentration of MgATP and is half-maximal at 30 and 100 μM MgATP, respectively (Dickinson et al., 1997; Löffler-Walz and Quast, 1998). If MgATP is necessary for opener binding, it also appears to be sufficient because, provided that MgATP is kept at physiological levels, the opener binding profile of homogenized cells matches that of intact cells (Löffler-Walz and Quast, 1998). Recently, these binding assays have been repeated on membrane preparations from cells overexpressing recombinant K_{ATP} channel subunits (Hambrock et al., 1998; Schwanstecher et al., 1998) in order to provide more direct evidence for the molecular identity of the opener's receptor.

Heterologous expression of SUR subunits endows COS-7 or HEK293 cells with MgATPdependent [^{3}H]P1075 binding sites which are absent from nontransfected cells or from cells expressing Kir6.1 or Kir6.2 subunits. The P1075 affinities for the postulated vascular and cardiac isoforms SUR2B (3–12 nM) and SUR2A (46 nM) are essentially the same as those measured in vascular (Bray and Quast, 1992) and cardiac (Atwal et al., 1998) myocytes. Moreover, the affinities of other openers and sulfonylureas for SUR2B are in good agreement with those obtained in vascular cells (Hambrock et al., 1998). Opener binding requires both Mg^{2+} and ATP acting through a single site having an affinity for MgATP of $\sim 5~\mu M$. Mn^{2+} can replace Mg^{2+}, but ADP and nonhydrolyzable ATP analogs cannot replace ATP, implying that binding is enhanced by hydrolysis of ATP (Schwanstecher et al., 1998). These results, obtained in cells expressing SUR2B alone, were partially reproduced without obvious differences in cells coexpressing SUR2B and Kir6.2 (Schwanstecher et al., 1998). Thus, the binding site for openers does not appear to be modulated by intersubunit allosteric interactions.

Taken together, these results strongly indicate that the SUR subunit of the K_{ATP} channel and the potassium channel opener receptor are the very same protein. The observed lack of cooperativity in the binding profile of openers indicates that their binding sites are identical and noninteracting. These findings point to the existence of a single opener site on SUR. This site appears to be distinct from the sulfonylurea binding site, but allosterically coupled to it.

Current evidence further suggests that an ATP hydrolysis step controls the accessibility of openers to their binding site. Although SUR possesses necessary sequence motifs within the two nucleotide-binding domains, NBD1 and NBD2, for such a function, it is still unknown whether this K_{ATP} channel subunit hydrolyzes ATP, like other ABC transporters (Senior and Gadsby, 1997). Moreover, these domains have been implicated in the effect of MgATP on opener binding because functional disruption of either NBD1 or NBD2, by mutations of key residues known to affect nucleotide binding and hydrolysis in other ABC transporters, greatly reduces binding of openers to SUR (Schwanstecher et al., 1998). The fact that MgATP needs to bind at the two domains to enable opener binding implies, however, that the SUR2B and SUR1 isoforms may function rather differently. Photolabeling with radiolabeled azido-ATP has shown that MgADP at submillimolar concentrations fully antagonizes ATP binding to SUR1 (Ueda et al., 1997). This contrasts with the observation that MgADP has little effect on ATP binding to SUR2B, because at millimolar concentrations it does not interfere with the stimulatory effect of 100 μM MgATP on opener binding to SUR2B (Schwanstecher et al., 1998). Further work is now needed to investigate whether the SUR1 and SUR2 isoforms of the sulfonylurea receptor are indeed regulated in distinct ways by nucleotides, despite nearly identical nucleotide-binding domains. Such differences may be at the origin of observed distinct pharmacological properties of SUR isoforms.

Having established that openers bind to SUR proteins, the signature sequences responsible for binding need to be identified. There are indications that two regions within SUR, namely, the C-region comprised of the last 42 amino acids of the protein and the TMD2 region between the two nucleotide-binding domains (Fig. 1), are tightly linked to the opener binding sites. The two splice variants SUR2A and SUR2B differ only in the 28 amino acids at the very C-region, yet SUR2A/Kir6.2 channels are significantly less sensitive to openers like diazoxide and nicorandil than SUR2B/Kir6.2 channels (Isomoto et al., 1996; Shindo et al., 1998). Binding assays have shown that openers have roughly a fivefold higher affinity for SUR2B than for SUR2A (Schwanstecher et al., 1998). However, recent studies using chimeras constructed of SUR2A and SUR1 have paradoxically demonstrated that the C-region is only a weak determinant of the opener sensitivity. For example, although SUR1 neither binds nor is activated by cromakalim, replacing the C-region of SUR2A with that of SUR1 does not reduce its binding affinity for cromakalim (Schwanstecher et al., 1998). Conversely, replacing the C-region of SUR1 with that of SUR2A does not render SUR1 any more sensitive to cromakalim in patch-clamp experiments (D'hahan et al., 1999a). On the other hand, the TMD2 region of SUR2A, when inserted into SUR1, is sufficient to confer to SUR1 the sensitivity that SUR2A exhibits toward cromakalim (D'hahan et al., 1999a). As the TMD2 region is predicted to span the membrane, it would fit the requirement for a binding site accessible to hydrophobic molecules, such as openers. The cytoplasmic C-region may have more of a regulatory role, perhaps through its interaction with nearby nucleotide-binding domains.

3.3. From Binding Sites to Channel Gating

Using the patch-clamp technique, it is, in principle, possible to study the functional consequences of drug application on channel gating. However, in the case of K_{ATP} channels, where the same modulators affect both drug binding and channel gating in a complex manner by interacting with distinct subunits, experimental results often have been difficult to interpret and sometimes contradictory. A case in point is that of nucleotides, which exist in Mg^{2+}-bound and Mg^{2+}-free forms. Nucleotides may interact directly with K_{ATP} channels in a form-dependent manner at apparently three allosterically coupled sites, the two nucleotide-binding domains of SUR and the inhibitory site of Kir6.2. They may further modify channels indirectly via kinases and phosphatases present in the microenvironment of the channel.

Although the very C-terminus of Kir6.2 is known to be expendable for physical association with SUR (Lorenz *et al.*, 1998; Lorenz and Terzic, 1999), overall there is little information on the structural basis underlying interactions between the SUR and Kir6.2 subunits. The open-state probability of the Kir6.2 pore is controlled by ATP binding at a site near the cytoplasmic entrance of the channel (Tucker *et al.*, 1998), where the physical gate is supposed to be located (Doyle *et al.*, 1998). If ATP binding affects the gate, it's probably also true that channel gating may influence the affinity of ATP for its binding site. Thus, it is likely that whether drug binding to SUR modifies channel activity by acting on the channel gate or by acting on the ATP binding site, a change in ATP affinity will be recorded. Thus, activation of K_{ATP} channels by pharmacological openers is accompanied by a reduction in ATP inhibition (Thuringer and Escande, 1989; Terzic *et al.*, 1995; Forestier *et al.*, 1996), and so is activation by MgADP (Nichols and Lederer, 1991), phospholipids (Baukrowitz *et al.*, 1998; Shyng and Nichols, 1998), or protons (Vivaudou and Forestier, 1995). Establishing a causative relationship, that activation by openers is due to relief of inhibition by ATP, may therefore not be entirely justified although it would be a reasonable explanation of the mechanism of action of some openers, like HOE-234 and ER-00153. However, it may not account for the ability of openers like levcromakalim or aprikalim to apparently elicit K_{ATP} channel activation in the absence of nucleotides in cardiac tissues (Terzic *et al.*, 1995). Although binding data suggest a single opener site, these functional differences point to possible variations in the way openers interact with that site to regulate channel activity.

Two other apparent contradictions exist between binding and functional data. First, there is an absolute requirement for MgATP to measure opener binding, but channel activation by some openers in cell-free patches is seen in the absence of nucleotides, as well as in the absence of magnesium ions (Forestier *et al.*, 1993, 1996). One explanation could be that channels may remain in their phosphorylated state(s) for the rather short duration of patch-clamp experiments. Second, the opener diazoxide was found to bind better to the cardiac SUR2A than to the pancreatic isoform SUR1 (Schwanstecher *et al.*, 1998). In contrast, it was shown to activate both native and recombinant SUR1/Kir6.2 pancreatic β-cell channels but have no effect on either native or recombinant SUR2A/Kir6.2 cardiac channels (Okuyama *et al.*, 1998). The recent discovery that cardiac channels are in fact sensitive to diazoxide provided MgADP is present (D'hahan *et al.*, 1999b) should partly resolve this discrepancy, as it implies that diazoxide does bind to the SUR1 and SUR2 isoforms. It further suggests that the transduction steps linking this binding and channel opening are regulated by the interaction of MgATP and MgADP with the nucleotide-binding domains, not only in SUR1 (Gribble *et al.*, 1997a; Shyng *et al.*, 1997), but also in SUR2.

4. CELLULAR PHARMACOLOGY OF CARDIAC K_{ATP} CHANNELS: FROM THE SARCOLEMMA TO MITOCHONDRIA

4.1. Action on the Sarcolemma

The majority of potassium channel openers preferentially activate K^+ channels in smooth muscle cells leading to membrane hyperpolarization. Thereby, the primary pharmacodynamic effect of potassium channel openers is relaxation of smooth muscle affecting the vasculature, bronchial network, and urinary tract (Quayle *et al.*, 1997; Yokoshiki *et al.*, 1998). Potassium channel openers also open K_{ATP} channels in the myocardium. Although such action requires higher concentrations of openers in the normally oxygenated heart, the opening effect becomes more pronounced under conditions of impaired cellular metabolism, as in ischemia (Escande *et al.*, 1988; Weiss and Venkatesh, 1993). The predominant outcome is shortening of the action potential duration, with diminished time available for Ca^{2+} influx. This can reduce the force of contraction but also preserves energy expenditure associated with maintenance of cellular Ca^{2+} homeostasis (Nichols and Lederer, 1991; Terzic *et al.*, 1995). Because the resting membrane potential of a cardiomyocyte is already near the equilibrium potential for K^+, potassium channel openers may produce only a limited hyperpolarization of the cardiac cell membrane. At the single-channel level, potassium channel openers act by enhancing the open-state probability, without affecting channel conductance. The increase in K^+ current through K_{ATP} channels, along with the modification in action potential duration produced by potassium channel openers, is antagonized by K_{ATP} channel blockers. Because K_{ATP} channels are normally closed within an intact cardiomyocyte, the action of potassium channel openers translates into a functional antagonism of the natural channel block produced by intracellular ATP.

Potassium channel openers improve the function of the myocardium following metabolic insult, prevent loss of total adenine nucleotides that could occur as a result of ischemia–reperfusion injury, and diminish infarct size (Grover, 1997; Gross and Fryer, 1999). This protective effect is independent of changes in collateral blood flow, suggesting a direct cytoprotective action on the cardiomyocyte. In fact, overexpression of cardiac K_{ATP} channel genes, Kir6.2 and SUR2A, in conjunction with a potassium channel opener, can enhance cytoprotection and maintain cellular Ca^{2+} homeostasis under stress (Jovanovic *et al.*, 1998b). Conversely, sulfonylurea drugs, by blocking K_{ATP} channels, may reduce cardioprotection (Brady and Terzic, 1998). Although it has been assumed that the action of potassium channel openers is mediated through sarcolemmal K_{ATP} channels, newer evidence indicates that their cardioprotective efficacy does not always correlate with the degree of action potential shortening, implying additional sites of drug action (Garlid *et al.* 1997; Jovanovic et al. 1998b; Liu *et al.* 1998; Gross and Fryer, 1999). This may include the mitochondrial K_{ATP} (mito K_{ATP}) channel.

4.2. Action on Mitochondria

The presence of K_{ATP} channels in the inner membrane of mitochondria was initially demonstrated by patch-clamp measurement of a K^+-selective, ATP-gated ion current in mitoplasts (Inoue et al., 1991). The presence of functional K_{ATP} channels in mitochondria was confirmed after isolation of an inner-membrane protein fraction which, when incorporated into lipid bilayers and liposomes, reconstituted a K^+ conductance with properties similar to those described for other K_{ATP} channels (Paucek et al., 1992). Although the molecular composition of this channel remains elusive, it may consist of the inwardly rectifying K^+ channel subunit, Kir6.1, responsible for K^+ permeance (Suzuki et al., 1997). It has been hypothesized that K^+ influx through this channel may contribute to the osmotic regulation of mitochondrial volume (Garlid et al., 1996). While the efficacy of K_{ATP} channel blockers to block mito K_{ATP} channels may be defined by the operative condition of the channel (Jaburek et al. 1998), potassium channel openers have been consistently found to promote K^+ flux in both intact mitochondria and reconstituted systems containing purified mito K_{ATP} channels (Paucek et al., 1992; Garlid et al., 1996). Potassium channel openers depolarize the mitochondrial membrane, increase the rate of mitochondrial respiration, and oxidize the mitochondrial matrix (Szewczyk, 1997; Holmuhamedov et al., 1998; Liu et al., 1998). This activation of mito K_{ATP} channels by potassium channel openers is associated with an increase in the number of surviving cardiac cells following ischemia, a delay in the onset of contracture, and an improved postischemic recovery of heart muscle (Garlid et al., 1997; Liu et al., 1998). The mechanism responsible for the cardioprotective effect of potassium channel openers, at the mitochondrial level, remains speculative (Gross and Fryer, 1999). However, because mitochondrial dysfunction, secondary to excessive accumulation of Ca^{2+}, is implicated in ischemia–reperfusion injury, it has been proposed that potassium channel openers act by dissipating mitochondrial membrane potential and reducing the driving force for Ca^{2+} accumulation (Holmuhamedov et al., 1998; Liu et al., 1998). In this regard, targeting mito K_{ATP} channels may provide a novel approach for maintaining mitochondrial Ca^{2+} homeostasis under stress conditions (Holmuhamedov et al., 1999).

5. PERSPECTIVES: FROM TRANSLATIONAL PHARMACOLOGY TO THERAPEUTICS

5.1. Priorities

As sensors of the cellular metabolic state, K_{ATP} channels are considered important targets in the modern pharmacotherapy of stress-associated conditions. Historically, the blockade of K_{ATP} channels was first exploited in clinical practice. This followed the discovery, over half a century ago, of the hypoglycemic property of sulfonylurea drugs. During the search for new antibiotics, sulfonamides were formulated. These included a derivative which, when tested in patients with typhoid fever, could produce hypoglycemia. The action of sulfonamides required an intact pancreas and was associated with a direct release of insulin. Today, sulfonylurea drugs are the mainstay of therapy in patients with non-insulin-dependent diabetes mellitus. It is generally accepted that their principal antidiabetic action relies on their occupancy of sulfonylurea SUR1 receptors expressed in pancreatic β-cells. This event leads to inhibition of K^+ flux through the K_{ATP} channel pore, membrane depolarization with opening of voltage-dependent Ca^{2+} channels, and Ca^{2+}-dependent release of insulin (Ashcroft, 1996). Although highly efficient as oral hypoglycemics, the use of sulfonylurea drugs has raised controversy owing to possible concomitant inhibition of cardioprotective myocardial K_{ATP} channels (Brady and Terzic, 1998). Certain epidemiological studies have demonstrated increased cardiovascular mortality in patients taking these agents, and recently sulfonylurea drug use has been associated with a higher risk of in-hospital mortality in diabetic patients undergoing direct angioplasty for acute myocardial infarction (Garratt et al., 1999). Therefore, development of sulfonylurea drugs that would specifically target pancreatic β-cells, without producing adverse effects on the myocardium, is warranted.

Although in clinical practice the experience with openers of K_{ATP} channels is more limited, it is recognized that potassium-channel-opening drugs may offer a unique therapeutic potential (Terzic, 1999). In cardiovascular medicine, exploiting the combined cardioprotective and vasodilatory features of potassium channel openers is particularly attractive. Potassium channel openers are currently being considered for applications in cardiovascular surgery to preserve donor transplant hearts and as novel cardioplegic agents or adjuncts in cardiopulmonary bypass to protect the myocardium during global surgical ischemia (Lopez et al., 1996; Kirsch et al., 1998). Also, because of their efficacy in protecting the ischemic cardiomyocyte and dilating stenotic blood vessels, potassium channel openers may be of value in the management of ischemic heart disease, coronary artery spasm, and hypertension. In addition, potassium channel openers regulate the repolarization phase of cardiac action potentials and are therefore candidate drugs for the treatment of congenital long QT syndrome, an ion channel disease characterized by aberrant repolarization leading to sudden cardiac death. Beyond cardiovascular medicine, potassium channel openers hold promise in a series of clinical conditions, such as asthma, bladder hyperactivity, myotonic dystrophy, cerebrovascular spasm, epilepsy, hyperinsulinemia, and alopecia. The effectiveness of openers in such diverse conditions is associated with their property to clamp membrane potential at values close to the equilibrium potential for K^+, thereby controlling membrane excitability. Notwithstanding, the mechanisms underlying the actions of potassium channel openers are only partially understood. In fact, openers have been proposed to have a multifaceted pharmacodynamic profile. This is the case in alopecia, where potassium channel openers may act to promote hair growth through a direct action on hair follicles, as well as by promoting vascularization of dermal papilla via stimulation of vascular endothelial growth factor expression (Lachgar et al., 1998).

The clinical pharmacology of K_{ATP} channel modulators is at a crossroads. Although specific indications for these agents have been identified, the use of tissue-specific drugs is lagging. Development of openers and blockers exhibiting high selectivity for targeted tissues, and possibly disease-dependent efficacy, is a top priority. With continuous refinement in the pharmacokinetic and pharmacodynamic properties of K_{ATP} channel modulators, along with progress in pharmacogenetics and pharmacogenomics, individualized patient therapy is the ultimate goal. As of now, the main aim is elucidating the precise role(s) of K_{ATP} channels in specific tissues. This is a necessary step to fully capitalize on the therapeutic potential of modulating such unique channels.

5.2. Challenges and Strategies

Indeed, major progress is being made toward unraveling the intimate physiology and pathophysiology of K_{ATP} channels. In particular, the discovery of disease-causing mutations in Kir6.2 and SUR1 in patients with familial persistent hyperinsulinemic hypoglycemia of infancy implicates K_{ATP} channel malfunction as a cause of human channelopathy (Thomas et al., 1995; Dunne et al., 1997; Nestorowicz et al., 1998). Also, polymorphisms in Kir6.2 have been incriminated as contributing to the polygenic etiology of type II diabetes mellitus (Hani et al., 1998). The recent generation of K_{ATP} channel-deficient transgenic animals, by genetic disruption of Kir6.2, directly demonstrates the critical role of this ion conductance in regulating insulin secretion, tissue response to this hormone, and cellular survival (Miki et al., 1997; 1998). Future targeted disruptions of SUR isoforms, and of the related Kir6.1 channel, may provide valuable information on the tissue-specific roles of the K_{ATP} channel. Progress is also being made in establishing gene therapy based upon delivery of K_{ATP} channel subunits. The proof of such principle has been recently obtained in somatic cell lines which acquire resistance to metabolic stress following delivery of recombinant Kir6.2 and SUR isoforms (Jovanovic et al., 1998a, b, 1999).

In conclusion, the therapeutic potential of K_{ATP} channel modulators remains largely untapped, mainly because currently available openers and blockers lack the specificity required for efficient targeting of diseased tissues. The problem is particularly acute for potassium channel openers, which continue to baffle chemists by their ability to produce the same functional effects despite having few recognizable common structural features and no obvious pharmacophore. As the drug target is now known to be the sulfonylurea receptor, as efficient assays have been developed to explore the structure–function relationship of channel regulation, and as more crystallographic data on ion channels and ABC transporters are made available, biologists and pharmacologists alike have the means to elucidate the functional mechanisms and identify the structural determinants of drug action. Already, potential sequence elements of the binding sites for blockers and openers have been identified within the transmembrane domains of SUR, and allosteric interactions between these sites and nucleotide-binding domains have been uncovered. Refining this information and correlating it with the different SUR isoforms and their tissue distribution should bring us closer to an understanding of how to confer more selectivity to existing drugs. With the identification of the molecular structure of subcellular K_{ATP} channels, such a strategy may further allow the development not only of tissue-specific modulators, but also of organelle-selective drugs capable of discriminating between plasmalemmal versus mitochondrial K_{ATP} channels. Thus, in years to come, K_{ATP} channel modulators are destined to be a major center of interest in molecular pharmacology, at a time when they are poised to demonstrate their full potential as therapeutic agents.

REFERENCES

Abraham, M. R., Jahangir, A., Alekseev, A. E., and Terzic, A., 1999, Channel opathies of inwardly rectifying potassium channels, *FASEB J.* **13:**1901–1910.

Aguilar-Bryan, L., and Bryan, J., 1999, Molecular biology of adenosine triphosphate-sensitive potassium channels, *Endocr. Rev.* **20:**101–135.

Aguilar-Bryan, L., Nichols, C., Wechsler, S., Clement, J., Boyd, A., Gonzalez, G., Herrerasosa, H., Nguy, K., Aguilar-Bryan, L., Clement, J. P., Gonzalez, G., Kunjilwar, K., Babenko, A., and Bryan, J., 1998, Toward understanding the assembly and structure of K_{ATP} channels, *Physiol. Rev.*78:227–245.

Alekseev, A. E., Brady, P., and Terzic, A., 1998, Ligand-insensitive state of cardiac ATP-sensitive K^+ channels --Basis for channel opening, *J. Gen. Physiol.* 111:381–394.

Amoroso, S., Schmid-Antomarchi, H., Fosset, M., and Lazdunski, M., 1990, Glucose, sulfonylureas and neurotransmitter release:Role of ATP-sensitive potassium channels, *Science* 247:852–854.

Arena, J. P., and Kass, R. S., 1989, Activation of ATP-sensitive K channels in heart cells by pinacidil:Dependence on ATP, *Am. J. Physiol.* 257:H2092–H2096.

Ashcroft, F. M., 1996, Mechanisms of the glycaemic effects of sulfonylureas, *Horm. Metab. Res.* 28:456–463.

Ashcroft, S. J. H., and Ashcroft, F. M., 1992, The sulfonylurea receptor, *Biochim. Biophys. Acta* 1175:45–49.

Ashcroft, F. M., and Gribble, F. M., 1998, Correlating structure and function in ATP-sensitive K^+ channels, *Trends Neurosci.* 21:288–294.

Atwal, K. S., Grover, G. J., Lodge, N., Normandin, D., Traeger, S. C., Sleph, P., Cohen, R., Bryson, C., and Dickinson, K., 1998, Binding of ATP-sensitive potassium channel (K_{ATP}) openers to cardiac membranes:Correlation of binding affinities with cardioprotective and smooth muscle relaxing potencies, *J. Med. Chem.* 41:271–275.

Babenko, A. P., Gonzalez, G., Aguilar-Bryan, L., and Bryan, J., 1998, Reconstituted human cardiac K_{ATP} channels:Functional identity with the native channels from the sarcolemma of human ventricular cells, *Circ. Res.* 83:1132–1143.

Baukrowitz, T., Schulte, U., Oliver, D., Herlitze, S., Krauter, T., Tucker, S. J., Ruppersberg, J. P., and Fakler, B., 1998, PIP_2 and PIP as determinants for ATP inhibition of K_{ATP} channels, *Science* 282:1141–1144.

Belles, B., Hescheler, J., and Trube, G., 1987, Changes of membrane currents in cardiac cells induced by long whole-cell recordings and tolbutamide, *Pflügers Arch.* 409:582–588.

Bernardi, H., Fosset, M., and Lazdunski, M., 1992, ATP/ADP binding sites are present in the sulfonylurea binding protein associated with brain ATP-sensitive K^+ channels, *Biochemistry* 31:6328–6332.

Brady, P. A. and Terzic, A., 1998, The sulfonylurea controversy:More questions from the heart, *J. Am. Coll. Cardiol.* 31:950–956.

Brady, P. A. Alekseev, A. E., and Terzic, A., 1998, Operative condition-dependent response of cardiac ATP sensitive K^+ channels toward sulfonylureas, *Circ. Res.* 82:272–278.

Bray, K. M., and Quast, U., 1992, A specific binding site for K^+ channel openers in rat aorta, *J. Biol. Chem.* 267:11689–11692.

Bryan, J., and Nelson, D., 1995, Cloning of the beta cell high-affinity sulfonylurea receptor:A regulator of insulin secretion, *Science* 268:423–426.

Bryan, J., and Aguilar-Bryan, L., 1997, The ABCs of ATP-sensitive potassium channels:More pieces of the puzzle, *Curr. Opin. Cell. Biol.* 9:553–559.

Chutkow, W. A., Simon, M., LeBeau, M., and Burant, C., 1996, Cloning, tissue expression, and chromosomal localization of SUR2, the putative drug-binding subunit of cardiac, skeletal muscle, and vascular K_{ATP} channels, *Diabetes Care*145:1439–1445.

Clement, J. P., Kunjilwar, K., Gonzalez, G., Schwanstecher, M., Panten, U., Aguilar-Bryan, L., and Bryan, J., 1997, Association and stoichiometry of K_{ATP} channel subunits, *Neuron* 18:827–838.

Daut, J., Maier-Rudolph, W., von Beckerath, N., Mehrke, G., Gunther, K., and Goedel-Meinen, L., 1990, Hypoxic dilation of coronary arteries is mediated by ATP-sensitive potassium channels, *Science* 247:1341–1344.

Davies, N. W., Standen, N. B., and Stanfield, P. R., 1991, ATP-dependent potassium channels of muscle cells:Their properties, regulation, and possible functions, *J. Bioenerg. Biomembr.* 23:509–535.

Decking, U. K., Arens, S., Schlieper, G., Schulze, K., and Schrader, J., 1997, Dissociation between adenosine release, MVO_2, and energy status in working guinea pig hearts, *Am. J. Physiol.* 272:H371–H381.

Deutsch, N., Klitzner, T. S., Lamp, S. T., and Weiss, J. N., 1991, Activation of cardiac ATP-sensitive K^+ current during hypoxia: Correlation with tissue ATP levels, *Am. J. Physiol.* 261:H671–H676.

de Weille, J. R., Schmid-Antomarchi, H., Fosset, M., and Lazdunski, M., 1989, Regulation of ATP-sensitive K^+ channels in insulinoma cells: Activation by somatostatin and protein kinase C and the role of cAMP, *Proc. Natl. Acad. Sci. USA.* 86:2971–2975.

D'hahan, N., Jacquet, H., Moreau, C., Catty, P., and Vivaudou, M., 1999a, A transmembrane domain of the sulfonylurea receptor mediates activation of ATP-sensitive K^+ channels by K^+ channel openers, *Mol. Pharmacol.* 56:308–315.

D'hahan, N., Moreau, C., Prost, A. L., Jacquet, H., Alekseev, A. E., Terzic, A., and Vivaudou, M., 1999b. Pharmacological plasticity of cardiac ATP-sensitive potassium channels towards diazoxide revealed by ADP, *Proc. Natl. Acad. Sci. U.S.A.* **96**:12162–12167.

Dickinson, K. E. J., Bryson, C. C., Cohen, R. B., Rogers, L., Green, D. W., and Atwal, K. S., 1997, Nucleotide regulation and characteristics of potassium channel opener binding to skeletal muscle membranes, *Mol. Pharmacol.* **52**:473–481.

Doyle, D. A., Cabral, J. M., Pfuetzner, R. A., Kuo, A., Gulbis, J. M., Cohen, S. L., Chait, B. T. and MacKinnon, R., 1998, The structure of the potassium channel: Molecular basis of K$^+$ conduction and selectivity, *Science* **280**:69-77.

Dunne, M. J., Bullett, M. J., Li, G., Wollheim, C. B., and Petersen, O. H., 1989, Galanin activates nucleotide-dependent K$^+$ channels in insulin-secreting cells via a pertussis toxin-sensitive G-protein, *EMBO J.* **8**:413–420.

Dunne, M. J., Kane, C., Shepherd, R. M., Sanchez, J. A., James, R .F., Johnson, P. R., Aynsley-Green, A., Lu, S., Clement, J. P., Lindley, K. J., Seino, S., and Aguilar-Bryan, L, 1997, Familial persistent hyperinsulinemic hypoglycemia of infancy and mutations in the sulfonylurea receptor. *N. Engl. J. Med.* **336**:703–706,

Dzeja, P. P., and Terzic, A., 1998, Phosphotransfer reactions in the regulation of ATP-sensitive K$^+$ channels, *FASEB J.* **12**:523–529.

Edwards, G., and Weston, A. H., 1993, The pharmacology of ATP-sensitive potassium channels,*Annu. Rev. Pharmacol. Toxicol.* **33**:597–637.

Elvir-Mairena, J. R., Jovanovic, A., Gomez, L. A., Alekseev, A. E., and Terzic, A., 1996, Reversal of the ATP-liganded state of ATP-sensitive K$^+$ channels by adenylate kinase activity, *J. Biol. Chem.* **271**:31903–31908.

Escande, D., Thuringer, D., Leguern, S., and Cavero, I., 1988, The potassium channel opener cromakalim (BRL 34915) activates ATP-dependent K$^+$ channels in isolated cardiac myocytes, *Biochem. Biophys. Res. Commun.* **154**:620–625.

Findlay, I., 1992, Effects of pH upon the inhibition by sulphonylurea drugs of ATP-sensitive K$^+$ channels in cardiac muscle, *J. Pharmacol. Exp. Ther.* **262**:71–79.

Forestier, C., Depresle, Y., and Vivaudou, M., 1993, Intracellular protons control the affinity of skeletal muscle ATP-sensitive K$^+$ channels for potassium-channel-openers, *FEBS Lett.*325:276–80.

Forestier, C., Pierrard, J., and Vivaudou, M., 1996, Mechanism of action of K$^+$ channel openers on skeletal muscle K$_{ATP}$ channels:Interactions with nucleotides and protons, *J. Gen. Physiol.* **107**:489-502.

Fosset, M., de Weille, J. R., Green, R. D., Schmid-Antomarchi, H., and Lazdunski, M., 1988, Antidiabetic sulfonylureas control action potential properties in heart cells via high-affinity receptors that are linked to ATP-dependent K$^+$ channels, *J. Biol. Chem.* **263**: 7933–7936.

Garlid, K. D., Paucek, P., Yarov-yarovoy, V., Sun, X., and Schindler, P.A., 1996, The mitochondrial K-ATP channel as a receptor for potassium channel openers, *J. Biol. Chem.* **271**:8796–8799.

Garlid, K. D., Paucek, P., Yarov-Yarovoy, V., Murray, H. N., Darbenzio, R. B., D'Alonzo, A. J., Lodge, N. J., Smith, M. A., and Grover, G. J., 1997, Cardioprotective effect of diazoxide and its interaction with mitochondrial ATP-sensitive K$^+$ channels—possible mechanism of cardioprotection, *Circ. Res.* **81**:1072–1082.

Garratt, K. N., Brady, P. A., Hassinger, N. L., Grill, D. E., Terzic, A., and Holmes, D. R., 1999, Sulfonylurea drugs increase early mortality in patients with diabetes mellitus after direct angioplasty for acute myocardial infarction, *J. Am. Coll. Cardiol.* **33**: 119–124.

Gribble, F. M., Tucker, S. J. and Ashcroft, F. M., 1997a, The essential role of the Walker A motifs of SUR1 in K$_{ATP}$ channel activation by MgADP and diazoxide, *EMBO J.* **16**: 1145–1152.

Gribble, F. M., Tucker, S. J., and Ashcroft, F. M., 1997b, The interaction of nucleotides with the tolbutamide block of cloned ATP-sensitive K$^+$ channel currents expressed in *Xenopus oocytes:* Areinterpretation, *J. Physiol. (London)* **504**:35–45.

Gribble, F., Tucker, S., Seino, S., and Ashcroft, F., 1998, Tissue specificity of sulfonylureas:Studies on cloned cardiac and ?-cell K$_{ATP}$ channels, *Diabetes* **47**:1412-1418.

Gribble, F. M., Ashfield, R., and Ashcroft, F. M., 1999, Identification of the high affinity tolbutamide site on the SUR1 subunit of the K$_{ATP}$ channel, *Biophys. J.* **76**:A14.

Gross, G J., and Fryer, R. M., 1999, Sarcolemmal versus mitochondrial ATP-sensitive K$^+$ channels and myocardial preconditioning, *Circ. Res.* **84**:973–979.

Grover, G. J., 1997, Pharmacology of ATP-sensitive potassium channel openers in models of myocardial ischemia and reperfusion, *Can. J. Physiol. Pharmacol.*75:309–315.

Hambrock, A., Loffler-Walz, C., Kurachi, Y., and Quast, U., 1998, Mg^{2+} and ATP dependence of K_{ATP} channel modulator binding to the recombinant sulphonylurea receptor, SUR2B, *Br. J. Pharmacol.* **125**:577–583.

Hani, E. H., Boutin, P., Durand, E., Inoue, H., Permutt, M. A., Velho, G., and Froguel, P., 1998, Missense mutations in the pancreatic islet beta cell inwardly rectifying K^+ channel gene (KIR6.2/BIR):A meta-analysis suggests a role in the polygenic basis of type II diabetes mellitus in Caucasians, *Diabetologia* **41**:1511–1515.

Heron, L., Virsolvy, A., Peyrollier, K., Gribble, F., LeCam, A., Ashcroft, F., and Bataille, D., 1998, Human α-endosulfine, a possible regulator of sulfonylurea-sensitive K_{ATP} channel:Molecular cloning, expression and biological properties, *Proc. Natl. Acad. Sci. USA* **95**:8387–8391.

Hoffman, F.J., Lenfers, J.B., Niemers, E., Pleiss, U., Scriabine, A., and Janis, R.A., 1993, High affinity binding of a potassium channel agonist to intact rat insulinoma cells, *Biochem. Biophys. Res. Commun.* **190**:551–558.

Holmuhamedov, E. L., Jovanovic, S., Dzeja, P. P., Jovanovic, A., and Terzic, A., 1998, Mitochondrial ATP-sensitive K^+ channels modulate cardiac mitochondrial function, *Am. J. Physiol.* **44**: H1567–H1576.

Holmuhamedov, E. L., Wang, L., and Terzic, A., 1999, ATP-sensitive K^+ channel openers prevent Ca^{2+} overload in rat cardiac mitochondria, *J. Physiol. (London)* **519**:347–360.

Inagaki, N., Gonoi, T., Clement, J. P., Namba, N., Inazawa, J., Gonzalez, G., Aguilar-Bryan, L., Seino, S., and Bryan, J., 1995, Reconstitution of I_{KATP}:An inward rectifier subunit plus the sulfonylurea receptor, *Science* **270**:1166–1170.

Inagaki, N., Gonoi, T., Clement, J., Wang, C. Z., Aguilar-Bryan, L., Bryan, J., and Seino, S., 1996, A family of sulfonylurea receptors determines the pharmacological properties of ATP-sensitive K^+ channels, *Neuron* **16**:1011–1017.

Inoue, I., Nagase, H., Kishi, K., and Higuti, T., 1991, ATP-sensitive K^+ channel in the mitochondrial inner membrane, *Nature* **352**:244–247.

Isomoto, S., Kondo, C., Yamada, M., Matsumoto, S., Higashiguchi, O., Horio, Y., Matsuzawa, Y., and Kurachi, Y., 1996, A novel sulfonylurea receptor forms with BIR (Kir6.2) a smooth muscle type K_{ATP} channel, *J. Biol. Chem.* **271**:24321–24324.

Jaburek, M., Yarov-Yarovoy, V., Paucek, P., and Garlid, K.D., 1998, State-dependent inhibition of the mitochondrial K_{ATP} channel by glyburide and 5-hydroxydecanoate, *J. Biol. Chem.* **273**:13578—13582.

John, S. A., Monck, J. R., Weiss, J .N., and Ribalet, B., 1998, The sulphonylurea receptor SUR1 regulates ATP-sensitive mouse Kir6.2 K^+ channels linked to the green fluorescent protein in human embryonic kidney cells (HEK293), *J. Physiol. (London)* **510**:333—345.

Jovanovic, A., and Terzic, A., 1995, Diadenosine-hexaphosphate is an inhibitory ligand of myocardial ATP-sensitive K^+ channels, *Eur. J. Pharmacol.* **286**:R1—R2.

Jovanovic, A., and Terzic, A., 1996, Diadenosine tetraphosphate-induced inhibition of ATP-sensitive K^+ channels in patches excised from ventricular myocytes, *Br. J. Pharmacol.* **17**:233—235.

Jovanovic, A., Alekseev, A.E. and Terzic, A., 1996, Cardiac ATP sensitive K^+ channel:A target for diadenosine 5',5"-P-1,P-5-pentaphosphate, *Naunyn-Schmiedeberg's Arch. Pharmacol* **353**:241—244.

Jovanovic, A., Alekseev, A. E., and Terzic, A., 1997, Intracellular diadenosine polyphosphates:A novel family of inhibitory ligands of the ATP-sensitive K^+ channel, *Biochem. Pharmacol.* **54**:219—225.

Jovanovic, A., Jovanovic, S., Carrasco, A. J., and Terzic, A., 1998a, Acquired resistance of a mammalian cell line to hypoxia-reoxygenation through cotransfection of Kir6.2 and SUR1 clones, *Lab. Invest.* **78**:1101—1107.

Jovanovic, A., Jovanovic, S., Lorenz, E., and Terzic, A., 1998b, Recombinant cardiac ATP-sensitive K^+ channel subunits confer resistance to chemical hypoxia-reoxygenation injury, *Circulation* **98**:1548—1555.

Jovanovic, A., Jovanovic, S., Mays, D. C., Lipsky, J. J., and Terzic, A., 1998c, Diadenosine 5',5'-P1,P5-pentaphosphate harbors the properties of a signaling molecule in the heart, *FEBS Lett.* **423**:314—318.

Jovanovic, N., Jovanovic, S., Jovanovic, A., and Terzic, A., 1999, Gene delivery of Kir6.2/SUR2A in conjunction with pinacidil handles intracellular Ca^{2+} homeostasis under metabolic stress. *FASEB J.* **13**:923—929.

Kaubisch, N., Hammer, R., Wollheim, C., Renold, A., and Offord, R., 1982, Specific receptors for sulfonylureas in brain and in a beta-cell tumor of the rat, *Biochem. Pharmacol.* **31**:1171—1174.

Keung, E. C., and Li, Q., 1991, Lactate activates ATP-sensitive potassium channels in guinea pig ventricular myocytes, *J. Clin. Invest.* **88**:1772—1777.

Kirsch, M., Baufreton, C., Fernandez, C., Brunet, S., Pasteau, F., Astier, A., and Loisance, D. Y., 1998, Preconditioning with cromakalim improves long-term myocardial preservation for heart transplantation, *Ann. Thorac. Surg.* **66**:417—424.

Lachgar, S., Charveron, M., Gall, Y., and Bonafe, J., 1998, Minoxidil upregulates the expression of vascular endothelial growth factor in human hair dermal papilla cells, *Br. J. Dermatol.* **138**:407—411.

Liu, Y., Sato, T., O'Rourke, B., and Marban, E., 1998, Mitochondrial ATP-dependent potassium channels: Novel effectors of cardioprotection?, *Circulation* **97**:2463—2469.

Löffler-Walz, C., and Quast, U., 1998, Binding of K_{ATP} channel modulators in rat cardiac membranes, *Br. J. Pharmacol.* **123**:1395—1402.

Lopez, J. R., Jahangir, R., Jahangir, A., Shen, W. K., and Terzic, A., 1996, Potassium channel openers prevent potassium-induced calcium loading of cardiac cells: Possible implications in cardioplegia, *J. Thorac. Cardiovasc. Surg.* **112**:820—831.

Lorenz, E., and Terzic, A., 1999, Physical association between recombinant cardiac ATP-sensitive K^+ channel subunits Kir6.2 and SUR2A, *J. Mol. Cell. Cardiol.* **31**:425—434.

Lorenz, E., Alekseev, A. E., Krapivinsky, G. B., Carrasco, A. J., Clapham, D. E., and Terzic, A., 1998, Evidence for direct physical association between a K^+ channel (Kir6.2) and an ATP-binding cassette protein (SUR1) which affects cellular distribution and kinetic behavior of an ATP-sensitive K^+ channel, *Mol. Cell. Biol.* **18**:1652—1659.

Martin, S., Yule, D., Dunne, M., and Petersen, O., 1989, Vasopressin directly closes ATP-sensitive potassium channels evoking membrane depolarization and an increase in the free intracellular Ca^{2+} concentration in insulin-secreting cells, *EMBO J.* **8**:3595—3599.

Miki, T., Tashiro, F., Iwanaga, T., Nagashima, K., Yoshitomi, H., Aihara, H., Nitta, Y., Gonoi, T., Inagaki, N., Miyazaki, J., and Seino, S., 1997, Abnormalities of pancreatic islets by targeted expression of a dominant-negative K_{ATP} channel, *Proc. Natl. Acad. Sci. U.S.A.* **94**:11969-11973.

Miki, T., Nagashima, K., Tashiro, F., Kotake, K., Yoshitomi, H., Tamamoto, A., Gonoi, T., Iwanaga, T., Miyazaki, J., and Seino, S., 1998, Defective insulin secretion and enhanced insulin action in K_{ATP} channel-deficient mice, *Proc. Natl. Acad. Sci. U.S.A.* **95**:10402—10406.

Nelson, M. T., Huang, Y., Brayden, J. E., Hescheler, J., and Standen, N. B., 1990, Arterial dilations in response to calcitonin gene-related peptide involve activation of K^+ channels, *Nature* **344**:770—773.

Nestorowicz, A., Glaser, B., Wilson, B., Shyng, S., Nichols, C., Stanley, C., Thornton, P., and Permutt, M., 1998, Genetic heterogeneity in familial hyperinsulinism, *Hum. Mol. Genet.* **7**:1119—1128.

Nichols, C. G., and Lederer, W. J., 1991, Adenosine triphosphate-sensitive potassium channels in the cardiovascular system, *Am. J. Physiol.* **261**:H1675—H1686.

Nichols, C. G., Shyng, S., Nestorowicz, A., Glaser, B., Clement, J., Gonzalez, G., Aguilar-Bryan, L., Permutt, M., and Bryan, J., 1996, Adenosine diphosphate as an intracellular regulator of insulin secretion, *Science* **272**:1785—1787.

Noma, A., 1983, ATP-regulated K^+ channels in cardiac muscle, *Nature* **305**:147—148.

Notsu, T., Tanaka, I., Takano, M., and Noma, A., 1992, Blockade of the ATP-sensitive K^+ channel by 5-hydroxydecanoate in guinea pig ventricular myocytes, *J. Pharmacol. Exp. Ther.* **260**:702—708.

Okuyama, Y., Yamada, M., Kondo, C., Satoh, E., Isomoto, S., Shindo, T., Horio, Y., Kitakaze, M., Hori, M., and Kurachi, Y., 1998, The effects of nucleotides and potassium channel openers on the SUR2A/Kir6.2 complex K^+ channel expressed in a mammalian cell line, HEK293T cells, *Pflügers Arch.* **435**:595—603.

Olson, L. K., Schroeder, W., Robertson, R P., Goldberg, N. D., and Walseth, T. F., 1996, Suppression of adenylate kinase catalyzed phosphotransfer precedes and is associated with glucose-induced insulin secretion in intact HIT-T15 cells, *J. Biol. Chem.* **271**:16544—16552.

O'Rourke, B., Ramza, B., and Marban, E., 1994, Oscillations of membrane current and excitability driven by metabolic oscillations in heart cells, *Science* **265**:962-966.

Paucek, P., Mironova, G., Mahdi, F., Beavis, A., Woldegiorgis, G., and Garlid, K .D., 1992, Reconstitution and partial purification of the glibenclamide-sensitive, ATP-dependent K^+ channel from rat liver and beef heart mitochondria, *J. Biol. Chem.* **267**:26062—26069.

Proks, P., and Ashcroft, F. M., 1997, Phentolamine block of K_{ATP} channels is mediated by Kir6.2, *Proc. Natl. Acad. Sci. U.S.A.* **94**:11716—11720.

Quast, U., Bray, K. M., Andres, H., Manley, P. W., Baumlin, Y., and Dosognc, J., 1993, Binding of the K^+ channel opener [3H]P1075 in rat isolated aorta:Relationship to functional effects of openers and blockers, *Mol. Pharmacol.* **43**:474—481.

Quayle, J. M., Nelson, M. T., and Standen, N. B., 1997, ATP-sensitive and inwardly rectifying potassium channels in smooth muscle, *Physiol. Rev.* **77**:1165—1232.

Repunte, V. P., Nakamura, H., Fujita, A., Horio, Y., Findlay, I., Pott L., and Kurachi, Y., 1999, Extracellular links in Kir subunits control the unitary conductance of SUR/Kir6.0 ion channels, *EMBO J.* **18**:3317—3324.

Ribalet, B., Mirell, C. J., Johnson, D. G., and Levin, S. R., 1996, Sulfonylurea binding to a low-affinity site inhibits the Na/K-ATPase and the K_{ATP} channel in insulin-secreting cells, *J. Gen. Physiol.* **107**:231—241.

Ripoll, C., Martin, F., Rovira, J. M., Pintor, J., MirasPortugal, M. T., and Soria, B., 1996, Diadenosine polyphosphates:A novel class of glucose-induced intracellular messengers in the pancreatic beta-cell, *Diabetes* **45**:1431—1434.

Sanchez, J. A., Gonoi, T., Inagaki, N., Katada, T., and Seino, S., 1998, Modulation of reconstituted ATP-sensitive K^+-channels by GTP-binding proteins in a mammalian cell line, *J. Physiol. (London)* **507**:315—324.

Schmid-Antomarchi, H., de Weille, J., Fosset, M., and Lazdunski, M., 1987, The receptor for antidiabetic sulfonylureas controls the activity of the ATP-modulated K^+ channel in insulin secreting cells, *J. Biol. Chem.* **262**:15840—15844.

Schwanstecher, M., Loser, S., Rietze, I., and Panten, U., 1991, Phosphate and thiophosphate group donating adenine and guanine nucleotides inhibit glibenclamide binding to membranes from pancreatic islets, *Naunyn-Schmiedeberg's Arch. Pharmacol.* **343**:83—89.

Schwanstecher, C., Dickel, C., and Panten, U., 1992, Cytosolic nucleotides enhance the tolbutamide sensitivity of the ATP-dependent K^+ channel in mouse pancreatic β cells by their combined actions at inhibitory and stimulatory receptors, *Mol. Pharmacol.* **41**:480—486.

Schwanstecher, M., Schwanstecher, C., Dickel, C., Chudziak, F., Moshiri, A., and Panten, U., 1994, Location of the sulphonylurea receptor at the cytoplasmic face of the beta-cell membrane, *Br. J. Pharmacol.* **113**:903—911.

Schwanstecher, M., Sieverding, C., Dorschner, H., Gross, I., Aguilar-Bryan, L., Schwanstecher, C., and Bryan, J., 1998, Potassium channel openers require ATP to bind to and act through sulfonylurea receptors, *EMBO J.* **17**:5529—5535.

Seino, S., 1999, ATP-sensitive potassium channels: A model of heteromultimeric potassium channel/receptor assemblies, *Annu. Rev. Physiol.* **61**:337—362.

Senior, A. E. and Gadsby, D. C., 1997, ATP hydrolysis cycles and mechanism in P-glycoprotein and CFTR, *Semin. Cancer Biol.* **8**:143—150.

Sheppard, D. N., and Welsh, M. J., 1992, Effect of ATP-sensitive K^+ channel regulators on cystic fibrosis transmembrane conductance regulator chloride currents, *J. Gen. Physiol.* **100**:573—591.

Shindo, T., Yamada, M., Isomoto, S., Horio, Y., and Kurachi, Y., 1998, SUR2 subtype (A and B)-dependent differential activation of the cloned ATP-sensitive K^+ channels by pinacidil and nicorandil, *Br. J. Pharmacol.* **124**:985—991.

Shyng, S. L., and Nichols, C. G., 1997, Octameric stochiometry of the K_{ATP} channel complex, *J. Gen. Physiol.* **110**:655—664.

Shyng, S. L. and Nichols, C. G., 1998, Membrane phospholipid control of nucleotide sensitivity of K_{ATP} channels, *Science* **282**:1138—1141.

Shyng, S. L., Ferrigni, T., and Nichols, C. G., 1997, Regulation of K_{ATP} channel activity by diazoxide and MgADP:Distinct functions of the two nucleotide binding folds of the sulfonylurea receptor, *J. Gen. Physiol.* **110**:643—654.

Spanswick, D., Smith, M. A., Groppi, V. E., Logan, S. D. and Ashford, M. L. J., 1997, Leptin inhibits hypothalamic neurons by activation of ATP-sensitive potassium channels, *Nature* **390**:521—525.

Standen, N. B., Quayle, J. M., Davies, N. W., Brayden, J. E., Huang, Y., and Nelson, M. T., 1989, Hyperpolarizing vasodilators activate ATP-sensitive K^+ channels in arterial smooth muscle, *Science* **245**:177—180.

Sturgess, N. C., Ashford, M., Cook, D., and Hales, C., 1985, The sulphonylurea receptor may be an ATP-sensitive potassium channel, *Lancet* **8453**:474—475.

Suzuki, M., Kotake, K., Fujikura, K., Inagaki, N., Suzuki, T., Gonoi, T., Seino, S., and Takata, K., 1997, Kir6.1:A possible subunit of ATP-sensitive K^+ channels in mitochondria, *Biochem. Biophys. Res. Commun.* **241**:693—697.

Szewczyk, A., 1997, Intracellular targets for antidiabetic sulfonylureas and potassium channel openers, *Biochem. Pharmacol.* **54**:961—965.

Terzic, A., 1999, New frontiers of cardioprotection. *Clin. Pharmacol. Ther.* **66**:105—109.

Terzic, A., Tung, R. T., Inanobe, A., Katada, T., and Kurachi, Y., 1994, G Proteins activate ATP-sensitive K^+ channels by antagonizing ATP-dependent gating, *Neuron* **12**:885—893.

Terzic, A., Jahangir, A., and Kurachi, Y., 1995, Cardiac ATP-sensitive K^+ channels:Regulation by intracellular nucleotides and K^+ channel-opening drugs, *Am. J. Physiol.* **38**:C525—C545.

Thomas, P., Cote, G., Wohllk, N., Haddad, B., Mathew, P., Rabl, W., Aguilar-Bryan, L., Gagel, R., and Bryan, J., 1995, Mutations in the sulfonylurea receptor gene in familial persistent hyperinsulinemic hypoglycemia of infancy, *Science* **268**:426—429.

Thuringer, D., and Escande, D., 1989, Apparent competition between ATP and the potassium channel opener RP 49356 on ATP-sensitive K^+ channels of cardiac myocytes, *Mol. Pharmacol.* **36**:897—902.

Trube, G., Rorsman, P., and Ohno-Shosaku, T., 1986, Opposite effects of tolbutamide and diazoxide on the ATP-dependent K^+ channel in mouse pancreatic beta-cells, *Pflügers Arch.* **407**:493—499.

Tucker, S. J., Gribble, F. M., Zhao, C., Trapp, S., and Ashcroft, F. M., 1997, Truncation of Kir6.2 produces ATP-sensitive K^+ channels in the absence of the sulphonylurea receptor, *Nature* **387**:179—183.

Tucker, S. J., Gribble, F. M., Proks, P., Trapp, S., Ryder, T.J., Haug, T., Reimann, F., and Ashcroft, F. M., 1998, Molecular determinants of K_{ATP} channel inhibition by ATP, *EMBO J.* **17**:3290—3296.

Ueda, K., Inagaki, N. and Seino, S., 1997, MgADP antagonism to Mg^{2+}-independent ATP binding of the sulfonylurea receptor SUR1, *J. Biol. Chem.* **272**:22983—22986.

Virsolvy-Vergine, A., Leray, H., Kuroki, S., Lupo, B., Dufour, M., and Bataille, D., 1992, Endosulfine, an endogenous peptidic ligand for the sulfonylurea receptor: Purification and partial characterization from ovine brain, *Proc. Natl. Acad. Sci. U.S.A.* **89**:6629—6633.

Vivaudou, M., and Forestier, C., 1995, Modification by protons of frog skeletal muscle K_{ATP} channels: Effects on ion conduction and nucleotide inhibition, *J. Physiol. (London)* **486**:629—645.

Vivaudou, M., Arnoult, C., and Villaz, M., 1991, Skeletal muscle ATP-sensitive potassium channels recorded from sarcolemmal blebs of split fibers: ATP inhibition is reduced by magnesium and ADP, *J. Membrane Biol.* **122**:165—175.

Weiss, J. N., and Venkatesh, N., 1993, Metabolic regulation of cardiac ATP-sensitive K^+ channels, *Cardiovasc. Drug Ther.* **7**:499—505.

Yamada, M., Isomoto, S., Matsumoto, S., Kondo, C., Shindo, T., Horio, Y., and Kurachi, Y., 1997, Sulphonylurea receptor 2B and Kir6.1 form a sulphonylurea-sensitive but ATP-insensitive K^+ channel, *J. Physiol. (London)* **499**:715—720.

Yao, X. Q., Chang, A.Y., Boulpaep, E. L., Segal, A. S., and Desir, G. V., 1996, Molecular cloning of a glibenclamide-sensitive, voltage-gated potassium channel expressed in rabbit kidney, *J. Clin. Invest.* **97**:2525—2533.

Yokoshiki, H., Sunagawa, M., Seki, T., and Sperelakis, N., 1998, ATP-sensitive K^+ channels in pancreatic, cardiac and vascular smooth muscle cells, *Am. J. Physiol.* **43**:C25—C37.

Zunkler, B. J., Lins, S., Ohno-Shosaku, T., Trube, G., and Panten, U., 1988, Cytosolic ADP enhances the sensitivity to tolbutamide of ATP-dependent K^+ channels from pancreatic β-cells, *FEBS Lett.* **239**:241—244.

Zunkler, B. J., Trube, G., and Panten, U., 1989, How do sulfonylureas approach their receptor in the β-cell membrane?, *Naunyn-Schmiedeberg's Arch. Pharmacol.* **340**:328—332.

Part IV

Potassium Channels in the Heart

Chapter 16

Overview: Molecular Physiology of Cardiac Potassium Channels

B. M. Heath, X. Wehrens, and R. S. Kass

1. INTRODUCTION

Potassium (K^+) channels regulate K^+ ion movement across the *cell* membrane and are important in maintaining the electrical activity in most excitable cells, because they control cellular resting potential and action potential duration. Action potentials recorded from cardiac cells are characterized by their long duration and slow repolarization, quite unlike action potentials found in other electrically excitable cells such as nerve and skeletal muscle. This prolonged depolarization is important in regulating the strength and duration of the contraction of the heart. Outward currents through K^+ channels play important roles in influencing the morphology of the action potential in the heart. For example, inward rectifier K^+ current (I_{Kir}) is important in controlling the resting membrane potential, whereas current through voltage-dependent K^+ (Kv) channels plays a major role in controlling the duration of the action potential in cardiac cells. Many K^+ channels are the physiological targets of neurotransmitters and hormones, which influence heart rate and contractility through their action on many different types of ion channels, including K^+ channels.

Insights into K^+ channel structure and function were made possible by the cloning of the *Shaker* gene from *Drosophila* (Kamb *et al.*, 1987), which codes for a variety of voltage-gated K^+ (Kv) channels. Typically, the α-subunit of a K^+ channel is homologous to only one of the four domains of Na^+ or Ca^{2+} channel α-subunits, consistent with the association of four distinct α-subunits to form a tetrameric Kv channel complex that represents the pore-forming structure (MacKinnon *et al.*, 1993). From the amino acid sequence, hydrophobicity analysis shows that each α-subunit consists of six putative transmembrane-spanning regions (S1–S6) with both N- and

B. M. Heath, X. Wehrens, and R. S. Kass • Department of Pharmacology, College of Physicians and Surgeons of Columbia University, New York, New York 10032.

Potassium Channels in Cardiovascular Biology, edited by Archer and Rusch. Kluwer Academic/Plenum Publishers, New York, 2001.

C-termini located intracellularly (see Chapters 2 and 18). Functions have been assigned to different parts of the α-subunit based on pharmacological and mutagenesis experiments. For example, in *Shaker* channels, the binding of the pore-blocking toxin charybdotoxin (CTX) is influenced by site-directed mutagenesis in the linker region between S5 and S6, consistent with the hypothesis that this portion of the channel contributes to the pore region. Further evidence for this concept is provided by the findings that (1) CTX binding also localizes to this region, (2) the external and internal binding sites for block by tetraethylammonium (TEA) are found at this site, and (3) the region between S5 and S6 confers single-channel conductance properties and channel selectivity to the Kv channel (MacKinnon and Yellen, 1990; MacKinnon and Miller, 1989; Miller, 1989). Thus, it seems likely that this linker region (H5 or P-region) forms part of the channel pore by folding back into the membrane as two antiparallel β-sheets, although other parts of the channel protein also appear to be involved in the formation of the pore (Lopez *et al.*, 1994). In this regard, the crystal structure of the bacterial K^+ channel KcsA has been resolved for the pore region by MacKinnon and colleagues (Doyle *et al.*, 1998a) and has provided a structural basis for conduction and selectivity of the Kv channel (see Chapters 2 and 18). The S4 transmembrane region, which contains positively charged amino acids at every third position, initially was shown to function as the voltage sensor for Na^+ channels, and, subsequently, mutagenesis experiments have shown this region to have the same function for K^+ channels (Isacoff *et al.*, 1990).

Inwardly rectifying K^+ (Kir) channels also are thought to be formed from tetramers, but the α-subunit of these channels is likely to consist of only two membrane-spanning domains homologous to the S5 and S6 regions of the α-subunits of voltage-dependent channels (Ho *et al.*, 1993; Kubo *et al.*, 1993). Although Kir channels have no voltage-sensing S4 region, there is a region, denoted M0 by Ho et al. (1993), that shows some homology to S4 and contains repeated charged residues (Fig. 1). The inward rectification of these channels is thought to result, in part, from voltage-dependent block by cytoplasmic Mg^{2+}.

Recently, Ketchum *et al.* (1995) cloned another family of Kir channels with a novel structure consisting of two pore regions and four transmembrane domains, named TWIK channels (for *tandem* of P-domains in a *weak inward* rectifying K^+ channel). The mRNA for this channel is widely expressed in human tissue, especially in the heart, and the proteins are thought to form channels through the coassembly of dimers (Lesage *et al.*, 1996a,b).

There is, therefore, an ever-increasing diversity of K^+ channels, not only with respect to the number of different channels that have been cloned, but also through alternate splicing of genes (Attali *et al.*, 1993), the heterotetramerization of α-subunits from the same and different subfamilies (Po *et al.*, 1993), and the modification of channel activity by association with additional non-pore-forming subunits. Many such accessory subunits, known as β-subunits, have been cloned (England *et al.*, 1995a,b; Heinemann *et al.*, 1995) and are believed to associate with the α-subunit of K^+ channels to modify their properties. For example, a β-subunit cloned from human atria, Kvβ3, accelerates the inactivation of Kv1.4 channels when coexpressed in *Xenopus* oocytes. Recently, additional accessory proteins have been cloned, including Kv5.1, Kv6.1, Kv8.1, and Kv9.1-2. Although these proteins are homologous to the pore-forming α-subunit, they cannot generate electrical activity by themselves. Instead, they regulate

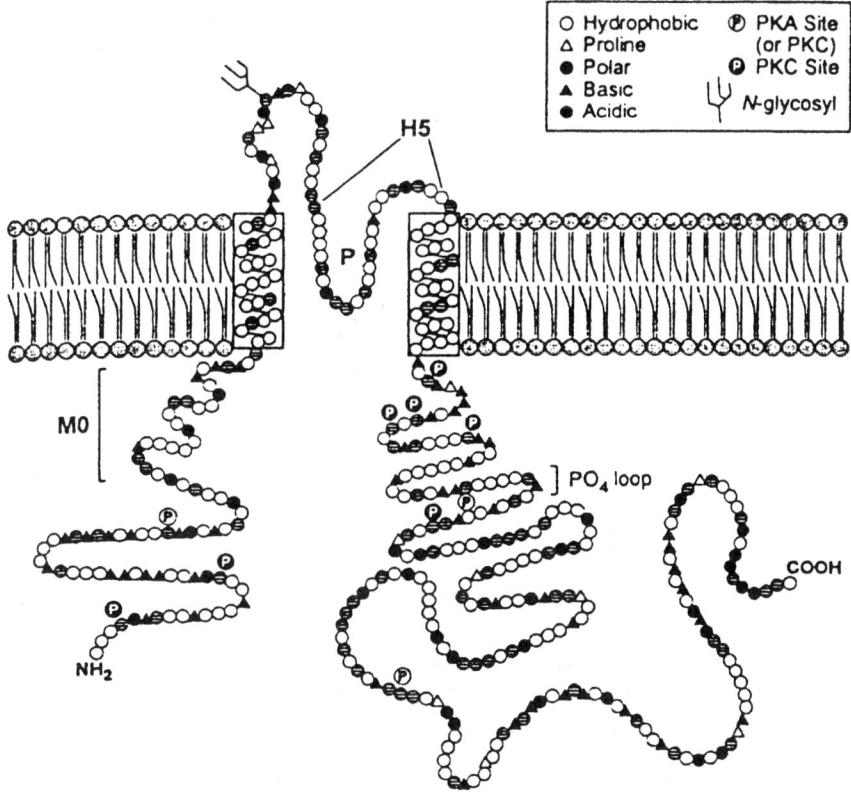

Figure 1. Structural analog for the predicted ROMK1 channel protein. The proposed pore-forming P-segment (P) of ROMK1 is located between membrane-spanning segments M1 and M2. The M0 segment is assumed not to span the membrane, so the N-terminus is depicted in a cytoplasmic location. PKA, Protein Kinase A; PKC, Protein Kinase C. (Reprinted, with permission from Ho *et al.*, 1993. Copyright 1993 Macmillan Magazines Limited.)

the function of Kv2.1 and Kv2.2 channels and channels from the Kv4 family. To date, these silent α-subunits have been found only in the brain, but such a mechanism of K^+ channel regulation also may exist in the heart.

2. VOLTAGE-DEPENDENT K^+ CHANNELS

At least 20 different functional voltage-gated K^+ (Kv) channels have been cloned to date. They are divided into six subfamilies, designated Kv1 (*Shaker*), Kv2 (*Shab*), Kv3 (*Shaw*), Kv4 (*Shal*), KvLQT1 (KCNQ1), and *ether-a-go-go* (EAG). The following section briefly describes the characteristics of the different Kv currents found in the heart and, where known, their molecular basis.

2.1. Transient Outward K$^+$ Current (I_{to})

I_{to} was first identified in sheep Purkinje fibers (Fozzard and Hiraoka, 1973) and later in single cells (Josephson *et al.*, 1984). Subsequent work has shown that I_{to} can be separated into two components: one that is voltage-dependent and insensitive to the cytosolic free calcium concentration ($[Ca^{2+}]_i$), and one that is activated by $[Ca^{2+}]_i$ (Kenyon and Sutko, 1987). These two currents also can be distinguished by their sensitivity to the Kv channel-blocking drug, 4-aminopyridine (4-AP). I_{to1}, the voltage-dependent current, is blocked by 4-AP whereas I_{to2}, the Ca^{2+}-dependent current, is insensitive to 4-AP block. It is now generally accepted that I_{to2} is carried predominantly by Cl^- ions (Zygmunt and Gibbons, 1992; Zygmunt, 1994; Zygmunt *et al.*, 1997), and therefore only I_{to} will be described further here as the apparently genuine I_{to} component.

I_{to} is found in both atrial and ventricular cells from a wide range of species, including those from the human, canine, cat, rat, ferret, rabbit, and mouse heart. It is characterized by a rapid activation and inactivation, and it is the inactivation process that is responsible for the transient *nature* of this current. I_{to} is important in the early phase of action potential repolarization, and because of its rapid activation, the initial phase of repolarization is usually characterized by a "notch" in cells expressing a prominent I_{to}, such as Purkinje fibers. Regional differences in the density of I_{to} across the wall of the heart have been observed. A much larger I_{to} density in epicardial and mid-myocardial cells compared to that in endocardial cells may contribute to the regional differences in action potential morphology across the left ventricular wall (Fedida and Giles, 1991).

Ribonuclease protection assays have revealed the presence of Kv1.2, Kv1.4, Kv1.5 (*Shaker* family), Kv2.1 (*Shab*), and Kv4.2 (*Shal*) in the adult rat atria and ventricle, which express a prominent I_{to} (Dixon and McKinnon, 1994; Dixon *et al.*, 1996; Wymore *et al.*, 1997). The molecular basis of the cardiac I_{to} is not entirely clear, but the most likely candidate genes are from the Kv4 family of Kv channels (Fig. 2). In the rat heart, I_{to} arises from a combination of Kv4.2 and Kv4.3, which are expressed at different levels: the levels of expression of Kv4.2 vary within the wall of the rat heart in the same manner as cardiac I_{to}, whereas levels of the Kv4.3 channel are constant (Dixon and McKinnon, 1994). In canine and human heart, I_{to} is thought to arise from Kv4.3 alone, and the pharmacology of this cloned channel is similar to the native I_{to}. In contrast, I_{to} in rabbit atria is thought to arise from a combination of three cloned channels: Kv4.2, Kv4.3, and Kv1.4 (Wang *et al.*, 1999).

The fast inactivation of channels such as I_{to} is thought to occur through "N-type" inactivation, a mechanism that involves the blocking of the pore from the inside by a cytoplasmic "ball" after the channel has opened. In some Kv channels, the amino terminus of the channel protein may form the blocking "ball" (Morales *et al.*, 1996; Hoshi *et al.*, 1990; Zagotta *et al.*, 1990). Cardiac I_{to} is blocked by 4-AP (5 mM) but is insensitive to the same blocking concentration of tetraethylammonium (TEA). The Kv4 family can be distinguished pharmacologically from Kv1.4 by its sensitivity to block by flecainide and heteropodatoxin peptides (Yeola and Snyders, 1997; Sanguinetti *et al.*, 1997).

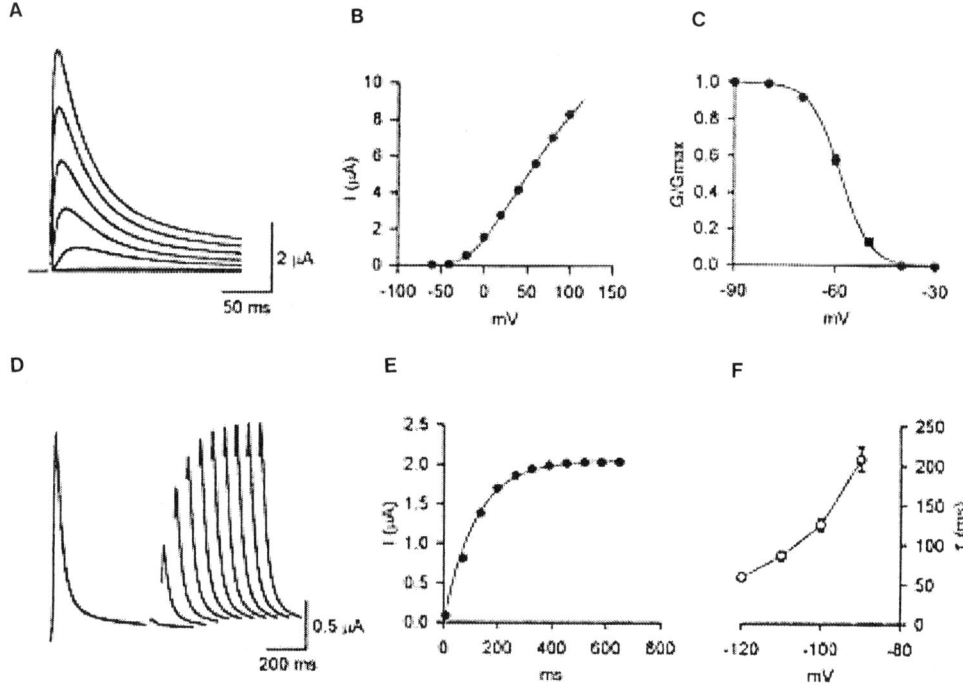

Figure 2. Kinetic properties of the rat Kv4.3 channel expressed in *Xenopus* oocytes resemble those of native I_{to}. (Reproduced, with permission, from Dixon *et al.*, 1996.)

2.2. Delayed Rectifier K$^+$ Currents (I_K)

The delayed rectifier K$^+$ current is a voltage- and time-dependent K$^+$ current originally described in sheep cardiac Purkinje fibers (Noble and Tsien, 1969). Two components of I_K have been separated on the basis of the activation kinetics: a rapidly activating current called I_{Kr} and a slowly activating component called I_{Ks}. I_{Kr} also can be separated pharmacologically by its sensitivity to block by class III antiarrhythmic drugs such as E-4031 and dofetilide (Fig. 3) (Carmeliet, 1992). The genes coding for these two channels have been recently cloned. I_{Kr} is thought to be formed by the EAG channel (Sanguinetti *et al.*, 1995), originally cloned by screening a hippocampal library (Warmke and Ganetzky, 1994) and later cloned from the human heart and named HERG (*human ether-a-go-go*-related *gene*) (Curran *et al.*, 1995). I_{Ks} is thought to arise from the coassembly of two proteins (Sanguinetti *et al.*, 1996a; Barhanin *et al.*, 1996): the minK protein, first cloned from the kidney (Takumi *et al.*, 1988), which is a small protein of about 130 amino acids with one putative transmembrane-spanning region, and KvLQT1 (KCNQ1), a recently cloned K$^+$ channel with the typical six transmembrane-spanning regions (Wang *et al.*, 1996). Coexpression of KvLQT1 with minK results in an increase in K$^+$ current and slowing of KvLQT1 kinetics (Fig. 4) and an

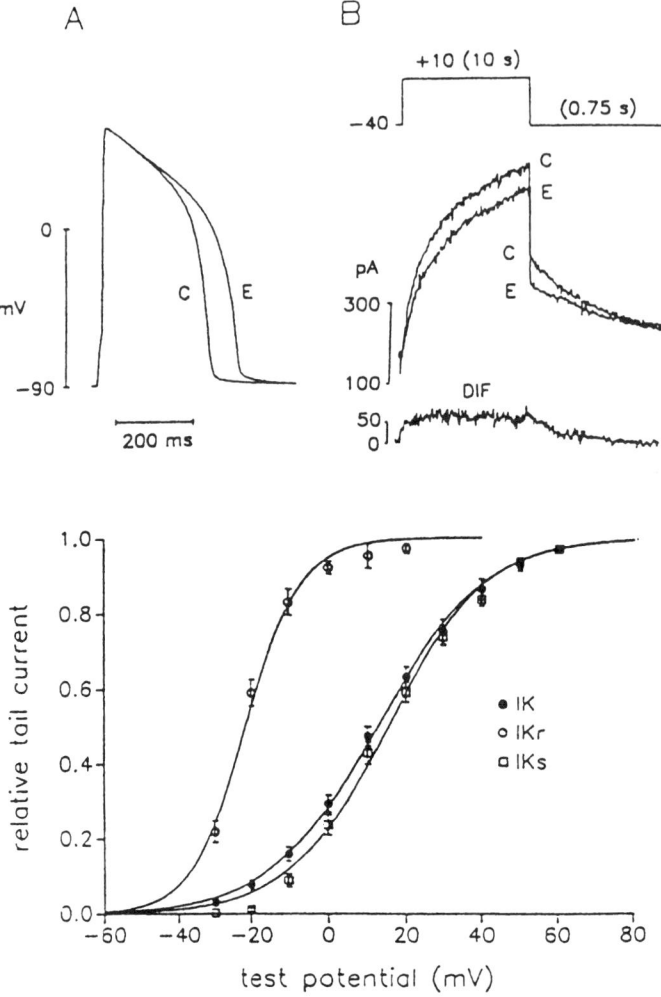

Figure 3. The two components of total delayed rectifier K$^+$ current are revealed by differential sensitivity to block by class III antiarrhythmic agents. *Top:* Action potentials (*left*) and membrane currents (*right*) recorded in the absence (C) and presence (E) of 5 μM E-4031. Difference currents (DIF) reveal the properties of the current blocked by E-4031, which is the rapidly activating component called I_{Kr}. *Bottom:* Voltage dependence of activation of total delayed rectifier current (I_K, ●), as well as the contributing components of I_{Kr} (○) and the slowly activating current, I_{Ks} (□). (Reproduced from Sanguinetti and Jurkiewicz, 1990, by copyright permission of the Rockefeller University Press.)

increase in the single-channel conductance (Pusch, 1998; Yang and Sigworth, 1998). Insight into the stoichiometry of the interaction between KvLQT1 and minK was provided by experiments with tandem fusion proteins of minK and KvLQT1, which were found to reproduce the predicted current (Wang *et al.*, 1998). The I_{Ks} channel formed from the KvLQT1 and minK proteins is unusual in that it is thought that both proteins contribute to the formation of the pore, even though the minK protein does not contain the classical P-region (Tai and Goldstein, 1998).

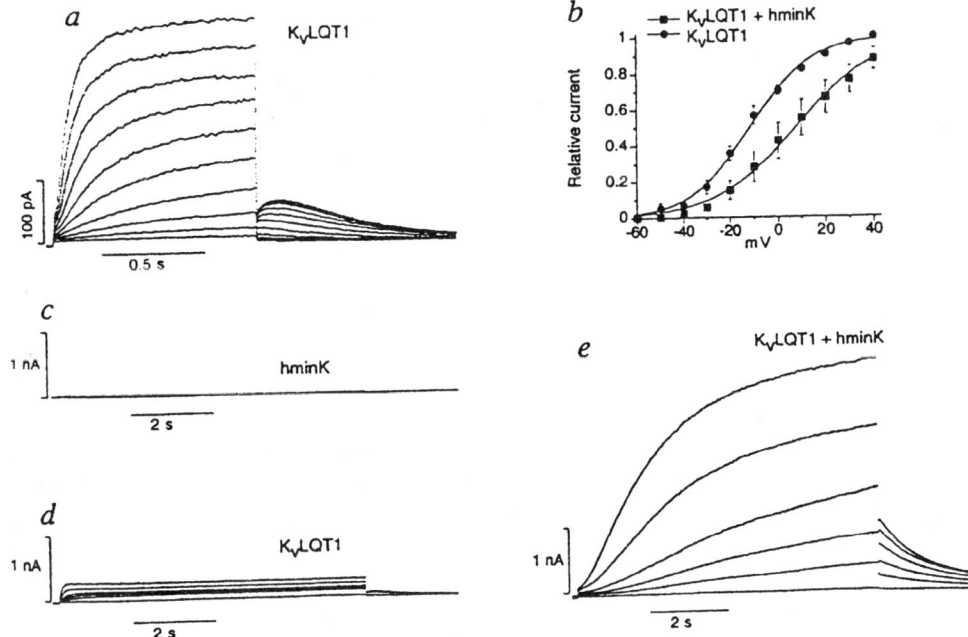

Figure 4. Coassembly of two subunits, KvLQT1 and minK, is needed to reconstitute the biophysical properties of I_{Ks}. (Reprinted with permission, from Sanguinetti *et al.*, 1996a. Copyright 1996 Macmillan Magazines Limited.)

I_{Kr} activates over a more negative range of voltages than I_{Ks}. In guinea pig cardiac myocytes, the half-maximal voltage ($V_{1/2}$) for I_{Kr} activation is -21.5 mV, whereas the $V_{1/2}$ value for I_{Ks} is $+15.7$ mV. Furthermore, I_{Ks} shows rectification at more positive potentials, characterized by a negative-slope conductance at voltages positive to about 0 mV (Sanguinetti and Jurkiewicz, 1990). More recent studies have shown that this apparent inward rectification is caused by rapid C-type inactivation and is abolished by mutation of one amino acid residue (S631A) in the outer mouth of the pore of HERG. Channels in which C-type inactivation is reduced by amino acid mutations show a greatly reduced sensitivity to block by class III antiarrhythmic drugs. Unlike C-type inactivation in many other channels, that in HERG appears to be unique in being voltage-dependent (Schonenherr and Heinemann, 1996; Smith *et al.*, 1996).

Cardiac I_{Ks} is regulated by the sympathetic nervous system. Stimulation of β-adrenoceptors enhances I_{Ks} through activation of cAMP-dependent protein kinase (PKA) and probable phosphorylation of the channel protein. Regulation of I_{Ks} also occurs in the absence of β-adrenergic stimulation. For example, elevation of $[Ca^{2+}]_i$ above 10 nM enhances I_{Ks} without altering its voltage sensitivity (Kass and Wiegers, 1982; Walsh and Kass, 1991). Noise analysis indicates that that elevated $[Ca^{2+}]_i$ increases the open-state probability of the channel without changing its unitary amplitude, estimated to be 0.21 pA at pCa 10 and 0. 19 pA at pCa 7 (Toshe, 1990). I_{Ks} also is regulated indirectly by $[Ca^{2+}]_i$ through activation of the Ca^{2+}-dependent protein kinase (PKC) (Toshe *et al.*, 1987). Thus, exposure of cardiac myocytes to phorbol esters such as 12-O-tetradecanoylphorbol 13-acetate enhances I_K recorded from guinea pig ventricular myocytes, and PKC activation also may be the mechanism by which α-adrenergic agonists such as phenylephrine enhance I_K (Walsh and Kass, 1988).

2.3. Ultrarapid K$^+$ Current (I_{Kur})

First described in human atrial myocytes, ultrarapid K$^+$ current (I_{Kur}) is a depolarization-induced sustained K$^+$ current that remains after I_{to} inactivation and can be distinguished from I_{to} on the basis of its voltage-dependent inactivation and much higher sensitivity to block by 4-AP (Wang *et al.*, 1993). Pharmacological block of this current by a low concentration of 4-AP (typically 50 μM) prolongs the cardiac action potential, demonstrating its role in the process of cardiac repolarization. It seems likely that I_{Kur} results from current through the Kv1.5 channel (Wang *et al.*, 1993), which has been cloned from rat and human heart (Tamkun *et al.*, 1991; Roberds and Tamkun, 1991). Expression of Kv1.5 results in a K$^+$ current with many of the biophysical and electrophysiological properties of I_{Kur} in the heart, including adrenergic modulation (Fedida *et al.*, 1993; Li *et al.*, 1996). The human I_{Kur} channel is thought to be localized in the atria rather than in the ventricles (Wang *et al.*, 1993), although in guinea pig, rat, and mouse, it also has been found in ventricular cells (Boyle and Nerbonne, 1991; Apkon and Nerbonne, 1991; Backx and Marban, 1993; Fiset *et al.*, 1997).

2.4. Potassium Channels and Long QT Syndrome

Mutations in certain ion channel genes are thought to give rise to some instances of the long QT syndrome (LQTS), an inherited cardiac disorder that causes syncope, seizures, and sudden death in otherwise healthy individuals (Moss and Robinson, 1992). Inherited forms of LQTS can result from mutations in at least four different genes, which have been mapped to chromosomes 11p15.5 (LQT1), 7q35–36 (LQT2), 3p21–24 (LQT3), and 4q25–27 (LQT4) (Kass and Davies, 1996). Two of these genes code for K$^+$ channels that are found in the heart, and a third codes for the minK protein. LQT1 is thought to underlie cardiac I_{Ks} (KvLQT1 or KCNQ1), and LQT2 is thought to contribute to I_{Kr} (HERG) and LQT5 (minK). New mutations in HERG and KvLQT1 channels continue to be discovered, with some 30 mutations in KvLQT1 (Wang *et al.*, 1996; Russell *et al.*, 1996; Vandenberg, 1987; Wollnik *et al.*, 1997; Shalaby *et al.*, 1997) and 13 in HERG (Curran *et al.*, 1995; Schulze-Bahr *et al.*, 1995; Benson *et al.*, 1996; Dausse *et al.*, 1996; Satler *et al.*, 1996; Chouabe *et al.*, 1997; Tanaka *et al.*, 1997; Sanguinetti *et al.*, 1996b) reported to date, together with two in the minK protein (Splawski *et al.*, 1997). It is thought likely that these mutations cause some dysfunction of the K$^+$ channels leading to abnormal repolarization in the heart.

3. INWARD RECTIFIER K$^+$ CHANNELS

Inwardly rectifying K$^+$ (Kir) channels represent a genetically diverse group of proteins that conduct K$^+$ more efficiently in the inward direction than in the outward direction. At least three kinds of Kir current have been described in the heart: the cardiac inward rectifier K^+ current (I_{K1}), the acetylcholine-sensitive K^+ current ($I_{K(ACh)}$), and the ATP-sensitive K^+ current (I_{KATP}). These channels play important roles in the heart by controlling membrane potential, influencing repolarization, and contributing to the vagal control of heart rate.

3.1. Cardiac Inward Rectifier K^+ Current (I_{K1})

The cardiac inward rectifier K^+ current (I_{K1}) is very important in maintaining the resting membrane potential and influencing the final phase of repolarization of the cardiac action potential. I_{K1} is activated by hyperpolarization of the membrane and is characterized by inward rectification that, in general, allows very little K^+ current to pass through the channel at positive potentials. Its high conductance at negative potentials allows cells to maintain a stable resting membrane potential. The rectification of I_{K1} channels can result from block of the channels by intracellular Mg^{2+} (Vandenberg, 1987; Raab-Graham et al., 1994) and polyamines (Nichols and Lopatin, 1997; Ficker et al., 1994; Lopatin et al., 1994). At least six subfamilies of inward rectifier K^+ channels (Kir1–6) have been cloned to date from various tissues, and it is thought that channels from the Kir2 subfamily (Kir 2.1, 2.2, and 2.3) compose the native cardiac I_{K1} channel (Ishii et al., 1994; Wible et al., 1995). The density of I_{K1} channels differs throughout the heart. The current is practically absent from pacemaker cells in the sinoatrial and atrioventricular nodes but is expressed in the atria, and its density is greatest in the ventricle and Purkinje system.

3.2. Acetylcholine-Activated K^+ Current ($I_{K(ACh)}$)

$I_{K(ACh)}$ channels are found in atrial and nodal cells and open in response to stimulation of muscarinic (m2) receptors in the heart. Release of acetylcholine from vagal nerves in the heart activates $I_{K(ACh)}$ channels, leading to a hyperpolarization of the myocardial *cell* membrane and a slowing of heart rate. These inwardly rectifying K^+ channels are believed to be directly coupled to muscarinic receptors by a fast membrane-delimited pathway involving the activation of a pertussis-toxin-sensitive G-protein; the released $G\beta\gamma$ subunits bind to and directly activate the channels (Logothetis et al., 1987).

Native $I_{K(ACh)}$ channels are believed to be formed as heteromultimers of two K^+ channel α-subunits—a G-protein-coupled inwardly rectifying K^+ channel, GIRK1 (Kir 3.1), and GIRK4 (Kir3.4)—in a 1:1 stoichiometry (Corey et al., 1998). The GIRK4 subunit appears to be essential in the processing and cell-surface localization of functional $I_{K(ACh)}$ channels and contains a $G\beta\gamma$ binding site essential for channel activation (Krapivinsky et al., 1998).

3.3. ATP-Sensitive K^+ Current (I_{KATP})

K_{ATP} channels were discovered originally in cardiac muscle and are characterized by an inhibition of channel opening when the ATP concentration at the cytoplasmic

Cell surface is increased (Noma, 1983). These channels play an important role in a variety of cellular responses by linking the metabolic status of the cell to its membrane potential, and they have now been found in various noncardiac tissues, including pancreatic β-cells, skeletal muscle, brain, and smooth muscle (Davis *et al.*, 1991).

In the heart, the activation of K_{ATP} channels has been implicated in the shortening of the action potential duration and the cellular loss of K^+ that occurs during various forms of metabolic stress, including ischemia, hypoxia, and inhibition of glycolysis or oxidative phosphorylation (Findlay, 1994). This physiological response reduces contractility, which may protect the myocardium while the function of the heart is impaired. However, during early ischemia, the opening of K_{ATP} channels and the electrophysiological effect of extracellular K^+ accumulation may lead to conditions that promote the induction of cardiac arrhythmias (Gasser and Vaughan-Jones, 1990; Wilde and Janse, 1994).

It has been shown that the K_{ATP} channel in the heart is a heteromer, formed from a complex of at least two subunits: the K^+ channel subunit Kir 6.2, a member of the inward rectifier family (Inagaki *et al.*, 1995a), and the sulfonylurea receptor (SUR2A), a member of the ATP-binding cassette family (Inagaki *et al.*, 1996). Coexpression of the cloned SUR2A and Kir6. 2 produces a channel with the primary characteristics of the cardiac K_{ATP} channel (Inagaki *et al.*, 1996) (Fig. 5).

The subunit of the K_{ATP} complex that confers ATP sensitivity is not clear. Kir6.2 associates with both SUR1 cloned from pancreatic β-cells (Chutkow *et al.*, 1996) and SUR2A cloned from cardiac cells to form K_{ATP} channels with contrasting properties. The channel formed with SUR2A has a lower sensitivity to ATP and the sulfonylurea glibenclamide and, unlike the channel formed with SUR1, is not activated by diazoxide. Therefore, it has been suggested that the SUR subunit confers the pharmacological and ATP-sensitive properties to the K_{ATP} channel. However, recent work showing that the channel demonstrates ATP sensitivity when Kir6.2 is expressed without SUR (as a truncated form) is consistent with the hypothesis that it is the Kir channel itself that is sensitive to the ATP concentration. Further support for this concept comes from the finding that coexpression of Kir6.1 with SUR2B, the smooth muscle isoform of SUR, fails to confer ATP sensitivity to the Kir6.1 channel (Tucker *et al.*, 1997; Yamada *et al.*, 1997).

4. BACKGROUND K$^+$ CURRENTS

A new family of Kir channels has been found in the heart that also may contribute to cardiac background K^+ currents. These channels, including TWIK-1, have a novel two-pore domain structure and four transmembrane regions. They are thought to dimerize to form functional channels with a weak Mg^{2+}-dependent rectification and sensitivity to block by barium. The messenger RNA for members of this family of K^+ channels has been found in both atrial and ventricular myocytes from the rat and human heart, with greater expression found in the ventricle than in the atria in the human cardiac myocytes. It is possible that these channels contribute to the background K^+ conductance in the heart (Attali *et al.*, 1993; Lesage *et al.*, 1996a,b).

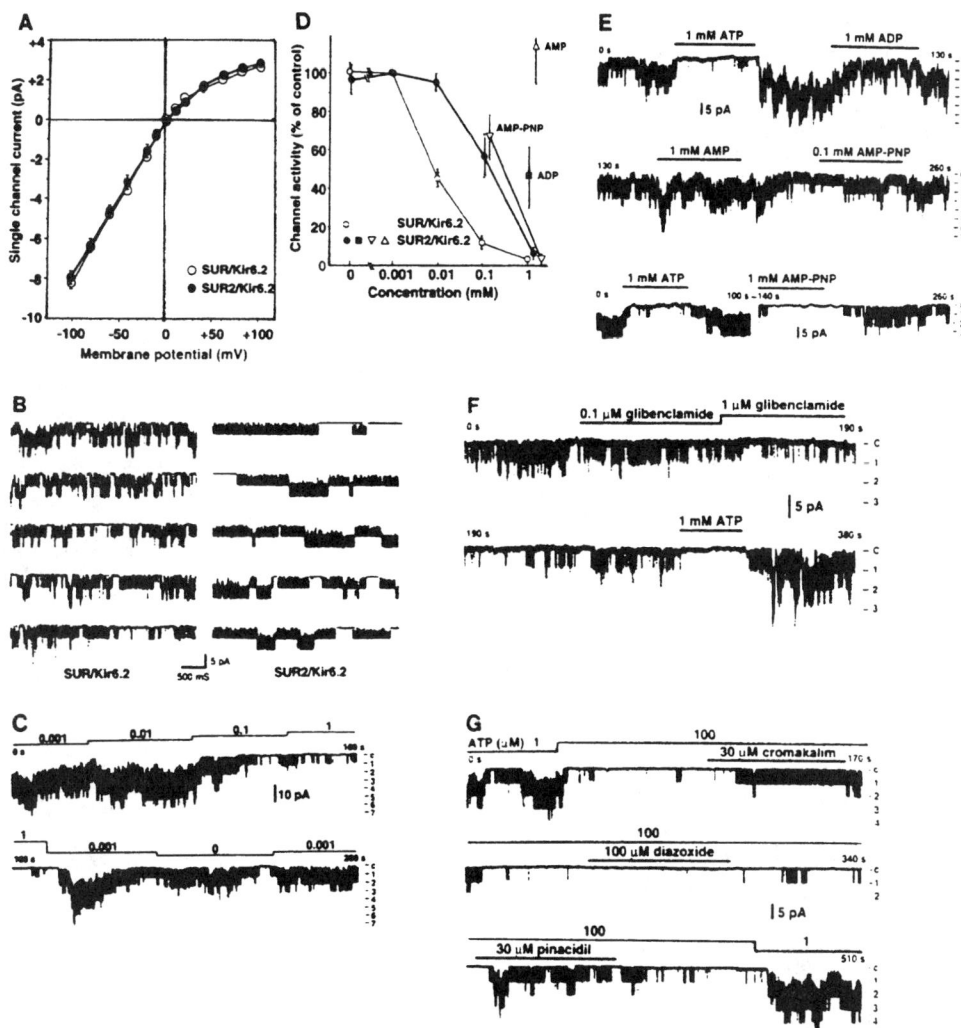

Figure 5. Coexpression of the inward rectifier subunit Kir 6.2 with the sulfonylurea receptor SUR2A encodes channel activity that demonstrates conductance, permeation, and ATP sensitivity of native I_{KATP} channels. (Reproduced, with permission, from Inagaki *et al.*, 1995b.)

5. SUMMARY

This chapter has presented a brief overview of the molecular and cellular properties of the major K$^+$ channel currents that are important in controlling the repolarization and action potential duration in the heart. Many of these K$^+$ channels are the targets of drugs for treating cardiac arrhythmias, since the prolongation of the cardiac action potential is considered an important antiarrhythmic intervention in the treatment of reentrant arrhythmias. However, pathological prolongation of the action potential duration, such as that found in long QT syndrome, can result from inherited disorders in K^+ channels or, in the acquired form, from excessive K^+ channel block by drugs. Therefore, a better understanding of the characteristics and pharmacology of these channels in both the atria and ventricles of the heart may lead to improved therapy for many cardiac diseases.

REFERENCES

Apkon, M., and Nerbonne, J. M., 1991, Characterization of two distinct depolarization-activated K^+ currents in isolated adult rat ventricular myocytes, *J. Gen. Physiol.* **97**:973–1011.

Attali, B., Lesage, F., Ziliani, P., Guillemare, E., Honore, E., Waldmann, R., Mattei, M. G., Lazdunski, M., and Barhanin, J., 1993, Multiple mRNA isoforms encoding the mouse cardiac Kv1.5 delayed rectifier K^+ channel, *J. Biol. Chem.* **268**:24283–24289.

Backx, P. H., and Marban, E., 1993, Background potassium current active during the plateau of the action potential in guinea pig ventricular myocytes, *Circ. Res.* **72**:890–900.

Barhanin, J., Lesage, F., Guillemare, E., Fink, M., Lazdunski, M., and Romey, G., 1996, $K_{(V)}$LQT1 AND IsK (MinK) proteins associate to form the I_{Ks} cardiac potassium current, *Nature* **384**:78–80.

Benson, D. W., Macrae, C. A., Vesely, M. R., Walsh, E. P., Seidman, J. G., Seidman, C. E., and Satler, C. A., 1996, Missense mutation in the pore region of HERG causes familial long QT syndrome, *Circulation* **93**:1791–1795.

Boyle, W. A., and Nerbonne, J. M., 1991, A novel type of depolarization-activated K^+ current in isolated adult rat atrial myocytes, *Am. J. Physiol.* **260**:H1236–H1247.

Carmeliet, E., 1992, Voltage- and time-dependent block of the delayed K^+ current in cardiac myocytes by dofetilide, *J. Pharmacol. Exp. Ther.* **262**:809–817.

Chouabe, C., Neyroud, N., Guicheney, P., Lazdunski, M., Romey, G., and Barhanin, J., 1997, Properties of KvLQT1 K^+ channel mutations in Romano–Ward and Jervell and Lange-Nielsen inherited cardiac arrhythmias, *EMBO J.* **16**:5472–5479.

Chutkow, W. A., Simon, M. C., Le Beau, M. M., and Burant, C. F., 1996, Cloning, tissue expression and chromosomal localization of SUR2, the putative drug-binding subunit of cardiac, skeletal muscle, and vascular K_{ATP} channels, *Diabetes* **45**:1439–1445.

Corey, S., Krapivinsky, G., Krapivinsky, L., and Clapham, D. E., 1998, Number and stoichiometry of subunits in the native atrial G-protein-gated K^+ channel, I_{KACh}, *J. Biol. Chem.* **273**:5271–5278.

Curran, M. E., Splawski, I., Timothy, K. W., Vincent, G. M., Green, E. D., and Keating, M. T., 1995, A molecular basis for cardiac arrhythmia: *HERG* mutations cause long QT syndrome, *Cell* **80**:795–803.

Dausse, E., Berthet, M., Denjoy, I., Andre-Fouet, X., Cruaud, C., Bennaceur, M., Faure, S., Coumel, P., Schwartz, K., and Guicheney, P., 1996, A mutation in HERG associated with notched T waves in long QT syndrome, *J. Mol. Cell. Cardiol.* **28**:1609–1615.

Davis, N. W., Standen, N. B., and Stanfield, P. R., 1991, ATP-dependent potassium channels of muscle cells: Their properties, regulation, and possible functions [Review], *J. Bioenerg. Biomembr.* **23**:509–535.

Dixon, J. E., and McKinnon, D., 1994, Quantitative analysis of potassium channel mRNA expression in atrial and ventricular muscle of rats, *Circ. Res.* **75**:252–260.

Dixon, J. E., Shi, W., Wang, H. S., McDonald, C., Yu, H., Wymore, R. S., and Cohen, I. S., and McKinnon, D., 1996, Role of the Kv4.3 K^+ channel in ventricular muscle: A molecular correlate for the transient outward current, *Circ. Res.* **79**:659–668 [published erratum appears in *Circ. Res.* **80**:147 (1997)].

Drewe, J. A., Verma, S., Frech, G., and Joho, R. H., 1992, Distinct spatial and temporal expression patterns of K^+ channel mRNAs from different subfamilies, *J. Neurosci.* **12**:538–548.

England, S. K., Uebele, V. N., Shear, H., Kodali, J., Bennett, P. B., and Tamkun, M. M., 1995a, Characterization of a voltage-gated K^+ channel β subunit expressed in human heart, *Proc. Natl. Acad. Sci. U.S.A.* **92**:6309–6313.

England, S. K., Uebele, V. N., Kodali, J., Bennett, P. B., and Tamkun, M. M., 1995b, A novel K^+ channel β-subunit (hKvβ1.3) is produced via alternative mRNA splicing, *J. Biol. Chem.* **270**:28531–28534.

Fedida, D., and Giles, W. R., 1991, Regional variations in action potentials and transient outward current in myocytes isolated from rabbit left ventricle, *J. Physiol. (London)* **442**:191–209.

Fedida, D., Wible, B., Wang, Z., Fermini, B., Faust, F., Nattel, S., and Brown, A. M., 1993, Identity of a novel delayed rectifier current from human heart with a cloned K^+ channel current, *Circ. Res.* **73**:210–216.

Ficker, E., Taglialatela, M., Wible, B. A., Henley, C. M., and Brown, A. M., 1994, Spermine and spermidine as gating molecules for inward rectifier K^+ channels, *Science* **266**:1068–1072.

Findlay, I., 1994, The ATP sensitive potassium channel of cardiac muscle and action potential shortening during metabolic stress, *Cardiovasc. Res.* **28**:760–761.

Fiset, C., Clark, R. B., Larsen, T. S., and Giles, W. R., 1997, A rapidly activating sustained K^+ current modulates repolarization and excitation–contraction coupling in adult mouse ventricle, *J. Physiol. (London)* **504**:557–563.

Fozzard, H. A., and Hiraoka, M., 1973, The positive dynamic current and its inactivation properties in cardiac Purkinje fibres, *J. Physiol. (London)* **234**:569–586.

Gasser, R. N. A., and Vaughan-Jones, R. D., 1990, Mechanism of potassium efflux and action potential shortening during ischaemia in isolated mammalian cardiac muscle, *J. Physiol. (London)* **431**:713–741.

Heinemann, S. H., Rettig, J., Wunder, F., and Pongs, O., 1995, Molecular and functional characterization of a rat brain Kvβ3 potassium channel subunit, *FEBS Lett.* **377**:383–389.

Ho, K., Nichols, C. G., Lederer, W. J., Lytton, J., Vassilev, P. M., Kanazirska, M. V., and Hebert, S. C., 1993, Cloning and expression of an inwardly rectifying ATP-regulated potassium channel, *Nature* **362**:31–38.

Hoshi, T., Zagotta, W. N., and Aldrich, R. W., 1990, Biophysical and molecular mechanisms of Shaker potassium channel inactivation [see comments], *Science* **250**:533–538.

Hugnot, J. P., Salinas, M., Lesage, F., Guillemare, E., De Weille, J., Heurteaux, C., Mattei, M. G., and Lazdunski, M., 1996, Kv8.1, a new neuronal potassium channel subunit with specific inhibitory properties towards Shab and Shaw channels, *EMBO J.* **15**:3322–3331.

Inagaki, N., Gonoi, T., Clement, J. P., Namba, N., Inazawa, J., Gonzalez, G., Aguilar-Bryan, L., Seino, S. and Bryan, J., 1995a, Reconstruction of I_{KATP}—an inward rectifier subunit plus the sulfonylurea receptor. *Science* **270**:1166–1170.

Inagaki, N., Tsuura, Y., Namba, N., Masuda, K., Gonoi, T., Horie, M., Seino, Y., Mizuta, M., and Seino, S., 1995b, Cloning and functional characterization of a novel ATP-sensitive potassium channel ubiquitously expressed in rat tissues, including pancreatic islets, pituitary, skeletal muscle, and heart, *J. Biol. Chem.* **270**:5691–5694.

Inagaki, N., Gonoi, T., Clement, J. P., Wang, C. Z., Aguilar-Bryan, L., Bryan, J., and Seino, S., 1996, A family of sulfonylurea receptors determines the pharmacological properties of ATP-sensitive K^+ channels, *Neuron* **16**:1011–1017.

Isacoff, E., Papazian, D., Timpe, L., Jan, Y. N., and Jan, L. Y., 1990, Molecular studies of voltage-gated potassium channels, *Cold Spring Harbor Symp. Quantit. Biol.* **55**:9–17.

Ishii, K., Yamagishi, T., and Taira, N., 1994, Cloning and functional expression of a cardiac inward rectifier K^+ channel, *FEBS Lett.* **338**:107–111.

Josephson, I. R., Sanchez-Chapula, J., and Brown, A. M., 1984, Early outward current in rat single ventricular cells, *Circ. Res.* **54**:157–162.

Kamb, A., Iverson, L. E., and Tanouye, M. A., 1987, Molecular characterization of Shaker, a *Drosophila* gene that encodes a potassium channel, *Cell* **50**:405–413.

Kass, R. S., and Davies, M. P., 1996, The roles of ion channels in an inherited heart disease: Molecular genetics of the long QT syndrome, *Cardiovasc. Res.* **32**:443–454.

Kass, R. S., and Wiegers, S. E., 1982, The ionic basis of concentration-related effects of noradrenaline on the action potential of calf cardiac Purkinje fibres, *J. Physiol. (London)* **322**:541–558.

Kenyon, J. L., and Sutko, J. L., 1987, Calcium- and voltage-activated plateau currents of cardiac Purkinje fibers, *J. Gen. Physiol.* **89**:921–958.

Ketchum, K. A., Joiner, W. J., Sellers, A. J., Kaczmarek, L. K., and Goldstein, S. A., 1995, A new family of outwardly rectifying potassium channel proteins with two pore domains in tandem, *Nature* **376**:690–695.

Krapivinsky, G., Krapivinsky, L., Wickman, K., and Clapham, D. E., 1995, G $\beta\gamma$ binds directly to the G protein-gated K^+ channel, I_{KACh}, *J. Biol. Chem.* **270**:29059–29062.

Krapivinsky, G., Kennedy, M. E., Nemec, J., Medina, I., Krapivinsky, L., and Clapham, D. E., 1998, Gβ binding to GIRK4 subunit is critical for G protein-gated K^+ channel activation, *J. Biol. Chem.* **273**:16946–16952.

Kubo, Y., Reuveny, E., Slesinger, P. A., Jan, Y. N., and Jan, L. Y., 1993, Primary structure and functional expression of a rat G-protein-coupled muscarinic potassium channel [see comments], *Nature* **364**:802–806.

Lesage, F., Guillemare, E., Fink, M., Duprat, F., Lazdunski, M., Romey, G., and Barhanin, J., 1996a, TWIK-1, a ubiquitous human weakly inward rectifying K^+ channel with a novel structure, *EMBO J.* **15**:1004–1011.

Lesage, F., Reyes, R., Fink, M., Duprat, F., Guillemare, E., and Lazdunski, M., 1996b, Dimerization of TWIK-1 K^+ channel subunits via a disulfide bridge, *EMBO J.* **15**:6400–6407.

Li, G. R., Feng, J., Wang, Z., Fermini, B., and Nattel, S., 1996, Adrenergic modulation of ultrarapid delayed rectifier K^+ current in human atrial myocytes, *Circ. Res.* **78**:903–915.

Logothetis, D. E., Kurachi, Y., Galper, J., Neer, E. J., and Clapham, D. E., 1987, The $\beta\gamma$ subunits of GTP-binding proteins activate the muscarinic K^+ channel in heart, *Nature* **325**:321–326.

Lopatin, A. N., Makhina, E. N., and Nichols, C. G., 1994, Potassium channel block by cytoplasmic polyamines as the mechanism of intrinsic rectification, *Nature* **372**:366–369.

Lopez, G. A., Jan, Y. N., and Jan, L. Y., 1994, Evidence that the S6 segment of the Shaker voltage-gated K^+ channel comprises part of the pore, *Nature* **367**:179–182.

MacKinnon, R., and Miller, C., 1989, Mutant potassium channels with altered binding of charybdotoxin, a pore-blocking peptide inhibitor, *Science* **245**:1382–1384.

MacKinnon, R., and Yellen, G., 1990, Mutations affecting TEA blockade and ion permeation in voltage activated K^+ channels, *Science* **250**:276–279.

MacKinnon, R., Aldrich, R. W., and Lee, A. W., 1993, Functional stoichiometry of Shaker potassium channel inactivation, *Science* **262**:757–759.

Miller, C., 1989, Genetic manipulation of ion channels: A new approach to structure and mechanism, *Neuron* **2**:1195–1205.

Morales, M. J., Wee, J. O., Wang, S., Strauss, H. C., and Rasmusson, R. L., 1996, The N-terminal domain of a K^+ channel β subunit increases the rate of C-type inactivation from the cytoplasmic side of the channel, *Proc. Natl. Acad. Sci. U.S.A.* **93**:15119–15123.

Moss, A. J., and Robinson, J. L., 1992, The long-QT syndrome: Genetic considerations, *Trends Cardiovasc. Med.* **2**:81–83.

Nichols, C. G., and Lopatin, A. N., 1997, Inward rectifier potassium channels, *Annu. Rev. Physiol.* **59**:171–191.

Noble, D., and Tsien, R. W., 1969, Outward membrane currents activated in the plateau range of potentials in cardiac Purkinje fibres, *J. Physiol. (London)* **200**:205–231.

Noma, A., 1983, ATP-regulated K^+ channels in cardiac muscle, *Nature* **305**:147–148.

Po, S., Roberds, S., Snyders, D. J., Tamkun, M. M., and Bennett, P. B., 1993, Heteromultimeric assembly of human potassium channels: Molecular basis of a transient outward current?, *Circ. Res.* **72**:1326–1336.

Pusch, M., 1998, Increase of the single-channel conductance of KvLQT1 potassium channels induced by the association with minK, *Pflügers Arch.* **437**:172–174.

Raab-Graham, K. F., Radeke, C. M., and Vandenberg, C. A., 1994, Molecular cloning and expression of a human heart inward rectifier potassium channel, *NeuroReport* **5**:2501–2505.

Roberds, S. L., and Tamkun, M. M., 1991, Cloning and tissue-specific expression of five voltage-gated potassium channel cDNAs expressed in rat heart, *Proc. Natl. Acad. Sci. U.S.A.* **88**:1798–1802.

Russell, M. W., Dick, M., Collins, F. S., and Brody, L. C., 1996, KVLQT1 mutations in three families with familial or sporadic long QT syndrome, *Hum. Mol. Genet.* **5**:1319–1324.

Sanguinetti, M. C., and Jurkiewicz, N. K., 1990, Two components of cardiac delayed rectifier K^+ current: Differential sensitivity to block by class III antiarrhythmic agents, *J. Gen. Physiol.* **96**:195–215.

Sanguinetti, M. C., Jiang, C., Curran, M. E., and Keating, M. T., 1995, A mechanistic link between an inherited and an acquired cardiac arrhythmia: HERG encodes the I_{Kr} potassium channel, *Cell* **81**:299–307.

Sanguinetti, M. C., Curran, M. E., Zou, A., Shen, J., Spector, P. S., and Keating, M. T., 1996a, Coassembly of $K_{(V)}LQT1$ and minK (IsK) proteins to form cardiac I_{Ks} potassium channels, *Nature* **384**:80–83.

Sanguinetti, M. C., Curran, M. E., Spector, P. S., and Keating, M. T., 1996b, Spectrum of HERG K^+ channel dysfunction in an inherited cardiac arrhythmia, *Proc. Natl. Acad. Sci. U.S.A.* **93**:8796–8796.

Sanguinetti, M. C., Johnson, J. H., Hammerland, L. G., Kelbaugh, P. R., Volkmann, R. A., Saccomano, N. A., and Mueller, A. L., 1997, Heteropodatoxins: Peptides isolated from spider venom that block Kv4.2 potassium channels, *Mol. Pharmacol.* **51**:491–498.

Satler, C. A., Walsh, E. P., Vesely, M. R., Plummer, M. H., Ginsburg, G. S., and Jacob, H. J., 1996, Novel missense mutation in the cyclic nucleotide-binding domain of HERG causes long QT syndrome, *Am. J. Med. Genet.* **65**:27–35.

Schonenherr, R., and Heinemann, S. H., 1996, Molecular determinants for activation and inactivation of HERG, a human inward rectifier potassium channel, *J. Physiol. (London)* **493**:5–42.

Schulze-Bahr, E., Haverkamp, W., and Funke, H., 1995, The long-QT syndrome [letter; comment], *N. Engl. J. Med.* **333**:1783–1784.

Shalaby, F. Y., Levesque, P. C., Yang, W. P., Little, W. A., Conder, M. L., Jenkins-West, T., and Blanar, M. A., 1997, Dominant-negative KvLQT1 mutations underlie the LQT1 form of long QT syndrome [see comments], *Circulation* **96**:1733–1736.

Slesinger, P. A., Jan, Y. N., and Jan, L. Y., 1993, The S4–S5 loop contributes to the ion-selective pore of potassium channels, *Neuron* **11**:739–749.

Smith, P. L., Baukrowitz, T., and Yellen, G., 1996, The inward rectification mechanism of the HERG cardiac potassium channel, *Nature* **379**:833–836.

Splawski, I., Tristani-Firouzi, M., Lehmann, M. H., Sanguinetti, M. C., and Keating, M. T., 1997, Mutations in the hminK gene cause long QT syndrome and suppress I_{Ks} function, *Nat. Genet.* **17**:338–340.

Tai, K. K., and Goldstein, S. N., 1998, The conduction pore of a cardiac potassium channel, *Nature* **391**:605–608.

Takumi, T., Ohkubo, H., and Nakanishi, S., 1988, Cloning of a membrane protein that induces a slow voltage-gated potassium current, *Science* **242**:1042–1045.

Tamkun, M. M., Knoth, K. M., Walbridge, J. A., Kroemer, H., Roden, D. M., and Glover, D. M., 1991, Molecular cloning and characterization of two voltage-gated K^+ channel cDNAs from human ventricle, *FASEB J.* **5**:331–337.

Tanaka, T., Nagai, R., Tomoike, H., Takata, S., Yano, K., Yabuta, K., Haneda, N., Nakano, O., Shibata, A., Sawayama, T., Kasai, H., Yazaki, Y., and Nakamura, Y., 1997, Four novel KVLQT1 and four novel HERG mutations in familial long-QT syndrome, *Circulation* **95**:565–567.

Toshe, N., 1990, Calcium-sensitive delayed rectifier potassium current in guinea pig ventricular cells, *Am. J. Physiol.* **258**:H1200–H1207.

Toshe, N., Kameyama, M., and Irasawa, H., 1987, Intracellular Ca and PKC modulate K current in guinea pig heart cells, *Am. J. Physiol.* **253**:H1321–H1324.

Tucker, S. J., Gribble, F. M., Zhao, C., Trapp, P. S., and Ashcroft, F. M., 1997, Truncation of Kir6.2 produces ATP-sensitive K^+ channels in the absence of the sulphonylurea receptor, *Nature* **387**:179–183.

Vandenberg, C. A., 1987, Inward rectification of a potassium channel in cardiac ventricular cells depends on internal magnesium ions, *Proc. Natl. Acad. Sci. U.S.A.* **84**:2560–2564.

Walsh, K. B., and Kass, R. S., 1988, Regulation of a heart potassium channel by protein kinase A and C, *Science* **242**:67–69.

Walsh, K. B., and Kass, R. S., 1991, Distinct voltage-dependent regulation of a heart-delayed I_K by protein kinases A and C, *Am. J. Physiol.* **261**:C1081–C1090.

Wang, Q., Curran, M. E., Splawski, I., Burn, T. C., Millholland, J. M., Vanraay, T. J., Shen, J., Timothy, K. W., Vincent, G. M., Dejager, T., Schwartz, P. J., Towbin, J. A., Moss, A. J., Atkinson, D. L., Landes, G. M., Connors, T. D., and Keating, M. T., 1996, Positional cloning of a novel potassium channel gene—KVLQT1 mutations cause cardiac arrhythmias, *Nature Genet.* **12**:17–23.

Wang, W., Xia, J., and Kass, R. S., 1998, MinK–KvLQT1 fusion proteins, evidence for multiple stoichiometries of the assembled IsK channel [In Process Citation], *J. Biol. Chem.* **273**:34069–34074.

Wang, Z., Fermini, B., and Nattel, S., 1993, Sustained depolarization-induced outward current in human atrial myocytes: Evidence for a novel delayed rectifier K^+ current similar to Kv1.5 cloned channel currents, *Circ. Res.* **73**:1061–1076.

Wang, Z., Feng, J., Shi, H., Pond, A., Nerbonne, J. M., and Nattel, S., 1999, Potential molecular basis of different physiological properties of the transient outward K^+ current in rabbit and human atrial myocytes [see comments], *Circ. Res.* **84**:551–561.

Warmke, J. W., and Ganetzky, B., 1994, A family of potassium channel genes related to *eag* in *Drosophila* and mammals, *Proc. Natl. Acad. Sci. U.S.A.* **91**:3438–3442.

Wible, B. A., De Biasi, M., Majumder, K., Taglialatela, M., and Brown, A. M., 1995, Cloning and functional expression of an inwardly rectifying K^+ channel from human atrium, *Circ. Res.* **76**:343–350.

Wilde, A. A., and Janse, M. J., 1994, Electrophysiological effects of ATP sensitive potassium channel modulation: Implications for arrhythmogenesis, *Cardiovasc. Res.* **28**:16–24.

Wollnik, B., Schroeder, B. C., Kubisch, C., Esperer, H. D., Wieacker, P., and Jentsch, T. J., 1997, Pathophysiological mechanisms of dominant and recessive KVLQT1 K$^+$ channel mutations found in inherited cardiac arrhythmias, *Hum. Mol. Genet.* **6:**1943–1949.

Wymore, R. S., Gintant, G. A., Wymore, R. T., Dixon, J. E., McKinnon, D., and Cohen, I. S., 1997, Tissue and species distribution of mRNA for the I_{Kr}-like K$^+$ channel, erg, *Circ. Res.* **80:**261–268.

Yamada, M., Isomoto, S., Matsumoto, S., Kondo, C., Shindo, T., Horio, Y., and Kurachi, Y., 1997, Sulphonylurea receptor 2B and Kir6.1 form a sulphonylurea-sensitive but ATP-insensitive K$^+$ channel, *J. Physiol. (London)* **499:**715–720.

Yang, Y., and Sigworth, F. J., 1998, Single-channel properties of I_{Ks} potassium channels, *J. Gen. Physiol.* **112:**665–678.

Yeola, S. W., and Snyders, D. J., 1997, Electrophysiological and pharmacological correspondence between Kv4.2 current and rat cardiac transient outward current, *Cardiovasc. Res.* **33:**540–547.

Zagotta, W. N., Hoshi, T., and Aldrich, R. W., 1990, Restoration of inactivation in mutants of Shaker potassium channels by a peptide derived from ShB, *Science* **250:**568–571.

Zagrovic, B., and Aldrich, R., 1999, For the latest information, tune to channel KcsA [comment], *Science* **285:**59–61.

Zygmunt, A. C., 1994, Intracellular calcium activates a chloride current in canine ventricular myocytes, *Am. J. Physiol.* **267:**H1984–H1995.

Zygmunt, A. C., and Gibbons, W. R., 1992, Properties of the calcium-activated chloride current in heart, *J. Gen. Physiol.* **99:**391–414.

Zygmunt, A. C., Goodrow, R. J., and Antzelevitch, C., 1997, Sodium effects on 4-aminopyridine-sensitive transient outward current in canine ventricular cells, *Am. J. Physiol.* **272:**H1–H11

Chapter 17

Molecular Mechanisms Controlling Functional Voltage-Gated K^+ Channel Diversity and Expression in the Mammalian Heart

Jeanne M. Nerbonne

1. INTRODUCTION

Depolarization-activated, Ca^{2+}-independent, outward K^+ currents contribute to determining the magnitude and the duration of action potentials in cardiac cells. Several distinct types of voltage-gated K^+ currents that serve these functions have been characterized (Anumonwo *et al.*, 1991; Barry and Nerbonne, 1996). Important among these are the transient outward currents, $I_{to,f}$ and $I_{to,s}$, and several components of delayed rectification, including I_{Kr} ($I_{K(rapid)}$), I_{Ks} ($I_{K(slow)}$), and I_{Kur} ($I_{K(ultrarapid)}$), and others. There are interspecies differences in the expression of the various voltage-gated, cardiac K^+ currents, as well as regional differences amongst myocytes from different regions of the heart within a single species. These distinct K^+ current expression patterns contribute to the variability in action potential waveforms recorded in different regions of the heart (Antzelevitch *et al.*, 1994; Barry and Nerbonne, 1996). Importantly, however, the time- and voltage-dependent properties of the various I_{to} and I_K currents in different cardiac cell types (and species) are quite similar, suggesting that the molecular identity of the K^+ channels underlying these currents is predictable. A number of voltage-gated K^+ channel (Kv) pore-forming (α) and accessory (β and minK) subunits are expressed in or have been cloned from the heart. A variety of experimental approaches have been exploited to define the relationships between these subunits and the functional voltage-gated K^+ channels in cardiac cells. Considerable progress has been made in defining these relationships, and all of the results obtained to date (discussed in detail in subsequent sections) suggest that distinct molecular

Jeanne M. Nerbonne • Department of Molecular Biology and Pharmacology, Washington University Medical School, St. Louis, Missouri 63110.

Potassium Channels in Cardiovascular Biology, edited by Archer and Rusch. Kluwer Academic/Plenum Publishers, New York, 2001.

entities underlie the electrophysiologically distinct types of voltage-gated K^+ currents/channels in myocardial cells.

It is well documented that there are marked and rather stereotypical changes in the densities and the properties of voltage-gated K^+ currents in the heart during normal development (Wetzel and Klitzner, 1996; Nerbonne, 1998). These changes in functional expression of voltage-gated K^+ channels reshape action potential waveforms (Wetzel and Klitzner, 1996) and explain variations in the sensitivity to antiarrhythmics in the developing heart, as compared with the mature heart (Abrahamsson et al., 1994). In addition, Kv channels are important downstream targets for the actions of a variety of endogenous transmitters, hormones, and exogenous drugs that modulate cardiac function (Gadsby, 1990; Anumonwo et al.,1991; Abrahamsson et al., 1994). Changes in the densities and functional properties of Kv currents also occur with myocardial disease in human hearts, as well as in various experimental models (Ten Eick et al., 1989, 1993; Bénitah et al., 1993; Beuckelmann et al., 1993; Näbauer et al., 1993; Tomita et al., 1994; Boyden and Jeck, 1995; Kääb et al., 1996; Mészáros et al., 1996; Potreau et al., 1996; Qin et al., 1996; Bailly et al., 1997; Freeman et al., 1997; Gomez et al., 1997; Shipsey et al., 1997; Van Wagoner et al., 1997; Yue et al., 1997; Näbauer and Kääb, 1998; Rozanski et al., 1998). These changes in channel function have important consequences, including the generation of life-threatening arrhythmias. In addition to the interest in defining the molecular correlates of functional cardiac K^+ channels, therefore, there is also considerable interest in understanding the molecular mechanisms controlling the regulation, modulation, and functional expression of these channels. Although there is evidence for transcriptional, translational, and posttranslational regulation of functional Kv channels in cardiac cells, the detailed molecular mechanisms and physiological relevance of these regulatory events are not well understood.

2. ELECTROPHYSIOLOGICAL DIVERSITY OF VOLTAGE-GATED K^+ CURRENTS

The amplitudes and the durations of action potentials in the mammalian heart are largely determined by voltage-gated K^+ channels (Fig. 1). Two broad classes of Kv channel currents have been distinguished based on differences in time- and voltage-dependent properties and pharmacological sensitivities: (1) rapidly activating and inactivating, 4-aminopyridine (4-AP)-sensitive K^+ currents, referred to as I_{to} (transient outward), and (2) delayed, slowly inactivating, and typically 4-AP-insensitive K^+ currents, referred to as I_K (delayed rectifiers) (Table 1). These K^+ current types serve distinct functional roles in action potential repolarization. I_{to} underlies the early phase of action potential repolarization (Fig. 1, portion 1 of the curve on the right-hand side) whereas I_K, which activates and inactivates somewhat more slowly, contributes to the late phase of membrane repolarization (portion 3 of the curve on the right-hand side of Fig. 1. Nevertheless, I_{to} and I_K are broad classifications, and each current is comprised of multiple components (Table 1). In addition, the properties and the densities of the various Kv currents (Table 1) vary among species and between different regions of the heart in the same species. These differences contribute to the variability in the waveforms of action potentials recorded in different cardiac cell types (Fig. 1).

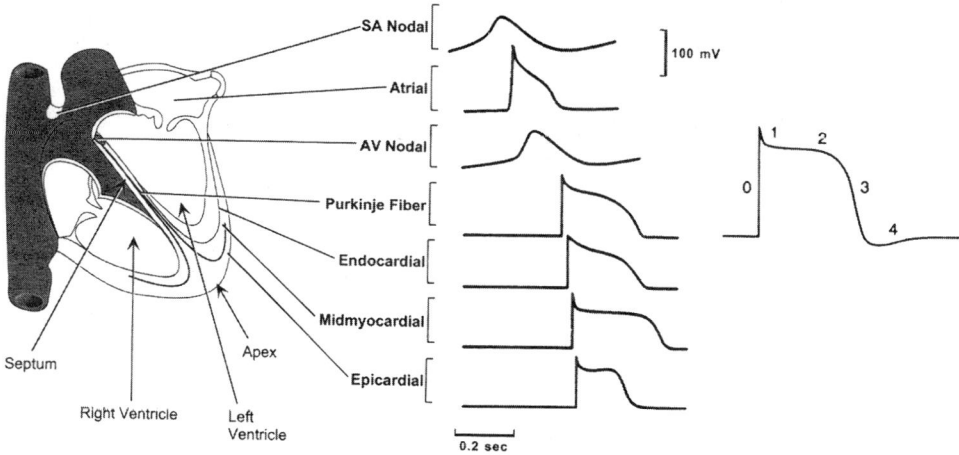

Figure 1. Action potential waveforms are variable in different regions of the heart. (A) Schematic representation of the heart and action potential waveforms recorded in different regions of the heart are illustrated; action potentials are displaced in time to reflect the temporal sequence of propagation through the heart. (B) A schematic of a ventricular action potential with the various phases numbered as follows: (0) depolarization; (1) early repolarization, which underlies the "notch"; (2) plateau; (3) late phase of repolarization; and, (4) afterhyperpolarization and subsequent return to the resting membrane potential.

2.1. Transient Outward K^+ Currents, I_{to}, in Cardiac Cells

Transient, outward currents, first described in cardiac Purkinje fibers, were originally thought to be carried primarily by Cl^- (Deck and Trautwein, 1964; Dudel *et al.*, 1967; Fozzard and Hiraoka, 1973). Subsequent work, however, distinguished two transient outward current components, referred to as I_{to1} and I_{to2} (Kenyon and Gibbons, 1979a,b; Coraboeuf and Carmeliet, 1982). I_{to1} is Ca^{2+}–independent and blocked by millimolar concentrations of 4-AP, whereas I_{to2} is 4-AP-insensitive and Ca^{2+}-dependent (Kenyon and Gibbons, 1979b; Coraboeuf and Carmeliet, 1982). Considerable evidence has now been accumulated demonstrating that I_{to2} is a Cl^--, not a K^+-, selective conductance pathway (Kenyon and Gibbons, 1979a,b; Zygmunt and Gibbons, 1991, 1992; Zygmunt, 1994), and, for this reason, it will not be discussed further here. Although cardiac transient outward K^+ currents have previously been referred to as I_{to}, I_{to1}, or I_t (Campbell *et al.*, 1995; Giles *et al.*, 1996; Barry and Nerbonne, 1996; Nerbonne, 1998), it has now been demonstrated that there are (at least) two distinct transient outward K^+ currents in cardiac cells and, in addition, that these currents are differentially distributed (Brahmajothi *et al.*, 1999; Wang *et al.*, 1999; Xu *et al.*, 1999a). These will be considered separately here and referred to as $I_{to,fast}$ ($I_{to,f}$) and $I_{to,slow}$ ($I_{to,s}$), as suggested by Xu *et al.* (1999a).

2.1.1. $I_{to,fast}$ ($I_{to,f}$)

$I_{to,f}$ has been characterized in considerable detail in cat (Furukawa *et al.*, 1990), dog (Litovsky and Antzelevitch, 1988; Tseng and Hoffman, 1989), ferret (Campbell *et*

Table 1
Voltage-Gated K$^+$ Currents/Channels in the Mammalian Heart[a]

Current	Activation	Inactivation	Blocker[b]	Tissue	Species
$I_{to,f}$	Fast	Fast	4-AP (mM), HaTX, HpTX	Atria	Dog, human, mouse, rabbit, rat
				Ventricle	Cat, dog, ferret, human, mouse, rat
$I_{to,s}$	Fast	Slow	4-AP (mM)	Atria	Rabbit
				Ventricle	Ferret, mouse, rabbit
				Node	Rabbit
I_{Kr}	Moderate	Fast	E-4031, dofetilide, lanthanum	Atria	Dog, guinea pig, human, rat
				Ventricle	Cat, dog, guinea pig, human, mouse, rabbit, rat
				Node	Rabbit
I_{Ks}	Very slow	Very, very slow	NE-10064, NE-10133	Atria	Dog, guinea pig, human
				Ventricle	Dog, guinea pig, human
				Node	Guinea pig
I_{Kur}	Fast	No	4-AP (μM)	Atria	Dog, human, rat
I_{Kp}	Fast	No	Ba^{2+}	Ventricle	Guinea pig
I_{K}	Slow	Slow	TEA (mM)	Ventricle	Rat
$I_{K,slow}$	Fast	Slow	4-AP (mM), TEA (mM)	Atria	Mouse
				Ventricle	Mouse
$I_{K,slow}$ ($I_{K,DTX}$)	Fast	Very slow	4-AP (mM), DTX	Atria	Human, rat
I_{ss}	Slow	No	TEA (mM), 4-AP (mM)	Atria	Mouse, rat
				Ventricle	Dog, human, mouse, rat

[a] Adapted from J. M. Nerbonne, *J. Neurobiol.* **37**:37–59 (1998).
[b] Abbreviations: 4-AP, 4-Aminopyridine; HaTX, hanatoxin; HpTX, heteropodatoxin; TEA, tetraethylammonium; DTX, dendrotoxin.

al., 1993), human (Wettwer *et al.*, 1993, 1994; Konarzewska *et al.*, 1995), mouse (Benndorf *et al.*, 1987; Benndorf and Nilius, 1988; Xu *et al.*, 1999a,b), and rat (Apkon and Nerbonne, 1991; Wettwer *et al.*, 1993) ventricular myocytes. $I_{to,f}$ is also evident in atrial cells from dog (Yue *et al.*, 1996a), rat (Boyle and Nerbonne, 1991, 1992), human (Escande *et al.*, 1985, 1987; Shibata *et al.*, 1989; Fermini *et al.*, 1992), mouse (Xu *et al.*, 1999b), rabbit (Giles and Imaizumi, 1988; Wang *et al.*, 1999), and guinea pig (Wang *et al.*, 1991) (Table 1). Although guinea pig ventricular myocytes reportedly lack $I_{to,f}$ (Sanguinetti and Jurkiewicz, 1990, 1991), a rapidly activating and inactivating ($I_{to,f}$-like) outward K^+ current is detectable in these cells when extracellular Ca^{2+} is removed (Inoue and Imanaga, 1993). This unexpected finding suggests that the functional expression of $I_{to,f}$ (at least in guinea pig ventricles) is highly regulated.

The time- and voltage-dependent properties of $I_{to,f}$ in various cardiac cell types (Table 1) are similar in that activation, inactivation, and recovery from steady-state inactivation are all rapid (Litovsky and Antzelevitch, 1988; Tseng and Hoffman, 1989; Apkon and Nerbonne, 1991; Boyle and Nerbonne, 1992; Campbell *et al.*, 1993; Wettwer *et al.*, 1993, 1994; Konarzewska *et al.*, 1995; Yue *et al.*, 1996a; Xu *et al.*, 1999a). Although not characterized extensively, the single-channel correlates of $I_{to,f}$ (unitary conductance = 20–27 pS) in different cells are also similar (Benndorf and Nilius, 1988; Nakayama and Irisawa, 1985). The rapid recovery from steady-state inactivation ($\tau \approx 30$–50 ms) has been exploited to distinguish between $I_{to,f}$ and $I_{to,s}$ (Fig. 2) in cardiac myocytes (Brahmajothi *et al.*, 1999; Wang *et al.*, 1999; Xu *et al.*, 1999a). It has also been demonstrated that $I_{to,f}$ is selectively attenuated by Heteropodatoxin-2 and -3 (Xu *et al.*, 1999a; Guo *et al.*, 1999) in ventricular myocytes from rat (Sanguinetti *et al.*, 1997), ferret (Brahmajothi *et al.*, 1999), and mouse (Xu *et al.*, 1999a; Guo *et al.*, 1999). Heteropodatoxin-2 and -3, isolated from the venom of the spider *Heteropoda venatoria* (Sanguinetti *et al.*, 1997), can be used to distinguish $I_{to,f}$ and $I_{to,s}$, particularly when these currents are expressed in the same cells (Brahmajothi *et al.*, 1999; Xu *et al.*, 1999a; Guo *et al.*, 1999). The fact that the properties of $I_{to,f}$ in different cardiac cells are similar (Table 1) led to speculations that the molecular correlates of functional $I_{to,f}$ channels in different cell types and species are also the same (Barry and Nerbonne, 1996). All available evidence suggests that members of the Kv4 subfamily of α-subunits underlie the functional cardiac $I_{to,f}$ current (see Section 4.1.1). Nevertheless, there are some subtle differences in the properties of the currents referred to as $I_{to,f}$ (Table 1) in different cells. For example, comparison of rat atrial and ventricular $I_{to,f}$ reveals significant differences in rates of inactivation and recovery from steady-state inactivation (Apkon and Nerbonne, 1991; Boyle and Nerbonne, 1992). These differences likely reflect the contribution of distinct Kv4 α-subunits to rat atrial and ventricular $I_{to,f}$ (Bou-Abboud and Nerbonne, 1999; see Section 4.1.1).

2.1.2. $I_{to,slow}$ ($I_{to,s}$)

It has long been recognized that there are marked differences in the properties of the currents referred to as I_{to} in some cells, most notably in rabbit atrial and ventricular myocytes (Giles and Imaizumi, 1988; Giles *et al.*, 1996). In contrast to the rapid inactivation of transient outward currents in most cells (which have now been referred to as $I_{to,f}$; see Table 1 and Section 2.1.1), inactivation of I_{to} (I_t) in rabbit atrial and

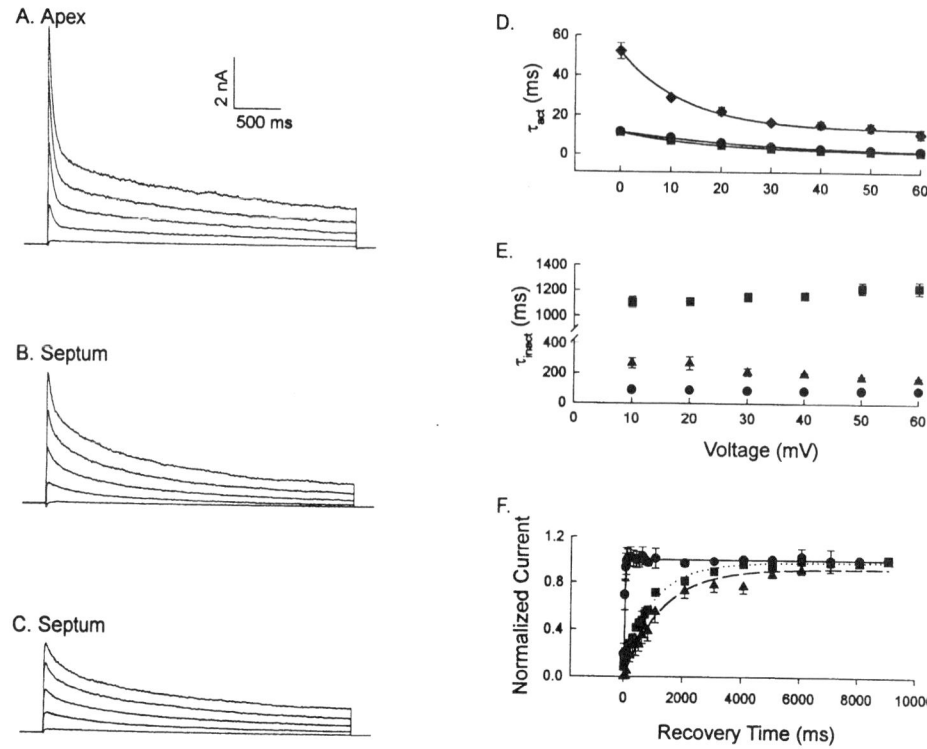

Figure 2. Distinct transient outward K^+ currents, $I_{to,f}$ and $I_{to,s}$, are expressed in mouse left ventricular myocytes. Whole-cell outward K^+ currents in isolated adult mouse left ventricular myocytes were evoked during 4.5 sec depolarizing voltage steps to potentials between -20 and $+60$ mV from a holding potential of -70 mV in 20 mV increments. The records displayed were obtained from cells isolated from the left ventricular apex (A) or septum (B, C). Note that the peak outward current amplitudes are larger in cells isolated from the apex (A), and the decay phases of the currents are slower in cells isolated from the septum (than apex). (D) The rates of activation of $I_{to,f}$ (●), $I_{to,s}$ (▲) and $I_{K,slow}$ (■) in mouse ventricular myocytes are similar, whereas I_{ss} (◆) activates more slowly. Mean $\pm$ SEM activation time constants, determined from single exponential fits to the rising phases of the separated current components, are plotted (as points) as a function of test potential, and the solid lines represent the best single exponential fits to the data. (E) The decay phases of the outward currents in records such as those in panels A–C were fitted to the sum of two or three exponentials, and mean $\pm$ SEM inactivation time constants for $I_{to,f}$ (●), $I_{to,s}$ (▲) and $I_{K,slow}$ (■) were determined and are plotted as a function of test potential; none of the inactivation time constants displays any appreciable voltage-dependence. (F) The rates of recovery of $I_{to,f}$, $I_{to,s}$ and $I_{K,slow}$ from steady-state inactivation are distinct. To determine the time constants of recovery from steady-state inactivation, cells were first depolarized to $+50$ mV for 5 sec (to inactivate the currents) and subsequently were hyperpolarized to -70 mV for varying times (to allow recovery) prior to test depolarizations to]50 mV (to assess the extent of recovery). The amplitudes of $I_{to,f}$, $I_{to,s}$ and $I_{K,slow}$ evoked at $+50$ mV following each recovery period were then determined, and normalized to the current amplitudes evoked following the 10 sec recovery period. Mean $\pm$ SEM normalized recovery data for $I_{to,f}$ (●), $I_{to,s}$ (▲) and $I_{K,slow}$ (■) are plotted. As is evident, $I_{to,f}$ recovers rapidly, whereas $I_{to,s}$ and $I_{K,slow}$ recover slowly, from steady-state inactivation (see Xu *et al.*, 1999a and Guo *et al.*, 1999).

ventricular myocytes is slow, and recovery from steady-state inactivation of I_{to} in these cells is very slow, with complete recovery requiring seconds (Giles and Imaizumi, 1988; Fedida and Giles, 1991; Wang *et al.*, 1999). In mouse ventricular myocytes, two distinct transient outward K^+ currents have been identified: $I_{to,f}$, as described earlier, and a second, more slowly decaying transient K^+ current referred to as $I_{to,slow}$ or $I_{to,s}$ (Xu *et al.*, 1999a). In addition, mouse ventricular $I_{to,s}$ recovers from steady-state inactivation much more slowly ($\tau_{recovery} \approx 1$ s) than $I_{to,f}$ ($\tau_{recovery} \approx 50$ ms) and is insensitive to the Heteropodatoxin-2 and -3 (Xu *et al.*, 1999a; Guo *et al.*, 1999). In addition, in mouse left ventricle, $I_{to,f}$ and $I_{to,s}$ are differentially distributed (Xu *et al.*, 1999a; Guo *et al.*, 1999). In cells isolated from the apex of the left ventricle, for example, only $I_{to,f}$ (Fig. 2) is present, whereas cells isolated from the left ventricular septum express either $I_{to,s}$ alone ($\sim 20\%$ of the cells) or both $I_{to,f}$ and $I_{to,s}$ ($\sim 80\%$ of the cells) (Xu *et al.*, 1999a; Guo *et al.*, 1999). In addition, $I_{to,f}$ density is significantly ($p < .001$) higher in myocytes derived from the apex versus those from the septum (Xu *et al.*, 1999a; Guo *et al.*, 1999). As discussed in Section 4.1, the molecular correlates of mouse ventricular $I_{to,s}$ and $I_{to,s}$ are distinct (Barry *et al.*, 1998; Guo *et al.*, 1999).

In myocytes from the ferret left ventricle, the rates of inactivation and recovery from steady-state inactivation are significantly slower in endocardial than in epicardial cells (Brahmajothi *et al.*, 1999). These observations were interpreted as suggesting that distinct K^+ channel subunits underlie I_{to} in ferret left ventricular epicardial and endocardial cells (see also below). Although experiments aimed at testing this hypothesis directly have not been completed to date, there is experimental evidence demonstrating regional differences in Kv α-subunit expression patterns in ferret heart (Brahmajothi *et al.*, 1999). Importantly, the time- and voltage-dependent properties and the pharmacological sensitivities of the transient outward K^+ currents in ferret epicardial and endocardial left ventricular myocytes are very similar to those of mouse ventricular $I_{to,f}$ and $I_{to,s}$, respectively. It seems appropriate, therefore, to also refer to the ferret ventricular K^+ currents by the same names (i.e., as $I_{to,f}$ and $I_{to,s}$) (Table 1). The properties of I_{to} (I_t) in rabbit myocytes are also similar to those of mouse ventricular $I_{to,s}$, suggesting that the transient currents in rabbit should also be termed $I_{to,s}$ (Table 1). Consistent with this suggestion, the molecular correlates of the slow transient K^+ current in rabbit atria and $I_{to,s}$ in mouse ventricle appear to be the same (Wang *et al.*, 1999; Guo *et al.*, 1999; see also Section 4.1.2).

2.2. Delayed Rectifier K^+ Currents/Channels, I_K, in Cardiac Cells

In addition to the transient outward currents, delayed rectifier K^+ currents, I_K, have been characterized extensively in myocytes isolated from canine (Tseng and Hoffman, 1989; Liu and Antzelevitch, 1995; Yue *et al.*, 1996a,b), feline (Follmer and Golatsky, 1990; Furakawa *et al.*, 1992), guinea pig (Hume and Uehara, 1985; Balser *et al.*, 1990; Horie *et al.*, 1990; Sanguinetti and Jurkiewicz, 1990, 1991; Walsh *et al.*, 1991; Anumonwo *et al.*, 1992; Freeman and Kass, 1993), human (Wang *et al.*, 1994; Li *et al.*, 1996), mouse (Fiset *et al.*, 1998; Zhou *et al.*, 1998; Xu *et al.*, 1999a,b; Guo *et al.*, 1999), rabbit (Shibasaki, 1987; Veldkamp *et al.*, 1993), and rat (Apkon and Nerbonne, 1991; Boyle and Nerbonne, 1992; Pond *et al.*, 2000) hearts. In most cells, multiple components of I_K (Table 1), with distinct kinetic and voltage-dependent properties and pharmacological sensitivities, are evident.

2.2.1. $I_{K,rapid}$ (I_{Kr}) and $I_{K,slow}$ (I_{Ks})

Two prominent components of I_K, I_{Kr} ($I_{K,rapid}$) and I_{Ks} ($I_{K,slow}$), have been distinguished in guinea pig atrial and ventricular myocytes (Sanguinetti and Jurkiewicz 1990, 1991, 1992). I_{Kr} activates rapidly, inactivates very rapidly, and displays marked inward rectification. In addition, I_{Kr} is selectively blocked by lanthanum and by several class III antiarrhythmics, including dofetilide, E-4031, and sotalol (Sanguinetti and Jurkiewicz, 1990, 1991). No inward rectification is evident for I_{Ks}, and this conductance pathway is selectively attenuated by several class III antiarrhythmics, such as NE-10064 and NE-10133, that do not appear to affect I_{Kr} appreciably (Busch et al., 1994). Similar currents have been described in atrial and ventricular myocytes isolated from human (Wang et al., 1993a, 1994; Li et al., 1996) and canine (Liu and Antzelevitch, 1995; Yue et al., 1996a) hearts and in rabbit ventricular cells (Veldkamp et al., 1993; Salata et al., 1996). In some myocardial cells, however, only I_{Kr} or I_{Ks} is expressed. In guinea pig nodal cells, for example, only I_{Ks} is evident (Anumonwo et. al., 1992), whereas in feline (Follmer and Colatsky, 1990) and rat (Pond et al., 2000) ventricular myocytes and in rat atrial (Pond et al., 2000) and rabbit nodal (Shibasaki, 1987) cells, only I_{Kr} is detected.

In symmetrical K^+, the single-channel conductances of I_{Kr} and I_{Ks} channels are 10–13 and 3–5 pS, respectively (Hume and Uehara, 1985; Shibasaki, 1987; Balser et al., 1990; Veldkamp et al., 1993). These observations led to suggestions that I_{Kr} and I_{Ks} reflect the expression of distinct molecular entities (Barry and Nerbonne, 1996; Deal et al., 1996). Considerable evidence has now been provided in support of this hypothesis (see Sections 4.2.1 and 4.2.2).

2.2.2. $I_{K,ultrarapid}$ (I_{Kur})

In guinea pig ventricular myocytes, a novel, 12–14-pS, K^+-selective channel that is distinct from I_{Kr} and I_{Ks} and referred to as I_{Kp} has also been reported (Yue and Marban, 1988). I_{Kp} activates very rapidly on depolarization, does not inactivate, and is sensitive to millimolar concentrations of Ba^{2+} (Backx and Marban, 1993). In rat (Boyle and Nerbonne, 1991, 1992), human (Wang et al., 1993a,b), and dog (Yue et al., 1996a,b) atrial myocytes, rapidly activating, noninactivating K^+ currents with properties similar to guinea pig I_{Kp} have also been described, although these currents were given different names: I_{ss} (steady-state), I_{sus} (sustained), or I_{Kur} (ultrarapid). The similarities in the properties of the currents characterized in rat, human, and canine atrial myocytes, together with experimental findings demonstrating that the molecular correlates of the rat and human currents are the same (Feng et al., 1997; Bou-Abboud and Nerbonne, 1999; see Section 4.2.3), suggest that these currents should all be referred to by the same name, I_{Kur} (Table 1). The fact that the properties of guinea pig I_{Kp} are very similar to those of the dog, human, and rat atrial currents further suggests that I_{Kp} should probably also be referred to as I_{Kur}.

2.2.3. Other Delayed Rectifiers

In several cardiac cell types, there are additional components of I_K with properties different from those of I_{Ks}, I_{Kr}, and I_{Kur} (Table 1), and these are assumed to reflect the expression of distinct molecular entities (Barry and Nerbonne, 1996; Nerbonne, 1998). In mouse ventricular and atrial myocytes, for example, two additional voltage-gated K^+ currents, referred to as $I_{K,slow}$ and I_{ss} (steady-state), have been identified (London *et al.*, 1998b; Fiset *et al.*, 1998; Zhou *et al.*, 1998; Xu *et al.*, 1999a,b). $I_{K,slow}$ is a rapidly activating and slowly inactivating K^+ current with kinetic and pharmacological properties distinct from those of $I_{to,f}$ and $I_{to,s}$ and I_{ss}. In addition, $I_{K,slow}$ is blocked effectively and selectively by micromolar concentrations of 4-AP (London *et al.* 1998b; Fiset *et al.*, 1998; Zhou *et al.*, 1998) which do not affect $I_{to,f}$ or $I_{to,s}$ in the same cells (Xu *et al.*, 1999a,b). In contrast, the current, I_{ss}, remaining at the end of long (up to 10 s) depolarizing voltage steps is slowly activating and 4-AP-insensitive (Xu *et al.*, 1999a,b). In contrast to the differential distribution of $I_{to,f}$ and $I_{to,s}$ (see Sections 2.1 and 2.3), however, $I_{K,slow}$ and I_{ss} appear to be expressed in all mouse ventricular and atrial myocytes (Xu *et al.*, 1999a,b; Guo *et al.*, 1999).

In rat ventricular and atrial myocytes, there are also novel delayed rectifier K^+ currents that have been referred to as I_K and $I_{K,slow}$, respectively (Table 1) (Apkon and Nerbonne, 1991; Boyle and Nerbonne, 1991, 1992; Xu *et al.*, 1996; Bou-Abboud and Nerbonne, 1999). In addition to differences in inactivation kinetics, I_K and $I_{K,slow}$ have distinct pharmacological properties: I_K, for example, is selectively blocked by tetraethylammonium (Apkon and Nerbonne, 1991), whereas $I_{K,slow}$ is blocked by millimolar concentrations of 4-AP (Boyle and Nerbonne, 1991, 1992) and nanomolar concentrations of dendrotoxin (Van Wagoner *et al.*, 1996). Interestingly, rat atrial $I_{K,slow}$ closely resembles the 4-AP- and dendrotoxin-sensitive, rapidly activating, slowly inactivating K^+ current I_D in mammalian central neurons (Nerbonne, 1998) and is distinct from $I_{to,f}$ and I_{Kur} in the same cells (Bou-Abboud and Nerbonne, 1999; see Section 4.2.3). However, rat atrial $I_{K,slow}$ and mouse (atrial and ventricular) $I_{K,slow}$ have very different properties, and these should not be considered identical K^+ conductance pathways. Rat atrial $I_{K,slow}$ is sensitive to dendrotoxin (Van Wagoner *et al.*, 1996), whereas the currents in mouse cardiac myocytes are not (Xu *et al.*, 1999a). Conversely, mouse $I_{K,slow}$ is blocked effectively by micromolar concentrations of 4-AP, and rat $I_{K,slow}$ is not (Boyle and Nerbonne, 1992; Fiset *et al.*, 1998; London *et al.*, 1998b; Xu *et al.*, 1999a). For this reason, it would seem appropriate to refer to the slowly inactivating current in rat atrial myocytes as $I_{K,DTX}$ (Table 1) to distinguish it from mouse $I_{K,slow}$. Once the α- and β-subunits contributing to the various types of channels are determined, it would be preferable to change the channel nomenclature to reflect the molecular composition(s) of the channels, rather than the kinetic or the pharmacological properties of the currents, as has been done in the past.

2.3. Regional Differences in Voltage-Gated K^+ Channel Expression

Although the properties of the currents referred to as $I_{to,f}$ in different cell types and species (Table 1) are similar, there is variability in $I_{to,f}$ densities among different myocardial cell types. In rat and human myocytes, for example, $I_{to,f}$ density is higher in atrial than in ventricular myocytes (Boyle and Nerbonne, 1991, 1992; Varro et al., 1993), whereas in the mouse, mean $I_{to,f}$ density is higher in ventricular than in atrial cells (Xu et al., 1999b). The density of $I_{to,f}$ also varies in different regions of the ventricle in canine (Litovsky and Antzelevitch, 1988; Liu et al., 1993), cat (Furukawa et al., 1992), ferret (Brahmajothi et al., 1999), human (Wettwer et al., 1994; Konarzewska et al., 1995), mouse (Xu et al., 1999a; Guo et al., 1999), and rat (Clark et al., 1993) heart. In canine left ventricle, for example, $I_{to,f}$ density is 5- to 6-fold higher in epicardial and midmyocardial than in endocardial cells (Liu et al., 1993). $I_{to,f}$ density is also reportedly higher in guinea pig atrial epicardium than endocardium (Wang et al., 1991). Although $I_{to,s}$ density in rabbit myocytes is significantly higher in atrial than in ventricular myocytes (Giles and Imaizumi, 1988), $I_{to,s}$ is not expressed in mouse atrial myocytes (Xu et al., 1999b). Similar to $I_{to,f}$ density, however, the density of $I_{to,s}$ does vary markedly in both mouse and ferret ventricles (Brahmajothi et al., 1999; Xu et al., 1999a; Guo et al., 1999). For example, $I_{to,s}$ is undetectable in ferret epicardial (Brahmajothi et al., 1999) and mouse apex (Xu et al., 1999a; Guo et al., 1999) left ventricular cells.

The densities of I_{Ks} and I_{Kr} vary in different myocardial cell types. In guinea pig, for example, I_{Kr} and I_{Ks} densities are 2-fold higher in atrial than in ventricular myocytes (Sanguinetti and Jurkiewicz, 1991). In dog heart, I_{Ks} density is higher in epicardial and endocardial cells than in M-cells (Liu and Antzelevitch, 1995). Although M-cells have not been identified in the guinea pig left ventricle, there are regional differences in I_{Kr} and I_{Ks} expression (Bryant et al., 1998; Main et al., 1998). In cells isolated from the guinea pig left ventricular free wall, for example, the density of I_{Kr} is reportedly higher in subepicardial than in midmyocardial or subendocardial myocytes (Main et al., 1998). In contrast, at the base of the left ventricle, both I_{Kr} and I_{Ks} densities are significantly lower in endocardial than in midmyocardial or epicardial cells. It seems certain that differences in the densities of the various Kv currents contribute to cell-type-specific and regional variations in action potential waveforms (Fig. 1) (Litovsky and Antzelevitch, 1988; Wang et al., 1991; Liu and Antzelevitch, 1995; Antzelevitch et al., 1994; Campbell et al., 1995; Giles et al., 1996; Barry and Nerbonne, 1996; Bryant et al., 1998; Main et al., 1998; Nerbonne, 1998).

3. MOLECULAR DIVERSITY OF Kv CHANNELS IN CARDIAC CELLS

3.1. Voltage-Gated K$^+$ Channel Pore-Forming α-Subunits

3.1.1. Homologous Voltage-Gated (Kv) Subfamilies of α-Subunits

The first Kv α-subunit was cloned from the Shaker locus in Drosophila (Papazian et al., 1987; Kamb et al., 1988; Pongs et al., 1988). Sequence analysis revealed a protein with six transmembrane domains and a region between the fifth and sixth transmembrane domains (Fig. 3A) that contributes to the K$^+$-selective pore (Jan and Jan, 1992; Pongs, 1992). Heterologous expression of Shaker in Xenopus oocytes reveals Kv channel currents (Fig. 3B) similar to the rapidly activating and inactivating I_{to} in wild-type Drosophila muscle and nerve (Timpe et al., 1988). Three homologous Kv

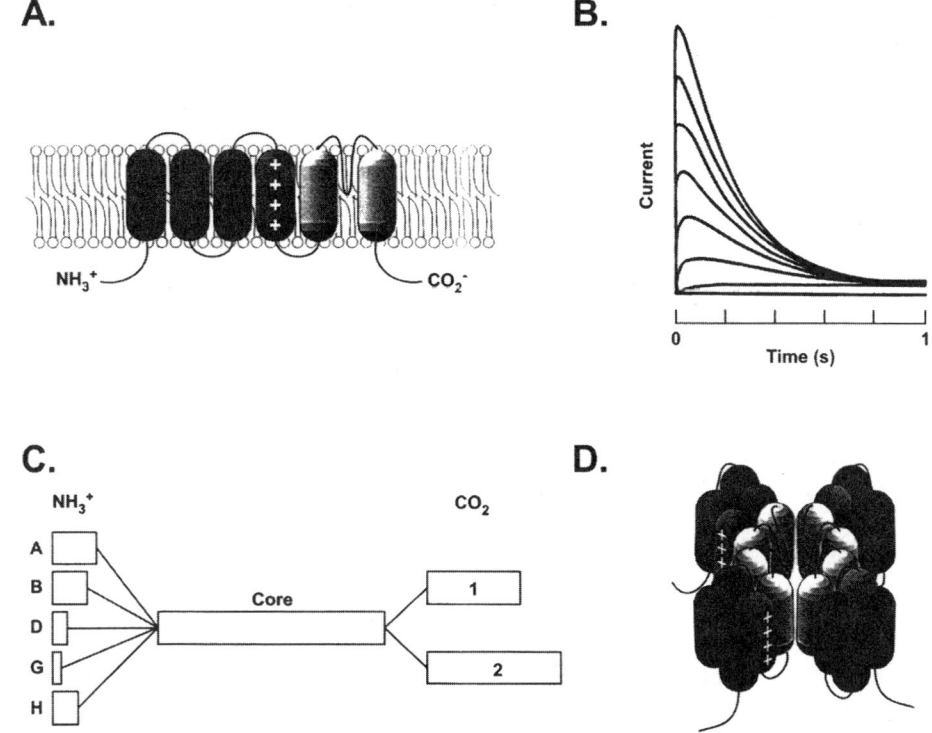

Figure 3. Structure, expression and assembly of voltage-gated K^+ channel pore forming (α) subunits. (A) *Shaker* is an integral membrane protein with six transmembrane domains, intracellular N- and C-termini and a positively charged S4, placing it in the S4 superfamily of voltage-gated ion channels (see Papazian *et al.*, 1987; Kamb *et al.*, 1988; Pongs *et al.*, 1988). (B) Schematic of outward K^+ current waveforms produced on heterologous expression of *Shaker* (see Timpe *et al.*, 1988). The currents activate and inactivate rapidly on membrane depolarization, and outward K^+ currents with distinct time- and voltage-dependent properties are also observed on heterologous expression of the *Shab*, *Shaw* or *Shal* (see Butler *et al.*, 1989; Wei *et al.*, 1990). (C) Alternative splicing of *Shaker* (as well as other Kv α subunits) gives rise to proteins with novel N- and C- termini and currents with distinct properties (see Schwarz *et al.*, 1988; Wei *et al.*, 1990). (D) Schematic of a voltage-gated K^+ channel illustrating four Kv α-subunits contributing to the K^+ selective pore (see MacKinnon, 1991).

α-subunit subfamilies, referred to as *Shab*, *Shaw*, and *Shal*, were subsequently cloned from *Drosophila* (Butler *et al.*, 1989; Wei *et al.*, 1990). A number of vertebrate homologs have been identified. K^+ channel α-subunit genes of the *Shaker*, *Shab*, *Shaw*, and *Shal* subfamilies (Chandy and Gutman, 1995) are referred to as Kv1.*x*, Kv2.*x*, Kv3.*x*, and Kv4.*x*, respectively (Table 2). In addition to the four subfamilies, the main mechanism for generating functional K^+ channel diversity in *Drosophila* is through alternative splicing of transcripts (Schwarz *et al.*, 1988) (Fig. 3C). Although alternative splicing also occurs for some vertebrate Kv α-subunits, considerably more diversity results from the presence of multiple members of each subfamily (Table 2).

The positively charged fourth transmembrane domain in the Kv α-subunits (Fig. 3A) is homologous to the corresponding region in voltage-gated Na^+ and Ca^{2+}

Table 2
Kvα Subunits and Voltage-Gated Myocardial K⁺ Currents/Channels[a]

Family	Subunit[b]	Activation	Inactivation	Blocker[b]	Endogenous current
Kv1					
(Shaker)	Kv 1.1				
	Kv 1.2	Fast	Very slow	4-AP, DTX	$I_{K,slow}$ (rat) ($I_{K,DTX}$)
	Kv 1.3				
	Kv 1.4	Fast	Slow	4-AP	$I_{to,s}$
	Kv 1.5	Fast	No	4-AP	I_{Kur} $I_{K,slow}$ (mouse)
	Kv 1.6				
	Kv 1.7	Fast	Fast	NTX	??
	Kv 1.8				
Kv2					
(Shab)	Kv 2.1	Slow	Very slow	TEA, HaTx	$I_{K,slow}$ (mouse)
	Kv 2.2	Slow	Very slow	TEA	??
Kv3					
(Shaw)	Kv 3.1				
	Kv 3.2				
	Kv 3.3				
	Kv 3.4				
Kv4					
(Shal)	Kv 4.1				
	Kv 4.2	Fast	Fast	4-AP, HaTx, HpTx	$I_{to,f}$
	Kv 4.3	Fast	Fast	4-AP, HpTx	$I_{to,f}$
Kv5–9					
eag family					
eag	eag				
elk	elk				
erg1	erg1	Moderate	Fast	E-4031	I_{Kr} (with mirp)
	erg2				
	erg3				
KvLQTl family					
KCNQ1	KvLQT1	Very slow	Very, very slow	NE-10064	I_{Ks} (with mink)
KCNQ2					
KCNQ3					

[a] Adapted from J. M. Nerbonne, *J. Neurobiol.* **37**:37–59 (1998).

[b] Boxes indicate cardiac expression.

[c] Abbreviations: 4-AP, 4-Aminapyridine; DTX, dendrotoxin; NTX, noxiustoxin; TEA, tetraethylammonium; HaTX, hanatoxin; HpTX, heteropodatoxin.

channels, placing these genes in the "S4" superfamily of voltage-gated channels (Jan and Jan, 1992; Pongs, 1992). In contrast to these other channels, Kv channels are comprised of four α-subunits (Fig. 3D) (MacKinnon, 1991). In addition to the multiplicity of Kv α-subunits and alternative splicing, further diversity could arise, therefore, through the formation of heteromultimeric channels between two or more types of Kv α-subunit proteins in the same subfamily (Covarrubias *et al.*, 1991). At present, it is not known if

Kv α-subfamily members coassemble *in vivo* to form functional heteromeric voltage-gated K$^+$ channels in the mammalian heart.

Several additional Kv α-subunit subfamilies, termed Kv5.*x* through Kv9.*x* (Table 2), have been identified (Drewe *et al.*, 1992; Hugnot *et al.*, 1996; Salinas *et al.*, 1997), although heterologous expression of these (Kv5.1, Kv6.1, Kv8.1, or Kv9.1) subunits alone does not yield a functional channel (Drewe *et al.*, 1992; Hugnot *et al.*, 1996; Castellano *et al.*, 1997; Salinas *et al.*, 1997). Interestingly, however, coinjection of any of these (Kv5.1, Kv6.1, Kv8.1, or Kv9.1) subunits with *Shab* (Kv2.*x*) subfamily members attenuates the amplitudes of the *Shab-* (Kv2.*x-*) induced currents (Salinas *et al.*, 1997). These observations, together with sequence similarities (with the Kv2.*x* subfamily), suggest that Kv5.1, Kv6.1, Kv8.1, and Kv9.1 may more appropriately be considered regulatory subunits of the Kv2.*x* subfamily (Castellano *et al.*, 1997), rather than distinct subfamilies (Salinas *et al.*, 1997). It has been reported that the Kv5.1 message is detectable in isolated ferret left ventricular myocytes (Brahmajothi *et al.*, 1996), although the role of this subunit in the generation of functional cardiac K$^+$ channels has not been explored directly to date. In addition, it has not been demonstrated experimentally that any of the other (Kv6, Kv8, or Kv9) subfamilies play a role in the generation of functional Kv channels in cardiac (or other) cells.

3.1.2. Ether-a-go-go-Related Gene Subfamily, ERG

Another subfamily of the S4 superfamily of voltage-gated K$^+$ channel α-subunit genes was revealed with the cloning of the *Drosophila ether-a-go-go* (*eag*) locus (Warmke *et al.*, 1991). Although the overall amino acid identity is only 10–15%, the S4 and pore regions of the Kv α-subunits and *eag* are homologous (Warmke *et al.*, 1991). Furthermore, the predicted membrane topology of the *eag* protein is similar to that of the Kv α-subunits (Fig. 3A). In addition, *eag* is also homologous to cyclic-nucleotide-gated channels, which are nonselective cation channels (Guy *et al.*, 1991). Subsequent homology screening of a human hippocampal cDNA library led to the cloning of human *eag*, as well as the *human eag-r*elated *g*ene, referred to as HERG (Warmke and Ganetzky, 1994). HERG was subsequently identified as the locus of mutations leading to one form of familial long QT syndrome, LQT2 (Curran *et al.*, 1995). Expression of HERG (human ERG1) in *Xenopus* oocytes reveals inwardly rectifying voltage-gated, K$^+$-selective currents (Sanguinetti *et al.*, 1995; Trudeau *et al.*, 1995) with properties similar to those of cardiac I_{Kr} (see Section 4.2.1). Additional members of this subfamily (ERG2 and ERG3) have been cloned from brain (Table 2), although these appear to be nervous system-specific and are not expressed in heart (Shi *et al.*, 1997). Alternatively processed forms of HERG1 and MERG1, the mouse homolog of HERG1, have been cloned from mouse and human heart cDNA libraries and are postulated to contribute to cardiac I_{Kr} (Lees-Miller *et al.*, 1997; London *et al.*, 1997, 1998c; Kupershmidt *et al.*, 1998). Biochemical evidence in support of this hypothesis, however, is lacking (Pond *et al.*, 2000; see Section 4.2.1).

3.1.3. KvLQT1 Subfamily

A positional cloning strategy led to the identification of another subfamily of voltage-gated K$^+$ channel α-subunits, and the first member of this subfamily was referred to as KvLQT1 (Wang *et al.*, 1996). Because mutations in this gene underlie another inherited form of long QT syndrome (LQT1), it is also referred to as KCNQ1 (Table 2). Heterologous expression of KvLQT1 reveals rapidly activating and noninactivating K^+ currents (Barhanin *et al.*, 1996; Sanguinetti *et al.*, 1996). However, coexpression of KvLQT1 with minK (I_{sK}) (Section 3.2.1) produces slowly activating K^+ currents that resemble cardiac I_{Ks} (Barhanin *et al.*, 1996; Sanguinetti *et al.*, 1996). This suggests that functional I_{Ks} channels are heteromers (Fig. 4A), comprised of minK and KvLQT1 (Barhanin *et al.*, 1996; Sanguinetti *et al.*, 1996; Section 4.2.1).

Additional members of the KvLQT (KCNQ) subfamily, termed KCNQ2 and KCNQ3, have been cloned from brain (Biervert *et al.*, 1998; Wang *et al.*, 1998; Schroeder *et al.*, 1998). Expression of either of these subunits in *Xenopus* oocytes produces slowly activating, noninactivating K^+-selective currents that deactivate very slowly on membrane repolarization (Schroeder *et al.*, 1998; Wang *et al.*, 1998). The unique kinetic and pharmacological properties of the expressed currents led Wang *et al.* (1998) to suggest that KCNQ2/KCNQ3 are the molecular correlates of functional neuronal M-channels. Interestingly, the densities of the currents are significantly higher

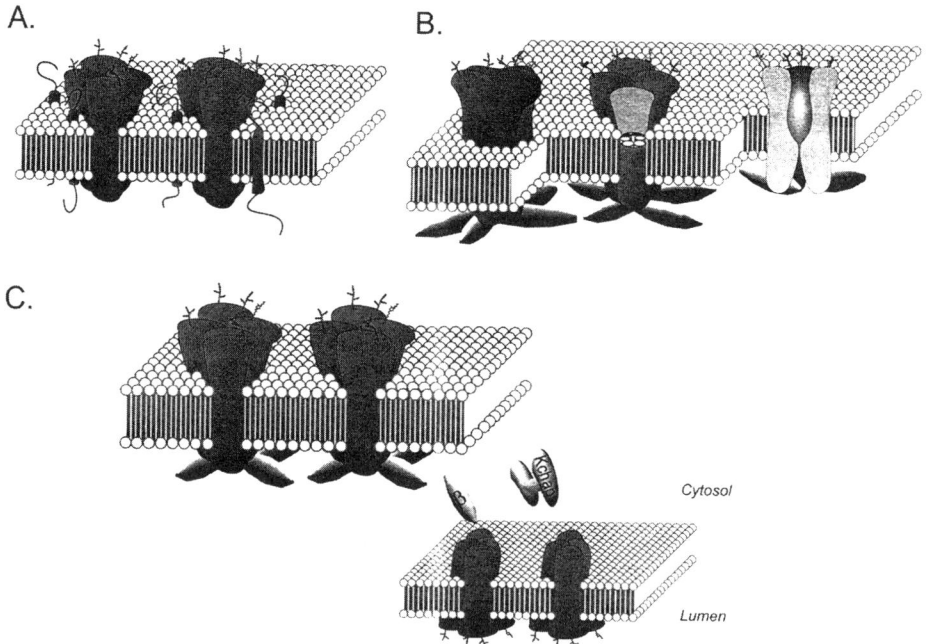

Figure 4. Accessory subunits and the formation of functional voltage-gated K^+ channels. Schematic representation of the putative membrane topology and subunit composition of voltage-gated K^+ channels with minK (A) or accessory β (B) subunits. Although the Kv α and β-subunits are assumed to associate in 1:1 ratios, the stoichiometry of minK- and KvLQT1-(or other α subunits) containing K^+ channels has not been defined. (C) Schematic representation of voltage-gated K^+ biogenesis illustrating channel biosynthesis, assembly and post-translational processing in the endoplasmic reticulum. Accessory β and regulatory KChAPs subunits appear to play roles in mediating the biochemical maturation of functional voltage-gated K^+ channels, as well as in regulating the cell surface cell expression of these channels (see Fink *et al.*, 1996; Shi *et al.*, 1996; Accili *et al.*, 1997a,b; Wible *et al.*, 1998).

when KCNQ2 and KCNQ3 are coexpressed than when either subunit is expressed alone. This implies that functional M-channels are also heteromultimeric (Wang *et al.*, 1998). It is also of interest to note that KCNQ2 and KCNQ3 have been identified as loci of mutations that lead to benign familial neonatal convulsions (Biervert *et al.*, 1998; Schroeder *et al.*, 1998). Neither KCNQ2 nor KCNQ3 contributes to the formation of functional K^+ channels in cardiac cells, however, because neither subunit appears to be expressed in the heart (Schroeder *et al.*, 1998; Wang *et al.*, 1998).

3.2. Accessory Subunits of Voltage-Gated K^+ Channels

3.2.1. Minimal K^+ Channel Subunit, I_{sK} or minK

Another mechanism for generating further functional voltage-gated K^+ channel diversity was identified with the molecular cloning of I_{sK} (or minK), which encodes a protein of 130 amino acids with a single membrane-spanning domain (Murai *et al.*, 1989; Folander *et al.*, 1991; Lesage *et al.*, 1992). Early studies revealed that heterologous expression of I_{sK} in *Xenopus* oocytes resulted in Kv currents (Folander *et al.*, 1990). The properties of the currents were affected by mutations in the transmembrane domain of I_{sK} (Goldstein and Miller, 1991; Takumi *et al.*, 1991; Wang *et al.*, 1996), suggesting that I_{sK} alone forms functional Kv channels. However, subsequent studies in cells other than *Xenopus* oocytes failed to reveal functional voltage-gated K^+ channels when I_{sK} was expressed alone. Resolution of this seeming controversy was obtained with the demonstration that the K^+ channels formed in oocytes injected with cRNA encoding I_{sK} reflect coassembly (of I_{sK}) with the *Xenopus* homolog of KvLQT1 (Sanguinetti *et al.*, 1996). Thus, I_{sK} does not, by itself, form functional Kv channels; rather, it coassembles with KvLQT1 to form functional cardiac I_{Ks} channels (Fig. 4A) (Barhanin *et al.*, 1996; Sanguinetti *et al.*, 1996; see Section 4.2.2).

3.2.2. Accessory β-Subunits

Another type of Kv channel accessory subunit, the low-molecular-weight (~ 45 kDa) accessory β-subunits, was initially identified biochemically (Muniz *et al.*, 1992) and subsequently cloned (Rettig *et al.*, 1994). Three homologous Kv β-subunits, now termed Kvβ1, Kvβ2, and Kvβ3, as well as alternatively spliced transcripts of Kvβ1.x, were cloned from heart and brain cDNA libraries (Castellino *et al.*, 1995; England *et al.*, 1995a,b; Majumder *et al.*, 1995; Morales *et al.*, 1995; Deal *et al.*, 1996). Kvβ1.2 and Kvβ1.3 are found in heart (Deal *et al.*, 1996). In contrast to I_{sK}, the amino acid sequences of the Kvβ-subunits suggest that they are cytosolic proteins that interact with the intracellular domains of the Kv1 α-subunits in assembled voltage-gated K^+ channels (Fig. 4B) (Rettig *et al.*, 1994; Castellino *et al.*, 1995; England *et al.*, 1995a,b; Majumder *et al.*, 1995; Morales *et al.*, 1995; Deal *et al.*, 1996). Functional studies in heterologous systems in which various Kv1.x α- and β-subunits have been expressed demonstrate clearly that β-subunit coexpression can affect the time- and voltage-dependent properties of the resulting currents (Castellino *et al.*, 1995; England *et al.*, 1995a,b; Majumder *et al.*, 1995; Morales *et al.*, 1995; Accili *et al.*, 1997a,b). Coexpression of accessory β-subunits can also increase the functional expression of Kv channels on the cell surface (Shi *et al.*, 1996; Accili *et al.*, 1997a,b). Because Kv α- and β-subunits coassemble in the endoplasmic reticulum (Nagaya and Papzian, 1997), the increase in functional channel expression suggests that β-subunits affect channel assembly, processing, and stability. β-Subunits may function as chaperone proteins (Fig. 4C). Homology screening of a brain cDNA library with a probe corresponding to Kvβ2 subsequently identified Kvβ4. This, subunit increases the functional expression of Kv2.x channels (Fink *et al.*, 1996). Although the Kvβ4 message is not detected in the heart (Fink *et al.*, 1996), there may be additional, Kv β-subunits that are selective for the cardiac Kv2.x subfamily.

Heterologous coexpression experiments have demonstrated that the effects of the Kv β-subunits are subfamily specific; i.e., Kvβ1, Kvβ2, and Kvβ3 interact only with the Kv1 subfamily of α-subunits (Nakahira *et al.*, 1996; Sewing *et al.*, 1996), whereas Kvβ4 is specific for the Kv2 subfamily (Fink *et al.*, 1996). By analogy to the specificity of the interactions between Kv α-subunits (Covarrubias *et al.*, 1991) and between Kv α- and β-subunits (Fink *et al.*, 1996; Nakahira *et al.*, 1996; Sewing *et al.*, 1996), it seems reasonable to suggest that there may be additional Kv β-subunit subfamilies, (i.e., specific for Kv3, ERG, etc). The role of the β-subunits in the generation of functional cardiac K^+ channels is unclear at present, primarily because it is unknown which Kv1.x α-subunit(s) and cardiac β-subunits are associated *in vivo*.

3.2.3. K^+ Channel Regulatory Proteins

A novel K^+ channel regulatory protein, KChAP (for voltage-gated K^+ channel accessory protein), was identified in heart using a yeast two-hybrid screen designed to identify cytoplasmic regulatory proteins of Kv α-subunit-containing K^+ channels (Wible *et al.*, 1998). Sequence analysis of KChAP revealed a 574-amino-acid-protein with no transmembrane domains and no homology to Kv α or β-subunits (Wible *et al.*, 1998). Coexpression of KChAP with Kv2.1 (or Kv2.2) in *Xenopus* oocytes, however, markedly increases functional Kv2.x-induced current densities without measurably affecting the time- or the voltage-dependent properties of the currents (Wible *et al.*, 1998). Similar to Kv β-subunits (see Section 3.2.2), therefore, KChAP influences functional Kv2.x channel cell-surface expression, suggesting that KChAP functions as a chaperone protein (Fig. 4C). KChAP interacts with the amino termini of Kv1.x α-subunits (as well as with the carboxy termini of Kvβ1.x subunits). However, KChAP coexpression had no effect on functional Kv1.5-induced currents in *Xenopus* oocytes (Wible *et al.*, 1998). These results suggest either that KChAP and Kv1.5 do not associate in oocytes or that the interaction between KChAP and Kv1.5 is ineffective in influencing channel function. Biochemical experiments, using antibodies directed against KChAPs and Kv α-subunits, will be required to define the *in vivo* partners of KChAP.

4. MOLECULAR CORRELATES OF FUNCTIONAL Kv CHANNELS

Considerable progress has been made in understanding the relationships between the various subunits and the functional voltage-gated K^+ channels expressed in cardiac cells (Table 2).

4.1. Molecular Correlates of Cardiac Transient Outward K^+ Currents (I_{to})

Heterologous expression of three of the Kv channel α-subunits cloned from heart and brain (Table 2), Kv1.4 (Tseng-Crank *et al.*, 1990; Roberds and Tamkun, 1991a; Tamkun *et al.*, 1991; Po *et al.*, 1992; Comer *et al.*, 1994), Kv4.2 (Baldwin *et al.*, 1991; Blair *et al.*, 1991; Roberds and Tamkun, 1991a), and Kv4.3 (Dixon *et al.*, 1996), reveals rapidly activating, 4-AP-sensitive K^+ currents that qualitatively resemble cardiac I_{to}. Because Kv1.4 was cloned first and from several species (Tseng-Crank *et al.*, 1990; Tamkun *et al.*, 1991; Po *et al.*, 1992; Comer *et al.*, 1994), this subunit was considered a likely candidate for the $I_{to,f}$, either as a homomultimer (Tamkun *et al.*, 1991; Tseng-Crank *et al.*, 1991; Murray *et al.*, 1994) or as a heteromultimer with other Kv1.*x* subunits (Po *et al.*, 1993; Roberds *et al.*, 1993) (Table 1; Section 2.1.1). Subsequently, the finding that mRNA levels for Kv4.2 vary through the thickness of the rat ventricular wall led Dixon and McKinnon (1994) to postulate that Kv4.2, rather than Kv1.4, underlies rat ventricular $I_{to,f}$. Consistent with this hypothesis, Kv4.2 protein expression in rat heart is high, whereas Kv1.4 is barely detectable (Barry *et al.*, 1995; Xu *et al.*, 1996). Experimental evidence demonstrating directly that Kv4 α-subunits underlie cardiac $I_{to,f}$ has recently been provided (see Section 4.1.1). Nonetheless, several studies also suggest a role for Kv1.4, specifically in the generation of $I_{to,s}$ (see Section 4.1.2).

4.1.1. Kv4.2/Kv4.3 α-Subunits Underlie $I_{to,f}$

Fiset *et al.* (1997) reported reductions in $I_{to,f}$ density ($\sim 50\%$) in rat ventricular myocytes exposed to antisense oligodeoxynucleotides (AsODNs) targeted against Kv4.2 or Kv4.3. In contrast, an AsODN targeted against the translation start site of Kv4.2, in rat atrial myocytes, attenuated $I_{to,f}$ selectively, whereas an AsODN against Kv4.3 was without effect (Bou-Abboud and Nerbonne, 1999). These findings suggest that the Kv4.2 and Kv4.3 subunits do not associate in rat atria (Bou-Abboud and Nerbonne, 1999). In addition, these observations suggest that both Kv4.2 and Kv4.3 contribute to rat ventricular $I_{to,f}$, whereas only Kv4.2 contributes to rat atrial $I_{to,f}$ (Apkon and Nerbonne, 1991; Boyle and Nerbonne, 1992, Fiset *et al.*, 1997; Bou-Abboud and Nerbonne, 1999). Nevertheless it remains to be demonstrated directly that Kv4.2 and Kv4.3 actually associate to form heteromeric $I_{to,f}$ channels in rat ventricles *in vivo*.

Reductions in rat ventricular $I_{to,f}$ density have also been observed in cells exposed to adenoviral constructs encoding a truncated Kv4.2 subunit (Kv4.2ST) that functions as a dominant negative (Johns *et al.*, 1997). Subsequently, Barry *et al.* (1998) produced a pore mutant of Kv4.2 (Kv4.2[W362F]) that also functions as a dominant negative and generated transgenic mice expressing this construct driven by the α-myosin heavy chain (α-MHC) promoter to direct cardiac-specific expression of the transgene. Expression of Kv4.2[W362F] results in the elimination of $I_{to,f}$ and increased action potential durations in mouse ventricular myocytes (Barry *et al.*, 1998). In addition, QT intervals are prolonged significantly in Kv4.2[W362F]-expressing animals, consistent with a defect in ventricular repolarization (Barry *et al.*, 1998). Subsequent studies revealed that $I_{to,f}$ is eliminated and action potential durations are increased in atrial myocytes isolated from Kv4.2[W362F]-expressing transgenics (Xu *et al.*, 1999b). Taken together, these results demonstrate that members of the Kv4 subfamily underlie $I_{to,f}$ in mouse and rat. Given that the properties of $I_{to,f}$ in other cells (Table 1) are similar, if not identical, to those of mouse and rat $I_{to,f}$, it seems reasonable to speculate that members of the Kv4 subfamily of α-subunits also underlie $I_{to,f}$ in other species (Table 2). In larger animals (such as dog and human), the candidate subunit is Kv4.3 because Kv4.2 appears not to be expressed (Dixon et al., 1996). Interestingly, two splice variants of Kv4.3 have been identified in human (Kong et al., 1998), as well as rat (Takimoto et al., 1996; Ohya et al., 1997), heart. The longer version of Kv4.3 contains a 19-amino-acid insert in the carboxy tail and is the more abundant isoform in heart at the message level (Takimoto et al., 1996; Ohya et al., 1997; Kong et al., 1998). However, the relative expression levels of the two Kv4.3 proteins and their contributions to the generation of functional cardiac $I_{to,f}$ channels have not been determined.

4.1.2. Kv1.4 Underlies $I_{to,s}$

As discussed in Section 2.1.2, the properties of transient outward K^+ currents in rabbit atrial and ventricular myocytes are distinct from those of $I_{to,f}$ in other species in that inactivation is slow and recovery from steady-state inactivation is very slow, proceeding with time constants in the range of 5–8 s (Clark et al., 1988; Giles and Imaizumi, 1988; Fermini et al., 1992). These differences, as well as the similarities between the transient K^+ currents in rabbit cells and the slow transient outward K^+ current ($I_{to,s}$) present in mouse myocytes isolated from the left ventricular septum (Xu et al., 1999a; Guo et al., 1999), led to the suggestion above (see Section 2.1.2) that the currents in rabbit myocytes should also be referred to as $I_{to,s}$ (Table 1). Interestingly, the properties of $I_{to,s}$ in rabbit cardiac myocytes are similar to those of heterologously expressed Kv1.4 (Tseng-Crank et al., 1990; Comer et al., 1994; Petersen and Nerbonne, 1999), an observation interpreted as suggesting that Kv1.4 could play a role in the generation of $I_{to,s}$ in the rabbit heart (Nerbonne, 1998). Support for this hypothesis is provided by experiments using AsODNs targeted against Kv1.4 (and other Kv α-subunits) in rabbit and human atrial myocytes. $I_{to,s}$ is attenuated in rabbit atrial cells exposed to the Kv1.4 AsODN, whereas human atrial $I_{to,f}$ is unaffected (Wang et al., 1999). In addition, $I_{to,s}$ is absent in myocytes isolated from the left ventricular septum of mice with a targeted deletion in Kv1.4 (Guo et al., 1999). Taken together, these results suggest that Kv1.4 indeed underlies $I_{to,s}$ in rabbit atrial (Wang et al., 1999) and mouse ventricular (Guo et al., 1999) myocytes. It has also been reported that the time- and voltage-dependent properties of the transient outward K^+ currents in ferret left ventricular epicardial and endocardial myocytes are distinct (Brahmajothi et al., 1999; see Section 2.1.2). In addition, observed regional differences in the expression of Kv1.4 and Kv4. 2/Kv4.3 have led to the suggestion that Kv1.4 and Kv4. 2/Kv4.3, respectively, underlie $I_{to,s}$ and $I_{to,f}$ in ferret left ventricular endocardial and epicardial myocytes (Brahmajothi et al., 1999).

4.2. Molecular Correlates of Cardiac Delayed Rectifiers

Considerable progress has been made in defining the molecular identities of several components of delayed rectifier currents (I_{Kr}, I_{Ks}, I_{Kur}, and $I_{K,slow}$) in mammalian cardiac myocytes (Table 2).

4.2.1. ERG1 Underlies I_{Kr}

As discussed in Section 3.1.2., ERG is the molecular correlate of functional cardiac I_{Kr} channels (Sanguinetti *et al.*, 1995). Interestingly, however, AsODNs targeted against I_{sK} (minK) attenuate I_{Kr} in AT-1 (an atrial tumor line) cells (Yang *et al.*, 1995). In addition, it has been shown that heterologously expressed HERG and I_{sK} coimmunoprecipitate (McDonald *et al.*, 1997), suggesting that functional I_{Kr} channels are multimers of ERG and I_{sK} (Yang *et al.*, 1995; McDonald *et al.*, 1997). It has not, however, been demonstrated directly that ERG and I_{sK} coassemble in mammalian cardiac cells. Alternatively processed forms of HERG1 and MERG1 with unique N- and C-termini have also been identified in mouse and human heart (Lees-Miller *et al.*, 1997; London *et al.*, 1997, 1998c; Kupershmidt *et al.*, 1998) and have been suggested to be important in the generation of functional I_{Kr} channels (Lees-Miller *et al.*, 1997; London *et al.*, 1997). However, immunoblots in rat and human atria and ventricles reveal only full-length ERG proteins, suggesting that alternatively spliced ERG variants do not contribute to I_{Kr} (Pond *et al.*, 2000).

4.2.2. KvLQT1 Underlies I_{Ks}

It appears that I_{sK} coassembles with KvLQT1 (Fig. 4A) to form functional cardiac I_{Ks} channels (Barhanin *et al.*, 1996; Sanguinetti *et al.*, 1996). The fact that mutations in the transmembrane domain of I_{sK} alter the properties of I_{Ks} channels further suggests that the transmembrane segment of I_{sK} contributes to the K^+-selective pore of I_{Ks} channels (Goldstein and Miller, 1991; Takumi *et al.*, 1991; K.-W. Wang *et al.*, 1996; Tai and Goldstein, 1998). However, direct biochemical evidence for coassembly of KvLQT1 and I_{sK} in the mammalian heart has not been provided to date, and the stoichiometry of functional I_{Ks} channels is unknown. In addition, the functional role of N-terminal splice variants of KvLQT1 in the generation of I_{Ks} channels *in vivo* remains to be determined. These splice variants exert a dominant-negative effect when coexpressed with full-length KvLQT1 (Jiang *et al.*, 1997).

4.2.3. Kv1.5 Underlies I_{Kur}

The time- and voltage-dependent properties and the pharmacological sensitivity of Kv1.5-induced K^+ currents in *Xenopus* oocytes are similar to those of the ultrarapid component of cardiac delayed rectification, I_{Kur} (Table 1) (Boyle and Nerbonne, 1992; Wang *et al.*, 1993a,b; Van Wagoner *et al.*, 1996; Yue *et al.*, 1996a,b). The similarities in the properties of cardiac I_{Kur} and the Kv1.5-induced currents and the fact that the Kv1.5 message (Tamkun *et al.*, 1991; Roberds *et al.*, 1993; Dixon and McKinnon 1994) and protein (Barry *et al.*, 1995; Mays *et al.*, 1995) are abundant in heart led to the hypothesis that Kv1.5 underlies I_{Kur} (Fedida *et al.*, 1993; Barry *et al.*, 1995; Wang *et al.*, 1993b; Mays *et al.*, 1995; Yue *et al.*, 1996b). I_{Kur} is selectively attenuated in human atrial myocytes exposed to AsODNs targeted against Kv1.5, whereas AsODNs against Kv1.4 were without effects on the K^+ currents (Feng *et al.*, 1997). In addition, rat atrial I_{Kur} is selectively reduced following treatment with Kv1.5 AsODNs, whereas AsODNs targeted against Kv1.2, Kv.2.1, Kv4.2, Kv4.3, and KvLQT1 did not alter I_{Kur} (Bou-

Abboud and Nerbonne, 1999). Similar approaches could be exploited to test the hypothesis that Kv1.5 also underlies I_{Kp} (I_{Kur}) in guinea pig ventricular myocytes (Yue and Marban, 1988; Backx and Marban, 1993), as well as I_{Kur} in other cells (Table 1).

4.2.4. Molecular Correlates of Other Cardiac Delayed Rectifiers

An *in vivo* dominant-negative strategy, involving cardiac-specific expression of a truncated Kv1.1 subunit (*Kv1.1N206Tag*), was developed and exploited by London *et al.* (1998b) in experiments focused on determining if Kv α-subunits of the Kv1 subfamily contribute to $I_{K,slow}$ in mouse ventricle. Electrophysiological recordings from ventricular myocytes isolated from *Kv1.1N206Tag*-expressing transgenic mice revealed that $I_{K,slow}$ is selectively attenuated in these cells, consistent with the hypothesis that a member of the Kv1 subfamily underlies this slowly inactivating current (London *et al.*, 1998b). Similar to the effects of Kv4.2[W362F], cardiac-specific expression of *Kv1.1N206Tag* markedly prolongs ventricular action potentials and QT intervals (London *et al.*, 1998b). Electrocardiographic recordings also revealed an increased frequency of premature ventricular beats, ventricular arrhythmias, and spontaneous ventricular tachyarrhythmias in the *Kv1.1N206Tag*-expressing animals (London *et al.*, 1998b). Interestingly, although action potential durations and QT intervals are prolonged to a greater extent in the Kv4.2[W362F]-expressing mice (Barry *et al.*, 1998) than in the *Kv1.1N206Tag* transgenic animals, spontaneous ventricular arrhythmias are not seen in the former (Barry *et al.*, 1998). This suggests that other factors, in addition to prolonged ventricular repolarization, play an important role in determining the propensity to develop and to sustain arrhythmias.

In experiments on rat atrial myocytes, it has been shown that I_{Kslow} ($I_{K,DTX}$) is selectively attenuated in cells exposed to AsODNs targeted against Kv1.2 (Bou-Abboud and Nerbonne, 1999). These results clearly demonstrate that $I_{K,slow}$ ($I_{K,DTX}$) in these cells is a unique molecular entity, distinct from $I_{to,f}$ and I_{Kur}. As noted above (see Section 4.2.3), it has also been demonstrated that an AsODN targeted against Kv1.5 specifically and selectively affected only I_{Kur} (Bou-Abboud and Nerbonne, 1999). The simplest interpretation of these findings is that Kv1.2 and Kv1.5 do not coassemble to form heteromultimeric K^+ channels in rat atria *in vivo*.

5. REGULATION OF FUNCTIONAL CARDIAC Kv CHANNELS

Changes in the functional expression and the properties of Kv channels occur during normal cardiac development and in the damaged or diseased heart, although little is known concerning the underlying molecular mechanisms involved. In addition, the properties and/or the functional expression of cardiac Kv channels can be modulated both *in vivo* (Shimoni *et al.*, 1992, 1995, 1997; Takimoto and Levitan, 1994; Drici *et al.*, 1996; Levitan *et al.*, 1996) and *in vitro* (Matsubara *et al.*, 1993; Mori *et al.*, 1993; Guo *et al.*, 1995, 1996, 1997b,c; Wickenden *et al.*, 1997) by a variety of physiological stimuli, including hormones, growth factors, and cell depolarization. There are multiple molecular mechanisms, including transcriptional, translational, and posttranslational modulation, that could be involved in the developmental and pathophysiological regulation of Kv channels.

5.1. Developmental Regulation of Cardiac K⁺ Channel Expression and Properties

In many species, marked changes in the waveforms of myocardial action potentials and in excitation–contraction coupling are seen during embryonic and postnatal development (Wetzel and Klitzner, 1996). Although the detailed molecular mechanisms underlying these changes have not been delineated (Wetzel and Klitzner, 1996; Nerbonne, 1998), it is clear that at least some of the age-related changes in action potential waveforms reflect developmental differences in the functional expression levels and/or in the properties of the voltage-gated cardiac K⁺ currents.

5.1.1. Development of Cardiac Transient Outward K⁺ Currents/Channels

During postnatal development in most species, ventricular action potentials shorten, phase 1 repolarization (Fig. 1) becomes more pronounced, and functional $I_{to,f}$ density is increased (Table 3) (Kilborn and Fedida, 1990; Jeck and Boyden, 1992; Nuss and Marban, 1994; Wahler *et al.*, 1994; Sánchez-Chapula *et al.*, 1994; Guo *et al.*, 1996, 1997a; Xu *et al.*, 1996; Shimoni *et al.*, 1997; Wang and Duff, 1997; Wickenden *et al.*, 1997). Age-dependent changes in atrial action potential waveforms (Escande *et al.*, 1985) and changes in the properties and densities of atrial $I_{to,f}$ also occur (Gross *et al.*, 1995; Crumb *et al.*, 1995). In neonatal canine ventricular myocytes, for example, the "notch" (Fig. 1) and the rapid (phase 1) repolarization that are typical in adult cells are not clearly evident (Jeck and Boyden, 1992). In addition, action potentials in neonatal

Table 3
Variations in Ventricular K⁺ Channel Expression during Development[a,b]

Current	Species	Fetal	Neonatal	Adult
$I_{to,f}$	Dog	ND	−	+ + + +
	Human	ND	+	+ + + +
	Mouse	+	+ +	+ + + +
	Rat	+	+ +	+ + + +
$I_{to,s}$	Rabbit[c]	ND	+ +	+ + + +
	Rat	ND	+	−
I_{Kr}	Mouse	+ +	+ +	−
	Rat	ND	+	+
I_{Ks}	Mouse	−	+ +	−
	Rat	ND	−	−
I_{Kur}	Rat	+ +	+	−
$I_{K,slow}$	Mouse	+	+ +	+ +
	Rat	+	+ +	+ +

[a] Adapted from J. M. Nerbonne, *J. Neurobiol.* **37**:37–59.
[b] Key: +, Detectable; + +, moderate density; + + + +, high density; −, not detected; ND, not determined.
[c] The properties of rabbit $I_{to,s}$ also change with age (Sánchez-Chapula *et al.*, 1994).

(day 0 to day 14) cells are insensitive to 4-AP, and voltage-clamp recordings do not reveal the presence of $I_{to,f}$ (Jeck and Boyden, 1992). In myocytes isolated from 2-month-old animals, $I_{to,f}$ is present and phase 1 repolarization is clearly evident (Jeck and Boyden, 1992). The density of $I_{to,f}$ is low in ventricular myocytes from neonatal mice (Nuss and Marban, 1994; Wang and Duff, 1997) and rats (Kilborn and Fedida, 1990; Wahler et al., 1994; Guo et al., 1996, 1997a; Xu et al., 1996; Shimoni et al., 1997; Wickenden et al., 1997) and increases 5- to 6-fold during postnatal development (Table 3). $I_{to,f}$ inactivation and recovery from inactivation are slower in rat ventricular myocytes at postnatal days 1–2 (Wickenden et al., 1997) than at postnatal day 5 or in adult cells (Xu et al., 1996). Indeed, the transient outward currents in postnatal day 1–2 rat ventricular cells more closely resemble $I_{to,s}$ than $I_{to,f}$, suggesting a developmental switch in the Kv α-subunits, from Kv1.4 (i.e., $I_{to,s}$) in early neonatal cells to Kv4.2/Kv4.3 ($I_{to,f}$) in more mature cells (Wickenden et al., 1997). However, neither $I_{to,s}$ nor $I_{to,f}$ is detectable in embryonic ventricular myocytes (Petersen and Nerbonne, 1999), further suggesting that the expression of Kv1.4 (and $I_{to,s}$) occurs transiently in early postnatal development and that the switch to Kv4.2/Kv4.3 (and $I_{to,f}$) must occur abruptly between postnatal days 2 and 5 (Xu et al., 1996; Guo et al., 1996, 1997a; Wickenden et al., 1997).

In rabbit ventricular myocytes, the density of the transient outward K^+ currents also increases and the kinetic properties of the currents change during postnatal development (Sánchez-Chapula et al., 1994). Specifically, the rate of recovery from steady-state inactivation is 10 times faster in neonatal cells (mean recovery time ≈ 100 ms) than in adult cells (mean recovery time ≈ 1300 ms) cells (Sánchez-Chapula et al., 1994). The slow recovery of the transient outward currents from steady-state inactivation underlies the marked broadening of action potentials seen in adult, but not neonatal, rabbit ventricular myocytes at high stimulation frequencies (Fermini et al., 1992; Sánchez-Chapula et al., 1994). It remains to be determined if the changes in the properties of the transient outward K^+ currents in rabbit ventricular cells reflect a switch in the Kv α-subunits underlying the currents [i.e., from Kv4.x ($I_{to,f}$) in the neonate to Kv1.4 ($I_{to,s}$) in the adult], changes in β-subunit composition, or post-translational processing of the same α-subunit(s).

5.1.2. Development of Cardiac Delayed Rectifier K^+ Currents/Channels

Marked changes in the functional expression of delayed rectifier K^+ currents are also evident during postnatal development (Table 3). Both I_{Kr} and I_{Ks}, for example, are prominent in fetal and/or neonatal mouse cardiac myocytes (Nuss and Marban, 1994; Davies et al., 1996; Wang and Duff, 1996; L. Wang et al., 1996), whereas these currents are difficult to detect in adult cells (Xu et al., 1999a,b; Guo et al., 1999). In the fetal mouse heart, the density of I_{Kr}, the predominant Kv current controlling action potential repolarization (Davies et al., 1996; Wang and Duff, 1996; L. Wang et al., 1996; Duff et al., 1997), decreases with age. In contrast, the density of I_{Ks} increases during late embryonic development (Davies et al., 1996) and subsequently decreases during postnatal development (Nuss and Marban, 1994; Davies et al., 1996; L. Wang et al., 1996). Because I_{Kr} and I_{Ks} are prominent repolarizing K^+ currents in adult cardiac cells in several species (Table 1), changes in the expression and/or the properties of these conductance pathways must be distinct from those observed in the mouse heart.

Although not reported in adult rat ventricular myocytes (Apkon and Nerbonne, 1991; Xu *et al.*, 1996), I_{Kur} is evident in neonatal rat ventricular myocytes *in vitro* (Guo *et al.*, 1997a,c,d), and these cells are immunoreactive for the Kv1.5 protein (Guo *et al.*, 1997a). The number of cells with detectable I_{Kur} and Kv1.5 protein decreases during postnatal development *in vitro*, in parallel with an increase in $I_{to,f}$ density (Guo *et al.*, 1997a,c,d). In contrast to the marked changes in $I_{to,f}$ and I_{Kur}, the density of I_K does not change significantly in rat ventricular cells after postnatal day 5 (Xu *et al.*, 1996; Shimoni *et al.*, 1997). There is, however, a small, but statistically significant difference in I_K density at postnatal days 1–2, compared with adult rat ventricular myocytes (Wickenden *et al.*, 1997).

5.2. Changes in Kv Current Expression in Myocardial Disease

Profound changes in myocardial membrane excitability and in excitation–contraction coupling are evident in a variety of myocardial disease states in humans, as well as in several animal models of human cardiac disease. Importantly, alterations in myocardial action potential waveforms, particularly if occurring in specific regions or cell types, can increase dispersion of repolarization and change the sequence and the spread of excitation through the myocardium, predisposing individuals to life-threatening arrhythmias. In addition, action potential prolongation *per se* can be arrhythmogenic, leading to afterdepolarizations and triggered activity. Although the detailed molecular mechanisms underlying changes in action potential waveforms in myocardial disease have not been delineated, it is clear that at least some of these reflect alterations in the characteristics and expression of functional Kv currents (Ten Eick *et al.*, 1989, 1993; Bénitah *et al.*, 1993; Beuckelmann *et al.*, 1993; Näbauer *et al.*, 1993; Tomita *et al.*, 1994; Boyden and Jeck, 1995; Gidh-Jain *et al.*, 1996; Kääb *et al.*, 1996, 1998; Mészáros *et al.*, 1996; Potreau *et al.*, 1996; Qin *et al.*, 1996; Bailly *et al.*, 1997; Freeman *et al.*, 1997; Gomez *et al.*, 1997; Shipsey *et al.*, 1997; Van Wagoner *et al.*, 1997; Wilde and Veldkamp, 1997; Yue *et al.*, 1997; Näbauer and Kääb, 1998; Rozanski *et al.*, 1998).

In several experimental models of cardiac hypertrophy, ventricular action potential durations are markedly increased (Ten Eick *et al.*, 1989, 1993; Bénitah *et al.*, 1993; Tomita *et al.*, 1994; Boyden and Jeck, 1995; Kääb *et al.*, 1996; Mészáros *et al.*, 1996; Qin *et al.*, 1996; Gomez *et al.*, 1997; Shipsey *et al.*, 1997; Rozanski *et al.*, 1998). Action potentials are also prolonged significantly in myocytes isolated from hypertrophied and failing human hearts (Beuckelmann *et al.*, 1993; Näbauer *et al.*, 1993; Bailly *et al.*, 1997; Näbauer and Kääb, 1998). Voltage-clamp recordings from myocytes isolated from hypertrophied rat hearts, induced by either pressure overload secondary to aortic constriction, or induced by myocardial infarction, reveal marked reductions in $I_{to,f}$ densities (Bénitah *et al.*, 1993; Tomita *et al.*, 1994; Jeck and Boyden, 1995; Qin *et al.*, 1996; Gomez *et al.*, 1997; Rozanski *et al.*, 1998). The density of $I_{to,f}$ is also reduced by catecholamine-induced hypertrophy of the rat heart (Mészáros *et al.*, 1996; Shipsey *et al.*, 1997), and, interestingly, the reduction in $I_{to,f}$ density and the resulting prolongation of the action potential are specific to the epicardium (Shipsey *et al.*, 1997) consistent with the transmural gradient of $I_{to,f}$ in rat left ventricles (Clark *et al.*, 1993). In contrast to the reduction in $I_{to,f}$ in hypertrophied rat ventricles, functional $I_{to,f}$ density is increased in right ventricular myocytes isolated from hypertrophied cat heart (Ten Eick et al., 1993). These unexpected results suggest that, at least in this model, effects on ionic currents in addition to $I_{to,f}$ likely play a role in the electrical remodeling that occurs with hypertrophy (Ten Eick et al., 1993; Boyden and Jeck, 1995; Näbauer and Kääb, 1998).

In myocytes isolated from hypertrophied and failing human hearts, however, $I_{to,f}$ density is reduced significantly (Beuckelmann et al., 1993; Näbauer et al., 1993; Bailly et al., 1997; Näbauer and Kääb, 1998). In addition, in a canine pacing-induced model of heart failure, action potential plateaus are elevated, action potential durations are prolonged significantly, and $I_{to,f}$ is attenuated compared with control dogs (Kääb et al., 1996). In fact, voltage-clamp recordings from left ventricular myocytes isolated from these animals suggest that the attenuation of ventricular $I_{to,f}$ is the predominant ionic mechanism underlying the observed changes in action potential waveforms associated with this model (Kääb et al., 1996). Importantly, the reduction in $I_{to,f}$ density, at least in the rat pressure-overload model of hypertrophy, cannot be attributed to changes in the properties of single $I_{to,f}$ channels because single-channel conductance and mean open-state probability are not altered (Tomita et al., 1994). Thus, the attenuation of $I_{to,f}$ reflects a decrease in the number of functional (cell surface) $I_{to,f}$ channels, due to a reduction in $I_{to,f}$ channel protein and/or to changes in the properties of at least some of the $I_{to,f}$ channels that render them nonfunctional. Support for the former hypothesis is provided by the demonstration that Kv4.3 message levels are decreased in failing human left ventricles (Kääb et al., 1998). Presumably, the reduction in Kv4.3 message is translated into a corresponding decrease in Kv4.3 protein, although this has not been demonstrated directly (Kääb et al., 1998). Reduced Kv4.2/4.3 protein expression has, however, been reported in hypertrophied rat left ventricular myocytes, after myocardial infarction (Gidh-Jain et al., 1996).

In contrast to the prolongation of action potentials in the hypertrophied and failing myocardium (Ten Eick et al., 1993; Boyden and Jeck, 1995; Näbauer and Kääb, 1998), acute and chronic atrial fibrillation are associated with shortening of the atrial refractory period, and action potential durations are reduced significantly (Boutjdir et al., 1986; Daoud et al., 1996). These observations led to suggestions that voltage-gated K^+ channel/current densities are likely increased in fibrillating atria. Whole-cell voltage-clamp recordings from atrial myocytes isolated from patients with chronic atrial fibrillation, however, have revealed marked reductions (rather than increases) in outward K^+ current densities (Van Wagoner et al., 1997). Both the peak and the plateau (measured at the end of 450-ms depolarizing voltage steps) outward K^+ currents are attenuated significantly, suggesting either that functional $I_{to,f}$ and I_{Kur} channel densities are reduced in atrial fibrillation or that the properties of single $I_{to,f}$ and I_{Kur} channels are altered, or both. Support for the former hypothesis is provided with the demonstration that Kv1.5 protein (which underlies I_{Kur}; see Section 4.2.3) expression is reduced in atrial membranes isolated from patients with chronic atrial fibrillation, whereas Kv2.1 expression levels are unaffected (Van Wagoner et al., 1997). In a canine pacing-induced model of atrial fibrillation, $I_{to,f}$ density is also reduced, and the attenuation of $I_{to,f}$ was directly related to the duration of pacing (Yue et al., 1997). The observation that neither the time- nor the voltage-dependent properties of the currents were affected suggests that functional $I_{to,f}$ density is reduced directly by electrical pacing (Yue *et al.*, 1997). Thus, understanding the ionic pathways affected in various myocardial disease states, as well as the underlying molecular mechanisms involved in mediating these changes, will be pivotal to the development of more effective treatment strategies.

5.3. Transcriptional Regulation of Functional Kv Channels

Transcriptional regulation of Kv channel α-subunit expression plays a role in controlling the number of functional K^+ channels in the adult and developing heart. In adult rats, for example, adrenalectomy results in a 6-fold reduction in ventricular Kv1.5 mRNA and protein expression levels (Takimoto and Levitan, 1994). In addition, when adrenalectomized animals are treated with dexamethasone, 50-fold and 20-fold increases in Kv1.5 message and protein levels, respectively, are observed (Takimoto and Levitan, 1994). Expression of Kv1.5 is also increased in response to cold stress (Levitan *et al.*, 1996) and decreased with hypertrophy (Matsubara *et al.*, 1993); Kv1.4 message levels, in contrast, are increased with hypertrophy (Matsubara *et al.*, 1993). The effects of adrenalectomy and dexamethasone on ventricular Kv1.5 appear to be specific, in that no effects on ventricular Kv1.4 or Kv2.1 or on atrial Kv1.5 expression were observed (Takimoto and Levitan, 1994). The functional consequences of the adrenalectomy-induced downregulation of Kv1.5, however, are unclear (Takimoto and Levitan, 1994; Levitan *et al.*, 1996). Indeed, given that Kv1.5 underlies I_{Kur} (Feng *et al.*, 1997; Bou-Abboud and Nerbonne, 1999; see Section 4.2.3) and that I_{Kur} is not evident in adult rat ventricular myocytes (Apkon and Nerbonne, 1991; Xu *et al.*, 1996; Guo *et al.*, 1997d), it seems unlikely that downregulation of Kv1.5 will affect voltage-gated K^+ currents in these cells. It would be interesting, however, to explore the possibility that I_{Kur} is evident in ventricular myocytes isolated from dexamethasone-treated adrenalectomized rats or from rats subjected to cold stress, representing situations in which Kv1.5 expression is increased (Takimoto and Levitan, 1994; Levitan *et al.*, 1996).

Steroid hormones have also been shown to affect the expression of myocardial Kv channel α-subunits (Drici *et al.*, 1996). Treatment of ovariectomized adult rabbits with either estradiol or dihydrotestosterone, for example, resulted in large (~ 10-fold) reductions in Kv1.5 mRNA levels, and electrocardiographic recordings revealed that QT intervals are prolonged in these animals, indicative of alterations in ventricular repolarization (Drici *et al.*, 1996). It is also important to note that steroids, notably progesterone and the antiestrogen tamoxifen, have been shown to produce reversible blockade of Kv channels, including I_{Kr}, in cardiac and other cells (Ehring *et al.*, 1998; Liu *et al.*, 1998). These observations suggest that studies focused on exploring the roles of steroid hormones in the regulation of voltage-gated K^+ currents in the mammalian heart may be complicated by nontranscriptional mechanisms. The expression of Kv1.5 in neonatal rat ventricular myocytes *in vitro* has also been shown to be increased by treatments that increase intracellular cAMP and by membrane depolarization (Matsubara *et al.*, 1993; Mori *et al.*, 1993; Guo *et al.*, 1997b). Although the molecular mechanisms involved in mediating these effects have not been delineated, the regulation of Kv1.5 mRNA levels by cAMP could be direct, given that the 5'-flanking region of the rat Kv1.5 gene contains a cAMP response element (Mori *et al.*, 1993). In addition, the effect of membrane depolarization on Kv1.5 expression is interesting in light of the evidence that functional K^+ channel density is affected by pacing (see Section 5.2) and suggests that electrical activity *per se* may play a role in the regulation of Kv channel α-subunit gene expression *in vivo*.

Electrophysiological studies have demonstrated that thyroid hormone levels affect transient outward K^+ currents in cardiac cells (Shimoni *et al.*, 1992, 1995, 1997; Wickenden *et al.*, 1997). In adult rabbit heart, for example, $I_{to,s}$ density is increased when plasma thyroid hormone levels are increased (Shimoni *et al.*, 1992). In left ventricular myocytes isolated from hypothyroid adult rats, $I_{to,f}$ amplitudes are reduced in epicardial myocytes, whereas $I_{to,f}$ in endocardial cells is unaffected (Shimoni *et al.*, 1995). Although no changes in $I_{to,f}$ densities are evident when adult animals are made hyperthyroid (Shimoni *et al.*, 1995), it has been shown that plasma thyroid hormone levels increase between postnatal days 5 and 20 in the rat (Shimoni *et al.*, 1997), a time course similar to that associated with the developmental increase in functional $I_{to,f}$ density (Kilborn and Fedida, 1990; Xu *et al.*, 1996; Shimoni *et al.*, 1997). In addition, the $I_{to,f}$ density and Kv4.2/Kv4.3 message levels are increased in neonatal animals by daily injections of thyroid hormone (Shimoni *et al.*, 1997). In thyroid-hormone-treated neonatal rat ventricular myocytes *in vitro*, Kv4.3 message levels are also elevated, whereas Kv1.4 decreases and Kv4.2 is unaffected (Wickenden *et al.*, 1997). Although these results suggest a correlation between Kv4.2 and Kv4.3 message levels and the increases in functional $I_{to,f}$ density in thyroid-hormone-treated neonatal rat ventricular cells, it remains to be demonstrated that Kv4.2/Kv4.3 protein levels are actually elevated in response to thyroid hormone treatment. In adult animals, changes in thyroid status reportedly have variable effects on the message levels of Kv α-subunits (notably Kv1.2, Kv1.4, Kv1.5, Kv2.1, Kv4.2, and Kv4.3), although the functional consequences of these changes in mRNA levels have not been determined (Abe *et al.*, 1998; Nishiyama *et al.*, 1998).

5.4. Posttranscriptional Regulation of Functional Kv Channels

The functional expression of myocardial Kv channels could also be regulated at the level of translation of the underlying K^+ channel pore-forming α and/or accessory (β and minK) subunits. Because Kv channels are tetrameric (MacKinnon, 1991) and heteromeric channels can be formed by coassembly of Kv α-subunits in the same subfamily (Covarrubias *et al.*, 1991), changes in subunit composition could alter functional K^+ channel expression levels or properties. Alternatively, changes in the contributions of accessory subunits (e.g., β-subunits or KChAPs) could lead to variations in channel density or function. It is possible, but unproven, that changes in functional K^+ current densities in the developing or diseased myocardium reflect differences in subunit composition.

Alternatively, functional Kv channel expression could be regulated at the level of posttranslational modifications of an unchanged collection of α- and β-subunits. Certainly, the properties and the densities of Kv currents produced by heterologous expression of Kv α-subunits can be modified by posttranslational processing (Drain *et al.*, 1994; Covarrubias *et al.*, 1994; Murray *et al.*, 1994; Fadool *et al.*, 1997; Murakoshi *et al.*, 1997; Nakamura *et al.*, 1997; Roeper *et al.*, 1997; Beck *et al.*, 1998). Activators of protein kinase C (PKC), for example, cause a small, transient increase, followed by a maintained decrease, in Kv1.4-induced currents in *Xenopus* oocytes (Murray *et al.*, 1994). Because Kv1.4 can be phosphorylated by PKC *in vitro*, these observations suggest that the changes in current densities reflect direct phosphorylation of Kv1.4 (Murray *et al.*, 1994). Activation of PKC also attenuates Kv4.2- and Kv4.3-induced K^+ currents in *Xenopus* oocytes (Nakamura *et al.*, 1997). Because PKC activation suppresses $I_{to,f}$ in rat ventricular myocytes (Apkon and Nerbonne, 1988), this suggests that phosphorylation of Kv4.2 and Kv4.3 may play a role in regulating the density of functional cardiac $I_{to,f}$ channels (Nakamura *et al.*, 1997).

In addition to increasing functional K^+ channel densities, current properties can also be altered by subunit phosphorylation (Drain *et al.*, 1994; Covarrubias *et al.*, 1994; Fadool *et al.*, 1997; Murakoshi *et al.*, 1997; Roeper *et al.*, 1997). Inactivation of *Shaker* D-induced K^+ currents in *Xenopus* oocytes, for example, is slowed by phosphatase treatment (Drain *et al.*, 1994). This slowing was reversed *in situ* upon addition of ATP and the catalytic subunit of protein kinase A (PKA), suggesting that direct phosphorylation of *Shaker* D or an endogenous (Xenopus) Kv1.x α-or β-subunit that associates with *Shaker* D regulates the rate of current inactivation (Drain *et al.*, 1994). In contrast, phosphorylation of Kv1.4, by Ca^{2+}/calmodulin-dependent kinase, slows inactivation and accelerates recovery, whereas dephosphorylation has the opposite effect (Roeper *et al.*, 1997). Similarly, phosphorylation of the rapidly inactivating Kv3.4 subunit by PKC eliminates fast inactivation (Covarrubias *et al.*, 1994; Beck *et al.*, 1998). Activators of PKA and PKC also affect ERG1 currents expressed in *Xenopus* oocytes (Barros *et al.*, 1998; Kiehn *et al.*, 1998). Interestingly, however, the effects of various kinases are distinct. Phosphorylation by PKA increases functional K^+ current densities (Kiehn *et al.*, 1998), whereas activation of PKC slows I_{Kr} activation and accelerates deactivation (Barros *et al.*, 1998). Heterologously expressed Kv1 α-subunits are also phosphorylated by tyrosine kinases, and both the kinetic properties and the densities of the currents are affected (Fadool *et al.*, 1997). Because many hormone and growth factor receptors are tyrosine kinases, these results suggest the interesting possibility that tyrosine phosphorylation may play a role in regulating Kv channel expression and/or functional properties in the heart.

There are other potential posttranslational mechanisms that could contribute to changes in the properties or the expression levels of voltage-gated myocardial K^+ channels. Glycosylation, for example, reportedly alters the kinetic properties of the inwardly rectifying K^+ channel α-subunit ROMK1 (Schwalbe *et al.*, 1995). In addition, in experiments on expressed Kv1.1 currents, glycosylation has been shown to influence the voltage dependence as well as the kinetics of current activation, without appreciably affecting cell-surface expression (Thornhill *et al.*, 1996). It has also been reported that ERG1 is glycosylated and that the extent of glycosylation varies in different cell types (Zhou *et al.*, 1998; Pond *et al.*, 2000). Importantly, glycosylation of ERG1 increases the amount of protein, and, therefore, the number of functional channels, on the cell surface (Zhou *et al.*, 1998). Although it is possible that the properties of ERG1 channels are also affected, this is difficult to determine, given that glycosylation appears to be required for cell-surface expression (Zhou *et al.*, 1998). Clearly, experiments focused on examining posttranslational processing of voltage-gated myocardial K^+ channel α- and β-subunits, determining the functional consequences of this processing, and delineating the molecular mechanisms controlling processing will be of considerable interest.

ACKNOWLEDGMENTS

The author thanks several of the past and present members of her laboratory who have made important contributions to our understanding of structural and functional myocardial K^+ channel diversity: Drs. Michael Apkon, Dianne M. Barry, Elias Bou-Abboud, Walter A. Boyle, Sylvain Brunet, Weinong Guo, Huilin Li, Amber L. Pond, and Haodong Xu. The author also thanks Andrew Benedict, Bridget Scheve, and Rebecca Hood for their expert technical assistance in all aspects of the research effort in the Nerbonne laboratory focused on understanding the molecular basis of structural and functional myocardial K^+ channel diversity. Preparation of this chapter and the work from the author's laboratory cited here has been supported by the National Heart, Lung and Blood Institute of the National Institutes of Health, by the Monsanto/Searle/Washington University Biomedical Research Agreement, and by the National Office and the Midwest Affiliate of the American Heart Association.

REFERENCES

Abe, A., Yamamoto, T., Isome, M., Ma, M. L., Yaoita, E., Kawasaki, K., Kihara, I., and Aizawa, Y., 1998, Thyroid hormone regulates expression of *Shaker*-related potassium channel mRNA in rat heart, *Biochem. Biophys. Res. Commun.* **245**:226–230.

Abrahamsson, C., Palmer, M., Ljung, B., Duker, G., Bäärnheilm, C., Carlsson, L., and Danielsson, B., 1994, Induction of rhythm abnormalities in the fetal rat heart: A tentative mechanism for the embryotoxic effect of the class III antiarrhythmic agent almokalant, *Cardiovasc. Res.* **28**:337–344.

Accili, E. A., Kiehn, J., Yang, Q., Wang, Z., Brown, A. M., and Wible, B. A., 1997a, Separable $Kv\beta$ subunit domains alter expression and gating of potassium channels, *J. Biol. Chem.* **272**:25824–25831.

Accili, E. A., Kiehn, J., Wible, B. A., and Brown, A. M., 1997b, Interactions among inactivating and noninactivating $Kv\beta$ subunits, and $Kv\alpha1.2$, produce potassium currents with intermediate inactivation, *J. Biol. Chem.* **272**:28232–28236.

Antzelevitch, C., Sicouri, S., Lukas, A., Nesterenko, V. V., Liu, D.-W., and Didiego, J. M., 1994, Regional differences in the electrophysiology of ventricular cells: Physiological implications, in: *Cardiac Electrophysiology: From Cell to Bedside* (D. P. Zipes, and J. Jalife, eds.), W. B. Saunders Co., Philadelphia, pp. 228–245.

Anumonwo, J. M. B., Freeman, L. C., Kwok, W. M., and Kass, R. S., 1991, Potassium channels in the heart: Electrophysiology and pharmacological regulation, *Cardiovasc. Drug Rev.* **9**:299–316.

Anumonwo, J. M. B., Freeman, L. C., Kwok, W. M., and Kass, R. S., 1992, Delayed rectification in single cells isolated from guinea pig sinoatrial node, *Am. J. Physiol.* **262**:H921–H925.

Apkon, M., and Nerbonne, J. M., 1988, α_1-Adrenergic agonists selectively suppress voltage-dependent K^+ currents in rat ventricular myocytes, *Proc. Natl. Acad. Sci., U.S.A.* **85**:8756–8760.

Apkon, M., and Nerbonne, J. M., 1991, Characterization of two distinct depolarization-activated K^+ currents in isolated adult rat ventricular myocytes, *J. Gen. Physiol.* **97**:973–101.

Backx, P. H., and Marban, E., 1993, Background potassium current active during the plateau of the action potential in guinea pig ventricular myocytes, *Circ. Res.* **72**:890–900.

Bailly, P., Bénitah, J. P., Mouchoniere, M., Vassort, G., and Lorente, P., 1997, Regional alteration of the transient outward current in human left ventricular septum during compensated hypertrophy, *Circulation* **96**:1266–1274.

Baldwin, T. J., Tsaur, M. -L., Lopez, G. A., Jan, Y. N., and Jan, L. Y., 1991, Characterization of a mammalian cDNA for an inactivating voltage-sensitive K^+ channel, *Neuron* **7**:471–483.

Balser, J. R., Bennett, P. B., and Roden, D. M., 1990, Time-dependent outward currents in guinea pig ventricular myocytes: Gating kinetics of the delayed rectifier, *J. Gen. Physiol.* **96**:835–863.

Barhanin, J., Lesage, F., Guillemare, E., Fink, M., Lazdunski, M., and Romey, G., 1996, KvLQT1 and ISK (minK) proteins associate to form the I_{Ks} cardiac potassium current, *Nature* **384**:78–80.

Barros, F., Gomezvarela, D., Viloria, C. G., Palomero, T., Giraldez, T., and Delapena, P., 1998, Modulation of human ERG K^+ channel gating by activation of a G-protein coupled receptor and protein kinase C, *J. Physiol. (London)* **511**:333–346.

Barry, D. M., and Nerbonne, J. M., 1996, Myocardial potassium channels: Electrophysiological and molecular diversity, *Annu. Rev. Physiol.* **58**:363–394.

Barry, D. M., Trimmer, J. S., Merlie, J. P., and Nerbonne, J. M., 1995, Differential expression of voltage-gated K^+ channel subunits in adult rat heart: Relationship to functional K^+ channels, *Circ. Res.* **77**:361–369.

Barry, D. M., Xu, H., Schuessler, R. B., and Nerbonne, J. M., 1998, Functional knockout of the transient outward current, long QT syndrome and cardiac remodelling in mice expressing a dominant negative Kv4 α subunit, *Circ. Res.* **83**:560–567.

Beck, E. J., Sorensen, R. G., Slater, S. J., and Covarrubias, M., 1998, Interactions between multiple phosphorylation sites in the inactivation particle of a K^+ channel: Insights into the molecular mechanism of protein kinase C action. *J. Gen. Physiol.* **112**:71–74.

Bénitah, J. P., Gomez, A. M., Bailly, P., Da Ponte, J. P., Berson, G., Delgado, C., and Lorente, P., 1993, Heterogeneity of the early outward current in ventricular cells isolated from normal and hypertrophied rat hearts, *J. Physiol. (London)* **469**:111–138.

Benndorf, K., and Nilius, B., 1988, Properties of an early outward current in single cells of the mouse ventricle, *Gen. Physiol. Biophys.* **7**:449–466.

Benndorf, K., Markwardt, F., and Nilius, B., 1987, Two types of transient outward currents in cardiac ventricular cells of mice, *Pflügers Arch.* **409**:641–643.

Beuckelmann, D. J., Näbauer, M., and Erdmann, E., 1993, Alterations in K^+ currents in isolated human ventricular myocytes from patients with terminal heart failure, *Circ. Res.* **73**:379–385.

Biervert, C., Schroeder, B. C., Kubisch, C., Berkoviv, S. F., Propping, P., Jentsch, T. J., and Steinlein, O. K., 1998, A potassium channel mutation in neonatal human epilepsy, *Science* **279**:403–405.

Blair, T. A., Roberds, S. L., Tamkun, M. M., and Hartshorne, R. P., 1991, Functional characterization of RK5, a voltage-gated K^+ channel cloned from the rat cardiovascular system, *FEBS Lett.* **295**:211–213.

Bou-Abboud, E., and Nerbonne, J. M., 1999, Molecular correlates of the Ca^{++}-independent depolarization-activated K^+ currents in adult rat atrial myocytes, *J. Physiol. (London)* **517**:407–420.

Boutjdir, M., Le Heuzey, J. Y., Lavergne, T., Chauvaud, S., Guize, L., Carpentier, A., and Peronneau, P., 1986, Inhomogeneity of cellular refractoriness in human atrium: Factor of arrhythmia? *Pacing Clin. Electrophysiol.* **9**:1095–1100.

Boyden, P. A., and Jeck, C. D., 1995, Ion channel function in disease, *Cardiovasc. Res.* **29**:312–318.

Boyle, W. A., and Nerbonne, J. M., 1991, A novel type of depolarization-activated K^+ current in isolated adult rat atrial myocytes, *Am. J. Physiol.* **260**:H1236–H1247.

Boyle, W. A., and Nerbonne, J. M., 1992, Two functionally distinct 4–aminopyridine-sensitive outward K^+ currents in adult rat atrial myocytes, *J. Gen. Physiol.* **100**:1047–1061.

Brahmajothi, M. V., Morales, M. J., Liu, S., Rasmusson, R. L., Campbell, D. L., and Strauss, H. C., 1996, *In situ* hybridization reveals extensive diversity of K^+ channel mRNA in isolated ferret cardiac myocytes, *Circ. Res.* **78**:1083–1089.

Brahmajothi, M. V., Campbell, D. L., Rasmusson, R. L., Morales, M. J., Nerbonne, J. M., and Strauss, H. C., 1999, Distinct transient outward potassium current (I_{to}) phenotypes and distribution of fast-inactivating potassium channel α subunits in ferret left ventricular myocytes, *J. Gen. Physiol.* **113**:581–600.

Bryant, S. M., Wan, X. P., Shipsey, S. J., and Hart, G., 1998, Regional differences in the delayed rectifier currents (I_{Kr} and I_{Ks}) contribute to the differences in action potential duration in basal left ventricular myocytes in guinea pig, *Cardiovasc. Res.* **40**:322–331.

Busch, A. E., Malloy, K., Groh, W. J., Varnum, M. D., Adelman, J. P., and Maylie, J., 1994, The novel class III antiarrhythmics NE-10064 and NE-10133 inhibit I_{sK} channels expressed in *Xenopus* oocytes and I_{Ks} in guinea pig ventricular myocytes, *Biochem. Biophys. Res. Commun.* **202**:265–270.

Butler, A., Wei, A., Baker, K., and Salkoff, L., 1989, A family of putative potassium channel genes in *Drosophila*, *Science* **243**:943–947.

Campbell, D. L., Rasmusson, R. L., Comer, M. B., and Strauss, H. C., 1993, The calcium-independent transient outward potassium current in isolated ferret right ventricular myocytes. I. Basic characterization and kinetic analysis, *J. Gen. Physiol.* **101**:571–601.

Campbell, D. L., Rasmusson, R. L., Comer, M. B., and Strauss, H. C., 1995, The cardiac calcium-independent transient outward potassium current: Kinetics, molecular properties, and role in ventricular repolarization, in: *Cardiac Electrophysiology: From Cell to Bedside*, 2nd ed. (D. P. Zipes and J. Jalife, eds.), W. B. Saunders Co., Philadelphia, pp. 83–96.

Castellano, A., Chiara, M. D., Mellstr, B., Molina, A., Monje, F., Naranjo, J. R., and Lopez-Barneo, J., 1997, Identification and functional characterization of a K^+ channel α subunit with regulatory properties specific to brain, *J. Neurosci.* **17**:4652–4661.

Castellino, R. C., Morales, M. J., Strauss, H. C., and Rasmusson, R. L., 1995, Time- and voltage-dependent modulation of a Kv1.4 channel by a β subunit (Kvβ3) cloned from ferret ventricle, *Am. J. Physiol.* **268**:H385–H391.

Chandy, K. G., and Gutman, G. A., 1995, Voltage-gated K^+ channel genes, in: *Handbook of Receptors and Channels* (R. A. North, ed.), CRC Press, Boca Raton, Florida, pp. 1–71.

Clark, R. B., Giles, W. R., and Imaizumi, Y., 1988, Properties of the transient outward current in rabbit atrial cells, *J. Physiol. (London)* **405**:147–168.

Clark, R. B., Bouchard, R. A., Salinas-Stefanson, E., Sanchez-Chalupa, J., and Giles, W. R., 1993, Heterogeneity of action potential waveforms and potassium currents in rat ventricle, *Cardiovasc. Res.* **27**:1795–1799.

Comer, M. B., Campbell, D. L., Rasmusson, R. L., Lamson, D. R., Morales, M. J., Zhang, Y., and Strauss, H. C., 1994, Cloning and characterization of an I_{to}-like potassium channel from ferret ventricle, *Am. J. Physiol.* **267**:H1388–H1395.

Coraboeuf, E., and Carmeliet, E., 1982, Existence of two transient outward currents in sheep Purkinje fibers, *Pflügers Arch.* **392**:352–359.

Covarrubias, M., Wei, A., and Salkoff, L., 1991, *Shaker, Shal, Shab,* and *Shaw* express independent K^+ current systems, *Neuron* **7**:763–773.

Covarrubias, M., Wei, A., Salkoff, L., and Vybas, T. B., 1994, Elimination of rapid potassium channel inactivation by phosphorylation of the inactivation gate, *Neuron* **13**:1403–1412.

Crumb, W. J., Jr., Pigott, J. D., and Clarkson, C. W., 1995, Comparison of the transient outward current in young and adult human atrial myocytes: Evidence for developmental changes, *Am. J. Physiol.* **268**:H1335–H1342.

Curran, M. E., Splawski, I., Timothy, K. W., Vincent, G. M., Green, E. D., and Keating, M. T., 1995, A molecular basis for cardiac arrhythmia: *herg* mutations cause long QT syndrome, *Cell* **80**:795–803.

Daoud, E. G., Bogun, F., Goyal, R., Harvey, M., Man, K. C., Strickberger, A., and Morady, F., 1996, Effect of atrial fibrillation on atrial refractoriness in humans, *Circulation* **94**:1600–1606.

Davies, M. P., An, R. H., Doevendans, P., Kubalak, S., Chien, K. R., and Kass, R. S., 1996, Developmental changes in ionic channel activity in the embryonic murine heart, *Circ. Res.* **78**:15–25.

Deal, K. K., England, S. K., and Tamkun, M. M., 1996, Molecular physiology of cardiac potassium channels, *Physiol. Rev.* **76**:49–67.

Deck, K. A., and Trautwein, W., 1964, Ionic currents in cardiac excitation. Pflügers Archives. *Eur. J. Physiol.* **280**:63–80.

Dixon, J. E., and McKinnon, D., 1994, Quantitative analysis of mRNA expression in atrial and ventricular muscle of rats, *Circ. Res.* **75**:252–260.

Dixon, J. E., Shi, W. S., Wang, H. S., McDonald, C., Yu, H., Wymore, R. S., Cohen, I. S., and McKinnon, D., 1996, The role of the Kv4.3 K^+ channel in ventricular muscle: A molecular correlate for the transient outward current, *Circ. Res.* **79**:659–668.

Drain, P., Dubin, A. E., and Aldrich, R. W., 1994, Regulation of *Shaker* K^+ channel inactivation gating by the cAMP-dependent protein kinase, *Neuron* **12**:1097–1109.

Drewe, J. A., Verma, S., Frech, G., and Joho, R. L., 1992, Distinct spatial and temporal expression patterns of K^+ channel mRNAs from different subfamilies, *J. Neurosci.* **12**:538–548.

Drici, M. D., Burklow, T. R., Haridasse, V., Glazer, R. I., and Woosley, R. L., 1996, Sex hormones prolong the QT interval and down regulate potassium channel expression in the rabbit heart, *Circulation* **94**:1471–1474.

Dudel, J., Peper, K., Rudel, R., Trautwein, W., 1967, The dynamic chloride component of membrane current in Purlinje fibers. *Pflügers Archiv. fur die Gesamte Physiologie des Menschen und der Tiere.* **295**(3):197–212.

Duff, H. J., Feng, Z. P., Wang, L., and Sheldon, R. S., 1997, Regulation of expression of the [^{3}H]dofetilide binding site associated with the delayed rectifier K$^+$ channel by dexamethasone in neonatal mouse ventricle, *J. Mol. Cell. Cardiol.* **29**:1959–1965.

Ehring, G. R., Kerschbaum, H. H., Eder, C., Neben, A. L., Fanger, C. M., Khoury, R. M., Negulescu, P., and Cahalan, M. D., 1998, A nongenomic mechanism for progesterone-mediated immunosuppression— inhibition of K$^+$ channels, Ca^{2+} signaling, and gene expression in T lymphocytes, *J. Exp. Med.* **188**:1593–1602.

England, S. K., Uebele, V. N., Shear, H., Kodali, K., Bennett, P. B., and Tamkun, M. M., 1995a, Characterization of a K$^+$ channel β subunit expressed in human heart, *Proc. Natl. Acad. Sci., U.S.A.* **92**:6309–6313.

England, S. K., Uebele, V. N., Kodali, K., Bennett, P. B., and Tamkun, M. M., 1995b, A novel K$^+$ channel β subunit (hKv?1. 3) is produced via alternative mRNA splicing, *J. Biol. Chem.* **270**:28531–28534.

Escande, D., Loisance, D., Planche, C., and Coraboeuf, E., 1985, Age-related changes of the action potential plateau shape in isolated human atrial fibers, *Am. J. Physiol.* **249**:H843–H850.

Escande, D., Coulombe, A., Faivre, J. F., Deroubaix, E., and Corabouef, E., 1987, Two types of transient outward currents in adult human atrial cells, *Am. J. Physiol.* **252**:H142–H148.

Fadool, D. A., Holmes, T. C., Berman, K., Dagan, D., and Levitan, I. B., 1997, Tyrosine phosphorylation modulates current amplitude and kinetics of a neuronal voltage-gated potassium channel, *J. Neurophysiol.* **78**:1563–1573.

Fedida, D., and Giles, W. R., 1991, Regional variations in action potentials and transient outward current in myocytes isolated from rabbit left ventricle, *J. Physiol. (London)* **442**:191–209.

Fedida, D., Wible, B., Wang, Z., Fermini, B., Faust, F., Nattel, S., and Brown, A. M., 1993, Identity of a novel delayed rectifier current from human heart with a cloned K$^+$ channel current, *Circ. Res.* **73**:210–216.

Feng, J., Wible, B., Li, G. R., Wang, Z., and Nattel, S., 1997, Antisense oligonucleotides directed against Kv1.5 mRNA specifically inhibit ultrarapid delayed rectifier K$^+$ current in cultured adult human atrial myocytes, *Circ. Res.* **80**:572–579.

Fermini, B., Wang, Z., Duan, D., and Nattel, S., 1992, Differences in rate dependence of the transient outward current in rabbit and human atrium, *Am. J. Physiol.* **263**:H1747–H1754.

Fink, M., Duprat, F., Lesage, F., Heurteaux, C., Romey, G., Barhanin, J., and Lazdunski, M., 1996, A new K$^+$ channel β subunit to specifically enhance Kv2.2 (CDRK) expression, *J. Biol. Chem.* **271**:26341–26348.

Fiset, C., Clark, R. B., Shimoni, Y., and Giles, W. R., 1997, Shal-type channels contribute to the Ca^{2+}-independent transient outward K$^+$ current in rat ventricle, *J. Physiol. (London)* **500**:51–64.

Fiset, C., Clark, R. B., Larsen, T. S., and Giles, W. R., 1998, A rapidly activating, sustained K$^+$ current modulates repolarization and excitation–contraction coupling in adult mouse ventricles, *J. Physiol. (London)* **504**:557–563.

Fisher, D. A., Dussault, J. H., Sack, J., and Chopra, I., 1977, Ontogenesis of hypothalamic–pituitary–thyroid function and metabolism in man, sheep and rat, *Rec. Prog. Horm. Res.* **33**:59–107.

Folander, K., Smith, J. S., Antanavage, J., Bennett, C., Stein, R. B., and Swanson, R., 1990, Cloning and expression of the delayed rectifier I$_{sK}$ channel from neonatal rat heart and diethylstilbestrol-primed rat uterus, *Proc. Natl. Acad. Sci., U.S.A.* **87**:2975–2979.

Follmer, C. H., and Colatsky, T. J., 1990, Block of delayed rectifier potassium current, I$_K$, by flecainide and E-4031 in cat ventricular myocytes, *Circulation* **82**:289–293.

Fozzard, H. A., and Hiraoka, M., 1973, The positive dynamic current and its inactivation properties in cardiac Purkinje fibres, *J. Physiol. (London)* **234**:569–586.

Freeman, L. C., and Kass, R. S., 1993, Delayed rectifier potassium channels in ventricle and sinoatrial node of guinea pig: Molecular and regulatory properties, *Cardiovasc. Drugs Ther.* **7**:627–635.

Freeman, L. C., Pacioretty, L. M., Moise, N. S., Kass, R. S., and Gilmour, R. F., Jr., 1997, Decreased density of I$_{to}$ in left ventricular myocytes from German shepherd dogs with inherited arrhythmias, *J. Cardiovasc. Electrophysiol.* **8**:872–883.

Furukawa, T., Myerburg, R. J., Furukawa, N., Bassett, A. L., and Kimura, S., 1990, Differences in transient outward currents of feline endocardial and epicardial myocytes, *Circ. Res.* **67**:1287–1291.

Furukawa, T., Kimura, S., Furukawa, N., Bassett, A. L., and Myerburg, R. J., 1992, Potassium rectifier currents differ in myocytes of endocardial and epicardial origin, *Circ. Res.* **70**:91–103.

Gadsby, D. C., 1990, Effects of β adrenergic catecholamines on membrane currents in cardiac cells, in: *Cardiac Electrophysiology: A Textbook* (M. R. Rosen, M. J. Jansen, and A. L. Wit, eds.), Futura Publishing Co., Mt. Kisco, New York, pp. 857–876.

Gidh-Jain, M., Huang, B., Jain, P., and El-Sherif, N., 1996, Differential expression of voltage-gated K^+ channel genes in left ventricular remodeled myocardium after experimental myocardial infarction, *Circ. Res.* **79:**669–675.

Giles, W. R., and Imaizumi, Y., 1988, Comparison of potassium currents in rabbit atrial and ventricular cells, *J. Physiol. (London)* **405:**123–145.

Giles, W. R., and Van Ginneken, A. C., 1985, A transient outward current in isolated cells from the crista terminalis of rabbit heart, *J. Physiol. (London)* **368:**243–264.

Giles, W. R., Clark, R. B., and Braun, A. P., 1996, Ca^{2+}-independent transient outward current in mammalian heart, in: *Molecular Physiology and Pharmacology of Cardiac Ion Channels and Transporters* (M. Morad, Y. Kurachi, A. Noma, and M. Hosada, eds.), Kluwer Press, Amsterdam, pp. 141–168.

Goldstein, S. A., and Miller, C., 1991, Site-specific mutations in a minimal voltage-gated K^+ channel alter ion selectivity and open channel block, *Neuron* **7:**403–408.

Gomez, A. M., Bénitah, J. -P., Henzel, D., Vinet, A., Lorente, P., and Delgado, C., 1997, Modulation of electrical heterogeneity by compensated hypertrophy in rat left ventricle, *Am. J. Physiol.* **272:**H1078–H1086.

Gross, G. J., Burke, R. P., and Castle, N. A., 1995, Characterization of transient outward current in young human myocytes, *Cardiovasc. Res.* **29:**112–117.

Guo, W., Kamiya, K., and Toyama, J., 1995, bFGF promotes functional expression of transient outward currents in cultured neonatal rat ventricular myocytes, *Pflügers Arch.* **430:**1015–1017.

Guo, W., Kamiya, K., and Toyama, J., 1996, Modulated expression of transient outward current in cultured neonatal rat ventricular myocytes: Comparison with development *in situ*, *Cardiovasc. Res.* **32:**524–533.

Guo, W., Kamiya, K., and Toyama, J., 1997a, Roles of the voltage-gated K^+ channel subunits, Kv1.5 and Kv1.4, in the developmental changes of K^+ currents in cultured neonatal rat ventricular cells, *Pflügers Arch.* **434:**206–208.

Guo, W., Kamiya, K., and Toyama, J., 1997b, Differential effects of chronic membrane depolarization on the K^+ channel activities in cultured rat ventricular cells, *Cardiovasc. Res.* **33:**139–146.

Guo, W., Kada, K., Kamiya, K., and Toyama, J., 1997c, IGF-I regulates K^+ channel expression of cultured neonatal rat ventricular myocytes, *Am. J. Physiol.* **272:**H2599–H2606.

Guo, W., Kamiya, K., Liu, W., and Toyama, J., 1997d, Developmental changes of the ultrarapid delayed rectifier K^+ current in rat ventricular myocytes, *Pflügers Arch.* **433:**442–445.

Guo, W., Kamiya, K., Hojo, M., Kodama, I., and Toyama, J., 1998, Regulation of Kv4.2 and Kv1.4 K^+ channel expression by myocardial hypertrophic factors in cultured newborn rat ventricular cells, *J. Mol. Cell. Cardiol.* **30:**1449–1455.

Guo, W., Xu, H., London, B., and Nerbonne, J. M., 1999, Molecular basis of transient outward K^+ diversity in mouse ventricular myocytes, *J. Physiol. London)*, submitted.

Guy, H. R., Durell, S. R., Warmke, J., Drysdale, R., Ganetzky, B., 1991, Similarities in amino acid sequences of Drosophila eag and cyclic nucleotide-gated channels. *Science* **254(5032):**730.

Hiraoka, M., and Kawano, S., 1989, Calcium-sensitive and insensitive transient outward current in rabbit ventricular myocytes, *J. Physiol. (London)* **410:**187–212.

Honoré, E., Attali, B., Romey, G., Heurteaux, C., Ricard, P., Lesage, F., and Lazdunski, M., 1991, Cloning, expression, pharmacology and regulation of a delayed rectifier K^+ channel in mouse heart, *EMBO J.* **10:**2805–2811.

Horie, M., Hayashi, S., and Kawai, C., 1990, Two types of delayed rectifying K^+ channels in atrial cells of guinea pig heart, *Jpn. J. Physiol.* **40:**479–490.

Hugnot, J. P., Salinas, M., Lesage, F., Guillemare, E., De Weile, J., Heurteaux, C., Mattei, M. G., and Lazdunski, M., 1996, Kv8.1, a new neuronal potassium channel subunit with specific inhibitory Fproperties towards *Shab* and *Shaw* channels, *EMBO J.* **15:**3322–3331.

Hume, J. R., and Uehara, A., 1985, Ionic basis of the different action potential configurations of single guinea-pig atrial and ventricular myocytes, *J. Physiol. (London)* **368:**525–544.

Inoue, M., and Imanaga, I., 1993, Masking of A-type K^+ channel in guinea pig cardiac cells by extracellular Ca^{++}, *Am. J. Physiol.* **264:**C1434–C1438.

Jan, L. Y., and Jan, Y. N., 1992, Structural elements involved in specific K^+ channel functions. *Annu. Rev. Physiol.* **54:**537–555.

Jeck, C. D., and Boyden, P. A., 1992, Age-related appearance of outward currents may contribute to developmental differences in ventricular repolarization, *Circ. Res.* **71:**1390–1403.

Jiang, M., Tseng-Crank, J., and Tseng, G. N., 1997, Suppression of slow delayed rectifier current by a truncated isoform of KvLQT1 cloned from normal human heart, *J. Biol. Chem.* **272**:24109–24112.

Johns, D. C., Nuss, H. B., and Marban, E., 1997, Suppression of neuronal and cardiac transient outward currents by viral gene transfer of dominant-negative Kv4.2 constructs, *J. Biol. Chem.* **272**: 31598–31603.

Kääb, S., Nuss, H. B., Chiamvimonvat, N., O'Rourke, B., Pak, P. H., Kass, D. A., Marban, E., and Tomaselli, G. F., 1996, Ionic mechanism of action potential prolongation in ventricular myocytes from dogs with pacing-induced heart failure, *Circ. Res.* **78**:262–273.

Kääb, S., Dixon, J., Duc, J., Ashen, D., Näbauer, M., Beuckelmann, D. J., Steinbeck, G., McKinnon, D., and Tomaselli, G. F., 1998, Molecular basis of transient outward potassium current downregulation in human heart failure—a decrease in Kv4.3 mRNA correlates with a reduction in current density, *Circulation* **98**:1383–1393.

Kalman, K., Nguyen, A., Tseng-Crank, J., Dukes, I. D., Chandy, G., Hustad, C. M., Copeland, N. G., Jenkins, N. A., Mohrenweiser, H., Brandriff, B., Cahalan, M., Gutman, G. A., and Chandy, K. G., 1998, Genomic organization, chromosomal localization, tissue distribution, and biophysical characterization of a novel mammalian *Shaker*-related voltage-gated potassium channel, Kv1.7, *J. Biol. Chem.* **273**:5851–5857.

Kamb, A., Tseng-Crank, J., and Tanouye, M. A., 1988, Multiple components of the *Drosophila Shaker* gene may contribute to potassium channel diversity, *Neuron* **1**:421–430.

Kenyon, J. L., and Gibbons, W. R., 1979a, Influence of chloride, potassium, and tetraethylammonium on the early outward current of sheep cardiac Purkinje fibers, *J. Gen. Physiol.* **73**:117–138.

Kenyon, J. L., and Gibbons, W. R., 1979b, 4-Aminopyridine and the early outward current of sheep cardiac Purkinje fibers, *J. Gen. Physiol.* **73**:139–157.

Kiehn, J., Karle, C., Thomas, D., Yao, X. Z., Brachman, J., and Kubler, W., 1998, HERG potassium channel activation is shifted by phorbol esters via protein kinase A-dependent pathways, *J. Biol. Chem.* **273**:25285–25291.

Kilborn, M. J., and Fedida, D. A., 1990, A study of the developmental changes in outward currents in rat ventricular myocytes, *J. Physiol. (London)* **430**:37–60.

Konarzewska, H., Peeters, G. A., and Sanguinetti, M. C., 1995, Repolarizing K^+ currents in nonfailing human hearts: Similarities between right-septal subendocardial and left epicardial ventricular myocytes, *Circulation* **92**:1179–1187.

Kong, W., Po, S., Yamagishi, T., Ashen, M. D., Stetten, G., and Tomaselli, G. F., 1998, Isolation and characterization of the human gene encoding I_{to}: Further diversity by alternative mRNA splicing, *Am. J. Physiol.* **275**:H1963–H1970.

Kupershmidt, S., Snyders, D., Raes, A., and Roden, D., 1998, A K^+ channel splice variant common in human heart lacks a C-terminal domain required for expression of rapidly activating delayed rectifier current, *J. Biol. Chem.* **273**:27231–27235.

Lees-Miller, J. P., Kondo, C., Wang, L., and Duff, H. J., 1997, Electrophysiological characterization of an alternatively processed ERG K^+ channel in mouse and human hearts, *Circ. Res.* **81**:719–726.

Lesage, F., Attali, B., Lazdunski, M., and Barhanin, J., 1992, I_{sK}, a slowly activating voltage-sensitive K^+ channel: Characterization of multiple cDNAs and gene organization in the mouse, *FEBS Lett.* **301**:168–172.

Levitan, E. S., Gealy, R., Trimmer, J. S., and Takimoto, K., 1995, Membrane depolarization inhibits Kv1.5 voltage-gated K^+ channel gene transcription and protein expression in pituitary cells, *J. Biol. Chem.* **270**:6036–6041.

Levitan, E. S., Hershman, K. M., Sherman, T. G., and Takimoto, K., 1996, Dexamethasone and stress upregulate Kv1.5 K^+ channel gene expression in rat ventricular myocytes, *Neuropharmacology* **35**:1001–1006.

Li, G. R., Feng, J., Yue, L., Carrier, M., and Nattel, S., 1996, Evidence for two components of delayed rectifier K^+ current in human ventricular myocytes, *Circ. Res.* **78**:689–696.

Litovsky, S. H., and Antzelevitch, C., 1988, Transient outward current prominent in canine ventricular epicardium but not endocardium, *Circ. Res.* **72**:1092–1103.

Liu, D. -W., and Antzelevitch, C., 1995, Characteristics of the delayed rectifier current (I_{Kr} and I_{Ks}) in canine ventricular epicardial, midmyocardial, and endocardial myocytes, *Circ. Res.* **76**:351–365.

Liu, D. -W., Gintant, G. A., and Antzelevitch, C., 1993, Ionic basis for electrophysiological distinctions among epicardial, midmyocardial and epicardial myocytes, *Circ. Res.* **72**:671–687.

Liu, X. K., Katchman, A., Ebert, S. N., and Woosley, R. L., 1998, The antiestrogen tamoxifen blocks the delayed rectifier potassium current, I_{Kr}, in rabbit ventricular myocytes, *J. Pharmacol. Exp. Ther.* **287**:877–883.

London, B., Trudeau, M. C., Newton, K. P., Beyer, A. K., Copeland, N. G., Gilbert, D. J., Jenkins, N. A., Satler, C. A., and Robertson, G. A., 1997, Two isoforms of the mouse ether-a-go-go-related gene coassemble to form channels with properties similar to the rapidly activating component of the cardiac delayed rectifier K$^+$ current, *Circ. Res.* **81**:870–878.

London, B., Wang, D. W., Hill, J. A., and Bennett, P. B., 1998a, The transient outward current in mice lacking the potassium channel gene Kv1.4, *J. Physiol. (London)* **81**:870–878.

London, B., Jeron, A., Zhou, J., Buckett, P., Han, X., Mitchell, G. F., and Koren, G., 1998b, Long QT and ventricular arrhythmias in transgenic mice expressing the N terminus and the first transmembrane segment of a voltage-gated potassium channel, *Proc. Natl. Acad. Sci., U.S.A.* **95**:2926–2931.

London, B., Aydar, E., Lewarchik, C. M., Seibel, J. S., January, C. T., and Robertson, G. A., 1998c, N- and C-terminal isoforms of HERG in the human heart, *Biophys. J.* **74**:A26.

MacKinnon, R., 1991, Determination of the subunit stoichiometry of a voltage-activated potassium channel, *Nature* **350**:232–235.

Main, M. C., Bryant, S. M., and Hart, G., 1998, Regional differences in action potential characteristics and membrane currents of guinea-pig left ventricular myocytes, *Exp. Physiol.* **83**:747–761.

Majumder, K., Debiasi, M., Wang, Z., and Wible, B., 1995, Molecular cloning and functional expression of a novel potassium channel β subunit from human atrium, *FEBS Lett.* **361**: 13–16.

Matsubara, H., Liman, E. R., Hess, P., and Koren, G., 1991, Pretranslational mechanisms determine the type of potassium channel expressed in the rat skeletal and cardiac muscles, *J. Biol. Chem.* **266**:13324–13328.

Matsubara, H., Suzuki, J., and Inada, M., 1993, *Shaker*-related potassium channel, Kv1.4, mRNA regulation in cultured rat heart myocytes and differential expression of Kv1.4 and Kv1.5 genes in myocardial development and hypertrophy, *J. Clin. Invest.* **92**:1659–1666.

Mays, D. J., Foose, J. M., Philipson, L. H., and Tamkun, M. M., 1995, Localization of the Kv1.5 K$^+$ channel protein in explanted cardiac tissue, *J. Clin. Invest.* **96**:282–292.

McDonald, T. V., Yu, Z., Ming, Z., Palma, E., Meyers, M. B., Wang, K. W., Goldstein, S. A., and Fishman, G. I., 1997, A minK–HERG complex regulates the cardiac potassium current $I_{(Kr)}$. *Nature* **388**:289–292.

Mészáros, J., Ryder, K. O., and Hart, G., 1996, Transient outward current in catecholamine-induced cardiac hypertrophy in the rat, *Am. J. Physiol.* **271**:H2360–H2367.

Morales, M. J., Castellino, R. C., Crews, A. L., Rasmusson, R. L., and Strauss, H. C., 1995, A novel β subunit increases the rate of inactivation of specific voltage-gated potassium channel α subunits, *J. Biol. Chem.* **270**:6272–6277.

Mori, Y., Matsubara, H., Folco, E., Siegel, A., and Koren, G., 1993, The transcription of a mammalian voltage-gated potassium channel is regulated by cAMP in a cell-specific manner, *J. Biol. Chem.* **268**:26482–26493.

Muniz, Z. M., Parcej, D. N., and Dolly, J. O., 1992, Characterization of monoclonal antibodies against voltage-dependent K$^+$ channels raised using α-dendrotoxin acceptors purified from bovine brain, *Biochemistry* **31**:12297–12303.

Murai, T., Kakizuka, A., Takumi, T., Ohkubo, H., and Nakanishi, S., 1989, Molecular cloning and sequence analysis of human genomic DNA encoding a novel membrane protein which exhibits slowly activating potassium channel activity, *Biochem. Biophys. Res. Commun.* **61**:176–181.

Murakoshi, H., Shi, G. Y., Scannevin, R. H., and Trimmer, J. S., 1997, Phosphorylation of the Kv2.1 K$^+$ channel alters voltage-dependent activation, *Mol. Pharmacol.* **52**:821–828.

Murray, K. T., Fahrig, S. A., Deal, K. K., Po, S. S., Hu, N. N., Snyders, D. J., Tamkun, M. M., and Bennett, P. B., 1994, Modulation of an inactivating human cardiac K$^+$ channel by protein kinase C, *Circ. Res.* **75**:999–1005.

Näbauer, M., and Kääb, M., 1998, Potassium channel down regulation in heart failure, *Cardiovasc. Res.* **37**:324–334.

Näbauer, M., Beuckelmann, D. J., and Erdmann, E., 1993, Characteristics of transient outward current in human ventricular myocytes from patients with terminal heart failure, *Circ. Res.* **73**:386–394.

Näbauer, M., Barth, A., and Kääb, S., 1998, A second calcium-independent transient outward current present in human ventricular myocardium, *Circulation* **98**:I-231.

Nagaya, N., and Papazian, D. M., 1997, Potassium channel α and β subunits assemble in the endoplasmic reticulum, *J. Biol. Chem.* **272**:3022–3027.

Nakahira, K., Shi, G., Rhodes, K. J., and Trimmer, J. S., 1996, Selective interaction of voltage-gated K$^+$ channel β subunits with α subunits, *J. Biol. Chem.* **271**:7084–7089.

Nakamura, K., and Iijima, T., 1994, Postnatal changes in mRNA expression of the K^+ channel in rat cardiac ventricles, *Jpn. J. Pharmacol.* **66**:489–492.

Nakamura, T. Y., Coetzee, W. T., Vega-Saenz De Miera, E., Artman, M., and Rudy, B., 1997, Modulation of Kv4 channels, key components of rat ventricular transient outward current, by PKC, *Am. J. Physiol.* **273**:H1775–H1786.

Nakayama, T., and Irisawa, H., 1985, Transient outward current carried by potassium and sodium in quiescent atrioventricular node cells of rabbits, *Circ. Res.* **57**:65–73.

Nerbonne, J. M., 1998, Regulation of voltage-gated K^+ channel expression in the developing mammalian myocardium, *J. Neurobiol.* **37**:37–59.

Nishiyama, A., Kambe, F., Kamiya, K., Seo, H., and Toyama, J., 1998, Effects of thyroid status on expression of voltage-gated potassium channels in rat left ventricle, *Cardiovasc. Res.* **40**:343–351.

Nuss, H. B., and Marban, E., 1994, Electrophysiological properties of neonatal mouse cardiac cells in primary culture, *J. Physiol. (London)* **479**:265–279.

Nuss, H. B., Johns, D. C., Kääb, S., Tomaselli, G. F., Kass, D., Lawrence, J. H., and Marban, E., 1996, Reversal of potassium channel deficiency in cells from failing hearts by adenoviral gene transfer: A prototype for gene therapy for disorders of cardiac excitability and contractility, *Gene Ther.* **3**:900–912.

Ohya, S., Tanaka, M., Oku, T., Asai, Y., Watanabe, M., Giles, W. R., and Imaizumi, Y., 1997, Molecular cloning and tissue distribution of an A-type K^+ channel α-subunit, Kv4.3 in the rat, *FEBS Lett.* **420**:47–53.

Papazian, D. M., Schwarz, T. L., Temple, B. L., Jan, Y. N., and Jan, L. Y., 1987, Cloning of genomic and complementary DNA from *Shaker*, a putative potassium channel gene from *Drosophila*, *Science* **237**:749–753.

Petersen, K. R., and Nerbonne, J. M., 1999, Expression environment determines K^+ current properties: Kv1 and Kv4 α subunit-induced K^+ currents in mammalian cell lines and cardiac myocytes, *Pflügers Arch.* **437**:381–392.

Po, S., Snyders, D. J., Baker, R., Tamkun, M. M., and Bennett, P. B., 1992, Functional expression of an inactivating potassium channel cloned from human heart, *Circ. Res.* **71**:732–736.

Po, S., Roberds, S., Snyders, D. J., Tamkun, M. M., and Bennett, P. B., 1993, Heteromultimeric assembly of human potassium channels: Molecular basis of a transient outward current?, *Circ. Res.* **72**:1326–1336.

Pond, A. L., Scheve, B. K., Benedict, A. T., Petrecca, K., Van Wagoner, D. R., Shrier, A., and Nerbonne, J. M., 2000, Expression of distinct *ERG* proteins in rat, mouse, and human heart: Relation to functional I_{Kr} channels, *J. Biol. Chem* **275**:5997–6006.

Pongs, O., 1992, Molecular biology of voltage-dependent potassium channels, *Physiol. Rev.* **72**:S69–S88.

Pongs, O., Kecskemethy, N., Muller, R., Krah-Jentgens, I., Baumann, A., Kiltz, H. H., Canal, I., Llamazares, S., and Ferrus, A., 1988, *Shaker* encodes a family of putative potassium channel proteins in the nervous system, *EMBO J.* **7**:1087–1096.

Potreau, D., Gomez, J. P., and Fares, N., 1996, Depressed transient outward current in single hypertrophied cardiomyocytes isolated from the right ventricle of ferret heart, *Cardiovasc. Res.* **30**:440–448.

Qin, D., Zhang, Z. -H., Boutjdir, M., Jain, P., and El-Sherif, N., 1996, Cellular and ionic basis of arrhythmias in post-infarction remodeled ventricular myocardium, *Circ. Res.* **79**:461–473.

Rettig, J., Heinemann, S. H., Wunder, F., Lorra, C., Parcej, D. N., Dolly, J. O., and Pongs, O., 1994, Inactivation properties of voltage-gated K^+ channels altered by presence of β-subunit, *Nature* **369**:289–294.

Roberds, S. L., and Tamkun, M. M., 1991a, Cloning and tissue-specific expression of five voltage-gated potassium channel cDNAS expressed in rat heart, *Proc. Natl. Acad. Sci. U.S.A.* **88**:1798–1802.

Roberds, S. L., and Tamkun, M. M., 1991b, Developmental expression of cloned cardiac potassium channels, *FEBS Lett.* **284**:152–154.

Roberds, S. L., Knoth, K. M., Po, S., Blair, T. A., Bennett, P. B., Hartshorne, R. P., Snyders, D. J., and Tamkun, M. M., 1993, Molecular biology of the voltage-gated potassium channels of the cardiovascular system, *J. Cardiovasc. Electrophysiol.* **4**:68–80.

Robertson, B., 1997, The real life of voltage-gated K^+ channels: More than model behaviour, *Trends Pharmacol. Sci.* **18**:474–483.

Roeper, J., Lorra, C., and Pongs, O., 1997, Frequency-dependent inactivation of mammalian A-type K^+ channel Kv 1.4 regulated by Ca^{2+}/calmodulin-dependent protein kinase, *J. Neurosci.* **17**:3379–3391.

Rosen, M. R., Cohen., I. R., Danilo, P., and Steinberg, S. P., 1998, The heart remembers, *Cardiovasc. Res.* **40**:469–482.

Rozanski, G. J., Xu, Z., Zhang, K., and Patel, K. P., 1998, Altered K$^+$ current of ventricular myocytes in rats with chronic myocardial infarction, *Am. J. Physiol.* **273**:259–265.

Salata, J. J., Jurkiewicz, N. K., Jow, B., Folander, K., Guinoso, P. J., Raynor, B., Swanson, R., and Fermini, B., 1996, I_K of rabbit ventricle is composed of two currents: Evidence for I_{Ks}, *Am. J. Physiol.* **271**:H2477–H2489.

Salinas, M., Duprat, F., Heurteaux, C., Hugnot, J. P., and Lazdunski, M., 1997, New modulatory α subunits for mammalian *Shab* K$^+$ channels, *J. Biol. Chem.* **272**:24371–24379.

Sánchez-Chapula, J., Elizalde, A., Navarro-Polanco, R., and Barajas, H., 1994, Differences in outward currents between neonatal and adult rabbit ventricular cells, *Am. J. Physiol.* **266**:H1184–H1194.

Sanguinetti, M. C., and Jurkiewicz, N. K., 1990, Two components of cardiac delayed rectifier K$^+$ current, *J. Gen. Physiol.* **96**:195–215.

Sanguinetti, M. C., and Jurkiewicz, N. K., 1991, Delayed rectifier outward K$^+$ current is composed of two currents in guinea pig atrial cells, *Am. J. Physiol.* **260**:H393–H399.

Sanguinetti, M. C., Jurkiewicz, N. K., 1992, Role of external Ca^{2+} and K$^+$ in gating of cardiac delayed rectifier K$^+$ currents, *Pflügers Archiv.–Eur. J. Physiol.* **420(2)**:180–186.

Sanguinetti, M. C., Jiang, C., Curran, M. E., and Keating, M. T., 1995, A mechanistic link between an inherited and an acquired cardiac arrhythmia: *HERG* encodes the I_{Kr} potassium channel, *Cell* **81**:299–307.

Sanguinetti, M. C, Curran, M. E., Zou, A., Shen, J., Spector, P. S., Atkinson, D. L., and Keating, M. T., 1996, Coassembly of KvLQT1 and minK (I_{sK}) proteins to form cardiac I_{Ks} potassium channel, *Nature* **384**:80–83.

Sanguinetti, M. C., Johnson, J. J., Hammerland, L. G., Kelbaugh, P. R., Volkman, R. A., Saccomano, N. A., and Mueller, A. L., 1997, Heteropodatoxins: Peptides isolated from spider venom that block Kv4.2 potassium channels, *Mol. Pharmacol.* **51**:491–498.

Schroeder, B. J., Kubisch, C., Stein, V., and Jentsch, T. J., 1998, Moderate loss of cyclic-AMP-modulated KCNQ2/KCNQ3 K$^+$ channels causes epilepsy, *Nature* **396**:687–690.

Schwalbe, R. A., Wang, Z., Wible, B., and Brown, A. M., 1995, Potassium channel structure and function as reported by a single glycosylation sequon, *J. Biol. Chem.* **270**:15336–15340.

Schwarz, T. L., Temple, B. L., Papazian, D. M., Jan, Y. N., and Jan, L. Y., 1988, Multiple potassium-channel components are produced by alternative splicing at the *Shaker* locus in *Drosophila*, *Nature* **331**:137–142.

Sewing, S., Roeper, J., and Pongs, O., 1996, Kvβ1 subunit binding specific for *Shaker*-related potassium channel α subunits, *Neuron* **16**:455–463.

Shi, G., Nakahira, K., Hammond, S., Rhodes, K. J., Schechter, L. E., and Trimmer, J. S., 1996, β subunits promote K$^+$ channel surface expression through effects early in biosynthesis, *Neuron* **16**:843–852.

Shi, W., Wymore, R. S., Wang, H. -S., Pan, Z., Cohen, I. S., McKinnon, D., and Dixon, J. E., 1997, Identification of two nervous system-specific members of the erg potassium channel gene family, *J. Neurosci.* **17**:9423–9432.

Shibasaki, T., 1987, Conductance and kinetics of delayed rectifier potassium channels in nodal cells of the rabbit heart, *J. Physiol.* **387**:227–250.

Shibata, E. F., Drury, T., Refsum, H., Aldrete, V., and Giles, W., 1989, Contributions of a transient outward current to repolarization in human atrium, *Am. J. Physiol.* **257**:H1773–H1781.

Shimoni, Y., Banno, H., and Clark, R. B., 1992, Hyperthyroidism selectively modifies a transient outward potassium current in rabbit ventricular and atrial myocytes, *J. Physiol. (London)* **457**:369–389.

Shimoni, Y., Firek, I., Severson, D., and Giles, W. R., 1994, Short term diabetes alters K$^+$ currents in rat ventricular myocytes, *Circ. Res.* **74**:620–628.

Shimoni, Y., Severson, D., and Giles, W. R., 1995, Thyroid status and diabetes modulate regional differences in potassium currents in rat ventricle, *J. Physiol. (London)* **488**:673–688.

Shimoni, Y., Fiset, C., Clark, R. B., Dixon, J. E., McKinnon, D., and Giles, W. R., 1997, Thyroid hormone regulates postnatal expression of transient K$^+$ channel isoforms in rat ventricle, *J. Physiol. (London)* **500**:65–73.

Shipsey, S. J., Bryant, S. M., and Hart, G., 1997, Effects of hypertrophy on regional action potential characteristics in the rat left ventricle: A cellular basis for T wave inversion, *Circulation* **96**:2061–2068.

Swartz, K. J., and MacKinnon, R., 1995, An inhibitor of the Kv2.1 potassium channel isolated from the venom of a Chilean tarantula, *Neuron* **15**:941–949.

Tai, K.-K., and Goldstein, S. A. N., 1998, The conduction pore of a cardiac potassium channel, *Nature* **391**:605–607.

Takimoto, K., and Levitan, E. S., 1994, Glucocorticoid induction of Kv1.5 K^+ channel gene expression in ventricle of rat heart, *Circ. Res.* **75:**1006–1013.

Takimoto, K., and Levitan, E. S., 1996, Altered K^+ channel subunit composition following hormone induction of Kv1.5 gene expression, *Biochemistry* **35:**14149–14156.

Takimoto, K., Gealy, R., and Levitan, E. S., 1995, Multiple protein kinases are required for basal Kv1.5 K^+ channel expression in GH3 clonal pituitary cells, *Biochim. Biophys. Acta* **1265:**22–28.

Takimoto, K., Li, D., Hershman, K. M., Li, P., Jackson, E. K., and Levitan, E. S., 1996, Decreased expression of Kv4.2 and novel Kv4.3 K^+ channel subunit mRNAs in ventricles of renovascular hypertensive rats, *Circ. Res.* **81:**533–539.

Takumi, T., Ohkubo, H., and Nakinishi, S., 1988, Cloning of a membrane protein that induces a slow voltage-gated potassium current, *Science* **242:**1042–1045.

Takum, T., Moriyoshi, K., Aramori, I., Ishii, T., Oiki, S., Okada, Y., Ohkubo, H., Nakanishi, S., 1991, Alteration of channel activities and gating by mutations of slow ISK potassium channel. *J. Biol. Chem.* **266(33):**22192–22198.

Tamkun, M. M., Knoth, K. M., Walbridge, J. A., Kroemer, H., Roden, D. M., and Glover, D. M., 1991, Molecular cloning and characterization of two voltage-gated K^+ channel cDNAs from human ventricle, *FASEB J.* **5:**331–337.

Ten Eick, R. E., Houser, J. R., and Bassett, A. L., 1989, Cardiac hypertrophy and altered cellular activity of the myocardium, in: *Physiology and Pathophysiology of the Heart*, 2nd. ed. (N. Speralakis, ed.), Martinus Nijhoff, Boston, pp. 573–594.

Ten Eick, R. E., Zhang, K., Harvey, R. D., and Bassett, A. L., 1993, Enhanced functional expression of transient outward current in hypertrophied feline myocytes, *Cardiovasc. Drugs Ther.* **7:**611–619.

Thornhill, W. B., Wu, M. B., Jiang, X., Wu, X., Morgan, P. T., and Margiotta, J. F., 1996, Expression of Kv1.1 delayed rectifier potassium channels in *L ec* mutant Chinese hamster ovary cell lines reveals a role for sialidation in channel function, *J. Biol. Chem.* **271:**19093–19098.

Timpe, L. C., Schwarz, T. L., Temple, B. L., Papazian, D. M., Jan, Y. N., and Jan, L. Y., 1988, Expression of functional potassium channels from *Shaker* cDNA in *Xenopus* oocytes, *Nature* **331:**143–145.

Tomita, F., Bassett, A. L., Myerburg, R. J., and Kimura, F., 1994, Diminished transient outward currents in rat hypertrophied ventricular myocytes, *Circ. Res.* **75:**296–303.

Trimmer, J. S., 1991, Immunological identification and characterization of a delayed rectifier K^+ channel polypeptide in rat brain, *Proc. Natl. Acad. Sci., U.S.A.* **88:**10764–10768.

Trudeau, M. C., Warmke, J. W., Ganetzky, B., and Robertson, G. A., 1995, *H-ERG*, a human inward rectifier with structural and functional homology to voltage-gated K^+ channels, *Science* **269:**92–95.

Tseng, G. -N., and Hoffman, B. F., 1989, Two components of transient outward current in canine ventricular myocytes, *Circ. Res.* **64:**633–647.

Tseng-Crank, J. C. L., Tseng, G. -N., Schwartz, A., and Tanouye, M. A., 1990, Molecular cloning and functional expression of a potassium channel cDNA isolated from a rat cardiac library, *FEBS Lett.* **268:**63–68.

Van Wagoner, D. R., Kirian, M., and Lamorgese, M., 1996, Phenylephrine suppresses outward K^+ currents in rat atrial myocytes, *Am. J. Physiol.* **271:**H937–H946.

Van Wagoner, D. R., Pond, A. L., McCarthy, P. M., Trimmer, J. S., and Nerbonne, J. M., 1997, Outward K^+ current densities and Kv1.5 expression are reduced in chronic human atrial fibrillation, *Circ. Res.* **80:**772–781.

Varro, A., Nanasi, P. P., and Lathrop, D. A., 1993, Potassium currents in isolated human atrial and ventricular cardiocytes, *Acta Physiol. Scand.* **149:**133–142.

Veldkamp, M. W., Van Ginneken, A. C. G., and Bouman, L. N., 1993, Single delayed rectifier channels in the membrane of rabbit ventricular myocytes, *Circ. Res.* **72:**865–878.

Wahler, G. M., Dollinger, S. J., Smith, J. M., and Flemal, K. I., 1994, Time course of postnatal changes in rat heart action potential and transient outward current is different, *Am. J. Physiol.* **267:**H1157–H1166.

Walsh, K. B., Arena, J. P., Kwok, W. M., and Freeman, l., 1991, Delayed-rectifier potassium channel activity in isolated membrane patches of guinea pig ventricular myocytes, *Am. J. Physiol.* **260:**H1390–H1393.

Wang, H. S., Pan, Z., Shi, W., Brown, B. S., Wymore, R. S., Cohen, I. S., Dixon, J. E., and McKinnon, D., 1998, KCNQ2 and KCNQ3 potassium channel subunits: Molecular correlates of the M-channel, *Science* **282:**1890–1893.

Wang, K. -W., Tai, K. -K., and Goldstein, S. A. N., 1996, MinK residues line a potassium channel pore, *Neuron* **16:**571–577.

Wang, L., and Duff, H. J., 1996, Identification and characteristics of delayed rectifier K$^+$ current in fetal mouse ventricular myocytes, *Am. J. Physiol.* **270**:H2088–H2093.

Wang, L., and Duff, H. J., 1997, Developmental changes in transient outward current in mouse ventricle, *Circ. Res.* **81**:120–127.

Wang, L., Feng, Z. -P., Kondo, C. S., Sheldon, R. S., and Duff, H. J., 1996, Developmental changes in the delayed rectifier K$^+$ channels in mouse heart, *Circ. Res.* **79**:79–85.

Wang, Q., Curran, M. E., Splawski, I., Burn, T. C., Millholland, J. M., Vanray, T. J., Shen, J., Timothy, K. W., Vincent, G. M., Dejager, T., Schwartz, P. J., Toubin, J. A., Moss, A. J., Atkinson, D. L., Landes, G. M., Connors, T. D., and Keating, M. T., 1996, Positional cloning of a novel potassium channel gene: KvLQT1 mutations cause cardiac arrhythmias, *Nat. Genet.* **12**:17–23.

Wang, Z., Fermini, B., and Nattel S., 1991, Repolarization differences between guinea pig atrial endocardium and epicardium, *Am. J. Physiol.* **260**:H1501–H1506.

Wang, Z., Fermini, B., and Nattel S., 1993a, Delayed rectifier outward current and repolarization in human atrial myocytes, *Circ. Res.* **73**:276–285.

Wang, Z., Fermini, B., and Nattel, S., 1993b, Sustained depolarization-induced outward current in human atrial myocytes: Evidence for a novel delayed rectifier K$^+$ current similar to Kv1.5 cloned channel currents, *Circ. Res.* **73**:1061–1076.

Wang, Z., Fermini, B., and Nattel, S., 1994, Rapid and slow components of delayed rectifier current in human atrial myocytes, *Cardiovasc. Res.* **28**:1540–1546.

Wang, Z., Feng, J., Shi, H., Pond, A. L., Nerbonne, J. M., and Nattel, S., 1999, Potential molecular basis of different physiological properties of transient outward K$^+$ current in rabbit and human atrial myocytes, *Circ. Res.* **84**:551–561.

Warmke, J. W., and Ganetzky, B., 1994, A family of potassium channel genes related to *eag* in *Drosophila* and mammals, *Proc. Natl. Acad. Sci. U.S.A.* **91**:3438–3442.

Warmke, J., Drysdale, R., and Ganetzky, B., 1991, A distinct potassium channel polypeptide encoded by the *Drosophila eag* locus, *Science* **252**:1560–1564.

Wei, A., Covarrubias, M., Butler, A., Baker, K., Pak, M., and Salkoff, L., 1990, K$^+$ current diversity is produced by an extended gene family conserved in *Drosophila* and mouse, *Science* **248**:599–603.

Wettwer, E., Amos, G., Gath, J., Zerkowski, H.-R., Reidemeister, J.-C., and Ravens, U., 1993, Transient outward current in human and rat ventricular myocytes, *Cardiovasc. Res.* **27**:1662–1669.

Wettwer, E., Amos, G., Posival, H., and Ravens, U., 1994, Transient outward current in human and ventricular myocytes of subepicardial and subendocardial origin, *Circ. Res.* **75**:473–482.

Wetzel, G. T., and Klitzner, T. S., 1996, Developmental cardiac electrophysiology: Recent advances in cellular physiology, *Cardiovasc. Res.* **31**:E52–E60.

Wible, B. A., Yang, Q., Kuryshev, Y. A., Accili, E. A., and Brown, A. M., 1998, Cloning and expression of a novel K$^+$ channel regulatory protein, KChAP, *J. Biol. Chem.* **273**:11745–11751.

Wickenden, A. D., Kaprielan, R., Parker, T. G., Jones, O. T., and Backx, P. H., 1997, Effects of development and thyroid hormone on K$^+$ currents and K$^+$ channel gene expression in rat ventricle, *J. Physiol. (London)* **504**:271–286.

Wilde, A. A. M., and Veldkamp, M. W., 1997, Ion channels, the QT interval and arrhythmias, *Pace* **20**:2048–2051.

Xu, X., and Best, P. M., 1991. Decreased transient outward K$^+$ current in ventricular myocytes form acromegalic rats. *Am. J. Physiol.* **260**:H935–H942.

Xu, H., Dixon, J. E., Barry, D. M., Trimmer, J. S., Merlie, J. P., McKinnon, D., and Nerbonne, J. M., 1996, Developmental analysis reveals mismatches in the expression of K$^+$ channel α-subunits and voltage-gated K$^+$ channel currents in rat ventricular myocytes, *J. Gen. Physiol.* **108**:405–419.

Xu, H., Guo, W., and Nerbonne, J. M., 1999a, Four kinetically distinct depolarization activated K$^+$ channel currents in adult mouse ventricular myocytes, *J. Gen. Physiol.* **113**:661–678.

Xu, H., Li, H., Barry, D. M., and Nerbonne, J. M., 1999b, Elimination of the transient outward current and action potential prolongation in atrial myocytes expressing a dominant negative Kv4 alpha subunits, *J. Physiol. (London)*, **519**:11–21.

Yang, T., Kupershmidt, S., and Roden, D., 1995, Anti-minK antisense decreases the amplitude of the cardiac delayed rectifier K$^+$ current, *Circ. Res.* **77**:1246–1253.

Yeola, S. W., and Snyders, D. J., 1997, Electrophysiological and pharmacological correspondence between Kv4.2 current and rat cardiac transient outward current. *Cardiovasc. Res.* **33**:540–547.

Yue, D. T., and Marban, E., 1988, A novel cardiac potassium channel that is active and conductive at depolarized potentials, *Pflügers Arch.* **413:**127–133.

Yue, L., Feng, J., Li, G. R., and Nattel, S., 1996a, Transient outward and delayed rectifier currents in canine atrium: properties and role of isolation methods, *Am. J. Physiol.* **270:**H2157–H2168.

Yue, L., Feng, J., Li, G. R., and Nattel, S., 1996b, Characterization of an ultrarapid delayed rectifier potassium channel involved in canine atrial repolarization, *J. Physiol. (London)* **496:**647–662.

Yue, L., Feng, J., Gaspo, R., Li, G. R., Wang, Z., and Nattel, S., 1997, Ionic remodelling underlying action potential changes in a canine model of atrial fibrillation, *Circ. Res.* **81:**512–525.

Zhou, Z., Gong, Q., Ye, B., Fan, Z., Makielski, J., Robertson, G. A., and January, C. T., 1998, Properties of HERG channels stably expressed in HEK 293 cells studied at physiological temperature, *Biophys. J.* **74:**230–241.

Zygmunt, A. C., 1994, Intracellular calcium activates a chloride current in canine ventricular myocytes, *Am. J. Physiol.* **267:**H1984–H1995.

Zygmunt, A. C., and Gibbons, W. R., 1991, Calcium-activated chloride current in rabbit ventricular myocytes, *Circ. Res.* **68:**424–437.

Zygmunt, A. C., and Gibbons, W. R., 1992, Properties of the calcium-activated chloride current in heart, *J. Gen. Physiol.* **99:**391–414.

Chapter 18

Voltage-Gated Potassium Channels in the Myocardium

Joanne T. Hulme, Jeffrey R. Martens,
Ricardo A. Navarro-Polanco, Atsushi Nishiyama,
and Michael M. Tamkun

1. INTRODUCTION

1.1. Cardiac Action Potential

The cardiac action potential results from the complex, but precisely controlled, movement of ions across the cell membrane. Although the shape of the action potential is known to vary between different regions of the heart (e.g., sinoatrial node vs. atrial vs. ventricle), the present discussion will focus solely on the ventricular myocardium. The upstroke of the action potential (phase 0) in ventricular cells results from the rapid influx of Na^+ through voltage-gated Na^+ channels (Fozzard, 1994). A subsequent rapid, but incomplete, repolarization is achieved in phase 1 due to K^+ efflux through transiently activated K^+ channels (I_{to}). This early outward current sets the initial plateau potential, thus influencing the behavior of subsequently activated ion channels and the duration of the action potential. The plateau of the action potential (phase 2) maintains the cell membrane potential in a relatively depolarized state for up to several hundred milliseconds. This plateau phase is characterized by minimal net ion flow owing to the balance of an inactivating inward Ca^{2+} current and K^+ efflux through slowly activating voltage-gated K^+ (Kv) channels. Consequently, this balance between inward and outward currents is an important determinant of action potential duration and cardiac contractility. Finally, the action potential is terminated by a rise in K^+

Joanne T. Hulme, Jeffrey R. Martens, Ricardo A. Navarro-Polanco, Atsushi Nishiyama • Department of Physiology, Colorado State University, Fort Collins, Colorado 80523. *Michael M. Tamkun* • Departments of Biochemistry and Molecular Biology, Colorado State University, Fort Collins, Colorado 80523.

Potassium Channels in Cardiovascular Biology, edited by Archer and Rusch. Kluwer Academic/Plenum Publishers, New York, 2001.

permeability (phase 3) that ultimately repolarizes the cell membrane back to the resting membrane potential level (phase 4).

As in most excitable cells, Kv channels in cardiac myocytes play an important role in determining the magnitude and duration of the action potential. Differences in the type(s) and levels of K^+ channel expression contribute to the heterogeneity of action potential configuration that exists in different regions of the heart. In addition, many distinct Kv currents have been identified in cardiac myocytes isolated from a number of different species (Terzic *et al.*, 1995; Coraboeuf and Nargeot, 1993; Takano and Noma, 1993; Anumonwo *et al.*, 1991; Antzelevitch *et al.*, 1991; Nichols and Lederer, 1991; Kaplan and Trout, 1969).

1.2. Diversity of Voltage-Gated K^+ Currents in Myocardial Cells

Kv currents in the heart were classified initially into two types, based on their differing time- and voltage-dependent properties and pharmacological sensitivities: (1) a transient outward current called I_{to} and (2) a slowly activating delayed rectifier current, I_K (Gintant *et al.*, 1991). More recent voltage-clamp studies of isolated cardiac myocytes from various species, have however, identified at least five distinct types of Kv currents in the myocardium (Z. Wang *et al.*, 1994, 1993; Apkon and Nerbonne, 1991; Sanguinetti and Jurkiewicz, 1991; Balser *et al.*, 1990; Tseng and Hoffman, 1989; Yue and Marban, 1988; Benndorf *et al.*, 1987), suggesting that the diversity of cardiac Kv currents is much more complex than originally anticipated.

1.2.1. Transient Outward K^+ Current (I_{to})

I_{to} is an important determinant of the early repolarization phase of the cardiac action potential. This current is expressed in many regions of the heart, including atrial and ventricular muscle, the sinoatrial (SA) and atrioventricular (AV) nodes, and Purkinje fibers of several species (for a review, see Campbell *et al.*, 1995). One notable exception, however, appears to be guinea pig ventricular myocytes, in which I_{to} is either absent or relatively small (Hume *et al.*, 1990; Hirano and Hiraoka, 1988).

In recent years, it has become increasingly clear that marked differences in the density of I_{to} exist in different regions of the heart. For example, I_{to} density is more prominent in the atrium than in the ventricle of the rat, rabbit, and human heart (Bond *et al.*, 1994; Varro *et al.*, 1993; Boyle and Nerbonne, 1992; Giles and Imaizumi, 1988). Furthermore, I_{to} density is reported to vary across the ventricular wall in a number of species, including dog (Liu *et al.*, 1993; Litovsky and Antzelevitch, 1988), cat (Furukawa *et al.*, 1990), rat (Clark *et al.*, 1993), rabbit (Fedida and Giles, 1991), and human (Wettwer *et al.*, 1993). For example, I_{to} is prominent in the subepicardium and underlies the characteristic "spike and dome" morphology, whereas in the subendocardium, I_{to} is less marked and the spike and dome morphology is less pronounced (Liu *et al.*, 1993; Antzelevitch *et al.*, 1991).

Transient outward currents, first described in sheep Purkinje fibers, were thought to be carried primarily by Cl^- (Fozzard and Hiraoka, 1973; Dudel *et al.*, 1967). However, subsequent studies have demonstrated that much of this current is carried predominantly by K^+ (Dukes and Morad, 1991; Apkon and Nerbonne, 1991; Tseng and Hoffman, 1989; Giles and Van Kinneken, 1985; Kenyon and Gibbons, 1979). Furthermore, reports of functional heterogeneity, as evidenced by differences in single-channel conductance and activation and inactivation kinetics, suggest the presence of more than one I_{to} channel protein. I_{to} is now thought to be composed of two components: (1) a rapidly activating and inactivating, Ca^{2+}-independent current (I_{to1}) that is carried by K^+ and is sensitive to block by $1-10$ μM 4-aminopyridine (4-AP) (Tseng and Hoffman, 1989; Coraboeuf and Carmeliet, 1982) and (2) a Ca^{2+}-activated Cl^- current that is typically smaller than I_{to1} (Zygmunt, 1994; Zygmunt and Gibbons, 1991).

1.2.2. Delayed Rectifier K^+ Current (I_K)

Like I_{to}, I_K was originally thought to represent a single current (Apkon and Nerbonne, 1991). However, it was later shown that the delayed rectifier current in the myocardium consists of at least two components, I_{Kr} (rapid) and I_{Ks} (slow) (Sanguinetti and Jurkiewicz, 1991, 1990). These two components have been identified in a number of species, including the guinea pig (Sanguinetti and Jurkiewicz, 1991; Balser et al., 1990; Sanguinetti and Jurkiewicz, 1990), dog (Blair et al., 1991; Benndorf, 1988), and rabbit (Salata et al., 1996). I_{Kr} activates faster and at more negative potentials than I_{Ks} and displays an apparent inward rectification at potentials positive to 0 mV (Sanguinetti and Jurkiewicz, 1991, 1990). This rectification is due to an ultrarapid inactivation as opposed to a pore-based rectification (Snyders and Chaudhary et al., 1996). Furthermore, single-channel studies have revealed single-channel conductances of 10–13 and 3–5 ρS for I_{Kr} and I_{Ks}, respectively (Veldkamp et al., 1993; Balser et al., 1990; Horie et al., 1990; Shibasaki, 1987). In addition, these individual currents can be separated based on their pharmacology: I_{Kr} is one of the few K^+ currents that is sensitive to dofetilide (Carmeliet, 1992), whereas I_{Ks} is blocked by other class III antiarrhythmics, such as NE-10133 (Busch et al., 1994).

The densities of I_{Kr} and I_{Ks} also vary in different regions of the heart. For example, in the guinea pig, the densities of I_{Kr} and I_{Ks} are approximately twofold higher in the atrium compared to the ventricle, thus contributing to the shorter action potential recorded in atrial myocytes (Sanguinetti and Jurkiewicz, 1991, 1990). In addition, Liu and Antzelevitch (1995) reported that the density of I_{Ks} is greater in subepicardial and subendocardial cells than in mid-wall myocardial (M) cells in the dog ventricle.

In human and rat atrial myocytes, a voltage-activated but noninactivating K^+ current has been described that is distinctly faster than I_{Kr} (Fedida et al., 1993; Z. Wang et al., 1993; Boyle and Nerbonne, 1991) and is commonly referred to as I_{Ksus} or I_{Kur}, for sustained or ultrarapid K^+ current, respectively. Another current, I_{Kp}, has been identified in guinea pig ventricle and termed a "plateau current" because it is active at depolarized potentials (Backx and Marban, 1992; Yue and Marban, 1988).

1.2.3. Limitations of Traditional Electrophysiological and Biochemical Approaches

The presence of multiple overlapping currents in cardiac myocytes complicates the study of individual K^+ channels. Traditional approaches that are used to eliminate all but the current of interest, including holding-potential modulation, ion substitution, and pharmacological dissection, have obvious limitations. For example, the divalent cations often used to block Ca^{2+} currents (Co^{2+}, Cd^{2+}, La^{3+}) also modify the gating properties of K^+ channels (Fan and Hiraoka, 1991). Furthermore, because several cloned K^+ channels contain putative extracellular protease cleavage sites, concerns that enzymatic dissociation of cardiac myocytes itself may inactivate some physiologically important currents while leaving others unaltered must be considered. The use of molecular cloning and cDNA expression technologies, however, has allowed single cloned isoforms to be expressed in a system of choice. This approach now makes it possible to study individual channels under relatively physiological conditions in the absence of contaminating currents.

2. OVERVIEW OF MOLECULAR APPROACHES TO THE STUDY OF VOLTAGE-GATED K^+ CHANNELS

2.1. Cloning Strategies

Potassium channel cDNA has been cloned using a variety of approaches. Cloning based on chromosomal location was used not only to isolate the first K^+ channel from *Drosophila* but also to isolate the human KvLQT1 channel gene responsible for one form of long QT syndrome (Q. Wang *et al.*, 1996). Without information on chromosomal location, K^+ channel genes were cloned by either (i) functional screening of cDNA libraries for channel activity (expression screening), (ii) purification of channel protein and use of nucleotide sequences derived from the primary structure to screen libraries by hybridization, or (iii) simple homology screening (Deal *et al.*, 1996). This last approach has generated the greatest number of clones.

2.1.1. The Drosophila Genetic Approach: Cloning-Based Chromosomal Location

The first cloned Kv channel was obtained using *Drosophila* genetics and chromosomal walking. Flies displaying the *Shaker* phenotype, which involves abnormal leg shaking in response to ether exposure, were found to be missing an I_{to}-like K^+ current from their leg and flight muscles (Salkoff, 1985). Thus, it was postulated that the defective gene responsible for the *Shaker* phenotype was a K^+ channel and that the locus of the *Shaker* mutation in the *Drosophila* genome marked the location of this channel. Isolation of this genetic locus involved extensive isolation and reisolation of genomic fragments, allowing investigators to "walk" down the *Drosophila* chromosome toward the *Shaker* defect. Using this approach, two independent groups cloned the first Kv channel (Pongs *et al.*, 1988; Tempel *et al.*, 1987). The original *Shaker* clone is one of the most studied proteins in terms of its structure–function relationship. Site-directed mutagenesis followed by functional expression and voltage-clamp analysis has identified sequence regions that are involved in ion permeation, gating, voltage sensing, and subunit assembly (Chandy and Gutman, 1995). These domains of the *Shaker* channel are illustrated in Fig. 1.

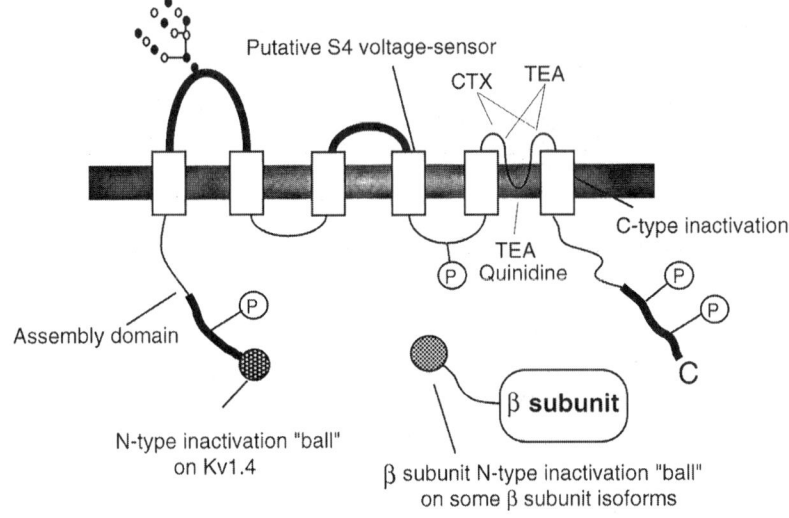

Putative S4 voltage-sensor
CTX   TEA
C-type inactivation
TEA
Quinidine
Assembly domain
N-type inactivation "ball"
on Kv1.4
β subunit
β subunit N-type inactivation "ball"
on some β subunit isoforms

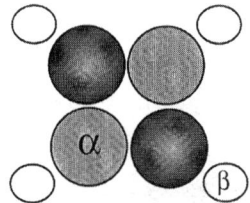

α subunits can form heterotetrameric channels

Figure 1. Postulated structure of the *Shaker*-like K⁺ channels present in the heart: Summary of the current ideas with regard to the structure and functionally important regions of *Shaker*-like K⁺ channels CTX, Charybdotoxin; TEA, tetraethylammonium.

Similar genetic approaches were taken by the Ganetzky laboratory to clone the K⁺ channels responsible for the *Slowpoke* (Butler *et al.*, 1993), *ether-a-go-go* (Warmke *et al.*, 1991), and *Hyperkinetic* (Chouinard *et al.*, 1995) phenotypes. These *Drosophila* genes encode the Ca^{2+}-activated K⁺ channel, a non-*Shaker*-like delayed rectifer K⁺ channel, and β-subunits that assemble with *Shaker* channels, respectively.

2.1.2. Homology Screening

Using the *Shaker* gene as a homology probe, three related *Drosophila* K⁺ channel genes were isolated (Butler *et al.*, 1989). These are referred to as the *Shaw*, *Shab*, and *Shal* channels, being named in part for the initials of the investigators who cloned them. Homology screening also identified numerous mammalian homologs for all the above-mentioned *Drosophila* K⁺ channel genes, homologs for at least 12 channels with high similarity to the *Shaker* gene were described (Chandy and Gutman, 1995). Under the accepted classification scheme, mammalian Kv1 family members are most similar to *Shaker*, while the Kv2, Kv3, and Kv4 families are most similar to *Shab*, *Shaw*, and *Shal*, respectively (Chandy and Gutman, 1995). To account for the diversity among mammals, the Kv5–Kv9 subfamilies have been added. In addition to the Kv1–Kv9 channels, there are mammalian homologs for the *Drosophila ether-a-go-go* (Warmke and Ganetzky, 1994), *Slowpoke* (Butler *et al.*, 1993), and *Hyperkinetic* (Chouinard *et al.*, 1995) channel proteins.

2.1.3. Expression Cloning

This cloning strategy utilizes heterologous expression to screen library clones for channel activity. Here the cDNA is divided into pools of perhaps thousands of individual clones, and cRNA is produced from each clone pool and injected into *Xenopus* oocytes. The oocytes are then voltage-clamped after 1–3 days of incubation. If current is detected in the cRNA-injected oocytes, the clone pool containing the channel cDNA is subdivided and the process repeated until cRNA derived from a single clone generates the desired current. This cloning approach was used to isolate cDNAs encoding rat minK and rat Kv2.1 channels (Frech *et al.*, 1989; Takumi *et al.*, 1988).

2.1.4. Cloning Based on K^+ Channel Protein Purification

This approach cloned mammalian β-subunits that were discovered because they copurified with Kv channel α-subunits (see Section 3.3). At the time, expression cloning was not an option because the functional effect of the β-subunit on the α-subunit current profile was unknown. Furthermore, without knowledge of chromosomal location or related homologs, positional cloning and homology screening could not be used. Rather, using protein purification methods, α/β complexes were purified from bovine brain, and the amino acid sequence of the β-subunit protein was obtained. Degenerate nucleotide probes then cloned the first two Kvβ subunits from a rat brain cDNA library.

2.2. Heterologous Expression Systems

Xenopus oocytes have been by far the expression system of choice owing to the ease of cRNA injection and the two-electrode voltage-clamp technique. For the most part, this system has been reliable in terms of producing physiologically meaningful results. However, there are at least two notable exceptions. The first is that several channels have appeared insensitive to pore-blocking drugs in the oocyte system, while being readily blocked when expressed in mammalian cells. For example, quinidine blocks the Kv1.5 delayed rectifier with a dissociation constant (K_d) of 6 μM in mouse L-cells, but this concentration has little effect following channel expression in oocytes. In this case, the hydrophobic yolk within the oocyte probably represents a major sink for any membrane-permeant compound. The second exception deals with the expression of cloned minK protein. Investigators were puzzled for years by the finding that minK only could be expressed reproducibly in oocytes, but not in mammalian cells. This result finally was explained by the finding that (1) minK is not a true channel but rather a β-subunit that is required for the proper expression of KvLQT1, and (2) oocytes express an endogenous KvLQT1 α-subunit that is silent until minK cRNA is injected (Barhanin *et al.*, 1996; Sanguinetti *et al.*, 1996b).

Although fibroblast-like tissue culture cells are perhaps the best expression system to use in terms of their pharmacology and fast voltage clamp, they too can suffer from endogenous subunit problems. For example, hamster embryonic kidney (HEK293) cells occasionally express endogenous delayed rectifier currents that resemble a number of cloned Kv channels. In addition, cell lines such as mouse L-cells and Chinese hamster ovary (CHO) cells express Kvβ subunits which can alter the function of heterologously expressed channels (Uebele *et al.*, 1996).

In summary, no heterologous expression system is perfect. However, mammalian cells may be more physiologically relevant, because the mammalian protein processing machinery is operating at a physiological temperature, whereas *Xenopus* oocytes are maintained at 18°C. However, the possibility of endogenous subunits remains a concern with all systems.

2.3. Relation of Cloned Channels with Native Currents

Although kinetic and pharmacological properties can be used to relate cloned channels to endogenous cardiac currents, this approach is limited by the functional and pharmacological similarities between cloned channels. In addition, differences between native and cloned currents may reflect differences in channel subunit composition or posttranslational modifications. Therefore, the field has emphasized the use of both antisense and transgenic knockout/dominant-negative technologies to relate cloned cardiac channels to native currents.

2.3.1. Antisense Studies

Typically, antisense studies are conducted by incubating short-term primary cultures of either rat or human cardiac myocytes with antisense oligonucleotides. This antisense approach is difficult in that statistically significant data can be elusive, given the variability in outward currents between cells and the partial downregulation that is achieved with antisense. Antisense often only decreases gene expression by 50%, and, for channel proteins with a long half-life, even 100% removal of mRNA may not translate into a detectable effect. Nonetheless, this approach was taken by Nattel and co-workers to demonstrate that Kv1.5 encodes I_{Kur} in human atrial myocytes (Feng *et al.*, 1997) and by Fiset and collaborators to demonstrate that Kv4.2 and 4.3 are components of I_{to} (Fiset *et al.*, 1997). This approach also has been used to suggest that minK functionally alters I_{Kr}, because minK antisense decreases I_{Kr} in a mouse atrial tumor (AT-1) cell line (Yang *et al.*, 1995).

2.3.2. *Transgenic Studies*

The classic transgenic knockout experiment offers, in theory, a more direct approach and the promise of allowing examination of whole animal physiology in the absence of the channel of interest. Analysis of mice lacking the Kv1.4 gene shows normal cardiovascular function and normal I_{to} currents in ventricular myocytes, indicating that this channel is unlikely to be a major component of I_{to}, at least in adult animals. However, the minK knockout prolongs the QT interval at slow heart rates as expected if this protein represents a component of the I_{Ks} current, as predicted from heterologous expression studies (Drici *et al.*, 1998). Interestingly, one result of this knockout was an effect on inner ear development, which explains why some patients with long QT syndromes (LQTS) are deaf. The Kv1.5 knockout has been less informative. These mice appear normal as evaluated by resting heart rate, PR and QT interval lengths, and QRS morphology on electrocardiogram. Although they do exhibit differential susceptibility to class III antiarrhythmic agents compared with wild-type animals voltage-clamp studies of isolated ventricular myocytes from the adult animals reveal only subtle differences in K^+ current function thus far (J. Hill, personal communication). Clearly, additional studies are warranted.

A related approach that has been used by several investigators involves overexpression of a dominant-negative channel protein in transgenic mice. Nerbonne and co-workers used this approach to support a role for Kv4.2 and Kv4.3 in I_{to} (Barry *et al.*, 1998). In contrast, other studies show that the overexpression of a dominant-negative α-subunit channel decreases both Kv1.5 protein and I_{Kur} in the ventricle, as well as inducing QT interval prolongation (Babila *et al.*, 1996). However, it is likely that this dominant-negative construct downregulates most, if not all, Kv1 channel family members. The discrepancy between the QT prolongation induced by dominant-negative α-subunit expression and the lack of QT prolongation observed with the true Kv1.5 knockout illustrates potential problems with both approaches. It is possible that the true knockout induces upregulation of a compensatory current; furthermore, the Kv1 dominant-negative approach may remove another channel in addition to Kv1.5.

3. FUNCTIONAL EXPRESSION OF CLONED Kv CHANNELS

To date, at least seven Kv channels have been cloned from the heart and functionally characterized using heterologous expression systems. As with native currents, these Kv channels can be divided into either delayed rectifiers or transient outward currents. The delayed rectifiers include Kv1.1, Kv1.2, Kv1.5, Kv2.1, KvLQT1, and h-erg (HERG), and the transient outward currents include Kv1.4, Kv4.2, and Kv4.3 (Barry and Nerbonne, 1996; Deal *et al.*, 1996; Dixon *et al.*, 1996). Figure 2 shows currents generated by expression of Kv1.2, Kv1.5, Kv2.1, and Kv4.2 clones in mouse L-cells. Among these four K^+ channels, the time to peak current ranges from 2 to 300 ms at $+40$ mV, with Kv4.2 activating most rapidly and Kv1.2 most slowly. Kv4.2 displays nearly complete inactivation, whereas Kv1.2 and Kv1.5 show varying degrees of slow inactivation. Direct correlation of the cloned K^+ channels with native cardiac K^+ currents is complicated by the marked diversity of K^+ channels, which arises in

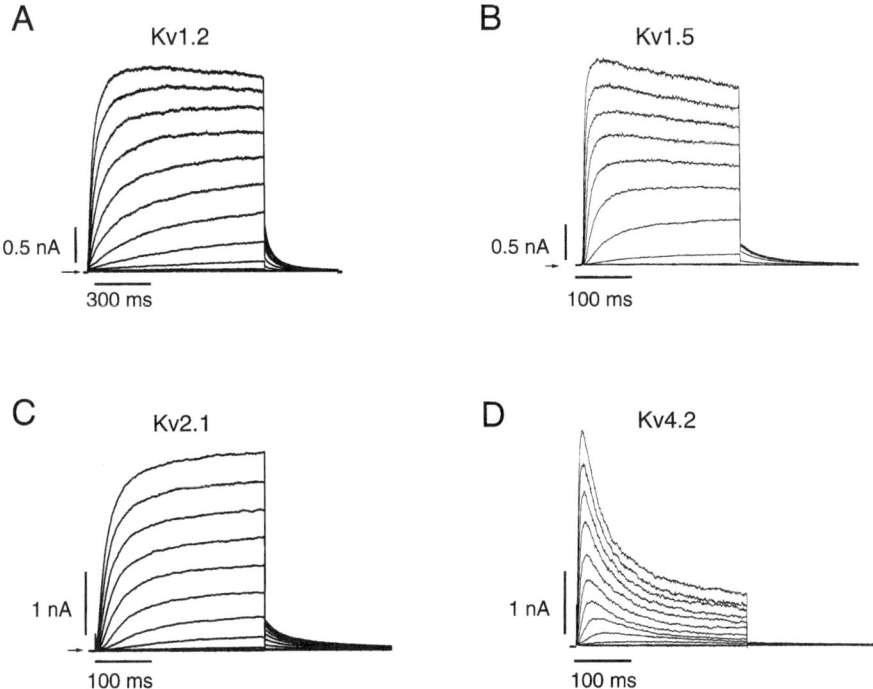

Figure 2. Currents produced by cloned cardiovascular K^+ channels expressed in mouse L-cells. Whole-cell voltage-clamp tracings were elicited in response to 10-mV step depolarizations from a holding potential of -80 to $+40$ mV; tails were recorded at -50 mV. No tail currents are visible for Kv4.2 because this current nearly fully inactivated during depolarization. All channels are from rat except for Kv1.5, which is from human. Cells were transfected using the lipofectamine method. Cells transfected without K^+ channel cDNA did not display voltage-gated currents under these recording conditions.

part from the large number of K^+ channel genes. Additional diversity is generated by alternative RNA splicing, the potential for heteromeric assembly of K^+ channel proteins, and the existence of accessory β-subunits (for a review, see Deal *et al.*, 1996). In the following sections we will summarize our current understanding of which cDNA clones represent the best molecular correlates of native Kv currents in the myocardium.

3.1. *Shaker*-Like Kv Channels with Delayed-Rectifier-Like Properties

3.1.1. *Kv1.2*

Kv1.2 was cloned from the rat heart and brain (Roberds and Tamkun, 1991; McKinnon, 1989). Northern analysis (Roberds and Tamkun, 1991) and ribonuclease protection assays have detected Kv1.2 mRNA in both fetal (Xu *et al.*, 1996; Roberds and Tamkun, 1991) and adult rat heart (Dixon and McKinnon, 1994), and Western analysis has confirmed protein expression of Kv1.2 in rat ventricular myocytes (Barry *et al.*, 1995). Developmental changes in the level of Kv1.2 mRNA and protein also have been reported. For example, Kv1.2 mRNA and protein levels increase approximately 10-fold between birth and adulthood (Xu *et al.*, 1996; Nakamura and Iijlma, 1994).

Heterologous expression of Kv1.2 in mouse L-cells produces K^+ currents that exhibit several notable differences compared to Kv1.2 currents recorded from *Xenopus* oocytes (see Fig. 2A; Murai *et al.*, 1989). For example, the activation kinetics of Kv1.2 current expressed in mouse L-cells are 10-fold slower than when the current is expressed in *Xenopus* oocytes. Additionally, expression of Kv1.2 in mouse L-cells shifts the voltage sensitivity of activation by ~25 mV in the depolarizing direction (M. M. Tamkun, unpublished data). Differences in protein processing between these two expression systems (Deal *et al.*, 1996), including differences in possible phosphorylation pathways and the presence of endogenous α- and β-subunits (Uebele *et al.*, 1996) capable of assembling with transfected α-subunits, may help to explain such discrepancies.

The observation that Kv1.2 mRNA levels are higher in the atria than in the ventricle (Dixon and McKinnon, 1994) led several investigators to propose that Kv1.2 was the molecular correlate of the atrial K^+ current, I_{Kur} (Paulmichl *et al.*, 1991). A major difference between Kv1.2 and the native current, however, is its sensitivity to block by the K^+ channel toxin dendrotoxin (DTX); Kv1.2 is sensitive to nanomolar concentrations of DTX (Chandy and Gutman, 1995; Grissmer *et al.*, 1994), whereas both the rat and the human atrial current are DTX-insensitive (Z. Wang *et al.*, 1993). Furthermore, the potential for heteromeric assembly of Kv α-subunits in the myocardium has complicated the correlation of cloned Kv channels with native currents. For example, it has been demonstrated that Kv1.2 and Kv1.5 α-subunits can assemble to form a functional heteromeric channel *in vitro* and that the presence of the Kv1.5 toxin-insensitive channel dominates the pharmacology of the heteromeric channel (Overturf *et al.*, 1994). Thus, the potential for heteromeric channel formation may provide one explanation for the pharmacological and electrophysiological discrepancies between cloned and native K^+ currents. Antisense and gene knockout strategies are necessary to define the functional role of Kv1.2 in the heart.

3.1.2. Kv1.5

Kv1.5 was cloned originally from rat heart and rat brain (Roberds and Tamkun, 1991; Swanson *et al.*, 1990). The human Kv1.5 (HK2) was subsequently cloned from ventricle (Tamkun *et al.*, 1991) and atrium (Grissmer *et al.*, 1994) and from an insulinoma cell line (Philipson *et al.*, 1991) and is only 86% identical to the rat channel. Most of the nonidentity occurs in the NH_2- and COOH-terminal regions, which have 64% and 75% identity, respectively. Canine and mouse Kv1.5 also have been cloned from colonic smooth muscle and heart, respectively (Attali *et al.*, 1993b; Overturf *et al.*, 1994), and show 85% overall identity with the human homolog. It is most likely that these differences represent simple interspecies variations as opposed to different K^+ channel isoforms. Of all the *Shaker*-like K^+ channels cloned to date, Kv1.5 represents the most cardiac-specific clone, based on mRNA expression (Dixon and McKinnon, 1994). Northern blot (Roberds and Tamkun, 1991) and ribonuclease protection analyses (Dixon and McKinnon, 1994) have revealed that the Kv1.5 transcript is equally distributed between the atria and ventricles in the rat heart. In contrast, in human heart, Northern blot analysis suggests that Kv1.5 is more abundant in the atrium than in ventricle (Tamkun *et al.*, 1994, 1991), although a polymerase chain reaction (PCR)-based analysis suggests that transcript levels are similar between human atrium and ventricle (Fedida *et al.*, 1993). More specifically, immunolocalization of Kv1.5 protein in both human and rat myocardium indicates that Kv1.5 is expressed equally in both atrial and ventricular myocytes (Mays *et al.*, 1995). Moreover, Kv1.5 is not evenly distributed over the myocyte surface, but is localized in regions of high density at the intercalated discs (Mays *et al.*, 1995).

Voltage-clamp analysis has detected outward currents in rat and human atrial myocytes that resemble Kv1.5 when expressed in heterologous systems (Z. Wang *et al.*, 1993; Boyle and Nerbonne, 1992). Both these currents are sensitive to block by low concentrations of 4-AP and quinidine (Z. Wang *et al.*, 1995) but are insensitive to block by tetraethylammonium (TEA). Consistent with the idea that Kv1.5 is the cloned isoform corresponding to the native current I_{Kur}, Nattel and co-workers demonstrated that antisense oligonucleotides directed against Kv1.5 specifically inhibit I_{Kur} in human atrial myocytes but have no measurable effects on current density in ventricular myocytes (Feng *et al.*, 1997). Although Kv1.5 protein has been found in human ventricular myocytes, whole-cell currents similar to I_{Kur} have not been detected in these cells. Given the potential for heteromeric assembly of Kv channel proteins within a given subfamily and the existence of ancillary β-subunits that may alter channel function, it is possible that Kv1.5 forms a homomeric channel in the atrium while perhaps forming a heteromeric channel in the ventricle. A number of β-subunits have been cloned from the heart (Morales *et al.*, 1995; Makita *et al.*, 1994) that alter the inactivation kinetics of Kv1.5 (Fig. 3A, B) and shift the threshold for channel activation to more hyperpolarized potentials (Uebele *et al.*, 1996). It is important to determine if these β-subunits colocalize with the Kv1.5 channel α-subunit in the human ventricle but not in the atrium.

Knockout mice lacking the Kv1.5 gene have been generated in an attempt to determine the physiological role of the Kv1.5 protein. However, the greatest effect reported in these mice has been the development of sinus bradycardia and sinus arrest, suggesting a role for Kv1.5 in repolarization of the sinus node. Surprisingly, voltage-clamp studies of isolated ventricular myocytes from adult mice have not revealed significant differences in outward current between knockout and control animals (J. Hill, personal communication), raising the possibility of a compensatory increase in a similar outward current.

3.1.3. Kv2.1

Using ribonuclease protection assays, Dixon and McKinnon (1994) demonstrated that the mRNA for Kv2.1 was readily detected in adult rat ventricular myocytes. However, the expression patterns of Kv2.1 mRNA and protein during development are quite distinct. For example, expression of Kv2.1 mRNA increases 3–5 fold between birth and adulthood, whereas Kv2.1 protein levels decrease (Xu *et al.*, 1996), suggesting that Kv2.1 protein expression is regulated posttranscriptionally. This finding clearly emphasizes the need for caution when attempting to draw conclusions about the expression of functional channels based on analysis of steady-state mRNA levels alone.

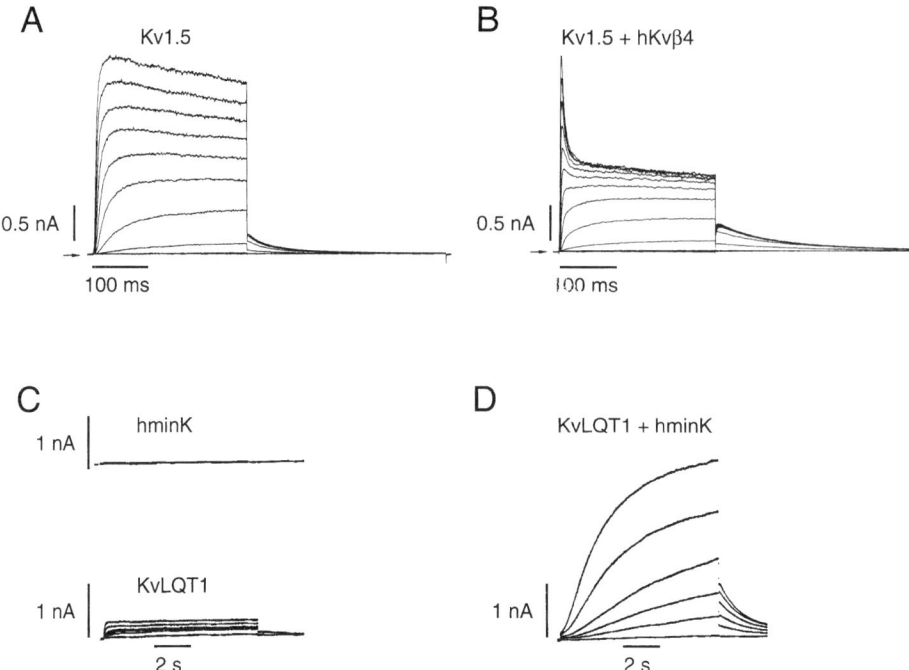

Figure 3. *Top panels*: Functional effects of human Kvβ1.3 (hKvβ4) on human Kv1.5 currents. Outward currents recorded from HEK293 cells transfected with either Kv1.5 alone (A) or with Kv1.5 and hKvβa.3 (B). Currents were elicited in response to 10-mV step depolarizations from a holding potential of −80 to +60 mV; tails were recorded at −50 mV. *Lower panels*: Currents recorded during 7.5-s pulses to −40, −20, −10, 0, +20, and +40 mV in cells transfected with hminK, KvLQT1, or KvLQT1 and hminK. Tail currents were measured at −70 mV (C) or −50 mV (D). (Adapted, with permission, from Sanguinetti *et al.*, 1996b. Copyright 1996 Macmillan Magazines Limited.)

The functional role of Kv2.1 in atrial and ventricular myocytes is unclear. Heterologous expression of Kv2.1 reveals slowly activating, TEA sensitive currents similar to I_K recorded in adult rat ventricular myocytes (Apkon and Nerbonne, 1991). The TEA-sensitivity of Kv2.1, however, is too great for this channel to be considered as a candidate for I_{Kur}. Furthermore, the human clone is insensitive to block by 4-AP whereas the human atrial current is sensitive (Z. Wang *et al.*, 1995; Albrecht *et al.*, 1993; Fedida *et al.*, 1993). Nerbonne and co-workers used a dominant-negative approach in transgenic mice to suggest that a truncated Kv2.1 subunit functionally alters I_{Ks} in mouse ventricular myocytes (Xu *et al.*, 1999). However, further studies are necessary to define the role, if any, of Kv2.1 in the myocardium.

A major frustration involves the attempts to immunolocalize Kv2.1 expression in the heart in order to answer the simple question of localization. Western analysis indicates that Kv2.1 protein is readily detectable in mouse heart, but these same antibodies do not stain any cardiac cell type (M. M. Tamkun, unpublished results). The best explanation for this negative result is that the antibody epitopes are masked in the native cell.

3.2. *Shaker*-Like Kv Channels with Rapid Inactivation (I_{to}-Like Channels)

3.2.1. *Kv1.4*

Kv1.4 was one of the first members of the *Shaker* K^+ channel gene family to be cloned from rat and human heart (Roberds and Tamkun, 1991), and it initially was thought to encode I_{to} (Tseng-Crank *et al.*, 1990; Stuhmer *et al.*, 1989). However, although Kv1.4 mRNA levels are abundant in the heart of several species (Dixon and McKinnon, 1994; Matsubara *et al.*, 1993; Roberds and Tamkun, 1991), Western blot analysis and immunohistochemistry have failed to detect Kv1.4 protein in adult rat ventricular myocytes (Barry *et al.*, 1995). Messenger RNA expression of Kv1.4, like many of the *Shaker*-like K^+ channels, also changes with development. For example, Kv1.4 mRNA appears to be greatest in the rat neonatal heart but subsequently decreases to very low levels in the adult heart (Wickenden *et al.*, 1997; Xu *et al.*, 1996; Dixon and McKinnon, 1994; Matsubara *et al.*, 1993; Roberds and Tamkun, 1991). In contrast, two K^+ channel genes, Kv4.2 and Kv4.3, also believed to encode I_{to}, are expressed at relatively low levels during the neonatal period but predominate in the adult heart (Shimoni *et al.*, 1997; Dixon *et al.*, 1996; Xu *et al.*, 1996; Barry *et al.*, 1995; Dixon and McKinnon, 1994; Roberds and Tamkun, 1991). Developmental regulation of these genes is associated with changes in action potential and I_{to} characteristics. For example, the marked shortening of the action potential that occurs during postnatal development is associated with an increase in the density of I_{to} (Wickenden *et al.*, 1997; Smith and Wahler, 1996; Wahler *et al.*, 1994; Kilborn and Fedida, 1990) and an increase in recovery from inactivation (Wickenden *et al.*, 1997). These data suggest that the predominant K^+ channel gene encoding I_{to} switches from Kv1.4 to Kv4.2/Kv4.3 during development, (Wickenden *et al.*, 1997).

Heterologous expression of Kv1.4 in *Xenopus* oocytes and mammalian cell lines produces rapidly activating and inactivating K^+ currents that are sensitive to block by 4-AP, similar to the native I_{to} channel (Po *et al.*, 1992; Tseng-Crank *et al.*, 1990; Stuhmer *et al.*, 1989). This finding led to the earlier suggestion that Kv1.4 may be responsible for I_{to} in rat ventricle. Recovery of the cloned channel from inactivation, however, is much slower than that of native I_{to}. Cloned Kv1.4 channels recover from inactivation with a time constant of 3–8 s (Bertoli *et al.*, 1994; Po et al., 1993, 1992), whereas the rat and human native I_{to} currents recover with time constant of ~ 50 ms (Nabauer *et al.*, 1993; Wettwer *et al.*, 1993). In addition, Kv1.4 is only weakly sensitive to flecainide ($K_d > 60$ μM) (Franqueza et al., 1999), whereas the native I_{to} current in human and rat heart is blocked with an estimated K_d of 3–5 μM (Z. Wang *et al.*, 1995; Slawsky and Castle, 1994). Using a gene targeting approach to engineer mice lacking Kv1.4, London *et al.* (1998) demonstrated that the electrophysiological and pharmacological properties of I_{to} in homozygous knockout mice are identical to those in wild-type mice. Nerbonne and co-workers used *in vitro* antisense combined with an *in vivo* transgenic approach to produce preliminary data suggesting that I_{to} may be differentially distributed in the mouse ventricle (Guo *et al.*, 1999). An $I_{to,fast}$ component was found predominantly in epicardial myocytes, but an additional $I_{to,slow}$ component was found in septal myocytes (Guo *et al.*, 1999). Furthermore, expression of a dominant-negative fragment of Kv1.4 in septal myocytes using an adenoviral vector decreased $I_{to,slow}$, suggesting that Kv1.4 may be the molecular correlate of $I_{to,slow}$ in the mouse septum (Guo *et al.*, 1999).

3.2.2. Kv4.2 and Kv4.3

Kv4.2 and Kv4.3 have emerged as strong molecular candidates for the cardiac I_{to} current (Barry *et al.*, 1995; Dixon *et al.*, 1996). Kv4.2 was cloned from both rat brain and heart (Baldwin *et al.*, 1992; Roberds and Tamkun, 1991), whereas Kv4.3 was cloned from the canine, human, and rat heart (Dixon *et al.*, 1996; Kong *et al.*, 1998). In the human myocardium, the Kv4.3 cDNA corresponds to a 637-amino-acid protein and is 99% identical to the rat homolog. It shares 95% amino acid identity in the transmembrane regions and overall 76% identity with the related human Kv4.2 channel (Kong *et al.*, 1998). Sequence analysis by reverse-transcription PCR (RT-PCR) of the COOH terminus of human Kv4.3 has revealed the existence of two alternatively spliced variants in ventricular muscle; the long splice variant has a 19-amino-acid insert that contains a consensus site for protein kinase C (PKC) phosphorylation after the S6 region (Kong *et al.*, 1998). Similar splice variants have also been described in the rat ventricle (Nishiyama *et al.*, 1998; Ohya *et al.*, 1997; Takimoto *et al.*, 1997).

Kv4.3 mRNA is abundantly expressed in the rat, dog, and human heart, whereas Kv4.2 mRNA has been detected only in the rat heart (Dixon *et al.*, 1996; Dixon and McKinnon, 1994; Baldwin *et al.*, 1992; Roberds and Tamkun, 1991). In addition, a gradient of Kv4.2 mRNA expression exists across the ventricular wall that follows an increase in I_{to} current as one moves from the endocardium to epicardium (Dixon and McKinnon, 1994; Clark *et al.*, 1993). The kinetics and 4-AP sensitivities ($K_d = 0.7-4$ μM) of Kv4.2 and Kv4.3, although not identical, closely resemble those of I_{to} (Dixon *et al.*, 1996). Direct manipulation of channel expression in cultured cardiac myocytes has supported a role for Kv4.2 and Kv4.3 in the etiology of I_{to}. Fiset *et al.* (1997) demonstrated that antisense oligonucleotides directed against Kv4.2 and Kv4.3 selectively decrease I_{to} in rat ventricular myocytes. Similarly, expression of a dominant-negative fragment of Kv4.2 in rat cardiac myocytes using an adenoviral vector decreases I_{to} (Johns *et al.*, 1997). These studies suggest that Kv4.2 and Kv4.3 are the molecular components of I_{to} in the rat heart. The question of whether rat I_{to} reflects the expression of homomers or heteromers of Kv4.2 and Kv4.3 remains unresolved and clearly warrants further investigation. On the basis of the correlation between mRNA level and I_{to} current density in human ventricular myocytes (Kaab *et al.*, 1998), Kv4.3 has emerged as the leading α-subunit K^+ channel gene encoding the cardiac I_{to} in dog and humans.

3.2.3. Are Other Shaker-Like Channels Expressed in the Heart?

Of all the *Shaker*-like K^+ channels cloned thus far, the ones described above have been identified as those expressed in the heart based on an extensive analysis of rat cardiac K^+ channel mRNA expression using ribonuclease protection assay studies by Dixon and McKinnon, 1994). Specifically, the Kv1.1, Kv1.3, Kv1.6, Kv2.2, and Kv4.1 channel mRNAs were found to be either very low in the myocardium or undetectable. Kv1.1 was cloned from rat aorta and found by Northern analysis to be expressed in the rat atrium (Roberds and Tamkun, 1991). However, because this Northern study included the atrial septum, which contains some neuronal cell bodies, and the ribonuclease protection assays specifically avoided this region, it is possible that Kv1.1

is solely of neuronal origin. After all, Kv1.1 was the first Kv channel cloned from the brain and is abundantly expressed in the central nervous system.

3.3. Accessory β-Subunits

More recently, another avenue of K^+ channel diversity was confirmed: the presence of β-subunits that modulate the functional properties of K^+ channel α-subunits expressed in a heterologous system. The idea that accessory subunits exist for K^+ channels is not new, because the inactivation kinetics, voltage dependence, and current amplitudes of voltage-gated Na^+ and Ca^{2+} channels are modulated by accessory β-subunits (Patton et al., 1994; De Waard et al., 1994; Makita et al., 1994; Bennett et al., 1993; Isom et al., 1992; Messner et al., 1986). Early purification studies of K^+ channel proteins, utilizing the knowledge that multiple K^+ channel isoforms bind the naturally occurring toxin DTX (Rehm and Lazdunski, 1988), identified a 39-kDa protein that consistently copurified with a 78-kDa DTX-binding channel of the Shaker-related Kv1 family (Dolly et al., 1994; Scott et al., 1994a,b; Parcej et al., 1992). The Kv2 subfamily also is associated with β-subunits, because smaller proteins also copurify from rat brain with the Kv2.1 protein using antibodies specific for this α-subunit (Trimmer, 1991). The Kv4 subfamily also may be associated with function-altering β-subunits. Rudy and co-workers showed that 2- to 4-Kilobase rat brain poly $(A)^+$-mRNA modifies the function of Kv4.2 expressed in Xenopus oocytes (Serodio et al., 1994). Although the brain mRNA may encode a channel-modifying enzyme, it is also possible that it encodes a β-subunit.

Rettig et al. (1994) isolated two full-length β-subunit cDNAs, now designated Kvβ1.1 and Kvβ2.1, from a rat brain cortex library. The Kvβ1.1 cDNA sequence corresponds to a 401-amino-acid protein, and the rat Kvβ2.1 sequence contains 367 amino acids. The rat Kvβ1 protein sequence differs markedly from the Kvβ2.1 sequence in the NH_2 terminus, whereas the remaining 329 COOH-terminal amino acids are 85% identical, suggesting that the NH_2-terminus is important for functional differences. Kvβ subunits are probably cytoplasmic proteins that associate with the cytoplasmic domain of Kvα subunits, because the Kvβ subunits show neither putative membrane-spanning domains nor potential NH_2-linked glycosylation sites indicative of membrane-delineated proteins. The rat Kvβ1.1 subunit confers rapid A-type inactivation on the Kv1.1 delayed rectifier channel, which appears to be mediated by an NH_2 terminal "inactivation ball" within the variable NH_2 terminus of Kvβ1.1 (Rettig et al., 1994). The rat Kvβ2.1 isoform lacks this inactivation domain and does not induce fast inactivation.

Simultaneous work by three independent groups indicated that cardiac K^+ channel diversity is further complicated by the presence of β-subunits. Two groups cloned a Kvβ1.2 subunit from human atrium (Majumder et al., 1995) and ventricle (England et al., 1995b), while a third group cloned the equivalent protein from ferret heart (Morales et al., 1995). A detailed analysis of the tissue-specific expression of this β-subunit in humans has not been performed; the ferret homolog is expressed most prominently in the following tissues: aorta > left ventricle > atrium = skeletal > brain > right ventricle > kidney. Another β-subunit, Kvβ1.3, also has been cloned from human heart (England et al., 1995a). Kvβ1.1, Kvβ1.2, and Kvβ1.3 represent splice variants derived from the same gene. These three proteins share a similar COOH-terminal sequence but possess unique NH_2 termini contributed by an isoform-defining specific exon. It is this NH_2-terminal region of Kvβ1.3 that confers fast inactivation onto the Kv1.5 delayed rectifier (see Fig. 3A,B).

Clearly, the discovery of functional K^+ channel β-subunits further complicates attempts to correlate cloned cardiac K^+ channels and endogenous myocyte currents. The same K^+ channel protein may be expressed in both atrium and ventricle, but by virtue of being associated with various β-subunit isoforms, it may yield vastly different electrophysiological characteristics.

3.4. Erg (H-erg and M-erg) Kv Channels

The *ether-a-go-go* (EAG) K^+ channel is not homologous to *Shaker* but was cloned from *Drosophila* using the same genetic approach of chromosomal walking (Bruggemann *et al.*, 1993; Warmke *et al.*, 1991). Whereas EAG-related channels share a common six-transmembrane-domain motif with *Shaker* (Ludwig et al.,1994; Warmke and Ganetzky, 1994), the homology between EAG and the Kv channels is very low (10–17% amino acid identity), indicating that EAG represents a novel family of K^+ channel genes. Mouse and human homologs of EAG [m-erg and h-erg (HERG), respectively] have been cloned from the brain (Warmke and Ganetzky, 1994) and the heart (London *et al.*, 1997; Lees-Miller *et al.*, 1997). Two splice variants of erg mRNA have been identified in mouse and human heart, designated m-ergB and h-ergB, respectively. M-ergA is the longest isoform, corresponding to a 1162-amino-acid protein that is 96% identical to the human homolog, h-ergA (London *et al.*, 1997). M-ergB is identical to m-ergA from S1 to the COOH terminal but has a considerably shorter 37-amino-acid NH_2-terminal domain that exhibits no significant homology to any previously cloned genes. Both RT-PCR and ribonuclease protection studies indicate that m-ergA and h-ergA are abundantly expressed in the brain, atrium, and ventricle, whereas m-ergB and h-ergB are abundant only in the heart (Lees-Miller *et al.*, 1997; London *et al.*, 1997). Erg message also has been reported to be abundantly expressed in the atrium and ventricle of ferret, guinea pig, rabbit, dog, and human (Brahmajothi *et al.*, 1997; Wymore *et al.*, 1997).

As shown in Fig. 4, heterologous expression of h-erg in *Xenopus* oocytes generates currents that exhibit many properties similar to those of the cardiac delayed rectifier, I_{Kr} (Sanguinetti *et al.*, 1995; Trudeau *et al.*, 1995). Properties shared by the h-erg current and I_{Kr} include strong inward rectification (see Fig. 4A,B) which is due to a rapid voltage-dependent inactivation (Shibasaki, 1987; Sanguinetti *et al.*, 1995; Trudeau *et al.*, 1995), block by methanesulfonanilides, such as E-4031 or dofetilde (Sanguinetti *et al.*, 1995; Trudeau *et al.*, 1995), and a single-channel conductance of 10–12 p5 (Zou *et al.*, 1997; Veldkamp *et al.*, 1993; Shibasaki, 1987). Although h-erg and I_{Kr} share many electrophysiological and pharmacological similarities, they are not identical. For example, the time constants of activation and deactivation of h-erg are 4–10 times slower than those of native I_{Kr} in guinea pig and mouse cardiac myocytes (Sanguinetti and Jurkiewicz, 1990), suggesting that h-erg may associate with other related α- or β-subunits to form the native channel. Indeed, simultaneous work by two independent groups (Lees-Miller *et al.*, 1997; London *et al.*, 1997) suggested that heteromeric assembly of physiologically occurring splice variants may form the channel responsible

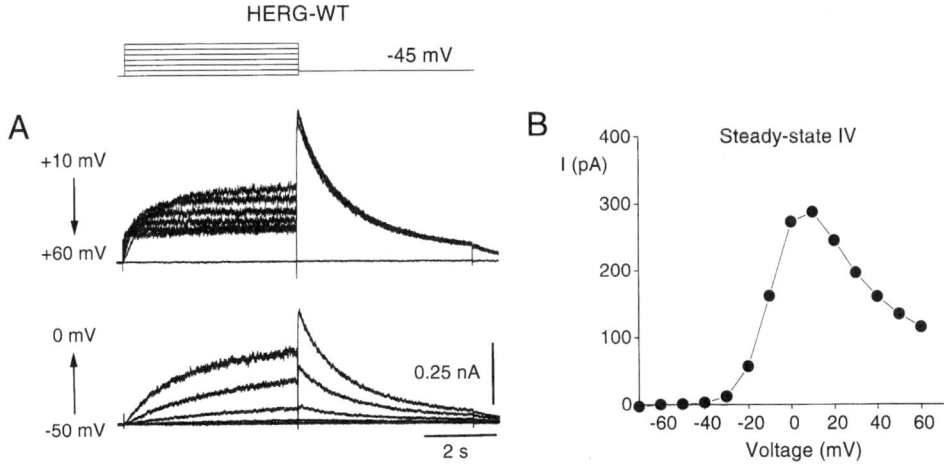

Figure 4. Functional expression of wild-type H-erg (HERG) in HEK293 cells. (A) Currents were elicited in response to 5-s depolarizing pulses in 10-mV steps from a holding potential of -60 mV to +60 mV; tails were recorded at −45 mV. (B) Plot of the steady-state current (measured at the end of the depolarizing pulses) against pulse potential. (Kindly provided by D. J. Snyders.)

for I_{Kr} in native myocytes. Robertson and co-workers demonstrated that coexpression of the cardiac-specific NH_2-terminal splice variant of h-erg, h-ergB (or the mouse homolog, m-ergB), with wild-type h-erg produces currents with activation and deactivation kinetics nearly identical to those of I_{Kr} (London *et al.*, 1997). More recently, Roden and colleagues reported that coexpression of a cardiac-specific COOH-terminal splice variant of h-erg with wild-type h-erg can modify the gating properties of h-erg and identified a 104-amino-acid COOH-terminal domain of HERG that appears to be essential for reconstitution of I_{Kr} in heterologous systems (Kupershmidt *et al.*, 1998). In fact, regions of the COOH terminus of rat EAG have been implicated in subunit–subunit interactions crucial for generating K^+ currents (Ludwig *et al.*, 1997).

Several reports also have suggested that minK (see Section 3.5) associates with h-erg to increase I_{Kr} (McDonald *et al.*, 1997; Yang *et al.*, 1995). A role for minK in I_{Kr} was initially suggested by the observation that antisense oligonucleotides directed against minK diminish I_{Kr} in an atrial tumor (AT-1) cell line (Yang *et al.*, 1995). Later, McDonald *et al.* (1997) demonstrated that minK physically associates with h-erg and that this heteromeric association may regulate I_{Kr}.

3.5. KvLQT1 and minK (I_{SK})

3.5.1. *KvLQT1*

KvLQT1 is a novel K^+ channel gene cloned by Keating and co-workers in 1996 (Q. Wang *et al.*, 1996). It is responsible for one form of the human LQTS and was found by identifying the locus responsible for this disease. Heterologous expression studies have confirmed that KvLQT1 encodes a voltage-dependent, rapidly activating and slowly deactivating outward K^+ current that does not resemble any of the known K^+ currents described in native myocytes (Sanguinetti *et al.*, 1995; Trudeau *et al.*, 1995). It is postulated, therefore, that KvLQT1 coassembles with other subunits to form a functional channel with biophysical and pharmacological properties similar to those of a cardiac K^+ current.

3.5.2. minK (I_{SK})

The unique, slowly activating minK clone was first discovered by expression cloning from rat kidney (KCNE1) (Takumi *et al.*, 1991; Murai *et al.*, 1989). Heterologous expression of minK in *Xenopus* oocytes revealed slowly activating K^+ currents that resembled the I_{Ks} current described in guinea pig ventricular myocytes (Sanguinetti and Jurkiewicz, 1990; Hume and Uehara, 1985) and human atrial myocytes (Z. Wang *et al.*, 1994). The protein has been detected immunohistochemically in adult guinea pig ventricular myocytes (Freeman and Kass, 1993), but the mRNA is extremely low in adult mouse (Felipe *et al.*, 1994) and rat heart (Folander *et al.*, 1990) relative to neonatal tissue.

The minK clone encodes a protein of only 130 amino acids with a single putative membrane-spanning domain that is not homologous to the S4 voltage-sensor or pore-forming sequence of other cloned K^+ channels. However, mutations in this region do alter gating kinetics (Takumi *et al.*, 1991). The orientation of minK in the membrane has not been confirmed, but consensus sites for glycosylation suggest that the NH_2 terminus may be extracellular, placing a PKC phosphorylation site and a COOH-terminal cysteine possibly involved in subunit interaction on the intracellular face (Philipson and Miller, 1992). The size and unusual structure of minK, with no homology to other cation channels, generated some doubt as to whether this protein alone forms functional channels (Attali *et al.*, 1993a; Lesage *et al.*, 1993). Arguments in favor of minK as a channel included the observation that mutagenesis of the transmembrane segment alters channel properties (Takumi *et al.*, 1991). In contrast, studies by Lazdunski and co-workers suggested that minK activates both endogenous K^+ and Cl^- currents in *Xenopus* oocytes (Attali *et al.*, 1993a). Furthermore, the expression of minK in mammalian cells has been extremely difficult to achieve (Lesage *et al.*, 1993). On the basis of these findings, Lazdunski and colleagues proposed that this protein was not a true channel but rather an activator of endogenous channel activity in *Xenopus* oocytes. In support of this hypothesis, this group demonstrated that KvLQT1, when coexpressed with minK, forms a channel with biophysical properties identical to those of the cardiac slowly activating delayed rectifier current, I_{Ks} (Barhanin *et al.*, 1996). Simultaneous work by another group (Sanguinetti *et al.*, 1996b) confirmed that KvLQT1 coassembles with minK to form the cardiac I_{Ks} channel and that *Xenopus* oocytes express their own form of KvLQT1. This endogenous KvLQT1 explains why minK expression succeeds only in the oocyte system. As shown in Fig. 3C, expression of KvLQT1 in CHO cells induces a voltage-dependent, rapidly activating and slowly inactivating outward K^+ current with biophysical properties distinct from those of any of the known cardiac K^+ currents, whereas minK does not produce any detectable current when expressed alone. In contrast, coexpression of KvLQT1 and minK produces a slowly activating delayed rectifier current that is much larger than the current in cells transfected with KvLQT1 alone (Fig. 3D) and which resembles the native cardiac I_{Ks}.

As mentioned in Section 2.3.2, the mouse minK gene has been deleted by two different groups (Kupershmidt *et al.*, 1999; Drici *et al.*, 1998). In both cases, cardiac myocytes isolated from minK $(-/-)$ animals lack I_{Ks}, as expected if minK plays a role in producing this current. In addition, I_{Kr} is significantly reduced and its deactivation slowed in the $-/-$ animals. This effect on the I_{Kr} current further supports a role for minK in modulating both I_{Ks} and I_{Kr}. A major difference between the two knockouts is that Kupershmidt *et al.* (1999) found no difference in the electrocardiograms between $(+/+)$ and $(-/-)$ animals at adult or neonatal stages, whereas Drici *et al.* (1998) reported action potential prolongation. The reason for this discrepancy is unclear.

It is important to realize that defects in three of the K^+ channel subunits discussed in this chapter, KvLQT1, h-erg (HERG), and minK, have been linked to three forms of inherited LQTS. The features of LQTS are discussed elsewhere in this book (see Chapters 35 and 36) and the reader is referred to these chapters for details. Briefly, LQTS is a disease characterized by an abnormally long action potential duration that predisposes many affected individuals to fatal cardiac arrhythmias, and, in several common forms, gene mutations in KvLQT1, h-erg (HERG), or minK produce a dominant-negative effect on I_{Ks} or I_{Kr} channel function. These human mutations confirm not only the importance of I_K components to cardiac function but also confirm the subunit composition of these currents. In this exciting area of research, K^+ channel molecular biology has interfaced perfectly with the human pathophysiology, a goal of all basic *science* in recent years.

4. SUMMARY AND CONCLUSIONS

During the last decade, tremendous progress has been made in the area of cardiac K^+ channel molecular physiology. As reviewed in this chapter, cDNA molecules have been cloned for all the major outward currents in the mammalian myocardium. Additionally, heterologous expression systems, antisense knockdowns, and transgenic mice have been developed as models to relate cloned cardiac channels to native currents. However, relating K^+ channel physiology to human pathophysiology has only been possible for the long QT syndrome, and the K^+ channel mechanisms of other potentially channel-based cardiovascular pathologies remain unknown.

So where will the cardiac K^+ channel field go next? Identification of new LQT mutations, although important, will not have the impact that they did two years ago. Determination of the exact subunit stoichiometry of cardiac K^+ channels, although also important, will not generate great excitement. Perhaps future research will focus on the regulation of cardiac K^+ channel function. All the K^+ channels reviewed here are likely to be well regulated by diverse signaling pathways. These channels also are very likely to be localized near, or within, carefully scaffolded arrays of signaling molecules. Elucidation of how channel function is coupled to the myocyte's internal and external environment, especially in the diseased myocardium, may prove the next challenge.

REFERENCES

Albrecht, B., Lorra, C., Stocker, M., and Pongs, O., 1993, Cloning and characterization of a human delayed rectifier potassium channel gene, *Receptors Channels* **1**:99–110.

Antzelevitch, C., Sicouri, S., Litovsky, S. H., Lukas, A., Krishnan, S. C., Di Diego, J. M., Gintant, G. A., and Liu, D. W., 1991, Heterogeneity within the ventricular wall: Electrophysiology and pharmacology of epicardial, endocardial, and M cells, *Circ. Res.* **69**:1427–1449.

Anumonwo, J. M. B., Freeman, L. C., Kwok, W. M., and Kass, R.S., 1991, Potassium channels in the heart: Electrophysiology and pharmacological regulation, *Cardiovasc. Drug Res.* **9**:299–316.

Apkon, M., and Nerbonne, J. M., 1991, Characterization of two distinct depolarization activated K^+ currents in isolated adult rat ventricular myocytes, *J. Gen. Physiol.* **97**:973–1011.

Attali, B., Guillemare, E., Lesage, F., Honore, E., Romey, G., Lazdunski, M., and Barhanin, J., 1993a, The protein I_{sK} is a dual activator of K^+ and Cl^- channels, *Nature* **365**:850–852.

Attali, B., Lesage, F., Ziliani, P., Guillemare, E., Honore, E., Waldmann, R., Hugnot, J. P., Mattei, M. G., Lazdunski, M., and Barhanin, J., 1993b, Multiple mRNA isoforms encoding the mouse cardiac Kv1.5 delayed rectifier K^+ channel, *J. Biol. Chem.* **268**:24283–24289.

Babila, T., Moscucci, A., Wang, H. Y., Weaver, F. E., and Koren, G., 1996, Assembly of mammalian voltage-gated potassium channels: Evidence for an important role of the first transmembrane segment, *Neuron* **12**:615–626.

Backx, P. H., and Marban, E., 1992, Background potassium conductance active during the plateau of the action potential in guinea pig ventricular myocytes, *Circ. Res.* **72**:890–900.

Baldwin, T. J., Tsaur, M. L., Lopez, G. A., Jan, Y. N., and Jan, L. Y., 1992, Characterization of a mammalian cDNA for an inactivating voltage-sensitive K^+ channel, *Neuron* **7**:471–483.

Balser, J. R., Bennett, P. B., and Roden, D., 1990, Time dependent outward current in guinea pig ventricular myocytes: Gating kinetics of the delayed rectifier, *J. Gen. Physiol.* **96**:835–863.

Barhanin, J., Lesage, F., Guillemare, E., Fink, M., Lazdunski, M., and Romey, G., 1996, $K_{(v)}$LQT1 and I_{sK} (minK) proteins associate to form the I_{sK} cardiac potassium current, *Nature* **384**:78–80.

Barry, D. M., and Nerbonne, J. M., 1996, Myocardial potassium channels: Electrophysiological and molecular diversity, *Annu. Rev. Physiol.* **58**:363–394.

Barry, D. M., Trimmer, J. S., Merlie, J. P., and Nerbonne, J. M., 1995, Differential expression of voltage-gated K^+ channel subunits in adult rat heart—relation to functional K^+ channels?, *Circ. Res.* **77**:361–369.

Barry, D. M., Xu, H. D., Schuessler, R. B., and Nerbonne, J. M., 1998, Functional knockout of the transient outward current, long-QT syndrome, and cardiac remodeling in mice expressing a dominant-negative Kv4 alpha subunit, *Circ. Res.* **83**:560–566.

Benndorf, K., 1988, Three types of single K channels contribute to the transient outward current in myocardial mouse cells, *Biomed. Biochim. Acta* **47**:401–416.

Benndorf, K., Markwardt, F., and Nilius, B., 1987, Two types of transient outward currents in cardiac ventricular cells of mice, *Pflügers Arch.* **413**:127–133.

Bennett, P. B., Makita, N., and George, A. L., 1993, A molecular basis for gating mode transitions in human skeletal muscle Na^+ channels, *FEBS Lett.* **326**:21–24.

Bertoli, A., Moran, O., and Conti, F., 1994, Activation and deactivation properties of rat brain channels of the *Shaker* -related subfamily, *Eur. J. Biophys.* **23**:379–384.

Blair, T. A., Roberds, S. L., Tamkun, M. M., and Hartshorne, R. P., 1991, Functional expression of RK5, a voltage-gated K^+ channel isolated from the rat cardiovascular system, *FEBS Lett.* **295**:211–213.

Bond, C. T., Pessia, M., Xia, X. M., Lagrutta, A., Kavanaugh, M. P., and Adelman, J. P., 1994, Cloning and expression of a family of inward rectifier potassium channels, *Receptors Channels* **2**:183–191.

Boyle, W. A., and Nerbonne, J. M., 1991, A novel type of depolarization-activated K^+ current in adult rat atrial myocytes, *Am. J. Physiol.* **260**:H1236–H1247.

Boyle, W. A., and Nerbonne, J. M., 1992, Two functionally distinct 4-aminopyridine-sensitive outward K^+ currents in rat atrial myocytes, *J. Gen. Physiol.* **100**:1041–1067.

Brahmajothi, M. V., Morales, M. J., Reimer, K. A., and Strauss, H. C., 1997, Regional localization of ERG, the channel protein responsible for the rapid component of the delayed rectifier, K^+ current in the ferret heart, *Circ. Res.* **81**:128–135.

Bruggemann, A., Pardo, L. A., Stuhmer, W., and Pongs, O., 1993, Ether-a-go-go encodes a voltage-gated channel permeable to K^+ and Ca^{2+} and modulated by cAMP, *Nature* **365**:445–448.

Busch, A. E., Malloy, K., Groh, W. J., Varnum, M. D., Adelman, J. P., and Maylie, J., 1994, The novel class III antiarrhythmics NE-10064 and NE-10133 inhibit Isk channels expressed in *Xenopus* oocytes and ISK in guinea pig myocytes, *Biochem. Biophys. Res. Commun.* **202**:265–270.

Butler, A., Wei, A., Baker, K., and Salkoff, L., 1989, A family of putative potassium channel genes in *Drosophila, Science* **243**:943–947.

Butler, A., Tsunoda, S., McCobb, D. P., Wei, A., and Salkoff, L., 1993, *mSlo*, a complex mouse gene encoding "maxi" calcium-activated potassium channels, *Science* **261**:221–224.

Campbell, D. L., Rasmusson, R. L., Comer, M. B., and Strauss, H. C., 1995, The cardiac calcium-independent transient outward potassium current: Kinetics, molecular properties and role in ventricular repolarization, in: *Cardiac Electrophysiology: From Cell to Bedside* (D. Zipes and J. Jalife, eds.), W.B. Saunders Co, Philadelphia, pp. 88–96.

Carmeliet, E., 1992, Voltage- and time-dependent block of the delayed K^+ current in cardiac myocytes by dofetilide, *J. Pharmacol. Exp. Ther.* **262**:809–817.

Chandy, K. G., and Gutman, G. A., 1995, Voltage-gated K^+ channels, in: *Handbook of Receptors and Channels: Ligand- and Voltage-Gated Ion Channels* (R. A. North, eds), CRC Press, Boca Raton, Florida, pp. 1–71.

Chouinard, S. W., Wilson, G. F., Schlimgen, A. K., and Ganetzky, B., 1995, A potassium channel beta subunit related to the aldo-keto reductase superfamily is encoded by the *Drosophila* Hyperkinetic locus, *Proc. Natl. Acad. Sci. U.S.A.* **92**:6763–6767.

Clark, R. B., Bouchard, R. A. A., Salinas-Stefanon, E., Sanchez-Chapula, J., and Giles, W. R., 1993, Heterogeneity of action potential waveforms and potassium currents in rat ventricle, *Cardiovasc. Res.* **27**:1795–1799.

Coraboeuf, E., and Carmeliet, E., 1982, Existence of two transient outward currents in sheep cardiac Purkinje fibers, *Pflügers Arch.* **392**:352–359.

Coraboeuf, E., and Nargeot, J., 1993, Electrophysiology of human cardiac cells, *Cardiovasc. Res.* **27**:1713–1725.

Deal, K. K., England, S. K., and Tamkun, M. M., 1996, Molecular physiology of cardiac potassium channels, *Physiol. Rev.* **76**:49–67.

De Waard, M., Pragnell, M., and Campbell, K. P., 1994, Ca^{2+} channel regulation by a conserved β subunit domain, *Neuron* **13**:495–503.

Dixon, J. E., and McKinnon, D., 1994, Quantitiative analysis of potassium channel mRNA expression in atrial and ventricular muscle of rats, *Circ. Res.* **75**:252–260.

Dixon, J. E., Shi, W. M., Wang, H. S., Mcdonald, C., Yu, H., Wymore, R. S., Cohen, I. S., and McKinnon, D., 1996, Role of the Kv4.3 K^+ channel in ventricular muscle—a molecular correlate for the transient outward current, *Circ. Res.* **79**:659–668.

Dolly, J. O., Rettig, J., Scott, V. E., Parcej, D. N., Wittkat, R., Sewing, S., and Pongs, O., 1994, Oligomeric and subunit structures of neuronal voltage-sensitive K^+ channels, *Biochem. Soc. Trans.* **22**:473–478.

Drici, M. D., Arrighi, I., Chouabe, C., Mann, J. R., Lazdunski, M., Romey, G., and Barhanin, J., 1998, Involvement of I_{sK}-associated K^+ channel in heart rate control of repolarization in a murine engineered model of Jervell and Lange-Nielsen syndrome, *Circ. Res.* **83**:95–102.

Dudel, J., Peper, K., Rudel, R., and Trautwein, W., 1967, The dynamic chloride component of membrane current in Purkinje fibers, *Pflügers Arch.* **295**:197–212.

Dukes, I. D., and Morad, M., 1991, The transient outward K^+ currents in rat ventricular myocytes: Evaluation of its Ca^{2+} and Na^+ dependence, *J. Physiol. (London)* **435**:395–420.

England, S. K., Uebele, V. N., Kodali, J., Bennett, P. B., and Tamkun, M. M., 1995a, A novel K^+ channel β-subunit (hKv $\beta1.3$) is produced via alternative mRNA splicings, *J. Biol. Chem.* **270**:28531–28534.

England, S. K., Uebele, V. N., Shear, H., Kodali, J., Bennett, P. B., and Tamkun, M. M., 1995b, Characterization of a voltage-gated K^+ channel β subunit expressed in human heart, *Proc. Natl. Acad. Sci. U.S.A.* **92**:6309–6313.

Fan, Z., and Hiraoka, M., 1991, Depression of delayed outward K^+ current by Co^{2+} in guinea pig ventricular myocytes, *Am. J. Physiol.* **261**:23–31.

Fedida, D., and Giles, W. R., 1991, Regional variations in action potentials and transient outward current in myocytes isolated from rabbit left ventricle, *J. Physiol. (London)* **442**:191–201.

Fedida, D., Wible, B., Wang, Z., Fermini, B., Faust, F., Nattel, S., and Brown, A. M., 1993, Identity of a novel delayed rectifier current from human heart with a cloned K^+ channel current, *Circ. Res.* **73**:210–216.

Felipe, A., Knittle, T. J., Doyle, K. L., Snyders, D. J., and Tamkun, M. M., 1994, Differential expression of I_{sk} mRNAs in mouse tissue during development and pregnancy, *Am. J. Physiol.* **267**:700–705.

Feng, J. L., Wible, B., Li, G. R., Wang, Z. G., and Nattel, S., 1997, Antisense oligodeoxynucleotides directed against Kv1.5 mRNA specifically inhibit ultrarapid delayed rectifier K^+ current in cultured adult human atrial myocytes, *Circ. Res.* **80**:572–579.

Fiset, C., Clark, R. B., Shimoni, Y., and Giles, W. R., 1997, Shal-type channels contribute to the Ca^{2+}-independent transient outward K^+ current in rat ventricle, *J. Physiol. (London)* **500**:51–64.

Folander, K., Smith, J. S., Antanavage, J., Bennett, C., Stein, R. B., and Swanson, R., 1990, Cloning and expression of the delayed-rectifier I_{sK} channel from neonatal rat heart and diethylstilbestrol-primed rat uterus, *Proc. Natl. Acad. Sci. U.S.A.* **87**:2975–2979.

Fozzard, H. A., 1994, Cardiac electrogenesis and the sodium channel, in: *Ion Channels in the Cardiovascular System: Function and Dysfunction* (P. M. Spooner and A. M. Brown, eds.), Futura Publishing Co., Armonk, New York, pp. 81–99.

Fozzard, H. A., and Hiraoka, M., 1973, The positive dynamic current and its inactivation properties in cardiac Purkinje fibres, *J. Physiol. (London)* **234**:569–586.

Franqueza, L., Valenzuela, J., Tamkun, M. M., Tamargo, J., and Snyders, D. J., 1999, Functional expression of an inactivating potassium channel (Kv4.3) in a mammalian cell line, *Cardiovasc. Res.* **41**:212–219.

Frech, G. C., Vandongen, A. M. J., Schuster, G., Brown, A. M., and Joho, R. H., 1989, A novel potassium channel with delayed rectifier properties isolated from rat brain by expression cloning, *Nature* **340**:642–645.

Freeman, L. C., and Kass, R. S., 1993, Expression of a minimal K^+ channel protein in mammalian cells and immunolocalization in guinea pig heart, *Circ. Res.* **73**:968–973.

Furukawa, T., Myerburg, R. J., Furukawa, N., Bassett, A. L., and Kimura, S., 1990, Differences in transient outward currents of feline endocardial and epicardial myocytes, *Circ. Res.* **67**:1287–1291.

Giles, W. R., and Imaizumi, Y., 1988, Comparison of potassium currents in rabbit atrial and ventricular cells, *J. Physiol. (London)* **405**:123–145.

Giles, W. R., and Van Kinneken., 1985, A transient outward current in cells from the crista terminalis of rabbit heart, *J. Physiol. (London)* **368**:243–264.

Gintant, G. A., Cohen, I. S., Datyner, N. B., and Kline, R. P., 1991, Time-dependent outward currents in the heart, in: *The Heart and Cardiovascular System: Scientific Foundations* (H. A. Fozzard, E. Haber, and R. B. Jennings, eds.), Raven Press, New York, pp. 1121–1169.

Grissmer, S., Nguyen, A. N., Aiyar, J., Hanson, D. C., Mather, R. J., Gutman, G. A., Karmilowicz, M. J., Auperin, D. D., and Chandy, K. G., 1994, Pharmacological characterization of five cloned voltage-gated K^+ channels, types Kv1.1, 1.2, 1.3, 1.5, and 3.1, stably expressed in mammalian cell lines, *Mol. Pharmacol.* **45**:1227–1234.

Guo, W., Xu, H., and Nerbonne, J. M., 1999, Molecular basis of transient outward K^+ current diversity in mouse ventricular myocytes, *Biophys. J.* **76**:A331.

Hirano, Y., and Hiraoka, M., 1988, Barium-induced automatic activity in isolated ventricular myocytes from guinea-pig hearts, *J. Physiol. (London)* **480**:449–463.

Horie, M., Hayashi, S., and Kawai, C., 1990, Two types of delayed rectifying K^+ channels in atrial cells of guinea pig heart, *Jpn. J. Physiol.* **40**:479–490.

Hume, J. R., and Uehara, A., 1985, Ionic basis of the different action potential configurations of single guinea-pig atrial and ventricular myocytes, *J. Physiol. (London)* **368**:525–544.

Hume, J. R., Uehara, A., Hadley, R. W., and Harvey, R. D., 1990, Comparison of K^+ channels in mammalian atrial and ventricular myocytes, *Prog. Clin. Biol. Res.* **334**:17–41.

Isom, L. L., De Jongh, K. S., Patton, D. E., Reber, B. F., Offord, J., Charbonneau, H., Walsh, K., Goldin, A. L., and Catterall, W. A., 1992, Primary structure and functional expression of the β_1 subunit of the rat brain sodium channel, *Science* **256**:839–842.

Johns, D. C., Nuss, H. B., and Marban, E., 1997, Suppression of neuronal and cardiac transient outward currents by viral gene transfer of dominant-negative Kv4.2 constructs, *J. Biol. Chem.* **272**:31598–31603.

Kaab, S., Dixon, J., Duc, J., Ashen, D., Nabauer, M., Beuckelmann, D. J., Steinbeck, G., McKinnon, D., and Tomaselli, G. F., 1998, Molecular basis of transient outward potassium current downregulation in human heart failure: A decrease in Kv4.3 mRNA correlates with a reduction in current density, *Circulation* **98**:1383–1393.

Kaplan, W. D., and Trout, W. E., 1969, The behavior of four neurological mutants of *Drosophila*, *Genetics* **61**:399–409.

Kenyon, J. L., and Gibbons, W. R., 1979, Influence of chloride, potassium, and tetraethylammonium ions on the early outward current of sheep Purkinje fibers, *J. Gen. Physiol.* **73**:117–138.

Kilborn, M. J., and Fedida, D., 1990, A study of the developmental changes in outward currents of rat ventricular myocytes, *J. Physiol. (London)* **430**:37–60.

Kong, W., Po, S., Yamagishi, T., Ashen, M. D., Stetten, G., and Tomaselli, G. F., 1998, Isolation and characterization of the human gene encoding I_{to}: Further diversity by alternative mRNA splicing, *Am. J. Physiol.* **275**:H1963–H1970.

Kupershmidt, S., Snyders, D. J., Raes, A., and Roden, D. M., 1998, A K$^+$ channel splice variant common in human heart lacks a C-terminal domain required for expression of rapidly activating delayed rectifier current, *J. Biol. Chem.* **273**:27231–27235.

Kupershmidt, S., Yang, T., Anderson, M. E., Wessels, A., Niswender, K. D., Magnuson, M. A., and Roden, D. M., 1999, Replacement by homologous recombination of the minK gene with lacZ reveals restriction of minK expression to the mouse cardiac conduction system, *Circ. Res.* **84**:146–152.

Lees-Miller, J. P., Kondo, C., Wang, L., and Duff, H. J., 1997, Electrophysiological characterization of an alternatively processed ERG K$^+$ channel in mouse and human hearts, *Circ. Res.* **81**:719–726.

Lesage, F., Attali, B., Lakey, J., Honore, E., Romey, G., Faurobert, E., Lazdunski, M., and Barhanin, J., 1993, Are *Xenopus* oocytes unique in displaying functional I_{sK} channel heterologous expression?, *Receptors Channels* **1**:143–152.

Litovsky, S. H., and Antzelevitch, C., 1988, Transient outward current prominent in canine ventricular epicardium but not endocardium, *Circ. Res.* **62**:116–126.

Liu, D. W., and Antzelevitch, C., 1995, Characteristics of the delayed rectifier current (I_{Kr} and I_{Ks}) in canine ventricular epicardial, midmyocardial, and endocardial myocytes, *Circ. Res.* **76**:351–365.

Liu, D. W., Gintant, G. A., and Antzelevitch, C., 1993, Ionic basis for electrophysiological distinctions among epicardial, midmyocardial, and epicardial myocytes, *Circ. Res.* **72**:671–687.

London, B., Trudeau, M. C., Newton, K. P., Beyer, A. K., Copeland, N. G., Gilbert, D. J., Jenkins, N. A., Satler, C. A., and Robertson, G. A., 1997, Two isoforms of the mouse ether-a-go-go-related gene coassemble to form channels with properties similar to the rapidly activating component of the cardiac delayed rectifier K$^+$ current, *Circ. Res.* **81**:870–878.

London, B., Wang, D. W., Hill, J. A., and Bennett, P. B., 1998, The transient outward current in mice lacking the potassium channel gene Kv1.4, *J. Physiol. (London)* **509**:171–182.

Ludwig, J., Terlau, H., Wunder, F., Bruggemann, A., Pardo, L. A., Marquardt, A., Stuhmer, W., and Pongs, O., 1994, Functional expression of a rat homologue of the voltage gated *ether a go-go* potassium channel reveals differences in selectivity and activation kinetics between the *Drosophila* channel and its mammalian counterpart, *EMBO J.* **13**:4451–4458.

Ludwig, J., Owen, D., and Pongs, O., 1997, Carboxy-terminal domain mediates assembly of the voltage-gated rat Ether-a-go-go potassium channel, *EMBO J.* **16**:6337–6345.

Majumder, K., De Biasi, M., Wang, Z., and Wible, B. A., 1995, Molecular cloning and functional expression of a novel potassium channel β-subunit from human atrium, *FEBS Lett.* **361**:13–16.

Makita, N., Bennett, P. B., and George, A. L., 1994, Voltage-gated Na$^+$ channel β_1 subunit mRNA expressed in adult human skeletal muscle, heart, and brain is encoded by a single gene, *J. Biol. Chem.* **269**:7571–7578.

Matsubara, H., Suzuki, J., and Inada, M., 1993, *Shaker*-related potassium channel, Kv1.4, mRNA regulation in cultured rat heart myocytes and differential expression of Kv1.4 and Kv1.5 genes in myocardial development and hypertrophy, *J. Clin. Invest.* **92**:1659–1666.

Mays, D. J., Foose, J. M., Philipson, L. H., and Tamkun, M. M., 1995, Localization of the Kv1.5 K$^+$ channel protein in explanted cardiac tissue, *J. Clin. Invest.* **96**:282–292.

McDonald, T. V., Yu, Z. H., Ming, Z., Palma, E., Meyers, M. B., Wang, K. W., Goldstein, S. A. N., and Fishman, G. I., 1997, A minK–HERG complex regulates the cardiac potassium current I_{Kr}, *Nature* **388**:289–292.

McKinnon, D., 1989, Isolation of a cDNA clone coding for a putative second potassium channel indicates the existence of a gene family, *J. Biol. Chem.* **264**:8230–8236.

Messner, D. J., Feller, D. J., Scheuer, T., and Catterall, W. A., 1986, Functional properties of rat brain sodium channels lacking the $\beta1$ or $\beta2$ subunit, *J. Biol. Chem.* **261**:14882–14890.

Morales, M. J., Castellino, R. C., Crews, A. L., Rasmusson, R. L., and Strauss, H. C., 1995, A novel β subunit increases rate of inactivation of specific voltage-gated potassium channel alpha subunits, *J. Biol. Chem.* **270**:6272–6277.

Murai, T., Kakizuka, A., Takumi, T., Ohkubo, H., and Nakanishi, S., 1989, Molecular cloning and sequence analysis of human genomic DNA encoding a novel membrane protein which exhibits a slowly activating potassium channel activity, *Biochem. Biophys. Res. Commun.* **61**:176–181.

Nabauer, M., Beuckelmann, D. J., and Erdmann, E., 1993, Characteristics of transient outward current in human ventricular myocytes from patients with terminal heart failure, *Circ. Res.* **73**:386–394.

Nakamura, K., and Iijima, T., 1994, Postnatal changes in mRNA expression of the K$^+$ channel in rat cardiac ventricles, *Jap. J. Pharmacol.* **66**(4):489–492.

Nichols, C. G., and Lederer, W. J., 1991, Adenosine triphosphate-sensitive potassium channels in the cardiovascular system, *Am. J. Physiol.* **261**:H1675–H1686.

Nishiyama, A., Kambe, F., Kamiya, K., Seo, H., and Toyama, J., 1998, Effects of thyroid status on the expression of voltage-gated K$^+$ channels in rat left ventricle, *Cardiovasc. Res.* **40**:343–351.

Ohya, S., Tanaka, M., Oku, T., Asai, Y., Watanabe, M., Giles, W. R., and Imaizumi, Y., 1997, Molecular cloning and tissue distribution of an alternatively spliced variant of an A-type K$^+$ channel alpha-subunit, Kv4.3 in the rat, *FEBS Lett.* **420**:47–53.

Overturf, K. E., Russell, S. N., Carl, A., Vogalis, F., Hart, P. J., Hume, J. R., Sanders, K. M., and Horowitz, B., 1994, Cloning and characterization of a Kv1.5 delayed rectifier K$^+$ channel from vascular and visceral smooth muscles, *Am. J. Physiol.* **267**:C1231–C1238.

Parcej, D. N., Scott, V. E. S., and Dolly, O., 1992, Oligomeric properties of α-dendrotoxin-sensitive potassium ion channels purified from bovine brain, *Biochemistry* **31**:11084–11088.

Patton, D. E., Isom, L. L., Catterall, W. A., and Goldin, A. L., 1994, The adult rat brain β_1 subunit modifies activation and inactivation gating of multiple sodium channel α subunits, *J. Biol. Chem.* **269**:17649–17655.

Paulmichl, M., Nasmith, P., Hellmiss, K., Reed, K., Boyle, W. A., Nerbonne, J. M., Peralta, E. G., and Clapham, D. E., 1991, Cloning and expression of a rat delayed rectifier potassium channel, *Proc. Natl. Acad, Sci. U.S.A.* **88**:7892–7895.

Philipson, L. H., and Miller, R. J., 1992, A small K$^+$ channel looms large, *Trends Pharmacol. Sci.* **13**:8-11.

Philipson, L. H., Hice, R. E., Schaeffer, K., Lamendola, J., Bell, G. I., Nelson, D. J., and Steiner, D. F., 1991, Sequence and functional expression in *Xenopus* oocytes of a human insulinoma and islet potassium channel, *Proc. Natl. Acad. Sci. U.S.A.* **88**:53–57.

Po, S. S., Snyders, D. J., Baker, R., Tamkun, M. M., and Bennett, P. B., 1992, Functional expression of an inactivating potassium channel cloned from human heart, *Circ. Res.* **71**:732–736.

Po, S. S., Roberds, S. L., Snyders, D. J., Tamkun, M. M., and Bennett, P. B., 1993, Heteromultimeric assembly of human potassium channels, *Circ. Res.* **72**:1326–1336.

Pongs, O., Kecskemthy, N., Muller, R., Krah-Jentgens, I., Baumann, A., Kiltz, H. H., Canal, I., Llamazares, S., and Ferrus, A., 1988, *Shaker* encodes a family of putative potassium channel proteins in the nervous system of *Drosophila*, *EMBO J.* **7**:1087–1096.

Rehm, H., and Lazdunski, M., 1988, Purification and subunit structure of a putative K$^+$ channel protein identified by its binding properties for dendrotoxin I, *Proc. Natl. Acad. Sci. U.S.A.* **85**:4919–4923.

Rettig, J., Heinemann, S. H., Wunder, F., Lorra, C., Parcej, D. N., Dolly, J. O., and Pongs, O., 1994, Inactivation properties of voltage-gated K$^+$ channels altered by presence of β-subunit, *Nature* **369**:289–294.

Roberds, S. L., and Tamkun, M. M., 1991, Cloning and tissue-specific expression of five voltage-gated potassium channel cDNAs expressed in rat heart, *Proc. Natl. Acad. Sci. U.S.A.* **88**:1798–1802.

Salata, J. J., Jurkiewicz, N. K., Jow, B., Folander, K., Guinosso, P. J. J., Raynor, B., Swanson, R., and Fermini, B., 1996, I_K of rabbit ventricle is composed of two currents: Evidence for I_{Ks}, *Am. J. Physiol.* **271**:H2477–H2489.

Salkoff, L., 1985, Development of ion channels in the flight muscles of *Drosophila, J. Physiol. (Paris)* **80**:275–282.

Sanguinetti, M. C., and Jurkiewicz, N. K., 1990, Two components of cardiac delayed rectifier K$^+$ current: Differential sensitivity to block by class III antiarrhythmic agents, *J. Gen. Physiol.* **96**:195–215.

Sanguinetti, M. C., and Jurkiewicz, N. K., 1991, Delayed rectifier outward K$^+$ current is composed of two currents in guinea pig atrial cells, *Am. J. Physiol.* **260**:393–399.

Sanguinetti, M. C., Jiang, C., Curran, M. E., and Keating, M. T., 1995, A mechanistic link between an inherited and an acquired cardiac arrhythmia: HERG encodes the I_{Kr} potassium channel, *Cell* **81**:299–307.

Sanguinetti, M. C., Curran, M. E., Zou, A., Shen, J., Spector, P. S., Atkinson, D. L., and Keating, M. T., 1996b, Coassembly of $K_{(v)}$LQT1 and minK (IsK) proteins to form cardiac I_{Ks} potassium channel, *Nature* **384**:80–83.

Scott, V. E., Muniz, Z. M., Sewing, S., Lichtinghagen, R., Parcej, D. N., Pongs, O., and Dolly, J. O., 1994a, Antibodies specific for distinct Kv subunits unveil a heterooligomeric basis for subtypes of α-dendrotoxin-sensitive K^+ channels in bovine brain, *Biochemistry* **33**:1617–1623.

Scott, V. E., Rettig, J., Parcej, D. N., Keen, J. N., Findlay, J. B., Pongs, O., and Dolly, J. O., 1994b, Primary structure of a β subunit of α-dendrotoxin-sensitive K^+ channels from bovine brain, *Proc. Natl. Acad. Sci. U.S.A.* **91**:1637–1641.

Serodio, P., Kentros, C., and Rudy, B., 1994, Identification of molecular components of A-type channels activating at subthreshold potentials, *J. Neurophysiol.* **72**:1516–1529.

Shibasaki, T., 1987, Conductance and kinetics of delayed rectifier potassium channels in nodal cells of the rabbit heart, *J. Physiol.* **387**:227–250.

Shimoni, Y., Fiset, C., Clark, R. B., Dixon, J. E., McKinnon, D., and Giles, W. R., 1997, Thyroid hormone regulates postnatal expression of transient K^+ channel isoforms in rat ventricle, *J. Physiol. (London)* **500**:65–73.

Slawsky, M. T., and Castle, N. A., 1994, K^+ channel blocking actions of flecainide compared with those of propafenone and quinidine in adult rat ventricular myocytes, *J. Pharmacol. Exp. Ther.* **269**:66–74.

Smith, J. M., and Wahler, G. M., 1996, ATP-sensitive potassium channels are altered in ventricular myocytes from diabetic rats, *Mol. Cell. Biochem.* **158**:43–51.

Snyders, D. J., and Chaudhary, A., 1996, High affinity open channel block by dofetilide of HERG expressed in a human cell line, *Mol. Pharmacol.* **49**:949–955.

Stuhmer, W., Ruppersberg, J. P., Schroter, K. H., Sakmann, B., Stocker, M., Giese, K. P., Perschke, A., Baumann, A., and Pongs, O., 1989, Molecular basis of functional diversity of voltage-gated potassium channels in mammalian brain, *EMBO J.* **8**:3235–3244.

Swanson, R., Marshall, J., Smith, J. S., Williams, J. B., Boyle, M. B., Folander, K., Luneau, C. J., Aantanavage, J., Oliva, C., Buhrow, S. A., Bennett, C., Stein, R. B., and Kaczmarek, L. K., 1990, Cloning and expression of cDNA and genomic clones encoding three delayed rectifier potassium channels in rat brain, *Neuron* **4**:929–939.

Takano, M. and Noma, A., 1993, The ATP-sensitive K^+ channel, *Prog. Neurobiol.* **41**:21–30.

Takimoto, K., Li, D. Q., Hershman, K. M., Li, P., Jackson, E. K., and Levitan, E. S., 1997, Decreased expression of Kv4.2 and novel Kv4.3 K^+ channel subunit mRNAs in ventricles of renovascular hypertensive rats, *Circ. Res.* **81**:533–539.

Takumi, T., Ohkubo, H., and Nakanishi, S., 1988, Cloning of a membrane protein that induces a slow voltage-gated potassium current, *Science* **242**:1042–1045.

Takumi, T., Moriyoshi, K., Aramori, I., Ishii, T., Oiki, S., Okada, Y., Ohkubo, H., and Nakanishi, S., 1991, Alteration of channel activities and gating by mutations of slow-I_{sk} potassium channel, *J. Biol. Chem.* **266**:22192–22198.

Tamkun, M. M., Knoth, K., Walbridge, J. A., Kroemer, H., Roden, D., and Glover, D., 1991, Molecular cloning and characterization of two voltage-gated K^+ channel cDNAs from human ventricle, *FASEB J.* **5**:331–337.

Tamkun, M. M., Bennett, P. B., and Snyders, D. J., 1994, Cloning and expression of human cardiac K^+ channels, in: *Cardiac Electrophysiology: From Cell to Bedside* (D. Zipes and J. Jalife, eds.), W.B. Saunders Co, Philadelphia, pp. 21–31.

Tempel, B. L., Papazian, D., Schwarz, T., Jan, Y. N., and Jan, L. Y., 1987, Sequence of a probable potassium channel component encoded at *Shaker* locus of *Drosophila*, *Science* **237**:770–775.

Terzic, A., Jahangir, A., and Kurachi, Y., 1995, Cardiac ATP-sensitive K^+ channels: Regulation by intracellular nucleotides and K^+ channel-opening drugs, *Am. J. Physiol.* **38**:C525–C545.

Trimmer, J. S., 1991, Immunological identification and characterization of a delayed rectifier K^+ channel polypeptide in rat brain, *Proc. Natl. Acad. Sci. U.S.A.* **88**:10764–10768.

Trudeau, M. C., Warmke, J. W., Ganetzky, B., and Robertson, G. A., 1995, H-ERG, a human inward rectifier in the voltage-gated potassium channel family, *Science* **269**:92–95.

Tseng, G. N., and Hoffman, B. F., 1989, Two components of transient outward current in canine ventricular myocytes, *Circ. Res.* **64**:633–647.

Tseng-Crank, J. C., Tseng, G. N., Schwartz, A., and Tanouye, M. A., 1990, Molecular cloning and functional expression of a potassium channel cDNA isolated from a rat cardiac library, *FEBS Lett.* **268**:63–68.

Uebele, V. N., England, S. K., Chaudhary, A., Tamkun, M. M., and Snyders, D. J., 1996, Functional differences in Kv1.5 currents expressed in mammalian cell lines are due to the presence of endogenous Kvβ2.1 subunits, *J. Biol. Chem.* **271**:2406–2412.

Varro, A., Nanasi, P. P., and Lathrop, D. A., 1993, Potassium currents in isolated human atrial and ventricular cardiocytes, *Acta Physiol. Scand.* **149**:133–142.

Veldkamp, M. W., van Ginneken, A. C., and Bouman, L. N., 1993, Single delayed rectifier channels in the membrane of rabbit ventricular myocytes, *Circ. Res.* **72**:865–878.

Wahler, G. M., Dollinger, S. J., Smith, J. M., Flemal, K. L., 1994 Time course of postnatal changes in rat heart action potential and in transient outward current is different. *Am. J. Physiol.* **267**:(3pt2): H1157–1166.

Wang, Q., Shen, J., Splawski, I., Atkinson, D., Li, Z., Robinson, J. L., Moss, A. J., Towbin, J. A., and Keating, M. T., 1995, SCN5A mutations associated with an inherited cardiac arrhythmia, long QT syndrome, *Cell* **80**:805–811.

Wang, Q., Curran, M. E., Splawski, I., Burn, T. C., Millholland, J. M., Vanraay, T. J., Shen, J., Timothy, K. W., Vincent, G. M., Dejager, T., Schwartz, P. J., Towbin, J. A., Moss, A. J., Atkinson, D. L., Landes, G. M., Connors, T. D., and Keating, M. T., 1996, Positional cloning of a novel potassium channel gene: KVLQT1 mutations cause cardiac arrhythmias, *Nat. Genet.* **12**:17–23.

Wang, Z., Fermini, B., and Nattel, S., 1993, Sustained depolarization-induced outward current in human atrial myocytes: Evidence for a novel delayed rectifier potassium current similar to Kv1.5 cloned channel currents, *Circ. Res.* **73**:1061–1076.

Wang, Z., Fermini, B., and Nattel, S., 1994, Rapid and slow components of delayed rectifier current in human atrial myocytes, *Cardiovasc. Res.* **28**:1540–1546.

Wang, Z., Fermini, B., and Nattel, S., 1995, Effects of flecainide, quinidine, and 4-aminopyridine on transient outward and ultrarapid delayed rectifier currents in human atrial myocytes, *J. Pharmacol. Exp. Ther.* **272**:184–196.

Warmke, J. W., and Ganetzky, B., 1994, A family of potassium channel genes related to *eag* in *Drosophila* and mammals, *Proc. Natl. Acad. Sci. U.S.A.* **91**:3438–3442.

Warmke, J., Drysdale, R., and Ganetzky, B., 1991, A distinct potassium channel polypeptide encoded by the *Drosophila eag* locus, *Science* **252**:1560–1562.

Wettwer, E., Amos, G. J., Gath, J., Zerkowski, H. R., Reidemeister, J. G., and Ravens, U., 1993, Transient outward current in human and rat ventricular myocytes, *Cardiovasc. Res.* **27**:1662–1669.

Wickenden, A. D., Kaprielian, R., Parker, T. G., Jones, O. T., and Backx, P. H., 1997, Effects of development and thyroid hormone on K^+ currents and K^+ channel gene expression in rat ventricle, *J. Physiol. (London)* **504**:271–286.

Wymore, R. S., Gintant, G. A., Wymore, R. T., Dixon, J. E., McKinnon, D., and Cohen, I. S., 1997, Tissue and species distribution of mRNA for the I_K-like K^+ channel, ERG, *Circ. Res.* **80**:261–268.

Xu, H., Dixon, J. E., Barry, D. M., Trimmer, J. S., Merlie, J. P., McKinnon, D., and Nerbonne, J. M., 1996, Developmental analysis reveals mismatches in the expression of K^+ channel alpha subunits and voltage-gated K^+ channel currents in rat ventricular myocytes, *J. Gen. Physiol.* **108**:405–419.

Xu, H., Barry, D. M., Li, H., and Nerbonne, J. M., 1999, Alteration of $I_{K,slow}$ and induction of triggered activity in ventricular myocytes isolated from transgenic mice expressing truncated Kv2.1 subunit, Kv2.1N216-Flag, *Biophys. J.* **76**:A327.

Yang, T., Kupershmidt, S., and Roden, D. M., 1995, Anti-minK antisense decreases the amplitude of the rapidly activating cardiac delayed rectifier K^+ current, *Circ. Res.* **77**:1246–1253.

Yue, D. T., and Marban, E., 1988, A novel cardiac potassium channel that is active and conductive at depolarized potentials, *Pflügers Arch.* **413**:127–133.

Zou, A. R., Curran, M. E., Keating, M. T., and Sanguinetti, M. C., 1997, Single HERG delayed rectifier K^+ channels expressed in *Xenopus* oocytes, *Am. J. Physiol.* **41**:H1309–H1314.

Zygmunt, A. C., 1994, Intracellular calcium activates a chloride current in canine ventricular myocytes, *Am. J. Physiol.* **267**:H1984–H1995.

Zygmunt, A. C., and Gibbons, W. R., 1991, Calcium-activated chloride current in rabbit ventricular myocytes, *Circ. Res.* **68**:424–437.

Chapter 19

Inward Rectifying and ATP-Sensitive K^+ Channels in the Ventricular Myocardium

Akikazu Fujita and Yoshihisa Kurachi

I. INTRODUCTION

In cardiac myocytes, such as Purkinje fibers and ventricular myocytes, the depolarizing phase (phase 0) of the action potential is mediated by a rapid influx of Na^+ ions through voltage-gated Na^+ channels. A subsequent rapid, but incomplete, repolarization in phase 1 is achieved by K^+ efflux through transiently activated K^+ channels. This early outward current sets the initial plateau potential, thus influencing the behavior of subsequently activated ion channels and the duration of the action potential. The plateau of the action potential (phase 2) is characterized by minimal net ion flow due to the balance between an inactivating Ca^{2+} inward current and K^+ efflux through slowly activating voltage-gated channels. In phase 3, the K^+ permeability increases with time and ultimately results in complete repolarization. In phase 4, the K^+ permeability sets the deep resting potential, which governs the time period between action potentials. Therefore, K^+ channels play essential roles in various phases of the cardiac action potential.

The K^+ channels in cardiac myocytes can be classified into two major categories: inward rectifying K^+ (Kir) channels and voltage-gated K^+ (Kv) channels. Cardiac Kir channels include classical Kir (I_{K1}) (Kurachi, 1985; Sakmann and Trube, 1984), G-protein-activated muscarinic K^+ (K_{ACh}) (Kurachi et al., 1986; Sakmann et al., 1983), Na^+-activated K^+ (K_{Na}) (Kameyama et al., 1984), and ATP-sensitive K^+ (K_{ATP}) channels (Noma, 1983). The Kv channels that have been identified electrophysiologically include A-type transient outward K^+ (I_{to}) channels (Coraboeuf and Carmeliet, 1982; Kenyon and Gibbons, 1979) and delayed rectifier K^+ (I_K) channels, including I_{Kur}, I_{Kr}, and I_{Ks} channels (Coraboeuf and Carmeliet, 1982; Kenyon and Gibbons, 1979). I_{Kur}, I_{Kr}, and I_{Ks} are, respectively, ultrarapidly, rapidly, and slowly activating

Akikazu Fujita and Yashihisa Kurachi • Department of Pharmacology II, Faculty of Medicine and Graduate School of Medicine, Osaka University, Suita, Osaka 565-0871, Japan.

Potassium Channels in Cardiovascular Biology, edited by Archer and Rusch. Kluwer Academic/Plenum Publishers, New York, 2001.

components of I_K. Kv channels are activated at depolarized membrane potentials during the action potential and initiate its repolarization, whereas Kir channels have diverse cellular functions, including roles in setting the resting membrane potential near the K^+ equilibrium potential (I_{K1} channel), acetylcholine-induced deceleration of the heartbeat (K_{ACh} channel), and metabolic impairment-induced shortening of the action potential (K_{ATP} channel and possibly K_{Na} channel). Kir channels are distributed in the different regions of the heart. Three of these Kir channels, I_{K1}, K_{ATP}, and K_{Na}, are detected in the ventricular myocardium.

Unlike Kv channels, the Kir channels are not regulated by the membrane potential across the cell membrane but by intracellular substances such as polyamines and Mg^{2+} ions (see Isomoto *et al.*, 1997a). These substances block the outward-going K^+ currents through Kir channels above the K^+ equilibrium potential, resulting in the inward rectifying property. Another feature common to Kir channels is that the channel conductance increases as the extracellular K^+ concentration ($[K^+]_o$) increases. Also, each of the Kir channel types is regulated by specific intracellular ligands, namely G-protein (K_{ACh}), ATP (K_{ATP}), and Na^+ (K_{Na}). These features are essential for the functional roles of the Kir channels in various tissues.

In 1993, a K^+-transport Kir channel, Kir1.1/ROMK (Ho *et al.*, 1993), and a classical Kir channel, Kir2.1/IRK1 (Kubo *et al.*, 1993a), were cloned by the expression cloning technique from the outer medulla of rat kidney and a mouse macrophage cell line, respectively. They have a common molecular motif in the primary structure, i.e., two putative membrane-spanning regions (M1 and M2) and one potential pore-forming loop (H5) [Fig.1B(a)]. Thus, the primary structure of these Kir channel subunits resembles that of the S5, H5, and S6 segments of Kv channels [Fig. 1A(a)] (Jan and Jan, 1994, 1992; Pongs, 1992). Recent studies have provided evidence that the Kir2.1 channel as well as a functional Kv channel can be tetrameric assemblies of these

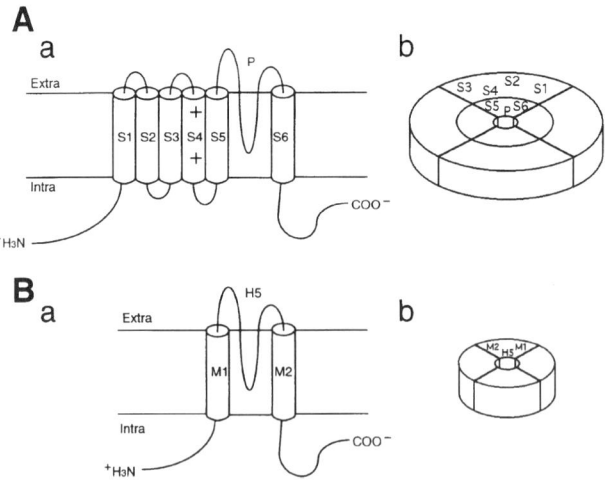

Figure 1. Hypothetical structures of a pore-forming subunit of an α-subunit of Kv channels (A) and Kir channels (B). Putative transmembrane segments are numbered. H5 and P are putative pore-forming regions. S4 may be the voltage-sensing segment.

subunits (Yang et al., 1995b; Jan and Jan, 1994; MacKinnon, 1991). This tetrameric structure also has been suggested for Kir3.0/GIRK channels (Inanobe et al., 1995; Corey et al., 1998). Thus, it seems likely that, as in the case of Kv channels, the Kir channel subunits are assembled to form homo- or heterotetrameric functional channels [Fig.1A(b) and 1B(b)]. In the Kv channels, the S4 region, which possesses the repeated positively charged amino acid residues, is presumed to be the voltage-sensor region of the channels. On the other hand, the Kir channel subunits do not have a segment corresponding to the S4 region in Kv channels. The blockade of outwardly flowing currents through Kir channels is caused by intracellular substances such as Mg^{2+} and polyamines, but not by intrinsic voltage-dependent gating of the channels (Yamada and Kurachi, 1995; Ficker et al., 1994; Fakler et al., 1994a; Lopatin et al., 1994). The cDNAs encoding one of the pore-forming subunits of K_{ATP} channels (Kir6.1/uK_{ATP}-1) and K_{ACh} channels (Kir3.1/GIRK1) also have been cloned (Inagaki et al., 1995b; Kubo et al., 1993b; Dascal et al., 1993). All of these channel subunits exhibit the same primary structure. So far, more than 10 cDNAs encoding Kir channel subunits in mammals have been isolated. These cloned Kir channel subunits are classified into four main groups: (1) the Kir2.0 subfamily, which represents classical K_{IR} channels (Takahashi et al., 1994; Morishige et al., 1994, 1993; Kubo et al., 1993a); (2) the Kir3.0 subfamily of G-protein-activated K^+ channels (Isomoto et al., 1996a; Lesage et al., 1994; Kubo et al., 1993b; Dascal et al., 1993); (3) the Kir1.0 and Kir4.0/K_{AB} subfamily of K^+-transport K^+ channels (Takumi et al., 1995; Bond et al., 1994; Ho et al., 1993); and (4) the Kir6.0 subfamily of ATP-sensitive K^+ channels (Inagaki et al., 1995a,b; Sakura et al., 1995). The molecular biology of the Kir channel subfamilies is reviewed in Chapter 5 of this book.

Recent progress in the molecular biology of Kir channels has enabled us to understand the structure–function relationships of channel biophysics, the physiological regulation, and the pharmacology of these channels at the molecular level. In this chapter, we will summarize the current understanding of the molecular properties and functional roles of Kir channels in the ventricular myocardium.

2. CLASSICAL INWARD RECTIFYING K^+ CHANNELS IN HEART: I_{K1}

2.1. Physiological Roles of I_{K1} Channels

The I_{K1} channel is a strong inward rectifying K^+ channel, which is constitutively active at all physiological potentials and is modulated by external K^+ and internal Mg^{2+}. The I_{K1} current is blocked by external Cs^+ and Ba^{2+}. The single-channel conductance is approximately proportional to the root value of $[K^+]_o$. This K^+ channel is distinct from other Kir channels such as K_{ACh}, K_{ATP}, and K_{Na} channels in its metabolic regulation and degree of rectification. The classical background Kir channel, I_{K1}, is not activated by transmitters or voltage but seems to be constitutively active, and its inward rectifying property is now believed to arise from a combination of blockade mediated by intracellular Mg^{2+} (Mg_i^{2+}) and blockade mediated by polyamines such as spermine and spermidine (see below), which contribute to the gating of strong inward rectification.

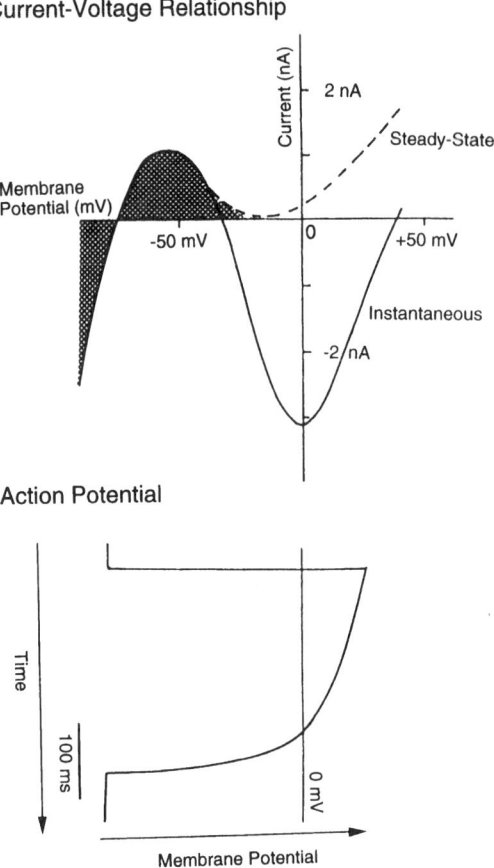

Figure 2. Current–voltage relationship and action potential in the ventricular cell. The current–voltage relationship in the voltage-clamp experiment for the instantaneous current is indicated by the solid curve, and the relationship for the steady-state current at 1 s by the dashed curve. The action potential is drawn sideways to correlate it to the above current–voltage relationship. Shaded area shows I_{K1} currents. (Reproduced, with permission from Isomoto *et al.*, 1997a.)

The Kir conductance in cardiac myocytes, including Purkinje fibers as well as ventricular and atrial tissues (Kurachi, 1985; Rougier *et al.*, 1968; Beeler and Reuter, 1970; McAllister and Noble, 1966), but not in nodal cells (Noma *et al.*, 1984), is the dominant component of the resting conductance of these tissues. It is defined as the time-independent background K$^+$ current, I_{K1}. The single-channel conductance and mean open time at ~ -80 mV of the I_{K1} channel in the ventricular myocytes are 40 pS and 100 ms, respectively (Kurachi, 1985). The contributions of I_{K1} to the action potential are illustrated in the current–voltage relationship in Fig. 2. The large inward conductance at membrane potentials below the K$^+$ equilibrium potential, E_K, and the relatively large outward conductance at those potentials just above E_K set the resting membrane potential close to E_K. The lack of outward conductance at plateau-level potentials caused by the strong inward rectification of I_{K1} prevents massive K$^+$ efflux during the action potential plateau, resulting in the maintenance of depolarization.

When repolarization is initiated by activation of the delayed rectifier Kv channels at the plateau potential, relatively large outward currents pass through the I_{K1} channel in the negative-slope region, resulting in the acceleration of repolarization. Thus, I_{K1} plays important roles in (1) setting the resting potential, (2) maintaining the plateau phase, and (3) causing rapid repolarization.

The rectification of I_{K1} channels results from block by intracellular Mg^{2+} and polyamines. It has been elucidated recently that the differences in inward rectification are attributable to the efficacy of these substances in blocking the outward-going K^+ currents at specific amino acid residues located in the M2 domain and carboxyl-terminal regions of cloned Kir channels (Yang et al., 1995a). For cloned strong inward rectifiers, which are believed to encode I_{K1} channels (Kir2.0 subfamily) and K_{ACh} channels (Kir3.0), one or two amino acid residues at these sites are negatively charged. For weak-inward rectifiers such as the K_{ATP} channel, the corresponding sites of the cloned channel subunits (K_{ATP} subfamily) contain only neutral residues. Indeed, substitution of a negatively charged amino acid by a neutral one in strong inward rectifiers reduces the degree of inward rectification by decreasing the apparent affinity for polyamines and Mg^{2+} ions (Stanfield et al., 1994; Wible et al., 1994).

2.2. Molecular Aspect of I_{K1} Channels

The features of the IRK subfamily (Kir2.1–2.3) closely resemble those of classical Kir channels. In our laboratory, three Kir2.0 subfamily members have been cloned from mouse brain cDNA libraries (Takahashi et al., 1994; Morishige et al., 1994, 1993). The amino acid sequence of mouse brain Kir2.1 shares 70% and 61% identity with Kir2.2 and Kir2.3, respectively. These three Kir2.0s encode proteins of 428, 427, and 445 amino acids, respectively. The amino acid sequence is well conserved at M1, M2, and H5. Especially, at H5 they differ from each other at only a single amino acid residue.

Xenopus oocytes injected with cRNAs derived from Kir2.0 subfamily members express strong inward rectifying K^+ currents. Hyperpolarizing voltage steps elicit rapidly activated large inward currents. These expressed currents are blocked by external Ba^{2+} and Cs^+ in a concentration- and voltage-dependent manner. The single-channel conductances of Kir2.1, Kir2.2, and Kir2.3 are ~ 22, ~ 34, and ~ 13 pS, respectively, with 150 mM $[K^+]_o$ (Fig. 3). The steady-state open-channel open-state probability (P_o) of the Kir2.2 channel decreases with hyperpolarization, whereas those of Kir2.1 and Kir2.3 remain constant. A detailed gating kinetic analysis strongly indicates that an increase of long closed gaps (> 200 ms) between clusters causes the prominent reduction of steady-state P_o at hyperpolarized potentials in the case of Kir2.2. The dominant cardiac I_{K1} is reported to have a single-channel conductance of 30–40 pS with 150 mM $[K^+]_o$ and to exhibit hyperpolarization-induced inactivation in the absence of blocking cations (Kurachi, 1985; McAllister and Noble, 1966). These properties are quite similar to those of Kir2.2 (Takahashi et al., 1994). Reverse-transcription polymerase chain reaction (RT-PCR) analyses of mRNAs obtained from isolated single ventricular and atrial myocytes show that these myocytes express the mRNA of only Kir2.2, but not those of Kir2.1 and Kir2.3 (Matsumoto et al., 2000). Because Kir2.0 subunits seem to assemble as homotetramers (Yang et al., 1995b), the I_{K1} channel protein is thought, at present, to be composed of homotetrameric Kir2.2 subunits.

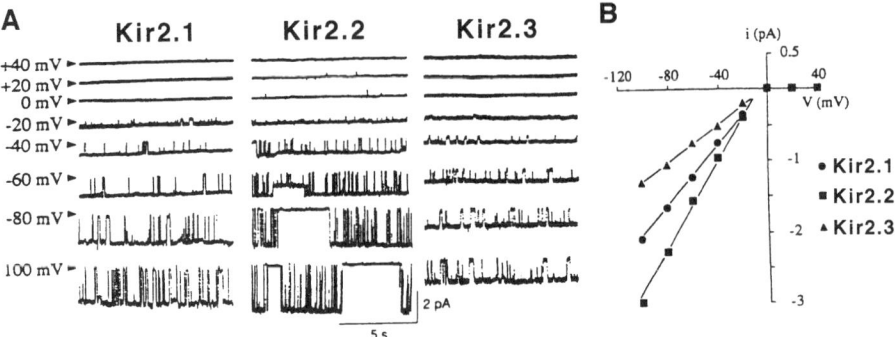

Figure 3. Single-channel recordings from cell-attached membrane patches of *Xenopus* oocytes expressing Kir2.0 channels. (A) Membrane current traces were recorded at the membrane potential values indicated to the left of each trace. Arrows to the right of certain traces indicate the patch current level recorded when all channels were closed. Each of these patches appeared to contain one channel. (B) Current–voltage relationships of the channel records shown in (A). (Reproduced, with permission from Isomoto *et al.*, 1997a).

2.3. Modulation of Kir2.0 Channels

Collins *et al.* (1996) reported that Kir2.3 is inhibited by ATP at physiological concentrations (K_D of 1.47 mM). This effect is antagonized by ADP, again in the physiological range, which implies that this channel is sensitive to the intracellular [ATP]/[ADP] ratio. Kir2.3 is, however, not a K_{ATP} channel, because K_{ATP} channels have weaker rectification and much higher sensitivity to ATP than Kir2.3. The inhibition of Kir2.3 currents by ATP does not require Mg^{2+} and is mimicked by nonhydrolyzable ATP analogs, indicating that hydrolysis of ATP is not a requisite step. These effects, not observed in Kir2.1, may be specific to Kir2.3. On the other hand, patch-damp studies by Fakler *et al.* (1994b) showed that Kir2.1-mediated currents, after rundown upon excision of the patch, can be partly restored by application of 1 mM ATP with 1 mM free Mg^{2+} but not by the nonhydrolyzable ATP analogs, suggesting that Kir2.1 is regulated by ATP hydrolysis, possibly by phosphatidylinositol bisphosphate (PIP_2).

3. ATP-SENSITIVE K$^+$ CHANNELS

3.1. Physiological Roles of K_{ATP} Channels

The ATP-sensitive K$^+$ (K_{ATP}) channel is a weakly inward rectifying K$^+$ channel that is inhibited by intracellular ATP (ATP_i) and activated by intracellular nucleoside diphosphates (NDP_i). Thus, it provides a link between cellular metabolism and excitability (Terzic *et al.*, 1995, 1994b). The cardiac K_{ATP} channel may be involved in the increase of K$^+$ efflux and shortening of the action potential duration in the ischemic heart (Nichols and Lederer, 1991; Faivre and Findlay, 1990; Findlay *et al.*, 1989). Both are major factors contributing to the electrophysiological abnormalities that predispose the heart to the development of reentrant arrhythmias (Wilde and Janse, 1994; Wilde, 1993; Gasser and Vaughan-Jones, 1992). On the other hand, opening of the cardiac K_{ATP} channel also has been implicated as a cardioprotective mechanism underlying ischemia-related preconditioning (Parratt and Kane, 1994; Yao *et al.*, 1993; Cole, 1993; Downey, 1992; Gross and Auchampach, 1992; Grover *et al.*, 1992).

3.2. Electrophysiological Properties of K_{ATP} Channels

In cell-attached and excised-patch recordings from human and guinea pig atrial myocytes, K_{ATP} channels are readily distinguished from I_{K1}, K_{Ach}, and K_{Na} channels based on their biophysical properties. The degree of inward rectification of K_{ATP} channels is much weaker as compared to that of I_{K1} or K_{ACh} channels. The weak inward rectification of K_{ATP} channels is also mediated by Mg^{2+} (Terzic *et al.*, 1995; Findlay, 1987a,b). With symmetrical ~ 150 mM K^+, the conductance of the unitary inward current through the K_{ATP} channels in cardiac myocytes is ~ 70–90 pS (Ashcroft and Ashcroft, 1990; Heidbuchel *et al.*, 1990; Horie *et al.*, 1987; Kakei and Noma, 1984; Trube and Heschler, 1984; Noma, 1983), which is larger than that for I_{K1} and K_{Ach} channels (~ 30–45 pS) (Kurachi *et al.*, 1986; Kurachi, 1985; Sakmann and Trube, 1984; Sakmann *et al.*, 1983) but smaller than that for the K_{Na} channel (~ 180–230 pS) (Wang *et al.*, 1991; Kameyama *et al.*, 1984). The K_{ATP} channels are active at every potential and essentially time- and voltage-independent. The mean open time of the K_{ATP} channel is ~ 1 ms, and longer open times may represent bursts of activity (Terzic *et al.*, 1995, 1994b; Takano and Noma, 1993; Nichols and Lederer, 1991; Heidbuchel *et al.*, 1990; Horie *et al.*, 1987; Kakei and Noma, 1984; Trube and Heschler, 1984).

3.3. Regulation of K_{ATP} Channels by Intracellular Nucleotides

The K_{ATP} channels are known to be regulated by various intracellular factors such as ATP_i (Fig. 4) and nucleotide diphosphates (NDPs). ATP_i is the main regulator of classical K_{ATP} channels and has two functions: closing the channels and maintaining channel activity in the presence of Mg^{2+} (Takano *et al.*, 1990; Ohno-Shosaku *et al.*, 1987; Findlay and Dunne, 1986; Trube and Heschler, 1984). The first action of ATP_i is referred to as the "ligand action," because it is assumed to require the binding of ATP_i to the K_{ATP} channel and persists as long as ATP_i is bound to the channel. Typically, K_{ATP} channels have a very low probability of being open at physiological concentrations of ATP. Half-maximum inhibition of the K_{ATP} channel in cardiac muscle cells is achieved by 20–30 μM ATP, and this value is independent of whether ATP is in the free acid form (ATP^{4-}) or is bound to Mg^{2+} (MgATP) (Terzic *et al.*, 1995; Ashcroft and Ashcroft, 1990; Findlay, 1988). In pancreatic β-cells, on the other hand, it has been shown that the ATP concentration at which half-maximum inhibition of the channel occurs is increased by Mg^{2+} from 4 to 26 μM ATP, suggesting that ATP^{4-} is a more potent inhibitor of the pancreatic K_{ATP} channel than MgATP (Ashcroft and Kakei,

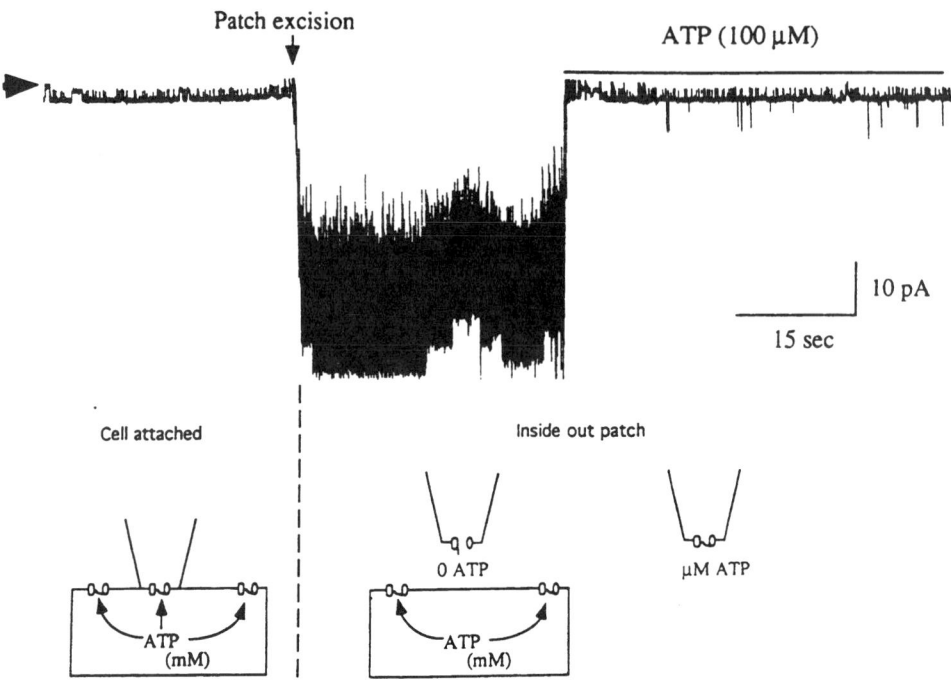

Figure 4. Inhibition of K_{ATP} channels by ATP_i. In the cell-attached configuration of the patch-clamp technique (holding potential -75 mV), no activity of the K_{ATP} channel is recorded with millimolar levels of ATP inside the cardiac cell. After patch excision in an ATP-free solution, several K_{ATP} channels immediately open. Application of micromolar concentrations of ATP to the cytosolic side of the inside-out patch inhibits K_{ATP} channel openings. Note that Kir current is recorded even in the presence of ATP. (Reproduced, with permission from Terzic *et al.*, 1995).

1989). Although information about the inhibition of K_{ATP} channels in smooth muscle by ATP_i is limited, it has been established that MgATP is less effective than ATP^{4-} (Nelson and Quayle, 1995; Kajioka *et al.*, 1991). Half-maximum inhibition of the K_{ATP} channel in the rat portal vein is achieved by 29 μM ATP^{4-}, whereas MgATP is ineffective.

The second action of ATP_i is referred to as "hydrolysis-dependent," because it apparently requires the hydrolysis of ATP in the presence of Mg^{2+} and can last for several tens of minutes after the removal of ATP_i. The effect of ATP_i on K_{ATP} channels depends on the state of the channel protein. When the channels are operative, ATP_i inhibits channel opening. When the channels are not operative, treatment with MgATP restores channel opening. Recently, it has been shown that PIP_2 added to the intracellular side of the membrane restores K_{ATP} channel activity after rundown, as described later.

NDPs are also major regulators of K_{ATP} channel activity. NDPs have two distinct actions: (1) attenuating ATP-induced channel inhibition by competing with the binding of ATP_i to the K_{ATP} channels and (2) permitting K_{ATP} channel opening even after rundown (Terzic *et al.*, 1994a; Beech *et al.*, 1993a; Tung and Kurachi, 1991; Dunne and

Petersen, 1991; Faivre and Findlay, 1989). The regulation of channel activity by nucleotides is modulated by several factors exogenous to the channel protein such as hormones, including galanine and somatostatin, which are known to inhibit insulin secretion by activation of G-proteins (de Weille *et al.*, 1989, 1988). Whereas the K_{ACh} channels are activated by $G_{\beta\gamma}$, the K_{ATP} channels are activated by a GTP-bound form of $G_{i\alpha}$, at least in the heart (Terzic *et al.*, 1994c; Ito *et al.*, 1992).

3.4. Pharmacological Regulation of K_{ATP} Channels

The K_{ATP} channels exhibit characteristic pharmacological properties: they are selectively inhibited by antidiabetic sulfonylurea derivatives (SUs), such as glibenclamide and tolbutamide, and activated by a certain class of vasorelaxant agents, such as pinacidil, levcromakalim, and nicorandil, which are collectively termed K^+ channel openers (KCOs). Details regarding the pharmacological profile of K_{ATP} channels in the heart are available elsewhere in this book. One interesting aspect is the tissue specificity of KCO action. For example, the K_{ATP} channels in cardiac myocytes are activated by pinacidil but not by diazoxide, whereas K_{ATP} channels in pancreatic β-cells are activated by diazoxide but only weakly by pinacidil. The K_{ATP} channels in smooth muscle cells are activated effectively by both of these compounds (Terzic *et al.*, 1995; Nelson and Quayle, 1995; Ashcroft and Ashcroft, 1990). Thus, properties of K_{ATP} channels vary among tissues, and this has led to the premise that this K^+ channel family may be composed of heterogeneous K^+ channel proteins (see below).

3.5. Molecular Aspect of K_{ATP} Channels

The molecular biology of K_{ATP} channels is reviewed in the introductory chapters of this book and so is only briefly presented here. So far, two Kir subunits belonging to the K_{ATP} subfamily, Kir6.1 and Kir6.2, have been isolated (Inagaki *et al.*, 1995a,b; Sakura *et al.*, 1995). The predicted amino acid sequences show that Kir6.1 and Kir6.2 have roughly 70% identity with each other and 40–50% identity with other members of the Kir channel family. The highly conserved motif in the H5 region, Gly-Tyr-Gly, found in all other members of the Kir channel family is Gly-Phe-Gly in both Kir6.1 and Kir6.2. Inagaki *et al.* (1995b) examined the distribution of their mRNAs by Northern blot analysis and showed that Kir6.1 is ubiquitously expressed in various tissues, whereas Kir6.2 is expressed in pancreas, heart, skeletal muscle, and brain. Whereas members of the Kir2.0, Kir3.0 or Kir1.1, Kir1.2/Kir4.1/K_{AB} subfamily can themselves function as Kir channels (Krapivinsky *et al.*, 1995; Jan and Jan, 1994; Ho *et al.*, 1993; Kubo *et al.*, 1993b; Jan and Jan, 1992; Pongs, 1992), both Kir6.1 and Kir6.2 appear to require coupling with sulfonylurea receptors (SURs) to form a functional K_{ATP} channel (Yamada *et al.*, 1997; Inagaki *et al.*, 1996, 1995a; Isomoto *et al.*, 1996b; Sakura *et al.*, 1995). Inagaki *et al.* (1996, 1995a) demonstrated that the pancreatic β-cell and cardiac K_{ATP} channels are complexes composed of Kir6.2 and SUR1 and SUR2A, respectively (Fig. 5). The stoichiometry of SURs and Kir subunits in the K_{ATP} channels was investigated by studying SUR1/Kir6.2 fusion constructs (Clement *et al.*, 1997;

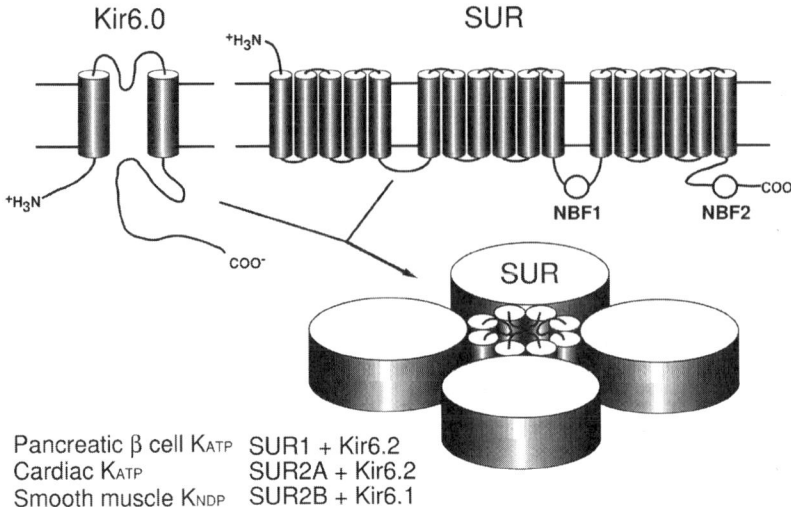

Pancreatic β cell K$_{ATP}$ SUR1 + Kir6.2
Cardiac K$_{ATP}$ SUR2A + Kir6.2
Smooth muscle K$_{NDP}$ SUR2B + Kir6.1

Figure 5. Molecular structure of ATP-sensitive K$^+$ channels. Pancreatic, cardiac, and skeletal muscle ATP-sensitive K$^+$ (K$_{ATP}$) channels are composed of two distinct subunits: a sulfonylurea receptor (SUR) and a K$^+$ (Kir) channel subunit, Kir6.2 (Inagaki *et al.*, 1995a, 1996; Sakura *et al.*, 1995; Isomoto, 1996; Aguilar-Bryan *et al.*, 1997). On the other hand, the complex of SUR2B and Kir6.1 corresponds to the "small-conductance K$_{ATP}$" or "nucleoside-diphosphate-dependent K$^+$ (K$_{NDP}$)" channel in vascular smooth muscle (Yamada *et al.*, 1997).

Inagaki *et al.*, 1997; Shyng and Nichols, 1997). These studies indicated that the K$_{ATP}$ channels are heterooctamers, consisting of four SURs interacting with four Kir subunits. Although Ammälä *et al.* (1996) demonstrated that Kir6.1 also couples to SUR1 and acquires sulfonylurea sensitivity, other features of K$_{ATP}$ channels, such as regulation by ATP$_i$ and NDPs and activation by K$^+$ channel openers, have been identified on the channel reconstituted from Kir6.1 and SUR1. Recently, we found that Kir6.1 could form a smooth muscle K$_{NDP}$ channel with another sulfonylurea receptor, SUR2B (see below) (Satoh *et al.*, 1998; Yamada *et al.*, 1997).

3.6. Molecular Heterogeneity of Sulfonylurea Receptors

The first sulfonylurea receptor, SUR1, was cloned from insulinoma cells by Aguilar-Bryan *et al.* (1995). Its protein is assumed to possess 17 potential transmembrane regions (Fig. 5) (Tusnady *et al.*, 1997), two nucleotide binding folds (NBFs) with Walker A and B consensus motifs, two N-linked glycosylation sites, and several protein kinase A- and C-dependent phosphorylation sites. Coexpression of hamster (ha)-SUR1 and mouse (m)-Kir6.2 elicits a K$_{ATP}$ conductance (I$_{KATP}$), which is inhibited by glibenclamide (half-maximal inhibition at 32 μM) and activated by diazoxide (half-maximal activation at 60 μM) (Inagaki *et al.*, 1995a). The single-channel conductance is 76 pS in symmetric 140 mM K$^+$ solution. ATP$_i$ inhibits the SUR1/Kir6.2 channel activity, with half-maximal inhibition at 10 μM. These properties are the same as those of the pancreatic β-cell K$_{ATP}$ channel (Dunne and Petersen, 1991; Garrino et al., 1989; Findlay et al., 1985; Cook and Hales, 1984). Northern blot analysis has revealed that SUR1 mRNA is expressed at a high level in pancreatic islets and at a low level in heart and skeletal muscle (Inagaki et al., 1995a; Aguilar-Bryan et al., 1995). Gene mapping data show that both Kir6.2 and SUR1 genes are clustered on human chromosome 11 at 11p15.1 (Inagaki et al., 1995a). Thus, SUR1 seems to form the pancreatic β-cell K$_{ATP}$ channel with Kir6.2. Mutations in the SUR1 protein have been shown to lead to a nonfunctional K$_{ATP}$ channel, resulting in persistent hyperinsulinemic hypoglycemia of infancy, a disease associated with unregulated insulin secretion (Kane et al., 1996; Thomas et al., 1995).

Two additional homologs of SUR1 have been isolated subsequently. SUR2 was first cloned from rat brain by Inagaki et al. (1996). We also have cloned a mouse homolog of SUR2 and a novel isoform of the sulfonylurea receptor from a mouse heart cDNA library (Isomoto et al., 1996b). Sequence analysis indicates that this novel member of the sulfonylurea receptors is essentially the same as mouse SUR2 except for 42 amino acid residues in the carboxyl-terminal end, suggesting that these two sulfonylurea receptors are formed by alternative splicing of a single gene. Based on these results, we proposed that the original SUR2 be renamed SUR2A and designated the third member SUR2B (Isomoto et al., 1996b). The amino acid sequence of rat (r)-SUR1 has 66% and 67% identity with those of m-SUR2A and m-SUR2B, respectively. A further splice variant of m-SUR2A, which has a deleted predicted 35 amino acid residues in the intracellularloop between the 11th and 12th membrane-spanning domains, also has been cloned, but its function is unclear (Chutkow et al., 1996). It is now designated SUR2C (Ashcroft and Gribble, 1998).

Coexpression of Kir6.2 and either SUR2A or SUR2B also elicits I_{KATP} with a single-channel conductance of $\sim$80 pS in symmetrical $\sim$145 mM K$^+$ solution. The r-SUR2A/m-Kir6.2 channel activity is inhibited by ATP$_i$ in a concentration-dependent manner with a half-maximal value of 100 μM and is only partially inhibited by 1 μM glibenclamide (Inagaki et al., 1996). In contrast to the ha-SUR1/m-Kir6.2 channel, the m-SUR2A/r-Kir6.2 channel is activated by pinacidil but not by diazoxide. These are the features of cardiac and skeletal muscle K$_{ATP}$ channels (Terzic et al., 1995, 1994b; Findlay, 1992; Nichols et al., 1991; Nichols and Lederer, 1991; Ashcroft and Ashcroft, 1990; Faivre and Findlay, 1990; Fosset et al., 1988). The m-SUR2B/r-Kir6.2 channel is activated by both pinacidil and diazoxide, suggesting that the features of this channel closely resemble the responses of smooth muscle to KCOs (Nelson and Quayle, 1995; Beech et al., 1993b; Lorenz et al., 1992; Kajioka et al., 1991; Standen et al., 1989). Thus, pharmacological properties of K$_{ATP}$ channels may be determined by sulfonylurea receptors. Because diazoxide activates K$_{ATP}$ channels containing SUR1 or SUR2B, but not SUR2A, the alternative splicing region between SUR2A and SUR2B may be a functional domain important for diazoxide activation of K$_{ATP}$ channels. Interestingly, the sequence of the last 42 amino acids of SUR2B exhibits 74% and 33% identity with those of the corresponding region of r-SUR1 and m-SUR2A, respectively. On the other hand, the functional domain for pinacidil may be in regions other than the carboxyl-terminal end, because pinacidil activates K$_{ATP}$ channels reconstituted from SUR2A or SUR2B but not from SUR1 (Isomoto et al., 1996b).

Northern blot analysis of rat tissues using the probe that may include a common nucleotide sequence for both isoforms of SUR2 reveals that mRNA for either SUR2A or SUR2B in rat is expressed at high levels in heart, skeletal muscle, and ovary, at moderate levels in brain, tongue, and pancreatic islets, at low levels in lung, testis, and adrenal gland, and at very low levels in stomach, colon, thyroid, and pituitary (Inagaki *et al.*, 1996). On the other hand, RT-PCR analysis shows that m-SUR2A mRNA is expressed in heart, skeletal muscle, cerebellum, eye, and urinary bladder, whereas m-SUR2B mRNA is ubiquitously distributed not only in these tissues but also in forebrain, lung, liver, pancreas, kidney, spleen, stomach, small intestine, colon, uterus, ovary, and fat tissue (Isomoto *et al.*, 1996b). *In situ* hybridization using a probe that is common to SUR2 isoforms shows that SUR2 isoforms are expressed in the parenchyma of the heart and skeletal muscle and in the vascular structures of various tissues (Chutkow *et al.*, 1996). Because SUR2A is believed to be the cardiac and skeletal muscle type sulfonylurea receptor, other isoforms of SUR2 may be expressed in the vascular structure. This supports the idea that SUR2B represents a component of the vascular smooth muscle type K_{ATP} channels. However, to clarify the distribution of each isoform of sulfonylurea receptors, further studies using probes specific for each isoform are needed.

SUR2B is thought to be one of the subunits reconstituting smooth muscle type "K_{ATP}" channels because of its pharmacological properties and tissue distribution. However, the SUR2B/Kir6.2 channel expresses I_{KATP} with a single-channel conductance of ~ 80 pS, which is distinct from the single-channel conductances of physiological smooth muscle "K_{ATP}" channels. In addition, mRNA for Kir6.2 is expressed in restricted tissues, whereas that of SUR2B is expressed in a variety of tissues. These findings suggest that SUR2B may reconstitute some types of "K_{ATP}" channels in smooth muscle cells by coupling with members of the Kir channel family other than Kir6.2. Recently, we demonstrated that coexpression of SUR2B and Kir6.1 forms a functional K^+ channel with the features of a smooth muscle K^+ channel, which is described as the K_{NDP} channel observed in rat portal vein (Satoh *et al.*, 1998; Yamada *et al.*, 1997). The m-SUR2B/m-Kir6.1 channel exhibits a single-channel conductance of ~ 33 pS in symmetrical ~ 145 mM K^+ solution and is activated by diazoxide, pinacidil, and nicorandil. Surprisingly, the channel does not spontaneously open on patch excision even in the absence of ATP_i. In excised patches, the channel activity is activated by NDPs and inhibited by sulfonylureas but not by ATP_i. ATP_i on its own at concentrations greater than 100 μM activates the channel. Thus, the SUR2B/Kir6.1 channel closely resembles a K_{NDP} channel in vascular smooth muscle in the following respects: (1) activation by internal nucleoside di- and triphosphates, (2) activation by K^+ channel openers and inhibition by sulfonylureas, (3) little inhibition by ATP_I, and (4) a single-channel conductance of 22 pS with 60 mM $[K^+]_o$ (Zhang and Bolton, 1996). It also is noteworthy that SUR2 and Kir6.1 genes are clustered in the distal region of mouse chromosome 6 (Isomoto *et al.*, 1997b).

3.7. Molecular Mechanism of K_{ATP} Channel Inhibition by Intracellular ATP

One of the hallmarks of the classical K_{ATP} channels is the inhibition of channel activity by micromolar concentrations of ATP_i (Terzic *et al.*, 1995; Ashcroft, 1988; Noma, 1983). Tucker *et al.* (1997) recently found that a Kir6.2 channel whose last 26 amino acids at the carboxy terminus are deleted (Kir6.2ΔC26) is functionally expressed in the absence of SUR. A charge-neutralization mutation on Lys185 of Kir6.2ΔC26 reduces the ATP_i sensitivity of the Kir6.2ΔC26 channel by a factor of ~ 40. The ATP_i sensitivity of the Kir6.2ΔC26 channel was increased by a factor of 5–8 by coexpression of SUR1. Thus, it is likely that the primary inhibitory ATP_i-binding site resides in Kir6.2, whereas SUR1 increases the ATP_i sensitivity of Kir6.2. Koster et al. (1998) showed that the complex of SUR1 and the Kir6.2 whose N-terminal 30 amino acids were deleted (Kir6.2ΔN30) exhibited ~ 10 times lower ATP_i sensitivity than the SUR1/Kir6.2 channel. Interestingly, the SUR1/Kir6.2ΔN30 channel also is less sensitive to intracellular ADP (ADPi) and tolbutamide than the SUR1/Kir6.2 channel. Therefore, Kir6.2 might interact with SUR1 through its N-terminus, and the low ATP_i sensitivity of the SUR1/Kir6.2ΔN30 channel might be due to impaired coupling between SUR1 and Kir6.2.

It is unknown how SUR1 increases the ATP_i sensitivity of Kir6.2. Gribble et al. (1997b) found that the ATP_i sensitivity of the SUR1/Kir6.2 channel was not modified by mutations on either or both of the two conserved lysine residues in the Walker A motifs in the first or the second NBF of SUR1 (K719A and K1384M, respectively). On the other hand, Ueda et al. (1997) demonstrated that SUR1 possesses two distinct ATP_i-binding sites with high and low affinities. The high-affinity binding site was saturated with 10 μM ATP_i in the absence of Mg_i^{2+}. Substitution of the conserved lysine residue in the Walker A motif (K719R and K719M) or the aspartate residue in the Walker B motif (D854N) in the first NBF abolished the high-affinity ATP_i-binding, whereas the corresponding mutations in the second NBF did not have any significant effect. Because Ueda et al. (1997) and Gribble et al. (1997b) used different mutations (K719R, K719M, or D854N vs. K719A), it is not clear whether the ATP_i binding found by Ueda et al. (1997) underlies the sensitization of Kir6.2 to ATP_i by SUR1.

No corresponding studies have been done with SUR2s. However, ATP_i inhibits the SUR2A/Kir6.2 and the Kir6.2ΔC26 channels with similar potencies (~ 100 μM) (Okuyama, et al., 1998; Tucker et al., 1997; Inagaki et al., 1996). Thus, SUR2A may not substantially enhance the ATP_i sensitivity of Kir6.2. However, the native cardiac K_{ATP} channel has been reported to be ~ 3–10 times more sensitive to ATP_i than the SUR2A/Kir6.2 channel. Thus, some unidentified factors in cardiac myocytes might sensitize the SUR2A/Kir6.2 channel to ATP_i in vivo. Indeed, various factors including intracellular polyvalent cations and actin polymerization have been suggested to affect the ATP_i sensitivity of the cardiac K_{ATP} channel (Hiraoka et al., 1996; Terzic and Kurachi, 1996; Deutsch et al., 1994).

The SUR2A/Kir6.2 channel is equally sensitive to Mg^{2+}-free and Mg^{2+}-bound ATP_i (Fig. 6A), whereas the SUR2B/Kir6.2 channel is more sensitive to Mg^{2+}-free than Mg^{2+}-bound ATP_i (Fig. 6B). This difference may be ascribed to the difference in the amino acid sequence of the C-terminus between SUR2A and SUR2B (Isomoto et al., 1996b). As stated previously, the sequence of the last 42 amino acids in the C-terminus of SUR2B is more similar to that of the corresponding part of SUR1 than of SUR2A, and the SUR1/Kir6.2 channel is more sensitive to Mg^{2+}-free than Mg^{2+}-bound ATP_i (Nichols et al., 1996). Therefore, the last 42 amino acids in the C-terminus of SURs may be involved in discrimination between Mg^{2+}-bound and Mg^{2+}-free ATP_i by K_{ATP} channels. This part is very close to the second NBF in the primary structure, and the second NBF is known to play a crucial role in NDP_i-induced activation of K_{ATP}

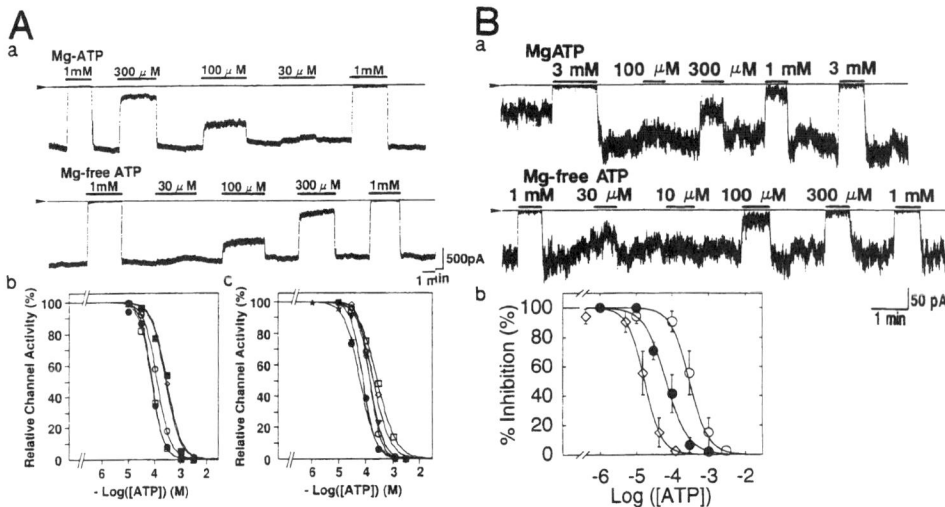

Figure 6. Inhibition of the SUR2A/Kir6.2 and SUR2B/Kir6.2 channels by intracellular ATP. (A) The SUR2A/Kir6.2 channel. (a) Effect of intracellular ATP in inside-out patch membranes in the presence (upper trace) or the absence (lower trace) of ~ 1.4 mM intracellular free Mg^{2+}. (b) and (c). Relationship between total ATP concentrations and channel activity in the presence (b) and the absence (c) of Mg^{2+}. The data are expressed as a percentage of the value obtained in the absence of ATP. Different symbols represent the data obtained from different patches. Curves represent the fit of each set of data to the Hill equation. (Adapted from Okuyama *et al.*, 1998.) (B) The SUR2B/Kir6.2 channel. (a) Effect of ATP in inside-out patch membranes in the presence (upper trace) or the absence (lower trace) of ~ 1.4 mM free Mg^{2+}. (b) Relationship between channel activity in the presence (open symbols) and the absence (filled symbols) of Mg^{2+} and total ATP concentration ($\bigcirc$, $\bullet$) or calculated free ATP concentration ($\diamondsuit$). Symbols and bars indicate the mean $\pm$ SEM ($n = 3$ for each point). Curves represent the best fit of each set of data to the Hill equation. (Reproduced from Okuyama *et al.*, 1998, copyright Springer-Verlag, and Isomoto *et al.*, 1996b, with permission).

channels (Gribble *et al.*, 1997b; Nichols *et al.*, 1996). Ueda *et al.* (1997) found in SUR1 that the binding of ADP_i to the second NBF potently antagonized the ATP_i binding to the first NBF of the same protein in the presence of Mg_i^{2+}. Therefore, the C-terminal tail of SURs might serve to regulate either ATP_i hydrolysis on or ADP_i binding to the second NBF in the presence of Mg_i^{2+}.

ATP_i inhibits both SUR2A/Kir6.2 and SUR2B/Kir6.2 channels with a Hill coefficient significantly larger than unity (~ 1.8). No cooperativity is detected, however, with the SUR1/Kir6.2 channel nor the Kir6.2ΔC26 channel (Gribble *et al.*, 1997a,b; Tucker *et al.*, 1997). Therefore, SUR2s, but not SUR1, may function to create the cooperative interaction between ATP_i and K_{ATP} channels through an unknown molecular mechanism.

3.8. Molecular Mechanism of Response to Intracellular Nucleoside Diphosphates

NDP_is such as UDP exhibit distinct effects on the cardiac K_{ATP} channel before and after rundown (Terzic *et al.*, 1995, 1994a). UDP antagonizes the inhibitory effect of ATP_i before rundown. After rundown, UDP restores the channel activity without

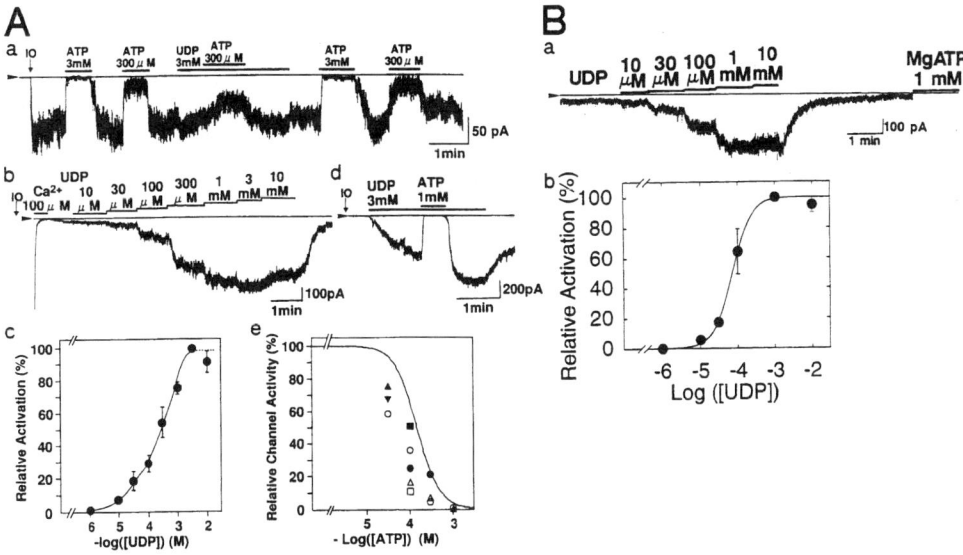

Figure 7. Effect of nucleoside diphosphates on the SUR2A/Kir6.2 and SUR2B/Kir6.2 channels. (A) The SUR2A/Kir6.2 channel. (a), (b), and (d). Inside-out patch recordings. (c) Relationship between UDP concentrations and channel activity after rundown. Channel activity is expressed as a percentage of the maximum activity induced by 3 mM UDP. Symbols and bars represent the mean $\pm$ SEM ($n = 3$–12 for each point). The curve is the best fit of the data to the Hill equation. (e) Concentration-dependent inhibitory effect of ATP on channel activity induced by 3 mM UDP after rundown in the presence of Mg^{2+}. Different symbols represent data from different patches. The curve represents the averaged inhibitory effect of ATP on the spontaneous channel activity in the absence of UDP. (Adapted from Okuyama *et al.*, 1998.) (B) The SUR2B/Kir6.2 channel. (a) An inside-out patch recording. (b) Relationship between UDP concentration and channel activity after rundown. Channel activity is expressed as a percentage of the maximum activity induced by 1 mM UDP. Symbols and bars represent the mean $\pm$ SD ($n = 3$ for each point). The curve is the best fit of the data to the Hill equation. (Reproduced, with permission, from Okuyama *et al.*, 1998. Copyright 1998 Springer-Verlag.)

attenuating the ATP_i sensitivity of the channels (Tung and Kurachi, 1991). The SUR2A/Kir6.2 channel mimics this dualistic response of the cardiac K_{ATP} channel to NDP_is (Fig. 7) (Okuyama *et al.*, 1998).

As shown in Fig. 7A(a), ATP_i exhibits a weaker inhibitory effect on spontaneous channel activity of the SUR2A/Kir6.2 channel in the presence than in the absence of UDP. Removal of UDP almost completely restores the channel's ATP_i sensitivity. Thus, UDP antagonizes the ATP_i-mediated inhibition of channel activity before rundown.

After rundown, UDP activates the channel in a concentration-dependent manner [Fig. 7A(b)], with a half-maximum effective concentration (EC_{50}) of 240 μM [Fig.7A)c)]. This effect is completely dependent on Mg_i^{2+} (not shown). ATP_i inhibits the channel activity induced by UDP [Fig. 7A(d)] in a concentration-dependent manner [as shown by the data points in Fig. 7(Ae)] as potently as it inhibits spontaneous activity in the absence of UDP [represented by the curve Fig. 7A(e), which is the average of the data shown in Fig. 6A(b)]. Thus, the SUR2A/Kir6.2 and native cardiac K_{ATP} channels respond to NDP_i in a very similar way.

UDP activates the post-rundown SUR2B/Kir6.2 channel in a concentration-dependent manner with an EC_{50} of 71.7 μM and a Hill coefficient of 1.74 (Fig. 7B). This response also is dependent on Mg_i^{2+} (not shown). UDP also antagonizes the inhibitory effect of ATP_i on the SUR2B/Kir6.2 channel before but not after rundown, as is the case for the SUR2A/Kir6.2 channel (unpublished observation). Overall, SUR2A/Kir6.2 and SUR2B/Kir6.2 channels have very similar responses to NDP_i.

Nichols *et al.* (1996) found that a human mutation (G1479R) in the second NBF of SUR1 that characterizes one form of persistent *hyper*insulinemic *hypo*glycemia of infancy (PHHI) abolishes the antagonizing effect of ADP_i on the ATP_i-induced inhibition of the SUR1/Kir6.2 channel. Tucker *et al.* (1997) demonstrated that $MgADP_i$ inhibits the Kir6.2ΔC26 channel but activates the SUR1/Kir6.2ΔC26 channel. Gribble *et al.* (1997b) showed that either the K719A or the K1384M mutation of rat SUR1 abolishes the stimulatory effects of ADP_i on the partial rundown of the SUR1/Kir6.2 channel both in the presence and in the absence of ATP_i. Furthermore, hydrolysis-resistant β-methylene-ADP fails to activate the SUR1/Kir6.2 channel. Thus, it is likely that hydrolysis of $MgADP_i$ (and probably also the other NDP_is) at the NBFs of SUR1 may be critically involved in the activating effect of the nucleotides.

Although no corresponding studies have been done in SUR2s, a similar mechanism may underlie the activating effect of NDP_i on the SUR2/Kir6.2 channel. The molecular mechanism responsible for the dualistic responses of the SUR2/Kir6.2 channels to NDP_i is, however, unknown.

3.9. Molecular Mechanism of Rundown

Rundown of channel activity in inside-out membrane patches is not necessarily a phenomenon specifically associated with K_{ATP} channels, as it can be seen in other Kir channels as well. Nevertheless, many investigators have been interested in the mechanism underlying the rundown of K_{ATP} channels, because K_{ATP} channels run down more prominently than other Kir channels (Terzic *et al.*, 1995; Ashcroft, 1988). The K_{ATP} channels can be reactivated after rundown with ATP_i in the presence, but not in the absence, of Mg_i^{2+} (Takano *et al.*, 1990; Ohno-Shosaku *et al.*, 1987; Findlay and Dunne, 1986). Nonhydrolyzable ATP analogs cannot mimic this effect of ATP_i even in the presence of Mg_i^{2+} (Takano *et al.*, 1990; Ohno-Shosaku *et al.*, 1987). Therefore, rundown/refreshment of K_{ATP} channel activity might be crucially related to hydrolysis of ATP_i or phosphorylation/dephosphorylation of K_{ATP} channels (Terzic *et al.*, 1995; Ashcroft, 1988).

Kir6.2ΔC26 channels also exhibit rundown and are reactivated with $MgATP_i$ (Tucker *et al.*, 1997), indicating that the rundown/refreshment is primarily associated with a certain functional alteration of Kir6.2. Hilgeman and Ball (1996) showed that PIP_2 added to the intracellular side of the membrane restores K_{ATP} channel activity after rundown. Recently, Huang *et al.* (1998) showed that various Kir channels, including homomeric Kir1.1, Kir2.1, and Kir3.2 and heteromeric Kir3.1/Kir3.4 channels, are reactivated by PIP_2 after rundown. Furthermore, the ATP_i-mediated restoration of activity was inhibited by antibodies against PIP_2. Thus, PIP_2 and its generation by ATP-dependent lipid kinases appear to be critically involved in spontaneous Kir channel activity. The dualistic behavior of K_{ATP} channels in response to NDP_is (Terzic et al., 1995, 1994a) may have to be reexamined from this respect.

More recently, it has been shown that PIP_2 shifts the dose–response curves for ATP inhibition of K_{ATP} channels toward higher concentrations. This result suggests that PIP_2 metabolism may regulate the ATP sensitivity of K_{ATP} channels (Shyng and Nichols, 1998; Baukrowitz et al., 1998).

4. Na^+-ACTIVATED K^+ CHANNELS

4.1. Physiological Roles and Electrophysiological Properties of K_{Na} Channels

The Na^+-activated K^+ (K_{Na}) channel was first described by Kameyama et al. (1984) in cardiac myocytes. A cDNA clone for this channel has not been isolated yet. This channel is activated by Na_i^+ at concentrations of >20 mM, shows an inward rectification property and has the highest conductance (>200 pS) among ion channels in the heart. In inside-out patches, relatively high concentrations of Na^+ ions are required for activation. From the concentration–open-state probability relationship, the Na^+ required for half-activation (K_D) of the K_{Na} channel was 66 mM with an estimated Hill coefficient of 2.8 (Kameyama et al., 1984). The threshold for activation of the channel is around 30 mM Na^+. This low sensitivity to activation by Na^+ raises questions about the role of the channel, because the physiological intracellular Na^+ concentration is estimated to be around 10 mM or less (Pike et al., 1990; Guarnieri, 1987; Kleber, 1983). Notably, higher concentrations of Na^+ ions occuring in pathological conditions, such as ischemia (Pike et al., 1990; Guarnieri, 1987), in digitalis intoxication (Eisner et al., 1984), and during perfusion with Ca^{2+}, Mg^{2+}-free medium (Rodrigo and Chapman, 1990) may reach the activation threshold. The sensitivity of the K_{Na} channel to Na^+ is greater in cell-attached patches than in excised membrane patches. Rodrigo (1993) suggested that the Na^+ sensitivity of the channel in inside-out patches is probably underestimated owing to the loss of some intracellular factor. In neurons, on the other hand, the current is thought to play a physiological role as a repolarizing current during the action potential and is believed to be activated by the inrushing Na^+ ions during the upstroke of the action potential (Dryer, 1991; Bader et al., 1985).

4.2. Pharmacological Regulation of K_{Na} Channels

Many drugs have been tested but none can be considered selective for the K_{Na} channel. The first drug described to block K_{Na} channels was the experimental drug R56865, known for its inhibition of the electrophysiological effects of digitalis intoxication (Vollmer et al., 1987). However, it also inhibits, with even higher potency, other currents such as the transient inward current (Leyssens and Carmeliet, 1991) and the slowly inactivating Na^+ current (Carmeliet and Tygat, 1991). Pharmacological block of the K_{Na} channel is also possible with other drugs (Luk and Carmeliet, 1990). The most efficient are flunarizine, amiodarone, nicardipine, verapamil, and tedisamil. K_{ATP} channel blockers such as glibenclamide or class I drugs (lidocaine) are not efficient blockers at therapeutic concentrations.

5. CONCLUSIONS

In this chapter, we have reviewed recent molecular biological aspects of inward rectifying K^+ channels in the ventricular myocardium. Further understanding at the molecular level of the Kir channels in the heart may enable us to clarify the roles of these channels in cardiac physiology and pathophysiology, which may allow further development of strategies and pharmacological agents to treat various cardiac diseases.

REFERENCES

Aguilar-Bryan, L., Nichols, C. G., Wechsler, S. W., Clement, J. P. I., Boyd, A. E. I., Gonzalez, G., Herrera-Sosa, H., Nguy, K., Bryan, J., and Nelson, D. A., 1995, Cloning of the β cell high-affinity sulfonylurea receptor: A regulator of insulin secretion, *Science* **268**:423–426.

Ammälä, C., Moorhouse, A., Gribble, F., Ashfield, R., Proks, P., Smith, P. A., Sakura, H., Coles, B., Ashcroft, S. J., and Ashcroft, F. M., 1996, Promiscuous coupling between the sulfonylurea receptor and inwardly rectifying potassium channels, *Nature* **379**:545–548.

Ashcroft, F. M., 1988, Adenosine 5′-triphosphate-sensitive potassium channels, *Annu. Rev. Neurosci.* **11**:97–118

Ashcroft, F. M., and Gribble, F. M., 1998, Correlating structure and function in ATP-sensitive K^+ channels, Trends Neurosci. **21**:288–294.

Ashcroft, F. M., and Kakei, M., 1989, ATP-sensitive K^+ channels in rat pancreatic β-cells: Modulation by ATP and Mg^{2+} ions, *J. Physiol. (London)* **416**:349–367.

Ashcroft, S. J. H., and Ashcroft, F. M., 1990, Properties and functions of ATP-sensitive K-channels, *Cell. Signal.* **2**:197–214.

Bader, C. R., Bernheim, L., and Bertrand, D., 1985, Sodium-activated potassium current in cultured avian neurons, *Nature* **317**:540–542.

Baukrowitz, T., Schulte, U., Oliver, D., Herlitze, S., Krauter, T., Tucker, S. J., Ruppersberg, J. P., and Fakler, B., 1998, PIP_2 and PIP as determinants for ATP inhibition of K_{ATP} channels, *Science* **282**:1141–1144.

Beech, D. J., Zhang, H., Nakao, K., and Bolton, T. B., 1993a, K channel activation by nucleotide diphosphates and its inhibition by glibenclamide in vascular smooth muscle cells, *Br. J. Pharmacol.* **110**:573–582.

Beech, D. L., Zhang, H., Nakao, K., and Bolton, T. B., 1993b, Single channel and whole-cell K-currents evoked by levcromakalim in smooth muscle cells from the rabbit portal vein, *Br. J. Pharmacol.* **110**:583–590.

Beeler, G. W., and Reuter, H., 1970, Voltage clamp experiments on ventricular myocardial fibers, *J. Physiol. (London)* **207**:165–190.

Bond, C. T., Ammälä-, C., Ashfield, R., Blair, T. A., Gribble, F., Khan, R. N., Lee, K., Proks, P., Rowe, I. C., Sakura, H., Ashford, M. J., Adelman, J. P., Ashcroft, F. M., 1995, Cloning and functional expression of the cDNA encoding an inwardly-rectifying potassium channel expressed in pancreatic beta-cells and in the brain. *FEBS Letters.* **367**(1):61-66.

Bryan, J., Aguilar-Bryan, L., 1997, The ABCs of ATP-sensitive potassium channels: more pieces of the puzzle. [Review] [44refs] *Current Opinion in Cell Biology* **9**(4):553-559.

Carmeliet, E., and Tygat, J., 1991, Agonistic and antagonistic effects of R 56865 on the Na^+ channel in cardiac cells, *Eur. J. Pharmacol.* **196**:53–60.

Chutkow, W. A., Simon, M. C., Beau, M. M. L., and Burant, C. F., 1996, Cloning, tissue expression, and chromosomal localization of SUR2, the putative drug-binding subunit of cardiac, skeletal muscle, and vascular K$_{ATP}$ channels, *Diabetes* **45:**1439–1445.

Clement, J. P. I., Kunjilwar, K., Gonzalez, G., Schwanstecher, M., Panten, U., Aguilar-Bryan, L., and Bryan, J., 1997, Association and stoichiometry of K$_{ATP}$ channel subunits, *Neuron* **18:**827–838.

Cole, W. P., 1993, ATP-sensitive K$^+$ channels in cardiac ischemia: An endogenous mechanism for protection of the heart, *Cardiovasc. Drugs Ther.* **7:**527–537.

Collins, A., German, M. S., Jan, Y. N., Jan, L. Y., and Zhao, B., 1996, A strongly inwardly rectifying K$^+$ channel that is sensitive to ATP, *J. Neurosci.* **16:**1 9.

Cook, D. L., and Hales, C. N., 1984, Intracellular ATP directly blocks K$^+$ channels in pancreatic β-cells, *Nature* **311:**271–273.

Coraboeuf, E., and Carmeliet, E., 1982, Existence of two transient outward currents in sheep cardiac Purkinje fibers, *Pflügers Arch.* **392:**352–359.

Corey, S., Krapivinsky, G., Krapivinsky, L., and Clapham, D. E., 1998, Number and stoichiometry of subunits in the native atrial G-protein-gated K$^+$ channel, I$_{KACh}$, *J. Biol. Chem.* **273:**5271–5278.

Dascal, N., Lim, N. F., Schreibmayer, W., Wang, W., Davidson, N., and Lester, H. A., 1993, Expression of an atrial G-protein-activated potassium channel in *Xenopus* oocytes, *Proc. Natl. Acad. Sci. USA* **90:**6569–6600.

Deutsch, N., Matsuoka, S., and Weis, J. N., 1994, Surface charge and properties of cardiac ATP-sensitive K$^+$ channels, *J. Gen. Physiol.* **104:**773–800.

de Weille, J. R., Schmid-Antomarchi, H., Fosset, M., and Lazdunski, M., 1988, ATP-sensitive K$^+$ channels that are blocked by hypoglycemia-inducing sulfonylureas in insulin-secreting cells are activated by galanin, a hyperglycemia-inducing hormone, *Proc. Natl. Acad. Sci. U.S.A.* **85:**1312–1316.

de Weille, J. R., Schmid-Antomarchi, H., Fosset, M., and Lazdunski, M., 1989, Regulation of ATP-sensitive K$^+$ channels in insulinoma cells: Activation by somatostatin and protein kinase C and role of cAMP, *Proc. Natl. Acad. Sci. U.S.A.* **86:**2971–2975.

Downey, J. M., 1992, Ischemic preconditioning: Nature's own cardioprotective intervention, *Trends Cardiovasc. Med.* **2:**170–176.

Dryer, S. E., 1991, Na$^+$-activated K$^+$ channels and voltage-evoked ionic currents in brain stem and parasympathetic neurons of the chick, *J. Physiol. (London)* **435:**513–532.

Dunne, W. J., and Petersen, O. H., 1991, Potassium selective ion channels in insulin secreting cells: Physiology, pharmacology and their role in insulin secreting cells, *Biochim. Biophys. Acta* **1071:**67–82.

Eisner, D. A., Lederer, W. J., and Vaughan-Jones, R. D., 1984, The quantitative relationship between twitch tension and intracellular sodium activity in sheep cardiac Purinje fibers, *J. Physiol. (London)* **355:**251–266.

Faivre, J.-F., and Findlay, I., 1989, Effects of tolbutamide, glibenclamide and diazoxide upon action potentials recorded from rat ventricular muscle, *Biochim. Biophys. Acta* **984:**1–5.

Faivre, J.-F., and Findlay, I., 1990, Action potential duration and activation of the ATP-sensitive K$^+$ current in isolated guinea pig ventricular myocytes, *Biochim. Biophys. Acta* **1029:**167–172.

Fakler, B., Brèndle, U., Glowatzki, E., Weidemann, S., Zenner, H. P., and Ruppersberg, J. P., 1994a, Strong voltage dependent inward rectification of inward rectifier K$^+$ channels is caused by intracellular spermine, *Cell* **80:**149–154.

Fakler, B., Brèndle, U., Glowatzki, E., Zenner, H.-P. and Ruppersberg, J. P., 1994b, Kir2.1 inward rectifier K$^+$ channels are regulated independently by protein kinases and ATP hydrolysis, *Neuron* **13:**1413–1420.

Ficker, E., Taglialatela, M., Wible, B. A., Henley, C. M., and Brown, A. M., 1994, Spermine and spermidine as gating molecules for inward rectifier K$^+$ channels, *Science* **266:**1068–1072.

Findlay, I., 1987a, ATP-sensitive K$^+$ channels in rat ventricular myocytes are blocked and inactivated by internal divalent cations, *Pflügers Arch.* **410:**313–320.

Findlay, I., 1987b, The effects of magnesium upon adenosine triphosphate-sensitive potassium channels in a rat insulin-secreting cell line, *J. Physiol. (London)* **391:**611–629.

Findlay, I., 1988, ATP^{4-} and ATP-Mg inhibit the ATP-sensitive K$^+$ channel of rat ventricular myocytes, *Pflügers. Arch.* **412:**37–41.

Findlay, I., Dunne, M. J., and Petersen, O. H., 1985, ATP-sensitive inward rectifier and voltage and calcium activated K$^+$ channels in cultured pancreatic islet cells, *J. Membr. Biol.* **88:**165–172.

Findlay, I. and Dunne, M. J., 1986, ATP maintains ATP-inhibited K$^+$ channels in an operational state, *Pflügers Arch.* **407:**238–240.

Findlay, I., 1992, Inhibition of ATP-sensitive K^+ channels in cardiac muscle by the sulfonylurea drug glibenclamide, *J. Pharmacol. Exp. Ther.* **261**:540–545.

Findlay, I., Deroubaix, E., Guiraudou, P., and Coraboeuf, E., 1989, Effects of activation of ATP-sensitive K^+ channels of mammalian ventricular myocytes, *Am. J. Physiol.* **257**:H1551–H1559.

Fosset, M., Eeille, J. R., Green, R. D., Schmid-Antomarchi, H., and Lazdunski, M., 1988, Antidiabetic sulfonylureas control action potential properties in heart cells via high affinity receptors that are linked to ATP-dependent K^+ channels, *J. Biol. Chem.* **263**:7933–7936.

Garrino, M. G., Plant, T. D., and Henquin, J. C., 1989, Effects of putative activators of K^+ channels in mouse pancreatic β-cells, *Br. J. Pharmacol.* **98**:957–965.

Gasser, R. N. A., and Vaughan-Jones, R. D., 1992, Mechanism of potassium efflux and action potential shortening during ischemia in isolated mammalian cardiac muscle, *J. Physiol. (London)* **431**:713–741.

Gribble, F. M., Ashfield, R., and Ashcroft, F. M., 1997a, Properties of cloned ATP sensitive K^+ currents expressed in *Xenopus* oocytes *J. Physiol. (London)* **498**:87–98.

Gribble, F. M., Tucker, S. J., and Ashcroft, F. M., 1997b, The essential role of the Walker A motifs of SUR1 in K-ATP channel activation by Mg-ADP and diazoxide, *EMBO J.* **16**:1145–1152.

Gross, G. J., and Auchampach, A., 1992, Role of ATP-dependent potassium channels in myocardial ischaemia, *Cardiovasc. Res.* **26**:1011–1016.

Grover, G. J., Sleph, P. G., and Dzwonczyk, S., 1992, Role of myocardial ATP-sensitive potassium channels in mediating preconditioning in the dog heart and their possible interaction with adenosine A1-receptor, *Circulation* **86**:1310–1316.

Guarnieri, T., 1987, Intracellular sodium–calcium dissociation in early contractile failure in hypoxic ferret papillary muscle, *J. Physiol. (London)* **388**:449–465.

Heidbuchel, H., Vereecke, J., and Carmeliet, E., 1990, Three different potassium channels in human atria: Contribution to the basal potassium conductance, *Circ. Res.* **66**:1277–1286.

Hilgeman, D. W., and Ball, R., 1996, Regulation of cardiac Na^+, Ca^{2+} exchanger and K_{ATP} potassium channels by PIP_2, *Science* **273**:956–959.

Hiraoka, M., Sawanobori, T., Kawano, S., Hirano, Y., and Furukawa, T., 1996, Functins of cardiac ion channels under normal and pathological conditions, *Jpn. Heart J.* **37**:693–707.

Ho, K., Nichols, C. G., Lederer, W. J., Lytton, J., Vassilev, P. M., Kanazirska, M. V., and Hebert, S. C., 1993, Cloning and expression of an inwardly rectifying ATP-regulated potassium channel, *Nature* **362**:31–38.

Horie, M., Irisawa, H., and Noma, A., 1987, Voltage-dependent magnesium block of adenosine-triphosphate-sensitive potassium channel in guinea pig ventricular cells, *J. Physiol. (London)* **387**:251–272.

Huang, C.-L., Feng, S., and Hilgeman, D. W., 1998, Direct activation of inward rectifier potassium channels by PIP_2 and its stabilization by $G_{\beta\gamma}$, *Nature* **391**:803–806.

Inagaki, N., Gonoi, T., Clement, J. P., IV, Namba, N., Inazawa, J., Gonzalez, G., Aguilar-Bryan, L., Seino, S., and Bryan, J., 1995a, Reconstitution of I_{KATP}: An inward rectifier subunit plus the sulfonylurea receptor, *Science* **270**:1166–1170.

Inagaki, N., Tsuura, Y., Namba, N., Masuda, K., Gonoi, T., Horie, M., Seino, Y., Mizuta, M., and Seino, S., 1995b, Cloning and functional characterization of a novel ATP-sensitive potassium channel ubiquitously expressed in rat tissues, including pancreatic islets, pituitary, skeletal muscle and heart, *J. Biol. Chem.* **270**:5691–5694.

Inagaki, N., Gonoi, T., Clement, J. P., IV, Wang, C.-Z., Aguilar-Bryan, L., Bryan, J., and Seino, S., 1996, A family of sulfonylurea receptors determines the pharmacological properties of ATP-sensitive K^+ channels, *Neuron* **16**:1011–1017.

Inagaki, N., Gonoi, T., and Seino, S., 1997, Subunit stoichiometry of the pancreatic β-cell ATP-sensitive K^+ channel, *FEBS Lett.* **409**:232–236.

Inanobe, A., Ito, H., Ito, M., Hosoya, Y., and Kurachi, Y., 1995, Immunological and physical characterization of the brain G protein-gated muscarinic potassium channel, *Biochem. Biophys. Res. Commun.* **217**:1238–1244.

Isomoto, S., Konda, C., Yamada, M., Matsumoto, S., Higashiguchi, O., Horio, Y., Matsuzawa, Y., Kurachi, Y., 1996, A novel sulfonylurea receptor forms with BIR (Kir6.2) a smooth muscle type ATP-sensitive K^+ channel. *J. Biol. Chem.* **271**(40):24321–24324.

Isomoto, S., Kondo, C., Takahashi, N., Matsumoto, S., Yamada, M., Takumi, T., Horio, Y., and Kurachi, Y., 1996a, A novel ubiquitously distributed isoform of GIRK2 (GIRK2B) enhances GIRK1 expression of the G-protein-gated K^+ current in *Xenopus* oocytes, *Biochem. Biophys. Res. Commun.* **218**:286–291.

Isomoto, S., Kondo, C., Yamada, M., and Matsumoto, S., 1996b, A novel sulfonylurea receptor forms with BIR (Kir6.2) a smooth muscle type ATP-sensitive K$^+$ channel, *J. Biol. Chem.* **271:**24321–24324.

Isomoto, S., Kondo, C., and Kurachi, Y., 1997a, Inwardly rectifying potassium channels: Their molecular heterogeneity and function, *Jpn. J. Physiol.* **47:**11–39.

Isomoto, S., Horio, Y., Matsumoto, S., Kondo, C., Yamada, M., Gilbert, D. J., Copeland, N. G., Jenkins, N. G., and Kurachi, Y., 1997b, Sur2 and Kcnj8 genes are tightly linked on the distal region of mouse chromosome 6, *Mamm. Genome* **8:**790–791.

Ito, H., Tung, R. T., Sugimoto, T., Kobayashi, I., Takahashi, K., Katada, T., Ui, M., and Kurachi, Y., 1992, On the mechanism of G protein $\beta\gamma$ subunit activation of the muscarinic K$^+$ channel in guinea pig atrial cell membrane: Comparison with the ATP-sensitive K$^+$ channel, *J. Gen. Physiol.* **99:**961–983.

Jan, L. Y., and Jan, Y. N., 1992, Structure elements involved in specific K$^+$ channel functions, *Annu. Rev. Physiol.* **54:**537–555.

Jan, L. Y., and Jan, Y. N., 1994, P$_o$tassium channels and their evolving gates, *Nature* **371:**119–122.

Kajioka, S., Kitamura, K., and Kuriyama, H., 1991, Guanosine diphosphate activates an adenosine 5′-triphosphate-sensitive K$^+$ channel in the rabbit portal vein, *J. Physiol. (London)* **444:**397–418.

Kakei, M., and Noma, A., 1984, Adenosine-5′-triphosphate-sensitive single potassium channel in the atrioventricular node cell of the rabbit heart, *J. Physiol. (London)* **352:**265–284.

Kameyama, M., Kakei, M., Sato, R., Shibasaki, T., Matsuda, H., and Irisawa, H., 1984, Intracellular Na$^+$ activates a K$^+$ channel in mammalian cardiac cells, *Nature* **309:**354–356.

Kamouchi, M., and Kitamura, K., 1994, Regulation of ATP-sensitive K$^+$ channels by ATP and nucleotide diphosphate in rabbit portal vein, *Am. J. Physiol.* **266:**H1687–H1698.

Kane, C., Shepherd, R. M., Squires, P. E., Johnson, P. R., James, R. F., Milla, P. J., Aynsley-Green, A., Lindley, K. J., and Dunne, M. J., 1996, Loss of functional K$_{ATP}$ channels in pancreatic β-cells causes persistent hyperinsulinemic hypoglycemia of infancy, *Nat. Med.* **2:**1344–1347.

Kenyon, J. L., and Gibbons, W. R., 1979, 4-Aminopyridine and the early outward current of sheep cardiac Purkinje fibers, *J. Gen. Physiol.* **73:**139–157.

Kleber, A. G., 1983, Resting membrane potential, extracellular potassium activity, and intracellular sodium activity during global ischemia in isolated perfused guinea pig hearts, *Circ. Res.* **52:**442–450.

Koster, J. C., Shyng, S. -L., Sha, Q., and Nichols, C. G., 1998, Involvement of the N terminus of Kir6.2 in regulating ATP-sensitivity of K$_{ATP}$ channels, *Biophys. J.* **74:**A230.

Krapivinsky, G., Gordon, E. A., Wickman, K., Velimirovic, B., Krapivinsky, L., and Clapham, D. E., 1995, The G-protein-gated atrial K$^+$ channel I$_{KACh}$ is a heteromultimer of two inwardly rectifying K$^+$-channel proteins, *Nature* **374:**135–141.

Krapivinsky, G., Medina, I., Eng, L., Krapivinsky, L., Yang, Y., and Clapham, D. E., 1998, A novel inward rectifier K$^+$ channel with unique pore properties, *Neuron* **20:**995–1005.

Kubo, Y., Baldwin, T. J., Jan, Y. N., and Jan, L. Y., 1993a, Primary structure and functional expression of a mouse inward rectifier potassium channel, *Nature* **362:**127–133.

Kubo, Y., Reuveny, E., Slesinger, P. A., Jan, Y. N., and Jan, L. Y., 1993b, Primary structure and functional expression of a rat G protein-coupled muscarinic potassium channel, *Nature* **364:**802–806.

Kurachi, Y., 1985, Voltage-dependent activation of the inward rectifier potassium channel in the ventricular cell membrane of guinea-pig heart, *J. Physiol. (London)* **366:**295–309.

Kurachi, Y., Nakajima, T., and Sugimoto, T., 1986, Acetylcholine activation of K$^+$ channels in cell-free membranes of atrial cells, *Am. J. Physiol.* **251:**H681–H684.

Lesage, F., Duprat, F., Fink, M., Guillemare, E., Coppola, T., Lazdunski, M., and Hugnot, J. P., 1994, Cloning provides evidence for a family of inward rectifier and G-protein coupled K$^+$ channels in the brain, *FEBS Lett.* **353:**37–42.

Leyssens, A., and Carmeliet, E., 1991, Block of transient inward current by R56865 in guinea-pig ventricular myocytes, *Eur. J. Pharmacol.* **196:**43–51.

Lopatin, A. N., Makhina, E. N., and Nichols, C. G., 1994, Potassium channel block by cytoplasmic polyamines as the mechanism of intrinsic rectification, *Nature* **372:**366–369.

Lorenz, J. N., Schnermann, J., Brosius, F. C., Briggs, J. P., and Furspan, P. B., 1992, Intracellular ATP can regulate afferent arteriolar tone via ATP-sensitive K$^+$ channels in the rabbit, *J. Clin. Invest.* **90:**733–740.

Luk, H.-N., and Carmeliet, E., 1990, Na$^+$-activated K$^+$ current in cardiac cells: Rectification, open probability, block and role in digitalis toxicity, *Pflügers. Arch.* **416:**766–768.

MacKinnon, R., 1991, Determination of the subunit stoichiometry of a voltage-activated potassium channel, *Nature* **350**:232–235.

Matsumoto, S., Isomoto, S., Inanobe, A., Hibino, H., Saito, H., Katayama, Y., Motone, M., Chachin, M., Yamada, M., Horio, Y., and Kurachi, Y., 2000, The cardiac inwardly rectifying background K^+ channel (I_{K1}) is composed of Kir2.2: RT-PCR analysis of single channel myocyte mRNA, *Med. J. Osaka Univ.* **44**:15–23.

McAllister, R. E., and Noble, D., 1966, The time and voltage dependence of the slow outward current in cardiac Purkinje fibers, *J. Physiol.* (*London*) **186**:632–662.

Morishige, K., Takahashi, N., Findlay, I., Koyama, H., Zanelli, J. S., Peterson, C., Jenkins, N. A., Copeland, N. G., and Kurachi, Y., 1993, Molecular cloning, functional expression and localization of an inward rectifier potassium channel in the mouse brain, *FEBS Lett.* **336**:375–380.

Morishige, K., Takahashi, N., Jahangir, A., Yamada, M., Koyama, H., Zanelli, J. S., and Kurachi, Y., 1994, Molecular cloning and functional expression of a novel brain-specific inward rectifer potassium channel, *FEBS Lett.* **346**:303–307.

Nelson, M. T., and Quayle, J. M., 1995, Physiological roles and properties of potassium channels in arterial smooth muscle, *Am. J. Physiol.* **268**:C799–C822.

Nichols, C. G., and Lederer, W. J., 1991, Adenosine trisphosphate-sensitive potassium channels in the cardiovascular system, *Am. J. Physiol.* **261**:H1675–H1686.

Nichols, C. G., Ripoll, C. and Lederer, W. J., 1991, ATP-sensitive K^+ channel modulation of the guinea pig ventricular action potential and contraction, *Circ. Res.* **68**:280–287.

Nichols, C. G., Shyng, S. -L., Nestorowicz, A., Glaser, B., Clement, J. P. I., Gonzalez, G., Aguilar-Bryan, L., Permutt, M. A., and Bryan, J., 1996, Adenosine diphosphate as an intracellular regulator of insulin secretion, *Science* **272**:1785–1787.

Noma, A., 1983, ATP-regulated K^+ channels in cardiac muscle, *Nature* **305**:147–148.

Noma, A., Nakayama, T., Kurachi, Y., Irisawa, H., 1984, Resting K conductances in pacemaker and non-pacemaker heart cells of the rabbit. *Jpn. J. Physiol.* **34**(2):245–254.

Ohno-Shosaku, T., Zfnkler, B. J., and Trube, G., 1987, Dual effect of ATP on K^+ currents of mouse pancreatic β-cells, *Pflügers Arch.* **408**:133–138.

Okuyama, Y., Yamada, M., Kondo, C., Satoh, E., Isomoto, S., Shindo, T., Horio, Y., Kitakaze, Y., Horio, M., and Kurachi, Y., 1998, The effects of nucleotides and potassium channel openers on the SUR2A/Kir6.2 complex K^+ channel expressed in a mammalian cell line, HEK293T cells, *Pflügers Arch.* **435**:595–603.

Parratt, J. R., and Kane, K. A., 1994, K_{ATP} channels in ischemic preconditioning, *Cardiovasc. Res.* **28**:783–787.

Pike, M. M., Kitakaze, M., and Marban, E., 1990, ^{23}Na-NMR measurements of intracellular sodium in intact perfused ferret hearts during ischemia and reperfusion, *Am. J. Physiol.* **259**:H1767–H1773.

Pongs, O., 1992, Molecular biology of voltage-dependent potassium channels, *Physiol. Rev.* **72**:61–88.

Rodrigo, G. C., 1993, The Na^+-dependence of Na^+ activated K^+ channels (I_{KNa}) in guinea pig ventricular myocytes is different in excised inside/out patches and cell attached patches, *Pflügers. Arch.* **422**:530–532.

Rodrigo, G. C., and Chapman, R. A., 1990, A sodium-activated potassium current in intact ventricular myocytes isolated from the guinea-pig heart, *Exp. Physiol.* **75**:839–842.

Rougier, O., Vassort, G., and Stèmpfri, R., 1968, Voltage clamp experiments on frog atrial muscle fibers with the sucrose gap techniqe, *Pflügers Arch.* **301**:91–108.

Sakmann, B., and Trube, G., 1984, Conductance properties of single inwardly rectifying potassium channels in ventricular cells from guinea-pig, *J. Physiol.* (*London*) **347**:641–657.

Sakmann, B., Noma, A., and Trautwein, W., 1983, Acetylcholine activation of single muscarinic K^+ channels in isolated pacemaker cells of mammalian heart, *Nature* **303**:250–253.

Sakura, H., Ammälä-, C., Smith, P. A., Gribble, F. M., and Aschcroft, F. M., 1995, Cloning and functional expression of the cDNA encoding a novel ATP-sensitive potassium channel subunit expressed in pancreatic β-cells, heart and skeletal muscle, *FEBS Lett.* **377**:338–344.

Satoh, E., Yamada, M., Kondo, C., Okuyama, Y., Isomoto, S., Horio, Y., and Kurachi, Y., 1998, Kir subunits determine the molecular mode of pinacidil activation of SUR/Kir6.0 complex potassium channels, *J. Phystol.* (*London*) **511**:663–674.

Shyng, S., and Nichols, C. G., 1997, Octameric stoichiometry of the K_{ATP} channel complex, *J. Gen. Physiol.* **110**:655–664.

Standen, N. B., Quayle, J. M., Davies, N. W., Brayden, J. E., Huang, Y., and Nelson, M. T., 1989, Hyperpolarizing vasodilators activate ATP-sensitive K^+ channels in atrial smooth muscle, *Science* **245**:177–180.

Stanfield, P. R., Davies, N. W., Shelton, P. A., Sutcliffe, M. J., Khan, I. A., Brammar, W. J., and Conley, E. C., 1994, A single aspartate residue is involved in both intrinsic gating and blockade by Mg^{2+} of the inward rectifier, IRK1, *J. Physiol. (London)* **478**:1–6.

Takahashi, N., Morishige, K., Jahangir, A., Yamada, M., Findlay, I., Koyama, H., and Kurachi, Y., 1994, Molecular cloning and functional expression of cDNA encoding a second class of inward rectifier potassium channels in the mouse brain, *J. Biol. Chem.* **269**:23274–23279.

Takano, M., and Noma, A., 1993, The ATP-sensitive K$^+$ channel, *Prog. Neurobiol.* **41**:21–30.

Takano, M., Qin, D., and Noma, A., 1990, ATP-dependent decay and recovery of K$^+$ channels nea-pig cardiac myocytes, *Am. J. Physiol.* **258**:H45–H50.

Takumi, T., Ishii, T., Horio, Y., Morishige, K., Takahashi, N., Yamada, M., Yamashita, T., and Koyama, H., 1995, A novel ATP-dependent inward rectifier potassium channel expressed predominantly in glial cells, *J. Biol. Chem.* **270**:16339–16346.

Terzic, A., and Kurachi, Y., 1996, Actin microfilament disrupters enhance K$_{ATP}$ channel opening in patches from guinea-pig cardiomyocytes, *J. Physiol. (London)* **492**:395–404.

Terzic, A., Findlay, I., Hosoya, Y., and Kurachi, Y., 1994a, Dualistic behavior of ATP-sensitive K$^+$ channels toward intracellular nucleoside diphosphates, *Neuron* **12**:1–20.

Terzic, A., Tung, R., and Kurachi, Y., 1994b, Nucleotide regulation of ATP sensitive potassium channels, *Cardiovasc. Res.* **28**:746–753.

Terzic, A., Tung, R. T., Inanobe, A., Katada, T., and Kurachi, Y., 1994c, G proteins activate ATP-sensitive K$^+$ channels by antagonizing ATP-dependent gating, *Neuron* **12**:885–893.

Terzic, A., Jahangir, A., and Kurachi, Y., 1995, Cardiac ATP-sensitive K$^+$ channels: Regulation by intracellular nucleotides and K$^+$ channel-opening drugs, *Am. J. Physiol.* **269**:C525–C545.

Thomas, P. M., Cote, G. J., Wohllk, N., Haddad, B., Mathew, P. M., Rabl, W., Aguilar-Bryan, L., Gagel, R. F., and Bryan, J., 1995, Mutations in the sulfonylurea receptor gene in familial persistent hyperinsulinemic hypoglycemia of infancy, *Science* **268**:426–428.

Trube, G., and Heschler, J., 1984, Inward-rectifying channels in isolated patches of the heart cell membrane: ATP-dependence and comparison with cell-attached patches, *Pflügers Arch.* **401**:178–184.

Tucker, S. J., Gribble, F. M., Zhao, C., Trapp, S., and Ashcroft, F. M., 1997, Truncation of Kir6.2 produces ATP-sensitive K$^+$ channels in the absence of the sulfonylurea receptor, *Nature* **387**:179–183.

Tung, R., and Kurachi, Y., 1991, On the mechanism of nucleotide diphosphate activation of the ATP-sensitive K$^+$ channel in ventricular cell of guinea-pig, *J. Physiol. (London)* **437**:239–256.

Tusnady, G. E., Bakos, E., Varadi, A., and Sarkadi, B., 1997, Membrane topology distinguishes a subfamily of the ATP-binding cassette (ABC) transporters, *FEBS Lett.* **402**:1–3.

Ueda, K., Inagaki, N., and Seino, S., 1997, MgADP antagonism to Mg^{2+} independent ATP binding of the sulfonylurea receptor SUR1, *J. Biol. Chem.* **272**:22983–22986.

Veldkamp, M. W., Vereecke, J., and Carmeliet, E., 1994, Effects of intracellular Na$^+$ and H$^+$ on the Na$^+$-activated K$^+$ channel in isolated patches from guinea pig ventricular myocytes, *Cardiovasc. Res.* **28**:1036–1041.

Vollmer, B., Meuter, C., and Janssen, P. A. J., 1987, R 56865 prevents electrical and mechanical signs of intoxication in guinea-pig papillary muscle, *Eur. J. Pharmacol.* **142**:137–140.

Wang, W., and Giebisch, G., 1991, Dual modulation of renal ATP-sensitive K$^+$ channel by protein kinase A and C, *Proc. Natl. Acad. Sci. U.S.A.* **88**:9722–9725.

Wang, Z., Kimitsuki, T., and Noma, A., 1991, Conductance properties of the Na$^+$-activated K$^+$ channel in guinea-pig ventricular cells, *J. Physiol. (London)* **433**:241-258.

Wible, B. A., Taglialatela, M., and Ficker, E., 1994, Gating of inwardly rectifying K$^+$ channels localized to a single negatively charged residue, *Nature* **371**:246–249.

Wilde, A. A. M., 1993, Role of ATP-sensitive K$^+$ channel current in ischemic arrhythmias, *Cardiovasc. Drugs Ther.* **7**:521–526.

Wilde, A. A. M. and Janse, M. J., 1994, Electrophysiological effects of ATP-sensitive potassium channel modulation: Implications for arrhythmogenesis, *Cardiovasc. Res.* **28**:16–24.

Yamada, M., and Kurachi, Y., 1995, Spermine gates inward-rectifying muscarinic but not ATP-sensitive K$^+$ channels in rabbit atrial myocytes, *J. Biol. Chem.* **270**:9289–9294.

Yamada, M., Isomoto, S., Matsumoto, S., Kondo, C., Shindo, T., Horio, Y., and Kurachi, Y., 1997, Sulfonylurea receptor 2B and Kir6.1 form a sulfonylurea-sensitive but ATP-insensitive K$^+$ channel, *J. Physiol. (London)* **499**:715–720.

Yang, J., Jan, Y. N., and Jan, L. Y., 1995a, Control of rectification and permeation by residues in two distinct domains in an inward rectifier K$^+$ channel, *Neuron* **14:**1047–1054.

Yang, J., Jan, Y. N., and Jan, L. Y., 1995b, Determination of the subunit stoichiometry of an inwardly rectifying potassium channel, *Neuron* **15:**1441–1447.

Yao, Z., Cavero, I., and Gross, G., 1993, Activation of cardiac K$_{ATP}$ channels: An endogenous protective mechanism during repetitive ischemia, *Am. J. Physiol.* **264:**H495–H504.

Zhang, H. L., and Bolton, T. B., 1996, Two types of ATP-sensitive potassium channels in rat portal vein smooth muscle cells, *Br. J. Pharmacol.* **118:**105–114.

Chapter 20

Cholinergic and Adrenergic Modulation of Cardiac K^+ Channels

Christopher Parker and David Fedida

1. INTRODUCTION

The modulation of cardiac K^+ channels by the cholinergic and adrenergic divisions of the autonomic nervous system is central to regulating chronotropy and inotropy in the heart. Interaction of drugs with autonomic receptors in cardiac tissues represents an important intervention in the control of arrhythmias, heart failure, hypertension, hypotension, and shock. The effects of adrenergic and cholinergic stimulation on cardiac K^+ currents have been well described. More recent advances in molecular biology have also allowed us to better understand the signal transduction and effector mechanisms underlying these effects at a subcellular level and can provide a unifying approach to understanding the seemingly diverse effects of autonomic stimulation on cardiac K^+ currents and function. Furthermore, there is substantial overlap between the effects of cholinergic and adrenergic responses. Finally, activation of one population of autonomic receptors may accentuate or antagonize the effects of another. This complex cross talk allows a finer control over cardiac K^+ channel function and may arise at any level of the signal transduction pathways, including receptors, G-proteins, second messengers, downstream effectors, or even the channel proteins themselves.

Gaskell (1886) was the first to report that vagal stimulation resulted in electrical changes within the heart. The classic experiments of Loewi and Navratil (1926) established the role of chemical synaptic transmission in the modulation of cardiac function. They showed that a diffusible substance released upon vagal stimulation led to the inhibition or slowing of atrioventricular conduction. Since that time, it has become apparent that the physiological mechanism underlying this vagal response involves a change in the membrane permeability to K^+ in both nodal pacemaker and atrial cells.

Christopher Parker • Department of Physiology, Queen's University, Kingston, Ontario, Canada K7L 3N6. *David Fedida* • Department of Physiology, University of British Columbia, Vancouver, British Columbia, Canada V6T 1Z3.

Potassium Channels in Cardiovascular Biology, edited by Archer and Rusch. Kluwer Academic/Plenum Publishers, New York, 2001.

Adrenoceptors (ARs) are abundant in both atrial and ventricular tissues, and are important positive regulators of chronotropy and inotropy. Classically, ARs have been divided into two broad classes, β and α. However, radioligand binding and pharmacological studies have subsequently identified multiple subtypes in each class. Many of these pharmacologically defined subtypes have also been cloned. β-ARs in the heart are composed of a heterogeneous population of β_1, β_2, and β_3 subtypes, and there is pharmacological evidence for a fourth subtype, which is not yet cloned (Kaumann, 1997; Kaumann and Molenaar, 1997). Myocardial α-ARs, conversely, appear to be primarily composed of the α_1 subtype (Graham et al., 1996).

The cloning and heterologous expression of K^+ channels that are the putative molecular correlates of native cardiac K^+ currents allows us to further characterize the signal transduction and regulatory pathways involved in the autonomic regulation of the heart. When the behavior of cloned K^+ channels is correlated with K^+ channel responses observed in intact cells, these molecular techniques are powerful tools for advancing our understanding of the complex mechanisms of cholinergic and adrenergic modulation of cardiac function.

2. CHOLINERGIC RESPONSES

It has long been established that the atria receive vagal innervation and that vagal stimulation results in negative chronotropy and inotropy. Although early anatomical studies failed to demonstrate parasympathetic innervation to the ventricles, it is now apparent that the ventricles indeed receive vagal input (Loffelholz and Pappano, 1985). Furthermore, the ventricles, like the atria, contain primarily M_2-type muscarinic cholinergic receptors (M_2-AChRs) (Michel and Whiting, 1987; Giraldo et al., 1988; Ehlert et al., 1989; Hoover and Neely, 1997). In atrial myocytes and nodal cells, parasympathetic stimulation and the release of the neurotransmitter acetylcholine (ACh) directly results in a specific increase in K^+ conductance (the muscarinic K^+ current, denoted $I_{K(ACh)}$). The role of vagal innervation in the ventricles is more subtle, and, in addition to activation of ventricular $I_{K(ACh)}$ (Koumi and Wasserstrom, 1994), cholinergic stimulation appears to involve antagonism of β-adrenergic stimulation at the level of the signal transduction pathways. This latter activity, referred to as "accentuated antagonism," may represent the most important cholinergically mediated effect on ventricular tissue. Accentuated antagonism has been proposed as a possible mechanism of the antiarrhythmic effects of muscarinic agonists or adenosine following myocardial ischemia (Rauch and Niroomand, 1991; Priori et al., 1993). These effects are considered in greater detail subsequently.

2.1. Effects of Cholinergic Stimulation on Cardiac K^+ Currents

2.1.1. Effects on $I_{K(ACh)}$

2.1.1a. Nodal and Atrial $I_{K(ACh)}$. The negative chronotropic response in nodal pacemakers and atrial myocytes in response to vagal stimulation results largely from the activation of a unique inwardly rectifying K^+ current, $I_{K(ACh)}$, that is distinct from both the ATP-sensitive K$^+$ current (I_{KATP}) and the strong inward rectifier (I_{K1}) current (Sakmann et al., 1983; Kurachi et al., 1986a,b). Along with cholinergic modulation of the pacemaker current (I_f) and the calcium current (I_{Ca}), activation of $I_{K(ACh)}$ contributes to negative chronotropy and slowed conduction velocity at the sinoatrial and atrioventricular nodes. $I_{K(ACh)}$ has been characterized electrophysiologically in a number of tissues, including rabbit nodal cells (Sakmann et al., 1983) and human (Sato et al., 1990; Heidbuchel et al., 1990), rabbit (Soejima and Noma, 1984), and guinea pig (Kurachi et al., 1986a,b) atrial myocytes. The gating properties of $I_{K(ACh)}$ in these tissues are similar, and, at the single-channel level, M$_2$-AChR stimulation results in the activation of a 40-pS K$^+$ conductance in symmetrical K$^+$ (Koumi and Wasserstrom, 1994). Activation of $I_{K(ACh)}$ requires the presence of GTP and is inhibited by pertussis toxin (PTX), consistent with the involvement of the PTX-sensitive inhibitory G-protein (Kurachi et al., 1986a,b; Koumi and Wasserstrom, 1994), G$_i$. Further studies suggested that activation of $I_{K(ACh)}$ involves a direct interaction of the K$_{(ACh)}$ channel with either G$_{\alpha i}$ (Yatani et al., 1987; Codina et al., 1987) or G$_{\beta \gamma}$ (Logothetis et al., 1987; Kurachi et al., 1989a). However, current evidence favors a more important role of G$_{\beta \gamma}$ (see Section 2.2).

Prolonged exposure to ACh results in a desensitization of the I$_{K(ACh)}$ response to ACh in both atrial myocytes and nodal cells, with a time-dependent multiphasic (i.e., fast, slow, and ultraslow) decrease in the magnitude of $I_{K(ACh)}$ (Jalife et al., 1980; Mubagwa and Carmeliet, 1983; Carmeliet and Mubagwa, 1986; Boyett and Roberts, 1987; Kurachi et al., 1987). Although beyond the scope of this chapter, it has been suggested that desensitization occurs due to modifications at various stages of the transduction pathway. There is evidence to suggest a role of phosphorylation of the M$_2$-AChR, perhaps by a G-protein-coupled receptor kinase (GRK), in mediating the slow phase of desensitization, possibly because phosphorylation reduces the agonist affinity for the receptor (Kwatra and Hosey, 1986; Kwatra et al., 1987; Shui et al., 1995). Downstream elements of the signal transduction pathway, including the K$_{(Ach)}$ channel itself, may also be modulated in mediating the fast phase of desensitization (Pappano and Mubagwa, 1991; Shui et al., 1997).

2.1.1*b*. *Ventricular* $I_{K(ACh)}$. At the level of the cell, isolated human, rat, cat, ferret, and guinea pig ventricular myocytes display $I_{K(ACh)}$ (McMorn et al., 1993; Koumi and Wasserstrom, 1994; Ito et al., 1995). Furthermore, at the level of the organ, anatomical studies have demonstrated that these ventricles receive vagal input (Loffelholz and Pappano, 1985). Cholinergic activation of $I_{K(ACh)}$ in ferret (Ito et al., 1995) and rat (McMorn et al., 1993) ventricular myocytes resulted in shortening of the action potential duration (APD), suggesting that $I_{K(ACh)}$ may be of physiological importance in these ventricular tissues. The single-channel conductance and gating kinetics of atrial and ventricular $I_{K(ACh)}$ are similar, which suggests a common molecular correlate. However, in human hearts, ventricular $I_{K(ACh)}$ is less sensitive to ACh than atrial $I_{K(ACh)}$ (K_d values being 30 nM and 130 nM, respectively) (Koumi and Wasserstrom, 1994). This may be explained by (i) a decreased M$_2$-AChR affinity for ACh in ventricular tissue or (ii) differences in atrial and ventricular signal transduction. Precisely where in the signal transduction pathway those differences lie is still unclear, although it has been suggested that they reside proximal to G-proteins, perhaps involving differences in the coupling of G-proteins to the muscarinic receptor (Koumi and Wasserstrom, 1994).

2.1.2. Effects on Other Cardiac K^+ Currents

2.1.2a. *Effects on* I_{KATP}. The cardiac I_{KATP}, which is suppressed by elevated intracellular ATP levels (ATP_i), provides a clear link between cellular metabolic activity and electrical excitability (Ashcroft and Rorsman, 1990; Nichols and Lederer, 1991; Noma, 1993; Terzic et al., 1995). During times of metabolic stress, such as hypoxia or ischemia, a drop in ATP_i and the subsequent disinhibition of I_{KATP} shortens the APD (Lederer et al., 1989). Under some experimental conditions, however, I_{KATP} can also be activated by cholinergic agonists (Wang and Lipsius, 1995a,b). However, in contrast to the direct $G_{\beta\gamma}$-mediated modulation of $I_{K(ACh)}$, stimulation of I_{KATP} by cholinergic agonists likely involves activated G_{zi} (Ito et al., 1992). Cholinergic effects on I_{KATP} may occur as a result of direct G-protein-mediated antagonism of ATP_i-dependent inhibitory gating of the channels (Terzic et al., 1994). However, in isolated perfused dog atria or ventricles, the glibenclamide-sensitive component of cholinergic-induced negative chronotropy and inotropy is small (Murakami et al., 1992, 1994), and experiments in isolated guinea pig atrium failed to demonstrate a reduction by glibenclamide of cholinergically induced negative inotropy (Urquhart et al., 1993). These observations suggest that under normal physiological conditions, cholinergic activation of I_{KATP} plays only a minor role, relative to stimulation of $I_{K(ACh)}$, in the net cardiac cholinergic response.

2.1.2b. *Effects on Components of the Delayed Rectifier* I_K. In many species, including human (Veldkamp et al., 1995), rabbit (Veldkamp et al., 1993), rat (Apkon and Nerbonne, 1991), and cat (Follmer and Colatsky, 1990), the I_K of ventricular myocytes is comprised of a single component. Conversely, in some other species and tissues, including human atrial (Wang et al., 1993a,b; Li et al., 1996a) and guinea pig atrial and ventricular myocytes (Sanguinetti and Jurkiewicz, 1990a, 1991), I_K appears to be a composite of currents that differ in time and voltage dependence, as well as in pharmacology (Sanguinetti and Jurkiewicz, 1990a, 1991; Li et al., 1996b). The best described of these components have been the slowly activating (I_{Ks}), rapidly activating (I_{Kr}), and ultrarapidly activating (I_{Kur}) delayed rectifier-type currents (Horie et al., 1990; Sanguinetti and Jurkiewicz, 1990b; Chinn, 1993; Liu and Antzelevitch, 1995; Yue et al., 1996a).

Cholinergic stimulation inhibits several of these I_K components. In guinea pig ventricular myocytes, carbachol reduces the amplitude of I_{Ks} (Harvey and Hume, 1989; Yazawa and Kameyama, 1990), although this effect was dependent upon previous β-AR stimulation. Therefore, the effects of cholinergic agonists on I_{Ks} may actually be secondary to antagonism of adenylate cyclase (AC) activation at the level of the signal transduction pathways (see Section 2.2.2). In isolated guinea pig sinoatrial nodal cells, however, it appears that cholinergic agonists can decrease the amplitude of I_{Ks} via a nonmuscarinic, noncholinergic, cAMP-independent pathway that remains to be fully elucidated (Freeman and Kass, 1995).

2.2. Signal Transduction Mechanisms

Cardiac cholinergic receptors are primarily of the M_2 subtype (Giraldo et al., 1988; Ehlert et al., 1989; Hoover and Neely, 1997) and are coupled to the PTX-sensitive class

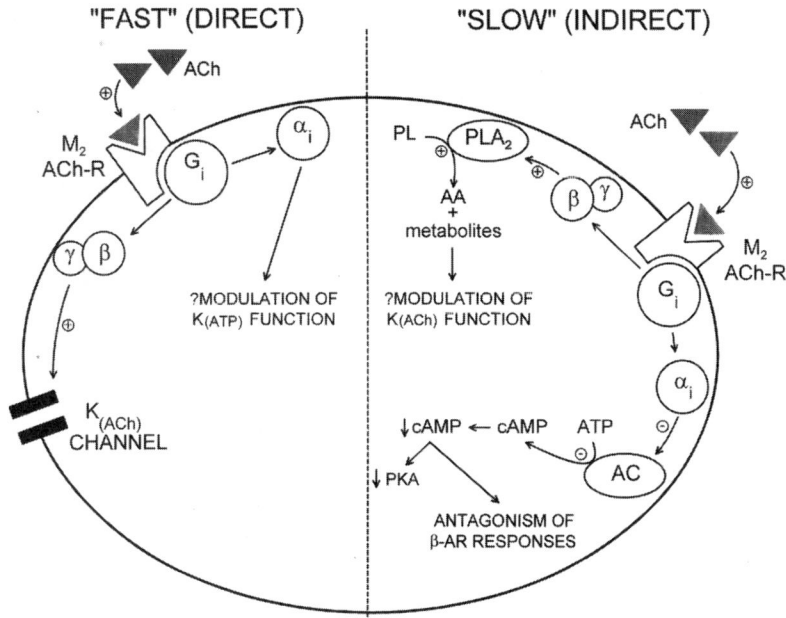

Figure 1. Generalized schematic of important signal transduction pathways and intracellular effects of cholinergic modulation of K$^+$ channels. ACh, Acetylcholine; PL, membrane phospholipids; PLA$_2$, phospholipase A$_2$; AA, arachidonic acid; AC, adenylate cyclase. Other abbreviations as described in the text.

of inhibitory G-proteins, G$_i$. Activation of M$_2$-AChRs in the heart results in both "fast" and "slow" responses (Wickman and Clapham, 1995). The fast, or direct responses, which develop quickly following stimulation of the M$_2$-AChR, are membrane-delimited, do not depend on cytoplasmic components, and involve the direct interaction of G-protein subunits with K$^+$ channels (Kurachi et al., 1992). In contrast, the slow, or indirect, responses are not membrane-delimited and are the result of the modulation of cytoplasmic signaling pathways, including cAMP signaling, by activated G-protein subunits (Wickman and Clapham, 1995). These pathways are illustrated in Fig. 1.

2.2.1. Direct Pathways

The K$_{(Ach)}$ channel was the first ion channel demonstrated to be directly activated by G-protein subunits (Pfaffinger et al., 1985; Kurachi et al., 1986a, 1986b; Logothetis et al., 1987; Breitwieser and Szabo, 1988). This obligatory role of the G-proteins was deduced from several lines of evidence. First, in cell-attached membrane patch recordings from rabbit atrial myocytes, extracellular administration of ACh failed to activate K$_{(Ach)}$ channels, whereas intracellular administration, via the pipette filling solution, resulted in channel activation (Soejima and Noma, 1984). Second, observations using excised inside-out patches from atrial myocytes demonstrated a need for GTP in linking receptor stimulation and $I_{K(ACh)}$ activation. These studies also implicated a requirement for a PTX-sensitive G-protein, such as G$_i$ (Pfaffinger et al., 1985; Kurachi et al.,

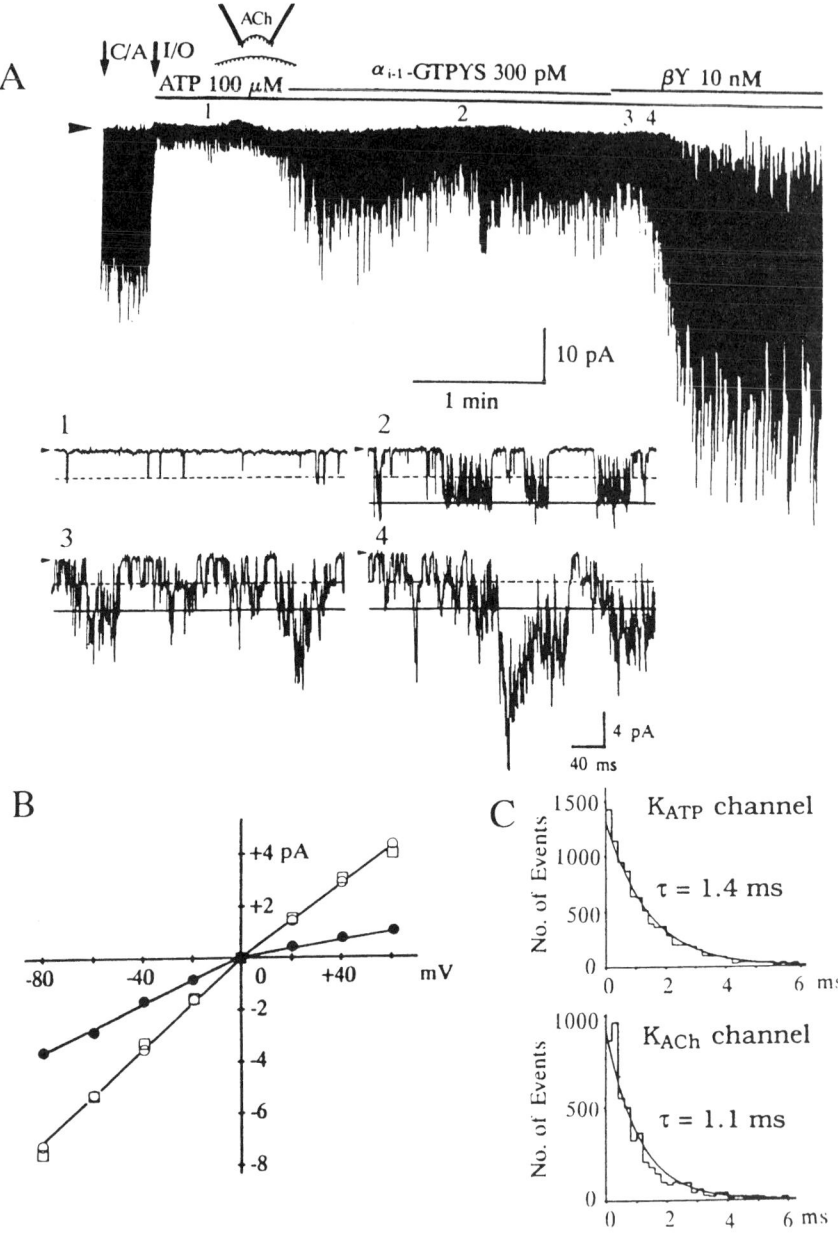

Figure 2. Differential effects of $G_{\alpha i}$ and $G_{\beta \gamma}$ on K_{ATP} and $K_{(Ach)}$ channels in inside-out patches from guinea pig atrial myocytes. (A) Current trace from a representative patch. The inside-out (I/O) patch was formed from the cell-attached configuration (C/A) as indicated by the arrows. The pipette filling solution continuously contained 100 μM ATP and 0.5 mM MgCl$_2$. The nonhydrolyzable analog $G_{\alpha i-1}$-guanosine-5'-O-(3-thiotriphosphate) ($G_{\alpha i-1}$-GTPγS), at a concentration of 300 pM, was applied to the internal side of the patch as indicated in the figure. This induced openings of a K$^+$ channel with a unitary conductance of $\sim$90 pS, consistent with K_{ATP}. Subsequently, $G_{\beta \gamma}$ (10 nM) was applied to the cytoplasmic side of the patch, which caused a dramatic increase of 45-pS $K_{(Ach)}$ channel openings in the same patch. Numbers above the current

1986a,b). The effects of cholinergic agonists on $I_{K(ACh)}$ could be prolonged by intracellular application of nonhydrolyzable GTP analogs such as guanosine 5'-O-(3-thiotriphosphate) (GTPγS) or prevented by application of guanosine 5'-O-(2-thiodiphosphate) (GDPβS), which inhibits G-protein activation (Kurachi *et al.*, 1986a, b). Furthermore, $I_{K(ACh)}$ could be activated by AlF$_4^-$ (Yatani and Brown, 1991; Kurachi *et al.*, 1992), which directly activates heterotrimeric G-proteins (Chabre, 1990). Third, application of purified G-protein subunits to the intracellular face of excised atrial inside-out patches activated $I_{K(ACh)}$. Although it has been reported that either G$_{\alpha i}$ (Yatani *et al.*, 1987; Codina *et al.*, 1987) or G$_{\beta \gamma}$ (Logothetis *et al.*, 1987; Kurachi *et al.*, 1989a) could activate the K$_{(ACh)}$ channel, current evidence favors a more important role of G$_{\beta \gamma}$. Application of purified bovine brain G$_{\beta \gamma}$ to excised inside-out membrane patches from atria activated $I_{K(ACh)}$ (Logothetis *et al.*, 1988; Ito *et al.*, 1992). Subsequent experiments using purified recombinant G$_{\beta \gamma}$ (Wickman *et al.*, 1994) appear to have excluded the possibility that the effects observed in the bovine preparations were due to contamination by preactivated G$_\alpha$ subunits or detergent (Ito *et al.*, 1991, 1992). The demonstration that G$_{\beta \gamma}$ activates the K$^+$ channel directly was the source of considerable controversy, as it had been previously accepted that downstream effects of G-proteins were mediated exclusively through the activated G$_\alpha$ subunits (Brown and Birnbaumer, 1990; Kurachi, 1995; Wickman and Clapham, 1995). G$_{\alpha i}$, however, appears to have a role in modulating the direct effects of M$_2$-AChR stimulation of the G$_{\beta \gamma}$-mediated gating of the K$_{(ATP)}$ channel. This was elegantly demonstrated by Ito *et al.* (1992), who showed that application of purified G$_{\beta \gamma}$ to inside-out atrial myocyte membrane patches resulted in the stimulation of $I_{K(ACh)}$, whereas application of activated G$_{\alpha i}$ to the same patches specifically stimulated I_{KATP}. These effects are illustrated in Fig. 2. In addition, because G$_{\alpha i}$ contains intrinsic GTPase activity, it is likely that G$_{\alpha i}$ has an important regulatory role in determining the kinetics of association of the G$_{\beta \gamma}$ subunits with K$^+$ channels by determining the rate at which the inactive G$_{\alpha i}$–ADP–G$_{\beta \gamma}$ heterotrimeric complex re-forms (Kurachi, 1995).

2.2.2. Indirect Pathways

Indirect pathways of M$_2$-AChR signal transduction involve the modulation of cytoplasmic signal transduction processes and second-messenger systems such as cAMP. Indirectly mediated effects develop more slowly than those involving direct gating by G-protein subunits and are not membrane-delimited. The coupling of the cardiac M$_2$-AChR to G$_i$ suggests that cholinergic stimulation can decrease cAMP levels

---

tracing indicate the location of each expanded current trace below. In the expanded current tracings, the dotted line is the first level of the K$_{(Ach)}$ channel, and the continuous line is that of the K$_{ATP}$ channel. The arrowhead at the start of each trace denotes zero current level. (B) Representative current–voltage relationships of G$_{\alpha i-1}$-induced K$_{ATP}$ channels in ventricular (□) and atrial (○) cell membrane, and G$_{\beta \gamma}$-induced K$_{(Ach)}$ channels in atrial cell membrane (●). (C) the Open-time histograms of representative G$_{\alpha i-1}$-induced K$_{ATP}$ channel openings (in ventricular membrane) and G$_{\beta \gamma}$-induced K$_{(Ach)}$ channel openings (in atrial membrane) obtained in response to voltage steps to −80 mV. The mean channel open times (τ) for the K$_{ATP}$ and the K$_{(Ach)}$ channels were 1.4 ms and 1.1 ms, respectively. (Reproduced from Ito *et al.*, 1992, by copyright permission of The Rockefeller University Press.)

secondary to a $G_{\alpha i}$-mediated inhibition of adenylate cyclase. The importance of this indirect pathway in regulating cardiac function has been best described in reference to cholinergic modulation of I_f and I_{Ca} (Chang *et al.*, 1987; Petit-Jacques *et al.*, 1993; Renaudon *et al.*, 1997; Accili *et al.*, 1998). Indirect pathways of M_2-AChR signal transduction have also been implicated in the ability of ACh to antagonize β-adrenergic inhibition of I_{K1} in guinea pig ventricular myocytes (Koumi *et al.*, 1995a, b). This mechanism may also be involved in the reduction of I_{Ks} amplitude by cholinergic agonists observed in guinea pig ventricular cells that have been exposed to β-adrenergic agonists (Harvey and Hume, 1989). Inhibition of AC by $G_{\alpha i}$, therefore, provides a point of convergence between the M_2-cholinergic and β-adrenergic pathways.

There are also reports that in bovine retinal rod cells, $G_{\beta\gamma}$ stimulates phospholipase A_2 (PLA_2) activity, resulting in the production of arachidonic acid (AA) and its metabolites as second messengers (Jelsema and Axelrod, 1987). This raised speculation that a similar mechanism may be operating in heart to modulate $I_{K(ACh)}$ (Kim *et al.*, 1989). However, direct application of AA derivatives only weakly stimulated $I_{K(ACh)}$ when compared with M_2-AChR activation. Furthermore, pharmacological inhibition of lipoxygenases and PLA_2 failed to prevent $I_{K(ACh)}$ activation by $G_{\beta\gamma}$ (Kurachi *et al.*, 1992). Current opinion, therefore, favors a modulatory role of lipid derivatives in $I_{K(ACh)}$ function and holds that $I_{K(ACh)}$ activation by direct binding of $G_{\beta\gamma}$ to the channel does not require generation of AA metabolites (Kurachi *et al.*, 1989b,c; Scherer *et al.*, 1993). However, modulation of $I_{K(ACh)}$ by AA derivatives may represent another potential convergence point between cholinergic and adrenergic pathways, since it is known that α_1-AR stimulation activates PLA_2 in atria (Kurachi *et al.*, 1989b,c).

2.3. Correlation with Cloned Channel Currents

$K_{(ACh)}$ channels are heterotetramers formed from the 1:1 stoichiometric association of GIRK1 and GIRK4 (Kir3.1 and Kir3.4/CIR, respectively) (Krapivinsky *et al.*, 1995; Barry and Nerbonne, 1996; Silverman *et al.*, 1996; Dascal, 1997), and heterotetrameric GIRK1/GIRK4 channels, when heterologously coexpressed, bind and are activated by purified $G_{\beta\gamma}$ in a manner similar to that observed for native $K_{(ACh)}$ channels (Reuveny *et al.*, 1994; Krapivinsky *et al.*, 1995; Nair *et al.*, 1995). This is illustrated by the data in Fig. 3. Furthermore, G-protein-binding domains on both GIRK1 and GIRK4 have been characterized (Huang *et al.*, 1995; Dascal, 1997), and experiments have also identified a domain on $G_{\beta\gamma}$ that is responsible for this interaction (Yan and Gautam, 1996).

Studies involving cloned GIRK1/GIRK4 channels have also sought to address the issue of the specificity of $G_{\beta\gamma}$-mediated activation of $I_{K(ACh)}$. Current evidence suggests that $G_{\beta\gamma}$ coupled to $G_{\alpha i2}$ or $G_{\alpha i3}$ is likely involved in the *in vivo* activation of $I_{K(ACh)}$, whereas $G_{\beta\gamma}$ coupled to other G_α subunits may not be involved (Dascal, 1997). The reason for this selectivity is unclear, since, in theory, activation of any heterotrimeric G-protein should liberate a pool of free $G_{\beta\gamma}$. In fact, it seems that, at high expression levels of receptors or G-proteins, $G_{\beta\gamma}$ liberated from $G_{\alpha s}$ can indeed activate homologous GIRK1 channels expressed in oocytes (Lim *et al.*, 1995). More recently, selective antagonism between GIRK1, $G_{\beta\gamma}$, and specific G_α subunits has been proposed

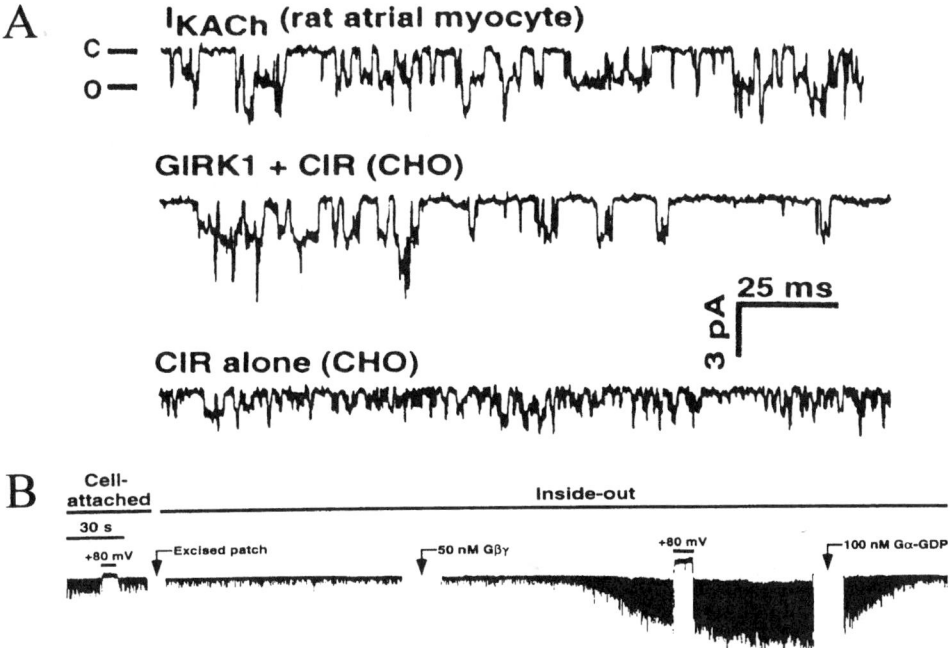

Figure 3. Channels produced by GIRK1 and CIR/GIRK4 coexpression in CHO cells are similar to native atrial $I_{K(ACh)}$ channels and are activated by purified $G_{\beta\gamma}$. (A) Comparison of inside-out patch recordings from rat atrial myocytes, CHO cells coexpressing GIRK1 and CIR/GIRK4, and CHO cells expressing CIR/GIRK4 alone, after stimulation by guanosine-5'-O-(3-thiotriphosphate) (GTPγS). The patches were stepped to a test potential of −80 mV. Open and closed levels of the channels are indicated by o and c , respectively. (B) Coexpression of GIRK1 and CIR/GIRK4 in CHO cells resulted in channels that were minimally activated after patch excision, but could be activated (in response to a test potential of −80 mV) by purified bovine brain $G_{\beta\gamma}$ (50 nM) applied to the intracellular face of the membrane. This effect could be reversed upon addition of 100 nM of purified bovine brain G_α-GDP to the cytoplasmic face. In these experiments, the pipette and bath solutions were identical and contained 140 mM K$^+$. (Reproduced and modified, with permission, from Krapivinsky *et al.*, 1995. Copyright 1995 Macmillan Magazines Ltd.)

as a mechanism underlying effector specificity (Schreibmayer *et al.*, 1996). This discussion is beyond the scope of this chapter but has been reviewed elsewhere (Dascal, 1997).

3. β-ADRENERGIC RESPONSES

3.1. Effects of β-AR Stimulation on Cardiac K$^+$ Currents

In the heart, under physiological conditions, activation of β-ARs is likely the most important mechanism of positive chronotropy and inotropy. As for M$_2$-AChRs, adrenergic pathways in the heart are tonically activated, and heart rate and contractile

Table 1
β-Adrenergic Effects on Native and Cloned Cardiac K^+ Currents

Current	Species	Effect of β-AR stimulation on current	Mechanism	Reference(s)
I_K	Guinea pig	↑	PKA	Yazawa and Kameyama, 1990; Harada and Iijima, 1994; Walsh and Kass, 1988; Walsh et al., 1989
			Direct coupling of I_K channel to β-AR via membrane-delimited G-protein pathway	Freeman et al., 1992
I_{Kur}	Human	↑	PKA	Li et al., 1996a
I_{ss}	Rat	↓	PKA	Scamps, 1996
I_{K1}	Guinea pig	↓	PKA	Koumi et al., 1995a,b
$I_{K(ACh)}$	Rat	↑[a]	?PKA	Kim, 1990
	Cloned GIRKI[b]	↑	Direct gating of GIRK by $G_{\beta\gamma}$ liberated in response to β-AR stimulation	Lim et al., 1995
I_{KATP}	Cat	↑	?Depletion of subsarcolemmal ATP secondary to enhanced production of cAMP	Schackow and Ten Eick, 1994
	Dog	↑	cAMP and cAMP-dependent effectors (?PKA)	Tseng and Hoffman, 1990

[a] Studies of GIRK1 do not support this finding in oocytes.
[b] Coexpression of GIRK1, β_2-AR, and $G_{\alpha s}/G_{\beta\gamma}$ in oocytes.

force can therefore be modulated by increasing or decreasing the stimulation of the ARs. Pharmacological intervention at the level of the β-ARs, and subsequent modulation of inotropy, chronotropy, and rhythmicity, is therefore commonly used clinically to modify cardiac function.

Conversely, β-AR stimulation can be arrhythmogenic and can increase the risk of supraventricular and ventricular arrhythmias. In addition to a β-AR-mediated modulation of I_{Ca}, and stimulation of arrhythmogenic transient inward currents, these effects are a result of modulation of K^+ conductances in the heart, including components of I_K, I_{K1}, $I_{K(ACh)}$, and I_{KATP}. The effects of β-AR stimulation on several native and cloned cardiac K^+ currents are summarized in Table 1 and are considered in greater detail below.

3.1.1. Effects on Components of the Delayed Rectifier, I_K

An increase in the magnitude of I_K in response to β-AR stimulation has been demonstrated in a number of mammalian cardiac preparations, including Purkinje fibers (Tsien et al., 1972; Bennett et al., 1986), isolated ventricular myocytes (Bennett and Begenisich, 1987; Yazawa and Kameyama, 1990; Harada and Iijima, 1994), and atrial pacemaker cells (Brown and DiFrancesco, 1980). More recently, it was shown that I_{Kur} in human atrial myocytes is increased by β-AR stimulation, and this may play a role in the genesis of atrial arrhythmias (Li et al., 1996a). I_{ss}, a delayed rectifier-like current found in isolated rat ventricular myocytes, is decreased by isoprenaline. β-AR inhibition of I_{ss} significantly prolongs APD (Scamps, 1996). Interestingly, the modulation of myocardial I_K by β-ARs is temperature-dependent (Walsh et al., 1989), unlike that of I_{Ca}, perhaps indicating differences in the underlying regulatory mechanisms or temperature-dependent differences in the kinetics of some required regulatory event, such as channel phosphorylation (Walsh et al., 1988; Walsh and Kass, 1988).

3.1.2. Effects on Other Cardiac K⁺ Currents

At least three other cardiac K⁺ currents are modulated by β-adrenergic stimulation. I_{K1}, recorded from guinea pig ventricular myocytes, is decreased by isoprenaline (Koumi et al., 1995a). At the single-channel level, this appears to result from a decreased open-state probability (Koumi et al., 1995b). In this tissue, the β-AR-mediated decrease in I_{K1} could be antagonized by ACh, which highlights the importance of cross talk between autonomic signaling pathways in the heart.

In one study of isolated rat ventricular myocytes, isoprenaline not only augmented the magnitude of $I_{K(ACh)}$ induced by ACh, but could also activate $I_{K(ACh)}$ in the absence of M_2-AChR activation. This was reflected at the single-channel level by a prolongation of both mean channel open and closed times in the presence of isoprenaline, with an overall increase in channel open-state probability (Kim, 1990). Although it was speculated that these effects were the result of direct $K_{(Ach)}$ channel phosphorylation, it is possible that β-AR stimulation results in the modification of other components of the signaling pathways involved, such as the M_2-AChR or G-proteins, which indirectly increase channel activity (Kim, 1990). In fact, studies of cloned $K_{(Ach)}$ channel subunits have suggested that $I_{K(ACh)}$ is likely not physiologically modulated by β-adrenergic stimulation (Lim et al., 1995).

In some cardiac tissues at least, I_{KATP} also appears to be stimulated by β-ARs. In isolated cat ventricular cells dialyzed with an ATP-free solution, application of isoprenaline resulted in the rapid (< 60 s) development of a glyburide-sensitive current, consistent with I_{KATP}. This current remained elevated even after the removal of the β-agonist (Schackow and Ten Eick, 1994). This contrasts with the relatively slow (10–25 min) development of I_{KATP} that occurs in cells dialyzed with an ATP-free solution in the absence of a β-agonist. Although some authors have proposed a role of cAMP and cAMP-dependent effectors in the β-adrenergic modulation of I_{KATP} (Tseng and Hoffman, 1990), others (Schackow and Ten Eick, 1994) have suggested that the augmentation of I_{KATP} occurs as a result of depletion of ATP from the subsarcolemmal surface in response to the activation of AC.

3.2. β-AR Receptor Subtypes Present in the Heart

β-ARs in mammalian heart are a heterogeneous population composed of several receptor subtypes. The existence of both cardiac β_1- and β_2-ARs has been demonstrated in a number of species using receptor-specific ligands and molecular techniques (Jones et al., 1989; Chevalier et al., 1991; Bohm and Lohse, 1994; Beau et al., 1995; Rodefeld et al., 1996; Xiao et al., 1998). The total number of β-ARs in human myocardium appears to be similar in both atrial and ventricular tissues, and although β_1-ARs predominate in terms of receptor density in both atrium and ventricle, the relative proportion of β_2-ARs is somewhat higher in atrial, as compared to ventricular, tissues (Kaumann et al., 1989; Brodde, 1991; Steinfath et al., 1992). Functionally, activation of β_1- or β_2-ARs results in increased chronotropy, and β-ARs are the most important receptor population in the human heart mediating positive inotropy under normal conditions. In human heart, myocardial relaxation is accelerated by stimulation of either β_1- or β_2-ARs (Del Monte et al., 1993; Kaumann et al., 1996), whereas in other animal cardiac tissues, only β_1-AR stimulation has this effect (Lemoine and Kaumann, 1991; Xiao et al., 1995).

There is also mounting evidence for the existence of β_3-ARs in myocardium, although their existence and function in human heart remain controversial. Earlier studies involving human myocardial tissues failed to detect either β_3-AR mRNA (Walsh and Kass, 1988; Berkowitz et al., 1995) or functional responses to β_3-AR-specific stimulation (Kaumann, 1989). Although Evans et al. (1996) found that β_3-AR mRNA was virtually undetectable in rat hearts, subsequent studies of isolated human ventricular tissue demonstrated not only the presence of β_3-AR mRNA but also showed that β_3-AR-selective stimulation results in a cardiodepressant effect and a shortening of APD (Gauthier et al., 1996). Although the physiological role of cardiac β_3-ARs is unclear, it has been speculated that since β_3-ARs are relatively insensitive to downregulation (Strosberg, 1993), they may become of increasing significance in disease states such as heart failure, when other β-ARs are downregulated (Gauthier et al., 1996).

Most recently, it has been suggested that a fourth β-AR subtype, the putative β_4-AR, exists in mammalian myocardium, including human (Kaumann, 1996; Kaumann and Molenaar, 1997) and rat (Kaumann and Molenaar, 1996) atrial tissues and human (Kaumann and Molenaar, 1997) and ferret (Lowe et al., 1998) ventricular preparations. Stimulation of this putative receptor subtype results in a positive inotropic and chronotropic response similar to that produced in response to β_1- or β_2-AR activation (Kaumann, 1997; Molenaar et al., 1997). In ferret ventricle, specific stimulation of the putative β_4-AR resulted in a prolongation of the plateau phase of the cardiac action potential and an overall shortening of the APD, suggesting that, in this tissue at least, β_4-ARs may have a physiological role (Lowe et al., 1998). As for the β_3-AR, however, the physiological relevance of the putative β_4-AR in other cardiac tissues, including human heart, is still unclear.

3.3. Signal Transduction Mechanisms

3.3.1. G-Proteins and Adenylate Cyclase

The important signal transduction pathways involved in the β-AR-mediated modulation of cardiac K^+ channels are illustrated in Fig. 4. In the heart, the coupling of β_1- and β_2-ARs to the stimulatory G-protein, G_s, has been clearly demonstrated.

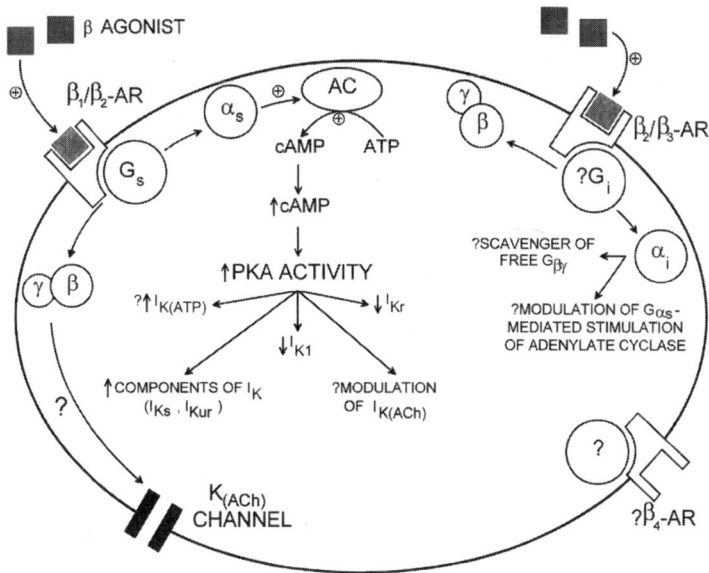

Figure 4. Schematic of intracellular effects of β-AR-mediated modulation of cardiac K$^+$ channels. AC, adenylate cyclase; PKA, protein kinase A. Other abbreviations as described in the text.

Activated $G_{\alpha s}$ directly interacts with the catalytic subunit of AC and results in the increased production of the second messenger cAMP from ATP by AC (Bristow *et al.*, 1990; Levitzki *et al.*, 1993). In human myocardial tissues, independent stimulation of either the β_1- or β_2-AR populations results in comparable intracellular increases in cAMP levels, despite the lower density of β_2-ARs (Kaumann *et al.*, 1989). This has been attributed to a tighter coupling between the β_2-AR and the G_s/AC system as compared to that between the β_1-AR and this system (Gille *et al.*, 1985; Green *et al.*, 1992; Levy *et al.*, 1993) and highlights the importance of the β_2-AR population in modulating cardiac function despite the lower myocardial tissue density of β_2-ARs.

In some cardiac tissues, β-ARs may couple not only with G_s, but with inhibitory G-proteins, such as G_i, as well, in a manner previously described in rat adipocytes (Chaudhry *et al.*, 1994). This has been demonstrated for the β_2-AR, but not the β_1-AR, in rat ventricular myocytes (Xiao *et al.*, 1995), where it was found that myocyte pretreatment with PTX enhanced contractility and the magnitude of I_{Ca} in response to β_2-adrenergic stimulation. Studies involving transgenic mice overexpressing the β_2-AR also indicate that both G_s and G_i may couple to this receptor population (Xiao *et al.*, 1999). Coupling of β_3-ARs to G_i in human ventricular tissue has also been suggested by the finding that PTX could largely abolish the negative inotropic effects of β_3-AR-selective stimulation (Gauthier *et al.*, 1996). The $G_{\beta\gamma}$ subunits of G_i can activate some isoforms of AC when heterologously expressed in COS cells (Federman *et al.*, 1992), which suggests the possibility of a further level of complexity in the interaction between the stimulatory G_s and the inhibitory G_i pathways, although the relevance of such an interaction in cardiac tissues is unknown.

Several studies have demonstrated the ability of M_2-AChR stimulation to antagonize β-adrenergic responses on ion currents in cardiac tissues (Koumi *et al.*, 1995a,b). Activation of G_i in response to M_2-AChR stimulation liberates a pool of free $G_{\beta\gamma}$ subunits, which are thought to scavenge activated $G_{\alpha s}$ that is formed in response to β-adrenergic stimulation, thereby competing with AC for the activated $G_{\alpha s}$ (Bristow *et al.*, 1990). The subsequent reduction in AC activity decreases intracellular cAMP concentrations and can explain the requirement for prior β-AR stimulation in order to observe cholinergically mediated inhibition of I_{Ks} and I_{Ca} in ventricular myocytes (Yazawa and Kameyama, 1990; Dascal, 1997) or the inhibition of I_f in Purkinje fibers (Chang *et al.*, 1987). Additionally, in guinea pig ventricular myocytes, ACh can reverse the stimulation of I_K after β-AR activation with isoprenaline but has no effect on I_K in the absence of the β-AR agonist (Harvey and Hume, 1989). Furthermore, cholinergic stimulation of M_2-AChRs antagonizes the β-AR-mediated reduction in I_{K1} (Koumi *et al.*, 1995a,b). AC, therefore, represents an important point of convergence between the adrenergic and cholinergic systems. The convergent role of AC in coupling neural input to I_{K1} attests to the intricacy and complexity of the signaling pathways that regulate K^+ channel function.

Regulation of cardiac K^+ channels may be mechanistically different from that of Ca^{2+} channels (Iijima *et al.*, 1990; Harada and Iijima, 1994), and signal transduction between β-ARs and cardiac ion channels is likely more complex than originally thought. For example, it appears that some cardiac ion channels, including Ca^{2+} (Brown *et al.*, 1989; Yatani and Brown, 1989) and Na^+ (Schubert *et al.*, 1989) channels, can be directly gated by G-protein subunits after β-AR activation. A direct interaction between G-protein subunits and I_K in guinea pig ventricular myocytes has also been suggested (Freeman *et al.*, 1992), and transfection of adult rat atrial myocytes with the β_1-subunit of a rat brain heterotrimeric G-protein resulted in a direct gating of $I_{K(ACh)}$ by $G_{\beta\gamma}$ in response to isoproterenol (Bender *et al.*, 1998). The mechanism of this cAMP-independent gating is analogous to that described previously for the cholinergically mediated direct interaction of $G_{i\beta\gamma}$ with $K_{(ACh)}$ channels. Bender *et al.* (1998) postulated that in native cardiac cells, promiscuous activation of $I_{K(ACh)}$ via $G_{\beta\gamma}$ liberated from G_s in response to β-AR stimulation is suppressed by sequestration of $G_{\beta\gamma}$ and that overexpression of $G_{\beta\gamma}$ may overcome this inhibition. Although the physiological relevance of this finding is unclear, $G_{s\beta\gamma}$-mediated activation may represent a further level of complexity in the β-adrenergic signaling pathway.

3.3.2. Role of cAMP-Dependent Protein Kinase

Protein kinase A (cAMP-dependent protein kinase; PKA) is a well-characterized effector of cAMP, and PKA activity in the heart increases in response to β-AR stimulation. Phosphorylation of cardiac K^+ channels by PKA has been implicated in the β-adrenergic effects on guinea pig I_K (Walsh *et al.*, 1989; Harvey and Hume, 1989; Yazawa and Kameyama, 1990; Walsh and Kass, 1991) and I_{K1} (Koumi *et al.*, 1995b), human I_{Kur} (Li *et al.*, 1996a), and rat I_{ss} (Scamps, 1996) and $I_{K(ACh)}$ (Kim, 1990). A PKA-dependent mechanism has also been proposed for the β-adrenergic augmentation of pinacidil-induced I_{KATP} in canine ventricular myocytes (Tseng and Hoffman, 1990). Conversely, it has been suggested that in feline ventricular myocytes, PKA is not involved in the stimulation of I_{KATP} amplitude in the absence of K_{ATP} channel openers (Schackow and Ten Eick, 1994). Although Kim (1990) demonstrated that PKA increased $I_{K(ACh)}$ in rat atrial myocytes, studies involving cloned K$^+$ channels expressed in oocytes have failed to confirm this finding (Lim *et al.*, 1995). It is therefore possible that PKA does not directly modulate the $K_{(Ach)}$ channels, but rather regulates other components of the signaling pathways that are not present in oocytes.

In guinea pig ventricular myocytes, I_{Ks} is increased in response to both PKA and protein kinase C (PKC) stimulation. These effects differ in terms of voltage and temperature dependence (Walsh and Kass, 1988, 1991) and suggest the existence of multiple regulatory sites on the K$^+$ channel that are differentially regulated by PKA and PKC. Furthermore, whereas PKA stimulation increases I_{Ks}, it decreases I_{Kr}. These and other issues have been further addressed through the use of cloned K$^+$ channels.

3.4. Correlation with Cloned Channel Currents

In guinea pig ventricular myocytes, β-adrenergic stimulation or direct activation of PKA is able to modulate both the I_{Ks} and the I_{Kr} components of the net delayed rectifier current. Native cardiac I_{Ks} is likely the result of the heteromultimeric expression of cloned IsK (minK) and KvLQT1, although expression of homomultimeric IsK channels results in a current with most of the characteristics of I_{Ks} (Wang and Goldstein, 1995; Tzounopoulos *et al.*, 1995; Tai *et al.*, 1997). Human, guinea pig, rabbit, rat, mouse, and cat IsK channels are modulated by PKA when heterologously expressed in *Xenopus* oocytes, and the resulting currents are increased after PKA activation (Kaczmarek and Blumenthal, 1997; Lo and Numann, 1998). Human IsK (hIsK) current is also independently modulated by PKC. Activation of PKC by high-dose phorbol ester resulted in a biphasic hIsK current response, with an initial increase in current, followed by a sustained decrease (Lo and Numann, 1998). Treatment with low-dose phorbol ester results in a monophasic increase in current. PKA and PKC modulation of IsK may be mutually exclusive, in that phosphorylation by either PKA or PKC appears to prevent subsequent modulation by the other. This is illustrated by the data in Fig. 5 and suggests the presence of multiple interacting phosphorylation sites on the IsK protein (Lo and Numann, 1998). Mechanistically, this may represent a novel form of channel regulation, in that IsK, and therefore I_{Ks}, may be modulated by either PKA (in response to β-adrenergic stimulation) or PKC (in response to α_1-adrenergic stimulation; see Section 4), but not by both concurrently. Under normal physiological conditions in human heart, β-adrenergic stimulation, and therefore PKA modulation, is likely to predominate, but α_1-AR-mediated activation of PKC may potentially be of increased importance in pathophysiological states such as heart failure, when the population of β-ARs is relatively downregulated.

In contrast to I_{Ks}, guinea pig ventricular I_{Kr} is decreased by PKA. I_{Kr} is thought to arise from the expression of the ether-a-go-go-related gene product, ERG, and heterologous expression of human ERG (HERG) in oocytes confirms that HERG current is reduced by PKA in a manner analogous to native I_{Kr} (Kiehn *et al.*, 1998).

Molecular correlates of $I_{K(ACh)}$ are also modulated by β-adrenergic stimulation. Coexpression of the GIRK1 clone with G_s and the β_2-AR in *Xenopus* oocytes has

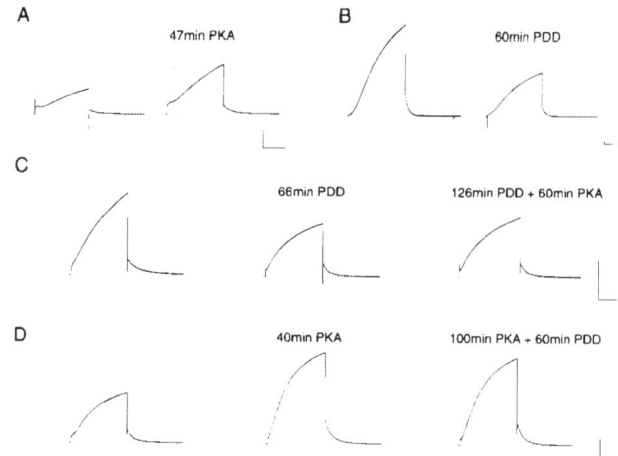

Figure 5. Effects of PKC and PKA activation on IsK (I_{Ks}) currents recorded from oocytes injected with cRNA for either human (hIsK) or cat (cIsK) IsK. (A) hIsK current from a representative oocyte was increased in the presence of 20 μM forskolin plus 500 μM 3-isobutyl-1-methylxanthine (IBMX) to activate PKA. (B) PKC activation by a 60-min application of 100 nM phorbol 12,13-didecanoate (PDD) resulted in a decrease in hIsK current. (C) Recordings from another oocyte show that a 60-min application of PDD decreased cIsK current and that subsequent application of PKA activators had little additional effect. (D) Application of PKA activators increased hIsK current 2-fold, but subsequent application of PDD for 60 min had no additional effect. Scale bars = 500 nA and 2 s. (Reproduced, with permission, from Lo and Numann, 1998.)

indicated that, in this expression system at least, $G_{\beta\gamma}$ liberated in response to $G_{\alpha s}$ activation can directly gate GIRK1 (Lim *et al.*, 1995). cAMP-dependent signaling did not appear to be involved, as direct elevation of cAMP levels had no effect on homomultimeric GIRK1 currents. Furthermore, in rat atrial myocytes transfected with a G-protein β-subunit cloned from rat brain, isoprenaline could increase the magnitude of $I_{K(ACh)}$ through a G_s-coupled mechanism that involved the direct interaction of $G_{\beta\gamma}$ with the $K_{(Ach)}$ channels (Bender *et al.*, 1998). These results differ from the previous finding that β-AR stimulation with isoproterenol activated native $I_{K(ACh)}$ through a PKA-dependent mechanism in rat atrial cells (Kim, 1990). It is feasible, however, that the stimulation of native $I_{K(ACh)}$ observed in response to PKA activation is the result of the regulation of other components of either the β-adrenergic or cholinergic pathways that are not expressed in oocytes. Alternatively, coexpression of GIRK4 may be required for regulation by PKA to occur. Therefore, the physiological relevance of $I_{K(ACh)}$ modulation by β-AR stimulation in intact cardiac tissues and the potential role of PKA in mediating this effect remain unresolved.

4. α_1-ADRENERGIC RESPONSES

Although experimental evidence unequivocally confirms the importance of β-AR stimulation in the regulation of inotropy and chronotropy, the positive inotropic response to α-AR stimulation is more variable, with both tissue and species dependence. In some tissues, the α-adrenergic positive inotropic response has been demonstrated to be as much as 60% of that of the β-AR-mediated effects (Hescheler et al., 1988; Tiaho et al., 1993). However, the role of α-AR modulation of cardiac function becomes of increasing importance pathophysiologically when the population of α-ARs is upregulated relative to the level of β-ARs, such as during heart failure (Bristow et al., 1988) or in response to ischemia (Corr et al., 1981; Molina-Viamonte et al., 1991). In addition, α-ARs have been implicated in the pathogenesis of cardiac dysfunction, including ventricular hypertrophy and ischemia-induced arrhythmias (Ikeda et al., 1991; Kurz et al., 1991).

4.1. Effects of α_1-AR Stimulation on Cardiac Function and K$^+$ Currents

The effects of α_1-AR stimulation on cardiac inotropy, and the effects on individual K$^+$ currents, have been described in detail (Terzic et al., 1993; Fedida, 1993). However, the transduction pathways involved in the α_1-adrenergic modulation of cardiac K$^+$ currents are not as clearly defined. There is general consensus on which signal transduction mechanisms are activated following α_1-AR stimulation, but it has been difficult to show that these pathways are indeed responsible for channel modulation. Only recently has experimental evidence reconciled subcellular signaling mechanisms with regulation of channel function.

Stimulation of α_1-ARs, either autonomically or pharmacologically, results in a positive inotropic effect in the majority of cardiac preparations studied, including human (Schumann et al., 1978; Bruckner et al., 1984; Schmitz et al., 1987), rat (Wenzel and Su, 1966; Fedida and Bouchard, 1992), cat (Hartmann et al., 1988), and rabbit (Benfey and Varma, 1967; Skomedal et al., 1990; Fedida et al., 1990; Jahnel et al., 1992) preparations. Conversely, in canine (Endoh et al., 1978), ferret (Endoh et al., 1989), or guinea pig (Hescheler et al., 1988; Koga et al., 1989) ventricle, there is a small or nonexistent α_1-adrenergic positive inotropic response. Indeed, some studies involving human heart have also failed to demonstrate a sizable positive inotropy in response to α_1-AR stimulation (Jakob et al., 1988). The variability of these effects is partially attributable to the substantial variability in α_1-AR density between species. α_1-AR density is high in rat, but lower in guinea pig and dog, which correlates well with the observed inotropic responses to α_1-AR stimulation. Conversely, feline cardiomyocytes display robust α_1-adrenergic responses despite a relatively low α_1-AR density (Mukherjee et al., 1983; Hartmann et al., 1988), suggesting that interspecies differences in β_1-AR density alone is not sufficient to explain the observed variability (Fedida, 1993). Therefore, differences in receptor subtypes, G-proteins, effector mechanisms, or ion channels themselves may also contribute to the diversity of α_1-adrenergic effects.

α_1-Adrenergic inotropy is mediated through several mechanisms, including modulation of I_{Ca} and an apparent sensitization of ventricular myofilaments to Ca^{2+} (Fedida, 1993). However, α_1-AR stimulation modulates several K$^+$ conductances in the heart, and this likely constitutes a major mechanism of positive inotropy. At least four membrane K$^+$ currents are reduced in response to α_1-AR stimulation. These include the transient outward current, I_{to}, and components of I_K (including I_{Kur}, I_{K1}, and $I_{K(Ach)}$.

Table 2
β_1-Adrenergic Effects on Native and Cloned Cardiac K^+ Currents

Current	Species tissue	Effect of β-AR stimulation on current	Mechanism	Reference(s)
I_{to}	Rabbit atrium	↓	PTX-insensitive G-protein	Fedida et al., 1989; Braun eta l., 1990
	Rat ventricle	↓	Likely PKC	Apkon and Nerbonne, 1988; Wang et al., 1991
	Cloned Kv4.2[a]	↓	PKC	Nakamura et al., 1997
I_K	Rat ventricle	↓		Ravens et al., 1989
I_{Kur}	Human atrium	↓	PKC	Li et al., 1996a
	Rat atrium	↓	Effects mimicked by DAG analog (∴ ?PKC)	Van Wagoner et al., 1996
	Cloned Kv1.5		Current decreased by PKC in oocytes or COS cells	Tseng et al., 1997; Vogalis et al., 1995
I_{Ks}	Guinea pig ventricle	↑	PKC	Tohse et al., 1987, 1992
		↑	Calmodulin-dependent mechanism, not involving phosphorylation by PKA, PKC, or CaMKII (?direct modulation by Ca^{2+}-calmodulin)	Nitta et al., 1994
	Cloned gpISK		PKC increases gpIsK current in oocytes[b]	Varnum et al., 1993
	Cloned hIsK, mIsK	↓	PKC	Honoré et al., 1991; Tseng et al., 1997
	Cloned hIsK[c]	↑	Ca^{2+}-calmodulin-dependent effectors	Tseng et al., 1992
I_{K1}	Rabbit atrium	↓	PTX-insensitive G-protein	Braun et al., 1992
	Rabbit ventricle	↓	PTX-insensitive G-protein	Fedida et al., 1991
	Canine Purkinje fibers	↓	Increased Na^+/K^+ pump current; PTX-sensitive pathway	Shah et al., 1988
$I_{K(ACh)}$	Rabbit atrium	↓	PTX-insensitive G-proatein	Braun et al., 1992

[a] Coexpression of Kv4.2 and $\alpha_{1A/c}$-AR in oocytes.
[b] Effects dependent on presence of asparagine residue at position 102 (mutation to serine results in gpIsK current that is decreased by PKC).
[c] Coexpression of hIsK and $\alpha_{1A/c}$-AR in oocytes.

However, at least one K^+ current, I_{Ks} recorded from guinea pig ventricle, is increased in response to α_1-AR stimulation (Tohse et al., 1992). The effects of α_1-adrenergic stimulation on native and cloned cardiac K^+ currents are summarized in Table 2.

α_1-Adrenergic blockade of K^+ currents in the heart has several important consequences. First, stimulation of cardiac α_1-ARs leads to a prolongation of the APD. This effect is observed in several cardiac preparations, including isolated rat ventricular myocytes (Apkon and Nerbonne, 1988; Ravens et al., 1989; Vogel and Terzic, 1989; Jahnel et al., 1991; Fedida and Bouchard, 1992), sinoatrial nodal cells (Satoh and Hashimoto, 1988), and Purkinje fibers (Giotti et al., 1973) and in muscle preparations from rabbit atrium (Miura and Inui, 1984; Fedida et al., 1989, 1990; Jahnel et al., 1992) or ventricle (Hescheler *et al.*, 1988; Jahnel *et al.*, 1992). Blockade of K$^+$ currents responsible for early (I_{to}) and late (I_K, I_{K1}) components of repolarization prolongs the plateau phase (phase 2) of the cardiac action potential. This delays the deactivation of L-type Ca^{2+} channels, allowing an increased influx of Ca^{2+}, which likely contributes to positive inotropy (Fedida, 1993). Second, although there appears to be little effect of α_1-AR stimulation on the sinoatrial node (Hewett and Rosen, 1985), α_1-AR-mediated modulation of cardiac K$^+$ conductances affects latent pacemaker activity (Sheridan, 1986; Terzic *et al.*, 1993). Therefore, this effect likely plays only a minor role physiologically, but it becomes of increased importance in the presence of cardiac disease states, including myocardial ischemia. Third, excessive α_1-adrenergic stimulation, either autonomically or pharmacologically, can be arrhythmogenic, and this may be synergistic with β-adrenergic proarrhythmic effects (Priori *et al.*, 1993). For example, phenylephrine induces delayed afterdepolarizations and an increase in triggered activity in hypoxic myocytes (Priori *et al.*, 1991). The potential role of α_1-adrenergic modulation of K$^+$ currents in this process is still unclear, but it has been suggested to contribute to the development of malignant arrhythmias and sudden death (Kurz *et al.*, 1991). The effects of α_1-AR stimulation on individual cardiac K$^+$ currents are considered in greater detail in the following sections.

4.1.1. Effects on I_{to}

I_{to} is the most prominent outward current in human ventricular myocytes (Wettwer *et al.*, 1993, 1994; Murray *et al.*, 1994; Chonmaitree and Heikkinen, 1997) and has also been described in rat (Josephson *et al.*, 1984; Apkon and Nerbonne, 1991; Wettwer *et al.*, 1993), rabbit (Jahnel *et al.*, 1994), cat (Schackow *et al.*, 1995), dog (Simurda *et al.*, 1989), ferret (Campbell *et al.*, 1993), and mouse (Benndorf *et al.*, 1987) ventricular cells. I_{to} is also prominent in atrial tissue from humans (Le Grand *et al.*, 1992; Mansourati and Le Grand, 1993; Crumb *et al.*, 1995), rabbits (Delpón *et al.*, 1992; Qi *et al.*, 1994), and dogs (Yue *et al.*, 1996a,b). In contrast, I_{to} is virtually absent in guinea pig cardiac preparations. The distribution of I_{to} correlates well to the tissues and species that exhibit positive inotropy to α_1-AR agonists. Furthermore, in rabbit (Endoh and Schumann, 1975) and rat (Capogrossi *et al.*, 1991) at least, the positive inotropic response to α_1-AR agonists disappears at stimulation frequencies above 2 Hz, where the rate-sensitive I_{to} is also largely inactivated. A similarly rate-dependent positive inotropy is observed in human subjects in response to α_1-adrenergic stimulation (Curiel *et al.*, 1989; Landzberg *et al.*, 1991).

I_{to} is largely responsible for the rapid but incomplete repolarization characteristic of phase 1 of the cardiac action potential and likely also helps to determine the duration of phase 2, and therefore the APD as a whole, by setting the initial plateau potential. Not surprisingly, α_1-adrenergic block of I_{to} has been demonstrated to prolong the APD in several preparations, including isolated rabbit atrial (Fedida *et al.*, 1989) and rat ventricular (Tohse *et al.*, 1990; Fedida and Bouchard, 1992) myocytes, as illustrated by the data in Fig. 6. However, in rat atrial myocytes, α_1-AR stimulation with 50 μM phenylephrine only weakly inhibits I_{to} (by $\sim 10\%$) (Ertl *et al.*, 1991; Van Wagoner *et al.*, 1996). This may either reflect an α_1-AR subtype-specific effect (Wang *et al.*, 1991) or tissue-specific differences in signal transduction coupling.

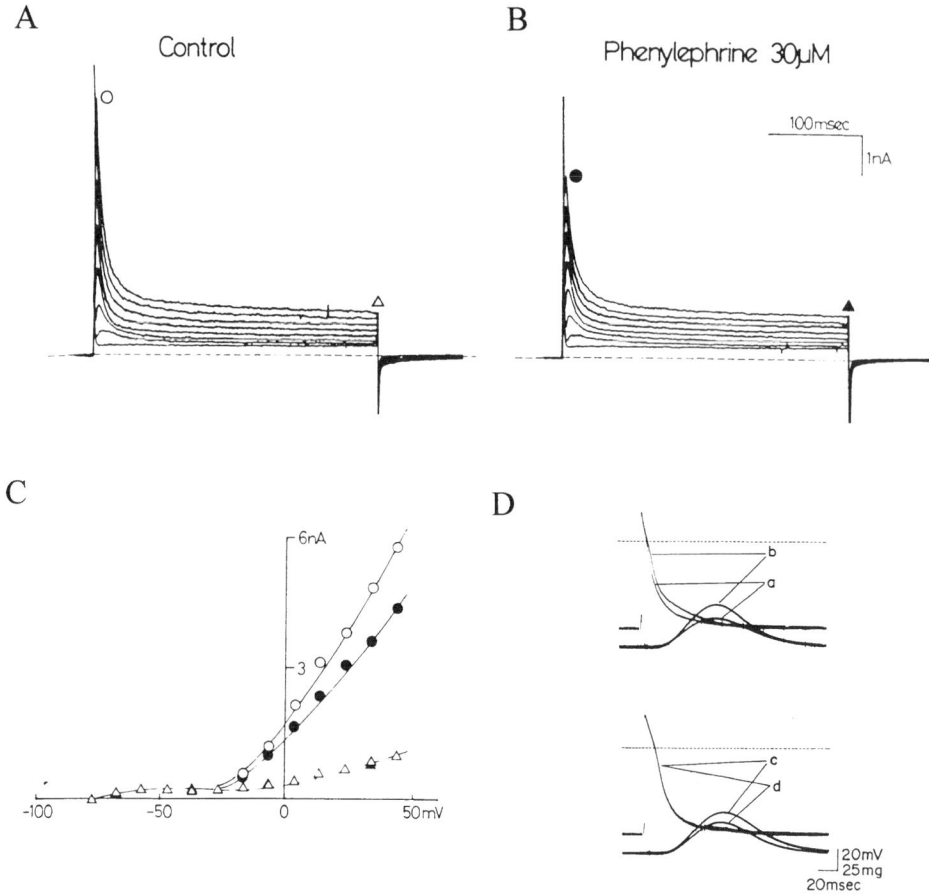

Figure 6. Effects of phenylephrine on I_{to} and on the action potential duration (APD) and contractile tension in isolated rat ventricular cells. (A) and (B) Control current recordings obtained in a Na^+-free external solution (A) and in the presence of 30 μM phenylephrine (B). I_{to} was elicited by test pulses of 300-ms duration from -27 mV to $+43$ mV in 10-mV increments. The holding potential was -77 mV. Phenylephrine decreased the peak outward current, relative to control, at every potential. The sustained current at the end of the pulse was not changed by phenylephrine. The dashed line indicates zero current level. (C) Current–voltage relationship from a representative myocyte from a holding potential of -77 mV. Open circles indicate the peak current (I_{to}) in control solution, and closed circles indicate I_{to} amplitude in the presence of 30 μM phenylephrine. Triangles represent the current magnitude at the end of the 300-ms test pulse in control solution ($\triangle$) and in the presence of 30 μM phenylephrine ($\blacktriangle$). (D) Changes in APD and contractile tension induced by 30 μM phenylephrine in the absence (*top*) and presence (*bottom*) of 1.5 mM 4-aminopyridine (4-AP) to block I_{to}. Propranolol (1 μM was continuously present. Superimposed action potentials and twitch curves are shown. In the absence of 4-AP, application of 30 μM phenylephrine prolonged the APD and resulted in an increase in contractile force (tracings labeled b) relative to control (tracings labeled a). However, in the presence of 4-AP, 30 μM phenylephrine did not change the APD and failed to produce a positive inotropic effect (tracings labeled d) relative to control (tracings labeled c). In fact, a negative inotropic response was instead observed, possibly due to attenuation of the positive inotropic component in the presence of 4-AP. (Reproduced and modified, with permission, from Tohse *et al.*, 1990. Copyright 1990 Springer-Verlag.)

4.1.2. Effects on Components of I_K

A number of components of I_K in different species are modulated by α_1-AR stimulation. In adult rat ventricular myocytes, I_K was reduced by the α-agonist phenylephrine (Ravens *et al.*, 1989). More recently, I_{Kur} in human (Li *et al.*, 1996a) and rat (Van Wagoner *et al.*, 1996) atrial cells was also demonstrated to be blocked by phenylephrine. Interestingly, in guinea pig ventricle, phenylephrine (10–30 μM) increased the magnitude of I_{Ks} (Tohse *et al.*, 1987, 1992) suggesting that the underlying signal transduction mechanisms or coupling of receptor to channel may be tissue-specific. This, coupled with the fact that guinea pig ventricular myocytes lack I_{to}, has been suggested as a plausible mechanism to explain why the APD is shortened, rather than lengthened, in guinea pig ventricular cells (Tohse *et al.*, 1992).

4.1.3. Effects on the Inwardly Rectifying I_{K1}

The inwardly rectifying I_{K1} has been well described in human (Heidbuchel *et al.*, 1990), guinea pig (Hume and Uehara, 1985), and rabbit (Giles and Imaizumi, 1988) atrial cells. It is also present in the ventricular myocytes of humans (Chonmaitree and Heikkinen, 1997), dogs (Tseng *et al.*, 1987), rats (Josephson and Brown, 1986), guinea pigs (Hume and Uehara, 1985) and rabbits (Giles and Imaizumi, 1988). I_{K1} plays a prominent role in determining the diastolic membrane potential and contributes to the repolarization characteristic of phase 3 of the cardiac action potential (Giles and Imaizumi, 1988). In rabbit ventricular myocytes, the α_1-AR agonist methoxamine, in the presence of propranolol to block β-ARs and 4-aminopyridine (4-AP) to block I_{to}, reduces the magnitude of I_{K1} in both the inward and outward directions (Fedida *et al.*, 1991). Methoxamine thereby depolarizes the diastolic membrane potential and prolongs the APD, in a dose-dependent, reversible manner. A similar reduction in I_{K1} in rabbit atrial myocytes has also been suggested (Braun *et al.*, 1992). It should be noted, however, that the prolongation of the APD as a result of I_{K1} block by α_1-AR agonists in these tissues was relatively minor when compared with the marked lengthening of the APD when 4-AP was omitted, reinforcing the important role of I_{to} in α_1-adrenergic modulation of cardiac function (Fedida *et al.*, 1991).

4.1.4. Effects on $I_{K(ACh)}$

$I_{K(ACh)}$, the inwardly rectifying K$^+$ current activated by muscarinic agonists as described above, is also regulated by α_1-AR stimulation. Braun *et al.* (1992) have demonstrated that in isolated rabbit atrial myocytes superfused with 4-AP, the APD is markedly shortened by ACh due to activation of $I_{K(ACh)}$. However, when 200 μM methoxamine was added, in the presence of ACh and 4-AP, the APD was actually lengthened relative to control, suggesting that, in this tissue at least, $I_{K(ACh)}$ is reduced by α_1-AR stimulation. Conversely, however, Kurachi *et al.* (1989b) demonstrated that in guinea pig atria, α_1-AR agonists activate $I_{K(ACh)}$. These species-dependent effects may be explained, in part, by differences in the signal transduction mechanisms (see Section 4.3).

4.2. α_1-AR Subtypes Present in the Heart

α-ARs in the mammalian myocardium are predominantly of the $\alpha 1$ subtype (Ruffolo and Hieble, 1994; Graham *et al.*, 1996). Based on receptor ligand studies, α_1-ARs were initially classified into α_{1A} and α_{1B} subtypes (Morrow and Creese, 1986). Subsequently, three genes encoding α_1-ARs were cloned and denoted $\alpha_{1a/d}$, α_{1b}, and α_{1c} (Graham *et al.*, 1996). This complicated the initial pharmacological classification and led to significant confusion regarding nomenclature. It is now accepted, however, that the cloned α_{1b} is identical to the pharmacologically defined α_{1B}, and it is now designated α_{1B} (Hieble *et al.*, 1995; Graham *et al.*, 1996). The $\alpha_{1a/d}$ clone (Schwinn *et al.*, 1990; Lomasney *et al.*, 1991; Perez *et al.*, 1991) was recognized as a novel subtype not previously identified by either pharmacological sensitivity or radioligand binding. The $\alpha_{1a/d}$ clone is now designated α_{1D} (Hieble *et al.*, 1995). The gene encoding α_{1c} corresponds to the pharmacologically defined α_{1A} (Laz *et al.*, 1994; Perez *et al.*, 1994; Rokosh *et al.*, 1994), and receptors of this subtype are therefore designated either as α_{1A} (Hieble *et al.*, 1995) or alternatively as $\alpha_{1A/c}$ (Graham *et al.*, 1996) to reflect the initial designation of the cloned channel. This latter convention will be used here. In both human and rat heart, both $\alpha_{1A/c}$ and α_{1B} subtypes are present (Graham *et al.*, 1996), with $\alpha_{1A/c}$ predominating in human heart (Price *et al.*, 1994a) and α_{1B} predominating in rat heart (Price *et al.*, 1994b). Although some studies have detected low expression of α_{1D}-ARs in human and rat myocardium (Stewart *et al.*, 1994; Price *et al.*, 1994a,b), other studies have failed to support this finding (Weinberg *et al.*, 1994), and the role of this receptor population, if present, remains unclear.

4.3. Signal Transduction Mechanisms

In contrast to the signal transduction mechanisms that underlie muscarinic and β-AR stimulation, the subcellular pathways and effector mechanisms involved in α_1-AR modulation of ion channel function remain somewhat of an enigma. Study of α_1-AR signal transduction is complicated by several factors, including the lack of specific pharmacological inhibitors for downstream signaling elements, such as protein kinases. α_1-ARs likely couple multiple G-proteins (so-called promiscuous coupling) with differing affinities, and each G-protein likely has multiple downstream effectors. In addition, expression of components of the α_1-adrenergic signaling pathways in different tissues varies with developmental age or in the presence of cardiac disease states (Del Balzo *et al.*, 1990; Kurz *et al.*, 1991; Anyukhovsky *et al.*, 1992; Rybin and Steinberg, 1994). However, a generalized schematic of the important signal transduction mechanisms involved in the α_1-adrenergic modulation of cardiac K^+ channels is presented in Fig. 7.

4.3.1. G-Proteins

In vitro expression systems have demonstrated that $\alpha_{1A/c}$-ARs can couple to G_q, G_{11}, and G_{14}, whereas α_{1B}-ARs can couple to G_q, G_{11}, G_{14}, and G_{16} (Graham *et al.*, 1996). Of these, however, only G_q and G_{11} are expressed in adult mammalian myocardium (Hansen *et al.*, 1995). α_{1B}-ARs may also couple to a novel high-molecular-weight (~ 74 kDa), PTX-insensitive G-protein, denoted G_h. G_h is distinct from the

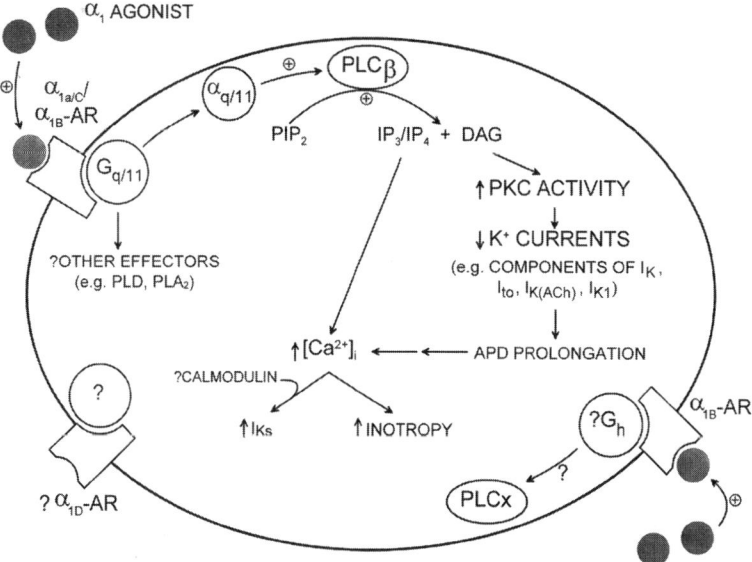

Figure 7. Schematic of α_1-AR-mediated effects on intracellular signal transduction mechanisms and cardiac K$^+$ currents. PLCβ, β-isoform of phospholipase C; PIP$_2$, phosphatidylinositol-4,5-bisphosphate; IP$_3$, inositol-1,4,5-trisphosphate; IP$_4$, inositol-1,3,4,5-tetrakisphosphate; DAG, diacylglycerol; PLD, phospholipase D; PLA$_2$, phospholipase A$_2$; PLCx, novel isoform of PLC; PKC, protein kinase C; APD, action potential duration. Other abbreviations as described in the text.

heterotrimeric G-proteins in that it exists as a heterodimer and possesses both transglutaminase and receptor-signaling functions (Im and Graham, 1990; Im *et al.*, 1990). G$_h$ has been detected in human, rat, and bovine heart (Nakaoka *et al.*, 1994), and α_1-adrenergic stimulation of rabbit atrial or ventricular myocytes activates a high-affinity, PTX-insensitive GTPase consistent with G$_h$ (Braun and Walsh, 1993). Further work, however, is required to elucidate the exact role, if any, of G$_h$ in modifying K$^+$ currents in these and other preparations.

In the majority of cardiac tissues studied, α_1-AR-mediated effects on K$^+$ conductances are transduced initially by members of the G$_{q/11}$ subfamily. In rabbit atrial myocytes, reduction of I_{to}, $I_{K(ACh)}$, and I_{K1} after α_1-AR stimulation was not inhibited by PTX, and internal dialysis of rabbit atrial myocytes with the nonhydrolyzable GTP analog GTPγS produced an irreversible block of I_{to} after α_1-AR stimulation. These observations support a role of G$_{q/11}$. However, in some cardiac tissues at least, such as canine Purkinje fibers, α_1-AR effects on I_{K1} appeared to be transduced via a PTX-sensitive G-protein (Shah *et al.*, 1988).

4.3.2. Role of Phospholipases (PLC/PLA$_2$)

The β-isozyme of phospholipase C (PLCβ) is a well-described effector of G$_{q/11}$, and its activation by G$_{q/11}$ results in the cleavage of the membrane phospholipid phosphatidylinositol 4,5-bisphosphate (PIP_2) to form the second messengers inositol-1,4,5-trisphosphate (IP_3), inositol 1,3,4,5-tetrakisphosphate (IP_4), and diacylglycerol (DAG). IP_3, IP_4, and potentially other inositol polyphosphates are generated in response to α_1-AR stimulation in rat heart (Scholz et al., 1988, 1992). Inositol phosphates interact with specific receptors on sarcoplasmic reticulum to release stored Ca^{2+}, thus increasing the intracellular Ca^{2+} concentration ($[Ca^{2+}]_i$) (Berridge, 1993). DAG is a potent activator of PKC, and DAG accumulation in cardiac tissues has been demonstrated in response to α_1-AR stimulation (Bordoni et al., 1991).

Interestingly, there is more recent evidence for the existence of a novel isoform of PLC that is expressed in heart (Das et al., 1993; Baek et al., 1993). This 69-kDa isoform, which has not yet been cloned, is activated by G_h (Das et al., 1993; Nakaoka et al., 1994) and may represent an important intracellular effector coupling α_1-ARs to PLC, via G_h, independent of $G_{q/11}$ activation.

In some tissues, PLA_2 is activated by α_1-ARs, resulting in the generation of free AA and its derivatives, including prostaglandins, leukotrienes, and thromboxanes (Axelrod et al., 1988). AA can activate PKC through a DAG-independent pathway that does not require membrane phospholipids and does not result in membrane translocation of PKC (Khan et al., 1992, 1995). In guinea pig atria, activation of $I_{K(ACh)}$ in response to α_1-AR stimulation is blocked by inhibitors of the AA/leukotriene pathways, but not by inhibitors of prostaglandin synthesis (Kurachi et al., 1989c). Thus, a modulatory role of AA or its derivatives on ion channel function has been proposed (Kurachi et al., 1989b; Yatani et al., 1990). The potential for couplings of phospholipase D (PLD) to α_1-ARs has also been suggested (Thompson et al., 1991; Terzic et al., 1993). PLD activation can activate PKC by generating phosphatidic acid, which is metabolized to DAG. However, it must be noted that direct evidence for a role of PLA_2 or PLD pathways in cardiac tissues is still lacking.

4.3.3. Role of Increased $[Ca^{2+}]$

In most tissues, PLC activation results in the formation of inositol derivatives, including IP_3 and IP_4, which release Ca^{2+} from sarcoplasmic stores (Berridge, 1993). However, the role of inositol derivatives in cardiac muscle may be somewhat different, and IP_3 may increase $[Ca^{2+}]_i$ levels by additional mechanisms (Zhu and Nosek, 1991; Kijima et al., 1993). Whereas some authors have demonstrated a direct release of Ca^{2+} from the cardiac sarcoplasmic reticulum in response to IP_3 (Kentish et al., 1990; Vites and Pappano, 1990; Gilbert et al., 1991), others have failed to support this finding and have proposed that IP_3 may merely act to facilitate Ca^{2+}-induced Ca^{2+} release (Movsesian et al., 1985; Fabiato, 1992).

Although the exact role of inositol phosphates in cardiac preparations remains controversial, it is widely agreed that $[Ca^{2+}]_i$ increases in response to α_1-AR stimulation. In addition to the inositol phosphate-related mechanisms described above, this increase may also result from delayed Ca^{2+} channel deactivation, secondary to K^+ current inhibition, which allows a greater influx of Ca^{2+} during the plateau phase of the cardiac action potential. This has been proposed as a prominent mechanism of positive inotropy in cardiac tissues (Fedida and Bouchard, 1992).

In isolated membrane patches from guinea pig ventricular myocytes, a component of I_K (consistent with I_{Ks}) was increased by elevation of $[Ca^{2+}]_i$ through a staurosporine-insensitive, calmodulin-dependent mechanism that did not involve phosphorylation of the channel by PKA, PKC, or calmodulin-dependent protein kinase type II (CaMKII) (Nitta et al., 1994). It was instead proposed that Ca^{2+}-calmodulin directly altered channel behavior. This effect was demonstrated at physiological concentrations of Ca^{2+}, which may suggest that the increase in $[Ca^{2+}]_i$ that occurs after α_1-AR stimulation may contribute to the shortening of the APD in guinea pig ventricle through an increase in the magnitude of repolarizing K$^+$ currents.

4.3.4. Role of PKC

PKC, activated by DAG in response to α_1-AR stimulation, is translocated to the plasma membrane, where it acts to phosphorylate a number of intracellular substrates, including ion channels (Henrich and Simpson, 1988; Mochly-Rosen et al., 1990; Talosi and Kranias, 1992). Alternatively, DAG-independent activation of PKC by AA does not involve membrane translocation and may allow a degree of substrate selection through a lack of membrane targeting.

There has, however, been conflicting evidence for a role of PKC in the modulation of cardiac K$^+$ conductances. It has been suggested that PKC activation contributes to the reduction of transient and sustained outward K$^+$ currents recorded from rat ventricular myocytes in response to α_1-AR stimulation (Apkon and Nerbonne, 1988). This is supported by the demonstration that activation of PKC by phorbol 12-myristate 13-acetate (PMA) reduces I_{to} in rat ventricular myocytes (Nakamura et al., 1997). Conversely, PKC activation was found to potentiate I_{Ks} in guinea pig ventricular myocytes. There have been conflicting reports that this increase is a result of a negative shift in the voltage dependence of activation, although such an effect was not observed in a study involving homomultimeric hIsK channels, the putative molecular correlate of I_{Ks}, expressed in Xenopus oocytes (Lo and Numann, 1998). These variations may reflect species-specific differences in the response to PKC or, alternatively, may indicate a role for KvLQT1 in mediating this effect.

PKC does not appear to be involved in the α_1-AR-mediated reduction of I_{to}, I_{K1}, or $I_{K(ACh)}$ from rabbit atrial myocytes (Braun et al., 1990, 1992). However, PKC activation is likely responsible for the reduction of I_{Kur} observed in human atrial myocytes after stimulation with phenylephrine (Li et al., 1996a). More recent work involving expression of cloned α_1-AR receptor subtypes, K$^+$ channels, and components of the putative signal transduction pathways has sought to clarify these results and is described in greater detail below.

4.4. Correlation with Cloned K$^+$ Channel Currents

Heterologous expression of cloned K$^+$ channels and α_1-AR subtypes has sought to further clarify the subcellular mechanisms of α_1-adrenergic signal transduction, and a growing body of evidence now supports a role of PKC in mediating α_1-AR-mediated effects on the molecular correlates of some cardiac K$^+$ currents. For example, phenylephrine reduces Kv4.2 current in *Xenopus* oocytes coexpressing Kv4.2 and the $\alpha_{1A/c}$-AR (Nakamura *et al.*, 1997). This effect was mimicked by activation of PKC by PMA and was largely prevented by the PKC inhibitor chelerythrine. The reduction of Kv4.2 and Kv4.3 currents in response to PKC activation by PMA is illustrated by the data in Fig. 8. Kv4.2 and Kv4.3 are thought to be key components of I_{to} in human and rat ventricular myocytes (Dixon *et al.*, 1996; Tsaur *et al.*, 1997; Nakamura *et al.*, 1997), and these findings are therefore in agreement with the demonstration that PKC is involved in the α_1-adrenergic reduction of native I_{to} in rat ventricular cells (Nakamura *et al.*, 1997).

Other cloned K^+ channels that are present in cardiac tissues (Barry and Nerbonne, 1996), including Kv1.2, Kv1.4, and Kv1.5, may also be modulated by PKC. Coexpression of Kv1.2 or Kv1.4 with the $\alpha_{1A/c}$-AR in *Xenopus* oocytes resulted in a slow reduction of current in response to phenylephrine, which was prevented by prazosin or the PKC inhibitor staurosporine, but which could be mimicked by the PKC activator 12-*O*-tetradecanoylphorbol 13-acetate (TPA) (Tseng *et al.*, 1997). The exact roles of Kv1.2 and Kv1.4 in cardiac tissues are uncertain, although it has been suggested that Kv1.2 may contribute to the delayed rectifier current in some cardiac tissues (Barry and Nerbonne, 1996). Phorbol esters, which activate PKC, reduced Kv1.5 currents expressed in *Xenopus* oocytes (Tseng *et al.*, 1997) or COS cells (Vogalis *et al.*, 1995). Kv1.5 likely underlies I_{Kur} in human and rat atrial myocytes (Barry and Nerbonne, 1996), and these results are therefore in agreement with the observation that activation of PKC is involved in the reduction of native I_{Kur} from atrial cells (Van Wagoner *et al.*, 1996; Li *et al.*, 1996a).

IsK (minK) contributes to cardiac I_{Ks} (Barry and Nerbonne, 1996; Tseng *et al.*, 1997), and phenylephrine could increase the magnitude of hIsK current when hIsK was coexpressed with the $\alpha_{1A/c}$-AR in *Xenopus* oocytes. Interestingly, TPA *decreased* the hIsK current, which is in agreement with results obtained in other studies that have used phorbol 12,13-didecanoate to activate PKC (Lo and Numann, 1998). However, the effects of phenylephrine could be largely antagonized by intracellular EGTA (30 μM) or the calmodulin antagonist W-7 (Tseng *et al.*, 1997). This supports the work of Nitta *et al.* (1994), which demonstrated that native I_{Ks} is stimulated by increases in $[Ca^{2+}]_i$ and may suggest that regulation of native I_{Ks} depends on the balance between inositol phosphates and DAG generated in response to α_1-AR stimulation (Tseng *et al.*, 1997). IP_3 would increase $[Ca^{2+}]_i$ and, through calmodulin, could regulate the activity of a number of intracellular effectors or, alternatively could directly modify channel behavior through an allosteric interaction (Nitta *et al.*, 1994). DAG, conversely, could activate PKC and inhibit I_{Ks}.

These results, however, do not account for the apparent increase in I_{Ks} magnitude that was observed in guinea pig ventricular myocytes in response to PKC activation. It has been suggested that the amino acid sequence surrounding PKC phosphorylation consensus sites on the IsK protein is a major determinant of whether phosphorylation of this channel by PKC results in an increase or a decrease in current magnitude (Busch *et al.*, 1992; Zhang *et al.*, 1994). The guinea pig IsK clone, whose current is increased by PKC activation (Varnum *et al.*, 1993; Zhang *et al.*, 1994), differs in this region when compared to the human, rat, or mouse IsK clones, whose currents are all decreased by PKC (Honoré *et al.*, 1991; Busch *et al.*, 1992; Tseng *et al.*, 1997). The amino acid residue at position 102 plays an important role in determining the effects of PKC. The guinea

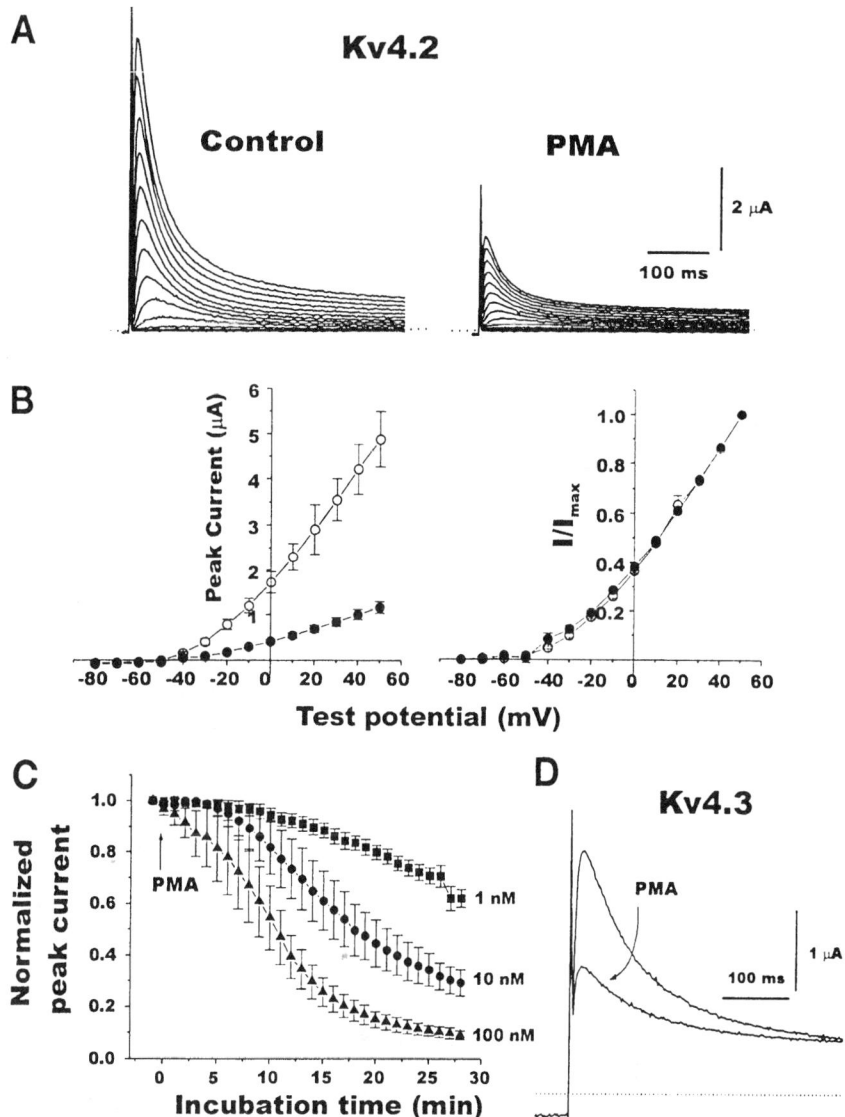

Figure 8. Effects of phorbol esters on Kv4.2 and Kv4.3 currents expressed in oocytes. (A) Kv4.2 currents before and 30 min after exposure to 10 n*M* phorbol 12-myristate 13-acetate (PMA), as indicated. Currents were elicited in response to a series of 900-ms depolarizing voltage steps from −80 mV to +50 mV, in 10-mV increments, from a holding potential of −110 mV, at a frequency of 0.07 Hz. Dotted lines indicate zero current level. PKC activation by PMA resulted in a decrease in Kv4.2 current at all test potentials studied. (B) Mean current–voltage relationship of peak Kv4.2 current amplitude (*left*) or normalized current–voltage relationship (*I*/*I*$_{max}$; *right*, normalized values obtained at +50 mV) before (○) and 30 min after application of 10 n*M* PMA (●). Currents were recorded in response to an identical voltage-step protocol to that described in (A). Data are expressed as means ±SE (*n* = 10 for each group). (C) Time course of effect of PMA (1, 10, and 100 n*M* as indicated) on Kv4.2 currents. Peak current amplitude was normalized to current amplitude before PMA application (time 0). Results are expressed as means ±SE (*n* = 6 for each group). (D) Effects of 10 n*M* PMA on Kv4.3 current, recorded at +20 mV from a holding potential of −110 mV, after 30-min exposure to PMA. Kv4.3 current was also reduced by PKC in response to PMA activation. (Reproduced, with permission, from Nakamura *et al.*, 1997)

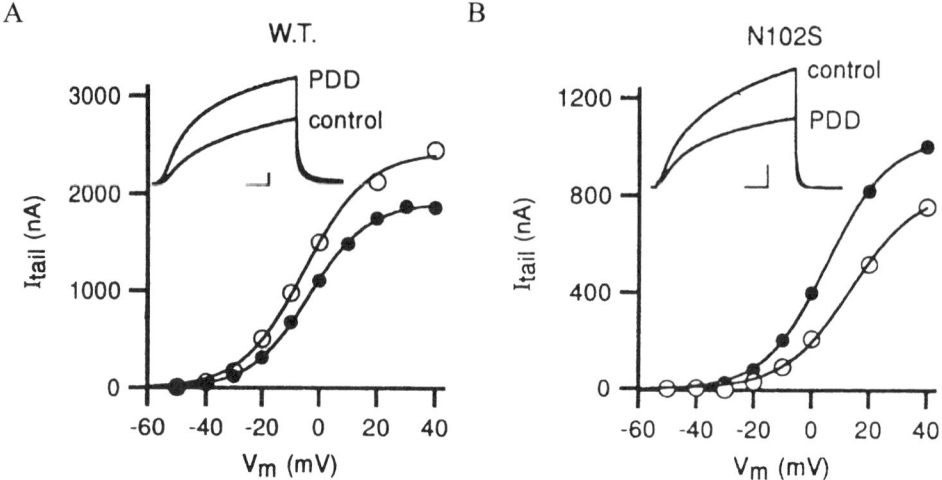

Figure 9. PKC regulation of wild-type and mutated guinea pig IsK current (gpIsK) expressed in oocytes. (A) Voltage dependence of activation of wild-type (W.T.) gpIsK, before (●) and after (○) 20-min exposure to 50 nM phorbol-12,13-didecanoate (PDD) to activate PKC. Continuous curves were drawn according to a Boltzmann function representing $I_{max}(V_m - E_{rev})/(1 + e^{-(V_m - V_{1/2})/k})$, where I_{max} is the maximal predicted current, V_m is the test potential, E_{rev} is the reversal potential, $V_{1/2}$ is the potential of half-maximal activation, and k is the slope. Control: I_{max} = 1917 nA, $V_{1/2}$ = −3.3 mV, k = 10.1 mV. Treated: I_{max} = 2429 nA, $V_{1/2}$ = −5.1 mV, k = 10.9 mV. Inset shows representative gpIsK current tracings elicited by 30-s steps to +20 mV (scale bars: 5 s, 500 nA) in control and in the presence of PDD, as indicated. PKC activation by PDD resulted in an increase in gpIsK current magnitude. (B) Voltage-dependence of activation of N102S (gpIsK in which the asparagine reside at position 102 was mutated to a serine) before (●) and after (○) exposure to 50 nM PDD as described in (A). Control: I_{max} = 1036 nA, $V_{1/2}$ = 5.3 mV, k = 10.6 mV. Treated: I_{max} = 900 nA, $V_{1/2}$ = 16.5 mV, k = 12.6 mV. Inset shows representative N102S current tracings recorded in response to a similar protocol as described in (A) Mutation of asparagine to serine at position 102 resulted in a current that was decreased by PKC activation by PDD (scale bars: 5 s, 500 nA). (Reproduced, with permission, from Varnum et al., 1993. Copyright 1993 National Academy of Sciences, U.S.A.)

pig clone has an asparagine residue at this position, compared with a serine in the human, rat, and mouse clones, and mutation of the guinea pig Asp_{102} to Ser_{102} results in a mutant channel that is downregulated by PKC stimulation (Varnum et al., 1993) (Fig. 9).

A more recent study of homomultimeric human IsK channels, however, has demonstrated both decreased *and* increased IsK current responses to PKC, which may be indicative of multiple phosphorylation sites that are preferentially phosphorylated by different PKC isozymes (Lo and Numann, 1998). It is important to recognize that native I_{Ks} is likely not homomultimeric, but rather the product of a heteromultimeric association between IsK and KvLQT1; IsK functions as an important regulator of this channel complex (Barhanin et al., 1996; Sanguinetti et al., 1996). Furthermore, IsK may also contribute to cardiac I_{Kr}, through an association with the ERG product (McDonald et al., 1997). Although it has been demonstrated that HERG, expressed in oocytes, is insensitive to PKC modulation (Kiehn et al., 1998), further work is required to clarify the effects of α_1-adrenergic stimulation on these heteromultimeric channels.

REFERENCES

Accili, E. A., Redaelli, G., and DiFrancesco, D., 1998, Two distinct pathways of muscarinic current responses in rabbit sino-atrial node myocytes, *Pflügers Arch.* **437**:164–167.

Anyukhovsky, E. P., Rybin, V. O., Nikashin, A. V., Budanova, O. P., and Rosen, M. R., 1992. Positive chronotropic responses induced by α_1-adrenergic stimulation of normal and "ischemic" Purkinje fibers have different receptor–effector coupling mechanisms, *Circ. Res.* **71**:526–534.

Apkon, M., and Nerbonne, J. M., 1988, α_1-Adrenergic agonists selectively suppress voltage-dependent K$^+$ currents in rat ventricular myocytes, *Proc. Natl. Acad. Sci. U.S.A.* **85**:8756–8760.

Apkon, M., and Nerbonne, J. M., 1991, Characterization of two distinct depolarization-activated K$^+$ currents in isolated adult rat ventricular myocytes, *J. Gen. Physiol.* **97**:973–1011.

Ashcroft, F. M., and Rorsman, P., 1990, ATP-sensitive K$^+$ channels: A link between B-cell metabolism and insulin secretion, *Biochem. Soc. Trans.* **18**:109–111.

Axelrod, J., Burch, R. M., and Jelsema, C. L. 1988, Receptor-mediated activation of phospholipase A2 via GTP-binding proteins: Arachidonic acid and its metabolites as second messengers, *Trends Neurosci.* **11**:117–123.

Baek, K. J., Das, D. K., Gray, C., Antar, S., Murugesan, G., and Im, M.-J., 1993, Evidence that the G$_h$ protein is a signal mediator from α_1-adrenoceptor to a phospholipase C. I. Identification of α_1-adrenoceptor-coupled G$_h$ family and purification of G$_{h7}$ from bovine heart. *J. of Biol. Chem.* **268**:27390–27397.

Barhanin, J., Lesage, F., Guillemare, E., Fink, M., Lazdunski, M., and Romey, G., 1996, KvLQT1 and IsK (minK) proteins associate to form the I_{Ks} cardiac potassium current, *Nature* **384**:78–80.

Barry, D. M., and Nerbonne, J. M., 1996, Myocardial potassium channels: Electrophysiological and molecular diversity, *Annu. Rev. Physiol.* **58**:363–394.

Beau, S. L., Hand, D. E., Scheussler, R. B., Bromberg, B. I., Kwon, B., Boineau, J. P., and Saffitz, J. E., 1995, Relative densities of muscarinic and beta-adrenergic receptors in the canine sinoatrial node and their relation to sites of pacemaker activity, *Circ. Res.* **77**:957–963.

Bender, K., Wellner-Kienitz, M.-C., Meyer, T., and Pott, L., 1998, Activation of muscarinic K$^+$ current by β-adrenergic receptors in cultured atrial myocytes transfected with β_1 subunit of heterotrimeric G-proteins, *FEBS Lett.* **439**:115–120.

Benfey, B. G., and Varma, D. R., 1967, Interactions of sympathomimetic drugs, propranolol, and phentolamine on atrial refractory period and contractility, *Br. J. Pharmacol.* **30**:603–611.

Benndorf, K., Markwardt, F., and Nilius, B., 1987, Two types of transient outward current in cardiac ventricular cells of mice, *Pflügers Arch.* **409**:641–643.

Bennett, P. B., and Begenisich, T. B., 1987, Catecholamines modulate the delayed rectifying potassium current (I_K) in guinea pig ventricular myocytes, *Pflügers Arch.* **410**:217–219.

Bennett, P., McKinney, L., Begenisich, T., and Kass, R. S., 1986, Adrenergic modulation of the delayed rectifier potassium channel in calf cardiac Purkinje fibers, *Biophys. J.* **49**:839–848.

Berkowitz, D. E., Nardone, N. A., Smiley, R. M., Price, D. T., Kreutter, D. K., Fremeau,R.T., and Schwinn, D. A., 1995, Distribution of beta 3-adrenoceptor mRNA in human tissues, *Eur. J. Pharmacol.* **289**:223–228.

Berridge, M. J., 1993, Inositol trisphosphate and calcium signalling, *Nature* **361**:315–325.

Bohm, M., and Lohse, M. J., 1994, Quantification of beta-adrenoceptors and beta-adrenoceptor kinase on protein and mRNA levels in heart failure, *Eur. Heart J.*, **15**:30–34.

Bordoni, A., Biagi, P. L., Rossi, C. A., and Hrelia, S., 1991, Alpha-1-stimulated phosphoinositide breakdown in cultured cardiomyocytes: Diacylglycerol production and composition in docosahexaenoic acid supplemented cells, *Biochem. Biophys. Res. Commun.* **174**:869–877.

Boyett, M. R., and Roberts, A., 1987, The fade of the response to acetylcholine at the rabbit isolated sino-atrial node. *J. Physiol. (London)* **393**:171–194.

Braun, A. P., and Walsh, M. P., 1993, Cardiac α_1-adrenoceptors stimulate a high-affinity GTPase activity in sarcolemmal membranes from rabbit atrial and ventricular myocytes, *Eur. J. Biochem.* **213**:57–65.

Braun, A. P., Fedida, D., Clark, R. B., and Giles, W. R., 1990, Intracellular mechanisms for α_1-adrenergic regulation of the transient outward current in rabbit atrial myocytes, *J. Physiol.)(London)* **431**:689–712.

Braun, A. P., Fedida, D., and Giles, W. R., 1992, Activation of α_1-adrenoceptors modulates the inwardly rectifiying potassium currents of mammalian atrial myocytes, *Pflügers Arch.* **421**:431–439.

Breitwieser, G. E. and Szabo, G., 1988, Mechanism of muscarinic receptor-induced K$^+$ channel activation as revealed by hydrolysis-resistant GTP analogues, *J. Gen. Physiol.* **91**:469–493.

Bristow, M. R., Minobe, W., Rasmussen, R., Hershberger, R. E., and Hoffman, B. B. 1988, Alpha-1 adrenergic receptors in the nonfailing and failing human heart, *J. Pharmacol. Exp. Ther.* **247:**1039–1045.

Bristow, M. R., Hershberger, R. E., Port, J. D., Gilbert, E. M., Sandoval, A., Rasmussen, R., Cates, A. E., and Feldman, A. M., 1990, β-Adrenergic pathways in nonfailing and failing human ventricular myocardium. *Circulation,* **82,** I-12–I-25.

Brodde, O.-E., 1991, β_1- and β_2-adrenoceptors in the human heart: Properties, function, and alterations in chronic heart failure, *Pharmacol. Rev.* **43:**203–242.

Brown, A. M., and Birnbaumer, L., 1990, Ionic channels and their regulation by G-protein subunits, *Annu. Rev. Physiol.* **52:**197–213.

Brown, H. and DiFrancesco, D., 1980, Voltage-clamp investigations of membrance currents underlying pace-maker activity in rabbit sino-atrial node, *J. Physiol. (London)* **308:**331–351.

Brown, A. M., Yatani, A., Imoto, Y., Codina, J., Mattera, R., and Birnbaumer, L., 1989, Direct G-protein regulation of Ca^{2+} channels, *Ann. N.Y. Acad. Sci.* **560:**373–386.

Bruckner, R., Meyer, W., Mugge, A., Schmitz, W., and Scholz, H., 1984, α-Adrenoceptor-mediated positive inotropic effect of phenylephrine in isolated human ventricular myocardium, *Eur. J. Pharmacol.,* **99:**345–347.

Busch, A. E., Varnum, M. D., North, R. A., and Adelman, J. P., 1992, An amino acid mutation in a potassium channel that prevents inhibition by protein kinase C, *Science* **255:**1705–1707.

Campbell, D. L., Rasmusson, R. L., Qu, Y., and Strauss, H. C., 1993, The calcium-independent transient outward potassium current in isolated ferret right ventricular myocytes. I. Basic characterization and kinetic analysis, *J. Gen. Physiol.* **101:**571–601.

Capogrossi, M. C., Kachadorian, W. A., Gambassi, G., Spurgeon, H. A., and Lakatta, E. G., 1991, Ca^{2+} dependence of α-adrenergic effects on the contractile properties and Ca^{2+} homeostasis of cardiac myocytes, *Circ. Res.* **69:**540–550.

Carmeliet, E. and Mubagwa, K., 1986, Desensitization of the acetylcholine-induced increase of potassium conductance in rabbit cardiac Purkinje fibres, *J. Physiol (London)* **371:**239–255.

Chabre, M., 1990, Aluminofluoride and beryllofluoride complexes: new phosphate analogs in enzymology, *Trends Biochem. Sci.* **15:**6–10.

Chang, F., Gao, J., Tromba, C., Cohen, I., and DiFrancesco, D., 1987, Acetylcholine reverses effects of β-agonists on pacemaker current in canine cardiac Purkinje fibers but has no direct action. A difference between primary and secondary pacemakers, *Circ. Res.* **66:**633–666.

Chaudhry, A., MacKenzie, R. G., Georgie, L. M., and Granneman, J. G., 1994, Differential interaction of β_1- and β_3-adrenergic receptors with G_i in rat adipocytes, *Cell. Signal.* **6:**457–465.

Chevalier, B., Mansier, P., Teiger, E., Callen-el Amrani, F., and Swynghedauw, B., 1991, Alterations in beta adrenergic and muscarinic receptors in aged rat heart. Effects of chronic administration of propranolol and atropine, *Mech. Ageing Dev.* **60:**215–224.

Chinn, K. (1993). Two delayed rectifiers in guinea pig ventricular myocytes distinguished by tail current kinetics. *J. Parmacol. Exp. Ther.* **264:**553–560.

Chonmaitree, T., and Heikkinen, T., 1997. Role of viruses in middle-ear disease, *Ann. N.Y. Acad. Sci.* **830:**143–157.

Codina, J., Yatani, A., Grenet, D., Brown, A. M., and Birnbaumer, L., 1987, The α subunit of the GTP binding protein G_k opens atrial potassium channels, *Science* **236:**442–445.

Corr, P. B., Shayman, J. A., Kramer, J. B., and Kipnis, R. J., 1981, Increased α-adrenergic receptors in ischemic cat myocardium. *J. Clin. Invest.* **67:**1232–1236.

Crumb, W. J., Jr., Pigott, J. D., and Clarkson, C. W., 1995, Comparison of I_{to} in young and adult human atrial myocytes: Evidence for developmental changes, *Am. J. Physiol. Heart Circ. Physiol.* **268:**H1335–H1342.

Curiel, R., Perez-Gonzalez, J., Brito, N., Zerpa, R., Tellez, D., Cabrera, J., Curiel, C., and Cubeddu, L., 1989, Positive inotropic effects mediated by α_1 adrenoceptors in intact human subjects, *J. Cardiovasc. Pharmacol.* **14:**603–615.

Das, T., Baek, K. J., Gray, C., and Im, M.-J., 1993, Evidence that the G_h protein is a signal mediator from α_1-adrenoceptor to a phospholipase C. II. Purification and characterization of a G_h-coupled 69-kDa phospholipase C and reconstitution of α_1-adrenoceptor, G_h family, and phospholipase C, *J. Biol. Chem.* **268:**27398–27405.

Dascal, N., 1997, Signalling via the G-protein-activated K^+ channels, *Cell. Signal.* **9:**551–573.

Del Balzo, U., Rosen, M. R., Malfatto, G., Kaplan, L. M., and Steinberg, S. F., 1990, Specific α_1-adrenergic receptor subtypes modulate catecholamine- induced increases and decreases in ventricular automaticity, *Circ. Res.* **67:**1535–1551.

Del Monte, F., Kaumann, A. J., Poole-Wilson, P. A., Wynne, D. G., Pepper, J., and Harding, S. E., 1993, Coexistence of functioning β_1- and β_2-adrenoceptors in single myocytes from human ventricle, *Circulation* **88:**854–863.

Delpón, E., Tamargo, J., and Sánchez-Chapula, J., 1992, Effects of imipramine on the transient outward current in rabbit atrial single cells, *Br. J. Pharmacol.* **106:**464–469.

Dixon, J. E., Shi, W. M., Wang, H. S., McDonald, C., Yu, H., Wymore, R. S., Cohen, I. S., and McKinnon, D., 1996, Role of the Kv4.3 K$^+$ channel in ventricular muscle—a molecular correlate for the transient outward current, *Circ. Res.*, **79:**659–668.

Ehlert, F. J., Delen, F. M., Yun, S. H., Friedman, D. J., and Self, D. W., 1989, Coupling of subtypes of the muscarinic receptor to adenylate cyclase in the corpus striatum and heart, *J. Pharm. Exp. Ther.* **251:**660–671.

Endoh, M. and Schumann, H. J., 1975, Frequency-dependence of the positive inotropic effect of methoxamine and naphazoline mediated by α-adrenoceptors in the isolated rabbit papillary muscle, *Naunyn-Schmiedeberg's Arch. Pharmacol.* **287:**377–389.

Endoh, M., Shimizu, T., and Yanagisawa, T., 1978, Characterization of adrenoceptors mediating positive inotropic responses in the ventricular myocardium of the dog, *Br. J. Pharmacol.* **64:**53–61.

Endoh, M., Hiramoto, T., and Kushida, H., 1989, Preponderance of β- over α-adrenoceptors in mediating the positive inotropic effect of phenylephrine in the ferret myocardium, *Naunyn-Schmiedeberg's Arch. Pharmacol.* **339:**362–366.

Ertl, R., Jahnel, U., Nawrath, H., Carmeliet, E., and Vereecke, J., 1991, Differential electrophysiologic and inotropic effects of phenylephrine in atrial and ventricular heart muscle preparations from rats, *Naunyn-Schmiedeberg's Arch. Pharmacol.* **344:**574–581.

Evans, B. A., Papaioannou, M., Bonazzi, V. R., and Summers, R. J., 1996, Expression of 3-adrenoceptor mRNA in rat tissues, *Br. J. Pharmacol.* **117:**210–216.

Fabiato, A., 1992, Two kinds of calcium-induced release of calcium from the sarcoplasmic reticulum of skinned cardiac cells, *Adv. Exp. Med. Biol.* **311:**245–262.

Federman, A. D., Conklin, B. R., Schrader, K. A., Reed, R. R., and Bourne, H. R., 1992,, Hormonal stimulation of adenylyl cyclase through G$_i$-protein beta gamma subunits, Nature **356:**159–161.

Fedida, D., 1993, Modulation of cardiac contractility by α_1 adrenoceptors, *Cardiovasc. Res.* **27:**1735–1742.

Fedida, D., and Bouchard, R. A., 1992, Mechanisms for the positive inotropic effect of α_1-adrenoceptor stimulation in rat cardiac myocytes, *Circ. Res.* **71:**673–688.

Fedida, D., Shimoni, Y., and Giles, W. R., 1989, A novel effect of norepinephrine on cardiac cells is mediated by α_1-adrenoceptors, *Am J. Physiol.* **256:**H1500–H1504.

Fedida, D., Shimoni, Y., and Giles, W. R., 1990, α-Adrenergic modulation of the transient outward current in rabbit atrial myocytes, *J. Physiol. (London)* **423:**257–277.

Fedida, D., Braun, A. P., and Giles, W. R., 1991, α_1-Adrenoceptors reduce background K$^+$ current in rabbit ventricular myocytes. *J. Physiol. (London)* **441:**673–684.

Follmer, C. H., and Colatsky, T. J., 1990, Block of delayed rectifier potassium current, I_K, by flecainide and E-4031 in cat ventricular myocytes, *Circulation* **82:**289–293.

Freeman, L. C., and Kass, R. S., 1995, Cholinergic inhibition of slow delayed-rectifier K$^+$ current in guinea pig sino-atrial node is not mediated by muscarinic receptors, *Mol. Pharmacol.* **47:**1248–1254.

Freeman, L. C., Kwok, W.-M., and Kass, R. S., 1992, Phosphorylation-independent regulation of cardiac I_K by guanine nucleotides and isoproterenol, *Am. J. Physiol. Heart Circ. Physiol.* **262:**H1298–H1302.

Gaskell, W. H., 1886, The electrical changes in the quiescent cardiac muscle which accompany stimulation of the vagus nerve, *J. Physiol. (London)* **7:**451–452.

Gauthier, C., Tavernier, G., Charpentier, F., Langin, D., and Le Marec, H., 1996, Functional β_3-adrenoceptor in the human heart. *J. Clin. Invest.* **98:**556–562.

Gilbert, J. C., Shirayama, T., and Pappano, A. J., 1991, Inositol trisphosphate promotes Na–Ca exchange current by releasing calcium from sarcoplasmic reticulum in cardiac myocytes, *Circ. Res.* **69:**1632–1639.

Giles, W. R., and Imaizumi, Y., 1988, Comparison of potassium currents in rabbit atrial and ventricular cells, *J. Physiol. (London)* **405:**123–145.

Gille, E., Lemoine, H., Ehle, B., and Kaumann, A. J., 1985, The affinity of ($-$)-propranolol for beta 1- and beta 2-adrenoceptors of human heart. Differential antagonism of the positive inotropic effects and adenylate cyclase stimulation by ($-$)-noradrenaline and ($-$)-adrenaline, *Naunyn-Schmiedeberg's Arch. Pharmacol.* **331:**60–70.

Giotti, A., Ledda, F., and Mannaioni, P. F., 1973, Effects of noradrenaline and isoprenaline, in combination with α- and β-receptor blocking substances, on the action potential of cardiac Purkinje fibres. *J. Physiol. (London)* **229**:99–113.

Giraldo, E., Martos, F., Gomez, A., Garcia, A., Vigano, M. A., Ladinsky, H., and Sanchez de La Cuesta, F., 1988, Characterization of muscarinic receptor subtypes in human tissues, *Life Sci.* **43**:1507–1515.

Graham, R. M., Perez, D. M., Hwa, J., and Piascik, M. T., 1996, α_1-Adrenergic receptor subtypes—molecular structure, function, and signaling, *Circ. Res.* **78**:737–749.

Green, S. A., Holt, B. D., and Liggett, S. B., 1992, Beta 1- and beta 2-adrenergic receptors display subtype-selective coupling to G_s, *Mol. Pharmacol.* **41**:889–893.

Hansen, C. A., Schroering, A. G., and Robishaw, J. D., 1995, Subunit expression of signal transducing G-proteins in cardiac tissue: Implications for phospholipase C-beta regulation, *J. Mol. Cell. Cardiol.* **27**:471–484.

Harada, K., and Iijima, T., 1994, Differential modulation by adenylate cyclase of Ca^{2+} and delayed K^+ current in ventricular myocytes, *Am. J. Physiol.* **266**:H1551–H1557.

Hartmann, H. A., Mazzocca, N. J., Kleimann, R. B., and Houser, S. R., 1988, Effects of phenylephrine on calcium current and contractility of feline ventricular myocytes, *Am. J. Physiol.* **25**:H1173–H1180.

Harvey, R. D., and Hume, J. R., 1989, Autonomic regulation of delayed rectifier K^+ current in mammalian heart involves G-proteins, *Am. J. Physiol.* **257**:H818–H823.

Heidbuchel, H., Callewaert, G., Vereecke, J., and Carmeliet, E., 1990, ATP-dependent activation of atrial muscarinic K^+ channels in the absence of agonist and G-nucleotides, *Pflügers Arch.* **416**:213–215.

Henrich, C. J. and Simpson, P. C., 1988, Differential acute and chronic response of protein kinase C in cultured neonatal rat heart myocytes to alpha-1-adrenergic and phorbol ester stimulation, *J. Mol. Cell. Cardiol.* **20**:1081–1085.

Hescheler, J., Nawrath, H., Tang, M., and Trautwein, W., 1988, Adrenoceptor-mediated changes of excitation and contraction in ventricular heart muscle from guinea-pigs and rabbits, *J. Physiol. (London)* **397**:657–670.

Hewett, K. W., and Rosen, M. R., 1985, Developmental changes in the rabbit sinus node action potential and its response to adrenergic agonists, *J. Pharmacol. Exp. Ther.* **235**:308–312.

Hieble, J. P., Bylund, D. B., Clarke, D. E., Eikenburg, D. C., Langer, S. Z., Lefkowitz, R. J., Minneman, K. P., and Ruffolo, R. R., Jr., 1995, International Union of Pharmacology. X. Recommendation for nomenclature of α_1-adrenoceptors: Consensus update, *Pharmacol. Rev.* **47**:267–270.

Honoré, E., Attali, B., Romey, G., Heurteaux, C., Ricard, P., Lesage, F., Lazdunski, M., and Barhanin, J., 1991, Cloning, expression, pharmacology and regulation of a delayed rectifier K^+ channel in mouse heart, *EMBO J.* **10**:2805–2811.

Hoover, D. B., and Neely, D. A., 1997, Differentiation of muscarinic receptors mediating negative chronotropic and vasoconstrictor responses to acetylcholine in isolated rat hearts, *J. Pharmacol. Exp. Ther.* **282**:1337–1344.

Horie, M., Hayashi, S., and Kawai, C., 1990, Two types of delayed rectifying K^+ channels in atrial cells of guinea pig heart, *Jpn. J. Physiol.* **40**:479–490.

Huang, C. L., Slesinger, P. A., Casey, P. J., Jan, Y. N., and Jan, L. Y., 1995, Evidence that direct binding of G beta gamma to the GIRK1 G-protein-gated inwardly rectifying K^+ channel is important for channel activation, *Neuron* **15**:1133–1143.

Hume, J. R., and Uehara, A., 1985, Ionic basis of the different action potential configurations of single guinea-pig atrial and ventricular myocytes, *J. Physiol. (London)* **368**:525–544.

Iijima, T., Imagawa, J.-I., and Taira, N., 1990, Differential modulation by *beta* adrenoceptors of inward calcium and delayed rectifier potassium current in single ventricular cells of guinea pig heart, *J. Pharmacol. Exp. Ther.* **254**:142–146.

Ikeda, U., Tsuruya, Y., and Yaginuma, T., 1991, Alpha$_1$-Adrenergic stimulation is coupled to cardiac myocyte hypertrophy, *Am. J. Physiol.* **260**:H953–H956.

Im, M.-J., and Graham, R. M., 1990, A novel guanine nucleotide-binding protein coupled to the α_1-adrenergic receptor. I. Identification by photolabeling of membrane and ternary complex preparations, *J. Biol. Chem.* **265**:18944–18951.

Im, M.-J., Riek, R. P., and Graham, R. M., 1990, A novel guanine nucleotide-binding protein coupled to the α_1-adrenergic receptor. II. Purification, characterization, and reconstitution, *J. Biol. Chem.* **265**:18952–18960.

Ito, H., Sugimoto, T., Kobayashi, I., Takahashi, K., Katada, T., Ui, M., and Kurachi, Y., 1991, On the mechanism of basal and agonist-induced activation of the G-protein-gated muscarinic K$^+$ channel in atrial myocytes of guinea pig heart, *J. Gen. Physiol.* **98**:517–533.

Ito, H., Tung, R. T., Sugimoto, T., Kobayashi, I., Takahashi, K., Katada, T., Ui, M., and Kurachi, Y., 1992, On the mechanism of G-protein $\beta\gamma$ subunit activation of the muscarinic K$^+$ channel in guinea pig atrial cell membrane. Comparison with the ATP-sensitive K$^+$ channel, *J. Gen. Physiol.* **99**:961–983.

Ito, H., Hosoya, Y., Inanobe, A., Tomoike, H., and Endoh, M., 1995, Acetylcholine and adenosine activate the G-protein-gated muscarinic K$^+$ channel in ferret ventricular myocytes, *Naunyn-Schmiedeberg's Arch Pharmacol.* **351**:610–617.

Jahnel, U., Nawrath, H., Carmeliet, E., and Vereecke, J., 1991, Depolarization-induced influx of sodium in response to phenylephrine in rat atrial heart muscle, *J. Physiol. (London)* **432**:621–637.

Jahnel, U., Kaufmann, B., Rombusch, M., and Nawrath, H., 1992, Contribution of both α- and β-adrenoceptors to the inotropic effects of catecholamines in the rabbit heart, *Naunyn-Schmiedeberg's Arch. Pharmacol.* **346**:665–672.

Jahnel, U., Klemm, P., and Nawrath, H., 1994, Different mechanisms of the inhibition of the transient outward current in rat ventricular myocytes, *Naunyn-Schmiedeberg's Arch. Pharmacol.* **349**:87–94.

Jakob, H., Nawrath, H., and Rupp, J., 1988, Adrenoceptor-mediated changes of action potential and force of contraction in isolated human ventricular heart muscle, *Br. J. Pharmacol.* **94**:584–590.

Jalife, J., Hamilton, A. J., and Moe, G. K., 1980, Desensitization of the cholinergic receptor at the sinoatrial cell of the kitten, *Am. J. Physiol.* **238**:H439–H448.

Jelsema, C. L., and Axelrod, J., 1987, Stimulation of phospholipase A$_2$ activity in bovine rod outer segments by the $\beta\gamma$ subunits of transducin and its inhibition by the α subunit, *Proc. Natl. Acad. Sci. U.S.A.* **84**:3623–3627.

Jones, C. R., Molenaar, P., and Summers, R. J., 1989, New views on human cardiac β-adrenoceptors, *J. Mol. Cell. Cardiol.* **21**:519–535.

Josephson, I. R., Brown, A. M., 1986, Inwardly rectifying single-channel and whole cell K$^+$ currents in rat venticular myocytes. *J. Membr. Biol.* **94**(1):19–35.

Josephson, I. R., Sanchez-Chapula, J., and Brown, A. M., 1984, Early outward current in rat single ventricular cells, *Circ. Res.* **54**:157–162.

Kaczmarek, L. K., and Blumenthal, E. M., 1997, Properties and regulation of the minK potassium channel protein, *Physiol. Rev.* **77**:627–641.

Kaumann, A. J., 1989, Is there a third heart β-adrenoceptor? *Trends Pharmacol. Sci.* **10**:316–320.

Kaumann, A. J., 1996, (−)-CGP 12177-induced increase of human atrial contraction through a putative third beta-adrenoceptor, *Br. J. Pharmacol.* **117**:93–98.

Kaumann, A. J., 1997, Four β-adrenoceptor subtypes in the mammalian heart, *Trends Pharmacol. Sci.* **18**:70–76.

Kaumann, A. J., and Molenaar, P., 1996, Differences between the third cardiac β-adrenoceptor and the colonic β_3-adrenoceptor in the rat, *Br. J. Pharmacol.* **118**:2085–2098.

Kaumann, A. J., and Molenaar, P., 1997, Modulation of human cardiac function through 4 β-adrenoceptor populations, *Naunyn-Schmiedeberg's Arch. Pharm.* **355**:667–681.

Kaumann, A. J., Hall, J. A., Murray, K. J., Wells, F. C., and Brown, M. J., 1989, A comparison of the effects of adrenaline and noradrenaline on human heart: The role of β_1- and β_2-adrenoceptors in the stimulation of adenylate cyclase and contractile force, *Eur. Heart J.* **10**:29–37.

Kaumann, A. J., Sanders, L., Lynham, J. A., Bartel, S., Kuschel, M., Karczewski, P., and Krause, E. G., 1996, β_2-adrenoceptor activation by zinterol causes protein phosphorylation, contractile effects and relaxant effects through a cAMP pathway in human atrium, *Mol. Cell. Biochem.* **163–164**:113–123.

Kentish, J. C., Barsotti, R. J., Lea, T. J., Mulligan, I. P., Patel, J. R., and Ferenczi, M. A., 1990, Calcium release from cardiac sarcoplasmic reticulum induced by photorelease of calcium or ins(1,4,5)P$_3$, *Am. J. Physiol.* **258**:H610–H615.

Khan, W. A., Blobe, G. C., and Hannun, Y. A., 1992, Activation of protein kinase C by oleic acid. Determination and analysis of inhibition by detergent micelles and physiologic membranes: Requirement for free oleate, *J. Biol. Chem.*, **267**:3605–3612.

Khan, W. A., Blobe, G. C., and Hannun, Y. A., 1995, Arachidonic acid and free fatty acids as second messengers and the role of protein kinase C, *Cell. Signal* **7**:171–184.

Kiehn, J., Karle, C., Thomas, D., Yao, X., Brachmann, J., and Kubler, W., 1998, HERG potassium channel activation is shifted by phorbol esters via protein kinase A-dependent pathways, *J. Biol. Chem.* **273**:25285–25291.

Kijima, Y., Saito, A., Jetton, T. L., Magnuson, M. A., and Fleischer, S., 1993, Different intracellular localization of inositol 1,4,5-trisphosphate and ryanodine receptors in cardiomyocytes. *J. Biol. Chem.* **268:**3499–3506.

Kim, D., 1990, β-adrenergic regulation of the muscarinic-gated K^+ channel via cyclic AMP-dependent protein kinase in atrial cells, *Circ. Res.* **67:**1292–1298.

Kim, D., Lewis, D. L., Graziadei, L., Neer, E. J., Bar-Sagi, D., and Clapham, D. E., 1989, G-protein βγ-subunits activate the cardiac muscarinic K^+-channel via phospholipase A_2, *Nature* **337:**557–560.

Koga, T., Shiraki, Y., and Sakai, K., 1989, Demonstration in Tupaia papillary muscle preparations of α-adrenoceptors meditaing positive inotropic effects: Comparison with guinea-pigs, *Br. J. Pharmacol.* **98:**552–556.

Koumi, S.-I. and Wasserstrom, J. A., 1994, Acetylcholine-sensitive muscarinic K^+ channels in mammalian ventricular myocytes, *Am. J. Physiol.* **266:**H1812–H1821.

Koumi, S.-I., Wasserstrom, J. A., and Ten Eick, R. E., 1995a, β-Adrenergic and cholinergic modulation of the inwardly rectifying K^+ current in guinea-pig ventricular myocytes, *J. Physiol. (London)* **486:**647–659.

Koumi, S.-I., Wasserstrom, J. A., and Ten Eick, R. E., 1995b, β-Adrenergic and cholinergic modulation of inward rectifier K^+ channel function and phosphorylation in guinea-pig ventricle, *J. Physiol. (London)* **486:**661–678.

Krapivinsky, G., Gordon, E. A., Wickman, K., Velimirovic, B., Krapivinsky, L., and Clapham, D. E., 1995, The G-protein-gated atrial K^+ channel I_{KACh} is a heteromultimer of two inwardly rectifying K^+-channel proteins, *Nature* **374:**135–141.

Kurachi, Y., 1995, G-protein regulation of cardiac muscarinic potassium channel, *Am. J. Physiol.* **269:**C821–C830.

Kurachi, Y., Nakajima, T., and Sugimoto, T., 1986a, Acetylcholine activation of K^+ channels in cell-free membrane of atrial cells, *Am. J. Physiol.* **251:**H681–H684.

Kurachi, Y., Nakajima, T., and Sugimoto, T., 1986b, On the mechanism of activation of muscarinic K^+ channels by adenosine in isolated atrial cells: Involvement of GTP-binding proteins. *Pflugers, Arch.* **407:**264–272.

Kurachi, Y., Nakajima, T., and Sugimoto, T., 1987, Short-term desensitization of muscarinic K^+ channel current in isolated atrial myocytes and possible role of GTP-binding proteins, *Pflügers. Arch.* **410:**227–233.

Kurachi, Y., Ito, H., Sugimoto, T., Katada, T., and Ui, M., 1989a, Activation of atrial muscarinic K^+ channels by low concentrations of βγ subunits of rat brain G-protein, *Pflügers Arch.* **413:**325–327.

Kurachi, Y., Ito, H., Sugimoto, T., Shimizu, T., Miki, I., and Ui, M., 1989b, Arachidonic acid metabolites as intracellular modulators of the G-protein-gated cardiac K^+ channel, *Nature* **337:**555–557.

Kurachi, Y., Ito, H., Sugimoto, T., Shimizu, T., Miki, I., and Ui, M., 1989c, α-Adrenergic activation of the muscarinic K^+ channel is mediated by arachidonic acid metabolites, *Pflugers Arch.* **414:**102–104.

Kurachi, Y., Tung, R. T., Ito, H., and Nakajima, T., 1992, G-protein activation of cardiac muscarinic K^+ channels, *Prog. Neurobiol.* **39:**229–246.

Kurz, T., Yamada, K. A., DaTorre, S. D., and Corr, P. B., 1991, $α_1$-adrenergic system and arrhythmias in ischemic heart disease, *Eur. Heart J.* **12:**88–98.

Kwatra, M. M. and Hosey, M. M., 1986, Phosphorylation of the cardiac muscarinic receptor in intact chick heart and its regulation by a muscarinic agonist, *J. Biol. Chem.* **261:**12429–12432.

Kwatra, M. M., Leung, E., Maan, A. C., McMahon, K. K., Ptasienski, J., Green, R. D., and Hosey, M. M., 1987, Correlation of agonist-induced phosphorylation of chick heart muscarinic receptors with receptor desensitization, *J. Biol. Chem.* **262:**16314–16321.

Landzberg, J. S., Parker, J. D., Gauthier, D. F., and Colucci, W. S., 1991, Effects of myocardial $α_1$-adrenergic receptor stimulation and blockade on contractility in humans, *Circulation* **84:**1608–1614.

Laz, T. M., Forray, C., Smith, K. E., Bard, J. A., Vaysse, P. J. J., Branchek, T. A., and Weinshank, R. L., 1994, The rat homologue of the bovine $α_{1c}$-adrenergic receptor shows the pharmacological properties of the classical $α_{1A}$ subtype, *Mol. Pharmacol.* **46:**414–422.

Lederer, W. J., Nichols, C. G., and Smith, G. L., 1989, The mechanism of early contractile failure of isolated rat ventricular myocytes subjected to complete metabolic inhibition, *J. Physiol. (London)* **413:**329–349.

Le Grand, B., Deroubaix, E., Couétil, J.-P., and Coraboeuf, E., 1992, Effects of atrionatriuretic factor on Ca^{2+} current and Ca_i-independent transient outward K^+ current in human atrial cells, *Pflügers Arch.* **421:**486–491.

Lemoine, H. and Kaumann, A. J., 1991, Regional differences of β_1- and β_2-adrenoceptor-mediated functions in feline heart. A β_2-adrenoceptor-mediated positive inotropic effect possibly unrelated to cyclic AMP, *Naunyn-Schmiedeberg's Arch. Pharmacol.* **344**:56–69.

Levitzki, A., Marbach, I., and Bar-Sinai, A., 1993, The signal transduction between β-receptors and adenylyl cyclase, *Life Sci.* **52**:2093–2100.

Levy, F. O., Zhu, X., Kaumann, A. J., and Birnbaumer, L., 1993, Efficacy of beta 1-adrenergic receptors is lower than that of β_2-adrenergic receptors, *Proc. Natl. Acad. Sci. U.S.A.* **90**:10798–10802.

Li, G. R., Feng, J. L., Wang, Z. G., Fermini, B., and Nattel, S., 1996a, Adrenergic modulation of ultrarapid delayed rectifier K$^+$ current in human atrial myocytes, *Circ. Res.* **78**:903–915.

Li, G. R., Feng, J. L., Yue, L. X., Carrier, M., and Nattel, S., 1996b, Evidence for two components of delayed rectifier K$^+$ current in human ventricular myocytes, *Circ. Res.* **78**:689–696.

Lim, N. F., Dascal, N., Labarca, C., Davidson, N., and Lester, H. A., 1995, A G-protein-gated K channel is activated via β_2-adrenergic receptors and G$\beta\gamma$ subunits in *Xenopus* oocytes, *J. Gen. Physiol.* **105**:421–439.

Liu, D.-W., and Antzelevitch, C., 1995, Characteristics of the delayed rectifier current (I_{Kr} and I_{Ks}) in canine ventricular epicardial, midmyocardial, and endocardial myocytes: A weaker I_{Ks} contributes to the longer action potential of the M cell, *Circ. Res.* **76**:351–365.

Lo, C. F., and Numann, R., 1998, Independent and exclusive modulation of cardiac delayed rectifying K$^+$ current by protein kinase C and protein kinase A, *Circ. Res.* **83**:995–1002.

Loewi, O. and Navratil, E., 1926, Uber humorale ubertragbarkeit der herznervenwirkung, *Pflügers Arch.* **189**:239–242.

Loffelholz, K. and Pappano, A. J., 1985, The parasympathetic neuroeffector junction of the heart, *Pharmacological Rev.* **37**:1–24.

Logothetis, D. E., Kurachi, Y., Galper, J., Neer, E. J., and Clapham, D. E., 1987, The beta gamma subunits of GTP-binding proteins activate the muscarinic K$^+$ channel in heart, *Nature* **325**:321–326.

Logothetis, D. E., Kim, D. H., Northup, J. K., Neer, E. J., and Clapham, D. E., 1988, Specificity of action of guanine nucleotide-binding regulatory protein subunits on the cardiac muscarinic K$^+$ channel, *Proc. Natl. Acad. Sci. U.S.A.* **85**:5814–5818.

Lomasney, J. W., Cotecchia, S., Lorenz, W., Leung, W.-Y., Schwinn, D. A., Yang-Feng, T. L., Brownstein, M., Lefkowitz, R. J., and Caron, M. G., 1991, Molecular cloning and expression of the cDNA for the α_{1A}-adrenergic receptor, *J. Biol. Chem.* **266**:6365–6369.

Lowe, M. D., Grace, A. A., Vandenberg, J. I., and Kaumann, A. J., 1998, Action potential shortening through the putative β_4-adrenoceptor in ferret ventricle: Comparison with β_1- and β_2-adrenoceptor-mediated effects, *Br. J. Pharmacol.* **124**:1341–1344.

Mansourati, J., and Le Grand, B., 1993, Transient outward current in young and adult diseased human atria, *Am. J. Physiol.* **265**:H1466–H1470.

McDonald, T. V., Yu, Z. H., Ming, Z., Palma, E., Meyers, M. B., Wang, K. W., Goldstein, S. A. N., and Fishman, G. I., 1997, A minK–HERG complex regulates the cardiac potassium current I_{Kr}, *Nature* **388**:289–292.

McMorn, S. O., Harrison, S. M., Zang, W.-J., Yu, X.-J., and Boyett, M. R., 1993, A direct negative inotropic effect of acetylcholine on rat ventricular myocytes, *Am. J. Physiol.* **265**:H1393–H1400.

Michel, A. D., and Whiting, R. L., 1987, Direct binding studies on ileal and cardiac muscarinic receptors, *Br. J. Pharmacol.* **92**:755–767.

Miura, Y., and Inui, J., 1984, Multiple effects of α-adrenoceptor stimulation on the action potential of the rabbit atrium, *Naunyn-Schmiedeberg's Arch. Pharmacol.* **325**:47–53.

Mochly-Rosen, D., Henrich, C. J., Cheever, L., Khaner, H., and Simpson, P. C., 1990, A protein kinase C isozyme is translocated to cytoskeletal elements on activation, *Cell. Regul.* **1**:693–706.

Molenaar, P., Sarsero, D., and Kaumann, A. J., 1997, Proposal for the interaction of non-conventional partial agonists and catecholamines with the 'putative beta 4-adrenoceptor' in mammalian heart, *Clin. Exp. Pharmacol. Physiol.* **24**:647–656.

Molina-Viamonte, V., Anyukhovsky, E. P., and Rosen, M. R., 1991, An α_1-adrenergic receptor subtype is responsible for delayed afterdepolarizations and triggered activity during simulated ischemia and reperfusion of isolated canine Purkinje fibers. *Circ.*, **84**:1732–1740.

Morrow, A. L. and Creese, I., 1986, Characterization of α_1-adrenergic receptor subtypes in rat brain: A reevaluation of [^{3}H]WB4104 and [^{3}H]prazosin binding, *Mol. Pharmacol.* **29**:321–330.

Movsesian, M. A., Thomas, A. P., Selak, M., and Williamson, J. R., 1985, Inositol trisphosphate does not release Ca^{2+} from permeabilized cardiac myocytes and sarcoplasmic reticulum, *FEBS Lett.* **185:**328–332.

Mubagwa, K., and Carmeliet, E., 1983, Effects of acetylcholine on electrophysiological properties of rabbit cardiac Purkinje fibers, *Circ. Res.* **53:**740–751.

Mukherjee, A., Haghani, Z., Brady, J., Bush, J., McBride, W., Buja, L.M ., and Willerson, J. T., 1983, Differences in myocardial α- and β-adrenergic receptor numbers in different species. *Am. J. Physiol.* **245:**H957–H961.

Murakami, M., Furukawa, Y., Karasawa, Y., Ren, L. M., Takayama, S., and Chiba, S., 1992, Inhibition by glibenclamide of negative chronotropic and inotropic responses to pinacidil, acetylcholine, and adenosine in the isolated dog heart, *J. Cardiovasc. Pharmacol.* **19:**618–624.

Murakami, M., Furukawa, Y., and Chiba, S., 1994, Effects of glibenclamide on negative cardiac responses to cholinergic and puringeric stimuli on cromokalim in the isolated dog heart, *Jpn. J. Pharmacol.* **65:**215–222.

Murray, K. T., Fahrig, S. A., Deal, K. K., Po, S. S., Hu, N. N., Snyders, D. J., Tamkun, M. M., and Bennett, P. B., 1994, Modulation of an inactivating human cardiac K^+ channel by protein kinase C, *Circ. Res.* **75:**999–1005.

Nair, L. A., Inglese, J., Stoffel, R., Koch, W. J., Lefkowitz, R. J., Kwatra, M. M., and Grant, A. O., 1995, Cardiac muscarinic potassium channel activity is attenuated by inhibitors of Gβγ, *Circ. Res.* **76:**832–838.

Nakamura, T. Y., Coetzee, W. A., De Miera, E. V. S., Artman, M., and Rudy, B., 1997, Modulation of Kv4 channels, key components of rat ventricular transient outward K^+ current, by PKC, *Am. J. Physiol.* **273:**H1775–H1786.

Nakaoka, H., Perez, D. M., Baek, K. J., Das, T., Husain, A., Misono, K., Im, M.-J., and Graham, R. M., 1994, G_h: A GTP-binding protein with transglutaminase activity and receptor signalling function, *Science* **264:**1593–1595.

Nichols, C. G., and Lederer, W. J., 1991, Adenosine triphosphate-sensitive potassium channels in the cardiovascular system. *Am. J. Physiol.* **261:**H1675–H1686.

Nitta, J., Furukawa, T., Marumo, F., Sawanobori, T., and Hiraoka, M., 1994, Subcellular mechanism for Ca^{2+}-dependent enhancement of delayed rectifier K^+ current in isolated membrane patches of guinea pig ventricular myocytes, *Circ. Res.* **74:**96–104.

Noma, A., 1993, Gating properties of ATP-sensitive K^+ channels in the heart, *Cardiovasc. Drugs Ther.* **7** (Suppl. **3**):515–520.

Pappano, A. J., and Mubagwa, K., 1991, Muscarinic agonist-induced actions on potassium and calcium channels in atrial myocytes: Differential desensitization, *Eur. Heart J.* **12:**70–75.

Perez, D. M., Piascik, M. T., and Graham, R. M., 1991, Solution-phase library screening for the identification of rare clones: Isolation of an $α_{1D}$-adrenergic receptor cDNA, *Mol. Pharmacol.* **40:**876–883.

Perez, D. M., Piascik, M. T., Malik, N., Gaivin, R., and Graham, R. M., 1994, Cloning, expression, and tissue distribution of the rat homolog of the bovine $α_{1C}$-adrenergic receptor provide evidence for its classification as the $α_{1A}$ subtype, *Mol. Pharmacol.* **46:**823–831.

Petit-Jacques, J., Bois, P., Bescond, J., and Lenfant, J., 1993, Mechanism of muscarinic control of the high-threshold calcium current in rabbit sino-atrial node myocytes, *Pflügers. Arch.* **423:**21–27.

Pfaffinger, P. J., Martin, J. M., Hunter, D. D., Nathanson, N. M., and Hille, B., 1985, GTP-binding proteins couple cardiac muscarinic receptors to a K channel, *Nature* **317:**536–538.

Price, D. T., Lefkowitz, R. J., Caron, M. G., Berkowitz, D., and Schwinn, D. A., 1994a, Localization of mRNA for three distinct $α_1$-adrenergic receptor subtypes in human tissues: Implications for human α-adrenergic physiology. *Mol. Pharmacol.,* **45:**171–175.

Price, D. T., Chari, R. S., Berkowitz, D. E., Meyers, W. C., and Schwinn, D. A., 1994b, Expression of $α_1$-adrenergic receptor subtype mRNA in rat tissues and human SK-N-MC neuronal cells: Implications for $α_1$-adrenergic receptor subtype classification, *Mol. Pharmacol.* **46:**221–226.

Priori, S. G., Yamada, K. A., and Corr, P. B., 1991, Influence of hypoxia on adrenergic modulation of triggered activity in isolated adult canine myocytes, *Circulation* **83:**248–259.

Priori, S. G., Napolitano, C., and Schwartz, P. J., 1993, Cardiac receptor activation and arrhythmogenesis, *Eur. Heart J.* **14:**20–26.

Qi, A., Yeung-Lai-Wah, J. A., Xiao, J., and Kerr, C. R., 1994, Regional differences in rabbit atrial repolarization: Importance of transient outward current, *Am. J. Physiol.* **266:**H643–H649.

Rauch, B., and Niroomand, F., 1991, Specific M_2-receptor activation: An alternative to treatment with beta-receptor blockers?, *Eur. Heart J.* **12:**76–82.

Ravens, U., Wang, X.-L., and Wettwer, E., 1989, Alpha adrenoceptor stimulation reduces outward currents in rat ventricular myocytes, *J. Pharmacol. Exp. Ther.* **250**:364–370.

Renaudon, B., Bois, P., Bescond, J., and Lenfant, J., 1997, Acetylcholine modulates I_f and $I_{K(ACh)}$ via different pathways in rabbit sino-atrial node cells, *J. Mol. Cell. Cardiol.* **29**:969–975.

Reuveny, E., Slesinger, P. A., Inglese, J., Morales, J. M., Iñiguez-Lluhi, J. A., Lefkowitz, R. J., Bourne, H. R., Jan, Y. N., and Jan, L. Y., 1994, Activation of the cloned muscarinic potassium channel by G-protein $\beta\gamma$ subunits, *Nature* **370**:143–146.

Rodefeld, M. D., Beau, S. L., Schuessler, R. B., Boineau, J. P., and Saffitz, J. E., 1996, Beta-adrenergic and muscarinic cholinergic receptor densities in the human sinoatrial node: Identification of a high β_2-adrenergic receptor density. *J. Cardiovasc. Electrophysiol.* **7**:1039–1049.

Rokosh, D. G., Bailey, B. A., Stewart, A. F., Karns, L. R., Long, C. S., and Simpson, P. C., 1994, Distribution of α_{1C}-adrenergic receptor mRNA in adult rat tissues by RNase protection assay and comparison with α_{1B} and α_{1D}, *Biochem. Biophys. Res. Commun.* **200**:1177–1184.

Ruffolo, R. R., Jr., and Hieble, J. P., 1994, α-Adrenoceptors, *Pharmacol. Ther.* **61**:1–64.

Rybin, V. O., and Steinberg, S. F., 1994, Protein kinase C isoform expression and regulation in the developing rat heart, *Circ. Res.* **74**:299–309.

Sakmann, B., Noma, A., and Trautwein, W., 1983, Acetylcholine activation of single muscarinic K⁺ channels in isolated pacemaker cells of the mammalian heart, *Nature* **303**:250–253.

Sanguinetti, M. C., and Jurkiewicz, N. K., 1990a, Two components of delayed rectifier K⁺ current, *J. Gen. Physiol.* **96**:195–215.

Sanguinetti, M. C., and Jurkiewicz, N. K., 1990b, Lanthanum blocks a specific component of I_K and screens membrane surface charge in cardiac cells, *Am. J. Physiol.* **259**:H1881–H1889.

Sanguinetti, M. C., and Jurkiewicz, N. K., 1991, Delayed rectifier outward K⁺ current is composed of two currents in guinea pig atrial cells, *Am. J. Physiol.* **260**:H393–H399.

Sanguinetti, M. C., Curran, M. E., Zou, A., Shen, J., Spector, P. S., Atkinson, D. L., and Keating, M. T., 1996, Coassembly of KvLQT1 and minK (IsK) proteins to form cardiac I_{Ks} potassium channel, *Nature* **384**:80–83.

Sato, R., Hisatome, I., Wasserstrom, J. A., Arentzen, C. E., and Singer, D. H., 1990, Acetylcholine-sensitive potassium channels in human atrial myocytes, *Am. J. Physiol.* **259**:H1730–H1735.

Satoh, H., and Hashimoto, K., 1988, Effect of α_1-adrenoceptor stimulation with methoxamine and phenylephrine on spontaneously beating rabbit sino-atrial node cells, *Naunyn-Schmiedeberg's Arch. Pharmacol.* **337**:415–422.

Scamps, F., 1996, Characterization of a β-adrenergically inhibited K⁺ current in rat cardiac venticular cells, *J. Physiol. (London)* **491**:81–97.

Schackow, T. E., and Ten Eick, R. E., 1994, Enhancement of ATP-sensitive potassium current in cat ventricular myocytes by β-adrenoreceptor stimulation, *J. Physiol. (London)* **474**:131–145.

Schackow, T. E., Decker, R. S., and Ten Eick, R. E., 1995, Electrophysiology of adult cat cardiac ventricular myocytes: Changes during primary culture, *Am. J. Physiol.* **268**:C1002–C1017.

Scherer, R. W., Lo, C. F., and Breitwieser, G. E., 1993, Leukotriene C_4 modulation of muscarinic K⁺ current activation in bullfrog atrial myocytes, *J. Gen. Physiol.* **102**:125–141.

Schmitz, W., Scholz, H., and Erdmann, E., 1987, Effects of α- and β-adrenergic agonists, phosphodiesterase inhibitors and adenosine on isolated human heart muscle preparations, *Trends Pharmacol. Sci.* **8**:447–450.

Scholz, J., Schaefer, B., Schmitz, W., Scholz, H., Steinfath, M., Lohse, M., Schwabe, U., and Puurunen, J., 1988, Alpha-1 adrenoceptor-mediated positive inotropic effect and inositol trisphosphate increase in mammalian heart, *J. Parmacol. Exp. Thera.* **245**:327 335.

Scholz, J., Troll, U., Sandig, P., Schmitz, W., Scholz, H., and Schulte am Esch, J., 1992, Existence and α_1-adrenergic stimulation of inositol polyphosphates in mammalian heart, *Mol. Pharmacol.* **42**:134–140.

Schreibmayer, W., Dessauer, C. W., Vorobiov, D., Gilman, A. G., Lester, H. A., Davidson, N., and Dascal, N., 1996, Inhibition of an inwardly rectifying K⁺ channel by G-protein α-subunits, *Nature* **380**:624–627.

Schubert, B., VanDongen, A. M. J., Kirsch, G. E., and Brown, A. M., 1989, β-adrenergic inhibition of cardiac sodium channels by dual G-protein pathways, *Science* **245**:516–519.

Schumann, H. J., Wagner, J., Knorr, A., Reidemeister, J. C., Sadony, V., and Schramm, G, 1978, Demonstration in human atrial preparations of α-adrenoceptors mediating positive inotropic effects, *Naunyn-Schmiedeberg's Arch. Pharmacol.* **302**:333–336.

Schwinn, D. A., Lomasney, J. W., Lorenz, W., Szklut, P. J., Fremeau, R. T. Jr., Yang-Feng, T. L., Caron, M. G., Lefkowitz, R. J., and Cotecchia, S, 1990, Molecular cloning and expression of the cDNA for a novel α_1-adrenergic receptor subtype, *J. Biol. Chem.* **265**:8183–8189.

Shah, A., Cohen, I. S., and Rosen, M. R., 1988, Stimulation of cardiac alpha receptors increases Na/K pump current and decreases G_K via a pertussis toxin-sensitive pathway, *Biophys. J.* **54**:219–225.

Sheridan, D. J., 1986, Alpha adrenoceptors and arrhythmias, *J. Mol. Cell. Cardiol.* **18**:59–68.

Shui, Z., Boyett, M. R., Zang, W. J., Haga, T., and Kameyama, K., 1995, Receptor kinase-dependent desensitization of the muscarinic K^+ current in rat atrial cells, *J. Physiol. (London)* **48**:359–366.

Shui, Z., Boyett, M. R., and Zang, W. J., 1997, ATP-dependent desensitization of the muscarinic K^+ channel in rat atrial cells, *J. Physiol. (London)* **505**:77–93.

Silverman, S. K., Lester, H. A., and Dougherty, D. A., 1996, Subunit stoichiometry of a heteromultimeric G-protein-coupled inward-rectifier K^+ channel, *J. Biol. Chem.* **271**:30524–30528.

Simurda, J., Simurdov, M., and Christe, G., 1989, Use-dependent effects of 4-aminopyridine on transient outward current in dog ventricular muscle, *Pflügers Arch.* **415**:244–246.

Skomedal, T., Aass, H., and Osnes, J.-B., 1990, Prazosin-sensitive component of the inotropic response to norepinephrine in rabbit heart, *J. Pharmacol. Exp. Ther.* **252**:853–858.

Soejima, M. and Noma, A., 1984, Mode of regulation of the ACh-sensitive K-channel by the muscarinic receptor in rabbit atrial cells, *Pflügers Arch.* **400**:424–431.

Steinfath, M., Lavicky, J., Schmitz, W., Scholz, H., Doring, V., and Kalmar, P., 1992, Regional distribution of beta 1- and beta 2-adrenoceptors in the failing and nonfailing human heart, *Eur. J. Clin. Pharmacol.* **42**:607–611.

Stewart, A. F. R., Rokosh, D. G., Bailey, B. A., Karns, L. R., Chang, K. C., Long, C. S., Kariya, K., and Simpson, P. C., 1994, Cloning of the rat α_{1C}-adrenergic receptor from cardiac myocytes: α_{1C}, α_{1B}, and α_{1D} mRNAs are present in cardiac myocytes but not in cardiac fibroblasts, *Circ. Res.* **75**:796–802.

Strosberg, A. D., 1993, Structure, function, and regulation of adrenergic receptors. *Protein Sci.* **2**:1198–1209.

Tai, K. K., Wang, K. W., and Goldstein, S. A. N., 1997, MinK potassium channels are heteromultimeric complexes, *J. Biol. Chem.* **272**:1654–1658.

Talosi, L. and Kranias, E. G., 1992, Effect of α-adrenergic stimulation on activation of protein kinase C and phosphorylation of proteins in intact rabbit hearts, *Circ. Res.* **70**:670–678.

Terzic, A., Pucat, M., Vassort, G., and Vogel, S. M., 1993, Cardiac α_1-adrenoceptors: An overview, *Pharmacolog. Rev.* **45**:147–175.

Terzic, A., Tung, R. T., Inanobe, A., Katada, T., and Kurachi, Y., 1994, G-proteins activate ATP-sensitive K^+ channels by antagonizing ATP-dependent gating, *Neuron* **12**:885–893.

Terzic, A., Jahangir, A., and Kurachi, Y., 1995, Cardiac ATP-sensitive K^+ channels: Regulation by intracellular nucleotides and K^+ channel-opening drugs. *Am. J. Physiol.* **269**:C525–C545.

Thompson, N. T., Bonser, R. W., and Garland, L. G., 1991, Receptor-coupled phospholipase D and its inhibition, *Trends Pharmacol. Sci.* **12**:404–408.

Tiaho, F., Nargeot, J., and Richard, S., 1993, Repriming of L-type calcium currents revealed during early whole-cell patch-clamp recordings in rat ventricular cells, *J. Physiol. (London)* **463**:367–389.

Tohse, N., Hattori, Y., Nakaya, H., and Kanno, M., 1987, Effects of α-adrenoceptor stimulation on electrophysiological properties and mechanics in rat papillary muscle, *Gen. Pharmacol.* **5**:539–546.

Tohse, N., Nakaya, H., Hattori, Y., Endou, M., and Kanno, M., 1990, Inhibitory effect mediated by α_1-adrenoceptors on transient outward current in isolated rat ventricular cells, *Pflügers Arch.* **415**:575–581.

Tohse, N., Nakaya, H., and Kanno, M., 1992, α_1-Adrenoceptor stimulation enhances the delayed rectifier K^+ current of guinea pig ventricular cells through the activation of protein kinase C, *Circ. Res.* **71**:1441–1446.

Tsaur, M. L., Chou, C. C., Shih, Y. H., and Wang, H. L., 1997, Cloning, expression and CNS distribution of Kv4.3, an A- type K^+ channel α subunit, *FEBS Lett.* **400**:215–220.

Tseng, G.-N., and Hoffman, B. F., 1990, Actions of pinacidil on membrane currents in canine ventricular myocytes and their modulation by intracellular ATP and cAMP, *Pflügers. Arch.* **415**:414–424.

Tseng, G.-N., Robinson, R. B., and Hoffman, B. F., 1987, Passive properties and membrane currents of canine ventricular myocytes, *J. Gen. Physiol.* **90**:671–701.

Tseng, G. N., Yao, J. A., and Tseng-Crank, J., 1997, Modulation of K channels by coexpressed human α_{1c}-adrenoceptor in *Xenopus* oocytes, *Am. J. Physiol.* **272**:H1275–H1286.

Tsien, R. W., Giles, W., and Greengard, P., 1972, Cyclic AMP mediates the effects of adrenaline on cardiac Purkinje fibres, *Nat. New. Biol.* **240**:181–183.

Tzounopoulos, T., Guy, H. R., Durell, S., Adelman, J. P., and Maylie, J., 1995, min K channels form by assembly of at least 14 subunits, *Proc. Natl. Acad. Sci. U.S.A.* **92**:9593–9597.

Urquhart, R. A., Ford, W. R., and Broadley, K. J., 1993, Potassium channel blockade of atrial negative inotropic responses to P$_1$-purinoceptor and muscarinic receptor agonists and to cromokalim, *J. Cardiovasc. Pharmacol.* **21**:279–288.

Van Wagoner, D. R., Kirian, M., and Lamorgese, M., 1996, Phenylephrine suppresses outward K$^+$ currents in rat atrial myocytes, *Am. J. Physiol.* **271**:H937–H946.

Varnum, M. D., Busch, A. E., Bond, C. T., Maylie, J., and Adelman, J. P., 1993, The min K channel underlies the cardiac potassium current I_{Ks} and mediates species-specific responses to protein kinase C, *Proc. Natl. Acad. Sci. U.S.A.* **90**:11528–11532.

Veldkamp, M. W., van Ginneken, A. C., and Bouman, L. N., 1993, Single delayed rectifier channels in the membrane of rabbit ventricular myocytes, *Circ. Res.* **72**:865–878.

Veldkamp, M. W., van Ginneken, A. C., Opthof, T., and Bouman, L. N., 1995, Delayed rectifier channels in human ventricular myocytes, *Circulation* **92**:3497–3504.

Vites, A.-M., and Pappano, A., 1990, Inositol 1,4,5-trisphosphate releases intracellular Ca^{2+} in permeabilized chick atria, *Am. J. Physiol.* **258**:H1745–H1752.

Vogel, S. M., and Terzic, A., 1989, α-Adrenergic regulation of action potentials in isolated rat cardiomyocytes, *Eur. J. Pharmacol.* **164**:231–239.

Vogalis, F., Ward, M., Horowitz, B. 1995, Suppression of two cloned smooth muscle-derived delayed rectifier potassium channels by cholinergic agonists and phorbol esters, *Mol. Pharmacol.* **48(6)**:1015–1023.

Walsh, K. B., and Kass, R. S., 1988, Regulation of a heart potassium channel by protein kinase A and C, *Science* **242**:67–69.

Walsh, K. B., and Kass, R. S., 1991, Distinct voltage-dependent regulation of a heart-delayed I_K by protein kinases A and C, *Am. J. Physiol.* **261**:C1081–C1090.

Walsh, K. B., Begenisich, T. B., and Kass, R. S., 1988, β-Adrenergic modulation in the heart; independent regulation of K and Ca channels, *Pflügers Arch.* **411**:232–234.

Walsh, K. B., Begenisich, T. B., and Kass, R. S., 1989, β-Adrenergic modulation of cardiac ion channels. Differential temperature sensitivity of potassium and calcium currents, *J. Gen. Physiol.* **93**:841–854.

Wang, K.-W., and Goldstein, S. A. N., 1995, Subunit composition of MinK potassium channels, *Neuron* **14**:1303–1309.

Wang, X.-L., Wettwer, E., Gross, G., and Ravens, U., 1991, Reduction of cardiac outward currents by *alpha*-1 adrenoceptor stimulation: A subtype-specific effect, *J. Pharmacol. Exp. Ther.* **259**:783–788.

Wang, Y. G., and Lipsius, S. L., 1995a, Acetylcholine activates a glibenclamide-sensitive K$^+$ current in cat atrial myocytes, *Am. J. Physiol.* **268**:H1322–H1334.

Wang, Y. G., and Lipsius, S. L., 1995b, β-Adrenergic stimulation induces acetylcholine to activate ATP-sensitive K$^+$ current in cat atrial myocytes, *Circ. Res.* **77**:565–575.

Wang, Z., Fermini, B., and Nattel, S., 1993a, Delayed rectifier outward current and repolarization in human atrial myocytes, *Circ. Res.* **73**:276–285.

Wang, Z., Fermini, B., and Nattel, S., 1993b, An ultra-rapidly activating delayed rectifier carries sustained outward current in human atrial myocytes, *Biophys. J.* **64**:A199.

Weinberg, D. H., Trivedi, P., Tan, C. P., Mitra, S., Perkins-Barrow, A., Borkowski, D., Strader, C. D., and Bayne, M., 1994, *Biochem. Biophys. Res. Commun.* **201**:1296–1304.

Wenzel, D. G., and Su, J. L., 1966, Interactions between sympathomimetic amines and blocking agents on the rat ventricle strip, *Arch. Int. Pharmacodyn. Ther.* **160**:379–389.

Wettwer, E., Amos, G., Gath, J., Zerkowski, H.-R., Reidemeister, J.-C., and Ravens, U., 1993, Transient outward current in human and rat ventricular myocytes, *Cardiovasc. Res.* **27**:1662–1669.

Wettwer, E., Amos, G. J., Posival, H., and Ravens, U., 1994, Transient outward current in human ventricular myocytes of subepicardial and subendocardial origin, *Circ. Res.* **75**:473–482.

Wickman, K., and Clapham, D. E., 1995, Ion channel regulation by G-proteins, *Physiol. Rev.* **75**:865–885.

Wickman, K. D., Iñiguez-Lluhi, J. A., Davenport, P. A., Taussig, R., Krapivinsky, G. B., Linder, M. E., Gilman, A. G., and Clapham, D. E., 1994, Recombinant G-protein βγ-subunits activate the muscarinic-gated atrial potassium channel, *Nature* **368**:255–257.

Xiao, R.-P., Ji, X., and Lakatta, E. G., 1995, Functional coupling of the β$_2$-adrenoceptor to a pertussis toxin-sensitive G-protein in cardiac myocytes, *Mol. Pharmacol.* **47**:322–329.

Xiao, R.-P., Tomhave, E. D., Wang, D. J., Ji, X., Boluyt, M. O., Cheng, H., Lakatta, E. G., and Koch, W. J., 1998, Age-related reductions in cardiac beta$_1$- and beta$_2$-adrenergic responses without changes in inhibitory G-proteins or receptor kinases, *J. Clin. Invest.* **101**:1273–1282.

Xiao, R.-P., Avdonin, P., Zhou, Y. Y., Cheng, H., Akhter, S. A., Eschenhagen, T., Lefkowitz, R. J., Koch, W. J., and Lakatta, E. G., 1999, Coupling of β_2-adrenoceptor to G_i proteins and its physiological relevance in murine cardiac myocytes, *Circ. Res.* **84:**43–52.

Yan, K. and Gautam, N., 1996, A domain on the G-protein β subunit interacts with both adenylyl cyclase 2 and the muscarinic atrial potassium channel, *J. Biol. Chem.* **271:**17597–17600.

Yatani, A., and Brown, A. M., 1989, Rapid beta-adrenergic modulation of cardiac calcium channel currents by a fast G-protein pathway, *Science* **245:**71–74.

Yatani, A., and Brown, A. M., 1991, Mechanism of fluoride activation of G-protein-gated muscarinic atrial K^+ channels, *J. Biol. Chem.* **266:**22872–22877.

Yatani, A., Codina, J., Brown, A. M., and Birnbaumer, L., 1987, Direct activation of mammalian atrial muscarinic potassium channels by GTP regulatory protein G_k, *Science* **235:**207–211.

Yatani, A., Okabe, K., Birnbaumer, L., and Brown, A. M., 1990, Detergents, dimeric $G\beta\gamma$, and eicosanoid pathways to muscarinic atrial K^+ channels, *Am. J. Physiol.* **258:**H1507–H1514.

Yazawa, K., and Kameyama, M., 1990, Mechanism of receptor-mediated modulation of the delayed outward potassium current in guinea-pig ventricular myocytes, *J. Physiol. (London)* **421:**135–150.

Yue, L. X., Feng, J. L., Li, G. R., and Nattel, S., 1996a, Characterization of an ultrarapid delayed rectifier potassium channel involved in canine atrial repolarization, *J. Physiol. (London)* **496:**647–662.

Yue, L. X., Feng, J. L., Li, G. R., and Nattel, S., 1996b, Transient outward and delayed rectifier currents in canine atrium: Properties and role of isolation methods, *Am. J. Physiol.* **270:**H2157–H2168.

Zhang, Z. J., Jurkiewicz, N. K., Folander, K., Lazarides, E., Salata, J. J., and Swanson, R., 1994, K^+ currents expressed from the guinea pig cardiac IsK protein are enhanced by activators of protein kinase C, *Proc. Natl. Acad. Sci. U.S.A.* **91:**1766–1770.

Zhu, Y., and Nosek, T. M., 1991, Inositol trisphosphate enhances Ca^{2+} oscillations but not Ca^{2+}-induced Ca^{2+} release from cardiac sarcoplasmic reticulum, *Pflügers Arch.* **418:**1–6.

Chapter 21

Cardiac K$^+$ Channel Expression and Function at Birth and in the Neonate

Fuhua Chen and Thomas S. Klitzner

1. INTRODUCTION

Potassium (K$^+$) current plays a role in early repolarization during phase 1 of the action potential, delayed repolarization during phase 3, and maintenance of the resting membrane potential. Changes in the action potential characteristics during development reflect, in part, developmental changes in K$^+$ channel expression and function (Fig. 1). The past decade has witnessed tremendous growth in our understanding of the structure and function of membrane K$^+$ channels in cardiac cells. This progress has resulted from several technical advances, including improvements in cardiac myocyte isolation, whole-cell voltage-clamp and patch-clamp recordings, and molecular biological techniques. During this period, characterization of cardiac ionic channel function has allowed a more detailed examination of the origin of the cardiac action potential and the relation between cardiac membrane potential and contractile function in the mature heart (Katz, 1993; Noble and Bett, 1993).

Although delineation of individual cardiac ion channel characteristics has not led to a precise replication of the action potential, the time course of the major inward and outward transsarcolemmal currents involved in the generation of the cardiac action potential has been described (Task Force of the Working Group on Arrhythmias of the European Society of Cardiology, 1991). This information has been used to develop a conceptual framework for understanding the pharmacology of antiarrhythmic drugs and their use in controlling clinically significant arrhythmias.

On the other hand, the characterization of developmental changes in cardiac K$^+$ channel activity has proceeded more slowly. The subcellular mechanisms responsible for developmental changes in action potential configuration and excitation–contraction coupling are only now being elucidated. This task has been made more difficult by

Fuhua Chen and Thomas S. Klitzner • Department of Pediatrics, Division of Cardiology, University of California at Los Angeles School of Medicine, Los Angeles, California 90095

Potassium Channels in Cardiovascular Biology, edited by Archer and Rusch. Kluwer Academic/ Plenum Publishers, New York, 2001.

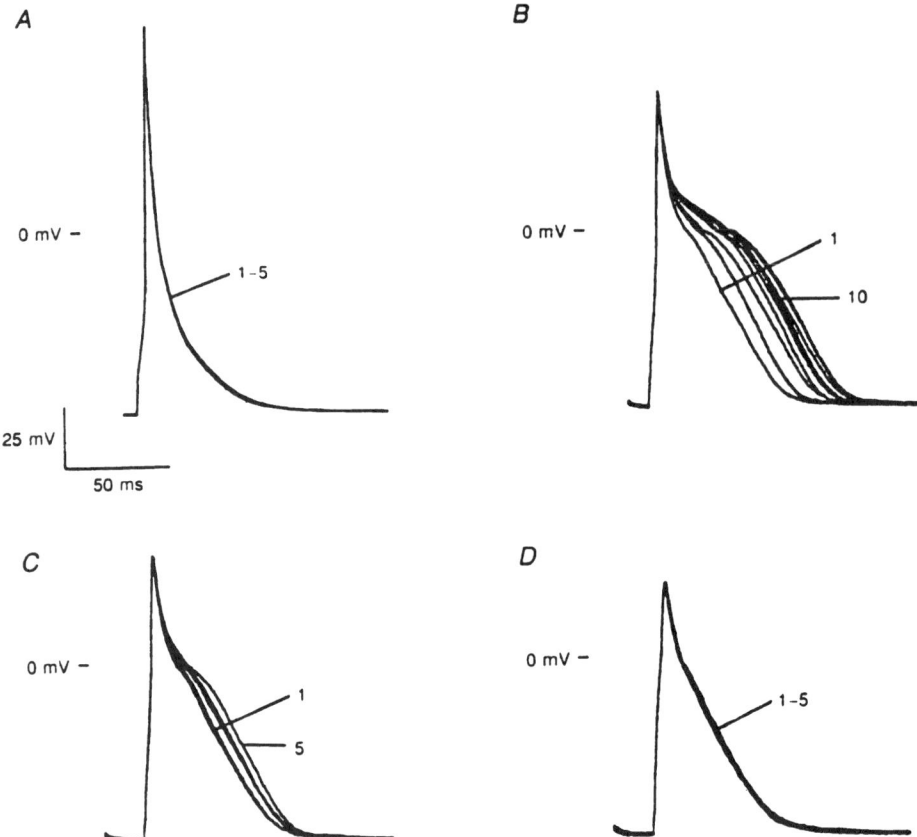

Figure 1. Action potential characteristics in cultured neonatal and adult rat ventricular myocytes. Representative action potentials recorded from an adult right ventricular myocyte stimulated at 1 Hz (A) and a neonatal ventricular myocyte maintained in culture for 24 h and stimulated at a frequency of 1 Hz (B), 0.2 Hz (C), and 0.067 Hz (D) are depicted. Action potentials were evoked after a rest period of 30 s. Sweep numbers are indicated. In (A), the post-rest adult action potential was short. Action potential duration did not lengthen with 1-Hz stimulation. In (B)–(D), the post-rest action potential in the neonatal cell was longer than in (A). Also, (B) and (C) show that the neonatal action potential lengthens with repetitive stimulation.

significant age- and species-dependent differences in cardiac electrophysiology. For example, the action potential in rat cardiac myocytes is very brief (Fig. 1). The short plateau phase of neonatal rat myocytes is nearly absent in adult rat heart (Fermini and Schanne, 1991; Wahler *et al.*, 1994). In contrast, the action potential of neonatal rabbit ventricular myocytes has a prominent plateau similar to that recorded from mature rabbit heart (Chen *et al.*, 1991, Osaka *et al.*, 1989). These age- and species-related differences also occur in other cardiac ionic currents such as the Ca^{2+} current. Mature myocytes utilize Ca^{2+} entry via sarcolemmal voltage-gated Ca^{2+} channels to trigger Ca^{2+} release from the sarcoplasmic reticulum. In the neonatal rat, voltage-gated Ca^{2+} channel currents are more prominent than in mature myocytes and most probably play a major role in cell contraction (Cohen and Lederer, 1988). In contrast, immature rabbit myocytes are relatively deficient in Ca^{2+} channel activity and can exhibit maximal contractions despite Ca^{2+} channel blockade.

Rather than endeavoring to provide a comprehensive survey of the ontogeny of the cardiac K$^+$ current and action potential, we have attempted to integrate recent descriptions of developmental changes in K$^+$ ionic currents. The physiological implications of these observations in terms of cardiac action potential configuration and excitation–contraction coupling are also discussed.

2. INWARD RECTIFIER K$^+$ CURRENT

By providing a prominent K$^+$ conductance at membrane potentials near the K$^+$ reversal potential (-60 to -100 mV, depending on cell type and species), inward rectifier K$^+$ (Kir) channels are thought to play a major role in maintaining resting membrane potential (Fig. 2). It has been demonstrated that the inward rectifier K$^+$ current (I_{KIR}) of ventricular cells increases markedly during development in both rat heart (Masuda and Sperelakis, 1993) and the embryonic chick heart (Josephson and Sperelakis, 1990). In rats, the I_{KIR} density of ventricular cells increases markedly during cardiac development between fetal day 12 and neonatal day 5; however, there is no further increase on neonatal day 10 (Masuda and Sperelakis, 1993). In the embryonic chick, the magnitude of I_{KIR} conductance in early myocytes is small but increases approximately fivefold in the older embryonic myocytes. Correspondingly, the density of Kir channels is greater on the surface membrane of the 17-day, as compared to the

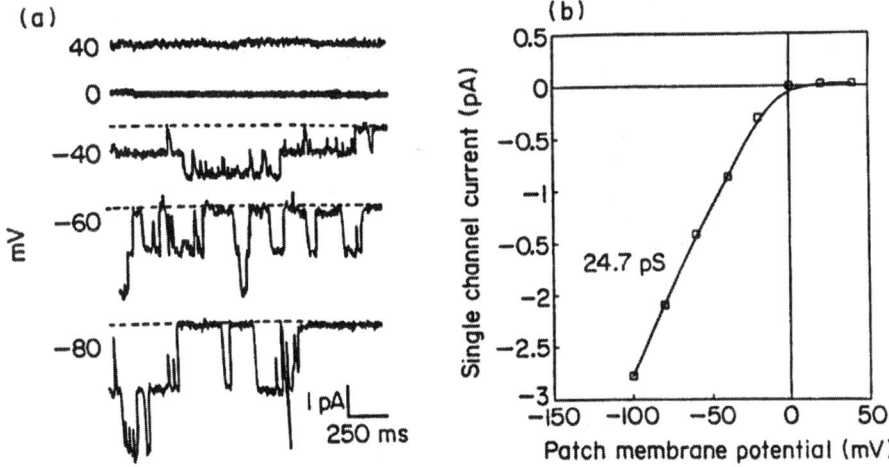

Figure 2. Single-channel currents recorded from an inside-out membrane patch of a neonatal rabbit ventricular cell. (a) Single-channel current recordings from a typical neonatal membrane patch. The patch membrane potential was held at the values given to the left of each trace. Inward current is shown as negative (downward). (b) Single-channel current–voltage (I–V) relation obtained from the record shown in (a). The slope of the linear portion of the I–V relation was determined by linear regression and yields a single-channel conductance of 24.7 pS. Both the patch pipette and bath solution contained 150 mM KCl.

3-day, embryonic myocyte. In addition, the single-channel conductance for 17-day myocytes is severalfold larger than for the 3-day myocytes. These results suggest that cardiac Kir channels increase in number and may also undergo structural alterations during development (Fig. 3) (Josephson and Sperelakis, 1990).

Therefore, the increase in I_{KIR} in the ventricular myocardium during development may be due to two factors: (1) an increase in the number of channel molecules and (2) an increase in single-channel conductance (Masuda and Sperelakis, 1993). The single-channel conductance in the early fetal rat is much less than that in the later fetus and the neonate. However, the mean open time of the Kir channel is longer in young fetal cells than in later fetal or neonatal cells. These observations suggest that the structure of the Kir channel in the heart changes dramatically during development. Perhaps this reflects a developmentally regulated change in channel isoform. Wahler (1992) found that ventricular myocytes from adult rats have two populations of Kir channels, classified by conductances of 30 and 42 pS. In contrast, neonatal cells exhibited only a single population of Kir channels with an average conductance of 30 pS. The appearance of the larger-conductance Kir channel may contribute to the reported developmental increase in the I_{KIR}. However, other researchers have found that these 42-pS channels may conduct the muscarinic K^+ current (Xie et al., 1997). It may be argued that the previously recorded "small 11-pS" Kir channel reported by Masuda and Sperelakis (1993) may be coincidentally a one-third sublevel of the 30-pS channel. We have also noted that sublevel conductance phenomena are very often associated with Kir channels, and shown that the smaller single-channel conductance of Kir channels in immature versus mature hearts may be related to the greater occurrence of sublevel conductance states in the immature hearts (Chen et al., 1992b).

An increase in I_{KIR} is likely to be the major factor contributing to the observed increase in resting membrane potential (i.e., hyperpolarization) that occurs during cardiac development. Development is also marked by a decrease in membrane resistivity and membrane time constant. An increase in functional Kir channel expression into myocardium may also account for the decrease in the permeability ratio of sodium to potassium (P_{Na}/P_K) that occurs during development, specifically resulting from an increase in the P_K (Sperelakis and Shigenobu, 1972). Therefore, these developmental changes in I_{KIR} may explain, in part, the less negative resting membrane potential in the neonate and decreased sensitivity of the neonatal membrane potential with respect to changes in the external K^+ concentration (Sperelakis and Shigenobu, 1972).

3. DELAYED RECTIFIER K^+ CURRENT

The delayed rectifier K^+ current, I_K, is one of the major repolarizing cardiac K^+ currents in the hearts of many species (Gintant et al., 1992) and is also the target for many class III antiarrhythmic agents in humans (Colastsky et al., 1990). Two components of I_K in cardiac myocytes, namely, the rapidly activating I_{Kr} and the slowly activating I_{Ks}, have been identified by the patch-clamp technique, based on their different activation kinetics, rectification properties, and pharmacological sensitivities (Sanguinetti and Jurkiewicz, 1990). Recently, Wang and Duff (1995) reported that I_{Kr}

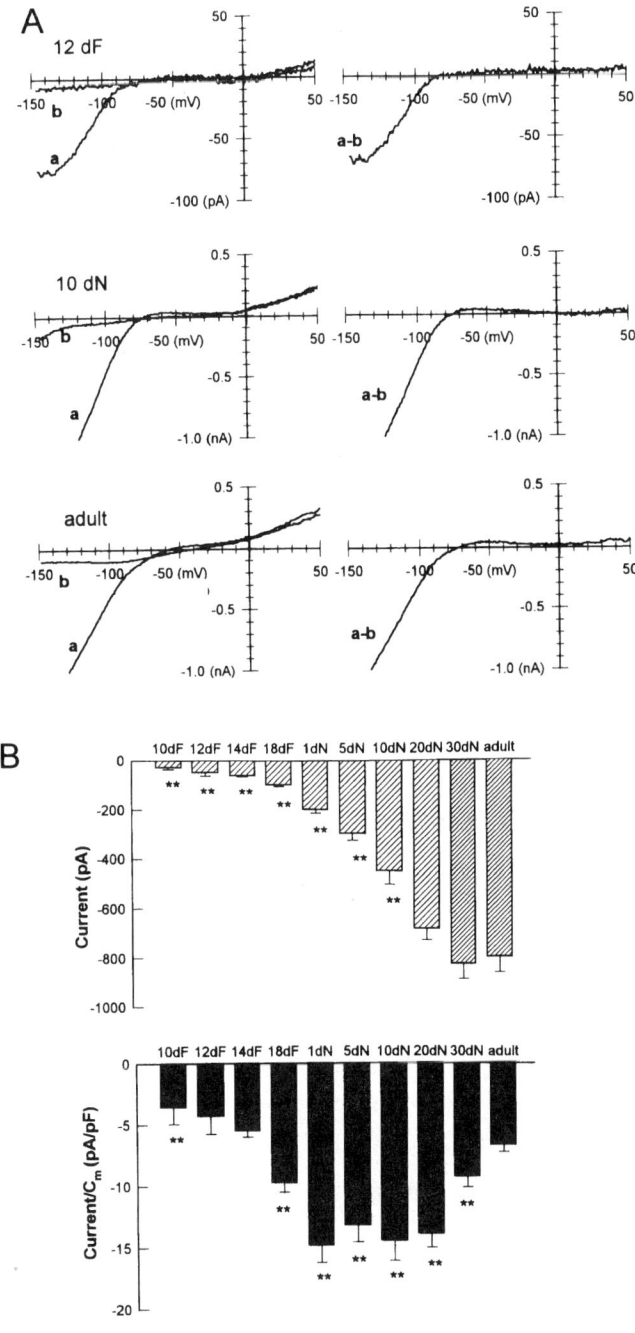

Figure 3. Developmental changes of I_{K1}. (A) $I-V$ curves recorded by applying ramp pulses from a holding potential of -50 mV to $+50$ mv. The rate of the ramp was ± 200 mV/S (see Xie *et al.*, 1997). *Top:* Fetal day 12 (12 dF), *middle:* neonatal day 10 (10 dN); *bottom:* adult. $I-V$ curves recorded in control Tyrose solution (5.4 mM K$^+$) (a) and K$^+$-free Tyrose solution (b) are shown on the left-hand side. The difference a $-$ b is shown at right. (B) Current amplitude (in picoamperes) measured at -110 mV (*top*) and current density normalized to the cellular membrane capacitance (C_m) (in picoamperes per picofarad)/(*bottom*) between fetal day 10 (10 dF) and neonatal day 30 (30 dN) and in adults. Values are means $\pm$ SE. **$P < 0.05$ compared with adults.

is the dominant repolarizing K^+ current in fetal mouse ventricular myocytes. However, there are species variations, and Benndorf and Nilius (1988) found that the transient outward K^+ current (I_{to}) is the dominant repolarizing K^+ current in adult mouse ventricular myocytes. Recently, two candidate genes that may encode I_{Kr} and I_{Ks} channel proteins have been identified, namely, *human ether-a-go-go-related gene* (*HERG*) and *minK* (Sanguinetti *et al.*, 1995; Takumi *et al.*, 1988; Folander *et al.*, 1990; Hausdorff *et al.*, 1991; see Chapters 17 and 36).

I_K has also been identified in neonatal mouse ventricular myocytes (Honoré *et al.*, 1991; Nuss and Marban, 1994). Nuss and Marban (1994) reported that I_K in a mixed population of day-1 and day-3 neonatal mouse ventricular myocytes consists of both I_{Kr} and I_{Ks}, but predominantly I_{Ks}. In another study, it was found that neither atrial nor ventricular mouse cells express I_{Ks} until just before birth (Davies *et al.*, 1996). Consistent with this finding, I_{Kr} appears to be the sole component of I_K in day-18 fetal mouse ventricular myocytes, whereas both I_{Kr} and I_{Ks} were observed in day-1 neonatal ventricular myocytes (Wang *et al.*, 1996). By day 3, I_{Ks} became the dominant component of I_K. In adult mouse ventricular myocytes, neither I_{Kr} nor I_{Ks} is observed.

Developmental changes in electrophysiological properties also occur in guinea pig ventricular cardiomyocytes. Kato *et al.* (1996) measured action potential duration in myocytes from fetal (45–55 days after conception), neonatal (1–5 days after birth), and adult (45–60 days after birth) guinea pigs. Action potential duration at 50% and 90% repolarization decreased between the fetal and neonatal periods and increased between neonate and adult (Kato *et al.*, 1996). The I_K current density at 0 and $+30$ mV was significantly smaller in fetal cells as compared with neonatal and adult cells. Of note, L-type Ca^{2+} current density at 0 and $+10$ mV was significantly smaller in fetal and neonatal cells than in the adult, although the voltage dependence and inactivation kinetics of the channel were similar between the three age groups. These authors concluded that the observed changes in action potential duration could be explained by changes in the balance between I_K and L-type Ca^{2+} current (Kato *et al.*, 1996).

4. ATP-SENSITIVE K^+ CURRENT

Near the resting membrane potential (~ -80 mV), K^+ is the dominant permeant cation in the ventricular myocardium (Katz, 1977) However, outward K^+ currents become quite small in the membrane potential range near the action potential plateau, owing to inward rectification of the I_{KIR} current–voltage relation (Josephson and Brown, 1986). During metabolic inhibition, large K^+ currents develop in the positive range of membrane potentials, and the K^+ current–voltage relation may become nearly linear (Isenberg *et al.*, 1983; Noma and Shibasaki, 1985). Under these conditions, increased outward current results in action potential shortening, thereby decreasing Ca^{2+} influx and developed tension. In addition, the increase in extracellular K^+ may result in a dispersion of refractoriness, precipitating arrhythmias and abnormal contractions during hypoxia or ischemia (Weiss and Shine, 1982). Recent studies suggest that increased K^+ current in the hypoxic and ischemic myocardium is related to the presence of ATP-sensitive K^+ (K_{ATP}) channels which are activated when internal ATP levels fall (Noma and Shibasaki, 1985; Noma, 1983; Kakei *et al.*, 1985).

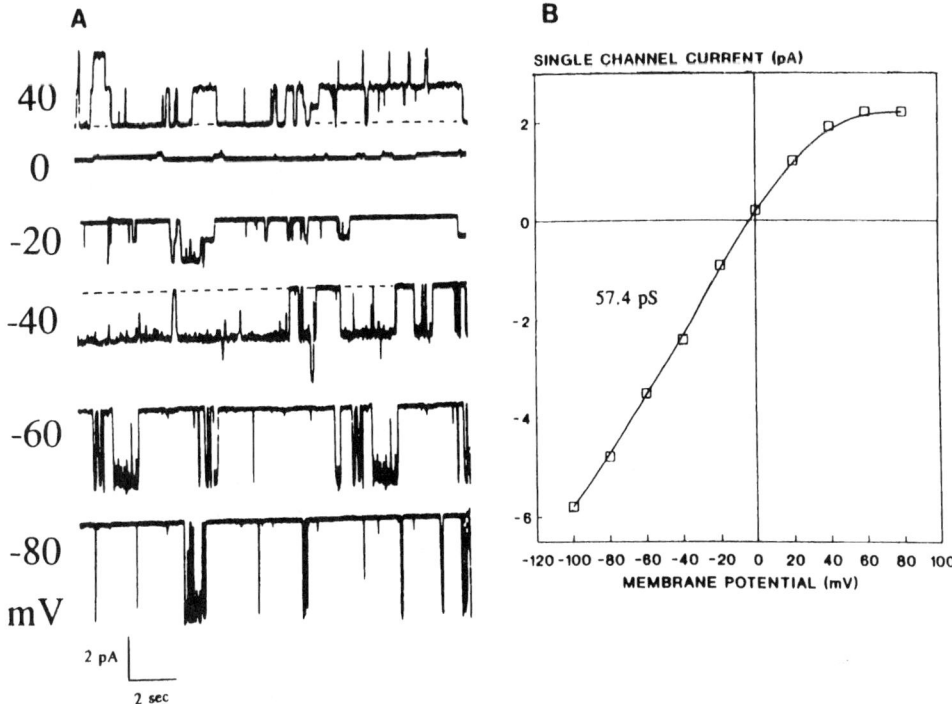

Figure 4. Voltage dependence of a typical single K$_{ATP}$ channel recorded from a rabbit neonatal myocyte. (A) Current traces recorded from a typical inside-out membrane patch. The membrane was held at the potentials indicated at the left of each trace. External and internal [K$^+$] were 150 mM, and the zero-current potential was approximately 0 mV. When the patch was held at potentials positive to the K$^+$ equilibrium potential, a significant outward current was seen. The magnitude of the outward current at +40 mV is almost equal to the magnitude of the inward current at −40 mV, indicating a fairly linear current–voltage relation in this membrane potential range. (B) The magnitude of the single-channel current (y-axis) plotted versus membrane potential (x-axis) for the experiment shown in (A). The single-channel current–voltage relation for membrane potentials negative to +20 mV can be approximated by a straight line yielding a single-channel conductance of 57.4 pS.

We have identified an ATP-sensitive K$^+$ channel in excised membrane patches from freshly isolated neonatal rabbit ventricular myocytes (Fig. 4) (Chen *et al.*, 1992a). Our results suggest that the K$_{ATP}$ channel in neonatal ventricular myocytes shares many of the properties of the K$_{ATP}$ channel reported in other preparations. The conductance of the K$_{ATP}$ channel in neonatal rabbit cardiac cells is more than twice that of the inwardly rectifying K$^+$ (Kir) channel, and the current–voltage relation shows little inward rectification when measured at physiological concentrations of K$^+$. We also found that in adult ventricular myocytes the single-channel conductance of the K$_{ATP}$ channel is significantly larger, and the channel density higher, than in neonatal cells (Chen *et al.*, 1992a).

In contrast to our findings in neonatal rabbits, Xie *et al.* (1997) observed no developmental changes in single-channel conductance or mean open times of either Kir or K$_{ATP}$ channels in rat ventricular myocytes (Fig. 5). However, these authors did find

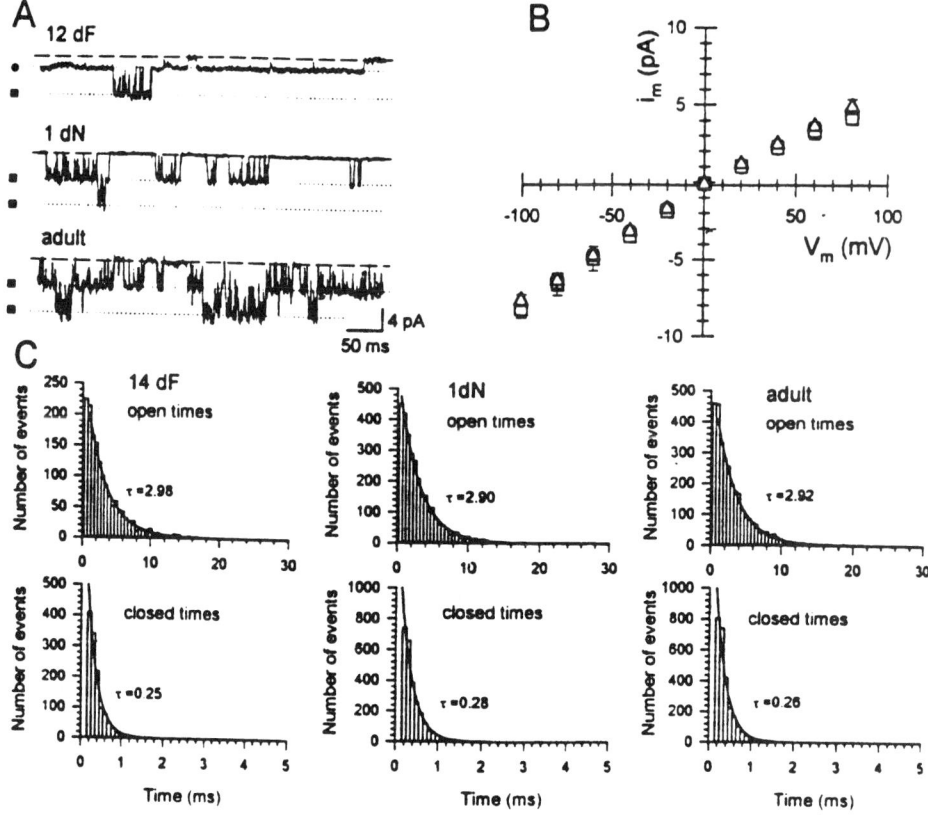

Figure 5. Single-channel characteristics of I_{KATP} in rat ventricular myocytes at different ages. (A) Representative inside-out patch recording at $-60\,mV$ in ATP-free solution obtained from ventricular cells of ages indicated [in days (d), F, fetal; N, neonatal] Dashed lines indicate current level with channel closed, and dotted lines indicate openings of I_{KI} (●) and I_{KATP} (■) channels. (B) I-V relationships for single K_{ATP} channels from 12dF (○), 1dN (□), and adult (Δ) cells. Values of SE not visible are within symbols. (C), Open-time (*top*) and closed-time (*bottom*) distributions obtained from 14dF (*left*), 1dN (*middle*), and adult (*right*) myocytes. Histograms were determined from data segments of 10-s duration. Distribution in all histograms was fitted with a single-exponential function. Time constants (τ) are shown above each curve.

that the open-state probability of the K_{ATP} channel was lower in the ventricular myocardium of fetuses, and the sensitivity to ATP was highest in 1-day neonates (Fig. 6). These results may reflect species-dependent differences in the K_{ATP} channel.

The physiological function of the ATP-sensitive K^+ current (I_{KATP}) early in development has also been explored. Davies *et al.* (1996) isolated embryonic atrial and ventricular cells derived from timed-pregnant female mice at various gestational periods and used patch-clamp procedures to investigate age- and chamber-specific expression of ionic channels. Their data indicate that K^+ channel expression undergoes significant changes during development. Neither atrial nor ventricular cells express I_{Ks} until just before birth. I_{KIR} activity, associated with determination of cellular resting potential, is not markedly apparent until late stages of embryogenesis. However, they found robust expression of the K_{ATP} channels at early and late stages of embryonic development, with

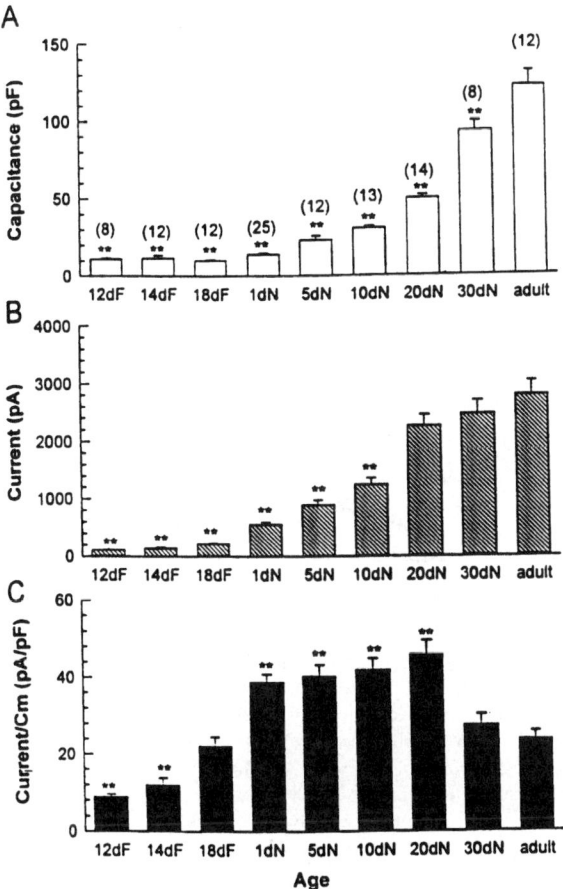

Figure 6. Developmental changes in cellular membrane capacitance (A) and current magnitude (B) or density (C) of I_{KATP} in rat ventricular myocytes. Numerals in parentheses indicate numbers of experiments, and vertical bars indicate SE. **$P < 0.05$ compared with adult cells, fetal day (dF), neonatal day (dN).

K_{ATP} activity in virtually every patch of membrane. The high level of expression of I_{KATP} in fetal heart raises questions about possible functional roles of this channel during the development of the heart. These authors argued that the low levels of Kir channel expression in the early stages of murine embryonic heart cells of both chambers suggest that cell resting membrane potential must depend on other ionic currents, such as I_{KATP} (Davies *et al.*, 1996). Further experiments are required to determine the relative physiological roles of I_{KATP} during murine fetal development.

5.   TRANSIENT OUTWARD K$^+$ CURRENT

The transient outward K$^+$ current, I_{to}, has been reported to increase during development of the rat heart. In neonatal rat ventricular cells, the density of I_{to} was

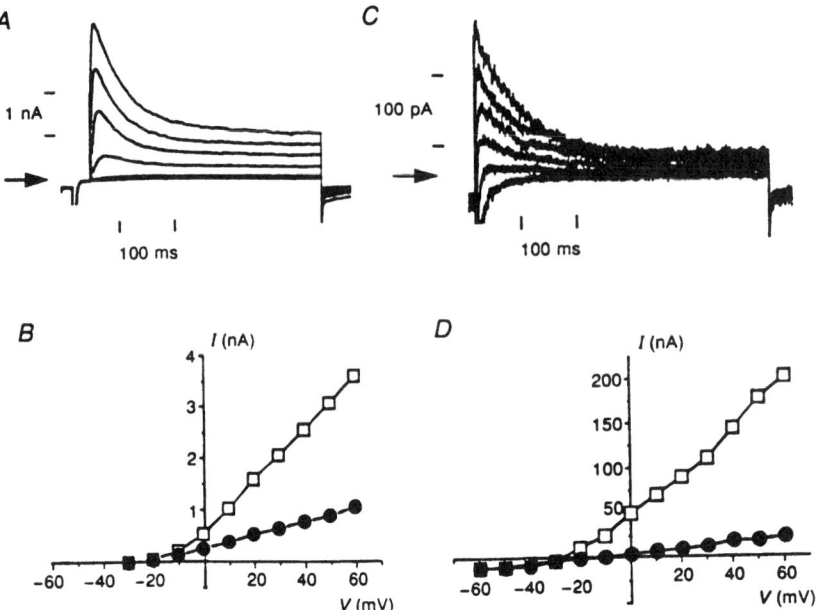

Figure 7. Typical outward currents recorded from an adult rat right ventricular myocyte and a cultured neonatal rat ventricular myocyte. Families of outward currents from a 125-pF freshly isolated adult rat right ventricular myocyte (A) and a 22-pF neonatal ventricular myocyte after 24 h in culture (C) are shown. Families were obtained by depolarizing the cells to a range of potentials (-30, -20, 0, 20, 40, and 60 mV for the adult cell and -40, -20, 0, 20, 40, and 60 mV for the neonatal cell) for 500 ms. Arrows represent the zero current level. Current–voltage relationship for the currents shown in (A) and (C) are shown in (B) and (D), respectively. □, Peak outward current induced by the depolarizing stimulus; ●, current remaining at the end of the 500-ms pulse. The difference between the peak and sustained current components is I_{to}.

reported to increase about 4-fold between day 1 and day 10 (Kilborn and Fedida, 1990). Wickenden *et al.* (1997) investigated action potential characteristics and I_{to} in cultured neonatal rat ventricular myocytes and adult rat hearts (Fig. 7). They found that in myocytes isolated from the right ventricle of adult rat heart, action potential duration was short and independent of rate of stimulation. Recovery from inactivation was best described by a single exponential. In contrast, action potential duration was prolonged and markedly rate-dependent in cultured neonatal rat ventricular myocytes. The current density of I_{to} measured in cultured ventricular myocytes from 1- to 2-day-old rats was larger than in the adults. These investigators found that recovery from inactivation for I_{to} was best described by the sum of two exponentials. They concluded that, in rat ventricular cells, postnatal development involves a shortening of action potential duration and an increase in the density of I_{to}. Furthermore, rat heart development is also associated with a loss of rate dependence of action potential duration and an acceleration in the rate of recovery of I_{to}.

Others have reported that the increase in I_{to} contributes to the abbreviation of the action potential that occurs in ventricular cells of the neonatal rat. It has been demonstrated that current density in adult myocytes is markedly larger than that in the neonatal period for both rat (Wahler *et al.*, 1994) and canine (Jeck and Boyden, 1992) ventricular cells. In addition, a substantial amount of I_{to} has been observed in early embryonic chick heart cells (Satoh and Sperelakis, 1994). In contrast, I_{to} has been described in adult but not neonatal canine myocytes. This lack of I_{to} in the neonatal heart appears to correlate with the absence of phase 1 early repolarization of the action potential, prolongation of action potential duration, and insensitivity of the action potential plateau to rapid pacing (Jeck and Boyden, 1992).

I_{to} has recently been described in human atrial cells (Crumb *et al.*, 1995). This study reported an increase in the prevalence and current density of I_{to} in patients older than 2.5 years of age as compared to patients less than 10 months of age. However, recovery of I_{to} from inactivation was more rapid in cells from older hearts.

6. MUSCARINIC K$^+$ CURRENT

The muscarinic K$^+$ channel is present mainly in supraventricular tissue and plays a major role in the negative chronotropic effect of parasympathetic stimulation. The muscarinic K$^+$ current ($I_{K(ACh)}$) is activated by both the neurotransmitter acetylcholine (ACh) and local hormones, such as adenosine and endothelin (Kim, 1991; Kurachi *et al.*, 1986).

Using whole-cell voltage-clamp techniques, Takano and Noma (1997) studied developmental changes in muscarinic K$^+$ current in fetal and neonatal rat heart. They found that the current density of ACh-induced K$^+$ current in ventricular cells increased on gestational day 12, reached a maximum on neonatal day 20, and then decreased in adult myocytes. Developmental changes in adenosine-induced K$^+$ current were shown to follow a similar time course, except for a remarkable decrease after neonatal day 10. The authors speculated that changes in $I_{K(ACh)}$ may also be related to the functional development of cardiac myocytes. In contrast, another group found that the $I_{K(ACh)}$ is present in the ventricle at all stages of development but at a much lower density than in the atrium (Xie *et al.*, 1997). It has been reported that adenosine receptors potently regulate mammalian heart rates via multiple effector systems at very early stages of prenatal development (Hofman *et al.*, 1997).

7. SUMMARY

While this overview of cardiac K$^+$ channel development does not suggest a single unifying theme as regards developmental changes in the expression or regulation of cardiac ion channels, some generalization is possible. For example, Ca^{2+} and K$^+$ channel currents become more prominent with development. By contrast, developmental changes in Na^+ currents are less dramatic, and Na^+–Ca^{2+} exchange currents appear to decrease with age. Concomitant changes in other physiological factors, such as local intracellular Ca^{2+} accumulation, increased resting membrane potential, and decreased heart rate in the mature heart may inhibit or augment channel activity. These developmental changes in ion channel expression and function are likely to have a significant effect on the generation of the action potential in individual myocytes.

Developmental changes in the characteristics of the action potential may, therefore, have a major influence on the initiation, propagation, and termination of autonomic, triggered, and reentrant arrhythmias. Progress in this area will provide an essential foundation for the evolution of a systematic approach to pediatric arrhythmias comparable to that being developed for arrhythmias in the mature heart.

REFERENCES

Benndorf, K., and Nilius, B., 1988, Properties of an early outward current in single cells of the mouse ventricle, *Gen. Physiol. Biophys.* **7**:449–466.

Chen, F., Wetzel, G. T., Friedman, W. F., and Klitzner, T. S., 1991, Single-channel recording of inwardly rectifying potassium currents in developing myocardium, *J. Mol. Cell. Cardiol.* **23**:259–267.

Chen, F., Wetzel, G. T., Friedman, W. F., Klitzner, T. S., 1992a, ATP—sensitive potassium channels in neonatal and adult rabbit ventricular myocytes, *Pediatr. Res.* **32**:230–235.

Chen. F., Wetzel, G. T., Friedman, W. F., and Klitzner, T. S., 1992b, Smaller single channel K^+ conductance in neonatal cardiac cells is due to decreased sub-state current amplitude, *Am. J. Cardiol.* **70**:561–568.

Cohen, N. M., and Lederer, W. J., 1988, Changes in the calcium current of rat heart ventricular myocytes during development, *J. Physiol. (London)* **406**:115–146.

Colastsky, J., Follmer, C. H., and Starmer, C. F., 1990, Channel specificity in antiarrhythmic drug action: Mechanism of potassium channel block and its role in suppressing and aggravating cardiac arrhythmias, *Circulation* **82**:2235–2242.

Crumb, W. J., Jr., Pigott, J. D., and Clarkson, C. W., 1995, Comparison of I_{to} in young and adult human atrial myocytes: Evidence for developmental changes, *Am. J. Physiol.* **268**:H1335–1342.

Davies, M. P., An, R. H., Doevendans, P., Kubalak, S., Chien, K. R., and Kass, R. S., 1996, Developmental changes in ionic channel activity in the embryonic murine heart, *Circ. Res.* **78**:15–25.

Fermini, B., and Schanne O. F., 1991, Determinants of action potential duration in neonatal rat ventricle cells, *Cardiovasc. Res.* **25**:235–243.

Folander, K., Smith, J. S., Antanavage, J., Bennett, C., Stein, R. B., and Swanson, R., 1990, Cloning and expression of the delayed-rectifier IsK channel from neonatal rat heart and diethylstilbestrol-primed rat uterus, *Proc. Natl. Acad. Sci. U.S.A.* **87**:2975–2979.

Gintant, G. A., Cohen, I. S., Datyner, N. B., and Klein, R. P., 1991, Time-dependent outward currents in heart, in: (H. A. Fozzard, E. Haber, R. B. Jennings, A. M. Katz, and A. E. Morgan, eds.), *The Heart and Cardiovascular System: Scientific Foundations*, 2nd ed. Raven Press, New York, pp. 1121–1169.

Hausdorff, S. F., Goldstein, S. A., Rushin, E. E., and Miller, C., 1991, Functional characterization of a minimal K^+ channel expressed from a synthetic gene, *Biochemistry* **30**:3341–3346.

Hofman, P. L., Hiatt, K., Yoder, M. C., and Rivkees, S. A., 1997, A_1 adenosine receptors potently regulate heart rate in mammalian embryos, *Am. J. Physiol.* **273**:R1374–R1380.

Honoré, E., Attali, B., Romey, G., Heurteaux, C., Ricard, P., Lesage, F., Lazdunski, M., and Barhanin, J., 1991, Cloning, expression, pharmacology and regulation of a delayed rectifier K^+ channel in mouse heart, *EMBO J.* **10**:2805–2811.

Isenberg, G., Vereecke, J., Van der Heyden, G., and Carmeliet, E., 1983, The shortening of the action potential by DNP in guinea-pig ventricular myocytes is mediated by an increase of a time-independent K conductance, *Pflügers Archi.* **397**:251–259.

Jeck, C. D., and Boyden, P. A., 1992, Age-related appearance of outward currents may contribute to developmental differences in ventricular repolarization, *Circ. Res.* **71**:1390–1403.

Josephson, I. R., and Brown, A. M., 1986, Inwardly rectifying single-channel and whole cell K^+ currents in rat ventricular myocytes, *J. Membr. Biol.* **94**:19–35.

Josephson, I. R., and Sperelakis, N., 1990, Developmental increases in the inwardly-rectifying K^+ current of embryonic chick ventricular myocytes, *Biochim. Biophys. Acta* **1052**:123–127.

Kakei, M., Noma, A., and Shibasaki, T., 1985, Properties of adenosine-triphosphate-regulated potassium channels in guinea-pig ventricular cells, *J. Physiol. (London)* **363**:441–462.

Kato, Y., Masumiya, H., Agata, N., Tanaka, H., and Shigenobu, K., 1996, Developmental changes in action potential and membrane currents in fetal, neonatal and adult guinea-pig ventricular myocytes, *J. Mol. Cell. Cardiol.* **28:**1515–1522.

Katz, A. M., 1977, *Physiology of the Heart*, Raven Press, New York, pp. 229–256.

Katz, A. M., 1993, Cardiac ion channels, *N. Engl. J. Med.* **328:**1244–1251.

Kilborn, M. J., and Fedida, D., 1990, A study of the developmental changes in outward currents of rat ventricular myocytes, *J. Physiol. (London)* **430:**37–60.

Kim, D., 1991, Endothelin activation of an inwardly rectifying K$^+$ current in atrial cells, *Circ. Res.* **69:**250–255.

Kurachi, Y., Nakajima, T., and Sugimoto, T.,1986, On the mechanism of activation of muscarinic K$^+$ channels by adenosine in isolated atrial cells: Involvement of GTP-binding proteins, *Pflügers Arch.* **407:**264–274.

Masuda, H., and Sperelakis, N., 1993, Inwardly rectifying potassium current in rat fetal and neonatal ventricular cardiomyocytes, *Am. J. Physiol.* **265:**H1107–H1111.

Noble, D., and Bett, G., 1993, Reconstructing the heart: A challenge for integrative physiology, *Cardiovasc. Res.* **27:**1701–1712.

Noma, A., 1983, ATP-regulated K$^+$ channels in cardiac muscle, *Nature* **305:**147–148.

Noma, A., and Shibasaki, T., 1985, Membrane current through adenosine-triphosphate-regulated potassium channels in guinea-pig ventricular cells, *J. Physiol. (London)* **363:**463–480.

Nuss, H. B., and Marban, E., 1994, Electrophysiological properties of neonatal mouse cardiac myocytes in primary culture, *J. Physiol. (London)* **479:**265–279.

Osaka, T., Ramza, B. M., Jan, R. C., and Joyner, R. W., 1989, Developmental changes in the electrophysiologic properties of rabbit papillary muscles, *Pediatr. Res.* **26:**543–547.

Sanguinetti, M. C., and Jurkiewicz, N. K., 1990, Two components of cardiac delayed rectifier K$^+$ current: Differential sensitivity to block by class III antiarrhythmic agents, *J. Gen. Physiol.* **96:**195–215.

Sanguinetti, M. C., Jiang, C., Curran, and M. E., and Keating, M.T., 1995, A mechanistic link between an inherited and an acquired cardiac arrhythmia: HERG encodes the I_{Kr} potassium channel, *Cell.* **81:**299–307.

Satoh, H., and Sperelakis, N., 1994, Identification of and developmental changes in transient outward current in embryonic chick cardiomyocytes. *J. Dev. Physiol.* **20:**149–154.

Sperelakis, N., and Shigenobu, K., 1972, Changes in membrane properties of chick embryonic hearts during development, *J. Gen. Physiol.* **60:**430–453.

Takano, M., and Noma, A., 1997, Development of muscarinic potassium current in fetal and neonatal rat heart, *Am. J. Physiol.* **272:**H1188–H1195.

Takumi, T., Ohkubo, H., and Nakanishi, S., 1988, Cloning of a membrane protein that induces a slow voltage-gated potassium current, *Science* **242:**1042–1045.

Task Force of the Working Group on Arrhythmias of the European Society of Cardiology, 1991, The Sicilian Gambit: A new approach to the classification of antiarrhythmic drugs based on their actions on arrhythmogenic mechanisms, *Circulation* **84:**1831–1851.

Wahler, G. M., 1992, Developmental increases in the inwardly rectifying potassium current of rat ventricular myocytes, *Am. J. Physiol.* **262:**C1266–C1272.

Wahler, G. M., Dollinger, S. J., Smith, J. M., and Flemal, K.L., 1994, Time course of postnatal changes in rat heart action potential and in transient outward current is different, *Am. J. Physiol.* **267:**H1157–H1166.

Wang, L., and Duff, H. J., 1995, The fast component of delayed rectifier K$^+$ current (I_{Kr}) in fetal mouse ventricular myocytes, *Biophys. J.* **68:**A36 (Abstract).

Wang, L., Feng, Z. P., Kondo, C. S., Sheldon, R. S., and Duff, H. J., 1996, Developmental changes in the delayed rectifier K$^+$ channels in mouse heart, *Circ. Res.* **79:**79–85.

Weiss, J., and Shine, K. I., 1982, Extracellular K$^+$ accumulation during myocardial ischemia in isolated rabbit, *Am. J. Physiol.* **242:**H619–H628.

Wickenden, A. D., Kaprielian, R., Parker, T. G., Jones, O. T., and Backx, P. H., 1997, Effects of development and thyroid hormone on K$^+$ currents and K$^+$ channel gene expression in rat ventricle. *J. Physiol. (London)* **504:**271–286.

Xie, L. H., Takano, M., and Noma, A., 1997, Development of inwardly rectifying K$^+$ channel family in rat ventricular myocytes, *Am. J. Physiol.* **272:**H1741–H1750.

Part V

Potassium Channels in Vascular Smooth Muscle

Chapter 22

Overview: Physiological Role of K^+ Channels in the Regulation of Vascular Tone

Joseph E. Brayden

1. INTRODUCTION

Calcium entry through voltage-dependent Ca^{2+} channels plays an important role in the contractile responses of vascular smooth muscle, particularly in the resistance vasculature. Thus, agents or interventions that open or close these Ca^{2+} channels have significant effects on smooth muscle contractile activity. Accordingly, regulation of smooth muscle membrane potential through changes in K^+ channel activity, and subsequent alterations in the activity of voltage-dependent Ca^{2+} channels, is a major mechanism of vasodilation and vasoconstriction, both in normal and in pathophysiological conditions. Activation of K^+ channels will result in hyperpolarization, closure of voltage-dependent Ca^{2+} channels, and vasodilation, whereas inhibition of K^+ channels will have the opposite effects. Several different types of K^+ channels are present in most vascular smooth muscle cells. Membrane potential and diameter are determined, in part, by the integrated activity of these K^+ channels, which are regulated by multiple dilator and constrictor signals in vascular smooth muscle. The objective of this chapter is to provide an overview of the evidence supporting the important functional roles of the major K^+ channels found in vascular smooth muscle.

2. BK_{Ca} CHANNELS: ROLE IN REGULATION OF MYOGENIC TONE

Large-conductance, Ca^{2+}-activated K^+ (BK_{Ca}) channels have been cloned from several tissue sources, and the biophysical characteristics of these channels have been extensively characterized (Nelson and Quayle, 1995). A prominent physiological role

Joseph E. Brayden • Department of Pharmacology, The University of Vermont, Burlington, Vermont. 05405-9998

Potassium Channels in Cardiovascular Biology, edited by Archer and Rusch. Kluwer Academic/Plenum Publishers, New York, 2001.

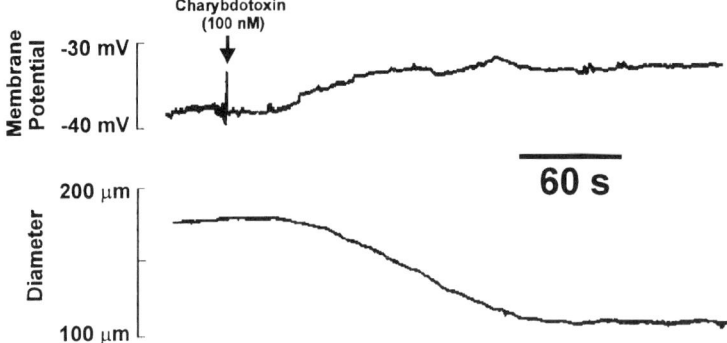

Figure 1. Charybdotoxin, a blocker of large-conductance, Ca^{2+}-sensitive K^+ channels, depolarizes and constricts rabbit cerebral arteries with myogenic tone. (Reproduced, with permission, from Brayden and Nelson, 1992. Copyright 1992 American Association for the Advancement of Science.)

for BK_{Ca} channels in vascular smooth muscle has been revealed whereby these channels play a negative-feedback role, opposing the depolarization, Ca^{2+} entry, and constriction that occur in cerebral arteries during the development of pressure-induced myogenic tone (Brayden and Nelson, 1992) (Fig. 1). Both depolarization and elevated cytosolic Ca^{2+} concentration, $[Ca^{2+}]_i$, increase BK_{Ca} channel activity, and the associated hyperpolarization prevents excessive vasoconstriction that occurs in the absence of this feedback system. Further advances in understanding the precise pathways and mechanisms by which this negative-feedback system is controlled have recently been made (Nelson et al., 1995; Jaggar et al., 1998; Perez et al., 1999). BK_{Ca} channels in vascular smooth muscle are activated by localized bursts of Ca^{2+} (Ca^{2+} sparks) derived from the sarcoplasmic reticulum (SR) just under the sarcolemma. These Ca^{2+}-release events do not change global Ca^{2+} concentrations directly. Ca^{2+} sparks do result in large local changes in $[Ca^{2+}]_i$. As a result, many nearby BK_{Ca} channels are activated, which is observed macroscopically in the form of spontaneous transient outward currents (STOCs). The net effect of an increased frequency of STOCs is hyperpolarization of the smooth muscle cell membrane, decreased Ca^{2+} entry through voltage-dependent Ca^{2+} channels, and decreased global $[Ca^{2+}]_i$. When the smooth muscle cells are depolarized, that is, when intravascular pressure is increased or in the presence of some vasoconstrictor agonists, Ca^{2+} entry through voltage-dependent Ca^{2+} channels is increased. This in turn increases the Ca^{2+} load of the sarcoplasmic reticulum with a subsequent increase in the amplitude and frequency of Ca^{2+} sparks and STOCs. Thus, the relationship between membrane depolarization, increased Ca^{2+} entry and activation of BK_{Ca} channels is similar to that originally proposed, but with the addition of another key regulatory site involving the sarcoplasmic reticulum, ryanodine-sensitive Ca^{2+} release channels, and Ca^{2+} sparks.

Several endogenous vasodilators may act, at least in part, through activation of BK_{Ca} channels. For instance, adenosine-induced relaxations of rat coronary arteries in vitro are inhibited by BK_{Ca} channel blockers such as tetraethylammonium (TEA; 1 mM) and iberiotoxin (Cabell et al., 1994). This finding is consistent with the observation that protein kinase A (PKA) can activate BK_{Ca} channels in coronary arteries

(Scornik *et al.*, 1993), since it is known that adenosine activates adenylyl cyclase in vascular smooth muscle. Interestingly, cGMP-dependent protein kinase (PKG) also can activate BK$_{Ca}$ channels in arterial smooth muscle (Robertson *et al.*, 1993; Taniguchi *et al.*, 1993). This could be an important mechanism by which the endothelium-derived vasodilator nitric oxide relaxes arteries, since nitric oxide activates PKG. Although PKA and PKG can directly activate BK$_{Ca}$ channels, some of the activation may also involve modulation of Ca^{2+} spark activity. Activators of PKA (adenosine, forskolin, cAMP) and of PKG (sodium nitroprusside, nicorandil) increase Ca^{2+} spark and STOC frequency, and this may contribute to their dilator effect (Porter *et al.*, 1998).

Another coronary endothelial factor known as endothelium-derived hyperpolarizing factor (EDHF), which is distinct from nitric oxide and thought to be an epoxyeicosatrienoic acid, also activates smooth muscle BK$_{Ca}$ channels (Campbell *et al.*, 1996), apparently via a G-protein-coupled pathway (Li and Campbell, 1997). Thus, both direct and indirect signaling pathways for activation of vascular smooth muscle BK$_{Ca}$ channels by endogenous vasodilators have been documented. Finally, nitric oxide may directly activate BK$_{Ca}$ channels (i.e., independent of guanylyl cyclase/cGMP/PKG activity) in aortic smooth muscle (Bolotina *et al.*, 1994).

Many vasoconstrictors depolarize vascular smooth muscle, and one possible mechanism underlying this response is the direct inhibition of BK$_{Ca}$ channels. Angiotensin II and the thromboxane A$_2$ agonist U46619 do, in fact, inhibit BK$_{Ca}$ channels from coronary artery smooth muscle (Toro *et al.*, 1990; Scornik and Toro, 1992) in lipid bilayers. Vasoconstrictors may also inhibit BK$_{Ca}$ channel activity via a PKC-dependent inhibition of Ca^{2+} spark and STOC activity (Bonev *et al.*, 1997).

3. SK$_{Ca}$ CHANNELS: POSSIBLE ROLE IN ACTIONS OF EDHF

Recent evidence from studies of intact arteries suggests that small-conductance, Ca^{2+}-activated K$^+$ (SK$_{Ca}$) channels are present in vascular smooth muscle and may be involved in endothelium-dependent vasodilator responses in some instances. The SK$_{Ca}$ channel has been characterized in skeletal muscle and other tissues and is a low-conductance, Ca^{2+}-sensitive K$^+$ channel that is voltage-independent and inhibited by apamin (K$_i \approx 0.3$ nM) (Garcia-Pascual *et al.*, 1991). In contrast to BK$_{Ca}$ channels, where Ca^{2+} activation occurs via interactions at a specific domain within the channel protein structure, the Ca^{2+} sensitivity of SK$_{Ca}$ channels is determined by tightly bound calmodulin (Xia *et al.*, 1998). Molecular evidence for the presence of SK$_{Ca}$ channels in vascular smooth muscle is limited. However, mRNA for an apamin-sensitive K$^+$ channel is expressed in porcine vascular smooth muscle (Sokol *et al.*, 1994), suggesting that the channel is indeed present in vascular smooth muscle. An apamin-sensitive K$_{Ca}$ channel, albeit of intermediate conductance (~ 70 pS), has been observed in renal vascular smooth muscle (Gebremedhin *et al.*, 1996).

Pharmacological data from studies of intact arteries suggest that apamin-sensitive K$^+$ channels are involved in the mechanism of relaxation of some arteries to EDHF. Apamin, as well as an elevated K$^+$ concentration in the bathing medium, blocked a nitroarginine-resistant, acetylcholine-induced decrease in perfusion pressure of an isolated mesenteric vascular bed preparation (Adeagbo and Triggle, 1993). Very similar results were obtained in isolated ring segments of bovine oviductal artery (Garcia-Pascual *et al.*, 1995). Direct measurements of EDHF-evoked hyperpolarization in rabbit mesenteric vascular smooth muscle also indicate a role for apamin-sensitive K^+ channels in this response (Murphy and Brayden, 1995a). Additional studies are required to confirm the presence and role of apamin-sensitive K^+ channels in the vascular response to EDHF and possibly to other vasodilator signals.

4. Kv CHANNELS: REGULATION OF MEMBRANE POTENTIAL AND INHIBITION BY AGONISTS

Kv channels are voltage-dependent K^+ channels that are activated by membrane depolarization. Kv channels have been observed in vascular smooth muscle (Clapp and Gurney, 1991; Gelband and Hume, 1992; Smirnov and Aaronson, 1992; Volk and Shibata, 1993; Robertson and Nelson, 1994) and other types of smooth muscle (Boyle *et al.*, 1992; Carl, 1995). These channels are inhibited by aminopyridines [4-aminopyridine(4-AP) and 3,4-diaminopyridine(3,4-DAP)] but not by blockers of several other K^+ channels. Single-channel conductances for the Kv channel fall into two ranges, ~ 5–8 pS (Volk and Shibata, 1993; Robertson and Nelson, 1994) and ~ 60–70 pS (Ishikawa *et al.*, 1993; Gelband *et al.*, 1993). Kv channels have been cloned. Two different cDNA clones, Kv1.5 and Kv1.2, both expressing 4-AP-sensitive Kv channels, have been isolated from canine smooth muscle (Hart *et al.*, 1993; Overturf *et al.*, 1994). These channel proteins contain several consensus sites for PKC-mediated phosphorylation within the cytosolic domains, suggesting the potential for channel regulation by phosphorylation.

Kv channels undergo inactivation following membrane depolarization. However, estimates of the degree of voltage-dependent activation and inactivation of these channels suggest that steady-state current through Kv channels should be sizable at physiological membrane potentials. Direct measurements of steady-state Kv currents in cerebral artery myocytes have verified this concept (Robertson and Nelson, 1994). Pharmacological and biophysical studies indicate that more than one subtype of Kv channel is present in some vascular smooth muscle cells (Clapp and Gurney, 1991; Robertson and Nelson, 1994). Kv channels are inhibited by intracellular Ca^{2+} and Mg^{2+} (Gelband *et al.*, 1993) and activated by intracellular ATP (Evans *et al.*, 1994).

Data from studies of intact tissues indicate several possible functional roles for Kv channels in vascular smooth muscle. Kv channels appear to play a significant role in setting the steady-state values of membrane potential in intact smooth muscle cells. Because these channels are activated by depolarization, they may serve in part to limit the membrane depolarization that occurs in response to vasoconstrictor stimuli (Nelson and Quayle, 1995). Inhibition of Kv channels depolarizes current-clamped single vascular smooth muscle cells (Yuan, 1995) and, in intact pressurized arteries, causes substantial, additional depolarization and contraction (Knot and Nelson, 1995) (Fig. 2), as would be predicted if Kv channels contribute significantly to the overall membrane conductance under these conditions. Several investigators have proposed that hypoxic pulmonary vasoconstriction involves inhibition of Kv channels (Post *et al.*, 1992; Yuan *et al.*, 1995; Smirnov *et al.*, 1994; Archer *et al.*, 1998). Recent evidence

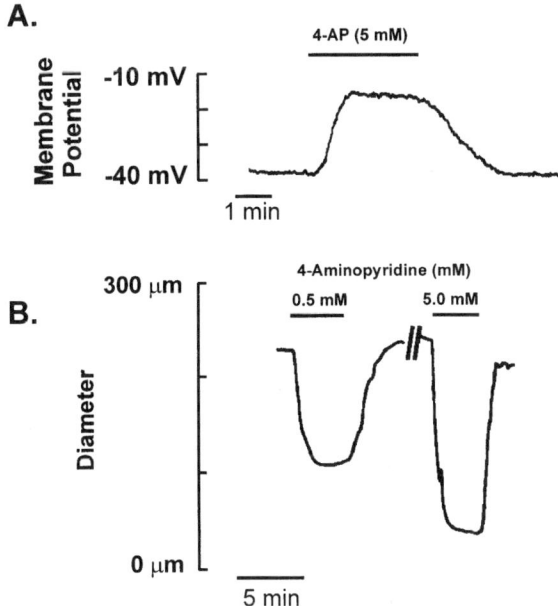

Figure 2. 4-Aminopyridine, (4-AP) a blocker of voltage-dependent K⁺ channels, depolarizes (A) and constricts (B) cerebral arteries with tone. (Reproduced, with permission from Knot and Nelson, 1995.)

indicates that Kv channels can be regulated by receptor ligands. Activation of β-adrenoceptors, adenylyl cyclase, and PKA increased the amplitude of Kv currents in rabbit portal vein smooth muscle cells (Aiello *et al.*, 1995). Nitric oxide activates Kv channels in rat pulmonary artery myocytes, and nitric oxide-induced relaxations of intact pulmonary arteries are inhibited by 4-AP (Yuan *et al.*, 1996). Histamine (Ishikawa *et al.*, 1993) and angiotensin II (Gelband and Hume, 1995; Clement-Chomienne *et al.*, 1996) inhibit Kv currents activated by voltage steps in isolated coronary and renal artery and portal vein myocytes. In canine renal artery myocytes, part of the inhibitory effect of angiotensin II could be ascribed to release of Ca^{2+} from intracellular stores, which then inhibits Kv channels. Activation of PKC results in inhibition of Kv channels in cerebral artery and portal vein smooth muscle (Cole *et al.*, 1996).

5. K_ATP CHANNELS: ACTIVATION BY PHARMACOLOGICAL AND ENDOGENOUS VASODILATORS

ATP-sensitive potassium (K_{ATP}) channels have been identified in numerous cell types, including vascular smooth muscle (Ashcroft and Ashcroft, 1990). A primary distinguishing feature of this channel is that it is inhibited by exposure of the intracellular face of the cell membrane to ATP ($K_i \cong 50–500\,\mu$M. Other factors, such as the ADP/ATP ratio (Lederer and Nichols, 1989) or intracellular pH (Davies *et al.*,

1992) can alter the sensitivity of the channel to inhibition by ATP. Cloning studies indicate that K_{ATP} channels are composed of at least two subunits, one a permeation subunit and one a regulatory protein that binds sulfonylurea agents (Inagaki *et al.*, 1995).

K_{ATP} channels are inhibited by several agents, most specifically by sulfonylurea compounds such as glibenclamide and tolbutamide (Ashcroft and Ashcroft, 1990; Quayle *et al.*, 1997). These channels were first directly identified in vascular smooth muscle with the use of the patch-clamp technique (Standen *et al.*, 1989) and have since been found in several different vascular smooth muscle types (Inoue *et al.*, 1989; Kajioka *et al.*, 1991; Lorenz *et al.*, 1992; Clapp and Gurney, 1992; Russell *et al.*, 1992).

K_{ATP} channels in vascular smooth muscle can be activated by a number of pharmacological and endogenous substances [for a review, see Nelson and Quayle (1995)]. Vascular responses to pharmacological vasodilators such as cromakalim, pinacidil, diazoxide, minoxidil, and nicorandil are well defined. A key feature that these substances have in common is their ability to increase K^+ conductance in vascular smooth muscle and thereby hyperpolarize the smooth muscle cell membrane and inhibit tone. The open-state probability of single K_{ATP} channels in vascular smooth muscle cells is enhanced by 10- to 100-fold by cromakalim, depending on the concentration of ATP to which the cells are exposed, and glibenclamide dramatically inhibits this effect (Standen *et al.*, 1989). Pharmacological K^+ channel openers increase the efflux of labeled K^+ or rubidium, and this effect is inhibited by K_{ATP} channel blockers (Quast and Cook, 1989). The K^+ channel openers also induce a glibenclamide-sensitive hyperpolarization of vascular smooth muscle cells, either in the intact tissue (Brayden *et al.*, 1991) or in isolated cells (Clapp and Gurney, 1992). In the presence of tone, isolated arteries are relaxed by these K^+ channel openers, and this response can be reversed by the application of glibenclamide (Standen *et al.*, 1989).

K_{ATP} channel activation is also an important mechanism of action of endogenous vasodilators. Calcitonin gene-related peptide (CGRP), a potent endogenous vasodilator found in perivascular nerves in many tissues, activates K_{ATP} channels in isolated vascular smooth muscle cells (Fig. 3) (Nelson *et al.*, 1990; Miyoshi and Nakaya, 1995). CGRP also hyperpolarizes the smooth muscle cells in intact arteries, and this action is abolished by glibenclamide but is not affected by inhibitors of K_{Ca} channels. The effects of CGRP are mediated by the adenylyl cyclase/cAMP/PKA second-messenger pathway

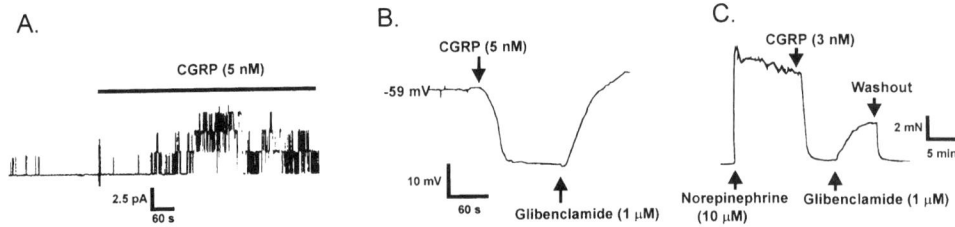

Figure 3. Calcitonin gene-related peptide (CGRP) activates K_{ATP} channels (A), hyperpolarizes mesenteric artery smooth muscle cells (B), and relaxes intact mesenteric artery (C). The latter response is partially reversed by glibenclamide. (Reproduced, with permission, from Nelson *et al.*, 1990. Copyright 1990 Macmillan Magazines Limited.)

(Quayle *et al.*, 1994), suggesting that phosphorylation of the K_{ATP} channel is an important activation pathway. In contrast to the actions of cromakalim, CGRP-induced vasodilation is only partially inhibited by glibenclamide. Thus, for this as well as some other endogenous dilators, membrane hyperpolarization due to activation of K_{ATP} channels may account for only part of the overall mechanism of dilation. Vasoactive intestinal polypeptide (VIP) is another potent endogenous vasodilator. The effects of K_{ATP} channel blockers on the hyperpolarizing and vasodilator actions of this peptide are comparable to those indicated above for CGRP (Standen *et al.*, 1989). Endothelial-derived nitric oxide relaxes vascular smooth muscle by several mechanisms, one of which is through activation of K^+ channels. Although much of the evidence indicates that K_{Ca} channels are the primary channels involved, activation of K_{ATP} by nitric oxide may also play a role (Murphy and Brayden, 1995b). At least in the rabbit, EDHF also seems to activate K_{ATP} channels (Brayden, 1990), whereas for other endogenous vasodilators, only part of the dilation induced by endothelium-dependent vasodilators is due to the membrane hyperpolarization. Prostacyclin is a third endothelial factor that appears to hyperpolarize and relax vascular smooth muscle, in this case primarily via activation of K_{ATP} channels (Jackson *et al.*, 1993; Murphy and Brayden, 1995b).

Adenosine has long been implicated as an important endogenous vasodilator in the coronary circulation. Activation of K_{ATP} channels plays a significant role in the coronary vasodilator response to adenosine in several species (Daut *et al.*, 1990; Belloni and Hintze, 1991; Merkel *et al.*, 1992). The specific adenosine receptor involved (A_1, A_2) seems to be tissue and species specific, but presumably activation of PKA signaling is involved (Kleppisch and Nelson, 1995). In coronary arteries and in the microcirculation, K_{ATP} channels also play a role in control of resting vascular tone. K_{ATP} channels are active under resting conditions and contribute to the maintenance of basal coronary vascular tone (Imamura *et al.*, 1992; Samaha *et al.*, 1992). Infusion of glibenclamide into the coronary circulation causes large increases in coronary vascular resistance that are reversed upon removal of the K_{ATP} channel blocker. In the microcirculation of the hamster cheek pouch and cremaster muscle, glibenclamide causes concentration-dependent vasoconstriction, indicating a tonic activation of K_{ATP} channels in these vascular beds (Jackson, 1993). Ongoing activity and contributions of K_{ATP} channels to resting membrane potential of pulmonary artery myocytes have been reported (Clapp and Gurney, 1992), although correlative studies of the effects of K_{ATP} channel blockers in intact pulmonary arteries have not been reported.

Part of the mechanism of action of vasoconstrictors may involve inhibition of K^+ channels and an associated depolarization. Endothelin (Miyoshi *et al.*, 1992), vasopressin (Wakatsuki *et al.*, 1992), and angiotensin II (Miyoshi and Nakaya, 1991) may act this way in part through inhibition of K_{ATP} channels. In light of evidence suggesting that K_{ATP} channels are tonically active in coronary artery smooth muscle cells, inhibition of these channels could represent an important mechanism of depolarization and constriction by a variety of agents in the coronary circulation.

Coronary, cerebral, and skeletal muscle arteries dilate in response to hypoxia. Daut *et al.* (1990) found that hypoxia induced a pronounced vasodilation in isolated, intact guinea pig hearts and that this response was blocked by glibenclamide. Metabolic inhibitors such as dinitrophenol, cyanide, or 2-deoxyglucose also induce a glibenclamide-sensitive vasodilation in this preparation. Thus, it seems likely that interventions that alter cellular metabolism and, therefore, production of ATP may be sufficient to activate K_{ATP} channels in coronary vascular smooth muscle cells and cause vasodilation. In these studies, adenosine did not appear to be a mediator of the hypoxic vasodilation, in that 8-phenyltheophylline, an adenosine receptor antagonist, inhibited the response to exogenous adenosine but had little or no effect on the dilator response to hypoxia. Reactive hyperemia in the coronary (Kanatsuka et al., 1992) and cerebral (Bari et al., 1998) circulations, which occurs following a brief period of ischemia, may involve activation of K_{ATP} channels. This may be related to release of adenosine during the period of reperfusion or to other mechanisms whereby arterial K_{ATP} channels are activated. The autoregulatory vasodilator response of small (<100 μm) epicardial arteries that occurs during mild or severe coronary stenosis or during coronary artery occlusion in vivo may also involve activation of K_{ATP} channels (Komaru et al., 1991). In the coronary (Ishizaka and Kuo, 1996) and cerebral circulations (Kinoshita and Katusic, 1997), acidosis induces a pH-dependent, but endothelium-independent, vasodilation that is reversed by glibenclamide.

6. Kir CHANNELS: REGULATION OF RESTING MEMBRANE POTENTIAL AND ROLE IN K⁺-INDUCED DILATIONS

Inward rectifier K^+ (Kir) channels were first identified in vascular smooth muscle using electrophysiological approaches (Hirst and Edwards, 1989). More recent studies have documented the presence of Kir currents in single smooth muscle cells isolated from cerebral (Quayle et al., 1993) and coronary (Robertson et al., 1996) arteries using the patch-clamp technique. Potassium currents through Kir channels in arterial smooth muscle, as in other cells, show inward rectification. Such currents are activated by hyperpolarization and by increases in extracellular K^+ in the concentration range between 7 and 15 mM. Kir channels in vascular smooth muscle are inhibited by low concentrations of barium ions, with a half-inhibition constant of 2.2 μM at -60 mV. Barium blockade of Kir channels increases with hyperpolarization (e-fold increase in sensitivity for a 24-mV hyperpolarization). Kir channels are also blocked by Cs^+ ions ($K_d = 1.6$ mM at -50 mV) but are relatively unaffected by other K^+ channel blockers (Quayle et al., 1993).

Studies of intact arteries have confirmed the presence and functional role of Kir channels in vascular smooth muscle. Elevations of extracellular K^+ from normal levels to only 7–10 mM causes a large hyperpolarization and vasodilation of cerebral arteries (Fig. 4) (Hirst and Edwards, 1989; Knot et al., 1996). These effects of K^+ are reversed by low concentrations of barium (<10 μM), suggesting a specific activation of Kir channels. This cerebral vasodilator mechanism may contribute to the vasodilation observed during increased neuronal activity, hypoxia, ischemia, or hypoglycemia, each of which is associated with elevations in extracellular K^+ ions (see Nelson and Quayle, 1995). Moreover, a recent report has proposed that K^+ released from endothelial cells may be a type of EDHF, which activates Kir channels on adjacent vascular smooth muscle cells (Edwards et al., 1998). In some arteries, the resting membrane potential (i.e., membrane potential in unstimulated arteries; -60 to -75 mV) is significantly depolarized by concentrations of barium ions

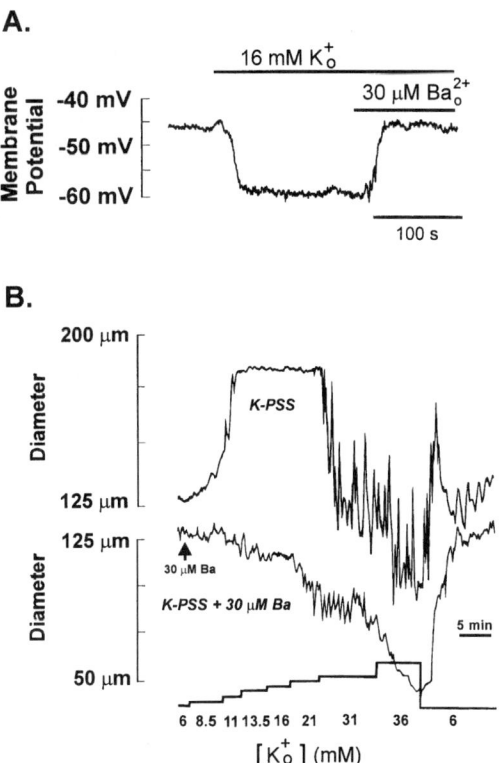

Figure 4. Elevation in extracellular K$^+$ induces a barium-sensitive hyperpolarization (A) and dilation of cerebral arteries with myogenic tone (B). (Reproduced, with permission, from Knot *et al.*, 1996.)

that should selectively inhibit Kir channels (Knot *et al.*, 1996). This suggests that Kir channels may make a significant contribution to the overall membrane K$^+$ conductance in unstimulated arteries.

7. SUMMARY

Potassium channel activity plays a predominant role in setting the membrane potential of vascular smooth muscle cells, which in turn regulates vascular tone and the local control of blood flow in virtually every organ system. A change in membrane potential via altered K$^+$ channel activity influences the activity of voltage-dependent Ca^{2+} channels, which results in changes in intracellular [Ca^{2+}] and altered vascular tone. Four major types of K$^+$ channels (K$_{Ca}$, Kv, Kir, and K$_{ATP}$) have been identified in vascular smooth muscle. High-conductance, Ca^{2+} activated K$^+$ (BK$_{Ca}$) channels act as negative-feedback regulators of intrinsic (myogenic) tone and can also be activated by substances that increase intracellular cyclic nucleotide levels (e.g., nitric oxide, adenosine, and vasoactive peptides). This may occur directly or through modulation of release of Ca^{2+} from intracellular storage sites (i.e., via altered Ca^{2+} "spark" activity). Small-conductance, Ca^{2+}-activated K^+ (SK_{Ca}) channels also occur in vascular smooth muscle and may be activated by endothelial factors in some vascular beds. Kv channels are activated at physiological membrane potentials (-50 to -30 mV). Inhibition of Kv channels by hypoxia or certain agonists may contribute to vasoconstriction by these interventions. Kir channels are present in resistance artery smooth muscle, mediate extracellular K^+-induced dilations, and may be involved in autoregulation of blood flow. ATP-sensitive K^+ (K_{ATP}) channels respond to altered cellular metabolism, are tonically active in some arterial smooth muscles, and are activated by agents such as adenosine, neuropeptides, and endothelial factors. In any given vascular bed, it is the distribution of K^+ channel types and the net activity of this array of K^+ channels which regulates membrane potential, Ca^{2+} entry, and vascular resistance and ultimately helps determine blood flow.

REFERENCES

Adeagbo, A. S., and Triggle, C. R., 1993, Varying extracellular $[K^+]$: A functional approach to separating EDHF- and EDNO-related mechanisms in perfused rat mesenteric arterial bed, *J. Cardiovasc. Pharmacol.* **21**:423–429.

Aiello, E. A., Walsh, M. P., and Cole, W. C., 1995, Phosphorylation by protein kinase A enhances delayed rectifier K^+ current in rabbit vascular smooth muscle cells, *Am. J. Physiol.* **268**:H926–H934.

Archer, S. L., Souil, E., Dinh-Xuan, A. T., Schremmer, B., Mercier, J. C., El Yaagoubi, A., Nguyen-Huu, L., Reeve, H. L., and Hampl, V., 1998, Molecular identification of the role of voltage-gated K^+ channels, Kv1.5 and Kv2.1, in hypoxic pulmonary vasoconstriction and control of resting membrane potential in rat pulmonary artery myocytes, *J. Clin. Invest.* **101**:2319–2330.

Ashcroft, S. J., and Ashcroft, F. M., 1990, Properties and functions of ATP-sensitive K-channels, *Cell. Signal.* **2**:197–214.

Bari, F., Louis, T. M., and Busija, D. W., 1998, Effects of ischemia on cerebral arteriolar dilation to arterial hypoxia in piglets, *Stroke* **29**:222–227.

Belloni, F. L., and Hintze, T. H., 1991, Glibenclamide attenuates adenosine-induced bradycardia and coronary vasodilatation, *Am. J. Physiol.* **261**:H720–H727.

Bolotina, V. M., Najibi, S., Palacino, J. J., Pagano, P. J., and Cohen, R. A., 1994, Nitric oxide directly activates calcium-dependent potassium channels in vascular smooth muscle, *Nature* **368**:850–853.

Bonev, A. D., Jaggar, J. H., Rubart, M., and Nelson, M. T., 1997, Activators of protein kinase C decrease Ca^{2+} spark frequency in smooth muscle cells from cerebral arteries, *Am. J. Physiol.* **273**:C2090–C2095.

Boyle, J. P., Tomasic, M., and Kotlikoff, M. I., 1992, Delayed rectifier potassium channels in canine and porcine airway smooth muscle cells, *J. Physiol. (London)* **447**:329–350.

Brayden, J. E., 1990, Membrane hyperpolarization is a mechanism of endothelium- dependent cerebral vasodilation, *Am. J. Physiol.* **259**:H668–H673.

Brayden, J. E., and Nelson, M. T., 1992, Regulation of arterial tone by activation of Ca^{2+} dependent potassium channels, *Science* **256**:532–535.

Brayden, J. E., Quayle, J. M., Standen, N. B., and Nelson, M. T., 1991, Role of potassium channels in the vascular response to endogenous and pharmacological vasodilators, *Blood Vessels* **28**:147–153.

Cabell, F., Weiss, D. S., and Price, J. M., 1994, Inhibition of adenosine-induced coronary vasodilation by block of large- conductance Ca^{2+}-activated K^+ channels, *Am. J. Physiol.* **267**:H1455–H1460.

Campbell, W. B., Gebremedhin, D., Pratt, P. F., and Harder, D. R., 1996, Identification of epoxyeicosatrienoic acids as endothelium-derived hyperpolarizing factors, *Circ. Res.* **78**:415–423.

Carl, A., 1995, Multiple components of delayed rectifier K^+ current in canine colonic smooth muscle, *J. Physiol. (London)* **484**:339–353.

Clapp, L. H., and Gurney, A. M., 1991, Outward currents in rabbit pulmonary artery cells dissociated with a new technique, *Exp. Physiol.* **76**:677–693.

Clapp, L. H., and Gurney, A. M., 1992, ATP-sensitive K$^+$ channels regulate resting potential of pulmonary arterial smooth muscle cells, *Am. J. Physiol.* **262**:H916–H920.

Clement-Chomienne, O., Walsh, M. P., and Cole, W. C., 1996, Angiotensin II activation of protein kinase C decreases delayed rectifier K$^+$ current in rabbit vascular myocytes, *J. Physiol. (London)* **495**:689–700.

Cole, W. C., Clement-Chomienne, O., and Aiello, E. A., 1996, Regulation of 4-aminopyridine-sensitive, delayed rectifier K$^+$ channels in vascular smooth muscle by phosphorylation, *Biochem. Cell Biol.* **74**:439–447.

Daut, J., Maier-Rudolph, W., von Beckerath, N., Mehrke, G., Gunther, K., and Goedel-Meinen, L., 1990, Hypoxic dilation of coronary arteries is mediated by ATP-sensitive potassium channels, *Science* **247**:1341–1344.

Davies, N. W., Standen, N. B., and Stanfield, P. R., 1992, The effect of intracellular pH on ATP-dependent potassium channels of frog skeletal muscle, *J. Physiol. (London)* **445**:549–568.

Edwards, G., Dora, K. A., Gardener, M. J., Garland, C. J., and Weston, A. H., 1998, K$^+$ is an endothelium-derived hyperpolarizing factor in rat arteries [see comments], *Nature* **396**:269–272.

Evans, A. M., Clapp, L. H., and Gurney, A. M., 1994, Augmentation by intracellular ATP of the delayed rectifier current independently of the glibenclamide-sensitive K-current in rabbit arterial myocytes, *Br. J. Pharmacol.* **111**:972–974.

Garcia-Pascual, A., Labadia, A., Jimenez, E., and Costa, G., 1995, Endothelium-dependent relaxation to acetylcholine in bovine oviductal arteries: Mediation by nitric oxide and changes in apamin-sensitive K$^+$ conductance, *Br. J. Pharmacol.* **115**:1221–1230.

Gebremedhin, D., Kaldunski, M., Jacobs, E. R., Harder, D. R., and Roman, R. J., 1996, Coexistence of two types of Ca^{2+}-activated K$^+$ channels in rat renal arterioles, *Am. J. Physiol.* **270**:F69–F81.

Gelband, C. H., and Hume, J. R., 1992, Ionic currents in single smooth muscle cells of the canine renal artery, *Circ. Res.* **71**:745–758.

Gelband, C. H., and Hume, J. R., 1995, [Ca^{2+}]$_i$ inhibition of K$^+$ channels in canine renal artery. Novel mechanism for agonist-induced membrane depolarization, *Circ. Res.* **77**:121–130.

Gelband, C. H., Ishikawa, T., Post, J. M., Keef, K. D., and Hume, J. R., 1993, Intracellular divalent cations block smooth muscle K$^+$ channels, *Circ. Res.* **73**:24–34.

Hart, P. J., Overturf, K. E., Russell, S. N., Carl, A., Hume, J. R., Sanders, K. M., and Horowitz, B., 1993, Cloning and expression of a Kv1.2 class delayed rectifier K$^+$ channel from canine colonic smooth muscle, *Proc. Natl. Acad. Sci. U.S.A.* **90**:9659–9663.

Hirst, G. D., and Edwards, F. R., 1989, Sympathetic neuroeffector transmission in arteries and arterioles, *Physiol. Rev.* **69**:546–604.

Imamura, Y., Tomoike, H., Narishige, T., Takahashi, T., Kasuya, H., and Takeshita, A., 1992, Glibenclamide decreases basal coronary blood flow in anesthetized dogs, *Am. J. Physiol.* **263**:H399–H404

Inagaki, N., Gonoi, T., Clement, J. P., Namba, N., Inazawa, J., Gonzalez, G., Aguilar-Bryan, L., Seino, S., and Bryan, J., 1995, Reconstitution of I$_{KATP}$: An inward rectifier subunit plus the sulfonylurea receptor [see comments], *Science* **270**:1166–1170.

Inoue, I., Nakaya, Y., Nakaya, S., and Mori, H., 1989, Extracellular Ca^{2+}-activated K channel in coronary artery smooth muscle cells and its role in vasodilation, *FEBS Lett.* **255**:281–284.

Ishikawa, T., Hume, J. R., and Keef, K. D., 1993, Modulation of K$^+$ and Ca^{2+} channels by histamine H$_1$-receptor stimulation in rabbit coronary artery cells, *J. Physiol. (London)* **468**:379–400.

Ishizaka, H., and Kuo, L., 1996, Acidosis-induced coronary arteriolar dilation is mediated by ATP-sensitive potassium channels in vascular smooth muscle, *Circ. Res.* **78**:50–57.

Jackson, W. F., 1993, Arteriolar tone is determined by activity of ATP-sensitive potassium channels, *Am. J. Physiol.* **265**:H1797–H1803.

Jackson, W. F., Konig, A., Dambacher, T., and Busse, R., 1993, Prostacyclin-induced vasodilation in rabbit heart is mediated by ATP-sensitive potassium channels, *Am. J. Physiol.* **264**:H238–H243.

Jaggar, J. H., Stevenson, A. S., and Nelson, M. T., 1998, Voltage dependence of Ca^{2+} sparks in intact cerebral arteries, *Am. J. Physiol.* **274**:C1755–C1761.

Kajioka, S., Kitamura, K., and Kuriyama, H., 1991, Guanosine diphosphate activates an adenosine 5'-triphosphate-sensitive K$^+$ channel in the rabbit portal vein, *J. Physiol. (London)* **444**:397–418.

Kanatsuka, H., Sekiguchi, N., Sato, K., Akai, K., Wang, Y., Komaru, T., Ashikawa, K., and Takishima, T., 1992, Microvascular sites and mechanisms responsible for reactive hyperemia in the coronary circulation of the beating canine heart, *Circ. Res.* **71**:912–922.

Kinoshita, H., and Katusic, Z. S., 1997, Role of potassium channels in relaxations of isolated canine basilar arteries to acidosis, *Stroke* **28**:433–437.

Kleppisch, T., and Nelson, M. T., 1995, Adenosine activates ATP-sensitive potassium channels in arterial myocytes via A$_2$ receptors and cAMP-dependent protein kinase, *Proc. Natl. Acad. Sci. U.S.A.* **92**:12441–12445.

Knot, H. J., and Nelson, M. T., 1995, Regulation of membrane potential and diameter by voltage-dependent K$^+$ channels in rabbit myogenic cerebral arteries, *Am. J. Physiol.* **269**:H348-H355.

Knot, H. J., Zimmermann, P. A., and Nelson, M. T., 1996, Extracellular K$^+$-induced hyperpolarizations and dilatations of rat coronary and cerebral arteries involve inward rectifier K$^+$ channels, *J. Physiol. (London)* **492**:419–430.

Komaru, T., Lamping, K. G., Eastham, C. L., and Dellsperger, K. C., 1991, Role of ATP-sensitive potassium channels in coronary microvascular autoregulatory responses, *Circ. Res.* **69**:1146–1151.

Lederer, W. J., and Nichols, C. G., 1989, Nucleotide modulation of the activity of rat heart ATP-sensitive K$^+$ channels in isolated membrane patches, *J. Physiol. (London)* **419**:193–211.

Li, P. L., and Campbell, W. B., 1997, Epoxyeicosatrienoic acids activate K$^+$ channels in coronary smooth muscle through a guanine nucleotide binding protein, *Circ. Res.* **80**:877–884.

Lorenz, J. N., Schnermann, J., Brosius, F. C., Briggs, J. P., and Furspan, P. B., 1992, Intracellular ATP can regulate afferent arteriolar tone via ATP-sensitive K$^+$ channels in the rabbit, *J. Clin. Invest.* **90**:733–740.

Merkel, L. A., Lappe, R. W., Rivera, L. M., Cox, B. F., and Perrone, M. H., 1992, Demonstration of vasorelaxant activity with an A1-selective adenosine agonist in porcine coronary artery: Involvement of potassium channels, *J. Pharmacol. Exp. Ther.* **260**:437–443.

Miyoshi, Y., and Nakaya, Y., 1991, Angiotensin II blocks ATP-sensitive K$^+$ channels in porcine coronary artery smooth muscle cells, *Biochem. Biophys. Res. Commun.* **181**:700–706.

Miyoshi, H., and Nakaya, Y., 1995, Calcitonin gene-related peptide activates the K$^+$ channels of vascular smooth muscle cells via adenylate cyclase, *Basic Res. Cardiol.* **90**:332–336.

Miyoshi, Y., Nakaya, Y., Wakatsuki, T., Nakaya, S., Fujino, K., Saito, K., and Inoue, I., 1992, Endothelin blocks ATP-sensitive K$^+$ channels and depolarizes smooth muscle cells of porcine coronary artery, *Circ. Res.* **70**:612–616.

Murphy, M. E., and Brayden, J. E., 1995a, Apamin-sensitive K$^+$ channels mediate an endothelium-dependent hyperpolarization in rabbit mesenteric arteries, *J. Physiol. (London)* **489**:723–734.

Murphy, M. E., and Brayden, J. E., 1995b, Nitric oxide hyperpolarizes rabbit mesenteric arteries via ATP-sensitive potassium channels, *J. Physiol. (London)* **486**:47–58.

Nelson, M. T., and Quayle, J. M., 1995, Physiological roles and properties of potassium channels in arterial smooth muscle, *Am. J. Physiol.* **268**:C799–C822.

Nelson, M. T., Huang, Y., Brayden, J. E., Hescheler, J., and Standen, N. B., 1990, Arterial dilations in response to calcitonin gene-related peptide involve activation of K$^+$ channels, *Nature* **344**:770–773.

Nelson, M. T., Cheng, H., Rubart, M., Santana, L. F., Bonev, A. D., Knot, H. J., and Lederer, W. J., 1995, Relaxation of arterial smooth muscle by Ca^{2+} sparks [see comments], *Science* **270**:633–637.

Overturf, K. E., Russell, S. N., Carl, A., Vogalis, F., Hart, P. J., Hume, J. R., Sanders, K. M., and Horowitz, B., 1994, Cloning and characterization of a Kv1.5 delayed rectifier K$^+$ channel from vascular and visceral smooth muscles, *Am. J. Physiol.* **267**:C1231–C1238

Perez, G., Bonev, A. D., Patlak, J. B., and Nelson, M. T., 1999, Functional coupling of ryanodine receptors to K$_{Ca}$ channels in smooth muscle cells from rat cerebral arteries, *J. Gen. Physiol.* **113**:385–388.

Porter, V. A., Bonev, A. D., Knot, H. J., Heppner, T. J., Stevenson, A. S., Kleppisch, T., Lederer, W. J., and Nelson, M. T., 1998, Frequency modulation of Ca^{2+} sparks is involved in regulation of arterial diameter by cyclic nucleotides, *Am. J. Physiol.* **274**:C1346–C1355.

Post, J. M., Hume, J. R., Archer, S. L., and Weir, E. K., 1992, Direct role for potassium channel inhibition in hypoxic pulmonary vasoconstriction, *Am. J. Physiol.* **262**:C882-C890.

Quast, U. and Cook, N. S., 1989, in vitro and in vivo comparison of two K$^+$ channel openers, diazoxide and cromakalim, and their inhibition by glibenclamide, *J. Pharmacol. Exp. Ther.* **250**:261–271.

Quayle, J. M., McCarron, J. G., Brayden, J. E., and Nelson, M. T., 1993, Inward rectifier K$^+$ currents in smooth muscle cells from rat resistance-sized cerebral arteries, *Am. J. Physiol.* **265**:C1363–C1370.

Quayle, J. M., Bonev, A. D., Brayden, J. E., and Nelson, M. T., 1994, Calcitonin gene-related peptide activated ATP-sensitive K$^+$ currents in rabbit arterial smooth muscle via protein kinase A, *J. Physiol. (London)* **475**:9-13.

Quayle, J. M., Nelson, M. T., and Standen, N. B., 1997, ATP-sensitive and inwardly rectifying potassium channels in smooth muscle, *Physiol. Rev.* **77**:1165–1232.

Robertson, B. E., and Nelson, M. T., 1994, Aminopyridine inhibition and voltage dependence of K$^+$ currents in smooth muscle cells from cerebral arteries, *Am. J. Physiol.* **267**:C1589–C1597.

Robertson, B. E., Schubert, R., Hescheler, J., and Nelson, M. T., 1993, cGMP-dependent protein kinase activates Ca-activated K channels in cerebral artery smooth muscle cells, *Am. J. Physiol.* **265**:C299–C303.

Robertson, B. E., Bonev, A. D., and Nelson, M. T., 1996, Inward rectifier K$^+$ currents in smooth muscle cells from rat coronary arteries: Block by Mg^{2+}, Ca^{2+} and Ba^{2+}, *Am. J. Physiol.* **271**:H696–H705.

Russell, S. N., Smirnov, S. V., and Aaronson, P. I., 1992, Effects of BRL 38227 on potassium currents in smooth muscle cells isolated from rabbit portal vein and human mesenteric artery, *Br. J. Pharmacol.* **105**:549–556.

Samaha, F. F., Heineman, F. W., Ince, C., Fleming, J., and Balaban, R. S., 1992, ATP-sensitive potassium channel is essential to maintain basal coronary vascular tone in vivo, *Am. J. Physiol.* **262**:C1220–C1227.

Scornik, F. S., and Toro, L., 1992, U46619, a thromboxane A$_2$ agonist, inhibits K$_{Ca}$ channel activity from pig coronary artery, *Am. J. Physiol.* **262**:C708–C713.

Scornik, F. S., Codina, J., Birnbaumer, L., and Toro, L., 1993, Modulation of coronary smooth muscle K$_{Ca}$ channels by G$_s\alpha$ independent of phosphorylation by protein kinase A, *Am. J. Physiol.* **265**:H1460–H1465.

Smirnov, S. V., and Aaronson, P. I., 1992, Ca^{2+}-activated and voltage-gated K$^+$ currents in smooth muscle cells isolated from human mesenteric arteries, *J. Physiol. (London)* **457**:431–454.

Smirnov, S. V., Robertson, T. P., Ward, J. P., and Aaronson, P. I., 1994, Chronic hypoxia is associated with reduced delayed rectifier K$^+$ current in rat pulmonary artery muscle cells, *Am. J. Physiol.* **266**:H365–H370.

Sakol, P. T., Hu, W., Yi, L., Toral, J., Chandra, M., Ziai, M. R., 1994, Cloning of an apamin binding protein of vascular smooth muscle, *J. Prot. Chem.* **13**:117–128.

Standen, N. B., Quayle, J. M., Davies, N. W., Brayden, J. E., Huang, Y., and Nelson, M. T., 1989, Hyperpolarizing vasodilators activate ATP-sensitive K$^+$ channels in arterial smooth muscle, *Science* **245**:177–180.

Taniguchi, J., Furukawa, K. I., and Shigekawa, M., 1993, Maxi K$^+$ channels are stimulated by cyclic guanosine monophosphate-dependent protein kinase in canine coronary artery smooth muscle cells, *Pflügers Arch.* **423**:167–172.

Toro, L., Amador, M., and Stefani, E., 1990, ANG II inhibits Ca^{2+} activated potassium channels from coronary smooth muscle in lipid bilayers, *Am. J. Physiol.* **258**:H912–H915.

Volk, K. A., and Shibata, E. F., 1993, Single delayed rectifier potassium channels from rabbit coronary artery myocytes, *Am. J. Physiol.* **264**:H1146–H1153.

Wakatsuki, T., Nakaya, Y., and Inoue, I., 1992, Vasopressin modulates K$^+$-channel activities of cultured smooth muscle cells from porcine coronary artery, *Am. J. Physiol.* **263**:H491–H496.

Xia, X. M., Fakler, B., Rivard, A., Wayman, G., Johnson-Pais, T., Keen, J. E., Ishii, T., Hirschberg, B., Bond, C. T., Lutsenko, S., Maylie, J., and Adelman, J. P., 1998, Mechanism of Ca^{2+} gating in small-conductance Ca^{2+} activated potassium channels, *Nature* **395**:503–507.

Yuan, X. J., 1995, Voltage-gated K$^+$ currents regulate resting membrane potential and [Ca^{2+}]$_i$ in pulmonary arterial myocytes, *Circ. Res.* **77**:370–378.

Yuan, X. J., Tod, M. L., Rubin, L. J., and Blaustein, M. P., 1995, Hypoxic and metabolic regulation of voltage-gated K$^+$ channels in rat pulmonary artery smooth muscle cells, *Exp. Physiol.* **80**:803–813.

Yuan, X. J., Tod, M. L., Rubin, L. J., and Blaustein, M. P., 1996, NO hyperpolarizes pulmonary artery smooth muscle cells and decreases the intracellular Ca^{2+} concentration by activating voltage-gated K$^+$ channels, *Proc. Natl. Acad. Sci. U.S.A.* **93**:10489–10494.

Chapter 23

Modulation of Vascular K^+ Channels by Extracellular Messengers

D. J. Beech, A. Cheong, R. Flemming, C. Guibert, and S. Z. Xu

1. INTRODUCTION

There can be no doubt from what is now a considerable volume of literature that a host of endogenous extracellular messengers activate or inhibit vascular K^+ channels directly or via intracellular or intercellular coupling mechanisms. It is also true, however, that some investigators have reported that K^+ channel blockers have no effect on the actions of vasodilators and vasoconstrictors that have been suggested by other authors to involve K^+ channels. Such apparent contradictions may be explained in some cases by the vasculature's heterogeneity—between species and vascular beds, and the location, size, and type of blood vessel in question. However, this is only one possible explanation. Blood vessels also seem to have a wealth of parallel or backup mechanisms. As an example, for illustration purposes, a vasodilatory extracellular messenger may simultaneously activate K^+ channels and suppress voltage-gated Ca^{2+} channels in vascular smooth muscle cells. Either effect may be sufficient for full vasodilation such that if one effect is prevented experimentally, there is no change in the size of the vasodilation. This is a parallel system. Alternatively, a backup system may exist whereby a mechanism that is not normally operative becomes essential in a certain condition, perhaps during or after ischemia. Developing this further, it is plausible (and there is supporting evidence) that vasoconstrictor and vasodilator mechanisms have commonality and thus interact in such a way that a vasodilatory mechanism is negated by one vasoconstrictor mechanism but not another. The combination of biological complexity, experimental variability, and the different experimental conditions used by investigators has, perhaps not surprisingly, produced a

D. J. Beech, A. Cheong, R. Flemming, C. Guibert, and S. Z. Xu • School of Biomedical Sciences, University of Leeds, Leeds LS2 9JT, United Kingdom.

Potassium Channels in Cardiovascular Biology, edited by Archer and Rusch. Kluwer Academic/Plenum Publishers, New York, 2001.

spectrum of conclusions from an essential functional role of K^+ channels to no role at all.

There are many types of K^+ channels and auxiliary subunits. In excess of 50 mammalian K^+ channel genes are now known. Not all types are expressed in blood vessels, but there is still plenty of diversity. To a large extent, the exact molecular identities of K^+ channel types in blood vessels are unknown. Indeed, this is an important area for further investigation. At this point in time, we must use a fairly broad classification of six types of K^+ channels in blood vessels: voltage-gated K^+ channels (K_v), which include the delayed rectifiers and the A-currents; ATP-sensitive K^+ channels (K_{ATP} channels); classical strong inward rectifiers (K_{ir} channels); large-conductance, Ca^{2+}-activated K^+ channels (BK_{Ca} channels); intermediate-conductance, Ca^{2+}-activated K^+ channels (IK_{Ca} channels); and small-conductance, Ca^{2+}-activated K^+ channels (SK_{Ca} channels). Kv channels, Kir channels, and K_{ATP} channels are expressed in vascular smooth muscle cells and endothelial cells. It seems, although it is not certain, that BK_{Ca} channels are predominantly in vascular smooth muscle cells, whereas IK_{Ca} channels and SK_{Ca} channels are predominantly in endothelial cells. This is probably too simplistic, however, because there is heterogeneity within the six classes of K^+ channels—different genes, splice variants, and auxiliary subunits—and thus, at a detailed molecular level, the complement of K^+ channels in endothelial cells may be substantially different from that in vascular smooth muscle cells. Additional types of K^+ channels may also be expressed, including apamin- and charybdotoxin-sensitive small-conductance K^+ channels (Gebremedhin *et al.*, 1996) and background K^+ channels (Prior *et al.*, 1998a).

Presently, there are only three pharmacological agents that have been widely used for which there remains good reason to believe that they are selective for a class of K^+ channel. These are iberiotoxin, which blocks BK_{Ca} channels, glibenclamide ($\leqslant 1$ μM), which blocks K_{ATP} channels, and apamin, which blocks SK_{Ca} channels. 4-Aminopyridine blocks some Kv channels (Grissmer *et al.*, 1994), but it also blocks K_{ATP} channels (Beech and Bolton, 1989) and Ca^{2+}-ATPase (Ishida and Honda, 1993) and may change intracellular pH because of the high concentrations commonly used (5–10 mM) and its alkaline pK_a. Extracellular Ba^{2+} (<0.1 mM) may be fairly selective for Kir channels, but K_{ATP} channels are related channels that are also Ba^{2+}-sensitive ($K_i < 1$ mM); for example, Kinoshita and Katusic (1997) observed that 1 mM Ba^{2+} abolished relaxation induced by the K_{ATP} channel opener cromakalim. The sensitivity of BK_{Ca} channels to intracellular Ba^{2+} (Benham *et al.*, 1985) may also be important in whole-tissue experiments because Ba^{2+} is permeant in voltage-gated Ca^{2+} channels and might accumulate inside cells. Furthermore, Ba^{2+} has been observed to cause endothelium-dependent nitric oxide-mediated vasorelaxation (Yamazaki *et al.*, 1998). Unfortunately, charybdotoxin is still commonly used as a selective inhibitor of BK_{Ca} channels, even though this toxin also blocks IK_{Ca} channels (Jensen *et al.*, 1998) and some types of Kv channels (e.g., Kv1.3; Spencer *et al.*, 1997). Charybdotoxin should not be used without a comparison to iberiotoxin.

It has been suggested that data from whole-tissue experiments using K^+ channel blockers are open to misinterpretation if the blocker increases the tone in a vessel in control conditions. The effect of a vasodilatory messenger could then be inhibited, not because a K^+ channel is an element of the signaling cascade, but because of functional antagonism as the vasoconstriction becomes supramaximal. Nevertheless, although basal constriction induced by K$^+$ channel blockers has been reported, there are also many reports where there was no vasoconstrictor effect of the blocker, perhaps in a different blood vessel or under different experimental conditions. Furthermore, there are reports where vasoconstriction in response to the K$^+$ channel blocker was offset experimentally. Sufficient reports now show effects of K$^+$ channel blockers on responses to vasodilatory messengers when functional antagonism was not a complication.

K$^+$ channels are most commonly implicated in vasodilator responses when experiments require the blood vessel to have myogenic tone or tone induced by a vasoconstrictor agonist. Therefore, in different studies, and even between experiments in the same study, the degree of contraction and the concentration of intracellular Ca^{2+} before the vasodilator is applied will vary. There is also likely to be variation in membrane potential, and some vasoconstrictor agents might inhibit certain K$^+$ channel types as part of their complement of procontractile mechanisms. For example, the initial (prevasodilator) intracellular Ca^{2+} concentration might be 300 nM in one study and 60 nM in another, or the initial membrane potential might be -20 mV in one study and -50 mV in another. This is likely to be of consequence because K$^+$ channel subtypes are differentially regulated by intracellular Ca^{2+} and membrane potential. Elevated Ca^{2+} levels increase and hyperpolarization decreases activity of BK$_{Ca}$ channels. Furthermore, Ca^{2+} and hyperpolarization decrease activity of Kv channels. Finally, Ca^{2+} inactivates and hyperpolarization has no effect on the activity of K$_{ATP}$ channels. Therefore, the use of different vasoconstrictors, the concentration of vasoconstrictor used, and the absence or presence of myogenic tone are variables that may influence whether one K$^+$ channel type or another is found to play a role in a vasodilatory mechanism. A biological purpose of activation of multiple K$^+$ channels by a vasodilatory messenger may be to avoid dependence on one K$^+$ channel type that might be suppressed by some vasoconstrictors or inhibited by a pathological condition. For example, Kv channels are thought to be suppressed and BK$_{Ca}$ channels potentiated in hypertension (Martens and Gelband, 1996; Liu *et al.*, 1998), whereas K$_{ATP}$ channel function is suppressed in diabetes (Mayhan and Faraci, 1993).

In vitro relaxation or contraction experiments primarily involve the use of isometric tension recording from rings of vessel (wire myograph) or, less commonly, diameter measurements from cannulated and pressurized vessels (pressure myograph) (Halpern and Kelley, 1991). The choice of method is important when investigating the involvement of K$^+$ channels because luminal pressure (pressure myograph) generates depolarization of the membrane potential of vascular smooth muscle cells into a range where voltage-gated Ca^{2+} channels play a pivotal role in governing Ca^{2+} influx. Therefore, stimulation of K$^+$ channels will induce hyperpolarization, closure of voltage-gated Ca^{2+} channels, and a reduction of Ca^{2+} influx. On the wire myograph, by contrast, the membrane potential is usually more hyperpolarized and may be negative to the effective threshold for activation of voltage-gated Ca^{2+} channels. Thus, membrane potential-independent mechanisms may predominate and K$^+$ channels may be found to have no role. Indeed, dissociation between the capacities of a vasodilator to cause hyperpolarization and relaxation has been observed when making recordings on a wire myograph (e.g., Plane *et al.*, 1998; Edwards *et al.*, 1998).

The subsequent sections survey, in some cases selectively, current knowledge about the roles played by K$^+$ channels in the actions of 20 or so extracellular messengers. The messengers include nitric oxide, oxygen, reactive oxygen species, potassium, protons, arachidonic acid, hydroxyeicosatetraenoic acids (HETEs), epoxyeicosatrienoic acids (EETs), prostacyclin, adenosine, calcitonin gene-related peptide (CGRP), β-adrenoceptor agonists, histamine, endothelins, and 17β-estradiol. The focus of this chapter is on the extracellular messengers that act directly on vascular smooth muscle cells, at least in part.

2. ROLE OF K$^+$ CHANNELS IN THE ACTIONS OF EXTRACELLULAR MESSENGERS ON BLOOD VESSELS

2.1. Nitric Oxide

High local concentrations of nitric oxide are produced in blood vessels (up to 0.25 μM; Cohen *et al.*, 1997) either basally or in response to a vasodilatory substance such as acetylcholine or bradykinin. An extensive literature on nitric oxide shows that it is produced by endothelial cells and nitrergic neurons and also within vascular smooth muscle cells by inducible nitric oxide synthase. Maintained levels of nitric oxide may also induce heme oxygenase and thus enhance endogenous carbon monoxide production (Durante *et al.*, 1997), and some effects of carbon monoxide are similar to those of nitric oxide (Wang and Wu, 1997). There is a considerable volume of literature on the subject of nitric oxide interaction with vascular K$^+$ channels—some of it positive, some of it negative. Although there is overwhelming evidence that nitric oxide does act on vascular K$^+$ channels, nitric oxide can also induce vasorelaxation by K$^+$ channel-independent mechanisms, and these are sometimes sufficient for a full vasodilatory response.

There is electrophysiological evidence from patch-clamp studies that BK$_{Ca}$ channels, Kv channels, and K$_{ATP}$ channels are activated by nitric oxide or nitric oxide donors in isolated vascular smooth muscle cells from various blood vessels (e.g., Williams *et al.*, 1988; Kubo *et al.*, 1994; Archer *et al.*, 1996; Yuan *et al.*, 1996; Bychkov *et al.*, 1997; Mistry and Garland, 1998). The literature is dominated by reports that BK$_{Ca}$ channels are sensitive to nitric oxide (e.g., Fig. 1), and there are relatively few reports of electrophysiological evidence that K$_{ATP}$ channels are modulated. BK$_{Ca}$ channels and K$_{ATP}$ channels are stimulated by cGMP-dependent phosphorylation (Taniguchi *et al.*, 1993; Kubo *et al.*, 1994) and thus are expected to be stimulated by nitric oxide simply because nitric oxide raises cGMP levels. There is also evidence for cGMP-independent effects of nitric oxide (e.g., Weisbrod *et al.*, 1998), and this includes the observation that BK$_{Ca}$ channels are stimulated even when guanylyl cyclase is inhibited (Bolotina *et al.*, 1994; Mistry and Garland, 1998; Sun *et al.*, 1998). This may not reflect a direct action of nitric oxide on BK$_{Ca}$ channels but rather inhibition of 20-HETE synthesis by nitric oxide and thus the loss of 20-HETE-induced inhibition of BK$_{Ca}$ channels (Sun *et al.*, 1998; see below).

In vitro and *in vivo* pharmacological studies have commonly shown that blockers of BK$_{Ca}$ channels, Kv channels, and K$_{ATP}$ channels attenuate or abolish nitric oxide donor-induced vasodilation or vascular hyperpolarization (e.g., Bari *et al.*, 1996; Kitazono *et al.*, 1997; Zhao *et al.*, 1997). Some studies reveal strong inhibition of nitric oxide-induced relaxation by iberiotoxin alone but not glibenclamide, indicating a role

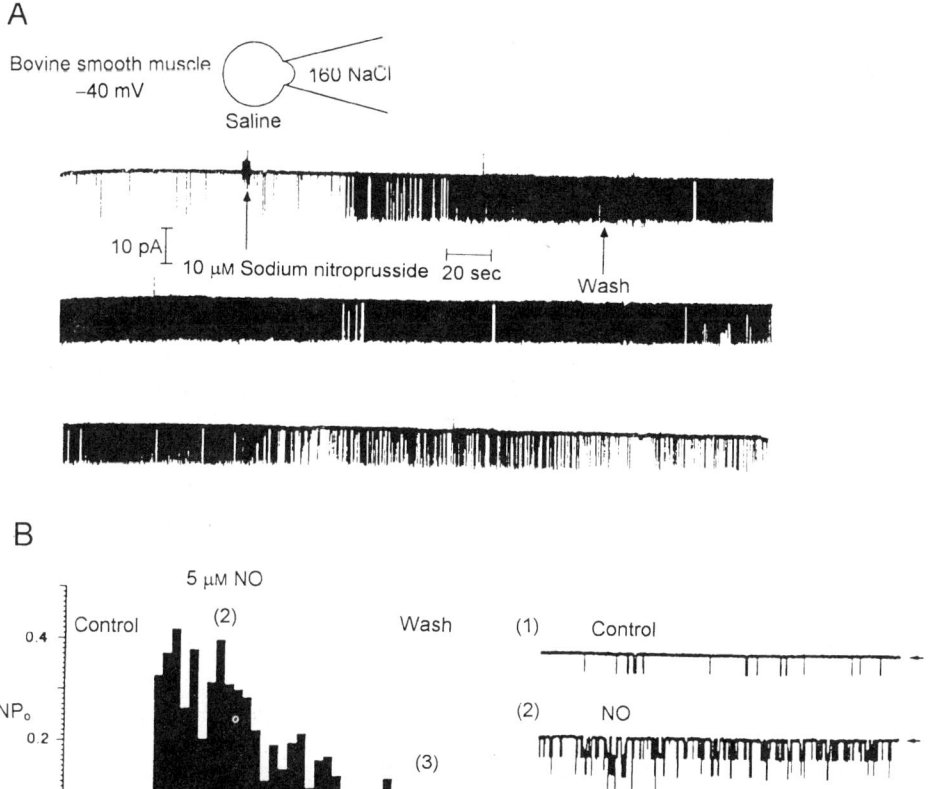

Figure 1. Stimulation of BK$_{Ca}$ channels by nitric oxide. (A) Effect of 10 μM sodium nitroprusside on BK$_{Ca}$ single-channel currents in a cell-attached patch. (Experiments were conducted on bovine aortic smooth muscle cells). (Reproduced from Williams *et al.*, 1988.) (B) Effect of 5 μM nitric oxide (NO) on BK$_{Ca}$ single-channel currents in an inside-out membrane patch. On the left is a histogram of NP$_o$ (number of channels × open probability) against time. On the right are original single-channel currents. Horizontal arrows mark the closed-channel current. Experiments were conducted on rabbit aortic smooth muscle cells. (Reprinted, with permission, from Bolotina *et al.*, 1994. Copyright 1994 Macmillan Magazines Limited.)

for BK$_{Ca}$, but not K$_{ATP}$, channels (e.g., Dong *et al.*, 1998). In contrast, others have found inhibition of nitric oxide-induced relaxation by glibenclamide alone but not by iberiotoxin or other blockers of BK$_{Ca}$ channels (e.g., Murphy and Brayden, 1995). The latter two studies used rabbit cerebral and mesenteric arteries, respectively. Thus, a vascular bed difference may be implicated. Alternatively, it is worth considering that Murphy and Brayden (1995) measured membrane potential on the wire myograph, the resting potential was quite negative (-56 mV), and a spasmogen was not used. These conditions favor K$_{ATP}$ channel activity. Dong *et al.*, (1998), by contrast, studied contraction on the wire myograph and induced contraction with histamine, which elevates intracellular Ca^{2+} levels and promotes activity of BK_{Ca} channels (see below). It is interesting to note that in a review of all types of smooth muscle, BK_{Ca} channels were the most commonly identified K^+ channel target for nitric oxide donors (Beech, 1997). Dong *et al.*, (1998) observed that nitric oxide-induced relaxation could also be inhibited by apamin in endothelium-denuded arteries, suggesting that nitric oxide might also activate SK_{Ca} channels in vascular smooth muscle cells. Such *in vitro* observations suggest a predominant role for Ca^{2+}-activated K^+ channels in vasorelaxation induced by nitric oxide. However, it is intriguing that in an *in vivo* study by Berg and Koteng (1997), 4-aminopyridine, the blocker of Kv channels, rather than iberiotoxin, inhibited the hypotensive response to sodium nitroprusside.

The above section is a positive description of the role of K^+ channels in nitric oxide action. There are also reports showing that block of BK_{Ca} channels, Kv channels, or K_{ATP} channels, can be without effect on nitric oxide or nitric oxide donor-induced vasorelaxation (e.g., Ishizaka and Kuo, 1996; Gidday *et al.*, 1996; Kinoshita and Katusic, 1997; Gambone *et al.*, 1997; Husken *et al.*, 1997; Deka *et al.*, 1997; Plane *et al.*, 1998). There are numerous reasons why K^+ channel involvement may not have been revealed (see Section 1) but it is interesting to consider the possibility that K^+ channel activation may not always have a direct vasodilatory role but instead may serve to facilitate the effect of a second vasodilator. This idea is supported by the observation that synergism between the nitric oxide- and prostacyclin-mediated components of bradykinin-induced vasodilation was abolished by glibenclamide, the inhibitor of K_{ATP} channels (Gambone *et al.*, 1997).

It has been suggested that K^+ channel involvement in the relaxant effect of nitric oxide donors is revealed if endogenous nitric oxide production is first inhibited (e.g., Murphy and Brayden, 1995). Clearly, if endogenous nitric oxide is already maximally activating K^+ channels, then exogenous nitric oxide will have no effect. However, nitric oxide donors are good K^+ channel-independent vasodilators against the backdrop of endogenous nitric oxide production, and thus other non-K^+ channel mechanisms are not maximally activated by endogenous nitric oxide. Thus, there is reason to suspect that endogenous nitric oxide has different, perhaps restricted or localized, effects compared with those resulting from flooding of the system with exogenous nitric oxide. It has also been observed that K^+ channel involvement in nitric oxide donor action is revealed after experimentally induced subarachnoid hemorrhage (Onoue and Katusic, 1998). The reason for this effect could be that intracellular Ca^{2+} levels are elevated and the membrane potential is more depolarized in vascular smooth muscles cells following hemorrhage, promoting activity of BK_{Ca} channels and Kv channels.

2.2. Oxygen and Reactive Oxygen Species

In most vascular beds, a decrease in oxygen tension (hypoxia or anoxia) induces vasodilation, and this can occur via a multitude of mechanisms, rather like the effect of an increase in nitric oxide concentration. There is electrophysiological evidence that in blood vessels, other than the pulmonary arteries, hypoxia increases activity of BK_{Ca} channels (Gebremedhin *et al.*, 1994) and K_{ATP} channels (Dart and Standen, 1995) in isolated vascular smooth muscle cells (Fig. 2). Hypoxic vasodilation in intact blood

A   B

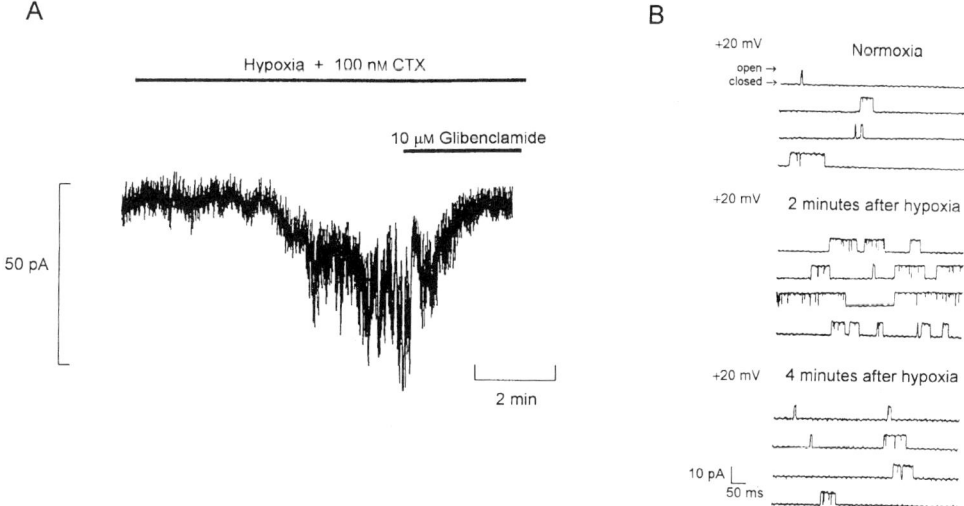

Figure 2. Stimulation of K$_{ATP}$ channels (A) and BK$_{Ca}$ channels (B) by hypoxia. (A) Induction of inward whole-cell K$^+$ current by hypoxia in the presence of 100 nM charybdotoxin (CTX), which blocks BK$_{Ca}$ channels. The bath solution contained 143 mM K$^+$, and the holding potential was -60 mV. The inward current was inhibited by 10 μM glibenclamide, which blocks K$_{ATP}$ channels. Experiments were conducted on coronary artery smooth muscle cells. (Reproduced from Dart and Standen, 1995.) (B) Transient stimulation of BK$_{Ca}$ single-channel currents by hypoxia, observed in an inside-out patch from a feline middle cerebral artery smooth muscle cell. (Reproduced from Gebremedhin et al., 1994.)

vessels is also inhibited by iberiotoxin or glibenclamide (von Beckerath et al., 1991; Armstead, 1998a). Hypoxia, however, produces complex cellular and tissue responses, and thus K$^+$ channel involvement in hypoxic responses may not reflect a direct effect of oxygen on a K$^+$ channel. Changes in proton or adenosine levels, for example, are known to affect K$^+$ channel activity (see below) and could mediate effects of hypoxia. Nitric oxide can also be released during hypoxia, although Armstead (1998a) observed that cerebral hypoxic vasodilation was iberiotoxin-sensitive whereas nitric oxide-induced vasodilation was not.

There is compelling evidence that acute and chronic hypoxia modulate Kv channels in vascular smooth muscle cells of pulmonary arteries and that these effects play primary roles in the specialized oxygen responses of the lung's blood supply (e.g., Yuan et al., 1993; Archer et al., 1996; Osipenko et al., 1997; Wang et al., 1997; Osipenko et al., 1998). Kv channels of non-pulmonary blood vessels have been found to be resistant to hypoxia (e.g., Yuan et al., 1993).

Reactive oxygen species are produced during several oxidative metabolic processes, and their production is increased during aging, inflammation, and cerebral vascular injury. Reactive oxygen species are vasoactive and affect K$^+$ channels. There is good agreement that hydrogen peroxide stimulates BK$_{Ca}$ channels of vascular smooth muscle cells (Krippeitdrews et al., 1995; Sobey et al., 1997; Barlow and White, 1998; Hayabuchi et al., 1998a). This effect appears to be crucial for bradykinin-induced endothelium-dependent vasodilation in the cerebral circulation (Fig. 3). Oxidizing agents, including NAD$^+$ and 5,5'-dithio-bis(2-nitrobenzoic acid), also stimulate BK$_{Ca}$ channels (Park et

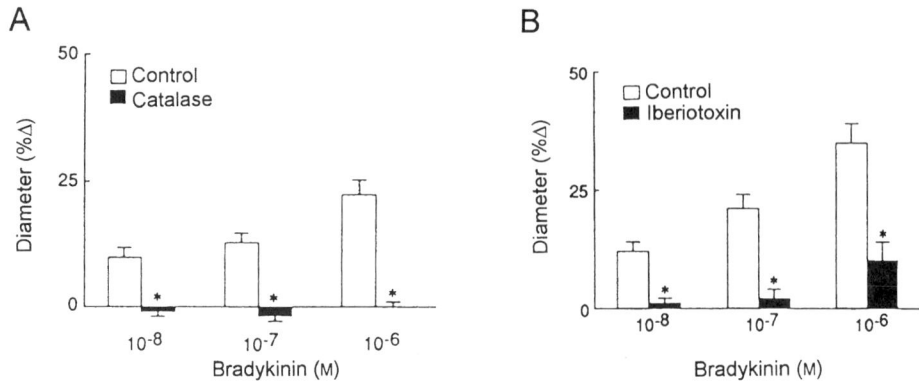

Figure 3. Functional role of reactive oxygen species in BK_{Ca} channel activation. (A) Inhibition of the vasodilatory response to bradykinin by 100 U/ml catalase (black bars), which degrades hydrogen peroxide. (B) Inhibition of the vasodilatory response to bradykinin by 50 nM iberiotoxin (black bars), which blocks BK_{Ca} channels. The vasodilatory responses were observed in rat arterioles in the parietal cortex viewed via a cranial window. Values are mean $\pm$SEM. *$p < 0.05$ vs control response. (Reproduced from Sobey *et al.*, 1997.)

al., 1995), as does the superoxide radical (Wei *et al.*, 1996). K_{ATP} channels are stimulated by hydrogen peroxide or peroxynitrite (Wei *et al.*, 1996), although, surprisingly, antioxidants appear to inhibit hydrogen peroxide- and peroxynitrite-induced vasodilation by directly blocking K_{ATP} channels (Wei *et al.*, 1998).

2.3. Protons

Extracellular acidosis occurs during conditions such as ischemia, and appears to form a component of the vasodilatory response to hypercapnia, and may modulate intracellular pH (Kontos *et al.*, 1977; Aalkjaer and Poston, 1996; Kitakaze *et al.*, 1997). Extravascular acidosis causes vasodilation in various vascular beds, including in the brain and heart. Although other intermediate messengers may contribute to the effect, it seems that there is an endothelium- and nitric oxide-independent effect of acid.

Ishizaka and Kuo (1996) observed that acidosis (e.g., a fall in pH from 7.4 to 7.0) induced pronounced arteriolar dilation that was strongly inhibited by glibenclamide or Ba^{2+} but was resistant to iberiotoxin or endothelium denudation (Fig. 4). Observations made by Faraci *et al.* (1994) and Hayabuchi *et al.* (1998b) also suggest a primary role for K_{ATP} channels in acid-induced vasodilation. Although Kinoshita and Katusic (1997) found only a small inhibitory effect of glibenclamide, raising the extracellular K^+ concentration to 20 mM almost abolished the acid-induced relaxation, suggesting that K^+ channels were involved but that they were mostly not K_{ATP} channels. The effect was resistant to 4-aminopyridine and charybdotoxin, suggesting that BK_{Ca} channels, IK_{Ca} channels, and at least some types of Kv channel were not responsible. The K^+ channels involved may be Kir channels because Okazaki *et al.*, (1998) observed an endothelium-

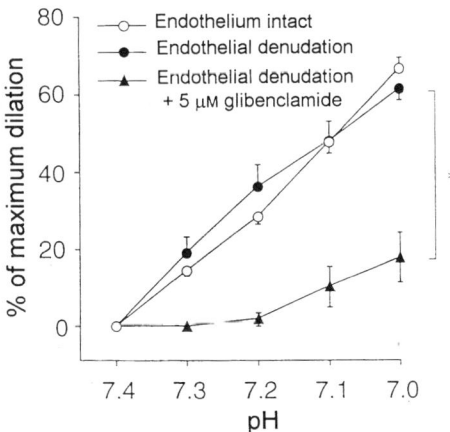

Figure 4. Role of vascular smooth muscle K_{ATP} channels in vasodilation induced by mild acidosis. Percentage dilation of porcine coronary arterioles is plotted against extracellular pH. The acid-induced vasodilation was blocked by 5 μM glibenclamide but not by removal of the endothelium. *$P < 0.05$. (Reproduced from Ishizaka and Kuo, 1996.)

dependent vasodilatory response to hypercapnia that was blocked by 0.3 mM Ba^{2+} but resistant to 4-aminopyridine, 1 mM tetraethylammonium (which blocks BK$_{Ca}$ channels), apamin, and glibenclamide.

Kv channels are modulated by extracellular and intracellular pH, although the functional role of the effects is not established. Berger *et al.*, (1998) observed differential effects of intracellular acidosis on Kv channels of coronary and pulmonary arterial smooth muscle cells. Acidosis increased the Kv channel current in coronary cells but suppressed it in pulmonary cells. It was suggested that the differential effects may help explain why intracellular acidosis relaxes coronary but constricts pulmonary arteries. Intriguingly, only the effect in pulmonary cells was abolished by dendrotoxin, which blocks Kv channels, including Kv1.1 and Kv1.2. It is puzzling that Ahn and Hume (1997) observed that Kv channel current was stimulated by intracellular acidosis in canine pulmonary arteries, the opposite effect to that observed by Berger *et al.* (1998) in the rat. Extracellular acid stimulates voltage-dependent K⁺ current in cerebral artery smooth muscle cells (Bonnet *et al.*, 1991) and suppresses Kv channel current in pulmonary artery smooth muscle cells (Ahn and Hume, 1997).

2.4. Potassium

It has long been recognized that extracellular K⁺ levels elevate during ischemia as K⁺ is released from the surrounding organ. Although an artificially high extracellular K⁺ concentration causes depolarization, modestly elevated levels of K⁺, which are relevant to the *in vivo* situation (e.g., 10–15 mM), paradoxically induce vasodilation, and this effect is likely to have a real function in reducing ischemic damage of the organ. Edwards *et al.* (1998) have recently proposed that modest elevations of the extracellular

K^+ concentration may additionally be a physiological intercellular signal. It was suggested that K^+ leaving through Ca^{2+}-activated K^+ channels (probably SK_{Ca} and IK_{Ca} channels) of endothelial cells acted as an endothelium-derived factor that hyperpolarized and relaxed vascular smooth muscle cells [i.e., K^+ is an endothelium-derived hyperpolarizing factor (EDHF)]. Intriguingly, it was possible to measure elevated levels of K^+ in response to acetylcholine in what seemed to be a myoendothelial junction (Fig. 5A). Clearly, if this is the case between endothelial cells and

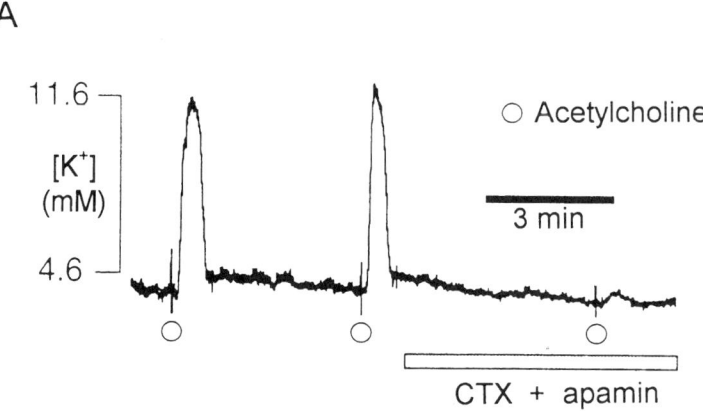

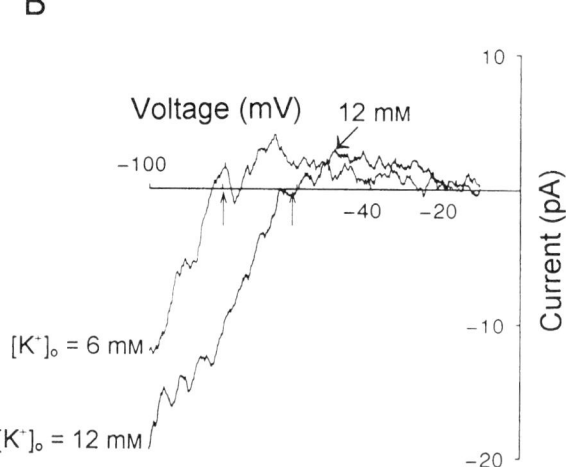

Figure 5. K^+ channels and K^+-induced vasodilation. (A) Acetylcholine (10 μM)-induced increase in myoendothelial K^+ concentration measured in endothelium-intact rat hepatic arteries using a K^+-sensitive microelectrode. The increase in K^+ concentration was prevented by a combination of 100 nM charybdotoxin (CTX) and 100 nM apamin, which are presumed to have blocked IK_{Ca} channels and SK_{Ca} channels respectively. (Reproduced from Edwards *et al.*, 1998.) (B) Ba^{2+}-sensitive whole-cell K^+ current recorded from a porcine small coronary artery smooth muscle cell in the presence of 6 mM or 12 mM extracellular K^+. The 12 mM K^+ record is the thicker line. Arrows indicate approximate reversal potentials. Currents are presumed to have been carried entirely by Kir channels. (Reproduced from Quayle *et al.*, 1996.)

smooth muscle cells, it can also be the case simply among the vascular smooth muscle cells of a blood vessel whenever K$^+$ channels are activated. Thus, K$^+$ channel-mediated hyperpolarization of vascular smooth muscle cells may be potentiated by extracellular K$^+$-induced vasodilatory mechanisms.

One mechanism that does mediate K$^+$-induced vasodilation is stimulation of Na$^+$–K$^+$–ATPase (Henderickx and Casteels, 1974; Webb and Bohr, 1978; Prior *et al.*, 1998b). However, this mechanism appears to be relatively ineffective once K$^+$ concentrations exceed 10 mM (McCarron and Halpern, 1990; Prior *et al.*, 1998b). Edwards and Hirst (1988) and Edwards *et al.* (1988) first demonstrated that Kir channels also mediate K$^+$-induced hyperpolarization and vasodilation. This mechanism is effective even at high levels of extracellular K$^+$ (McCarron and Halpern, 1990; Knot *et al.*, 1996; Edwards *et al.*, 1998). It is also restricted to small arteries and arterioles where specialized vascular smooth muscle cells express Kir channels—highly Ba^{2+}-sensitive and strongly rectifying inward rectifier K$^+$ channels (Edwards *et al.*, 1988; Quayle *et al.*, 1996; Prior *et al.*, 1998b). The conductance of the Kir channel, like that of other K$^+$ channels, increases when extracellular K$^+$ levels are raised. However, the outward current through Kir channels, unlike other K$^+$ channels, exhibits a strong voltage-dependent blocking effect of intracellular Mg^{2+} or polyamines that generates a bell-shaped current–voltage relationship. Changes in the extracellular K$^+$ concentration shift the peak of this bell-shaped curve toward less negative voltages, thus increasing the outward current through Kir channels at voltages close to the membrane potential of pressurized arteries. Outward current through Kir channels is very small in isolated vascular smooth muscle cells, but the increase in Ba^{2+}-sensitive outward current has been measured in response to an elevation of extracellular K$^+$ concentration from 6 mM to 12 mM (Fig. 5B).

2.5. Arachidonic Acid, Prostaglandins, HETEs, and EETs

Arachidonic acid is the parent substance for a host of biologically active derivatives produced by the actions of cyclooxygenase, lipoxygenase, or cytochrome P-450-dependent monooxygenases (McGiff, 1991). Prostaglandins such as prostacyclin resulting from cyclooxygenase activity and HETEs and EETs resulting from cytochrome P-450-dependent activity may be the most important derivations in relation to vascular K$^+$ channels. However, arachidonic acid itself is biologically active and could be a signaling molecule in its own right. Arachidonic acid modulates K$^+$ channels in isolated smooth muscle cells, activating BK$_{Ca}$ channels (Kirber *et al.*, 1992) and some Kv channels (Smirnov and Aaronson, 1996), and activating K$_{ATP}$ channels in the presence of intracellular ATP but inhibiting them in the absence of ATP (Xu and Lee, 1996). Although these effects are expected to mediate vasodilation, arachidonic acid-induced vasodilation has been found to be endothelium-dependent (Adeagbo and Malik, 1991; Zou *et al.*, 1996; Muira and Gutterman, 1998).

Prostaglandins are formed via the cyclooxygenase pathway. One of the prostaglandins produced is the potent vasodilator prostacyclin, which is now widely recognized as an endothelium-derived vasodilator. Various investigators have suggested that K$_{ATP}$ channels mediate the effect of prostacyclin because the vasodilation is sensitive to inhibition by glibenclamide in coronary arteries (Jackson *et al.*, 1993; Bouchard *et al.*, 1994), aorta (Bouchard *et al.*, 1994), pulmonary arteries (Dumas *et al.*, 1997; Gambone *et al.*, 1997), and rat tail artery (Schubert *et al.*, 1997). In contrast, Clapp *et al.*, (1998) and Dong *et al.*, (1998) found no effect of glibenclamide on prostacyclin-induced vasodilation. Clapp *et al.*, (1998) suggested instead a role for BK_{Ca} channels, in agreement with Schubert *et al.*, (1996), and Dong *et al.*, (1998) found that prostacyclin-induced vasodilation was inhibited by dendrotoxin, 4-aminopyridine, iberiotoxin, and apamin. Thus, it seems that a variety of K^+ channels can be involved in the action of prostacyclin. This should not be surprising given that prostacyclin receptors couple positively to adenylyl cyclase and many K^+ channels are stimulated by cAMP-dependent phosphorylation. Vasodilatory effects of other prostaglandins may also involve K^+ channels: actions of PGE_1, PGE_2, and PGD_2 are associated with K_{ATP} channels (Bouchard *et al.*, 1994; Ney and Feelisch, 1995).

The cytochrome P-450 4A enzyme appears to be the primary enzyme producing HETEs and EETs in blood vessels. It is expressed primarily in small arteries and arterioles (Gebremedhin *et al.*, 1998; Sun *et al.*, 1998) and is in the endothelial rather than the smooth muscle cell layer (Imig *et al.*, 1996), fueling supposition that HETEs and EETs are endothelium-derived relaxing and contracting factors. Furthermore, arachidonic acid-induced vasodilation in human coronary arterioles is endothelium-dependent and blocked by 17-octadecynoic acid, an inhibitor of cytochrome P-450 enzymes (Muira and Gutterman, 1998). 20-HETE inhibits the activity of BK_{Ca} channels in vascular smooth muscle cells (Fig. 6A) and induces vasoconstriction (Ma *et al.*, 1993; Zou *et al.*, 1996; Sun *et al.*, 1998). In contrast, 19(R)-HETE is vasodilatory (Ma *et al.*, 1993). Some EETs, for example, 11,12-EET and 14,15-EET, stimulate BK_{Ca} channels in vascular smooth muscle cells (Fig. 6B) and induce hyperpolarization and endothelium-independent vasodilation (Gebremedhin *et al.*, 1992; Hu and Kim, 1993; Oltman *et al.*, 1998; Eckman *et al.*, 1998). Intriguingly, glibenclamide was found to prevent the stimulatory action of 5,6-EET on BK_{Ca} channels (Hu and Kim, 1993), suggesting involvement of the sulfonylurea receptor, which is a subunit of K_{ATP} channels.

2.6. Acetylcholine, Bradykinin, and Substance P

Some extracellular messengers produce effects by acting on endothelial cells without any effect on vascular smooth muscle. Examples include the vasodilator effects of acetylcholine, bradykinin, and substance P. These messengers elevate the intracellular Ca^{2+} concentration, and this activates Ca^{2+}-dependent K^+ channels in the endothelium. A variety of vasodilators may then be released as Ca^{2+} levels rise, including nitric oxide and prostacyclin and, perhaps, EETs and potassium. Actions of these secondary (intercellular) messengers may also involve K^+ channels, as described above.

2.7. Adenosine

In most vascular beds, adenosine acts via A_{2A} or A_{2B} receptors on vascular smooth muscle cells or the endothelium to produce vasodilation. In the cerebral circulation, for

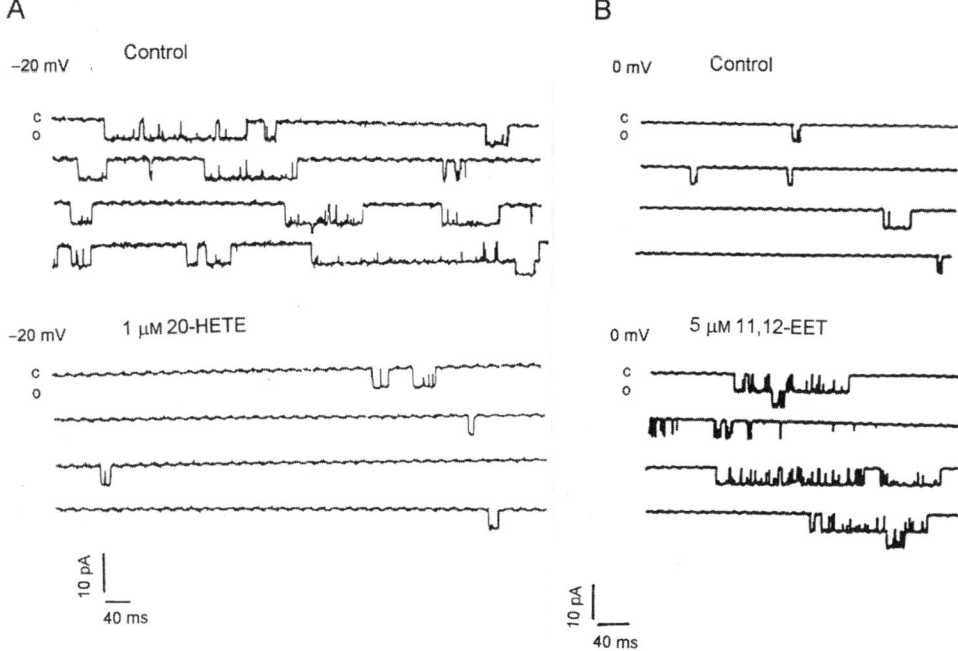

Figure 6. Opposite effects of a hydroxyeicosatetraenoic acid (HETE) and an epoxyeicosatrienoic acid (EET) on BK$_{Ca}$ channels. (A) Inhibition of BK$_{Ca}$ single-channel currents by 1 μM 20-HETE in a cell-attached patch on a canine renal arcuate artery smooth muscle cell. (Reproduced from Ma *et al.*, 1993.) (B) Stimulation of BK$_{Ca}$ single-channel currents by 5 μM 11,12-EET in a cell-attached patch on a feline middle cerebral artery smooth muscle cell. (Reproduced from Gebremedhin *et al.*, 1992.) The letters "c" and "o" to the left of the traces designate the closed and the open state, respectively.

example, hypoxic vasodilation is attenuated by adenosine receptor blockade, and adenosine is released from neurons (and possibly other cell types) when there is hypoxia, ischemia, hypotension, or epilepsy (Berne *et al.*, 1974; Morii *et al.*, 1987; Meno *et al.*, 1991; Edvinsson *et al.*, 1993; Pelligrino *et al.*, 1995). The small vessels of the kidney, primarily the afferent arteriole, are an exception in terms of adenosine's vasoactive properties because adenosine acts on A$_1$ receptors to induce vasoconstriction (Holz and Steinhausen, 1987; Churchill and Bidani, 1990). A$_1$ receptors have also been reported to mediate dilation in rat diaphragmatic arterioles and rabbit femoral artery (Danialou *et al.*, 1997; Sakai *et al.*, 1998).

Adenosine-induced vasodilation may involve K$^+$ channels either because of the release of a factor from endothelial cells or because A$_2$ receptors expressed in smooth muscle cells couple to K$^+$ channels, presumably via the second messenger cAMP. BK$_{Ca}$ channels, K$_{ATP}$ channels, and Kv channels are activated by cAMP in patch-clamp experiments on single smooth muscle cells (e.g., Quayle *et al.*, 1994; for additional references, see Beech, 1997). Furthermore, particularly in the coronary circulation, K$^+$ channel inhibitors have been shown to attenuate vasodilator effects of adenosine, and there are electrophysiological data showing stimulation of K$_{ATP}$ channels by adenosine, or adenosine receptor agonists, in isolated vascular smooth muscle cells (Dart and

Standen, 1993; Kleppisch and Nelson, 1995b). Adenosine-induced dilation of coronary vessels from various species, hamster cheek pouch arterioles, and the retinal microcirculation is sensitive to inhibition by glibenclamide, suggesting a role for K_{ATP} channels (von Beckerath *et al.*, 1991, Jackson *et al.*, 1993; Jiang and Collins, 1994; Randall, 1995; Kuo and Chancellor, 1995; Gidday *et al.*, 1996). In contrast, adenosine receptor-mediated vasodilation is resistant to glibenclamide in other arteries (e.g., Taguchi *et al.*, 1994; Tabrizchi and Lupichuk, 1995; Prior *et al.*, 1998c). Merkel *et al.* (1992) observed that A_2 receptor-mediated vasorelaxation was also resistant to glibenclamide in porcine coronary artery but found an effect mediated by an atypical adenosine receptor that was glibenclamide-sensitive. This atypical effect was not observed in renal arcuate artery (Prior *et al.*, 1998c). In coronary vessels from the dog and rat, adenosine-induced dilation is inhibited by tetraethylammonium ions and iberiotoxin, indicating a role for BK_{Ca} channels (Cabell *et al.*, 1994; Price *et al.*, 1996).

Sakai and Saito (1998) observed interactions between various endogenous vasodepressors, including adenosine. Adenosine, for example, augmented responses to vasoactive intestinal polypeptide (VIP) or CGRP, but not to acetylcholine. The responses were inhibited by glibenclamide, suggesting involvement of K_{ATP} channels. Similarly, Duncker *et al.* (1995) have proposed that K_{ATP} channel activation and adenosine interact synergistically in producing vasodilation in normal hearts during exercise.

2.8. Calcitonin Gene-Related Peptide

CGRP is produced by alternative processing of mRNA from the calcitonin gene. It is released from nerves that are widely distributed in the cardiovascular system (Bell and McDermott, 1996) and is a potent vasodilator (Brain *et al.*, 1985; Kitazono *et al.*, 1993). Several electrophysiological studies show that CGRP stimulates K_{ATP} channels in isolated vascular smooth muscle cells (Quayle *et al.*, 1994; Miyoshi and Nakaya, 1995; Kleppisch and Nelson, 1995a; Wellman *et al.*, 1998). CGRP-induced vasodilation is also inhibited by glibenclamide in cranial window experiments on the rat (Kitazono *et al.*, 1993). K_{ATP} channels are stimulated as a result of elevated cAMP levels (Quayle *et al.*, 1994), and thus it might be expected that other types of K^+ channels can also be involved in the action of CGRP. Indeed Miyoshi and Nakaya (1995) observed that CGRP stimulated BK_{Ca} channel activity in smooth muscle cells cultured from pig coronary artery.

2.9. 17β-Estradiol

Estrogens reduce peripheral vascular resistance, and this appears to contribute to the premenopausal cardiovascular protection observed in females. Wellman *et al.* (1996), for example, observed that coronary arteries from female rats were less constricted than those from males, a sex difference attributed to high circulating levels of 17β-estradiol.

17β-Estradiol stimulates BK_{Ca} channels in endothelial cells and vascular smooth muscle cells (Fig. 7), and this effect appears to occur as a result of elevated levels of

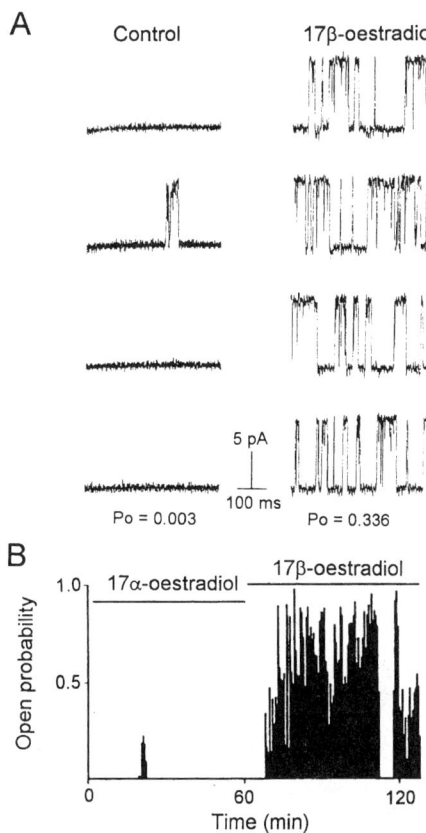

Figure 7. Stimulation of BK$_{Ca}$ channels by 17β-estradiol. (A) BK$_{Ca}$ single-channel currents recorded from a cell-attached patch before (control) and after exposure to 10 μM 17β-estradiol (40 mins). The lower current level in each trace is the closed-channel current. (B) Lack of effect of 17α-estradiol and effect of 17β-estradiol on BK$_{Ca}$ channel activity plotted as open probability (Po). (Reproduced from White *et al.*, 1995.)

nitric oxide (Rusko and van Breemen, 1995; White *et al.*, 1995; Wellman *et al.*, 1996; Darkow *et al.*, 1997). The enhanced generation of nitric oxide may occur as a result of endothelial nitric oxide synthase activity (Wellman *et al.*, 1996) or induced nitric oxide synthase activity in vascular smooth muscle cells (Darkow *et al.*, 1997).

2.10. Histamine

Histamine is released by immunological stimuli and also by nonimmunological stimuli such as ischemia, surgery, and drugs. *In vivo*, histamine causes a decrease in peripheral vascular resistance and a fall in systemic blood pressure, but in isolated vessels, vasoconstriction or vasodilation may be observed.

Histamine-induced vasoconstriction is generally associated with H$_1$ receptors on vascular smooth muscle cells (Van de Voorde *et al.*, 1994). However, similar to other

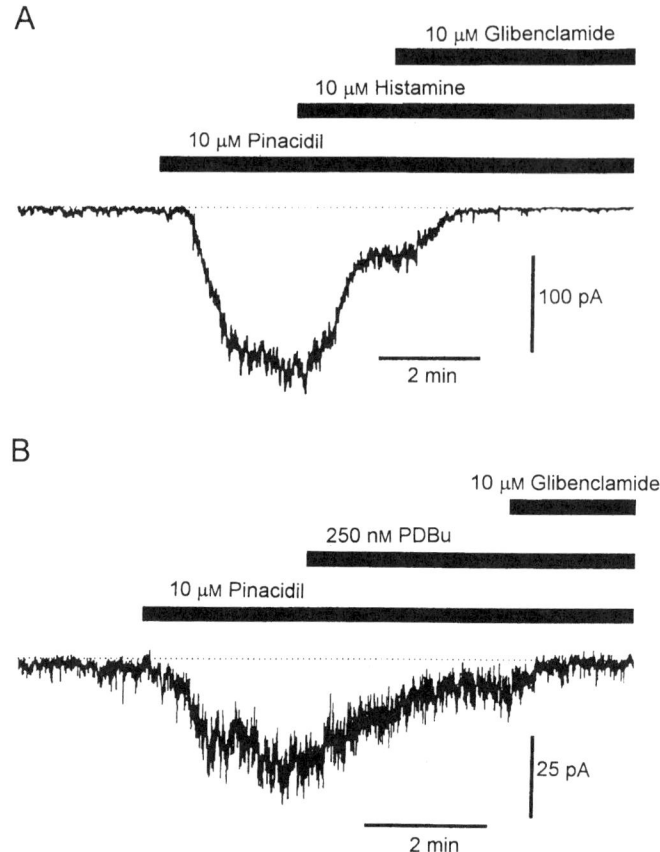

Figure 8. Inhibition of K_{ATP} channels by histamine or a phorbol ester in smooth muscle cells from rabbit basilar artery. Inward whole-cell K_{ATP} channel currents were activated by 10 μM pinacidil, which is a synthetic drug. The induced current was inhibited by histamine (10 μM) (A) or 250 nM phorbol 12,13-dibutyrate (PDBu) (B), which activates protein kinase C. Residual inward current was blocked by 10 μM glibenclamide, which blocks K_{ATP} channels. (Reproduced from Kleppisch and Nelson, 1995a.)

phospholipase C-coupled receptors, histamine stimulation also can result in the activation of BK_{Ca} channels in vascular smooth muscle cells (Hashemzadeh-Gargari and Rembold, 1992; Ishikawa *et al.*, 1993; Helliwell *et al.*, 1994). This may occur as periodic oscillations of whole-cell K^+ current (Desilets *et al.*, 1989; Wang and Large, 1993; Kang *et al.*, 1995), or high concentrations of histamine may induce a brief transient activation of BK_{Ca} channels (Neliat *et al.*, 1989). Histamine also inhibits K_{ATP} channel activity in vascular smooth muscle cells (Fig. 8A) (Bonev and Nelson 1996; Kleppisch and Nelson 1995a) and Kir channel activity in venous endothelial cells (Nilius *et al.*, 1993). Inhibition of K_{ATP} channels occurs via protein kinase C (Fig. 8B), an effect commonly associated with agonists acting on phospholipase C-coupled receptors (see also below).

Histamine-induced vasodilation can be mediated by H$_1$, H$_2$, or H$_3$ receptors. Adeagbo and Oriowo (1998) investigated the relative roles of histamine receptor subtypes and K$^+$ channels in the rat mesenteric bed. H$_1$ receptors predominantly mediated endothelium-dependent vasodilation. In contrast, endothelium-independent hyperpolarization was mediated by H$_2$ and H$_3$ receptors and was inhibited by 80 mM K$^+$ or 1 μM dequalinium, suggesting that SK$_{Ca}$ channels in vascular smooth muscle cells are involved. SK$_{Ca}$ channels are also activated by histamine in human umbilical vein endothelial cells (Muraki *et al.*, 1997).

2.11. Adrenoceptor Agonists and 5-Hydroxytryptamine

There is compelling evidence that activation of β-adrenoceptors leads to stimulation of K$_{ATP}$ channels and that this effect plays a functional role in adrenaline-induced vasodilation mediated by β-adrenoceptors (Nakashima and Vanhoutte, 1995; Randall and McCulloch, 1995). There are, however, reports that do not support such a functional role of K$_{ATP}$ channels (Sheridan *et al.*, 1997; Huang and Kwok, 1997) and data reported by Husken *et al.* (1997) suggest that impaired ATP metabolism may be required before the effect is revealed. Like K$_{ATP}$ channels, BK$_{Ca}$ channels and Kv channels are also susceptible to stimulation by cAMP-dependent phosphorylation and thus may mediate β-adrenoceptor-mediated vasodilation. Indeed, investigations of vascular smooth muscle cells from basilar artery and portal vein support the view that β-adrenoceptor activation also enhances BK$_{Ca}$ channel and Kv channel currents (Song and Simard, 1995; Aiello *et al.*, 1995, 1998). Noradrenaline acting at α_1-adrenoceptors, like many vasoconstrictors that act on phospholipase C-coupled receptors, induces transient activation of BK$_{Ca}$ channels in vascular smooth muscle cells (Benham and Bolton, 1986; Byrne and Large, 1988; Hogg *et al.*, 1994).

5-Hydroxytryatamine (5-HT) causes vasoconstriction or vasodilation. For example, it increases precapillary and postcapillary resistances and pulmonary capillary pressure and reduces resistance in coronary arteries of the rat (Barman, 1997a,b). 5-HT inhibits K$_{ATP}$ channel activity of vascular smooth muscle cells isolated from rabbit cerebral artery (Kleppisch and Nelson, 1995a).

2.12. Endothelins

Of the three endothelin peptides, endothelin-1 (ET-1) is the most potent in regulating vascular tone and often causes profound and long-lasting vasoconstriction. ET-1 was initially identified as an endothelium-derived contracting factor but is now thought to be released from a range of cell types, including astrocytes, neurons, smooth muscle cells, and endothelial cells (Pluta *et al.*, 1997). ET-1 does seem to be released in physiological conditions but is primarily associated with vascular abnormality, disease, or injury relating to hemorrhage, hypertension, vasospasm, or ischemic stroke (Rubanyi and Polokoff, 1994; see also references in Guibert and Beech, 1999).

ET-1 does affect BK$_{Ca}$ channels. Stimulatory effects are expected because ET-1, acting at the ETA receptor, is a vasoconstrictor that releases Ca^{2+} from the sarcoplasmic reticulum, induces Ca^{2+} influx via receptor-operated channels, and causes depolarization (Guibert and Beech, 1999). A transient ET-1-induced stimulation of BK_{Ca} channels has been reported for vascular smooth muscle cells of aorta and pulmonary, coronary, mesenteric, and renal arteries (Van Renterghem *et al.*, 1988; Klöckner and Isenberg, 1991; Gordienko *et al.*, 1994; Salter *et al.*, 1995; Salter and Kozlowski, 1996, 1998; Peng *et al.*, 1998). A small sustained stimulation of BK_{Ca} channels observed by Hill *et al.*, (1997) might be expected, given that sustained depolarization and Ca^{2+} influx are associated with ET-1-induced vasoconstriction, although other researchers focusing on BK_{Ca} channels have not observed such an effect. This may be related to the fact that ET-1 can also have a sustained inhibitory effect on BK_{Ca} channels. Some investigators have observed a bell-shaped (stimulation followed by inhibition) concentration-dependent effect of ET-1 or ET-3 on what seems to be current carried by BK_{Ca} channels (Hu *et al.*, 1991, 1997; Peng *et al.*, 1998). Others have also observed a sustained inhibitory effect (Minami *et al.*, 1995; Salter and Kozlowski, 1998). If there is a net inhibitory effect of ET-1 on BK_{Ca} channels, this may enhance the depolarization associated with ET-1-induced vasoconstriction. ET-1 induces sustained inhibition of K_{ATP} channels (Miyoshi *et al.*, 1992) and Kv channels (Salter and Kozlowski, 1996, 1998; Shimoda *et al.*, 1998), and these effects will also promote depolarization.

2.13. Other Extracellular Messengers Linked with K^+ Channels

The number of extracellular messengers known to modulate vascular K^+ channels continues to increase. The list of 20 or more above is not exhaustive. K^+ channels may also be involved in vasodilation induced by insulin (McKay and Hester, 1996), opioids (Armstead, 1997, 1998b), interleukin-1β (Takizawa *et al.*, 1997), endotoxin (Hall *et al.*, 1996; Price *et al.*, 1997; Hoang and Mathers, 1998), neuropeptide Y (Xiong and Cheung, 1994), cannabinoids (Zygmunt *et al.*, 1997; Chataigneau *et al.*, 1998), and atrial natriuretic factor (Kubo *et al.*, 1994). Furthermore, vasodilation in response to a fall in luminal pressure may involve activation of K_{ATP} channels (Toyoda *et al.*, 1997), and constrictor effects of thromboxane A_2, angiotensin II, and melatonin may result from inhibition of BK_{Ca} channels or K_{ATP} channels (Toro *et al.*, 1990; Scornik and Toro, 1992; Kubo *et al.*, 1997; Geary *et al.*, 1997, 1998).

3. CONCLUSIONS

BK_{Ca} channels and K_{ATP} channels are the K^+ channels most commonly reported to be involved in the actions of extracellular messengers. Kv channels have been less readily studied because of the lack of a highly selective blocker, but there are strong indications that these channels do play an important role and that subtypes are selectively expressed and regulated in the vasculature. IK_{Ca} and SK_{Ca} channels are also present and are activated by vasodilators, notably in endothelial cells, and Kir channels contribute to K^+-induced vasodilation and may be involved in acid-induced vasodilation. Therefore, all six classes of K$^+$ channels participate in responses of blood vessels to extracellular messengers.

As a generalization, it is not appropriate to think of one K$^+$ channel subtype as having an exclusive and specialized role in mediating the effect of a particular extracellular messenger. Nevertheless, BK$_{Ca}$ channels seem to be particularly important in responses to nitric oxide, reactive oxygen species, and HETEs and EETs, whereas K$_{ATP}$ channels are more associated with factors that change during ischemia, such as oxygen tension, adenosine levels, and pH. Agonists acting on cAMP-coupled receptors (e.g., prostacyclin, CGRP, and adenosine) can be expected to activate several K$^+$ channel types, including BK$_{Ca}$ channels, K$_{ATP}$ channels, and Kv channels. However, as with other agonists that are linked to broad-spectrum second messengers, there are non-K$^+$ channel mechanisms that are also activated, and so the agonist response may persist even when K$^+$ channels are blocked.

Agonists acting at phospholipase C-coupled receptors (e.g., ET-1, angiotensin II, and noradrenaline) commonly elicit transient activation of BK$_{Ca}$ channels (because there is Ca^{2+} release from the sarcoplasmic reticulum), and this may be followed by broad suppression of activity in K$_{ATP}$ channels, Kv channels, and possibly BK$_{Ca}$ channels. Therefore, such spasmogenic agonists will, depending on their concentration, repress K$^+$ channel activity and make it less available as a mechanism to induce vasodilation. This will be of consequence experimentally, perhaps hiding the role of K$^+$ channels; we speculate that spasmogenic agonist concentrations are not sufficiently high *in vivo* to prevent K$^+$ channel opening, and thus *in vitro* experiments using high concentrations of a vasoconstrictor may give an artificially negative view of the role of K$^+$ channels. However, this does not rule out a functional role for K$^+$ channel suppression by vasoconstrictors, especially the rather pronounced inhibition of K$_{ATP}$ channels. It is also worth considering that K$^+$ channels may have a more striking role in blood vessels exposed to luminal pressure, compared with blood vessels studied by the more common and classical approach of isometric tension recording, because of the depolarization induced by pressure and the consequent enhanced dependence of contraction on membrane potential and Ca^{2+} influx through voltage-gated Ca^{2+} channels.

On balance, it seems that K$^+$ channels are a major vasodilatory mechanism underlying responses to a large number of extracellular messengers. Negative reports in the literature reflect the important contribution of other (non-K$^+$ channel) mechanisms, but it is also quite probable that experimental conditions can bias an investigator away from concluding that K$^+$ channels are involved. Many messengers appear to couple to several K$^+$ channels, perhaps to protect against a situation in which one K$^+$ channel type is suppressed, either acutely or as a consequence of disease. K$^+$ channels are a key part of the vasculature and will continue to be a fruitful area for further investigation.

ACKNOWLEDGMENTS

We are grateful to the Wellcome Trust and the British Heart Foundation for supporting the research of the authors of this chapter.

REFERENCES

Aalkjaer, C. and Poston, L., 1996, Effects of pH on vascular tension: Which are the important mechanisms? *J. Vasc. Res.* **33**(5):347–359.

Adeagbo, A. S. O., and Malik, K. U., 1991, Contribution of K^+ channels to arachidonic acid-induced endothelium-dependent vasodilation in rat isolated perfused mesenteric arteries, *J. Pharmacol. Exp. Ther.* **258**:452–458.

Adeagbo, A. S. O., and Oriowo, M. A., 1998, Histamine receptor subtypes mediating hyperpolarization in the isolated, perfused rat mesenteric pre-arteriolar bed, *Eur. J. Pharmacol.* **347**(2–3):237–244.

Ahn, S. D., and Hume, J. R., 1997, pH regulation of voltage-dependent K^+ channels in canine pulmonary arterial smooth muscle cells, *Pflügers Arch.* **433**:758–765.

Aiello, E. A., Walsh, M. P., and Cole, W. C., 1995, Phosphorylation by protein kinase A enhances delayed rectifier K^+ current in rabbit vascular smooth muscle cells, *Am. J. Physiol.* **268**:H926–H934.

Aiello, E. A., Malcolm, A. T., Walsh, M. P., and Cole, W. C., 1998, Beta-adrenoceptor activation and PKA regulate delayed rectifier K^+ channels of vascular smooth muscle cells, *Am. J. Physiol.* **275**:H448–H459.

Archer, S. L., Huang, M. C., Reeve, H. L., Hampi, V., Tolarova, S., Michelakis, E., and Weir, E. K., 1996, Differential distribution of electrophysiologically distinct mysocytes in conduit and resistance arteries determines their responses to nitric oxide and hypoxia, *Circ. Res.* **78**:431–442.

Armstead, W. M., 1997, Role of activation of calcium-sensitive K^+ channels and cAMP in opioid-induced pial artery dilation, *Brain Res.* **747**:252–258.

Armstead, W. M., 1998a, Contribution of K_{Ca} channels activation to hypoxic cerebrovasodialtion does not involve NO, *Brain Res.* **799**:44–48.

Armstead, W. M., 1998b, Relationship among NO, the K_{ATP} channel, and opioids in hypoxic pial artery dilation in *Am. J. Physiol.* **275**:H988–H994.

Bari, F., Errico, R. A., Louis, T. M., and Busija, D. W., 1996, Interaction between ATP-sensitive K^+ channels and nitric oxide on pial arterioles in piglets, *J. Cereb. Blood Flow Metab.* **16**:1158–1164.

Barlow, R. S., and White, R. E., 1998, Hydrogen peroxide relaxes porcine coronary arteries by stimulating BK_{Ca} channel activity, *Am. J. Physiol.* **275**:H1283–H1289.

Barman, S. A., 1997a, Pulmonary vasoreactivity to serotonin during hypoxia is modulated by ATP-sensitive potassium channels *J. Appl. Physiol.* **83**:569–574.

Barman, S. A., 1997b, Role of calcium-activated potassium channels and cyclic nucleotides on pulmonary vasoreactivity to serotonin, *Am. J. Physiol.* **273**:L142–L147.

Beech, D. J., 1997, Actions of neurotransmitters and other messengers on Ca^{2+} channels and K^+ channels in smooth muscle cells, *Pharmacol. Ther.* **73**:91–119.

Beech, D. J. and Bolton, T. B., 1989, Properties of the cromakalim-induced potassium conductance in smooth muscle cells isolated from the rabbit portal vein, *Br. J. Pharmacol.* **98**:851–864.

Bell, D., and McDermott, B. J., 1996, Calcitonin gene-related peptide in the cardiovascular system: Characterisation of receptor populations and their (patho)physiological significance *Pharmacol. Rev.* **48**:253–288.

Benham, C. D., and Bolton, T. B., 1986, Spontaneous transient outward currents in single visceral and vascular smooth muscle cells of the rabbit, *J. Physiol, (London)* **381**:385–406.

Benham, C. D., Bolton, T. B., Lang, R. J., and Takewaki, T., 1985, The mechanism of action of Ba^{2+} and TEA on single Ca^{2+}-activated K^+ channels in arterial and intestinal smooth muscle cell membranes, *Pflügers Arch.* **403**:120–127.

Berg, T., and Koteng, O, 1997, Signalling pathways in bradykinin- and nitric oxide-induced hypotension in the normotensive rat; role of K^+-channels, *Br. J. Pharmacol.* **121**:1113–1120.

Berger, M. G., Vandier, C., Bonnet, P., Jackson, W. F., and Rusch, N. J., 1998, Intracellular acidosis differentially regualtes Kv channels in coronary and pulmnary vascular muscle, *Am. J. Physiol.* **275**:H1351–H1359.

Berne, R. M., Rubio, R., amd Curnish, R. R., 1974, Release of adenosine from ischemic brain: Effect on cerebral vascular resistance and incorporation into cerebral adenine nucleotides, *Circ. Res.* **35**:262–271.

Bolotina, V. M., Najibi, S., Palacino, J. J., Pagano, P. J., and Cohen, R. A., 1994, Nitric oxide directly activates calcium-dependent potassium channels in vascular smooth muscle, *Nature* **368**:850–853.

Bonev, A. D., and Nelson, M. T., 1996, Vasoconstrictors inhibit ATP-sensitive K^+ channels in arterial smooth muscle through protein kinase C, *J. Gen. Physiol.* **108**:315–323.

Bonnet, P., Rusch, N. J., and Harder, D. R., 1991, Characterization of an outward K$^+$ current in freshly dispersed cerebral arterial muscle cells, *Pflügers Arch.* **418**:292–296.

Bouchard, J.-F., Dumont, E., and Lamontagne, D., 1994, Evidence that prostaglandins I$_2$, E$_2$, and D$_2$ may activate ATP-sensitive potassium channels in the isolated rat heart, *Cardiovasc. Res.* **28**:901–905.

Brain, S. D., Williams, T. J., Tippins, J. R., Morris, H. R., and MacIntyre, I., 1985, Calcitonin gene-related peptide is a potent vasodilator, *Nature* **313**:54–56.

Bychkov, R., Gollash, M., Steinke, T., Ried, C., Luft, F. C., and Haller, H., 1997, Calcium-activated potassium channels and nitrate-induced vasodilation in human coronary arteries, *J. Pharmacol. Exp. Ther.* **285**:293–298.

Byrne, N. G., and Large, W. A., 1988, Membrane ionic mechanisms activated by noradrenaline in cells isolated from the rabbit portal vein, *J. Physiol, (London)* **404**:557–573.

Cabell, F., Weiss, D. S., and Price, J. M., 1994, Inhibition of adenosine-induced coronary vasodilation by block of large-conductance Ca^{2+}-activated K$^+$ channels, *Am. J. Physiol.* **267**:H1455–H1460.

Chataigneau, T., Félétou, M., Thollon, C., Villeneuve, N., Vilaine, J.-P., Duhault, J., and Vanhoutte, P. M., 1998, Cannabinoid CB$_1$ receptor and endothelium-dependent hyperpolarization in guinea-pig carotid, rat mesenteric and porcine coronary arteries, *Br. J. Pharmacol.* **123**:968–974.

Churchill, P. C. and Bidani, A. K.,1990, Adenosine and renal function, in: *Adenosine and Adenosine Receptors* (M. Williams, ed.), The Humana Press, Clifton, New Jersey, pp 335–380.

Clapp, L. H., Turcato, S., Hall, S., and Baloch, M., 1998, Evidence that Ca^{++}-activated K$^+$ channels play a major role in mediating the vascular effects of iloprost and cicaprost, *Eur. J. Pharmacol.* **356**:215–224.

Cohen, R. A., Plane, F., Najibi, S., Huk, I., Malinski, T., and Garland, C. J., 1997, Nitric oxide is the mediator of both endothelium-dependent relaxation and hyperpolarization of the rabbit carotid artery, *Proc. Natl. Acad. Sci. U.S.A.* **94**:4193–4198.

Danialou, G., Vicaut, E., Sambe, A. , Aubier, M., and Boczkowski, J., 1997, Predominant role of A$_1$ adenosine receptors in mediating adenosine induced vasodilation of rat diaphragmatic arterioles: Involvement of nitric oxide and the ATP-dependent K$^+$ channels, *Br. J. Pharmacol.* **121**:1355–1363.

Darkow, D. J., Lu, L., and White, R. E., 1997, Estrogen relaxation of coronary artery smooth muscle is mediated by nitric oxide and cGMP, *Am. J. Physiol.* **272**:H2765–H2773.

Dart, C., and Standen, N. B., 1993, Adenosine-activated potassium current in smooth muscle cells isolated from the pig coronary artery, *J. Physiol, (London)* **471**:767–786.

Dart, C., and Standen, N. B., 1995, Activation of ATP-dependent K$^+$ channels by hypoxia in smooth muscle cells isolated from the pig coronary artery, *J. Physiol, (London)* **483**:29–30.

Deka, D. K., Raviprakash, V., and Mishra, S. K., 1997, K(ATP) channels do not mediate vasodilation by 3-morpholinosydnonimine in goat coronary artery, *Eur. J. Pharmacol.* **330**(2):157–164.

Desilets, M., Driska, S. P., and Baumgarten, C. M., 1989, Current fluctuations and oscillations in smooth muscle cells from hog carotid artery: Role of the sarcoplasmic reticulum, *Circ Res.* **65**:708–722

Dong, H., Waldorn, G. J., Cole, W. C., and Triggle C. R., 1998, Roles of calcium-activated and voltage-gated delayed rectifier potassium channels in endothelium-dependent vasorelaxation of the rabbit middle cerebral artery, *Br. J. Pharmacol.* **123**:821–832.

Dumas, M., Dumas, J.-P., Rochette, L., Advenier, C., and Guidicelli, J.-F., 1997, Role of potassium channels and nitirc oxide in the effects of iloprost and prostaglandin E1 on hypoxic vasoconstriction in the isolated perfused lung of the rat, *Br. J. Pharmacol.* **120**:405–410.

Duncker, D. J., van Zon, N. S., Pavek, T. J., Herrlinger, S. K., and Bache, R. J., 1995, Endogeneous adenosine mediates coronary vasodilation during exercise after K$_{ATP}$ blockade, in *J. Clin. Invest.* **95**:285–295.

Durante, W., Kroll, M. H., Christodoulides, N., Peyton, K. J., and Schafer, A. I., 1997, Nitric oxide induces heme oxygenase-1 gene expression and carbon monoxide production in vascular smooth muscle cells, *Circ. Res.* **80**:557–564.

Eckman, D. M., Hopkins, N., McBride, C., and Keef, K. D., 1998, Endothelium-dependent relaxation and hyperpolarization in guinea-pig coronary artery: Role of epoxyeicosatrienoic acid, *Br. J. Pharmacol.* **124**:181–189.

Edvinsson, L., MacKenzie, E. T., and McCulloch, J., 1993, *Cerebral Blood Flow and Metabolism*, Raven Press, New York.

Edwards, F. R., and Hirst, G. D. S., 1988, Inward rectification in submucosal arterioles of guinea-pig ileum in *J. Physiol, (London)* **404**:437–454.

Edwards, F. R., Hirst, G. D. S., and Silverberg, G. D., 1988, Inward rectification in rat cerebral arterioles; involvement of potassium ions in autoregulation, *J. Physiol, (London)* **404**:455–466.

Edwards, G., Dora, K. A., Gardener, M. J., Garland, C. J., and Weston, A. H., 1998, K$^+$ is an endothelium-derived hyperpolarizing factor in rat arteries, *Nature* **396**:269–272.

Faraci, F. M., Breese, K. R., and Heistad, D. D., 1994, Cerebral vasodilation during hypercapnia: Role of glibenclamide-sensitive potassium channels and nitric oxide, *Stroke* **25**:1679–1683.

Fukamo, Y., Toki, Y., Numaguchi, Y., Nakashima, Y., Mukawa, H., Matsui, H., Okumura, K., and Ito, T., 1998, Nitroglycerin-induced aortic relaxation mediated by calcium-activated potassium channel is markedly diminished in hypertensive rats, *Life Sci.* **63**:1047–1055.

Gambone, L. M., Murray, P. A., and Flavahan, N. A., 1997, Synergistic interaction between endothelium-derived NO and prostacyclin in pulmonary artery: Role for K$_{ATP}^+$ channels, *Br. J. Pharmacol.* **121**:271–279.

Geary, G. G., Krause, D. N., and Duckles, S. P., 1997, Melatonin directly constricts rat cerebral arteries through modulation of potassium channels, *Am. J. Physiol.* **273**:H1530–H1536.

Geary, G. G., Duckles, S. P. and Krause, D. N., 1998, Effect of melatonin in the rat tail artery: Role of K$^+$ channels and endothelial factors, *Br. J. Pharmacol.* **123**:1533–1540.

Gebremedhin, D., Ma, Y.-M., Flack, J. R., Roman, R. J., vanRollins, M., and Harder, D. R., 1992, Mechanism of action of cerebral epoxyeicosatrienoic acids on cerebral arterial smooth muscle, *Am. J. Physiol.* **263**:H519–H525.

Gebremedhin, D., Bonnet, P., Greene A. S., England, S. K., Rusch, N. J., Lombard, J. H., and Harder, D. R., 1994, Hypoxia increases the activity of Ca^{2+}-sensitive K$^+$ channels in cat cerebral arterial muscle cell membranes, *Pflügers Arch.* **428**:621–630.

Gebremedhin, D., Kaldunski, M., Jacobs, E. R., Harder, D. R., and Roman, R. J., 1996, Coexistence of two types of Ca^{2+}-activated K$^+$ channels in rat renal arterioles, *Am. J. Physiol.* **270**:F69–F81.

Gebremedhin, D., Lange, A. R., Narayan, J., Aebly, M. R., Jacobs, E. R., and Harder, D. R., 1998, Cat cerebral arterial smooth muscle cells express cytochrome P450 4A$_2$ enzyme and produce the vasoconstrictor 20-HETE which enhances L-type Ca^{2+} current, *J. Physiol, (London)* **507**:771–781.

Gidday, J. M., Maceren, R. G., Shah, A. R., Meier, J. A., and Zhu Y., 1996, K$_{ATP}$ channels mediate adenosine-induced hyperemia in retina, *Invest. Ophthalmol. Visual Sci.* **37**:2624–2633.

Gordienko, D. V., Clausen, C., and Goligorsky, M. S., 1994, Ionic currents and endothelin signaling in smooth muscle cells from rat renal resistance arteries, *Am. J. Physiol.* **266**:25–41.

Grissmer, S., Nguyen, A. N., Aiyar, J., Hanson, D. C., Mather, R. J., Gutman, G. A., Karmilowicz, M. J., Auperin, D. D., and Chandy, K. G., 1994, Pharmacological characterisation of five cloned voltage-gated K$^+$ channels, types Kv1.1, 1.2, 1.3, 1.5, and 3.1, stably expressed in mammalian cell lines, *Mol. Pharmacol.* **45**:1227–1234.

Guibert, C., and Beech, D. J.,1999, Positive and negative coupling of the endothelin ETA receptor to Ca^{2+}-permeable channels in rabbit cerebral cortex arterioles, *J. Physiol, (London)* **514**:843–856.

Hall, S., Turcato, S., and Clapp, L., 1996, Abnormal activation of K$^+$ channels underlies relaxation to bacterial lipopolysaccharide in rat aorta, *Biochem. Biophys. Res. Commun.* **224**:184–190.

Halpern, W., and Kelley, M., 1991, In vitro methodology for resistance arteries, *Blood Vessels* **28**:245–251.

Hashemzadeh-Gargari, H., and Rembold, C. M., 1992, Histamine activates whole cell K$^+$ currents in swine carotid arterial smooth muscle cell, *Comp. Biochem. Physiol. C* **102**(1):33–37.

Hayabuchi, Y., Nakaya, Y., Matsuoka, S., and Kuroda, Y., 1998a, Hydrogen peroxide-induced vascular relaxation in porcine coronary arteries is mediated by Ca^{2+}-activated K$^+$ channels, *Heart Vessels* **13**:9–17.

Hayabuchi, Y., Nakaya, Y., Matsuoka, S., and Kuroda, Y., 1998b, Effect of acidosis on Ca^{2+}-activated K$^+$ channels in cultured porcine coronary artery smooth muscle cells, *Pflügers Arch.* **436**:509–514.

Helliwell, R. M., Wang, Q., Hogg, R. C., and Large, W. A., 1994, Synergistic action of histamine and adenosine triphosphate on the response to noradrenaline in rabbit pulmonary artery smooth muscle cells, *Pflügers Arch.* **426**:433–439.

Henderickx, H., and Casteels, R., 1974, Electrogenic sodium pump in arterial smooth muscle cells, *Pflügers Arch.* **346**:299–306.

Hill, C. E., Kirton, A., Wu, D. D., and Vanner, S. J., 1997, Role of maxi-K$^+$ channels in endothelin-induced vasoconstriction of mesenteric and submucosal arterioles, *Am. J. Physiol.* **273**: G1087–G1093.

Hoang, L. M., and Mathers, D. A., 1998, Internally applied endotoxin and the activation of BK channels in cerebral artery smooth muscle via a nitric oxide-like pathway, *Br. J. Pharmacol.* **123**:5–12.

Hogg, R. C., Wang, Q., and Large, W. A., 1994, Effects of Cl channel blockers on Ca-activated chloride and potassium currents in smooth muscle cells from rabbit portal vein, *Br. J. Pharmacol.* **111**:1333–1341.

Holz, F. G., and Steinhausen, M., 1987, Renovascular effects of adenosine receptor agonists, *Renal Physiol* **10**:272–282.

Hu, S., and Kim, H. S., 1993, Activation of K$^+$ channel in vascular smooth muscles by cytochrome P450 metabolites of arachidonic acid, *Eur. J. Pharmacol.* **230**:215–221.

Hu, S. L., Kim, H. S., and Jeng, A. Y., 1991, Dual action of endothelin-1 on the Ca^{2+}-activated K$^+$ channel in smooth-muscle cells of porcine coronary-artery, *Eur. J. Pharmacol.* **194**:31–36.

Hu, S., Kim, H. S., Savage, P., and Jeng, A. Y., 1997, Activation of BK$_{(Ca)}$ channel via endothelin ET$_{(A)}$ receptors in porcine coronary artery smooth muscle cells, *Eur. J. Pharmacol* **324**:277–282.

Huang, Y., and Kwok, K. H., 1997, Effects of putative K$^+$ channel blockers on β-adrenoceptor-mediated vasorelaxation of rat mesenteric artery, *J. Cardiovasc. Pharmacol.* **29**:515–519.

Husken, B. C., Pfaffendorf, M., and van Zwieten, P. A., 1997, Contribution of ATP-sensitive potassium channels to β-adrenoceptor-mediated responses, *Naunyn-Schmiedebergs Arch. Pharmacol.* **355**:97–102.

Imig, J. D., Zou, A.-P., Stec, D. E., Harder, D. R., Flack, J. R., and Roman, R. J., 1996, Formation and actions of 20-hydroxyeicosatetraenoic acid in rat renal arterioles, *Am. J. Physiol.* **270**:R217–R227.

Ishida, Y., and Honda, H., 1993, Inhibitory action of 4-aminopyridine on Ca^{2+}-ATPase of the mammalian sarcoplasmic reticulum, *J. Biol. Chem.* **268**:4021–4024.

Ishikawa, T., Hume, J. R., and Keef, K. D., 1993, Modulation of K$^+$ and Ca^{2+} channels by histamine H1-receptor stimulation in rabbit coronary artery cells, *J. Physiol, (London)* **468**:379–400.

Ishizaka, H., and Kuo, L., 1996, Acidosis-induced coronary arteriolar dilation is mediated by ATP-sensitive potassium channels in vascular smooth muscle, *Circ. Res.* **78**:50–57.

Jackson, W. F., Konig, A., Dambacher, T., and Busse, R., 1993, Prostacyclin-induced vasodilation in rabbit heart is mediated by ATP-sensitive potassium channels, *Am. J. Physiol.* **264**:H238–H243.

Jensen, B. S., Strobaek, D., Christophersen, P., Jorgensen, T. D., Hansen, C., Silahtaroglu, A., Olesen, S.-P., and Ahring, P. K., 1998, Characterisation of the cloned human intermediate-conductance Ca^{2+}-activated K$^+$ channel, *Am. J. Physiol.* **275**:C848–C856.

Jiang, C., and Collins, P., 1994, Inhibition of hypoxia-induced relaxation of rabbit isolated coronary arteries by N^G-monomethyl-L-arginine but not glibenclamide, *Br. J. Pharmacol.* **111**:711–716.

Kang, T. M., So, I., and Kim, K. W., 1995, Caffeine- and histamine-induced oscillations of K$_{Ca}$ current in single smooth muscle cells of rabbit cerebral artery, *Pflügers Arch.* **431**:91–100.

Kinoshita, H., and Katusic, Z. S., 1997, Role of potassium channels in relaxations of isolated canine basilar arteries to acidosis, *Stroke* **28**:433–438.

Kirber, M. T., Ordway, R. W., Clapp, L. H., Walsh, J. V. J., and Singer, J. J., 1992, Both membrane stretch and fatty acids directly activate large conductance Ca^{++}-activated K$^+$ channels in vascular smooth muscle cells, *FEBS Lett.* **297**:24–28.

Kitakaze, M, Takshima, S., Minamino, T., Node, K, Shinozaki, Y., Mori, H., and Hori, M., 1997, Temporary acidosis during reperfusion limits myocardial infarct size in dogs, *Am. J. Physiol.* **272**:H2071–H2078.

Kitazono, T., Faraci, F. M., and Heistad, D. D., 1993, Role of ATP-sensitive K$^+$ channels in CGRP-induced dilation *in vivo*, *Am. J. Physiol.* **265**:H581–H585.

Kitazono, T., Ibayashi, S., Nagao, T., Fujii, K., and Fujishima, M., 1997, Role of Ca^{2+}-activated K$^+$ channels in acetylcholine-induced dilatation of the basilar artery *in vivo*, *Br. J. Pharmacol.* **120**:102–106.

Kleppisch, T., and Nelson, M. T., 1995a, ATP-sensitive K$^+$ currents in cerebral arterial smooth muscle: Pharmacological and hormonal modulation, *Am. J. Physiol.* **269**:H1634–H1640.

Kleppisch, T., and Nelson, M. T., 1995b, Adenosine activates ATP-sensitive potassium channels in arterial myocytes via A$_2$ receptors and cAMP-dependent protein kinase, *Proc. Natl. Acad. Sci., U.S.A.* **92**:12441–12445.

Klöckner, U., and Isenberg, G., 1991, Endothelin depolarizes myocytes from porcine coronary and human mesenteric arteries through a Ca-activated chloride current, *Pflügers Arch.* **418**:168–175.

Knot, H. J., Zimmerman, P. A., and Nelson, M. T., 1996, Extracellular K$^+$-induced hyperpolarizations and dilatations of rat coronary and cerebral arteries involve inward rectifer K$^+$ channels, *J. Physiol, (London)* **492**:419–430.

Kontos, H. A., Raper, A. J., and Patterson, J. L., 1977, Analysis of vasoactivity of local pH, PCO$_2$ and bicarbonate on pial vessels, *Stroke* **8**:258–360.

Krippeitdrews, P., Haberland, C., Fingerle, J., Drews, G., and Lang, F., 1995, Effect of H$_2$O$_2$ on membrane potential and [Ca^{2+}]$_i$ of cultured rat arterial smooth-muscle cells, *Biochem. Biophys. Res. Commun.* **209**:139–145.

Kubo, M., Nakaya, Y., Matsuoka, S., Saito, K., and Kuroda, Y., 1994, Atrial natriuretic factor and isosorbide dinitrate modulate the gating of ATP-sensitive K^+ channels in cultured vascular smooth muscle cells, *Circ. Res.* **74:**471–476.

Kubo, M., Quayle, J. M., and Standen, N. B., 1997, Angiotensin II inhibition through ATP-sensitive K^+ currents in rat smooth muscle cells through protein kinase C, *J. Physiol, (London)* **503:**489–496.

Kuo, L., and Chancellor, J. D., 1995, Adenosine potentiates flow-induced dilation of coronary arterioles by activating K_{ATP} channels in endothelium, *Am. J. Physiol.* **269:**H541–H549.

Ma, Y.-H., Gebremedhin, D., Schwartzman, M. L., Falck, J. R., Clark, J. E., Masters, B. S., Harder, D. R., and Roman, R. J., 1993, 20-Hydroxyeicosatetranoic acid is an endogenous vasoconstrictor of canine renal arcuate arteries, *Circ. Res.* **72:**126–130.

Martens, J. R., and Gelband, G. H., 1996, Alterations in rat interlobar artery membrane potential and K^+ channels in genetic and nongenetic hypertension, *Circ. Res.* **79:**295–301.

Mayhan, W. G., and Faraci, F. M., 1993, Responses of cerebral arterioles in diabetic rats to activation of ATP-sensitive potassium channels, *Am. J. Physiol.* **265:**H152–H157.

McCarron, J. G., and Halpern, W., 1990, Potassium dilates rat cerebral arteries by two independent mechanisms, *Am. J. Physiol.* **259:**H902–H908.

McGiff, J. C., 1991, Cytochrome P450 metabolism of arachidonic acid, *Annu. Rev. Pharmacol. Toxicol.* **31:**339–369.

McKay, M. K., and Hester, R. L., 1996, Role of nitric oxide, adenosine, and ATP-sensitive potassium channels in insulin-induced vasodilation, *Hypertension* **28:**202–208.

Meno, J. R., Ngai, A. C., Ibayashi, S., and Winn, H. R., 1991, Adenosine release and changes in pial arteriolar diameter during transient cerebral ischaemia and reperfusion, *J. Cereb. Blood Flow Metab.* **11:**986–993.

Liu, Y., Hudetz, A. G., Knaus, H.-G., Rusch, N. J. 1998, Increased expression of Ca^{2+}-sensitive K^+ channels in the cerebral microcirculation of genetically hypertensive rates, *Circ. Res.* **82:**729–737.

Minami, K., Hirata, Y., Tokumura, A., Nagaya, Y., and Fukuzawa, K., 1995, Protein-kinase C-independent inhibition of the Ca^{2+}-activated K^+ channel by angiotensin-II and endothelin-1, *Biochem. Pharmacol.* **49:**1051–1056.

Mistry, D. K., and Garland, C. J., 1998, Nitric oxide (NO)-induced activation of large conductance Ca^{2+}-dependent K^+ channels (BK_{Ca}) in smooth muscle cells isolated from the rat mesenteric artery, *Br. J. Pharmacol.* **124:**1131–1140.

Miyoshi, H., and Nakaya, Y., 1995, Calcitonin gene-related peptide activates the K^+ channels of vascular smooth muscle cells via adenylate cyclase, *Basic Res. Cardiol.* **90:**332–336.

Miyoshi, Y., Nakaya, Y., Wakatsuki, T., Nakaya, S., Fujino, K., Saito, K., and Inoue, I., 1992, Endothelin blocks ATP-sensitive K^+ channels and depolarizes smooth muscle cells of porcine coronary artery, *Circ. Res.* **70:**612–616.

Morii, S., Ngai, A.C., Ko, A. R., and Winn, H. R., 1987, Role of adenosine in regulation of cerebral blood flow: Effects of theophylline during normoxia and hypoxia, *Am. J. Physiol.* **253:**H165–H175.

Muira, H., and Gutterman, D. D., 1998, Human coronary arteriolar dilation to arachidonic acid depends on cytochrome P450 monooxygenase and Ca^{++}-activated K^+ channels, *Circ. Res.* **83:**501–507.

Muraki, K., Imaizumi, Y., Ohya, S., Sato, K., Takii, T., Onozaki, K., and Watanabe, M., 1997, Apamin-sensitive Ca^{2+}-dependent K^+ current and hyperpolarization in human endothelial cells, *Biochem. Biophys. Res. Commun.* **236:**340–343.

Murphy, M. E., and Brayden, J. E., 1995, Nitric oxide hyperpolarises rabbit mesenteric arteries via ATP-sensitive potassium channels, *J. Physiol, (London)* **486:**47–58.

Nakashima, M., and Vanhoutte, P. M., 1995, Isoproterenol causes hyperpolarization through opening of ATP-sensitive potassium channels in vascular smooth muscle of the canine saphenous vein, *J. Pharmacol. Exp. Ther.* **272:**379–384

Neliat, G., Masson, F., and Gargouil, Y. M., 1989, Modulation of the spontaneous transient outward currents by histamine in single vascular smooth muscle cells, *Pflügers Arch.* **414** (Suppl. 1):S186–S187.

Ney, P., and Feelisch, M., 1995, Vasodilator effects of PGE_1 in the coronary and systemic circulation of the rat are mediated by ATP-sensitive potassium (K^+) channels, *Agents Actions*—Suppl. **45:**71–76.

Nilius, B., Schwarz, G., and Droogmans, G., 1993, Modulation by histamine of an inwardly rectifying potassium channel in human endothelial cells, *J.Physiol. (London)* **472:**359–371.

Okazaki, K., Endou, M., and Okumura, F., 1998, Involvement of barium-sensitive K^+ channels in endothelium-dependent vasodilation produced by hypercapnia in rat mesenteric vascular beds, *Br. J. Pharmacol.* **125:**168–185.

Oltman, C. I., Weitraub, N. L., VanRollins, M., and Dellsperger, K. C., 1998, Epoxyeicosatrienoic acids and dihydroxyeicosatrienoic acids are potent vasodilators in the canine coronary microcirculation, *Circ. Res.* **83**:932–939.

Onoue, H., and Katusic, Z. S., 1998, Subarachnoid hemorrhage and the role of potassium channels in relaxations of canine basilar artery to nitrovasodilators, *J. Cereb. Blood Flow Metab.* **18**:186–195.

Osipenko, O. N., Evans, A. M., and Gurney, A. M., 1997, Regulation of the resting potential of rabbit pulmonary artery myocytes by a low threshold, O$_2$-sensing potassium current, *Br. J. Phamacol.* **120**:1461–1470.

Park, M. K., Lee, S., H., Lee, S. J., Ho, W. K., and Earm, Y. E., 1995, Different modulation of Ca-activated K channels by the intracellular redox potential in pulmonary and ear arterial smooth muscle cells of the rabbit, *Pflügers Arch.* **430**:308–314.

Pelligrino, D. A., Wang, Q., Koenig, H. M., and Albrecht, R. F., 1995, Role of nitric oxide, adenosine, N-methyl-D-aspartate, and neuronal activation in hypoxia-induced pial arteriolar dilation in rats, *Brain Res.* **704**:61–70.

Peng, W., Michael, J. R., Hoidal, J. R., Karwande, S. V., and Farrukh, I. S., 1998, ET-1 modulates K$_{Ca}$-channel activity and arterial tension in normoxic and hypoxic human pulmonary vasculature, *Am.J. Physiol.* **275**:L729–L739.

Plane, F., Wiley, K. E., Jeremy, J. Y., Cohen, R. A., and Garland, C. J., 1998, Evidence that different mechanisms underlie smooth muscle relaxation to nitric oxide donors in the rabbit isolated carotid artery, *Br. J. Pharmacol.* **123**:1351–1358.

Pluta, R. M., Boock, R. J., Afshar, J. K., Clouse, K., Bacic, M., Ehrenreich, H., and Oldfield, E. H., 1997, Source and cause of endothelin-1 release into cerebrospinal fluid after subarachnoid hemorrhage, *J. Neurosurg.* **87**:287–293.

Price, J. M., Cabell, J. F., and Hellerman, A., 1996, Inhibition of cAMP mediated relaxation in rat coronary vessels by block of Ca^{++} activated K$^+$ channels, *Life Sci.* **58**:2225–2232.

Price, J. M., Baker, C. H., and Bond, R. F., 1997, Calcium-activated potassium channel-mediated arteriolar relaxation during endotoxic shock, *Shock* **7**:294–299.

Prior, H. M., Yates, M. S., and Beech, D. J.,1998a, Functions of large conductance (BK$_{Ca}$), delayed rectifier (K$_V$) and background K$^+$ channels in the control of membrane potential in rabbit renal arcuate artery, *J. Physiol. (London)* **511**:159–169.

Prior, H. M., Webster, N., Quinn, K. V., Beech, D. J., and Yates, M. S., 1998b, K$^+$-induced dilation in a small renal artery: No role for inward rectifier K$^+$ channels, *Cardiovasc. Res.* **37**:780–790.

Prior, H. M., Yates, M. S., and Beech, D. J., 1998c, Role of K$^+$ channels in A$_{2A}$ adenosine receptor-mediated dilation of the pressurised renal arcuate artery, *Br. J. Pharmacol.* **126**:494–500.

Quayle, J. M., Bonev, A. D., Brayden, J. E., and Nelson, M. T., 1994, Calcitonin gene-related peptide activated ATP-sensitive K$^+$ currents in rabbit arterial smooth muscle via protein kinase A, *J. Physiol. (London)* **475**:9–13.

Quayle, J. M., Dart, C., and Standen, N. B., 1996, The properties and distribution of inward rectifier potassium currents in pig coronary arterial smooth muscle, *J. Physiol. (London)* **494**:715–726.

Randall, M. D.,1995, The involvement of ATP-sensitive potassium channels and adenosine in the regulation of coronary flow in the isolated perfused rat heart, *Br. J. Pharmacol.* **116**:3068–3074.

Randall, M. D., and McCulloch, A. I., 1995, The involvement of ATP-sensitive potassium channels in β-adrenoceptor-mediated vasorelaxation in the rat isolated mesenteric arterial bed, *Br. J. Pharmacol.* **115**:607–612.

Rubanyi, G. M., and Polokoff, M. A., 1994, Endothelins: Molecular biology, biochemistry, pharmacology, physiology, and pathophysiology, *Pharmacol. Rev.* **46**:325–415.

Rusko, J., Li, L., and van Breemen, C., 1995, 17β-Estradiol stimulation of endothelial K$^+$ channels, *Biochem. Biophys. Res. Commun.* **214**:367–372.

Sakai, Y., and Saito, K., 1998, Reciprocal interactions among neuropeptides and adenosine in the cardiovascular system of rats: A role of K$_{(ATP)}$ channels, *Eur. J. Pharmacol.* **345**:279–284.

Sakai, Y., Yoshikawa, N., Akima, M., and Saito, K., 1998, Role for adenosine A$_1$ and A$_2$ receptors in femoral vasodilation induced by intra-arterial adenosine in rabbits, *Eur. J. Pharmacol.* **353**:257–264.

Salter, K. J., and Kozlowski, R. Z., 1996, Endothelin receptor coupling to potassium and chloride channels in isolated rat pulmonary arterial myocytes, *J. Pharmacol. Exp. Ther.* **279**:1053–1062.

Salter, K. J., and Kozlowski, R. Z., 1998, Differential electrophysiological actions of endothelin-1 on Cl$^-$ and K$^+$ currents in myocytes isolated from aorta, basilar and pulmonary artery, *J. Pharmacol. Exp. Ther.* **284**:1122–1131.

Salter, K. J., Turner, J. L., Albawarni, S., Clapp, L. H., and Kozlowski, R. Z., 1995, Ca^{2+}-activated Cl^- and K^+ channels and their modulation by endothelin-1 in rat pulmonary arterial smooth-muscle cells, *Exp. Physiol.* **80:**815–824.

Schubert, R., Serebyakov, V. N., Engel, H., and Hopp, H.-H., 1996, Iloprost activates K_{Ca} channels of vascular smooth muscle cells: Role of cAMP-dependent protein kinase, *Am. J. Physiol. (London)* **271:**C1203–C1211.

Schubert, R., Serebyrakov, V. N., Mewes, H., and Hopp, H.-H., 1997, Iloprost dilates rat small arteries: Role of K_{ATP}-channel activation by cAMP-dependent protein kinase, *Am. J. Physiol. (London)* **272:**H1147–H1156.

Scornik, F. S., and Toro, L., 1992, U46619, a thromboxane A_2 agonist, inhibits K_{ca} channel activity from pig coronary artery, *Am. J. Physiol.* **262:**C708–C713.

Sheridan, B. C., McIntyre, R. C., Jr., Meldrum, D. R., and Fullerton, D. A., 1997, K_{ATP} channels contribute to beta- and adenosine receptor-mediated pulmonary vasorelaxation, *Am. J. Physiol.* **273:**L950–L956.

Shimoda, L. A., Sylvester, J. T., and Sham, J. S. K., 1998, Inhibition of voltage-gated K^+ current in rat intrapulmonary arterial myocytes by endothelin-1, *Am. J. Physiol.* **274:**L842–L853.

Smirnov, S. V., and Aaronson, P. I., 1996, Modulatory effects of arachidonic acid on the delayed rectifier K^+ current in rat pulmonary arterial myocytes: Structural aspects and involvement of protein kinase C, *Circ. Res.* **79:**20–31.

Sobey, C. G., Heistad, D. D., and Faraci, F. M., 1997, Mechanisms of bradykinin-induced cerebral vasodilatation in rats: Evidence that reactive oxygen species activate K^+ channels, *Stroke* **28:**2990–2295.

Song, Y., and Simard, J. M., 1995, β-Adrenoceptor stimulation activates large-conductance Ca^{2+}-activated K^+ channels in smooth muscle cells from basilar artery of guinea pig, *Pflügers Arch.* **430:**984–993.

Spencer, R. H., Sokolov, Y., Li, H., Takenas, B., Milici, A. J., Aiyar, J., Nguyen, A., Park, H., Jap, B. K., Hall, J. E., Gutman, G. A., and Chandy, K. G., 1997, Purification, visualisation, and biophysical characterisation of Kv1.3 tetramers, *J. Biol. Chem.* **272:**2389–2395.

Sun, C.-W., Alonso-Galicia, M., Taheri, M. R., Flack, J. R., Harder, D. R., and Roman, R. J., 1998, Nitric oxide–20-hydroxyeicosatetraenoic acid interaction in the regulation of K^+ channel activity and vascular tone in renal arterioles, *Circ. Res.* **83:**1069–1079.

Tabrizchi, R., and Lupichuk, S. M., 1995, Vasodilation produced by adenosine in isolated rat perfused mesenteric artery: A role for endothelium, *Naunyn-Schmiedeberg's Arch. Pharmacol.* **352:**412–418.

Taguchi, H., Heistad, D. D., Kitazono, T., and Faraci, F. M., 1994, ATP-sensitive K^+ channels mediate dilatation of cerebral arterioles during hypoxia, *Circ. Res.* **74:**1005–1008.

Takizawa, S., Ozaki, H., and Karaki, H., 1997, Interleukin-1β-induced, nitric oxide-dependent and -independent inhibition of vascular smooth muscle contraction, *Eur. J. Pharmacol.* **330:**143–150.

Taniguchi, J., Furukawa, K.-I., and Shigekawa, M., 1993, Maxi K^+ channels are stimulated by cyclic guanosine monophosphate-dependent protein kinase in canine coronary artery smooth muscle cells, *Pflügers Arch.* **423:**167–172.

Toro, L., Amador, M., and Stefani, E., 1990, ANG II inhibits calcium-activated potassium channels from coronary smooth muscle in lipid bilayers, *Am. J. Physiol.* **258:**H912–H915.

Toyoda, K., Fujii, K., Ibayashi, S., Kitazono, T., Nagao, T., and Fujishma, M., 1997, Role of ATP-sensitive potassium channels in brain stem circulation during hypotension, *Am. J. Physiol.* **273:**H1342–H1346.

Van de Voorde, J., Brochez, V., and Vanheel, B., 1994, Heterogenous effects of histamine on isolated rat coronary arteries, *Eur. J. Pharmacol.* **271:**17–23.

Van Renterghem, C., Vigne, P., Barhanin, J., Schmid-Alliana, A., Frelin, C., and Lazdunski, M., 1988, Molecular mechanisms of action of the vasoconstrictor peptide endothelin, *Biochem. Biophys. Res. Commun.* **157:**977–985.

von Beckerath, N., Cyrys, C., Dischner, A., and Daut, A., 1991, Hypoxic vasodilation in isolated, perfused guinea-pig heart: An analysis of the underlying mechanisms, *J. Physiol. (London)* **442:**297–319.

Wang, J., Juhaszova, M., Rubin, L. J., and Yuan X.-J., 1997, Hypoxia inhibits gene expression of voltage-gated K^+ channel α subunits in pulmonary artery smooth muscle cells, *J. Clin. Invest.* **100:**2347–2353.

Wang, Q., and Large, W. A., 1993, Action of histamine on single smooth muscle cells dispersed from the rabbit pulmonary artery, *J. Physiol. (London)* **468:**125–139.

Wang, R., and Wu, L., 1997, The chemical modification of K_{Ca} channels by carbon monoxide in vascular smooth muscle cells, *J. Biol. Chem.* **273:**8222–8226.

Webb, R. C., and Bohr, D. F., 1978, Potassium-induced relaxation as an indicator of Na^+,K^+-ATPase activity in vascular smooth muscle, *Blood Vessels* **15:**198–207.

Wei, E. P., Kontos, H., A., and Beckman, J. S.,1996, Mechanisms of cerebral vasodilatation by superoxide, hydrogen peroxide, and peroxynitrite, *Am. J. Physiol.* **271:**H1262–H1266.

Wei, E. P., Kontos, H. A., and Beckman, J. S., 1998, Antioxidants inhibit ATP-sensitive potassium channels in cerebral arterioles, *Stroke* **29:**817–823.

Weisbrod, R. M., Grisworld, M. C., Yaghoubi, M., Komalavilas, P., Linclon, T. M., and Cohen, R. A., 1998, Evidence that additional mechanisms to cyclic GMP mediate the decrease in intracellular calcium and relaxation of rabbit aortic smooth muscle to nitric oxide, *Br. J. Pharmacol.* **125:**1695–1707.

Wellman, G. C, Bonev, A. D., Nelson, M. T., and Brayden, J. E., 1996, Gender differences in coronary artery diameter involve estrogen, nitric oxide, and Ca^{++}-dependent K$^+$ channels, *Circ. Res.* **79:**1024–1030.

Wellman, G. C., Quayle, J. M., and Standen, N. B., 1998, ATP-sensitive K$^+$ channel activation by calcitonin gene-related peptide and protein kinase A in pig coronary arterial smooth muscle, *J. Physiol. (London)* **507:**117–129.

White, R. E., Darkow, D. J., and Falvo Lang, J. L., 1995, Estrogen relaxes coronary arteries by opening BK$_{Ca}$ channels through a cGMP-dependent mechanism, *Circ. Res.* **77:**936–942.

Williams, D. L., Jr., Katz, G. M., Roy-Contancin, L., and Reuben, J. P., 1988, Guanosine 5'-monophosphate modulates gating of high-conductance Ca^{2+}-activated K$^+$ channels in vascular smooth muscle cells, *Proc. Natl. Acad. Sci., U.S.A.* **85:**9360–9364.

Xiong, Z., and Cheung, D. W., 1994, Neuropeptide Y inhibits Ca^{2+}-activated K$^+$ channels in vascular smooth muscle cells from the rat tail artery, *Pflügers Arch.* **429:**280–284.

Xu, X., and Lee, K. S., 1996, Dual effects of arachidonic acid on ATP-sensitive K$^+$ current of coronary smooth muscle cells, *Am. J. Physiol.* **270:**H1957–H1962.

Yamazaki, J., Sato, K., Ohara, F., and Nagao, T., 1998, Direct activation of endothelial NO pathway by Ba^{2+} in canine coronary artery, *Br. J. Pharmacol.* **124:**1149–1158.

Yuan, X.-J., Goldman, W. F., Tod, M. L., Rubin, L. J., and Blaustein, M. P., 1993, Hypoxia reduces potassium currents in cultured rat pulmonary but not mesenteric arterial myocytes, *Am. J. Physiol.* **264:**L116–L123.

Yuan, X.-J., Goldman, W. F., Tod, M. L., Rubin, L. J., and Blaustein, M. P., 1996, NO hyperpolarizes pulmonary artery smooth muscle cells and decreases the intracellular Ca^{2+} concentration by activating voltage-gated K$^+$ channels, *Proc. Natl. Acad. Sci. U.S.A.* **93:**10489–10494.

Zhao, Y.-J., Wang, J., Rubin, L. J., and Yuan, X.-J., 1997, Inhibition of K$_V$ and K$_{Ca}$ channels antagonizes NO-induced relaxation in pulmonary artery, *Am. J. Physiol.* **272:**H904–H912.

Zou, A.-P., Fleming, J. T., Falck, J. R., Jacobs, E. R., Gebremedhin, D., Harder, D. R., and Roman, R. J., 1996, Stereospecific effects of epoxyeicosatrienoic acids on renal vascular tone and K$^+$-channel activity, *Am. J. Physiol.* **270:**F822–F832

Zygmunt, P. M., Edwards, G., Weston, A. H., Larsson, B., and Hogestatt, E. D., 1997, Involvement of voltage-depedendent potassium channels in the EDHF-mediated ralxation of rat hepatic artery, *Br. J. Pharmacol.* **121:**141–149.

Chapter 24

Delayed Rectifier K$^+$ Channels of Vascular Smooth Muscle: Characterization, Function, and Regulation by Phosphorylation

W. C. Cole and M. P. Walsh

1. INTRODUCTION

Regulation of membrane potential in vascular smooth muscle cells is critical for determining the level of tone in the arterial wall and, as a result, vessel diameter and peripheral vascular resistance. K$^+$ channels play a critical role in setting the basal level of membrane potential and in electrical and mechanical responses to changes of intraluminal pressure and vasoactive agonists. Voltage-activated K$^+$ current (Kv) is an important component of outward K$^+$ conductance of smooth muscle and may be divided into two varieties based on distinct inactivation characteristics: (i) rapidly inactivating, transient outward (K$_{TO}$) or A-type K$^+$ current, which exhibits rapid activation and fast, N-type inactivation kinetics, and (ii) delayed rectifier (K$_{DR}$) current, which displays no, or only slow, C-type inactivation. In the past five years, it has become apparent that vascular K$_{DR}$ channels (i) are regulated by endogenous vasoactive molecules via phosphotransferase reactions involving serine/threonine kinases and (ii) play an important role in the control of electrical and mechanical activity, arterial diameter, and peripheral resistance. This chapter reviews our current understanding of the biophysical and pharmacological propertics of vascular K$_{DR}$ channels, the signal transduction pathways involved in their regulation, and their possible roles in the control of tone and arterial diameter.

W. C. Cole ● Smooth Muscle Research Group., Department of Pharmacology and Therapeutics, Faculty of Medicine, University of Calgary, Calgary, Alberta, Canada T2N 4N1. *M. P. Walsh* ● Smooth Muscle Research Group, Department of Biochemistry and Molecular Biology, Faculty of Medicine, University of Calgary, Calgary, Alberta, Canada T2N 4N1.

Potassium Channels in Cardiovascular Biology, edited by Archer and Rusch. Kluwer Academic/Plenum Publishers, New York, 2001.

2. PROPERTIES OF VASCULAR K_{DR} CHANNELS

Macroscopic currents due to K_{DR} channel activity have been recorded from isolated myocytes of all varieties of smooth muscle. For example, vascular (Gollasch *et al.*, 1996; Jackson *et al.*, 1996; Halliday *et al.*, 1995; Knot and Nelson, 1995; Leblanc *et al.*, 1994; Bolzon *et al.*, 1993; Boyle *et al.*, 1992; Gelband and Hume, 1992; Smirnov and Aaronson, 1992; Volk *et al.*, 1991; Beech and Bolton, 1989; Okabe *et al.*, 1987), gastrointestinal (Carl, 1995; Xiong *et al.*, 1995; Akbarali, 1993; Thornbury *et al.*, 1992), urogenital (Zhang *et al.*, 1993), and airway (Snetkov *et al.*, 1995; Fleischmann *et al.*, 1993; Boyle *et al.*, 1992; Kotlikoff, 1990) smooth muscles all exhibit voltage-activated, time-dependent current positive to approximately -50 mV (due to K_{DR} channels). However, some cell-to-cell variability in K_{DR} current density has been reported: myocytes isolated from small pulmonary arteries were reported to possess larger K_{DR} currents than cells from large vessels (Archer, 1996; Archer *et al.*, 1996), and in human bronchial smooth muscle, little or no K_{DR} current was observed in some myocytes, although the same tissue samples also yielded myocytes with robust K_{DR} currents (Snetkov *et al.*, 1995). Whether these differences in current density reflect a differential expression of K_{DR} channels or lack of availability due to phosphorylation or dephosphorylation (see below) remains to be determined.

Whole-cell K_{DR} currents have the following properties: (i) voltage-dependent activation positive to approximately -50 mV at $37°C$, (ii) slow, incomplete C-type inactivation which leaves a component of noninactivating, steady-state current of varied amplitude, (iii) slowly deactivating tail currents, and (iv) inhibition by 4-aminopyridine (4-AP) in the range of $0.3-5$ mM. Inactivation of K_{DR} occurs over the range of -80 to 0 mV, with variable values for half-maximal availability depending on the source of myocyte. Despite the fact that K_{DR} channels exhibit voltage-dependent inactivation, the presence of overlapping activation and availability curves for K_{DR} current in smooth muscle, as shown in Fig. 1, indicates that K_{DR} channels contribute sustained outward "window" current between approximately -45 and 0 mV (Aiello *et al.*, 1995; Leblanc *et al.*, 1994; Thornbury *et al.*, 1992; Volk *et al.*, 1991; Beech and Bolton, 1989). Moreover, single K_{DR} channel activity can be recorded from isolated cells at voltages consistent with membrane potential recorded in intact tissues (Aiello *et al.*, 1998). These data indicate the potential of K_{DR} channels to contribute to steady-state control of membrane potential in vascular smooth muscle.

Native whole-cell K_{DR} current may not be due to a single-channel type; rather, two or more channel types with distinct properties are likely present, as suggested for canine colonic myocytes by Carl (1995). This view is supported by findings that K_{DR} currents of different smooth muscles show variations in:

(i) pharmacology: Varied IC_{50} values are reported for inhibition of the channels by 4-AP and tetraethylammonium ion (TEA^+). Moreover, 4-AP-resistant components have also been described, and these currents appear to inactivate over a more negative voltage range compared to 4-AP-sensitive current (Carl, 1995; Leblanc *et al.*, 1994; Gelband and Hume, 1992; Thornbury *et al.*, 1992; Beech and Bolton, 1989; Okabe *et al.*, 1987).

(ii) voltage dependence and extent of inactivation: Values for half-maximal inactivation range from -60 to -15 mV (Aiello *et al.*, 1995; Boyle *et al.*, 1992;

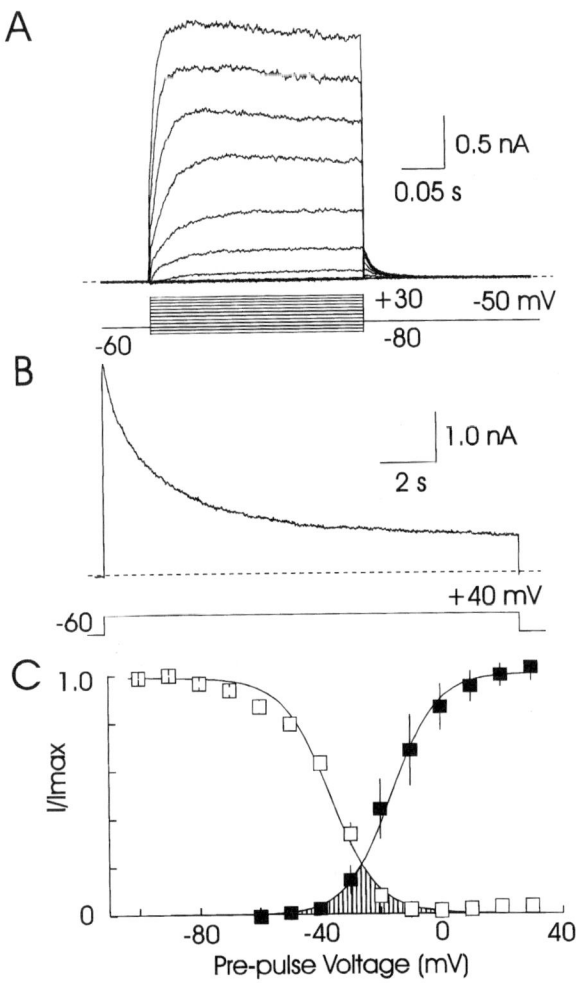

Figure 1. Characteristics of whole-cell K_{DR} current of rabbit portal vein smooth muscle cell. (A) Family of K_{DR} currents recorded at 22°C from a freshly dispersed myocyte via standard whole-cell patch-clamp method using a pipette solution containing 10 mM BAPTA to buffer intracellular Ca^{2+} and minimize contamination by BK_{Ca} channel activity. Note that activation of current occurs positive to approximately -40 mV and that slow, C-type inactivation occurs positive to -10 mV. (B) Decay of K_{DR} current during C-type inactivation. The decay requires seconds and is best fitted by a double-exponential function with time constants of approximately 0.5 and 3 s. (C) Steady-state availability and activation of rabbit portal vein K_{DR} current. Note that there is a range of voltages (indicated by the hatching) over which steady-state channel activity will occur owing to the presence of "window" current that represents the overlapping of steady-state activation with inactivation. Half-maximal values for inactivation and activation were obtained at approximately -35 and -15 mV, respectively.

Smirnov and Aaronson, 1992; Thornbury *et al.*, 1992; Volk *et al.*, 1991; Beech and Bolton, 1989; Kotlikoff, 1990).

(iii) kinetics of inactivation and deactivation: The rate of inactivation is more rapid in portal vein compared to coronary, cerebral, and mesenteric arterial smooth muscle cells (David Sontag, Mircea Iftinca, Frances Plane, and

William Cole, unpublished observations). Slow and fast deactivation is apparent for 4-AP-sensitive and 4-AP-resistant components, respectively (Carl, 1995).

(iv) single-channel conductance: K_{DR} channels having a conductance of 5–20 pS are present in coronary artery (Volk and Shibata, 1993), portal vein (Aiello *et al.*, 1998; Beech and Bolton, 1989), trachea (Boyle *et al.*, 1992), and circular muscle of the colon (Koh *et al.*, 1996). However, larger channels of 25–70-pS unitary conductance were also detected in myocytes from pulmonary (Yuan *et al.*, 1996; Post *et al.*, 1995), coronary (Ishikawa *et al.*, 1993), cerebral (Bonnet *et al.*, 1991), and renal conduit arteries (Gelband and Hume, 1995). The properties of the whole-cell currents due to these channels have not been fully characterized. For example, it is unknown whether they exhibit inactivation at positive potentials, but they are inhibited by elevated $[Ca^{2+}]_i$ (Gelband and Hume, 1995; Ishikawa *et al.*, 1993).

The view that multiple components of K_{DR} current are present in vascular smooth muscle cells is further supported by data showing the expression of several different mRNAs encoding Kv channel pore-forming α- and modulatory β-subunits (see below).

Ensemble averaged currents resulting from the activation of 5–20-pS K_{DR} channels during depolarizing voltage steps exhibit a time course of inactivation that has similar kinetics to that of whole-cell K_{DR} current (Aiello *et al.*, 1998; Koh *et al.*, 1996; Volk and Shibata, 1993; Boyle *et al.*, 1992). These data support the view that the small-conductance K_{DR} channels underlie the inactivating component of whole-cell K_{DR} current. Whether the large-conductance K_{DR} channels of conduit arteries also inactivate at positive voltages has not been established. Additionally, a novel noninactivating Kv current distinct from K_{TO} and K_{DR} currents was recently identified in pulmonary arterial myocytes by Gurney and co-workers (Osipenko *et al.*, 1997; Evans *et al.*, 1996). This conductance was sensitive to 4-AP but activated with very slow kinetics and over a more negative voltage range (between -80 and -60 mV) compared to that typically described for whole-cell K_{DR} currents of vascular myocytes.

K_{DR} currents are sensitive to a wide variety of agents in addition to the classical K^+ channel-blocking drugs, 4-AP and TEA^+. For example, the bradycardic agent tedisamil (Pfunder and Kreye, 1992), the antidepressant fluoxetine (Farrugia, 1996), the appetite suppressant drugs fenfluramine and dexfenfluramine (Weir *et al.*, 1996), the psychedelic drug phencyclidine (Halliday *et al.*, 1995), and cytochrome P-450 inhibitors (Waldron et al.,1999; Edwards *et al.*, 1996; Yuan *et al.*, 1995) all reduce current amplitude. The inhibition of K_{DR} current by the cytochrome P-450 inhibitors appears to be unrelated to reduced synthesis of an epoxyeicosatrienoic acid metabolite of cytochrome P-450 (Waldron *et al.*, 1999). This conclusion is based on the inconsistent effects of various types of P-450 inhibitors on K^+ channel function. For example, clotrimazole depressed K_{DR} current of rabbit portal vein via a direct open block mechanism, characterized by enhanced inactivation during depolarizing pulses and crossover of tail currents. In contrast, the P-450 inhibitor ketoconazole did not alter the rate of channel inactivation or deactivation. Furthermore, the irreversible P-450 inhibitor 1-aminobenzotriazole was without effect on K_{DR} current of the portal vein (Waldron *et al.*, 1999) but depressed K_{DR} current of pulmonary artery in a reversible manner (Yuan *et al.*, 1995). The finding of reversible channel inhibition is unexpected if an irreversible inhibition of cytochrome P-450 is involved in the mechanism of K^+ channel inhibition. Thus, there is uncertainty as to the relationship between K^+ channel regulation and the function of the P-450 system.

3. MOLECULAR IDENTIFY OF VASCULAR K_{DR} CHANNELS

The past 10 years have seen an explosion of interest in identification of the molecular basis of ion channels, and Chapter 3 in this book deals with current knowledge concerning the molecular identity of Kv channels, including K_{DR} channels. Several criteria must be met to definitively identify a specific channel subunit as a component of the native current, including

(i) the following [based on Tamkun *et al.* (1995)]: Northern blot evidence for the expression of mRNA encoding the channel combined with immunocytological, immunohistochemical, or Western blot evidence of the expression of the channel protein in the cell or tissue of interest,

(ii) similar biophysical properties of the cloned channel and the native current,

(iii) similar pharmacology,

(iv) affinity purification of the native channel to identify the presence of the channel protein, as well as associated α- (heterotetramers) and β-subunits,

(v) suppression or alteration in properties of the native current using an antisense oligonucleotide approach or isoform-specific antibodies raised against the cloned channel subunit, and

(vi) modification of the native current by deletion of the gene encoding the cloned channel in a transgenic model.

To date, criteria (i), (ii), (iii), and (v) have been used to identify Kv1.5 as a component of the delayed rectifier K^+ current in vascular smooth muscle, but the evidence for other Kv channel subunits is less developed.

cDNAs encoding K_{DR} channels have been isolated from a variety of smooth muscle preparations. However, our understanding of the molecular basis of native K_{DR} current remains limited; the only statement that can be made with certainty is that native K_{DR} channels are not homomultimers of a single variety of pore-forming α-subunit. Native K_{DR} channels of vascular myocytes likely consist of α- and β-subunits, as suggested for other cell types, the former constituting the pore-forming functional channel and the latter having a modulatory role (Rettig *et al.*, 1992). Molecular biological and immunocytological approaches have been employed to demonstrate the presence of several Kv channel α- and β-subunits in a variety of vascular and nonvascular smooth muscle cells. For example, cDNA clones encoding Kv1.2, Kv1.5, and Kv2.2 were isolated from canine colonic smooth muscle (Schmalz *et al.*, 1998; Overturf *et al.*, 1994; Hart *et al.*, 1993), Kv1.5 was cloned from rabbit vascular smooth muscle (Clément-Chomienne *et al.*, 1999a), and mRNA encoding Kv1.5 was shown to be present in human airway smooth muscle (Adda *et al.*, 1996). Northern analysis and/or immunolocalization studies have shown that Kv1.5 is expressed differentially along the canine gastrointestinal tract, being absent from longitudinal muscles of the fundus, duodenum, and colon but abundant in circular muscles from these regions, as well as in the uterus and large conduit arteries and portal vein (Overturf *et al.*, 1994). This subunit is also present in rat aorta (Roberds and Tamkun, 1991), coronary arteries of rats and humans (Mays *et al.*, 1995; Wang *et al.*, 1994), and rat pulmonary artery (Yuan *et al.*, 1998). Interestingly, prolonged hypoxia reduced native K_{DR} current amplitude of pulmonary vascular myocytes (Smirnov *et al.*, 1994), as well as the expression of Kv1.5 and Kv1.2 subunits (Wang *et al.*, 1997), providing functional evidence for a contribution of these subunits to native K_{DR} current. Clément-Chomienne et al. (1999a) compared the biophysical and pharmacological properties of native K_{DR} channels and Kv1.5 channels isolated from rabbit portal vein and expressed in mammalian cells. The whole-cell and single-channel currents were virtually identical with the exception of differences in the rate and voltage sensitivity of inactivation (Clément-Chomienne *et al.*, 1999a), as well as the state dependence of block by 4-AP; native channels demonstrated both open and closed (resting) state inhibition, but Kv1.5 expressed in mammalian L or HEK293 cells was affected by 4-AP only after activation (Clément-Chomienne *et al.*, 1999b). These data provide strong evidence that the native K_{DR} channel must be a heteromultimer.

As noted above, there are several lines of evidence indicating that native K_{DR} current is composed of more than one component and has varied biophysical properties in myocytes derived from different vessels. It would seem likely that a differential expression of α- and β-subunits contributes to this diversity. The Kv channel α-subunits that are expressed in vascular myocytes, in addition to Kv1.5, include Kv1.1, Kv1.2, Kv1.3, Kv1.6, Kv2.1, and Kv2.2, as well as the electrically silent subunit Kv9.3, based on the presence of mRNA encoding these proteins and, in the case of Kv1.2 and Kv2.1, immunoblot data obtained with the use of isoform-specific antibodies (Yuan *et al.*, 1998; Patel *et al.*, 1997; Schmalz *et al.*, 1998; Roberds and Tamkum, 1991). Modulatory β-subunits, which associate with Kv channel α-subunits (Rhodes *et al.*, 1996), are also expressed in vascular myocytes: for example, rat pulmonary artery (Yuan *et al.*, 1998) and rabbit portal vein (Kevin Thorneloe, Michael Walsh, and William Cole, unpublished observations) express mRNAs encoding $Kv\beta1.1$, $Kv\beta1.2$, $Kv\beta1.3$, $Kv\beta2.1$, and/or $Kv\beta4$. cDNAs encoding Kv1.1, Kv1.2, Kv1.5, Kv1.6, and Kv2.1 and Kv2.2, when expressed in heterologous cells, all yield delayed rectifier-like currents similar to native K_{DR} current, and the formation of heteromultimers involving members of the Kv1 or the Kv2 family is known to occur (Blaine and Ribera, 1998; Rhodes *et al.*, 1996). Kv1.5 and other members of the Kv1 family show a similar sensitivity to 4-AP as native K_{DR} current (IC_{50}s in the submillimolar range) and may contribute to this component as heteromultimers. In contrast, Kv2.1 channel currents are suppressed by 4-AP at a concentration ($IC_{50} > 18$ mM) considerably greater than that required for inhibition of Kv1 channels, but similar to that at which the 4-AP-resistant, noninactivating component of K_{DR} current of vascular myocytes is inhibited.

4. REGULATION OF VASCULAR K_{DR} CHANNELS BY PHOSPHOTRANSFERASE REACTIONS

Regulation of vascular tone by vasoactive agonists is well known to involve the modulation of ion channels, including K^+ channels, and the literature contains many examples of decreased and increased K^+ channel open-state probabilities in the presence of vasoconstrictors and vasodilators, respectively (Kuriyama *et al.*, 1995; Nelson and Quayle, 1995; Nelson *et al.*, 1990). Enhanced K^+ channel activity leading to relaxation involves phosphorylation by cAMP-dependent (PKA) or cGMP-dependent (PKG) protein kinase (Kuriyama *et al.*, 1995; Nelson and Quayle, 1995; Walsh, 1994). For example, activation of β-adrenoceptors, adenosine A_2 receptors, or prostaglandin I_2 receptors influences vascular K^+ channel activity via PKA, and nitric oxide released by nerves or the endothelium acts via PKG. The specific K^+ channel affected depends on the agonist and the vessel: BK_{Ca}, K_{ATP}, and K_{DR} channels are all known to exhibit increased activity following vasodilator treatment, and there are several instances in which multiple channels are involved (Cole and Clément-Chomienne, 1999; Waldron and Cole, 1999).

Regulation of smooth muscle K_{DR} channels by PKA and PKG is suggested by the following evidence. K_{DR} channels of vascular (Aiello *et al.*, 1998, 1995) and gastrointestinal (Koh *et al.*, 1996; Shuttleworth *et al.*, 1996) smooth muscles are regulated by PKA activated in response to β-adrenoceptor and vasoactive intestinal polypeptide (VIP) receptor occupancy, respectively. Figure 2 illustrates that hyperpolarization of rabbit portal vein induced by isoprenaline can be prevented or reversed by 4-AP treatment. A similar inhibition of the response of colonic smooth muscle to VIP was obtained with 4-AP (Du *et al.*, 1994). The evidence supporting a role for PKA in the regulation of K_{DR} of these tissues is as follows:

(i) Activation of β-adrenoceptors and VIP receptors increased whole-cell K_{DR} currents and single-channel open state probability, and this effect was mimicked by stimulation of adenylyl cyclase by forskolin and prevented by an appropriate receptor antagonist (Aiello *et al.*, 1998, 1995; Koh *et al.*, 1996; Shuttleworth *et al.*, 1996).

(ii) Isoprenaline had no effect on vascular K_{DR} current in the presence of the peptide inhibitor of PKA, PKI (Aiello *et al.*, 1995).

(iii) Isoprenaline and VIP enhanced the open state probability of inactivating, 15-pS and 20-pS K_{DR} channels of portal vein and colonic myocytes, respectively (Aiello *et al.*, 1998; Koh *et al.*, 1996; Shuttleworth *et al.*, 1996).

(iv) Application of the catalytic subunit of PKA in the presence of ATP increased the open state probability of K_{DR} channels of inside-out membrane patches excised from either cell type.

(v) Inhibition of basal K_{DR} current was observed upon dialysis of vascular myocytes with pipette solutions containing PKI (Aiello *et al.*, 1995; Edwards *et al.*, 1993).

Activation of 4-AP-sensitive K_{DR} channels by a similar signaling pathway involving adenylyl cyclase, cAMP, and PKA causes relaxation of rabbit middle cerebral arteries in response to endothelium-derived prostaglandin I_2 (Dong *et al.*, 1998). Relaxation of pig coronary artery due to prostanoid release from the endothelium also appears to involve 4-AP-sensitive K_{DR} channels (Nishiyama *et al.*, 1998; Shimizu and Paul, 1997).

Cyclic GMP and phosphorylation by PKG are thought to mediate the actions of endogenous ligands, such as nitric oxide and atrial natriuretic factor, as well as clinically employed organic nitrates, such as sodium nitroprusside. Nitric oxide is synthesized by Ca^{2+}-dependent nitric oxide synthase of endothelial cells and released

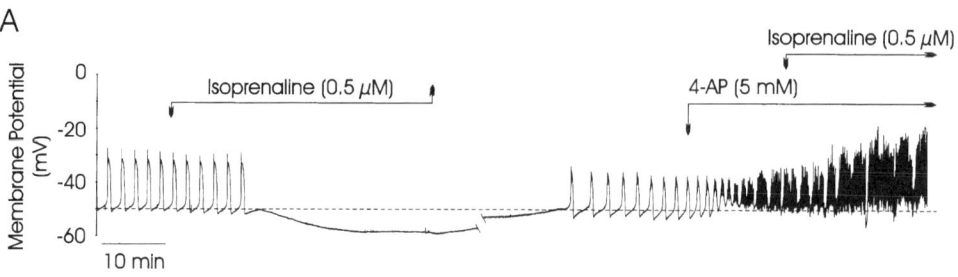

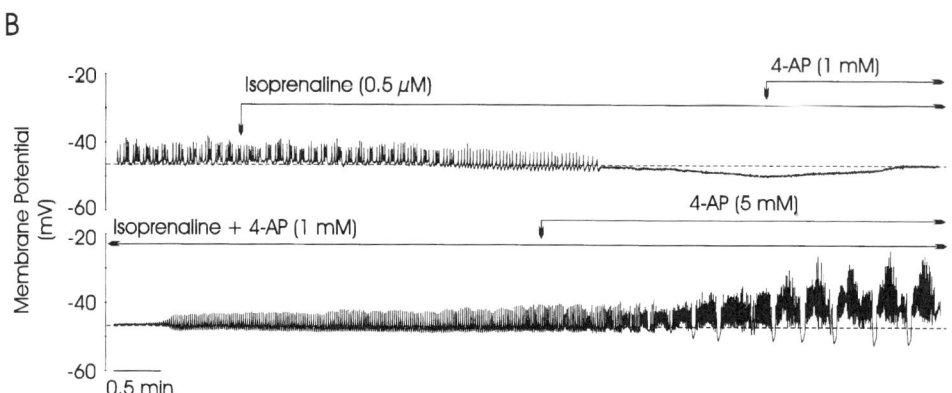

Figure 2. Involvement of K_{DR} channels in isoprenaline-induced hyperpolarization of rabbit portal vein. (A) Isoprenaline is shown to inhibit slow waves and spikes and to produce reversible hyperpolarization of membrane potential recorded by sharp microelectrode technique, but not after application of 4-aminopyridine (4-AP) and block of K_{DR} channels. (B) Inhibition of activity and hyperpolarization due to isoprenaline are shown to be reversed in a dose-dependent fashion by 4-AP. Records in (A) and (B) were made using Krebs physiological saline solution containing wortmannin (5 μM), tetrodotoxin (1 μM), atropine (1 μM), and prazosin (10 μM). (Mircea Iftinca, Stefan Sigurdsson, and William Cole, unpublished observations.)

in response to shear stress and blood-borne factors, such as bradykinin, to induce smooth muscle relaxation and arterial dilation. Depending on the vascular bed, nitric oxide (Brayden, 1990; Tare et al. 1990), atrial natriuretic factor (Taniguchi *et al.*, 1993), and organic nitrates (Karashima *et al.*, 1982) elicit hyperpolarization and relaxation due to K^+ channel activation. The identity of the K^+ channel(s) affected by nitric oxide, cGMP, and PKG appears to vary in different vascular smooth muscles. For example, nitric oxide and elevated cGMP relax cerebral arteries and noncerebral vessels, such as the pulmonary artery (Mikawa *et al.*, 1997; Yuan *et al.*, 1996; Taguchi *et al.*, 1995; Hamaguchi *et al.*, 1992). However, BK_{Ca} inhibition by charybdotoxin or iberiotoxin had no effect on the response of cerebral arteries, but in noncerebral vessels, both blockers were found to suppress nitric oxide and/or cGMP-induced relaxations (Mikawa *et al.*, 1997; Yuan *et al.*, 1996; Taguchi *et al.*, 1995; Hamaguchi *et al.*, 1992). Moreover, the vasodilatory response of some vessels, such as pulmonary arteries, may

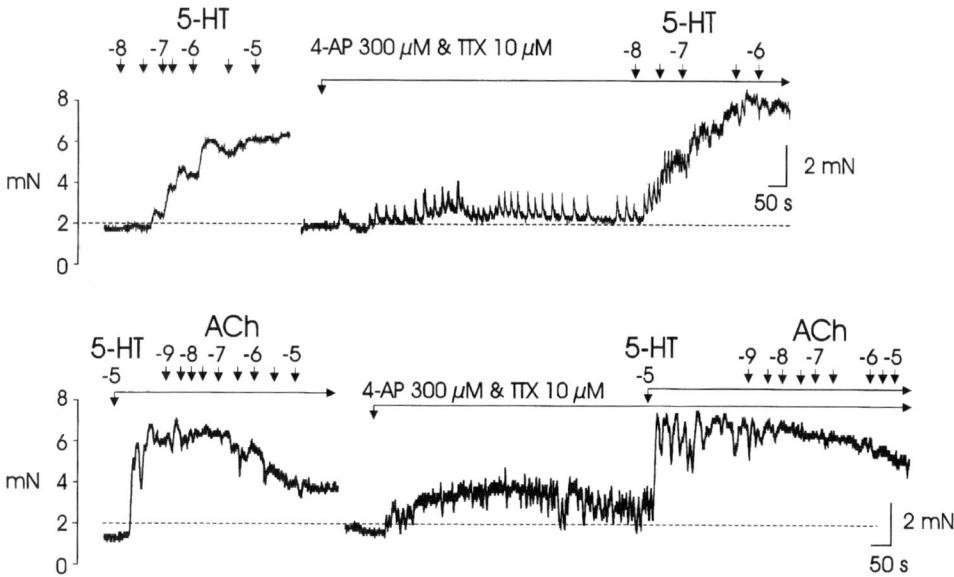

Figure 3. Effect of inhibition of K_{DR} channels on contractile behavior of rat basilar arteries. (A) Representative effect of 4-AP treatment on dose-dependent contraction of basilar artery induced by serotonin (5-HT; numbers represent log concentrations, with intervening unlabeled arrows being half-log concentrations). Note the spontaneous oscillations of tone in the presence of 4-aminopyridine (4-AP) and greater response to 5-HT at each concentration applied, indicative of increased reactivity in the presence of reduced K_{DR} conductance. (B) Representative effect of inhibition of K_{DR} channels on endothelium-dependent relaxation of a basilar artery preconstricted with 5-HT induced by acetylcholine (ACh) in the absence and presence of 4-AP. Note the spontaneous oscillations in tone in the presence of 4-AP and reduced relaxation at each concentration of ACh, indicative of block of endothelium-dependent relaxation. (Tracy Allen, Christopher Sobey, and William Cole, unpublished observations.)

involve both K_{DR} and BK_{Ca} channels, based on inhibition of the response by 4-AP and charybdotoxin (Yuan *et al.*, 1996; Zhao *et al.*, 1997). Recent data indicate that vasodilation of rat basilar artery *in vivo* in response to endothelial nitric oxide release following treatment with acetylcholine involves 4-AP-sensitive K_{DR} channels (Sobey and Faraci, 1999). Figure 3 shows the ability of 4-AP to suppress relaxation of these arteries *in vitro*, in response to acetylcholine treatment (Tracy Allen, Christopher Sobey, and William Cole, unpublished observations). The identity of the 4-AP-sensitive conductance involved in this response is unknown. Pulmonary arteries are also relaxed by nitric oxide via a mechanism apparently involving K_{DR} channels (Zhao *et al.*, 1997; Yuan *et al.*, 1996). In this case, a 4-AP-sensitive (5 mM), 40-pS channel sensitive to block by 4-AP (5 mM) was postulated to mediate the response to nitric oxide. In contrast, Archer *et al.* (1996) reported that NO-induced pulmonary vasodilatation was mediated in large part by activation of BK_{Ca} channels.

Several agonists are known to contract smooth muscles by a mechanism involving Ca^{2+}/phospholipid-dependent protein kinase C (PKC) activation and sustained Ca^{2+}

influx (Walsh, 1994). At least five PKC isoforms are known to be expressed in smooth muscle tissues, including the Ca^{2+}-dependent isoforms, PKCα and PKCβ, as well as the Ca^{2+}-independent isoforms, PKCδ, PKCϵ, and PKCζ (Walsh, 1994). Inhibition of K_{DR} current via a signal transduction pathway involving PKC activation by vasoconstrictors has been demonstrated for vascular myocytes of rabbit portal vein (Aiello *et al.*, 1996; Clément-Chomienne *et al.*, 1996) and rat pulmonary arteries (Salter *et al.*, 1998; Shimoda *et al.*, 1998; Salter and Kozlowski, 1996). Angiotensin II, a contractile agonist known to activate PKC in vascular smooth muscle, depressed K_{DR} current of portal vein myocytes (Clément-Chomienne *et al.*, 1996). The involvement of AT_1 receptors, diacylglycerol (DAG), and PKC activation in the signal transduction pathway activated by angiotensin II is supported by the following:

(i) The decrease in K_{DR} current was blocked by losartan, an angiotensin AT_1 receptor-specific antagonist.

(ii) Treatment with direct activators of PKC, such as the phorbol ester phorbol 12,13-dibutyrate and an active DAG analog, 1,2-dioctanoyl-*sn*-glycerol (1,2-diC$_8$), inhibited K_{DR} current, but the inactive DAG analog 1,3-dioctanoyl-*sn*-glycerol (1,3-diC$_8$) was without effect. Figure 4 illustrates the effect of phorbol ester treatment on steady-state K_{DR} current of a rabbit coronary arterial myocyte. Note that the inhibition of K_{DR} current occurs throughout the range of voltage over which the channels are active, including the range of -60 to approximately -30 mV which is consistent with the level of membrane potential recorded from intact coronary arteries. These data are consistent,

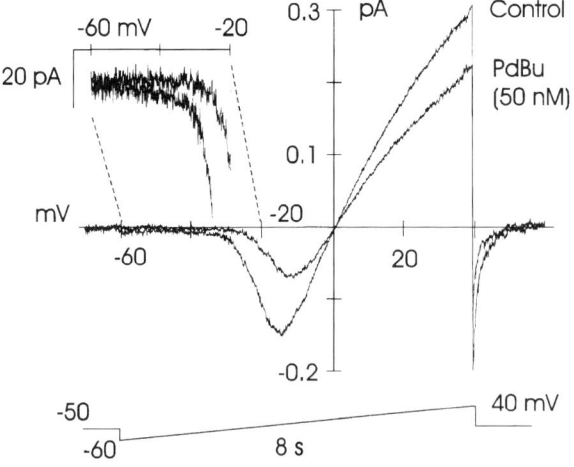

Figure 4. Reduction in K_{DR} current of rabbit coronary arterial myocyte by activation of protein kinase C (PKC). Records shown are representative ramp protocol-induced quasi-steady-state K_{DR} currents recorded in the absence and presence of the PKC activator phorbol 12,13-dibutyrate (PdBu) using symmetrical 140 mM KCl whole-cell patch clamp recording conditions. Note the inhibition of current over the physiologically relevant range of voltage between approximately -50 and -20 mV in the presence of PKC activator. (David Sontag and William Cole, unpublished observations.)

therefore, with the possibility that reduced K_{DR} activity contributes to depolarization during PKC activation by vasoconstrictor agonists.

(iii) Depression of K_{DR} current by 1,2-diC$_8$ and angiotensin II was blocked by pretreatment with the selective inhibitors of PKC, calphostin C and chelerythrine.

The isoforms of PKC present in rabbit portal vein myocytes were identified in immunoblots using antibodies to several PKC isoenzymes, including α, β, γ, δ, ε, η, and ζ, in the absence or presence of competing peptide antigen. The presence of the α, ε, and ζ isoforms was identified, but it was concluded that PKCα, a Ca^{2+}-dependent isoform, was likely not involved because the whole-cell voltage-clamp experiments were performed with high intracellular Ca^{2+} chelation. Similarly, PKCζ was not involved since it is not sensitive to DAG, phorbol esters, or chelerythrine. The inhibition of K_{DR} current by angiotensin II was therefore attributed to PKCε, but the involvement of the Ca^{2+}-dependent isoform PKCα in the control of K_{DR} channels at physiological $[Ca^{2+}]_i$ was not excluded. Subsequent experiments on rat pulmonary arterial myocytes have shown that a similar pathway for regulating K_{DR} current involving DAG and PKC is activated by endothelin-1 stimulation of endothelin B (ET$_B$) receptors (Shimoda et al., 1998). In addition, evidence suggesting that Ca^{2+}-dependent or Ca^{2+}-independent PKC isoforms will inhibit K_{DR} channels was obtained through the use of perforated-patch and standard whole-cell voltage-clamp techniques, the latter employing a pipette solution with a high concentration of a Ca^{2+} chelator (Shimoda et al., 1998).

Attempts to assess the role of phosphotransferase reactions mediated by other kinases, such as tyrosine kinases and calmodulin-dependent kinase II, have been compromised by direct channel inhibition by the compounds employed to modulate the activities of these enzymes. For example, Smirnov and Aaronson (1995) found that the tyrosine kinase inhibitors genistein and ST 638 caused a direct block of K_{DR} current of rat pulmonary myocytes. Similarly, the calmodulin-dependent kinase II inhibitor KN-93 and its inactive analog KN-92 produced open-state block of K_{DR} current of coronary arterial myocytes (Leblanc and Chartier, 1998). Future studies will need to employ different pharmacological tools to determine whether these kinases are involved in the regulation of vascular K_{DR} currents.

5. ROLE OF K_{DR} CHANNELS IN CONTROL OF VASCULAR TONE

K_{DR} currents are widely recognized to contribute to repolarization of action potentials in the heart and neurons, and a similar function has been proposed for K_{DR} channels of smooth muscle. However, in contrast to the cells of excitable smooth muscle tissues, such as the portal vein, myocytes of small arteries do not normally fire action potentials (Nelson and Quayle, 1995). Yet K_{DR} current is present, and in some instances at a similar density to that reported for vascular tissues that exhibit spikes. For example, K_{DR} current density of myocytes of large coronary arteries is identical to that of portal vein smooth muscle cells (~ 40 pA/pF; Clément-Chomienne et al., 1996; Leblanc et al., 1994; Volk et al., 1991). This suggests that K_{DR} channels must, in addition to controlling action potential repolarization, function in other capacities in

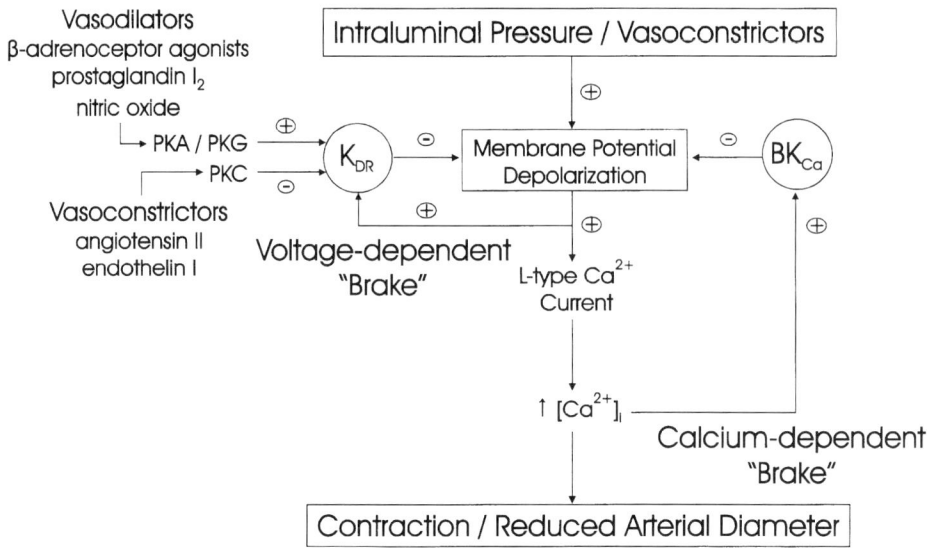

Figure 5. Model for role of K_{DR} channels in control of vascular tone and arterial diameter. See text for details.

the control of electrical and mechanical activity in vascular smooth muscle. Figure 5 summarizes the possible role of K_{DR} channels in the regulation of vascular smooth muscle tone.

A variety of lines of evidence indicate that changes in the open state probability of K^+ channels, including K_{DR} channels, play a critical role in the control of arterial smooth muscle membrane potential and tone. Arterial diameter in small resistance vessels is well known to be dependent on the level of membrane potential of the vascular smooth muscle cells (Neild and Keef, 1985; Brayden and Nelson, 1992; Nelson and Quayle, 1995): Membrane potential values between -60 and -30 mV are recorded *in vitro* when vessels are subjected to intraluminal pressures between 20 and 140 mm Hg (Knot and Nelson, 1995; Nelson and Quayle, 1995). Development of tone by resistance arteries and arterioles in response to increased intraluminal pressure, referred to as the myogenic response, depends on depolarization due to the activation of inward nonselective cation current, voltage-dependent activation of L-type Ca^{2+} channels, and increased intracellular $[Ca^{2+}]$ (Meininger and Davis, 1992; D'Angelo and Meininger, 1994). Figure 5 shows the cascade of events following an increase in intraluminal pressure leading to smooth muscle constriction. Treatment of pressurized vessels with blockers of L-type Ca^{2+} channels, such as 1,4-dihydropyridines, results in relaxation, implying the presence of sustained, steady-state Ca^{2+} influx during the myogenic response (Nelson *et al.*, 1990; Meininger and Davis, 1992; D'Angelo and Meininger, 1994). Moreover, patch-clamp experiments performed with freshly dispersed, isolated vascular myocytes maintained in solutions with a physiologically relevant Ca^{2+} concentration and temperature reveal steady-state L-type Ca^{2+} channel activity at voltages consistent with the level of membrane potential of myocytes in pressurized arteries (Nelson *et al.*, 1990). Functionally significant L-type Ca^{2+} channel activation and Ca^{2+} influx occur when membrane potential is positive to approximately -50 to -45 mV, and changes of only a few millivolts lead to substantial variations in arterial diameter (Nelson *et al.*, 1990). Control of arterial diameter is attained by low-amplitude, tonic changes in membrane potential within a limited voltage range to elicit graded changes in $[Ca^{2+}]$ and tone. K^+ channel activity plays a critical role in the myogenic response by regulating the extent of depolarization. When activated by myogenic depolarization, L-type Ca^{2+} current has the potential to initiate a regenerative influx of Ca^{2+}. If this occurred, then graded, sustained changes in arterial diameter in response to increased intraluminal pressure would not be possible.

There is evidence that at least four different types of K^+ channels can contribute to the control of membrane potential in vascular smooth muscle, depending on the vessel and the physiological condition. These include:

(i) Kv channels, including K_{DR} channels: 4-AP treatment has been shown to cause depolarization, increased tone, or reduced arterial diameter, as well as depolarization and increased tone of airway (Fleischmann *et al.*, 1993) and gastrointestinal smooth muscle (Du *et al.*, 1994). Figure 3 illustrates the effect of 4-AP treatment on the tone of rat basilar artery tone. The concentration of 4-AP employed, 300 μM, is consistent with the IC_{50} for 4-AP block of Kv1 channels (Grissmer *et al.*, 1994), including Kv1.5 cloned from vascular smooth muscle and expressed in mammalian cell types (Clément-Chomienne *et al.*, 1999). These data provide evidence for a contribution of 4-AP-sensitive K^+ channels to control of membrane potential and tone of smooth muscle.

(ii) BK_{Ca} channels: The participation of BK_{Ca} channel activity in response to focal Ca^{2+} release from the sarcoplasmic reticulum (Ca^{2+} sparks) in the control of membrane potential has received considerable attention (Perez *et al.*, 1999; Nelson *et al.*, 1995; Brayden and Nelson, 1992).

(iii) K_{ATP} channels: Inhibition of K_{ATP} channels by glibenclamide has variable effects; some vessels, such as renal arterioles, are unaffected (Loutzenhiser and Parker, 1994), but in others, such as coronary arteries, glibenclamide induced vasoconstriction, implying a contribution of these channels to resting K^+ conductance (Imamura *et al.*, 1992).

(iv) Inward rectifier K^+ (K_{IR}) channels: K_{IR} current density was reported to be greater in smaller arteries (Quayle *et al.*, 1993), suggesting that these channels may have a greater influence on membrane potential with decreasing arterial diameter. There is strong evidence that this conductance contributes to arterial dilation associated with elevated external K^+ concentration (up to $10-15$ mM; Knot *et al.*, 1996), as occurs during cerebral ischemia and reactive hyperemia.

To date, evidence has been obtained that K_{DR} and BK_{Ca} channels play a role in the basal regulation of the myogenic response (Fig. 5).

Figure 5 shows that activation of K_{DR} current constitutes a voltage-dependent regulatory mechanism which acts in parallel to a Ca^{2+}-dependent mechanism provided by BK_{Ca} channels to regulate the myogenic response. Ca^{2+}-dependent activation of BK_{Ca} channels clearly plays a role in limiting the extent of Ca^{2+} influx during the myogenic response (Sturek *et al.*, 1991; Brayden and Nelson, 1992; Loutzenhiser and Parker, 1994; Nelson *et al.*, 1995). BK_{Ca} channel activity appears to be more important as intraluminal pressure rises: selective inhibition of BK_{Ca} by iberiotoxin or TEA^+ caused depolarization in mesenteric vessels pressurized to >60 mm Hg; however, these blockers had little influence on arterial diameter when intraluminal pressure was <60 mm Hg (Brayden and Nelson, 1992; Nelson et al., 1995). Moreover, graded myogenic responses have been shown to occur in the presence of BK_{Ca} channels inhibitors (Loutzenhiser and Parker, 1994). A second pathway for control of the myogenic response is provided by K_{DR} channel activation, which acts as a voltage-dependent "brake" to limit myogenic depolarization (Fig. 5). Experiments on pressurized cerebral arteries (Knot and Nelson, 1995) provide clear evidence of enhanced depolarization and myogenic reactivity in the presence of 4-AP. Figure 5 also shows that K_{DR} channel activation can also influence the control of vascular tone by vasoactive agonists. Agonists such as serotonin, α-adrenoceptor agonists, and endothelin are known to activate Ca^{2+} channels, nonselective cation channels, and Cl^- channels of vascular myocytes (Nelson et al., 1990; Kuriyama et al., 1995). Activation of K_{DR} channels in response to depolarization-induced vasoconstrictor likely contributes to regulating the response. For example, Fig. 3 shows that contractions of the rat basilar artery in response to serotonin treatment are enhanced in the presence of 4-AP, consistent with a role for K_{DR} channels in limiting the response of the tissue to the vasoconstrictor (Tracy Allen, Christopher Sobey, and William Cole, unpublished observations).

Direct modulation of K_{DR} channel activity may also be involved in the control of membrane potential and myogenic reactivity by vasoactive agonists (Fig. 5). As described above, K_{DR} channel activity is modulated by vasoactive agonists via phosphorylation by PKA, PKG, and PKC. Increased K_{DR} current due to PKA and PKG activation will offset larger changes in inward current amplitude during myogenic depolarization. This may be expected to shift the myogenic response to the right, such that greater changes in intraluminal pressure are required to produce an equivalent level of depolarization and constriction. On the other hand, a PKC-mediated inhibition of K_{DR} current and decline in the voltage-dependent "brake" provided by this conductance would be expected to enhance myogenic reactivity. Activation of PKC by vasoconstrictors is known to enhance myogenic reactivity (Liu et al., 1994), and it is possible that a direct suppression of K_{DR} channels may be involved.

6. SUMMARY

This chapter reviews our current understanding of the properties, regulation, and role of K_{DR} channels of vascular smooth muscle. It is apparent that whole-cell K_{DR} currents are due to the presence of multiple types of K^+ channel subtypes and that there is evidence for a varied expression of K_{DR} channels in different smooth muscle tissues. The molecular basis for this diversity is not established, but it is likely that tissue-dependent expression of different Kv channel α- and β-subunits accounts for the variability in properties of K_{DR} currents. K_{DR} channels of smooth muscle are regulated by phosphotransferase reactions involving serine/threonine kinases, including PKA, PKG, and PKC. K_{DR} channels are active at voltages consistent with the level of membrane potential recorded from intact vessels and likely contribute to the regulation of depolarization during the myogenic response to increased intraluminal pressure and exposure to vasoconstrictors, as well as inducing hyperpolarization in response to vasodilator treatment. Major goals for the future include the identification of the specific K$_{DR}$ channels involved in these responses, determination of the specific α- and β-subunits comprising K$_{DR}$ channels in different tissues, and characterization of the mechanism by which channel activity is modulated by phosphorylation, i.c., including the identification of the residues on the α- and/or β-subunits that are involved.

ACKNOWLEDGMENTS

W.C.C. is a Senior Scholar and M.P.W. is a Medical Scientist of the Alberta Heritage Foundation for Medical Research. The work was supported by grants from the Medical Research Council of Canada (MT-13505 and MT-10569).

REFERENCES

Adda, S., Fleischmann, B. K., Freedman, B. D., Yu, M., Hay, D. W., and Kotlikoff, M. I., 1996, Expression and function of voltage-dependent potassium channel genes in human airway smooth muscle, *J. Biol. Chem.* **271**:13239–13243.

Aiello, E. A., Walsh, M. P., and Cole, W., 1995, Phosphorylation by protein kinase A enhances delayed rectifier K$^+$ current in rabbit vascular smooth muscle cells, *Am. J. Physiol.* **268**:H926–H934.

Aiello, E. A., Clément-Chomienne, O., Sontag, D. P., Walsh, M. P., and Cole, W. C., 1996, Protein kinase C inhibits delayed rectifier K$^+$ current in rabbit vascular smooth muscle cells, *Am. J. Physiol.* **271**:H109–H119.

Aiello, E. A., Malcolm, A. T., Walsh, M. P., and Cole, W. C., 1998, β-Adrenoceptor activation and PKA regulate delayed rectifier K$^+$ channels of vascular smooth muscle cells, *Am. J. Physiol.* **275**:H448–H459.

Akbarali, H. I., 1993, K$^+$ currents in rabbit esophageal muscularis mucosae, *Am. J. Physiol.* **264**:G1001–G1007.

Archer, S. L., 1996, Diversity of phenotype and function of vascular smooth muscle cells, *J. Lab. Clin. Med.* **127**:524–529.

Archer, S. L., Huang, J. M. C. Reeve, H. L., Hampl, V., Tolarova, S., Michelakis, E., and Weir, E. K., 1996, Differential distribution of electrophysiologically distinct myocytes in conduit and resistance arteries determines their response to nitric oxide and hypoxia, *Circ. Res.* **78**:431–442.

Beech, D. J., and Bolton, T. B., 1989, Two components of potassium current activated by depolarization of single smooth muscle cells from the rabbit portal vein, *J. Physiol. (London)* **418**:293–309.

Bolzon, B. J., Xiong, Z., and Cheung, D. W., 1993, Membrane rectification in single smooth muscle cells from the rat tail artery, *Pflügers Arch.* **425**:482–490.

Bonnet, P., Rusch, N. J., and Harder, D. R., 1991, Characterization of an outward K$^+$ current in freshly dispersed cerebral arterial muscle cells, *Pflügers Arch.* **418**:292–296.

Boyle, J. P., Tomasic, M., and Kotlikoff, M. I., 1992, Delayed rectifier potassium channels in canine and porcine airway smooth muscle cells, *J. Physiol. (London)* **447**:329–350.

Brayden, J. E., 1990, Membrane hyperpolarization is a mechanism of endothelium-dependent cerebral vasodilation, *Am. J. Physiol.* **259**:H668–H673.

Brayden, J. E., and Nelson, M. T., 1992, Regulation of arterial tone by activation of calcium-dependent potassium channels, *Science* **256**:532–535.

Carl, A., 1995, Multiple components of delayed rectifier K$^+$ current in canine colonic smooth muscle, *J. Physiol. (London)* **484**:339–353.

Clément-Chomienne, O., Walsh, M. P., and Cole, W. C., 1996, Angiotensin II activation of protein kinase C decreases delayed rectifier K$^+$ current in rabbit vascular myocytes, *J. Physiol. (London)* **495**:689–700.

Clément-Chomienne, O., Ishii, K., Walsh, M. P., and Cole, W. C., 1999a, Identification, cloning and expression of rabbit vascular smooth muscle Kv1.5 and comparison with native delayed rectifier K^+ current, *J. Physiol. (London)* **515**:653–667.

Clément-Chomienne, O., Walsh, M. P., and Cole, W. C., 1999b, State-dependent channel block by 4-aminopyridine: Comparison of native delayed rectifier and cloned Kv1.5 currents of rabbit vascular myocytes, *Biophys. J.* **76**:A188.

Cole, W. C., and Clément-Chomienne, O., 2000, K^+ channels in smooth muscle: Structural and functional diversity, in: *A View of Smooth Muscle* (L. Barr, ed.), in press.

D'Angelo, G., and Meininger, G. A., 1994, Transduction mechanism involved in the regulation of myogenic activity, *Hypertension* **23**:1096–1105.

Dong, H., Waldron, G. J., Cole, W. C., Triggle, C. R. 1998, Roles of calcium-activated and voltage-gated delayed rectifier potassium channels in endothelium-dependent vaso relaxation of the rabbit middle cerebral artery. *Br. J. Pharmacol.* **123**(5):821–832.

Du, C., Carl, A., Smith, T. K., Sanders, K. M., and Keef, K. D., 1994, Mechanism of cyclic AMP-induced hyperpolarization in canine colon, *J. Pharmacol. Exp. Ther.* **268**:208–215.

Edwards, G., Ibbotson, T., and Weston, A. H., 1993, Levcromakalim may induce a voltage-independent K-current in rat portal veins by modifying the gating properties of the delayed rectifier, *Br. J. Pharmacol.* **110**:1037–1048.

Edwards, G., Zygmunt, P. M., Hogestatt, E. D., and Weston, A. H., 1996, Effects of cytochrome P-450 inhibitors on potassium currents and mechanical activity in rat portal vein, *Br. J. Pharmacol.* **119**:691–701.

Evans, A. M., Osipenko, O. N., and Gurney, A. M., 1996, Properties of a novel potassium current that is active at the resting potential in rabbit pulmonary smooth muscle cells, *J. Physiol. (London)* **496**:407–420.

Farrugia, G., 1996, Modulation of ionic currents in isolated canine and human jejunal circular smooth muscle cells by fluoxetine, *Gastroenterology* **110**:1438–1445.

Fleischmann, B. K., Washabau, R. J., and Kotlikoff, M. I., 1993, Control of resting membrane potential by delayed rectifier potassium currents in ferret airway smooth muscle cells, *J. Physiol. (London)* **469**:625–638.

Gelband, C. H., and Hume, J. R., 1992, Ionic currents in single smooth muscle cells of the canine renal artery, *Circ. Res.* **71**:745–758.

Gelband, C. H., and Hume, J. R., 1995, $[Ca^{2+}]_i$ inhibition of K^+ channels in canine renal artery: Novel mechanism for agonist-induced membrane depolarization, *Circ. Res.* **77**:121–122.

Gollasch, M., Ried, C., Bychkov, R., Luft, F. C., and Haller, H., 1996, K^+ currents in human coronary artery vascular smooth muscle cells, *Circ. Res.* **78**:676–688.

Grissmer, S., Nguyen, A. N., Aiyar, J., Hanson, D. C., Mather, R. J., Gutman, G. A., Karmilowicz, M. J., Auperin, D. D., Chandy, K. G. 1994, Pharmacological characterization of five cloned voltage-gated K^+ channels, types Kv1.1, 1.2, 1.3, 1.5, and 3.1, stably expressed in mammalian cell lines, *Mol. Pharmacol.* **45**(6):1227–1234.

Halliday, F. C., Aaronson, P. I., Evans, A. M., and Gurney, A. M., 1995, The pharmacological properties of K^+ currents from rabbit isolated aortic smooth muscle cells, *Br. J. Pharmacol.* **116**:3139–3148.

Hamaguchi, M., Ishibashi, T., and Imai, S., 1992, Involvement of charybdotoxin-sensitive K^+ channel in the relaxation of bovine tracheal smooth muscle by glyceryltrinitrate and sodium nitroprusside, *J. Pharmacol. Exp. Ther.* **262**:263–270.

Hart, P. J., Overturf, K. E., Russell, S. N., Carl, A., Hume, J. R., Sanders, K. M., and Horowitz, B., 1993, Cloning and expression of a $K_v1.2$ class delayed rectifier K^+ channel from canine colonic smooth muscle, *Proc. Natl. Acad. Sci. U.S.A.* **90**:9659–9663.

Imamura, Y., Tomoike, H., Narishige, T., Takahashi, T., Kasuya, H., and Takeshita, A., 1992, Glibenclamide decreases basal coronary blood flow in anesthetized dogs, *Am. J. Physiol.* **263**:H399–H404.

Ishikawa, T., Hume, J. R., and Keef, K. D., 1993, Modulation of K^+ and Ca^{2+} channels by histamine H_1-receptor stimulation in rabbit coronary artery cells, *J. Physiol. (London)* **468**:379–400.

Jackson, W. F., Huebner, J. M., and Rusch, N. J., 1996, Enzymatic isolation and characterization of single vascular smooth muscle cells from cremasteric arterioles, *Microcirculation* **3**:313–328.

Karashima, T., Itoh, T., and Kuriyama, H., 1982, Effects of 2-nicotinamidoethyl nitrate on smooth muscle cells of the guinea-pig mesenteric and portal veins, *J. Pharmacol. Exp. Ther.* **221**:472–480.

Knot, H. J., and Nelson, M. T., 1995, Regulation of membrane potential and diameter by voltage-dependent K^+ channels in rabbit myogenic cerebral arteries, *Am. J. Physiol.* **269**:H348–H355.

Knot, H. J., Zimmermann, P. A., and Nelson, M. T., 1996, Extracellular K$^+$-induced hyperpolarizations and dilatations of rat coronary and cerebral arteries involve inward rectifier K$^+$ channels, *J. Physiol. (London)* **492:**419–430.

Koh, S. D., Sanders, K. M., and Carl, A., 1996, Regulation of smooth muscle delayed rectifier K$^+$ channels by protein kinase A, *Pflügers Arch.* **432:**401–412.

Kotlikoff, M. I., 1990, Potassium currents in canine airway smooth muscle cells, *Am. J. Physiol.* **259:**L384–L395.

Kuriyama, H., Kitamura, K., and Nabata, H., 1995, Pharmacological and physiological significance of ion channels and factors that modulate them in vascular tissues, *Pharmacol. Rev.* **47:**387–573.

Leblanc, N., and Chartier, D., 1998, Specific inhibitors of calmodulin-dependent protein kinase are potent blockers of voltage-dependent K$^+$ channels in vascular myocytes, *Biophys. J.* **74:**A111.

Leblanc, N., Wan, X., and Leung, P. M., 1994, Physiological role of Ca^{2+}-activated and voltage-dependent K$^+$ currents in rabbit coronary myocytes, *Am. J. Physiol.* **266:**C1523–C1537.

Liu, J., Hill, M. A., and Meininger, G. A., 1994, Mechanisms of myogenic enhancement by norepinephrine, *Am. J. Physiol.* **266:**H440–H446.

Loutzenhiser, R. D., and Parker, M. J., 1994, Hypoxia inhibits myogenic reactivity of renal afferent arterioles by activating ATP-sensitive potassium channels, *Circ. Res.* **74:**861–869.

Mays, D. J., Foose, J. M., Philipson, L. H., and Tamkun, M. M., 1995, Localization of the K$_v$1.5 K$^+$ channel protein in explanted cardiac tissue, *J. Clin. Invest.* **96:**282–292.

Meininger, G. A., and Davis, M. J., 1992, Cellular mechanisms involved in the vascular myogenic response, *Am. J. Physiol.* **263:**H647–H659.

Mikawa, K., Kume, H., and Takagi, K., 1997, Effects of BK$_{Ca}$ channels on the reduction of cytosolic Ca^{2+} in cGMP-induced relaxation of guinea-pig trachea, *Clin. Exp. Pharmacol. Physiol.* **24:**175–181.

Neild, T. O., and Keef, K., 1985, Measurements of the membrane potential of arterial smooth muscle in anesthetized animals and its relationship to changes in artery diameter, *Microvasc. Res.* **30**(1):19–28.

Nelson, M. T., and Quayle, J. M., 1995, Physiological roles and properties of potassium channels in arterial smooth muscle, *Am. J. Physiol.* **268:**C799–C822.

Nelson, M. T., Patlak, J. B., Worley, J. F., and Standen, N. B., 1990, Calcium channels, potassium channels, and voltage dependence of arterial smooth muscle tone, *Am. J. Physiol.* **259:**C3–C18.

Nelson, M. T., Cheng, H., Rubart, M., Santana, L. F., Bonev, A. D., Knot, H. J., and Lederer, W. J., 1995, Relaxation of arterial smooth muscle by calcium sparks, *Science* **270:**633–637.

Nishiyama, M., Hashitani, H., Kukuta, H., Yamomoto, Y., and Suzuki, H., 1998, Potassium channels activated in the endothelium-dependent hyperpolarization in guinea-pig coronary artery, *J. Physiol. (London)* **510:**455–465.

Okabe, K., Kitamura, K., and Kuriyama, H., 1987, Features of 4-aminopyridine sensitive outward current observed in single smooth muscle cells from the rabbit pulmonary artery, *Pflügers Arch.* **409:**561–568.

Osipenko, O. N., Evans, A. M., and Gurney, A. M., 1997, Regulation of resting potential of rabbit pulmonary artery myocytes by a low threshold, O$_2$-sensing potassium current, *Br. J. Pharmacol.* **120:**1461–1470.

Overturf, K. E., Russell, S. N., Carl, A., Vogalis, F., Hart, P. J., Hume, J. R., Sanders, K. M., and Horowitz, B., 1994, Cloning and characterization of a K$_v$1.5 delayed rectifier K$^+$ channel from vascular and visceral smooth muscles, *Am. J. Physiol.* **267:**C1231–C1238.

Patel, A. J., Lazdunski, M., and Honore, E., 1997, Kv2.1/Kv9.3, a novel ATP-dependent delayed-rectifier K$^+$ channel in oxygen-sensitive pulmonary artery myocytes, *EMBO J.* **16:**6615–6625.

Perez, G. J., Bonev, A. D., Patlak, J. B., and Nelson, M. T., 1999, Functional coupling of ryanodine receptors to KCa channels in smooth muscle cells from rat cerebral arteries, *J. Gen. Physiol.* **113:**229–238.

Pfrunder, D., and Kreye, V. A., 1992, Tedisamil inhibits the delayed rectifier K$^+$ current in single smooth muscle cells of the guinea-pig portal vein, *Pflügers Arch.* **421:**22–25.

Post, J. M., Gelband, C. H., and Hume, J. R., 1995, [Ca^{2+}]$_i$ inhibition of K$^+$ channels in canine pulmonary artery: Novel mechanism for hypoxia-induced membrane depolarization, *Circ. Res.* **77:**131–139.

Quayle, J. M., McCarron, J. G., Brayden, J. E., and Nelson, M. T., 1993, Inward rectifier K$^+$ currents in smooth muscle cells from rat resistance-sized cerebral arteries, *Am. J. Physiol.* **265:**C1363–C1370.

Rettig, J., Heinemann, S. H., Wunder, F., Lorra, C., Parcei, D. N., Dolly, J. O., and Pongs, O., 1992, Inactivation properties of voltage-gated K$^+$ channels altered by presence of β-subunit, *Nature* **369:**289–294.

Rhodes, K. J., Monaghan, M. M., Barrezueta, N. X., Nawoschik, S., Bekele-Arcuri, Z, Matos, M. F., Nakahira, K., Schechter, L. E., and Trimmer, J. S., 1996, Voltage-gated K$^+$ channel β subunits: Expression and distribution of Kvβ1 and Kvβ2 in adult rat brain, *J. Neurosci.* **16:**4846–4860.

Roberds. S. L., and Tamkun, M. M., 1991, Cloning and tissue-specific expression of five voltage-gated potassium channel cDNAs expressed in rat heart, *Proc. Natl. Acad. Sci. U.S.A.* **88**:1798–1802.

Salter, K. J., and Kozlowski, R. Z., 1996, Endothelin receptor coupling to potassium and chloride channels in isolated rat pulmonary arterial myocytes, *J. Pharmacol. Exp. Ther.* **279**:1053–1062.

Salter, K. J., Wilson, C. M., Kato, K., and Kozlowski, R. Z., 1998, Endothelin-1, delayed rectifier K channels, and pulmonary arterial smooth muscle, *J. Cardiovasc. Pharmacol.* **31**(Suppl 1):S81–S83.

Schmalz, F., Kinsella, J., Koh, S. D., Vogalis, F., Schneider, A., Flynn, E. R., Kenyon, J. L., and Horowitz, B., 1998, Molecular identification of a component of delayed rectifier current in gastrointestinal smooth muscles, *Am. J. Physiol.* **274**:G901–G911.

Shimizu, S., and Paul, R. J., 1997, The endothelium-dependent, substance P relaxation of porcine coronary arteries resistant to nitric oxide synthesis inhibition is partially mediated by 4-aminopyridine-sensitive voltage-dependent K^+ channels, *Endothelium* **5**:287–295.

Shimoda, L. A., Sylvester, J. T., and Sham, J. S., 1998, Inhibition of voltage-gated K^+ current in rat intrapulmonary arterial myocytes by endothelin-1, *Am. J. Physiol.* **274**:L842–L853.

Shuttleworth, C. W., Koh, S. D., Bayginov, O., and Sanders, K. M., 1996, Activation of delayed rectifier potassium channels in canine proximal colon by vasoactive intestinal peptide, *J. Physiol. (London)* **493**:651–663.

Smirnov, S. V., and Aaronson, P. I., 1992, Ca^{2+}-activated and voltage-gated K^+ currents in smooth muscle cells isolated from human mesenteric arteries, *J. Physiol. (London)* **457**:431–454.

Smirnov, S. V., and Aaronson, P. I., 1995, Inhibition of vascular smooth muscle cell K^+ currents by tyrosine kinase inhibitors genistein and ST 638, *Circ. Res.* **76**:310–316.

Smirnov, S. V., Robertson, T. P., Ward, J. P., and Aaronson, P. I., 1994, Chronic hypoxia is associated with reduced delayed rectifier K^+ current in rat pulmonary artery muscle cells, *Am. J. Physiol.* **266**:H365–H370.

Snetkov, V. A., Hirst, S. J., Twort, C. H. C., and Ward, J. P. T., 1995, Potassium currents in human freshly isolated bronchial smooth muscle cells, *Br. J. Pharmacol.* **115**:1117–1125.

Sobey, C. G., and Faraci, F. M., 1999, Role of voltage-dependent K^+ channels and soluble guanylate cyclase in dilator responses of the basilar artery to nitric oxide, *Br. J. Pharmacol.* **126**(6):1437–1443.

Sturek, M., Stehno-Bittel, L., and Obye, P., 1991, Modulation of ion channels by calcium release in coronary artery smooth muscle, in: *Ion Channels of Vascular Smooth Muscle Cells and Endothelial Cells* (N. Sperelakis, and H. Kuriyama, eds.), Elsevier, New York pp. 65–80.

Taguchi, H., Heistad, D. D., Kitazono, T., and Faraci, F. M., 1995, Dilatation of cerebral arterioles in response to activation of adenylate cyclase is dependent on activation of Ca^{2+}-dependent K^+ channels, *Circ. Res.* **76**:1057–1062.

Tamkun, M. M., Bennett, P. B., and Snyders, D. J., 1995, Cloning and expression of human cardiac potassium channels, in: *Cardiac Electrophysiology. From Cell to Bedside* (D. P. Zipes and J. Jalife, eds.), Sanders, Philadelphia pp. 21–31.

Taniguchi, J., Furukawa, K. I., and Shigekawa, M., 1993, Maxi K^+ channels are stimulated by cyclic guanosine monophosphate-dependent protein kinase in canine coronary artery smooth muscle cells, *Pflügers Arch.* **423**:167–172.

Tare, M., Parkington, H. C., Coleman, H. A., Neild, T. O., and Dusting, G. J., 1990, Hyperpolarization and relaxation of arterial smooth muscle caused by nitric oxide derived from the endothelium, *Nature* **346**:69–71.

Thornbury, K. D., Ward, S. M., and Sanders, K. M., 1992, Participation of fast-acting, voltage-dependent K currents in electrical slow waves of colonic circular muscle, *Am. J. Physiol.* **263**:C226–C236.

Volk, K. A., Shibata, E. F., 1993, Single delayed rectifier potassium channels from rabbit coronary artery myocytes. *Am. J. Physiol.* **264**(4pt 2):H1146–1153.

Volk, K. A., Matsuda, J. J., and Shibata, E. F., 1991, A voltage-dependent potassium current in rabbit coronary artery smooth muscle cells, *J. Physiol. (London)* **439**:751–768.

Waldron, G. J., and Cole, W. C., 1999, Activation of vascular smooth muscle K^+ channels by endothelium-derived relaxing factors, *Clin. Exp. Pharmacol. Physiol.* **26**:180–184.

Waldron, G. J., Iftinca, M., Clément-Chomienne, O., Triggle, C. R., and Cole, W. C., 1999, Direct block of voltage-gated K^+ current of vascular myocytes by clotrimazole not involving inhibition of cytochrome P450, *Biophys. J.* **76**:A188.

Walsh, M. P., 1994, Regulation of vascular smooth muscle tone, *Can. J. Physiol. Pharmacol.* **72**:919–936.

Wang, H., Mori, Y., and Koren, G., 1994, Expression of Kv1.5 in the rat tissues and the mechanism of cell specific regulation of Kv1.5 gene by cAMP, *Biophys. J.* **66**:207A.

Wang, J., Juhaszova, M., Rubin, L. J., and Yuan, X. J., 1997, Hypoxia inhibits gene expression of voltage-gated K$^+$ channel alpha subunits in pulmonary artery smooth muscle cells, *J. Clin. Invest.* **100**:2347–2353.

Weir, E. K., Reeve, H. L., Huang, J. M. C., Michelakis, E., Nelson, D. P., Hampl, V., and Archer, S. L., 1996, Anorexic agents aminorex, fenfluramine, and dexfenfluramine inhibit potassium current in rat pulmonary vascular smooth muscle and cause pulmonary vasoconstriction, *Circulation* **94**:2216–2220.

Xiong, Z., Sperelakis, N., Noffsinger, A., and Fenoglio-Preiser, A., 1995, Potassium currents in rat colonic smooth muscle cells and changes during development and aging, *Pflügers Arch.* **430**:563–572.

Yuan, X.-J., 1995, Voltage-gated K$^+$ currents regulate resting membrane potential and [Ca^{2+}]$_i$ in pulmonary arterial myocytes, *Circ. Res.* **77**:370–378.

Yuan, X.-J., Tod, M. L., Rubin, L. J., and Blaustein, M. P., 1995, Inhibition of cytochrome P-450 reduces voltage-gated K$^+$ currents in pulmonary arterial myocytes, *Am. J. Physiol.* **268**:C259–C270.

Yuan, X.-J., Tod, M. L., Rubin, L. J., and Blaustein, M. P., 1996, NO hyperpolarizes pulmonary artery smooth muscle cells and decreases the intracellular Ca^{2+} concentration by activating voltage-gated K$^+$ channels, *Proc. Natl. Acad. Sci. U.S.A.* **93**:10489–10494.

Yuan, X. J., Wang, J., Juhaszova, M., Golovina, V. A., and Rubin, L. J., 1998, Molecular basis and function of voltage-gated K$^+$ channels in pulmonary arterial smooth muscle cells, *Am. J. Physiol.* **274**:L621–L635.

Zhang, L., Bonev, A. D., Nelson, M. T., and Mawe, G. M., 1993, Ionic basis of the action potential of guinea pig gallbladder smooth muscle cells, *Am. J. Physiol.* **265**:C1552–C1561.

Zhao, Y. J., Wang, J., Rubin, L. J., and Yuan, X. J., 1997, Inhibition of K$_{(V)}$ and K$_{(Ca)}$ channels antagonizes NO-induced relaxation in pulmonary artery, *Am. J. Physiol.* **272**:H904–H912.

Chapter 25

Potassium Channels in the Circulation of Skeletal Muscle

William F. Jackson

1. INTRODUCTION

Skeletal muscle represents 40% of body mass and contributes substantially in the regulation of blood pressure, whole body metabolism, and cardiovascular homeostasis (Rowell, 1986). The membrane potential (E_m) of vascular muscle cells in the walls of small arteries and arterioles that supply blood to skeletal muscles regulates the tone of these vessels. Thus, as in other tissues, K^+ channels play a central role in the physiology and pathophysiology of the circulation in skeletal muscle. The focus of this chapter will be the function of ATP-sensitive (K_{ATP}), large-conductance, Ca^{2+}-activated (BK_{Ca}), voltage-gated (Kv), and inward rectifier (K_{IR}) channels in the regulation of the arterial blood supply of skeletal muscle, with particular emphasis on the roles played by these channels in the microcirculation of this tissue. The reader is referred to other chapters in this volume for detailed information about the molecular biology, pharmacology, and pathophysiology of these channels as well as related topics.

2. ATP-SENSITIVE K^+ CHANNELS IN THE SKELETAL MUSCLE CIRCULATION

2.1. Evidence for K_{ATP} Channels in Skeletal Muscle Arteries and Arterioles

K_{ATP} channel agonists, such as pinacidil and cromakalim, dilate arterioles in skeletal muscle (Tateishi and Faber, 1995a; Saito *et al.*, 1996; Crijns *et al.*, 1998; Struijker Boudier *et al.*, 1992; Jackson, 1993; De Wit *et al.* 1994). In addition,

William F. Jackson • Department of Biological Sciences, Western Michigan University, Kalamazoo, Michigan 49008.

Potassium Channels in Cardiovascular Biology, edited by Archer and Rusch. Kluwer Academic/Plenum Publishers, New York, 2001.

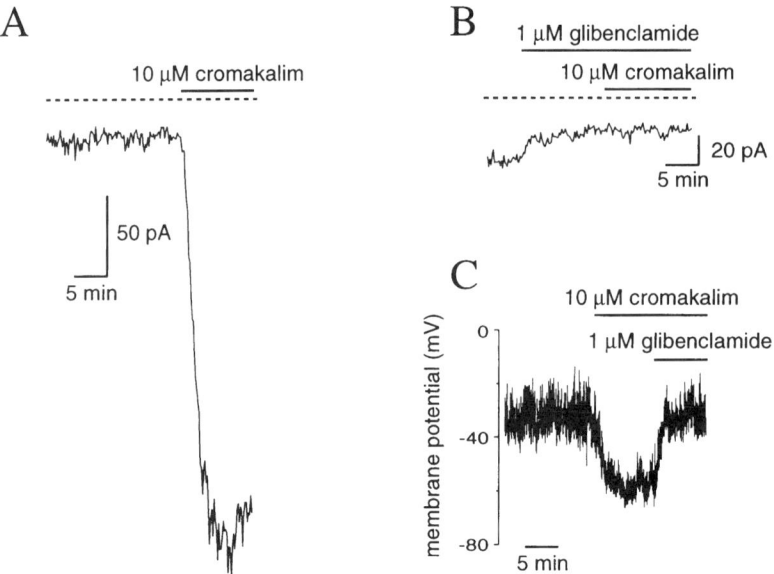

Figure 1. Cromakalim activates glibenclamide-sensitive currents and hyperpolarizes arteriolar muscle cells. Perforated-patch recordings of whole-cell holding currents (A and B) or membrane potential (C) from single hamster cremasteric arteriolar muscle cells are shown. Pipette solutions in all three panels contained 145 mM K^+ [see Jackson *et al.* 1997 for details]. In panels A and B, cells were superfused with solutions containing 145 mM K^+ and voltage-clamped at -60 mV; in panel C, cells were superfused with solutions containing 5 mM K^+. The dashed lines in panels A and B represent the zero-current level. The K_{ATP} channel agonist cromakalim produces a large increase in holding current (A), glibenclamide decreases resting current and prevents effects of cromakalim (B), and cromakalim causes hyperpolarization which can be reversed by glibenclamide (C).

cromakalim hyperpolarizes or increases glibenclamide-sensitive K^+ currents in single vascular muscle cells isolated from cremasteric arterioles (Fig. 1). These data suggest that recruitable K_{ATP} channels are present in the vascular muscle cells of skeletal muscle arterioles, although single K_{ATP} channel currents have yet to be reported from arteriolar muscle cells.

The molecular identity of K_{ATP} channels in the skeletal muscle circulation has not been established. Vascular muscle cells may express the SUR2B isoform of the sulfonylurea receptor subunit (Chapter 15 by A. Terzic and M. Vivaudou (Part III). However, the K^+ channel to which SUR2B is coupled in vascular smooth muscle membranes remains unknown. Expression of SUR2B with Kir6.2 channels forms an ATP-sensitive K^+ channel that has some characteristics of the K_{ATP} channels reported in vascular smooth muscle (Isomoto *et al.*, 1996). However, the single-channel conductance of the SUR2B/Kir6.2 channel (~ 70 pS) falls midway between the conductances of the large-conductance (>100 pS) and small/medium-conductance (<50 pS) K_{ATP} channels reported in native vascular smooth muscle cells (Quayle *et al.*, 1997). These and other data suggest that we have yet to identify the molecular components of endogenous vascular K_{ATP} channels.

2.2. Resting Membrane Potential, Tone, and K$_{ATP}$ Channels

Early studies of vascular smooth muscle cells from conduit arteries suggested that K$_{ATP}$ channels were closed under resting conditions (Nelson and Quayle, 1995; Quayle et al., 1997). These data fit with the known sensitivity of these channels to block by ATP ($K_i = 10–100$ μM) in vascular muscle cells and suggest that the channels would be closed at intracellular ATP concentrations of 3–5 mM. However, subsequent studies indicate that K$_{ATP}$ channels may be open in skeletal muscle resistance arteries and arterioles under resting conditions and thus contribute to the resting E_m and tone of these vessels. For example, glibenclamide causes concentration-dependent constriction of arterioles in cremaster muscles in vivo, with an EC$_{50}$ of approximately 100 nM (Jackson, 1998a). Other investigations support the hypothesis that K$_{ATP}$ channels are active in the microcirculation of skeletal muscle under resting conditions (Saito et al., 1996; Tateishi and Faber, 1995a; Vanelli and Hussain, 1994a; Kosmas et al., 1995; Vanelli et al., 1994).

Recent studies of isolated arterioles and arteriolar muscle cells further indicate that active K$_{ATP}$ channels reside in vascular muscle cells. Glibenclamide has been demonstrated to constrict isolated, cannulated rat cremasteric arterioles, in vitro (Tateishi and Faber, 1995a), suggesting that open K$_{ATP}$ channels are intrinsic to the vascular wall. Similarly, glibenclamide inhibits currents around the resting membrane potential (-30 to -60 mV) (Fig. 1B), decreases whole-cell membrane conductance in voltage-clamp experiments, and causes depolarization under current-clamp conditions in single vascular muscle cells, isolated from rat or hamster cremasteric arterioles (Jackson et al., 1997; Jackson, 1998a). Although these observations seem to be difficult to reconcile with the previously identified ATP sensitivity of these channels, the existence of endogenous activators of these channels, the compartmentalization of ATP in vascular muscle cells, the interaction between ATP and phosphatidylinositol 4, 5-bisphosphate (PIP$_2$) in the regulation of Kir6 channels (Shyng and Nichols, 1998; Baukrowitz et al., 1998), or a unique type of K$_{ATP}$ channel expressed in the microcirculation may permit channel activation at physiological ATP concentrations. Because only a few channels need to open to significantly influence vascular muscle cell E_m, even small changes in open-state probability may have profound effects on vascular tone.

2.3. Vasodilators and K$_{ATP}$ Channels

K$_{ATP}$ channels are the targets for agents such as cromakalim, pinacidil, and minoxidil sulfate, the so-called "potassium channel openers" (see Chapter Terzic and Vivaudou). In addition, many receptor-mediated dilators that stimulate cAMP formation cause glibenclamide-sensitive vasodilation including isoproterenol (Jackson, 1993; Randall and McCulloch, 1995; Nakashima and Vanhoutte, 1995), prostacyclin and its stable analogs (Jackson et al., 1993; Jackson, 1993; Schubert et al., 1997; Corriu et al., 1996), adenosine (Jackson, 1993; Kleppisch and Nelson, 1995a), and calcitonin gene-related peptide (CGRP) (Nelson et al., 1990; Quayle et al., 1994). The cAMP-dependent activation of protein kinase A (PKA), and presumably phosphorylation of K$_{ATP}$ channels, may be involved in these responses (Kleppisch and Nelson, 1995a; Quayle et *al.*, 1997; Nelson and Quayle, 1995; Quayle *et al.*, 1994; Zhang *et al.*, 1994). Both sulfonylurea receptors (SURs) and Kir6 subunits contain PKA consensus sequences, consistent with this hypothesis (Quayle *et al.*, 1997; Babenko *et al.*, 1998; Aguilar-Bryan *et al.*, 1998).

Vasodilator agents such as adenosine, isoproterenol, and prostacyclin also may cause glibenclamide-sensitive vasodilation independent from cAMP. Studies in the hamster cheek pouch (Jackson, 1993) and in the rat mesentery (Randall and McCulloch, 1995) have demonstrated that glibenclamide inhibits dilation to agents such as isoproterenol but has no effect on dilation induced by the cAMP analog dibutyryl-cAMP. These data are difficult to reconcile with the cAMP–PKA hypothesis presented above and suggest a role for K_{ATP} channels in agonist-induced vasodilation independent from cAMP.

Hypoxia, another vasodilator influence, activates K_{ATP} channels in vascular muscle cells isolated from porcine coronary arteries (Dart and Standen, 1995). Similar findings have yet to be reported for cells isolated from skeletal muscle arteries or arterioles. Instead, functional studies have provided compelling evidence that myogenic tone of vascular muscle cells in small systemic arteries and arterioles is insensitive to changes in the partial pressure of oxygen, PO_2, between 150 and 10 mm Hg (Jackson, 1987; Duling, 1974; Messina *et al.*, 1992; Fredricks *et al.*, 1994; Tateishi and Faber, 1995b; Busse *et al.* 1983, 1984). More severe reductions in PO_2 (6 mm Hg) may result in glibenclamide-sensitive dilation of rat cremasteric arterioles (Tateishi and Faber, 1995a).

Hypoxia may act on endothelial cells lining small arteries to stimulate the release of prostacyclin or other prostaglandins (Busse *et al.*, 1983, 1984; Messina *et al.*, 1992; Fredricks *et al.*, 1994) that then open K_{ATP} channels in vascular smooth muscle cells, as has recently been reported in rat middle cerebral arteries (Lombard *et al.*, 1999). In addition, Marshall and colleagues (Marshall *et al.*, 1993; Bryan and Marshall, 1999) proposed that adenosine released from skeletal muscle and from endothelial cells may activate both endothelial and vascular smooth muscle cell K_{ATP} channels to mediate vasodilation in the hind limbs of rats during systemic hypoxia.

Glibenclamide has been demonstrated to inhibit functional hyperemia (Thomas *et al.*, 1997; Saito *et al.*, 1996; Vanelli *et al.*, 1994), reactive hyperemia (Vanelli and Hussain, 1994a; Banitt *et al.*, 1996; Minkes *et al.*, 1995), and the decreased vascular resistance associated with reductions in blood flow (Vallet *et al.*, 1995) in several skeletal muscle models. These observations support a role for K_{ATP} channels in the local regulation of blood flow in skeletal muscle. In addition, activation of K_{ATP} channels during skeletal muscle contraction may play a role in the decrease in sympathetic nerve-induced vasoconstriction that occurs during muscle activity (Thomas *et al.*, 1997). The location of the K_{ATP} channels involved in these functional responses and the signaling pathways involved in their regulation have not been adequately explored.

2.4. Vasoconstrictors and K_{ATP} Channels

Vasoconstrictors such as norepinephrine (Fig. 2), the α_1-adrenergic agonist phenylephrine (Bonev and Nelson, 1996), histamine (Bonev and Nelson, 1996; Klep-

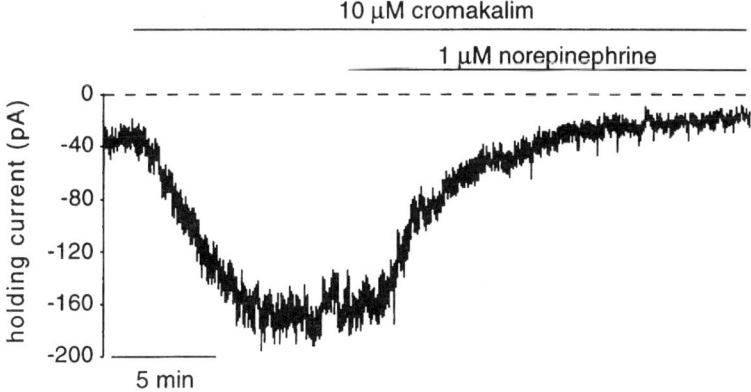

Figure 2. Norepinephrine inhibits K$_{ATP}$ currents in arteriolar muscle cells. Digitized record of holding current recorded as in Fig. 1 is shown. Cromakalim stimulates a large increase in current. Norepinephrine completely reversed this effect of cromakalim.

pisch and Nelson, 1995b), serotonin (Bonev and Nelson, 1996; Kleppisch and Nelson, 1995b), neuropeptide Y (Bonev and Nelson, 1996), angiotensin II (Miyoshi and Nakaya, 1991), vasopressin (Wakatsuki *et al.*, 1992a), and endothelin (Miyoshi *et al.*, 1992), all have been demonstrated to inhibit currents through K$_{ATP}$ channels. Nelson and colleagues have proposed that this inhibition is mediated by protein kinase C (PKC) (Bonev and Nelson, 1996; Quayle *et al.*, 1997; Nelson and Quayle, 1995). In addition, recent evidence suggests that K$_{ATP}$ channels also are sensitive to changes in membrane PIP$_2$ levels, which apparently modulate the ATP sensitivity of the Kir6 channels (Baukrowitz *et al.*, 1998; Shyng and Nichols, 1998). Thus, it is possible that agonists which activate phospholipase C may close K$_{ATP}$ channels by reducing local PIP$_2$ levels and increasing the ATP sensitivity of the channels.

While there is considerable evidence indicating that vasoconstrictor agents inhibit K$_{ATP}$ currents, the functional significance of these observations has not been thoroughly studied. Tateishi and Faber (1995a) suggested that α_{2D}-adrenergic receptor-mediated contraction of rat cremasteric arterioles involves closure of K$_{ATP}$ channels, because glibenclamide inhibits vasoconstriction induced by the α_{2D}-adrenergic agonist UK14,304, and, conversely, that the constriction induced by UK14,304 inhibits glibenclamide-induced increases in tone. In contrast, activation of $\alpha_{1A/D}$-adrenergic receptors with phenylephrine did not appear to interact with the effects of glibenclamide. However, vasoconstriction of rat hind limb induced by the α_{2D}-adrenergic agonist UK14,304 is unaffected by glibenclamide, suggesting that these findings are not universal (Thomas *et al.*, 1997).

The apparent lack of effect of α_1-adrenergic agonists on K$_{ATP}$ channels in the studies of Tateishi and Faber (1995a) is another point of confusion because the α_1-adrenergic agonist phenylephrine inhibits currents through K$_{ATP}$ channels in rabbit mesenteric artery myocytes by a PKC-dependent pathway (Bonev and Nelson, 1996). Thus, it is unclear why α_1-adrenergic receptor activation did not affect K$_{ATP}$ channels in the studies by Tateishi and Faber (1995a). Further research, involving direct measurement of the effects of adrenergic agonists on K$_{ATP}$ channels, as well as clarification of the signal transduction pathways involved in the effects of α_1- and α_2-adrenergic receptor activation on these currents, will be required to resolve these issues.

3. BK$_{Ca}$ CHANNELS IN THE SKELETAL MUSCLE CIRCULATION

3.1. Evidence for BK$_{Ca}$ Channels in Skeletal Muscle Arteries and Arterioles

Patch-clamp studies of cremasteric arteriolar muscle cells indicate that there is a high density of large-conductance K$^+$ channels (242 pS with symmetrical 140 mM K$^+$ gradient) that are both Ca^{2+}- and voltage-sensitive and which can be blocked by iberiotoxin (Jackson and Blair, 1998). These observations suggest that arteriolar muscle cells from skeletal muscle express BK$_{Ca}$ channels, as has been reported in other vascular beds (see other chapters in this volume).

3.2. Resting Membrane Potential, Tone, and BK$_{Ca}$ Channels

Inhibitors of BK$_{Ca}$ channels, such as iberiotoxin and tetraethylammonium (TEA), depolarize cerebral vascular smooth muscle cells and constrict small, myogenically active arteries from a number of vascular beds (see Chapters Broyden and Nelson). These data suggest that BK$_{Ca}$ channels may be active under resting conditions in isolated arteries and thereby contribute to resting E_m and tone. Furthermore, *in vivo* studies of canine diaphragm have demonstrated that the BK$_{Ca}$ channel blocker iberiotoxin decreases resting blood flow when infused into the blood perfusing this muscle (Vanelli *et al.*, 1994; Vanelli and Hussain, 1994b). However, these *in vivo* data must be interpreted cautiously, because the site of action of the K$^+$ channel blockers was not established so that vascular smooth muscle, endothelial, or neural BK$_{Ca}$ channels could have been affected.

In contrast to the reports cited above, a number of studies have failed to implicate BK$_{Ca}$ channels in the regulation of resting E_m and vascular tone in skeletal muscle. Patch-clamp studies of rat and hamster cremasteric arteriolar muscle cells have failed to find any effect of iberiotoxin on the resting E_m level or on whole-cell currents at physiological E_m (-90 to 0 mV) (Jackson *et al.*, 1997), even when cells are dialyzed with solutions containing 300 nM free Ca^{2+} (Jackson, 1998a). Similarly, neither iberiotoxin nor TEA affects the resting diameter of cremasteric arterioles, *in vivo*, despite substantial resting, myogenic tone (Jackson and Blair, 1998; Loeb *et al.*, 1997). A lack of effect of TEA on resting blood flow to feline hind limb also has been reported (Champion and Kadowitz, 1997). These data suggest that BK$_{Ca}$ channels may not contribute to resting E_m and tone in resistance arteries and arterioles in skeletal muscle.

The lack of apparent activity of BK$_{Ca}$ channels in the studies outlined above could be due to a low expression of the channels, a low voltage sensitivity (reduced slope of the voltage activation relationship), low Ca^{2+} sensitivity (reduced slope of the calcium activation relationship), or a change in the Ca^{2+} set point (Ca^{2+} threshold for voltage activation) of the channels expressed (Jackson and Blair, 1998). These hypotheses were recently tested in cremasteric arteriolar muscle cells (Jackson and Blair, 1998), which display a lack of BK$_{Ca}$ channel currents in whole-cell studies (Jackson et al., 1997; Jackson, 1998a). In this preparation, a high density of iberiotoxin-sensitive BK$_{Ca}$ channels were observed in inside-out membrane patches with typical voltage and calcium sensitivity (Fig. 3; Jackson and Blair, 1998). However, the channels had a relatively high (9 μM) calcium set point, which represents the concentration of Ca^{2+} required for half-maximal activation of the channels at 0 mV and reflects the threshold Ca^{2+} concentration required for physiological activation of the channels (Carl et al., 1996; Jackson and Blair, 1998). This value is about 10-fold higher than the value of 1 μM reported for vascular muscle cells isolated from larger arteries and other smooth muscles (Benham et al., 1986; Albarwani et al., 1994; Inoue et al., 1985; Carl et al., 1996; McManus, 1991). Thus, even in the presence of focal release of Ca^{2+} from internal stores (Ca^{2+} sparks) that may provide peak Ca^{2+} concentrations on the order of 300 nM (Nelson et al., 1995; Knot et al., 1998), the BK$_{Ca}$ channels in arteriolar muscle cells in the skeletal muscle microcirculation will likely remain silent at physiological E_m unless activated by other influences. The mechanism(s) responsible for the high calcium set point in cremasteric arteriolar muscle cells remains unknown, but it might result from the expression of spliced variants of the BK$_{Ca}$ α-subunit or from differential coupling of the α-subunits to ancillary regulatory subunits (Lagrutta et al., 1994; Tanaka et al., 1997; Jan and Jan, 1997; McCobb et al., 1995).

3.3. Vasodilators and BK$_{Ca}$ Channels

The role played by BK$_{Ca}$ channels in the mechanism of action of vasodilators remains controversial. Studies have shown that BK$_{Ca}$ channels can be activated by vasodilators that stimulate formation of cAMP or cGMP (Nelson and Quayle, 1995), and blockers of these channels have been demonstrated to inhibit cAMP- and cGMP-related vasodilation of coronary and peripheral arteries, in vitro (Schubert et al., 1997; Cabell et al., 1994; Bruch et al., 1997; Carrier et al., 1997). In contrast, blockers of BK$_{Ca}$ channels have no effect on vasodilation of canine hind limb in response to dilators thought to act via cAMP or cGMP (Champion and Kadowitz, 1997). Thus, BK$_{Ca}$ channels may not be involved in the mechanism of action of vasodilators that act through these second messengers in skeletal muscle circulation.

BK$_{Ca}$ channels may be the target for the actions of carbon monoxide (CO) produced endogenously from metabolism of heme groups by the heme oxygenases (Wang and Wu, 1997; Wang et al., 1997a, b; Abraham et al., 1996). In rat tail artery, vasodilation induced by CO is inhibited by charybdotoxin (Wang et al., 1997a), and CO directly activates BK$_{Ca}$ channels in inside-out membrane patches from the same preparation, independently of changes in cGMP (Wang and Wu, 1997; Wang et al., 1997b). These findings raise the possibility that CO may represent another gaseous signaling molecule that can modulate BK$_{Ca}$ channel activity.

Relatively few functional studies have been performed to determine the role played by BK$_{Ca}$ channels in the regulation of blood flow. Active hyperemia in canine diaphragm has been shown to be partially inhibited by iberiotoxin, particularly at low

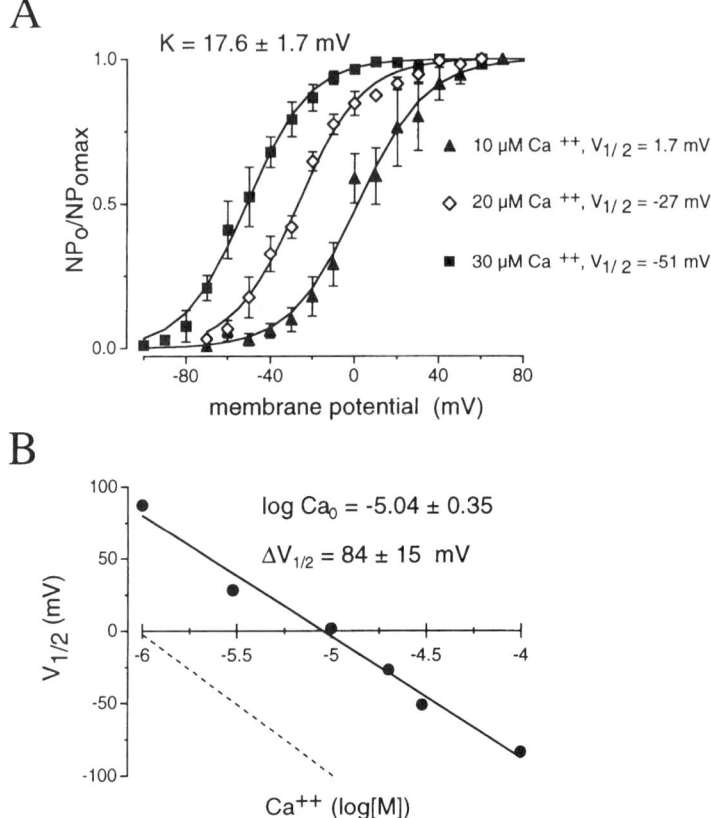

Figure 3. BK_{Ca} channels in arteriolar muscle cells have a high calcium set point. (A) Voltage–activation curves at three Ca^{2+} concentrations in inside-out patches of hamster cremasteric arteriolar muscle cells [see Jackson and Blair (1998) for detailed methods]. Data are mean channel activities $\pm SE$ ($n = 5$–10) expressed relative to the maximum channel activity observed. The solid lines represent best-fit curves to the Boltzmann equation $NP_o/$maximum $NP_o = 1/\{1 + \exp[(V_{1/2} - V_m)/K]\}$, where $V_{1/2}$ is the membrane potential required for half-maximal activation of the channels, V_m is the membrane potential, and K is the logarithmic voltage sensitivity (V required for an e-fold increase in activity). (Data from Jackson and Blair, 1998.) (B) Estimation of $\Delta V_{1/2}$ (an index of calcium sensitivity) and Ca_0 (the calcium set point) for cremasteric arteriolar muscle BK_{Ca} channels. $V_{1/2}$ values were estimated from data similar to those presented in panel (A) [see Jackson and Blair (1998) for more details]. Analysis of variance indicated a significant regression ($P < 0.05$). From the regression analysis, $\Delta V_{1/2}$ (the change in $V_{1/2}$ for a 10-fold change in Ca^{2+}, equal to the absolute value of the slope of the line), log Ca_0 [the log (Ca^{2+}) axis intercept, corresponding to log (Ca^{2+} set point)], and their 95% confidence intervals were estimated as shown in the figure. The dashed line represents the line of best fit through data from previous studies of BK_{Ca} channels in vascular muscle membranes from large vessels (see Benham et al. 1986; Albarwani et al.1994; Inoue et al., 1985; Carl et al., 1996; McManus, 1991). The slope of this line (an index of the $\Delta V_{1/2}$) was similar to that determined from arteriolar muscle. However, the Ca^{2+}-axis intercept was shifted to the left. This indicates that the Ca^{2+} set point of BK_{Ca} channels from these large vessels was approximately 10-fold lower than that observed in arteriolar muscle cells. [See Jackson and Blair (1998) for more information.]

levels of activity (Vanelli et al., 1994). The role played by BK$_{Ca}$ channels in reactive hyperemia is unclear. In canine diaphragm, iberiotoxin (Vanelli and Hussain, 1994b) inhibits a portion of the hyperemia induced by vascular occlusion, supporting a role for BK$_{Ca}$ channels in this response. In contrast, studies in the cat hind limb found no effect of TEA on reactive hyperemia (Champion and Kadowitz, 1997).

3.4. Vasoconstrictors and BK$_{Ca}$ Channels

BK$_{Ca}$ channels have been reported to play either a positive- or negative-feedback role in the mechanism of action of vasoconstrictors (Toro et al., 1990; Scornik and Toro, 1992; Lange et al., 1997; Wesselman et al., 1997; Jackson and Blair, 1998; Nelson et al., 1995; Nelson and Quayle, 1995; Brayden and Nelson, 1992; Rusch and Liu, 1997; Berczi et al., 1992; Hashemzadeh-Gargari and Rembold, 1992; Wakatsuki et al., 1992a; Ganitkevich and Isenberg, 1990). As with K$_{ATP}$ channels discussed earlier, activation of PKC, a common step in the signaling cascade of many vasoconstrictors, appears to inhibit BK$_{Ca}$ channels in some vascular muscle cells (Lange et al., 1997; Minami et al., 1993). The functional significance of this inhibition has not been well studied. However, in small mesenteric arteries, pharmacological blockade of BK$_{Ca}$ channels with charybdotoxin inhibits pressure-induced depolarization and myogenic reactivity (Wesselman et al., 1997), suggesting that closure of BK$_{Ca}$ channels plays a role in pressure-induced vasoconstriction in some systemic arteries.

In contrast, a number of studies suggest that BK$_{Ca}$ channels are activated during agonist- and pressure-induced vasoconstriction (Wakatsuki et al., 1992b; Jackson and Blair, 1998; Wakatsuki et al., 1992a; Salter and Kozlowski, 1998; Tsien and Rink, 1980; Berczi et al., 1992; Nelson et al., 1995; Brayden and Nelson, 1992; Wesselman et al., 1997; Champion and Kadowitz, 1997). For example, in hamster cremasteric arteriolar muscle cells, norepinephrine-induced contraction is correlated with a large, reversible increase in the opening of single BK$_{Ca}$ channels in cell-attached patches (Fig. 4B). This agonist-induced increase in K$^+$ efflux apparently acts as a negative-feedback mechanism to limit norepinephrine-induced contraction, because blockade of BK$_{Ca}$ channels with iberiotoxin augments contraction of single cells in response to this catecholamine (Fig. 4A). Thus, although the activation of PKC might cause channel inhibition, the increase in intracellular Ca^{2+}, as well as the membrane depolarization induced by vasoconstrictors, appears be sufficient to activate BK$_{Ca}$ channels and limit the membrane depolarization. Future studies in which E_m, subsarcolemmal Ca^{2+}, BK$_{Ca}$ channel activity, and vascular smooth muscle tone are measured simultaneously during application of vasoconstrictors or elevated pressure will be required to test this hypothesis.

4. Kv CHANNELS IN SKELETAL MUSCLE CIRCULATION

4.1. Evidence for Kv Channels in Skeletal Muscle Arteries and Arterioles

4-Aminopyridine, a relatively nonspecific inhibitor of Kv channels, inhibits outward, voltage-dependent K$^+$ currents in cremasteric arteriolar muscle cells, providing pharmacological evidence for the expression of Kv channels in the circulation of

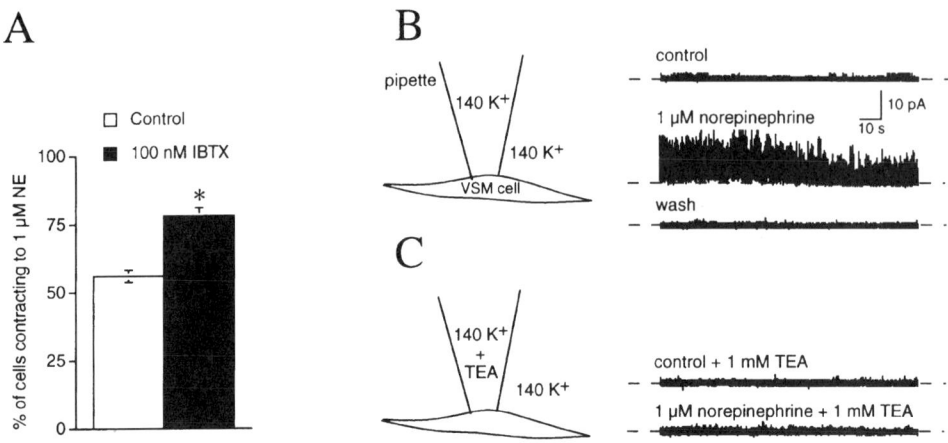

Figure 4. BK_{Ca} channels limit vasoconstriction in arteriolar muscle cells. (A) Mean $\pm SE$ ($n = 5$) percentage of single hamster cremasteric arteriolar muscle cells responding to norepinephrine (NE) in the absence and presence of 100 nM iberiotoxin (IBTX) (Data replotted from Jackson and Blair, 1998). (B) Single-channel currents through cell-attached patches of hamster cremasteric arteriolar muscle cells. Both the bath and the pipette contained 140 mM K^+, and the patch was held at -40 mV. The dashed lines represent the zero-current level. Norepinephrine (1 μM) caused cell contraction and an increase in the activity of a large-conductance channel. This was reversed by washout of the norepinephrine as shown. Similar results have been obtained in 6 cells. VSM, Vascular smooth muscle. (C) Experiments similar to those described in (B), but with 1 mM tetraethylammonium (TEA) in the pipette to block BK_{Ca} channels.

skeletal muscle (Jackson *et al.*, 1997). The specific Kv channel genes expressed in this vascular bed have yet to be examined. However, based on studies in other vascular smooth muscle cells, it is likely that several Kv channel subtypes will contribute to K^+ currents in skeletal muscle blood vessels.

4.2. Resting Membrane Potential, Tone, and Kv Channels

Blockade of Kv channels in cremasteric arteriolar muscle cells with 4-aminopyridine (1–3 mM) inhibits K^+ currents, reduces membrane conductance around the resting E_m, and depolarizes these cells (Fig. 5A, B; Jackson *et al.*, 1997; Jackson, 1998a). This Kv channel blocker also causes constriction of isolated, cannulated cremasteric arterioles (Fig. 5C). These data concur with previous studies in cerebral (Knot and Nelson, 1995; Robertson and Nelson, 1994) and coronary arteries (Leblanc *et al.*, 1994; Rusch *et al.*, 1996) which suggest that Kv channels importantly regulate resting E_m and tone of small arteries and arterioles.

4.3. Vasodilators and Kv Channels

The role played by Kv channels in the mechanism of action of vasodilators in the circulation of skeletal muscle has not been investigated. However, several studies have

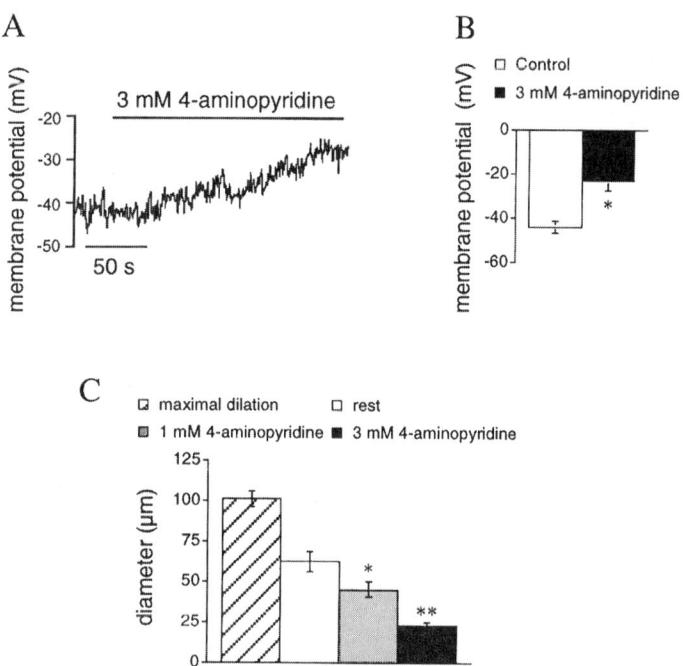

Figure 5. 4-Aminopyridine depolarizes arteriolar muscle cells and constricts cannulated arterioles *in vitro:* Evidence that Kv channels regulate resting membrane potential. (A) 4-Aminopyridine depolarizes current-clamped single rat cremasteric arteriolar muscle cells (see (Jackson *et al.*, 1997) for details). (B) Mean $\pm$ SE ($n = 8$) for experiments similar to those in (A) (data replotted from Jackson *et al.*, 1997). (C) Mean diameter $\pm$ SE ($n = 8$) of cannulated, second-order hamster cremasteric arterioles. Arterioles were isolated, cannulated with glass pipettes, and pressurized to 80 mm Hg [see [Jackson (1996) for methods]. The cross-hatched bar represents the maximum diameter of the arterioles in the absence of Ca^{2+}, demonstrating that these vessels had substantial resting tone. 4-Aminopyridine (1–3 mM) caused concentration-dependent constriction of the arterioles (*different from rest, **different from rest and 1 mM 4-aminopyridine; $P < 0.05$).

suggested that Kv channels may be activated by vasodilator stimuli. Agents that activate PKA have been reported to open Kv channels in rabbit vascular muscle cells (Aiello *et al.*, 1994, 1995, 1998). Kv channels also have been proposed to mediate endothelium-dependent vasodilation in several vascular beds (Yuan *et al.*, 1996; Eckman *et al.*, 1998; Shimizu and Paul, 1997; Hendriks *et al.*, 1993). Acidosis may also activate Kv channels in coronary vascular muscle cells (Berger *et al.*, 1998), although the functional significance of this effect has not been investigated. Hypoxia also may activate Kv channels in ductus arteriosus (Weir *et al.*, 1997), a topic that will be addressed in Chapter 30. Thus, Kv channels may participate in the mechanism of action of several classes of vasodilators and vasodilator stimuli, but their role in mediating the vasodilator reactivity of skeletal muscle arteries and arterioles has not been assessed. The discovery and application of blockers more selective than 4-aminopyridine for inhibition of vascular Kv channels would facilitate such studies in the future.

4.4. Vasoconstrictors and Kv channels

Given the voltage dependence of Kv channels, one would expect that, similarly to BK_{Ca} channels discussed above, these channels should be activated by vasoconstrictors that depolarize vascular smooth muscle cells. Thus, Kv channels could potentially limit the degree of depolarization, and hence vasoconstriction, in a negative-feedback manner. However, several lines of evidence suggest that Kv channel closure may help to mediate, rather than buffer, vasoconstrictor-induced depolarization and constriction. Activation of PKC appears to inhibit Kv channels (Aiello et al., 1996), as it does K_{ATP} and BK_{Ca} channels (see above). In addition, increases in intracellular Ca^{2+} ion concentration also inhibit Kv channel activity (Gelband et al., 1993; Ishikawa et al., 1993). Thus, stimulation of PKC and increases in intracellular Ca^{2+} would provide a positive-feedback signal that would promote depolarization and constriction for vasoactive agents that activate PKC and/or raise intracellular calcium. Finally, voltage-dependent inactivation of these channels may further reduce their open-state probability during prolonged stimulation. Functional studies in skeletal muscle supporting these hypotheses have yet to be performed.

5. K_{IR} CHANNELS IN THE SKELETAL MUSCLE CIRCULATION

5.1. Evidence for Kir Channels in Skeletal Muscle Arterioles

Perforated-patch recordings of whole-cell currents from single cremasteric arteriolar muscle cells have identified inwardly rectifying K^+ currents that are activated by elevated extracellular K^+ and inhibited by micromolar concentrations of Ba^{2+} (Jackson, 1998b). These data are consistent with studies from other tissues and suggest that vascular muscle cells in the walls of arterioles in skeletal muscle express K_{IR} channels (see Chapters 5, 19, and 22). Single-channel recordings and molecular studies of the K_{IR} channels expressed in vascular smooth muscle cells have yet to be performed to determine their phenotypic properties and subunit sequence.

5.2. Resting Membrane Potential, Tone, and Kir Channels

Very little is known about the role played by K_{IR} channels in the regulation of resting E_m and tone in skeletal muscle vascular beds. Preliminary studies suggest that 50 μM Ba^{2+} causes constriction of rat cremasteric arterioles in vivo (Loeb et al., 1997), although the site of action of Ba^{2+} was not determined. In contrast, studies of single hamster cremasteric arteriolar muscle cells found no effect of Ba^{2+} on currents measured at the resting E_m level (Jackson, 1998b). These data are inconsistent with the hypothesis that K_{IR} channels are active under resting conditions in these cells, and hence the role of K_{IR} channels in regulating resting E_m in skeletal muscle arterioles remains to be established.

5.3. Vasodilators and Kir Channels

In cerebral, coronary, and skeletal muscle vascular beds, elevated extracellular K$^+$ causes vasodilation that is associated with hyperpolarization of the vascular smooth muscle membrane (Sparks, 1980; Quayle *et al.*, 1997; Knot *et al.*, 1996; Nelson and Quayle, 1995). Two main events have been proposed to explain this K$^+$-induced hyperpolarization: activation of Na$^+$–K$^+$ ATPase and activation of K$_{IR}$ channels. Early studies showed that K$^+$–induced vasodilation could be inhibited by ouabain, suggesting that the Na$^+$–K$^+$ ATPase might be involved in this process (Johansson and Somlyo, 1980; Sparks, 1980). However, more recent evidence suggests that K$_{IR}$ channels mediate K$^+$-induced vasodilation in cerebral and coronary resistance arteries (Edwards *et al.*, 1988; Knot *et al.*, 1996; Salter and Kozlowski, 1998; Quayle *et al.*, 1993; Nelson and Quayle, 1995, Quayle *et al.*, 1997). Premliminary data support a role for K$_{IR}$ channels in K$^+$-induced dilation of arterioles in cremaster muscle (Loeb *et al.*, 1997). However, previous studies in this preparation have shown that K$^+$-induced dilation can be inhibited by millimolar concentrations of ouabain (Lombard and Stekiel, 1995). Thus, the role played by K$_{IR}$ channels in K$^+$-induced vasodilation of skeletal muscle arterioles remains unclear.

6. CONCLUSION AND QUESTIONS FOR THE FUTURE

Potassium channels represent the dominant ion-conducting pathway in the membrane of vascular muscle cells, including those in the skeletal muscle arteries and arterioles. From the information presented above and in other chapters in this volume, it should be clear that K$^+$ channels play a central role in the regulation of resting E_m, and vascular tone, and hence in the regulation of blood pressure and blood flow. Studies demonstrating the modulation of K$^+$ channel function by vasodilator and vasoconstrictor agents, acting through a variety of signal transduction pathways, suggest that these ion channels are likely involved in the local metabolic, myogenic, hormonal, and neural regulation of vascular resistance. However, despite the rapid increase in knowledge about K$^+$ channels that has occurred over the past two decades, a number of large questions still remain. Which K$^+$ channel genes are expressed in skeletal muscle resistance arteries and arterioles? What is the functional role of different K$^+$ channels *in vivo*? What role do K$^+$ channels play in conducted vasodilation in the microcirculation? In the next few years, a combination of molecular biology approaches, patch-clamp studies, and research into ionic regulation of integrated vascular systems should begin to provide answers to some of these questions about K$^+$ channels in the circulation of skeletal muscle.

REFERENCES

Abraham, N. G., Drummond, G. S., Lutton, J. D. and Kappas, A., 1996, The biological significance and physiological role of heme oxygnase, *Cell. Physiol. Biochem.* **6**:129–168.

Aguilar-Bryan, L., Clement, J. P., Gonzalez, G., Kunjilwar, K., Babenko, A., and Bryan, J., 1998, Toward understanding the assembly and structure of K_{ATP} channels, *Physiol.Rev.* **78**:227–245.

Aiello, E. A., Walsh, M. P., and Cole, W. C., 1994, Isoproterenol and forskolin increase and PKI inhibits delayed rectifier K^+ current in vascular myocytes isolated from rabbit coronary artery and portal vein, *Can. J. Physiol. Pharmacol.* **72**:47.

Aiello, E. A., Walsh, M. P., and Cole, W. C., 1995, Phosphorylation by protein kinase A enhances delayed rectifier K^+ current in rabbit vascular smooth muscle cells, *Am. J. Physiol. Heart Circ. Physiol.* **268**:H926–H934.

Aiello, E. A., Clement-Chomienne, O., Sontag, D. P., Walsh, M. P., and Cole, W. C., 1996, Protein kinase C inhibits delayed rectifier K^+ current in rabbit vascular smooth muscle cells, *Am. J. Physiol.* **271**:H109–H119.

Aiello, E. A., Malcolm, A. T., Walsh, M. P., and Cole, W. C., 1998, Beta-adrenoceptor activation and PKA regulate delayed rectifier K^+ channels of vascular smooth muscle cells, *Am. J. Physiol.* **275**:H448–H459.

Albarwani, S., Robertson, B. E., Nye, P. C. G., and Kozlowski, R. Z., 1994, Biophysical properties of Ca^{2+}- and Mg-ATP-activated K^+ channels in pulmonary arterial smooth muscle cells isolated from the rat, *Pflügers Arch.* **428**:446–454.

Babenko, A. P., Aguilar-Bryan, L., and Bryan, J., 1998, A view of $SUR/K_{IR}6.x$, K_{ATP} channels, *Annu. Rev. Physiol.* **60**:667–687.

Banitt, P. F., Smits, P., Williams, S. B., Ganz, P., and Creager, M. A., 1996, Activation of ATP-sensitive potassium channels contributes to reactive hyperemia in humans, *Am. J. Physiol.* **271**:H1594–H1598.

Baukrowitz, T., Schulte, U., Oliver, D., Herlitze, S., Krauter, T., Tucker, S. J., Ruppersberg, J. P., and Fakler, B., 1998, PIP_2 and PIP as determinants for ATP inhibition of K_{ATP} channels, *Science* **282**:1141–1144.

Benham, C. D., Bolton, T. B., Lang, R. J., and Takewaki, T., 1986, Calcium-activated potassium channels in single smooth muscle cells of rabbit jejunum and guinea-pig mesenteric artery, *J. Physiol. (London).* **371**:45–67.

Berczi, V., Stekiel, W. J., Contney, S. J., and Rusch, N. J., 1992, Pressure-induced activation of membrane K^+ current in rat saphenous artery, *Hypertension* **19**:725–729.

Berger, M. G., Vandier, C., Bonnet, P., Jackson, W. F., and Rusch, N. J., 1998, Intracellular acidosis differentially regulates K_v channels in coronary and pulomonary vascular muscle, *Am. J. Physiol* **275**:H1351–H1359.

Bonev, A. D., and Nelson, M. T., 1996, Vasoconstrictors inhibit ATP-sensitive K^+ channels in arterial smooth muscle through protein kinase C, *J. Gen. Physiol.* **108**:315–323.

Brayden, J. E., and Nelson, M. T., 1992, Regulation of arterial tone by activation of calcium-dependent potassium channels, *Science* **256**:532–535.

Bruch, L., Bychkov, R., Kastner, A., Bulow, T., Ried, C., Gollasch, M., Baumann, G., Luft, F. C., and Haller, H., 1997, Pituitary adenylate-cyclase-activating peptides relax human coronary arteries by activating K_{ATP} and K_{Ca} channels in smooth muscle cells, *J. Vasc. Res.* **34**:11–18.

Bryan, P. T. and Marshall, J. M., 1999, Cellular mechanisms by which adenosine induces vasodilatation in rat skeletal muscle: Significance for systemic hypoxia, *J. Physiol. (London)* **514**:163–175.

Busse, R., Pohl, U., Kellner, C., and Klemm, U., 1983, Endothelial cells are involved in the vasodilatory response to hypoxia, *Pflügers Arch.* **397**:78–80.

Busse, R., Forstermann, U., Matsuda, H., and Pohl, U., 1984, The role of prostaglandins in the endothelium-mediated vasodilatory response to hypoxia, *Pflügers Arch.* **401**:77–83.

Cabell, F., Weiss, D. S., and Price, J. M., 1994, Inhibition of adenosine-induced coronary vasodilation by block of large-conductance Ca^{2+}-activated K^+ channels, *Am. J. Physiol.* **267**:H1455–H1460.

Carl, A., Lee, H. K., and Sanders, K. M., 1996, Regulation of ion channels in smooth muscles by calcium, *Am. J. Physiol.* **271**:C9–C34.

Carrier, G. O., Fuchs, L. C., Winecoff, A. P., Giulumian, A. D., and White, R. E., 1997, Nitrovasodilators relax mesenteric microvessels by cGMP-induced stimulation of Ca-activated K channels, *Am. J. Physiol.* **273**:H76–H84.

Champion, H. C., and Kadowitz, P. J., 1997, Vasodilator responses to acetylcholine, bradykinin, and substance P are mediated by a TEA-sensitive mechanism, *Am. J. Physiol.* **273**:R414–R422.

Corriu, C., Feletou, M., Canet, E., and Vanhoutte, P.M., 1996, Endothelium-derived factors and hyperpolarization of the carotid artery of the guinea-pig, *Br. J. Pharmacol.* **119**:959–964.

Crijns, F. R., Struijker Boudier, H. A., and Wolffenbuttel, B. H., 1998, Arteriolar reactivity in conscious diabetic rats: Influence of aminoguanidine treatment, *Diabetes* **7**:918–923.

Dart, C., and Standen, N. B., 1995, Activation of ATP-dependent K$^+$ channels by hypoxia in smooth muscle cells isolated from the pig coronary artery, *J. Physiol. London.* **483**:29–39.

De Wit, C., Von Bismarck, P., and Pohl, U., 1994, Synergistic action of vasodilators that increase cGMP and cAMP in the hamster cremaster microcirculation, *Cardiovasc. Res.* **28**:1513–1518.

Duling, B. R., 1974, Oxygen sensitivity of vascular smooth muscle. II. *In vivo* studies, *Am. J. Physiol.* **227**:42–49.

Eckman, D. M., Hopkins, N., McBride, C., and Keef, K. D., 1998, Endothelium-dependent relaxation and hyperpolarization in guinea-pig coronary artery: Role of epoxyeicosatrienoic acid, *Br. J. Pharmacol.* **124**:181–189.

Edwards, F. R., Hirst, G. D. S., and Silverberg, G. D., 1988, Inward rectification in rat cerebral arterioles; involvement of potassium ions in autoregulation, *J. Physiol. (London)* **404**:455–466.

Fredricks, K. T., Liu, Y., and Lombard, J. H., 1994, Response of extraparenchymal resistance arteries of rat skeletal muscle to reduced PO$_2$, *Am. J. Physiol.* **267**:H706–H715.

Ganitkevich, V., and Isenberg, G., 1990, Isolated guinea pig coronary smooth muscle cells: Acetylcholine induces hyperpolarization due to sarcoplasmic reticulum calcium release activating potassium channels, *Circ. Res.* **67**:525–528.

Gelband, (c) H., Ishikawa, T., Post, J. M., Keef, K. D., and Hume, J. R., 1993, Intracellular divalent cations block smooth muscle K$^+$ channels, *Circ. Res.* **73**:24–34.

Hashemzadeh-Gargari, H., and Rembold, C. M., 1992, Histamine activates whole cell K$^+$ currents in swine carotid arterial smooth muscle cell, *Comp. Biochem. Physiol. C* **102C**:33–37.

Hendriks, M. G., Pfaffendorf, M., and Van Zwieten, P. A., 1993, The role of nitric oxide and potassium channels in endothelium-dependent vasodilation in SHR, *Blood Pressure* **2**:233–243.

Inoue, R., Kitamura, K., and Kuriyama, H., 1985, Two Ca-dependent K-channels classified by the application of tetraethylammonium distribute to smooth muscle membranes of the rabbit portal vein, *Pflügers Arch.* **405**:173–179.

Ishikawa, T., Hume, J. R., and Keef, K. D., 1993, Modulation of K$^+$ and Ca^{2+} channels by histamine H1-receptor stimulation in rabbit coronary artery cells, *J. Physiol. (London)* **468**:379–400.

Isomoto, S., Kondo, C., Yamada, M., Matsumoto, S., Higashiguchi, O., Horio, Y., Matsuzawa, Y., and Kurachi, Y., 1996, A novel sulfonylurea receptor forms with BIR (Kir6.2) a smooth muscle type ATP-sensitive K$^+$ channel, *J. Biol. Chem.* **271**:24321–24324.

Jackson, W. F., 1987, Arteriolar oxygen reactivity: Where is the sensor? *Am. J. Physiol.* **253**:H1120–H1126.

Jackson, W. F., 1993, Arteriolar tone is determined by activity of ATP-sensitive potassium channels, *Am. J. Physiol.* **265**:H1797–H1803.

Jackson, W. F., 1996, Rp diastereomeric analogs of cAMP inhibit both cAMP- and cGMP-induced dilation of hamster mesenteric small arteries, *Pharmacology* **52**:226–234.

Jackson, W. F., 1998a, Potassium channels and regulation of the microcirculation, *Microcirculation* **5**:85-90.

Jackson, W. F., 1998b, Potassium increases barium-sensitive potassium conductance of cremasteric arteriolar muscle cells, *FASEB J.* **12**(4):A14 (Abstract).

Jackson, W. F., 1999, Glibenclamide-sensitive oxygen reactivity in the absence of evidence for activation of ATP-sensitive K$^+$ channels, *FASEB J.* **13**(4):A31 (Abstract)

Jackson, W. F., and Blair, K. L., 1998, Characterization and function of Ca^{++}-activated K$^+$ channels in hamster cremasteric arteriolar muscle cells, *Am. J. Physiol.* **274**:H27–H34.

Jackson, W. F., König, A., Dambacher, T., and Busse, R., 1993, Prostacyclin-induced vasodilation in the rabbit heart is mediated by ATP-sensitive potassium channels, *Am. J. Physiol.* **264**:H238–H243.

Jackson, W. F., Huebner, J. M., and Rusch, N. J., 1997, Enzymatic isolation and characterization of single vascular smooth muscle cells from cremasteric arterioles, *Microcirculation* **4**(1):35–50.

Jan, L. Y., and Jan, Y. N., 1997, Ways and means for left shifts in the MaxiK channel, *Proc. Natl. Acad. Sci. U.S.A.* **94**:13383–13385.

Johansson, B., and Somlyo, A. P., 1980, Electrophysiology and excitation–contraction coupling, in: *Handbook of Physiology, Section 2: The Cardiovascular System, Volume II, Vascular Smooth Muscle*, (D. F. Bohr, A. P. Somlyo, and H. V. Sparks, eds.), American Physiological Society, Bethesda, Maryalnd, pp. 301–323.

Kleppisch, T. and Nelson, M. T., 1995a, Adenosine activates ATP-sensitive potassium channels in arterial myocytes via A2 receptors and cAMP-dependent protein kinase, *Proc. Natl. Acad. Sci. U.S.A.* **92**:12441–12445.

Kleppisch, T., and Nelson, M. T., 1995b, ATP-sensitive K^+ currents in cerebral arterial smooth muscle: Pharmacological and hormonal modulation, *Am. J. Physiol.* **269**:H1634–H1640.

Knot, H. J., and Nelson, M. T., 1995, Regulation of membrane potential and diameter by voltage-dependent K^+ channels in rabbit myogenic cerebral arteries, *Am. J. Physiol.* **269**:H348–H355.

Knot, H. J., Zimmermann, P. A., and Nelson, M. T., 1996, Extracellular K^+-induced hyperpolarizations and dilatations of rat coronary and cerebral arteries involve inward rectifier K^+ channels, *J. Physiol. (London)* **492**:419–430.

Knot, H. J., Standen, N. B., and Nelson, M. T., 1998, Ryanodine receptors regulate arterial diameter and wall $[Ca^{2+}]$ in cerebral arteries of rat via Ca^{2+}-dependent K^+ channels, *J. Physiol. (London)* **508**:211–221.

Kosmas, E. N., Levy, R. D., and Hussain, S. N., 1995, Acute effects of glyburide on the regulation of peripheral blood flow in normal humans, *Eur. J. Pharmacol.* **274**:193–199.

Lagrutta, A., Shen, K.-Z., North, R. A., and Adelman, J. P., 1994, Functional differences among alternatively spliced variants of *Slowpoke*, a *Drosophila* calcium-activated potassium channel, *J. Biol. Chem.* **269**:20347–20351.

Lange, A., Gebremedhin, D., Narayanan, J., and Harder, D., 1997, 20-Hydroxyeicosatetraenoic acid-induced vasoconstriction and inhibition of potassium current in cerebral vascular smooth muscle is dependent on activation of protein kinase C, *J. Biol. Chem.* **272**:27345–27352.

Leblanc, N., Wan, X., and Leung, P. M., 1994, Physiological role of Ca^{2+}-activated and voltage-dependent K^+ currents in rabbit coronary myocytes, *Am. J. Physiol.* **266**:C1523–C1537.

Loeb, A. L., Gödény, I., and Longnecker, D. E., 1997, Functional evidence for inward-rectifier potassium channels in rat cremaster muscle arterioles in vivo, *Microcirculation* **4(1)**:160 (Abstract).

Lombard, J. H., and Stekiel, W. J., 1995, Responses of cremasteric arterioles of spontaneously hypertensive rats to changes in extracellular K^+ concentration, *Microcirculation* **2**:355–362.

Lombard, J. H., Liu, Y., Fredricks, K. T., Bizub, D. M., and Rusch, N. J., 1999, Electrical and mechanical responses of rat middle cerebral arteries to reduced PO_2 and prostacyclin, *Am. J. Physiol.* **276**:H509–H516.

Marshall, J. M., Thomas, T., and Turner, L., 1993, A link between adenosine, ATP-sensitive K^+ channels, potassium and muscle vasodilatation in the rat in systemic hypoxia, *J. Physiol. (London)* **472**:1–9.

McCobb, D. P., Fowler, N. L., Featherstone, T., Lingle, C. J., Saito, M., Krause, J. E., and Salkoff, L., 1995, A human calcium-activated potassium channel gene expressed in vascular smooth muscle, *Am. J. Physiol.* **269**:H767–H777.

McManus, O. B., 1991, Calcium-activated potassium channels: Regulation by calcium, *J. Bioenerg. Biomembr.* **23**:537–560.

Messina, E. J., Sun, D., Koller, A., Wolin, M. S., and Kaley, G., 1992, Role of endothelium-derived prostaglandins in hypoxia-elicited arteriolar dilation in rat skeletal muscle, *Circ. Res.* **71**:790–796.

Minami, K., Fukuzawa, K., and Nakaya, Y., 1993, Protein kinase C inhibits the Ca^{2+}-activated K^+ channel of cultured porcine coronary artery smooth muscle cells, *Biochem. Biophys. Res. Commun.* **190**:263–269.

Minkes, R. K., Santiago, J. A., McMahon, T. J., and Kadowitz, P. J., 1995, Role of K_{ATP}^+ channels and EDRF in reactive hyperemia in the hindquarters vascular bed of cats, *Am. J. Physiol.* **269**:H1704–H1712.

Miyoshi, Y., and Nakaya, Y., 1991, Angiotensin II blocks ATP-sensitive K^+ channels in porcine coronary artery smooth muscle cells, *Biochem. Biophys. Res. Commun.* **181**:700–706.

Miyoshi, Y., Nakaya, Y., Wakatsuki, T., Nakaya, S., Fujino, K., Saito, K., and Inoue, I., 1992, Endothelin blocks ATP-sensitive K^+ channels and depolarizes smooth muscle cells of porcine coronary artery, *Circ. Res.* **70**:612–616.

Nakashima, M., and Vanhoutte, P. M., 1995, Isoproterenol causes hyperpolarization through opening of ATP-sensitive potassium channels in vascular smooth muscle of the canine saphenous vein, *J. Pharmacol. Exp. Ther.* **272**:379–384.

Nelson, M. T., and Quayle, J. M., 1995, Physiological roles and properties of potassium channels in arterial smooth muscle, *Am. J. Physiol.* **268**:C799–C822.

Nelson, M. T., Huang, Y., Brayden, J. E., Hescheler, J., and Standen, N. B., 1990, Arterial dilations in response to calcitonin gene-related peptide involve activation of K^+ channels, *Nature* **344**:770–773.

Nelson, M. T., Cheng, H., Rubart, M., Santana, L. F., Bonev, A. D., Knot, H. J., and Lederer, W.J., 1995, Relaxation of arterial smooth muscle by calcium sparks [see comments], *Science* **270**:633–637.

Quayle, J. M., McCarron, J. G., Brayden, J. E., and Nelson, M. T., 1993, Inward rectifier K^+ currents in smooth muscle cells from rat resistance-sized cerebral arteries, *Am. J. Physiol.* **265**:C1363–C1370.

Quayle, J. M., Bonev, A. D., Brayden, J. E., and Nelson, M. T., 1994, Calcitonin gene-related peptide activated ATP-sensitive K⁺ currents in rabbit arterial smooth muscle via protein kinase A, *J. Physiol. (London)* **475**:9–13.

Quayle, J. M., Nelson, M. T., and Standen, N. B., 1997, ATP-sensitive and inwardly rectifying potassium channels in smooth muscle, *Physiol. Rev.* **77**:1165–1232.

Randall, M. D., and McCulloch, A. I., 1995, The involvement of ATP-sensitive potassium channels in beta-adrenoceptor-mediated vasorelaxation in the rat isolated mesenteric arterial bed, *Br. J. Pharmacol.* **115**:607–612.

Robertson, B. E., and Nelson, M. T., 1994, Aminopyridine inhibition and voltage dependence of K⁺ currents in smooth muscle cells from cerebral arteries, *Am. J. Physiol.* **267**:C1589–C1597.

Rowell, L. B., 1986, *Human Circulation*, Oxford University Press, New York, pp. 96-116.

Rusch, N. J., and Liu, Y., 1997, Potassium channels in hypertension: Homeostatic pathways to buffer arterial contraction, *J. Lab. Clin. Med.* **130**:245–251.

Rusch, N. J., Gauthier-Rein, K., Liu, Y., Vandier, C., and Jackson, W. F., 1996, Regulation of membrane potential and diameter in small coronary arteries of human left ventricle by two K⁺ channel types, *FASEB J.* **10**(3):A571 (Abstract).

Saito, Y., McKay, M., Eraslan, A., and Hester, R. L., 1996, Functional hyperemia in striated muscle is reduced following blockade of ATP-sensitive potassium channels, *Am. J. Physiol.* **270**(5, Part 2):H1649–H1654.

Salter, K. J., and Kozlowski, R. Z., 1998, Differential electrophysiological actions of endothelin-1 on Cl⁻ and K⁺ currents in myocytes isolated from aorta, basilar and pulmonary artery, *J. Pharmacol. Exp. Ther.* **284**:1122-1131.

Schubert, R., Serebryakov, V. N., Mewes, H., and Hopp, H. H., 1997, Iloprost dilates rat small arteries: Role of K(ATP)- and K(Ca)-channel activation by cAMP-dependent protein kinase, *Am. J. Physiol.* **272**:H1147–H1156.

Scornik, F. S., and Toro, L., 1992, U46619, a thromboxane A₂ agonist, inhibits K_{Ca} channel activity from pig coronary artery, *Am. J. Physiol.* **262**:C708–C713.

Shimizu, S., and Paul, R. J., 1997, The endothelium-dependent, substance P relaxation of porcine coronary arteries resistant to nitric oxide synthesis inhibition is partially mediated by 4-aminopyridine-sensitive voltage-dependent K⁺ channels, *Endothelium* **5**:287–295.

Shyng, S., and Nichols, C. G., 1998, Membrane phospholipid control of nucleotide sensitivity of KATP channels, *Science* **282**:1138–1141.

Sparks, H. V., 1980, Effect of local metabolic factors on vascular smooth muscle, in: *Handbook of Physiology, Section sec. 2: The Cardiovascular System, Volume II, Microcirculation, Part 2* (D. F. Bohr, A. P. Somlyo, and H. V. Sparks, (eds.), American Physiological Society, Bethesda, Maryland, pp. 181–309.

Struijker Boudier, H. A. J., Messing, M. W. J., and Van Essen, H., 1992, Preferential small arteriolar vasodilatation by the potassium channel opener, BRL 38227, in conscious spontaneously hypertensive rats, *Eur. J. Pharmacol.* **218**:191–193.

Tanaka, Y., Meera, P., Song, M., Knaus, H. G., and Toro, L., 1997, Molecular constituents of maxi KCa channels in human coronary smooth muscle: Predominant $\alpha + \beta$ subunit complexes, *J. Physiol. (London)* **502**:545–557.

Tateishi, J., and Faber, J. E., 1995a, ATP-sensitive K⁺ channels mediate α_{2D}-adrenergic receptor contraction of arteriolar smooth muscle and reversal of contraction by hypoxia, *Circ. Res.* **76**:53–63.

Tateishi, J., and Faber, J. E., 1995b, Inhibition of arteriole α_2- but not α_1-adrenoceptor constriction by acidosis and hypoxia in vitro, *Am. J. Physiol.* **268**:H2068–H2076.

Thomas, G. D., Hansen, J., and Victor, R. G., 1997, ATP-sensitive potassium channels mediate contraction-induced attenuation of sympathetic vasoconstriction in rat skeletal muscle, *J. Clin. Invest.* **99**:2602–2609.

Toro, L., Amador, M., and Stefani, E., 1990, ANG II inhibits calcium-activated potassium channels from coronary smooth muscle in lipid bilayers, *Am. J. Physiol.* **258**:H912–H915.

Tsien, R. Y., and Rink, T. J., 1980, Neutral carrier ion-selective microelectrodes for measurement of intracellular free calcium, *Biochim. Biophys. Acta* **599**:623–638.

Vallet, B., Curtis, S. E., Guery, B., Mangalaboyi, J., Menager, P., Cain, S.M., Chopin, C., and Dupuis, B.A., 1995, ATP-sensitive K⁺ channel blockade impairs O₂ extraction during progressive ischemia in pig hindlimb, *J. Appl. Physiol.* **79**:2035–2042.

Vanelli, G., and Hussain, S. N., 1994a, Effects of potassium channel blockers on basal vascular tone and reactive hyperemia of canine diaphragm, *Am. J. Physiol.* **266**:H43–H51.

Vanelli, G. and Hussain, S. N. A., 1994b, Effects of potassium channel blockers on basal vascular tone and reactive hyperemia of canine diaphragm, *Am. J. Physiol.* **266**:H43–H51.

Vanelli, G., Chang, H. Y., Gatensby, A. G., and Hussain, S. N. A., 1994, Contribution of potassium channels to active hyperemia of the canine diaphragm, *J. Appl. Physiol.* **76**:1098–1105.

Wakatsuki, T., Nakaya, Y., and Inoue, I., 1992a, Vasopressin modulates K^+-channel activities of cultured smooth muscle cells from porcine coronary artery, *Am. J. Physiol.* **263**:H491–H496.

Wakatsuki, T., Nakaya, Y., Miyoshi, Y., Xiao-Rong, Z., Nomura, M., Saito, K., and Inoue, I., 1992b, Effects of vasopressin on ATP-sensitive and Ca^{2+}-activated K^+ channels of coronary arterial smooth muscle cells, *Jpn. J. Pharmacol.* **58** (Suppl. 2):339P.

Wang, R., and Wu, L., 1997, The chemical modification of K_{Ca} channels by carbon monoxide in vascular smooth muscle cells, *J. Biol. Chem.* **272**:8222–8226.

Wang, R., Wang, Z., and Wu, L., 1997b, Carbon monoxide-induced vasorelaxation and the underlying mechanisms, *Br. J. Pharmacol.* **121**:927–934.

Wang, R., Wu, L., and Wang, Z., 1997, The direct effect of carbon monoxide on K_{Ca} channels in vascular smooth muscle cells, *Pflügers. Arch.* **434**:285–291.

Weir, E. K., Reeve, H. L., Cornfield, D. N., Tristani-Firouzi, M., Peterson, D. A., and Archer, S. L., 1997, Diversity of response in vascular smooth muscle cells to changes in oxygen tension, *Kidney Int.* **51**:462–466.

Wesselman, J. P., Schubert, R., VanBavel, E. D., Nilsson, H., and Mulvany, M. J., 1997, K_{Ca}-channel blockade prevents sustained pressure-induced depolarization in rat mesenteric small arteries, *Am. J. Physiol.* **72**:H2241–H2249.

Yuan, X. J., Tod, M. L., Rubin, L. J., and Blaustein, M. P., 1996, NO hyperpolarizes pulmonary artery smooth muscle cells and decreases the intracellular Ca^{2+} concentration by activating voltage-gated K^+ channels, *Proc. Natl. Acad. Sci. U.S.A.* **93**:10489–10494.

Zhang, L., Bonev, A. D., Mawe, G. M., and Nelson, M. T., 1994, Protein kinase A mediates activation of ATP-sensitive K^+ currents by CGRP in gallbladder smooth muscle, *Am. J. Physiol.* **267** G494–G499.

Chapter 26

Regulation of Cerebral Artery Diameter by Potassium Channels

George C. Wellman and Mark T. Nelson

1. INTRODUCTION

One remarkable feature of the cerebral circulation is the maintenance of near-constant blood flow to the brain, even under the most extreme circumstances. Diameter regulation of cerebral resistance arteries to ensure this constant flow of oxygen and other nutrients to the brain involves the integration of a multitude of neural, humoral, endothelial, metabolic, and physical factors. Further complexity in the study of cerebral artery regulation arises when one considers that although total cerebral blood flow (accounting for approximately 20% of the entire blood flow in adult humans) remains constant, blood flow patterns within the brain can shift dramatically with changing electrical activity patterns. Although the factors regulating cerebral arterial diameter are both diverse and complex in nature, most, if not all, of these vasoactive stimuli ultimately act to control the contractile state of cerebral vascular smooth muscle cells via changes in the global intracellular free Ca^{2+} concentrations ($[Ca^{2+}]_i$) of these cells. These fluctuations in $[Ca^{2+}]_i$ largely mirror membrane potential changes in these cells, due to changes in the open-state probablity of voltage-sensitive Ca^{2+} channels, the major Ca^{2+} influx pathway in this tissue (see Nelson *et al.*, 1990 for a review).

   In vascular smooth muscle, the efflux rate of positively charged potassium ions through potassium-selective ion (K^+) channels contributes a large part of total cellular conductance and, thus, membrane potential (Hirst and Edwards, 1989). Given that plasmalemmal K^+ channels are a key determinant of membrane potential and thus global $[Ca^{2+}]_i$ in cerebral vascular smooth muscle, it is not surprising that modulation of K^+ channels represents a common and important underlying mechanism in the maintenance and regulation of cerebral blood flow (Faraci and Heistad, 1998; Nelson

George C. Wellman and Mark Nelson ● Department of Pharmacology, University of Vermont, Burlington, Vermont, 05405.

Potassium Channels in Cardiovascular Biology, edited by Archer and Rusch. Kluwer Academic/Plenum Publishers, New York, 2001.

et al., 1990). Four distinct types of K^+-selective ion channels have been described in cerebral arterial smooth muscle: (1) classical strong inwardly rectifying (Kir) K^+ channels, (2) ATP-sensitive (K_{ATP}) K^+ channels, (3) voltage-dependent (Kv) K^+ channels, and (4) Ca^{2+}-activated (BK_{Ca}) K^+ channels. This chapter will provide a brief biophysical and molecular description of each class of K^+ channel reported in cerebral vascular smooth muscle. However, the main focus here will be to emphasize the distinct features of these ion channels that allow physiological factors to selectively target these proteins, and thus regulate cerebral artery diameter.

2. STRONG INWARDLY RECTIFYING POTASSIUM (Kir) CHANNELS

Strong or "classical" inwardly rectifying K^+ (Kir) channels derive their name from their ability to pass inward current more readily than outward current. When the membrane potential is negative to the potassium equilibrium potential (E_K), the driving force for the movement of K^+ is in the inward direction and K^+ readily passes through Kir channels. However, at membrane potentials positive to E_K, the outward flow of K^+ ions through these channels is less. The mechanism underlying the inward rectification of this channel is blocking of the channel pore by positively charged cytoplasmic proteins and ions (e.g., spermine and magnesium) (Nichols and Lopatin, 1997). While this biophysical property of inward rectification is an accurate and suitable description of this channel, it is important to remember that, *in vivo*, the membrane potential of cerebral arterial myocytes (-60 to -35 mV) is always positive to E_K, and the membrane resistance of these cells is quite high (~ 10 MΩ). Thus, it is the relatively small efflux of K^+ through this channel that is physiologically relevant in the regulation of cerebral artery diameter (Nelson and Quayle, 1995; Quayle *et al.*, 1997).

2.1. Molecular Structure of Kir Channels in Cerebral Artery Myocytes

Evidence for the existence of Kir channels in the cerebral vasculature was first provided by Hirst and colleagues, who measured inwardly rectifying K^+ currents in intact voltage-clamped cerebral artery segments (Edwards *et al.*, 1988; Quayle *et al* 1993) subsequently demonstrated that these currents did indeed originate in vascular smooth muscle cells by recording membrane currents characteristic of Kir in vascular smooth muscle cells isolated from cerebral arteries (Fig. 1). The biophysical and pharmacological properties of these currents in cerebral vascular smooth muscle (Nelson and Quayle, 1995; Quayle *et al.*, 1997) are similar to those of Kir currents found in cardiac and skeletal muscle and are also consistent with these channels being composed of products from the Kir2.0 gene family. The inward rectifier potassium channel (Kir) gene family is currently divided into six subfamilies (Kir1.0 through Kir6.0) which share approximately 50% homology, predominately in the highly conserved pore (P) region (Doupnik *et al.*, 1995). Recently, the use of reverse-transcription polymerase chain reaction (RT-PCR) techniques on RNA from isolated cerebral vascular smooth muscle cells revealed transcripts for Kir2.1, but not Kir2.2 or

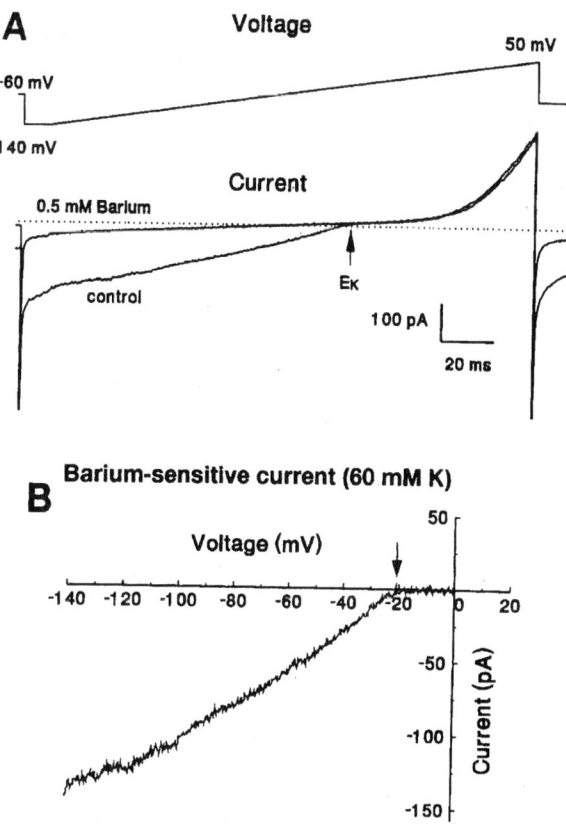

Figure 1.   Inward rectifying K$^+$ currents in cerebral arteries. (A) Membrane current recorded from a cell in response to a voltage ramp from -140 to $+50$ mV. Extracellular solution contained 60 mM K$^+$; intracellular solution contained 140 mM K$^+$. The arrow indicates the K$^+$ equilibrium potential (E_K). Two current traces are superimposed, one in control soution and the other after the cell was perfused with a solution containing 0.5 mM barium. Dotted line is zero-current level. (B) Current–voltage relationship of 0.5 mM barium-sensitive current. (Reproduced, with permission, from Quayle *et al.*, 1993.)

Kir2.3 (Bradley *et al.*, 1999). Kir2.1 was cloned from arterial smooth muscle and expressed in *Xenopus* oocytes. This expressed current exhibited properties very similar to those of the native Kir currents found in vascular smooth muscle (Bradley *et al.*, 1999) (Fig. 2).

    Cerebral arteries from genetically altered "knockout" mice have been used to confirm the molecular identity of the inward rectifier K$^+$ channel in cerebral vascular smooth muscle and the obligatory role of this channel in K$^+$-induced cerebral artery dilations. Kir currents and K$^+$-induced dilations (see Section 2.4) were absent in cerebral arteries of "knockout" mice in which the Kir2.1 gene was ablated (Eckman *et al.*, 1999). These data suggest that expression of the Kir2.1 gene is necessary for inward rectifier K$^+$ currents and K$^+$-induced dilations in small-diameter cerebral arteries.

A

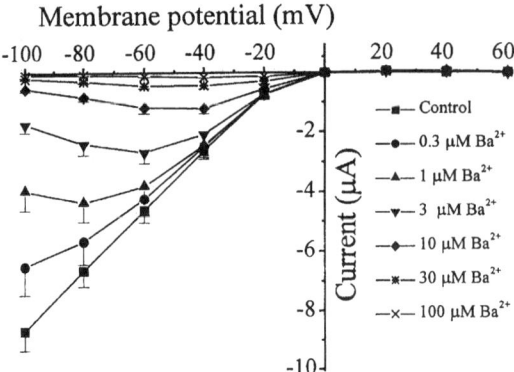

B

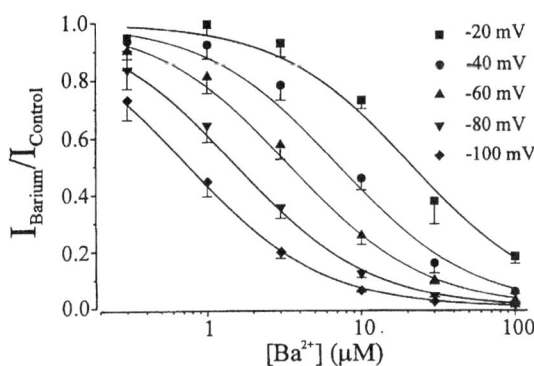

Figure 2. Barium inhibition of inward rectifier K$^+$ currents in *Xenopus* oocytes injected with RNA encoding Kir 2.1 cloned from vascular smooth muscle cells. (A) Current–voltage relationship demonstrating the effect of barium on membrane current. (B) Relationship between external barium concentration and the fractional inhibition of inward current at -20, -40, -60, -80, and -100 mV. (Reproduced, with permission, from Bradley *et al.*, 1999.)

2.2. Selective Inhibition of Channels by Micromolar Ba^{2+}: A Functional Fingerprint for the Role of Kir Channels in the Regulation of Cerebral Blood Flow

Consistent with the features of cloned Kir2.1 channels, inward rectifier K$^+$ currents from cerebral vascular smooth muscle cells exhibit a high-affinity block by external barium ions. Block of Kir currents by Ba^{2+} is highly voltage-dependent, with the dissociation constant (K_i) for Ba^{2+} decreasing approximately e-fold per 24-mV hyperpolarization, suggesting that Ba^{2+} binds to a blocking site within the channel pore (Fig. 2). At physiological membrane potentials, Kir channels are 20- to 1000-fold more sensitive to Ba^{2+} ($K_i \approx 5 \mu M$) than other K^+ channels found in vascular smooth muscle. In addition, highly selective blockers of other K^+ channels (e.g., glibenclamide and iberiotoxin; see following sections) have little, if any, affect on Kir currents. Thus, the use of low micromolar concentrations of Ba^{2+}, in concert with highly selective inhibitors of other K^+ channels, provides an effective means to examine the role of Kir channels in the regulation of cerebral artery diameter.

2.3. Voltage Dependence of Kir Activity: Contribution of Kir Channels to Resting Membrane Potential and Indirect Augmentation of the Action of Vasoactive Compounds

Membrane hyperpolarization and/or elevation of external K^+ will increase the activity of Kir channels. Therefore, fluctuations of external K^+, by even 1 mM, or slight changes in membrane potential will alter Kir channel conductance and its impact on cerebral artery function. In vascular smooth muscle, K^+ conductance (permeability) is a major determinant of membrane potential in unstimulated cells. Kir channels contribute significantly to this resting K^+ conductance in the cerebral vasculature (Hirst and Edwards, 1989; Johnson et al., 1998). From this intermediate level of channel activity, increases in Kir channel activity lead to membrane hyperpolarization and dilation, whereas decreases in Kir channel activity cause membrane depolarization and vasoconstriction.

Unlike the channel activity of other K^+ channels found in cerebral artery myocytes, Kir channel activity is increased at more negative membrane potentials and decreased by membrane depolarization. This voltage dependence of Kir channel activity will lead to an augmentation of the membrane potential (i.e., changes in $[Ca^{2+}]_i$ and arterial diameter) produced by vasoactive agents. For example, membrane depolarization caused by vasoconstrictors will tend to decrease Kir activity and lead to further membrane potential depolarization, whereas membrane hyperpolarization (produced by vasodilators) will tend to increase Kir activity and lead to additional membrane hyperpolarization. Thus, although these vasoactive agents do not directly alter Kir activity, secondary effects on Kir gating due to alterations in membrane potential enhance the efficacy of these stimuli (Fig. 3).

2.4. Modulation of Kir Gating by Extracellular K^+: Coupling of Neuronal Metabolic Activity to Cerebral Blood Flow

A second unique feature of Kir channels compared to other K^+ channels in cerebral vascular smooth muscle, is the action of extracellular K^+ ($[K^+]_o$) on the gating (activity) of this channel. Modest increases in $[K^+]_o$ (in the range of 3–20 mM) enhance K^+ efflux through Kir channels over the range of physiologically relevant membrane potentials, leading to membrane hyperpolarization and cerebral arterial dilation (Johnson et al., 1998; Knot et al., 1996). These K^+-induced dilations are

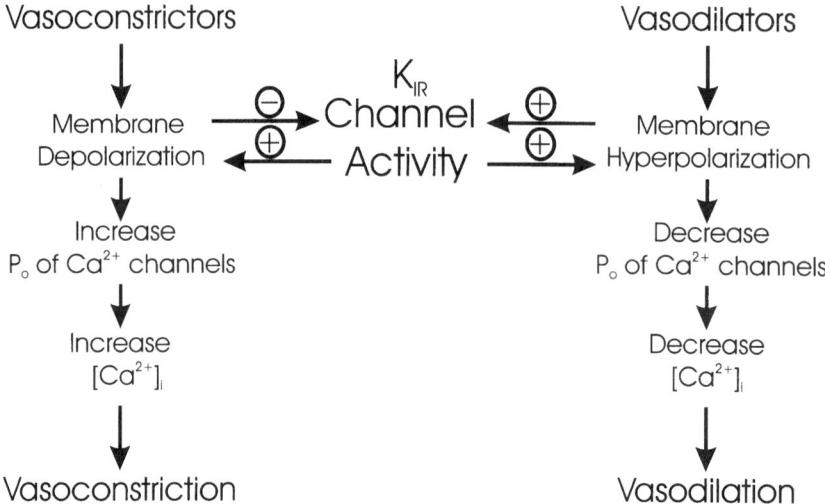

Figure 3. Augmentation of the effects of vasoactive compounds due to the voltage-dependent gating of Kir channels in arterial smooth muscle. Vasoconstrictors that cause membrane depolarization will decrease Kir channel activity, which leads to further membrane depolarization (and constriction). Conversely, vasodilators causing membrane hyperpolarization will increase Kir activity, leading to an additional membrane hyperpolarization (and dilation).

selectively inhibited by low-micromolar Ba^{2+}. This paradoxical increase in K^+ efflux despite a reduction in the K^+ electrochemical gradient can be explained by the unique ability of elevated $[K^+]_o$ to increase the open-state probability of Kir channels. This modulation of Kir activity by $[K^+]_o$ represents an important phenomenon coupling the metabolic activity of the brain to cerebral blood flow. Localized concentrations of extracellular K^+ in cerebrospinal fluid can range between 3 and 10 mM, depending upon neuronal activity (Somjen, 1979; Paulson and Newman, 1987). This localized K^+ increase in the proximity of the vasculature supplying areas of increased neuronal activity leads to localized cerebral arterial vasodilation, thus matching blood flow to demand. Increases in $[K^+]_o$ can also lead to more global cerebral arterial dilation under pathological conditions. For example, both ischemia and hypoglycemia cause K^+ efflux from neuronal tissue, leading to cerebral arterial vasodilation in an attempt to increased blood flow and limit neuronal damage.

3. ATP-SENSITIVE POTASSIUM (K_{ATP}) CHANNELS

As their name implies, one identifying feature of K_{ATP} channels is their sensitivity to intracellular concentrations of ATP. Inhibition of K_{ATP} channels by ATP ($K_i \approx 10 \, \mu M$) occurs at concentrations of this nucleotide (~ 5–10 mM) normally found in vascular smooth muscle, consistent with observations that selective inhibitors of K_{ATP} channels have little impact on cerebral blood flow under resting conditions

(Faraci and Heistad, 1998). However, this is not to say that K$_{ATP}$ channels do not significantly contribute to the regulation of cerebral blood flow. Indeed, as detailed below, K$_{ATP}$ channels are important targets for metabolic, neuronal, and synthetic stimuli at work under both physiological and pathophysiological conditions.

3.1. Molecular Distinction of Vascular K$_{ATP}$ Channels

Unlike the Kir channels observed in cerebral arteries, K$_{ATP}$ channels in vascular smooth muscle exhibit distinct differences when compared to K$_{ATP}$ channels found in other muscle types (e.g., cardiac and skeletal muscle). Rapid progress is being made with respect to defining and understanding the differing molecular structure of these K$_{ATP}$ channels. Functional K$_{ATP}$ channels are now thought to be composed of at least two distinct moieties: four channel subunits from the Kir6.0 gene family coexpressed with four sulfonylurea receptor proteins (SURs) (Babenko et al., 1998). As discussed in detail elsewhere in this book, the majority of the tissue-specific distinctions in K$_{ATP}$ channel activity can be accounted for by variations in SUR subtype expression or differences in the functional impact of K$_{ATP}$ channel activation.

3.2. Pharmacology of K$_{ATP}$ Channels in Cerebral Vascular Smooth Muscle

Most evidence for a functional role of K$_{ATP}$ channels in the cerebral vasculature relies on the inhibition of this channel by sulfonylurea compounds such as glibenclamide. Glibenclamide effectively inhibits whole-cell K$_{ATP}$ currents in vascular smooth muscle with a half-inhibition (K_i) of these currents at 100 nM (Quayle et al., 1995; Wellman et al., 1996). Although the potency of glibenclamide is somewhat less for vascular K$_{ATP}$ channels compared to cardiac and pancreatic K$_{ATP}$ isoforms ($K_i \approx 2\,nM$), the specificity of this compound for vascular K$_{ATP}$ channels compared to other vascular K$^+$ channels is still great (Nelson and Quayle, 1995; Quayle et al., 1997).

K$_{ATP}$ channels in cerebral vascular smooth muscle are also targets for a wide variety of naturally occurring and synthetic compounds (Kleppisch and Nelson, 1995) that activate these channels, presumably via a direct interaction with the SUR subunits of these channels. These synthetic activators of K$_{ATP}$ channels are potent vasodilators and represent a group of chemically diverse structures (Wellman and Quayle, 1997) termed K$^+$ channel openers. Consistent with involvement of K$_{ATP}$ channels, the membrane hyperpolarization and dilation of cerebral arteries in vitro caused by K$^+$ channel activators are inhibited by concentrations of glibeclamide selective for K$_{ATP}$ channels. Glibenclamide can also inhibit cerebral arterial dilations to these compounds in vivo (Faraci and Heistad, 1998). A number of these compounds have been in use as antihypertensive agents (e.g., minoxidil sulfate, pinacidil, and diazoxide). Despite a great deal of initial hope, adverse side effects have limited the clinical utility of these compounds in the treatment of hypertension. For example, diazoxide has been used clinically in the management of acute hypertensive crisis, but long term use of this compound results in hypoglycemia (due to activation of K$_{ATP}$ channels in pancreatic β-cells). Minoxidil is also a potent vasodilator that has limited use in the treatment of some severe forms of hypertension. In addition to side effects such as headaches and edema that are common to all vasodilators, minoxidil use is also associated with cardiotoxicity and hirsutism. Minoxidil is also the active compound in topical hair-growth treatments; its mechanism of action is unclear but may involve vasodilation and increased blood flow to the scalp. The use of pinacidil in the treatment of hypertension has been approved in most countries. Clinical data on this compound suggest that although headaches and edema are prevalent among users, these adverse side effects can be eliminated by coadministration of diuretics and β-adrenergic receptor blockers.

3.3. Modulation by Endogenous Compounds: Role of Second Messengers and Protein Phophorylation

Cerebral arterial dilations and membrane hyperpolarizations in response to a number of endogenously occurring compounds are also inhibited by glibenclamide, suggesting involvement of K_{ATP} channels. The list of physiological activators of K_{ATP} channels includes neurotransmitters such as calcitonin gene-related peptide (CGRP), vasoactive intestinal polypeptide (VIP), and opioids, as well as α-adrenergic receptor agonists (Faraci and Heistad, 1998). Other compounds such as adensosine, adrenomedullin, and prostacyclin that are produced locally by cerebral vascular smooth muscle and endothelial cells also cause glibenclamide-sensitive cerebral arterial dilations. A common feature of many of the endogenous K_{ATP} channel activators listed above is their activation of membrane receptors coupled to G_s, a G-protein linked to stimulation of adenylyl cyclase. Increased production of cAMP due to activation of adenylyl cyclase in turn leads to enhanced activity of cAMP-dependent protein kinase (PKA) (Stryer, 1995). Consistent with involvement of PKA in the dilations induced by these compounds, both forskolin (a synthetic activator of adenylyl cyclase) and membrane-permeable anologs of cAMP (e.g., Sp-cAMPS) produce glibenclamide-sensitive cerebral artery dilations (Nelson, 1992; Quayle *et al.*, 1997) (Fig. 4).

Use of the patch-clamp technique to measure membrane K^+ currents in isolated vascular smooth muscle cells has provided compelling direct evidence that the cAMP/PKA second-messenger pathway is a potent mechanism for K_{ATP} channel activation (Kleppisch and Nelson, 1995; Quayle *et al.*, 1994). For example, CGRP, adenosine, and isoproterenol all activate glibenclamide-sensitive K_{ATP} currents in a manner similar to forskolin. A similar current is activated by Sp-cAMPS or cell dialysis with the activated catalytic subunit of PKA in vascular smooth muscle (Fig. 4). Further, inhibitors of PKA completely abolished activation of K_{ATP} currents by these agents. It is unclear whether these phosphorylation events reflect phosphorylation of the Kir or SUR subunits, as both proteins appear to have a number of PKA phosphorylation sites (Aguilar-Bryan *et al.*, 1998).

In a manner opposite to the effects of PKA, protein kinase C (PKC) activation decreases K_{ATP} currents in vascular smooth muscle cells, including those of cerebral arteries (Bonev and Nelson, 1996; Kleppisch and Nelson, 1995). For example, activators of PKC such as analogs of diacylglycerol and phorbol esters inhibit K_{ATP}

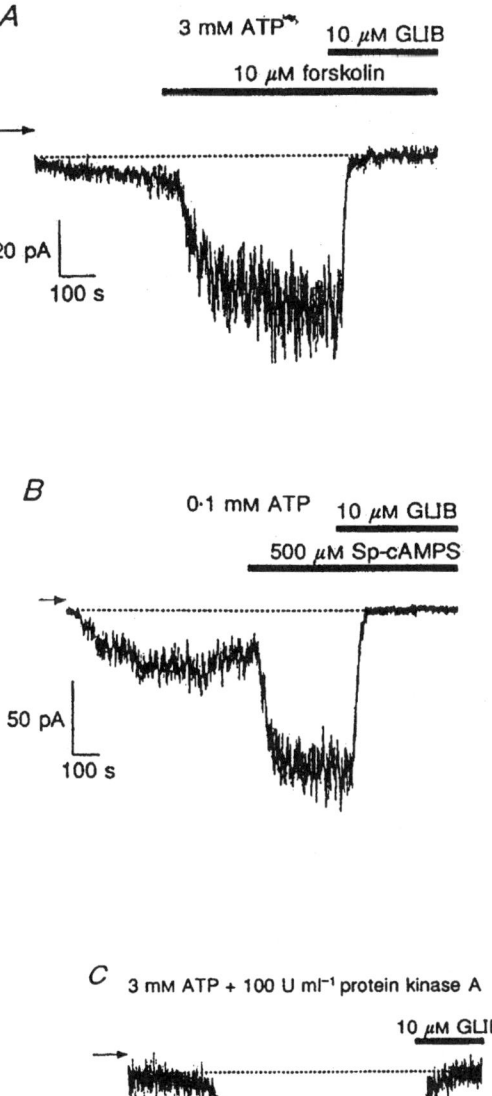

Figure 4. Activators of cAMP-dependent protein kinase (PKA) increase K_{ATP} currents in vascular smooth muscle cells. (A) Addition of forskolin (an activator of adenylyl cyclase) caused a glibenclamide-sensitive current in a vascular smooth muscle cell dialyzed with 3 mM ATP. (B) Activation of PKA by Sp-cAMPS also enhanced K_{ATP} currents in vascular smooth muscle (note the increased basal current in this cell dialyzed with a lower concentration of ATP). (C) Demonstration that cell dialysis with the catalytic subunit of PKA also increased K_{ATP} currents in vascular smooth muscle. (Reproduced, with permission, from Quayle *et al.*, 1994.)

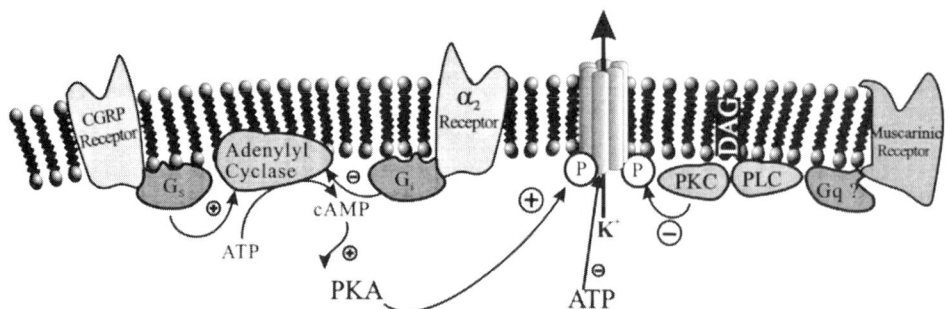

Figure 5. Model illustrating second-messenger pathways leading to the modulation of K_{ATP} channels in cerebral vascular smooth muscle.

currents activated by either synthetic K^+ channel openers (e.g., pinacidil and lev-cromakalim) or receptor-mediated K_{ATP} current activation in vascular smooth muscles. PKC inhibition of K_{ATP} currents may be an important mechanism with regard to the action of vasoconstrictor substances on the cerebral circulation. Neurotransmitters such as neuropeptide Y (NPY), serotonin (5HT), and histamine all constrict cerebral arteries and inhibit K_{ATP} currents. The inhibition of K_{ATP} currents by these vasoconstrictors was also abolished both by inhibitors of phospholipase C and inhibitors of PKC. The signal transduction pathways linking receptor activation to increased cAMP and PKA, decreased PKC, and the activity of K_{ATP} channels are illustrated in Fig. 5.

3.4. K_{ATP} Channel Activity and Changes in Cellular Metabolic State

An important function of activation of K_{ATP} currents in the cerebral vasculature is to increase blood flow during times of metabolic compromise (Faraci and Heistad, 1998; Quayle et al., 1997). In addition to the neuronal release of K^+ and sensory neurotransmitters (e.g., CGRP and VIP) described above, evidence now suggests that multiple metabolic signals converge to activate K_{ATP} channels located on cerebral arterial myocytes in an attempt to match cerebral blood flow to the local needs of the surrounding tissue. Several of these factors are intracellular signals originating within the vascular smooth muscle during times of metabolic compromise and are thought to have direct effects on K_{ATP} channels to increase their open-state probability. These metabolic signals include hypoxia, hypercapnia, and hypoglycemia as well as elevated intracellular levels of ADP and adenosine. These factors all increase vascular K_{ATP} currents (Nelson and Quayle, 1995; Quayle et al., 1997) and cause glibenclamide-sensitive cerebral arterial vasodilations (Faraci and Heistad, 1998; Faraci and Sobey, 1998). In addition, during ischemic conditions, surrounding tissue (neurons, astrocytes, and vascular endothelial cells) can release additional factors such as prostacylin and adenosine, which can activate K_{ATP} currents through a variety of mechanisms, including direct effects on channel subunits as well as activation of PKA.

4. Kv CHANNELS

Kv channels comprise a large family of ion channel proteins that are activated by membrane depolarization and are common in all types of vascular and nonvascular smooth muscle. In cerebral vascular smooth muscle, Kv currents have been measured in cells isolated from several species (e.g., Bonnet *et al.*, 1991; Robertson and Nelson, 1994). It is believed that membrane hyperpolarization associated with increased Kv activity acts as a negative-feedback mechanism to limit the depolarization and Ca^{2+} entry that occur in response to vasoconstrictor stimuli. Kv channels may also act as targets for a number of cerebral arterial vasodilators.

4.1. Kv Subtypes in Cerebral Arterial Smooth Muscle

The molecular structure of Kv channels is somewhat different from that of the Kir gene family products which make up the channel subunits of Kir and K$_{ATP}$ (Jan and Jan, 1997). Functional channels from Kir and Kv gene products each contain four channel subunits, and each individual Kir subunit contains two membrane-spanning regions. In contrast, members of the Kv gene family express pore-forming α-subunit proteins with six hydrophobic membrane-spanning domains (S1–S6). The voltage sensitivity of Kv channels is thought to arise from the S4 region, which is highly charged, containing basic amino acids (lysine or arginine) at every third residue. In addition, at least four different mammalian β-subunits have been cloned, and these are thought to play a modulatory role in Kv channel kinetics.

At this point, it is unclear how many different subtypes of Kv channels are expressed in cerebral vascular smooth muscle, and a detailed discussion of the plethora of molecular evidence regarding Kv channels is beyond the scope of this chapter. It suffices to say that the difficulty in dissecting distinct species of Kv channels arises not only from the lack of selective pharmacological inhibitors (discussed below), but also from the complexity that may exist in the expression of these channels. For example, it has been suggested that different Kv α-subunits may complex to form heteromultimers, and the possibility that Kv subunits may associate with different accessory β-subunits has also been proposed (Jan and Jan, 1997; Robertson, 1997).

4.2. Identification of Kv Currents in Cerebral Vascular Smooth Muscle

Kv currents recorded in response to depolarizing voltage steps exhibit both voltage-dependent activation and voltage-dependent inactivation (which occurs at a slower rate). The increased K$^+$ efflux through Kv channels in response to membrane depolarization occurs as a result of both an increase in the open-state probability of these channels and an increase in the single-channel conductance due to an increase in the driving force for K$^+$ efflux. The voltage dependence of Kv channel activation and inactivation for Kv currents obtained in smooth muscle cells isolated from cerebral arteries is depicted in Fig. 6. The physiologically relevant Kv current at a given

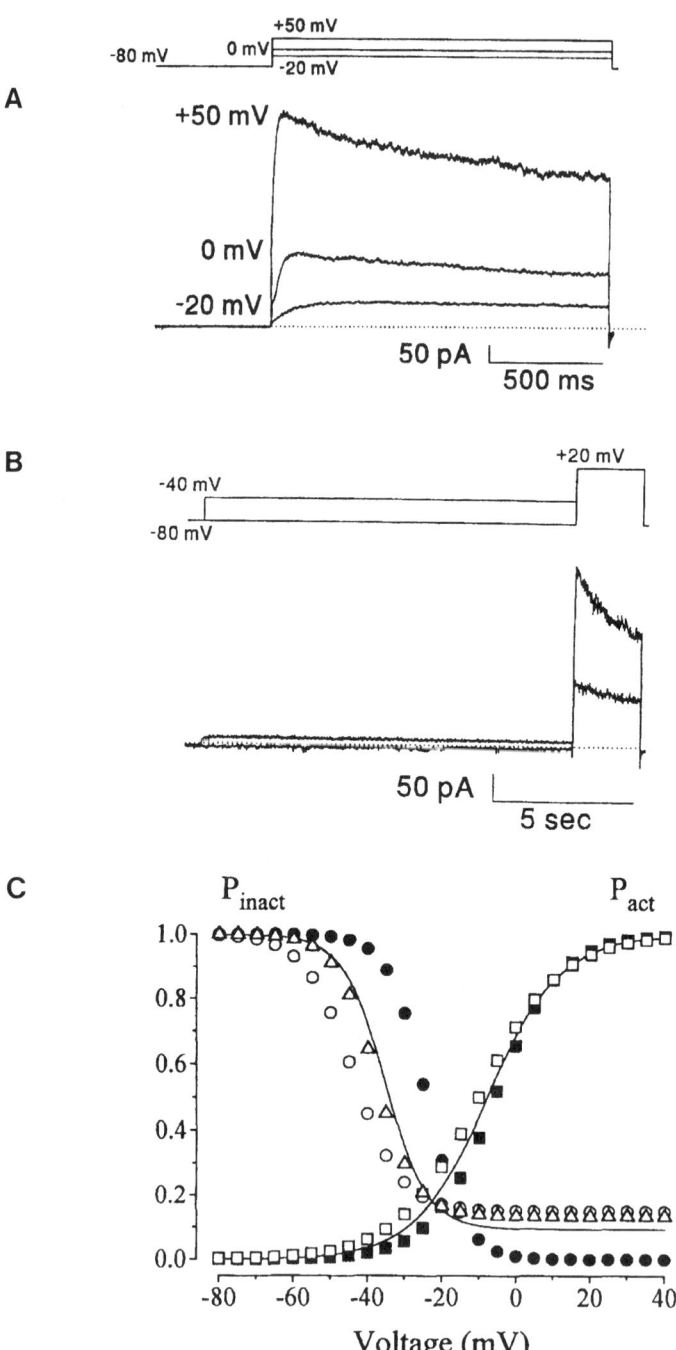

Figure 6. Voltage-dependent activation and inactivation of Kv channels in cerebral vascular smooth muscle cells. (A) Outward current traces elicited by membrane depolarization steps from a holding potential of −80 mV. These records were obtained in the presence of TEA$^+$ (1 mM) and nimodipine (1 μM) to inhibit Ca^{2+}-activated K$^+$ channels (B) Original record illustrating the voltage dependence of Kv channel inactivation. Outward Kv currents elicited by step depolarizations to +20 mV were decreased when the holding potential of the cell was changed from −80 mV to −40 mV. (C) Steady-state Kv activation and inactivation curves obtained from arterial smooth muscle. (Modified from Robertson and Nelson, 1994, and Quayle *et al.*, 1995, with permission.)

membrane potential should be represented by the steady-state balance between the activation and inactivation of these currents. During steady state, the probability that a Kv channel will be open (P_o) at a given membrane voltage should be equal to the product of the probability of the channel being activated (P_{act}) and the probability that the channel is not inactivated ($P_{1-inact}$). Using this approach, Nelson and Quayle have predicted that significant current should occur via Kv channels at membrane potentials in the physiological range (~ -35 to -40 mV) for cerebral arteries (Nelson, 1992). Consistent with this prediction, steady-state Kv currents were measured in cerebral arterial myocytes voltage-clamped at -40 mV (Robertson and Nelson, 1994).

4.3. Pharmacology of Kv Currents in Cells Isolated from Vascular Smooth Muscle

One inhibitor of vascular Kv currents that has been widely described in the literature is 4-aminopyridine (4-AP). In isolated arterial myocytes (including cerebral arterial myocytes), the K_i for 4-AP is approximately 0.5 mM. It is interesting to note that increasing the concentration of 4-AP above 5–10 mM produces no further inhibition of Kv currents. This observation could suggest the existence of multiple Kv subtypes in vascular smooth muscle. Several other compounds also produce partial inhibition of vascular Kv currents, including phencyclidine ($K_i \approx 30\,\mu M$), tedisamil ($K_i \approx 3\,\mu M$), quinidine ($K_i \approx 30\,\mu M$), and tetraethylammonium ions (TEA$^+$; $K_i > 5$mM). Selective inhibitors of other vascular K^+ channels such as glibenclamide (K_{ATP}), low-micromolar Ba^{2+} (Kir), and iberiotoxin (BK$_{Ca}$) do not inhibit Kv channels (Nelson and Quayle, 1995).

4.4. Physiological Contribution of Kv Channels to the Regulation of Cerebral Artery Diameter

Increases in intravascular pressure cause graded membrane depolarization, increased Ca^{2+} influx, and constriction of cerebral arteries (Brayden and Wellman, 1989; Knot and Nelson, 1998; Nelson et al., 1990). This pressure-induced depolarization has the potential of a feed-forward system (depolarization leading to increased Ca^{2+} influx through voltage-dependent Ca^{2+} channels, which would cause further membrane depolarization). Substantial evidence from the cerebral circulation suggests that activation of vascular Kv channels and the resulting K^+ efflux act as one negative-feedback pathway in response to pressure-induced depolarizations (Fig. 7); BK$_{Ca}$ channels, described below, are another. Consistent with this hypothesis, inhibition of Kv currents (by 4-AP) produced greater effects on cerebral arterial diameter at elevated intravascular pressures (Knot and Nelson, 1995). A similar negative-feedback role for Kv currents in response to vasoconstrictors acting via membrane depolarization is likely.

Phosphorylation and activation of Kv channels via the cAMP/PKA pathway (Aiello et al., 1995; Cole et al., 1996) may be involved in the vasodilation produced by a number of physiological mediators, including substances produced and released by the cerebral vascular endothelium (Dong et al., 1998). Inhibition of Kv channels by activators of PKC has also been described in vascular smooth muscle (Cole et al., 1996) and may play a role in the regulation of cerebral arterial diameter.

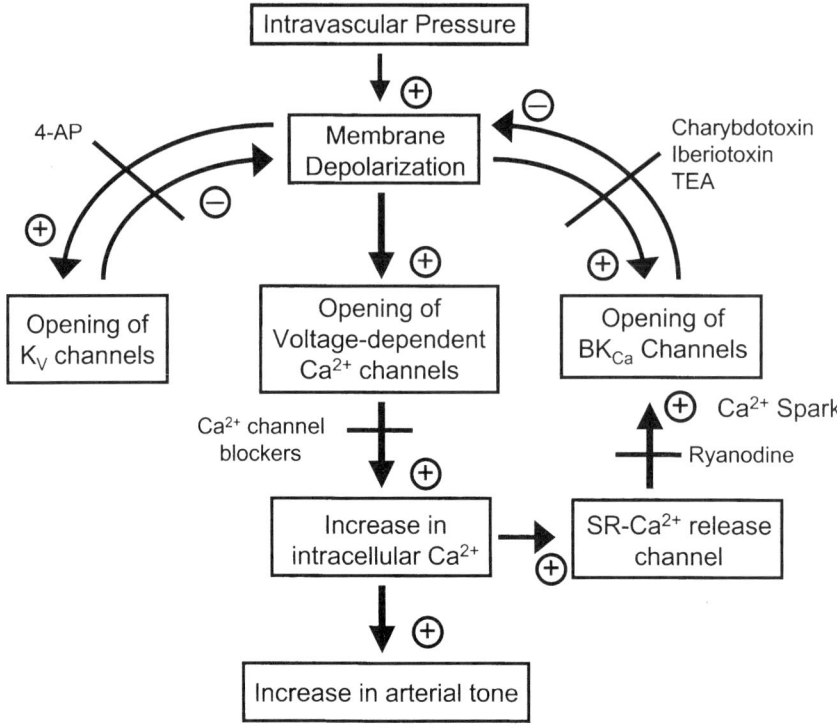

Figure 7. Flow diagram of the proposed role of Kv and BK_{Ca} channels in the negative-feedback regulation of pressure-induced membrane depolarization and contriction of cerebral arteries.

5. BK_{Ca} CHANNELS

Given the high density, large conductance, and activation characteristics of BK_{Ca} channels located in cerebral vascular smooth muscle, it is not surprising that BK_{Ca} channels are thought to play a major role in the regulation of cerebral blood flow. Increased BK_{Ca} activity occurs in response to both membrane depolarization and elevation of intracellular Ca^{2+}. Recent exciting new evidence suggests that the majority of BK_{Ca} activity in cerebral vascular smooth muscle occurs in response to very localized, transient, and high increases in subsarcolemmal Ca^{2+}.

5.1. Molecular Structure of BK_{Ca} Channels

Each of the four identical α-subunits that combine to form functional BK_{Ca} channels contain a total of 10 hydrophobic regions, S1–S10. Six of these hydrophobic segments, thought to be transmembrane loops (S1–S6), share a high degree of homology with other voltage-dependent (Kv) channels and, in all likelihood, encompass the voltage-sensing region of these ion channels. The additional S7–S10 regions present in the BK_{Ca} channel α-subunit are likely to act as the sensor to intracellular Ca^{2+}. Many splice variants of this gene exist, which are likely to account for the slight differences in BK_{Ca} channel properties observed in different tissues. In addition, it is now believed that modulatory β-subunits combine with the pore-forming α-subunits to further influence BK_{Ca} channel function. Although not described in cerebral vascular smooth muscle, the molecular structures of intermediate-conductance (IK_{Ca}) and small-conductance (SK_{Ca}) Ca^{2+}-activated channels have recently been elucidated. Unlike BK_{Ca} channels, IK_{Ca} and SK_{Ca} channels appear to derive their Ca^{2+} sensitivity by their close association with the Ca^{2+}-binding protein calmodulin (Vergara *et al.*, 1998).

5.2. Modulation of BK_{Ca} Channels in Cerebral Vascular Smooth Muscle by Ca^{2+} Sparks

Recent advances in vascular biology have led to the identification of elemental Ca^{2+} release events (Ca^{2+} sparks) from internal Ca^{2+} stores located in arterial smooth muscle (Nelson *et al.*, 1995). Ca^{2+} sparks arise from the activation of several, tightly clustered ryanodine-sensitive Ca^{2+} release channels (RyRs) located in the sarcoplasmic reticulum (SR) membrane (Gollasch *et al.*, 1998; Nelson *et al.*, 1995). These elemental Ca^{2+} events, although displaying many similarities to those observed in other muscle types, may functionally be very different. In vascular smooth muscle, Ca^{2+} sparks paradoxically lead to decreased global $[Ca^{2+}]_i$ and relaxation of arterial smooth muscle (Knot *et al.*, 1998; Nelson *et al.*, 1995; Porter *et al.*, 1998). In cerebral vascular smooth muscle, these localized increases in $[Ca^{2+}]_i$ are small, representing less than 1% of the surface area of the cell. Furthermore, in vascular smooth muscle (VSM), RyRs do not appear to be recruited in large numbers, and there is no coordinated release and fusing of Ca^{2+} sparks to increase global $[Ca^{2+}]_i$. Rather than directly altering global $[Ca^{2+}]_i$, each asynchronous Ca^{2+} spark activates an estimated 20–100 BK_{Ca} channels located on the plasma membrane to cause an outward K$^+$ current previously referred to as "spontaneous transient outward currents" or "STOCs." Recent simultaneous measurements have verified the causal relationship between Ca^{2+} sparks and STOCs (Fig. 8; Perez *et al.*, 1999). These Ca^{2+} sparks and STOCs exhibit similar temporal characteristics and are abolished by agents that either block sarcoplasmic reticulum (SR) Ca^{2+} release channels or deplete SR Ca^{2+} stores. Enhanced BK_{Ca} activity in the form of STOCs would cause membrane hyperpolarization and reduce Ca^{2+} influx by decreasing the open-state probability of dihydropyridine-sensitive L-type Ca^{2+} channels. The Ca^{2+} spark/STOC pathway may therefore act as an important negative-feedback system to limit vasoconstriction caused by stimuli such as increased intravascular pressure (Brayden and Nelson, 1992; Nelson *et al.*, 1995). It would also be predicted that increasing the Ca^{2+} spark frequency in vascular smooth muscle would promote vasodilation through increased BK_{Ca} channel activity and may represent a previously unknown mechanism of action for vasodilators.

5.3. Modulation of Ca^{2+} Spark/STOC Activity in Cerebral Arteries

Nitric oxide, a potent vasodilator, produced by both the vascular endothelium and clinically used nitrovasodilators, cause the activation of guanylyl cyclase, increased

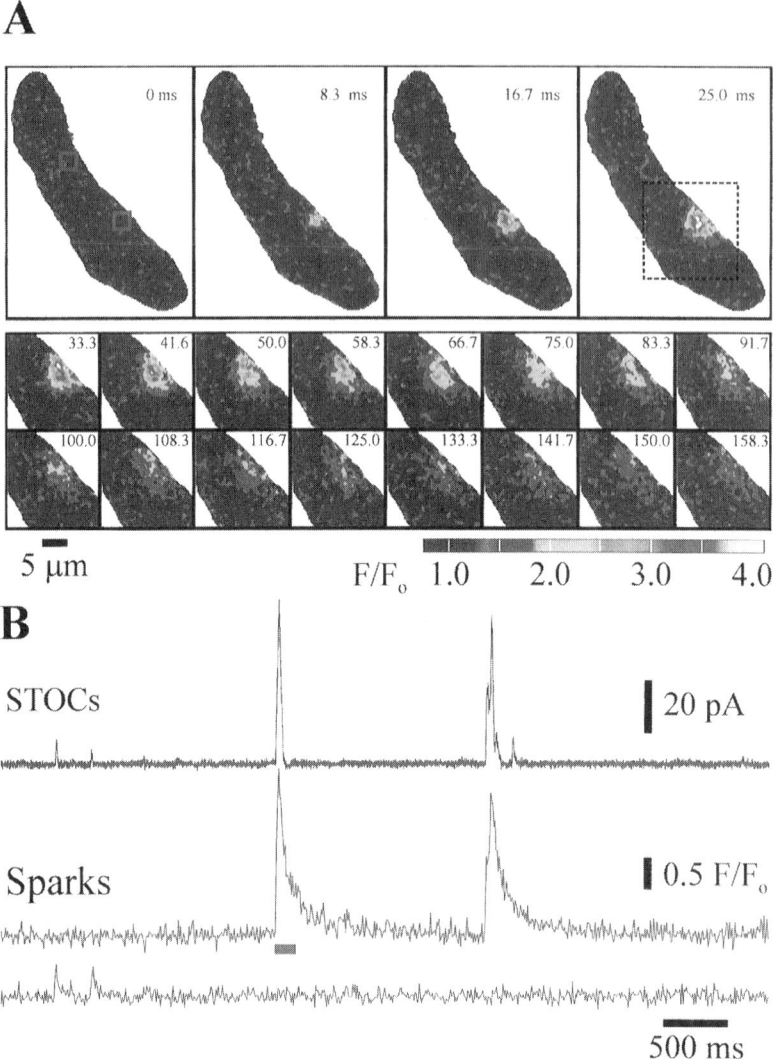

Figure 8. Ca^{2+} sparks generate BK_{Ca} currents (STOCs) in myocytes from rat basilar cerebral artery. (A) Original sequence of two-dimensional confocal images obtained every 8.33 ms of an entire smooth muscle cell (*top*), followed by subsequent images of the region of interest (dotted box) illustrating the time course of the fractional increase in fluorescence (F/F_o) and decay of a typical Ca^{2+} spark. The images are color-coded as indicated by the bar. (B) Simultaneous STOC (pA) and spark (F/F_o) measurements, at -40 mV, illustrating temporal association. Blue, Current; red and green, the F/F_o average of the red and green boxes (2.2 μm/side) indicated in (A), respectively. The pink bar indicates the segment of the trace illustrated in (A). (Reproduced from Perez *et al.*, 1999, by copyright of The Rockefeller University Press.)

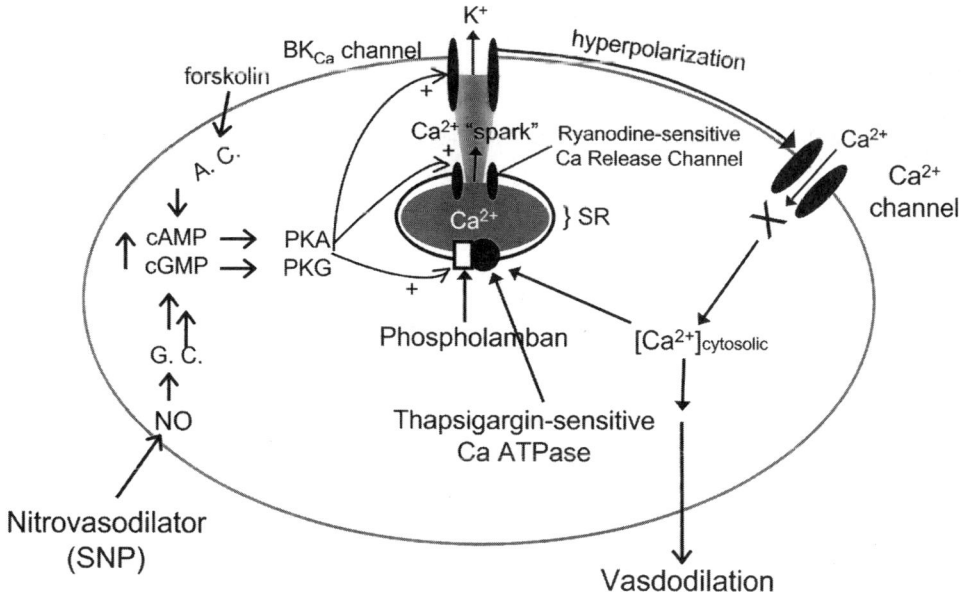

Figure 9. Proposed mechanism(s) for activators of cAMP-dependent protein kinase (PKA) and cGMP-dependent protein kinase (PKG) on Ca^{2+} sparks and BK$_{Ca}$ channels in arterial smooth muscle. Adenylyl cyclase (A. C.) produces cAMP when activated by forskolin or membrane receptors linked to G$_s$ proteins, which in turns stimulates PKA. Activation of PKA has a direct effect on the BK$_{Ca}$ channel to increase its open state probability. However, the predominant effect of PKA is on the sarcoplasmic reticulum (SR) to increase Ca^{2+} spark frequency. Possible targets of PKA and PKG are the BK$_{Ca}$ channel, RyR receptors, and the Ca^{2+}-ATPase (phospholamban). Combined effects of increased spark activity and direct effects on the BK$_{Ca}$ channel lead to significantly increased activity of the BK$_{Ca}$ channel. Resulting hyperpolarization closes voltage-dependent Ca^{2+} channels, leading to reduced global intracellular Ca^{2+} and vasodilation.

accumulation of cGMP, and stimulation of cGMP-dependent protein kinase (PKG). Many other endogenous vasodilators acting through membrane receptors coupled to G$_s$ proteins (e.g., CGRP, adenosine, and β-adrenoceptor agonists) cause increased adenylyl cyclase activity, elevated levels of cAMP, and stimulation of cAMP-dependent protein kinase (PKA). Several common mechanisms have been proposed for vasodilations mediated by PKG and PKA, including increased Ca^{2+} uptake into the SR and activation of BK$_{Ca}$ channels (Lincoln *et al.*, 1994; Robertson *et al.*, 1993; Taguchi *et al.*, 1995). It is possible that these two events are related through an increase in the frequency of Ca^{2+} sparks (Fig. 9). This increase in Ca^{2+} spark frequency may be a consequence of increased SR Ca^{2+} load due to enhanced SR Ca^{2+}-ATPase activity arising from the phosphorylation of the SR Ca^{2+}-ATPase regulatory protein, phospholamban, or a direct effect of these kinases on the ryanodine receptor (Fig. 9). Consistent with this hypothesis, parallel increases in Ca^{2+} spark and STOC frequency were observed in isolated cerebral arterial myocytes in response to compounds known to activate both PKA and PKG (Porter *et al.*, 1998). The parallel increases in Ca^{2+} spark and STOC frequency observed in this study likely represent a causal relationship between the two phenomena, as agents such as ryanodine and thapsigargin abolish both events. Further, cerebral arterial dilations to PKA activation are reduced by inhibitors of Ca^{2+} sparks (e.g., ryanodine). Additionally, an increase in STOC amplitude via a direct action on plasmalemmal BK_{Ca} channels by PKA and PKG represents a synergistic pathway to increase these outward potassium currents.

It has recently been demonstrated that activators of PKC decrease Ca^{2+} spark frequency in isolated cells from cerebral arteries (Bonev et $al.$, 1997). For example, activators of PKC, phorbol 12-myristate 13-acetate (PMA) and 1,2-sn-dioctanoyl-glycerol (DOG), decreased Ca^{2+} spark frequency by $>60\%$. However, 4α-PMA, an inactive phorbol ester analog, had no effect on Ca^{2+} spark frequency. PMA slightly decreased Ca^{2+} spark amplitude, with no effect on spatial spread or rate of decay. Consistent with 1:1 functional coupling between Ca^{2+} sparks and STOCs, these PKC activators also decreased STOC frequency. These effects of PKC activation on the Ca^{2+} spark/STOC pathway are independent of both changes in Ca^{2+} entry and SR Ca^{2+} content, suggesting that direct PKC phosphorylation of SR ryanodine receptors may underlie the decrease in spark frequency. It should also be noted that PKC has a direct inhibitory effect on the BK_{Ca} channel. This direct effect on the BK_{Ca} channel would amplify the effects of PKC on the Ca^{2+} spark/STOC pathway in cerebral artery myocytes.

6. SUMMARY

Potassium channels described in cerebral vascular smooth muscle (Kir, K_{ATP}, Kv, and BK_{Ca}) represent a heterogeneous group of proteins that play a pivotal role in the control of blood flow to the brain. The distinct molecular and biophysical properties of these K^+ channels allow for a wide array of physiological and pathophysiological signals to target these proteins. The net effect of these factors on membrane K^+ conductance, and ultimately membrane potential, dictates Ca^{2+} entry and thus the contractile state of cerebral vascular smooth muscle.

REFERENCES

Aguilar-Bryan, L., Clement, J. P. Gonzalez, G., Kunjilwar, K. Babenko, A., and Bryan, J., 1998, Toward understanding the assembly and structure of K_{ATP} channels, $Physiol.$ $Rev.$ **78**:227–245.

Aiello, E.A., M. P., Walsh, M. P., and Cole, W. C., 1995, Phosphorylation by protein kinase A enhances delayed rectifier K^+ current in rabbit vascular smooth muscle cells, $Am.$ $J.$ $Physiol.$ **268**:H926–H934.

Babenko, A. P., Aguilar-Bryan, L., and Bryan, J., 1998, A view of sur/Kir6.X, K_{ATP} channels, $Annu.$ $Rev.$ $Physiol.$ **60**:667–687.

Bonev, A. D., and Nelson, M. T., 1996, Vasoconstrictors inhibit ATP-sensitive K^+ channels in arterial smooth muscle through protein kinase C, $J.$ $Gen.$ $Physiol.$ **108**:315–323.

Bonev, A. D., Jaggar, J. H., Rubart, M., and Nelson, M. T., 1997, Activators of protein kinase C decrease Ca^{2+} spark frequency in smooth muscle cells from cerebral arteries, $Am.$ $J.$ $Physiol.$ **273**:C2090–C2095.

Bonnet, P., Rusch, N. J. and Harder, D. R., 1991, Characterization of an outward K^+ current in freshly dispersed cerebral arterial muscle cells, $Pflügers$ $Arch.$ **418**:292–296.

Bradley, K. K., Jaggar, J. H., Bonev, A. D., Heppner, T. J., Flynn, E. R., Nelson, M. T., and Horowitz, B., 1999, Kir2.1 encodes the inward rectifier potassium channel in rat arterial smooth muscle cells, $J.$ $Physiol.$ ($London$) **515**:639–651.

Brayden, J. E. and Nelson, M. T., 1992, Regulation of arterial tone by activation of calcium-dependent potassium channels, *Science* **256**:532-535.

Brayden, J. E., and Wellman, G. C., 1989, Endothelium-dependent dilation of feline cerebral arteries: Role of membrane potential and cyclic nucleotides, *J. Cereb. Blood Flow Metab.* **9**:256–263.

Cole, W. C., Clement-Chomienne, O., and Aiello, E. A., 1996, Regulation of 4-aminopyridine-sensitive, delayed rectifier K$^+$ channels in vascular smooth muscle by phosphorylation, *Biochem. Cell Biol.* **74**:439–447.

Dong, H., Waldron, G. J., Cole, W. C., and Triggle, C. R., 1998, Roles of calcium-activated and voltage-gated delayed rectifier potassium channels in endothelium-dependent vasorelaxation of the rabbit middle cerebral artery, *Br. J. Pharmacol.* **123**:821–832.

Doupnik, C. A., Dadidson, N., and Lester, H. A., 1995, The inward rectifier potassium channel family, *Curr. Opin. Neurobiol.* **5**:268–277.

Eckman, D. M., Wellman, G. C., Zaritsky, J. J., Schwarz, T. L., and Nelson, M. T., 1999, Expression of Kir 2.1 potassium channels is obligatory for strong inwardly rectifying K$^+$ currents and K$^+$-induced dilations in mouse cerebral arteries, *FASEB J.* **13**:A98.

Edwards, F. R., Hirst, G. D., and Silverberg, G. D., 1988, Inward rectification in rat cerebral arterioles; involvement of potassium ions in autoregulation, *J. Physiol. (London)* **404**:455-466.

Faraci, F. M., and Heistad, D. D., 1998, Regulation of the cerebral circulation: Role of endothelium and potassium channels, *Physiol. Rev.* **78**:53–97.

Faraci, F. M., and Sobey, C. G., 1998, Role of potassium channels in regulation of cerebral vascular tone, *J. Cereb. Blood Flow Metab.* **18**:1047–1063.

Gollasch, M., Wellman, G. C., Knot, H. J., Jaggar, J. H., Damon, D. H., Bonev, A. D., and Nelson, M. T., 1998, Ontogeny of local sarcoplasmic reticulum Ca^{2+} signals in cerebral arteries: Ca^{2+} sparks as elementary physiological events, *Circ. Res.* **83**:1104–1114.

Hirst, G. D., and Edwards, F. R., 1989, Sympathetic neuroeffector transmission in arteries and arterioles, *Physiol. Rev.* **69**:546–604.

Jan, L. Y., and Jan, Y. N., 1997, Cloned potassium channels from eukaryotes and prokaryotes, *Annu. Rev. Neurosci.* **20**:91–123.

Johnson, T. D., Marrelli, S. P., Steenberg, M. L., Childres, W. F., and Bryan, R. M. J., 1998, Inward rectifier potassium channels in the rat middle cerebral artery, *Am. J. Physiol.* **274**:R541–R547.

Kleppisch, T., and Nelson, M. T., 1995, ATP-sensitive K$^+$ currents in cerebral arterial smooth muscle: Pharmacological and hormonal modulation, *Am. J. Physiol.* **269**:H1634–H1640.

Knot, H. J., and Nelson, M. T., 1995, Regulation of membrane potential and diameter by voltage-dependent K$^+$ channels in rabbit myogenic cerebral arteries, *Am. J. Physiol.* **269**:H348–H355.

Knot, H. J., and Nelson, M. T., 1998a, Regulation of arterial diameter and wall [Ca^{2+}] in cerebral arteries of rat by membrane potential and intravascular pressure, *J. Physiol. (London)* **508**:199–209.

Knot, H. J., Zimmermann, P. A., and Nelson, M. T., 1996, Extracellular K$^+$-induced hyperpolarizations and dilatations of rat coronary and cerebral arteries involve inward rectifier K$^+$ channels, *J. Physiol. (London)* **492**:419–430.

Knot, H. J., Standen, N. B., and Nelson, M. T., 1998, Ryanodine receptors regulate arterial diameter and wall [Ca^{2+}] in cerebral arteries of rat via Ca^{2+}-dependent K$^+$ channels, *J. Physiol. (London)* **508**:211–221.

Lincoln, T. M., Komalavilas, P., and Cornwell, T. L., 1994, Pleiotropic regulation of vascular smooth muscle tone by cyclic GMP- dependent protein kinase, *Hypertension* **23**:1141–1147.

Nelson, M. T., 1992, Regulation of arterial tone by potassium channels, *Jpn. J. Pharmacol.* **58** (Suppl. 2):238P-242P.

Nelson, M. T., and Quayle, J. M., 1995, Physiological roles and properties of potassium channels in arterial smooth muscle, *Am. J. Physiol.* **268**:C799–C822.

Nelson, M. T., Patlak, J. B., Worley, J. F., and Standen, N. B., 1990, Calcium channels, potassium channels, and voltage dependence of arterial smooth muscle tone, *Am. J. Physiol.* **259**:C3–C18.

Nelson, M. T., Cheng, H., Rubart, M., Santana, L. F., Bonev, A. D., Knot, H. J., and Lederer, W. J., 1995, Relaxation of arterial smooth muscle by calcium sparks, *Science* **270**:633–637.

Nichols, C. G., and Lopatin, A. N., 1997, Inward rectifier potassium channels, *Annu. Rev. Physiol.* **59**:171–191.

Paulson, O. B., and Newman, E. A., 1987, Does the release of potassium from astrocyte endfeet regulate cerebral blood flow?, *Science* **237**:896–898.

Perez, G. J., Bonev, A. D., Patlak, J. B., and Nelson, M. T., 1999, Functional coupling of ryanodine receptors to K$_{Ca}$ channels in smooth muscle cells from rat cerebral arteries, *J. Gen. Physiol.* **113**:229–238.

Porter, V. A., Bonev, A., Knot, H. J., Heppner, T. J., Stevenson, A. S., Kleppisch, T., Lederer, M. R., and Nelson, M. T., 1998, Freqency modulation of Ca^{2+} sparks is involved in regulation of arterial diameter by cyclic nucleotides. *Am. J. Physiol.* **274:**C1346-C1355.

Quayle, J. M., McCarron, J. G., Brayden, J. E., and Nelson, M. T., 1993, Inward rectifier K^+ currents in smooth muscle cells from rat resistance-sized cerebral arteries, *Am. J. Physiol.* **265:**C1363–C1370.

Quayle, J. M., Bonev, A. D., Brayden, J. E., and Nelson, M. T., 1994, Calcitonin gene-related peptide activated ATP-sensitive K^+ currents in rabbit arterial smooth muscle via protein kinase A, *J. Physiol. (London)* **475:**9–13.

Quayle, J. M., Bonev, A. D., Brayden, J. E., and Nelson, M. T., 1995, Pharmacology of ATP-sensitive K^+ currents in smooth muscle cells from rabbit mesenteric artery, *Am. J. Physiol.* **269:**C1112–C1118.

Quayle, J. M., Nelson, M. T., and Standen, N. B., 1997, ATP-sensitive and inwardly rectifying potassium channels in smooth muscle, *Physiol. Rev.* **77:**1165–1232.

Robertson, B., 1997, The real life of voltage-gated K^+ channels: More than model behaviour, *Trends. Pharmacol. Sci.* **18:**474–483.

Robertson, B. E., and Nelson, M. T., 1994, Aminopyridine inhibition and voltage dependence of K^+ currents in smooth muscle cells from cerebral arteries, *Am. J. Physiol.* **267:**C1589–C1597.

Robertson, B. E., Schubert, R., Hescheler, J., and Nelson, M. T., 1993, cGMP-dependent protein kinase activates Ca-activated K channels in cerebral artery smooth muscle cells, *Am. J. Physiol.* **265:**C299–C303.

Somjen, G. G., 1979, Extracellular potassium in the mammalian central nervous system, *Annu. Rev. Physiol.* **41:**159–177.

Stryer L., 1995, *Biochemistry.* W. H. Freeman & Co., New York, pp. 325–360.

Taguchi, H., Heistad, D. D., Kitazono, T., and Faraci, F. M., 1995, Dilatation of cerebral arterioles in response to activation of adenylate cyclase is dependent on activation of Ca^{2+}-dependent K^+ channels, *Circ. Res.* **76:**1057–1062.

Vergara, C., Latorre, R., Marrion, N. V., and Adelman, J. P., 1998, Calcium-activated potassium channels, *Curr. Opin. Neurobiol.* **8:**321–329.

Wellman, G. C., and Quayle, J. M., 1997, ATP-sensitive potassium channels: Molecular structure and therapeutic potential in smooth muscle, *ID Research Alert* **2:**75–83.

Wellman, G. C., Quayle, J. M., and Standen, N. B., 1996, Evidence against the association of the sulphonylurea receptor with endogenous Kir family members other than K_{ATP} in coronary vascular smooth muscle, *Pflügers Arch.* **432:**355–357.

Chapter 27

The Role of Potassium Channels in the Control of the Pulmonary Circulation

Stephen Archer

1. INTRODUCTION

Potassium (K^+) channels serve many functions in controlling vascular tone in the pulmonary circulation. At the moment of birth, K^+ channels in pulmonary artery smooth muscle cells contribute to the transition of the pulmonary circulation from a high-resistance, low-flow circuit to a low-resistance, high flow circuit (Cornfield *et al.*, 1996; Reeve *et al.*, 1998; see Chapter 30). K^+ channels in pulmonary arterial smooth muscle cells also are intimately involved in the regional control of blood flow within the lung in response to regional variation in oxygenation by a mechanism called hypoxic pulmonary vasoconstriction (Archer *et al.*, 1986, 1993b; Hasunuma *et al.*, 1991a; Post *et al.*, 1992; Weir *et al.*, 1998; Yuan *et al.*, 1993) (Figs. 1 and 2). The pulmonary arterial smooth muscle cell's K^+ channels play a further role in amplifying (e.g., endothelin) (Salter *et al.*, 1998) or inhibiting [e.g. nitric oxide (NO)] (Archer *et al.*, 1994) vasoconstriction. Additionally, impaired expression of K^+ channels in the arterial media may contribute to the genesis of pulmonary hypertension (PHT) (see Chapter 40). Finally, K^+ channels are targets for drugs that may cause vascular diseases, such as the anorexigens aminorex or fenfluramine, which have caused outbreaks of PHT (see Chapter 41).

Of course, pulmonary vascular tone results from more than just the function of K^+ channels. It reflects the influence of innervation and rheology, plus the constrictor and dilator products of the endothelium. However, this chapter will take a reductionist approach to the pulmonary circulation and focus primarily on the contribution of K^+ channels in the plasma membrane of pulmonary arterial smooth muscle cells to the regulation of vascular tone in this circuit. The important roles of the many modulators of the pulmonary circulation other than K^+ channels are not discussed but are well

Stephen Archer ● Cardiology Division, Department of Medicine, University of Alberta, Edmonton, Alberta, Canada T6G 2B7.

Potassium Channels in Cardiovascular Biology, edited by Archer and Rusch. Kluwer Academic/Plenum Publishers, New York, 2001.

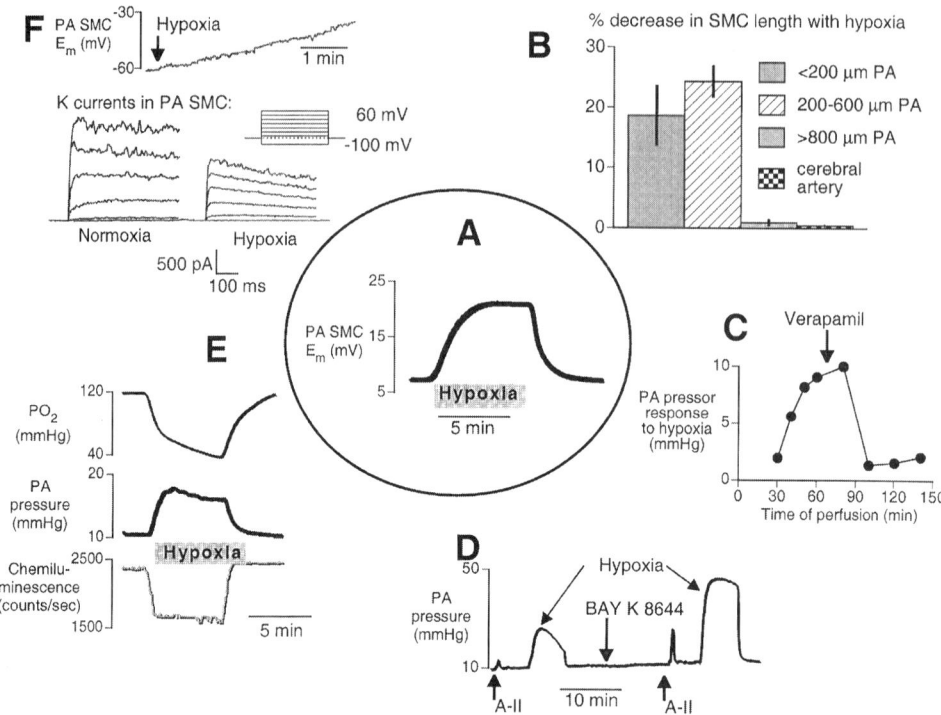

Figure 1. The mechanism of hypoxic pulmonary vasoconstriction: "A Tale of Two Channels." (A) Hypoxic pulmonary vasoconstriction in the isolated perfused rat lung model: The hypoxic response is preserved *ex vivo*, despite the absence of blood and innervation. (B) Isolated smooth muscle cells (SMC) from small pulmonary arteries (PA) contract in response to hypoxia — the work of Jane Madden. (C) Hypoxic pulmonary vasoconstriction is inhibited by the L-type Ca^{2+} channel blocker verapamil in the isolated rat lung — the work of Ivan McMurtry. (D) The Ca^{2+} channel agonist BAY K8644 enhances hypoxic pulmonary vasoconstriction in the isolated rat lung — the work of Ivan McMurtry. (E) Chemiluminescence from the isolated perfused rat lung is reduced by hypoxia prior to the rise in pulmonary artery pressure. (F) Hypoxia inhibits I_K and depolarizes E_m in freshly dispersed canine pulmonary arterial smooth muscle cells. (Reproduced, with permission, from Weir and Archer, 1995.)

described elsewhere (Weir *et al.*, 1992, 1994), nor is the importance of endothelial K^+ channels considered in this chapter (see Chapter 31). Rather, this chapter will briefly review pulmonary vascular physiology and then indicate the role of K^+ channels in the control of membrane potential (E_m), the mechanism of hypoxic pulmonary vasoconstriction, and the modulation of vascular tone by NO and endothelin-1.

2. SYNOPSIS OF PULMONARY VASCULAR PHYSIOLOGY — ADULT AND NEONATAL

The pulmonary circulation is engineered by evolution to convey the entire cardiac output of deoxygenated blood from the right heart, through a huge capillary bed, where it is oxygenated, to the left heart. Blood transits the pulmonary circuit in seconds at

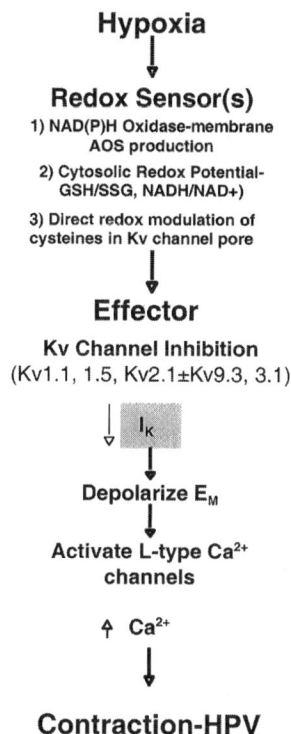

Figure 2. Proposed sensor and effector pathways of hypoxic pulmonary vasoconstriction (HPV). Hypoxic pulmonary vasoconstriction may be initiated by inhibition of a redox pathway (possibly the mitochondrial electron transport chain or NADPH oxidase). This leads to a reduced state (evident in the cytosolic redox pool) which inhibits several types of Kv channels in pulmonary arterial smooth muscle cells. The membrane depolarization leads to activation of the voltage-gated Ca^{2+} channel, Ca^{2+} influx, and vasoconstriction.

pressures which are only 25% of those found in the systemic circulation. Whereas systemic systolic/diastolic pressures are 120/80 mm Hg, systolic/diastolic pressures in the pulmonary artery are approximately 30/20 mm Hg. This low pulmonary vascular resistance results in part from a unique vascular architecture. Large, proximal elastic arteries serve as conduits to muscular, resistance arteries. The resistance arteries, where much of the regulation of regional blood flow occurs, have much less smooth muscle than similar-sized arteries in other organs. This arrangement permits the pulmonary circulation's thin-walled arterioles and capillaries to extract oxygen from the atmosphere and eliminate the carbon dioxide produced by aerobic metabolism in the body.

The pulmonary circulation does not begin life as a low-resistance/high-flow vascular bed, quite the contrary. *In utero*, the fetus's need for systemic oxygenation is met by the maternal circulation through the placenta. The liquid-filled fetal lung receives little of the cardiac output because the thick-walled and constricted arteries result in high pulmonary vascular resistance. Most of the fetus's systemic venous return to the right ventricle is shunted away from the pulmonary circulation through a large artery, the ductus arteriosus, to the aorta. The pulmonary arteries of the fetus are thick-walled and resemble systemic arteries in adults. At the moment of birth, the mechanical act of inspiration, augmented by the resulting rise in inspired oxygen tension (FiO_2), causes a dramatic vasodilation in the fetal pulmonary circulation. Over a period of days to weeks, vasodilation permits the transition to a low-resistance circulation with thin-walled arteries. This transition must occur for survival as an air-breathing animal.

3. HYPOXIC PULMONARY VASOCONSTRICTION

Hypoxic pulmonary vasoconstriction matches ventilation to perfusion by adjusting the segmental pulmonary vascular resistance in response to changes in local alveolar O_2 content (Bradford and Dean, 1894; von Euler and Liljestrand, 1946). This mechanism optimizes systemic oxygen saturation in conditions of regional hypoventilation, such as pneumonia and atelectasis. Consider the patient with left upper lobe pneumonia. In the absence of hypoxic pulmonary vasoconstriction (e.g., if it were inhibited by a vasodilator drug, such as nifedipine), the pneumonic lobe, containing hypoxic gas, would remain perfused. The perfusion of this hypoxic segment would add hypoxemic blood to an otherwise O_2-rich pulmonary venous return, constituting a "shunt." However, in normal subjects, hypoxic pulmonary vasoconstriction causes the blood to be diverted away from the pneumonic lobe to other, better ventilated lobes.

Since its modern description (von Euler and Liljestrand, 1946), many features of hypoxic pulmonary vasoconstriction have been elucidated (Weir and Archer, 1995). Hypoxic pulmonary vasoconstriction is more a response to airway partial pressure of oxygen (PO_2) than to mixed venous PO_2 and thus is strongest in the resistance arteries which are near the gas-exchange surfaces. Although the strength of hypoxic pulmonary vasoconstriction is modulated by many products of the endothelium (Archer et al., 1989b; Brashers et al., 1988; Hasunuma et al., 1991b; Robertson et al., 1990) and the activity of the autonomic nervous system, it exists in isolated lungs, free of innervation and blood (McMurtry et al., 1976). Hypoxic pulmonary vasoconstriction can even be demonstrated in isolated arteries and smooth muscle cells (Madden et al., 1985, 1992) (Fig. 1). The hypoxic response is strongest in small muscular pulmonary arteries, those third to fifth-generation vessels which are $\sim 200 \, \mu$m in diameter (Kato and Staub, 1966; Shirai et al., 1991, 1986) (Figs. 1B and 3). This permits regional control of blood flow without elevation of global pulmonary vascular resistance and the consequent increase in right-heart work. When the lung faces global hypoxia, as occurs with ascent to altitude, the entire circulation becomes hypoxic and pulmonary vascular resistance rises. If persistent, this may lead to chronic hypoxic pulmonary hypertension.

4. VASODILATOR AND VASOCONSTRICTOR AGONISTS

With the exception of the constrictor response to hypoxia, which is unique to the pulmonary circulation, most other agonists affect vascular tone in the pulmonary circulation much in the same way as in the systemic arterial tree. Furthermore, the signal transduction pathways used by these vasoconstrictor or vasodilator substances largely mirror those found in systemic vascular beds. Thus, this chapter will deal only with the dilator NO and the constrictor endothelin-1, which have a significant effect on

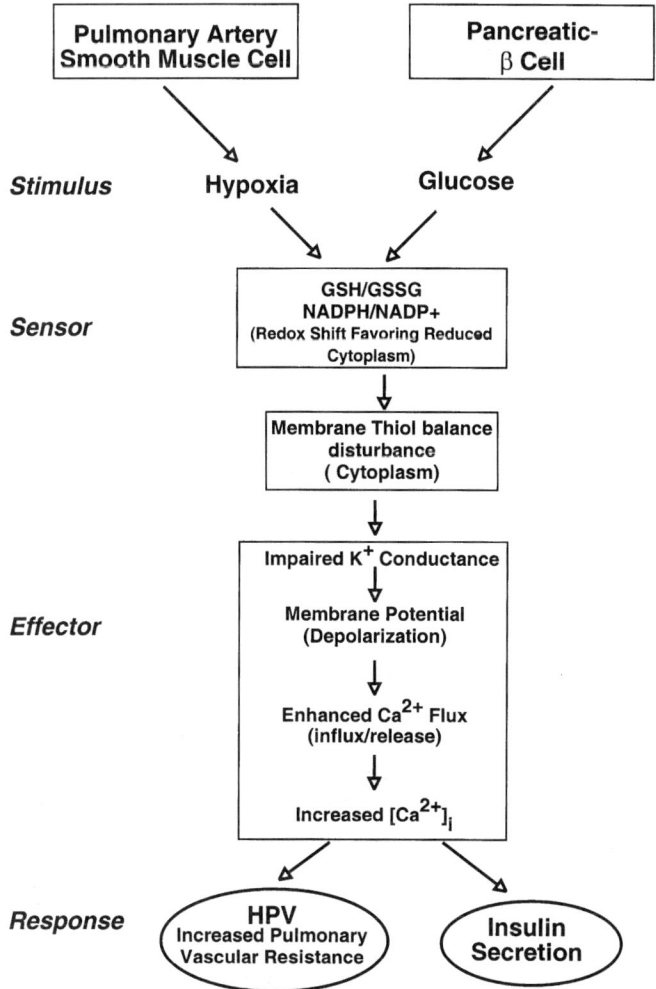

Figure 3. The "redox hypothesis" for regulation of pulmonary vascular tone This figure suggests a common redox mechanism for control of E_m in pulmonary arterial smooth muscle cells and pancreatic β-cells. In this model, changes in redox chemistry induced by alveolar hypoxia and blood glucose, respectively, impair K$^+$ conductance, leading to membrane depolarization and either hypoxic pulmonary vasoconstriction (HPV) or insulin secretion. (Reproduced, with permission, from Archer *et al.*, 1986.)

pulmonary vascular resistance, at least in part, through their ability to promote the opening and closing of K$^+$ channels.

5. PULMONARY VASODILATION AND VASCONSTRICTION—THE OPENING AND CLOSING OF K$^+$ CHANNELS

In the pulmonary circulation, as in other vascular tissues, pharmacological inhibitors of K$^+$ channels are used to characterize the contribution of a class of K$^+$

channels to the control of vascular tone. Using this approach, high-conductance Ca^{2+}-sensitive (BK_{Ca}), voltage-gated (Kv), or ATP-sensitive (K_{ATP}) channels have been studied. However, none of the inhibitors is completely specific. In the pulmonary circulation, drugs that inhibit whole-cell K^+ current (I_K) tend to be vasoconstrictors, whereas K^+ channel openers are vasodilators. The link between vascular tone and the K^+ channel is usually assumed to be the E_m (Fig. 2). Because K^+ channels set the E_m in smooth muscle cells, they control the open-state probability of the voltage-gated, L-type Ca^{2+} channel and thus regulate Ca^{2+} influx. Depolarization results in increased opening of the Ca^{2+} channel, and the resultant rise in cytosolic Ca^{2+}, $[Ca^{2+}]_i$, activates the contractile apparatus, leading to vasoconstriction (Archer et al., 1986; Nelson et al., 1990; Post et al., 1992).

What drugs are available to inhibit K^+ channels? Charybdotoxin (100 nM) is a peptide component of scorpion venom that is a relatively selective BK_{Ca} channel inhibitor (Garcia et al., 1991, 1995), although it also inhibits several Shaker K^+ channels (Garcia et al., 1991; Garcia-Calvo et al., 1994). Charybdotoxin blocks the external channel by interaction of basic amino acids with fixed negative charges in the channel. It is similarly effective in binding the channel in both open and closed conditions (see Chapter 13) Iberiotoxin is a more selective inhibitor of BK_{Ca} channels (Candia et al., 1992). Tetraethylammonium (TEA) is a cationic blocker that interferes with the outer pore of the BK_{Ca} channel and competes with charybdotoxin for this aspect of the channel (Sugg et al., 1990). At low concentrations (1–5 mM), TEA is a preferential BK_{Ca} inhibitor (Gelband and Hume, 1992), though at higher concentrations it blocks other types of K^+ channels. 4-Aminopyridine (4-AP) inhibits Kv channels by blocking the channel pore (Bouchard and Fedida, 1995). K_{ATP} channels are inhibited by glibenclamide (Clapp and Gurney, 1991; Clapp et al., 1993), which is also nonspecific at concentrations beyond 1 μM. For example, in the pulmonary circulation, glibenclamide may inhibit thromboxane A_2 receptors (Kaye et al., 1997). Drugs which open K_{ATP} channels cause vasodilation (Chang et al., 1992; Triggle et al., 1992). The K_{ATP} channel opener aprikalim induces relaxation of the pulmonary artery (Magnon et al., 1994). However, this does not indicate that the E_m is set by K_{ATP} channels. Rather, the opening of most classes of K^+ channels will permit K^+ efflux along its ionic concentration gradient, and the E_m level of the smooth muscle cell will correspondingly move toward the K^+ equilibrium potential of approximately −90 mV.

6. PHARMACOLOGICAL EVIDENCE FOR A ROLE FOR K^+ CHANNELS IN DETERMINING PULMONARY VASCULAR RESISTANCE

The first evidence that K^+ channels contribute to the regulation of pulmonary vascular resistance was the observation that some K^+ channel blockers cause pulmonary vasoconstriction. In 1980, TEA was reported to contract rabbit main pulmonary arteries and also to produce membrane depolarization (Haeusler and Thorens, 1980). In the same year, it was demonstrated that 4-AP potentiates KCl-induced contractions of isolated pulmonary arteries (Hara et al., 1980). This suggests that a basal efflux of K^+, via tonically active TEA- and 4-AP-sensitive K^+ channels, maintains the smooth muscle cell in a relatively hyperpolarized state. Conversely, inhibition of these K^+

channels depolarizes and constricts the pulmonary artery. The rudiments of this concept were first enunciated in 1986 as part of the "redox theory," which attempted to explain both hypoxic pulmonary vasoconstriction and insulin secretion by the pancreatic β-cell through a redox regulation of membrane potential (Archer *et al.*, 1986) (Fig. 3). Initially, this theory was vague as to the physical correlate of K$^+$ conductance. Subsequently, it became evident that reduction–oxidation chemistry was regulating, by means which still remain controversial.

The Kv inhibitor 4-AP increases normoxic pulmonary vascular resistance in the isolated perfused rat lung (Hasunuma *et al.*, 1991a) whereas TEA has little effect and glibenclamide has no effect on normoxic or hypoxic pulmonary vascular resistance (Hasunuma *et al.*, 1991a). The fact that Kv channel blockers cause vasoconstriction suggests that a basal outward I_K tonically contributes to the low pulmonary vascular resistance. This basal level of K$^+$ channel activation and smooth muscle cell polarization appears to make the pulmonary bed a normoxically dilated bed, a modification of the hypothesis of "normoxic pulmonary vasodilatation" initially proposed by Weir (1978).

7. PULMONARY ARTERIAL SMOOTH MUSCLE CELLS EXPRESS SEVERAL K$^+$ CHANNEL TYPES

Pharmacological and electrophysiological studies suggest that pulmonary arterial smooth muscle cells contain various types of K$^+$ channels, including BK$_{Ca}$ (Archer *et al.*, 1994, 1996), Kv (previously referred to as delayed rectifier channels) (Archer *et al.*, 1996; Post *et al.*, 1995; Sheehan *et al.*, 1994; Yuan, 1995), and K$_{ATP}$ channels (Clapp and Gurney, 1991; Clapp *et al.*, 1993; Wiener *et al.*, 1991).

Kv channels are largely responsible for setting the resting E_m in pulmonary arterial smooth muscle cells (Archer *et al.*, 1996; Post *et al.*, 1992; Sheehan *et al.*, 1994; Smirnov *et al.*, 1994; Yuan, 1995) (Fig. 4), although chloride channels may also contribute (Zhao *et al.*, 1998). The resting E_m of pulmonary arterial smooth muscle cells from resistance arteries is approximately -50 mV (Archer *et al.*, 1996; Post *et al.*, 1992; Sheehan *et al.*, 1994; Smirnov *et al.*, 1994; Yuan, 1995). At this E_m the L-type Ca^{2+} channels are maintained primarily in their closed state, promoting relaxation. Conversely, Kv channel blockers cause pulmonary vasoconstriction largely through their ability to depolarize E_m and thus open L-type Ca^{2+} channels (McMurtry, 1985; McMurtry *et al.*, 1976; Tolins *et al.*, 1986).

BK$_{Ca}$ channels are ubiquitous in vascular smooth muscle cells, including pulmonary arterial smooth muscle cells (Archer *et al.*, 1986, 1994; Farrukh *et al.*, 1998) (Fig. 5). Although their properties in different vascular cell types vary, all members of this family are activated by increases in $[Ca^{2+}]_i$ and are also voltage-dependent (see Chapters 4, 13, and 42). In pulmonary arterial smooth muscle cells, BK$_{Ca}$ channels therefore tend to act as a brake on vasoconstriction by restoring E_m toward basal levels.

K$_{ATP}$ channels open when intracellular ATP levels fall (see Chapter 15). Although K$_{ATP}$ channels are expressed in the plasma membranes of healthy, normoxic pulmonary arterial smooth muscle cells (Clapp *et al.*, 1993), their effects on vascular tone are most evident when there is hypertension (Pinheiro and Malik, 1992) or anoxia (Wiener *et al.*,

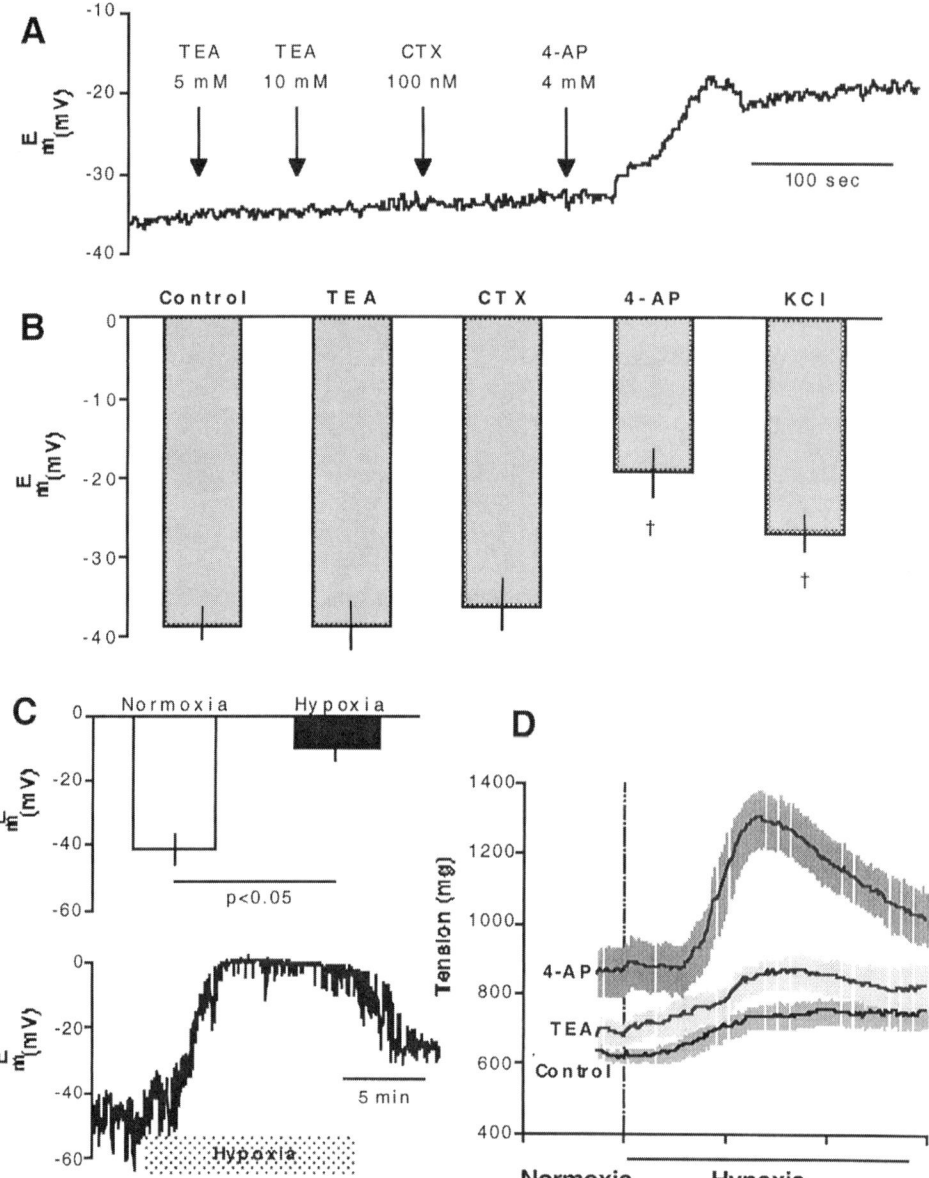

Figure 4. Contribution of Kv channels to resting membrane potential (E_m) in pulmonary arterial smooth muscle cells. (A) Representative trace of E_m in smooth muscle cells from rat resistance pulmonary arteries: note that 4-aminopyridine (4-AP), but not tetraethylammonium (TEA) or charybdotoxin (CTX), depolarizes the membrane. (B) Effects of TEA, CTX, 4-AP, and KCl on E_m. Data are plotted as means $\pm$ SEM; †$p < 0.0001$, value differs from control. (C) Representative recording (*bottom*) and E_m data (*top*; means $\pm$ SEM, $n = 5$) showing that hypoxia depolarizes Kv cells from rat resistance pulmonary arteries. Cells were studied using amphotericin-perforated whole-cell current-clamp technique. (D) Effects of TEA, 4-AP, and hypoxia on tone in isolated rat resistance pulmonary artery rings. All rings started from a resting tension of ~600 mg. 4-AP (10 mM) constricted the rings whereas TEA (10 mM) had no effect on basal tone. Hypoxia caused a monotonic constriction of resistance rings, which was enhanced by 4-AP but not TEA. Data are plotted as means $\pm$ SEM; *$P < 0.0001$, curve differs from control. (Reproduced with permission from Archer *et al.*, 1996.)

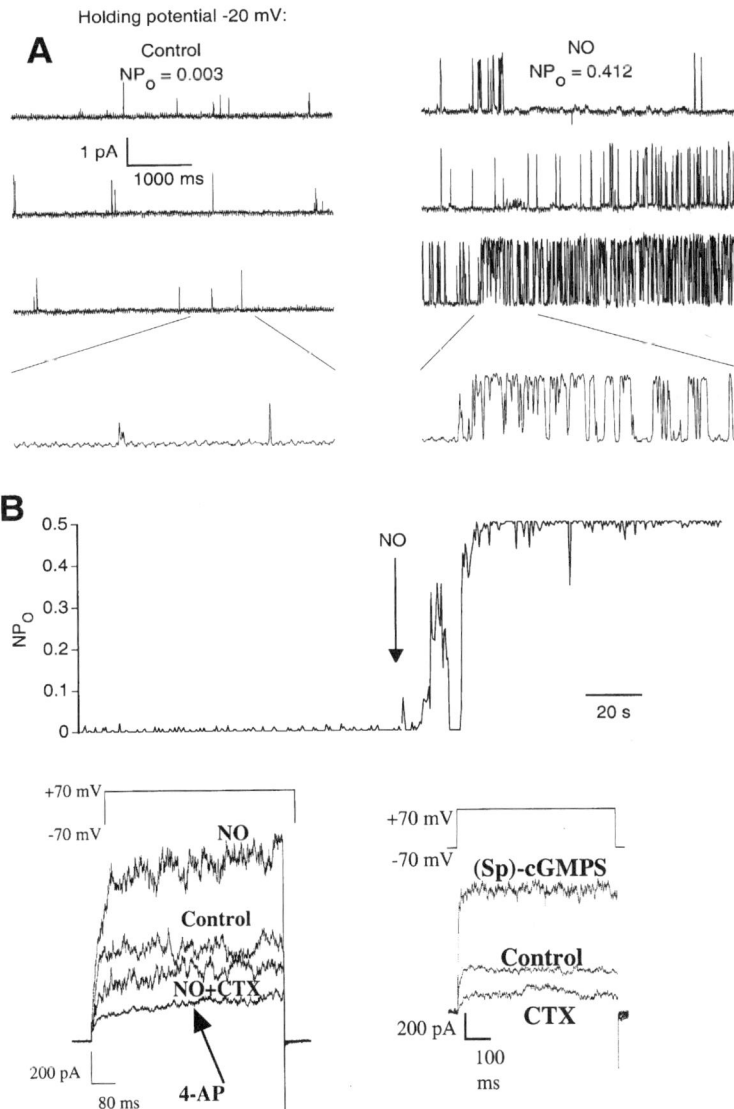

Figure 5. NO activates BK$_{Ca}$ channels by a cGMP protein phosphokinase-dependent mechanism. (A) Control data are shown in the left panel and the effects of NO on unitary BK$_{Ca}$ currents are shown in the right panel. Traces are sequential. The bottom trace in each panel shows the identified segment in greater detail to illustrate the duration of opening. Note that the holding potential is −20 mV. The BK$_{Ca}$ channel was not active until NO was applied to the cell (right panel). (B) Time course of open probability of BK$_{Ca}$ channel before and after NO application (indicated by arrow). *Bottom left*: Representative whole-cell currents produced by depolarization to +70 mV in freshly dispersed rat pulmonary arterial smooth muscle cells. Note the spontaneous spiking of the control current, an indicator of the opening of BK$_{Ca}$ channels. NO increases I_K, and this response is inhibited by charybdotoxin (CTX). 4-AP does reduce some of the basal current. *Bottom*: Representative tracing showing I_K produced by depolarization to +70 mV immediately (control) and 20 min after membrane rupture with a patch pipette containing (Sp)-cGMPS, a specific protein phosphokinase G activator. (Sp)-cGMPS increased I_K, and CTX reversed this effect, implying that it was mediated by BK$_{Ca}$ channel activation. (Reproduced, with permission, from Archer *et al.*, 1994. Copyright 1994 National Academy of Sciences, U.S.A.)

1991). Clapp and Gurney (1992) found that with physiological levels of ATP in the patch pipette (1 mM), pulmonary arterial smooth muscle cells had a mean E_m of -55 mV, as measured using current-clamp techniques in dialyzed cells. When ATP was omitted from the patch pipette solution, the cell hyperpolarized to -70 mV as K_{ATP} channels were permitted to open. Under these conditions, glibenclamide elicits a depolarizing response. However, K_{ATP} current is unlikely to be involved in maintaining the low pressure of the normoxic pulmonary vasculature because glibenclamide does not increase pulmonary vascular resistance in normal lungs (Archer et al., 1996; Yuan, 1995). Nonetheless, K_{ATP} channels are activated during anoxia, and this hyperpolarizes pulmonary arterial smooth muscle cells, reducing the constrictor response and moderating anoxic pulmonary vasoconstriction (Wiener et al., 1991). Drugs that open K_{ATP} channels, such as minoxidil or pinacidil (Chang et al., 1992; Mellemkjaer et al., 1989; Weir et al., 1976), or activate BK_{Ca} channels, such as dehydroepiandrosterone and NO, are also pulmonary vasodilators (Archer et al., 1994; Farrukh et al., 1998).

8. "ELECTRICAL REMODELING" OF THE PULMONARY CIRCULATION

The expression of K^+ channels in pulmonary arterial smooth muscle cells is highly plastic, and protein levels can change rapidly in response to numerous physiological (e.g., development, changes in PO_2) and pathological (e.g., hypertension) stimuli. For example, in the fetal pulmonary circulation, the BK_{Ca} channel appears to contribute to resting E_m (Cornfield et al., 1996; Reeve et al., 1998) whereas in the adult pulmonary arterial smooth muscle cells, E_m is largely determined by Kv channels (Archer et al., 1996; Post et al., 1995; Sheehan et al., 1994; Yuan, 1995). This important developmental change in the pattern of K^+ channel expression indicates the potential of the pulmonary circulation to undergo dynamic "electrical remodeling" (see Chapter 30). Perhaps this is not surprising given the relatively short half-life of some Kv channels (approximately 4 h) for kv 1.5 mRNA and the fact that membrane depolarization can selectively depress the synthesis of certain Kv channels (Levitan et al., 1995). There is evidence that K^+ channel expression is inhibited in several forms of PHT, including PHT induced experimentally by exposure to chronic hypoxia (Osipenko et al., 1998; Smirnov et al., 1994) and human primary PHT (see Chapter 40). However, it remains uncertain whether the relationship between impaired expression of Kv channels and development of PHT is cause or effect.

9. K^+ CHANNEL INHIBITION CONTRIBUTES TO ENDOTHELIN-INDUCED VASOCONSTRICTION

Endothelin-1 is a pulmonary vasoconstrictor (Chatfield et al., 1991; Ivy et al., 1994; MacLean et al., 1994; Oparil et al., 1995). Circulating endothelin-1 levels are increased in humans with PHT (Stewart et al., 1991). Expression of endothelin-1 in the endothelium is also increased in patients with the rare, often fatal disease, primary PHT (Giaid et al., 1993). It is possible that the local production of endothelin-1 may contribute to the elevation of pulmonary vascular resistance under these conditions, and Oparil *et al.*, (1995) have hypothesized that endothelin-1 may be intimately involved with eliciting hypoxic pulmonary vasoconstriction. Recently, it has been shown that endothelin-1 causes contraction, at least in part, by inhibition of I_K in pulmonary arterial smooth muscle cells (Salter and Kozlowski, 1998; Salter *et al.*, 1998; Shimoda *et al.*, 1998). Although the precise channels involved have not been identified, there is consensus that they are Kv channels, possibly including Kv1.5 (Salter *et al.*, 1998). It would not be unexpected if vasoconstrictors such as angiotensin II and endothelin-1 exerted part of their action by inhibiting K$^+$ channels, given the ability of these constrictors to enhance the activity of certain protein phosphokinase C isoforms (Damron *et al.*, 1993; Furukawa *et al.*, 1992; Kasemsri and Armstead, 1997; Takenaka *et al.*, 1993). Substances that activate protein phosphokinase C tend to inhibit K$^+$ channels (Damron, *et al.*, 1993; Kasemsri and Armstead, 1997). Protein phosphokinase C activators also promote opening of L-type Ca2$^+$ channels (Furukawa *et al.*, 1992; Takenaka *et al.*, 1993). Both effects enhance vasoconstriction. Activation of other kinases, such as protein phosphokinase G (Archer *et al.*, 1994), causes vasodilation by phosphorylating and activating BK$_{Ca}$ channels.

10. NO and cGMP ARE BK$_{Ca}$ CHANNEL OPENERS IN PULMONARY ARTERIAL SMOOTH MUSCLE CELLS

NO is a short-lived radical, originally described as endothelium-derived relaxing factor (EDRF) (Furchgott and Zawadzki, 1980). NO is produced by the vascular endothelium, including pulmonary arterial endothelium (Archer and Cowan, 1991). This lipophilic substance rapidly diffuses to pulmonary arterial smooth muscle cells, where it activates guanylate cyclase, thereby increasing levels of the second messenger cGMP (Ignarro *et al.*, 1986; Murad, 1986). Basal NO production contributes to low pulmonary vascular resistance. Rats treated with NO synthase inhibitors (Hampl *et al.*, 1993; Hampl and Mejsnar, 1986) and mice in which the gene for endothelial NO synthase has been disrupted (Steudel *et al.*, 1997) develop mild PHT. The ability of NO to cause pulmonary vasodilation depends, in part, on the cGMP-mediated activation of BK$_{Ca}$ channels. Much of NO's effect on the BK$_{Ca}$ channels is mediated by a cGMP-dependent protein kinase. The first direct evidence that K$^+$ channel activation contributed to the vasodilator activity of NO and cGMP in the pulmonary circulation was the observation that increasing cGMP levels in pulmonary arterial smooth muscle cells increased I_K and hyperpolarized the membrane. Both the relaxation of rat pulmonary artery rings and the increase in I_K were attenuated by charybdotoxin (Fig. 5) (Archer *et al.*, 1994, 1996). This vasodilator mechanism for NO is shared by cerebral (Robertson *et al.*, 1993) and coronary arteries (Tanaguchi *et al.*, 1993a). This pathway may also contribute to NO-induced vasodilation in human pulmonary arteries (Fig. 6). The NO–protein phosphokinase G–BK$_{Ca}$ cascade for vasodilation is developmentally conserved and seems to be crucial for mediating the fall in pulmonary vascular resistance that occurs at birth in response to increased oxygen levels (Cornfield *et al.*, 1996) (see Chapter 30).

There is controversy as to whether the effect of NO on K$^+$ channels occurs directly (Bolotina *et al.*, 1994) or through guanylate cyclase and cGMP-dependent protein

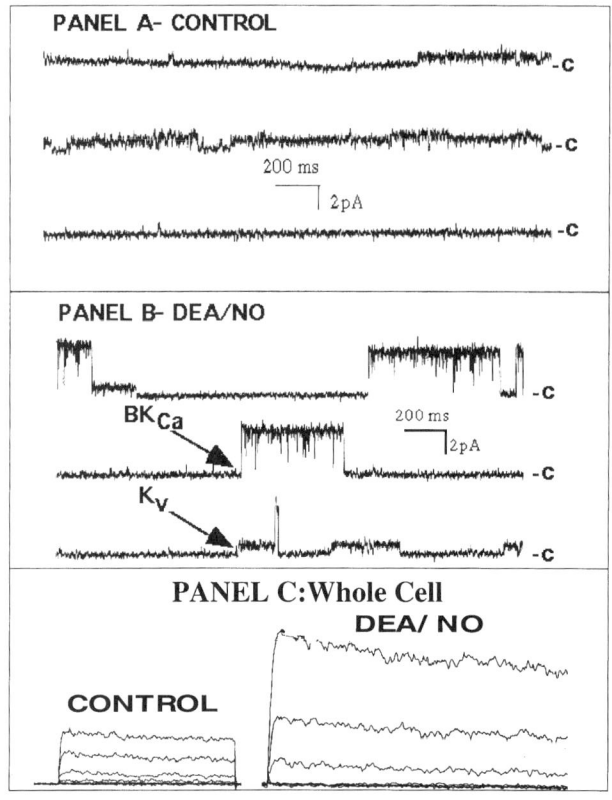

Figure 6. NO activates K$^+$ channels in human pulmonary arterial smooth muscle cells from resistance arteries. (A) Cell-attached patches of human pulmonary arterial smooth muscle cells from resistance arteries at $+20$ mV patch potential in drug-free solution display frequent opening of Kv channels, whereas BK$_{Ca}$ channels rarely open. (B) The NO donor diethylamine/NO (DEA/NO) increases BK$_{Ca}$ channel activity, whereas the Kv$_{DR}$ channel open-state probability is unchanged. (C) DEA/NO increases I_K in whole-cell recordings from enzymatically dispersed human pulmonary arterial smooth muscle cells derived from primary cultures of cells obtained from a patient undergoing lobectomy. Membrane potential was held at -70 mV and increased to $+70$ mV by 10-mV depolarization steps of 500-ms duration.

phosphokinase (Archer *et al.*, 1994; Robertson *et al.*, 1993; Tanaguchi *et al.*, 1993b). Both scenarios likely occur, although the cGMP mechanism appears to be more common and perhaps occurs at lower doses of NO. Direct channel activation by NO may occur via redox modulation of key sulfhydryl groups in the channel, as has been demonstrated to occur with diamide, oxidized glutathione, and other oxidants (Reeve *et al.*, 1995). However, the evidence supporting cGMP-dependent mechanisms is robust, being derived from the independent work of many laboratories on various types of arteries. First, inhibitors of guanylate cyclase diminish the vasodilator effects of NO. Second, the direct activation of protein phosphokinase G mimics the effects of NO on the BK$_{Ca}$ channel. The involvement of protein phosphokinase G in channel opening suggests that this mechanism should depend on phosphorylation, the hallmark of kinase activation. Third, inhibition of endogenous cGMP (type V) phosphodiesterase (using zaprinast) or inhibition of endogenous phosphatases [with okadaic acid, an inhibitor of protein phosphatases 1 and 2A (Ashizawa, 1989)] also activates BK$_{Ca}$ currents in pulmonary arterial smooth muscle cells. This suggests that enhancing BK$_{Ca}$ channel phosphorylation may be a useful strategy in the development of new vasodilator therapies (Hampl *et al.*, 1995). In fact, NO is not the only agent that causes pulmonary vasodilation by activating BK$_{Ca}$ channels. A variety of hormones, including dehydroepiandrosterone, also act via this mechanism (Farrukh *et al.*, 1998). In addition to its action as an activator of BK$_{Ca}$ channels, NO can also increase Kv current in pulmonary arterial smooth muscle cells (Zhao *et al.*, 1997).

11. HYPOXIC PULMONARY VASOCONSTRICTION: "A TALE OF TWO CHANNELS"

A major advance in our understanding of hypoxic pulmonary vasoconstriction resulted from the use of patch-clamp techniques to directly assess the role of K$^+$ channels (Post *et al.*, 1992). We now recognize that hypoxic pulmonary vasoconstriction results from the integrated action of Ca^{2+} and K$^+$ channels in the pulmonary arterial smooth muscle cell's plasmalemma (Fig. 1) (Weir and Archer, 1995). Hypoxic pulmonary vasoconstriction relies on Ca^{2+} influx through L-type, voltage-gated Ca^{2+} channels. Inhibitors of voltage-gated Ca^{2+} channels, including nifedipine, reduce hypoxic pulmonary vasoconstriction (Fig. 1C) (McMurtry *et al.*, 1976) whereas the response is augmented by the dihydropyridine agonist BAY K8644 (Fig. 1D) (McMurtry, 1985; Tolins *et al.*, 1986). As discussed earlier, the open-state probability of the L-type Ca^{2+} channel is regulated by E_m, which, in turn, is determined by the activity of various K$^+$ channels. The L-type Ca^{2+} channels are mainly inactive at resting E_m and progressively activate at more positive voltages (Franco-Obregon and Lopez-Barneo, 1996). Hypoxia induces membrane depolarization (Fig. 4C) and increases [Ca^{2+}]$_i$ by inhibiting Kv channels in pulmonary arterial smooth muscle cells (Archer *et al.*, 1993b, 1998; Post *et al.*, 1992; Yuan *et al.*, 1993, 1998). This mechanism is similar to the response to hypoxia in another O$_2$-sensitive, cardiovascular tissue, the carotid body. The E_m of smooth muscle cells in the ductus arteriosus (Tristani-Firouzi *et al.*, 1996) is also controlled by O$_2$-responsive K$^+$ channels—although in this case it is oxygenation rather than deoxygenation that inhibits the Kv channel. In each O$_2$-sensitive tissue, a change in PO$_2$ inhibits I_K, depolarizes E_m, activates L-type Ca^{2+} channels, and culminates in a tissue-specific effect, including hypoxic pulmonary vasoconstriction in the pulmonary circulation, neural discharge from the carotid body, and constriction of the ductus arteriosus (Weir and Archer, 1995).

Hypoxia does not inhibit I_K in systemic arterial smooth muscle cells (Post *et al.*, 1992), perhaps explaining the lack of hypoxic vasoconstriction in these arteries (Madden *et al.*, 1992; Vadula *et al.*, 1993). Indeed, hypoxia dilates systemic arteries (Hampl *et al.*, 1994), in part by activating K$_{ATP}$ channels (Dart and Standen, 1995; Daut *et al.*, 1990; Gasser *et al.*, 1993); see Chapter 38) or BK$_{Ca}$ channels (Gebremedhin *et al.*, 1994); see Chapter 42).

12. REDOX REGULATION OF PULMONARY VASCULAR K$^+$ CHANNELS

Although this chapter does not directly deal with O$_2$ sensing, it is important to mention that K$^+$ channels in pulmonary arterial smooth muscle cells are redox-regulated. Many K$^+$ channels contain cysteines and other sulfhydryl-rich amino acids at critical functional sites. Oxidation and reduction of these channels may open and close them, respectively (Archer *et al.*, 1986; Duprat *et al.*, 1995; Lee *et al.*, 1994a; Park *et al.*, 1995a,b; Reeve *et al.*, 1995; Weir *et al.*, 1998). There remains controversy as to whether PO$_2$-responsive K$^+$ channels are directly redox-regulated independent of cytosolic factors (Conforti and Millhorn, 1997; Jiang and Haddad, 1994), or whether a pathway involving some cytosolic redox messenger system exists. Candidate "sensors" include the reduced/oxidized redox couples glutathione (GSH/GSSG) or nicotinamide adenine dinucleotide (NADH/NAD$^+$) (Fig. 3) (Archer *et al.*, 1993a, 1986, 1993b, 1989a; Reeve *et al.*, 1995; Weir and Archer, 1995). The ratios of these freely diffusible couples are dynamically modulated by hypoxia and inhibitors of complex 1 of the mitochondrial electron transport chain, including rotenone (Archer *et al.*, 1993a,b). We have suggested that the low resting pulmonary vascular resistance in the pulmonary circulation results from a basal, tonic oxidation of K$^+$ channels by endogenous production of activated O$_2$ species, such as radicals and peroxides (Figs. 2 and 3) (Archer *et al.*, 1986). The unresolved debate regarding a direct versus an indirect mechanism for redox gating of K$^+$ channels is reminiscent of the NO controversy, discussed earlier. Ultimately, both direct and indirect redox sensor mechanisms may exist. Perhaps one mechanism is more commonly used under physiological conditions whereas the other occurs with more severe redox shifts, as in ischemia–reperfusion or in response to certain drugs. These potential redox regulatory systems can be studied *in vitro* (Peterson *et al.*, 1994, 1996).

13. MOLECULAR IDENTIFICATION OF Kv CHANNELS THAT SET E_m AND RESPOND TO HYPOXIA IN PULMONARY ARTERIAL SMOOTH MUSCLE CELLS

There is a gradual movement from a pharmacological system for classifying K$^+$ channels to a molecular scheme. This is possible because many K$^+$ channels have now been cloned. Kv channels in mammals (Kv1, Kv2, Kv3, and Kv4) are closely homologous to those cloned from *Drosophila* (*Shaker*, *Shab*, *Shaw*, and *Shal*, respectively) (Chandy and Gutman, 1993; Salkoff and Jegla, 1995; Xu *et al.*, 1995), see Chapters 1 and 18). Pulmonary arterial smooth muscle cells contain all the *Shaker* Kv channels except for Kv1.4 and also display the *Shab* channel Kv2.1 (Archer *et al.*, 1998) (Fig. 7).

Dissection of the individual K$^+$ channel types that contribute to I_K has traditionally rested on the use of pharmacological inhibitors. Classical assessment of Kv channel identity, using pharmacology and biophysics, is not precise enough to allow differentiation among Kv channel subtypes (e.g., Kv1.1 vs. Kv1.5) (Chandy and Gutman, 1993). This is particularly true in mammalian cells, in which many types of K$^+$ channels are simultaneously active, creating an ensemble current (Fig. 8). Expression of these cloned

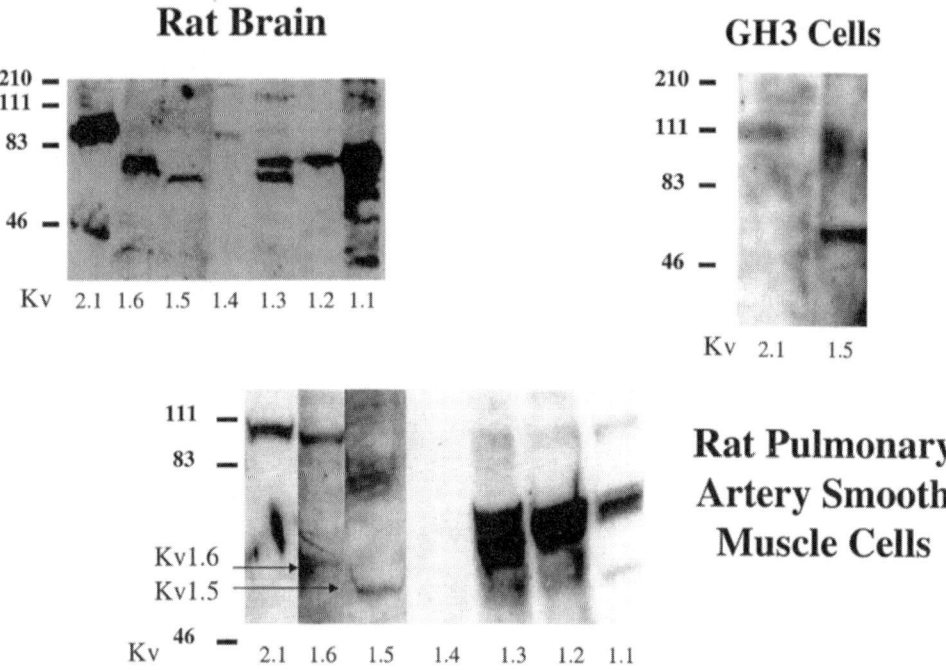

Figure 7. Kv2.1 and all Kv1 channels (except Kv1.4) are present in pulmonary arterial smooth muscle cells. The specificity of the antibodies was assessed in a clonal pituitary line (GH3) known to express abundant Kv1.5 but little Kv2.1 channel protein. The sensitivity of anti-Kv antibodies was also assessed in a more complex tissue that is known to express most Kv channel types, rat brain. All the Kv channel types were present in brain, with molecular weights as follows: Kv1.1, a broad band centered at 83 kDa; Kv1.2, 80 kDa; Kv3.1, two bands at ~66 and 80 kDa; Kv1.4, 90 kDa; Kv1.5, 66 kDa; Kv1.6, a broad band at 75–80 kDa, Kv2.1, a broad band at ~110 kDa.. Rat pulmonary arterial smooth muscle cells contain most Kv1 family channels as well as Kv2.1. Molecular weights are indicated on the vertical axis. (Reproduced, with permission, from Archer *et al.*, 1998.)

K$^+$ channels in *Xenopus* oocytes and mammalian expression systems [e.g., Chinese hamster ovary (CHO) cells], has defined their pure attributes (Grissmer *et al.*, 1994), albeit without necessarily revealing the important modulatory effects of β-subunits and heterotetramer formation with active or silent α-subunits. The properties of these clonal Kv channels can be compared with the profile of the O$_2$-inhibited I_K in pulmonary arterial smooth muscle cells (Archer *et al.*, 1998), allowing one to construct a relatively short list of which K$^+$ channels may underlie a particular current. It appears that several types of K$^+$ channel, as indicated by distinct single-channel conductances, are inhibited by hypoxia (Fig. 9A). This indicates the need for a molecular approach to the identification of the K$^+$ channel(s) involved in hypoxic pulmonary vasoconstriction. In this regard, the candidate current is slowly inactivating and voltage-gated and is inhibited by 4-AP, but not by charybdotoxin or glibenclamide (Archer *et al.*, 1996; Post *et al.*, 1995; Sheehan *et al.*, 1994; Yuan, 1995). Thus, inward rectifiers (e.g., GIRK1) and rapidly inactivating channels (e.g., Kv1.4 and Kv4.3) are unlikely candidates because of their discordant current morphology (Archer, 1996; Archer *et al.*, 1998). Kv1.1, Kv1.5, Kv2.1, and Kv3.1 are on the short list of candidate Kv channels because, similarly to

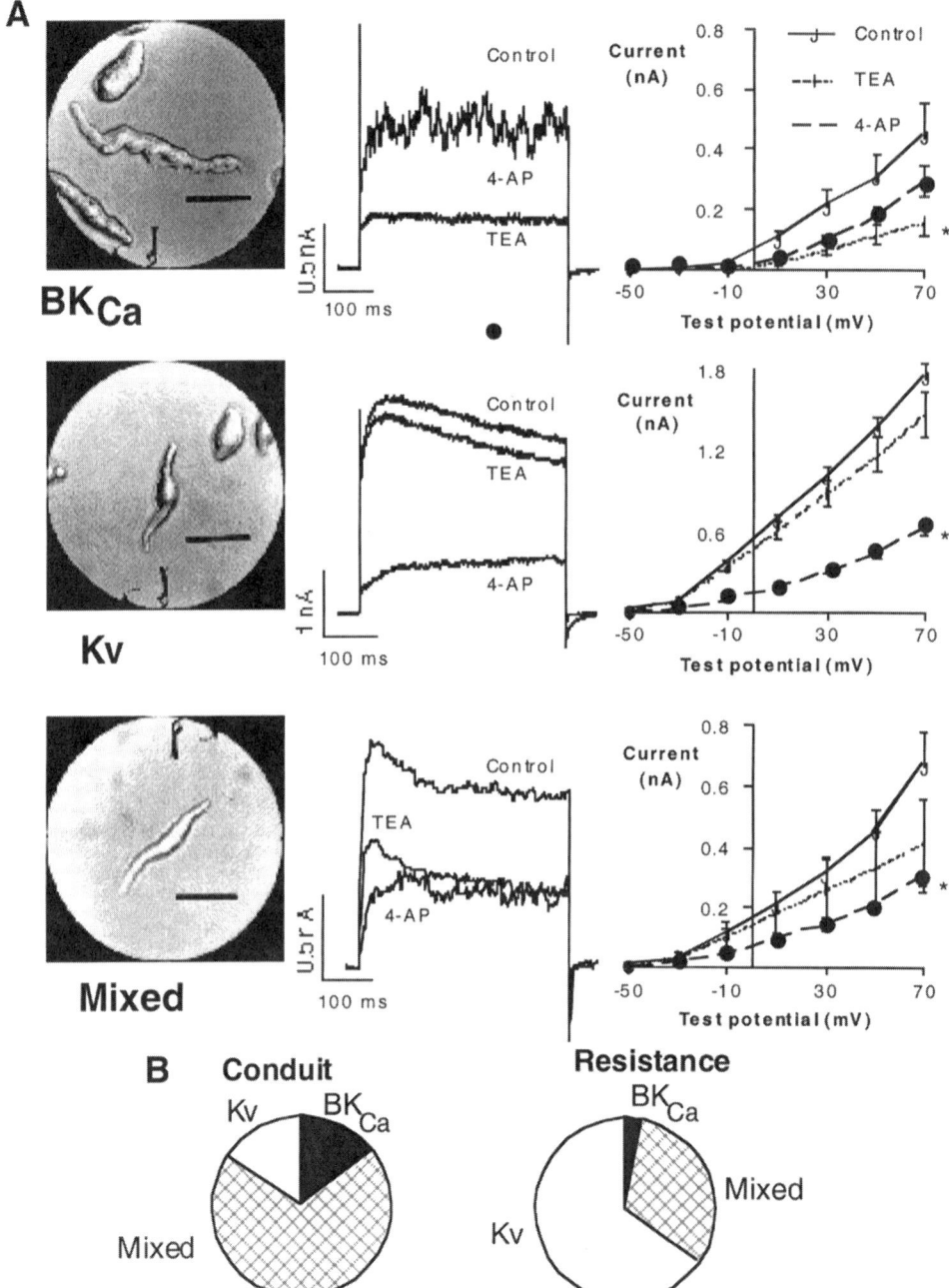

Figure 8. Diversity of K$^+$ channel expression in rat pulmonary artery. Three distinct cell phenotypes exist in large, conduit pulmonary artery, and these phenotypes correspond to the cells identified by current density and whole-cell pharmacology as expressing primarily BK$_{Ca}$, Kv, and mixed channel populations. (A) *Left column*: Note that the cell expressing primarily BK$_{Ca}$ current is larger and the cell expressing predominantly Kv current is smaller with a discrete perinuclear bulge. The cell showing mixed channel types has an intermediate morphology. The scale bar represents 25 μ m. *Center column*: Morphology of the macroscopic

the hypoxia-inhibited K$^+$ channel, they are voltage-gated, slowly inactivating, and inhibited by 4-AP, but not by charybdotoxin (Bouchard and Fedida, 1995; Grissmer *et al.*, 1994). Several other members of the Kv1 family are less likely candidates because they are inhibited by charybdotoxin; these include Kv1.2 and Kv1.3 (Grissmer *et al.*, 1994) and Kv1.6 (Garcia *et al.*, 1994)).

Recently, we have focused on two candidate channels, Kv2.1 and Kv1.5, inhibition of which may contribute to hypoxic pulmonary vasoconstriction. Kv1.5 is a member of the *Shaker* family that is found in vascular myocytes (Archer *et al.*, 1998; Overturf *et al.*, 1994; Yuan *et al.*, 1998), including pulmonary arterial smooth muscle cells. Kv1.5 is also expressed in bronchial smooth muscle, where it may control the E_m and modulate smooth muscle tone (Adda *et al.*, 1996). Kv1.5 is inhibited by 4-AP and has a single-channel conductance of roughly 17 pS (Bouchard and Fedida, 1995). Its turnover in the plasma membrane is rapid, measured in hours (Takimoto *et al.*, 1993), and thus, Kv1.5 may be dynamically up- or downregulated under a variety of pathophysiological conditions, including hypertension. Kv1.5 is also susceptible to redox regulation (Duprat *et al.*, 1995).

The second candidate channel, Kv2.1, is sensitive to both TEA and 4-AP (Shi *et al.*, 1994; Taglialatela *et al.*, 1993; Trimmer, 1993). Kv2.1 is found in rat pulmonary arterial smooth muscle cells (Archer *et al.*, 1998; Patel *et al.*, 1997; Yuan *et al.*, 1998). Phosphorylation of the cytoplasmic region of the carboxy terminus of Kv2.1 shifts the activation threshold of the channel to a more positive range (Murakoshi *et al.*, 1997). Kv2.1 is also redox-modulated. Its amino terminus participates in transitions leading to activation through interactions involving reduced cysteines that can be modulated from the cytoplasmic compartment (Pascual *et al.*, 1997), consistent with the redox theory (Fig. 3) (Archer, *et al.*, 1986, 1993b). The inhibition of Kv2.1 may account, in part, for the ability of 4-AP to inhibit I_K and depolarize pulmonary arterial smooth muscle cells (Fig. 9B,C).

14. "IMMUNOELECTROPHARMACOLOGY"

We recently devised a form of "immunoelectropharmacology" to assess whether Kv1.5 and Kv2.1 are involved in the control of E_m and whether their inhibition by hypoxia initiates vasoconstriction. The term immunoelectropharmacology indicates

---

currents obtained by conventional whole-cell technique (example shown is obtained by stepped depolarization from -70 to $+70$ mV). Note that the small, noisy, spiky current in BK$_{Ca}$ cells is inhibited by TEA, but not by 4-AP. In contrast, the large, smooth-profile current in Kv cells is much more sensitive to 4-AP than TEA. In mixed cells, 4-AP inhibits approximately half the total current, leaving the noisy, TEA-inhibitable portion intact. *Right column.* Current–voltage curves (mean $\pm$ SEM; $n = 7$ BK$_{Ca}$, 5 mixed, and 20 Kv cells, studied using conventional whole-cell technique). Current was elicited by 10-mV depolarization steps from -70 to $+70$ mV. *$P < 0.05$, curve differs from control. (B) The prevalence of the three cell types (identified by their morphology on light microscopy) was studied by light microscopy in 542 cells from conduit pulmonary arteries of 6 rats and 386 cells from resistance pulmonary arteries of 5 rats. Note the relatively greater occurrence of BK$_{Ca}$ cells in conduit arteries and the greater prevalence of Kv cells in resistance arteries. Mixed cell types are found in great numbers throughout the circulation. (Reproduced, with permission, from Archer *et al.*, 1996; *note that in the original publication Kv cells were called "delayed rectifier-predominant" or K_{DR} cells*).

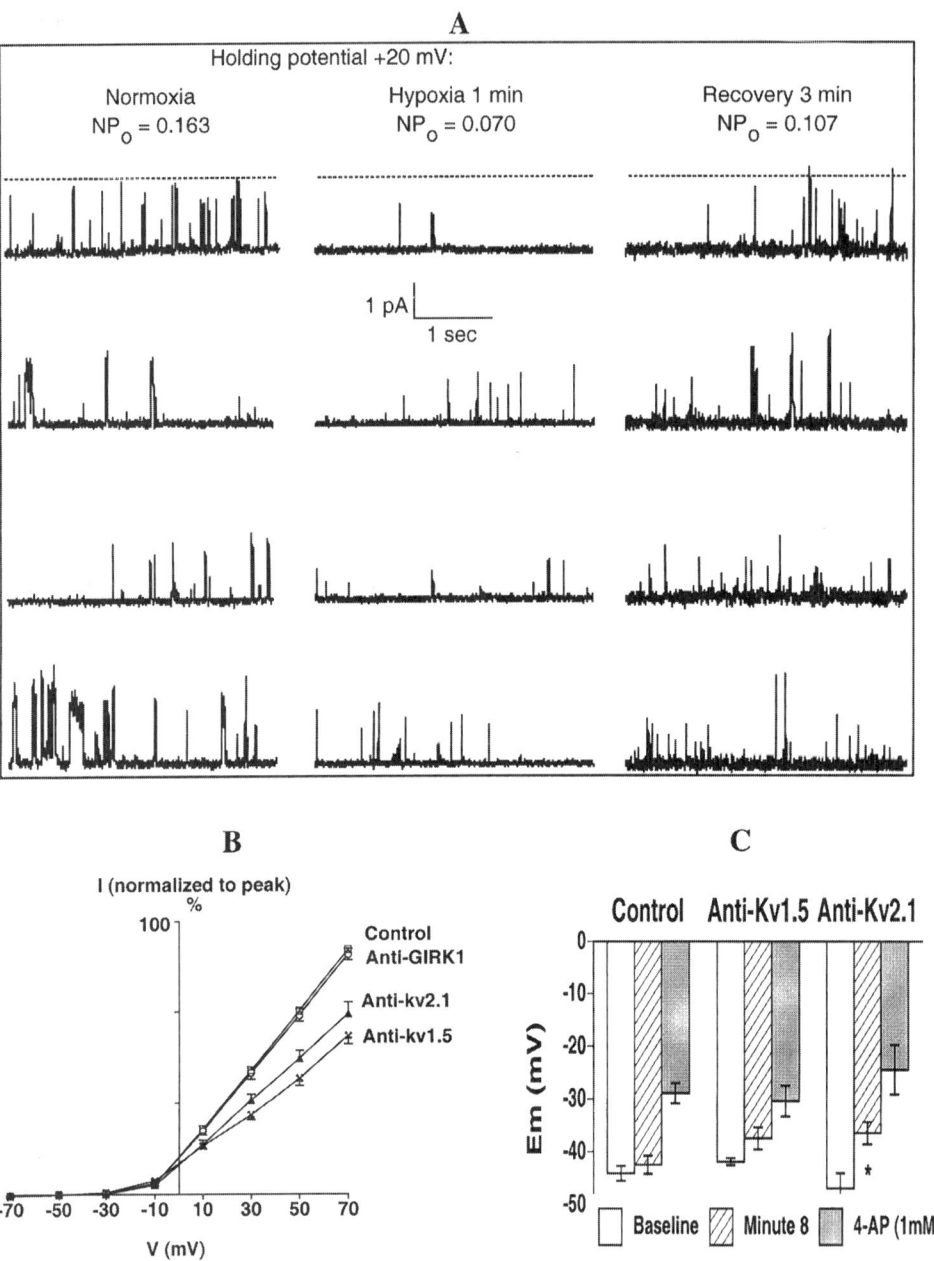

Figure 9. Molecular identification of hypoxia-inhibited K$^+$ channels. (A) Representative single-channel traces from a cell-attached patch of a rat pulmonary arterial smooth muscle cells (holding potential, +20 mV). The pipette contained TEA and niflumic acid to inhibit BK$_{Ca}$ and chloride channels, respectively. The dotted lines (upper panels) indicate the amplitude of the opening of a 37-pS Kv channel. The activity of this channel decreases in hypoxia. Note that there are several smaller channel amplitudes evident in this patch that are also hypoxia sensitive but which have not been fully characterized. Hypoxia decreases the open-state probability (NP$_o$) for all channels within 1 min, and this effect partially reverses within 3 min of the return to normoxia. (Reproduced, with permission, from Archer *et al.*, 1996.) (B) Kv1.5 and Kv2.1 contribute to the

that antibodies are used to functionally inhibit the channels against which they are targeted. The concept is that the specificity of the antibody for a clonal Kv channel will exceed that which could be achieved pharmacologically. In patch-clamp experiments, intracellular administration of an anti-Kv2.1 antibody, directed against the intracellular, carboxy terminus of the Kv2.1 α-subunit, partially inhibits I_K and depolarizes E_m (Fig. 9). Anti-Kv2.1 antibodies also elevate resting tension and diminish 4-AP-induced vasoconstriction in membrane-permeabilized pulmonary artery rings. Likewise, anti-Kv1.5 inhibits I_K, and, although it does not significantly depolarize the pulmonary arterial smooth muscle cells, it reduces the constriction caused by hypoxia and 4-AP (Archer *et al.*, 1998). Thus, Kv2. 1 may be an important determinant of resting E_m in pulmonary arterial smooth muscle cells from resistance arteries, and inhibition of both Kv2.1 and Kv1.5 may contribute to the initiation of hypoxic pulmonary vasoconstriction. The results of this work also demonstrate that, at least some types of anti-Kv antibodies can be useful tools to dissect the single-channel components of I_K in mammalian cells.

15. NEW CONCEPTS IN THE MOLECULAR REGULATION OF Kv CHANNELS

Four new ideas should be considered when assessing the role of K$^+$ channels in the pulmonary circulation. They introduce a degree of conceptual complexity, which may provide the theoretical basis for the observed interspecies and interorgan variability in K$^+$ channel function. These recent discoveries belie a "one K$^+$ channel, one current profile" hypothesis. However, these concepts may explain how similar K$^+$ channel building blocks (α-subunits) are differentially assembled or modulated to achieve diverse functions, including neurotransmission, blood pressure regulation, and cardiac pacemaking, in life forms ranging from prokaryotes to mammals (see Chapter 1).

15.1. Heterotetramers

The pore-forming structure of K$^+$ channels is tetrameric; it can be a homotetramer (e.g., all four α-subunits are Kv1.5) or a heterotetramer, composed of diverse α-subunits

---

pulmonary arterial smooth muscle cell's I_K. Intracellular administration of antibodies against Kv2.1 or Kv1.5 via the patch pipette partially inhibited I_K in rat pulmonary arterial smooth muscle cells. Anti-GIRK1 was used as a control, as this channel is not expressed in the pulmonary artery. Thus, the absence of an effect of anti-GIRK1 on current indicates that the presence of a nonspecific antibody in the patch pipette is not a cause of spurious current inhibition. The use of immunoelectropharmacology may help to attribute components of I_K to particular Kv channels. (Reproduced, with permission, from Archer *et al.*, 1998.) (C) Kv2.1 contributes to E_m in pulmonary arterial smooth muscle cells. Intracellular administration of anti-Kv2.1, but not anti-Kv1.5, depolarizes E_m in rat pulmonary arterial smooth muscle cells from resistance arteries. Anti-Kv2.1 antibodies depolarize E_m, indicating a significant contribution of Kv2.1 to resting E_m in these cells. The additional depolarization produced by 4-AP presumably reflects the involvement of multiple types of Kv channels, in addition to Kv2.1, in setting E_m. (Reproduced, with permission, from Archer *et al.*, 1998.)

from the same gene family (e.g., two Kv1.5 and two Kv1.1 α-subunits). The heterotetrameric channels may have kinetics and pharmacological sensitivities that are hybrids of those of their components. Alternatively, one subunit may determine the overall channel function, as occurs with dominant-negative channels produced by site-directed mutagenesis. This tetrameric diversity is one way in which Kv channel function can be varied from tissue to tissue while still using the same α-subunits as "building blocks" (Babila et al., 1994; Lee et al., 1994b; Russell et al., 1994; Yu et al., 1996). There is not yet evidence that heterotetramer formation in pulmonary arterial smooth muscle cells serves an important physiological role in vivo, and direct examination of this possibility is required.

15.2. β-Subunits

Three β-subunit genes (Kvβ1.1, Kvβ2, and Kvβ3) are expressed in pulmonary arterial smooth muscle cells (Yuan et al., 1998). However, their function is unknown. In expression systems, Kvβ1.1 induces inactivation of all Kv1 channels, except Kv1.6 (Heinemann et al., 1996). Kvβ2 shifts the activation threshold of Kv1.5 by approximately -10 mV but has no effect on channel deactivation (Heinemann et al., 1996). Tissue-specific differences in expression of Kvβ subunits could confer tissue-specific electrophysiological traits. However, the importance of Kvβ subunits to physiology or pathophysiology in vivo remains unproven.

15.3. Electrically Silent Kv Channels

Kv channels in families Kv5–9 do not conduct current themselves but, when associated with other functional Kv channel proteins, form heterotetramers with altered activation and inactivation properties (Patel et al., 1997; Salinas et al., 1997). For example, Patel et al. (1997) identified Kv2.1 and a novel Shab-like subunit, Kv9.3, in rat pulmonary arterial smooth muscle cells. Kv9.3 encodes an electrically silent subunit that associates with Kv2.1 and modulates its biophysical properties. The Kv2.1/9.3 heterotetramer activates in the voltage range of the resting E_m of pulmonary artery myocytes (Patel et al., 1997) whereas Kv2.1 alone activates at more positive potentials. In pulmonary arterial smooth muscle cells, Kv9.3 may confer O_2-responsiveness to Kv2.1 and contribute to its physiological importance; however, further study is required to clarify and confirm its regulatory function.

15.4. Regional Diversity of K^+ Channel Expression

Although the net response of the pulmonary circulation to acute hypoxia is to increase pulmonary vascular resistance, not all parts of the pulmonary vasculature behave in the same way. Proximal pulmonary arteries respond to hypoxia much like the aorta, with a biphasic change in tone. A small initial constriction is followed by a sustained relaxation of tension below baseline (Bennie *et al.*, 1991; Jin *et al.*, 1992). In contrast, the small, muscular, distal pulmonary arteries respond to hypoxia with a monophasic constriction (Fig. 4) (Archer *et al.*, 1996). How is it that the response of the pulmonary circulation can differ between these vascular segments? The finding that hypoxia inhibits Kv channels (Weir and Archer, 1995) while activating BK$_{Ca}$ channels (Gebremedhin, *et al.*, 1994) led us to assess whether the regional variability in response to hypoxia might be the consequence of regional variability in expression of Kv and BK$_{Ca}$ channels. Because hypoxia tends to activate BK$_{Ca}$ current by raising $[Ca^{2+}]_i$, in smooth muscle cells, cells expressing a preponderance of BK$_{Ca}$ channels would be expected to hyperpolarize and relax in response to hypoxia. In contrast, cells expressing primarily Kv channels would tend to constrict.

In this respect, freshly dispersed rat pulmonary arterial smooth muscle cells manifest three phenotypically and electrophysiologically distinct cell populations, coexisting side by side (Fig. 8). (Archer *et al.*, 1996; Michelakis *et al.*, 1997). Moreover, a patch-clamp "survey" reveals differential distribution of cells with distinctly different electrophysiological profiles along the length of the pulmonary arterial circulation. In conduit pulmonary arteries, we found cells that functionally, predominantly express BK$_{Ca}$ channels and, consequently, display a small, noisy I_K which is largely inhibited by low-dose TEA (5 mM), but not 4-AP. Conversely, the resistance arteries are enriched in cells that primarily manifest Kv channels and thus have high current density, greater sensitivity to low-dose 4-AP (2 mM), and minimal response to TEA (Archer *et al.*, 1996; Michelakis *et al.*, 1997) (Fig. 8). A third cell type, found in both conduit and resistance pulmonary arteries, has intermediate current density and its I_K is inhibited equally by 4-AP and TEA. These three electrophysiological cell types have a morphological correlate evident by light microscopy (Fig. 8) (Archer *et al.*, 1996).

As predicted, the BK$_{Ca}$-predominant cells respond to hypoxia with an increase in I_K whereas the current in Kv-predominant cells is inhibited by hypoxia (Archer *et al.*, 1996). The consequence of these regional differences in K$^+$ channel expression in the pulmonary circulation is manifest in regional variations in vascular tone, with preferential localization of hypoxic pulmonary vasoconstriction to resistance arteries. Teleologically, this would permit regional ventilation/perfusion matching without an obligatory increase in global pulmonary vascular resistance, which would otherwise occur if the entire arterial tree constricted homogeneously in response to hypoxia. Thus, the pulmonary arteries are mosaics of smooth muscle cell types which gradually vary their electrical characteristics as one progresses geographically from conduit to resistance sites. The genetic or environmental basis for this diversity remains uncertain.

16. CHRONIC HYPOXIC PULMONARY HYPERTENSION

K$^+$ channels are also involved in the chronic response to hypoxia. Chronic hypoxic pulmonary hypertension is paradoxically associated with diminished intensity of acute hypoxic pulmonary vasoconstriction (Isaacson *et al.*, 1994; McMurtry *et al.*, 1978). This attenuation of hypoxic pulmonary vasoconstriction is specific; other vasoconstrictor stimuli, for example, angiotensin II, cause enhanced constriction in chronically hypoxic lungs. It may be the result of diminished sensitivity of a pulmonary vascular O_2 sensor (Archer *et al.*, 1993b), diminished expression of hypoxia-responsive K^+ channels, or both (see Chapter 40). Indeed, pulmonary arterial smooth muscle cells from chronically hypoxic rats are somewhat depolarized (Smirnov and Aaronson, 1994), much as occurs in fetal pulmonary arterial smooth muscle cells, which also are exposed to chronic hypoxia (Cornfield *et al.*, 1996). In the case of these fetal cells, it recently has been found that Kv function increases at birth with the improved oxygenation (Reeve *et al.*, 1998).

Several groups have found that chronic hypoxic pulmonary hypertension is associated with inhibition of I_K in pulmonary arterial smooth muscle cells (Osipenko *et al.*, 1998; Smirnov *et al.*, 1994). On the basis of the pharmacology of the inhibited current, Smirnov *et al.* (1994) suggested that it is conducted by "delayed rectifier channels," now more commonly called Kv channels. This finding has been confirmed (Osipenko *et al.*, 1998), and the mechanism by which it occurs is a subject of active investigation. One interesting observation which may explain the loss of I_K in chronic hypoxic pulmonary hypertension is that sustained depolarization itself can lead to inhibition of the synthesis of Kv channels (Levitan *et al.*, 1995). Systemic hypertension and the associated left ventricular hypertrophy also are associated with the selective downregulation of specific Kv channels, such as Kv4.3, in the rat myocardium (Takimoto *et al.*, 1997; see Chapter 37). Moreover, Wang *et al.* (1997) recently used a model in which pulmonary arterial smooth muscle cells were cultured for several days under conditions of low PO_2, and, in this *in vitro* model, Kv1.5 channel expression was also inhibited by hypoxia.

While it seems highly likely that certain Kv channels are underexpressed in chronic hypoxic pulmonary hypertension, this may not be the sole mechanism for the impaired hypoxic pulmonary vasoconstriction. Chronic hypoxia may cause changes in redox chemistry that promote a "reduced" state in the pulmonary arterial smooth muscle cells. Specifically, in chronic hypoxia, cytosolic levels of reduced GSH are increased and production of activated O_2 species (radicals and peroxides) is diminished (unpublished data). An acute increase in reduced electron donors is postulated to be a sensor for hypoxic pulmonary vasoconstriction (Figs. 2 and 3) (Archer *et al.*, 1986, 1993b; Reeve, *et al.*, 1995). Rats exposed to chronic hypoxia may thus have impaired acute hypoxic pulmonary vasoconstriction because the normal small redox shift caused by acute hypoxia is masked in the already highly reduced environment evoked by chronic hypoxia. Certainly, animals that are genetically adapted to life at high altitude, such as the yak, have thin-walled pulmonary arteries and display minimal hypoxic pulmonary vasoconstriction (Durmowicz *et al.*, 1993). Further work is required to sort out the relative contributions of impaired K^+ channel expression and altered K^+ channel function to the impaired acute hypoxic pulmonary vasoconstriction seen in chronic hypoxic pulmonary hypertension.

REFERENCES

Adda, S., Fleischmann, B., Freedman, B., Yu, M., Hay, D. W., and Kotlikoff, M., 1996, Expression and function of voltage dependent potassium channel genes in human airway smooth muscle, *J. Biol Chem.* **271**:13239–13243.

Archer, S. L., 1996, Diversity of phenotype and function of vascular smooth muscle cells, *J. Lab. Clin. Med.* **127**:524–529.

Archer, S. L., and Cowan, N. J., 1991, Measurement of endothelial cytosolic calcium concentration and nitric oxide production reveals discrete mechanisms of endothelium-dependent pulmonary vasodilation, *Circ. Res.* **68**:1569–1581.

Archer, S., Will, J., and Weir, E., 1986, Redox status in the control of pulmonary vascular tone, *Herz* **11**:127–141.

Archer, S. L., Nelson, D. P., and Weir, E. K., 1989a, Simultaneous measurement of oxygen radicals and pulmonary vascular reactivity in the isolated rat lung, *J. Appl. Physiol.* **67**:1903–1911.

Archer, S. L., Tolins, J. P., Reye, L., and Weir, E. K., 1989b, Hypoxic pulmonary vasoconstriction is enhanced by inhibition of the synthesis of an endothelium derived relaxing factor, *Biochem. Biophys. Res. Commun.* **164**:1198–1205.

Archer, S. L., Huang, J., Post, J., Hume, J., and Weir, E., 1993a, t-Butyl hydroperoxide and gluta-thione modulate an outward K$^+$ current in rat pulmonary vascular smooth muscle cells, *Circulation* **88**:I-143.

Archer, S. L., Huang, J., Henry, T., Peterson, D., and Weir, E. K., 1993b, A redox based oxygen sensor in rat pulmonary vasculature, *Circ. Res.* **73**:1100–1112.

Archer, S. L., Huang, J. M. C., Hampl, V., Nelson, D. P., Shultz, P. J., and Weir, E. K., 1994, Nitric oxide and cGMP cause vasorelaxation by activation of a charybdotoxin-sensitive K channel by cGMP-dependent protein kinase, *Proc. Natl. Acad. Sci. U.S.A.* **91**:7583–7587.

Archer, S. L., Huang, J. M. C., Reeve, H. L., Hampl, V., Tolarova, S., Michelakis, E. D., and Weir, E. K., 1996, Differential distribution of electrophysiologically distinct myocytes in conduit and resistance arteries determines their response to nitric oxide and hypoxia., *Circ. Res.* **78**:431–442.

Archer, S. L., Souil, E., Dinh-Xuan, A. T., Schremmer, B., Mercier, J. C., El Yaagoubi, A., Nguyen-Huu, L., Reeve, H. L., and Hampl, V., 1998, Molecular identification of the role of voltage-gated K$^+$ channels, Kv1.5 and Kv2.1, in hypoxic pulmonary vasoconstriction and control of resting membrane potential in rat pulmonary artery myocytes, *J. Clin. Invest.* **101**:2319–2330.

Ashizawa, N., 1989, Relaxing action of okadaic acid, a black sponge toxin on arterial smooth muscle, *Biochem. Biophys. Res. Commun.* **162**:971–976.

Babila, T., Moscucci, A., Wang, H., Weaver, F. E., and Koren, G., 1994, Assembly of mammalian voltage-gated potassium channels: Evidence for an important role of the first transmembrane segment, *Neuron* **12**:615–626.

Bennie, R. E., Packer, C. S., Powell, D. R., Jin, N., and Rhoades, R. A., 1991, Biphasic contractile response of pulmonary artery to hypoxia, *Am. J. Physiol.* **261**:L156–L163.

Bolotina, V. M., Najibi, S., Pagano, P. J., and Cohen, R. A., 1994, Nitric oxide directly activates calcium-dependent potassium channels in vascular smooth muscle, *Nature* **368**:850–853.

Bouchard, R., and Fedida, D., 1995, Closed- and open-state binding of 4-aminopyridine to the cloned human potassium channel Kv 1.5, *J. Pharmacol. Exp. Ther.* **275**:864–876.

Bradford, J., and Dean, H., 1894, The pulmonary circulation, *J. Physiol. (London)* **16**:34–96.

Brashers, V. L., Peach, M. J., and Rose, C. E., 1988, Augmentation of hypoxic pulmonary vasoconstriction in the isolated perfused rat lung by in vitro antagonists of endothelium-dependent relaxation, *J. Clin. Invest.* **82**:1495–1502.

Candia, S., Garcia, M. L., and Latorre, R., 1992, Mode of action of iberiotoxin, a potent blocker of the large conductance Ca$^{(2+)}$-activated K$^+$ channel, *Biophys. J.* **63**:583–590.

Chandy, K. G., and Gutman, G. A., 1993, Nomenclature for mammalian potassium channel genes, *Trends Pharmacol.* **14**:434–440.

Chang, J. K., Moore, P., Fineman, J. R., Soifer, S. J., and Heymann, M. A., 1992, K$^+$ channel pulmonary vasodilation in fetal lambs: Role of endothelium-derived nitric oxide, *J. Appl. Physiol.* **73**:188–194.

Chatfield, B. A., McMurtry, I. F., Hall, S. L., and Abman, S. H., 1991, Hemodynamic effects of endothelin-1 in ovine fetal pulmonary circulation, *Am. J. Physiol.* **261**:R182–R187.

Clapp, L., and Gurney, A., 1991, Outward currents in rabbit pulmonary artery cells disassociated with a new technique, *Exp. Physiol.* **76**:667–693.

Clapp, L. H., and Gurney, A. M., 1992, ATP-sensitive K$^+$ channels regulate resting potential of pulmonary arterial smooth muscle cells, *Am. J. Physiol.* **262**:H916–H920.

Clapp, L. H., Davey, R., and Gurney, A. M., 1993, ATP-sensitive K$^+$ channels mediate vasodilation produced by lemakalim in rabbit pulmonary artery, *Am. J. Physiol.* **264**:H1907–H1915.

Conforti, L., and Millhorn, D. E., 1997, Selective inhibition of a slow-inactivating voltage-dependent K$^+$ channel in rat PC12 cells by hypoxia, *J. Physiol. (London)* **502**:293–305.

Cornfield, D. N., Reeve, H. L., Tolarova, S., Weir, E. K., and Archer, S. L., 1996, Oxygen causes fetal pulmonary vasodilation through activation of a calcium-dependent potassium channel, *Proc. Natl. Acad. Sci. U.S.A.* **93**:8089–8094.

Damron, D. S., Van Wagoner, D. R., Moravec, C. S., and Bond, M., 1993, Arachidonic acid and endothelin potentiate Ca^{2+} transients in rat cardiac myocytes via inhibition of distinct K^+ channels, *J. Biol. Chem.* **268**:27335–27344.

Dart, C., and Standen, N. B., 1995, Activation of ATP-dependent K^+ channels by hypoxia in smooth muscle cells isolated from the pig coronary artery, *J. Physiol. (London)* **483**:29–39.

Daut, J., Maier-Rudolph, W., von Beckerath, N., Mehrke, G., Gunther, K., and Goedel-Meinen, L., 1990, Hypoxic dilatation of coronary arteries is mediated by ATP-sensitive potassium channels, *Science* **247**:1341–1344.

Duprat, F., Guillcmare, E., Romey, G., Fink, M., Lesage, F., Lazdunski, M., and Honore, E., 1995, Susceptibility of cloned K^+ channels to reactive oxygen species, *Proc. Natl. Acad. Sci. U.S.A.* **92**:11796–11800.

Durmowicz, A. G., Hofmeister, S., Kadyraliev, T. K., Aldashev, A. A., and Stenmark, K. R., 1993, Functional and structural adaptation of the yak pulmonary circulation to residence at high altitude, *J. Appl. Physiol.* **74**:2276–2285.

Farrukh, I. S., Peng, W., Orlinska, U., and Hoidal, J. R., 1998, Effect of dehydroepiandrosterone on hypoxic pulmonary vasoconstriction: A $Ca^{(2+)}$-activated $K^{(+)}$-channel opener, *Am. J. Physiol.* **274**:L186–L195.

Franco-Obregon, A., and Lopez-Barneo, J., 1996, Differential oxygen sensitivity of calcium channels in rabbit smooth muscle cells of conduit and resistance pulmonary arteries, *J. Physiol. (London)* **491**:511–518.

Furchgott, R. F., and Zawadzki, J. V., 1980, The obligatory role of endothelial cells in the relaxation of arterial smooth muscle by acetylcholine, *Nature* **288**:373–376.

Furukawa, T., Ito, H., Nitta, J., Tsujino, M., Adachi, S., Hiroe, M., Marumo, F., Sawanobori, T., and Hiraoka, M., 1992, Endothelin-1 enhances calcium entry through T-type calcium channels in cultured neonatal rat ventricular myocytes, *Circ. Res.* **71**:1242–1253.

Garcia, M. L., Galvez, A., Garcia-Calvo, M., King, V. F., Vazquez, J., and Kaczorowski, G. J., 1991, Use of toxins to study potassium channels, *J. Bioenerg. Biomembr.* **23**:615–646.

Garcia, M. L., Garcia-Calvo, M., Hidalgo, P., Lee, A., and McKinnon, R., 1994, Purification and characterization of three inhibitors of voltage-depenent K^+ channels from *Leiurus quinquestriatus var. hebraeus* venom, *Biochemistry* **33**:6834–6839.

Garcia, M. L., Knaus, H.-G., Munujos, P., Slaughter, R. S., and Kaczorowski, G. J., 1995, Charybdotoxin and its effects on potassium channels, *Am. J. Physiol.* **269**:C1–C10.

Garcia-Calvo, M., Knaus, H. G., McManus, O. B., Giangiacomo, K. M., Kaczorowski, G. J., and Garcia, M. L., 1994, Purification and reconstitution of the high-conductance, calcium-activated potassium channel from tracheal smooth muscle, *J. Biol. Chem.* **269**:676–682.

Gasser, R., Klein, W., and Kickenweiz, E., 1993, Vasodilative response to hypoxia and simulated ischemia is mediated by ATP-sensitive K^+ channels in guinea pig thoracic aorta, *Angiology* **44**:228–243.

Gebremedhin, D., Bonnet, P., Greene, A. S., England, S. K., Rusch, N. J., Lombard, J. H., and Harder, D. R., 1994, Hypoxia increases the activity of Ca^{2+}-sensitive K^+ channels in cat cerebral arterial muscle cell membranes, *Pflügers Arch.* **428**:621–630.

Gelband, C. H., and Hume, J. R., 1992, Ionic currents in single smooth muscle cells of the canine renal artery, *Circ. Res.* **71**:745–758.

Giaid, A., Yanagisawa, M., Langleben, D., Michel, R. P., Levy, R., Shennib, H., Kimura, S., Masaki, T., Duguid, W. P., and Stewart, D. J., 1993, Expression of endothelin-1 in the lungs of patients with pulmonary hypertension, *N. Engl. J. Med.* **328**:1732–1739.

Grissmer, S., Nguyen, A. N., Aiyar, J., Hanson, D. C., Mather, R. J., Gutman, G. A., Karmilowicz, M. J., Auperin, D. D., and Chandy, K. G., 1994, Pharmacological characterization of five cloned voltage-gated K^+ channels, types Kv 1.1, 1.2, 1.3, 1.5, and 3.1, stably expressed in mammalian cell lines, *J. Pharmacol. Exp. Ther.* **45**:1227–1234.

Haeusler, G., and Thorens, S., 1980, Effects of tetraethylammonium chloride on contractile, membrane and cable properties of rabbit artery muscle, *J. Physiol. (London)* **303**:203–224.

Hampl, V., and Mejsnar, J., 1986, Thermogenic effect of noradrenaline in skeletal muscle perfused in vitro, *Physiol. Bohemoslov.* **35**:353–354.

Hampl, V., Archer, S. L., Nelson, D. P., and Weir, E. K., 1993, Chronic EDRF inhibition and hypoxia: Effects on the pulmonary circulation and systemic blood pressure, *J. Appl. Physiol.* **75**:1748–1757.

Hampl, V., Weir, E. K., and Archer, S. L., 1994, Endothelium-derived nitric oxide is less important for basal tone regulation in the pulmonary than the renal circulation of the adult rat, *J. Vasc. Med. Biol.* **5**:22–30.

Hampl, V., Huang, J. M. C., Weir, E. K., and Archer, S. L., 1995, Activation of the cGMP-dependent protein kinase mimics the stimulatory effect of nitric oxide and cGMP on calcium-gated potassium channels, *Physiol. Res.* **44**:39–44.

Hara, Y., Kitamura, K., and Kuriyama, H., 1980, Actions of 4-aminopyridine on vascular smooth muscle tissues of the guinea-pig, *Br. J. Pharmacol.* **68**:99–106.

Hasunuma, K., Rodman, D., and McMurtry, I., 1991a, Effects of K$^+$ channel blockers on vascular tone in the perfused rat lung, *Am. Rev. Respir. Dis.* **144**:884–887.

Hasunuma, K., Yamaguchi, T., Rodman, D., O'Brien, R., and McMurtry, I., 1991b, Effects of inhibitors of EDRF and EDHF on vasoreactivity of perfused rat lungs, *Am. J. Physiol.* **260**: L97–L104.

Heinemann, S. H., Rettig, J., Graack, H. R., and Pongs, O., 1996, Functional characterization of Kv channel beta-subunits from rat brain, *J. Physiol. (London)* **493**:625–633.

Ignarro, L. J., Harbison, R. G., Wood, K. S., and Kadowitz, P. J., 1986, Activation of purified soluble guanylate cyclase by endothelium-derived relaxing factor from intrapulmonary artery and vein: Stimulation by acetylcholine, bradykinin and arachidonic acid, *J. Pharmacol. Exp. Ther.* **237**:893–900.

Isaacson, T. C., Hampl, V., Weir, E. K., Nelson, D. P., and Archer, S. L., 1994, Increased endothelium-derived nitric oxide in hypertensive pulmonary circulation of chronically hypoxic rats, *J. Appl. Physiol.* **76**:933–940.

Ivy, D. D., Kinsella, J. P., and Abman, S. H., 1994, Physiologic characterization of endothelin A and B receptor activity in the ovine fetal pulmonary circulation, *J. Clin. Invest.* **93**:2141–2148.

Jiang, C., and Haddad, G. G., 1994, Oxygen deprivation inhibits a K$^+$ channel independently of cytosolic factors in rat central neurons, *J. Physiol. (London)* **481**:15–26.

Jin, N., Packer, C., and Rhoades, R., 1992, Pulmonary arterial hypoxic contraction: Signal transduction, *Am. J. Physiol.* **263**:L73–L78.

Kasemsri, T., and Armstead, W. M., 1997, Endothelin impairs ATP-sensitive K$^+$ channel function after brain injury, *Am. J. Physiol.* **273**:H2639–H2647.

Kato, M., and Staub, N., 1966, Response of small pulmonary arteries to unilobar alveolar hypoxia and hypercapnia, *Circ. Res.* **19**:426–440.

Kaye, A. D., Nossaman, B. D., Santiago, J. A., DeWitt, B. J., Ibrahim, I. N., and Kadowitz, P. J., 1997, Differential effects of glibenclamide on responses to thromboxane A2 mimic, U46619, in the pulmonary and hindquarters vascular beds of the cat, *Eur. J. Pharmacol.* **340**:187–193.

Lee, S., Park, M., So, I., and Earm, Y., 1994a, NADH and NAD modulates Ca^{2+}-activated K$^+$ channels in small pulmonary arterial smooth muscle cells of the rabbit, *Pflügers Arch.* **427**:378–380.

Lee, T. E., Philipson, L. H., Kuznetsov, A., and Nelson, D. J., 1994b, Structural determinant for assembly of mammalian K$^+$ channels, *Biophys. J.* **66**:667–673.

Levitan, E. S., Gealy, R., Trimmer, J. S., and Takimoto, K., 1995, Membrane depolarization inhibits Kv1.5 voltage-gated K$^+$ channel gene transcription and protein expression in pituitary cells, *J. Biol. Chem.* **270**:6036–6041.

MacLean, M. R., McCulloch, K. M., and Baird, M., 1994, Endothelin ET$_A$- and ET$_B$-receptor-mediated vasoconstriction in rat pulmonary arteries and arterioles, *J. Cardiovasc. Pharmacol.* **23**:838–845.

Madden, J., Dawson, C., and Harder, D., 1985, Hypoxia-induced activation in small isolated pulmonary arteries from the cat, *J. Appl. Physiol.* **59**:113–118.

Madden, J., Vadula, M., and Kurup, V., 1992, Effects of hypoxia and other vasoactive agents on pulmonary and cerebral artery smooth muscle cells, *Am. J. Physiol.* **263**:L384–L393.

Magnon, M., Durand, I., and Cavero, I., 1994, The contribution of guanylate cyclase stimulation and K$^+$ channel opening to nicorandil-induced vasorelaxation depends on the conduit vessel and on the nature of the spasmogen, *J. Pharmacol. Exp. Ther.* **268**:1411–1418.

McMurtry, I., 1985, BAY K8644 potentiates and A23187 inhibits hypoxic vasoconstriction in rat lungs, *Am. J. Physiol.* **249**:H741–H746.

McMurtry, I., Davidson, B., Reeves, J., and Grover, R., 1976, Inhibition of hypoxic pulmonary vasoconstriction by calcium antagonists in isolated rat lungs, *Circ. Res.* **38**:99–104.

McMurtry, I. F., Petrun, M. D., and Reeves, J. T., 1978, Lungs from chronically hypoxic rats have decreased pressor response to acute hypoxia, *Am. J. Physiol.* **235**:H104–H109.

Mellemkjaer, S., Nielsen-Kudsk, J., Nielsen, C., and Siggaard, C., 1989, A comparison of the relaxant effects of pinacidil in guinea-pig trachea, aorta and pulmonary artery, *Eur. J. Pharmacol.* **167**:275–280.

Michelakis, E., Reeve, H., Huang, J., Tolarova, S., Nelson, D., Weir, E., and Archer, S. L., 1997, Potassium channel diversity in vascular smooth muscle cells, *Can. J. Physiol. Pharmacol.* **75**:889–897.

Murad, F., 1986, Cyclic guanosine monophosphate as a mediator of vasodilatation, *J. Clin. Invest.* **78**:1–5.

Murakoshi, H., Shi, G., Scannevin, R. H., and Trimmer, J. S., 1997, Phosphorylation of the Kv2.1 K$^+$ channel alters voltage-dependent activation, *Mol. Pharmacol.* **52**:821–828.

Nelson, M., Patlak, J., Worley, J., and Standen, N., 1990, Calcium channels, potassium channels, and voltage dependence of arterial muscle tone, *Am. J. Physiol.* **259**:C3–C18.

Oparil, S., Chen, S. J., Meng, Q. C., Elton, T. S., Yano, M., and Chen, Y. F., 1995, Endothelin-A receptor antagonist prevents acute hypoxia-induced pulmonary hypertension in the rat, *Am. J. Physiol.* **12**:L95–L100.

Osipenko, O. N., Alexander, D., MacLean, M. R., and Gurney, A. M., 1998, Influence of chronic hypoxia on the contributions of non-inactivating and delayed rectifier K currents to the resting potential and tone of rat pulmonary artery smooth muscle, *Br. J. Pharmacol.* **124**:1335–1337.

Overturf, K. E., Russell, S. N., Carl, A., Vogalis, F., Hart, P. J., Hume, J. R., Sanders, K. M., and Horowitz, B., 1994, Cloning and characterization of a Kv1.5 delayed rectifier K$^+$ channel from vascular and visceral smooth muscles, *Am. J. Physiol.* **267**:C1231–C1238.

Park, M., Lee, S., Lee, S., Ho, W. and Earm, Y., 1995a, Different modulation of Ca-activated K channels by the intracellular redox potential in pulmonary and ear arterial smooth muscle cells of the rabbit, *Eur. J. Physiol.* **430**:308–314.

Park, M. K., Lee, S. H., Ho, W. K., and Earm, Y. E., 1995b, Redox agents as a link between hypoxia and the responses of ionic channels in rabbit pulmonary vascular smooth muscle, *Exp. Physiol.* **80**:835–842.

Pascual, J. M., Shieh, C. C., Kirsch, G. E., and Brown, A. M., 1997, Contribution of the NH$_2$ terminus of Kv2.1 to channel activation, *Am. J. Physiol.* **273**:C1849–C1858.

Patel, A. J., Lazdunski, M., and Honore, E., 1997, Kv2.1/Kv9.3, a novel ATP-dependent delayed-rectifier K$^+$ channel in oxygen-sensitive pulmonary artery myocytes, *EMBO J.* **16**:6615–6625.

Peterson, D., Archer, S., and Weir, E. K., 1994, Superoxide reduction of a disulfide: A model of intracellular redox modulation?, *Biochem. Biophys. Res. Commun.* **200**:1586–1591.

Peterson, D. A., Peterson, D. C., Archer, S. L. and Weir, E. K., 1996, Accelerated disulfide reduction with polyunsaturated fatty acids: A mechanism of ionic channel modulation?, *Redox Report* **2**:263–265.

Pinheiro, J. M., and Malik, A. B., 1992, K$_{ATP}$-channel activation causes marked vasodilation in the hypertensive neonatal pig lung, *Am. J. Physiol.* **263**:H1532–H1536.

Post, J., Hume, J., Archer, S., and Weir, E., 1992, Direct role for potassium channel inhibition in hypoxic pulmonary vasoconstriction, *Am. J. Physiol.* **262**:C882–C890.

Post, J. M., Gelband, C. H., and Hume, J. R., 1995, [Ca^{2+}]$_i$ inhibition of K$^+$ channels in canine pulmonary artery: Novel mechanism for hypoxia-induced membrane depolarization, *Circ. Res.* **77**:131–139.

Reeve, H. L., Weir, E. K., Nelson, D. P., Peterson, D. A., and Archer, S. L., 1995, Opposing effects of oxidants and antioxidants on K$^+$ channel activity and tone in vascular tissue, *Exp. Physiol.* **80**:825–834.

Reeve, H. L., Weir, E. K., Archer, S. L., and Cornfield, D. N., 1998, A maturational shift in pulmonary K$^+$ channels, from Ca^{2+} sensitive to voltage dependent, *Am. J. Physiol.* **275**:L1019–L1025.

Robertson, B. E., Warren, J. B., and Nye, P. C. G., 1990, Inhibition of nitric oxide synthesis potentiates hypoxic vasoconstriction in isolated rat lungs, *Exp. Physiol.* **75**:255–257.

Robertson, B. E., Schubert, R., Hescheler, J., and Nelson, M., 1993, cGMP-dependent protein kinase activates Ca-activated K channels in cerebral artery smooth muscle cells, *Am. J. Physiol.* **265**:C299–C303.

Russell, S. N., Overturf, K. E., and Horowitz, B., 1994, Heterotetramer formation and charybdotoxin sensitivity of two K$^+$ channels cloned from smooth muscle, *Am. J. Physiol.* **267**:C1729–C1733.

Salinas, M., Duprat, F., Heurteaux, C., Hugnot, J. P., and Lazdunski, M., 1997, New modulatory alpha subunits for mammalian *Shab* K$^+$ channels, *J. Biol. Chem.* **272**:24371–24379.

Salkoff, L., and Jegla, T., 1995, Surfing the DNA databases for K$^+$ channels nets yet more diversity, *Neuron* **15**:489–492.

Salter, K. J., and Kozlowski, R. Z., 1998, Differential electrophysiological actions of endothelin-1 on Cl$^-$ and K$^+$ currents in myocytes isolated from aorta, basilar and pulmonary artery, *J. Pharmacol. Exp. Ther.* **284**:1122–1131.

Salter, K. J., Wilson, C. M., Kato, K., and Kozlowski, R. Z., 1998, Endothelin-1, delayed rectifier K channels, and pulmonary arterial smooth muscle, *J. Cardiovasc. Pharmacol.* **31**:S81–S83.

Sheehan, D., Sylvester, J., and Sham, J., 1994, 4-Aminopyridine sensitive, delayed rectifier potassium (K$^+$)

current is the major K$^+$ current controlling resting membrane potential in porcine pulmonary vascular smooth muscle, *Am. J. Respir Crit. Care. Med.* **149**:A293.

Shi, G., Kleinklaus, A. K., Marrion, N. V., and Trimmer, J. S., 1994, Properties of Kv2.1 K$^+$ channels expressed in transfected mammalian cells, *J. Biol. Chem.* **269**:23204–23211.

Shimoda, L. A., Sylvester, J. T., and Sham, J. S., 1998, Inhibition of voltage-gated K$^+$ current in rat intrapulmonary arterial myocytes by endothelin-1, *Am. J. Physiol.* **274**:L842–L853.

Shirai, M., Sada, K., and Ninomiya, I., 1986, Effects of regional alveolar hypoxia and hypercapnia on small pulmonary vessels in cats, *J. Appl. Physiol.* **61**:440–448.

Shirai, M., Ninomiya, I., and Sada, K., 1991, Constrictor response of small pulmonary arteries to acute pulmonary hypertension during left atrial pressure elevation, *Jpn. J. Physiol.* **41**:129–142.

Smirnov, S. V., and Aaronson, P. I., 1994, Alteration of the transmembrane K$^+$ gradient during development of delayed rectifier in isolated rat pulmonary arterial cells, *J. Gen. Physiol.* **104**:241–264.

Smirnov, S., Robertson, T., Ward, J., and Aaronson, P., 1994, Chronic hypoxia is associated with reduced delayed rectifier K$^+$ current in rat pulmonary artery muscle cells, *Am. J. Physiol.* **266**:H365–H370.

Steudel, W., Ichinose, F., Huang, P. L., Hurford, W. E., Jones, R. C., Bevan, J. A., Fishman, M. C., and Zapol, W. M., 1997, Pulmonary vasoconstriction and hypertension in mice with targeted disruption of the endothelial nitric oxide synthase (NOS 3) gene, *Circ. Res.* **81**:34–41.

Stewart, D., Levy, R., Cernacek, P., and Langlehben, D., 1991, Increased plasma endothelin-1 in pulmonary hypertension: Marker or mediator of disease?, *Ann. Int. Med.* **114**:464–469.

Sugg, E. E., Garcia, M. L., Reuben, J. P., Patchett, A. A., and Kaczorowski, G. J., 1990, Synthesis and structural characterization of charybdotoxin, a potent peptidyl inhibitor of the high conductance Ca^{2+}-activated K$^+$ channel, *J. Biol. Chem.* **265**:18745–18748.

Taglialatela, M., Drewe, J. A., and Brown, A. M., 1993, Barium blockade of a clonal potassium channel and its regulation by a critical pore residue, *Mol. Pharmacol.* **44**:180–190.

Takenaka, T., Forster, H., and Epstein, M., 1993, Protein kinase C and calcium channel activation as determinants of renal vasoconstriction by angiotensin II and endothelin, *Circ. Res.* **73**:743–750.

Takimoto, K., Fomina, A. F., Gealy, R., Trimmer, J. S., and Levitan, E. S., 1993, Dexamethasone rapidly induces Kv1.5 K$^+$ channel gene transcription and expression in clonal pituitary cells, *Neuron* **11**:359–369.

Takimoto, K., Li, D., Hershman, K. M., Li, P., Jackson, E. K., and Levitan, E. S., 1997, Decreased expression of Kv4.2 and novel Kv4.3 K$^+$ channel subunit mRNAs in ventricles of renovascular hypertensive rats, *Circ. Res.* **81**:533–539.

Tanaguchi, J., Furukawa, K., and Shigekawa, M., 1993a, Maxi K$^+$ channels are stimulated by cyclic guanosine monophosphate-dependent protein kinase in canine coronary artery smooth muscle cells, *Pflügers Arch.* **423**:167–172.

Tanaguchi, J., Furukawa, K.-I., and Shigekawa, M., 1993b, Maxi K$^+$ channels are stimulated by cyclic guanosine monophosphate-dependent protein kinase in canine coronary artery smooth muscle cells, *Pflügers Arch.* **423**:167–172.

Tolins, M., Weir, E., Chesler, E., Nelson, D., and From, A., 1986, Pulmonary vascular tone is increased by a voltage-dependent calcium channel potentiator, *J. Appl. Physiol.* **60**:942–948.

Triggle, C. R., Li, Y. Q., and Wyse, D. G., 1992, Comparative effects of the K$^+$ channel openers, pinacidil and cromakalim on vascular tone: Sensitivity to glyburide and calcium, *Proc. West. Pharmacol. Soc.* **35**:97–102.

Trimmer, J. S., 1993, Expression of Kv2.1 delayed rectifier K$^+$ channel isoforms in the developing rat brain, *FEBS Lett.* **324**:205–210.

Tristani-Firouzi, M., Reeve, H. L., Tolarova, S., Weir, E. K., and Archer, S. L., 1996, Oxygen-induced constriction of the rabbit ductus arteriosus occurs via inhibition of a 4-aminopyridine-sensitive potassium channel, *J. Clin. Invest.* **98**:1959–1965.

Vadula, M. S., Kleinman, J. G., and Madden, J. A., 1993, Effect of hypoxia and norepinephrine on cytoplasmic free Ca^{2+} in pulmonary and cerebral arterial myocytes, *Am. J. Physiol.* **265**:L591–L597.

von Euler, U., and Liljestrand, G., 1946, Observations on the pulmonary arterial blood pressure in the cat, *Acta Physiol. Scand.* **12**:301–320.

Wang, J., Juhaszova, M., Rubin, L. J., and Yuan, X.-J., 1997, Hypoxia inhibits gene expression of voltage-gated K$^+$ channel α subunits in pulmonary artery smooth muscle cells, *J. Clin. Invest.* **100**:2347–2353.

Weir, E., 1978, Does normoxic pulmonary vasodilatation rather than hypoxic vasoconstriction account for the pulmonary pressor response to hypoxia, *Lancet* **i:**476–477.

Weir, E., Chidsey, C., Weil, J., and Grover, R., 1976, Minoxidil reduces pulmonary vascular resistance in dogs and cattle, *J. Lab. Clin. Invest.* **88:**885–894.

Weir, E. K., and Archer, S. L., 1995, The mechanism of acute hypoxic pulmonary vasoconstriction: The tale of two channels, *FASEB J.* **9:**183–189.

Weir, F. K., Archer, S. L., and Reeves, J. T., 1992, *The Diagnosis and Treatment of Pulmonary Hypertension*, Futura Press, Mount Kisco, New York.

Weir, E. K., Archer, S. L., and Rubin, L., 1994, Pulmonary hypertension, in: *Cardiovascular Medicine* (J. Cohn and J. T. Willerson, eds.), Churchill Livingstone, New York, pp 1495–1524.

Weir, E. K., Reeve, H. L., Peterson, D. A., Michelakis, E. D., Nelson, D. P., and Archer, S. L., 1998, Pulmonary vasoconstriction, oxygen sensing, and the role of ion channels: Thomas A. Neff lecture, *Chest* **114:**17S–22S.

Wiener, C. M., Dunn, A., and Sylvester, J. T., 1991, ATP-dependent K^+ channels modulate vasoconstrictor responses to severe hypoxia in isolated ferret lungs, *J. Clin. Invest.* **88:**500–504.

Xu, J., Yu, W., Jan, Y. N., Jan, L. Y., and Li, M., 1995, Assembly of voltage-gated potassium channels: Conserved hydrophilic motifs determine subfamily-specific interactions between the α-subunits, *J. Biol. Chem.* **270:**24761–24768.

Yu, W., Xu, J., and Li, M., 1996, NAB domain is essential for the subunit assembly of both αα and αβ complexes of Shaker-like potassium channels, *Neuron* **16:**441–453.

Yuan, X.-J., 1995, Voltage gated K^+ currents regulate resting membrane potential and $[Ca^{2+}]_i$ in pulmonary artery myocytes, *Circ. Res.* **77:**370–378.

Yuan, X.-J., Goldman, W., Tod, M., Rubin, L., and Blaustein, M., 1993, Hypoxia reduces potassium currents in cultured rat pulmonary but not mesenteric arterial myocytes, *Am. J. Physiol.* **264:**L116–L123.

Yuan, X. J., Wang, J., Juhaszova, M., Golovina, V. A., and Rubin, L. J., 1998, Molecular basis and function of voltage-gated K^+ channels in pulmonary arterial smooth muscle cells, *Am. J. Physiol.* **274:**L621–L635.

Zhao, Y. J., Wang, J., Rubin, L. J., and Yuan, X. J., 1997, Inhibition of $K_{(V)}$ and $K_{(Ca)}$ channels antagonizes NO-induced relaxation in pulmonary artery, *Am. J. Physiol.* **272:**H904–H912.

Zhao, Y. J., Wang, J., Rubin, L. J., and Yuan, X. J., 1998, Roles of K^+ and Cl⁻ channels in cAMP-induced pulmonary vasodilation, *Exp. Lung. Res.* **24:**71–83.

Chapter 28

Potassium Channels in the Renal Circulation

James D. Stockand and Steven C. Sansom

1. INTRODUCTION

The renal artery and its main tributaries, the arcuate and interlobular arteries, are responsible for the vascular supply to the most richly perfused organ of the human body. Because the kidney establishes a large portion of total peripheral resistance, it is one of the essential vascular beds controlling blood pressure. While rapid adjustments in tone of the main renal arteries can significantly modulate acute blood pressure, the renal arterioles regulate blood pressure on a more long-term basis by modulating urinary excretion in response to changing blood volume (Fig. 1).

1.1. Structure/Function of the Renal Corpuscle

The renal corpuscle has a unique structure which includes a capillary bracketed by resistance arterioles allowing control of glomerular filtration rate (GFR) somewhat independent of mean arterial pressure (MAP). Whereas MAP varies in response to systemic requirements, GFR is maintained relatively constant by intrinsic negative-feedback loops, localized entirely within the renal cortex. Such loops functionally couple arteriolar plasma flow with urinary flow in the tubule at the macula densa (tubuloglomerular feedback). Extrinsic humoral signals, such as atrial natriuretic factor (ANF), maintain salt and water homeostasis by influencing arteriolar diameter and resistance to modulate GFR. Thus, the vasculature of the renal corpuscle is the effector site for regulation of GFR by both intrinsic and extrinsic signals (Fig. 2).

James D. Stockand ● The Center for Cellular and Molecular Signaling, Department of Physiology, Emory University School of Medicine, Atlanta, Georgia 30322. *Steven C. Sansom* ● Department of Physiology and Biophysics, University of Nebraska Medical Center, Omaha, Nebraska 68198-4575.

Potassium Channels in Cardiovascular Biology, edited by Archer and Rusch. Kluwer Academic/Plenum Publishers, New York, 2001.

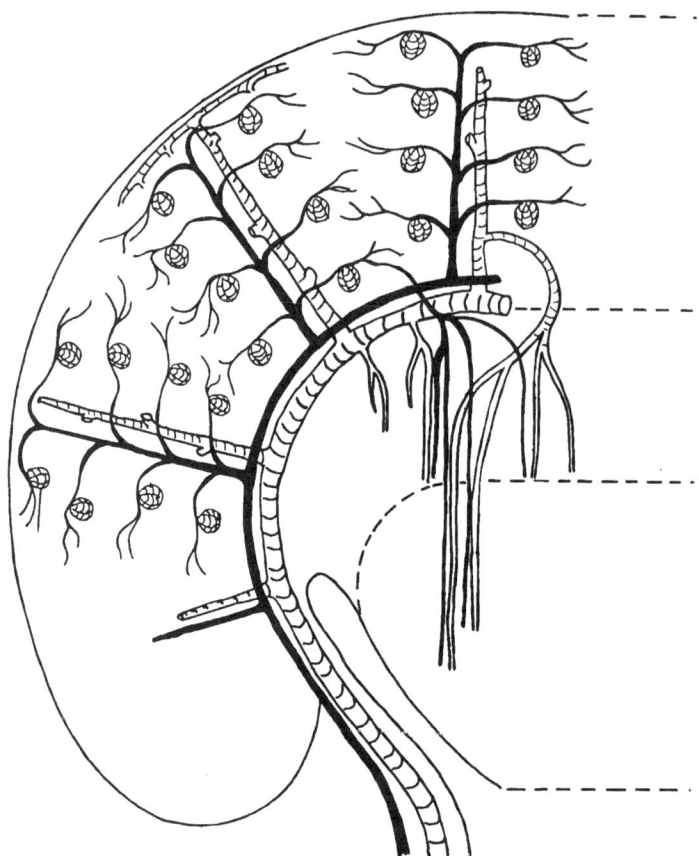

Figure 1. Illustration showing the organization of the renal vasculature. The renal artery enters the hilus and branches into the arcuate and interlobular arteries (arteries indicated by black lines). The afferent arterioles, formed off the interlobular arteries, enter into the capillary network of the glomerulus before forming the efferent arterioles. The renal tubular system runs in parallel. (Modified from Koeppen, B. M., and Stanton, B. A., *Renal Physiology*, 2nd ed., Mosby, St. Louis, 1997 Fig. 2.2, from Koushanpour, E., and Kriz, W., *Renal Physiology: Principles, Structure, and Function*, 2nd ed., Springer-Verlag, New York, 1986, and from Kriz, W., and Bankir, L., A standard nomenclature for structure of the kidney, *Am. J. Physiol.* **254:**, F1, 1988).

1.2. Roles of K^+ Channels

The plasmalemmal K^+ channels in renal vessels set the membrane potential, a role similar to that served by K^+ channels in vascular smooth muscle cells from other vascular beds (see Chapter 22). Furthermore, these K^+ channels are targets for the final action of relaxants, such as nitric oxide (NO), which govern local renal resistance and blood volume by regulating the filtered load entering Bowman's space. Intrarenal systems of autoregulation, such as tubuloglomerular feedback, designed to maintain constant urine flow, involve changes in the ionic conductances in the plasma membrane of arteriolar smooth muscle cells. Studies suggest that during rapid rates of urine

$$GFR = K_f \times P_{uf}$$

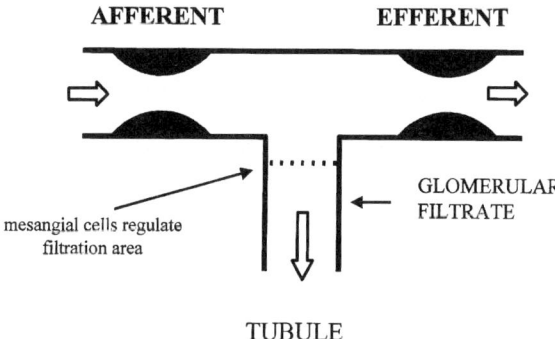

Figure 2. Illustration of roles of mesangial cells and afferent and efferent arterioles in regulation of glomerular filtration rate (GFR). Blood flow from the interlobar arteries enters the afferent arterioles. The filtration pressure (P_{uf}) is governed by the resistances of the afferent and efferent arterioles. The filtration constant, K_f, is proportional to the capillary surface area, which is partially controlled by mesangial cells.

delivery, the macula densa releases a constricting factor that inhibits K^+ channels in the afferent arteriole, resulting in contraction and a decrease in GFR (Harder *et al.*, 1995). With decreased tubular urine flow, a relaxing factor originating from the macula densa may elevate GFR by affecting K^+ channels in the afferent arterioles. In addition to intrinsic regulation of the ion channel activities of arteriolar smooth muscle, it is apparent that extrinsic humoral signals, such as ANF, act to control renal filtration rates by affecting the K^+ channels of glomerular mesangial cells.

Because K^+ channel activity in the renal microvasculature sets the normal tone of the glomerular microcirculation, these channels will be subjected to dysfunctional regulation in diseased states such as diabetes or hypertension. NO and arachidonic acid and its metabolites are a few of the important signaling agents that may be involved in the pathophysiological regulation of K^+ channel activity in afferent arterioles and mesangial cells.

Owing to the difficulty in isolating renal afferent and efferent arterioles, few studies have characterized directly the biophysical properties of the ion channels in these cells. Pharmacological studies using K^+ channel blockers and activators in conjunction with measurements of renal blood flow and arteriolar diameter have suggested that renal arteries and arterioles contain a complement of K^+ channels similar to those found in other vascular smooth muscle cells. These pharmacological studies have stressed the important roles played by K^+ channels in the homeostatic control of renal function and blood pressure.

2. K^+ CHANNELS IN RENAL ARTERIES

In most electrophysiological studies of the renal arteries, the smooth muscle cells were freshly extracted and experiments were performed within 12 hours of isolation.

Using these freshly dispersed cells, investigators have provided strong pharmacological and electrophysiological evidence for the presence of at least two types of K^+ channels in renal arteries, the large-conductance, Ca^{2+}-activated K^+ (BK_{Ca}) channels and 4-aminopyridine-sensitive delayed rectifier K^+ (K_{DR}) channels, now more commonly referred to as voltage-gated K^+ (Kv) channels. Only pharmacological evidence supports a role for ATP-sensitive K^+ (K_{ATP}) channels. Although data from the arcuate and interlobular arteries will be included in this section, there may be some differences in the roles of K^+ channels between these arterial segments.

2.1. BK_{Ca} Channels

Gelband and Hume (1992) provided the first electrophysiological evidence for BK_{Ca} channels in the renal vasculature in isolated vascular smooth muscle cells of the main renal artery of dogs. In whole-cell recordings, a component of a predominant outward, K^+ tail current was activated at $+10$ mV and was sensitive to tetraethylammonium (TEA) and charybdotoxin (CTX). Cell-attached patches revealed a large, K^+-selective channel with a unitary conductance of 232 pS (Fig. 3). Like BK_{Ca} channels of other vascular smooth muscle cells, this channel is inhibited by TEA and CTX and

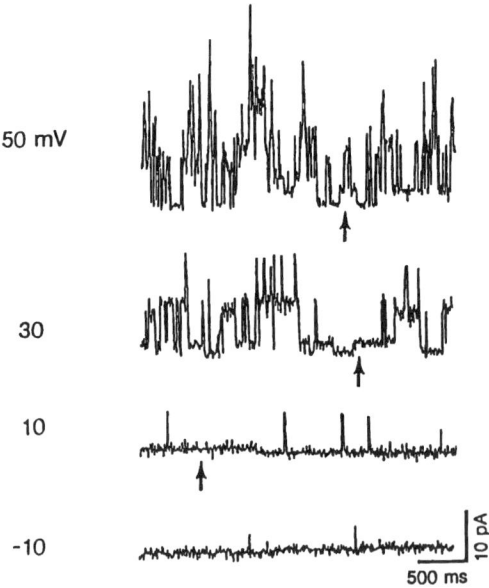

Figure 3. Original single-channel current tracings demonstrating both BK_{Ca} and Kv current in a cell-attached patch from isolated smooth muscle of the main renal artery from dog. The slope conductances of BK_{Ca} and Kv currents are 130 pS and 57 pS, respectively. Studies were performed with 5.4 mM KCl in the patch pipette and at depolarizing potentials between $+20$ and $+80$ mV, respectively. The lower-amplitude Kv current is indicated by the arrows. The large openings reflect activity of the BK_{Ca} channels. (Reproduced, with permission, from Gelband and Hume, 1992.)

is modulated by intracellular Ca^{2+} concentration ($[Ca^{2+}]_i$) and voltage.

Whole-cell currents consistent with BK_{Ca} channels also are found in arcuate and interlobular arteries isolated from rat kidney (Gordienko et al., 1994). Typical of BK_{Ca} channels, these outwardly directed currents are best demonstrated at depolarizing potentials (between $+30$ and $+50$ mV) and are inhibited by TEA. These BK_{Ca} currents are increased in response to bath application of caffeine, which stimulates Ca^{2+} release from ryanodine-sensitive Ca^{2+} channels of the sarcoplasmic reticulum. These results led to the hypothesis that the ability of BK_{Ca} channels to open in response to elevations of $[Ca^{2+}]_i$ might serve as a feedback mechanism to counteract the depolarizing effects of contractile agonists (Gordienko et al., 1994).

2.2. Kv Channels

Electrophysiological studies support the presence of K_{DR} channels, subsequently referred to as Kv channels, in isolated smooth muscle from the main renal artery of the dog (Gelband and Hume, 1992). Kv current is identified as the part of the whole-cell current which activates at -30 to -40 mV and is blocked by 4-aminopyridine (4-AP). In single-channel experiments, a 4-AP-sensitive, K^+-selective channel with a conductance of 104 pS may represent this Kv channel.

Unlike BK_{Ca} channels, Kv channels are normally active in cell-attached patches close to the resting membrane potential. Thus, the functional role of Kv channels is clearly different from that of BK_{Ca} channels. A study using smooth muscle cells from canine renal artery has shown that Kv channels are inhibited when $[Ca^{2+}]_i$ is elevated by adding angiotensin II (ANGII) or caffeine to the bathing solution (Gelband et al., 1993). Because these inhibitory effects of ANGII and caffeine are blocked by 4-AP, it was proposed that ANGII depolarizes the membrane potential by Ca^{2+}-mediated inhibition of Kv channels. More direct, single-channel experiments, using the inside-out patch configuration, show that these channels are inhibited by elevating $[Ca^{2+}]_i$ to 600 nM (Gelband et al., 1993). Therefore, Kv channels may be partly responsible for setting the resting membrane potential in renal arterial smooth muscle cells, and Ca^{2+}-mediated inactivation of these Kv channels may be a major mechanism for the depolarizing response to contractile agonists.

2.3. K_{ATP} Channels

The role of K_{ATP} channels in renal arteries is not as clearly defined as that of BK_{Ca} and Kv channels. Several investigators have shown that K_{ATP} channel openers (pinacidil, cromakalim, and lemakalim) reduce renal vascular resistance in anesthetized dogs (Tamaki et al., 1991; Hayashi et al., 1990) and counteract contraction of isolated renal arteries from a variety of species (Videbaek et al., 1988; Miura et al., 1995; Toda et al., 1985). Miura et al. (1995) found that glibenclamide completely blocked vasodilation of dog renal arteries induced by the K_{ATP} channel opener lemakalim. Furthermore, Wilson and Cooper (1989) found that the relaxing effects of cromakalim on rabbit renal arteries are reversed by glibenclamide. However, a study by Gojkovic and Kazic (1994) demonstrated that although pinacidil relaxes both precontracted rabbit mesenteric and renal arteries, glibenclamide only reverses this effect in the mesenteric arteries. This study raises the question of whether pinacidil-induced vasodilation occurs exclusively through activation of K_{ATP} channels. Despite the abundant pharmacological evidence supporting their existence, there have been no single-channel data describing K_{ATP} channels in smooth muscle cells from either the arcuate, interlobular, or main renal arteries.

2.4. Other K^+ Channels in Renal Arteries

Other K^+ channels in renal arteries have been described in studies using electrophysiological methods. Using the whole-cell patch-clamp technique, Gordienko *et al.* (1994) described a 4-AP-sensitive, transient outward current (I_{to}) in smooth muscle cells from interlobular and arcuate arteries of rat kidneys. I_{to} was activated at potentials positive to -20 mV and was Ca^{2+}-insensitive and refractory to block by TEA. In terms of its transient activating and inactivating nature, this current was similar to A-type currents described in human mesenteric artery and other muscle cells.

In renal arcuate arteries, Roman *et al.* (1993) described a 117-pS channel in cell-attached patches at -20 mV. Although a pharmacological profile was not performed, the single-channel conductance and relatively high open-state probability at resting membrane potentials in the cell-attached mode were similar to those described for Kv channels.

2.5. Regulation of K^+ Channels by NO and Metabolites of Arachidonic Acid

Radermacher *et al.* (1990) found that sodium nitroprusside (SNP), an NO donor, induced a decrease in renal vascular resistance of the perfused rat kidney. Yukimura *et al.* (1992) found that intrarenal arterial infusion of N^G-nitro-L-arginine (LNNA), an NO synthase inhibitor, also markedly decreases renal blood flow in dogs. These results suggest that NO plays a significant role in the regulation of the renal circulation. More recently, Alonso-Galicia *et al.* (1998) showed that application of the NO donor SNP increases cGMP in renal arteries and that the vasodilator response is inhibited by blockers of guanylyl cyclase and cGMP-dependent protein kinase. These results suggest that the mechanism of relaxation by NO is associated with cGMP-dependent protein phosphorylation.

Other studies also suggest that NO relaxes renal arteries, in part, by leading to the activation of specific K^+-selective channels. The effects of acetylcholine, which relaxes renal arteries by releasing endothelial NO, are inhibited by TEA and 4-AP, but not by glibenclamide. In agreement with these findings, Miura *et al.* (1995) found that adrenomedullin, which elevates NO levels, is an effective relaxant of renal arteries in the presence of glibenclamide. Thus, NO probably leads to stimulation of both BK_{Ca} (TEA-sensitive) and Kv (4-AP-sensitive) channels, but not K_{ATP} channels. To date, however, electrophysiological studies have not revealed the specific K^+ channels activated by NO in renal arteries.

Metabolic products of arachidonic acid (AA) regulate tone in many vascular beds, including the renal arteries. Ma *et al.* (1993) showed that the P-450 metabolite of AA, 20-hydroxyeicosatetraenoic acid (20-HETE), produced a dose-dependent (from 0.01 to 1 μM) constriction of canine renal arcuate arteries. Using the patch-clamp technique, it was demonstrated that 20-HETE significantly reduces the open-state probability of a 117-pS channel, recorded in the cell-attached mode. This suggests that AA production of 20-HETE could have a role in mediating the myogenic response in the renal circulation.

The finding that 11, 12-epoxyeicosatrienoic acid (11, 12-EET) causes vasodilation of preconstricted, perfused rabbit kidneys suggests that this metabolic product of AA may activate a K^+-selective channel in the renal vasculature (Carroll *et al.*, 1992). Zou *et al.* (1996b) tested this hypothesis by studying the direct effects of 11, 12-EET on BK_{Ca} channels of rat renal arteries. Extracellular application of the *R, S*-isomer of 11, 12-EET stereospecifically activated BK_{Ca} channels in cell-attached patches of renal arcuate and interlobular arteries, in a dose-dependent manner. However, 11, 12-EET failed to activate BK_{Ca} channels when applied to the intracellular aspect of the patch, in the inside-out patch configuration. This suggests that an intracellular mediator, removed when the patch was pulled from the cell surface, is required for these AA metabolites to cause channel activation. Moreover, it is not certain if 11, 12-EET represents an endogenous, physiological regulator of BK_{Ca} channels.

3. K^+ CHANNELS OF RENAL ARTERIOLES

The separate regulation of afferent and efferent arteriolar tone is essential for renal autoregulatory mechanisms, such as the myogenic response and tubuloglomerular feedback. In the afferent arteriole, K^+ channels regulate cell excitability. In contrast, the role of K^+ channels in the efferent arteriole remains unclear. However, it is possible that K^+ channel heterogeneity accounts for at least part of the differential regulation of these two vascular beds.

The preponderance of information describing K^+ channels in renal arterioles is derived from studies utilizing pharmacological agents in combination with a variety of methods for quantifying vessel diameter and microcirculatory responses. Preparations such as the perfused juxtamedullary nephron, the hydronephrotic kidney, and isolated perfused microvessels have enabled the direct determination of the responses of both afferent and efferent arterioles to pharmacological activators and blockers of K^+ channels. These preparations have allowed integration of knowledge derived from the *in vitro* electrophysiological experiments with *in vivo* and *ex vivo* studies of renal hemodynamics, GFR, and MAP. More recently, the patch-clamp technique has been applied to the afferent arterioles to characterize the properties of single K^+ channels. Electrophysiological and pharmacological experiments utilizing freshly microdissected, preglomerular arterioles from rats have identified BK_{Ca} and K_{ATP} channels in these small vessels. Both classes of channels function to counteract contraction by attenuating membrane depolarization necessary for entry of Ca^{2+} into the cell via voltage-gated Ca^{2+} channels.

3.1. BK_{Ca} Channels

BK_{Ca} channels have been characterized in excised, inside-out patches from smooth muscle cells isolated from renal afferent arterioles of rats (Ma et al., 1993; Gebremedhin et al., 1996) and rabbits (Lorenz et al., 1992). These arteriolar BK_{Ca} channels have large conductances (250–300 pS) in symmetrical KCl and are Ca^{2+}-sensitive and voltage-gated. Zou et al. (1996a) showed that these BK_{Ca} channels are inhibited by application on the intracellular surface of $BaCl_2$ (1 mM) or by external application of TEA (1 mM) and CTX. The open-state probability of BK_{Ca} channels was less than 5% in cell-attached patches at a membrane holding potential of $+40$ mV. This low activity at depolarizing potentials would suggest that BK_{Ca} channels contribute minimally to determining the resting membrane potential. Consistent with this view is the finding that 1 mM TEA only slightly reduces the diameter of afferent arterioles (Carmines et al., 1999).

As in other vascular beds, the role of BK_{Ca} channels in the afferent arteriole may be as a feedback inhibitor of contraction. However, contrary to this notion, application of TEA does not accentuate agonist-induced reduction in diameter of rat afferent arterioles (Carmines et al., 1999). The failure of BK_{Ca} channel blockers to augment the effects of contractile agonists is also observed for mesangial cells (see Section 4). This suggests that BK_{Ca} channels do not normally play a role as feedback inhibitors of contraction in renal arterioles. However, vasorelaxants, such as NO and ANF, do enhance the feedback response of BK_{Ca} in glomerular mesangial cells (Stockand and Sansom, 1996a; see below). Although it is not known if the feedback response to ANGII is similarly enhanced in renal arterioles by these vasorelaxants, this possibility is suggested by the finding that the BK_{Ca} channel opener NS1619 significantly attenuates the contractile effects of ANGII in these microvessels (Carmines et al., 1999). Therefore, it is possible that BK_{Ca} channels in afferent arterioles, as in mesangial cells, are significant feedback regulators only in the presence of agents that lead to stimulation of these channels.

3.2. K_{ATP} Channels

Binding of radiolabeled P1075, an activator of K_{ATP} channels, to the renal corpuscle vasculature isolated from the rat supports the notion that this smooth muscle contains ATP-sensitive K^+ channels (Metzger and Quast, 1999). Specific binding localizes primarily to afferent arteriolar endings adherent to the glomeruli, and other K_{ATP} channel openers and blockers compete for these binding sites. The finding that binding of labeled P1075 decreases upon treatment of the renal tissue with collagenase suggests that K_{ATP} channels are located on the plasma membranes, rather than at intracellular sites.

Although patch-clamp studies have not directly measured K_{ATP} currents, the presence of these channels in afferent arterioles is suggested by pharmacological studies. A contribution by K_{ATP} current to the resting membrane potential is unlikely because glibenclamide does not affect the diameter of afferent arterioles from normal rats (Ikenaga et al., 1996). However, although basal arteriolar diameter is unaffected by

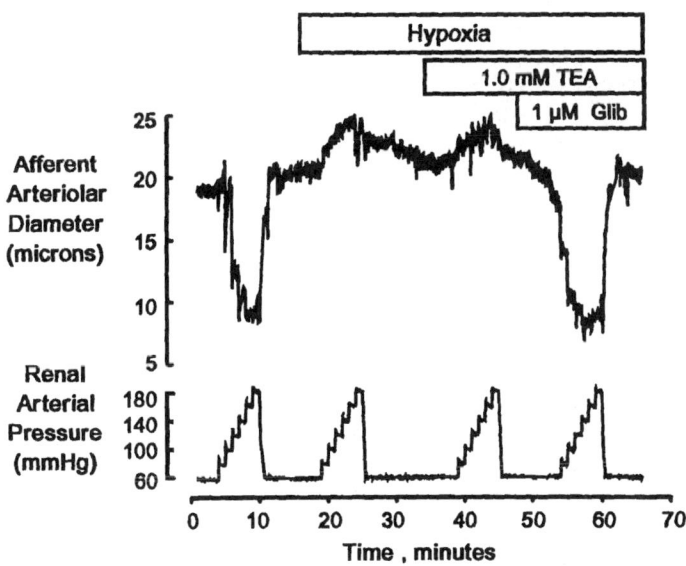

Figure 4. Hypoxia reduces myogenic response by activating K_{ATP} channels. As shown in these original tracings, the afferent arteriolar diameter normally constricts upon increasing renal arterial pressure. This myogenic response is inoperative during hypoxia and is not restored by blocking BK_{Ca} channels with tetrathylammonium (TEA). However, inhibiting K_{ATP} channels with glibenclamide (Glib) restores pressure-induced vasoconstriction. These results suggest that K_{ATP} channels, but not BK_{Ca} channels are activated by hypoxia in myogenically active afferent arterioles and act to buffer myogenic vasoconstriction. (Reproduced, with permission, from Loutzenhiser and Parker, 1994.)

K_{ATP} channel openers or blockers, constriction of afferent arterioles by ANGII is attenuated by pinacidil (Reslerova and Loutzenhiser, 1995). The pinacidil effect is reversed by glibenclamide, suggesting that openers of K_{ATP} channels may counteract ANGII-induced constriction of afferent arterioles.

Other studies suggest that the role of K_{ATP} channels may involve regulating the metabolic state of the preglomerular arteriole. Loutzenhiser and Parker (1994) showed that during hypoxia, glibenclamide, but not TEA, restores myogenic contraction to afferent arterioles in response to increased renal perfusion pressure. This demonstrates that K_{ATP}, but not BK_{Ca}, channels are probable effector sites for hypoxia-induced relaxation and is consistent with the findings of Lorenz et al. (1992), who reported that 2-deoxy-D-glucose, an inhibitor of glycolysis, significantly dilates preconstricted afferent arterioles. Thus, K_{ATP} channels, although quiescent at rest, may activate under conditions of hypoxia or metabolic stress to partially restore homeostasis (Fig. 4).

3.3. Other K$^+$-Selective Channels

Gebremedhin et al. (1996) supplied both whole-cell and single-channel evidence for the existence of apamin-sensitive Ca^{2+}-activated K^+ channels in preglomerular arterioles of the rat. Biophysically, these 68-pS channels are similar to BK_{Ca} channels;

pharmacologically, however, the 68-pS channel is inhibited by 50 nM apamin, a known blocker of small Ca^{2+}-activated K^+ (SK_{Ca}) channels in other cells. In whole-cell recordings, apamin decreased outward current by approximately 25–30% at command potentials greater than 0 mV. This absence of activity at negative potentials and the requirement of SK_{Ca} channels for Ca^{2+} for activation indicates that the 68-pS channel has no role in setting the resting membrane potential. The fact that the 68-pS channel was never observed in a patch in the absence of BK_{Ca} channels suggests that it may be related to or is a substate conductance of the BK_{Ca} channel.

Although several types of K^+ channels have been reported in afferent arterioles, none of these channels are active at resting membrane potentials. Indirect evidence for the existence of an inward rectifying K^+ channel (Kir) in preglomerular arterioles was provided by Chilton and Loutzenhiser (1998). Using the *in vitro* perfused hydronephrotic rat kidney preparation, they showed that an increase in bath K^+ from 5 to 15 mM caused a Ba^{2+}-sensitive vasodilation of the afferent arterioles. It was concluded that Kir channels are involved in setting the resting potential, since these channels are known to activate in response to elevations in external K^+ concentrations. However, as for K_{ATP} channels, no single-channel data have yet definitively established the existence of Kir channels in afferent arteries.

3.4. K^+ Channels of Efferent Arterioles

Because of technical difficulties in isolating efferent arterioles, there have been very few pharmacological studies and no direct patch-clamp experiments providing single-channel data in these segments. Carmines *et al.* (1993) used fura-2 fluorescence to show that, unlike afferent arterioles, efferent arterioles do not contain voltage-operated Ca^{2+} channels. It may therefore also be expected that the K^+ channel profile in efferent arterioles would differ from that in the afferent arterioles. However, Reslerova and Loutzenhiser (1995) showed that pinacidil, as in other vascular smooth muscle cells, hyperpolarizes the cell membrane and induces vasodilation of efferent arterioles from the rat. These actions of pinacidil on the efferent arteriole are reversed by glibenclamide and attenuated in the presence of 45 mM extracellular KCl, suggesting a requirement for K^+ efflux during K_{ATP}-mediated relaxation. However, it is a well-accepted view that K^+ channel openers accomplish vasodilation by hyperpolarizing the membrane, thereby inactivating voltage-operated Ca^{2+} channels. Hence, it is difficult to reconcile the reported effects of pinacidil with the previous studies that found no evidence for voltage-operated Ca^{2+} channels in efferent arterioles. Additional studies are needed to explore alternative explanations. For example, opening K^+ channels might reduce $[Ca^{2+}]_i$ by inhibiting release from internal inositol triphosphate-sensitive stores. Other mechanisms that do not involve voltage-operated Ca^{2+} channels may exist.

3.5. Regulation of Renal Arteriolar K^+ Channels

A few studies have utilized patch-clamp methods to show that specific K^+ channels in afferent arterioles are regulated by hormonal relaxing agents. Some of the agents, such as ANF, are systemically released; but most are locally produced paracrine factors, such as NO and metabolites of AA. These local agents, produced by a variety of glomerular cells, such as the endothelium, mesangium, and macula densa, are important regulators of the myogenic response and tubuloglomerular feedback.

NO, a paracrine agent that dilates smooth muscle of most, if not all, vascular beds, is a component of both normal and pathological regulation of the glomerular microcirculation. In the kidney, NO is produced from mesangial, epithelial (macula densa), and endothelial cells. In renal vascular smooth muscle, the vasodilator effects of bradykinin, acetylcholine, thrombin, and adenosine triphosphate all result from activation of NO synthase in renal vascular endothelial cells. In addition, NO is produced by invading macrophages during immune responses, such as those associated with numerous glomerulonephropathies.

Resting intrarenal NO levels regulate renal blood flow by establishing afferent arteriolar tone. The mechanism of afferent relaxation by NO was recently investigated by Alonso-Galicia et al. (1998). This group found that the vasodilator response of preconstricted rat afferent vessels to NO is partially blocked by KT-5823, an inhibitor of cGMP-dependent protein phosphokinase (PKG) (Alonso-Galicia et al., 1998). Because PKG activates BK_{Ca} channels in other tissues (see Chapter 27), these results suggest that NO relaxes afferent arterioles, in part by activating BK_{Ca} channels. However, a large portion of the total dilator response to SNP, an NO donor, is blocked by inhibiting the metabolism of AA by the cytochrome P-450 pathway. These results indicate that 20-HETE is involved in the action of NO on BK_{Ca} channels.

As first shown by Briggs et al. (1982) and Keeler and Azzarolo (1983), atrial-derived ANF modulates renal hemodynamics. ANF increases GFR by preferentially relaxing afferent, as compared to efferent, arteriolar smooth muscle cells, thereby increasing filtration pressure (Graray et al., 1985; Keeler and Azzarolo, 1983). ANF-induced salt and water excretion is blocked by inhibiting ANF's renal hemodynamic actions (Myers et al., 1975). It was shown that ANF increases intracellular cGMP in smooth muscle of afferent arterioles, which contain ANF receptors with a cytoplasmic guanylyl cyclase domain. However, it has not been determined whether PKG activates K^+ channels in afferent arterioles as shown for mesangial cells. As reviewed by Navar et al. (1996), AA has been reported to be a likely paracrine factor released from macula densa cells during tubuloglomerular feedback. The intrarenal paracrine factor 20-HETE reduces the diameter of preconstricted afferent arteriolar vessels of the rat and decreases the activity of the BK_{Ca} channels (Zou et al., 1996a). Zou and colleagues further demonstrated that 17-octadecynoic acid, an inhibitor of the P-450 pathway, activates arteriolar BK_{Ca} channels at resting membrane potentials. These results suggest that endogenous metabolic products of AA can modulate arteriolar tone by suppressing basal BK_{Ca} channel activity. However, the nature of the interaction between 20-HETE and BK_{Ca} channels is uncertain.

4. K^+ CHANNELS IN MESANGIAL CELLS

Mesangial cells regulate renal filtration rate by surrounding the glomerular capillaries and controlling the surface area available for filtration. Similarly to vascular smooth muscle cells, mesangial cells contract in response to agonists such as ANGII and endothelin-1 and relax in response to NO and ANF. As in most vascular smooth muscle cells, K^+-selective channels have a major role in the regulation of relaxation. Mesangial electrophysiology has been performed on cells cultured from rats, humans, and H-$2K^b$-tsA58 transgenic mice. Unlike isolated vascular smooth muscle cells, which may change phenotype within one generation in culture, the human mesangial cells maintain their phenotype for at least 10 generations in culture. In rat and human mesangial cells, patch-clamp studies have provided evidence for K_{ATP}, BK_{Ca}, intermediate-conductance Ca^{2+}-sensitive K^+ (IK_{Ca}), and SK_{Ca} channels.

Apamin and glibenclamide, but not iberiotoxin, initiate contraction of human mesangial cells (Sansom and Stockand, 1996). Moreover, only the iberiotoxin-sensitive channels are involved in the relaxation of mesangial cells in response to NO and ANF. These studies led to the conclusion that BK_{Ca} channels are inactive in unstimulated mesangial cells but are involved in relaxation induced by NO and ANF.

4.1. Mesangial Cell BK_{Ca} Channels

Several studies using human mesangial cells in culture have provided a wealth of information regarding BK_{Ca} channels and their roles in the regulation of mesangial contraction (Sansom *et al.*, 1997; Stockand and Sansom, 1996a, b). BK_{Ca} channels in human mesangial cells have a unitary conductance of 205 pS in a symmetrical KCl solution (140 mM K^+ intra- and extracellularly). They are voltage- and Ca^{2+}-dependent and are inhibited by scorpion toxins (Stockand and Sansom, 1994). At resting membrane potentials, mesangial BK_{Ca} channels are nearly completely quiescent. Upon addition of contractile agonists such as ANGII, BK_{Ca} channels are activated, in a negative-control feedback manner, by elevations in $[Ca^{2+}]_i$ (Stockand and Sansom, 1994).

4.2. Mesangial Cell K_{ATP} Channels

Barber *et al.* (1997) presented evidence for K_{ATP} channels in H-$2K^b$-tsA58 transgenic mice. In whole-cell patch-clamp experiments, outward currents were found that were inhibited by ATP and glibenclamide and stimulated by cromakalim. In human mesangial cells, induction of contraction by glibenclamide suggested that K_{ATP} channels were significantly active at resting potentials. However, these cells were grown in Waymouth's medium, which has a glucose concentration of 500 mg/100 ml. It is not known if K_{ATP} would be active at resting potentials under normal glucose conditions (see Section 5). As for other smooth muscle cells in the renal vasculature, definitive single-channel data consistent with the presence of K_{ATP} channels have been elusive.

4.3. Other K^+ Channels

Matsunaga *et al.* (1991) described an IK_{Ca} channel with a unitary conductance of 40 pS in cultured rat mesangial cells. In excised patches, this channel's open-state probability is enhanced by rises in $[Ca^{2+}]_i$ and by depolarization. IK_{Ca} channels are infrequently observed in cell-attached patches under basal conditions but are activated by arginine vasopressin and ANGII, presumably because these constrictor agonists elevate $[Ca^{2+}]_i$. It was therefore proposed that IK_{Ca} channels, like BK_{Ca} channels, may provide a homeostatic feedback regulatory mechanism that repolarizes the cell membrane after contraction.

Evidence in support of the presence of SK_{Ca} channels in mesangial cells is derived from studies using glomerular mesangial cells from $H\text{-}2K^b\text{-}tsA58$ transgenic mice. In whole-cell electrophysiological experiments, Barber *et al.* (1997) found that increases in $[Ca^{2+}]_i$ stimulate an apamin-sensitive K^+ current. In inside-out patches of cultured human mesangial cells, a K^+ channel with a unitary conductance of 9 pS, consistent with a SK_{Ca} channel, is also found (Sansom and Stockand, 1996). This conductance is similar to that of SK_{Ca} channels in other cell types; however, a pharmacological profile of the 9-pS channel was not performed, and, therefore, it could not be determined if the 9-pS channel was the same as that described in the mouse.

4.4. Regulation by Vasorelaxants

In mesangial cells, the feedback response of BK_{Ca} channels (i.e., the relaxant effect of these channels on vascular tone) is normally weak in the absence of a vasodilator. However, in the presence of NO or ANF, the feedback response is substantially increased (Stockand and Sansom, 1996a). In cell-attached patches, NO, ANF, and their second messenger cGMP, lead to the activation of BK_{Ca} channels and significantly increase the ability of these channels to influence membrane potential (hyperpolarization). As shown in inside-out patches, the mechanism of activation of BK_{Ca} channels by both NO and ANF is specifically mediated by PKG, which may directly activate BK_{Ca} channels by phosphorylation.

That PKG activation of BK_{Ca} is enzymatically reversible is shown in cell-attached patches by using inhibitors of protein phosphatases. Normally, BK_{Ca} channels are only activated transiently in cell-attached patches by NO, ANF, and cGMP. However, okadaic acid, a potent inhibitor of protein phosphatase 2A (PP2A) and protein phosphatase 1 (PP1), enhances and sustains the effects of these agents on BK_{Ca} channels (Sansom *et al.*, 1997). Thus, a protein phosphatase dephosphorylates and inactivates BK_{Ca} channels. Inside-out patches reveal that PP2A, but not PP1, reverses the activating effects of PKG. It is concluded that the activity of BK_{Ca} channels is controlled by a phosphorylation/dephosphorylation cycle. In the relaxed state in mesangial cells, the dephosphorylation of the channel by endogenous PP2A predominates, resulting in minimal basal activity. Relaxing agents such as NO induce an increase in BK_{Ca} activity by shifting the balance in favor of the phosphorylation limb of the cycle (Fig. 5).

The feedback activation of BK_{Ca} channels by ANGII is assumed to depend on the inositol triphosphate-induced elevation of $[Ca^{2+}]_i$, which then increases the activity of the channel. Although the time course of activation of BK_{Ca} channels generally parallels the ANGII-induced increase of $[Ca^{2+}]_i$, evidence now suggests that a Ca^{2+}-stimulated intermediary may be involved in channel activation. KCl-induced depolarization of the mesangial cell membrane induces a transient elevation of $[Ca^{2+}]_i$, which returns to

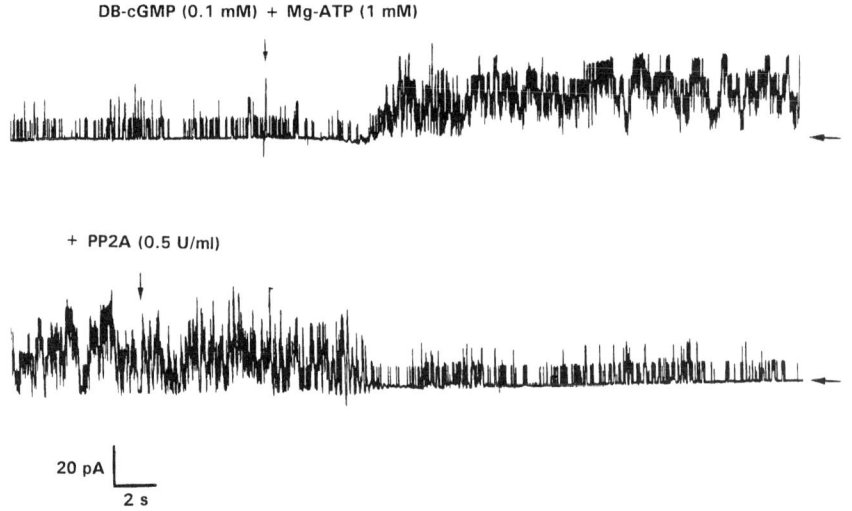

Figure 5. Tracings of BK_{Ca} in an inside-out patch of a human mesangial cell in culture. Addition of dibutyryl-cGMP (DB-cGMP) plus MgATP to the cytosolic side of the inside-out patch increases the open-state probability of the BK_{Ca} channel. Subsequent addition of protein phosphatase 2A (PP2A) reduces the open-state probability back to baseline. (Reproduced, with permission, from Sansom *et al.*, 1997.)

near baseline within 60 seconds. Although the onset of enhanced BK_{Ca} activity parallels the time course of the rise in $[Ca^{2+}]_i$, the increased open-state probability is sustained well beyond the time at which $[Ca^{2+}]_i$ returns to baseline (Hall *et al.*, 1998). Thus, some mediator, probably calmodulin-dependent protein kinase (CAMKII), prolongs the activation of BK_{Ca} channels. Patch-clamp studies show that specific CAMKII inhibitors, such as KN-62, totally inhibit the feedback activation of BK_{Ca} channels in response to agents that elevate $[Ca^{2+}]_i$. However, further investigations will be necessary to determine the details of the role of CAMKII in the feedback response of BK_{Ca}.

5. PATHOPHYSIOLOGICAL REGULATION OF K⁺ CHANNELS IN THE RENAL VASCULATURE

The renal vasculature often reacts in compensatory fashion to disorders of systemic volume and pressure regulation. In the early stages of diabetes mellitus, renal filtration rates approach levels of 50–100% greater than normal values. This state of "diabetic hyperfiltration" is the result of greater filtration pressure at the glomerulus and a failure of mesangial cells to compensate by contracting and reducing glomerular surface filtration area. Carmines *et al.* (1996) recently addressed this issue by determining the effects of diabetes on K_{ATP} channels in afferent arterioles in studies using the *in vitro* blood-perfused juxtamedullary nephron technique. Although glibenclamide did not affect the diameter of afferent arterioles from normal rats, in streptozotocin-induced diabetic rats, glibenclamide caused a concentration-dependent decrease in afferent arteriolar diameter. Moreover, pinacidil caused a significantly greater increase in diameter of the arterioles from diabetic rats, compared with control animals. These results suggested that K_{ATP} channels are both more active and upregulated in the cell membranes of diabetic rats. It was postulated that the decrease in glucose utilization in arteriolar smooth muscle leads to a decrease in ATP levels, resulting in activation of K_{ATP} and a dilation of afferent arterioles.

Other data suggest that diabetic hyperfiltration may be the result of dysfunctional regulation of BK_{Ca} channels in mesangial cells. Increased activity of BK_{Ca} channels would result in an exaggerated feedback response, manifest as an attenuated constrictor response to agonists, such as ANGII. Mesangial BK_{Ca} channels are activated by AA in a dose-dependent manner at concentrations as low as 10 nM (Stockand et al., 1998). There is evidence that the enhanced osmolality and growth factor production during diabetes will elevate levels of AA in the cell (Sansom et al., 1999). Accumulating concentrations of AA could therefore enhance the feedback response to a contractile agonist in diabetes, which would favor the relaxation of mesangial cells (Sansom et al., 1999; Stockand et al., 1998) and contribute to diabetic hyperfiltration.

In many vascular beds, K^+ channels play significant roles in adapting to increased perfusion pressure and tension on the vascular wall. Martens and Gelband (1996) used spontaneously hypertensive (SHR) rats plus deoxycorticosterone acetate (DOCA)-treated rat models to study hypertensive-induced alterations in Kv and BK_{Ca} currents of smooth muscle in interlobar arteries. It was found that the cells from hypertensive rats are depolarized and Kv current is greatly depressed, versus normotensive controls. The mechanism for the inhibition of Kv current during hypertension is not understood. The enhanced activity of BK_{Ca} channels in the renal smooth muscle cells of these hypertensive models is consistent with findings for other vascular smooth muscle cells exposed to high blood pressure (England et al., 1993) and is probably a homeostatic response that acts to buffer the elevated myogenic tone of hypertension. In support of this hypothesis, basal and peak ANGII-induced $[Ca^{2+}]_i$ are greater in smooth muscle cells from SHR and DOCA rats than in normotensive controls. However, the basal $[Ca^{2+}]_i$ is increased only marginally in these hypertensive myocytes (Martens and Gelband 1996). As identified by England et al, (1993), BK_{Ca} channel open state probability would not be significantly enhanced at the membrane potentials and $[Ca^{2+}]_i$ observed. Thus, it is likely that other variables are involved in the enhanced activity of BK_{Ca} channels in vascular smooth muscle membranes from hypertensive rats (see Chapter 42).

6. SUMMARY AND CONCLUSIONS

In renal arteries, pharmacological and electrophysiological experiments indicate the presence of at least three types of K^+-selective channels in the plasmalemmal membranes. The BK_{Ca} channels are inactive at resting membrane potentials but may be feedback regulators of agonist-induced elevations in $[Ca^{2+}]_i$. Kv channels are active at more negative membrane potentials and may be inhibited during the depolarizing response to contractile agonists. The existence of K_{ATP} channels in renal arteries is well

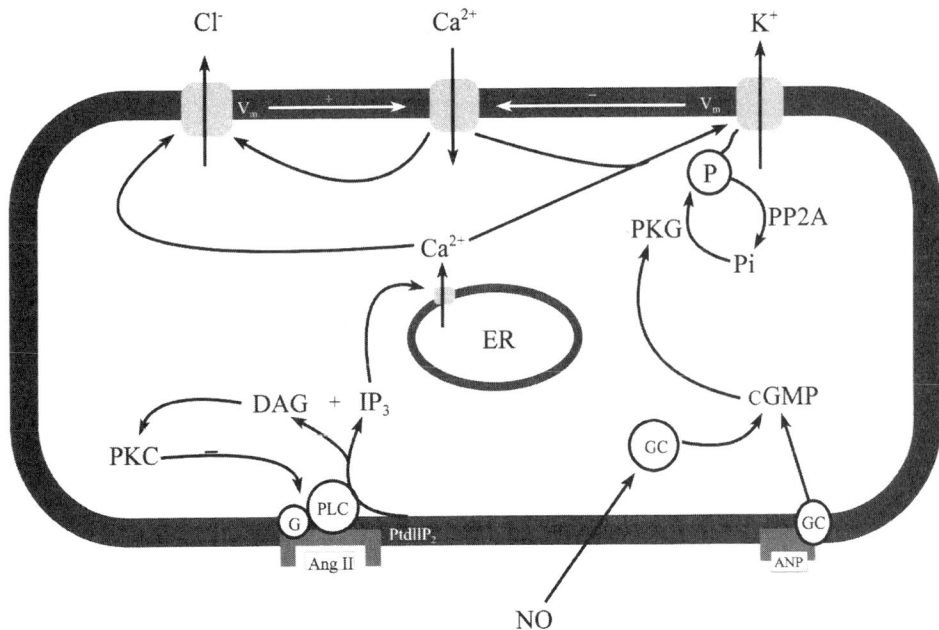

Figure 6. Cell model depicting the signal transduction pathways for the activation of BK_{Ca} channels by NO and atrial natriuretic factor (ANF). In the absence of a vasorelaxant, BK_{Ca} channels are maintained in a predominantly quiescent, dephosphorylated state by PP2A. Upon addition of relaxing agent, cGMP is elevated intracellularly and activates PKG, which either phosphorylates an intermediary or directly phosphorylates the BK_{Ca} channels, leading to increased channel opening. (Reproduced, with permission, from Stockand and Sansom, 1998.)

supported by pharmacological studies; however, these observations have not yet been verified in single-channel experiments. Pharmacological evidence suggests that NO leads to activation of BK_{Ca} and Kv channels, but not K_{ATP} channels.

The afferent and efferent arterioles have the important function of regulating the filtration pressure at the glomerulus. The smooth muscle cells of these arterioles clearly have differentiated electrophysiological mechanisms for excitation–contraction coupling. However, it is difficult to determine what role the K^+ channels have in the electrical heterogeneity of these cell types. Excellent electrophysiological evidence was provided by Zou *et al.* (1996a) for the existence of BK_{Ca} channels in the afferent arterioles and for the inhibition of these channels by 20-HETE. Pharmacological evidence shows that K_{ATP} channels are also present in preglomerular arterioles. K_{ATP} channels are quiescent at rest but are probably activated to relieve the hypoxia or metabolic stress induced by myogenic constriction. K_{ATP} channels of afferent arterioles may be activated during the high-glucose conditions of diabetes and contribute to the condition of "diabetic hyperfiltration."

Glomerular mesangial cells are smooth-muscle-like cells that surround the capillaries and regulate GFR. Mesangial tone, regulated by vascular contracting and relaxing agents, is partially governed by the K^+-selective channels in the plasmalemmal membrane. BK_{Ca} channels have a role in mesangial cells as feedback regulators of contraction. Thus, vasoconstrictor agonists activate BK_{Ca} channels by means of an inositol triphosphate-induced elevation of $[Ca^{2+}]_i$. Conversely, the vasodilators NO and ANF activate BK_{Ca} channels via the second messenger cGMP, which activates PKG, leading to channel phosphorylation (Fig. 6). Activation of BK_{Ca} channels by PKG is opposed by PP2A, which favors dephosphorylation of the channel, leading to quiescence at resting membrane potentials. Evidence from immortalized mouse mesangial cells supports the existence of IK_{Ca} and SK_{Ca} channels in mesangial cells. These channels are also likely involved in the feedback response. In whole-cell patches, outward currents that are increased by cromakalim and inhibited by glibenclamide suggest the presence of K_{ATP} channels.

In conclusion, BK_{Ca} channels have been found in all smooth muscle cells of the renal vasculature in which the patch-clamp technique has been successfully applied. Because they are activated only by large depolarizing potentials and increases in $[Ca^{2+}]_i$, BK_{Ca} channels will be activated late in the repolarizing recovery stage after an agonist-induced contraction. Although not confirmed by any single-channel data, an abundance of pharmacological studies show that K_{ATP} channels also are present in all smooth muscle cells of the renal vasculature. K_{ATP} channels may only play a role in pathological conditions such as hypoxia or other conditions characterized by deficient metabolic supplies.

REFERENCES

Alonso-Galicia, M., Sun, C.-W., Falck, J. R., Harder, D. R., and Roman, R. J., 1998, Contribution of 20-HETE to the vasodilator actions of nitric oxide in renal arteries, *Am. J. Physiol.* **275**:70–78.

Barber, R. D., Woolf, A. S., and Henderson, R. M., 1997, Potassium conductances and proliferation in conditionally immortalized renal glomerular mesangial cells from the *H-2K^b*-tsA58 transgenic mouse, *Biochim. Biophys. Acta* **1355**:191–203.

Breitwieser, G. E., 1996, Mechanisms of K^+ channel regulation, *J. Membr. Biol.* **152**:1–11.

Briggs, J. P., Steipe, B., Schubert, G., Schnermann, J., 1982, Micropuncture Studies of the renal effects of atrial natriuretic substances. *Pflüger Arch.* **395**:271276.

Carmines, P. K., Fowler, B. C., and Bell, P. D., 1993, Segmentally distinct effects of depolarization on intracellular $[Ca^{2+}]$ in renal arterioles, *Am. J. Physiol.* **265**:F677–F685.

Carmines, P. K., Fallet, R. W., Bast, J. P., Ishii, N., Fujiwara, K., and Sansom, S. C., 1999, Impact of Ca^{2+}-activated K^+ channels on afferent arteriolar responses to depolarizing agonists, *FASEB J.* Vol. 13, A717.

Carroll, M. A., Garcia, M. P., Falck, J. R., and McGiff, J. C., 1992, Cyclooxygenase dependency of the renovascular actions of cytochrome P-450 derived arachidonic acid metabolites, *J. Pharmacol. Exp. Ther.* **260**:104–109.

Chilton, L., and Loutzenhiser, R., 1998, Potassium-induced dilation of pre-glomerular arterioles: Evidence of the inward rectifier K channel (KIR) in the renal microcirculation, *J. Am. Soc. Nephrol* **9**:336A (Abstract).

England, S. K., Wooldridge, T. A., Stekiel, W. J., and Rusch, N. J., 1993, Enhanced single-channel K^+ current in arterial membranes from genetically hypertensive rats, *Am. J. Physiol.* **264**:H1337–H1345.

Gebremedhin, D., Kaldunski, M., Jacobs, E. R., Harder, D. R., and Roman, R. J., 1996, Coexistence of two types of Ca^{2+}-activated K^+ channels in rat renal arterioles, *Am. J. Physiol.* **270**:F69–F81.

Gelband, C. H., and Hume, J. R., 1992, Ionic currents in single smooth muscle cells of the canine renal artery, *Circ. Res.* **71**:745–758.

Gelband, C. H., Ishikawa, T., Post, J. M., Keef, K. D., and Hume, J. R., 1993, Intracellular divalent cations block smooth muscle K^+ channels, *Circ. Res.* **73**:24–34.

Gojkovic, L., and Kazic, T., 1994, A comparison of the relaxant effects of pinacidil in rabbit renal and mesenteric artery, *Gen. Pharmacol.* **25**:1711–1717.

Gordienko, D. V., Clausen, C., and Goligorsky, M. S., 1994, Ionic currents and endothelin signaling in smooth muscle cells from rat renal resistance arteries, *Am. J. Physiol.* **266**:25–41.

Graray, R., Hannaert, P., Rodrigue, F., Dunham, B., Marche, P., Genest, J., Braquet, P., Bianchi, C., Cantin, M., and Meyer, P., 1985, Atrial natriuretic factor inhibits Ca-dependent K fluxes in cultured vascular smooth muscle, *J. Hypertens.* **3**:S297–S298.

Hall, D., Carmines, P. K., and Sansom, S. C., 1998, Role of Ca^{2+}/calmodulin-dependent protein kinase (CAMKII) in the angiotensin II (ANGII)-induced feedback activation of BK_{Ca} in human glomerular mesangial cells, *J. Am. Soc. Nephrol* **9**:35A (Abstract).

Harder, D. R., Campbell, W. B., and Roman, R. J., 1995, Role of cytochrome P-450 enzymes and metabolites of arachidonic acid in the control of vascular tone, *J. Vasc. Res.* **32**:79–92.

Hayashi, K., Matsumura, Y., Yoshida, Y., Ohyama, T., Hisaki, K., Suzuki, Y., and Morimoto, S., 1990, Effects of BRL 34915 (cromakalim) on renal hemodynamics and function in anesthetized dogs, *J. Pharmacol. Exp. Ther.* **252**:1240–1245.

Ikenaga, H., Bast, J. P., Fallet, R. W., and Carmines, P. K., 1996, Role of ATP-sensitive K^+ channels in the renal afferent arteriolar dilation accompanying the early stage of IDDM in the rat, *J. Am. Soc. Nephrol* **7**:1582 (Abstract).

Ikenaga, H., Bast, J. P., Fallet, R. W., Carmines, P. K., 2000, Exaggerated impact of ATP-sensitive K^+ channels on afferent arteriolar diameter in diabetes mellitus. *J. Am. Soc. Nephrol.* **11**(7):11991207.

Keeler, R., and Azzarolo, A. M., 1983, Effects of atrial natriuretic factor on renal handling of water and electrolytes in rats, *Can. J. Physiol. Pharmacol.* **61**:996–1002.

Lorenz, J. N., Schnermann, J., Brosius, F. C., Briggs, J. P., and Furspan, P. B., 1992, Intracellular ATP can regulate afferent arteriolar tone via ATP-sensitive K^+ channels in the rabbit, *J. Clin. Invest.* **90**:733–740.

Loutzenhiser, R. D., and Parker, M. J., 1994, Hypoxia inhibits myogenic reactivity of renal afferent arterioles by activating ATP-sensitive K^+ channels, *Circ. Res.* **74**:861–869.

Ma, Y., Gebremedhin, D., Schwartzman, M. L., Falck, J. R., Clark, J. E., Masters, B. S., Harder, D. R., and Roman, R. J., 1993, 20-Hydroxyeicosatetraenoic acid is an endogenous vasoconstrictor of canine renal arcuate arteries, *Circ. Res.* **72**:126–136.

Martens, J. R., and Gelband, C. H., 1996, Alterations in rat interlobar artery membrane potential and K^+ channels in genetic and nongenetic hypertension, *Circ. Res.* **79**:295–301.

Matsunaga, H., Yamashita, N., Miyajima, Y., Okuda, T., Chang, H., Ogata, E., and Kurokawa, K., 1991, Ion channel activities of cultured rat mesangial cells, *Am. J. Physiol.* **261**:F808–F814.

Metzger, F., and Quast, U., 1999, Binding of ^{3}H-P1075, an opener of ATP-sensitive K channels, to rat glomerular preparations, *Naunyn-Schmiedeberg's Arch. Pharmacol.* **354**:452–459.

Miura, K., Ebara, T., Okumura, M., Matsuura, T., Kim, S., Yukimura, T., and Iwao, H., 1995, Attenuation of adrenomedullin-induced renal vasodilatation by N^G-nitro L-arginine but not glibenclamide, *Br. J. Pharmacol.* **115**:917–924.

Myers, B. D., Deen, W. M., and Brenner, B. M., 1975, Effects of norepinephrine and angiotensin II on the determinants of glomerular ultrafiltration and proximal tubule fluid reabsorption in the rat, *Circ. Res.* **37**:101–110.

Navar, L. G., Inscho, E. W., Majid, D. S. A., Imig, J. D., Harrison-Bernard, L. M., and Mitchell, K. D., 1996, Paracrine regulation of the renal microcirculation, *Physiol. Rev.* **76**:425–536.

Radermacher, J., Forstermann, K., and Frohlich, J. C., 1990, Endothelium-derived relaxing factor influences renal vascular resistance, *Am. J. Physiol.* **259**:F9–F17.

Reslerova, M., and Loutzenhiser, R., 1995, Divergent mechanisms of ATP-sensitive K^+ channel-induced vasodilation in renal afferent and efferent arterioles—evidence of L-type Ca^{2+} channel-dependent and -independent actions of pinacidil, *Circ. Res.* **77**:1114–1120.

Sansom, S. C., and Stockand, J. D., 1996, Physiological role of large, Ca-activated K channels in human glomerular mesangial cells. *Clin. Exp. Pharmacol. Physiol.* **23**:76–82.

Sansom, S. C., Stockand, J. D., Hall, D., and Williams, B., 1997, Regulation of large calcium-activated potassium channels by protein phosphatase 2A, *J. Biol. Chem.* **272**:9902–9906.

Sansom, S. C., Mehta, P., and Hall, D. A., 1999, Potentiating effects of hyper-omolality and epidermal growth factor on the release of arachidonic acid in human glomerular mesangial cells, *Diabetes Res. Clin. Practice.*

Stockand, J. D. and Sansom, S. C., 1994, Large Ca^{2+}-activated K^+ channels responsive to angiotensin II in cultured human mesangial cells, *Am. J. Physiol.* **267**:C1080–C1086.

Stockand, J. D., and Sansom, S. C., 1996a, Role of large, Ca-activated K channels in regulation by nitroprusside and ANP of mesangial contraction, *Am. J. Physiol.* **270**:C1773–C1779.

Stockand, J. D., and Sansom, S. C., 1996b, Mechanism of activation by cGMP-dependent protein kinase of large Ca^{2+}-activated K^+ channels in mesangial cells, *Am. J. Physiol.* **271**:C1669–C1677.

Stockand, J. D., and Sansom, S. C., 1998, Glomerular mesangial cells: Electrophysiology and regulation of contraction, *Physiol. Rev.* **78**:723–744.

Stockand, J. D., Silverman, M., Hall, D., Derr, T., Kubacak, B., and Sansom, S. C., 1998, Arachidonic acid potentiates the feedback response of mesangial BK_{Ca} channels to angiotensin II, *Am. J. Physiol.* **274**:F658–F664.

Tamaki, T., Hasui, K., Shoji, T., Fukui, K., Iwao, H., and Abe, Y., 1991, Effects of cromakalim on renal hemodynamics and function in dogs: Comparison with nicardipine, *J. Cardiovasc. Pharmacol.* **17**:305–309.

Toda, N., Nakajima, S., Miyazaki, M., and Ueda, M., 1985, Vasodilatation induced by pinacidil in dogs: Comparison with hydralazine and nifedipine, *J. Cardiovasc. Pharmacol.* **7**:1118–1126.

Videbaek, L. M., Aalkjaer, C., and Mulvany, M. J., 1988, Pinacidil opens K^+-selective channels causing hyperpolarization and relaxation of noradrenaline contractions in rat mesenteric resistance vessels, *Br. J. Pharmacol.* **95**:103–108.

Williams, S. B., Cusco, J. A., Roddy, M. A., Johnstone, M. T., and Creager, M. A., 1996, Impaired nitric oxide-mediated vasodilation in patients with non-insulin-dependent diabetes mellitus, *J. Am. Coll. Cardiol.* **27**:567–574.

Wilson, C., and Cooper, S. M., 1989, Effect of cromakalim on contractions in rabbit isolated renal artery in the presence and absence of extracellular Ca^{2+}, *Br. J. Pharmacol.* **98**:1303–1311.

Yukimura, T., Yamashita, Y., Miura, K., Okumura, M., Yamanaka, S., and Yamamoto, K., 1992, Renal effects of the nitric oxide synthase inhibitor, L-N^G-nitroarginine, in dogs, *Am. J. Hypertens.* **5**:484–487.

Zou, A., Fleming, J. T., Falck, J. R., Jacobs, E. R., Gebremedhin, D., Harder, D. R., and Roman, R. J., 1996a, 20-HETE is an endogenous inhibitor of the large-conductance Ca^{2+}-activated K^+ channel in renal arteries, *Am. J. Physiol.* **270**:R228–R237.

Zou, A. P., Fleming, J. T., Falck, J. R. Jacobs, E. R., Gebremedhin, D., Harder, D. R., and Roman, R. J., 1996b, Stereospecific effects of epoxyeicosatrienoic acids on renal vascular tone and K^+-channel activity, *Am. J. Physiol.* **270**:F822–F832.

Chapter 29

Potassium Channels in the Coronary Circulation

Maik Gollasch

1. INTRODUCTION

The heart consumes more oxygen per gram of tissue than any other organ of the body. One of the most striking features of the coronary circulation is the close relationship between the requirement for oxygen and metabolic substrates and the magnitude of coronary blood flow. The mechanisms underlying this interaction are still poorly understood but seem to involve adaptive regulatory processes in the coronary macro- and microvasculature. Daut *et al.* (1990) first demonstrated that hypoxic vasodilation in isolated, perfused guinea pig hearts is prevented by glibenclamide (Fig. 1), suggesting a role for adenosine triphosphate-sensitive K^+ (K_{ATP}) channels, in the regulation of adaptive processes in the coronary macro- and microvasculature and making the investigation of K_{ATP} channels in the coronary circulation clinically relevant.

Large coronary arteries and veins form separate anatomical and functional entities, whereas the microvasculature is an integral component of the tissue it supplies, and its function varies according to the activity of the surrounding tissue. There is evidence that coronary resistance arteries and the surrounding myocardium form a functional unit ("myoendothelial regulatory unit") with multiple electrochemical interactions between coronary endothelial cells, smooth muscle cells, perivascular nerves, and cardiomyocytes (Daut *et al.*, 1994). However, the microvasculature *in situ* is rather inaccessible for electrophysiological studies. This is one of the reasons why our knowledge about plasmalemmal ion channels and the regulation of coronary blood flow is still limited. Moreover, recent data have indicated that K^+ channels are diverse and that their distribution varies in different vascular beds (Albarwani *et al.*, 1995). Many studies of K^+ channels have involved the investigation of nonhuman coronary arteries, but human tissue may differ in many important ways from nonhuman tissue. Technical advances, such as the patch-clamp technique, as well as measurements of contractile properties of large and small coronary arteries have enabled several groups

Maik Gollasch • Franz Volhard Clinic at the Max Delbrück Center for Molecular Medicine, Charité University Hospitals, Humboldt University Berlin, D-13125 Berlin, Germany.

Potassium Channels in Cardiovascular Biology, edited by Archer and Rusch. Kluwer Academic/Plenum Publishers, New York, 2001.

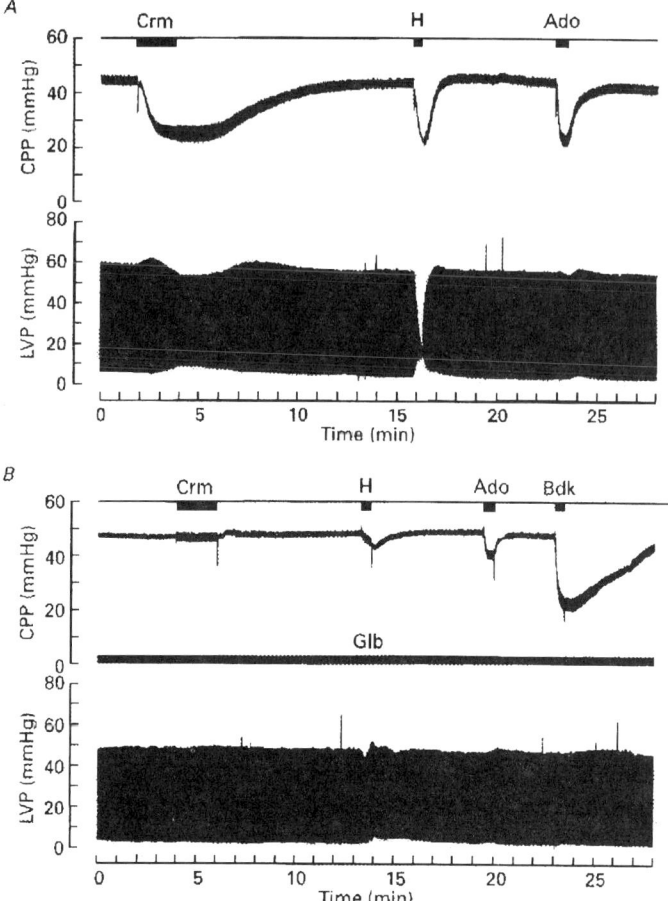

Figure 1. K_{ATP} channels and coronary circulation in beating heart. Coronary perfusion pressure (CPP) and isovolumetric left ventricular pressure (LVP) were recorded in an isolated perfused guinea pig heart. Cromakalim (Crm, 500 nM), hypoxia (H), adenosine (Ado, 1 μM), and bradykinin (Bdk, 1 nM) were applied as indicated by the horizontal boxes. Recordings were made under control conditions (A) or in the presence of glibenclamide (Glb, 1 μM) (B) Blockade of K_{ATP} channels by glibenclamide substantially inhibited coronary vasodilation (seen as a fall in coronary perfusion pressure) to cromakalim, hypoxia, and adenosine. (Reproduced, with permission, from von Beckerath *et al.*, 1991.)

to characterize K^+ channels in human coronary arteries and to investigate their role in the control of membrane potential and arterial diameter. The elucidation of K^+ channels in human coronary arteries may provide new insight into the control of coronary blood flow in the human heart and also may have future therapeutic implications.

Several avenues of recent research on K^+ channels in human coronary arterial smooth muscle cells are discussed in this chapter. Evidence that four types of K^+ channels (Kv, BK_{Ca}, Kir, and K_{ATP}) serve unique functions in the regulation of coronary arterial smooth muscle membrane potential is discussed. The channels integrate a variety of vasoactive signals to dilate or constrict coronary arteries through regulation of the membrane potential in arterial smooth muscle. I will emphasize that the role of K$^+$ channels in differentiation and proliferation of vascular smooth muscle cells is uncertain. Studies of K$^+$ channels in other types of nonhuman, noncoronary vascular smooth muscle (mesenteric artery, cerebral artery, urinary bladder, trachea, gastrointestinal cells, veins, cultured cells) and non-smooth-muscle cells are not discussed unless similar material from coronary arterial smooth muscle is not available.

2. VOLTAGE-DEPENDENT Ca^{2+} CHANNELS, MEMBRANE POTENTIAL AND REGULATION OF CORONARY MYOGENIC TONE

Resistance-sized coronary arteries develop inherent (myogenic) tone in response to increases in transmural pressure. The development of myogenic tone is associated with a graded depolarization of the smooth muscle cell membrane and an increase in the *global* intracellular Ca^{2+} concentration, $[Ca^{2+}]_i$, in the arterial wall (Harder, 1984; Nelson et al., 1990; Siegel et al., 1992; Brayden and Nelson, 1992; Knot and Nelson, 1995). For example, physiological pressure depolarizes smooth muscle cells of human coronary arterioles (diameter, $78 \pm 11 \mu m$) to about -40 mV at 60 mm Hg and constricts the arteries by about 38% (Knot et al., 1998; see also Miller et al., 1997; Miura and Gutterman, 1998; Miller et al., 1998; Wellman et al., 1996, Brayden and Nelson, 1992; Knot and Nelson, 1995). Under these conditions, human coronary and other arterial smooth muscle cells have constant or slowly changing membrane potentials, between -55 and -35 mV (Knot et al., 1998; Brayden and Nelson, 1992; Harder, 1984; Harder et al., 1987; Nelson et al., 1990; Hirst and Edwards, 1989; Neild and Keef, 1985). The relationship between membrane potential and *global* $[Ca^{2+}]_i$ in the arterial wall is very steep, so that even membrane potential changes of a few millivolts may cause significant changes in *global* $[Ca^{2+}]_i$ and blood vessel diameter (Brayden and Nelson, 1992; Nelson et al., 1990; Knot and Nelson, 1995; Knot et al., 1996a). Conversely, myogenic tone in these arteries is abolished by the membrane hyperpolarization caused by K$^+$ channel openers, by the removal of external Ca^{2+}, and by long-lasting ("L-type") Ca^{2+} channel-blocking drugs (Kuo et al., 1990; Eckman et al., 1994; Daut et al., 1994; Wellman et al., 1996; Gollasch and Nelson, 1997; Rusch and Liu, 1997). These observations suggest that voltage-dependent, L-type Ca^{2+} channels play an important role in controlling steady-state Ca^{2+} entry. Similar results have been obtained in large, epicardial coronary arteries of humans, suggesting a major role of voltage-dependent Ca^{2+} channels in controlling steady-state Ca^{2+} entry in human conduit coronary arteries (Nakashima et al., 1993; Gollasch et al., 1995, 1996a; Bruch et al., 1997; Godfraind et al., 1992).

Recently, single voltage-dependent Ca^{2+} channels have been studied in arterial smooth muscle using physiological low concentrations of charge carrier (1.25–2.0 mM Ca^{2+}) and at steady-state membrane potentials that occur in arterial smooth muscle (Gollasch et al., 1992; Rubart et al., 1996, Gollasch et al., 1996b; Gollasch and Nelson, 1997). The plots in Fig 2A–C demonstrate important features of voltage-dependent Ca^{2+} channels including a steep voltage dependence of channel openings at steady-state membrane potentials between -60 and -30 mV and a relatively high permeation rate

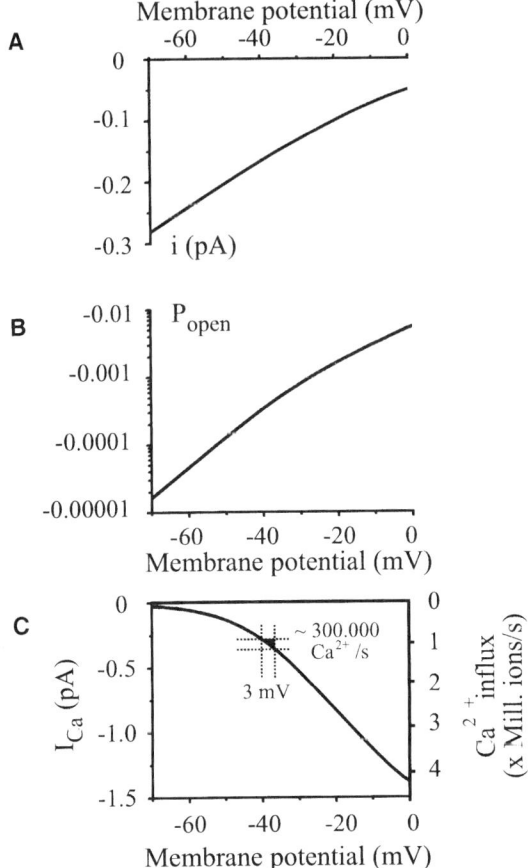

Figure 2. Relationship between voltage-dependent Ca^{2+} channel properties and Ca^{2+} entry in a voltage range that occurs in arterial smooth muscle. (A) Hypothetical relationship between the single-channel current amplitude (i) and voltage (V_m). This relationship was generated from the Goldman–Hodgkin–Katz equation: $i = (PV_m)\{[Ca_o - Ca_i \exp(zV_mF/RT)]/[1 - \exp(zV_mF/RT)]\}$, where P is $P_{Ca}z^2F^2/RT$ (F is Faraday's constant, $z = 2$, R is the gas constant, and $T = 293.16\,K$) and Ca_o (1.25 mM) and Ca_i ($10^{-7}\,M$) are external and internal Ca^{2+} concentrations, respectively. P is a constant and was scaled so that $i = -0.17\,pA$ at -40 mV. This would approximate unitary current at near physiological conditions (Gollasch and Nelson, 1997). (B) Hypothetical steady-state voltage dependence of open-state probability (P_{open}) generated from activation and inactivation curves. Steady-state P_{open} was generated by the following equation: $P_{open} = P_{act}(1 - P_{inact})P'_{funct}$, where $P_{act} = 1/\{1 + \exp[(V_{0.5} - V_m)/k_S]\}$ and $P_{inact} = (1 - C)/\{1 + \exp[(V_h - V_m)/k]\}$. For this example, midpoints of activation ($V_{0.5}$) and inactivation (V_h) curves, steepness factors of activation (k_S) and inactivation (k) factors, and a noninactivating component (C) were similar to those previously reported for voltage-dependent activation and a fast inactivation process of Ca^{2+} channels in arterial smooth muscle and were 6.2 mV, -24.2 mV, 9.5 mV, 9.6 mV, and 0.27, respectively (Rubart *et al.*, 1996). P'_{funct} was 0.05, because maximal P_{open} was <1 and P_{open} is reduced by an additional, slow inactivation process at steady-state membrane potentials between -40 mV and -20 mV in arterial smooth muscle, decreasing P_{open} by approximately 6-fold at all of these potentials (Rubart *et al.*, 1996). (C) Hypothetical steady-state whole-cell Ca^{2+} current (I_{Ca})–voltage relationship. I_{Ca} is the product of i [from (A)] and steady-state P_{open} [from (B)] multiplied by the number of channels (N); the value of N is 5000, which would be appropriate also for small myogenic arteries, including coronary arteries. (Reproduced with permission, from Gollasch and Nelson, 1997.)

at physiological Ca^{2+} concentrations, being about 10^6 Ca^{2+} ions/s at -50 mV. The relationships provide strong support for the idea that a small number of open L-type, but not T-type, Ca^{2+} channels supplies the steady-state Ca^{2+} influx (~ 0.5 pA) needed to maintain a constricted state in arterial smooth muscle (Fig. 2C). The *global* [Ca^{2+}]$_i$ regulates contractile force of arterial smooth muscle by mechanisms involving myosin light chain phosphorylation by a kinase activated by the Ca^{2+}-calmodulin complex. The phosphorylation leads to actin–myosin interaction, and hence to force development (Morano, 1992; Ruegg and Pfitzer, 1991; Rembold and Murphy, 1993; Somlyo and Somlyo, 1994). The close and positive correlation between the open-state probability of L-type Ca^{2+} channels, [Ca^{2+}]; and force development predicts that any physiological or pharmacological agent that alters membrane potential will significantly change *global* intracellular [Ca^{2+}]$_i$, blood vessel diameter, and contractile force.

L-type Ca^{2+} channels in vascular smooth muscle cells are affected by three main classes of calcium channel antagonist drugs, including dihydropyridines (e.g., nifedipine), phenylalkylamines (e.g., verapamil), and benzothiazepines (e.g., diltiazem) (Worley *et al.*, 1986, 1991; Kuga *et al.*, 1996; Hering *et al.*, 1993; Klöckner and Isenberg, 1991; Bean, 1991; Cai *et al.*, 1997). Recently, farnesol, a natural and endogenous metabolite present in all mammalian cells, has been found to block L-type channels in vascular smooth muscle by targeting a previously unrecognized regulatory site on the pore-forming α1-subunit. This constitutes a novel mechanism of channel block (Luft *et al.*, 1999).

3. EFFECT OF MEMBRANE POTENTIAL ON Ca^{2+} INFLUX THROUGH VOLTAGE-DEPENDENT Ca^{2+} CHANNELS

As illustrated in Fig. 2C, the relationship between Ca^{2+} influx through voltage-dependent Ca^{2+} channels and membrane potential is very steep in the physiological range of arterial smooth muscle membrane potential (e.g., between -60 and -30 mV), in accordance with the hypothesis that the membrane potential of smooth muscle cells primarily regulates muscle contractility through alterations in Ca^{2+} influx through voltage-dependent, L-type Ca^{2+} channels (Nelson *et al.*, 1990; Daut *et al.*, 1994). Membrane potential also may regulate Ca^{2+} entry through Na$^+$/Ca^{2+} exchange as well as affect intracellular Ca^{2+} release through the voltage dependence of inositol trisphosphate production (Ganitkevich and Isenberg, 1993; Itoh *et al.*, 1992; Ganitkevich and Isenberg, 1996; Nelson *et al.*, 1990). This assumption predicts that any physiological or pharmacological agent that alters membrane potential will significantly change blood vessel diameter.

Because the resting input resistance of arterial smooth muscle cells is high, i.e., on the order of ~ 10 GΩ, very few K$^+$ channels need to be open to contribute to the cell's membrane conductance. Based on a simple parallel-conductance model, Nelson and Quayle (1995) have calculated that increasing the membrane K$^+$ conductance by 15 pS (e.g., increasing the open probability of Kv channels from 0.003 to 0.005 and assuming a Kv channel conductance of about 25 pS at -60 mV) would hyperpolarize the cell membrane by 3 mV. Fiigure 2C illustrates that a membrane hyperpolarization of 3 mV in the membrane potentials range between -50 mV and -30 mV that occur in arterial smooth muscle cells of pressurized, myogenic coronary arteries would significantly impact Ca^{2+} influx (Wellman *et al.*, 1996; Knot *et al.*, 1998).

4. OUTWARDLY RECTIFYING K⁺ CHANNELS

In human coronary arterial smooth muscle cells, the outward K^+ current activated by depolarization is primarily comprised of two components (Fig. 3). One component is carried by BK_{Ca} channels, and the other by channels whose characteristics resemble those of Kv (delayed rectifier) channels in other types of smooth muscle cells (Gollasch *et al.*, 1996a). In the human cells, current activated at potentials negative to $+20$ mV is mainly Kv current and is blocked by 4-aminopyridine and possibly also by high

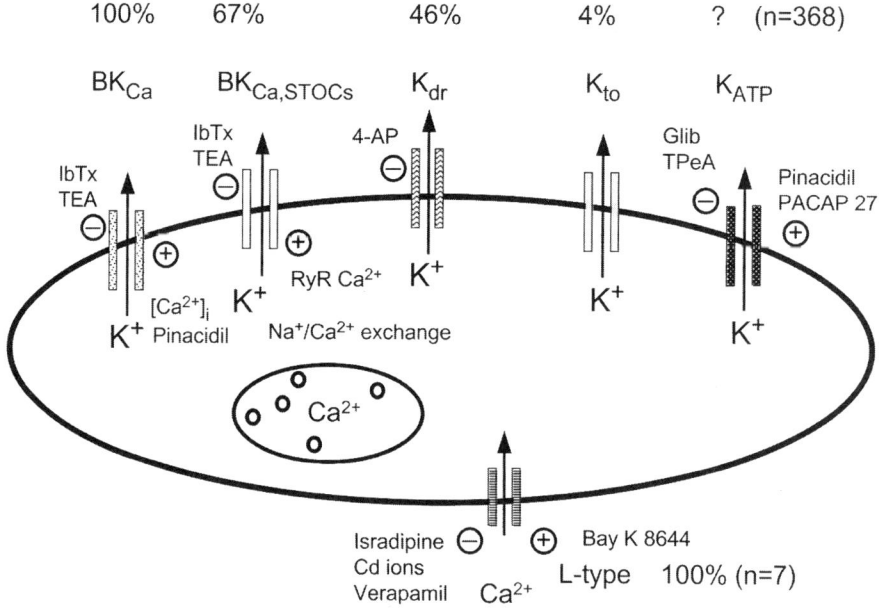

Figure 3. K^+ channels identified in smooth muscle cells from large conduit coronary arteries of humans. Despite similar microscopic appearance and passive electrical properties, the K^+ channels displayed functional heterogeneity. The most prominent K^+ currents in human coronary arterial smooth muscle cells were currents through BK_{Ca} channels (expressed in 100% of 368 cells) and Kv "delayed rectifier" channels (K_{dr}; 46%) and spontaneous transient outward K^+ currents (STOCs; 67%) (data from Gollasch *et al.*, 1996a). Transient outward K^+ (K_{to}) currents were observed only in a small fraction of cells (4%). Inwardly rectifying K^+ (Kir) currents were not observed in smooth muscle cells from large conduit coronary arteries of humans. The ATP-dependent K^+ (K_{ATP}) channel current was small under physiological conditions but increased markedly when the cells were activated by pinacidil or pituitary adenylate cyclase activating peptides (PACAP 27). L-type Ca^{2+} channels were observed in all cells tested (Ried, 1997). Abbreviations: 4-AP, 4-Aminopyridine; TEA, tetraethylammonium ions; TpeA, tetrapentylammonium ions; Glib, glibenclamide; IbTx, iberiotoxin; RyR, ryanodine receptors; L-type, dihydropyridine-sensitive voltage-dependent Ca^{2+} channel. $+$, Stimulation; $-$, inhibition.

concentrations of tetraethylammonium (TEA) ions. The current activated at more positive potentials consists of both BK$_{Ca}$ and Kv currents. The BK$_{Ca}$ component is blocked by low concentrations of TEA, as well as iberiotoxin (100 and 300 nM), whereas the Kv component is blocked by 4-aminopyridine. The BK$_{Ca}$ component is sensitive to the internal Ca^{2+} concentration. Direct measurement of K$^+$ channel unitary current amplitudes underlying the TEA- and iberiotoxin-sensitive current reveals a large unitary conductance of $\sim$150 pS between -10 and $+30$ mV (K$_o$/K$_i$, 6/130 mM) (Bychkov $et\ al.$, 1997a), similar to values reported for BK$_{Ca}$ channels in other preparations.

4.1. Ca^{2+}-Activated K$^+$ Channels

BK$_{Ca}$ channels have been described in almost all vascular smooth muscle cell types studied so far, including coronary arteries of guinea pig (Ganitkevich and Isenberg, 1990), dog (Buljubasic $et\ al.$, 1993), and rabbit (Ishikawa $et\ al.$, 1993, Leblanc $et\ al.$, 1994). These K$^+$ channels have often been called "big" or "maxi" K$_{Ca}$ channels. In arterial smooth muscle cells, estimates of BK$_{Ca}$ channel density ranges from 1000 to 10,000 channels per cell (Nelson and Quayle, 1995). Single-channel experiments reveal that at least three BK$_{Ca}$ channel subtypes (K$_L$, K$_S$, K$_M$) may exist in vascular smooth muscle cells (Inoue $et\ al.$, 1986). For example, in human coronary arteries, the K_i value for inhibiting BK$_{Ca}$ current with TEA (K_i of 150 μM at $+80$ mV) is 6.3-fold lower than the K_i observed in human mesenteric arteries (K_i at $+80$ mV, 850 μM) (Smirnov and Aaronson, 1992). In both vascular tissues, both currents are completely blocked by 100 nM iberiotoxin. The BK$_{Ca}$ channel activity in human coronary artery smooth muscle cells is unaffected by a dose of glibenclamide (3 μM) that blocks K$_{ATP}$ channels (Bychkov $et\ al.$, 1997a). Although the basic properties of BK$_{Ca}$ channels in human coronary arteries are not different from those described in arteries from other species (Ishikawa $et\ al.$, 1993; Leblanc $et\ al.$, 1994; Langton $et\ al.$, 1991, Gelband and Hume, 1992), the data indicate that there are differences in TEA sensitivity between BK$_{Ca}$ channels in human coronary arteries and in human mesenteric arteries. Furthermore, because both TEA-inhibition dose–response curves were well fitted by a Langmuir equation with a Hill coefficient of 1, both preparations may express only a single dominant population of different BK$_{Ca}$ channel subtypes. This finding may represent an important difference between human coronary smooth muscle cells and those of the rabbit, which show K_i values for TEA between 0.3 and 1 mM at $+60$ mV and a Hill coefficient unequal to one (Smirnov and Aaronson, 1992). Whether different BK$_{Ca}$ channels in human coronary and mesenteric smooth muscle cells have different functions in the regulation of vascular tone or smooth muscle cell proliferation remains to be determined. Because 4-aminopyridine ($K_i > 1$ mM), glibenclamide ($K_i > 10$ μM), and external Ba^{2+} ($K_i > 10$ mM) have little effect on arterial smooth muscle BK$_{Ca}$ channels (Nelson and Quayle, 1995), pharmacological approaches may be helpful in delineating the functional role of BK$_{Ca}$ channels from those of other K$^+$ channel types.

However, our group has observed that human and porcine coronary artery BK$_{Ca}$ channels can be stimulated by pinacidil. This stimulatory effect is not inhibited by glibenclamide (Bychkov $et\ al.$, 1997a). Stimulation of BK$_{Ca}$ channels by pinacidil and other synthetic agents also has been reported in rat portal vein and rabbit aorta. However, in these preparations the activation of BK_{Ca} channels by pinacidil and other agents is sensitive not only to the BK_{Ca} channel blocker charybdotoxin, but also to the K_{ATP} channel blocker glibenclamide (Hu et al., 1990; Gelband and McCullough, 1993). Activation of these channels also may mediate vasorelaxation induced by "potassium channel openers" such as cromakalim and P1075 (a pinacidil analog) in porcine coronary arteries (Balwierczak et al., 1995). The reason for the discrepancy in the glibenclamide sensitivity is unclear, but it may be explained by the presence of different BK_{Ca} ($BK_{Ca,ATP}$) channel subtypes in different tissue preparations. Activation of aortic and basilar artery BK_{Ca} channels by pinacidil and other synthetic "potassium channel openers" also has been observed in excised membrane patches (Hermsmeyer, 1988; Stockbridge et al., 1991) and in bilayer membranes (Gelband and McCullough, 1993). Although these results indicate that BK_{Ca} channels may be directly activated by "potassium channel openers," our data do not rule out an indirect effect of pinacidil on BK_{Ca} channels in human and porcine coronary artery vascular smooth muscle cells, since most of the experiments were performed using the perforated-patch configuration with nystatin. A possible direct activation of human coronary BK_{Ca} channels by pinacidil remains to be shown in future experiments, for example, by testing the effects of pinacidil on coronary arterial BK_{Ca} channels in excised membrane patches. Interestingly, Tanaka et al. (1997) reported that most native BK_{Ca} channels in human coronary artery smooth muscle cells are composed of a central, pore-forming α-subunit and an auxiliary β-subunit. The high Ca^{2+} sensitivity of the channels results from the coupling of the β-subunit to the α-subunit. Additional regulatory subunits and splice variants still may be discovered. Hence, significant functional and pharmacological heterogeneity of BK_{Ca} channels in different smooth muscle cell types may result from the expression of multiple isoforms of the pore-forming α-subunit and accessory subunits.

4.2. Kv Channels

Voltage-gated, "delayed rectifying" K^+(Kv) channels have been described in coronary vascular smooth muscle cells from rabbit (Ishikawa et al., 1993; Leblanc et al., 1994; Volk et al., 1991). The channels are strongly voltage-dependent, and their steady-state activation is described by a Boltzmann function of membrane potential. During prolonged depolarization, the Kv current inactivates slowly, with a time constant measured in seconds. In human coronary arterial smooth muscle cells, the 4-aminopyridine-sensitive component of outward current has characteristics similar to the basic properties of Kv currents reported in smooth muscle cells from different preparations (Leblanc et al., 1994; Volk et al., 1991; Beech and Bolton, 1989a; Robertson and Nelson, 1994; Gelband and Hume, 1992; Cole et al., 1996). In this respect, the Kv current is relatively insensitive to TEA at concentrations around 10 mM, and the whole-cell current shows relatively little noise, consistent with a small single-channel conductance (Volk et al., 1991; Ishikawa et al., 1993; Gollasch et al., 1996a). In an approximately physiological K^+ gradient, a single-channel conductance of 7.3 pS has been reported in porcine coronary arteries (Volk and Shibata, 1993).

However, a larger single-channel conductance of 70 pS exists in rabbit coronary arteries (140 mM/140 mM K$^+$) (Ishikawa *et al.*, 1993). The K_i value for 4-aminopyridine block of Kv current in human coronary vascular smooth muscle cells is 1.02 mM at $+20$ mV (Gollasch *et al.*, 1996a), which is close to the K_i value for sensitivity of Kv current to 4-aminopyridine in human mesenteric arteries (K_i of 1.04 mM at $+20$ mV) (Smirnov and Aaronson, 1992). However, marked differences are observed in the kinetic properties of Kv currents in the two preparations. First, Kv currents in human mesenteric vascular smooth muscle cells inactivated more rapidly than those in human coronary arterial cells. In mesenteric arteries, Kv current elicited at $+80$ mV inactivated by about 50% within 300 ms whereas, in human coronary arteries, this value was about 10%. Second, half-maximal inactivation ($V_{0.5}$) in human mesenteric arteries occurred at -38.0 and -29.7 mV for transient and sustained current components, respectively, and the currents increased as much as e-fold per 5.5-mV (k) and 6.2-mV depolarization, respectively. However, in human coronary arteries, half-maximal inactivation was observed at the more positive membrane potential of -26.0 mV, and Kv current increased e-fold per 12.1 mV (k). These differences may indicate that diverse populations of Kv channels are expressed in different human vascular beds. The inactivation parameters of human coronary Kv currents are also different from those reported for Kv channels in many other nonhuman vascular preparations, including coronary arteries (Leblanc *et al.*, 1994; Buljubasic *et al.*, 1993; Beech and Bolton, 1989a; Robertson and Nelson, 1994; Hume and Leblanc, 1989; Okabe *et al.*, 1987). Berger *et al.* (1998) presented evidence that Kv channels in rat coronary and pulmonary arterial smooth muscle cells may respond differentially to acidosis, suggesting the presence of distinct subtypes of Kv channels in these two vascular beds.

Despite similar microscopic appearance and passive electrical properties, 4-aminopyridine-sensitive Kv channels are not expressed in all smooth muscle cells isolated from human coronary arteries. The reason for this finding is unclear. One possible explanation is the existence of heterogeneous coronary vascular smooth muscle cells. In agreement with this suggestion are morphological and biochemical studies which demonstrate the heterogeneity of vascular smooth muscle cells in the arterial wall of pulmonary arteries (Frid *et al.*, 1994; Archer, 1996). Alternatively, some cells may not express functional channels because of different metabolic states induced by the cell isolation procedure or because of channel rundown after the isolation procedure. In this context, it should be noted that 4-aminopyridine-sensitive Kv channels were detected by Xu and Lee (1994) in canine coronary arteries, but not by Buljubasic *et al.* (1993), who studied the same preparation. In porcine coronary arteries, the number of Kv channels per cell is estimated to be about 5000 (Volk and Shibata, 1993).

Activation of Kv channels by membrane depolarization, as may occur in response to pressurization or hormonal vasoconstrictors, may limit membrane depolarization. Indeed, inhibition of Kv channels by 4-aminopyridine depolarizes and constricts many arteries (Hara *et al.*, 1980; Knot and Nelson, 1995; Ishikawa *et al.*, 1997). In addition, Kv channels may serve as targets for hormonal vasoconstrictors and vasodilators. In rabbit coronary arteries, a 4-aminopyridine-sensitive K$^+$ current is inhibited by activation of the histamine H1 receptor (Ishikawa *et al.*, 1993). In bovine coronary arteries, prostaglandin I$_2$ activates Kv channels (Li *et al.*, 1997). A more precise understanding of the properties and functional role of Kv channels in large and small coronary arteries remains an important goal for future research.

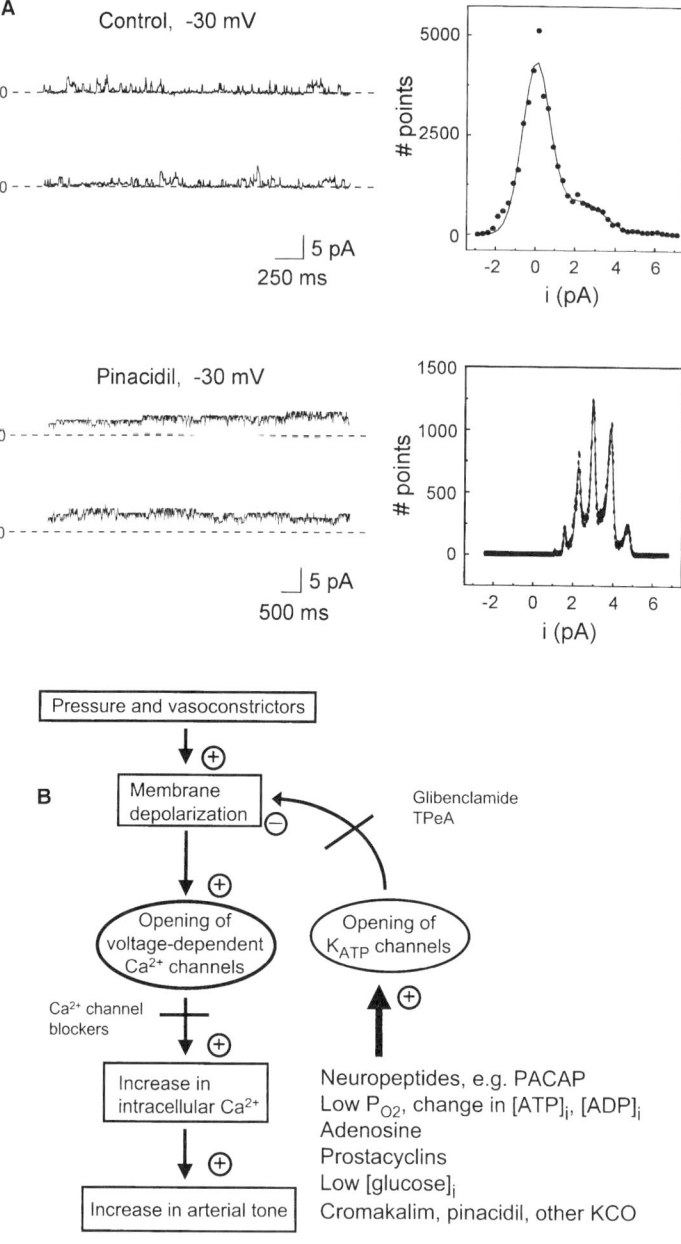

Figure 4. K_{ATP} channels and the regulation of arterial tone in coronary arteries. (A) Steady-state whole-cell recordings in a human coronary arterial smooth muscle cell made before (*upper panel, left*) and after application of pinacidil (1 μM) (*lower panel, left*), showing activation of single K_{ATP} channels by pinacidil. The dashed lines indicate the zero current levels. Corresponding histograms of current amplitudes are shown in the right part of the panels. The histograms were fitted by sums of Gaussian distributions with centers of 0.02 pA and 2.17 pA (upper panel) and 1.60, 2.26, 2.99, 3.87, and 4.73 pA (lower panel). Holding potential, -30 mV. (Reproduced with permission, from Bychkov *et al.*, 1997a.) (B) Flow diagram of proposed role of K_{ATP} channels in the regulation of myogenic tone in coronary arteries. Elevation of intravascular pressure and vasoconstrictors in small coronary arteries depolarizes the arterial smooth muscle cells, causing Ca^{2+}

4.3. Transient Outward K$^+$ Channels

A voltage-dependent transient outward K$^+$ current (K$_{to}$) has been observed only in a very small fraction of smooth muscle cells isolated from human coronary arteries (Gollasch *et al.*, 1996a). This current activates at potentials positive to 0 mV and inactivates very rapidly ($\tau = 65$ ms at $+50$ mV). The time constant of K$_{to}$ current decay decreases at more positive pulse potentials. The current has characteristics of the fast, transient K$^+$ current ('A'-type current) observed in neurons, and its properties (activation positive to -65 mV; $\tau \cong 65$ ms at $+20$ mV) resemble those of a 4-aminopyridine-sensitive, voltage-dependent, outward K$^+$ current (I_{fo}) described in portal vein vascular smooth muscle cells (Beech and Bolton, 1989b). The time constant of I_{fo} decay, however, increases proportionally to the amplitude of the voltage step. K$_{to}$ current apparently is not prominent in most vascular smooth muscle cells and may only be present in human coronary arterial smooth muscle cells. Its functional role remains unknown.

5. VOLTAGE-INDEPENDENT K$_{ATP}$ CHANNELS

Pinacidil and PACAP-27, at concentrations that induce vasorelaxation of human coronary arteries (Gollasch *et al.*, 1995; Bruch *et al.*, 1997), activate a K$^+$ current that shares several properties with the K$_{ATP}$ current activated by pinacidil and other K$^+$ channel openers in noncoronary vascular preparations (Fig. 4) (Zhang and Bolton, 1995; Standen *et al.*, 1989) and in porcine coronary arteries (Silberberg and van Breemen, 1990; Dart and Standen, 1993, 1995; Gollasch *et al.*, 1995). This K$^+$ current shows little voltage sensitivity and shifts the reversal potential of the entire transmembrane current to membrane potentials near the equilibrium potential for potassium (E_K) (Gollasch *et al.*, 1995, 1996a; Bruch *et al.*, 1997). The pinacidil (1 μM)-induced K$^+$ currents are blocked by glibenclamide (3 μM) but are not inhibited by iberiotoxin (100–300 nM). The K_i value for glibenclamide block of arterial smooth muscle K$_{ATP}$ channels is between 20 and 200 nM (Nelson and Quayle, 1995). 4-Aminopyridine (K_i, 0.2 mM), external Ba^{2+} (K_i at -80 mV, 100 μM), and TEA (K_i, 7 mM) have little effect on arterial smooth muscle K$_{ATP}$ channels; iberiotoxin (100 nM) and charybdotoxin (100 nM) have no effect on these channels (Nelson and Quayle, 1995). In human coronary smooth muscle cells, the density of the functional K$_{ATP}$ channels estimated from noise

---

influx through voltage-dependent L-type Ca^{2+} channels, an elevation of [Ca^{2+}]$_i$, and contraction of vascular smooth muscle cells. Myogenic tone depends on external Ca^{2+} and can be blocked by Ca^{2+} channel blockers. Note that activation of coronary arterial K$_{ATP}$ channels limits myogenic tone and the effects of depolarizing vasoconstrictors. K$_{ATP}$ channels in coronary artery vascular smooth muscle cells can be activated by neuropeptides (e.g., PACAP), hypoxia (low P_{O_2}, low intracellular concentrations of glucose, changes in intracellular levels of [ATP] [ADP], and other nucleotide diphosphates, adenosine, prostacyclin, and synthetic K$^+$ channel openers (KCO) such as cromakalim and pinacidil). The effects of K$_{ATP}$ channels on membrane potential and arterial tone can be blocked by glibenclamide or tetrapentylammonium ions (TPeA). (Modified, with permission, from Daut *et al.*, 1994.)

analysis, in the presence of pinacidil, is 150 channels per cell. A single-channel conductance of about 17 pS has been recorded at physiological membrane potentials (between -80 mV and -30 mV) and K^+ gradients (6 mM/130 mM) (Bychkov et al., 1997b). K_{ATP} channels with a small to medium conductance of 15–50 pS have been found in smooth muscle cells from porcine coronary arteries (Dart and Standen, 1993, 1995; Ottolia and Toro, 1996) and in cultured coronary smooth muscle artery cells (Miyoshi and Nakaya, 1991, 1993; Miyoshi et al., 1992; Wakatsuki et al., 1992), under symmetrical, high-K^+ conditions. This suggests that coronary arterial smooth muscle cells do not exhibit large-conductance K_{ATP} channels. In contrast, large-conductance K_{ATP} channels ($\sim$130 pS, measured under symmetrical, high-K^+ conditions) have been found in mesenteric arteries, renal arteries, and aorta (Quayle, 1997).

K_{ATP} channels in vascular smooth muscle are inhibited by intracellular ATP with half-inhibition occurring in the range of 10–200 μM in excised membrane patches (for a review, see Quayle et al., 1997). The K_i value is increased in the presence of intracellular Mg^{2+} and other regulatory factors, including nucleoside diphosphates, extracellular acidification or pH, and intracellular lactate. However, information about the channel regulation by intracellular ATP and cell metabolism is limited. In this regard, hypoxic solutions (with an oxygen partial pressure of 25–40 mm Hg) may activate glibenclamide-sensitive, whole-cell currents in porcine coronary arterial smooth muscle cells, studied using the perforated-patch technique (Dart and Standen, 1995). Furthermore, blockade of K_{ATP} channels by glibenclamide substantially inhibits coronary vasodilation, seen as a fall in coronary perfusion pressure in response to cromakalim, hypoxia, and adenosine in guinea pig heart (Fig. 1). Glibenclamide also partially inhibits relaxation of porcine coronary arteries induced by metabolic inhibition (Gollasch et al., 1995).

The open-state probability of K_{ATP} channels in coronary arterial smooth muscle cells is presumably quite low in the absence of activators of the channel and in the presence of physiological intracellular concentrations of ATP in the millimolar range. Regardless, recent evidence supports the idea that K_{ATP} channels play a role in the maintenance of tone in the coronary circulation under physiological (normoxic) conditions (Eckman et al., 1992; Imamura et al., 1992; Samaha et al., 1992; Daut et al., 1994). This suggestion is based on the observations that glibenclamide depolarizes and constricts coronary arteries or increases the resistance of the coronary vascular bed in some preparations. For example. a glibenclamide-sensitive K^+ (K_{ATP}) current is found in terminal arterioles from guinea pig heart, and inhibition of this K_{ATP} current induces a marked depolarization (Klieber and Daut, 1994). K_{ATP} channels of coronary arterial smooth muscle cells may also be regulated by phosphorylation via second messengers (Quayle et al., 1994; Bonev and Nelson, 1996; Wellman et al., 1996). In the intact heart, the open-state probability of K_{ATP} channels may be controlled by autacoids such as adenosine (Dart and Standen, 1993) and prostaglandins (Jackson et al., 1993; Bouchard et al., 1994; Parkington et al., 1995) and by vasoactive peptides such as calcitonin-gene-related peptide (CGRP) (Quayle et al., 1994; Wellman et al., 1998), vasoactive intestinal peptide (VIP) (Standen et al., 1989), atrial natriuretic peptide (ANP) (Kubo et al., 1994), or PACAP (Bruch et al., 1997, 1998) in the perivascular space. Moreover, the open-state probability of the channels may be regulated by nitric oxide (Kubo et al., 1994), endothelin (Miyoshi et al., 1992), and β2-receptor activation (Miyoshi and Nakaya, 1993) and, as mentioned above, by internal ATP and nucleoside diphosphates, extracellular pH, and intracellular lactate (for a review, see Quayle *et al.*, 1997). It is likely that all of these factors contribute to some extent to the activation of K$_{ATP}$ channels in the coronary circulation *in vivo*.

Findings that define the properties of coronary K$_{ATP}$ channels may have therapeutic implications. For instance, cardiac and coronary K$_{ATP}$ channels apparently operate with very low activity under normal metabolic conditions. Instead, they are activated when the oxygen supply, and consequently the intracellular high-energy-phosphate values, fall below critical levels (Daut *et al.*, 1990). Thus, the opening of K$_{ATP}$ channels may be considered as an "emergency" response to prevent energy failure and to preserve the viability of the tissue during ischemic episodes. Indeed, recent studies suggest that pinacidil may have beneficial effects in the ischemic myocardium. Pinacidil and other "K$^+$ channel openers" are viewed as exogenous "ischemic preconditioners," which enable the heart to survive during limited periods of ischemia by opening K$_{ATP}$ channels. Blockade of K$_{ATP}$ channels with glibenclamide interrupts this process in several animal species, including rabbits, dogs, and pigs (Parratt and Kane, 1994). In humans, glibenclamide at oral doses sufficient to treat type II diabetes mellitus prevented the beneficial effects of preconditioning (Tomai *et al.*, 1994). The cardiovascular mortality was threefold higher in diabetics treated with tolbutamide, another sulfonylurea that blocks K$_{ATP}$ channels, compared to those treated with insulin (Prout *et al.*, 1972). Thus, opening of K$_{ATP}$ channels appears to be a necessary link in the chain of events leading to cardioprotection and "preconditioning" initiated by endogenous signals, such as factors from heart muscle (adenosine) or perivascular nerves (CGRP, PACAP-27) that regulate smooth muscle membrane potential. These factors could conceivably place the heart in a state of "preconditioning" by opening K$_{ATP}$ channels. The present data demonstrating that PACAP-27 and pinacidil activate coronary K$_{ATP}$ channels in humans support this view.

6. INWARD RECTIFIER K$^+$ CHANNELS

Inward rectifier K$^+$ (Kir) currents have been identified in small, resistance-sized coronary arteries (diameter, $\sim$200 μm) (Fig. 5) (Knot *et al.*, 1996a, 1998), but not in large conduit coronary arteries of humans (Gollasch *et al.*, 1996a). Kir currents also have been identified in small, resistance-sized coronary arteries (diameter, $\sim$200 μm) of rats (Robertson *et al.*, 1996). In the pig coronary circulation, the relative density of Kir channels in smooth muscle cells increases with decrease in vessel diameter. Thus, cells isolated from the left anterior descending artery (diameter, $\sim$2 mm) show currents with little or no inward rectification, whereas cells from the fourth-order branches (diameter, 100–300 μm) show currents with marked inward rectification (Quayle *et al.*, 1993, 1997). This distribution is strikingly similar to distributions described in other arterial smooth muscle cells, supporting the view that Kir channels are most strongly expressed in smooth muscle cells in small, resistance vessels rather than in larger arteries. Current–voltage relationships of Kir currents show inward rectification; that is, the conductance is higher for inward than for outward currents (Fig. 5A). External Ba^{2+} is an effective blocker of Kir currents (K_i, $\cong 2\,\mu M$ at -60 mV) (Quayle *et al.*, 1993; Robertson *et al.*, 1996). External Ca^{2+} and Mg^{2+} at physiological concentrations also

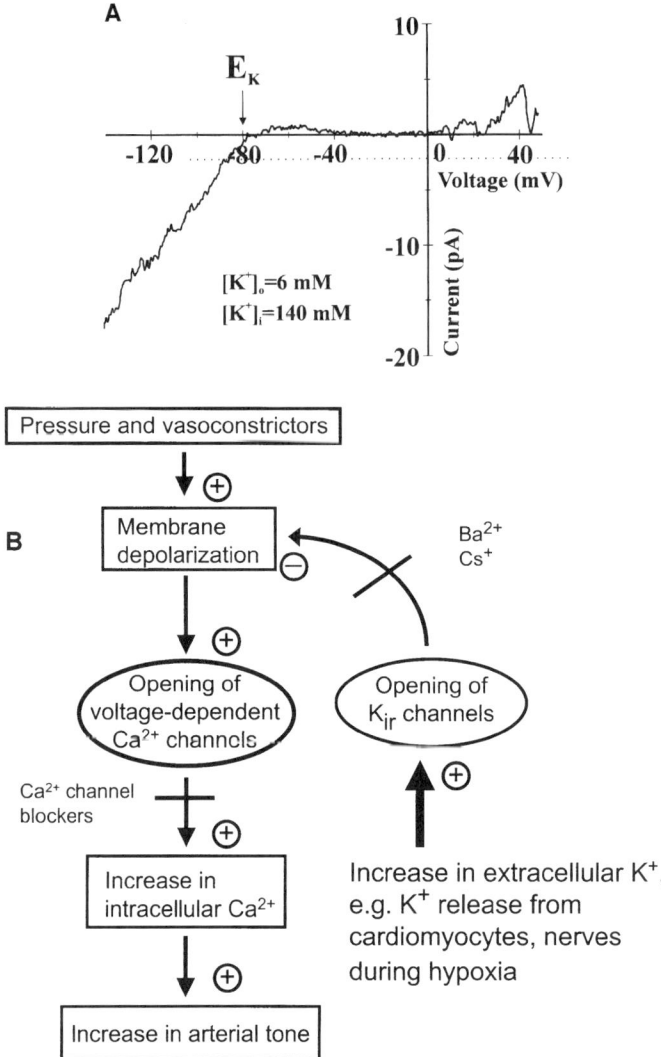

Figure 5. Kir channels and the regulation of arterial tone in small, resistance-sized coronary arteries. (A) Kir current in a smooth muscle cell from a small human coronary artery. E_K, Equilibrium potential for potassium. (Reproduced, with permission, from Knot *et al.*, 1996b.) (B) Flow diagram of proposed role of Kir channels in the regulation of myogenic tone of small coronary arteries (data from Knot *et al.*, 1998). Note that coronary arterial Kir channels contribute to the resting membrane conductance of coronary arterial smooth muscle cells and are activated by elevation of extracellular K^+ in the range of 10–16 mM. Activation of coronary arterial Kir channels limits myogenic tone and the effects of depolarizing vasoconstrictors. The effects of Kir channels on membrane potential and arterial tone are blocked by Ba^{2+} or Cs^+.

partially block rat coronary Kir currents ($\sim 50\%$ reduction of Kir current at 5 mM Ca^{2+} or Mg^{2+}) (Robertson *et al.*, 1996). External Cs^+ is also an effective blocker of Kir currents ($K_i \cong 3 \mu M$ at -60 to -50 mV) (Robertson *et al.*, 1996). 4-Aminopyridine (1 mM), glibenclamide (10 μM), TEA (1 mM), and charybdotoxin (100 nM) have little effect on arterial smooth muscle Kir currents (Quayle *et al.*, 1993; Robertson *et al.*, 1996).

The function of Kir currents in arterial smooth muscle cells is not completely understood. It has been suggested that Kir currents make a major contribution to the cell resting membrane conductance in small arteries and play a significant role in the vasodilator action of K$^+$ released locally from active tissue, for example, in heart, from cardiomyocytes and perivascular nerves (Fig. 5b). The extracellular K$^+$ concentration in the myocardium may reach 10–16 mM during ischemia, and even such small increases in K$^+$ are associated with significant dilation in the coronary circulation (for reviews, see Nelson and Quayle, 1995; Quayle *et al.*, 1997). In pressurized resistance-sized rat coronary arteries, hyperpolarizations and dilations to 16 mM K$^+$ are blocked by Ba^{2+} with a K_i in the range of 3–8 μM, very close to that seen for Kir channels at comparable membrane potential (Knot *et al.*, 1996b). Similarly, in resistance-sized human coronary arteries (diameter, $\sim$50 μm at 60 mm Hg), raising external K$^+$ from 6 to $\sim$16 mM causes a significant membrane hyperpolarization from about -40 mV to -62 mV and leads to vasorelaxation. These effects are also blocked by 30 μM Ba^{2+} (Knot *et al.*, 1998). Thus, by mediating the vasodilation in response to increased external K$^+$, Kir channels may play an important role in linking coronary blood flow in small coronary arteries to the metabolic demands of the heart.

7. STOCs

Spontaneous transient outward currents (STOCs) are observed in a majority of vascular smooth muscle cells from human coronary arteries (Fig. 6A) (Gollasch *et al.*, 1996a, Bychkov *et al.*, 1997b; 1998). STOCs with a similar duration are observed in vascular smooth muscle cells from guinea pig coronary artery, rabbit ear artery, dog carotid artery, and rat cerebral artery. However, STOCs have not been described in other smooth muscle cells from humans (Nelson *et al.*, 1995; Ganitkevich and Isenberg, 1990; Benham and Bolton, 1986). STOCs may reflect the summation of K$^+$ currents through 20–100 iberiotoxin-sensitive, BK$_{Ca}$ channels that have been activated by local and transient Ca^{2+} release ("Ca^{2+} sparks") through ryanodine receptors (RyRs) in the sarcoplasmic reticulum (Benham and Bolton, 1986, Bolton and Lim, 1989). It is possible that the Ca^{2+} spark-activated BK$_{Ca}$ channels represent a clustered population of maxi Ca^{2+}-activated K$^+$ channels (BK$_{Ca,STOCs}$). Nelson *et al.* (1995) proposed that STOCs control the diameter of small, myogenically active cerebral arteries. Recently, it has been proposed that the coordinated opening of a number of tightly clustered RyRs is required to cause a calcium spark (Gollasch *et al.*, 1998a). The BK$_{Ca}$ channels that are activated by Ca^{2+} sparks (e.g., STOCs) appear to regulate the *global* [Ca^{2+}]$_i$ in the arterial wall and hence the diameter of pressurized, myogenic cerebral arteries (Gollasch *et al.*, 1998a). Figure 6B illustrates a model based on the latter hypothesis. Specifically, more than one RyR is required for the occurrence of a Ca^{2+} spark, and it is postulated that this spark is the elementary, physiological, Ca^{2+} release event that regulates the membrane potential and diameter of pressurized small, resistance-sized arteries (Gollasch *et al.*, 1998a). In contrast, the open-state probability of BK$_{Ca}$ channels that are not activated by Ca^{2+} sparks, (such as those activated by subsarcolemmal

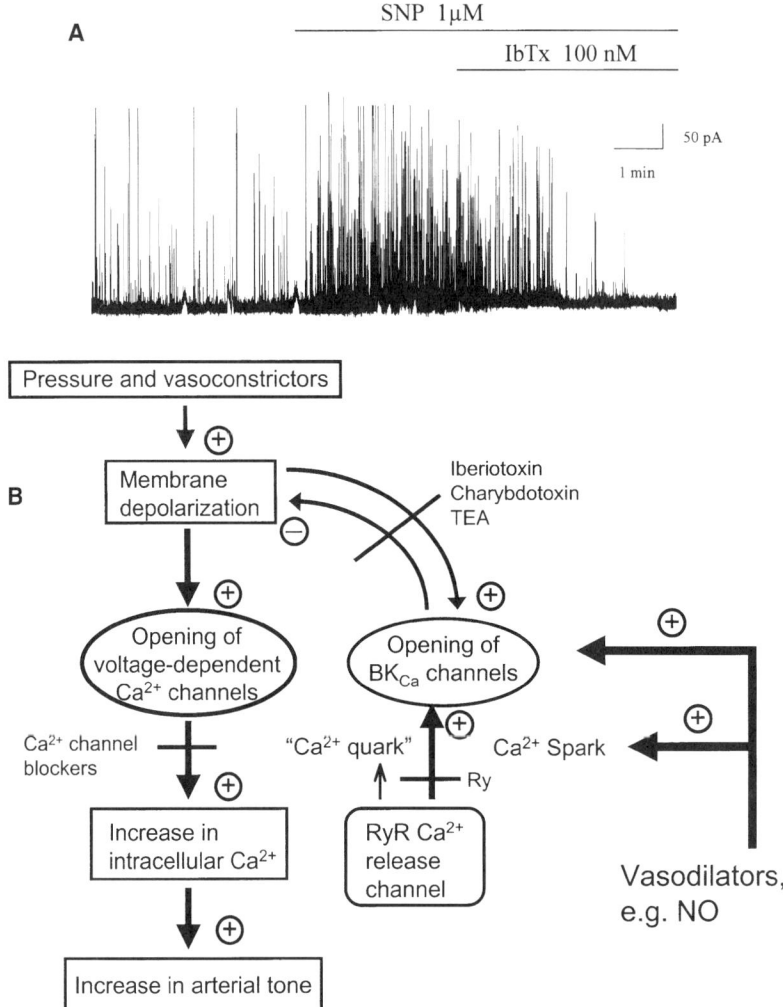

Figure 6. BK$_{Ca}$ channels and the regulation of vascular tone in coronary arteries. (A) Stimulation of spontaneous transient BK$_{Ca}$ currents (STOCs) by sodium nitroprusside (SNP) in human coronary artery vascular smooth muscle cells. Shown are superimposed current recordings before and after SNP (1 μM). The STOCs are blocked by 100 nM iberiotoxin (IbTX, 100 nM). (Reproduced, with permission, from Bychkov *et al.*, 1998.) (B) Flow diagram of proposed role of BK$_{Ca}$ channels in the regulation of myogenic tone of coronary arteries. Note that coronary arterial BK$_{Ca}$ channels limit myogenic tone and the effects of depolarizing vasoconstrictors if they are activated by elementary Ca^{2+} release events (Ca^{2+} sparks) through multiple ryanodine receptors (RyR) of the sarcoplasmic reticulum. BK$_{Ca}$ channels that are not activated by Ca^{2+} sparks (e.g., Ca^{2+} quarks) do not significantly regulate membrane potential and myogenic tone. Both BK$_{Ca}$ channels and Ca^{2+} sparks may represent targets for vasodilators, e.g., nitric oxide, in coronary arteries. The effects of BK$_{Ca}$ channels on membrane potential and arterial tone are blocked by iberiotoxin, charybdotoxin, tetraethylammonium (TEA), or ryanodine (Ry). (Modified from Daut *et al.*, 1994; Nelson *et al.*, 1995; and Gollasch *et al.*, 1998a.)

[Ca^{2+}] or by smaller Ca^{2+} release events of single RyRs termed "Ca^{2+}quarks" or "fundamental events" (Berridge, 1997; Lipp and Niggli, 1998), is too low to regulate the membrane potential of vascular smooth muscle cells and arterial myogenic tone (Gollasch et al., 1998a). Although the physiological meaning of STOCs for the regulation of large epicardial arteries is presently unclear, the striking similarity of STOCs observed in different smooth muscle preparations suggests that these phenomena may be common in arterial smooth muscle cells, including coronary arteries of humans. Ca^{2+} sparks have also been recorded in smooth muscle cells from rat coronary arteries (Jaggar et al., 1998). In contrast to Ca^{2+} sparks in cerebral arteries (Nelson et al., 1995; Jaggar et al., 1998), however, Ca^{2+} sparks in human coronary arterial smooth muscle cells seem to be regulated relatively independently of Ca^{2+} influx through voltage-dependent Ca^{2+} channels (Bychkov et al., 1998), possibly through a functional linkage between the sarcolemmal dihydropyridine receptor and the sarcoplasmic reticulum RyR (either directly or via an additional spanning protein), as reported at negative membrane potentials in cardiomyocytes (Satoh et al., 1998). Furthermore, there is experimental evidence that the generation of STOCs in human coronary arterial smooth muscle cells is regulated by Na$^+$/Ca^{2+} exchange activity (Bychkov et al., 1998), indicating the complex regulatory mechanisms of Ca^{2+} sparks in these cells.

Recently, it was shown that nitrovasodilators can activate BK$_{Ca}$ currents (Taniguchi et al., 1993; Robertson et al., 1993; Archer et al., 1994) and STOCs in smooth muscle cells from human coronary arteries (Fig. 6A) (Bychkov et al., 1998). Activation of these channels contributes to the vasorelaxant action of these drugs (Bychkov et al., 1998). Thus, Ca^{2+} sparks may serve as novel targets for drugs and vasoactive hormones to regulate coronary arterial tone (Fig. 6; Bychkov et al., 1998; Porter et al., 1998; Jaggar et al., 1998). These findings may have unique clinical significance for the development of novel antianginal and antihypertensive drugs targeting the function of K$^+$ channels.

8. ION CHANNELS AND DIFFERENTIATION OF VASCULAR SMOOTH MUSCLE CELLS

Arterial smooth muscle cells normally exist in a quiescent, differentiated state in the blood vessel wall. These cells express a unique repertoire of contractile proteins, ion channels, receptors, and signaling molecules that are acquired during differentiation of the cells and are necessary for their contractile function. One of the major features of chronic vascular diseases and atherosclerotic plaque development is the dedifferentiation of intimal vascular smooth muscle cells, with phenotypic changes and the partial loss of their differentiated properties (Benditt and Benditt, 1973; Campbell and Campbell, 1994). The dedifferentiation of the cells seems to be a prerequisite for their subsequent migration and proliferation in the intima (Ross, 1995). The dedifferentiation shows a high degree of plasticity and seems to be reversible in vitro (Schwartz et al., 1990, 1995). Although this phenomenon and its relevance for the pathogenesis of chronic vascular disease have been known for some time, only a few markers of vascular smooth muscle cell differentiation have been described so far. Dedifferentiation of human and experimental vascular smooth muscle cells is associated with a decreased expression of a number of smooth muscle (SM) contractile and cytoskeletal proteins, including SM α-actin, SM myosin heavy chain, calponin, SM-22 α, h-caldesmon, vinculin, and 20-kDa myosin light chains and an increased expression of the nonmuscle variants of these proteins. Most of them are markers for the early stages of differentiation (Owens, 1995). Although ion channels have been suggested to represent logical candidates for use as vascular smooth muscle cell differentiation markers (Owens, 1995), very little information has been available as to the differential expression of ion channels in vascular smooth muscle cells.

Recently, it has been shown that the expression of L-type Ca^{2+} channel α_{1C} (and β_2) subunits and functional dihydropyridine-sensitive Ca^{2+} channels in vascular smooth muscle cells is highly correlated with and can be induced together with the expression of the SM specific proteins necessary for specialized smooth muscle functions, i.e., with SM α-actin and SM myosin (Gollasch $et.$ $al.$, 1998b). The data indicate that ion channels may represent novel sensitive markers of late states of vascular smooth muscle cell differentiation. These findings open new questions about the role of intracellular calcium ions and ion channels in the process of differentiation. One could speculate that the expression of the dihydropyridine-sensitive Ca^{2+} channel itself plays a major role in the differentiation process. Indeed, the intracellular Ca^{2+} concentration and Ca^{2+} influx through dihydropyridine-sensitive Ca^{2+} channels are important for differentiation of skeletal muscle cells (Luo et $al.$, 1994) and neuronal cells (MacVicar, 1987; Reber and Reuter, 1991; Spitzer et $al.$, 1993); however, similar information for smooth muscle is lacking. The regulation of the expression of ion channels (e.g., Ca^{2+} channels and K^+ channels) that regulate Ca^{2+} influx during the differentiation of vascular smooth muscle cells may have physiological importance for normal smooth muscle function. Ion channel expression may also influence smooth muscle behavior in pathophysiological conditions, such as formation and evolution of atherosclerotic plaques in arteries.

The role of Ca^{2+} channels, K^+ channels, and membrane potential in the proliferation of vascular smooth muscle cells is unclear. Numerous studies report that the growth and proliferation of vascular smooth muscle cells are dependent on Ca^{2+} influx, and thus are blocked or reduced by Ca^{2+} removal or conventional Ca^{2+} antagonists (dihydropyridines, verapamil). However, the results are widely variable, and often suspiciously high doses of Ca^{2+} antagonists are required, much higher than those needed to modify contraction and selectively target L-type channels (Ko et $al.$, 1992; Munro et $al.$, 1994; Block and Buhler, 1992; Kruse et $al.$, 1994). Moreover, the effect of membrane potential changes on proliferation of vascular smooth muscle cells has not been studied so far, making the interpretation of the results difficult. For example, dihydropyridines at high concentrations may block other Ca^{2+}-selective ion channels important for the sustained Ca^{2+} influx necessary for proliferation. Moreover, the recent demonstration that the expression of L-type Ca^{2+} channels is downregulated during dedifferentiation of vascular smooth muscle cells (Gollasch et $al.$, 1998b) suggests that other Ca^{2+}-permeable ion channels may be involved in the control of Ca^{2+} influx necessary for maintenance of proliferation and regulation of gene expression. The identification of ion channels, including K^+ channels, regulating Ca^{2+} influx in dedifferentiated cells may provide novel therapeutic strategies to influence differentiation and proliferation properties of arterial vascular smooth muscle cells in chronic vascular diseases and atherosclerotic plaque development, making this topic of ion channel research clinically relevant.

Acknowledgments

The experiments cited here obviously represent the work of a large number of scientists, and I would like to thank those in our laboratory in Berlin, Germany, and in Dr. Mark T. Nelson's laboratory at the University of Vermont. The author is very grateful to Drs. Hermann Haller and Friedrich C. Luft for continuous support and for critically reading the manuscript. I would like to thank Dr. Matthias Löhn and Michael Furstenau for helpful discussions and Dr. Rudolf Schubert for critical comments on the manuscript. I acknowledge the past members of our patch-clamp laboratory whose work is cited here including Drs. Rostislav Bychkov, Christian Ried, Ulrich C. Luft, Tobias Steinke, Michael Liebold, Susan Scholze, and Friederike Behrendt. This work was supported by the Deutsche Forschungsgemeinschaft and Alexander von Humboldt-Stiftung.

REFERENCES

Albarwani S., Heinert G., Turner J. L., and Kozlowski R. Z., 1995, Differential K$^+$ channel distribution in smooth muscle cells isolated from the pulmonary arterial tree of the rat, *Biochem. Biophys. Res. Commun.* **208:**183–189.

Archer S. L., 1996, Diversity of phenotype and function of vascular smooth muscle cells, *J. Lab. Clin. Med.* **127:**524–529

Archer, S. L., Huang, J. M. C., Hampl, V., Nelson, D. P., Shultz, P. J., and Weir, E. K., 1994, Nitric oxide and cGMP cause vasorelaxation by activation of a charybdotoxin-sensitive K channel by cGMP-dependent protein kinase. *Proc. Natl. Acad. Sci. U.S.A.* **91:**7583–7587.

Balwierczak, J. L., Krulan, C. M., Kim, H. S., DelGrande, D., Weiss, G. B., and Hu, S., 1995, Evidence that BK$_{Ca}$ channel activation contributes to K$^+$ channel opener induced relaxation of the porcine coronary artery, *Naunyn-Schmiedeberg's Arch. Pharmacol.* **352:**213–221.

Bean B. P., 1991, Pharmacology of calcium channels in cardiac muscle, vascular muscle, and neurons. *Am. J. Hypertens.* **4:**406S–411S.

Beech D. J, and Bolton T. B., 1989a, Two components of potassium current activated by depolarization of single smooth muscle cells from rabbit portal vein, *J. Physiol.* **418:**293–309.

Beech D. J, and Bolton T. B., 1989b, A voltage-dependent outward current with fast kinetics in single smooth muscle cells isolated from rabbit portal vein, *J. Physiol.* **412:**397–414.

Benditt E. P, and Benditt J. M., 1973, Evidence for a monoclonal origin of human atherosclerotic plaques, *Proc. Natl. Acad. Sci. U.S.A.* **70:**1753–1756.

Benham C. D., and Bolton T. B., 1986, Spontaneous transient outward currents in single visceral and vascular smooth muscle cells of the rabbit, *J. Physiol.* **381:**385–406.

Berger M. G., Vandier C., Bonnet P., Jackson W. F., and Rusch N. J., 1998, Intracellular acidosis differentially regulates Kv channels in coronary and pulmonary vascular muscle, *Am. J. Physiol.* **275:**H1351–H1359.

Berridge, M. J., 1997, Elementary and global aspects of calcium signaling, *J. Physiol. (London)* **499:** 290–306.

Block L. H., and Buhler F. R., 1992, Atherosclerosis, cell motility, calcium, and calcium-channel blockers, *J. Cardiovasc. Pharmacol.* **19:**S1–S3.

Bolton T. B., and Lim S. P., 1989, Properties of calcium stores and transient outward currents in single smooth muscle cells of rabbit intestine, *J. Physiol.* **409:**385–401.

Bonev A. D., and Nelson M. T., 1996, Vasoconstrictors inhibit ATP-sensitive K$^+$ channels in arterial smooth muscle through protein kinase C, *J. Gen. Physiol.* **108:**315–323.

Bouchard J. F., Dumont E., and Lamontagne D., 1994, Evidence that prostaglandins I$_2$, E$_2$, and D$_2$ may activate ATP-sensitive potassium channels in the isolated rat heart, *Cardiovasc. Res.* **28:** 901–905.

Brayden J. E., and Nelson M. T., 1992, Regulation of arterial tone by activation of calcium-dependent potassium channels, *Science* **256:**532–535.

Bruch L., Bychkov, R., Kästner, A., Blow, T., Ried, C., Gollasch, M., Baumann, G., Luft, F. C., and Haller, H., 1997, Pituitary adenylate-cyclase-activating peptides relax human coronary arteries by activating K(ATP) and K(Ca) channels in smooth muscle cells, *J. Vasc. Res.* **34**:11–18..

Bruch, L., Rubel, S., Kästner, A., Gellert, K., Gollasch, M., and Witt, C., 1998, Pituitary adenylate cyclase activating peptides relax human pulmonary arteries by activating K_{ATP} and KCa channels, *Thorax* **53**:586–587

Buljubasic, N., Marijic, J., Kampine, J. P., and Bosnjak, Z. J., 1993, Calcium-sensitive potassium current in isolated canine coronary smooth muscle cells. *Can. J. Physiol. Pharmacol.* **72**:189–198.

Bychkov, R., Gollasch, M., Ried, C., Luft, F.C., and Haller, H., 1997a, Effects of pinacidil on Ca^{2+}-activated and ATP-dependent K^+ channels in human coronary artery vascular smooth muscle cells, *Am. J. Physiol.* **273**:C161–C171.

Bychkov, R., Gollasch, M., Ried, C., Luft, F.C., and Haller, H., 1997b, Regulation of spontaneous transient outward potassium currents in human coronary arteries, *Circulation* **95**:503-510.

Bychkov, R., Gollasch, M., Steinke, T., Ried, C., Luft, F. C., and Haller, H., 1998, Calcium-activated potassium channels and nitrate-induced vasodilation of human coronary arteries, *J. Pharmacol. Exp. Ther.* **285**:293–298.

Cai, D., Mulle, J. G., and Yue, D. T., 1997, Inhibition of recombinant Ca^{2+} channels by benzothiazepines and phenylalkylamines: Class-specific pharmacology and underlying molecular mechanisms, *Mol. Pharmcol.* **51**:872–881.

Campbell, J. H., and Campbell, G. R., 1994, Cell biology of atherosclerosis, *J Hypertens Suppl.* **12** (10):S129–S132.

Cole, W. C., Clement-Chmienne, O., and Aillo, E. A., 1996, Regulation of 4-aminopyridine-sensitive, delayed rectifier K^+ channels in vascular smooth muscle by phosphorylation, *Biochem. Cell. Biol.* **74**:439–447.

Dart, C., and Standen, N. B., 1993, Adenosine-activated potassium current in smooth muscle cells isolated from the pig coronary artery, *J. Physiol.* **471**:767–786.

Dart, C., and Standen, N. B., 1995, Activation of ATP-dependent K^+ channels by hypoxia in smooth muscle cells isolated from the pig coronary artery, *J. Physiol.* **483**:29–39.

Daut, J., Maier-Rudolph, W., von Beckerath, N., Mehrke, G., Günther, K., and Goedel-Meinen, L., 1990, Hypoxic dilation of coronary arteries is mediated by ATP-sensitive potassium channels, *Science*, **247**:1341–1344.

Daut, J., Standen, N. B., and Nelson, M. T., 1994, The role of the membrane potential of endothelial and smooth muscle cells in the regulation of coronary blood flow, *J Cardiovasc. Electrophysiol.* **5** (2):154–181.

Eckman, D. M., Frankovich, J. D., Keef, K. D., 1992, Comparison of the actions of acetylcholine and BRL 38227 in the guinea-pig coronary artery. *Br. J. Pharmacol.* **106**(1):9–16.

Eckman, D. M., Weinert, J. S., Buxton, I. L., and Keef, K. D., 1994, Cyclic GMP-independent relaxation and hyperpolarization with acetylcholine in guinea-pig coronary artery, *Br. J. Pharmacol.* **111**:1053–1060.

Frid, M. G., Moiseeva, E. P., and Stenmark, K. R., 1994, Multiple phenotypically distinct smooth muscle cell populations exist in the adult and developing bovine pulmonary arterial media in vivo, *Circ. Res.* **75**:669–681.

Ganitkevich, V. Y., and Isenberg, G., 1990, Isolated guinea pig coronary smooth muscle cells. Acetylcholine induces hyperpolarization due to sarcoplasmic reticulum calcium release activating potassium channels, *Circ. Res.* **67**:525–528.

Ganitkevich, Vya, and Isenberg, G., 1993, Membrane potential modulates inositol 1,4,5-trisphosphate-mediated Ca^{2+} transients in guinea-pig coronary myocytes, *J Physiol (London)* **470**:35–44.

Ganitkevich, V. Y., and Isenberg, G., 1996, Effect of membrane potential on the initiation of acetylcholine-induced Ca^{2+} transients in isolated guinea pig coronary myocytes, *Circ. Res.* **78**:717–723.

Gelband, C. H., and Hume, J. R., 1992, Ionic currents in single smooth muscle cells of the canine renal artery, *Circ. Res.* **71**:745–758.

Gelband, C. H., and McCullough, J. R., 1993, Modulation of rabbit aortic Ca^{2+}-activated K^+ channels by pinacidil, cromakalim, and glibenclamide, *Am. J. Physiol.* **264**:C1119–C1127.

Godfraind, T., Dessy, C., and Salomone, S., 1992, A comparison of the potency of selective L-type calcium channel blockers in human coronary and internal mammary arteries exposed to serotonin. *J. Pharmacol. Exp. Ther.* **263**:112–122.

Gollasch, M., and Nelson, M. T., 1997, Voltage-dependent Ca^{2+} channels in arterial smooth muscle cells, *Kidney Blood Pressure Res.* **20**:355–371.

Gollasch, M., Haller, H., Schultz, G., and Hescheler, J., 1991, Thyrotropin-releasing hormone induces opposite effects on Ca^{2+} channel currents in pituitary cells by two pathways, *Proc. Natl. Acad. Sci. U.S.A.* **88**:10262–10266.

Gollasch, M., Hescheler, J., Quayle, J. M., Patlak, J. B., and Nelson, M. T., 1992, Single calcium channel currents of arterial smooth muscle at physiological calcium concentrations. *Am. J. Physiol.* **263**(5):C948–C952.

Gollasch, M., Kleuss, C., Hescheler, J., Wittig, B., and Schultz, G., 1993, G$_{i2}$ and protein kinase C are required for thyrotropin-releasing hormone-induced stimulation of voltage-dependent Ca^{2+} channels in pituitary GH$_3$ cells, *Proc. Natl. Acad. Sci. U.S.A.* **90**:6265–6269.

Gollasch, M., Bychkov, R., Ried. C., Behrendt, F., Scholze, S., Luft, F. C., and Haller, H., 1995, Pinacidil relaxes porcine and human coronary arteries by activating ATP-dependent potassium channels in smooth muscle cells, *J. Pharmacol. Exp. Ther.* **275**:681–692.

Gollasch, M., Ried, C., Bychkov, R., Luft, F. C., and Haller, H., 1996a, K$^+$ currents in human coronary artery vascular smooth muscle cells, *Circ. Res.* **78**:676–688.

Gollasch, M., Ried, C., Liebold, M., Haller, H., Hofmann, F., and Luft, F. C., 1996b, High permeation of L-type Ca^{2+} channels at physiological [Ca^{2+}]: Homogeneity and dependence on the α_1-subunit, *Am. J. Physiol.* **271**:C842–C850.

Gollasch, M., Wellman, G. C., Knot, H. J., Jaggar, J. H., Damon D. H., Bonev, A. D., and Nelson, M. T., 1998a, Ontogeny of local SR calcium signals in cerebral arteries: Ca^{2+} sparks as elementary physiological events, *Circ. Res.* **83**:1104–1114.

Gollasch, M., Haase, H., Ried, C., Lindschau, C., Miethke, A., Morano, I., Luft, F. C., and Haller, H., 1998b, Expression of L-type calcium channels depends on the differentiated state of vascular smooth muscle cells, *FASEB J.* **12**:593–601.

Hara, Y., Kitamura, K., and Kuriyama, H., 1980, Actions of 4-aminopyridine on vascular smooth muscle tissues of guinea pig, *Br. J. Pharmacol.* **68**:99–106.

Harder, D. R., 1984, Pressure-dependent membrane depolarization in cat middle cerebral artery. *Circ. Res.* **55**:197–202.

Harder, D. R., Gilbert, R., and Lombard, J. H., 1987, Vascular muscle cell depolarization and activation in renal arteries on elevation of transmural pressure. *Am. J. Physiol.* **253**(4):F778–F781.

Hering, S., Hughes, A. D., Timin, E. N., and Bolton, T. B., 1993, Modulation of calcium channels in arterial smooth muscle cells by dihydropyridine enantiomers. *J. Gen. Physiol.* **101**:393–410.

Hermsmeyer, R. K., 1988, Pinacidil actions on ion channels in vascular muscle. *J. Cardiovasc. Pharmacol.* **12**(Suppl. II):S17–S22.

Hirst, G. D., and Edwards, F. R., 1989, Sympathetic neuroeffector transmission in arteries and arterioles, *Physiol. Rev.* **69**:546–604.

Hu, S., Kim, H. S., Okolie, P., and Weiss, G. B., 1990, Alterations by glyburide of effects of BRL 34915 and P 1060 on contraction, ^{86}Rb efflux and the maxi-K$^+$ channel in rat portal vein, *J. Pharmacol. Exp. Ther.* **253**:771–777.

Hume, J. R., and Leblanc, N., 1989, Macroscopic K$^+$ currents in single smooth muscle cells of the rabbit portal vein, *J. Physiol.* **413**:49–73.

Imamura, Y., Tomoike, H., Narishige, T., Takahashi, T., Kasuya, H., and Takshita, A., 1992, Glibenclamide decreases basal coronary blood flow in anesthetized dogs, *Am. J. Physiol.* **263**:H399–H404.

Inoue, R., Okabe, K., Kitamura, K., and Kuriyama, H., 1986, A newly identified Ca^{2+} dependent K$^+$ channel in the smooth muscle membrane of single cells dispersed from the rabbit portal vein, *Pflügers Arch.* **406**:138–143.

Ishikawa, T., Hume, J. R., and Keef, K. D., 1993, Modulation of K$^+$ and Ca^{2+} channels by histamine H$_1$-receptor stimulation in rabbit coronary artery cells, *J. Physiol.* **468**:379–400.

Ishikawa, T., Eckman, D. M., and Keef, K. D., 1997, Characterization of delayed rectifier K$^+$ currents in rabbit coronary artery cells near resting membrane potential, *Can. J. Physiol. Pharmacol.* **75**:1116–1122.

Itoh, T., Seki, N., Suzuki, S., Ito, S., Kajikuri, J., and Kuriyama, H., 1992, Membrane hyperpolarization inhibits agonist-induced synthesis of inositol 1,4,5-trisphosphate in rabbit mesenteric artery, *J. Physiol. (London)* **451**:307–328.

Jackson, W. F., Konig, A., Dambacher, T., and Busse, R., 1993, Prostacyclin-induced vasodilation in rabbit heart is mediated by ATP-sensitive potassium channels, *Am. J. Physiol.* **264**:H238–H243.

Jaggar, J. H., Wellman, G. C., Heppner, T. J., Porter, V. A., Perez, G. J., Knot, H. J., Gollasch, M., Kleppisch, T., Rubart, M., Stevenson, A.S., Lederer, W.J., Bonev, A.D., and Nelson, M.T., 1998, Ca^{2+} channels, ryanodine receptors, and Ca^{2+}-activated K^+ channels: A functional unit for regulating arterial tone, *Acta Scand. Physiol.* **164**:577–588.

Klieber, H. G., and Daut, J., 1994, A glibenclamide-sensitive potassium conductance in terminal arterioles isolated from giunea pig heart, *Cardiovasc. Res.* **28**:823–830.

Klückner, U., and Isenberg, G., 1991, Myocytes isolated from porcine coronary arteries: Reduction of currents through L-type Ca-channels by verapamil-type Ca-antagonists, *J. Physiol. Pharmacol.* **42**:163–179.

Knot, H. J., and Nelson, M. T., 1995, Regulation of membrane potential and diameter by voltage-dependent K^+ channels in rabbit myogenic cerebral arteries. *Am. J. Physiol.* **269**(1):H348–H355.

Knot, H. J., Zimmermann, P. A., and Nelson, M. T., 1996a, Extracellular K^+-induced hyperpolarizations and dilatations of rat coronary and cerebral arteries involve inward rectifier K^+ channels, *J. Physiol. (London)* **492**(Part 2): 419–430.

Knot, H. J., Brayden, J. B., and Nelson, M. T., 1996b, Calcium and potassium channels, in: *Biochemistry of Smooth Muscle Contraction* (M. Barany, ed.) Academic Press, San Diego, pp. 203–219.

Knot, H. J., Bonev, A. D., Mulieri, L. A., LeWinter, M. M., and Nelson, M. T., 1998, Functional role of inward rectifier K^+ (KIR) channels in coronary resistance arteries from humans, *Circulation* **98**:I489.

Ko, Y. D., Sachinidis, A., Graack, G. H., Appenheimer, M., Wieczorek, A.J., Dusing, R., and Vetter, H., 1992, Inhibition of angiotensin II and platelet-derived growth factor-induced vascular smooth muscle cell proliferation by calcium entry blockers, *Clin. Invest.* **70**(2):113–117.

Kruse, H. J., Bauriedel, G., Heimerl, J., Hofling, B., and Weber, P. C., 1994, Role of L-type calcium channels on stimulated calcium influx and on proliferative activity of human coronary smooth muscle cells, *J. Cardiovasc. Pharmacol.* **24**:328–335.

Kubo, M., Nakaya, Y., Matsuoka, S., Saito, K., and Kuroda, Y., 1994, Atrial natriuretic factor and isosorbide dinitrate modulate the gating of ATP-sensitive K^+ channels in cultured vascular smooth muscle cells, *Circ. Res.* **74**:471–476.

Kuga, T., Kobayashi, S., Hirakawa, Y., Kanaide, H., and Takeshita, A., 1996, Cell cycle-dependent expression of L- and T-type Ca^{2+} currents in rat aortic smooth muscle cells in primary culture, *Circ. Res.* **79**:14–19.

Kuo, L., Chilian, W. M., and Davis, M. J., 1990, Coronary arteriolar myogenic response is independent of endothelium, *Circ. Res.* **66**:860–866.

Langton, P. D., Nelson, M. T., Huang, Y., and Standen, N. B., 1991, Block of calcium-activated potassium channels in mammalian arterial myocytes by tetraethylammonium ions, *Am. J. Physiol.* **260**:H927–H934.

Leblanc, N., Wan, X., and Leung, P. M., 1994, Physiological role of Ca^{2+}-activated and voltage-dependent K^+ currents in rabbit coronary myocytes, *Am. J. Physiol.* **266**:C1523–C1537.

Li, P. L., Zou, A. P., and Campbell, W. B., 1997, Regulation of potassium channels in coronary arterial smooth muscle by endothelium-derived vasodilators, *Hypertension* **29**:262–267.

Lipp, P., and Niggli, E., 1998, Fundamental calcium release events revealed by two-photon excitation photolysis of caged calcium in guinea-pig cardiac myocytes, *J. Physiol.* **508**:801–809.

Luft, U. C., Bychkov, R., Gollasch, M., Rollet, J. B., Hofmann, F., Haller, H., and Luft, F. C., 1999, Farnesol blocks L-type Ca^{2+} channels targeting the α1C-subunit, *Arterioscler. Thromb. Vasc. Biol.* in press.

Luft, U. C., Bychkov, R., Gollasch, M., Gross, V., Roullet, J. B., McCarron, D. A., Ried, C., Yagil, Y., Yagil, H., Hofmann, F., Haller, H., Luft, F. C., 1999, Farnesol blocks L-type Ca^{2+} channels targeting the α1c-subunit. *Anterioscler. Throm. Vasc. Biol.* **19**:959–966.

Luo, Z., Fuentes, M. E., and Taylor, P., 1994, Regulation of acetylcholinesterase mRNA stability by calcium during differentiation from myoblasts to myotubes, *J. Biol. Chem.* **269**:27216–27223.

MacVicar, B. A., 1987, Morphological differentiation of cultured astrocytes is blocked by cadmium or cobalt, *Brain Res.* **420**:175–177.

Miller, F. J., Dellsperger, K. C., and Gutterman, D. D., 1997, Myogenic constriction of human coronary arterioles, *Am. J. Physiol.* **273**:H257–H264.

Miller, F. J., Dellsperger, K. C., and Gutterman, D. D., 1998, Pharmacologic activation of the human coronary microcirculation in vitro: Endothelium-dependent dilation and differential responses to acetylcholine, *Cardiovasc. Res.* **38**:744–750.

Miura, H., and Gutterman, D. D., 1998, Human coronary arteriolar dilation to arachidonic acid depends on cytochrome P-450 monooxygenase and Ca^{2+}-activated K^+ channels, *Circ. Res.* **83**:501–507.

Miyoshi, Y., and Nakaya, Y., 1991, Angiotensin II blocks ATP-sensitive K^+ channels in porcine coronary artery smooth muscle cells, *Biochem. Biophys. Res. Commun.* **181**:700–706.

Miyoshi, H., and Nakaya, Y., 1993, Activation of ATP-sensitive K^+ channels by cyclic AMP-dependent protein kinase in cultured smooth muscle cells of porcine coronary artery, *Biochem. Biophys. Res. Commun.* **193**:240–247.

Miyoshi, Y., Nakaya, Y., Wakatsuki, T., Nakaya, S., Fujino, K., Saito, K., and Inoue, I., 1992, Endothelin blocks ATP-sensitive K$^+$ channels and depolarizes smooth muscle cells of porcine coronary artery, *Circ. Res.* **70**:612 616.

Morano, I. L., 1992, Molecular biology of smooth muscle, *J. Hypertens.* **10**(5):411–416.

Munro, E., Patel, M., Chan, P., Betteridge, L., Gallagher, K., Schachter, M., Wolfe, J., and Sever P., 1994, Effect of calcium channel blockers on the growth of human vascular smooth muscle cells derived from saphenous vein and vascular graft stenoses, *J. Cardiovasc. Pharmacol.* **23**:779–784.

Nakashima, M., Momboult, J. V., Taylor, A. A., and Vanhoutte, P. M., 1993, Endothelium-dependent hyperpolarization caused by bradykinin in human coronary arteries, *J. Clin. Invest.* **92**:2867–2871.

Neild, T. O., and Keef, K., 1985, Measurements of the membrane potential of arterial smooth muscle in anesthetized animals and its relationship to changes in artery diameter, *Microvasc. Res.* **30**:19–28.

Nelson, M. T., and Quayle, J. M., 1995, Physiological roles and properties of potassium channels in arterial smooth muscle, *Am. J. Physiol.* **268**(4):C799–C822.

Nelson, M. T., Patlak, J. B., Worley, J. F., and Standen, N. B., 1990, Calcium channels, potassium channels, and voltage dependence of arterial smooth muscle tone, *Am. J. Physiol.* **259**(1):C3–C18.

Nelson, M. T., Cheng, H., Rubart, M., Santana, L.F., Bonev, A. D., Knot, H. J., and Lederer, W.J., 1995, Relaxation of arterial smooth muscle by calcium sparks, *Science* **270**:633–637.

Okabe, K., Kitamura, K., and Kuriyama, H., 1987, Features of 4-aminopyridine sensitive outward current observed in single smooth muscle cells from the rabbit pulmonary artery, *Pflügers Arch.* **409**:561–568.

Ottolia, M., and Toro, L., 1996, Reconstitution in lipid bilayers of an ATP-sensitive K$^+$ channel from pig coronary smooth muscle, *J. Membr. Biol.* **153**:203–209.

Owens, G. K., 1995, Regulation of differentiation of vascular smooth muscle cells, *Physiol. Rev.* **75**:487–517.

Parkington, H. C., Tonta, M. A., Coleman, H. A., and Tare, M., 1995, Role of membrane potential in endothelium-dependent relaxation of guinea-pig coronary arterial smooth muscle, *J. Physiol.* **484**:469480.

Parratt, J. R., and Kane, K. A., 1994, K$_{ATP}$ channels in ischaemic preconditioning, *Cardiovasc. Res.* **28**:783–787.

Porter, V. A., Bonev, A. D., Knot, H. J., Heppner, T. J., Stevenson, A. S., Kleppisch T., Lederer, W. J., and Nelson, M. T., 1998, Frequency modulation of Ca^{2+} sparks is involved in regulation of arterial diameter by cyclic nucleotides, *Am. J. Physiol.* **274**:C1346–C1355.

Prout, T. E., Knatterud, G. L., Meinert, C. L., and Klimt, C. R., 1972, The UGDP controversy. Clinical trials versus clinical implication, *Diabetes* **21**:1035–1040.

Quayle, J. M., McCarron, J. G., Brayden, J. E., and Nelson, M.T., 1993, Inward rectifier K$^+$ currents in smooth muscle cells from rat resistance-sized cerebral arteries, *Am. J. Physiol.* **265**:C1363–C1370.

Quayle, J. M., Bonev, A. D., Brayden, J. E., and Nelson, M. T., 1994, Calcitonin-gene related peptide activated ATP-sensitive K$^+$ currents in rabbit arterial smooth muscle via protein kinase A, *J. Physiol.* **475**:9–13.

Quayle, J. M., Nelson, M. T., and Standen, N. B., 1997, ATP-sensitive and inwardly rectifying potassium channels in smooth muscle, *Physiol. Rev.* **77**:1165–1232.

Reber, B. F., and Reuter, H., 1991, Dependence of cytosolic calcium in differentiating rat pheochromocytoma cells on calcium channels and intracellular stores, *J. Physiol. (London)* **435**:145–162.

Rembold, C. M., and Murphy, R. A., 1993, Models of the mechanism for crossbridge attachment in smooth muscle, *J. Muscle Res. Cell Motil.* **14**(3):325–334.

Ried, C., 1997, Kalzium- und Kaliumkanäle in humanen koronararteriellen Gefäßmuskelzellen, Doctoral thesis, Humboldt-Universität zu Berlin, Berlin, Germany.

Robertson, B. E., and Nelson, M. T., 1994, Aminopyridine inhibition and voltage dependence of K$^+$ currents in smooth muscle cells from cerebral arteries, *Am. J. Physiol.* **267**:C1589–C1597.

Robertson, B. E., Schubert, R., Hescheler, J., and Nelson, M. T., 1993, cGMP-dependent protein kinase activates Ca-activated K channels in cerebral artery smooth muscle cells, *Am. J. Physiol.* **265**:C299–C303.

Robertson, B. E., Bonev, A. D., and Nelson, M. T., 1996, Inward rectifier K$^+$ currents in smooth muscle cells from rat coronary arteries: Block by Mg^{2+}, Ca^{2+}, and Ba^{2+}, *Am. J. Physiol.* **271**:H696–H705.

Ross, R., 1995, Growth regulatory mechanisms and formation of the lesions of atherosclerosis, *Ann. N.Y. Acad. Sci.* **748**:1–4.

Rubart, M., Patlak, J. B., and Nelson, M. T., 1996, Ca^{2+} currents in cerebral artery smooth muscle cells of rat at physiological Ca^{2+} concentrations, *J. Gen. Physiol.* **107**:459–472.

Ruegg, J. C., and Pfitzer, G., 1991, Contractile protein interactions in smooth muscle, *Blood Vessels* **28**(1–3):159–163.

Rusch, N. J., and Liu, Y., 1997, Potassium channels in hypertension: Homeostatic pathways to buffer arterial contraction, *J. Lab. Clin. Med.* **130**:245–251.

Samaha, F. F., Heineman, F. W., Ince, C., Fleming, J., and Balaban, R. S., 1992, ATP-sensitive potassium channel is essential to maintain basal coronary vascular tone in vivo, *Am. J. Physiol.* **262**:C1220–C1227.

Satoh, H., Katoh, H., Velez, P., Fill, M., and Bers, D. M., 1998, Bay K 8644 increases resting Ca^{2+} spark frequency in ferret ventricular myocytes independent of Ca influx. Contrast with caffeine and ryanodine effects, *Circ. Res.* **83**:1192–1204.

Schubert, R., and Mulvany M. J., 1999, The myogenic response. Established facts and attractive hypotheses, *Clin. Sci.* **96**:313–326.

Schwartz, S. M., Heimark, R. L., and Majesky, M. W., 1990, Developmental mechanisms underlying pathology of arteries, *Physiol. Rev.* **70**:1177–1209.

Schwartz, S. M., Majesky, M. W., and Murry, C. E., 1995, The intima: Development and monoclonal responses to injury, *Atherosclerosis* **118**:S125–S140.

Siegel, G., Emden, J., Wenzel, K., Mironneau, J., and Stock, G., 1992, Potassium channel activation in vascular smooth muscle, *Adv. Exp. Med. Biol.* **311**:53–72.

Silberberg, S. D., and van Breemen, C., 1990, An ATP, calcium and voltage sensitive potassium channel in porcine coronary artery smooth muscle cells, *Biochem. Biophys. Res. Commun.* **172**:517–522.

Smirnov, S. V., and Aaronson P. I., 1992, Ca^{2+}-activated and voltage-gated K^+ currents in smooth muscle cells isolated from mesenteric arteries, *J. Physiol.* **457**:431–454.

Somlyo, A. P., and Somlyo, A. V., 1994, Smooth muscle: excitation–contraction coupling, contractile regulation, and the cross-bridge cycle, *Alcohol Clin. Exp. Res.* **18**(1):138–143.

Spitzer, N. C., Debaca, R. C., Allen, K. A., and Holliday, J., 1993, Calcium dependence of differentiation of GABA immunoreactivity in spinal neurons, *J. Comp. Neurol.* **337**(1):168–175.

Standen, N. B., Quayle, J. M., Davies, N. W., Brayden, J. E., Huang, Y., and Nelson, M. T., 1989, Hyperpolarizing vasodilators activate ATP-sensitive K^+ channels in arterial smooth muscle, *Science* **245**:177–180.

Stockbridge, N., Zhang, H., and Weir, B., 1991, Effects of K^+ channel agonists cromakalim and pinacidil on rat basilar artery smooth muscle cells are mediated by Ca^{2+} activated K^+ channels, Biochem. *Biophys. Res. Commun.* **181**:172–178.

Tanaka, Y., Meera, P., Song, M., Knaus, H. G., and Toro, L., 1997, Molecular constituents of maxi K_{Ca} channels in human coronary smooth muscle: Predominant alpha + beta subunit complexes, *J. Physiol.* **502**:545–557.

Taniguchi, J., Furukawa, K.-I., and Shigekawa, M., 1993, Maxi K^+ channels are stimulated by cyclic guanosine monophosphate-dependent protein kinase in canine coronary artery smooth muscle cells, *Pflügers Arch.* **423**:167–172.

Tomai, F., Crea, F., Gaspardone, A., Versaci, F., De Paulis, R., Penta de Peppo, A., Chiariello, L., and Gioffre, P. A., 1994, Blockade of ATP-sensitive potassium channels prevents myocardial preconditioning in man, *Circulation*, **90**:700–705.

Volk, K. A., and Shibata, E. F., 1993, Single delayed rectifier potassium channels from rabbit coronary artery myocytes, *Am. J. Physiol.* **264**:H1146–H1153.

Volk, K. A., Matsuda, J. J., and Shibata, E. F., 1991, A voltage-dependent potassium current in rabbit coronary artery smooth muscle cells, *J. Physiol.* **439**:751–768.

Von Beckerath, N., Cyrys, S., Dischner, A., and Daut, J., 1991, Hypoxic vasodilation in isolated, perfused guinea-pig heart: An analysis of the underlying mechanisms, *J. Physiol.* **442**:297–319.

Wakatsuki, T., Nakaya, Y., and Inoue, I., 1992, Vasopressin modulates K^+–channel activities of cultured smooth muscle cells from porcine coronary artery, *Am. J. Physiol.* **263**:H491–H496.

Wellman, G. C., Bonev, A. D., Nelson, M. T., and Brayden, J. E., 1996, Gender differences in coronary artery diameter involve estrogen, nitric oxide, and Ca^{2+}-dependent K^+ channels, *Circ. Res.* **79**:1024–1030.

Wellman, G. C., Quayle, J. M., and Standen, N. B., 1998, ATP-sensitive K^+ channel activation by calcitonin-gene related peptide and protein kinase A in pig coronary arterial smooth muscle, *J. Physiol.* **507**:117–129.

Worley, J. F. III, Deitmer, J. W., and Nelson, M. T., 1986, Single nisoldipine-sensitive calcium channels in smooth muscle cells isolated from rabbit mesenteric artery, *Proc. Natl. Acad. Sci. U.S.A.* **83**:5746–5750.

Worley, J. F. III, Quayle, J. M., Standen, N. B., and Nelson, M. T., 1991, Regulation of single calcium channels in cerebral arteries by voltage, serotonin, and dihydropyridines, *Am. J. Physiol.* **261:**H1951–H1960.

Xu, X., and Lee, K.S., 1994, Characterization of the ATP-inhibited K^+ current in canine coronary smooth muscle cells, *Pflügers Arch.* **427:**110–120.

Yang, J., Ellinor, P. T., Sather, W. A., Zhang, J. F., and Tsien, R. W., 1993, Molecular determinants of Ca^{2+} selectivity and ion permeation in L-type Ca^{2+} channels, *Nature* **366:**158–161.

Yatani, A., Bahinski, A., Mikala, G., Yamamoto, S., and Schwartz, A., 1994, Single amino acid substitutions within the ion permeation pathway alter single-channel conductance of the human L-type cardiac Ca^{2+} channel, *Circ. Res.* **75:**315–323.

Zhang, H., and Bolton, T.B., 1995, Activation by intracellular GDP, metabolic inhibition and pinacidil of a glibenclamide-sensitive K-channel in smooth muscle cells of rat mesenteric artery, *Br. J. Pharmacol.* **114:**662–672.

Chapter 30

Vascular K$^+$ Channel Expression and Function at Birth and in the Neonate

Helen L. Reeve and David N. Cornfield

1. INTRODUCTION

The fetal pulmonary circulation is a unique vascular bed which undergoes dramatic changes during the transition from fetus to neonate. Since the developing fetus has no need of its lungs for oxygen exchange, the pulmonary vasculature remains constricted and is characterized by high pulmonary vascular resistance (PVR) and minimal blood flow. Blood enters the right side of the heart and is directly shunted away into the systemic circulation via the ductus arteriosus (DA), a dilated blood vessel connecting the main pulmonary artery (PA) and the descending aorta. At birth, as the lungs become the primary site of oxygen exchange, two important changes occur in the pulmonary circulation. First, within 24 hours of birth, the pulmonary arterial bed dilates, resulting in an 8 to 10-fold increase in blood flow to the lungs and a decrease in pulmonary pressure to half systemic levels (Cassin *et al.*, 1964; Emmanouilides *et al.*, 1964; Dawes *et al.*, 1953). Second, the DA constricts to remove the right-to-left shunt pathway (Heymann and Rudolph, 1975). The closure of the DA occurs in two stages, with an initial O_2-dependent functional closure (characterized by constriction of the ductus) followed by intimal and medial necrosis and fibrosis (the anatomic phase). The time course over which these two phases occur is strongly species-dependent; in the normal human neonate, functional closure is generally complete within 48 hours of birth, with full anatomic closure occurring by 2–3 weeks of age (Eldridge and Hultgre, 1955). Although much of the decrease in PVR that occurs at birth is due to the mechanical process of lung ventilation (Dawes *et al.*, 1953), there is also an O_2-dependent factor in the fall in PVR. In the DA, functional closure is entirely dependent on O_2 (Kennedy and Clark, 1941, 1942), suggesting that these two opposing events (PA

Helen L. Reeve • Departments of Medicine and Physiology, University of Minnesota, Minneapolis, Minnesota 55455. *David N. Cornfield* • Department of Pediatrics, University of Minnesota, Minneapolis, Minnesota 55455.

Potassium Channels in Cardiovascular Biology, edited by Archer and Rusch. Kluwer Academic/Plenum Publishers, New York, 2001.

dilation versus DA constriction) can be induced by the same increase in oxygen partial pressure (PO_2) that occurs at birth. Emerging data suggest a role for K^+ channels in both these processes.

2. DILATION OF THE FETAL PULMONARY VASCULATURE

Successful dilation of the pulmonary circulation at birth requires a combination of factors: (1) mechanical ventilation of the lung (Teitel *et al.*, 1987), (2) increased pulmonary PO_2 (Assali *et al.*, 1968), and (3) increased shear stress (Abman and Accurso, 1989). The failure of these processes to occur results in persistent pulmonary hypertension of the newborn, a common cause of mortality and morbidity (see below). The role of mechanical ventilation was demonstrated in 1953 when Dawes *et al.* (1953) showed that ventilation of the fetal lamb lung with nitrogen was sufficient to induce partial dilation of the pulmonary circulation. More recently, the importance of endothelium-dependent vasoactive mediators such as nitric oxide (NO), thromboxanes, and prostaglandins has emerged (Abman *et al.*, 1990; Cornfield *et al.*, 1992a; Cassin, 1987; Soifer *et al.*, 1985). Pretreatment of animals with inhibitors of NO synthase can reduce the postnatal adaptation of the pulmonary circulation by preventing the release of endogenous NO (Fineman *et al.*, 1994; Abman *et al.*, 1990). Although NO appears to be an important mediator of the dilation induced by both the increase in PO_2 and the increase in shear stress, the mechanism by which it causes dilation is incompletely understood. Because the NO is primarily produced by the endothelium (Shaul and Wells, 1994), it must in some way affect the smooth muscle of the pulmonary arteries to initiate pulmonary vasodilation. It is known that exogenous NO can increase levels of smooth muscle cGMP via stimulation of guanylate cyclase (Archer *et al.*, 1994; Kukovetz *et al.*, 1991; Gold *et al.*, 1990; Arnold *et al.*, 1977); however the mechanism by which this rise in cGMP results in smooth muscle relaxation remains speculative.

3. K^+ CHANNELS IN THE FETAL AND NEONATAL PULMONARY ARTERY

One of the mechanisms by which blood vessel tone can be controlled is via modulation of K^+ channel activity in the membranes of the smooth muscle cells. Inhibition of a K^+ channel open at the resting membrane potential (E_m) of the cell will result in a buildup of positive charge on the inside of the cell membrane and thus cause membrane depolarization. Depolarization will change E_m to a value at which voltage-gated Ca^{2+} channels open, and then the resultant influx of Ca^{2+} ions will initiate constriction via the Ca^{2+}-calmodulin system. Conversely, activation of K^+ channels results in efflux of K^+ ions from the cell, leaving a negative charge on the inside of the membrane—i.e., hyperpolarization. This hyperpolarization will inactivate the voltage-dependent Ca^{2+} channels, resulting in smooth muscle relaxation and blood vessel dilation. The E_m recorded from smooth muscle cells taken from hypoxic fetal pulmonary arteries is significantly depolarized (approximately -20 to $-30\,mV$; Evans *et al.*, 1998; Reeve *et al.*, 1998; Cornfield *et al.*, 1996), as might be expected from a constricted, high-resistance vessel. Hypoxia can raise intracellular Ca^{2+} in fetal pulmonary artery smooth muscle cells (PASMC), suggesting why basal Ca^{2+} might be high in the developing fetus (Cornfield et al., 1993, 1994). Following birth, the E_m becomes more hyperpolarized, consistent with smooth muscle relaxation and blood vessel dilation (Evans et al., 1998; Reeve et al., 1998).

The presence of different classes of K^+ channels has been demonstrated in the perinatal pulmonary vasculature with the use of both pharmacological and electrophysiological techniques (Evans et al., 1998; Reeve et al., 1998; Boels et al., 1997; Theis et al., 1997; Cornfield et al., 1996; Tristani-Firouzi et al., 1996a; Chang et al., 1992; Cornfield et al., 1992b; Pinheiro and Malik, 1992). Furthermore, the K^+ channel population appears to be dynamic, with changes in expression occurring both during development and following the transition to neonate and adult stages (Reeve et al., 1998; Boels et al., 1997).

3.1. ATP-Dependent K^+ Channels

The presence of ATP-dependent K^+ channels (K_{ATP} channels) has been reported in the developing fetal PA of several species, including lamb and pig (Boels et al., 1997; Theis et al., 1997; Cornfield et al., 1996, 1992b). Whereas these channels do not appear to control resting E_m (Theis et al., 1997; Cornfield et al., 1996, 1992b) or basal tone, their activation causes significant and prolonged vasodilation of preconstricted arteries. Increased dilation to activators of K_{ATP} channels, such as levcromakalim, occurs during the transition from fetus to neonate, suggesting upregulation of these channels with maturation. Studies in porcine conduit and resistance vessels suggest that this upregulation may be dependent on the branch of the pulmonary tree studied, with maturation-dependent increases in relaxation to levcromakalim observed only in conduit rings (Boels et al., 1997). The cellular location of K_{ATP} channels within the pulmonary vasculature is controversial. Studies on isolated lamb arteries show that the relaxation response to activators of K_{ATP} channels can be prevented by endothelial denudation, suggesting that the K_{ATP} channels are located entirely in the endothelium (Theis et al., 1997). In contrast, in isolated fetal pig arteries, removal of the endothelium attenuates, but does not completely abolish, levcromakalim-induced relaxation, indicating the presence of K_{ATP} channels in both the pulmonary endothelial and smooth muscle cells (Boels et al., 1997). Patch-clamp studies of the K^+ channel activity in fetal PASMCs show K_{ATP} channels opening after several minutes of cell dialysis with ATP-free solution (Evans et al., 1998). However, glibenclamide, a specific blocker of K_{ATP} channels, has no effect on either E_m (Cornfield et al., 1996) or intracellular Ca^{2+} (Cornfield et al., 1994) recorded from fetal PASMCs, indicating that these channels, though present, are not open at resting E_m.

3.2. Large-Conductance, Ca^{2+}-Dependent K^+ Channels

Large-conductance, Ca^{2+}-dependent K^+ channels (BK_{Ca} channels) are also present in the fetal and neonatal pulmonary vasculature. Studies indicate that isolated arteries from these vasculatures constrict to BK_{Ca} channel blockers, such as tet-

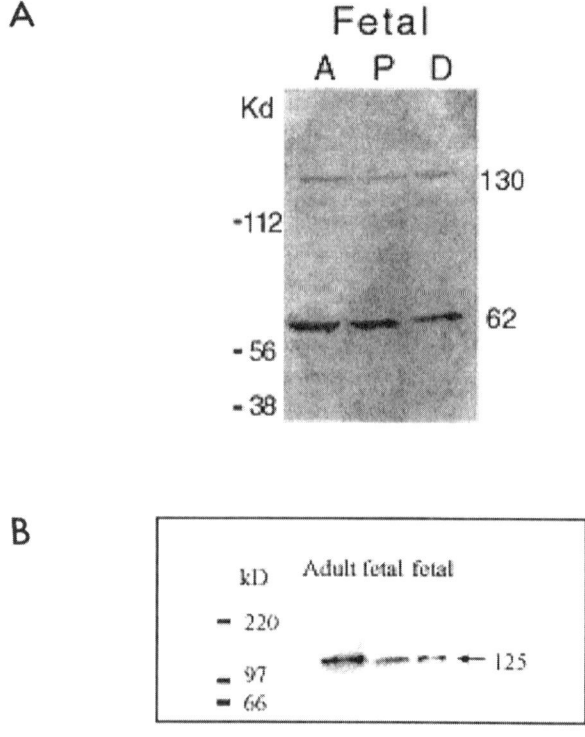

Figure 1. Western immunoblot analysis of the expression of BK_{Ca} α-subunit protein in fetal and adult pulmonary smooth muscle membrane. (A) Protein extract from the fetal (mean gestational age = 139 days) pulmonary vasculature. Each lane was loaded with 20 μg of protein, and subjected to SDS-PAGE to size separate proteins. The protein expresses both a 62-kDa fragment and a 135-kDa band. The 135-kDa band is consistent with a full-length BK_{Ca} α-subunit, and the 62-kDa band is consistent with an α-subunit fragment. The antibody used is directed against amino acids 455–477 of the α-subunit. These data support the notion that the BK_{Ca} channel is most abundantly expressed in the fetal and neonatal pulmonary circulation. A, Aorta; P, proximal pulmonary artery; D, distal pulmonary artery. (B) Western blot of the Kv2.1 channel in distal pulmonary artery from adult and fetal (mean gestational age = 140 days) sheep. The intensity of the band indicates increasing channel protein levels of the Kv2.1 with postnatal maturation. The 125 kDa band is consistent with the predicted molecular mass of the BK_{Ca} α-subunit.

raethylammonium (TEA) and charybdotoxin (CTX), whereas recent molecular studies also show the presence of both the RNA and protein for the BK_{Ca} channel in fetal pulmonary arteries (Fig. 1A). Electrophysiological studies using single smooth muscle cells from the fetal lamb indicate the presence of spontaneously transient outward currents (STOCs) (Reeve *et al.*, 1998; Cornfield *et al.*, 1996). STOCs result from the activation of BK_{Ca} channels by quantal release of intracellular Ca^{2+} from ryanodine-sensitive stores in the sarcoplasmic reticulum (Nelson *et al.*, 1995). Consistent with this, chelation of intracellular Ca^{2+} by including high concentrations of BAPTA in the patch pipette completely abolishes STOC activity (Reeve *et al.*, 1998). Fetal PASMCs are depolarized by exposure to TEA or CTX, but not to 4-aminopyridine (4-AP) or glibenclamide (Reeve *et al.*, 1998; Cornfield *et al.*, 1996). CTX is primarily a blocker of BK_{Ca} channels (Miller *et al.*, 1990) but also inhibits several voltage-dependent K^+ (Kv) channels (Kv) (Grissmer *et al.*, 1994), making its use a less specific way of demonstrating the presence of BK_{Ca} channels than originally believed. However, the depolarizing effect of TEA, coupled with the lack of effect of the Kv channel blocker 4-AP (Grissmer *et al.*, 1994), suggests that the channels controlling resting E_m in the constricted fetal pulmonary vasculature are BK_{Ca} channels. Concordant with this, CTX also increases intracellular Ca^{2+} in fetal PASMCs (Cornfield *et al.*, 1994). Interestingly, infusion of iberiotoxin (IBTX), low-dose TEA, or CTX into the whole, near-term, fetal lamb has no significant effect on basal PVR, as would be expected if BK_{Ca} channels were controlling resting tone (Storme *et al.*, 1999; Cornfield *et al.*, 1996). High doses of TEA have been shown to increase PVR during infusion, but a nonspecific effect of the drug cannot be discounted because 4-AP also produced an increase in resistance in some studies (Storme *et al.*, 1999) but not in others (Saqueton *et al.*, 1998). The reason for these discrepancies is unknown. The presence of endogenous vasodilators such as NO and dilator prostaglandins in the developing fetal circulation may make it difficult to produce significant constriction with the low concentrations of K^+ channel blockers that are specific for K^+ channel subtypes. Indeed, preliminary patch-clamp data suggest that the blockade of single BK_{Ca} channels by CTX can be overcome by NO.

In contrast to the findings in fetal PASMC, Kv channels appear to control resting E_m and hence tone in the adult pulmonary vasculature (Archer *et al.*, 1996; Yuan, 1995). These channels are also present in the fetal pulmonary vasculature, but in this developing bed their expression is diminished and their role in modulating vascular tone is limited. 4-AP-sensitive currents have been recorded in fetal smooth muscle cells from both pig and lamb pulmonary arteries (Evans *et al.*, 1998; Reeve *et al.*, 1998). In the lamb, they account for a very small percentage of total whole-cell current and, because of the STOC activity observed in these cells, are only clearly observed following chelation of intracellular Ca^{2+} (Reeve *et al.*, 1998). Maturational studies suggest that Kv channel expression increases following birth. The 4-AP-sensitive component of the whole-cell current becomes predominant as the fetal PASMC makes the transition to an adult smooth muscle cell whose E_m is determined by Kv channels (Reeve *et al.*, 1998). Because long-term exposure to hypoxia is known to downregulate the activity of Kv channels and depolarize resting E_m (Smirnov *et al.*, 1994; Wang *et al.*, 1997), it is possible that the hypoxic environment of the developing fetus maintains Kv channel expression low and accounts for the depolarized E_m recorded from these cells (Evans *et al.*, 1998; Reeve *et al.*, 1998; Cornfield *et al.*, 1996). Following birth, resting E_m becomes more hyperpolarized as the PA relaxes in the normoxic PO_2. Oxygenation may allow gradual upregulation of Kv channel expression. Preliminary molecular studies of the changes in expression of one of the Kv channel family, Kv2.1, suggest that this may be true. Kv2.1 protein appears to be minimally expressed in the resistance arteries of the fetal lamb, with expression increasing in the neonate and adult animals (Fig. 1B).

These studies suggest that the perinatal pulmonary vasculature has a dynamic K^+ channel population. In the developing fetus, the BK_{Ca} channel appears to be an important regulator of pulmonary vascular tone, whereas Kv channel activity increases following birth.

4. CONSTRICTION OF THE DUCTUS ARTERIOSUS

The successful transition from fetus to neonate requires not only the dilation of the pulmonary vasculature but constriction of the DA. Unlike pulmonary dilation, ductal constriction is an entirely O_2-dependent process that cannot be induced by ventilation of the fetal lungs with nitrogen (Kennedy and Clark, 1941, 1942). This O_2 dependency may explain the higher incidence of patent ductus (where the ductus fails to close at birth) at high altitude, where O_2 levels are low (Alzamora, 1953). Although the O_2 response has been known to be intrinsic to the ductus since the 1960s (Kovalcik, 1963), the full mechanism of functional closure still remains controversial (Smith, 1998). It does seem clear, however, that two separate mechanisms are involved in the closure: removal of the dilator influence of prostaglandins and active constriction of the DA smooth muscle (Coceani and Olley, 1988). Fay (1971) showed, using ductal strips, that the O_2-dependent constriction did not require the presence of either the adventitia or the endothelium and was consequently a function of the DA smooth muscle cells (DASMC). The importance of a rise in intracellular Ca^{2+} in DASMC has also been demonstrated. Removing extracellular Ca^{2+} by chelation with EGTA prevents O_2-induced constriction of DA rings, and normal constriction can be recovered by returning Ca^{2+} to the bathing medium (Kolvalcik, 1963). At present, the cyclooxygenase inhibitor indomethacin is still the most common medical treatment for patent ductus, indicating an important dilator role of prostaglandins (Gersony, 1983). In the rabbit, this dilation appears to be mediated through the prostaglandin EP_4 receptor (Smith et al., 1994), and EP_4-deficient mice show less ductal constriction at birth, demonstrating an active role for vasodilatory prostaglandins in counteracting constriction (Nguyen et al., 1997). Despite the importance of prostaglandins, the closure of the DA is mediated by more than just the withdrawal of a dilator influence (Smith, 1998), with the increase in PO_2 at birth likely to be a major factor controlling the mechanism of active ductal constriction.

5. K^+ CHANNELS IN THE DUCTUS ARTERIOSUS

The presence of K^+ channels in DASMC has been demonstrated by both pharmacological and, more recently, electrophysiological studies. Nakanishi et al. (1993) demonstrated that hypoxic rabbit DA rings could be constricted by glibenclamide and that O_2-constricted rings could be dilated by cromakalim, an opener of K_{ATP} channels. More recently, Tristani-Firouzi et al. (1996b) showed that fetal rabbit rings were more likely to constrict to low concentrations of the Kv channel blocker 4-AP than to glibenclamide, indicating that a Kv channel may be controlling resting E_m and hence tone. This constriction occurred independent of the endothelium and could not be blocked by cyclooxygenase inhibition, although it was obliterated by inhibition of the L-type Ca^{2+} channel (Tristani-Firouzi et al., 1996b). Tristani-Firouzi et al. (1996b) also showed that both 4-AP and O_2 inhibited Kv channels and led to similar and nonadditive membrane depolarization in DASMC. The presence of K_{ATP} channels has also been confirmed in patch-clamp studies. Although it is unlikely that these channels are open under resting conditions (glibenclamide has no effect on whole-cell currents

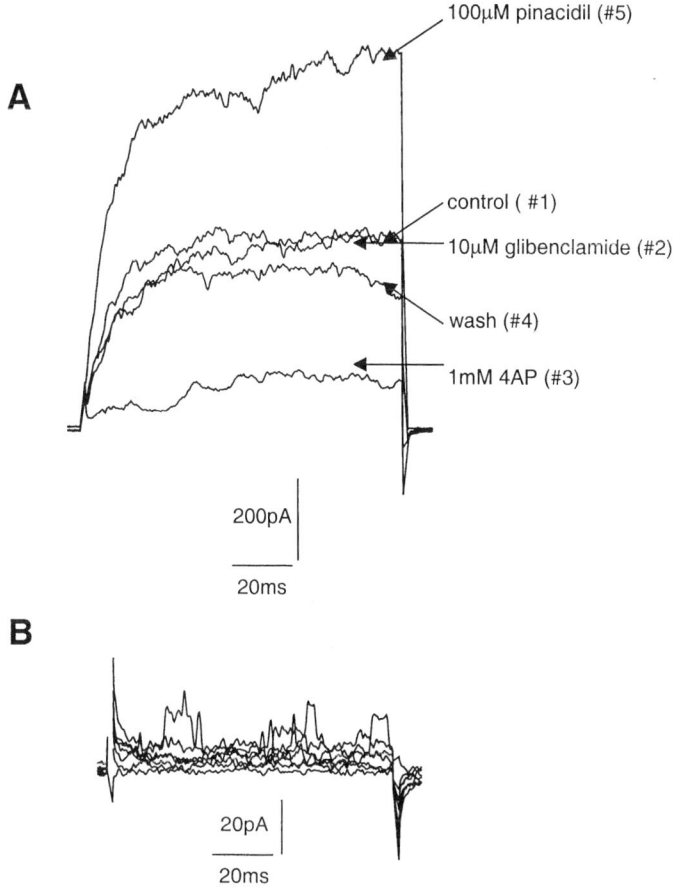

Figure 2. Pharmacology and kinetics of whole-cell currents recorded from ductus arteriosus (DA) smooth muscle cells. (A) Actual current traces recorded from a hypoxic DA smooth muscle cell by stepping from $-70\,mV$ to $+50\,mV$. Numbers in parentheses indicate order in which drugs were added. (B) Actual currents recorded by stepping from a holding potential of $-10\,mV$ to $+50\,mV$ in $+10$-mV increments. Note significantly reduced scale from (A).

recorded from DA SMC; see Fig. 2A), a glibenclamide-sensitive, voltage-independent current can be induced by exposure of cells to pinacidil, a K_{ATP} channel opener (Fig. 2A). Thus, while the presence of the channels is clear, their functional role is, as yet, unknown.

While insensitive to glibenclamide, whole-cell currents recorded from a holding potential of $-70\,mV$ in hypoxic DA SMC are inhibited by both 4-AP (Fig. 2A) and TEA, indicating the presence of both Kv and BK_{Ca} channels, as originally shown by Tristani-Firouzi et al. (1996b). By holding the cell at $-10\,mV$ to inactivate Kv channels, the 4-AP-sensitive component of the current can be completely abolished, indicating that these channels are open at more negative potentials and therefore are likely to be open at resting E_m (Fig. 2B). E_m recordings from hypoxic DA SMC also indicate that

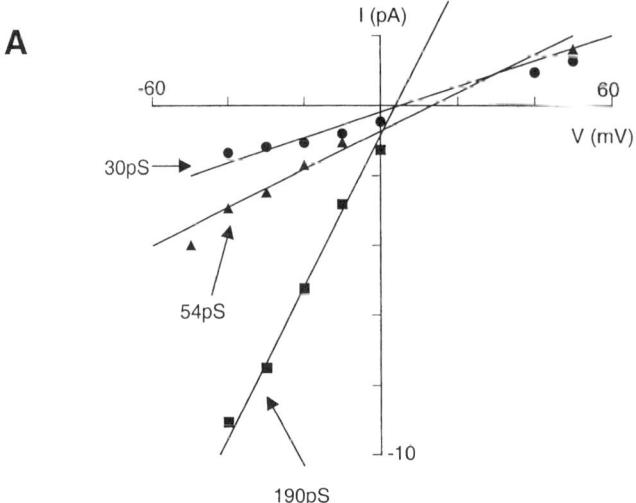

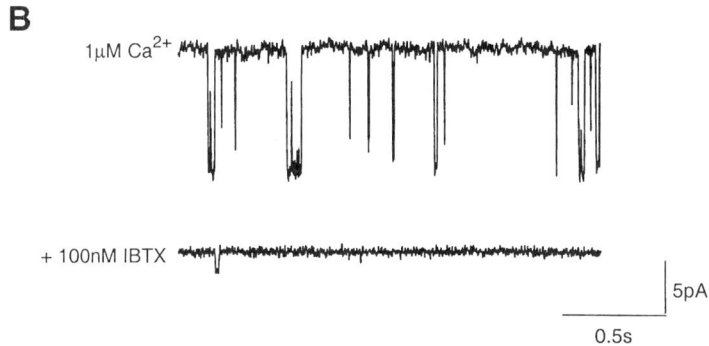

Figure 3. (A) Single-channel conductances recorded from ductus arteriosus smooth muscle cells (DASMC) using the inside-out configuration of the patch-clamp technique. (B) Recording of single-channel activity in an inside-out patch in $1 \mu M$ external Ca^{2+} (upper trace) and following exposure to $100 nM$ iberiotoxin (IBTX; lower trace).

cells can be depolarized only by exposure to 4-AP and not by exposure to TEA or glibenclamide (Tristani-Firouzi *et al.*, 1996b).

Single-channel studies from DASMC indicate the presence of several channels having different conductances, ranging from 30 pS to 150–190 pS (Fig. 3A). The large-conductance channel (150–190 pS) is Ca^{2+}-sensitive and is inhibited by IBTX in inside-out patches, indicating it to be a BK_{Ca} channel (Fig. 3B). An intermediate-conductance channel (58 pS), recorded using the cell-attached patch-clamp configuration, has been shown to be 4-AP-sensitive and so likely is also from the Kv family. Further studies are necessary to fully identify and categorize these single channels. Thus, like fetal PASMC, DASMC have Kv, K_{ATP}, and BK_{Ca} channels; however, in the DASMC, unlike the fetal PASMC, the Kv channels control resting E_m and hence tone.

6. MODULATION OF K$^+$ CHANNELS IN THE TRANSITION FROM FETUS TO NEONATE

6.1. The Fetal Pulmonary Artery, K$^+$ Channels, and Ventilation

Several critical physiological stimuli are required for successful transition of the fetal pulmonary vasculature at birth, including ventilation, shear stress, NO, and O_2. Emerging data suggest that changes in K$^+$ channel activity may play an important role in the responses to all of these stimuli. Dawes *et al.* (1953) demonstrated that partial dilation of the pulmonary vasculature could be achieved without the need for O_2, simply by the mechanical process of ventilation. More recently, Tristani-Firouzi *et al.* (1996a) showed that K$^+$ channel blockers could modify this ventilatory response. In the instrumented fetal lamb, ventilation with 10% O_2 increases left-PA blood flow by a mechanism that is significantly attenuated by infusion of TEA, but not of glibenclamide (Fig. 4). Although this low level of inspired O_2 could theoretically have sufficiently raised fetal PO_2 to produce the O_2-induced activation of BK_{Ca} channels (Cornfield *et al.*, 1996), blood gas measurements during the study did not indicate any change in PO_2 (Tristani-Firouzi *et al.*, 1996a).

6.2. The Fetal Pulmonary Artery, K$^+$ Channels, and Shear Stress

The increase in shear stress that occurs at birth along with the sudden increase in pulmonary blood flow contributes to the total vasodilatory response of the neonatal

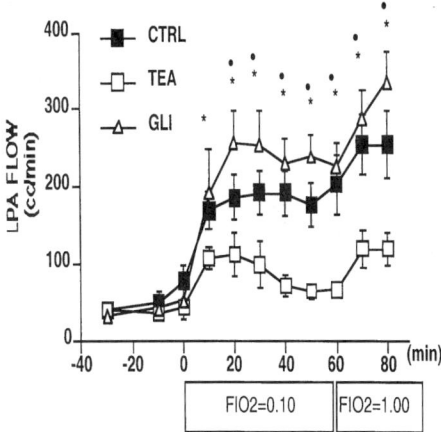

Figure 4. Effect of intrapulmonary infusion of tetraethylammonium (TEA; □) and glibenclamide (GLI; △) on left pulmonary artery blood flow (LPA flow) in response to sequential ventilation with low and high O_2. LPA flow in TEA group was attenuated compared to that in control group (CTRL; ■) in response to both low and high O_2. LPA flow did not differ in CTRL and GLI groups. FIO_2, Fractional inspired O_2 concentration. * $p < 0.01$, all groups compared to baseline value; • $p < 0.01$, CTRL vs. TEA. (From Tristani-Firouzi, M., Martin, E. B., Tolarova, S., Weir, E. K., Archer, S. L., Cornfield, D. N., 1996, Ventilation-induced pulmonary vasodilation at birth is modulated by potassium channel activity, *Am. J. Physiol. Heart. Circ. Physiol.* **271**:H2353–H2359.)

pulmonary vasculature. This shear stress response can be studied experimentally by partially compressing the DA with an occluder. This acutely increases pulmonary blood flow and PA pressure and thus elevates shear stress (Abman and Accurso, 1989). Over the first hour following DA compression, PVR falls as a result of a dilator response induced, at least in part, by elevated shear stress. This response is blocked by inhibition of NO synthase, suggesting a critical role of endogenous NO (Cornfield *et al.*, 1992a). However, more recent data indicate that NO may mediate shear stress-induced vasodilation through its ability to open K^+ channels. High doses of TEA and 4-AP can prevent the shear stress-mediated decrease in PVR (Storme *et al.*, 1999). Whereas the study by Tristani-Firouzi *et al.* (1996a) indicates that the BK_{Ca} channel may be important in the mechanical ventilatory response of the pulmonary vasculature, with shear stress it is possible that there is also a role for the Kv channel.

6.3. The Fetal Pulmonary Artery, K^+ Channels, and Oxygen

Whole-animal data, using the instrumented fetal lamb, indicate that the decrease in PVR that occurs in the lamb in response to increased O_2, can be attenuated by infusion of the animal with IBTX or TEA but not 4-AP or glibenclamide (Fig. 5). This suggests that O_2-induced pulmonary vasodilation may occur partially through the activation of a BK_{Ca} channel. Electrophysiological studies of hypoxic fetal PASMC also indicate a role for K^+ channels in O_2-induced pulmonary vasodilation. Exposure of these cells to acute normoxia greatly increases K^+ channel activity and causes significant hyperpolarization of E_m, both of which can be completely reversed by CTX

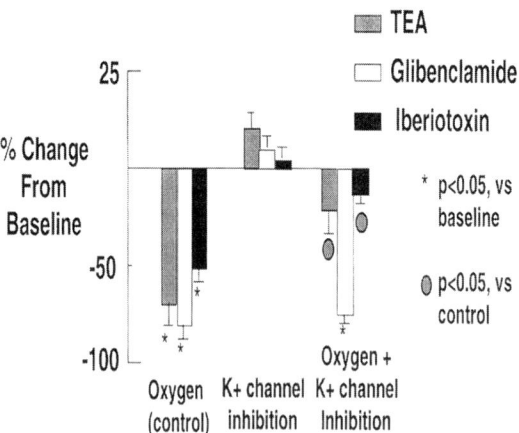

Figure 5. Effect of K^+ channel inhibition on O_2-induced fetal pulmonary vasodilation. In each of the control periods, O_2 caused a decrease in total pulmonary resistance (TPR) ($p < 0.05$). K^+ channel inhibition had no effect on basal TPR. Tetraethylammonium (TEA) blocked ($p < 0.05$) and glibenclamide had no effect on O_2-induced fetal pulmonary vasodilation. Iberiotoxin, a specific BK_{Ca} channel blocker, attenuated O_2-induced fetal pulmonary vasodilation ($p < 0.05$). (From Saqueton, C. B., Miller, R. B., Porter, V. A., Milla, C. E., Cornfield, D. N., 1999, NO causes perinatal pulmonary vasodilation through K^+ channel activation and intracellular Ca^{2+} release. *Am. J. Physiol. Lung. Cell. Moll. Cell. Physiol.* **275**:L925–L932.)

(Cornfield *et al.*, 1996). Because CTX can block Kv channels (Grissmer *et al.*, 1994) as well as BK_{Ca} channels, this observation alone does not definitively identify the BK_{Ca} channel as the O_2-sensitive channel. However, coupled with whole-animal data showing attenuation of O_2-induced decrease in PVR with IBTX (Cornfield *et al.*, 1996) but not 4-AP (Saqueton *et al.*, 1998), it is likely that increased BK_{Ca} channel activity is an important part of the O_2 dilator mechanism. A second study did not find an increase in whole-cell current during development from fetus to neonate, while still observing a similar maturation-dependent hyperpolarization of E_m (Evans *et al.*, 1998). The reason for this discrepancy is unknown, but it may be due to the longer cell preparation time used, which meant that fetal cells were maintained normoxic for several hours before study. Exposures of this length to normoxia may have already modified K^+ channel activity in these cells, as we have recently demonstrated (Reeve *et al.*, 1998).

6.4. The Fetal Pulmonary Artery, K^+ Channels, and NO

The mechanism by which K^+ channels in fetal PASMC are activated by O_2 and membrane hyperpolarization and vasodilation is unclear, but likely involves NO. Release of endothelial NO is both developmentally regulated and sensitive to changes in PO_2, with a gradual increase in NO production occurring from late in the third trimester through the first 4 weeks of life (Shaul *et al.*, 1993). Treatment of fetal lambs with inhibitors of NO synthase, such as N^G-nitro-L-arginine (L-NA), attenuates O_2- and shear stress-induced pulmonary vasodilation (Tiktinsky and Morin, 1993; McQuestion *et al.*, 1993; Cornfield *et al.*, 1992a). This suggests that NO and K^+ channel activation by O_2 are likely to be intimately linked. Furthermore, treatment of fetal lambs with inhibitors of either guanylate cyclase (LY 83583) or cGMP-dependent protein kinase (KT 5823) significantly attenuates O_2-dependent vasodilation. KT 5823 also prevents normoxic activation of K^+ channels. NO increases whole-cell K^+ current in fetal PASMC, but the mechanism by which this occurs has yet to be determined (Cornfield *et al.*, 1996). Nelson *et al.* (1995) proposed an elegant pathway by which arterial dilation could occur to counteract myogenic constrictions in adult systemic arteries. They proposed that quantal releases of Ca^{2+} from ryanodine-sensitive stores in the sarcoplasmic reticulum—known as Ca^{2+} sparks—activated BK_{Ca} channels located in the membrane, in close proximity to the stores. A myogenic constriction that caused an increase in intracellular Ca^{2+} via membrane depolarization and Ca^{2+} influx could therefore be self-limiting because the rise in Ca^{2+} would fill sarcoplasmic reticulum stores, thereby increasing spark activity, and thus promote BK_{Ca} channel activation, which would cause vasodilation. The activity of the BK_{Ca} channels regulated by these sparks is characteristic of STOCs, such as those previously recorded from the fetal PASMC (Reeve *et al.*, 1998; Cornfield *et al.*, 1996). Spark and STOC activity has been shown to be increased by elevating levels of cGMP (with sodium nitroprusside) or cAMP (with forskolin) (Porter *et al.*, 1998). This suggests that in the fetal PA, the rise in PO_2 at birth may increase release of endothelial NO, which in turn activates guanylate cyclase and increases cGMP levels. This rise in cGMP may increase Ca^{2+} spark activity and hence activate BK_{Ca} channels. The STOCs previously recorded from hypoxic cells may indicate a basal level of regulatory spark activity

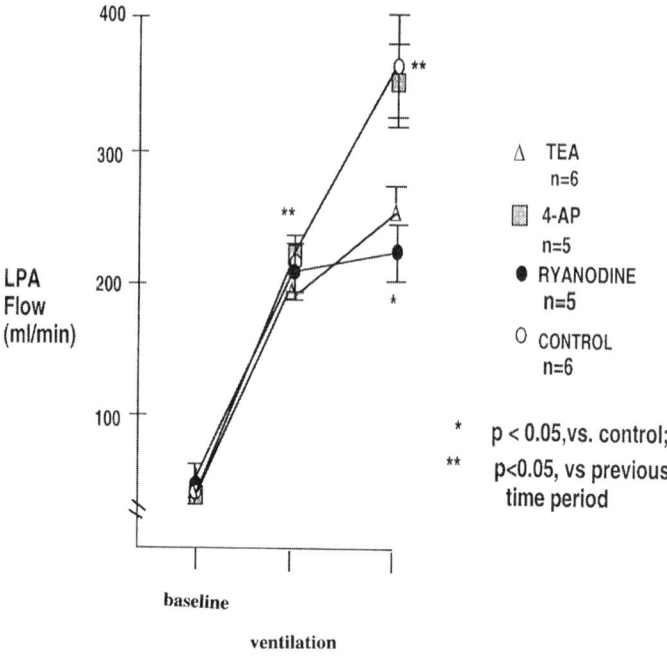

Figure 6. Effect of K$^+$ channel blockade with tetracthylammonium (TEA) or 4-aminopyridine (4-AP) or ryanodine treatment on nitric oxide (NO)-induced perinatal pulmonary vasodilation. In the presence of pharamcologic blockade of endogenous NO production, administration of inhaled NO caused a significant increase in left pulmonary artery blood flow) (LPA flow) in the control and 4-AP-treated animals. Following TEA treatment, the increase in LPA flow caused by administration of inhaled NO was significantly attenuated. Ryanodine treatment blocked the pulmonary vasodilation caused by inhaled NO.

necessary to maintain tone in the constricted, high-Ca^{2+} environment of the fetus. Although this potential mechanism is not yet fully resolved, preliminary data show that NO-induced fetal pulmonary vasodilation can be inhibited by ryanodine (Fig. 6), which prevents the release of Ca^{2+} stores necessary for spark generation. A schematic representation of the mechanism by which O$_2$, NO, and K$^+$ channels may interact to cause pulmonary vasodilation is shown in Fig. 7.

6.5. The Ductus Arteriosus, K$^+$ Channels, and Oxygen

Constriction of the DA with high levels of KCl attenuates O$_2$ induced constriction, suggesting that a portion of functional closure occurs by a membrane-dependent response. Furthermore, O$_2$-induced constriction of the DA is associated with smooth muscle cell depolarization (Roulet and Coburn, 1981). More recent studies indicate that there may be a role for K$^+$ channels in the O$_2$-induced constriction of the DA. Constriction of DA rings with the Kv channel blocker 4-AP prevents additional

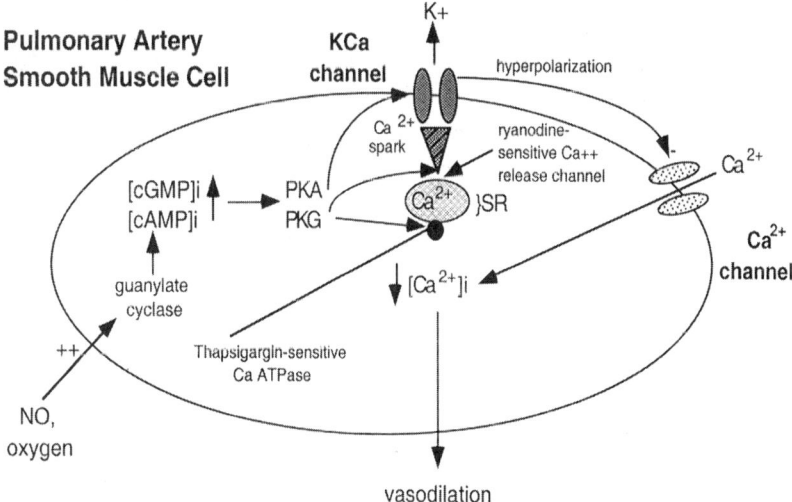

Figure 7. Proposed scheme for regulation of pulmonary arterial tone. Entry of extracellular Ca^{2+} can cause local release of Ca^{2+}, a Ca^{2+} spark, from an internal, ryanodine-sensitive store to cause elevation of intracellular Ca^{2+} in the region of the BK_{Ca} channel. Activation of the BK_{Ca} channel causes K^+ efflux, membrane hyperpolarization, and closure of the voltage-operated Ca^{2+} channel, a decrease in intracellular Ca^{2+}, and vasodilation. Vasodilator stimuli, such as NO or O_2, may act through cyclic-nucleotide-dependent kinases (PKA, PKG) to affect ryanodine receptor function and Ca^{2+} pumps on internal stores. Whether vasoconstrictors and vasodilators act on the Ca^{2+}-STOC pathway is unknown. (From Cornfield, D. N., Reeve, H. L., Tolarova, S., Weir, E. K., Archer, S. L., 1996, Oxygen causes fetal pulmonary vasodilation through activation of a calcium-dependent K^+ channel, *Proc. Natl. Acad. Sci.* **93**:80089–8094.)

constriction with O_2, whereas TEA and glibenclamide do not change the response to O_2 (Tristani-Firouzi *et al.*, 1996b). Whole-cell K^+ current recorded from DA SMC maintained in normoxia is significantly reduced from that recorded from hypoxic cells (Fig. 8A). Recordings of E_m indicate that O_2 can acutely depolarize hypoxic DA SMC (Tristani-Firouzi *et al.*, 1996b), whereas longer exposure to O_2 permanently shifts E_m to a depolarized level. Studies of changes in intracellular Ca^{2+} have also shown that both 4-AP and O_2 increase intracellular Ca^{2+} and that O_2-induced constriction can be prevented by L-type Ca^{2+} channel blockers (Takizawa *et al.*, 1994; Tristani-Firouzi *et al.*, 1996b). Together, these data indicate that the increase in O_2 that occurs at birth may initiate ductal constriction, at least in part, through inhibition of a Kv channel, membrane depolarizations and subsequent influx of Ca^{2+} through L-type Ca^{2+} channels.

The mechanism by which the K^+ channels in the DA smooth muscle may sense changes in O_2 is unknown but may involve redox modulation, as has been proposed in the adult pulmonary circulation (Weir and Archer, 1995). Indeed, there is substantial evidence that K^+ channel activity can be altered by redox changes (Reeve *et al.*, 1995; Archer *et al.*, 1993; Ruppersberg *et al.*, 1991; Meury and Robin, 1990). One route by which changes in cytosolic redox status may occur is through modification of mitochondrial oxidative phosphorylation. During this process, reducing equivalents, such as NADH, are oxidized to form ATP molecules. As a by-product of this process,

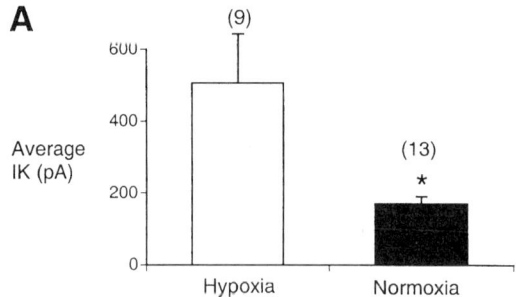

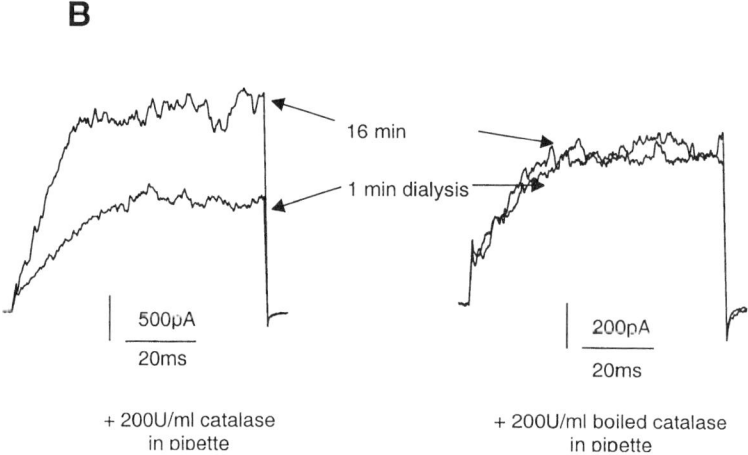

Figure 8. (A) Average K^+ currents (IK) recorded at $+50\,\text{mV}$ from hypoxic and normoxic ductus arteriosus smooth muscle cells (DASMC) maintained in hypoxia or normoxia as indicated. Currents are significantly reduced when cells are maintained in normoxia. $*P < 0.05$. (B) Effect of intracellular catalase on DASMC K^+ channel activity. Whole-cell currents were recorded from normoxic DASMC by stepping from $-70\,\text{mV}$ to $+50\,\text{mV}$ with either $200\,\text{U}$ catalase/ml in the patch pipette (*left*) or an equal concentration of inactive catalase in the pipette (*right*). Traces were recorded following 1-min and 16-min cell dialysis as indicated.

reactive O_2 species including superoxide and hydrogen peroxide (H_2O_2) are formed, in proportion to the level of cellular O_2. Although it is known that exogenous H_2O_2 can change the activity of K^+ channels (Filipovic and Reeves, 1997; Sobey *et al.*, 1997; Szabo *et al.*, 1997; Vega-Saenz de Miera and Rudy, 1992), the effect of endogenous H_2O_2 is more difficult to study. Preliminary patch-clamp studies in DA SMC may, however, provide evidence that altering endogenous H_2O_2 levels can affect K^+ channel activity. The enzyme responsible for the cellular degradation of H_2O_2 to H_2O and O_2 is catalase. In normoxic cells, inclusion of catalase in the patch-clamp pipette gradually increases whole-cell outward K^+ current, over a period of 20 min (Fig. 8B). This increase is not mimicked by dialysis of the cells for the same time without catalase in the pipette or by dialysis with a pipette solution containing boiled catalase (Fig. 8B). Na citrate (catalase buffer), or a millimolar-equivalent concentration of albumin. These findings suggest that endogenous H_2O_2 levels inhibit K^+ channel activity and that this inhibition is lost when H_2O_2 is removed by catalase. Although these data do not remove the possibility that loss of reactive O_2 species further down the reaction pathway from H_2O_2 (such as the hydroxyl radical) may account for the changes in K^+ current, they do provide the first evidence that modulating endogenous H_2O_2 levels can alter channel activity. It might be speculated that during fetal development, when PO_2 is low, H_2O_2 levels are concomitantly low and K^+ channels in the DA are open, the membrane is hyperpolarized, and the DA is dilated; at birth, the rise in H_2O_2 that occurs as a result of the rise in O_2 inhibits K^+ channel activity, resulting in membrane depolarization and constriction.

7. CLINICAL IMPLICATIONS

The failure of the fetal pulmonary circulation to dilate at birth results in persistent pulmonary hypertension of the newborn (PPHN), a condition with a high rate of mortality and morbidity (Roberts and Shaul, 1993). The reason for the unsuccessful transition of the fetal pulmonary vasculature is unknown, making prevention or treatment difficult. As discussed above, endothelial-derived NO is an important modulator of the perinatal pulmonary vasculature (Tiktinsky and Morin, 1993; Cornfield et al., 1992a; Abman et al., 1990), and inhaled NO has been shown to improve oxygenation in some infants with PPHN (Neonatal Inhaled Nitric Oxide Study Group, 1997; Roberts et al., 1997). However, it is by no means a successful therapy for all infants. Recent studies indicate that many infants do not respond at all to NO and that some others do not sustain their response (Neonatal Inhaled Nitric Oxide Study Group, 1997; Roberts et al., 1997; Steinhorn et al., 1997; Barefield et al., 1996). The gradual unraveling of the pathway by which neonatal pulmonary vasodilation occurs may allow the development of other, more successful therapeutic strategies.

7.1. K^+ Channel Agonists

Because it ultimately appears that activation of a K^+ channel causes relaxation of the PASMC and thus dilation of pulmonary arteries (Storme et al., 1999; Reeve et al., 1998; Cornfield et al., 1996; Tristani-Firouzi et al., 1996a), it might be hypothesized that K^+ channel agonists could provide a potential therapeutic strategy. Recent data suggest that BK_{Ca} channel activity may be decreased in arteries from lambs with experimental PPHN. Because it appears that the channel controlling resting E_m in the fetus is a BK_{Ca} channel and that this channel is also activated by O_2 (Reeve et al., 1998; Cornfield et al., 1996), decreased BK_{Ca} expression may, in part, explain the lack of dilation at birth in these infants. It would seem reasonable that an increase in the activity of these channels by exogenous activators might compensate for their decreased expression and hence induce dilation. BK_{Ca} channel agonists such as NS 004 and NS 1619 (Olesen et al., 1994; McManus et al., 1993) are available, but their use is compromised by the ubiquitous distribution of BK_{Ca} channels throughout the entire vasculature. To avoid significant systemic side effects, they would have to be locally delivered and maintained within the pulmonary circulation. Although K_{ATP} channels do not control resting tone in the fetus or neonate, their activation can induce dilation of arteries (Boels et al., 1997) or reduce PVR in lungs from pigs with experimental PPHN (Pinheiro and Malik, 1992). These data also suggest that K_{ATP} agonists, such as levcromakalim, may have a potential therapeutic role. Unfortunately, K_{ATP} channels are ubiquitously distributed throughout the vasculature, making systemic side effects almost inevitable with the use of K_{ATP} agonists. Since at present there are no data to suggest the existence of distinct types of K^+ channels in the perinatal pulmonary circulation versus other vascular beds, nonselective K^+ channel activation is not a viable therapeutic approach to treating PPHN. A more successful approach may be to find a step in the dilator pathway that leads to channel activation (and hence dilation) that is specific for the pulmonary circulation, which would permit an increase activity of K^+ channels primarily in the vascular bed of interest.

7.2. Phosphodiesterase Inhibitors

It appears likely that the endothelial NO liberated at birth by oxygenation increases the levels of smooth muscle cGMP, which in turn activates the BK_{Ca} channel, probably through increased Ca^{2+} spark activity (see above). As well as decreased BK_{Ca} channel activity, newborn lambs with experimental PPHN also appear to have decreased cGMP production (Steinhorn et al., 1995). Therefore, augmentation of cGMP levels might also provide a pathway by which perinatal pulmonary vasodilation could be induced. cGMP is normally hydrolyzed to its inactive metabolite by a family of phosphodiesterases (Murad, 1986). Consequently, inhibition of phosphodiesterase activity will enhance survival of cGMP and hence prolong vasodilation. There are several families of phosphodiesterases which have varying affinities for cGMP over cAMP (Soderling et al., 1998; Beavo, 1995). One family of phosphodiesterases, the PDE5 family, may be specifically expressed in the lung and platelets (Rabe et al., 1994; Thomas et al., 1990), making it an excellent target for a localized effect on the pulmonary circulation. Several PDE5 inhibitors have been investigated, including zaprinast and dipyridamole. However, both these drugs cause an unwanted adverse effect, namely significant systemic hypotension (Ziegler et al., 1998; Thusu et al., 1995; Mlczoch et al., 1977). This hypotension is likely due to inhibition of other phosphodiesterase isoenzymes present in the systemic vasculature. Recent data with a new inhibitor, E4021, appear to be more encouraging. E4021 has little effect on other phosphodiesterase isoenzymes, and preliminary studies in adult pigs (Saeki et al., 1995) and rats (Cohen et al., 1996) suggest that it may selectively dilate the pulmonary circulation without any associated systemic hypotension. The use of E4021 in the perinatal pulmonary vasculature is likely to provide clinically relevant data.

In conclusion, greater understanding of perinatal pulmonary vasodilation will provide a better opportunity to evaluate potential clinical interventions. The role of K^+ channels in this process is clear, but at present their ubiquitous nature complicates modulation of K^+ channel activity as a therapeutic modality. However, it is possible that, in the future, molecular studies may provide a key to specific ways in which fetal pulmonary K^+ channels can be modulated independently of channels found in the systemic vasculature.

REFERENCES

Abman, S. H., and Accurso, F. J., 1989, Acute effects of partial compression of ductus arteriosus on fetal pulmonary circulation, *Am. J. Physiol.* **257**:H626–H634.

Abman, S. H., Chatfield, B. A., Hall, S. L., and McMurtry, I. F., 1990, Role of endothelium-derived relaxing factor during transition of the pulmonary circulation at birth, *Am. J. Physiol.* **259**:H1921–H1927.

Alzamora, V., 1953, On the possible influence of great altitudes on the determination of certain cardiovascular anomalies, *Pediatrics* **12**:259–262.

Archer, S., Huang, J., Henry, T., Peterson, D., and Weir, E. K., 1993, A redox based O_2 sensor in rat pulmonary vasculature, *Circ. Res.* **73**:1100–1112.

Archer, S. L., Huang, J., Hampl, V., Nelson, D. P., Schultz, P. J., and Weir, E. K., 1994, Nitric oxide and cGMP cause vasorelaxation by activation of a charybdotoxin-sensitive K channel by cGMP-dependent protein kinase. *Proc Natl. Acad. Sci. U.S.A.* **91**:7583–7587.

Archer, S. L., Huang, J. M. C., Reeve, H. L., Hampl, V., Tolarova, S., Michelakis, E., and Weir, E. K., 1996, Differential distribution of electrophysiologically distinct myocytes in conduit and resistance arteries determines their response to nitric oxide and hypoxia, *Circ. Res.* **78**:431–442.

Arnold, W. P., Mittal, C. K., Katsuki, S., and Murad, F., 1977, NO activates guanylate cyclase and increases guanosine 3′,5′-cyclic monophosphate levels in various tissue preparations, *Proc. Natl. Acad. Sci. U.S.A.* **74**:3203–3207.

Assali, N. S., Kirschbaum, T. H., and Dilts, P. V., 1968, Effects of hyperbaric oxygen on utero placental and fetal circulation, *Circ. Res.* **22**:573–588.

Barefield, E. S., Karle, V. A., and Carlo, W. A., 1996, Inhaled nitric oxide in term infants with hypoxemic respiratory failure, *J. Pediatr.* **129**:279–286.

Beavo, J. A., 1995, Cyclic nucleotide phosphodiesterases: Functional implications of multiple isoforms, *Physiol. Rev.* **75**:725–748.

Boels, P. J., Gao, B., Deutsch, J., and Haworth, S. G., 1997, ATP-dependent K^+ channel activation in isolated normal and hypertensive newborn and adult porcine pulmonary vessels, *Pediatr. Res.* **42**:317–326.

Cassin, S., 1987, Role of prostaglandins, thromboxanes and leukotrienes in the control of the pulmonary circulation in the fetus and newborn, *Semin. Perinatol.* **11**:53–63.

Cassin, S., Dawes, G. S., Mott, J. C., Ross, B. B., and Strang, L. B., 1964, The vascular resistance of the fetal and newly ventilated lung of the lamb, *J. Physiol. (London)* **171**:61–79.

Chang, J. K., Moore, P., Fineman, J. R., Soifer, S. J., and Heymann, M. A., 1992, K^+ channel pulmonary vasodilation in fetal lambs: Role of endothelium-derived nitric oxide, *J. Appl. Physiol.* **73**:188–194.

Coceani, F., and Olley, P. M., 1988, Eicosanoids in the fetal and transitional pulmonary circulation, *Chest* **93**:112S–117S.

Cohen, A. H., Hanson, K., Morris, K., Fouty, B., McMurtry, I. F., Clarke, W., and Rodman, D. M., 1996, Inhibition of cyclic 3′-5′-guanosine monophosphate-specific phosphodiesterase selectively vasodilates the pulmonary circulation in chronically hypoxic rats, *J. Clin. Invest.* **97**:172–179.

Cornfield, D. N., Chatfield, B. A., McQuestion, J. A., McMurtry, I. F., and Abman S. H., 1992a, Effects of birth-related stimuli on L-arginine-dependent pulmonary vasodilation in ovine fetus, *Am. J. Physiol.* **262**:H1474–H1481.

Cornfield, D. N., McQuestion, J. A., McMurtry, I. F., Rodman, D. M., and Abman, S. H., 1992b, Role of ATP-sensitive potassium channels in ovine fetal pulmonary vascular tone, *Am. J. Physiol.* **263**:H1363–H1368.

Cornfield, D. N., Stevens. T., McMurtry, I. F., Abman, S. H., and Rodman, D. M., 1993, Acute hypoxia increases cytosolic calcium in fetal pulmonary artery smooth muscle cells, *Am. J. Physiol.* **265**:L53–L56.

Cornfield, D. N., Stevens, T., McMurtry, I. F., Abman, S. H., and Rodman, D. M., 1994, Acute hypoxia causes membrane depolarization and calcium influx in fetal pulmonary artery smooth muscle cells. *Am. J. Physiol.* **266**:L468–L475.

Cornfield, D. N., Reeve, H. L., Tolarova, S., Weir, E. K., and Archer, S., 1996, Oxygen causes fetal pulmonary vasodilation through activation of a calcium-dependent potassium channel, *Proc. Natl. Acad. Sci. U.S.A.* **93**:8089–8094.

Dawes, G. S., Mott, J. C., Widdicombe, J. G., and Wyatt, D. G., 1953, Changes in the lungs of the newborn lamb, *J. Physiol. (London)* **121**:141–162.

Eldridge, F. L., and Hultgre, H. N., 1955, The physiologic closure of the ductus arteriosus in the newborn infant, *J. Clin. Invest.* **34**:987–996.

Emmanouilides, G., Moss, A., Duffie, E., and Adams, F., 1964, Pulmonary arterial pressure changes in human infants from birth to 3 days of age, *J. Pediatr.* **65**:327–333.

Evans, A. M., Osipenko, O. N., Haworth, S. G., and Gurney, A. M., 1998, Resting potentials and potassium currents during development of pulmonary artery smooth muscle cells, *Am. J. Physiol.* **275**:H887–H899.

Fay, F. S., 1971, Guinea pig ductus arteriosus I: Cellular and metabolic basis for oxygen sensitivity, *Am. J. Physiol.* **221**:470–479.

Filipovic, D. M., and Reeves, W. B., 1997, Hydrogen peroxide activates glibenclamide-sensitive K^+ channels in LLC-PK1 cells, *Am. J. Physiol.* **272**:C737–C743.

Fineman, J. R., Wong, J., Morin, F. C., Wild, L. M., and Soifer, S. J., 1994, Chronic nitric oxide inhibition in utero produces persistent pulmonary hypertension in newborn lambs, *J. Clin. Invest.* **93**:2675–2683.

Gersony, W. M., Peckman, G. J., Ellison, R. C., Miettinen, O. S., Nadas, O. S., 1983, Effects of indomethacin in premature infants with patent ductus arteriosus: Results of a national collaborative study, *J. Pediatr.* **102**:895–906.

Gold, M. E., Wood, K. S., Byrns, R. E., Buga, G. M., and Ignarro, L. J., 1990, L-Arginine-dependent vascular smooth muscle and cGMP formation, *Am. J. Physiol.* **259**:H1813–H1821.

Grissmer, S., Nguyen, A. N., Aiyar, J., Hanson, D. C., Mather, R. J., Gutman, G. A., Karmilowicz, M. J., Auperin, D. D., and Chandy, K. G., 1994, Pharmacological characterization of five cloned voltage-gated K^+ channels stably expressed in mammalian cell lines, *Mol. Pharmacol.* **45**:1227–1234.

Heymann, M. A., and Rudolph, A. M., 1975, Control of the ductus arteriosus, *Physiol. Rev.* **55**:62–78.

Kennedy, J. A, and Clark, S. L., 1941, Observations on the ductus arteriosus of the guinea pig in relation to its method of closure, *Anat. Rec.* **79**:349.

Kennedy, J. A, and Clark, S. L., 1942, Observations on the physiological reactions of the ductus arteriosus, *Am. J. Physiol.* **136**:140.

Kovalcik, V., 1963, The response of the isolated ductus arteriosus to oxygen and anoxia, *J. Physiol. (London)* **169**:185–197.

Kukovetz, W. R., Holzmann, S., and Schmidt, K., 1991, Cellular mechanisms of action of therapeutic nitric oxide donors. *Eur. Heart J.* **12**:E16–E24.

McManus, O. B, Harris, G. H., Giangiacomo, K. M., Feigenbaum, P., Reuben, J. P., Addy, M. E., Burka, J. F., Kaczorowski, G. J., and Garcia, M. L., 1993, An activator of calcium-dependent potassium channels isolated from a medicinal herb, *Biochemistry* **32**:6128–6133.

McQuestion, J. A., Cornfield, D. N., McMurtry, I. F., and Abman, S. H., 1993, Effects of oxygen and exogenous L-arginine on EDRF activity in fetal pulmonary circulation, *Am. J. Physiol.* **264**:H865–H871.

Meury, J., and Robin, A., 1990, Glutathione-gated K^+ channels of *Escherichia coli* carry out K^+ efflux controlled by the redox status of the cell, *Arch. Microbiol.* **154**:475–482.

Miller, C., Moczydlowski, E., Latorre, R., and Phillips, M., 1990, Charybdotoxin, a protein inhibitor of single Ca^{2+}-activated K^+ channels from mammalian skeletal muscle, *Nature* **313**:316-318.

Mlczoch, J., Weir, E. K., and Grover R. F., 1977, Inhibition of hypoxic pulmonary vasoconstriction by dipyridamole is not platelet mediated, *Can. J. Physiol. Pharmacol.* **55**:448–451.

Murad, F., 1986, Cyclic guanosine monophosphate as a mediator of vasodilation, *J. Clin. Invest.* **78**:1–5.

Nakanishi, T., Gu, H., Hagiward, N., and Momma, K., 1993, Mechanism of oxygen-induced contraction of ductus arteriosus isolated from fetal rabbit, *Circ. Res.* **72**:1218–1228.

Nelson, M. T., Cheng, H., Rubart, M., Santana, L. F., Bonev, A. D., Knot, H. J., and Lederer, W. J., 1995, Relaxation of arterial smooth muscle by calcium sparks, *Science* **270**:633–637.

Neonatal Inhaled Nitric Oxide Study Group, 1997, Inhaled nitric oxide in full-term and nearly full-term infants with hypoxic respiratory failure, *N. Engl. J. Med.* **336**:605–610.

Nguyen, M., Camenisch, T., Snouwaert, J. N., Hicks, E., Coffman, T. M., Anderson, P. A. W., Malouf, N. N., and Koller, B. H., 1997, The prostaglandin receptor EP4 triggers remodeling of the cardiovascular system at birth, *Nature.* **390**:78–81.

Olesen, S. P., Munch, E., Moldt, P., and Drejer, J., 1994, Selective activation of Ca^{2+}-dependent K^+ channels in cerebellar granule cells, *Eur. J. Pharmacol.* **251**:53–59.

Pinheiro, J. M., and Malik, A. B., 1992, K^+ ATP-channel activation causes marked vasodilation in the hypertensive neonatal pig lung, *Am. J. Physiol.*, **263**:H1532–H1536.

Porter, V. A., Bonev, A. D., Knot, H. J., Heppner, T. J., Stevenson, A. S., Kleppisch, T., Lederer, W. J., and Nelson, M. T., 1998, Frequency modulation of Ca^{2+} sparks is involved in regulation of arterial diameter by cyclic nucleotides, *Am. J. Physiol.* **274**:C1346–C1355.

Rabe, K. F., Tenor, H., Dent, G., Schudt, C., Nakashima, M., and Magnussen, H., 1994, Identification of PDE isozymes in human pulmonary artery and effect of selective PDE inhibitors, *Am. J. Physiol.* **266**:L536–L543.

Reeve, H. L., Weir, E. K., Nelson, D. A, Peterson, D. P., and Archer, S. L., 1995, Opposing effects of oxidants and antioxidants on K^+ channel activity and tone in vascular tissue, *Exp. Physiol.* **80**:825–834.

Reeve H. L., Weir, E. K., Archer, S. L., and Cornfield, D. N., 1998, A maturational shift in pulmonary K^+ channels from Ca^{2+}-sensitive to voltage-dependent, *Am. J. Physiol.* 275:L1019–L1025.

Roberts, J. D., and Shaul, P. W., 1993, Advances in the treatment of persistent pulmonary hypertension of the newborn, *Pediatr. Clin. North Am.* **40**:983–1004.

Roberts, J. D., Fineman, J., Morin, F. C., III, Shaul, P. W., Rimar, S., Schreiber, M. D., Polin, R. A., Thusu, K. G., Zayek, M., Zwass, M. S., Zellers, T. M., Wylam, M. E., Gross, I., Zapol, W. M., and Heymann, M. A., 1997, Inhaled nitric oxide gas improves oxygenation in PPHN, *N. Engl. J. Med.* **336**:605–610.

Roulet, M. J., and Coburn, R. F., 1981, Oxygen-induced contraction in the guinea pig neonatal ductus arteriosus, *Circ. Res.* **49**:997–1002.

Ruppersberg, J., Stocker, M., Pongs, O., Heinemann, S., Frank, R., and Koenen, M., 1991, Regulation of fast inactivation of cloned mammalian $I_K(A)$ channels by cysteine oxidation, *Nature* **352**:711–714.

Saeki, T., Adachi, H., Takase, Y., Yoshitake, S., Souda, S., and Saito, I., 1995, A selective type V phosphodiesterase inhibitor, E4021, dilates porcine large coronary artery, *J. Pharmacol. Exp. Ther.* **272**:825–831.

Saqueton, C. B., Miller, R. B., Porter, V. A., and Cornfield, D. N., 1998, Inhaled nitric oxide causes pulmonary dilation through ryanodine-sensitive activation of a K_{Ca} channel, *Pediatr. Res.* **43**:296A.

Shaul, P. W., and Wells, L. B., 1994, Oxygen modulates nitric oxide production selectively in fetal pulmonary endothelial cells, *Am. J. Respir. Cell. Mol. Biol.* **11**:432–438.

Shaul, P. W., Farrar, M. A., and Magnuss, R. R., 1993, Pulmonary endothelial nitric oxide production is developmentally regulated in the fetus and newborn, *Am. J. Physiol.* **65**:H1056–1063.

Smirnov, S. V., Robertson, T. P., Ward, J. P. T., and Arronson, P. I., 1994, Chronic hypoxia is associated with reduced delayed rectifier K^+ current in rat pulmonary artery muscle cells, *Am. J. Physiol.* **266**:762–767.

Smith, G. C., 1998, The pharmacology of the ductus arteriosus, *Pharm. Rev.* **50**:35-58.

Smith G. C., Coleman, R. A., and McGrath, J. C., 1994, Characterization of dilator prostanoid receptors in the fetal rabbit ductus arteriosus, *J. Pharmacol. Exp. Ther.* **271**:390–396.

Sobey, C. G., Heistad, D. D., and Faraci, F., 1997, Mechanisms of bradykinin-induced cerebral vasodilatation in rats: Evidence that reactive oxygen species activate K^+ channels, *Stroke* **28**:2290–2294.

Soderling, S. H., Bayuga, S. J., and Beavo, J. A., 1998, Identification and characterization of a novel family of cyclic nucleotide phosphodiesterases, *J. Biol. Chem*, **273**:15553–15558.

Soifer, S. J., Loitz, R. D., Roman, C., and Heymann, M. A., 1985, Leukotriene end organ antagonists increase pulmonary blood flow in fetal lambs, *Am. J. Physiol.* **249**:H570–H576.

Steinhorn, R. H., Russell, J. A., Morin, F., III, 1995, Disruption of cyclic GMP production in pulmonary arteries isolated from fetal lambs with pulmonary hypertension, *Am. J. Physiol.* **268**:H1483–H1489.

Steinhorn, R. H., Cox, P. N., Fineman, J. R., Finer, N. N., Rosenberg, E. M., Silver, M. M., Tyebkhan, J., Zwass, M. S., and Morin, F. C., III, 1997, Inhaled nitric oxide enhances oxygenation but not survival in infants with alveolar capillary dysplasia, *J. Pediatr.* **130**:417–422.

Storme, L., Rairigh, R. L., Parker, T. A., Cornfield, D. N., Kinsella, J. P., and Abman, S. H., 1999, K^+-channel blockade inhibits shear stress-induced pulmonary vasodilation in the ovine fetus, *Am. J. Physiol.* **276**:L220–L228.

Szabo, I., Nilius, B., Zhang, X., Busch, A. E., Gulbins, E., Suessbrich, H., and Lang, F., 1997, Inhibitory effects of oxidants on n-type K^+ channels in *Xenopus* oocytes, *Pflügers Arch.* **433**:626–632.

Takizawa, T., Oda, T., Arishima. K., Ymamoto, M., Masaoka, T., Somiya, H., Akahori, F., and Shito, K., 1994, A calcium channel blocker, verapamil, inhibits the spontaneous closure of the ductus arteriosus in newborn rats, *J. Toxicol. Soc.* **19**:171–174.

Teitel, D. F., Iwamoto, H. S., and Rudolph, A. M., 1987, Effects of birth related events on central blood flow patterns, *Pediatr. Res.* **22**:557–566.

Theis, J. G., Liu, Y., and Coceani, F., 1997, ATP-gated potassium channel activity of pulmonary resistance vessels in the lamb, *Can. J. Physiol. Pharmacol.* **75**:1241–1248.

Thomas, M. K., Francis, S. H., and Corbine, J. D., 1990, Characterization of a purified bovine lung cGMP-binding cGMP phosphodiesterase, *J. Biol. Chem.* **265**:24964–24970.

Thusu, K. G., Morin, F. C., III, Russell, J. A., and Steinhorn, R. H., 1995, The cGMP phosphodiesterase inhibitor zaprinast enhances the effect of nitric oxide, *Am. J. Respir. Crit. Care Med.* **152:**1605–1610.

Tiktinsky, M. H., and Morin, F. C., III, 1993, Increasing oxygen tension dilates fetal pulmonary circulation via endothelium derived relaxing factor, *Am. J. Physiol.* **265:**H376–H380.

Tristani-Firouzi, M., Martin, E. B., Tolarova, S., Weir, E. K., Archer, S. L., and Cornfield, D. N., 1996a, Ventilation-induced pulmonary vasodilation at birth is modulated by potassium channel activity, *Am. J. Physiol.* **271:**H2353–H2539.

Tristani-Firouzi, M., Reeve, H. L., Tolarova, S., Weir, E. K., and Archer, S. L., 1996b, Oxygen-induced constriction of rabbit ductus arteriosus occurs via inhibition of a 4-aminopyridine-voltage-sensitive potassium channel, *J. Clin. Invest.* **98:**1959–1965.

Vega-Saenz de Miera, E., and Rudy, B., 1992, Modulation of K^+ channels by hydrogen peroxide. *Biochem. Biophys. Res. Commun.* **186:**1681–1687.

Wang, J., Juhaszova, M., Rubin, L. J., and Yuan, X.-J., 1997, Hypoxia inhibits gene expression of voltage-gated K^+ channel α subunits in pulmonary artery smooth muscle cells, *J. Clin. Invest.* **100:**2347–2353.

Weir, E. K., and Archer, S. L., 1995, The mechanism of acute hypoxic pulmonary vasoconstriction: The tale of two channels, *FASEB J.* **9:**183–189.

Yuan, X.-J., 1995, Voltage-gated K^+ currents regulate resting membrane potential and $[Ca^{2+}]_i$ in pulmonary arterial myocytes, *Circ. Res.* **77:**370–378.

Ziegler, J. W., Ivy, D. D., Fox, J. J., Kinsella, J. P., Clarke, W. R., and Abman, S. H., 1998, Dipyridamole potentiates pulmonary vasodilation induced by acetylcholine and nitric oxide in the ovine fetus. *Am. J. Respir. Crit. Care Med.* **157:**1104–1110.

Part VI

Potassium Channels in the Endothelium

Chapter 31

Overview: Potassium Channels in Vascular Endothelial Cells

Guy Droogmans and Bernd Nilius

1. INTRODUCTION

Vascular endothelium appears to be a unique organ that not only responds to numerous hormonal and chemical signals but also senses changes in physical parameters, such as changes in blood flow through shear stress and changes in blood pressure through stretch. The endothelium integrates these signals and responds to them by regulating the production and release of vasoactive substances that play a role in blood pressure regulation and vascular growth. These regulatory substances include prostaglandins, endothelium-derived relaxing factor (EDRF or NO) and endothelium-derived hyperpolarizing factor (EDHF), endothelin, natriuretic peptides, small signaling molecules, such as substance P, ATP, growth factors, steroids, and even larger proteins, such as receptors and proteins involved in the blood clotting cascade (for reviews, see Inagami *et al.*, 1995; Nilius and Casteels, 1996). In addition to this endocrine function, endothelial cells either prevent or trigger blood clotting in response to various signals and exert thrombolytic as well as thrombogenic activity. As antigen-presenting cells, they are also involved in immune responses. Their ability to modulate cell–cell contacts controls the permeability of the blood–tissue interface. Finally, they initiate angiogenesis and vessel repair.

2. K$^+$ CHANNELS OF ENDOTHELIUM

2.1. Inward Rectifier K$^+$ Channels

The most prominent K$^+$ channel in resting endothelial cells is the inward rectifier K$^+$ channel (Kir), which conducts inward currents at potentials more negative than the

Guy Droogmans and Bernd Nilius • Laboratorium voor Fysiologie, KU Leuven, Campus Gasthuisberg, B-3000 Leuven, Belgium.
Potassium Channels in Cardiovascular Biology, edited by Archer and Rusch. Kluwer Academic/Plenum Publishers, New York, 2001.

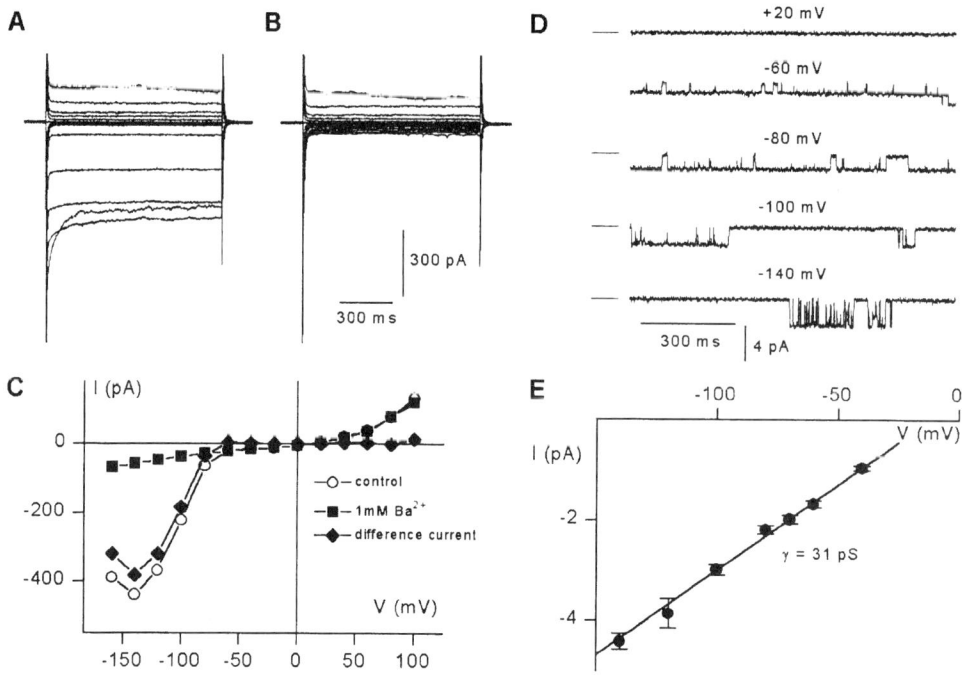

Figure 1.   Whole-cell and single-channel Kir currents in calf pulmonary artery endothelial (CPAE) cells. (A) and (B) Current traces recorded in CPAE under control conditions (A) and after application of 1 mM Ba^{2+} (B) at various potentials applied from a holding potential of 0 mV. Large inward currents are present at negative potentials, and a rapid inactivation is prominent at the most negative potentials. (C) I–V curves corresponding to the recordings in panels A ($\bigcirc$) and B ($\blacksquare$) and the Ba^{2+}-blocked difference current measured at the end of the applied voltage step ($\blacklozenge$). Notice the pronounced inward rectification of the current reversing near -80 mV. (D) Single-channel recordings in cell-attached mode at symmetrical K$^+$ concentrations. (E) Corresponding I–V curve reconstructed from the amplitudes of the measured single-channel currents. (Reproduced, with permission, from Kamouchi *et al.*, 1997b.)

K$^+$ equilibrium potential but permits much smaller currents at potentials positive to that potential (Fig. 1). The channel appears to be more abundant in cultured than in intact endothelial cells. In intact endocardial endothelial cells, it is mainly confined to the luminal side of the endothelium (Manabe *et al.*, 1995b). However, in monolayers of cultured bovine aortic endothelial cells, it appears to be randomly distributed (Colden-Stanfield *et al.*, 1992).

   The conductance of Kir channels, together with the basal activity of the volume-regulated anion channel and a background nonselective cation current, determines the resting membrane potential of endothelial cells (Campbell *et al.*, 1991; Fransen and Sys, 1997; Voets *et al.*, 1996), which ranges between -10 and   70 mV, depending on the cell type. The resting potential seems to be more negative in macrovascular than in microvascular cells (Daut *et al.*, 1994; Vargas *et al.*, 1994; Zunkler *et al.*, 1995) and also in electrically coupled confluent cells (Vargas *et al.*, 1994; Zunkler *et al.*, 1995). In bovine pulmonary endothelial cells, blocking the Kir channel by micromolar concentrations of Ba^{2+} results in a depolarization of cells with a negative resting potential but has no significant effect in less polarized cells, in which the resting potential is mainly determined by the other conductances (Nilius *et al.*, 1997a; Voets *et al.*, 1996).

2.1.1. Biophysical Properties

The single-channel conductance of endothelial Kir channels ranges from 23 to 30 pS in symmetrical K$^+$ solutions (Fig. 1; Elam and Lansman, 1995; Inazu *et al.*, 1994; Nilius *et al.*, 1993; Olesen and Bundgaard, 1993; Pennefather and DeCoursey, 1994; Silver and DeCoursey, 1990; Silver *et al.*, 1994; Takeda *et al.*, 1987) but is smaller at lower extracellular K$^+$ concentrations (Nilius and Riemann, 1990; Pennefather and DeCoursey, 1994; Silver *et al.*, 1994; Zunkler *et al.*, 1995). Its permeation profile is $P_K > P_{Rb} > P_{Cs}$ (Pennefather and DeCoursey, 1994; Silver *et al.*, 1994). The voltage dependence of current gating shifts along with the equilibrium potential for potassium, E_K, as is characteristic of Kir channels in other cells.

Extracellular barium (Ba^{2+}), tetraethylammonium (TEA), tetrabutylammonium (TBA), and cesium (Cs$^+$) block Kir channels (Nilius and Droogmans, 1995; Pasyk *et al.*, 1992; Revest and Abbott, 1992; von Beckerath *et al.*, 1996). Millimolar concentrations of Ca^{2+}, Sr^{2+}, Mg^{2+}, and Mn^{2+} also block the Kir channel in human capillary endothelial cells (Jow and Numann, 1998). The block by Sr^{2+} is voltage-dependent, in contrast with that by the other divalent cations, consistent with binding of Sr^{2+} to a site within the voltage field. At negative potentials, the channel shows time-dependent inactivation, which is largely due to block by extracellular Mg^{2+}, and this effect is antagonized by extracellular K$^+$ (Elam and Lansman, 1995). A mechanism was proposed in which Mg^{2+} binds to the closed channel during hyperpolarization and prevents it from opening until it is occupied by K$^+$.

Inward rectification is attributed to a time- and voltage-dependent block by intracellular Mg^{2+}. However, Kir channels in bovine pulmonary artery endothelial cells still exhibit voltage- and time-dependent gating and small outward currents in nominal Mg^{2+}-free internal pipette solutions (Silver and DeCoursey, 1990). Rectification may therefore represent an intrinsic gating property of Kir channels in these cells.

2.1.2. Modulation

Shear evokes reversible hyperpolarization in current-clamped endothelial cells and enhanced large inward and small outward whole-cell K$^+$ currents, which are blocked by Cs$^+$ in voltage-clamped endothelial cells. In luminal cell-attached patches, shear stress also enhances the open-state probability of the Kir channel (Olesen, 1993). Kir channels in inside-out patches also are activated by increases in the intracellular calcium concentration [Ca^{2+}]$_i$ from 10^{-7} to 10^{-6} M (Jacobs *et al.*, 1995).

Administration of ATP to the cytosolic surface of inside-out patches reversibly activates the K$^+$ channel within seconds (Olesen and Bundgaard, 1993). Also, intracellular ATP, but not its nonhydrolyzable analogs adenosine 5′-*O*-(3-thiotriphosphate) (ATPγS) and adenylyl imidodiphosphate (AMP-PNP), prevents rundown of the current in the whole-cell mode (Kamouchi *et al.*, 1997b). The phosphatase inhibitor okadaic acid also prevents rundown of the current, but protamine, an activator of protein phosphatase 2A (PP2A), enhances the rate of rundown. Phosphorylation of the channel molecule therefore seems to be essential for maintaining its activity, its rundown probably being due to dephosphorylation by PP2A (Kamouchi et al., 1997b).

Vasoactive agonists, such as angiotensin II, arginine vasopressin, vasoactive intestinal polypeptide (VIP), endothelin-1 (ET-1), and histamine, inhibit the Kir channels in capillary and macrovascular endothelial cells (Hoyer et al., 1991; Nilius et al., 1993; Pasyk et al., 1992; Zhang et al., 1994). Also, application of guanosine 5'-O-(3-thiotriphosphate) (GTPγS) via the patch pipette inhibits the Kir channel, both in control cells and in cells pretreated with pertussis toxin (PTX) (Hoyer et al., 1991; Kamouchi et al., 1997b), suggesting that activation of a PTX-insensitive G-protein may mediate these inhibitory actions, probably by modulating protein phosphatase 2A (PP2A).

2.1.3. Molecular Identity

Forsyth et al. (1997) isolated a 5.1-kilobase cDNA encoding a Kir channel from a bovine aortic endothelial cell library. Expression in $Xenopus$ oocytes showed that it is a K^+-specific strong inward rectifier channel, sensitive to extracellular Ba^{2+}, Cs^+, and a variety of antiarrhythmic agents. Sequence analysis revealed that it is a member of the Kir2.1 family of inward rectifier K^+ channels. Data from reverse-transcription polymerase chain reaction (RT-PCR) analysis are consistent with the previous findings (Kamouchi et al., 1997b). Although several potential phosphorylation sites have been demonstrated in the amino sequence of the Kir2 family, we failed to show any effect of protein kinase A (PKA) or protein kinase C (PKC) activation on the current (Kamouchi et al., 1997b).

2.2. Ca^{2+}-Activated K^+ Channels

In many cases, the initial response of endothelial cells to diverse stimuli involves an elevation of cytosolic Ca^{2+} and activation of Ca^{2+}-dependent enzymes, including nitric oxide (NO) synthase and phospholipase A_2. The magnitude of the influx of Ca^{2+} depends on its electrochemical gradient, which is modulated by the membrane potential. The influx of Ca^{2+} in stimulated cells has a depolarizing effect on the membrane potential, which is compensated by the activation of Ca^{2+}-dependent K^+ channels (K_{Ca}). The open-state probability of these channels is increased by the elevation of cytosolic Ca^{2+} to cause membrane hyperpolarization. Three subtypes of these channels can be differentiated according to conductance: BK_{Ca} or maxi-K channels, IK_{Ca} or intermediate-conductance channels, and SK_{Ca} or small-conductance channels. All three subtypes have been observed in endothelial cells. These channels appear to be more or less symmetrically distributed between the luminal and abluminal endothelial cell surface (Colden-Stanfield et al., 1992; Manabe et al., 1995b). However, the distribution over different endothelial tissues is rather puzzling. For example, freshly isolated rabbit aortic endothelial cells express mainly BK_{Ca} channels (Rusko et al., 1992), but the same cells in the rat contain two types of SK_{Ca} channels (Marchenko and Sage, 1996). Cultured endothelial cells from bovine aorta have been reported to contain primarily IK_{Ca} channels, although in another study both IK_{Ca} and BK_{Ca} channels appeared to be present.

2.2.1. BK$_{Ca}$ Channels

Channels having conductances between 165 and 220 pS have been reported in various endothelial cell types (Baron *et al.*, 1996; Hoyer *et al.*, 1994; Ling and Oneill, 1992; Nilius and Riemann, 1990; Rusko *et al.*, 1995, 1992). Their open-state probability depends on both intracellular Ca^{2+} and voltage, and they are blocked by TEA, charybdotoxin, *d*-tubocurarine (Baron *et al.*, 1996; Daut *et al.*, 1994; Rusko *et al.*, 1992), and intracellular alkalinization (Thuringer *et al.*, 1991).

It has been reported recently that NO directly activates BK$_{Ca}$ channels in vascular smooth muscle (Bolotina *et al.*, 1994) and bovine adrenal chromaffin cells (Chen *et al.*, 1998). However, in cultured endothelial cells, the NO donor *S*-nitrosocysteine neither directly activated BK$_{Ca}$ channels nor modulated BK$_{Ca}$ channels that had been activated by increasing [Ca^{2+}]$_i$ (Haburcak *et al.*, 1997). Epoxyeicosatrienoic acids (EETs), which have been described as possible candidates for EDHFs, exerted an autocrine effect on BK$_{Ca}$ channels recorded in inside-out patches of primary cultured pig coronary artery endothelial cells (Baron *et al.*, 1997).

The role of BK$_{Ca}$ channels in modulating agonist-induced changes in [Ca^{2+}]$_i$ was assessed by comparing the effects of ATP stimulation in cultured bovine pulmonary artery endothelial cells, which have no endogenous BK$_{Ca}$ channels, and in the same cells expressing hslo, the human homolog of BK$_{Ca}$ (Kamouchi *et al.*, 1997a). ATP depolarized nontransfected cells but hyperpolarized transfected cells. As a consequence, the ATP-induced sustained rise in [Ca^{2+}]$_i$, which is linked to Ca^{2+} influx, was more pronounced in the transfected cells (Fig. 2).

2.2.2. IK$_{Ca}$ Channels

Ca^{2+}-activated K$^+$ channels of intermediate conductance (30 and 80 pS in symmetrical K$^+$ and 15 pS at physiological extracellular K$^+$) have been observed in various endothelial cells (Ling and O'Neill, 1992; Manabe *et al.*, 1995a; Sauvé *et al.*, 1988, 1990; Vaca *et al.*, 1992; Van Renterghem *et al.*, 1995). Charybdotoxin, as well as quinine, and TBA are efficient blockers of these IK$_{Ca}$ channels (Daut *et al.*, 1994; Van Renterghem *et al.*, 1995). Noxiustoxin, a component of *Centruroides* scorpion toxin (Colden-Stanfield *et al.*, 1990; Vaca *et al.*, 1993), and NS1619 (Cai *et al.*, 1998), an activator of BK$_{Ca}$ channels, are also potent inhibitors of IK$_{Ca}$ channels.

Cultured bovine aortic endothelial cells (Vaca *et al.*, 1992) manifest 30-pS IK$_{Ca}$ channels. GTP, together with Mg^{2+}, as well as GTPγS stimulates these channels, whereas guanosine 5'-*O*-(2-thiodiphosphate) (GDPβS) reverses the stimulatory effects. In some preparations, IK$_{Ca}$ channels with a conductance between 40 and 80 pS are activated by the inositol trisphosphate (1,4,5-IP$_3$)-sensitive Ca^{2+} release caused by endothelium-dependent vasodilators, such as bradykinin, acetylcholine, and ATP (Sauvé *et al.*, 1990). In brain microvascular endothelial cells, ET-1 and ET-3 activate IK$_{Ca}$ channels via endothelin-A receptors (Van Renterghem *et al.*, 1995), without affecting Kir channels. Thus, the activity of IK$_{Ca}$ channels appears to be sensitive to receptor-mediated events and responsive to G-protein regulation.

The hydrophilic oxidative reagents 5,5'-dithio-bis(2-nitrobenzoic acid) and thimerosal reduce IK$_{Ca}$ channel activity in bovine aortic endothelial cells recorded in

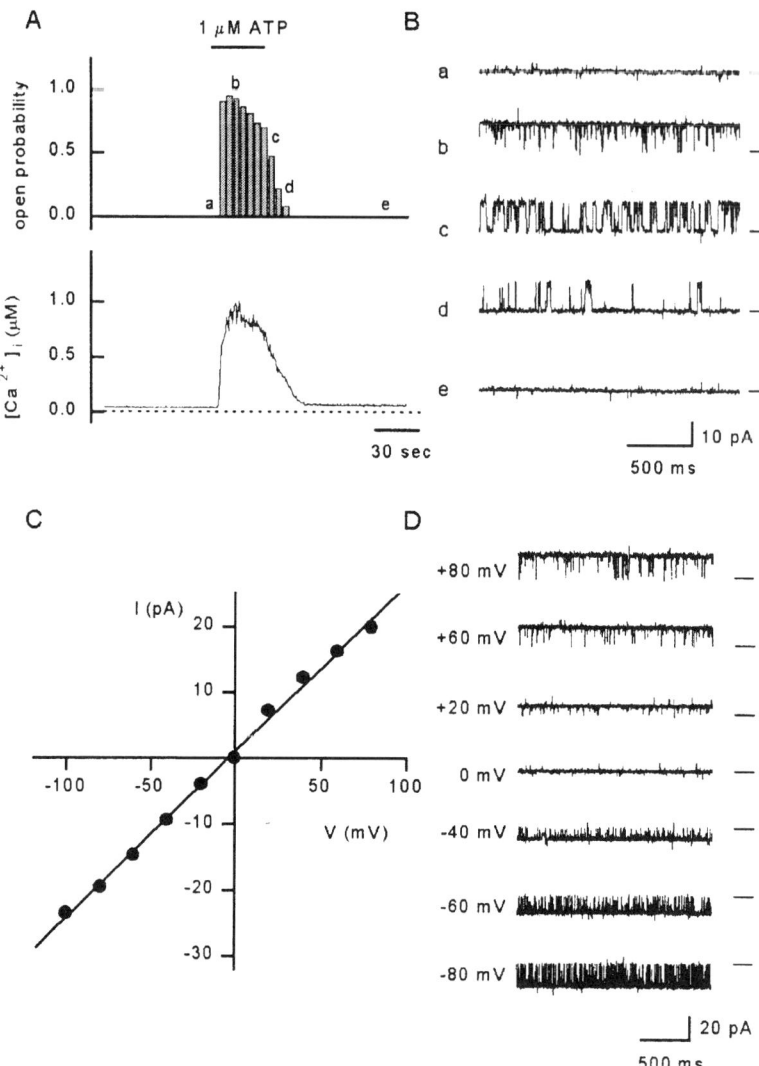

Figure 2. Single-channel currents through BK_{Ca} channels evoked by ATP in cultured bovine pulmonary artery endothelial (CPAE) cells transfected with *hslo*. (A) Change in open-state probability (*top*) of the BK_{Ca} channel during the Ca^{2+} transient (*bottom*) induced by 1 μM ATP in calf cells transiently transfected with *hslo*. Values for the open-state probability were calculated from the traces in (B). (B) Currents recorded in cell-attached mode at the time points labeled a–e in (A) at a holding potential of +40 mV and at symmetrical K^+ concentrations. (C) and (D) $I–V$ curve and unitary hslo current at various potentials in inside-out patches. (Reproduced, with permission, from Kamouchi *et al.*, 1997a.)

inside-out patch configuration. This inhibition occurs with no modification of the channel unitary conductance and is partly restored by the sulfhydryl (SH) reducing agents dithiothreitol or reduced glutathione. The lipid-soluble oxidative agent 4,4'-dithiodipyridine was less potent, suggesting that critical SH groups localized at the inner face of the cell membrane may be involved in channel gating (Cai and Sauv, 1997).

The biophysical and pharmacological profiles of these channels in endothelial cells are consistent with those of the recently cloned hIK channel (Ishii *et al.*, 1997; Jensen *et al.*, 1998).

2.2.3. SK$_{Ca}$ *Channels*

SK$_{Ca}$ channels with a conductance of about 10 pS in asymmetrical K$^+$ concentrations also have been observed in endothelial cells (Groschner *et al.*, 1992; Muraki *et al.*, 1997; Sakai, 1990). These channels lack voltage dependence and are blocked by extracellular TBA and apamin (Groschner *et al.*, 1992). Recently, two types of SK$_{Ca}$ channels were identified in excised patches of rat aortic endothelium; the conductances of these channels were 18 and 9 pS, respectively, in symmetrical 150 mM K$^+$ and 6.7 pS and 2.8 pS, respectively, at physiological extracellular K$^+$ concentrations (Marchenko and Sage, 1996). The lower-amplitude current was completely blocked by 10 nM apamin and 100 μM *d*-tubocurarine, whereas the 18-pS channel was insensitive to apamin but inhibited by charybdotoxin at concentrations greater than 50 nM.

2.3. ATP-Sensitive K$^+$ Channels

Intracellular dialysis of ATP, application of glucose-free NaCN solutions, or the K$_{ATP}$ channel opener pinacidil increases whole-cell and single-channel K$^+$ current in endothelial cells from rat aorta and brain microvessels. These effects are reversibly blocked by the sulfonylurea drug glibenclamide or by ATP in inside-out patches (Janigro *et al.*, 1993). In rabbit aortic endothelial cells, unitary currents with a conductance of 25 pS were evoked in response to lowering intracellular ATP concentration or application of the K$^+$ channel activator levcromakalim. Applying glibenclamide or increasing the ATP concentration inhibited channel activity in inside-out patches (Katnik and Adams, 1995, 1997).

The K$^+$ channel openers HOE 234, diazoxide, and pinacidil hyperpolarize freshly isolated guinea pig coronary capillaries to a membrane potential (E_m) close to E_K, as measured with the voltage-sensitive dye bis-oxonol. Similar hyperpolarization can be achieved by substituting L-glucose for D-glucose in the superfusing solution. This hyperpolarization is reversed by the subsequent addition of glibenclamide (Langheinrich and Daut, 1997). Indirect evidence also has been presented for a role of endothelial K$_{ATP}$ channels in the flow- and shear-stress-mediated vasodilation of coronary microvessels (Kuo and Chancellor, 1995) and rabbit aorta (Hutcheson and Griffith, 1994). In this regard, coronary arterial occlusion has been shown to increase osmolarity in the myocardial interstitium. Furthermore, intracoronary injection of hyperosmolar solutions reduces coronary vascular resistance. The endothelium-dependent vasodilation of coronary microvessels in response to an abluminal increase in osmolarity is significantly attenuated by glibenclamide, but not by low concentrations of BaCl$_2$ that inhibit the Kir channels or by iberiotoxin, an inhibitor of BK$_{Ca}$ channels (Ishizaka and Kuo, 1997).

2.4. Other K⁺ Channels

Besides these classical K^+ channels, a number of poorly characterized channels that apparently do not belong to any of the cloned subfamilies of K^+ channels have been observed in endothelial cells. For example, an inwardly rectifying K^+ channel with a large conductance of 170 pS has been described in aortic endothelial cells (Graier *et al.*, 1993). It is activated by isoprenaline, adenosine, forskolin, and membrane-permeable analogs of cAMP whereas it is inhibited by PKA inhibitors. Hoyer *et al.* (1997) reported the presence of a K^+-selective, stretch-activated channel in intact and isolated rat aortic endothelium that showed K^+:Na^+ permeability ratio of the endothelium 10.9:1. Compared with Wistar Kyoto (WKY) rats, spontaneously hypertensive rats (SHR) exhibited a 4.4-fold higher current density. In the corneal endothelium, the two major K^+ currents represented are an anion- and temperature-stimulated channel blocked by Cs^+ but not by most other K^+ channel blockers and a K^+ current similar to the A-currents in excitable cells. The latter current is blocked by both 4-aminopyridine and quinidine (Rae and Watsky, 1996).

3. FUNCTIONAL ASPECTS OF K⁺ CHANNELS

As sumarized in Table 1, the major and most obvious role of these channels is the modulation of E_m and the electrochemical gradient for Ca^{2+} influx, which affects endothelial functions regulated by changes in $[Ca^{2+}]_i$. The role of changes in intracellular Ca^{2+} in the synthesis of vasoactive substances, such as NO and prostag-

Table 1
K⁺ Channels in Endothelial Cells

Channel type	Single-channel conductance (pS)	Molecular identify	Modulation[a]	Blockers[b]
Inward rectifier (Kir)	23–30	Kir2.1	Shear stress ↑, Intracellular ATP ↑, PP2A ↓, angiotensin II, VIP, ET-1, and histamine ↓, GTPγS ↓	Ba^{2+}, Cs^+, TEA, TBA
Ca²⁺activated				
BK$_{Ca}$	165–220	hlk?	Intracellular alkalinization ↓	TEA, charybdotoxin, *d*-tubocurarine
IK$_{Ca}$	30–80		GTPγS ↑, GDPβS ↓	Charybdotoxin, quinine, TBA, noxiustoxin
SK$_{Ca}$	10			TBA, apamin
ATP-dependent (K$_{ATP}$)	25		Pinacidil ↑, ATP ↓	Glibenclamide

[a] Abbreviations PPA2, Phosphatase SA; VIP, vasoactive intestinal polypeptide; ET-1, endothelin-1; GTPγS, guanosine 5'-O-(3-thiotriphosphate); GDPβS, guanolsine 5'-O-(2-thiodiphosphate).
[b] Abbreviations: TEA, Tetraethylammonium; TBA, tetrabutylammonium.

landins, is well documented (for a review, see Nilius *et al.*, 1997b). Furthermore, the permeability of microvessels depends on changes in intracellular Ca^{2+} and appears to be modulated by changes in E_m (He and Curry, 1994). Although the single-channel events have not been characterized, the flow-evoked, endothelial release of ATP that contributes to vasodilation in the pulmonary vascular bed can be inhibited by glibenclamide, an indication that K_{ATP} channels may be involved in this release mechanism (Hassessian *et al.*, 1993).

K⁺ channels also appear to be involved in the regulatory volume decrease induced by hypotonic cell swelling of endothelial cells (Nilius *et al.*, 1995; Perry and O'Neill, 1993). In human umbilical vein endothelial cells, this regulatory volume decrease does not occur, because the activation of the volume-regulated anion current by cell swelling is not accompanied by a concomitant activation of an outward K⁺ current (De Smet *et al.*, 1994). Finally, these channels may also play a role in transcellular transport of K⁺ through the endothelial cell layer (Colden-Stanfield *et al.*, 1992; Manabe *et al.*, 1995b), in electrical communication between endothelium and smooth muscle (Beny and Connat, 1992; Beny and Pacicca, 1994), and in capillary sensing and communication (Tyml *et al.*, 1997).

REFERENCES

Baron, A., Frieden, M., Chabaud, F., and Beny, J. L., 1996, Ca^{2+}-dependent non-selective cation and potassium channels activated by bradykinin in pig coronary artery endothelial cells, *J. Physiol. (London)* **493:**691–706.

Baron, A., Frieden, M., and Beny, J. L., 1997, Epoxyeicosatrienoic acids activate a high-conductance, Ca^{2+}-dependent K⁺ channel on pig coronary artery endothelial cells, *J. Physiol. (London)* **504:**537–543.

Beny, J. L., and Connat, J. L., 1992, An electron-microscopic study of smooth muscle cell dye coupling in the pig coronary arteries. Role of gap junctions, *Circ. Res.* **70:**49–55.

Beny, J. L., and Pacicca, C., 1994, Bidirectional electrical communication between smooth muscle and endothelial cells in the pig coronary artery, *Am. J. Physiol.* **266:**H1465–H1472.

Bolotina, V. M., Najibi, S., Palacino, J. J., Pagano, P. J., and Cohen, R. A., 1994, Nitric oxide directly activates calcium-dependent potassium channels in vascular smooth muscle, *Nature* **368:**850–853.

Cai, S., and Sauvé, R., 1997, Effects of thiol-modifying agents on a K_{Ca} channel of intermediate conductance in bovine aortic endothelial cells, *J. Membr. Biol.* **158:**147–158.

Cai, S., Garneau, L., and Sauv, R., 1998, Single-channel characterization of the pharmacological properties of the K_{Ca} channel of intermediate conductance in bovine aortic endothelial cells, *J. Membr. Biol.* **163:**147–158.

Campbell, D. L., Strauss, H. C., and Whorton, A. R., 1991, Voltage dependence of bovine pulmonary artery endothelial cell function, *J. Mol. Cell. Cardiol.* **23**(S1):133–144.

Chen, C.-H., Houchi, H., Ohnaka, M., Sakamoto, S., Niwa, Y., and Nakaya, Y., 1998, Nitric oxide activates Ca^{2+}-activated K⁺ channels in cultured bovine adrenal chromaffin cells, *Neurosci. Lett.* **248:**127–129.

Colden-Stanfield, M., Schilling, W. P., Possani, L. D., and Kunze, D. L., 1990, Bradykinin-induced potassium current in cultured bovine aortic endothelial cells, *J. Membr. Biol.* **116:**227–238.

Colden-Stanfield, M., Cramer, E. B., and Gallin, E. K., 1992, Comparison of apical and basal surfaces of confluent endothelial cells—patch-clamp and viral studies, *Am. J. Physiol.* **263:**C573–C583.

Daut, J., Standen, N. B., and Nelson, M. T., 1994, The role of the membrane potential of endothelial and smooth muscle cells in the regulation of coronary blood flow, *J. Cardiovasc. Electrophysiol.* **5:**154–181.

De Smet, P., Oike, M., Droogmans, G., Van Driessche, W., and Nilius, B., 1994, Responses of endothelial cells to hypotonic solutions: Lack of regulatory volume decrease, *Pflügers Arch.* **428:**94–96.

Elam, T. R., and Lansman, J. B., 1995, The role of Mg^{2+} in the inactivation of inwardly rectifying K⁺ channels in aortic endothelial cells, *J. Gen. Physiol.* **105:**463–484.

Forsyth, S. E., Hoger, A., and Hoger, J. H., 1997, Molecular cloning and expression of a bovine endothelial inward rectifier potassium channel, *FEBS Lett.* **409**:277–282.

Fransen, P., and Sys, S. U., 1997, K^+ and Cl^- contribute to resting membrane conductance of cultured porcine endocardial endothelial cells, *Am. J. Physiol.* **272**:H1770–H1779.

Graier, W. F., Kukovetz, W. R., and Groschner, K., 1993, Cyclic AMP enhances agonist-induced Ca^{2+} entry into endothelial cells by activation of potassium channels and membrane hyperpolarization, *Biochem. J.* **291**:263–267.

Groschner, K., Graier, W. F., and Kukovetz, W. R., 1992, Activation of a small-conductance Ca-dependent K^+ channel contributes to bradykinin-induced stimulation of nitric oxide synthesis in pig aortic endothelial cells, *Biochim. Biophys. Acta* **1137**:162–170.

Haburcak, M., Wei, L., Viana, F., Prenen, J., Droogmans, G., and Nilius, B., 1997, Calcium-activated potassium channels in cultured human endothelial cells are not directly modulated by nitric oxide, *Cell Calcium* **21**:291–300.

Hassessian, H., Bodin, P., and Burnstock, G., 1993, Blockade by glibenclamide of the flow-evoked endothelial release of ATP that contributes to vasodilatation in the pulmonary vascular bed of the rat, *Br. J. Pharmacol.* **109**:466–472.

He, P., and Curry, F. E., 1994, Endothelial cell hyperpolarization increases $[Ca^{2+}]_i$ and venular microvessel permeability, *J. Appl. Physiol.* **76**:2288–2297.

Hoyer, J., Popp, R., Meyer, J., Galla, H. J., and Gögelein, H., 1991, Angiotensin II, vasopressin and GTPγS inhibit inward-rectifying K^+ channels in porcine cerebral capillary endothelial cells, *J. Membr. Biol.* **123**:55–62.

Hoyer, J., Distler, A., Haase, W., and Ggelein, H., 1994, Ca^{2+} influx through stretch-activated cation channels activates maxi K^+ channels in porcine endocardial endothelium, *Proc. Natl. Acad. Sci. U.S.A.* **91**:2367–2371.

Hoyer, J., Kohler, R., and Distler, A., 1997, Mechanosensitive cation channels in aortic endothelium of normotensive and hypertensive rats, *Hypertension* **30**:112–119.

Hutcheson, I. R., and Griffith, T. M., 1994, Heterogeneous populations of K^+ channels mediate EDRF release to flow but not agonists in rabbit aorta, *Am. J. Physiol.* **266**:H590–H596.

Inagami, T., Naruse, M., and Hoover, R., 1995, Endothelium: As an endocrine organ, *Annu. Rev. Physiol.* **57**:171–189.

Inazu, M., Zhang, H., and Daniel, E. E., 1994, Properties of the LP-805-induced potassium currents in cultured bovine pulmonary artery endothelial cells, *J. Pharmacol. Exp. Ther.* **268**:403–408.

Ishii, T. M., Silvia, C., Hirschberg, B., Bond, C. T., Adelman, J. P., and Maylie, J., 1997, A human intermediate conductance calcium activated potassium channel, *Proc. Natl. Acad. Sci. U.S.A.* **94**:11651–11656.

Ishizaka, H., and Kuo, L., 1997, Endothelial ATP-sensitive potassium channels mediate coronary microvascular dilation to hyperosmolarity, *Am. J. Physiol.* **273**:H104–H112.

Jacobs, E. R., Cheliakine, C., Gebremedhin, D., Birks, E. K., Davies, P. F., and Harder, D. R., 1995, Shear activated channels in cell-attached patches of cultured bovine aortic endothelial cells, *Pflügers Arch.* **431**:129–131.

Janigro, D., West, G. A., Gordon, E. L., and Winn, H. R., 1993, ATP-sensitive K^+ channels in rat aorta and brain microvascular endothelial cells, *Am. J. Physiol.* **265**:C812–C821.

Jensen, B. S., Størbræk, D., Christophersen, P., Jrgensen, T. D., Hansen, C., Silahtaroglu, A., Olesen, S.-P., and Ahring, P. K., 1998, Characterization of the cloned human intermediate-conductance Ca^{2+}-activated K^+ channel., *Am. J. Physiol.* **275**:C848–C856.

Jow, F., and Numann, R., 1998, Divalent ion block of inward rectifier current in human capillary endothelial cells and effects on resting membrane potential, *J. Physiol. (London)* **512**:119–128.

Kamouchi, M., Trouet, D., De Greef, C., Droogmans, G., Eggermont, J., and Nilius, B., 1997a, Functional effects of expression of *hslo* Ca^{2+} activated K^+ channels in cultured macrovascular endothelial cells, *Cell Calcium* **22**:497–506.

Kamouchi, M., Van Den Bremt, K., Eggermont, J., Droogmans, G., and Nilius, B., 1997b, Modulation of inwardly rectifying potassium channels in cultured bovine pulmonary artery endothelial cells, *J. Physiol. (London)* **504**:545–556.

Katnik, C., and Adams, D. J., 1995, An ATP-sensitive potassium conductance in rabbit arterial endothelial cells, *J. Physiol. (London)* **485**:595–606.

Katnik, C., and Adams, D. J., 1997, Characterization of ATP-sensitive potassium channels in freshly dissociated rabbit aortic endothelial cells, *Am. J. Physiol.* **272**:H2507–H2511.

Kuo, L., and Chancellor, J. D., 1995, Adenosine potentiates flow-induced dilation of coronary arterioles by activating K$_{ATP}$ channels in endothelium, *Am. J. Physiol.* **269:**H541–H549.

Langheinrich, U., and Daut, J., 1997, Hyperpolarization of isolated capillaries from guinea-pig heart induced by K$^+$ channel openers and glucose deprivation, *J. Physiol. (London)* **502:**397–408.

Ling, B. N., and O'Neill, W. C., 1992, Ca^{2+}-dependent and Ca^{2+}-permeable ion channels in aortic endothelial cells, *Am. J. Physiol.* **263:**H1827–H1838.

Manabe, K., Ito, H., Matsuda, H., and Noma, A., 1995a, Hyperpolarization induced by vasoactive substances in intact guinea-pig endocardial endothelial cells, *J. Physiol. (London)* **484:**25–40.

Manabe, K., Ito, H., Matsuda, H., Noma, A., and Shibata, Y., 1995b, Classification of ion channels in the luminal and abluminal membranes of guinea-pig endocardial endothelial cells, *J. Physiol. (London)* **484:**41–52.

Marchenko, S. M., and Sage, S. O., 1996, Calcium-activated potassium channels in the endothelium of intact rat aorta, *J. Physiol. (London)* **492:**53–60.

Muraki, K., Imaizumi, Y., Ohya, S., Sato, K., Takii, T., Onozaki, K., and Watanabe, M., 1997, Apamin-sensitive Ca^{2+}-dependent K$^+$ current and hyperpolarization in human endothelial cells, *Biochem. Biophys. Res. Commun.* **236:**340–343.

Nilius, B., and Casteels, R., 1996. Biology of the vascular wall and its interaction with migratory and blood cells in: *Comprehensive Human Physiology*, Vol. 2 (R. Greger and U. Windhorst, eds.), Springer-Verlag, Berlin, pp. 1981–1994.

Nilius, B., and Droogmans, G., 1995, Ion channels of endothelial cells, in: *Physiology and Pathophysiology of the Heart* (N. Sperelakis ed.), Kluwer Academic Publishers, Boston, pp. 961–973.

Nilius, B., and Riemann, D., 1990, Ion channels in human endothelial cells, *Gen. Physiol. Biophys.* **9:**89–112.

Nilius, B., Schwarz, G. and Droogmans, G., 1993, Modulation by histamine of an inwardly rectifying potassium channel in human endothelial cells, *J. Physiol. (London)* **472:**359–371.

Nilius, B., Sehrer, J., De Smet, P., Van Driesche, W., and Droogmans, G., 1995, Volume regulation in a toad epithelial cell line: Role of coactivation of K$^+$ and Cl$^-$ channels, *J. Physiol. (London)* **487:**367–378.

Nilius, B., Prenen, J., Kamouchi, M., Viana, F., Voets, T., and Droogmans, G., 1997a, Inhibition by mibefradil, a novel calcium channel antagonist, of Ca^{2+}- and volume-activated Cl$^-$ channels in macrovascular endothelial cells, *Br. J. Pharmacol.* **121:**547–555.

Nilius, B., Viana, F., and Droogmans, G., 1997b, Ion channels in vascular endothelium, *Annu. Rev. Physiol.* **59:**145–170.

Olesen, S. P., and Bundgaard, M., 1993, ATP-dependent closure and reactivation of inward rectifier K$^+$ channels in endothelial cells, *Circ. Res.* **73:**492–495.

Pasyk, E., Mao, Y. K., Ahmad, S., Shen, S. H., and Daniel, E. E., 1992, An endothelial cell-line contains functional vasoactive intestinal polypeptide receptors: They control inwardly rectifying K$^+$ channels, *Eur. J. Pharmacol.* **212:**209–214.

Pennefather, P. S., and DeCoursey, T. E., 1994, A scheme to account for the effects of Rb$^+$ and K$^+$ on inward rectifier K channels of bovine artery endothelial cells, *J. Gen. Physiol.* **103:**549–581.

Perry, P. B., and O'Neill, W. C., 1993, Swelling-activated K-fluxes in vascular endothelial cells-volume regulation via K–Cl cotransport and K-channels, *Am. J. Physiol.* **265:**C763–C769.

Rae, J. L., and Watsky, M. A., 1996, Ionic channels in corneal endothelium, *Am. J. Physiol.* **270:**C975–C989.

Revest, P. A,. and Abbott, N. J., 1992, Membrane ion channels of endothelial cells, *Trends Pharmacol. Sci.* **13:**404–407.

Rusko, J., Tanzi, F., Vanbreemen, C., and Adams, D. J., 1992, Calcium-activated potassium channels in native endothelial cells from rabbit aorta—conductance, Ca^{2+} sensitivity and block, *J. Physiol. (London)* **455:**601–621.

Rusko, J., Li, L., and van Breemen, C., 1995, 17-β-Estradiol stimulation of endothelial K$^+$ channels, *Biochem. Biophys. Res. Commun.* **214:**367–372.

Sakai, T., 1990, Acetylcholine induces Ca-dependent K currents in rabbit endothelial cells, *Jpn. J. Pharmacol.* **53:**235–246.

Sauvé, R., Parent, L., Simoneau, C., and Roy, G., 1988, External ATP triggers a biphasic activation process of a calcium-dependent K$^+$ channel in cultured bovine aortic endothelial cells, *Pflügers Arch.* **412:**469–481.

Sauvé, R., Chahine, M., Tremblay, J., and Hamet, P., 1990, Single-channel analysis of the electrical response of bovine aortic endothelial cells to bradykinin stimulation: Contribution of a Ca^{2+}-dependent K$^+$ channel, *J. Hypertens.* **8:**S193–S201.

Silver, M. R., and DeCoursey, T. E., 1990, Intrinsic gating of inward rectifier in bovine pulmonary artery endothelial cells in the presence or absence of internal Mg^{2+}, J. Gen. Physiol. **96**:109–133.

Silver, M. R., Shapiro, M. S., and DeCoursey, T. E., 1994, Effects of external Rb] on inward rectifier K^+ channels of bovine pulmonary artery endothelial cells, J. Gen. Physiol. **103**:519–548.

Takeda, K., Schini, V., and Stoeckel, H., 1987, Voltage-activated potassium, but not calcium currents in cultured bovine aortic endothelial cells, Pflügers Arch. **410**:385–393.

Thuringer, D., Diarra, A., and Sauvé, R., 1991, Modulation by extracellular pH of bradykinin-evoked activation of Ca^{2+}-activated K^+ channels in endothelial cells, Am. J. Physiol. **261**:H656–H666.

Tyml, K., Song, H., Munoz, P., and Ouellette, Y., 1997, Evidence for K^+ channels involvement in capillary sensing and for bidirectionality in capillary communication, Microvasc. Res. **53**:245–253.

Vaca, L., Schilling, W. P., and Kunze, D. L., 1992, G-protein-mediated regulation of a Ca^{2+}-dependent K^+ channel in cultured vascular endothelial cells, Pflügers Arch. **422**:66–74.

Vaca, L., Gurrola, G. B., Possani, L. D., and Kunze, D. L., 1993, Blockade of a K_{Ca} channel with synthetic peptides from noxiustoxin—a K^+ channel blocker, J. Membr. Biol. **134**:123–129.

Van Renterghem, C., Vigne, P., and Frelin, C., 1995, A charybdotoxin-sensitive, Ca^{2+}-activated K^+ channel with inward rectifying properties in brain microvascular endothelial cells: Properties and activation by endothelins, J. Neurochem. **65**:1274–1281.

Vargas, F. F., Caviedes, P. F., and Grant, D. S., 1994, Electrophysiological characteristics of cultured human umbilical vein endothelial cells, Microvasc. Res. **47**:153–165.

Voets, T., Droogmans, G., and Nilius, B., 1996, Ionic currents in non-stimulated endothelial cells from bovine pulmonary artery, J. Physiol. (London) **497**:95–107.

von Beckerath, N., Dittrich, M., Klieber, H. G., and Daut, J., 1996, Inwardly rectifying K^+ channels in freshly dissociated coronary endothelial cells from guinea-pig heart, J. Physiol. (London) **491**:357–365.

Zhang, H., Inazu, M., Weir, B., Buchanan, M., and Daniel, E., 1994, Cyclopiazonic acid stimulates Ca^{2+} influx through non-specific cation channels in endothelial cells, Eur. J. Pharmacol. **251**:119–125.

Zunkler, B. J., Henning, B., Grafe, M., Hildebrandt, A. G., and Fleck, E., 1995, Electrophysiological properties of human coronary endothelial cells, Basic Res. Cardiol. **90**:435–442.

Chapter 32

Single-Channel Properties of Ca^{2+}-Activated K^+ Channels in the Vascular Endothelium

Stewart O. Sage and Sergey M. Marchenko

1. INTRODUCTION

1.1. Cytosolic Ca^{2+} in Endothelial Function

The vascular endothelium performs a number of its functions through the release of several physiologically active mediators (Furchgott and Vanhoutte, 1989). These include nitric oxide (NO) and prostaglandin I_2 (PGI_2), which are involved in the regulation of vascular tone and the inhibition of platelet activity. The generation and release of both of these mediators is influenced by the cytosolic calcium concentration ($[Ca^{2+}]_i$) in the endothelial cells (Moncada *et al.*, 1991; Hallam *et al.*, 1988). Thus, endothelium-dependent vasodilators act, at least in part, through the modulation of endothelial $[Ca^{2+}]_i$. The $[Ca^{2+}]_i$ also influences endothelial barrier properties (Rotrosen and Gallin, 1986).

1.2. Generation of Endothelial Ca^{2+} Signals

Endothelium-dependent vasodilators such as acetylcholine, bradykinin, and histamine elevate $[Ca^{2+}]_i$ by employing Ca^{2+} from two sources. An initial elevation in $[Ca^{2+}]_i$ is achieved through the release of Ca^{2+} from intracellular stores via the second messenger inositol 1,4,5-trisphosphate (Freay *et al.*, 1989). Continued stimulation requires the entry of Ca^{2+} across the plasma membrane. The endothelium of most vessels appears to lack voltage-operated Ca^{2+} channels, although these have been

Stewart O. Sage ● Department of Physiology, University of Cambridge, Cambridge CB2 3EG, United Kingdom. *Sergey M. Marchenko* ● The Bogomoletz Institute of Physiology, Ukrainian Academy of Sciences, Kiev 24, 252601 GSP, Ukraine.

Potassium Channels in Cardiovascular Biology, edited by Archer and Rusch. Kluwer Academic/Plenum Publishers, New York, 2001.

reported in cultured microvascular endothelial cells (Bossu et al., 1989). As in many other cell types, a major mechanism for Ca^{2+} entry appears to be activated following depletion of intracellular Ca^{2+} stores. Store-dependent (or "capacitative") calcium entry has been demonstrated in cultured endothelial cells (Jacob, 1990; Hallam et al., 1989) and in native endothelium in situ in excised blood vessels (Usachev et al., 1995). Other receptor-mediated mechanisms for Ca^{2+} entry have also been proposed (Johns et al., 1987).

Agonist-evoked Ca^{2+} signals in endothelium vary in form. Some consist of an initial peak followed by a subsequent plateau above basal levels, reflecting the release of Ca^{2+} from intracellular stores and subsequent Ca^{2+} entry across the plasma membrane. Others are of a more complex nature, consisting of repetitive spikes or oscillations in $[Ca^{2+}]_i$. The spiking observed in single cultured endothelial cells can continue for a time in the absence of external Ca^{2+}, suggesting that it is due to the periodic release of Ca^{2+} from intracellular stores (Jacob et al., 1988). The synchronized oscillations in $[Ca^{2+}]_i$ that have been reported in cultured endothelial monolayers (Sage et al., 1989; Laskey et al., 1992) and in native endothelium in situ in excised blood vessels (Usachev et al., 1995) are critically dependent on external Ca^{2+}. This suggests that these synchronized oscillations may arise from modulation of Ca^{2+} entry.

1.3. Importance of Membrane Potential in Endothelial Ca^{2+} Signaling

A change in the membrane potential E_m alters the driving force for Ca^{2+} entry into cells. Depolarizing cultured endothelial cells at rest has been shown to reduce the $[Ca^{2+}]_i$, presumably by modulating Ca^{2+} entry through leak pathways (Cannell and Sage, 1989). Depolarization reduces agonist-evoked rises in $[Ca^{2+}]_i$ in cultured endothelial cells (Laskey et al., 1990; Lückhoff and Busse, 1990) and native endothelium in situ (Usachev et al., 1995). Agonist-evoked spikes in $[Ca^{2+}]_i$ in cultured endothelial cells cease when the cells are depolarised (Jacob et al., 1988).

1.4. Vasodilator-Evoked Changes in Endothelial Membrane Potential

Endothelium-dependent vasodilators such as acetylcholine evoke complex electrical responses in native endothelium in situ in excised blood vessels. In rat aorta, these responses consist of a rapid hyperpolarization which may be maintained, slowly decline toward the resting potential, or be followed by a pronounced depolarization beyond the resting potential (Marchenko and Sage, 1993). In some cases, oscillations occur in the depolarization phase.

Agonist-evoked hyperpolarization also has been reported in cultured endothelial cells from various sources (e.g., Chen and Cheung, 1992; Merke et al., 1991; Sauvé et al., 1988; Olesen et al., 1988; Busse et al., 1988; Colden-Stanfield et al., 1987). Other authors working with cultured endothelium have reported agonist-evoked hyperpolarization followed by depolarization accompanied by small oscillations in E_m (Merke and Daut, 1990) or depolarization alone (Bregestovski et al., 1988; Johns et al., 1987). The early vasodilator-evoked hyperpolarization is generated by a Ca^{2+}-activated K^+

conductance, which forms the subject of this chapter. The subsequent depolarization appears to be due to the activation of nonselective cation channels (for a review, see Nilius *et al.*, 1997).

2. VASODILATOR-EVOKED ENDOTHELIAL HYPERPOLARISATION

2.1. The Role of a K$^+$ Conductance

In the endothelium of intact rat aorta, the mean resting E_m was found to be -58 mV (Marchenko and Sage, 1993). Acetylcholine (2 μM) evoked a mean hyperpolarization of 19 mV (Marchenko and Sage, 1994), taking the E_m close to the estimated K$^+$ equilibrium potential, of -79.8 mV under the conditions used (Marchenko and Sage, 1993). This suggested that the hyperpolarization resulted from the activation of a K$^+$ conductance, an idea supported by the finding that hyperpolarization of the endothelium to a potential below the K$^+$ equilibrium potential by current injection converted the initial hyperpolarization into a depolarization (Marchenko and Sage, 1993).

Studies with cultured endothelium also support the hypothesis that vasodilator-evoked hyperpolarization is due to activation of a K$^+$ conductance. For example, in coronary artery endothelium from guinea pig, the magnitude of the bradykinin-evoked hyperpolarization was found to vary with the external K$^+$ concentration, decreasing and eventually reversing as external K$^+$ was elevated (Merke and Daut, 1990).

2.2. Evidence for a Ca^{2+}-Activated K$^+$ Conductance

Several pieces of evidence suggest that in the endothelium of intact rat aorta an elevation of $[Ca^{2+}]_i$ and subsequent activation of a Ca^{2+}-activated K$^+$ conductance underlies vasodilator-evoked hyperpolarizations. Removal of extracellular Ca^{2+} was found to reduce the duration of acetylcholine-evoked hyperpolarization, with the response becoming consistently transient in nature (Fig. 1A; Marchenko and Sage, 1993). Increasing the Ca^{2+} buffering power by loading the preparation with the Ca^{2+} chelator BAPTA reduced the amplitude of acetylcholine-evoked hyperpolarizations in a dose-dependent manner (Marchenko and Sage, 1994). Inhibitors of the Ca^{2+}-ATPase of the endoplasmic reticulum, such as thapsigargin, as well as the Ca^{2+} ionophore ionomycin, which all evoke receptor-independent rises in $[Ca^{2+}]_i$, mimicked vasodilator-evoked hyperpolarizations of the endothelium (Fig. 1B; Marchenko and Sage, 1994). Furthermore, simultaneous recordings of E_m and of $[Ca^{2+}]_i$ from the intact vessel preparation showed that the hyperpolarizations evoked by acetylcholine (Fig. 2) or ATP were temporally coincident with the rise and initial peak in $[Ca^{2+}]_i$ evoked by these agonists (Usachev *et al.*, 1995). Comparable observations have been reported in cultured endothelium. For example, the duration of vasodilator-evoked hyperpolarizations is reduced by the removal of extracellular Ca^{2+} in endothelium from a number of sources (Merke *et al.*, 1991; Merke and Daut, 1990).

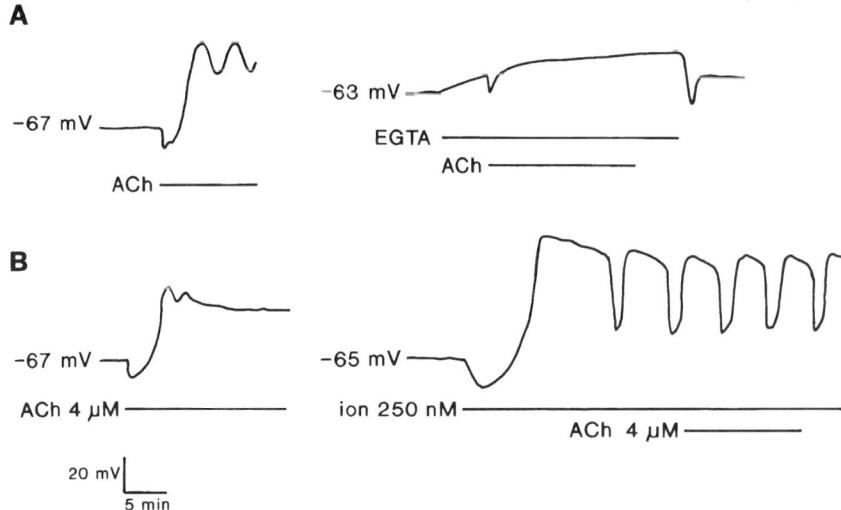

Figure 1. Importance of extracellular Ca^{2+} for the electrical response evoked by vasodilators in endothelium. (A) The membrane potential (E_m) of endothelium of intact excised rat aorta was measured under current clamp using the patch-clamp technique with amphotericin B-permeabilized patches. A typical control response evoked by 2 μM acetylcholine (ACh) in the presence of 1 mM external Ca^{2+} is shown on the left. The response after removal of external Ca^{2+} (in the presence of 2 mM EGTA) is shown on the right. In the absence of external Ca^{2+}, the ACh-evoked hyperpolarization becomes more transient and the depolarization phase is abolished. (Reproduced from Marchenko and Sage, 1993.) (B) A response evoked by 4 μM ACh in the presence of 1 mM external Ca^{2+} in the endothelium of intact excised rat aorta is shown on the left. The Ca^{2+} ionophore ionomycin (ion; 250 nM) evoked a similar response in the same preparation (*right*). After addition of ionomycin, the subsequent addition of ACh was without effect. (Reproduced, with permission, from Marchenko and Sage, 1994.)

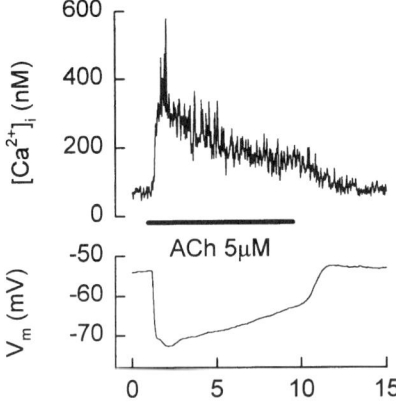

Figure 2. Relationship between $[Ca^{2+}]_i$ and E_m in endothelium of intact excised rat aorta. The upper panel shows the rise in $[Ca^{2+}]_i$ evoked by 5 μM acetylcholine (ACh), as monitored with the fluorescent indicator fura-2. The lower panel shows a simultaneous recording of membrane potential. The agonist-evoked rise in $[Ca^{2+}]_i$ is temporally coincident with the onset of membrane hyperpolarization. (Reproduced from Usachev *et al.*, 1995.)

3. CLASSIFICATION OF Ca^{2+}-ACTIVATED K^+ CHANNELS

Calcium-activated potassium channels (K_{Ca} channels) occur widely in mammalian cells and are classified into three groups on the basis of conductance and pharmacology. Large-conductance (100–250 pS) K_{Ca} channels (BK_{Ca} or maxi-K channels) are inhibited by nanomolar concentrations of charybdotoxin (Miller et al., 1985; Barrett et al., 1982) and are selectively blocked by iberiotoxin (Galves et al., 1990). Small-conductance (4–14 pS) K_{Ca} channels (SK_{Ca} channels) are inhibited by nanomolar concentrations of apamin (Blatz and Magelby, 1986). Intermediate-conductance (18–60 pS) K_{Ca} channels (IK_{Ca} channels) are, like BK_{Ca} channels, inhibited by charybdotoxin in the nanomolar range (Ewald et al., 1985; Grygorczyk and Schwarz, 1983).

4. Ca^{2+}-ACTIVATED K^+ CHANNELS IN CULTURED ENDOTHELIAL CELLS

All three classes of K_{Ca} channels have been found in cultured endothelial cells. BK_{Ca} channels were first reported in cultured bovine aortic endothelial cells (Fichtner et al., 1987). These BK_{Ca} channels appear to be of low density, being found in <4% of excised patches. They show a unitary conductance of 150 pS in physiological ion gradients (asymmetrical Na^+/K^+). A later report confirmed the presence of BK_{Ca} channels in cell-attached patches in a similar bovine aortic preparation, with channel conductance estimated at 165 pS in symmetrical K^+ (Ling and O'Neill, 1992). A BK_{Ca} channel with a conductance of 285 pS in symmetrical K^+ solutions also has been reported in cultured porcine coronary artery endothelium (Baron et al., 1996) and a low-density expression of BK_{Ca} channels has been found in cultured porcine aortic endothelial cells (Groschner et al., 1992).

Although BK_{Ca} channels are only found in 2.5% of membrane patches excised from cultured porcine aortic endothelium, the same preparation also possesses SK_{Ca} channels in 30% of patches (Groschner et al., 1992). The SK_{Ca} channels have a conductance of 9 pS in asymmetrical solutions, are insensitive to E_m, and are activated by Ca^{2+}, with the open-state probability (Po) rising from 0.05 to 0.64 as the $[Ca^{2+}]$ at the cytosolic membrane face increases from 10^{-7} to 10^{-6} M (Groschner et al., 1992). A Ca^{2+}-activated K^+ current has been reported in cultured human umbilical vein endothelial cells (Muraki et al., 1997). Although these authors did not perform single-channel analysis, they found that this current was inhibited by low concentrations of apamin but was unaffected by iberiotoxin. This pharmacology suggests the presence of SK_{Ca}, but not BK_{Ca}, channels in these cells.

Some workers have reported the presence of IK_{Ca} channels in cultured bovine aortic endothelial cells, along with BK_{Ca} channels (Ling and O'Neill, 1992). These IK_{Ca} channels have a conductance of 40 pS in symmetrical K^+ solutions. Other workers have reported only a 40-pS IK_{Ca} channel in this cell type (Sauvé et al., 1988). This current has been extensively characterized pharmacologically (Cai et al., 1998). A 40-pS IK_{Ca} channel has also been reported in cultured rat brain capillary endothelium (Van Renterghem et al., 1995).

It is difficult to draw clear conclusions regarding the relative levels of expression of the three major classes of K_{Ca} channels in endothelial cells from the few and disparate studies reported in the literature. The reports discussed previously cover the range of expression combinations and permutations, from each channel occurring alone to all three being simultaneously present. Some differences in the expression of K_{Ca} channels might be expected between endothelial cells from different species or between endothelium derived from different parts of the vascular tree. However, there is also disagreement between studies conducted using endothelial cells from ostensibly the same source. Thus, bovine aortic endothelial cells in culture have been reported to possess IK_{Ca} or BK_{Ca} channels alone (Sauvé et al., 1988; Fichtner et al., 1987) or in combination (Ling and O'Neill, 1992). Failure to detect the presence of BK_{Ca} channels may be due to their expression at low densities; however, the apparent variability of expression of IK_{Ca} channels in different cultures of the same cell type is more difficult to explain.

A major concern when working with cultured endothelial cells is that the protein expression profile *in vitro* does not reflect that *in vivo*. Although we are unaware of any systematic investigations of possible changes in endothelial ion channel expression in cells under culture, there are clearly documented changes in the transcription of some other genes. For example, freshly isolated bovine aortic endothelium contains the mRNA for M_1, M_2, and M_3 muscarinic receptor subtypes, whereas only the mRNA for the M_2 receptor has been identified in these same cells following passage in culture (Tracey and Peach, 1992). The presence of blood flow over the endothelium appears to have pronounced effects on endothelial morphology and function. The cytoskeleton undergoes major changes in the absence of the differentiating effects of hemodynamic stresses (Davies, 1995). Cytoskeletal changes may, in turn, influence the function of some, ion channels, particularly mechanosensitive ones (Marchenko and Sage, 1997). There are also a number of reports of changes in endothelial gene expression when cultured endothelial cells are subjected to shear stress. These include evidence for shear-evoked changes in the expression of the GTP-binding protein, $G_{\alpha i}$ (Redmond et al., 1988), of constitutive nitric oxide synthase (Ziegler et al., 1998) and of transcript 2 of the platelet-activating factor (PAF) receptor (Okahara et al., 1998). It is therefore quite possible that the expression of K_{Ca} channels by endothelium alters in culture or in the absence of shear stress. Hence, it is important to compare findings obtained in cultured cells, freshly isolated endothelial cells, and intact native endothelial preparations.

5. Ca^{2+}-ACTIVATED K^+ CHANNELS IN FRESHLY ISOLATED ENDOTHELIAL CELLS

The use of freshly isolated cells overcomes some of the concerns associated with the use of cultured endothelium. However, the need for enzymatic digestion and the removal of the endothelium from the influence of other cells in the vessel wall still leaves open some questions as to how representative the responses of these cells are of endothelial cells *in vivo*. Regardless, at least one can be certain that protein expression is unaltered. In freshly isolated endothelial cells from porcine coronary artery, only SK_{Ca} channels were observed (Davis and Sharma, 1994). These channels have a conductance for outward current of 13 pS with 150 mM K^+ at the cytosolic membrane face and 212 mM K$^+$ at the outer membrane face. This finding contrasts with those from the work of others on cultured endothelial cells from this source, in which BK$_{Ca}$, but not SK$_{Ca}$, channels were observed (Baron *et al.*, 1996). Failure to find BK$_{Ca}$ channels in the native cells may be due to their low density (e.g., Fichtner *et al.*, 1987). Alternatively, BK$_{Ca}$ channel expression may only occur in culture, not *in vivo*. Failure to observe SK$_{Ca}$ channels in the cultured cells might be due to inhibition of channel expression in the absence of some undefined condition(s) present only *in vivo*. Molecular expression studies are needed to address these questions.

Finally, BK$_{Ca}$ channels have been observed in freshly isolated rabbit aortic endothelial cells (Rusko *et al.*, 1992). These channels have a unitary conductance of 220 pS in symmetrical K$^+$ solutions and appeared to be responsible for the generation of spontaneous transient outward currents in these cells (Rusko *et al.*, 1995, 1992). Unfortunately, data from cultured cells from this source are not available for comparison.

6. Ca^{2+}-ACTIVATED K$^+$ CHANNELS IN THE ENDOTHELIUM OF INTACT RAT AORTA

6.1. Single-Channel Properties

We have extensively studied the electrical responses evoked by endothelium-dependent vasodilators in endothelium *in situ* in intact excised rat aorta (Marchenko and Sage, 1993, 1994). In this preparation, all vasodilators tested evoke a rapid initial hyperpolarization of the endothelium (Fig. 1A) which is K$^+$-dependent (Marchenko and Sage, 1993, 1994). Simultaneous recordings show that this hyperpolarization is coincident with an agonist-evoked rise in [Ca^{2+}]$_i$ (Fig. 2; Usachev *et al.*, 1995). These and other observations (see Section 1) suggest the presence of K$_{Ca}$ channels, and we have identified two such channels in membrane patches excised from the luminal face of intact rat aorta (Fig. 3; Marchenko and Sage, 1996). The activity of both channels is independent of voltage, and their linear single-channel current–voltage relationships reverse at -98.5 mV in asymmetrical solutions (K$^+$ at the cytosolic membrane face, Na$^+$ at the external face), close to the calculated equilibrium potential for K$^+$. The slope conductances of the channels were 6.7 pS and 2.8 pS in asymmetrical solutions and 18 pS and 9.1 pS in symmetrical K$^+$ solutions (Marchenko and Sage, 1996).

The nomenclature for these channels is based on their pharmacology. The larger (6.7 pS) channel is reversibly inhibited by charybdotoxin (Fig. 4) with an EC$_{50}$ of 137 nM but is unaffected by apamin at concentrations of up to 1 μM, so this channel is named the charybdotoxin-sensitive or K$_{Ch}$ channel. The smaller (2.8 pS) channel is unaffected by charybdotoxin at concentrations of up to 1 μM but is inhibited by apamin at concentrations greater than 1 nM (Fig. 4) with complete block at 10 nM. The smaller channels are thus referred to as apamin-sensitive or K$_{Ap}$ channels. Both channels are activated by a rise in [Ca^{2+}] at the cytosolic membrane face (Fig. 5). K$_{Ch}$ channels show essentially no activity when [Ca^{2+}] is less than 100 nM. The open-state probability rises sharply as [Ca^{2+}]$_i$ increases over the submicromolar range, reaching a Po of ~0.95 at 1 μM and an EC$_{50}$ of 340 nM. In contrast, K$_{Ap}$ channels are only

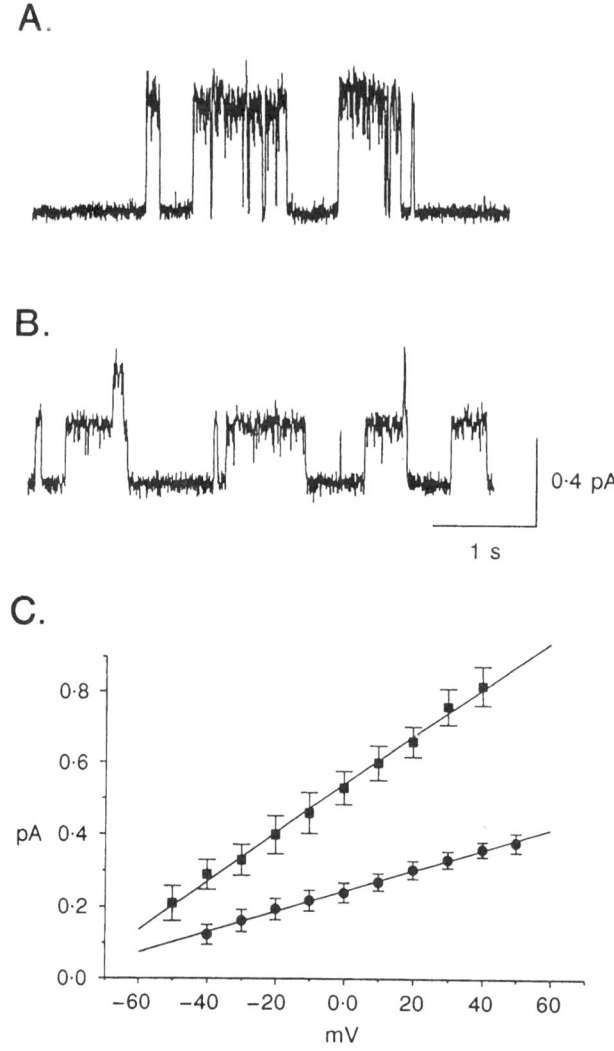

Figure 3. Two types of K_{Ca} channel in outside-out patches excised from endothelium of intact rat aorta. K_{Ch} (IK_{Ca}) channels (A) and K_{Ap} (SK_{Ca}) channels (B) were both recorded at an E_m of 0 mV with a K^+-containing pipette solution and a Na^+-enriched bath solution. The current–voltage relationships under these conditions for K_{Ch} (■) and K_{Ap} (●) channels are shown in (C). (Reproduced from Marchenko and Sage, 1996).

activated by $[Ca^{2+}]_i$ greater than 500 nM, with Po reaching ~0.2 at a $[Ca^{2+}]_i$ of 1 μM.

As well as showing different sensitivities to apamin and charybdotoxin, the two channels have different responses to d-tubocurarine (Fig. 6). K_{Ap} channels demonstrate a flickering block with d-tubocurarine at concentrations greater than 5 μM, with total block occurring at 100 μM. K_{Ch} channels are unaffected by d-tubocurarine over this

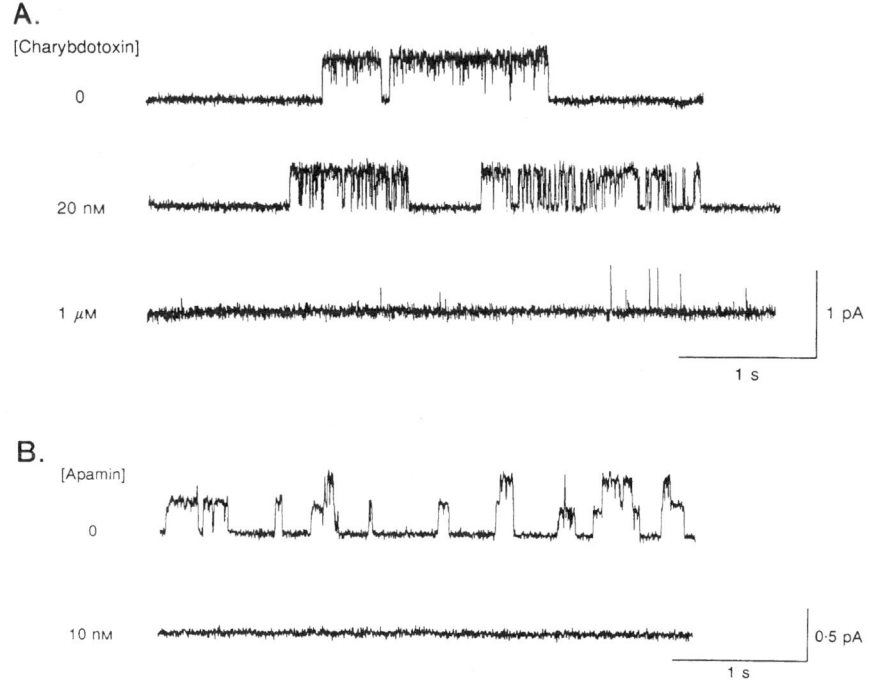

Figure 4. Effects of charybdotoxin and apamin on K$_{Ch}$ and K$_{Ap}$ channels. (A) Charybdotoxin blocked K$_{Ch}$ channels at concentrations greater than 50 nM, with complete block at about 1 μM. (B) Apamin blocked K$_{Ap}$ channels at concentrations greater than 1 nM, with complete block at 10 nM. Concentrations of the inhibitors are shown to the left of the traces, which are recordings from outside-out patches. The effects of both inhibitors were fully reversible (not shown). (Reproduced from Marchenko and Sage, 1996).

concentration range; only at concentrations of $\geqslant 1$ mM is the unitary current amplitude of the K$_{Ch}$ channel reduced. K$_{Ap}$ channels appear to be present at a much higher density in rat aortic endothelium, being found in 40% of 231 patches studied. In contrast, K$_{Ch}$ channels were found in only 5% of patches, and just 2.5% of patches possessed both channel types.

The conductance and apamin sensitivity of the K$_{Ap}$ channel allows it to be classified as an SK$_{Ca}$ channel. However, assigning the K$_{Ch}$ channel to one of the three established classes of K$_{Ca}$ channel is more difficult. The conductance of this channel falls within the range of SK$_{Ca}$ and IK$_{Ca}$ channels. Lack of sensitivity to apamin means that K$_{Ch}$ channels cannot be described as SK$_{Ca}$ channels. However, although they are inhibited by charybdotoxin, they are several hundred times less sensitive to this agent than other channels classified as IK$_{Ca}$ (Castle *et al.*, 1989). A similar channel with a relatively low conductance (35 pS) and a low sensitivity to charybdotoxin has been reported in *Aplysia* neurons (Hermann and Erxleben, 1987). These channels thus fall outside the presently defined classes of K$_{Ca}$ channels.

Although BK$_{Ca}$ channels have been reported in some studies of cultured or freshly isolated endothelial cells of aortic origin (Ling and O'Neill, 1992; Rusko *et al.*, 1992; Fichtner *et al.*, 1987), we found no evidence for BK$_{Ca}$ channels in native rat aortic

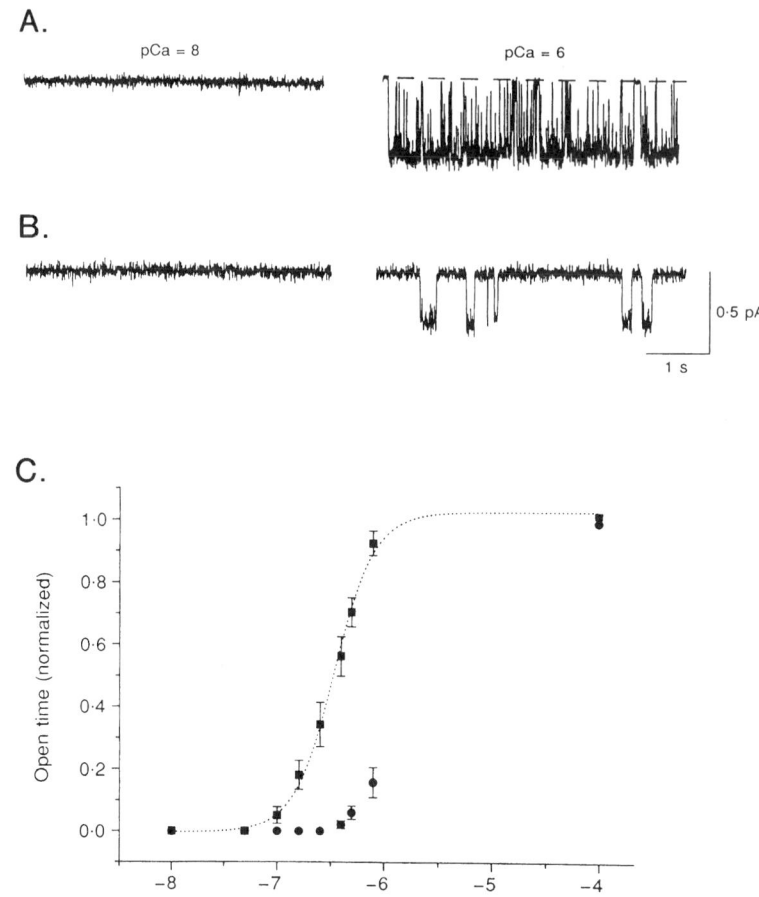

Figure 5. Ca^{2+}-dependent activation of K_{Ch} and K_{Ap} channels in inside-out patches from endothelium of intact rat aorta. (A) K_{Ch} channels were silent at a $[Ca^{2+}]_i$ of 10 nM (pCa = 8) and almost completely activated when $[Ca^{2+}]_i$ was increased to 1 μM (pCa = 6) at the cytosolic face of the membrane. (B) K_{Ap} channels were silent at a $[Ca^{2+}]_i$ of 10 nM and partially activated when $[Ca^{2+}]_i$ was increased to 1 μM at the cytosolic face of the membrane. (C) The dependence of the normalized open times for K_{Ch} (■) and K_{Ap} (●) channels on $[Ca^{2+}]_i$ at the cytosolic membrane face. (Reproduced from Marchenko and Sage, 1996).

endothelium. The failure to detect BK_{Ca} channels in patch-clamp studies might be ascribed to a low channel density. However, the absence of BK_{Ca} channels is supported by the low charybdotoxin sensitivity of the hyperpolarization evoked by acetylcholine (ACh) in rat aorta (Marchenko and Sage, 1996). The pharmacology of vasodilator-evoked responses of other rat vessels also suggests a similar lack of BK_{Ca} channels. For example, ACh-evoked hyperpolarization in rat hepatic artery is inhibited by the combined presence of charybdotoxin and apamin, but not by the combination of iberiotoxin and apamin (Edwards *et al.*, 1998). Vasodilator-evoked relaxations of rat hepatic and mesenteric artery are similarly insensitive to iberiotoxin (Zygmunt and Högestätt, 1996; Waldron and Garland, 1994).

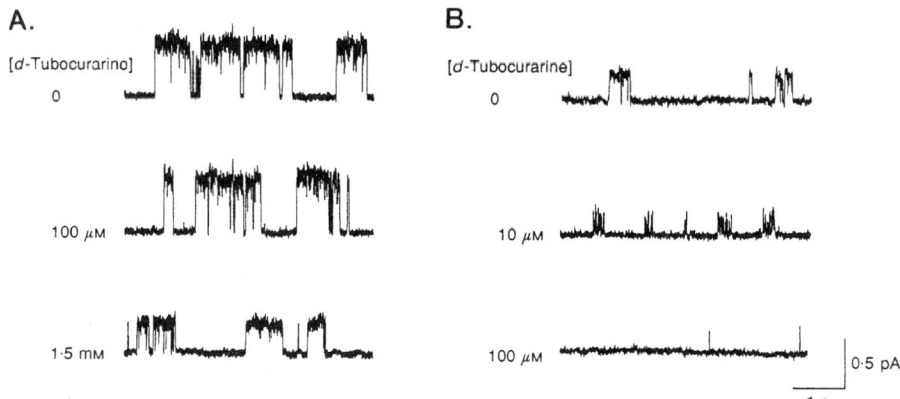

Figure 6. Effects of *d*-tubocurarine on Ca^{2+}-activated K^+ channels in inside-out patches of endothelium from intact rat aorta. (A) *d*-Tubocurarine at 100 μM did not affect K_{Ch} channels current, but at 1.5 mM reduced the single-channel current amplitude. (B) *d*-Tubocurarine at 10 μM induced burst activity of K_{Ap} channels and at 100 μM completely blocked them. Inhibitor concentrations are shown to the left of the traces. Both effects were reversible (not shown). These data further demonstrate the pharmacological difference between the two channels. (Reproduced from Marchenko and Sage, 1996.)

6.2. Physiological Role of K_{Ap} and K_{Ch} Channels

The distinct pharmacology of the two K_{Ca} channels found in the endothelium of intact rat aorta allowed us to assess their physiological roles. Apamin at concentrations of up to 1 μM was without effect on the resting E_m. Charybdotoxin at concentrations of up to 1 μM was similarly without effect on E_m in 88% of aortic endothelial cells, whereas 1 μM charybdotoxin caused a small depolarization of 1–2 mV in the remaining 12% of cells (Marchenko and Sage, 1996). These results suggest that neither type of K_{Ca} channel makes a significant contribution to the resting E_m. This is in accordance with expectations based on the $[Ca^{2+}]$ sensitivity of the channels, with the Po of both channels being about zero at a $[Ca^{2+}]_i$ of 100 nM. The resting $[Ca^{2+}]_i$ in the endothelium of intact rat aorta is estimated to be 95 nM (Usachev *et al.*, 1995).

Notably, ACh-evoked hyperpolarizations in intact rat aorta are unaffected by 1 μM apamin (Fig. 7), a concentration which totally inhibits activity of the K_{Ap} channels (Marchenko and Sage, 1996). This suggests that the K_{Ap} channels are not significantly involved in the ACh-evoked hyperpolarization. In contrast, the ACh-evoked hyperpolarization is inhibited by high (>20 nM) concentrations of charybdotoxin (Fig. 7B) and by high (>1 mM) concentrations of *d*-tubocurarine (Marchenko and Sage, 1996). These observations suggest that K_{Ch} channels are involved in, and perhaps solely responsible for, the vasodilator-evoked hyperpolarization. The role of the K_{Ap} channels is presently unclear.

6.3. Other Studies in Intact Preparations

We are aware of only one other study to date that has measured single K_{Ca} channels in native endothelium. (Hoyer *et al.*, 1994) reported that a BK_{Ca} channel is

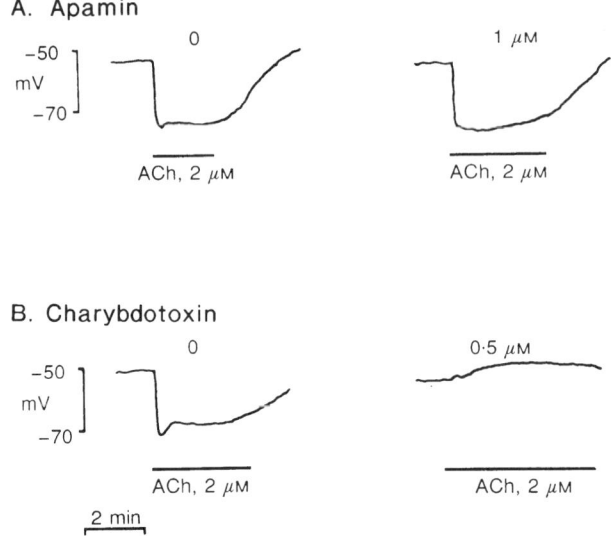

Figure 7. Effects of apamin or charybdotoxin on acetylcholine-evoked hyperpolarizations of the endothelium of intact rat aorta. Control responses evoked by 2 μM acetylcholine (ACh) are shown on the left. (A) Apamin at concentrations up to 1 μM was without effect on the ACh-evoked hyperpolarization. (B) Charybdotoxin at concentrations greater than 50 nM reduced ACh-evoked hyperpolarization and abolished it at 0.5 μM These results suggest that K_{Ch} but not K_{Ap} channels contribute to the vasodilator-evoked hyperpolarization. (Reproduced from Marchenko and Sage, 1996.)

present in excised patches of porcine endocardium from the right atrium. This channel has a conductance of 192 pS in symmetrical K^+ solutions and 157 pS in physiological ion gradients. The activity of the channel increases with depolarization and an increase in $[Ca^{2+}]$; at the cytosolic membrane face. At an E_m of -30 mV, Po is half-maximal at a $[Ca^{2+}]$; of 5.3 μM.

7. CONCLUSIONS

Single-channel studies of K_{Ca} channels in endothelial cells have yielded conflicting results concerning the distribution of different classes of these channels in vascular endothelium. The differences between various reports may reflect, at least in part, genuine differences in K_{Ca} channel expression in the endothelium of different parts of the vascular tree within a given species, as well as differences between species. Systematic comparative studies are needed to address these issues. Conflicting findings between reports using different cultured preparations of the same origin, and between those using cultured preparations and those using freshly isolated or native endothelium *in situ*, suggest that changes which occur after endothelial cells are removed from their normal environment *in vivo* may contribute to the confusing picture in the literature. A study comparing the expression and function of K_{Ca} channels in the same endothelial cell type measured *in situ* (in the vessel), in freshly isolated cells, and in cultured cells is needed to clarify the physiologically relevant level of channel expression.

Work on cultured endothelial cells clearly indicates the potential of this cell type to express K$_{Ca}$ channels in each of the three presently recognized classes. The limited data from studies using freshly isolated endothelial cells or native endothelium *in situ* tend to support this view, although as yet no single endothelial cell type has been found to contain all three K$_{Ca}$ channel subtypes—SK$_{Ca}$, IK$_{Ca}$, and BK$_{Ca}$ channels. Our own work on native rat aortic endothelium *in situ* indicates the presence of SK$_{Ca}$ (K$_{Ap}$) and IK$_{Ca}$ (K$_{Ch}$) channels, although the IK$_{Ca}$ channels differ from those classically described, by virtue of being less sensitive to charybdotoxin. In contrast, BK$_{Ca}$ channels do not appear to be present in native rat aortic endothelium. Future studies of native endothelial cells in other vascular preparations are awaited with interest.

Although the precise types of K$_{Ca}$ channels present in endothelium remains controversial, their functional importance appears certain. Vasodilator-evoked hyperpolarization mediated by K$_{Ca}$ channels influences the driving force for Ca^{2+} entry, and thus [Ca^{2+}]$_i$, and hence modulates the release of vasoactive substances from the endothelium. K$_{Ca}$ channels also may play an important role in signaling within the vessel wall, since it has been suggested that K$^+$ efflux through these channels may evoke hyperpolarization of the underlying smooth muscle (Edwards *et al.*, 1998). These latter findings in rat hepatic and mesenteric arteries suggest that the K$^+$ ions which leave the endothelium during activation of K$_{Ca}$ channels locally elevate the extracellular K$^+$ concentration in the myoendothelial space, which activate the Na$^+$–K$^+$-ATPase and inward rectifier K$^+$ (Kir) channels in the smooth muscle, thereby to induce hyperpolarization. Thus, K$^+$ ions may be the hitherto elusive endothelium-derived hyperpolarizing factor, EDHF, which appears to be involved in the relaxation of many blood vessels in response to endothelium-dependent vasodilators (Chen *et al.*, 1988; Feletou and Vanhoutte, 1988).

ACKNOWLEDGMENTS

The authors' work has been supported by the British Heart Foundation and the Wellcome Trust (grant no. 042948).

REFERENCES

Baron, A., Frieden, M., Chabaud, F., and Bény, J.-L., 1996, Ca^{2+}-dependent nonselective cation and potassium channels activated by bradykinin in pig coronary artery endothelial cells, *J. Physiol. (London)* **49**:699–706.

Barrett, J. N., Magelby, K. L., and Pallotta, B. S., 1982, Properties of single calcium-activated potassium channels in cultured rat muscle, *J. Physiol. (London)* **331**:211–230.

Blatz, A. L., and Magelby, K. L., 1986, Single apamin-blocked Ca^{2+}-activated K$^+$ channels in cultured rat muscle, *Nature* **323**:718–720.

Bossu, J. L., Elhamdani, A., and Feltz, A., 1989, Voltage-dependent calcium entry in confluent bovine capillary endothelial cells, *FEBS Lett* **299**:239–242.

Bregestovski, P., Bakhramov, A., Danilov, S., Moldobaeva, A., and Takeda, K., 1988, Histamine-induced inward currents in cultured endothelial cells from human umbilical vein, *Br. J. Pharmacol.* **95:**429–436.

Busse, R., Fichtner, H., Lückhoff, A., and Kohlhardt, M., 1988, Hyperpolarisation and increased free calcium in acetylcholine-stimulated endothelial cells, *Am. J. Physiol.* **255:**H965–H969.

Cai, S., Garneau, L., and Sauvé, R., 1998, Single-channel characterisation of the pharmacological properties of the $K(Ca^{2+})$ channel of intermediate conductance in bovine aortic endothelial cells, *J. Membr. Biol.* **163:**147–158.

Cannell, M. B., and Sage, S. O., 1989, Bradykinin-evoked changes in cytosolic calcium and membrane currents in cultured bovine pulmonary artery endothelial cells, *J. Physiol. (London)* **419:**555–568.

Castle, N. A., Haylett, D. G., and Jenkinson, D. H., 1989, Toxins in the characterisation of potassium channels, *Trends Neurosci* **12:**59–65.

Chen, G., and Cheung, D. W., 1992, Characterisation of acetylcholine-induced membrane hyperpolarization in endothelial cells, *Circ. Res.* **70:**257–263.

Chen, G., Suzuki, H., and Weston, A. H., 1988, Acetylcholine releases endothelium-derived hyperpolarising factor and EDRF from rat blood vessels, *Br. J. Pharmacol.* **95:**1165–1174.

Colden-Stanfield, M., Schilling, W. P., Ritchie, A. K., Eskin, S. G., Navarro, L. T., and Kunze, D.L., 1987, Bradykinin-induced increases in cytosolic calcium and ionic currents in cultured bovine aortic endothelial cells, *Circ. Res.* **61:**632–640.

Davies, P. F., 1995,. Flow-mediated endothelial mechanotransduction, *Phys. Rev.* 75: 519-560.

Davis, M. J., and Sharma, N. R., 1994, Mechanism of substance P-induced hyperpolarization of porcine coronary artery endothelial cells, *Am. J. Physiol.* **266:**H156–H164.

Edwards, G., Dora, K. A., Gardener, M. J., Garland, C. J., and Weston, A. H., 1998, K^+ is an endothelium-derived hyperpolarising factor in rat arteries, *Nature* **396:**269–272.

Ewald, D. A., Williams, A., and Levitan, I. B., 1985, Modulation of single Ca^{2+}-activated K^+ channel activity by protein phosphorylation, *Nature* **315:**503–506.

Feletou, M., and Vanhoutte, P. M., 1988, Endothelium-dependent hyperpolarization of canine coronary smooth muscle, *Br. J. Pharmacol.* **93:**515–524.

Fichtner, H., Frobe, U., Busse, R., and Kohldardt, M., 1987, Single nonselective cation channels and Ca^{2+}-activated K^+ channels in aortic endothelial cells, *J. Memb. Biol.* **98:**125–133.

Freay, A., Johns, A., Adams, D. J., Ryan, U. S., and van Breemen, C., 1989, Bradykinin and inositol 1,4,5-trisphosphate-stimulated calcium release from intracellular stores in cultured bovine endothelial cells, *Pflügers Arch.* **414:**377–384.

Furchgott, R. F., and Vanhoutte, P. M., 1989, Endothelium-derived relaxing and contracting factors, *FASEB J.* **3:**2007–2018.

Galves, A., Gimenez-Gallego, G., Reuben, J. P., Roy-Contancin, L., Feigenbaum, P., Kaczorowski, G. J., and Garcia, M. L, 1990, Purification and characterisation of a unique, potent, peptidyl probe for the high conductance calcium-activated potassium channel from venom of the scorpion *Buthus tamulus, J. Biol. Chem.* **265:**11083–11090.

Groschner, K., Graier, W. F., and Kukovetz, W. R., 1992, Activation of a small conductance Ca^{2+}-dependent K^+ channel contributes to bradykinin-induced stimulation of nitric oxide synthesis in pig aortic endothelial cells, *Biochim. Biophys. Acta* **1137:**162–170.

Grygorczyk, R., and Schwarz, W., 1983, Properties of the Ca^{2+}-activated K^+ conductance of human red cells as revealed by the patch clamp technique, *Cell Calcium* **4:**499–510.

Hallam, T. J., Pearson, J. D., and Needham, L. A., 1988, Thrombin-stimulated elevation of human endothelial cell cytoplasmic free calcium concentration causes prostacyclin production, *Biochem. J.* **251:**243-249.

Hallam, T. J., Jacob, R., and Merritt, J. E., 1989, Influx of bivalent cations can be independent of receptor stimulation in human endothelial cells, *Biochem. J.* **259:**125–129.

Hermann, A., and Erxleben, C., 1987, Charybdotoxin selectively blocks small Ca^{2+}-activated K^+ channels in *Aplysia* neurons, *J. Gen. Physiol.* **90:**27–47.

Hoyer, J., Distler, A., Haase, W., and Gogelen, H., 1994, Ca^{2+} influx through stretch-activated cation channels activates maxi K^+ channels in porcine endocardial endothelium, *Proc. Natl. Acad. Sci. U.S.A.* **91:**2367–2371.

Jacob, R., 1990, Agonist-stimulated divalent cation entry into single cultured human umbilical vein endothelial cells, *J. Physiol. (London)* **421:**55–77.

Jacob, R., Merritt, J. E., Hallam, T. J., and Rink, T. J., 1988, Repetitive spikes in cytoplasmic calcium evoked by histamine in human endothelial cells, *Nature* **335:**40–44.

Johns, A., Lategan, T. W., Lodge, N. J., Ryan, U. S., van Breemen, C., and Adams, D. J., 1987, Calcium entry through receptor-operated channels in bovine pulmonary artery endothelial cells, *Tissue Cell* **19**:733–745.

Laskey, R. E., Adams, D. J., Johns, A., Rubanyi, G. M., and van Breemen, C., 1990, Membrane potential and Na$^+$–K$^+$ pump activity modulate resting and bradykinin-stimulated changes in cytosolic free calcium in cultured endothelial cells from bovine atria, *J. Biol. Chem.* **265**:2613–2619.

Laskey, R. E., Adams, D. J., Cannell, M. B., and van Breemen, C., 1992, Calcium-entry dependent oscillations of cytoplasmic calcium concentration in cultured endothelial cell monolayers, *Proc. Natl. Acad. Sci. U.S.A.* **89**:1690–1994.

Ling, B. N., and O'Neill, W. C., 1992, Ca^{2+}-dependent and Ca^{2+}-permeable ion channels in aortic endothelial cells, *Am. J. Physiol.* **263**:H1827–H1838.

Lückhoff, A., and Busse, R., 1990, Calcium influx into endothelial cells and formation of endothelium-derived relaxing factor is controlled by the membrane potential, *Pflügers Arch.* **416**:305–311.

Marchenko, S. M., and Sage, S. O., 1993, Electrical properties of resting and acetylcholine-stimulated endothelium in intact rat aorta, *J. Physiol. (London)* **462**:735–751.

Marchenko, S. M., and Sage, S. O., 1994, Mechanism of acetylcholine action on membrane potential of endothelium of intact rat aorta, *Am. J. Physiol.* **266**:H2388–H2395.

Marchenko, S. M., and Sage, S. O., 1996, Calcium-activated potassium channels in the endothelium of intact rat aorta, *J. Physiol. (London)* **492**:53–60.

Marchenko, S. M., and Sage, S. O., 1997, A novel mechanosensitive cationic channel from the endothelium of rat aorta, *J. Physiol. (London)* **498**:419–425.

Merke, G., and Daut, J., 1990, The electrical response of cultured guinea-pig coronary endothelial cells to endothelium-dependent vasodilators, *J. Physiol. (London)* **430**:251–272.

Merke, G., Pohl, U., and Daut, J., 1991, Effects of vasoactive agonists on the membrane potential of cultured bovine aortic and guinea pig coronary endothelium, *J. Physiol. (London)* **439**:277–299.

Miller, C., Moczydlowski, E., Latorre, R., and Phillips, M., 1985, Charybdotoxin, an inhibitor of single Ca^{2+}-activated K$^+$ channels from mammalian skeletal muscle, *Nature* **313**:316–318.

Moncada, S., Palmer, M. J., and Higgs, E. A., 1991, Nitric oxide: Physiology, pathophysiology and pharmacology, *Pharmacol. Rev.* **43**:109–142.

Muraki, K., Imaizumi, Y., Ohya, S., Sato, K. Takii, T., Onozaki, K., and Watanabe, M., 1997, Apamin-sensitive Ca^{2+}-dependent K$^+$ current and hyperpolarization in human endothelial cells, *Biochem. Biophys. Res. Commun.* **236**:340–343.

Nilius, B., Viana, F., and Droogmans, G., 1997, Ion channels in vascular endothelium, *Annu. Rev. Physiol.* **59**:145–170.

Okahara, K, Sun, B, Kawasaki, T., Monden, and M., Kambayashi, J., 1998, Expression of platelet-activating factor receptor transcript-2 is induced by shear stress in HUVEC, *Prostaglandins Other Lipid Mediato.* **55**:323–329.

Olesen, S.-P., Davies, P. F., and Clapham, D. E., 1988, Muscarinic-activated K$^+$ current in bovine aortic endothelial cells, *Circ. Res.* **62**:1059–1064.

Redmond, E. M., Cahill, P. A., and Sitzman, J. V., 1998, Flow-mediated regulation of G-protein expression in cocultured vascular smooth muscle and endothelial cells, *Arterioscler. Thromb. Vasc. Biol.* **18**:75–83.

Rotrosen, D., and Gallin, J. I., 1986, Histamine type I occupancy increases endothelial cytosolic calcium, reduces F-actin and promotes albumin diffusion across cultured endothelial monolayers, *J. Cell. Biol.* **103**:2379–2387.

Rusko, J., Tanzi, F., van Breemen, C., and Adams, D. J., 1992, Calcium-activated potassium channels in native endothelial cells from rabbit aorta. Conductance, Ca^{2+} sensitivity and block, *J. Physiol. (London)* **455**:601–621.

Rusko, J., Vanslooten, G., and Adams, D. J., 1995, Caffeine-evoked, calcium-sensitive membrane currents in rabbit aortic endothelial cells, *Br. J. Pharmacol.* **115**:133–141.

Sage, S. O., Adams, D. J., and van Breemen, C., 1989, Synchronised oscillations in cytosolic free calcium concentration in confluent bradykinin-stimulated bovine pulmonary artery endothelial cell monolayers, *J. Biol. Chem.* **264**:6–9.

Sauvé, L., Parent, R., Simoneau, C., and Roy, G., 1988, External ATP triggers a biphasic activation process of a calcium-dependent K$^+$ current in cultured bovine aortic endothelial cells, *Pflügers Arch.* **412**:469–481.

Tracey, W. R., and Peach, M. J., 1992, Differential muscarinic receptor mRNA expression by freshly isolated and cultured bovine endothelial cells, *Circ. Res.* **70**:234–240.

Usachev, Y. M., Marchenko, S. M., and Sage, S. O., 1995, Cytosolic calcium concentration in resting and stimulated endothelium of excised intact rat aorta, *J. Physiol. (London)* **489:**309–317.

Van Renterghem, C., Vigne, P., and Frelin, C., 1995, A charybdotoxin-sensitive, Ca^{2+}-activated K^+ channel in cultured vascular endothelial cells, *J. Neurochem.* **65:**1274–1281.

Ziegler, T., Bouzourene K., Harrison, V. J., Brunner, H. R., and Hayoz, D., 1998, Influence of oscillatory and unidirectional flow environments on the expression of endothelin and nitric oxide synthase in cultured endothelial cells, *Arterioscler. Thromb. Vasc. Biol.* **18:**686–692.

Zymunt, P. M. and Högestätt, E. D., 1996, Role of potassium channels in endothelium-dependent relaxation resistant to nitroarginine in the rat hepatic artery, *Br. J. Pharmacol.* **117:**1600–1606.

Chapter 33

Endothelial Cell K$^+$ Channels, Membrane Potential and the Release of Vasoactive Factors from the Vascular Endothelium

Christopher R. Triggle

1. INTRODUCTION

Potassium channels play an important role in the regulation of the membrane potential (E_m) of endothelial cells and thereby modulate the entry of extracellular Ca^{2+} (Adams, 1994; Himmel *et al.*, 1993; Adams *et al.*, 1989). Ca^{2+} entry in concert with intracellular Ca^{2+} release is important for the synthesis of a number of endothelium-derived vasoactive factors. Thus, the synthesis of the endothelium-derived relaxing factor (EDRF), nitric oxide (NO), and of prostacyclin (PGI$_2$) requires, respectively, the Ca^{2+}-calmodulin-dependent activation of the constitutive endothelial cell nitric oxide synthase (eNOS) and the Ca^{2+}-dependent activation of phospholipase A$_2$ (Pollock *et al.*, 1991; Bredt and Snyder, 1990; Carter *et al.*, 1988; Hallam *et al.*, 1988). Similarly, the synthesis of the vasoconstrictor peptide endothelin-1 (ET-1) requires the mobilization of intracellular Ca^{2+} and the activation of protein kinase C (Yanagisawa *et al.*, 1989).

Indirect evidence indicates that endothelial cells possess receptor-operated and/or store-operated Ca^{2+} channels as well as stretch-activated channels that allow Ca^{2+} to permeate (Setoguchi *et al.*, 1997), and the consensus is that they lack voltage-gated L-type Ca^{2+} channels (Vanhoutte, 1988; Johns *et al.*, 1987). Of interest, however, is the report that voltage-gated R-type Ca^{2+} channels that are insensitive to nifedipine but sensitive to another 1,4-dihydropyridine, isradipine, may play a role in mediating a sustained increase in the intracellular Ca^{2+} concentration ([Ca^{2+}]$_i$) in human and canine aortic endothelial cells (Bkaily *et al.*, 1993). The presence of R-type Ca^{2+} channels in endothelium and their role in the regulation of Ca^{2+} entry may also explain the findings by Adeagbo and Triggle (1991) which described a role for a Ca^{2+} channel

Christopher R. Triggle ● Department of Pharmacology and Therapeutics, University of Calgary, Calgary, Alberta, Canada T2N 4N1.

Potassium Channels in Cardiovascular Biology, edited by Archer and Rusch. Kluwer Academic/Plenum Publishers, New York, 2001.

with properties distinct from an L-type channel in EDRF release from the mesenteric arterial vascular bed. Nonetheless, the opening of endothelium K^+ channels and resulting hyperpolarization of the endothelial cell will increase the electrochemical driving force for Ca^{2+} entry and thus may be required for the sustained intracellular Ca^{2+} signal and increased EDRF synthesis. Hyperpolarization of the underlying vascular smooth muscle (VSM) cells via low-resistance electrical myoendothelial coupling may play a more direct role in the regulation of VSM E_m (see Chaytor et al., 1998). A role for endothelial cell chloride (Cl^-) channels in receptor-mediated activation [i.e., subsequent to acetylcholine (ACh), bradykinin (BK), or substance P (SP) or purinoreceptor activation] and an increase in intracellular Ca^{2+} should also be considered (Ohashi et al., 1999; Nilius et al., 1997; Hosoki and Iijima, 1994; Nilius and Riemann, 1990). However, the dependence of E_m on extracellular K^+ [but not Cl^-] indicates that the predominant ionic permeability of the plasma membrane is provided by K^+ channels (Laskey et al., 1990; Johns et al., 1987).

Much of the data in the literature is obtained from studies of endothelial cells derived from large conduit or muscular arteries or veins, for instance, human umbilical vein endothelial cells (HUVEC), bovine aorta endothelial cells (BAEC), rat aorta endothelial cells (RAEC), and porcine coronary artery endothelial cells (PCAEC), with comparatively little data from endothelial cells derived from resistance vessels. The hypothesis that endothelial cells represent a homogeneous population has not been proven, and, indeed, evidence suggests otherwise. Thus, the extrapolation of data derived from endothelial cells of a particular vascular bed (and species) may be inappropriate. Indeed, the vascular endothelium demonstrates not only an anatomical heterogeneity but also a molecular and functional heterogeneity (see Woodley and Barclay, 1994; McCarthy et al., 1991). For instance, it is a common assumption that von Willebrand's factor is a specific marker of endothelial cells, whereas it is, in fact, a rather poor marker of microvascular endothelium such as in renal glomerular capillaries and in the lymphatics (Harach et al., 1983; Mukai et al., 1980). This being the case, it is perhaps reasonable to expect comparable heterogeneity with respect to the nature and actions of endothelium-derived vasoactive factors.

2. ENDOTHELIAL CELL MEMBRANE POTENTIAL

Although it is important not to make inappropriate generalizations, a resting E_m of between -40 and -55 mV has been reported in the intact endothelium from RAEC and PCAEC (Marchenko and Sage, 1994; Sharma and Davis, 1994). Values ranging from -20 to -80 mV have also been reported from endothelial cells derived from a variety of vascular beds in different species (von der Weid and Beny, 1992; Adams et al., 1989; Bregestovski et al., 1988; Daut et al., 1987; Johns et al., 1987; Takeda et al., 1987; Larson et al., 1983; Northover, 1980). K^+ channel activation and subsequent hyperpolarization can be mediated chemically, for instance, subsequent to ACh, BK, or SP receptor activation, or mechanically, via stretch-operated mechanisms, or facilitated by the effects of shear stress.

At least five different types of K^+ channels have been implicated in the regulation of resting E_m (and hence the synthesis and release of endothelium-derived vasoactive factors): (i) Ca^{2+}-activated K$^+$ (K$_{Ca}$) channels, (ii) inward rectifying K$^+$ (Kir channels, (iii) ATP-sensitive K$^+$ (K$_{ATP}$) channels, (iv) voltage-sensitive K$^+$ (Kv) channels, and (v) ACh-gated K$^+$ ($I_{K(ACh)}$) channels.

2.1. Ca^{2+}-Activated K$^+$ Channels

Probably the best-described K$^+$ channels in endothelial cells are the Ca^{2+}-activated K$^+$ (K$_{Ca}$) channels (Nilius et al., 1997; Adams, 1994; Himmel et al., 1993; Adams et al., 1989). Evidence derived from patch-clamp studies suggests a heterogeneous population of such channels. Data supporting an intermediate-conductance (approximately 40 pS) and charybdotoxin (ChTX)-sensitive K$^+$ channel have been presented (Olesen and Bundgaard, 1993). There also is evidence for a high-conductance, Ca^{2+}-activated K$^+$ channel (BK$_{Ca}$ channel) in rabbit aorta endothelial cells that is blocked by ChTX but is insensitive to 3,4-diaminopyridine, Ba^{2+}, and apamin (Rusko et al., 1992). Additionally, an apamin-sensitive, small-conductance K$_{Ca}$ channel (SK$_{Ca}$ channel) has been reported in endothelial cells cultured from pig aorta (Groschner et al., 1992). Using a glass microelectrode technique Ohashi et al. (1999) recently demonstrated that ACh activates ChTX- and apamin-sensitive channels in rabbit aortic valve endothelial cells. Additionally, an apamin-insensitive, but d-tubocurarine (dTC)-sensitive K$_{Ca}$ channel has also been reported in pig coronary artery endothelial cells (von der Weid and Beny, 1992). It should, however, be noted that von der Weid and Beny (1992) did not compare the blocking effects of dTC with those of a combination of apamin and ChTX. The dTC-sensitive K$^+$ channel described by von der Weid and Beny (1992) may be the same as, or related to, the channel described by others as having properties of an intermediate-conductance, Ca^{2+}-activated K$^+$ (IK$_{Ca}$) channel (see Cai et al., 1998). Also of interest is the report by Kohler et al. (1996) that describing the pharmacology of K$^+$ channels cloned from rat and human brain. These channels demonstrate differential sensitivity to apamin and dTC (the human clone, hSK1, was insensitive to apamin but sensitive to block by dTC) but possess some of the characteristics of SK$_{Ca}$ channels. They differ from the BK$_{Ca}$ channels in their greater sensitivity to Ca^{2+} and their voltage-independent inactivation. In addition, a K$^+$ channel in rat endothelial cells from hepatic artery and mesenteric arterioles has been described that is inhibited by a combination of apamin and ChTX, but not apamin and iberiotoxin (IBTX) (Edwards et al., 1998). Fransen et al. (1998) recently demonstrated, using fura-2 microfluorometry and perforated-patch whole-cell recordings, that both ACh and caffeine produce an initial increase in [Ca^{2+}]$_i$ in rabbit arterial endothelial cells. This increase was accompanied by a transient outward current. The caffeine-induced hyperpolarization was shown to be due to a dose-dependent increase in [Ca^{2+}]$_i$ and the activation of Ca^{2+}-sensitive K$^+$ and Cl$^-$ conductances. ACh has been shown to produce a two-component hyperpolarization (followed by a depolarization) in rabbit aortic valve endothelial cells (RAVEC) that is sensitive to ChTX (transient hyperpolarization) and ChTX and apamin (sustained hyperpolarization) but not to glibenclamide or Ba^{2+} (Ohashi et al., 1999).

ChTX is known to block certain voltage-dependent K$^+$ (Kv) channels as well as BK$_{Ca}$ channels (Nelson and Quayle, 1995; Cook and Quast, 1990) whereas IBTX is a highly selective and potent blocker of BK_{Ca} channels (Galvez et al., 1990). These data can be interpreted, therefore, to indicate the presence of a novel K^+ channel that has recognition sites for both apamin and ChTX (see Zygmunt et al., 1997).

Evidence from single-channel recordings also has been presented for a voltage-independent K_{Ca} channel of intermediate conductance that is present in BAEC and is inhibited by ChTX and also by tetraethylammonium (TEA) (IC_{50}, 23 mM) or dTC (IC_{50}, 4.4 mM) (Cai et al., 1998). Paradoxically, although insensitive to IBTX, the channel described was inhibited by NS1619, an activator of maxi-K_{Ca} channels, and also has some features attributed to the voltage-gated Kv1.2 K^+ channel. It is unclear at this time whether it is the same as, or related to, the channel described by Edwards et al. (1998). A channel with similar properties to that described by Cai et al. (1998) [inhibited by high concentrations of ChTX (> 50 nM) or dTC (> 1 mM)] has also been described in the endothelium of the intact rat aorta (Marchenko and Sage, 1996). A K^+ channel(s) that is sensitive to ChTX and the K_{Ca} blocker tetrabutylammonium (TBA), but not IBTX or apamin, also seems to be involved in the release/action of endothelium-derived hyperpolarizing factor (EDHF) in the rabbit carotid artery (Dong et al., 1997).

2.2. Inward Rectifying K⁺ Channels

Kir channels have consistently been reported in both freshly dispersed and cultured endothelial cells (Adams et al., 1989). A Kir current that can be activated by shear stress has been described in BAEC (Olesen et al. 1988a). It has also been shown, using E_m-sensitive fluorescent dyes, that shear stress produces endothelial cell hyperpolarization (Nakache and Gaub, 1988). In HUVEC, Kir channels may be activated by histamine and thrombin (Nilius and Riemann, 1990). The channel described in the latter studies is reported to have a single-channel conductance of 27 pS (when measured in symmetrical K^+, 140 mM).

2.3. ATP-Sensitive K⁺ Channel

A K_{ATP} channel has also been described in endothelial cells of the rat aorta (Janigro et al., 1992) and in rabbit thoracic aorta and pulmonary artery endothelial cells (Katnik and Adams, 1997, 1995). Another study, however, demonstrated that the inhibition of K_{ATP} channels in rat aorta endothelial cells does not inhibit ACh-mediated vasorelaxation, suggesting that the K_{ATP} channels, at least in this preparation, may not play a role in the regulation of NO synthesis and release (Demirel et al., 1994).

2.4. Voltage-Sensitive K⁺ Channels

Kv channels have not been observed in native rabbit arterial endothelial cells (David Adams, personal communication). However, Takeda et al. (1987) described two populations of cultured BAEC, one population with a K$^+$ current blocked by low concentrations of Ba^{2+} and possessing the expected properties of Kir channels, and a second population of cells with depolarization-activated outward currents. The latter, seen in cells from older primary and also secondary cultures, possessed the properties of the fast-inactivating or transient A-type K$^+$ current that has been described in neurons. Definitive data supporting the existence of such a channel in native endothelial cells have not been reported. Chen and Cheung (1992), however, have described a component of the ACh-mediated hyperpolarization of guinea pig coronary artery endothelial cells that is sensitive to low concentrations (0.5 mM) of 4-aminopyridine (4-AP) and thus appears to have the expected pharmacology of A-type Kv channels.

2.5. ACh-Gated K$^+$ Channels

ACh-mediated vasodilation has been linked to the activation of an $I_{K(ACh)}$ in cultured endothelial BAEC and RAEC (Olesen et $al.$, 1988b; Busse et $al.$, 1988) which results in a transient hyperpolarization, with the amplitude dependent on extracellular K$^+$ concentration and ACh. Because ACh did not gate Kir in the BAEC used by Olesen et $al.$ (1988b), it was concluded that an observed muscarinic receptor-gated K$^+$ current results from activation of a distinct channel (K$_{ACh}$). Furthermore, unlike the cardiac muscarinic receptor-gated K$^+$ channel, stimulation of this channel was not dependent on a pertussis-toxin-sensitive GTP-binding protein. The long-lasting hyperpolarization of the BAEC that was initiated by ACh led Olesen et $al.$ (1988b) to speculate that this endothelial cell hyperpolarization, via myoendothelial cell gap junctions, may lead to hyperpolarization of VSM cells. This conclusion supports the earlier report from Segal and Duling (1986) that there is a bidirectional conductance of an ACh-mediated vasodilatory response in microvessels that appears to be independent of either EDRF or blood flow.

3. NATURE OF ENDOTHELIUM-DERIVED RELAXING FACTORS

The role of the endothelium in the production of NO and PGI$_2$ is well established (Palmer et $al.$, 1987; Furchgott and Zawadski, 1980; Moncada et $al.$, 1976), and it is also well accepted that both NO and PGI$_2$ can play an important role in endothelium-dependent vasorelaxation. There is, however, considerable evidence that a factor(s) other than NO or PGI$_2$ contributes to endothelium-dependent vasorelaxation and, notably, vascular smooth muscle hyperpolarization. This factor, endothelium-derived hyperpolarizing factor (EDHF), has been the subject of considerable interest and is the focus of several recent reviews (Triggle et $al.$, 1999; Faraci and Heistad, 1998; Quilley et $al.$, 1997; Vanhoutte, 1996; Mombouli and Vanhoutte, 1997; Garland et $al.$, 1996; Waldron et $al.$, 1996). As stated by Mombouli and Vanhoutte (1997), the term "EDHF" should be limited to describing a non-NO/PGI$_2$ endothelial cell-derived vasorelaxant factor that hyperpolarizes vascular smooth muscle cells. It should be noted, however, that NO and PGI$_2$ can also elicit hyperpolarization in some blood vessels. There are a number of such reports of NO- and PGI$_2$-mediated hyperpolarization [see reports by

Dong *et al.* (1998); Mistry and Garland (1998), Mombouli and Vanhoutte (1997), Weidelt *et al.* (1997), Waldron *et al.* (1996). and Bolotina *et al.* (1994)]. Thus, the term EDHF should be used to refer to an additional factor distinct from NO and PGI_2.

3.1. Epoxyeicosatrienoic Acids

Candidate molecules for EDHF include the epoxyeicosatrienoic acids (EETs), which are derived from arachidonic acid via the action of cytochrome P-450. EETs and arachidonic acid can activate K^+ channels (McGiff, 1991; Ordway *et al.*, 1989). In addition, an EDHF with the pharmacological properties of a cytochrome P-450 arachidonic acid metabolite is synthesized by both native and cultured porcine endothelial cells and hyperpolarizes cultured smooth muscle cells downstream. Hyperpolarization in this bioassay using rat aorta results from the activation of TBA- and ChTX-sensitive BK_{Ca} channels (Popp *et al.*, 1996a). A limitation of many of the studies that imply an EET as the EDHF is the use of nonspecific inhibitors of cytochrome P-450. In addition, several studies indicate that a number of widely used cytochrome P-450 inhibitors have nonspecific inhibitory effects on K^+ channels in smooth muscle (Vanheel and van de Voorde, 1997). The role of metabolites of arachidonic acid via the cytochrome P-450 pathway in mediating endothelium-dependent hyperpolarization is controversial (see Chataigneau *et al.*, 1998a). However, endothelium-derived EETs (5,6- and 11,12-EETs) may play an important role in the regulation of the endothelial cell membrane potential. The release of EETs may lead to hyperpolarization and an increase in the driving force for Ca^{2+} entry (Hoebel *et al.*, 1997; Gravier *et al.*, 1995). Thus, treatment of endothelial cells with thiopentane sodium (to inhibit the cytochrome P-450) inhibits bradykinin-induced hyperpolarization of porcine aorta endothelial cells by about 50% (Hoebel *et al.*, 1997). In such a model, agonist (bradykinin)-induced release of inositol 1,4,5-trisphosphate (IP_3) and subsequent depletion of intracellular Ca^{2+} stores increase cytochrome P-450 activity, thereby linking Ca^{2+} store depletion to Ca^{2+} entry. Of particular interest, Gravier *et al.* (1995) have also reported that 5,6-EET can directly trigger the so-called store-operated Ca^{2+} entry pathway, and store-operated Ca^{2+} channels have been demonstrated in endothelial cells (Parekh and Penner, 1997; Vaca, 1996; Vaca and Kunze, 1995, 1994, 1993).

3.2. Cannabinoids

There has been considerable recent interest in the hypothesis that EDHF may be an endogenous cannabinoid. Anandamide (arachidonyl ethanolamine amide), the endogenous cannabinoid receptor agonist, induces a relaxation of rat mesenteric arteries that is independent of the release of endothelial-derived autacoid (Randall *et al.*, 1996). The cannabinoid CB1 receptor antagonist SR141716A inhibits the endothelium-dependent, NO- and PGI_2-independent, relaxation following treatment with carbachol or stimulation with the Ca^{2+} ionophores A23187 and calcimycin. Cannabinoids do not influence the endothelium-independent relaxation to sodium nitroprusside, levcromakalim or calcitonin gene-related peptide (CGRP) (Randall *et al.*, 1996). A number of subsequent studies indicate that whereas anandamide does cause concentration-dependent vasorelaxation in some vessels and will evoke hyperpolarization, it does not possess the same profile of activity as EDHF, at least in the vessels studied to date (Chataigneau *et al.*, 1998b; Plane *et al.*, 1997; White and Hiley, 1997; Zygmunt *et al.*, 1997). The vasorelaxant effect of anandamide has been reported to be endothelium-independent (White and Hiley, 1997). Anandamide has also been reported to increase intracellular Ca^{2+} from a caffeine-sensitive pool in a HUVEC line and, presumably, may thus stimulate the synthesis and release of EDRF and EDHF (Mombouli *et al.*, 1999). The controversy over anandamide, however, continues (Harris *et al.*, 1999).

3.3. Gap Junctions

Segal and Duling (1986) reported bidirectional conductance of an ACh-mediated vasodilation in microvessels. However, the role of myoendothelial cell gap junctions in facilitating electronic coupling (hyperpolarization) between endothelial and vascular smooth muscle cells remains controversial. On the one hand, dye coupling between endothelial and smooth muscle cells in large porcine coronary arteries is sparse, but, on the other hand, such coupling is abundant in coronary arterioles (Beny and Pacicca, 1994; Daut *et al.*, 1994). Again, one must recognize the concerns about extrapolation of data derived from one blood vessel to another. Nevertheless, these data raise the possibility that myoendothelial cell gap junctions may play an important role in endothelium-dependent hyperpolarization of vascular tissue, at least in the microcirculation.

In large conduit vessels, it can be argued (Bauersachs *et al.*, 1996) that a signal propagated from endothelial to smooth muscle cells would likely be attenuated simply because of the considerably greater mass of the smooth muscle cells in the vessel. In addition, the electrical coupling between endothelial and vascular smooth muscle cells may occur only in the retrograde direction, that is, from smooth muscle cells to the endothelium (Beny and Pacicca, 1994). Weidelt *et al.* (1997) have also shown that ACh-induced hyperpolarization of vascular smooth muscle cell is blocked by N^6-nitro-L-argininc methyl ester (L-NAME) in rat mesenteric arterioles, thus suggesting chemical transmission via NO. Furthermore, a study by von der Weid and Beny (1992) with the endothelium-intact pig coronary artery demonstrates a large concentration difference for ionophore-induced (A23187) hyperpolarization of endothelial versus vascular smooth muscle. At low concentrations, A23187 induces endothelial, but not vascular smooth muscle cell, hyperpolarization. In addition, 1 mM dTC blocked the A23187-mediated endothelial cell, but not vascular smooth muscle cell, hyperpolarization. These data argue against a role for myoendothelial gap junctions in mediating endothelium-to-vascular smooth muscle cell hyperpolarization and against the role of endothelial cell hyperpolarization as a prerequisite for the release of EDRF or EDHF.

Recently, Chaytor *et al.* (1998) used a GAP-27 peptide showing sequence homology to 11 residues within the extracellular loop of connexin 43 to investigate the electrical coupling between endothelial and vascular smooth muscle cells. The authors reported that the NO-independent component of ACh-induced relaxation in the rabbit aorta and superior mesenteric artery can be inhibited by GAP-27, presumably because of the disruption of the cell-to-cell junctions. The interpretation is that the EDHF-mediated response likely can be explained on the basis of electrical coupling and/or diffusion of a low-molecular-weight mediator through gap junctions. A similar conclusion was reached in a study of the NO-independent endothelium-dependent relaxation to ACh in the rabbit iliac artery, in which 18-α-glycyrrhetinic acid was used to disrupt gap junctions (Taylor *et al.*, 1998). These studies have refueled the controversy as to whether gap junctions represent the pathway for a component of smooth muscle hyperpolarization not only in resistance vessels, but also in at least some conduit blood vessels. Hutcheson *et al.* (1999) have recently reported a differential inhibition by GAP-27 peptide of ACh, but not A23187-mediated, L-NAME-resistant, endothelium-dependent relaxation of rabbit mesenteric artery. These data suggest that ACh mediates endothelial-dependent VSM cell hyperpolarization by means of gap junction communication, whereas hyperpolarization mediated by A23187 involves chemical transmission in the extracellular space. The inference from these results is that there is a colocalization of endothelium-dependent vasodilator receptors (i.e., muscarinic receptors K^+), channels, and gap junction proteins.

3.4. K^+ as an EDHF

The recent report by Edwards (1998) provides evidence that in the rat hepatic artery and mesenteric small resistance arteries, the endothelium-dependent, non-NO/PGI_2 vasorelaxation and hyperpolarization response is mediated by a small increase in extracellular K^+ (5–10 mM). Using a combination of tension measurements, whole-cell patch-clamp recordings, and K^+-sensitive microelectrode studies, Edwards *et al.* found that both EDHF- and K^+-induced hyperpolarization of vascular smooth muscle is inhibited by a combination of ouabain and a low concentration of barium (Ba^{2+}, 30 μM). This suggested that the EDHF is the K^+ ion itself and that its targets on the myocytes are Na^+–K^+ ATPase and a Kir channel, respectively. Furthermore, EDHF-induced hyperpolarization, in contrast to hyperpolarization induced by the addition of 5–10 mM K^+, was inhibited by a combination of ChTX and apamin, but not IbTX and apamin. Ba^{2+}, at 30 μM, provides specific inhibition of Kir channels (Knot *et al.*, 1996; Wellman *et al.*, 1996). As already noted, IBTX is selective for BK_{Ca} channels whereas ChTX not only inhibits BK_{Ca} channels but also inhibits IK_{Ca} and certain *Shaker* Kv channels (Nelson and Quayle, 1995; Cook and Quast, 1990; Galvez *et al.*, 1990).

Collectively, these data suggest that a small increase in extracellular K^+ represents EDHF in rat hepatic artery and mesenteric arterioles and that the hyperpolarization of vascular smooth muscle is mediated by the activation of both Kir channels and the ouabain-sensitive Na^+–K^+-ATPase in the vascular smooth muscle plasmalemma. On the basis of the lack of effect of IBTX, it may be inferred that BK_{Ca} channels are not involved—neither those on endothelial nor those on vascular smooth muscle cell membranes. Furthermore, the data suggest that the increase in K^+ efflux from endothelial cells results from the activation of apamin and ChTX-sensitive endothelial

K$^+$ channels subsequent to the elevation of intracellular Ca^{2+} following agonist (ACh) stimulation of the endothelial cell. Edwards *et al.* (1998) also noted that K$^+$ alone may not completely explain the EDHF response, at least in rat mesenteric arterioles, because a combination of Ba^{2+} and ouabain did not completely inhibit relaxation to ACh, and they suggested that myoendothelial gap junctions may mediate the Ba^{2+}/ouabain-insensitive response. In support of this hypothesis, data were presented indicating that the K$^+$-induced vasorelaxation, in the presence of Ba^{2+} and ouabain, was markedly greater in the presence of the endothelium.

Interestingly, the data from Edwards *et al.* (1998) appear to be in conflict with those reported by Okazaki *et al.* (1998), who investigated EDHF-mediated hyperpolarization in the rat mesenteric vascular bed. Okazaki *et al.* reported that an increase in CO$_2$ produces an EDHF-mediated hyperpolarization and that endothelial cell hyperpolarization results from activation of a Kir channel in the endothelial cell. Although both investigations involved studies of the rat mesenteric arterial system, only that of Edwards *et al.* (1998) included direct measurements of vascular smooth muscle cell E_m.

Edwards *et al.* (1998) were not the first to report that ouabain selectively blocks EDHF and not NO (Félétou and Vanhoutte, 1988; Vanhoutte *et al.*, 1996). However, given the numerous reports describing EDHFs with variable pharmacological properties (see Vanhoutte and Félétou, 1996; Triggle *et al.*, 1999), it is unlikely that K$^+$ alone, or together with gap junctions, can be responsible for the action of the multiple EDHFs that have been described. For instance, Hammarström *et al.* (1999) recently reported in a study of guinea pig mammary and coronary arteries that depolarization of the tissue differentially affected the EDHF responses in the coronary and mammary arteries, thus suggesting that the nature of EDHF may not be the same in the two arteries. Furthermore, Hashitani and Suzuki (1997) reported that both SP and ACh produce endothelium-dependent hyperpolarization of guinea pig submucosal arterioles. Because smooth muscle cell hyperpolarization was detected in solutions containing Ba^{2+}, Kir channels in VSM cells are unlikely to be involved in EDHF-mediated events in these arterioles. Also, as previously stated, Kir channels have been consistently detected in endothelial cells, thus raising the possibility that they may be involved in the regulation of NO/EDHF synthesis/release. Edwards *et al.* (1998), however, postulated that Kir channels located on vascular smooth muscle cell membranes are responsible for the hyperpolarization of vascular smooth muscle cells that is mediated by EDHF. However, studies of Kir channels in VSM have been limited, and these channels have thus far only been described in arterial and resistance vessels, not in myocytes from conduit vessels (Quayle *et al.*, 1997). Data from Hirst's laboratory (Edwards and Hirst, 1988; Edwards *et al.*, 1988) included glass-microelectrode measurements of a barium-sensitive Kir current in both guinea pig ileal submucosal and rat cerebral arterioles studied under voltage-clamp conditions. These two key papers provide convincing evidence for the existence of Kir channels in arterioles. However, as already noted, Kir channels have been consistently shown to be present in endothelial cells. Given that myoendothelial gap junctions may play an important role in endothelial–VSM cell communication, notably in resistance vessels, it remains uncertain whether Kir channels in vascular smooth muscle cells actually contribute to the effects of EDRF/EDHF.

There are relatively few studies that have demonstrated the existence of Kir channels in single vascular smooth muscle cells. Whole-cell Kir currents have been reported in rat cerebral resistance arteries (Quayle *et al.*, 1993), small-diameter porcine coronary arteries (Quayle *et al.*, 1996) and rat coronary arteries (Robertson *et al.*, 1996). The study by Quayle *et al.* (1996) is of particular interest because it demonstrated much larger density of Kir channels in the smaller (fourth-order) branches of the guinea pig left anterior descending coronary arteries than in conduit segments. The differential expression of Kir channels in resistance vessels helps explain the ability of such vessels to respond to changes in flow, pressure, and local metabolites, whereas larger vessels are generally less responsive or are insensitive (Jones *et al.*, 1995). Thus, a small increase in extracellular K^+, with the K^+ originating from the opening of endothelial cell K^+ channels may act as an EDHF in resistance vessels, but is unlikely to be of significance in larger arterioles and arteries. Quignard *et al.* (1999) have also recently reported that barium-sensitive Kir and ouabain-sensitive Na^+/K^+-ATPase are not involved in the EDHF-mediated hyperpolarization and relaxation of guinea pig carotid and porcine coronary arteries. In addition, a recent report from Chataigneau *et al.* (1999) indicates that aorta, carotid, coronary, and mesenteric arteries from either eNOS $(+/+)$ mice or the homozygous eNOS knockout mice (eNOS $-/-$) do not seem to release a non-NO/PGI_2 endothelium-derived product in response to ACh. Furthermore, ACh, BK, and SP all failed to induce endothelium-dependent hyperpolarization in coronary arteries from either eNOS $(+/+)$ or eNOS $(-/-)$ mice. These data clearly make questionable the hypothesis that K^+ and/or gap junctions are universal contributors to the EDHF response. It should be noted, however, that under somewhat different experimental conditions, Triggle *et al.* (1998) have reported the existence of an unidentified EDHF in both control and eNOS knockout mice.

4. Ca^{2+} DEPENDENCY OF THE SYNTHESIS AND ACTION OF ENDOTHELIUM-DERIVED VASOACTIVE FACTORS

The Ca^{2+} dependency of the synthesis of ET-1, NO, and PGI_2 has been noted in the introduction to this chapter. It is well established that endothelium-dependent hyperpolarization of vascular smooth muscle cells is Ca^{2+}-dependent, thus implying a role for Ca^{2+} in the synthesis and release and/or activity of EDHFs. The calcium ionophore calcimycin (A23187) elicits an endothelium-dependent hyperpolarization of canine coronary, rabbit carotid, and femoral arteries (Plane *et al.*, 1995; Nagao *et al.*, 1992; Chen and Suzuki, 1990). Furthermore, in the rabbit femoral artery (Plane *et al.*, 1995), simultaneous tension and E_m measurements indicated that the endothelium-dependent hyperpolarization of vascular smooth muscle was insensitive to NO synthase inhibition but was inhibited by raised extracellular K^+, thus implying a role for a non-NO EDHF. Of additional interest is the report that calmodulin antagonists can inhibit ACh- and BK-induced hyperpolarization of rabbit femoral vascular smooth muscle (Plane *et al.*, 1995) and canine coronary artery (Illiano *et al.*, 1992). K^+ channel blockers, such as TEA, can block both agonist-induced increases in $[Ca^{2+}]_i$ in endothelial cells (Demirel *et al.*, 1994), and the NO/PGI_2-independent relaxation of VSM (Hwa *et al.*, 1994, Waldron and Garland, 1994).

Although an increase in endothelial $[Ca^{2+}]_i$ is a requirement for the synthesis of NO, PGI$_2$, and EDHFs, there are quantitative differences in the rise in $[Ca^{2+}]_i$ required to elicit the synthesis of each type of vasodilator. For example, Parsaee et al. (1992) reported differential sensitivities of PGI$_2$ versus NO synthesis to increases in $[Ca^{2+}]_i$ in bovine aortic endothelial cells. A threshold for $[Ca^{2+}]_i$ of 350 nM was reported for PGI$_2$ release, whereas the threshold required for NO was less than 200 nM. Similarly, Bauersachs et al. (1996) reported that NO released by the NO donor C87-3786 reduced BK-mediated release of EDHF by reducing BK-induced increases in Ca^{2+} levels in HUVEC. In addition, Bauersachs et al. (1996) found that the hyperpolarization induced in cultured rat aortic VSM cells by the release of EDHF from a luminally perfused porcine coronary artery was reduced by C87-3786. In a study of ACh-induced endothelium-dependent hyperpolarization in rat mesenteric superior arteries, Fukao et al. (1997) concluded that EDHF release is dependent upon ACh-mediated phospholipase C activation. This led to IP$_3$ production and the release of intracellular Ca^{2+}. Ca^{2+} entry was also stimulated via a Ni^{2+}- and Mn^{2+}- sensitive pathway, which was insensitive to nifedipine. Thus, differing sources of activator Ca^{2+} as well as different intracellular Ca^{2+} thresholds, may be required for NO versus EDHF synthesis. It appears that the requirement for an increase in $[Ca^{2+}]_i$ is less for NO than for EDHF synthesis. Since both NO and PGI$_2$ have been reported to inhibit the release of EDHF (Yajima et al., 1999; Thorin et al., 1998; Bauersachs et al., 1996; Popp et al., 1996b), there likely exists a complex interrelationship between cellular processes regulating NO, PGI$_2$, and EDHF synthesis and release.

Rusko et al. (1992) demonstrated that BK, ACh, and adenosine diphosphate (ADP) increase the open-state probability of K$_{Ca}$ channels in native endothelial cells from rabbit aorta. The sustained increase in open-state probability was dependent upon the presence of extracellular Ca^{2+} and was attenuated by the Ca^{2+}-entry blocker Ni^{2+}, as well as the K$^+$ channel blocker TEA. These observations are in accordance with the interrelationship depicted in Fig. 1 between endothelial cell receptor activation, Ca^{2+} entry via receptor-operated channels or nonselective cation channels (NSCCs), intracellular Ca^{2+} release, and the activation of K$_{Ca}$ channels. It is conceivable that multiple cellular compartments may exist which differentially express and/or regulate the enzymes responsible for NO, PGI$_2$, or EDHF synthesis. Thus, eNOS may be activated by at least two independent signaling pathways, with the shear stress response being insensitive to calmodulin antagonists but modulated by tyrosine kinases, whereas agonist-induced activation is Ca^{2+}-dependent (Busse and Fleming, 1998; Ayajiki et al., 1996; Kuchan and Frangos, 1994). eNOS is associated with endothelial cell caveolae, therefore indicating that endothelial cell NO production and likely the processes involved in the Ca^{2+}-dependent regulation of this enzyme are also colocalized to the caveolae (Shaul and Anderson, 1998; Shaul et al., 1996). Furthermore, differences between agonist-mediated (BK-mediated) and mechanical (stretch-mediated) EDHF release have also been demonstrated (Popp et al., 1998). Stretch has been shown to alter cytoskeletal membrane proteins (Busse and Fleming, 1998; Hutcheson and Griffith, 1996) that serve to open endothelial cell K$^+$ channels and activate MAP kinases (Takahashi et al., 1997; Davies, 1995; Hutcheson and Griffith, 1994). However, the endothelial cell K$^+$ channels activated by changes in flow may not be the same as those activated by chemical stimulation (Hutcheson and Griffith, 1994).

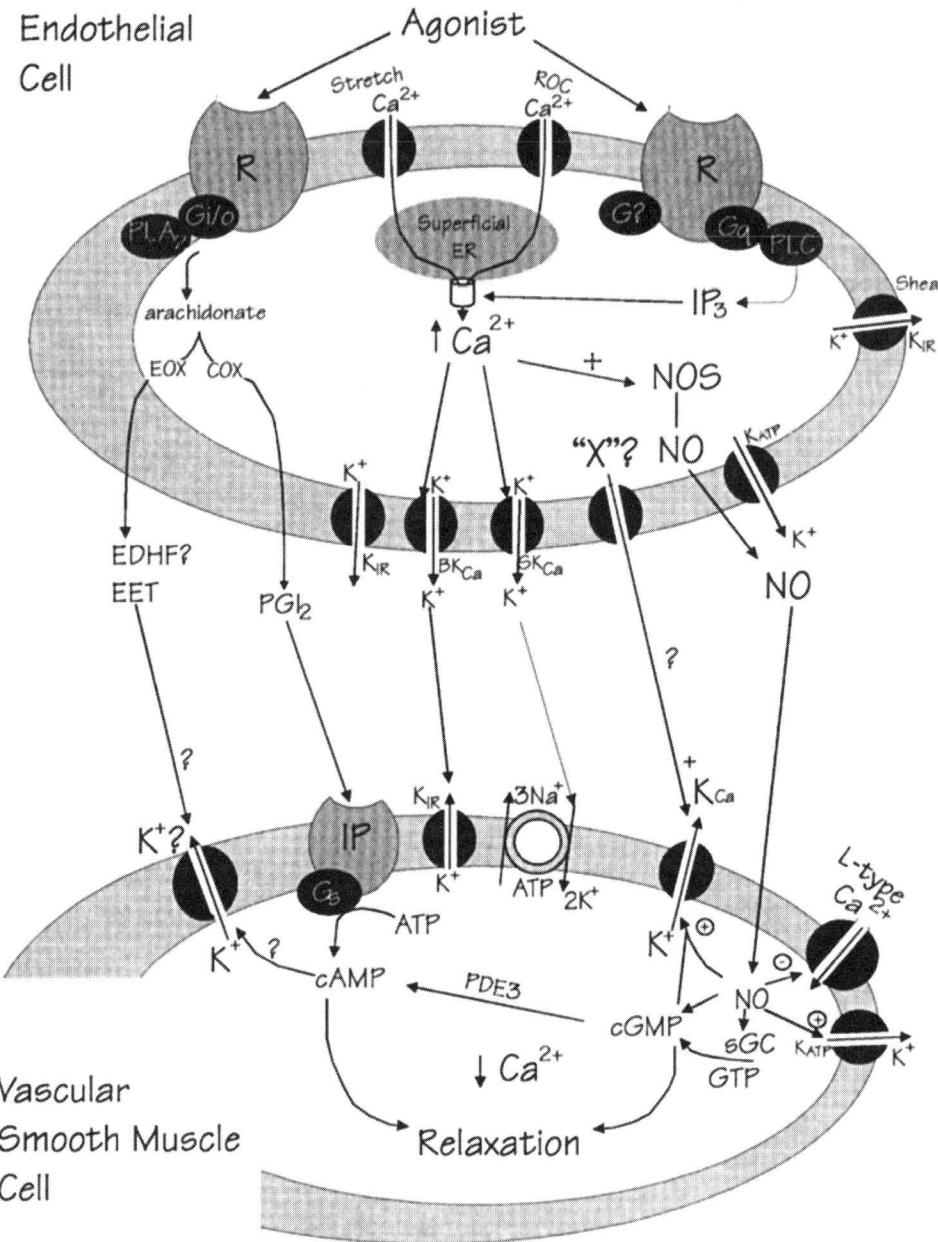

Figure 1. Pathways for the synthesis/release and cellular actions of endothelium-derived relaxing factors. The relationship between the endothelial cell and the vascular smooth muscle (VSM) cell and the regulation of the synthesis/release of nitric oxide (NO), prostacyclin (PGI$_2$), and epoxyeicosanoids (EETs), as well as the release of K$^+$ and a hypothesized factor "X" from the endothelial cell, are illustrated. Depicted on the endothelial cell are G-protein (Gi/o or Gq)-coupled receptors (R) that regulate phospholipase A$_2$ (PLA$_2$) or phospholipase C (PLC). PLA$_2$ regulates the mobilization of arachidonate (arachidonic acid) from phospholipids, and epoxygenase (EOX) and cyclooxygenase (COX) regulate the production of EETs and PGI$_2$,

5. NATURE OF ENDOTHELIAL K$^+$ CHANNELS INVOLVED IN THE SYNTHESIS AND RELEASE OF EDRF AND EDHF

One limitation of studies using intact tissues is that it is not possible to determine whether the inhibitory effects of K$^+$ channel blockers involve actions on the endothelial cell or the VSM cell or both. Table 1 summarizes some of the studies concerning endothelial cell K$^+$ channels and the release of vasoactive factors. There are very few studies that directly relate endothelial cell K$^+$ channel activation to the release and quantitative measurement of EDRF and EDHF, or their cellular effects. The study by Demirel *et al.* (1994) is one of the few that provides direct evidence for the role of endothelial K$^+$ channels in the synthesis and release of endothelial-derived factors. Using a perfusion–superfusion bioassay protocol with a segment of endothelium-intact rabbit thoracic aorta as the donor segment, and an endothelium-denuded rabbit thoracic aorta segment as the recipient tissue, these investigators demonstrated ACh-mediated production and bioactivity of EDRF. Spectrofluorometric measurements of fura-2-acetylmethylester (AM) loaded tissues also allowed measurements of ACh-induced changes in endothelial cell [Ca^{2+}]$_i$ to be monitored. Finally, membrane currents were recorded from rabbit thoracic aorta endothelial cells using the whole-cell patch-clamp configuration. These authors demonstrated that TEA, at a concentration appropriate for selective inhibition of K$_{Ca}$ channels, inhibited ACh-induced increases in endothelial cell [Ca^{2+}]$_i$ and also inhibited ACh-mediated relaxation of the recipient aortic vessel when perfused directly on the donor vessel, but not on the recipient vessel. The K$_{ATP}$ channel blocker glibenclamide had no effect. These data indicate that endothelial cell K$_{Ca}$ channels, but not K$_{ATP}$ channels, are important for the synthesis and release of EDRFs and that TEA blocks the ACh-induced increase in outward K$^+$ currents in rabbit aortic endothelial cells. Although the study did not demonstrate whether the EDRF was solely NO, nonetheless it did clearly demonstrate the important role of K$_{Ca}$ channels in the synthesis/release of EDRF(s). It should also be noted that Marchenko and Sage (1996) have reported that muscarinic receptor activation by ACh leads to the activation of ChTX- and dTC-sensitive K$_{Ca}$ channels in the endothelium of intact rat aorta; however, the VSM effects of any released EDRF/EDHF were not measured. Using the glass-microelectrode technique, Chen and Cheung (1992) studied ACh-mediated hyperpolarization in freshly dispersed endothelial cells from the guinea

---

respectively. Ca^{2+} influx into the endothelial cell may be regulated via stretch-activated and/or receptor-operated (ROC) cation channels, as well as a store-operated influx mechanism linked with the superficial endoplasmic reticulum (ER). Intracellular Ca^{2+} release may also be regulated by inositol trisphosphate (IP$_3$). The putative role of K$^+$ channels (Kir, inward rectifying; BK$_{Ca}$, large-conductance, Ca^{2+}-activated; SK$_{Ca}$, small-conductance, Ca^{2+}-activated; K$_{ATP}$, ATP sensitive) in regulation of shear- and chemical-mediated endothelial cell activation of EDRF/EDHF synthesis is also illustrated. The well-characterized actions of NO and PGI$_2$ are depicted as acting via the soluble guanylyl cyclase and adenylate cyclase systems, respectively, with PGI$_2$ activating adenylate cyclase via the IP prostanoid receptor. However, possible direct effects of NO on calcium-activated K$^+$ (K$_{Ca}$) and L-type voltage-gated Ca^{2+} (L-type Ca^{2+}) channels are also illustrated. Also noted is the interrelationship between cGMP and cAMP via cGMP-mediated inhibition of type 3 phosphodiesterase (PDE3). EDHF, which may reflect a small increase (5 mM) in extracellular K$^+$, is assumed to mediate VSM cell hyperpolarization via the activation of Kir and/or K$_{Ca}$ as well as the Na$^+$–K$^+$-ATPase.

TABLE 1
Endothelial K$^+$ Channels involved in Release of Nitric Oxide and EDHF

K$^+$ channel	Evidence supportive of ($+$)/against ($-$) role in NO release	Evidence supportive of ($+$)/against ($-$) role in EDHF release
K$_{Ca}$	($-$) Dong et al., 1998[a,b] ($+$) Demirel et al., 1994[d,e] ($+$) Groschner et al., 1992[g,h] ($\pm$) von der Weid and Beny, 1992[a,i]	($+$) Doughty et al., 1999[a,c] ($+$) Edwards et al., 1998[d,c,f] ($-$) Okazaki et al., 1998[a,c] ($+$) Zygmunt et al., 1998[a,j] ($+$) White and Hiley, 1997[a,c] ($+$) Demirel et al., 1994[d,e] ($+$) Hecker et al., 1994[a,i,j] ($+$) Adeagbo and Triggle, 1993[a,c] ($\pm$) von der Weid and Beny, 1992[a,i]
K$_{ATP}$	( ) White and Hiley, 1998[a,c] ($+$) Katnik and Adams, 1995, 1997[a,e,k] ($+$) Murphy and Brayden 1995[a,l]	($-$) Okazaki et al., 1998[a,c] ($+$) White and Hiley, 1997, 1998[a,c] ($-$) Murphy and Brayden 1995[a,l] ($-$) Fulton et al., 1994[a,m] ($-$) Hecker et al., 1994[a,b,i] ($-$) Adeagbo and Triggle, 1993[a,c] ($+$) Brayden, 1990[a,b]
Kir	($-$) Edwards et al., 1998[c,d,f] ($-$) Okazaki et al., 1998[a,c]	($-$) Edwards et al., 1998[c,d,f] ($+$) Okazaki et al., 1998[a,c] ($-$) White and Hiley, 1997[a,c]
Kv	($+$) Chen and Cheung, 1992[n,o],	($-$) Okazaki et al., 1998[a,c] ($+$) White and Hiley, 1997[a,c] ($+$) Chen and Cheung, 1992[n,o] ($-$) Adeagbo and Triggle, 1993[a,c]

[a] Indirect data reflected by reduction of NO/EDHF vasorelaxation.
[b] Rabbit middle cerebral artery.
[c] Rat mesenteric arterioles/vascular bed.
[d] Direct evidence.
[e] Rabbit thoracic and abdominal aorta.
[f] Rat hepatic artery.
[g] Measurement of changes in cGMP.
[h] Pig aorta.
[i] Pig coronary artery.
[j] Bovine coronary artery.
[k] Rabbit pulmonary artery.
[l] Rabbit mesenteric arterioles.
[m] Rat coronary artery (Langendorff).
[n] Indirect evidence with no measure of EDRF/EDHF effect or release.
[o] Guinea pig coronary artery.

pig coronary artery. Their data illustrate that ACh induced a long-lasting (>25 min) hyperpolarization that was inhibited by ChTX, TEA (1–5 mM), and 4-AP (0.5–1 mM). These K$^+$ channel blockers produced a rightward shift of the ACh dose–response curve. Chen and Cheung (1992) also demonstrated that apamin and glibenclamide had no effect on the ACh-mediated hyperpolarization in guinea pig coronary endothelial cells. Furthermore, sodium nitroprusside had no effect, thus indicating that endothelial-derived NO is unlikely to influence endothelial cell E_m. The sensitivity of the ACh-mediated hyperpolarization to 4-AP (0.5 mM) that Chen and Cheung (1992) reported suggests that a Kv channel, in addition to the K_{Ca} channel, may be involved in ACh-mediated endothelial cell hyperpolarization. A similar conclusion was also reached by Takeda *et al.* (1987) in their study of cultured bovine aortic endothelial cells. However, definitive data for the presence of Kv channels in native endothelial cells are lacking.

The recent publication of Edwards *et al.* (1998) implicating K$^+$ as an EDHF (see Section 3.4) also provides indirect support for the involvement of endothelial K_{Ca} channels in regulating the release of EDHF (K$^+$). Doughty *et al.* (1999) provided convincing evidence of a role for endothelial cell K_{Ca} channels in the regulation of EDHF release. They compared the effects of intraluminally applied versus superfused ChTX and apamin on ACh-induced relaxation of isometrically mounted, pressurized, third-order rat mesenteric arterioles. Apamin and ChTX abolished ACh-induced relaxation when applied intraluminally, but not when administered by superfusion. These data establish that, at least in rat mesenteric vessels, the site of action of apamin and ChTX is most likely the endothelial cell. It is, however, possible that endothelial K_{Ca} channels are not universally involved in the regulation of EDRF synthesis and release. Thus, von der Weid and Beny (1992) reported that although dTC significantly reduces (65–80%) both A23187- and SP-induced hyperpolarization of pig coronary artery endothelial cells, the respective relaxations are not inhibited. These data make questionable the role of endothelial cell hyperpolarization as a prerequisite for the release of EDRF and EDHF and/or VSM cell relaxation. However, the response to SP in the presence of nitroarginine alone was reduced, suggesting a role for a dTC-sensitive K$^+$ channel(s) in mediating the release or effects of an EDHF. Conversely, Cheung *et al.* (1999) reported that, based on their studies with rat mesenteric arteries, EDHF is more effective in inducing vascular smooth muscle cell hyperpolarization than relaxation, with the opposite being apparent for NO. In addition, Dong *et al.* (1998) recently reported that ACh-mediated endothelium-dependent relaxation of the rabbit middle cerebral artery is entirely mediated by NO and PGI$_2$, with no evidence for an EDHF contribution. In their study, Dong *et al.* also demonstrated that a nitroarginine-sensitive relaxation of this artery was observed in the presence of a high external concentration of K$^+$. These data would seem to argue against a role for endothelial K$^+$ channels in the regulation of, at least, NO synthesis and release and would support the hypothesis that intracellular Ca^{2+} release is critical for eNOS activation and NO production. White and Hiley (1998) reached a similar conclusion based on their myograph study of rat mesenteric arteries. In this study, the K_{ATP} channel activators levcromakalim and pinacidil reduced the NO-independent component of the endothelium-dependent relaxation response to carbachol. However, these channel openers did not inhibit relaxation in the absence of an inhibitor of NO synthesis. Furthermore, these K_{ATP} channel activators failed to reduce the portion of the relaxation response to A23187 that was NO-independent. Calcimycin directly elevates [Ca^{2+}]$_i$ in endothelial cells by a mechanism that is not dependent on the electrochemical gradient of Ca^{2+}. The absence of an effect of K_{ATP} channel activators on A-23187-induced effects supports the argument that their inhibitory effect on the L-NAME-resistant carbachol relaxation is due to an action at the level of the endothelial cell's synthesis/release of EDHF. Other investigators, however, have concluded that K_{ATP} channels on VSM cells are involved in NO-mediated hyperpolarization (Murphy and Brayden, 1995) and that apamin, but not K_{ATP} channel blockers, inhibits VSM cell hyperpolarization mediated by ACh-induced release of EDHF. These latter data from rabbit mesenteric resistance arteries can also be interpreted to indicate that endothelial cell K_{ATP} channels are not involved in the synthesis and release of EDHF and again reflect the evidence for a high degree of apparent species and tissue differences that exists in the literature. Relevant to the question concerning tissue heterogeneity and protocol differences are data recently reported by Hammarström *et al.* (1999) which indicate that in a study of guinea pig coronary and mammary arteries, the EDHF component of the response to ACh was more readily seen in depolarized arteries than in tissues at rest. This study underlies the importance of measurements of both vessel tone/diameter and membrane potential.

6. CONCLUSIONS

It is clear that the literature is not consistent regarding which endothelial cell K^+ channel(s) are involved in the regulation of synthesis/release of endothelium-derived NO and EDHF(s) (see Table 1). In part, this reflects the complexity of tissue and species differences, but it also certainly reflects that a large proportion of the literature describes indirect evidence obtained by different protocols. Few studies have clearly shown measurements of both changes in endothelial K^+ channel activity and, concurrently, the release and/or cellular actions of NO or EDHF. Table 1 lists just a few of the studies published on this topic.

A review of Table 1 illustrates some of the difficulties in interpreting the literature. Thus, many studies (too many to include in Table 1) have provided indirect data as to the actions of various K^+ channel blockers on the effects of NO and EDHF on VSM cell function. In many such studies, however, a change in vascular tone was the end point, and studies were performed on endothelium-intact tissue. As a consequence, an endothelial cell effect of the K^+ channel blocker, to prevent synthesis and release of the relaxant factor, is a potentially confusing influence. Several studies, such as those of Demirel *et al.* (1994) and Edwards *et al.* (1998), indicate an important role for endothelial K_{Ca} channels in the synthesis and release of EDRFs. These data are certainly very consistent with a more extensive literature that indicates that endothelium-dependent vasodilators (ACh, BK, SP, ADP) activate endothelial K_{Ca} channels. Despite extensive evidence for the presence of Kir channels in endothelial cells, there is very little supportive data for the role of Kir channels in regulating the release of NO or EDHF. One report consistent with a role for Kir channels in endothelium-dependent vasodilation is that of Okazaki *et al.* (1998). These authors described a barium-sensitive, endothelium-dependent vasodilator response of the perfused mesenteric vascular bed to hypercapnia. In the case of Kv channels in endothelial cells, evidence has been presented for both their expression and function. Takeda *et al.* (1987) described an A-type current in cultured bovine aortic endothelial cells. Furthermore, a 4-AP-sensitive K^+ current has been reported to contribute to ACh-mediated endothelial cell hyperpolarization (Chen and Cheung, 1992).

Several studies have also provided data that seemingly disassociate endothelial cell hyperpolarization from the release and vascular effects of endothelial-derived factors. For instance, von der Weid and Beny (1992) have shown a lack of effect of high extracellular K^+ concentrations on the synthesis and release of NO (also see Dong *et al.*, 1998). Such apparent contradictions clearly require a new experimental approach to analyze the relationship between NO/EDHF synthesis and release and the role of endothelium K$^+$ channels. A lack of knowledge of the nature of EDHFs and the means of reproducibly assaying EDHF and NO at the single-cell level has hindered such studies.

7. DIRECTIONS FOR FUTURE INVESTIGATION

As reflected in Table 1, there is a paucity of data that clearly correlate endothelial cell K$^+$ channel(s) activation to the release of NO, PGI$_2$, and EDHF. It is recognized that the generation of such data is not a simple matter, and, furthermore, it is likely that tissue heterogeneity may complicate any conclusions concerning a universally acceptable cellular mechanism. Of importance is the need to focus on native endothelial cell function and to employ not only measures of endothelial cell K$^+$ current but also semiqualitative and quantitative functional or chemical measurement of the vasoactive factor(s) that are released. Microelectrode detector systems for NO are available (Malinski and Taha, 1992), and changes in extracellular K$^+$ concentration can be measured (Edwards *et al.*, 1998). Bioassays have been extensively used for NO and PGI$_2$ and, with some limitations, for EDHF. Future studies need to focus not only on the identification of the endothelial cell K$^+$ channel currents initiated by chemical versus mechanical activation. Measurements of production of endothelial cell-derived vasoactive factors, such as NO, PGI$_2$, and EDHF, by individual cells are also required. There is much work that needs to be accomplished to identify the role of endothelial cell K$^+$ channels in both the regulation of endothelial cell E_m and the synthesis and release of endothelial vasoactive factors. Are the same K$^+$ channels involved in the synthesis and release of NO as are involved in the PGI$_2$ or EDHF pathways? How does NO or PGI$_2$ inhibit the synthesis and release of EDHF? Furthermore, is there a cell colocalization, or compartmentalization, of the receptors for endothelium-dependent vasodilators, endothelial cell ion channels, gap junction proteins, and the synthesis of the endothelial cell-derived vasodilator substances (see Hutcheson *et al.*, 1999), and how do these "compartments" interact and interrelate with respect to the synthesis of NO, PGI$_2$, and EDHF?

ACKNOWLEDGMENTS

The critical input from my colleagues, Drs. William C. Cole and Rodger Loutzenhiser, in the Smooth Muscle Research Group at the University of Calgary is gratefully acknowledged as is the valuable assistance of Mrs. Elizabeth Groves in preparing the manuscript.

REFERENCES

Adams, D. J., 1994, Ionic channels in vascular endothelial cells, *Trends Cardiovasc. Med.* **4**:18–26.
Adams, D. J., Barakeh, J., Laskey, R., and van Breemen, C., 1989, Ion channels and regulation of intracellular calcium in vascular endothelial cells, *FASEB J.* **3**:2389–2400.

Adeagbo, A. S. O., and Triggle, C. R., 1991, Effects of some inorganic divalent cations and protein kinase C inhibitors on endothelium-dependent vasorelaxation in rat isolated aorta and mesenteric arteries, *J. Cardiovasc. Pharmacol.* **18:**511–521.

Adeagbo, A. S., Triggle, C. R., 1993, Varying extracellular [K$^+$]: a functional approach to separating EDHF- and EDNA-related mechanisms in perfused rat mesenteric arterial bed, *J. Cardiovasc. Pharmacol.* **21**(3): 423–429.

Ayajiki, K., Kindermann, M., Hecker, M., Fleming, I., and Busse, R., 1996, Intracellular pH and tyrosine phosphorylation but not calcium determine shear stress-induced nitric oxide production in native endothelial cells, *Circ. Res.* **78:**750–758.

Bauersachs, J., Popp, R., Hecker, M., Sauer, E., Fleming, I., and Busse, R., 1996, Nitric oxide attenuates the release of endothelium-derived hyperpolarizing factor, *Circulation* **94:**3341–3346.

Beny, J. L., and Pacicca, C., 1994, Bidirectional electrical communication between smooth muscle and endothelial cells in the pig coronary artery, *Am. J. Physiol.* **260:**H1454–1472.

Bkaily, G., D'Orléans-Juste, P., Naik, R., Perodin, J., Stankova, J., Abdulnour, E., and Rola-Pleszczynski, M., 1993, PAI activation of a voltage-gated R-type Ca^{2+} channels in human and canine aortic endothelial cells, *Br. J. Pharmacol.* **110:**519–520.

Bolotina, V. M., Najibi, S., Palacino, J. J., Pagano, P. J., and Cohen, R. A., 1994, Nitric oxide directly activates calcium-dependent potassium channels in vascular smooth muscles, *Nature* **368:**850–853.

Brayden, J. E., 1990, Membrane hyperpolarization is a mechanism of endothelium-dependent cerebral vasodilation, *Am. J. Physiol.* **259:**H668–H673.

Bredt, D. S., and Snyder, S. H., 1990, Isolation of nitric oxide synthase, a calmodulin-requiring enzyme, *Proc. Natl. Acad. Sci. U.S.A.* **87:**682–685.

Bregestovski, P., Bakhramov, A., Danilov, S., Moldobaeva, A., and Takeda, K., 1988, Histamine-induced inward currents in cultured endothelial cells from human umbilical vein, *Br. J. Pharmacol.* **95:**429–436.

Busse, R., and Fleming, I., 1998, Pulsatile stretch and shear stress: Physical stimuli determining the production of endothelium-derived relaxing factors, *J. Vasc. Res.* **35:**73–84.

Busse, R., Eichtner, H., Lucknoff, A., and Kohlhardt, M., 1988, Hyperpolarization and increased free calcium in acetylcholine-stimulated endothelial cells, *Am. J. Physiol.* **255:**H965–H969.

Cai, S., Garneau, L., and Sauve, R., 1998, Single-channel characterization of the pharmacological properties of the K (Ca^{2+}) channel of intermediate conductance in bovine aortic endothelial cells, *J. Membr. Biol.* **163:**147–158.

Carter, T. D., Hallam, T. J., Cussack, N. J., and Pearson, J. D., 1988, Regulation of P$_{2Y}$-purinoceptor-mediated prostacyclin release from human endothelial cells by cytoplasmic calcium concentration, *Br. J. Pharmacol.* **95:**429–436.

Chataigneau, T., Feletou, M., Duhault, J., and Vanhoutte, P. M., 1998a, Epoxyeicosatrienoic acids, potassium channel blockers and endothelium-dependent hyperpolarization in the guinea pig carotid artery, *Br. J. Pharmacol.* **123:**574–580.

Chataigneau, T., Feletou, M., Thollon, C., Villeneuve, N., Vilaine, J-P., Duhault, J., and Vanhoutte, P. M., 1998b, Cannabinoid CB$_1$ receptor and endothelium-dependent hyperpolarization in guinea-pig carotid, rat mesenteric and porcine coronary arteries, *Br. J. Pharmacol.* **123:**968–974.

Chataigneau, T., Feletou, M., Huang, P. L., Fishman, M. C., Duhault, J., and Vanhoutte, P. M., 1999, Acetylcholine-induced relaxation in blood vessels from endothelial nitric oxide synthase knockout mice, *Br. J. Pharmacol.* **126:**219–226.

Chaytor, A. T., Evans, W. H., and Griffith, T. M., 1998, Central role of heterocellular gap junctional communication in endothelium-dependent relaxation of rabbit arteries, *J. Physiol.* (*London*) **508:**561–573.

Chen, G., and Cheung, D. W., 1992, Characterization of acetylcholine-induced membrane hyperpolarization in endothelial cells, *Circ. Res.* **70:**257–263.

Chen, G. and Suzuki, H., 1990, Calcium-dependency of the endothelium-dependent hyperpolarization in smooth muscle cells of the rabbit carotid artery, *J. Physiol.* (*London*) **421:**521–534.

Cheung, D. W., Chen G., MacKay, M. J., and Burnette, E., 1999, Regulation of vascular tone by endothelium-derived hyperpolarizing factor, *Clin. Exp. Pharmacol. Physiol.* **26:**172–175.

Cook, N. S., and Quast, U., 1990, Potassium channel pharmacology in: *Potassium Channels: Structure, Classification, Function and Therapeutic Potential* (N. S. Cook, ed.), Halstead Press, New York, pp. 181–255.

Daut, J., Mehrke, G., Nees, S., and Newman, W. H., 1987, Passive electrical properties and electrogenic sodium transport of cultured guinea-pig coronary endothelial cells, *J. Physiol.* (*London*) **402:**237–254.

Daut, J., Standen, N. B., and Nelson, M. T., 1994, The role of the membrane potential of endothelial and smooth muscle cells in the regulation of coronary blood flow, *J. Cardiovasc. Electrophysiol.* **5**:154–181.

Davies, P. F., 1995, Flow-mediated signal transduction in endothelial cells In: *Flow Dependent Regulation of Vascular Function* (J. A. Bevan, G. Kaley, and G. M. Rubanyi eds.), Oxford University Press, New York, pp. 46–61.

Demirel, E., Rusko, J., Laskey, R. E., Adams, D. J., and Van Breemen, C., 1994, TEA inhibits Ach-induced EDRF release: Endothelial Ca^{2+}-dependent K$^+$ channels contribute to vascular tone, *Am. J. Physiol.* **267**:H1135–H1141.

Dong, H., Waldron, G. J., Galipeau, D., Cole, W. C., and Triggle, C. R., 1997, NO/PGI$_2$-independent vasorelaxation and the cytochrome P-450 pathway in rabbit carotid artery, *Br. J. Pharmacol.* **120**:695–701.

Dong, H., Waldron, G. J., Cole, W. C., and Triggle, C. R., 1998, Roles of calcium-activated and voltage-gated rectifier potassium channels in endothelium-dependent vasorelaxation of the rabbit middle cerebral artery, *Br. J. Pharmacol.* **123**:821–832.

Doughty, J. M., Plane, F., and Langton, P. D., 1999, Charybdotoxin and apamin block EDHF in rat mesenteric artery if selectively applied to the endothelium, *Am. J. Physiol.* **276**:H1107–H1112.

Edwards, F. R., and Hirst, G. D. S., 1988, Inward rectification in sub-mucosal arterioles of guinea-pig ileum, *J. Physiol. (London)* **404**:437–454.

Edwards, F. R., Hirst, G. D. S., and Silverberg, G. D., 1988, Inward rectification of rat cerebral arterioles: Involvement of potassium ions in autoregulation, *J. Physiol. (London)* **404**:455–566.

Edwards, G., Dora, K. A., Gardener, M. J., Garland, C. J., and Weston, A. H., 1998, K$^+$ is an endothelium-derived hyperpolarizing factor in rat arteries, *Nature* **296**:269–272.

Faraci, F. M. and Heistad, D. D., 1998, Regulation of the cerebral circulation: Role of endothelium and potassium channels, *Physiol. Rev.* **78**:53–97.

Félétou, M., and Vanhoutte, P. M., 1988, Endothelium-dependent hyperpolarization of canine coronary smooth muscle, *Br. J. Pharmacol.* **93**:515–524.

Fransen, P., Katnik, C., and Adams, D. J., 1998, ACh- and caffeine-induced Ca^{2+} mobilization and current activation in rabbit arterial endothelial cells, *Am. J. Physiol.* **275**:H1748–H1758.

Fukao, M., Hattori, Y., Kanno, M., Sakuma, I., and Kitabatake, A., 1997, Sources of Ca^{2+} in relation to generation of acetylcholine-induced endothelium-dependent hyperpolarization in rat mesenteric artery, *Br. J. Pharmacol.* **120**:1328–1334.

Fulton, D., McGiff, J. C., and Quilley, J., 1994, Role of K$^+$ channels in the vasodilator response to bradykinin in the rat heart, *Br. J. Pharmacol.* **113**:954–958.

Furchgott, R. F., and Zawadski, J. V., 1980, The obligatory role of endothelial cells in the relaxation of arterial smooth muscle by acetylcholine, *Nature* **288**:373–375.

Galvez, A., Gimenez-Gallego, G., Reuben, J. P., Roy-Contancin, L. Feigenbaum, P., Kaczorowski, G. J., and Garcia, M. L., 1990, Purification and characterization of a unique, potent, peptidyl probe for the high conductance calcium-activated potassium channel from the venom of the scorpion *Buthus tamulus*, *J. Biol. Chem.* **265**:11083–11090.

Garland, C. J., Plane, F., Kemp, B. J. K., and Cocks, T. K., 1996, Endothelium-dependent hyperpolarization: A role in the control of vascular tone, *Trends Pharmacol. Sci.* **16**:23–30.

Gravier, W. F., Simecek, S., and Sturek, M., 1995, Cytochrome P450 mono-oxygenase-regulated signalling of endothelial Ca^{2+} entry, *J. Physiol. (London)* **483**:259–274.

Groschner, K., Gravier, W. F., and Kuskovetz, W.R., 1992, Activation of a small conductance Ca^{2+}-dependent K$^+$ channel contributes to bradykinin-induced stimulation of nitric oxide synthesis in pig aortic endothelial cells, *Biochim. Biophys. Acta* **1137**:162–170.

Hallam, T. J., Pearson, J. D., and Needham, L., 1988, Thrombin-stimulated elevation of endothelial cell cytoplasmic-free calcium concentration causes prostacyclin-production, *Biochem. J.* **257**:243–249.

Hammarström, A. K. M., Parkington, H. C., Tare, M., and Coleman, H. A., 1999, Endothelium-dependent hyperpolarization in resting and depolarized mammary and coronary arteries of guinea pigs, *Br. J. Pharmacol.* **126**:421–428.

Harach, H. R., Jasani, B., and Williams, E. D., 1983, Factor VIII as a marker of endothelial cells in follicular carcinoma of the thyroid, *J. Clin. Pathol.* **36**:1050–1054.

Harris, D., Kendall, D. A., and Randall, M. D., 1999, Characterization of cannabinoid receptors coupled to vasorelaxation by endothelium-derived hyperpolarizing factor, *Naunyn-Schmiedeberg's Arch. Pharmacol.* **359**:48–52.

Hashitani, H., and Suzuki, H., 1997, K$^+$ channels which contribute to the acetylcholine-induced hyperpolarization in smooth muscle of the guinea-pig submucosal arteriole, *J. Physiol. (London)* **501**:319–329.

Hecker, M., Bara, A. T., Bauersachs, J., and Busse, R., 1994, Characterization of endothelium-derived hyperpolarizing factor as a cytochrome P-450-derived arachidonic acid metabolite in mammals, *J. Physiol. (London)* **481**:407–414.

Himmel, H. M., Whorton, A. R., and Strauss, H. C., 1993, Intracellular calcium, currents, and stimulus–response coupling in endothelial cells, *Hypertension* **21**:112–127.

Hoebel, B. G., Kostner, G. M., and Gravier, W. F., 1997, Activation of microsomal cytochrome P450 mono-oxygenase by Ca^{2+}-store depletion and its contribution to Ca^{2+} entry in porcine aortic endothelial cells, *Br. J. Pharmacol.* **121**:1579 1588.

Hosoki, E., and Iijima, T., 1994, Chloride-sensitive Ca^{2+} entry by histamine and ATP in human aortic endothelial cells, *Eur. J. Pharmacol.* **266**:213–218.

Hutcheson, I. R., and Griffith, T. M., 1994, Heterogenous population of K$^+$ channels mediate EDRF release to flow but not agonists in rabbit aorta, *Am. J. Physiol.* **266**:H590–H596.

Hutcheson, I. R., and Griffith, T. M., 1996, Mechanotransductions through the endothelial cytoskeleton: Mediation of flow but not agonist-induced EDRF release, *Br. J. Pharmacol.* **118**:720 726.

Hutcheson, I. R., Chaytor, A. T., Evans, W. H., and Griffith, T. M., 1999, Nitric oxide-independent relaxations to acetylcholine and A23187 involve different routes of heterocellular communication. Role of gap junctions and phospholipase A$_2$, *Circ. Res.* **84**:53–63.

Hwa, J. J., Ghibaudi, L., Williams, P., and Chatterjee, M., 1994, Comparison of acetylcholine-dependent relaxation in large and small arteries of rat mesenteric vascular bed, *Am. J. Physiol.* **266**:H952–H958.

Illiano, S., Nagao, T., and Vanhoutte, P. M., 1992, Calmidazolium, a calmodulin inhibitor, inhibits endothelium-dependent relaxations resistant to nitro-L-arginine in the canine coronary artery, *Br. J. Pharmacol.* **107**:387–392.

Janigro, D., West, G. A., Gordon, E. L., and Winn, H. R., 1992, ATP-sensitive potassium channels in rat aorta and brain microvascular endothelial cells, *Am. J. Physiol.* **265**:C812–C821.

Johns, A., Lateyan, T. W., Lodge, N. J., Ryan, U. S., van Breemen, C., and Adams, D. J., 1987, Calcium entry through receptor-operated channels in bovine pulmonary artery endothelial cells, *Tissue Cell* **19**:733–745.

Jones, C. J. H., Kuo, L., Davis, M. J., and Chilian, W. M., 1995, Regulation of coronary blood flow: Coordination of heterogenous control mechanisms in vascular microdomains, *Cardiovasc. Res.* **29**:585–596.

Katnik, C., and Adams, D. J., 1995, An ATP-sensitive potassium conductance in rabbit arterial endothelial cells, *J. Physiol. (London)* **485**:595–606.

Katnik, C., and Adams, D. J., 1997, Characterization of ATP-sensitive potassium channels in freshly dissociated rabbit aortic endothelial cells, *Am. J. Physiol.* **272**:H2507–H2511.

Knot, H. J., Zimmermann, P. A., and Nelson, M. T., 1996, Extracellular K$^+$-induced hyperpolarization and dilations of rat coronary and cerebral arteries involve inward rectifier channels, *J. Physiol. (London)* **492**:419–430.

Kohler, M., Hirschberg, B., Bond, C. T., Kinzie, J. M., Marrrion, N. V., Maylie, J., and Adelman, J. P., 1996, Small conductance Ca^{2+}-activated potassium channels from mammalian brain, *Science* **273**:1709–1714.

Kuchan, M. J., and Frangos, J. A., 1994, Role of calcium and calmodulin in flow-induced nitric oxide production in endothelial cells, *Am. J. Physiol.* **266**:C628–C636.

Larson, D. M., Kam, E. Y., and Sheridan, J. D., 1983, Junctional transfer in cultured vascular endothelium: I Electrical coupling, *J. Membr. Biol.* **74**:103–113.

Laskey, R. E., Adams, D. J., Johns, A., Rubanyi, G. M., and van Breemen, C., 1990, Regulation of [Ca^{2+}]$_i$ in endothelial cells by membrane potential, in: *Endothelium-Derived Relaxing Factors* (G. M. Rubanyi, and P. M. Vanhoute, eds.), Karger, Basel, pp. 128-135.

Malinski, T., and Taha, Z., 1992, Nitric oxide release from a single cell measured in situ by a porphyrinic-based microsensor, *Nature*, **358**:676–678.

Marchenko, S. M., and Sage, S. O., 1994, Mechanism of acetylcholine action on membrane potential of endothelium of intact rat aorta, *Am. J. Physiol.* **266**:H2388–H2395.

Marchenko, S. M., and Sage, S. O., 1996, Calcium-activated potassium channels in the endothelium of intact rat aorta, *J. Physiol. (London)* **492**:53–60.

McCarthy, S. A., Kuzy, I., Gatter, K. C., and Bicknell, R., 1991, Heterogeneity of the endothelial cell and its role in organ preference of tumor metastasis, *Trends, Pharmacol. Sci.* **12**:462–467.

McGiff, J. C., 1991, Cytochrome P-450 metabolism of arachidonic acid, *Annu. Rev. Pharmacol. Toxicol.* **31:**339–369.

Mistry, D. K., and Garland, C. J., 1998, Nitric oxide (NO)-induced activation of large conductance Ca^{2+}-dependent K$^+$ channels (BK$_{Ca}$) in smooth muscle cells isolated from the rat mesenteric artery, *Br. J. Pharmacol.* **124:**1131–1140.

Mombouli, J-V., and Vanhoutte, P. M., 1997, Endothelium-derived hyperpolarizing factor(s): updating the unknown, *Trends, Pharmacol. Sci.* **18:**252–256.

Mombouli, J-V., Schaeffer, G., Holzmann, S., Kostner, G. M., and Graier, W. F., 1999, Anandamide-induced mobilization of cytosolic Ca^{2+} in endothelial cells, *Br. J. Pharmacol.* **126:**1593–1600.

Moncada, S., Gryglewski, R. J., Bunting, S., and Vane, J. R., 1976, An enzyme isolated from arteries transforms prostaglandin endoperoxides to an unstable substance that inhibits platelet aggregation, *Nature* **263:**663–665.

Mukai, K., Rosai, J., and Burgdorf, W. H., 1980, Localization of factor VIII-related antigen in vascular endothelial cells using an immunoperoxidase method, *Am. J. Surg. Pathol.* **4:**273–276.

Murphy, M. E., and Brayden, J. E., 1995, Nitric oxide hyperpolarizes rabbit mesenteric arteries via ATP-sensitive potassium channels, *J. Physiol.* **486:**47–58.

Nagao, T., Illiano, S., and Vanhoutte, P. M., 1992, Calmodulin antagonists inhibit endothelium-dependent hyperpolarization in canine coronary artery, *Br. J. Pharmacol.* **197:**282–286.

Nakache, M., and Gaub, H. E., 1988, Hydrodynamic hyperpolarization of endothelial cells, *Proc. Natl. Acad. Sci. U.S.A.* **85:**1841–1843.

Nelson, M. T., and Quayle, J. M., 1995, Physiological roles and properties of potassium channels in arterial smooth muscle, *Am. J. Physiol.* **268:**C794–C822.

Nilius, B., and Riemann, D., 1990, Ion channels in human endothelial cells, *Gen. Physiol. Biophys.* **9:**89–112.

Nilius, B., Viana, F., and Droogmans, G., 1997, Ion channels in vascular endothelium, *Ann. Rev. Physiol.* **59:**145–170.

Northover, B. J., 1980, The membrane potential of vascular endothelial cells, *Adv. Microcirc.* **9:**135–160.

Ohashi, M., Satoh, K., and Itoh, T., 1999, Acetylcholine-induced membrane potential changes in endothelial cells of rabbit aortic valve, *Br. J. Pharmacol.* **126:**19–26.

Okazaki, K., Endou, M., and Okamura, F., 1998, Involvement of barium-sensitive K$^+$ channels in endothelium-dependent vasodilation produced by hypercapnia in rat mesenteric vascular beds, *Br. J. Pharmacol.* **125:**168–174.

Olesen, S. P., and Bundgaard, M., 1993, ATP-dependent closure and reactivation of inward rectifier K$^+$ channels in endothelial cells, *Circ. Res.* **73:**492–495.

Olesen, S. P., Clapham, D. E., and Davies, P. F., 1988a, Haemodynamic shear stress activates a K$^+$ current in vascular endothelial cells, *Nature* **331:**168–170.

Olesen, S. P., Davies, P. F., and Clapham, D. E., 1988b, Muscarinic-activated K$^+$ current in bovine aortic endothelial cells, *Circ. Res.* **62:**1059–1064.

Ordway, R. W., Walsh, J. V., and Singer, J. J., 1989, Arachidonic acid and other fatty acids directly activate potassium channels in smooth muscle cells, *Science* **244:**1176–1179.

Palmer, R. M. J., Ferrige, A. G., and Moncada, S., 1987, Nitric oxide release accounts for biological activity of endothelium-derived relaxing factor, *Nature* **327:**524–526.

Parekh, A. B., and Penner, R., 1997, Store-depletion and calcium influx, *Physiol. Rev.* **77:**901–930.

Parsaee, H., Ewan, J. R., Joseph, S., and MacDermott, J., 1992, Differential sensitivities of the prostacyclin and nitric oxide biosynthetic pathways to cystolic calcium in bovine aortic endothelial cells, *Br. J. Pharmacol.* **107:**1013–1019.

Plane, F., Pearson, T., and Garland C. J., 1995, Multiple pathways underlying endothelium-dependent relaxation in the rabbit isolated femoral artery, *Br. J. Pharmacol.* **335:**31–38.

Plane, F., Holland, M., Waldron, G. J., Garland, C. J., and Boyle, J. P., 1997, Evidence that anandamide and EDHF act via different mechanisms in rat isolated mesenteric arteries, *Br. J. Pharmacol.* **121:**1509–1511.

Pollock, J. S., Fostermann, U., Mitchell, J. A., Warner, T. D., Schmidt, H. H. H. W., Nakane, M., and Murad, F., 1991, Purification and characterization of particulate endothelium-derived relaxing factor synthase from cultured and native bovine aortic endothelial cells, *Proc. Natl. Acad. Sci. U.S.A.* **88:**10480–10484.

Popp, R., Bauersachs, J., Hecker, M., Fleming, I., and Busse, R., 1996a, A transferable β-naphthoflavone-inducible hyperpolarizing factor is synthesized by native and cultured porcine coronary endothelial cells, *J. Physiol.* **497:**699–709.

Popp, R., Bauersachs, J., Sauer, E., Hecker, M., Fleming, I., and Busse, R., 1996b, The cytochrome P450 monooxygenase pathway and nitric oxide-independent relaxations, in: *Endothelium-Derived Hyperpolarizing Factor* (P. M. Vanhoutte, ed.), Harwood Academic Publishers, Amsterdam, pp. 115–127

Popp, R., Fleming, I., and Busse, R., 1998, Pulsatile stretch in coronary arteries elicits release of endothelium-derived hyperpolarizing factor: A modulator of arterial compliance, *Circ. Res.* **82**:696–703.

Quayle, J. M., McCarrron, J. G., Brayden, J. E., and Nelson, M. T., 1993, Inward rectifier K^+ currents in smooth muscle cells from rat resistance-sized cerebral arteries, *Am. J. Physiol.* **265**:C1363–C1370.

Quayle, J. M., Dart, C., and Standen, N. B., 1996, The properties and distribution of inward rectifier potassium currents in pig coronary arterial smooth muscle, *J. Physiol.* **494**:715–726.

Quayle, J. M., Nelson, M. T., and Standen, N. B., 1997, ATP-sensitive and inwardly rectifying potassium channels in smooth muscle, *Physiol. Rev.* **77**:1165–1232.

Quignard, J-F., Félwtou, M., Thollon, C., Vilaine, J.-P, Duhault, J., and Vanhoutte, P. M., 1999, Potassium ions and endothelium-derived hyperpolarizing factor in guinea pig carotid and porcine coronary arteries, *Br. J. Pharmacol.* **127**:27–34.

Quilley, J., Fulton, D., and McGiff, J. C., 1997, Commentary: Hyperpolarizing factors, *Biochem. Pharmacol.* **54**:1059–1070.

Randall, M. D., Alexander, S. P. H., Bennett, T., Boyd, E. A., Fry, J. R., Gardiner, S.M., Kemp, P. A., McCulloch, A. I., and Kendall, D. A., 1996, An endogenous cannabinoid as an endothelium-derived vasorelaxant, *Biochem. Biophys. Res. Commun.* **229**:114–120.

Robertson, B. E., Bonev, A. D., and Nelson, M. T., 1996, Inward rectifier K^+ currents in smooth muscle cells from rat coronary arteries: Block by Mg^{2+}, Ca^{2+} and Ba^{2+}, *Am. J. Physiol.* **271**:H696–H705.

Rusko, J., Tanzi, F., Van Breemen, C., and Adams, D. J., 1992, Calcium-activated potassium channels in native endothelial cells from rabbit aorta: Conductance, Ca^{2+} sensitivity and block, *J. Physiol.* **455**:601–621.

Segal, S. S., and Duling, B. R., 1986, Flow control among microvessels coordinated by intercellular conduction, *Science* **234**:868–870.

Setoguchi, M., Ohya, Y., Abe, I., and Fujishima, M., 1997, Stretch-activated whole-cell currents in smooth muscle cells from mesenteric resistance artery of guinea pig, *J. Physiol.* **501**:343–353.

Sharma, N. R., and Davis, M. J., 1994, Mechanism of substance P-induced hyperpolarization of porcine coronary artery endothelial cells, *Am. J. Physiol.* **266**:H156–H164.

Shaul, P. W., and Anderson, R. G. W. 1998, Role of plasmalemmal caveolae in signal transduction, *Proc. Nat. Acad. Sci. U.S.A.* **275**:845–851.

Shaul, P. W., Smart, E. J., Robinson, L. J., German, Z., Yuhanna, I. S., Ying, Y., Anderson, R. G., and Michel, T. 1996, Acylation targets endothelial nitric-oxide synthase to plasmalemmal caveolae, *J. Biol. Chem.* **271**:6518–6522.

Takahashi, M., Ishida, T., Traub, O., Corson, M. A., and Berk, B. C., 1997, Mechanotransduction in endothelial cells: Temporal signalling events in response to shear stress, *J. Vasc. Res.* **34**:212–219.

Takeda, K., Schini, V., and Stoeckel, H., 1987, Voltage-activated potassium, but not calcium currents, in cultured bovine aortic endothelial cells, *Pflügers Arch.* **410**:385–393.

Taylor, H. J., Chaytor, A. T., Evans, W. H., and Griffith, T. M., 1998, Inhibition of the gap junctional component of endothlium-dependent relaxations in rabbit iliac artery by 18α-glycyrrhetinic acid, *Br. J. Pharmacol.* **125**:1–3.

Thorin, E., Huang, P. L., Fishman, M. C., and Bevan, J. A., 1998, Nitric oxide inhibits α_2-adrenoceptor-mediated endothelium-dependent vasodilation, *Circ. Res.* **82**:1323–1329.

Triggle, C. R., Ding, H., Lovren, F., Kubes, P., and Waldron, G. J., 1998, Endothelium-dependent vascular relaxation in eNOS knockout mice, *Pharmacol. Toxicol.* **83**(Suppl.1):99.

Triggle, C. R., Dong, H., Waldron, G. J., and Cole, W. C., 1999, Endothelium-derived hyperpolarizing factor(s): Species and tissue heterogeneity, *Clin. Exp. Pharmacol. Physiol.* **26**:176–179.

Vaca, L., 1996, Calmodulin inhibits calcium influx current in vascular endothelium, *FEBS Lett.* **300**:289–293.

Vaca, L., and Kunze, D. L., 1993, Depletion and refilling of intracellular Ca^{2+} stores induces oscillations of Ca^{2+} current, *Am. J. Physiol.* **267**:C920–C925.

Vaca, L., and Kunze, D. L., 1994, Depletion of intracellular Ca^{2+} stores activates a Ca^{2+} selective channel in vascular endothelium, *Am. J. Physiol.* **267**:C733–C738.

Vaca, L., and Kunze, D. L., 1995, IP_3 activated Ca^{2+} channels in the plasma membrane of cultured vascular endothelial cells, *Am. J. Physiol.* **269**:C733–C738.

Vanheel, B., and van de Voorde, J., 1997, Evidence against the involvement of cytochrome P450 metabolites in endothelium-dependent hyperpolarization of the rat main mesenteric artery, *J. Physiol. (London)* **501**:331–341.

Vanhoutte, P. M., 1988, Vascular endothelium and Ca^{2+} antagonists, *J. Cardiovasc. Pharmacol.* **12**(Suppl. 6):521–528.

Vanhoutte, P. M., 1996, *Endothelium-Derived Hyperpolarizing Factor*, Harwood Academic Publishers, Amsterdam.

Vanhoutte, P. M., and Félétou, M. 1996, Conclusion: Existence of multiple endothelium-derived hyperpolarizing factor, in: *Endothelium-Derived Hyperpolarizing Factor* (P. M. Vanhoutte, ed.), Harwood Academic Publishers, Amsterdam, pp. ????

Vanhoutte, P. M., Félétou, M., Boulanger, C.M., Hoffner, U. and Rubanyi, G.M., 1996, Existence of multiple endothelium-derived relaxing factors, in: *Endothelium-Derived Hyperpolarizing Factor* (P. M. Vanhoutte, ed.), Harwood Academic Publishers, Amsterdam, pp. 88–111.

von der Weid, P-Y. and Beny, J.L., 1992, Effect of Ca^{2+} ionophores on membrane potential on pig coronary artery endothelial cells, *Am. J. Physiol.* **262**:H1823–1831.

Waldron, G. J., and Garland, C. J., 1994, Effect of potassium channel blockers on L-NAME-insensitive relaxations in rat small mesenteric artery, *Can. J. Physiol. Pharmacol.* **72** (Suppl. 1):26.

Waldron, G. J., Dong, H., Cole, W. C., and Triggle, C. R., 1996, Endothelium-dependent hyperpolarization of vascular smooth muscle: role for a non-nitric oxide synthase product, *Acta. Pharma. Sin.* **17**:3–7.

Weidelt, T., Boldt, W., and Markward, F., 1997, Acetylcholine-induced K$^+$ currents in smooth muscle cells of intact rat small arteries, *J. Physiol. (London)* **500**:617–630.

Wellman, G. C., Quayle, J. M., and Standen, N. B., 1996, Evidence against the association of the sulphonylurea receptor with endogenous Kir family members other than K$_{ATP}$ in coronary vascular smooth muscle, *Pflügers Arch.* **432**:355–357.

White, R., and Hiley, C. R., 1997, A comparison of EDHF-mediated and anandamide-induced relaxations in the rat isolated mesenteric artery, *Br. J. Pharmacol.* **122**:1573–1584.

White, R., and Hiley, C. R., 1998, Effects of K$^+$ channel openers on relaxations to nitric oxide and endothelium-derived hyperpolarizing factor in rat mesenteric artery, *Eur. J. Pharmacol.* **357**:41–51.

Woodley, N., and Barclay, J. K., 1994, Cultured endothelial cells from distinct vacular areas show differential responses to agonists, *Can. J. Physiol. Pharmacol.* **73**:1007–1012.

Yajima, K., Nishiyama, M., Yamamoto, Y., and Suzuki, H., 1999, Inhibition of endothelium-dependent hyperpolarization by endothelial prostanoids in guinea-pig coronary artery, *Br. J. Pharmacol.* **126**:1–10.

Yanagisawa, M., Inoue, A., Takuwa, Y., Mitsui, Y., Kobayashi, M., and Masaki, T., 1989, The human preproendothelin-1 gene: Possible regulation by endothelial phosphoinositide turnover signaling, *J. Cardiovasc. Pharmacol.* **13** (Suppl. 5):S13–17.

Zygmunt, P. M., Edwards, G., Weston, A. H., Larsson, B., and Hoegestatt, E. D., 1997, Involvement of voltage-dependent potassium channels in the EDHF-mediated relaxation of rat hepatic artery, *Br. J. Pharmacol.* **121**:141–149.

Zygmunt, P. M., Plane, F., Paulsson, M., Garland, C. J., and Högestatt, E. D., 1998, Interactions between endothelium-derived relaxing factors in the rat hepatic artery: Focus on regulation of EDHF, *Br. J. Pharmacol.* **124**:992–1000.

Chapter 34

Activation of Vascular Smooth Muscle K$^+$ Channels by Endothelium-Derived Factors

Michel Félétou and Paul M. Vanhoutte

1. ENDOTHELIUM-DERIVED VASODILATORS

Endothelial cells synthesize and release vasoactive mediators in response to various neurohumoral substances (e.g., acetylcholine, ATP, bradykinin, thrombin) and physical stimuli (e.g., shear stress exerted by the flowing blood) (Furchgott and Vanhoutte, 1989). Nitric oxide (NO) produced by the L-arginine–NO synthase pathway and prostacyclin produced from arachidonic acid by cyclooxygenase have been identified as potent endothelium-derived vasodilators (Moncada *et al.*, 1976; Moncada and Vane, 1979; Furchgott and Zawadzki, 1980; Palmer *et al.*, 1987, 1988). However, not all endothelium-dependent relaxations can be fully explained by the release of either NO or prostacyclin. Indeed, another unidentified substance(s) which hyperpolarizes the underlying vascular smooth muscle cells, termed endothelium-derived hyperpolarizing factor (EDHF), may contribute to endothelium-dependent relaxations (Furchgott and Vanhoutte, 1989; Komori and Vanhoutte, 1990; Félétou and Vanhoutte, 1996a; Mombouli and Vanhoutte, 1997).

1.1. Existence of a Third Pathway

De Mey *et al.* (1982) proposed the existence of a third pathway because, in canine femoral arteries, the endothelium-dependent relaxations to acetylcholine obtained in the presence of indomethacin were abolished by mepacrine whereas the responses to thrombin and ATP were unaffected. Bolton *et al.* (1984) showed that endothelium-dependent relaxations are accompanied by hyperpolarization of the cell membrane of

Michel Félétou ● Département de Diabétologie, Institut de Recherches Servier, 92150 Suresnes, France. *Paul M. Vanhoutte* ● Institut de Recherches Internationales Servier, 92410 Courbevoie, France.

Potassium Channels in Cardiovascular Biology, edited by Archer and Rusch. Kluwer Academic/Plenum Publishers, New York, 2001.

vascular smooth muscle cells. This initial observation has been confirmed in various blood vessels from different species (Félétou and Vanhoutte, 1985, 1988; Komori and Suzuki, 1987; Taylor et al., 1988; Chen et al., 1988), including humans (Nakashima et al., 1993; Petersson et al., 1995). When L-arginine analogs became available as specific inhibitors of the production of NO (Rees et al., 1989), it became obvious that endothelium-dependent relaxations and endothelium-dependent hyperpolarizations could be more or less resistant to inhibitors of cyclooxygenase and NO synthase (Fig. 1; Bény and Brunet, 1988; Richard et al., 1990; Cowan and Cohen, 1991; Mugge et al., 1991; Hasunuma et al., 1991; Illiano et al., 1992; Nagao et al., 1992a; Nagao and Vanhoutte, 1992a; Pacicca et al., 1992; Suzuki et al., 1992). Furthermore, endothelium-dependent responses, which are resistant to inhibitors of NO synthase and cyclo-oxygenase, are observed without an increase in intracellular levels of cyclic nucleotides (cGMP and cAMP) in the smooth muscle cells (Taylor et al., 1988; Cowan and Cohen, 1991; Mombouli et al., 1992; Zygmunt et al., 1994a; Garcia-Pascual et al., 1995, Hayabuchi et al., 1998).

Endothelium-dependent hyperpolarizations resistant to inhibitors of NO synthase and cyclooxygenase inhibition can be attributed to the release of a diffusible substance. Indeed, its existence has been demonstrated, using conventional intracellular microelectrode or patch-clamp techniques, under bioassay conditions whereby the source of EDHF was either native vascular segments or cultured endothelial cells (Félétou and Vanhoutte, 1988; Kauser et al., 1989; Chen et al., 1991; Plane et al., 1995; Mombouli et al., 1996; Popp et al., 1996; Harder et al., 1996; Fukuta et al., 1996; Hayabuchi et al., 1998). The technical difficulties in demonstrating the diffusible nature of EDHF can be explained by a very short half-life of the substance, its preferential abluminal release (Kauser and Rubanyi, 1992), the simultaneous release of a hypothetical endothelium-derived depolarizing factor (Mombouli et al., 1996; Corriu et al., 1996a), or a combination of these.

Endothelium-dependent hyperpolarization could also involve electrical coupling through myoendothelial junctions (Busse et al., 1988; Davies et al., 1988). Indeed, substances which produce endothelium-dependent hyperpolarization of vascular smooth muscle cells also hyperpolarize endothelial cells, with the same time course (Bény, 1990). Gap junctions couple smooth muscle and endothelial cells, and depolarization or hyperpolarization can be conducted from smooth muscle to endothelial cells (Marchenko and Sage, 1994; Bény and Chabaud, 1996). However, in most blood vessels, electrical propagation from endothelial to smooth muscle cells does not seem to occur (Bény, 1990; Bény and Chabaud, 1996). Conflicting results have been obtained with the nonspecific gap-junction uncoupler heptanol. It inhibits EDHF responses in porcine coronary artery (Kühberger et al., 1994) but does not affect these responses in rat hepatic artery (Zygmunt and Högestätt, 1996). More specific blockers of gap junctions, 18β-glycyrrhetinic acid and GAP-27 (a peptide which possesses a conserved sequence homology with a portion of connexin), inhibit EDHF-like responses in rabbit arteries (Chaytor et al., 1998; Yamamoto et al., 1998; Taylor et al., 1998). At present, this mechanism needs to be further explored to better understand its potential contribution to endothelium-dependent hyperpolarizations.

Endothelium-dependent hyperpolarizations require an increase in endothelial intracellular calcium and the subsequent activation of calmodulin (Chen and Suzuki, 1990; Nagao et al., 1992b; Illiano et al., 1992). The calcium and calmodulin depend-

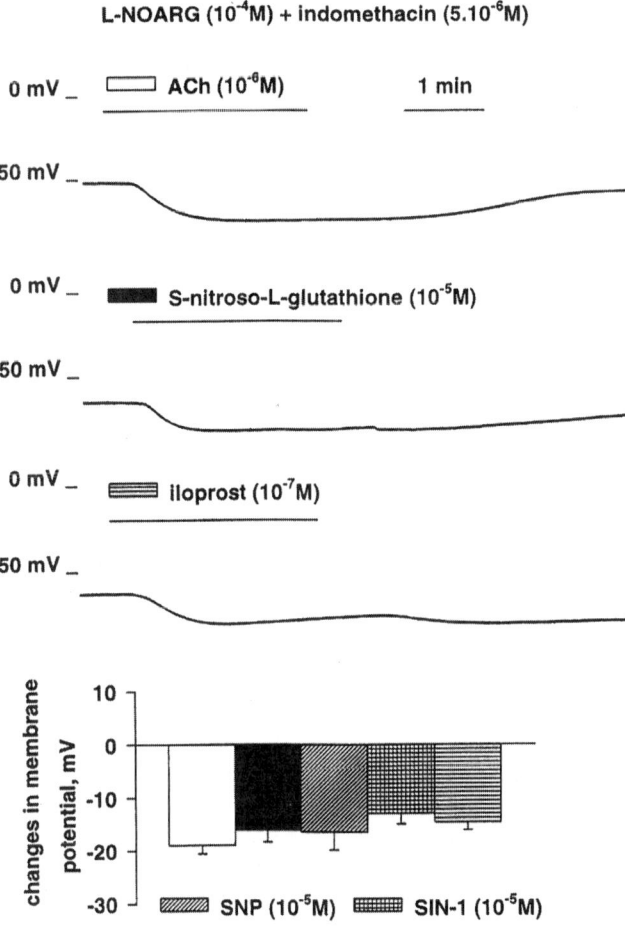

Figure 1. Hyperpolarizations produced by acetylcholine (ACh; 1 μM), by three nitrovasodilators—S-nitroso-L-glutathione, sodium nitroprusside (SNP), and SIN-1 (10 μM)—and by a stable analog of prostacyclin iloprost (0.1 μM), in guinea pig isolated carotid arteries. The hyperpolarization produced by acetylcholine is endothelium-dependent whereas the hyperpolarizations in response to nitrovasodilators or to iloprost can be observed in blood vessels without endothelium. An inhibitor of cyclooxygenase, indomethacin (5 μM) and an inhibitor of NO synthase, N^G-nitro-L-arginine (L-NOARG; 100 μM), were present throughout, suggesting that the hyperpolarization produced by acetylcholine is due to EDHF. The bar graph at the bottom of the figure shows the means ± SEM of the hyperpolarizing responses. These microelectrode experiments show that, in the guinea pig carotid artery, NO, prostacyclin, and EDHF are endothelium-derived factors that can provoke smooth muscle cell hyperpolarization. (Modified from Corriu et al. 1996a, with permission.)

encies of responses mediated by EDHF are similar to those observed with endothelial NO-dependent relaxations, although the former appear to be more sensitive to calmodulin blockers than the latter (Bredt and Snyder, 1990; Illiano et al., 1992).

To date, EDHF has not been identified with certainty, and multiple EDHFs may exist depending on the vascular bed and the species studied (Vanhoutte and Félétou, 1996).

1.2. Other Vasoactive Substances Released by Endothelial Cells

Endothelial cells can release various other vasoactive substances such as adenine nucleotides (AMP, ADP, and ATP) and adenosine (Shryock et al., 1988; Shinozuka et al., 1994), metabolites of arachidonic acid through the cytochrome P-450 or the lipoxygenase pathways, including epoxyeicosatrienoic acids (Campbell et al., 1996; Quilley et al., 1997; Hayabuchi et al., 1998), trihydroxyeicosatrienoic acids (Pfister et al., 1999), and 12-hydroxyeicosatetraenoic acid (De Mey et al., 1982; Weintraub et al., 1999), as well as anandamide, presumed to be the endogenous ligand for the cannabinoid CB_1 receptor (Devane et al., 1992; Di Marzo et al., 1994; Randall et al., 1996), and small molecules such as carbon monoxide (Coceani and Kelsey, 1997), hydroxyl radicals (Rosenblum, 1987; Prasad and Bharadwaj, 1996), hydrogen peroxide (Bény and von der Weid, 1991), and K^+ ions (Edwards et al., 1998). All these molecules are putative EDHFs as they are produced by the endothelial cells and induce hyperpolarization of the smooth muscle cells. However, in most of the tissues studied, none of these mediators can, alone, fully account for the observed endothelium-dependent hyperpolarizations.

2. ENDOTHELIUM-DERIVED MEDIATORS AND VASCULAR SMOOTH MUSCLE K^+ CHANNELS

2.1. Prostacyclin

Prostacyclin is the principal metabolite of arachidonic acid produced by cyclooxygenase in most blood vessels (Moncada et al., 1976), the endothelium being the major site of its synthesis (Moncada et al., 1977). Prostacyclin is generally described as an endothelium-dependent vasodilator but can also evoke contractile responses in vascular beds such as the rabbit aorta (Borda et al., 1983) or the human coronary and umbilical arteries (Pomerantz et al., 1978; Davis et al., 1980). These contractions are linked either to the activation of thromboxane receptors (Zhao and Wang, 1996) or to the release of endothelium-derived contracting factors (Adeagbo and Malik, 1990).

Relaxations caused by prostacyclin involve the stimulation of specific cell-surface receptors (IP receptors) and activation of adenylate cyclase, leading to an elevation of intracellular cAMP. Prostacyclin or its stable analogs (iloprost or cicaprost) hyperpolarize vascular smooth muscle cells from various species such as the rat, guinea pig, rabbit, sheep, and dog. In most of the blood vessels, these hyperpolarizations involve the opening of ATP-sensitive K^+ channels (K_{ATP}) and are blocked by sulfonylureas, such as glibenclamide (Figs. 1 and 2; Siegel et al., 1987; Jackson et al., 1993; Parkington et al., 1993, 1995; Murphy and Brayden 1995a; Corriu et al., 1996a). It is not yet completely clear whether the opening of the K_{ATP} channel is dependent on intracellular cAMP accumulation or on the activation of a G-protein (Wise and Jones, 1996). Other activators of adenylate cyclase, such as β-adrenergic agonists, hyperpolarize vascular smooth muscle cells by opening K_{ATP} channels (Nakashima and Vanhoutte 1995), and permeant analogs of cAMP also produce hyperpolarization (Parkington et al., 1993),

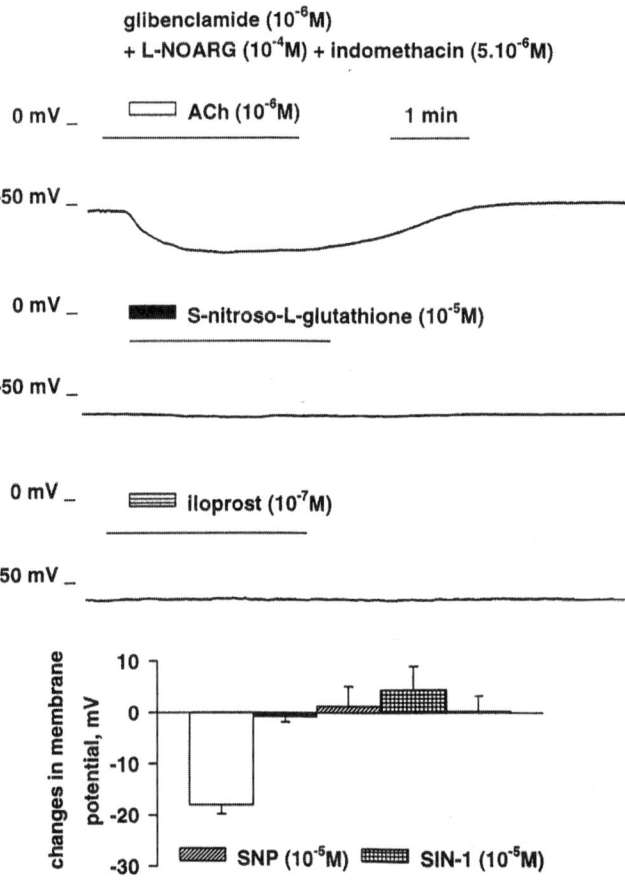

Figure 2. Effect of glibenclamide (1 μM), a K_{ATP} channel inhibitor, on the hyperpolarizations produced by acetylcholine (ACh; 1 μM), by three nitrovasodilators—S-nitroso-L-glutathione, sodium nitroprusside (SNP), and SIN-1 (10 μM)—and by a stable analog of prostacyclin, iloprost (0.1 μM), in guinea pig isolated carotid arteries. An inhibitor of cyclooxygenase, indomethacin (5 μM) and an inhibitor of NO synthase N^G-nitro-L-arginine (L-NOARG; 100 μM), were present throughout. Glibenclamide abolishes the hyperpolarizations to NO donors and to iloprost without affecting the endothelium-dependent hyperpolarization to acetylcholine. The bar graph at the bottom of the figure shows the means $\pm$ SEM of the responses. These microelectrode experiments indicate that, in the guinea pig carotid artery, the endothelium-dependent hyperpolarization produced by acetylcholine does not involve NO and prostacyclin release. (Modified from Corriu *et al.* 1996a, with permission.)

suggesting that cyclic-nucleotide-dependent protein kinases are involved in the hyperpolarization produced by prostacyclin. However, in the tail artery of the rat, prostacyclin and iloprost activate not only K_{ATP} but also large-conductance, Ca^{2+}-activated K^+ (BK_{Ca}) channels by a mechanism involving protein kinase A-dependent channel phosphorylation (Schubert *et al.*, 1996, 1997). In contrast, relaxations in the guinea pig aorta produced by iloprost and cicaprost involve BK_{Ca}, but not K_{ATP}, channels (Clapp *et al.*, 1998). Similarly, in the isolated smooth muscle of the rat portal vein, iloprost opens a K^+ conductance which possesses the characteristics of BK_{Ca} current (Siegel *et al.*, 1992). In patch-clamp studies on the isolated smooth muscle cells of the bovine coronary artery, prostacyclin opens a 4-aminopyridine-sensitive, "delayed rectifier" or voltage-gated K^+ (Kv) channel without affecting BK_{Ca} channels (Li et al., 1997). In the rat hepatic artery, the K^+ channel involved in the relaxation and hyperpolarization to iloprost has not been identified but is neither of K_{Ca} nor K_{ATP} channel origin (Zygmunt et al., 1998). In contrast, in the isolated coronary artery of the rat, prostacyclin and iloprost do not provoke hyperpolarization (Parkington et al., 1993).

Characterization of the K^+ conductances purely by pharmacological means may be misleading. For example, sulfonylurea inhibitors of K_{ATP} channels, such as glibenclamide and, to a lesser extent, tolbutamide, inhibit thromboxane receptors (Cocks et al., 1990; Zhang et al., 1991). It is not known whether sulfonylureas interact with the IP receptor. The inhibitor of Kv channels 4-aminopyridine (Nelson and Quayle, 1995) also inhibits cromakalim-activated K^+ channels (Beech and Bolton, 1989). In the guinea pig coronary artery, the identity of the K^+ channel involved in the slow hyperpolarization produced by endogenous prostaglandin is unresolved because the hyperpolarization is not blocked by either glibenclamide or 4-aminopyridine (Nishiyama et al., 1998).

The hyperpolarization produced by prostacyclin, and possibly the subtype of K^+ channel which is activated, may depend on the state of the vascular smooth muscle cells. In isolated guinea pig coronary arteries, the ability of endogenous or exogenous prostacyclin (or NO) to hyperpolarize the membrane depends upon the stretch exerted on the smooth muscle. The amplitude of these responses reaches a maximum when the tissues are stretched to the equivalent of 50 mm Hg, whereas no hyperpolarization is observed in unstretched preparations, although the resting membrane potential (E_m) of the vascular smooth muscle cell is not affected by the stretching process (Parkington et al., 1993). Furthermore, the ability of prostacyclin (or NO) to produce hyperpolarization of smooth muscle cells depends on the E_m of the latter. In blood vessels such as the uterine artery of the guinea pig, NO and prostacyclin do not provoke hyperpolarization unless the tissue is depolarized by an agonist (Tare et al., 1990; Parkington et al., 1993, 1996). The contribution of the hyperpolarization in the relaxation to prostacyclin can be very significant in some tissues (rabbit coronary artery: Jackson et al., 1993; rat tail artery: Schubert et al., 1997) whereas in others, such as the guinea pig coronary artery, the blockade of the hyperpolarization does not affect the relaxation produced by iloprost (Parkington et al., 1995). Finally, interactions between NO, prostacyclin, and K^+ channels can confuse the issue. Opening of K^+ channels can stimulate prostaglandin production (Siegel et al., 1990) in the porcine retinal and choroidal microcirculation; NO can release prostacyclin by opening endothelial BK_{Ca} channels (Hardy et al., 1998); and, in some vascular beds, such as the porcine coronary artery, prostacyclin releases NO from the endothelial cells (Shimokawa et al., 1988).

2.2. Nitric Oxide

The principal physiological action of NO is associated with the activation of cytosolic soluble guanylate cyclase and the consequent formation of cGMP (Moncada et al., 1991), but endothelial NO has many other targets on smooth muscle cells, including K^+ channels (Cohen and Vanhoutte, 1995).

Hyperpolarization of vascular smooth muscle in response to sodium nitroprusside or nitroglycerin in rabbit pulmonary and portal vein was reported early (Haeusler and Thorens, 1976; Ito *et al.*, 1978). This hyperpolarizing effect of nitrovasodilators or of authentic NO was confirmed in other vascular beds from various species such as mesenteric, coronary, and carotid arteries of the guinea pig, as well as in the aorta, tail, and mesenteric artery of the rat (Fig. 1; Tare *et al.*, 1990; Parkington *et al.*, 1993; Corriu *et al.*, 1996a; Garland and Plane, 1996). In the mesenteric artery of the rat, NO no longer produces hyperpolarization when cells have been contracted and depolarized (Garland and Plane, 1996). Conversely, in carotid and femoral arteries of the rabbit, uterine arteries of the guinea pig, or mesenteric arteries of the dog, NO and nitrovasodilators do not produce hyperpolarization in resting tissue but repolarize smooth muscle cells previously depolarized by an agonist (Tare *et al.*, 1990; Plane *et al.*, 1995; Cohen *et al.*, 1997). In some tissues such as the mesenteric artery of the rabbit, hyperpolarization in response to nitrovasodilators, requires that endogenous production of NO is suppressed either by endothelium removal or by a NO synthase inhibitor (Murphy and Brayden, 1995b). Finally, in some blood vessels such as the canine and porcine coronary arteries, the hepatic artery and the portal vein of the rat, and the basilar artery of the rabbit, NO and nitrovasodilators do not influence the E_m (Ito *et al.*, 1980a,b; Komori *et al.*, 1988; Félétou *et al.*, 1989, Félétou and Vanhoutte, 1996b; Brayden and Murphy, 1996; Garland and Plane, 1996; Zygmunt *et al.*, 1998).

In experiments on vascular smooth muscle involving the measurement of E_m with an intracellular microelectrode, the hyperpolarization produced by NO or NO donors in the coronary, and carotid arteries and mesenteric lymphatic vessels of the guinea pig (Parkington *et al.*, 1993; Corriu *et al.*, 1996a; von der Weid, 1998), in the mesenteric artery of the rat (Garland and Plane, 1996), and in the mesenteric and the femoral artery of the rabbit (Murphy and Brayden 1995b; Plane *et al.*, 1995) is sensitive to glibenclamide, implicating K_{ATP} channels (Fig. 2).

In electrophysiological experiments involving different configurations of the patch-clamp technique, the K$^+$ channels activated by nitrovasodilators [exogenous authentic NO or NO produced by either the constitutive endothelial cell nitric oxide synthase (eNOS) or the inducible iNOS] have been further characterized. Activation of K_{ATP} channels by NO has been demonstrated in isolated smooth muscle cells of the porcine coronary artery (Miyoshi *et al.*, 1994) and in the guinea pig carotid artery (Fig. 3; Quignard *et al.*, 1999a). In the porcine coronary artery, NO activates BK_{Ca} channels (Miyoshi and Nakaya, 1994) whereas in the carotid artery of the guinea pig, the nitrovosodilator SIN-1 or a permeant analog of cGMP, dibutyryl cGMP, does not significantly modify the open-state probability or the mean open time of BK_{Ca} channels (Fig. 4; Quignard *et al.*, 1999a). However, NO activates BK_{Ca} channels in isolated smooth muscle cells of cerebral and carotid arteries and in the aorta of the rabbit (Fig. 4; Robertson *et al.*, 1993; Bolotina *et al.*, 1994; Quignard *et al.*, 1999a), of pulmonary, coronary, mesenteric, and cerebral arteries of the rat (Archer *et al.*, 1994; Wellman *et al.*, 1996; Carrier *et al.*, 1997; Mistry and Garland, 1998; Hoang and Mathers, 1998), and of human pulmonary and coronary arteries (Peng *et al.*, 1996; Bychkov *et al.*, 1998). In the canine coronary artery (Taniguchi *et al.*, 1993), BK_{Ca} channels are stimulated by cGMP-dependent protein kinase. In bovine coronary artery, NO activates both BK_{Ca} and Kv channels (Li *et al.*, 1997). In most of the tissues mentioned above, the activation of BK_{Ca} channels was dependent upon cGMP-dependent protein kinase. However, NO

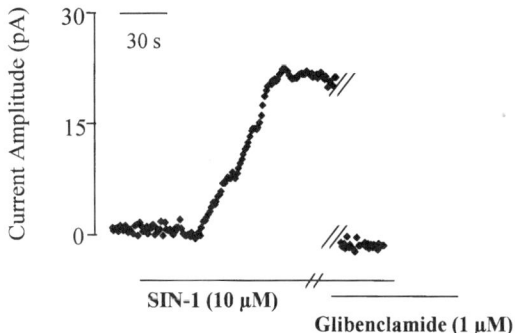

Figure 3. Effects of SIN-1 (10 μM) on K_{ATP} current amplitude in freshly isolated vascular smooth muscle cells of the guinea pig carotid artery. SIN-1 induced an increase in whole-cell K^+ current (intracellular solution with a low ATP concentration; holding potential = 0 mV). The effect of SIN-1 was abolished by the addition of glibenclamide (1 μM). These patch-clamp experiments confirm previous microelectrode experiments showing that, in the smooth muscle cells of the guinea pig carotid artery, NO activates K_{ATP} channels. (Modified from Quignard *et al.* 1999a. Copyright OPA N.V. Permission granted by Gordon and Breach Publishers.)

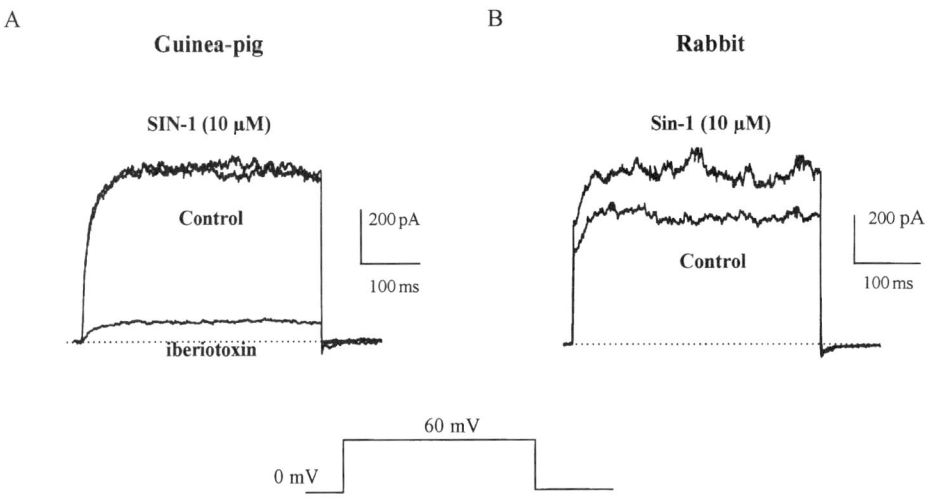

Figure 4. Effects of SIN-1 (10 μM) on whole-cell K^+ currents in freshly dissociated vascular smooth muscle cells of guinea pig (A) and rabbit (B) carotid arteries (holding potential = 0 mV; intracellular calcium concentration = 0.5 μM). SIN-1 induced an increase in BK_{Ca} current in smooth muscle cells of rabbit carotid arteries but did not affect this current in smooth muscle cells of the guinea pig. The recordings in the guinea pig smooth muscle cell demonstrates the sensitivity of the outward currents to iberiotoxin, a specific blocker of BK_{Ca} channels. These patch-clamp experiments show that NO can also activate BK_{Ca} channels and that the population of K^+ channels activated is species-dependent. (Modified from Quignard *et al.*, 1999a. Copyright OPA N.V. Permission granted by Gordon and Breach Publishers.)

produces a cGMP-independent activation of BK_{Ca} channels (direct effect) in smooth muscle cells from the rabbit aorta (Bolotina et al., 1994) and the mesenteric artery of the rat (Mistry and Garland, 1998) and in BK_{Ca} channels reconstituted into planar lipid bilayers (Shin et al., 1997). Similarly, in a study of the rat mesenteric artery with the single-microelectrode voltage-clamp method, NO produced a cGMP-independent activation of BK_{Ca} and K_{ATP} channels (Weidelt et al., 1997).

In functional studies performed in isolated or perfused pulmonary, mesenteric, coronary, and basilar arteries and the aorta of the rat (Archer et al., 1994; Nossaman et al., 1997; Zhao et al., 1997; Plane et al., 1996; Wellman et al., 1996; Price and Hellerman, 1997; Kitazono et al., 1997; Taguchi et al., 1996), mesenteric and carotid arteries and aorta of the rabbit (Khan et al., 1993; Najibi and Cohen, 1995; Plane et al., 1998; Bolotina et al., 1994), pulmonary artery of the guinea pig (Bialecki and Stinson-Fisher, 1995), and coronary and cerebral arteries of the dog (Pataricza et al., 1995; Node et al., 1997, Onoue and Katuzic, 1997) as well as in human, porcine, and bovine coronary arteries (Bychkov et al., 1998; Darkow et al., 1997; Li et al., 1998), the relaxation or the vasodilation produced by NO or NO donors was more or less sensitive to BK_{Ca} channel blockers (tetraethylammonium, charybdotoxin, or iberiotoxin), suggesting that these channels participate in the relaxation of vascular smooth muscle. In addition to BK_{Ca} channel activation, a 4-aminopyridine-sensitive relaxation, suggestive of the involvement of Kv channels, is observed in the pulmonary artery of the rat (Zhao et al., 1997). Apamin and glibenclamide impair NO relaxation in the rat mesenteric artery, suggestive of SK_{Ca} and K_{ATP} channel involvement, respectively (Plane et al., 1996), and an apamin-sensitive effect in lamb coronary arteries has been reported (Simonsen et al., 1997). In the aorta and carotid artery of the guinea pig, the relaxation to NO donors does not involve BK_{Ca} channels, because their blockers do not produce any inhibition (Bialecki and Stinson-Fisher, 1995). In these functional studies, a cGMP-independent activation of BK_{Ca} channels has been reported in the mesenteric and pulmonary arteries of the rat (Plane et al., 1996; Zhao et al., 1997) and in the carotid artery of the rabbit (Plane et al., 1998), confirming the hypothesis that direct activation of BK_{Ca} channels by NO might be an important mechanism in smooth muscle relaxation. Alternatively, a cGMP-independent but indirect activation of BK_{Ca} channels by NO can be produced. In rat renal arterioles, NO evokes a cGMP-independent inhibition of cytochrome P-450 monooxygenase, leading to a decrease in the endogenous production of 20-hydroxyeicosatetraenoic acid, a contractile substance that inhibits the opening of BK_{Ca} channels in preglomerular and cerebral arteries (Zou et al., 1996; Alonso-Galicia et al., 1997).

When these various studies are taken together (measurements of E_m by intracellular microelectrodes, ionic conductance with the use of the patch-clamp technique, and changes in tension), the effects of NO on vascular smooth muscle K$^+$ channels appear complex. This can be explained by various different mechanisms. First, a very heterogeneous population of K$^+$ channels is expressed in smooth muscle cells, and this heterogeneity can be demonstrated even in the same blood vessel. Archer et al., (1996) have shown that in the pulmonary arteries of the rat, various subpopulations of smooth muscle cells could be identified according to the relative distribution of the BK_{Ca} and Kv channels. NO relaxes conduit pulmonary arteries which possess predominantly a subpopulation of smooth muscle cells expressing preponderantly K_{Ca} channels. In the guinea pig carotid artery, the nitrovasodilator SIN-1 produced smooth muscle hyperpolarization by opening glibenclamide-sensitive K^+ channels (Corriu et al., 1996a). The opening of K_{ATP} channels by SIN-1 has been confirmed in isolated smooth muscle cells of the same artery studied with the patch-clamp technique using the whole-cell configuration (Quignard et al., 1999a). In the carotid artery of the rabbit, SIN-1 produces repolarization that is partially sensitive to iberiotoxin, and in isolated smooth muscle cells, the NO donors open BK_{Ca} channels (J.-F. Quignard, unpublished observations), indicating that the various electrophysiological techniques used produce consistent results and that this heterogeneity could be linked to intrinsic differences in the tissues studied (Fig. 4). Second, differences in protocols may explain why in most functional studies NO-induced relaxations are sensitive to BK_{Ca} channel blockers, whereas few studies measuring changes in E_m with intracellular microelectrodes have shown such a phenomenon. In functional studies, tissues are contracted (e.g., depolarized) in order to observe relaxations. In contrast, most of the electrophysiological studies with intracellular microelectrodes involve tissues at rest. BK_{Ca} channels are activated when intracellular Ca^{2+} concentration increases, a phenomenon that is voltage-dependent. The approximate threshold activation potential of BK_{Ca} channels with a low intracellular Ca^{2+} concentration (~ 100 nM) is -20 mV (Edwards and Weston, 1995), a E_m much more positive than that recorded in resting arteries (-50 to -65 mV). This may explain why in the canine coronary artery, nitroglycerin, NO, or sodium nitroprusside does not hyperpolarize resting tissue (Ito et al., 1980a; Komori et al., 1988; Félétou and Vanhoutte, 1996b), whereas in functional studies nitroglycerin induces a relaxation that is sensitive to iberiotoxin (Pataricza et al., 1995) and, in isolated smooth muscle cells of the same artery, BK_{Ca} channels are stimulated by a cGMP-dependent protein kinase, as observed in patch-clamp experiments (Taniguchi et al., 1993). Third, hyperpolarization of the smooth muscle cells may not be the predominant mechanism of relaxation. In the coronary artery of the guinea pig (Parkington et al., 1995) as well as in the basilar and femoral arteries of the rabbit (Plane and Garland, 1994; Plane et al., 1995), NO evokes glibenclamide-sensitive hyperpolarizations which do not contribute to the relaxation. Alternatively, in order to observe the contribution of hyperpolarization to the relaxation, prior inhibition of other pathways is required, as reported in the mesenteric artery of the rat (Plane et al., 1996). Fourth, different populations of K^+ channels could be activated by endogenous endothelial NO, exogenous authentic NO, or NO-releasing nitrovasodilators. In the rabbit isolated carotid artery, endogenous endothelial NO or exogenous authentic NO activates a cGMP-dependent K^+ conductance (which has not been fully characterized) and a cGMP-independent, charybdotoxin-sensitive conductance whereas nitrovasodilators activate only the former conductance (Plane et al., 1998).

In anesthetized pigs, the vasodilation produced by an NO donor is potently inhibited by iberiotoxin, suggesting that activation of vascular BK_{Ca} channels is an important component in the mechanism of vasodilation (Zanzinger et al., 1996). However, in anesthetized rats, 4-aminopyridine inhibits sodium nitroprusside-induced hypotension whereas iberiotoxin does not, suggesting that the mechanism involves activation of Kv and not BK_{Ca} channels (Berg and Koteng, 1997). These results should be interpreted with caution because the specificity and the sites of action of these drugs have not been established in vivo with certainty.

2.3. EDHF

In certain cases, endothelium-dependent hyperpolarizations are resistant to inhibitors of NO synthase and cyclooxygenase. They have been attributed to the release of an unidentified substance, EDHF (Furchgott and Vanhoutte, 1989; Komori and Vanhoutte, 1990; Félétou and Vanhoutte, 1996a; Mombouli and Vanhoutte, 1997).

The activation of a K$^+$ conductance as the mechanism of EDHF-mediated responses has been suggested by the following findings. The amplitude of the hyperpolarization is inversely related to the extracellular concentration of K$^+$ ions, and it disappears in K$^+$ concentrations higher than 25 mM (Chen and Suzuki, 1989a; Nagao and Vanhoutte, 1992b; Corriu et al., 1996b, Quignard et al., 1999b). Nonselective inhibitors of BK$_{Ca}$ channels, such as tetraethylammonium or tetrabutylammonium, prevent the hyperpolarization (Chen et al., 1991; Nagao and Vanhoutte, 1992b; Van de Voorde et al., 1992). Endothelium-dependent hyperpolarizations are associated with an increase in rubidium efflux (Taylor et al., 1988, Chen et al., 1988) and a decrease in membrane resistance, suggesting that the hyperpolarization is due to the opening and not to the closing of a conductance [such as chloride or nonspecific cationic conductances (Bolton et al., 1984; Chen and Suzuki, 1989a,b)]. Unfortunately, this last line of evidence should be interpreted with caution because the experiments were not performed in the presence of a combination of inhibitors of NO synthase and cyclooxygenase. Thus, opening of K$^+$ channels by concomitant NO and/or prostacyclin release cannot be ruled out.

The K$^+$ conductance involved in the EDHF-mediated responses does not involve K$_{ATP}$ channels, and its exact nature is still unknown and may depend on the tissue (Fig. 2). In mesenteric arteries of the rabbit, responses attributed to EDHF are fully blocked by apamin (Murphy and Brayden, 1995a), whereas in rat mesenteric and porcine and bovine coronary and oviductal arteries, they are only partially blocked by the toxin, suggesting that, in these species, small-conductance, Ca^{2+}-dependent K$^+$ channels (SK$_{Ca}$ channels) are involved (Adeagbo and Malik, 1991; Adeagbo and Triggle, 1993; Hecker et al., 1994; Garcia-Pascual et al., 1995). In rat mesenteric and hepatic arteries (Garland and Plane, 1996; Zygmunt and Högestätt, 1996; Chen and Cheung, 1997; Zygmunt et al., 1998) as well as in guinea pig carotid, coronary, basilar, and submucosal arteries (Corriu et al., 1996a; Hashitani and Suzuki, 1997; Petersson et al., 1997; Chataigneau et al., 1998a; Yamanaka et al., 1998), the combination of charybdotoxin (an inhibitor of BK$_{Ca}$, IK$_{Ca}$, and some Kv channels) plus apamin is required to abolish the hyperpolarization that is resistant to inhibitors of NO synthase and cyclooxygenase. Each blocker individually is either ineffective (charybdotoxin) or only partially effective (apamin). This combination of the two toxins appears to be specific because the hyperpolarizations produced by NO donors and prostacyclin analogs are unaffected (Fig. 5; Corriu et al., 1996a). However, the combination of iberiotoxin (a specific inhibitor of BK$_{Ca}$ channels) plus apamin did not mimic the effects of charybdotoxin and apamin (Zygmunt and Högestätt, 1996; Petersson et al., 1997; Zygmunt et al., 1997a; Chataigneau et al., 1998a; Yamanaka et al., 1998), indicating that BK$_{Ca}$ channels are not involved in the endothelium-dependent hyperpolarizations of the guinea pig carotid, coronary, and basilar arteries and rat mesenteric artery (Fig. 6). In contrast, in the guinea pig coronary artery, the effect of apamin is mimicked by scyllatoxin, a

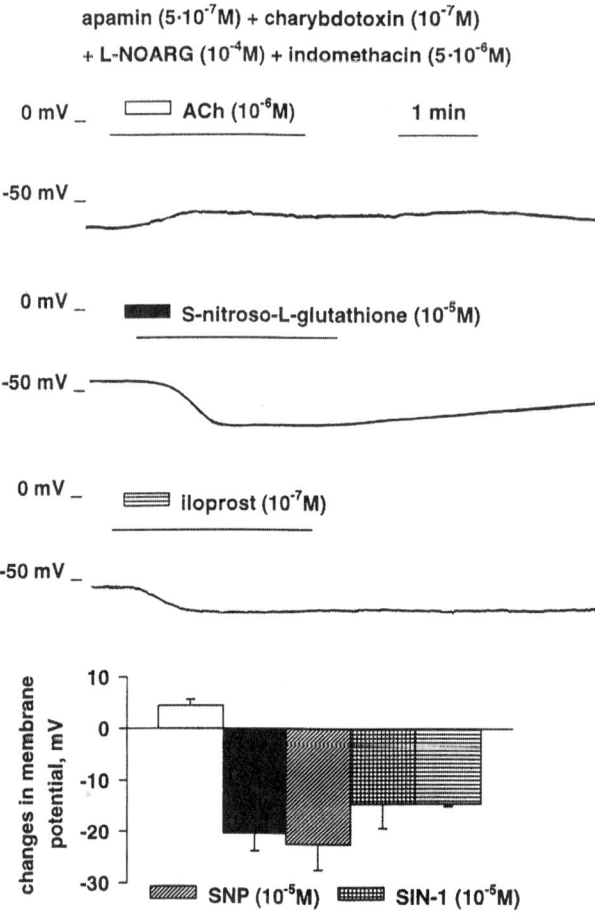

Figure 5. Effect of the combination of two toxins — apamin (0.5 μM), an SK_{Ca} channel inhibitor, plus charybdotoxin (0.1 μM), a nonspecific K_{Ca} channel inhibitor — on the hyperpolarizations produced by acetylcholine (ACh, 1 μM), by three nitrovasodilators — S-nitroso-L-glutathione, sodium nitroprusside (SNP), and SIN-1 (10 μM) — and by a stable analog of prostacyclin, iloprost (0.1 μM), in guinea pig isolated carotid arteries. An inhibitor of cyclooxygenase, indomethacin (5 μM), and an inhibitor of NO synthase, N^G-nitro-L-arginine (L-NOARG, 100 μM), were present throughout. The combination of the two toxins abolishes the endothelium-dependent hyperpolarization to acetylcholine without affecting the hyperpolarizations to the NO donor or iloprost. The bar graph at the bottom of the figure shows the means $\pm$ SEM of the responses. These microelectrode experiments further confirm that, in the guinea pig carotid artery, the endothelium-dependent hyperpolarization produced by acetylcholine does not involve NO and prostacyclin release. (Modified from Corriu *et al.*, 1996a, with permission.)

structurally different SK_{Ca} channel inhibitor, suggesting that SK_{Ca} channels play a role in these endothelium-dependent hyperpolarizations (Fig. 6; Corriu *et al.*, 1996a; Chataigneau *et al.*, 1998a). Similarly, in human blood vessels including coronary, cerebral, renal, omental, and subcutaneous arteries, responses attributed to EDHF are independent of K_{ATP} channel activation and are minimally affected or unaffected by charybdotoxin, iberiotoxin, or apamin individually but are blocked either by non-

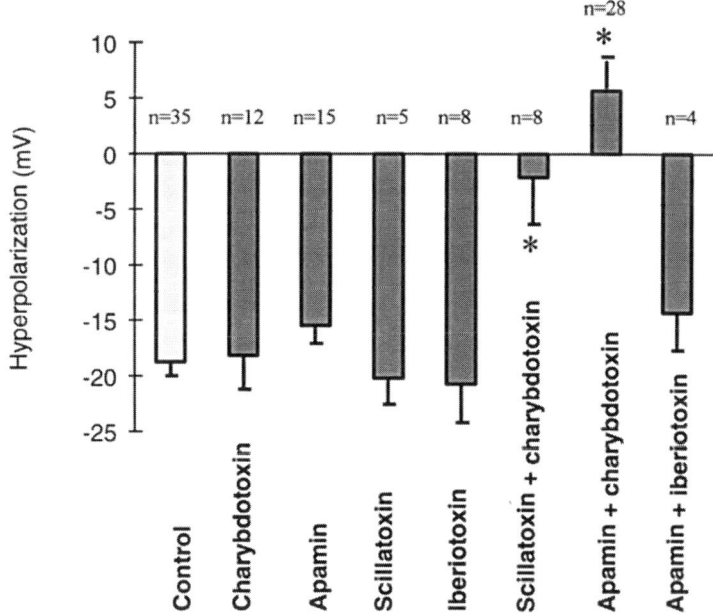

Figure 6. Effects of K$^+$ channel blockers on the endothelium-dependent hyperpolarization evoked by acetylcholine (1 μM) in the guinea pig isolated internal carotid artery [in the presence of N^G-nitro-L-arginine (100 μM) and indomethacin (5 μM)]. Data are shown as means $\pm$ SEM. Asterisks indicate a statistically significant difference from control values; n represents the number of cells studied. Apamin (0.5 μM) and scyllatoxin (50 nM), two specific inhibitors of SK$_{Ca}$ channels, iberiotoxin (30 nM), a specific inhibitor of BK$_{Ca}$ channels, and charybdotoxin (0.1 μM), a nonspecific inhibitor of K$_{Ca}$ channels, individually do not significantly affect the endothelium-dependent hyperpolarization to acetylcholine. The combination of either apamin plus charybdotoxin or scyllatoxin plus charybdotoxin inhibits the hyperpolarization whereas the combination of iberiotoxin plus apamin is ineffective. These results indicate that, in the guinea pig carotid artery, BK$_{Ca}$ channels are not involved in the endothelium-dependent hyperpolarization to acetylcholine whereas SK$_{Ca}$ channels or subunits of SK$_{Ca}$ channels are likely to be involved. The site of action of the combination of these toxins (smooth muscle versus endothelial cells) remains to be determined. (Modified from Chataigneau et al., 1998a, with permission.)

specific K$^+$ channel inhibitors (tetrabutylammonium or tetraethylammonium) or by the combination of charybdotoxin plus apamin (Nakashima et al., 1993; Kessler et al., 1996; Pascoal and Umans, 1996; Ohlmann et al., 1997; Petersson et al., 1997; Urakami-Harasawa et al., 1997; Wallerstedt and Bodelsson, 1997).

Additionally, in the guinea pig coronary and carotid arteries, 4-aminopyridine fully inhibits the endothelium-dependent hyperpolarization, whereas in the porcine coronary artery, it partially reduces the endothelium-dependent relaxation attributed to EDHF (Eckman et al., 1998; Quignard et al., 1999c; Shimizu and Paul, 1998). In the guinea pig carotid artery, this effect occurs in the same concentration range as that required to inhibit Kv channels in isolated smooth muscle cells (Fig. 7; Quignard et al., 1999c). However, although charybdotoxin can inhibit Kv channels (Chandy and Gutman, 1995), the effects of 4-aminopyridine and charybdotoxin appear to be unrelated because charybdotoxin, at the concentration used, did not inhibit Kv channels in isolated guinea pig carotid artery smooth muscle cells. Furthermore, the combination of apamin

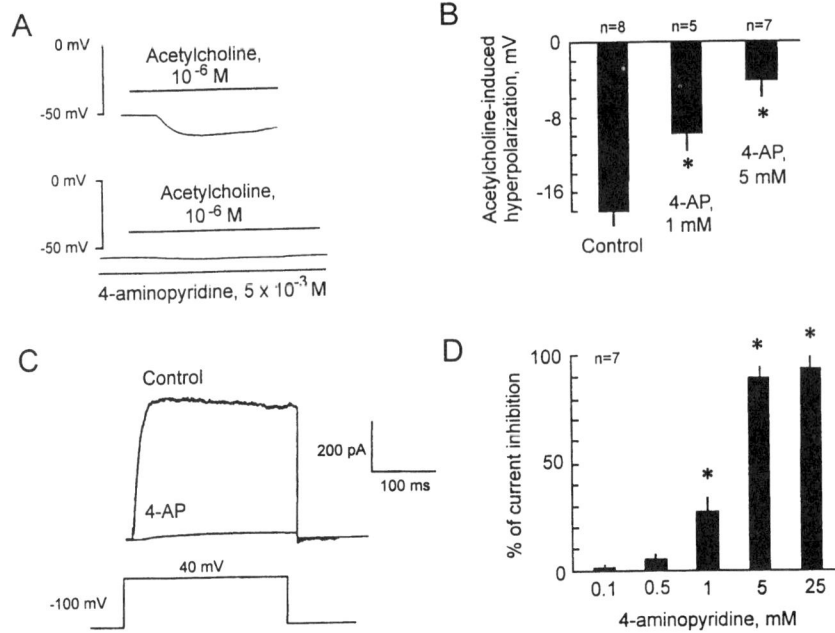

Figure 7. Effects of 4-aminopyridine (4-AP), a Kv channel blocker, on the guinea pig carotid artery. (A) 4-AP (5 mM) inhibits endothelium-dependent hyperpolarization evoked by acetylcholine (1 μM) in the guinea pig isolated internal carotid artery. (B) 4-AP induces a concentration-dependent inhibition of the acetylcholine-induced hyperpolarization. Hyperpolarization (mV) is shown as mean ± SEM. (C) 4-AP (5 mM) inhibits whole-cell K^+ current in freshly dissociated vascular smooth muscle cells of the guinea pig carotid artery (Ca^{2+}-free intracellular solution; holding potential = −100 mV, step depolarization to 40 mV). (D) 4-AP (0.1–25 mM) produces a concentration-dependent inhibition of the K^+ current. Data are shown as means ± SEM. Asterisks indicate a statistical difference from control values; n indicates the number of cells studied. These microelectrode and patch-clamp experiments indicate that, in the guinea pig carotid artery, Kv channels are involved in the endothelium-dependent hyperpolarization produced by acetylcholine. In freshly dissociated smooth muscle cells, a specific conductance sensitive to the combination of charybdotoxin plus apamin could not be observed (data not shown), but a Kv conductance sensitive to 4-AP was recorded. (Modified from Quignard et al., 1999c. Copyright OPA N.V. Permission granted by Gordon and Breach Publishers.)

plus 4-aminopyridine was not more effective than 4-aminopyridine alone, in contrast to the combination of apamin plus charybdotoxin (Quignard et al., 1999c). In the hepatic artery of the rat (Zygmunt et al., 1997a) and in the basilar artery of the guinea pig (Petersson et al., 1997), the effects of another Kv channel inhibitor, ciclazindol, and apamin were additive whereas 4-aminopyridine was ineffective, possibly because it was present at too low a concentration. In the porcine coronary artery, the inhibitory effect of 4-aminopyridine is potentiated by the addition of either apamin or charybdotoxin (Shimizu and Paul, 1998). In the perfused kidney of the rat, the vasodilation attributed to EDHF is blocked by charybdotoxin but is unaffected by either iberiotoxin or leiurotoxin, a SK_{Ca} channel inhibitor (Rapacon et al., 1996).

None of the experiments reported above can determine whether or not the site of action of the various K^+ channel inhibitors studied is at the level of the smooth muscle cells (e.g., inhibition of the target of EDHF) or the endothelial cells (e.g., inhibition of EDHF synthesis and/or release). In isolated vascular smooth muscle cells of the rat hepatic artery and in the guinea pig carotid artery (two vascular preparations in which EDHF inhibition requires the combination of apamin plus charybdotoxin), a K$^+$ conductance specifically sensitive to the combination of these two toxins (without involvement of BK$_{Ca}$ channels) was not detected (Zygmunt et al., 1997a; Quignard et al., 1999c). This indicates that either the site of action of the two toxins is not the smooth muscle cells, and therefore most likely is the endothelial cell, or that the presence of EDHF is required to activate this conductance in the smooth muscle cells. In the hepatic artery of the rat, the combination of charybdotoxin plus apamin inhibits the hyperpolarization of the endothelial cells produced by acetylcholine (Edwards et al., 1998). In the endothelial cells of the guinea pig coronary artery, the combination of the two toxins does not affect the acetylcholine-induced increase in intracellular free calcium concentration (Yamanaka et al., 1998). Thus, although charybdotoxin and apamin may act on the endothelial cells, it is not yet known whether this endothelial effect is responsible for the inhibition of the EDHF-mediated responses.

Endothelium-dependent hyperpolarizations sensitive to ouabain have been described in some vascular tissues such as the canine coronary artery (Félétou and Vanhoutte, 1988), the cerebral artery of the rabbit (Brayden, 1990), the porcine coronary artery (Olanrewaju et al., 1997), and the hepatic artery of the rat (Edwards et al., 1998). The concentration used or the duration of the incubation required to observe the inhibition suggest that ouabain may exert pharmacological effects other than the inhibition of the Na$^+$/K$^+$ pump.

2.4. Other Identified Endothelium-Derived Hyperpolarizing Factors

2.4.1. Epoxyeicosatrienoic Acids

EDHF may be a short-lived metabolite of arachidonic acid produced through the cytochrome P-450 monooxygenase pathway (Rubanyi and Vanhoutte, 1987; Komori and Vanhoutte, 1990). Inhibitors of this pathway inhibit endothelium-dependent vasodilator responses to acetylcholine, bradykinin, or arachidonic acid, which are resistant to inhibitors of NO synthase and cyclooxygenase. This inhibition has been observed in the perfused heart and kidney of the rat and in isolated human renal artery as well as in porcine, bovine, and human coronary arteries (Rosolowski and Campbell, 1993; Bauersachs et al., 1994; Hecker et al., 1994; Fulton et al., 1992, 1995; Kessler et al., 1996; Graier et al., 1996; Miura and Gutterman, 1998). Muscarinic agonists induce not only endothelium-dependent relaxation and hyperpolarization of bovine coronary arterial smooth muscle but also the release of epoxyeicosatrienoic acids from bovine coronary arterial endothelial cells (Rosolowski and Campbell, 1996). These responses are inhibited by metyrapone and miconazole (Campbell et al., 1996). Epoxyeicosatrienoic acids relax blood vessels such as cat cerebral and bovine, guinea pig, rat, and canine coronary arteries (Gebremedhin et al., 1992; Campbell et al., 1996; Graier et al., 1996; Eckman et al., 1998; Fulton et al., 1998; Oltman et al., 1998), hyperpolarize coronary arterial smooth muscle cells (Campbell et al., 1996; Eckman et al., 1998), and increase the open-state probability of BK$_{Ca}$ channels sensitive to tetraethylammonium, charybdotoxin, or iberiotoxin (Gebremedhin *et al.*, 1992; Hu and Kim, 1993; Campbell *et al.*, 1996; Li *et al.*, 1997; Gebremedhin *et al.*, 1998; Hayabuchi *et al.*, 1998). In bovine isolated coronary artery smooth muscle cells, 11,12-epoxyeicosatrienoic acid activates BK_{Ca} channels in a cAMP- and c-GMP-independent manner through a guanine nucleotide binding protein, Gsα, suggesting the possible existence of specific receptors to epoxyeicosatrienoic acids on the membrane of these vascular smooth muscle cells (Li and Campbell, 1998). Finally, bradykinin stimulates the release, from diverse types of endothelial cells, of a transferable factor that activates BK_{Ca} channels and hyperpolarizes vascular smooth muscle cells (Popp *et al.*, 1996; Gebremedhin *et al.*, 1998; Hayabuchi *et al.*, 1998). Taken in conjunction, these observations support the hypothesis that epoxyeicosatrienoic acids act as EDHFs in some vascular beds.

However, cytochrome P-450 inhibitors, studied at high concentrations, are notoriously unspecific and can inhibit hyperpolarizations induced by K^+ channel openers such as levcromakalim (Graier *et al.*, 1996; Eckman *et al.*, 1995; Edwards *et al.*, 1996; Vanhoutte and Félétou, 1996). In other studies involving blood vessels from humans (coronary and omental arteries), rats, guinea pigs, dogs, and pigs, chemically unrelated inhibitors of cytochrome P-450 do not inhibit the EDHF responses or produce a nonspecific inhibition (Corriu *et al.*, 1996b; Graier *et al.*, 1996; Zygmunt *et al.*, 1996; Fukao *et al.*, 1997; Ohlmann *et al.*, 1997; Urakami-Harasawa *et al.*, 1997; Wallerstedt and Bodelsson, 1997; Chataigneau *et al.*, 1998a). In guinea pig carotid arteries, epoxyeicosatrienoic acids do not produce relaxation or hyperpolarization (Fig. 8; Chataigneau *et al.*, 1998a). In the coronary artery smooth muscle of the same species,

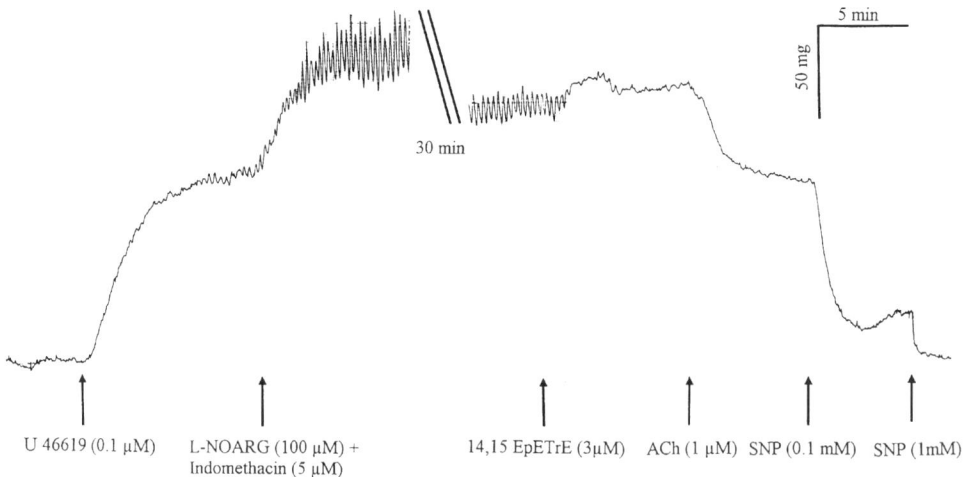

Figure 8. Original trace showing the changes of tension in the isolated guinea pig internal carotid artery with endothelium. U 46619, a stable analog of thromboxane A_2 (0.1 μM), produces a contraction, and the addition of N^G-nitro-L-arginine (L-NOARG; 100 μM) plus indomethacin (5 μM) produces a further increase in tension. 14,15 EpETrE (3 μM), an epoxyeicosatrienoic acid, does not produce a relaxation whereas acetylcholine (ACh; 1 μM) and sodium nitroprusside (SNP; 0.1 and 1 mM) relax the isolated blood vessels. Similar results were obtained with other epoxyeicosatrienoic acids, indicating that, in the guinea pig carotid artery, epoxyeicosatrienoic acids are unlikely to be EDHF. (Modified from Chataigneau *et al.*, 1998a, with permission.)

a high concentration of 11,12-epoxyeicosatrienoic acid produces hyperpolarization and relaxation which are sensitive to iberiotoxin but unaffected by 4-aminopyridine. However, the hyperpolarization and relaxation produced by acetylcholine are sensitive to the latter but unaffected by the former inhibitor (Eckman et al., 1998), indicating that the EDHF response induced by acetylcholine does not involve the release of epoxy-eicosatrienoic acid.

The discrepancies between the various studies presented can be explained in different ways. Without considering the nonspecific effects of cytochrome P-450 inhibitors, already mentioned, activation of cytochrome P-450 in human endothelial cells appears to be a more general requirement for increasing the intracellular calcium concentration and thus the release of endothelium-derived factors such as NO and EDHF (Graier et al., 1996). The particular agonist inducing endothelium-dependent hyperpolarization could be crucial in determining the mechanism of relaxation used (e.g., in human arteries: bradykinin versus arachidonic acid (Urakami-Harasawa et al., 1997; Miura and Gutterman, 1998)). Finally, epoxyeicosatrienoic acids produce effects which are not only tissue- and/or species-specific (Campbell et al., 1996; Chaigneau et al., 1998a) but can also differ markedly depending on the size of the vessel studied (Oltman et al., 1998). Thus, the choice of the agonist and the type of tissue studied must be considered in the interpretation of results of studies investigating the effects of inhibitors of cytochrome P-450 on EDHF-mediated responses.

2.4.2. Anandamide

Anandamide, a derivative of arachidonic acid, is thought to be an endogenous ligand for the cannabinoid CB$_1$ receptor (Devane et al., 1992; Di Marzo et al., 1994). In the isolated and perfused mesenteric and coronary arterial bed of the rat, anandamide induces a dilation which mimics responses to EDHF and which is blocked by the combination of charybdotoxin and apamin (Randall et al., 1996, 1997; Randall and Kendall, 1997, 1998). However, in the kidney of the rat the dilation caused by anandamide is due to the release of NO (Deutsch et al., 1997). In isolated blood vessels from various species (pig, guinea pig, rat), anandamide does not produce hyperpolarization or, if it does so, the underlying mechanism differs from that of EDHF-mediated responses. Indeed, some of these responses to anandamide are endothelium-dependent (Zygmunt et al., 1997b, Chaigneau et al., 1998b). Finally, CB$_1$ receptor antagonists do not inhibit endothelium-dependent hyperpolarization. These observations do not support the suggestion that an endogenous cannabinoid is the major mediator of endothelium-dependent hyperpolarizations (Fig. 9; Chaigneau et al., 1998b; Plane et al., 1997; Pratt et al., 1998; Zygmunt et al., 1997b; White and Hiley, 1997). Nevertheless, considering that anandamide can be produced and released by endothelial cells, it can relax vascular smooth muscle cells by a mechanism that involve K$^+$ channel activation (White and Hiley, 1998).

2.4.3. Adenosine

Adenine nucleotides (AMP, ADP, and ATP) and adenosine are released by endothelial cells (Shryock et al., 1988; Shinozuka et al., 1994). Although endothelial

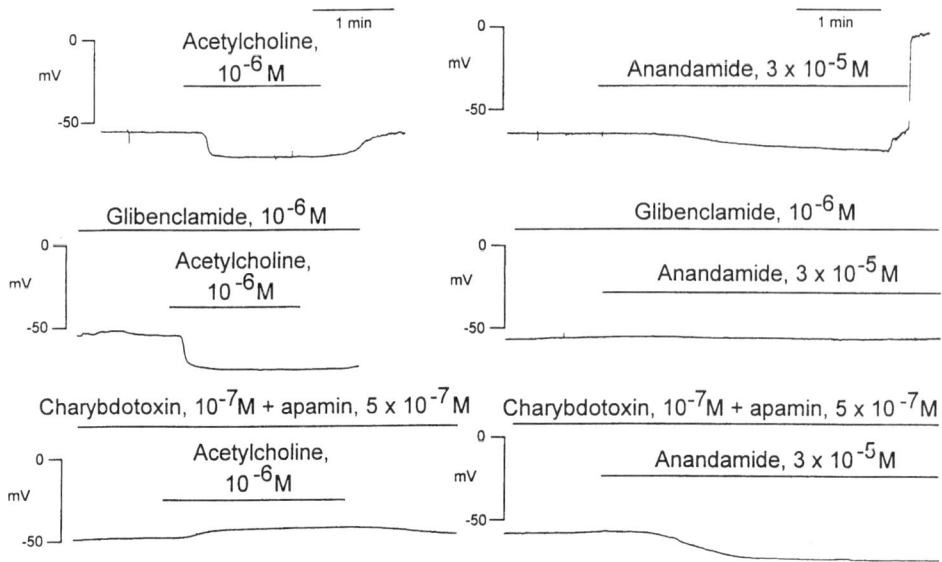

Figure 9. Effects of glibenclamide (1 μM) and the combination of charybdotoxin (0.1 μM) plus apamin (0.5 μM) [in the presence of N^G-nitro-L-arginine (100 μM) and indomethacin (5 μM)] on endothelium-dependent hyperpolarizations elicited by acetylcholine (ACh; 1 μM) (*left*) and anandamide (30 μM) (*right*) in the rat isolated mesenteric artery. The endothelium-dependent hyperpolarization to acetylcholine in the rat mesenteric artery is resistant to glibenclamide but fully blocked by the combination of charybdotoxin plus apamin as in the guinea pig carotid artery (Figs. 2 and 5). In contrast, the hyperpolarization induced by anandamide is blocked by glibenclamide but is unaffected by the combination of the two toxins. These microelectrode experiments indicate that, in the rat mesenteric artery, the endothelium-dependent hyperpolarization produced by acetylcholine is not mediated by anandamide. (Modified from Chataigneau *et al.*, 1998b, with permission.)

cells release mainly ATP, the nucleotide is rapidly transformed into ADP, AMP, and adenosine by ectonucleotidases. The release of adenosine by endothelial cells can be stimulated by chemical and physical stimuli such as hypoxia (Nees *et al.*, 1980) or increases in flow (Milner *et al.*, 1990), by neuromediators (including transmural nerve stimulation; Sedaa *et al.*, 1990), and by drugs such as α_1-adrenergic agonists (Shinozuka *et al.*, 1994). Adenosine induces relaxation and hyperpolarization of vascular and nonvascular smooth muscle, including the human coronary artery (Imai and Takeda, 1967; Herlihy *et al.*, 1976; Olanrewaju *et al.*, 1995). In most of the studies, the hyperpolarization of the vascular smooth muscle cells involves K_{ATP} channels. In rabbit mesenteric and guinea pig coronary arteries, adenosine stimulates an A_2 adenosine receptor subtype coupled to adenylate cyclase. The ensuing accumulation of cAMP activates protein kinase A, which then increases the activity of K_{ATP} channels (Kleppisch and Nelson, 1995; Mutafova-Yambolieva and Keef, 1997). In canine coronary, rat pulmonary, and porcine retinal arteries, the activation of K_{ATP} channels appears to be independent of the accumulation of cAMP (Akatsuka *et al.*, 1994; Gidday *et al.*, 1996; Sheridan *et al.*, 1997). In porcine coronary arteries, the receptor involved is an adenosine A_1 receptor subtype (Dart and Standen, 1993). However, in canine epicardial artery, the relaxation produced by adenosine may involve activation of BK$_{Ca}$, rather than K$_{ATP}$, channels (Makujina *et al.*, 1994; Cabell *et al.*, 1994). In certain blood vessels, including human coronary arteries, adenosine can also activate endothelial receptors and release NO by a mechanism which may involve activation of endothelial K$_{ATP}$ channels (Kuo and Chancellor, 1995; Olanrewaju *et al.*, 1995; Smits *et al.*, 1995).

2.4.4. Carbon Monoxide

The predominant biological source of carbon monoxide is the degradation of heme by heme oxygenase, an enzyme which could be inducible (HO-1) or constitutive (HO-2) (for a review, see Wang, 1998). Some arterial endothelial cells express HO-2 (Wang, 1998), and the expression of HO-1 in endothelial cells has been demonstrated in the ductus arteriosus of the lamb (Coceani and Kelsey, 1997) and in the rat thoracic aorta subjected to hypoxia (Caudill *et al.*, 1998). Carbon monoxide relaxes and hyperpolarizes vascular and nonvascular smooth muscles (Furchgott and Jonathiandan, 1991; Wang *et al.*, 1997a; Wang, 1998) by activating soluble guanylate cyclase (Furchgott and Jonathiandan, 1991; Wang *et al.*, 1997a; Wang, 1998) and, in the rat tail artery, by directly opening BK$_{Ca}$ channels (Wang *et al.*, 1997b). The increase in open-state probability of BK$_{Ca}$ channels induced by carbon monoxide is linked to the establishment of a hydrogen bond with the imidazole group of some histidyl residues (Wang and Wu, 1997). Additionally, at least in the ductus arteriosus, inhibition of cytochrome P-450 by carbon monoxide may suppress the synthesis of endogenous vasoconstrictors and lift the K$^+$ channel inhibition produced by these cytochrome P-450 derivatives (Coceani *et al.*, 1996a,b).

Heme oxygenase is inhibited by zinc protoporphyrin IX, and, although this inhibitor lacks specificity as it can also inhibit guanylate cyclase, NO synthase, and K$_{ATP}$ channels, especially if exposed to light (Zygmunt *et al.*, 1994b; Wang, 1998). It does not inhibit endothelium-dependent hyperpolarizations in the rat hepatic artery or in the guinea pig carotid artery (Zygmunt *et al.*, 1994b; Zygmunt and Högestätt, 1996; M. Félétou and C. Corriu, unpublished observations). Furthermore, endothelium-dependent hyperpolarizations are not affected by oxyhemoglobin, an active scavenger of carbon monoxide in a variety of blood vessels, suggesting that EDHF is not carbon monoxide.

2.4.5. Hydrogen Peroxide

Hydrogen peroxide can be produced by endothelial cells, spontaneously or in response to bradykinin either directly or as a by-product of the release of superoxide anion (Sundquist, 1991; Heinzel *et al.*, 1992). In the isolated rabbit aorta and in canine and porcine coronary arteries, hydrogen peroxide, but not superoxide anion or hydroxyl radical, produces relaxation and/or hyperpolarization (Needelman *et al.*, 1973; Rubanyi and Vanhoutte, 1986; Beny and von der Weid, 1991). In isolated smooth muscle cells from the rat aorta and the porcine coronary artery, the relaxation and/or hyperpolarization produced by hydrogen peroxide has been attributed to the opening of K$_{Ca}$ channels, a phenomenon which in the latter tissue involves a lipoxygenase metabolite of arachidonic acid (Krippeit-Drews *et al.*, 1995; Barlow and White, 1998).

However, in porcine coronary arteries, the hyperpolarization produced by hydrogen peroxide and the endothelium-dependent hyperpolarization to either substance P or bradykinin do not have the same time course. Whereas the former was sensitive to catalase, the latter was not, indicating that EDHF and hydrogen peroxide are two distinct molecules (Beny and von der Weid, 1991).

2.4.6. K^+ Ions

In certain vascular beds such as the coronary and cerebral arteries of the rat, increasing the extracellular concentration of K^+ ions (from 6 to 16 mM) relaxes the blood vessels and hyperpolarizes the smooth muscle cells up to 14 mV (Edwards et al., 1988; McCarron and Halpern, 1990; Knot et al., 1996). However, this phenomenon is usually observed in small, but not in large, arteries. For instance, K^+ induces a hyperpolarization in the small cerebral artery of the rat whereas it depolarizes larger cerebral arteries (Edwards et al., 1988). These hyperpolarizations induced by K^+ are inhibited by low concentrations of barium ($< 100 \mu M$), a specific inhibitor of inward rectifying K^+ (Kir) channels at these low concentrations (Nelson and Quayle, 1995). Kir channels are voltage-dependent K^+ channels whose open-state probability decreases with depolarization. The open-state probability is increased by a modest rise in extracellular K^+ concentration (Faraci and Heistad, 1998). The level of expression of Kir channels in vascular smooth muscle cells is inversely related to the size of the blood vessels. Thus, the expression of Kir channels is preponderant in smaller blood vessels. This explains the different effects of extracellular K^+ in small and large blood vessels (Edwards et al., 1988). A rise in extracellular K^+ may hyperpolarize and relax vascular smooth muscle cells by activating the Na^+/K^+ pump, without involvement of Kir channels (Prior et al., 1998).

Agonists that produce endothelium-dependent relaxations generate K^+ efflux from vascular endothelial cells (Gordon and Martin, 1983). Edwards et al. (1998) have suggested that, in the rat hepatic artery, K^+ ions are EDHF. The stimulation by acetylcholine of endothelial receptors opens charybdotoxin- and apamin-sensitive K^+ conductance on the endothelial cell membrane, leading to a K^+ efflux and accumulation in the intercellular space. The rise in K^+ activates Kir channels and the Na^+/K^+ pump in smooth muscle cells, provoking hyperpolarization and relaxation that are sensitive to the combination of barium and ouabain (Edwards et al., 1998). However, this cannot be generalized to every vascular bed (Vanhoutte, 1998). In guinea pig carotid and porcine coronary arteries, the endothelium-dependent hyperpolarizations are not affected by the combination of barium plus ouabain, and K^+ does not produce hyperpolarization (Fig. 10). Furthermore, in isolated smooth muscle cells of the guinea pig carotid artery, very few cells express the Kir channel (Quignard et al., 1999b). These results suggest that K^+ is not EDHF in these two blood vessels.

2.4.7. Others

Immunolocalization of numerous neuropeptides, including endothelin, vasoactive intestinal peptide (VIP), substance P, calcitonin gene-related peptide (CGRP), and arginine vasopressin, has been demonstrated in endothelial cells (Cai et al., 1993). VIP

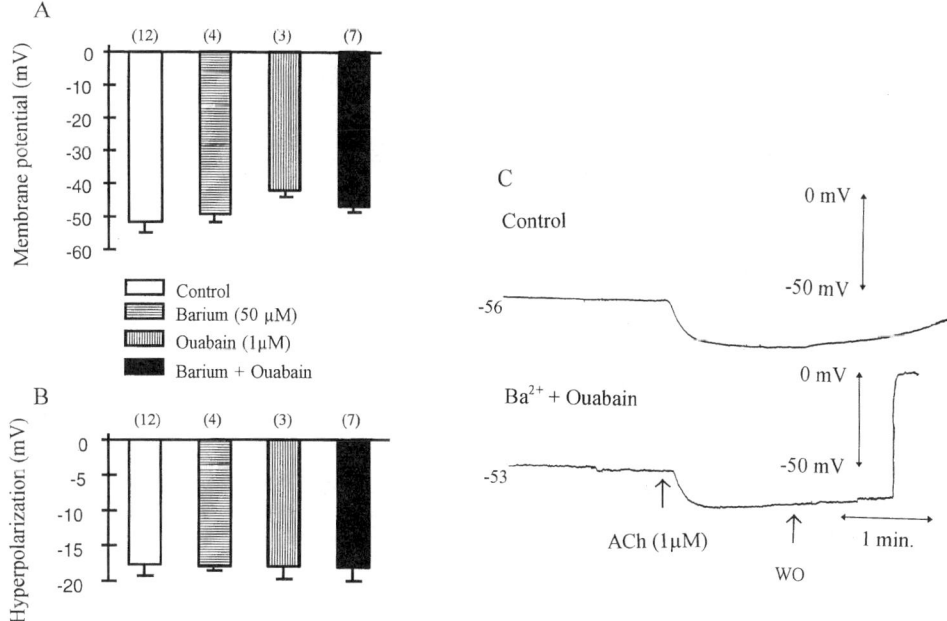

Figure 10. Effects of the Kir channel blocker barium (50 μM) and the Na$^+$/K$^+$ pump inhibitor ouabain (1 μM) on acetylcholine (1 μM)-induced endothelium-dependent hyperpolarization in the guinea pig isolated internal carotid artery with endothelium. The experiments were performed in the presence of N^G-nitro-L-arginine (100 μM), indomethacin (5 μM), and carboxy PTIO (10 μM; a NO scavenger). Neither barium nor ouabain altered the resting membrane potential (A) or hyperpolarization elicited by acetylcholine (B). Data are shown as Means $\pm$ SEM; numbers in parenthesis indicate the number of experiments. The upper and lower traces in panel C show the endothelium-dependent hyperpolarizations elicited by acetylcholine (1 μM) in control condition and in the presence of the combination of barium (50 μM) plus ouabain (1 μM, respectively. The combination of barium plus ouabain does not affect the endothelium-dependent hyperpolarization to acetylcholine. These microelectrode experiments indicate that, in the guinea pig carotid artery, K$^+$ ions are not mediating the EDHF response. (Modified from Quignard *et al.*, 1999b, with permission.)

and CGRP produce direct and endothelium-dependent relaxation of vascular smooth muscle. VIP stimulates adenylate cyclase in rat mesenteric artery and aorta (Ganz *et al.*, 1986), and in rabbit cerebral arteries it produces hyperpolarization by opening K$_{ATP}$ channels (Standen *et al.*, 1989). In smooth muscle cells of the porcine coronary artery, VIP activates BK$_{Ca}$ and Kv channels (Kawasaki *et al.*, 1997). Similarly to endothelin, CGRP is stored in Weibel–Palade bodies (Ozaka *et al.*, 1997). CGRP opens K$_{ATP}$ channels in blood vessels, such as the rabbit mesenteric artery (Nelson *et al.*, 1990) and the human mammary artery (Luu *et al.*, 1997). In porcine coronary artery, CGRP activates the adenylate cyclase–cAMP–protein kinase A pathway leading to K$_{ATP}$ and BK$_{Ca}$ channel activation (Miyoshi and Nakaya, 1995; Wellman *et al.*, 1998).

C-type natriuretic peptide (CNP) is also released by endothelial cells (Nazario *et al.*, 1995) and produces relaxation and hyperpolarization of vascular smooth muscle in arteries and veins. In porcine coronary artery and canine femoral vein, CNP induces accumulation of cGMP and opens BK$_{Ca}$ channels (Wei *et al.*, 1994; Banks *et al.*, 1996).

Finally, L-citrulline, the by-product of NO biosynthesis by NO synthase, is able to relax the rabbit aorta by a cGMP-dependent mechanism which may involve K^+ channel activation (Ruiz and Tejerina, 1998).

3. CONCLUSION

Endothelial cells are able to synthesize and release numerous vasoactive substances. The regulation of the opening and closure of K^+ channels by the release of endothelium-derived factors or directly through myoendothelial gap junctions is a key element in the control of the underlying vascular tube.

REFERENCES

Adeagbo, A. S. O., and Malik, K. U., 1990, Mechanism of vascular actions of prostacyclin in the rat isolated perfused mesenteric arteries, *J. Pharmacol. Exp. Ther.* **252**:26–34.

Adeagbo, A. S. O., and Malik, K. U., 1991, Contribution of K^+ channels to arachidonic acid-induced endothelium-dependent vasodilation in rat isolated perfused mesenteric arteries, *J. Pharmacol. Exp. Ther.* **258**:452–458.

Adeagbo, A. S. O., and Triggle, C. R., 1993, Varying extracellular $[K^+]$: A functional approach to separating EDHF- and EDNO-related mechanisms in perfused rat mesenteric arterial bed, *J. Cardiovasc. Pharmacol.* **21**:423–429.

Akatsuka, Y., Egashira, K., Katsuda, Y., Narishige, T., Ueno, H., Shimokawa, H., and Takeshita, A., 1994, ATP-sensitive potassium channels are involved in adenosine A_2 receptor mediated coronary vasodilatation, *Cardiovasc. Res.* **28**:906–911.

Alonso-Galicia, M., Drummond, H. A., Reddy, K. K., Falck, J. R., and Roman, R. J., 1997, Inhibition of 20-HETE production contributes to the vascular responses to nitric oxide, *Hypertension* **29**:320–325.

Archer, S. L., Huang, J. M. C., Hampl, V., Nelson, D. P., Shultz, P. J., and Weir, E. K., 1994, Nitric oxide and cyclic-GMP cause vasorelaxation by activation of a charybdotoxin-sensitive K channel by cyclic-GMP-dependent protein kinase, *Proc. Natl. Acad. Sci. U.S.A.* **91**:7583–7587.

Archer, S. L., Huang, J. M. C., Reeve, H. L., Hampl, V., Tolarova S., Michelakis, E., and Weir, E. K., 1996, Differential distribution of electrophysiologically distinct myocytes in conduit and resistance arteries determines their response to nitric oxide and hypoxia, *Circ. Res.* **78**:431–442.

Banks, M., Wei, C.-M., Kim, C. H., Burnett, J. C., and Miller, V. M., 1996, Mechanism of relaxations to C-type natriuretic peptide in veins, *Am. J. Physiol.* **271**:H1907–H1911.

Barlow, R. S., and White, R. E., 1998, Hydrogen peroxide relaxes porcine coronary arteries by stimulating BK_{Ca} channel activity, *Am. J. Physiol.* **44**:H1283–H1289.

Bauersachs, J., Hecker, M., and Busse, R., 1994, Display of the characteristics of endothelium-derived hyperpolarizing factor by a cytochrome P450-derived arachidonic acid metabolite in the coronary microcirculation, *Br. J. Pharmacol.* **113**:1548–1553.

Beech, D., and Bolton, T. B., 1989, Properties of the cromakalim-induced potassium conductance in smooth muscle cells isolated from the rabbit portal vein, *Br. J. Pharmacol.* **98**:851–854.

Bény, J.-L., 1990, Endothelial and smooth muscle cells hyperpolarized by bradykinin are not dye coupled. *Am. J. Physiol.* **258**:H836–H841.

Bény, J.-L., and Brunet, P. C., 1988 Electrophysiological and mechanical effects of substance P and acetylcholine on rabbit aorta, *J. Physiol. (London)* **398**:277–289.

Bény, J. L., and Chabaud, F., 1996, Kinins and endothelium-dependent hyperpolarization in porcine coronary arteries, in: *Endothelium-Derived Hyperpolarizing Factor*, Vol. 1 (P. M. Vanhoutte, ed.), Harwood Academic Publishers, Amsterdam, pp. 41–50.

Bény, J. L., and von der Weid, P. Y., 1991, Hydrogen peroxide, an endogenous smooth muscle cell hyperpolarizing factor, *Biochem. Biophys. Res. Commun.* **176**:378–384.

Berg, T., and Koteng, O., 1997, Signalling pathway in bradykinin- and nitric oxide-induced hypotension in the normotensive rat; role of K$^+$-channels, *Br. J. Pharmacol.* **121**:1113–1120.

Bialecki, R. A., and Stinson-Fisher, C., 1995, K$_{C_a}$ channel antagonists reduce NO donor-mediated relaxation of vascular and tracheal smooth muscle, *Am. J. Physiol.* **12**:L152–L159.

Bolotina, V. M., Najibi, S., Palacino, J. J., Pagano, P. J., and Cohen, R. A., 1994, Nitric oxide directly activates calcium-dependent potassium channels in vascular smooth muscle cells, *Nature* **368**:850–853.

Bolton, T. B., Lang, R. J., and Takewaki, T., 1984, Mechanism of action of noradrenaline and carbachol on smooth muscle of guinea-pig anterior mesenteric artery, *J. Physiol.* **351**:549–572.

Borda, E. S., Sterin-Borda, L., Gimeno, M. F., Lazzari, M. A., and Gimeno, A. C., 1983, The stimulatory effect of prostacyclin (PGI$_2$) on isolated rabbit and rat aorta is probably associated to the generation of a thromboxane A$_2$ (TXA$_2$) "like material," *Arch. Int. Pharmacodyn. Ther.* **261**:79-89.

Brayden, J. E.,1990, Membrane hyperpolarisation is a mechanism of endothelium-dependent cerebral vasodilation, *Am. J. Physiol.* **259**:H668–H673.

Brayden, J. E., and Murphy, M. E., 1996, Potassium channels activated by endothelium-derived factors in mesenteric and cerebral resistance arteries, in: *Endothelium-Derived Hyperpolarizing Factor*, Vol. 1 (P. M. Vanhoutte, ed.), Harwood Academic Publishers, Amsterdam, pp. 137–142.

Bredt, D. S., and Snyder, S. H., 1990, Isolation of nitric oxide synthase, a calmodulin-requiring enzyme, *Proc. Natl. Acad. Sci. U.S.A.* **87**:682–685.

Brunet, P. C., and Beny, J.-L. 1989. Substance P and bradykinin hyperpolarize pig coronary artery endothelial cells in primary culture, *Blood Vessels* **26**:228–234.

Busse, R., Fichtner, H., Luckhoff, A., and Kohlhardt, M., 1988, Hyperpolarisation and increased free calcium in acetylcholine-stimulated endothelial cells, *Am. J. Physiol.* **255**:H965–H969.

Bychkov, R., Gollasch, M., Steinke, T., Ried, C., Luft, F. C., and Haller, H., 1998, Calcium-activated potassium channels and nitrate-induced vasodilation in human coronary arteries, *J. Pharmacol. Exp. Ther.* **285**:293–298.

Cabell, F., Weiss, D. S., and Price, J. M., 1994, Inhibition of adenosine-induced coronary vasodilation by block of large-conductance Ca^{2+}-activated K$^+$ channels, *Am. J. Physiol.* **36**:H1455–H1460.

Cai, W. Q., Bodin, P., Loesch, A., Sexton, A., and Burnstock, G., 1993, Endothelium of human umbilical blood vessels — Ultrastructural immunolocalization of neuropeptides, *J. Vasc. Res.* **30**:348–355.

Campbell, W. B., Gebremedhin, D., Pratt, P. F., Harder, D. R., 1996, Identification of epoxyeicosatrienoic acids as endothelium-derived hyperpolarizing factor, *Circ. Res.* **78**:415–423.

Carrier, G. O., Fuchs, L. C., Winecoff, A. P., Giulumian, A. D., and White, R. E., 1997, Nitrovasodilators relax mesenteric microvessels by cyclic-GMP-induced stimulation of Ca-activated K channels, *Am. J. Physiol.* **42**:H76–H84.

Caudill, T. K., Resta, T. C., Kanagy, N. L., and Walker, B. R., 1998, Role of endothelial carbon monoxide in attenuated vasoreactivity following chronic hypoxia, *Am. J. Physiol.* **44**:1025–1030.

Chandy, K. G., and Gutman, G. A., 1995, Voltage-gated potassium channel genes, in *Ligand- and Voltage Gated Ion Channels* (R. A. North, ed.), CRC Press, Boca Raton, pp. 2–71.

Chataigneau T., Félétou M., Duhault J., and Vanhoutte P. M., 1998a, Epoxyeicosatrienoic acids, potassium channel blockers and endothelium-dependent hyperpolarisation in the guinea-pig carotid artery, *Br. J. Pharmacol.* **123**:574–580.

Chataigneau T., Félétou, M., Thollon, C.,. Villeneuve, N., Vilaine, J.-P., Duhault, J., and Vanhoutte., P. M., 1998b, Cannabinoid CB$_1$ receptor and endothelium-dependent hyperpolarisation in guinea-pig carotid, rat mesenteric and porcine coronary arteries, *Br. J. Pharmacol.* **123**:968–974.

Chaytor, A. Y., Evens, W. H., and Griffith T. M., 1998, Central role of heterocellular gap junction communication in endothelium-dependent relaxations of rabbit arteries. *J. Physiol. (London)* **508**: 561–573.

Chen, G., and Cheung, D. W., 1997, Effects of K$^+$ channel blockers on Ach-induced hyperpolarization and relaxation in mesenteric arteries, *Am. J. Physiol.* **41**:H2306–H2312.

Chen, G., and Suzuki, H., 1989a, Some electrical properties of the endothelium-dependent hyperpolarisation recorded from rat arterial smooth muscle cells, *J. Physiol.* **410**:91–106.

Chen, G., and Suzuki, H., 1989b, Direct and indirect action of acetylcholine and histamine on intrapulmonary artery and vein smooth muscles of the rat, *Jpn. J. Physiol.* **39**:51–65.

Chen, G., and Suzuki, H., 1990, Calcium dependency of the endothelium-dependent hyperpolarisation in smooth muscle cells of the rabbit carotid artery, *J. Physiol.* **421**:521–534.

Chen, G., Suzuki, H., and Weston, A. H., 1988, Acetylcholine releases endothelium-derived hyperpolarizing factor and EDRF from rat blood vessels, *Br. J. Pharmacol.* **95**:1165–1174.

Chen, G., Yamamoto, Y., Miwa, K., and Suzuki, H., 1991, Hyperpolarisation of arterial smooth muscle induced by endothelial humoral substances, *Am. J. Physiol.* **260:**H1888–H1892.

Clapp, L. H., Turcato, S., Hall, S., and Baloch, M., 1998, Evidence that Ca^{2+}-activated K^+ channels play a major role in mediating the vascular effects of iloprost and cicaprost, *Eur. J. Pharmacol.* **356:**215–224.

Coceani, F., and Kelsey, L., 1997, Carbon monoxide formation in the ductus arteriosus in the lamb: Implications for the regulation of muscle tone, *Br. J. Pharmacol.* **120:**599–608.

Coceani, F., Kelsey, L., and Seidlitz, E., 1996a, Carbon monoxide-induced relaxation of the ductus arteriosus in the lamb: Evidence against the prime role of guanylyl cyclase, *Br. J. Pharmacol.* **118:**1689–1696.

Coceani, F., Kelsey, L., Seidlitz, E., and Korzekwa, K., 1996b, Inhibition of the contraction of the ductus arteriosus to oxygen by 1-aminobenzotriazole, a mechanism-based inactivation of cytochrome P-450, *Br. J. Pharmacol.* **117:**1586–1592.

Cocks, T. M., King, S. J., and Angus, J. A., 1990, Glibenclamide is a competitive inhibitor of the thromboxane A_2 receptor in dog coronary artery in vitro, *Br. J. Pharmacol.* **100:**375–378.

Cohen, R. A., and Vanhoutte, P. M., 1995, Endothelium-dependent hyperpolarization—beyond nitric oxide and cyclic GMP, *Circulation* **92:**3337–3349.

Cohen, R. A., Plane, F., Najibi, S., Huk, I., Malinski, T., and Garland, C. J., 1997, Nitric oxide is the mediator of both endothelium-dependent relaxation and hyperpolarisation of the rabbit carotid artery, *Proc. Natl. Acad. Sci. U.S.A.* **94:**4193–4198.

Corriu, C., Félétou, M., Canet, E., and Vanhoutte, P. M. 1996a, Endothelium-derived factors and hyperpolarisations of the isolated carotid artery of the guinea-pig, *Br. J. Pharmacol.* **119:**959-964.

Corriu, C., Félétou, M., Canet, E., and Vanhoutte, P. M., 1996b, Inhibitors of the cytochrome P450-monooxygenase and endothelium-dependent hyperpolarisations in the guinea-pig isolated carotid artery, *Br. J. Pharmacol.* **117:**607–610.

Cowan, C. L., and Cohen, R,A., 1991, Two mechanisms mediate relaxation by bradykinin of pig coronary artery: NO-dependent and independent responses, *Am. J. Physiol.* **261:**H830–H835.

Darkow, D. J., Lu, L., and White, R. E., 1997, Estrogen relaxation of coronary artery smooth muscle is mediated by nitric oxide and cyclic-GMP, *Am. J. Physiol.* **41:**H2765–H2773.

Dart, C., and Standen, N. B., 1993, Adenosine-activated potassium current in smooth muscle cells isolated from the pig coronary artery, *J. Physiol. (London)* **471:**767–786.

Davies, P. F., Oleson, S. P., Clapham, D. E., Morel, E. M, and Schoen, F. J., 1988, Endothelial communication: State of the art lecture, *Hypertension* **11:**563–572.

Davis, K., Grinsburg, R., Bristow, M., and Harrison, D. C., 1980, Biphasic action of prostacyclin in the human coronary artery, *Clin. Res.* **28:**165A.

De Mey, J. G., Claeys, M., and Vanhoutte, P. M., 1982, Endothelium-dependent inhibitory effects of acetylcholine, adenosine triphosphate, thrombin and arachidonic acid in the canine femoral artery, *J. Pharmacol. Exp. Ther.* **222:**166–173.

Deutsch, D. G., Goligorsky, M. S., Schmid, P. C., Krebsbach, R. J., Schmid, H. H. O., Das, S. K., Dey, S. K., Arreaza, G., Thorup, C., Stefano, G., and Moore, L. C., 1997, Production and physiological actions of anandamide in the vasculature of the rat kidney, *J. Clin. Invest.* **100:**1538–1546.

Devane, W. A., Hanus, L., Breuer, A., Pertwee, R. G., Stevenson, L. A., Griffin, G., Gibson, D., Mandelbaum, A., Etinger, A., and Mechoulam, R.,1992, Isolation and structure of a brain constituent that binds to the cannabinoid receptor, *Science* **258:**1946–1949.

Di Marzo, V., Fontana, A., Cadas, H., Schinelli, S., Cimino, G., Schwartz, J. C., and Piomelli, D., 1994, Formation and inactivation of endogenous cannabinoid anandamide in central neurons, *Nature* **372:**686–691.

Eckman, D. M., Hopkins, N. O., and Keef, K. D., 1995, Effects of inhibitors of cytochrome P450 pathway on relaxation and hyperpolarisation induced with acetylcholine and lemakalim. *Circulation* **92:**I-751.

Eckman, D. M., Hopkins, N. O., McBride, C., and Keef, K. D., 1998, Endothelium-dependent relaxation and hyperpolarization in guinea-pig coronary artery: Role of epoxyeicosatrienoic acid, *Br. J. Pharmacol.* **124:**181–189.

Edwards, F. R., Hirst, G. D., and Silverberg, G. D., 1988, Inward rectification in rat cerebral arterioles, involvement of potassium ions in autoregulation, *J. Physiol. (London)* **404:**455–466.

Edwards, G., and Weston, A. H., 1995, Potassium channels in the regulation of vascular smooth muscle tone, in: *Pharmacologicol Control of Calcium and Potassium Homeostasis: Biological, Therapeutical and Clinical Aspects* (T. Godfraind, G. Mancia, M. P. Abbracchio, L. Aguilar-Bryan, and S. Govoni, eds.), Kluwer Academic Press, Dordrecht, The Netherlands, pp. 85–93.

Edwards, G., Zygmunt, P. M., Högestät, E. D., and Weston, A. H., 1996, Effects of cytochrome P450 inhibitors on potassium currents in mechanical activity in rat portal vein, *Br. J. Pharmacol.*, **119**:691–701.

Edwards, G., Dora, K. A., Gardener, M. J., Garland, C. J., and Weston, A. H., 1998, K$^+$ is an endothelium-derived hyperpolarizing factor in rat arteries, *Nature* **396**:269–272.

Faraci, F. M., and Heistad, D. D., 1998, Regulation of the cerebral circulation: Role of endothelium and potassium channel, *Physiol. Rev.* **78**:54–75.

Félétou, M., and Vanhoutte, P. M., 1985, Endothelium-derived relaxing factor(s) hyperpolarize(s) coronary smooth muscle, *Physiologist* **48**:325.

Félétou, M., and Vanhoutte, P. M., 1988, Endothelium-dependent hyperpolarisation of canine coronary smooth muscle, *Br. J. Pharmacol.* **93**:515–524.

Félétou, M., and Vanhoutte, P. M., 1996a, Endothelium-derived hyperpolarizing factor, *Clin. Exp. Pharmacol. Physiol.* **23**:1082–1090.

Félétou M., and Vanhoutte, P. M., 1996b, Biossay of endothelium-derived hyperpolarizing factor in canine arteries, in: *Endothelium-Derived Hyperpolarizing Factor*, Vol. 1 (P. M. Vanhoutte, ed.), Harwood Academic Publishers, Amsterdam, pp. 25–32.

Félétou, M., Hoeffner, U., and Vanhoutte, P. M., 1989, Endothelium-dependent relaxing factors do not affect the smooth muscle of portal-mesenteric vein, *Blood Vessels* **26**:21–32.

Fukao, M., Hattori, Y., Kanno, M., Sakuma, I., and Kitabatake, A., 1997, Evidence against a role of cytochrome P450-derived arachidonic acid metabolites in endothelium-dependent hyperpolarisation by acetylcholine in rat isolated mesenteric artery, *Br. J. Pharmacol.*, **120**:439–446.

Fukuta, H., Miwa, K., Hozumi, T., Yamamoto, Y., and Suzuki, H., 1996, Reduction by EDHF of the intracellular calcium concentration in vascular smooth muscle, in *Endothelium-Derived Hyperpolarizing Factor*, Vol. 1 (P. M. Vanhoutte, ed.), Harwood Academic Publishers, Amsterdam, pp. 143–153.

Fulton, D, McGiff, J. C., and Quilley, J., 1992, Contribution of NO and cytochrome P$_{450}$ to the vasodilator effect of bradykinin in the rat kidney, *Br. J. Pharmacol.* **107**:722–725.

Fulton, D., Mahboudi, K., Mcgiff, J. C., and Quilley, J., 1995, Cytochrome P450-dependent effects of bradykinin in the rat heart, *Br. J. Pharmacol.* **114**:99–102.

Fulton, D, McGiff, J. C., and Quilley, J., 1998, Pharmacological evaluation of an epoxide as the putative hyperpolarizing factor mediating the nitric oxide-independent vasodilator effect of bradykinin in the rat heart, *J. Pharmacol. Exp. Ther.* **287**:497–503.

Furchgott, R. F., and Jonathiandan, D., 1991, Endothelium-dependent and -independent vasodilatation involving cyclic GMP: Relaxation induced by nitric oxide, carbon monoxide and light, *Blood Vessels* **28**:52–61.

Furchgott, R. F., and Vanhoutte, P. M., 1989, Endothelium-derived relaxing and contracting factors, *FASEB J.* **3**:2007–2018.

Furchgott, R. F., and Zawadzki, J. V., 1980, The obligatory role of the endothelial cells in the relaxation of arterial smooth muscle by acetylcholine, *Nature* **288**:373–376.

Ganz, P., Sandbrock, A. W., Landis, S. C., Leopold, J., Gimbrone, M. A., and Alexander, R. W., 1986, Vasoactive intestinal peptide: Vasodilatation and cyclic AMP generation, *Am. J. Physiol.* **250**:H755–H760.

Garcia-Pascual, A., Labadia, A., Jimenez, E., and Costa, G., 1995, Endothelium-dependent relaxation to acetylcholine in bovine oviductal arteries: Mediation by nitric oxide and changes in apamin-sensitive K$^+$ conductance, *Br. J. Pharmacol.* **115**:1221–1230.

Garland, C. J., and Plane, F., 1996, Relative importance of endothelium-derived hyperpolarizing factor for the relaxation of vascular smooth muscle in different arterial beds, in: *Endothelium-Derived Hyperpolarizing Factor*, Vol. 1 (P. M. Vanhoutte, ed.), Harwood Academic Publishers, Amsterdam, pp. 173–179.

Gebremedhin, D., Ma, Y. H., Falck, J. R., Roman, R. J., VanRollins, M., and Harder, D. R., 1992, Mechanism of action of cerebral epoxyeicosatrienoic acids on cerebral arterial smooth muscle, *Am. J. Physiol.* **263**:H519–H525.

Gebremedhin, D., Harder, D. R., Pratt, P. F., and Campbell, W. B., 1998, Bioassay of an endothelium-derived hyperpolarizing factor from bovine coronary arteries: Role of a cytochrome P450 metabolite, *J. Vasc. Res.* **35**:274–284.

Gidday, J. M., Maceren, R. G., Shah, A. R., Meier, J. A., and Zhu, Y., 1996, K$_{ATP}$ channels mediate adenosine-induced hyperemia in retina, *Invest. Ophthalmol. Visual Sci* **37**:2624–2633.

Gordon, J. L., and Martin, W., 1983, Endothelium-dependent relaxation of the pig aorta: Relationship to stimulation of ^{86}Rb efflux from endothelial cells, *Br. J. Pharmacol.* **79**:531–541.

Graier, W. F., Simecek, S., and Sturek, M., 1995b, Cytochrome P450 mono-oxygenase-regulated signalling of Ca^{2+} entry in human and bovine endothelial cells, *J. Physiol. (London)* **482:**259–274.

Graier, W. F., Holzmann, S., Hoebel, B. G., Kukovetz, W. R., and Kostner, G. M., 1996, Mechanisms of L-N^G-nitroarginine/indomethacin-resistant relaxation in bovine and porcine coronary arteries, *Br. J. Pharmacol.* **119:**1177–1186.

Haeusler, G., and Thorens, 1976, The pharmacology of vaso-active antihypertensives, in: *Vascular Neuroeffector Mechanisms* (J. A. Bevan *et al.*, eds.), Kargel, Basel, Switzerland, pp. 232–241.

Harder, D. R., Campbell, W. B., Gebremedhin, D., and Pratt, P. F., 1996, Biossay of a cytochrome P450-dependent endothelial-derived hyperpolarizing factor from bovine coronary arteries, in: *Endothelium-Derived Hyperpolarizing Factor*, Vol. 1 (P. M. Vanhoutte, ed.), Harwood Academic Publishers, Amsterdam, pp. 73–81.

Hardy, P., Abran, D., Hou, X., Lahaie, I., Peri, K. G., Asselin, P., Varma, D. R., and Chemtob, S., 1998, A major role for prostacyclin in nitric oxide-induced ocular vasorelaxation in the piglet, *Circ. Res.* **83:**721–729.

Hashitani, H., and Suzuki, H., 1997, K^+ channels which contribute to the acetylcholine-induced hyperpolarization in smooth muscle of the guinea-pig submucosal arterioles, *J. Physiol. (London)* **501:**319–329.

Hasunuma, K., Yamaguchi, T., Rodman, D., O'Brien, R., and McMurtry, I., 1991, Effects of inhibitors of EDRF and EDHF on vasoreactivity of perfused rat lungs, *Am. J. Physiol.* **260:**L97–L104.

Hayabuchi, Y., Nakaya, Y., Matsukoa, S., and Kuroda, Y., 1998, Endothelium-derived hyperpolarizing factor activates Ca^{2+}-activated K^+ channels in porcine coronary artery smooth muscle cells, *J. Cardiovasc. Pharmacol.* **32:**642–649.

Hecker, M., Bara, A. T., Bauersachs, J., and Busse, R., 1994, Characterization of endothelium-derived hyperpolarizing factor as a cytochrome P_{450}-derived arachidonic acid metabolite in mammals, *J. Physiol.* **481:**407–414.

Heinzel, B., John, M., Klatt, P., Böhme, E., and Mayer, B., 1992, Ca^{2+}/calmodulin-dependent formation of hydrogen peroxide by brain nitric oxide synthase, *Biochem. J.* **281:**627–630.

Herlihy, J. T., Bockman, E. L., Berne, R. M., and Rubio, R., 1976, Adenosine relaxation of isolated vascular smooth muscle, *Am. J. Physiol.* **239:**1239–1243.

Hoang, L. M., and Mathers, D. A., 1998, Internally applied endotoxins and the activation of BK channels in cerebral artery smooth muscle via a nitric oxide-like pathway, *Br. J. Pharmacol.* **123:**5–12.

Hu, S., and Kim, H. S., 1993, Activation of K^+ channel in vascular smooth muscles by cytochrome P450 metabolites of arachidonic acid, *Eur. J. Pharmacol.* **230:**215–221.

Illiano, S. C., Nagao, T., and Vanhoutte, P. M., 1992, Calmidazolium, a calmodulin inhibitor, inhibits endothelium-dependent relaxations resistant to nitro-L-arginine in the canine coronary artery, *Br. J. Pharmacol.* **107:**387–392.

Imai, S., and Takeda, K., 1967, Effects of vasodilators upon the isolated taenia coli of the guinea-pig, *J. Pharmacol. Exp. Ther.* **156:**557–564.

Ito, Y., Suzuki, H., and Kuriyama, K., 1978, Effects of sodium nitroprusside on smooth muscle cells of rabbit pulmonary artery and portal vein, *J. Pharmacol. Exp. Ther.* **207:**1022–1031.

Ito, Y., Kitamura, K., and Kuriyama, K., 1980a, Actions of nitroglycerin on the membrane and mechanical properties of smooth muscle cells of the coronary artery of the pig, *Br. J. Pharmacol.* **70:**197–204.

Ito, Y., Kitamura, K., and Kuriyama, K., 1980b, Nitroglycerin and catecholamine actions on smooth muscle cells of the canine coronary artery, *J. Physiol. (London)* **309:**171–183.

Jackson, W. F., Konig, A., Dambacher, T., and Busse, R., 1993, Prostacyclin-induced vasodilation in rabbit heart is mediated by ATP-sensitive potassium channels, *Am. J. Physiol.* **264:**H238–H243.

Kauser, K., and Rubanyi, G. M., 1992, Bradykinin-induced nitro-L-arginine-insensitive endothelium-dependent relaxation of porcine coronary artery is not mediated by bioassayable substances, *J. Cardiovasc. Pharmacol.* **20:**S101–S104.

Kauser, K., Stekiel, W. J., Rubanyi, G. M., and Harder, D. R., 1989, Mechanism of action of EDRF on pressurized arteries:Effect on K^+ conductance, *Circ. Res.* **65:**199–204.

Kawasaki, J., Kobayashi, S., Miyagi, Y., Nishimura, J., Fujishima, M., and Kanaide, H., 1997, The mechanisms of the relaxation induced by vasoactive intestinal peptide in the porcine coronary artery, *Br. J. Pharmacol.* **121:**977–985.

Kessler, P., Lischke, V., and Hecker, M., 1996, Etomidate and thiopental inhibit the release of endothelium-derived-hyperpolarizing factor in the human renal artery, *Anesthesiology* **84:**1485–1488.

Khan, S. A., Mathews, S. R., and Meisheri, K. D., 1993, Role of calcium-activated K$^+$ channels in vasodilatation induced by nitroglycerin, acetylcholine and nitric oxide, *J. Pharmacol. Exp. Ther.* **267**:1327–1335.

Kitazono, T., Ibayashi, S., Nagao, T., Fujii, K., and Fujishima, M., 1997, Role of Ca^{2+}-activated K$^+$ channels in acetylcholine-induced dilatation of the basilar artery in vivo, *Br. J. Pharmacol.* **120**:102–106.

Kleppisch T., and Nelson, M. T., 1995, Adenosine activates ATP-sensitive potassium channels in arterial myocytes via A2 receptor and c-AMP-dependent protein kinase, *Proc. Natl. Acad. Sci. U.S.A.* **92**:12441–12445.

Knot, H. J., Zimmermann, P. A., and Nelson M. T.,1996, Extracellular potassium-induced hyperpolarization and dilatations of rat coronary and cerebral arteries involve inward rectifier potassium channels, *J. Physiol. (London)* **492**:419–430.

Komori, K., and Suzuki, H., 1987, Electrical responses of smooth muscle cells during cholinergic vasodilation in the rabbit saphenous artery, *Circ. Res.* **61**:586–593.

Komori, K., and Vanhoutte P. M., 1990, Endothelium-derived hyperpolarizing factor, *Blood Vessels* **27**:238–245.

Komori, K., Lorenz, R. R., and Vanhoutte, P. M., 1988, Nitric oxide, ACh and electrical and mechanical properties of canine arterial smooth muscle, *Am. J. Physiol.* **255**:H207-H212.

Krippeit-Drews, P., Haberland, C., Fingerle, J., Drews, G., and Lang, F., 1995, Effects of H$_2$O$_2$ on membrane potential and [Ca^{2+}]$_i$ of cultured rat arterial smooth muscle cells, *Biochem. Biophys. Res. Commun.* **209**:139–145.

Kühberger, E., Groschner, K., Kukovetz, W. R., and Brunner, F., 1994, The role of myoendothelial cell contact in non-nitric oxide-, non-prostanoid-mediated endothelium-dependent relaxation of porcine coronary artery, *Br. J. Pharmacol.* **113**:1289–1294.

Kuo, L., and Chancellor, J. D., 1995, Adenosine potentiates flow-induced dilation of coronary arterioles by activating K$_{ATP}$ channels in endothelium, *Am. J. Physiol.* **38**:H541–H549.

Li, P. L., and Campbell, W. B., 1998, Epoxyeicosatrienoic acids activate K$^+$ channels in coronary smooth muscle through a guanine nucleotide binding protein, *Circ. Res.* **80**:877–884.

Li, P. L., Zou, A. P., and Campbell, W. B., 1997, Regulation of potassium channels in coronary arterial smooth muscle by endothelium-derived vasodilators, *Hypertension* **29**:262–267.

Li, P. L., Jin, M. W., and Campbell, W. B., 1998, Effect of selective inhibition of soluble guanylate cyclase on the K$_{Ca}$ channel activity in coronary smooth muscle, *Hypertension* **31**:303–308.

Luu, T. N., Dashwood, M. R., Tadjkarimi, S., Chester, A. H., and Yacoub, M. H., 1997, ATP-sensitive potassium channels mediate vasodilatation by calcitonin gene related peptide in human internal mammary but not gastroepiploic arteries, *Eur. J. Clin. Invest.* **27**:960–966.

Makujina, S. R., Olanrewaju, H. A., and Mustafa, S. J., 1994, Evidence against K$_{ATP}$ channel involvement in adenosine receptor-mediated dilation of epicardial vessels, *Am. J. Physiol.* **267**:H716–H724.

Marchenko, S. M., and Sage, S. O., 1994, Smooth muscle cells affect endothelial membrane potential in rat aorta. *Am. J. Physiol.* **267**:H804–H811.

McCarron, J. G., and Halpern, W., 1990, Potassium dilates rat cerebral arteries by two independent mechanisms, *Am. J. Physiol.* **259**:H902–H908.

Milner, P., Kirkpatrick, K. A., Ralevic, V., Toothill, V., and Burnstock, G., 1990, Endothelial cells cultured from human umbilical vein release ATP and acetylcholine in response to increased flow, *Proc. R. Soc. London B.* **241**:245–248.

Mistry, D. K., and Garland, C. J., 1998, Nitric oxide (NO)-induced activation of large conductance Ca^{2+}-dependent K$^+$ channels (BK$_{Ca}$) in smooth muscle cells isolated from the rat mesenteric artery, *Br. J. Pharmacol.* **124**:1131–1140.

Miura, H., and Gutterman, D. D., 1998, Human coronary arteriolar dilation to arachidonic acid depends on cytochrome P450 monooxygenase and Ca^{2+}-activated K$^+$ channels, *Circ. Res.* **83**:501–507.

Miyoshi, H., and Nakaya, Y., 1994, Endotoxin-induced non endothelial nitric oxide activates the Ca^{2+}-activated K$^+$ channel in cultured vascular smooth muscle cells, *J. Mol. Cell. Cardiol.* **26**:1487–1495.

Miyoshi, H., and Nakaya, Y., 1995, Calcitonin gene related peptide activates the K$^+$ channels of vascular smooth muscle cells via adenylate cyclase, *Basic Res. Cardiol.* **90**:332–336.

Miyoshi, H., Nakaya, Y., and Moritoki, H., 1994, Nonendothelial-derived nitric oxide activates the ATP-sensitive K channel of vascular smooth muscle cells, *FEBS Lett.* **345**:47–49.

Mombouli, J.-V., and Vanhoutte, P. M., 1997, Endothelium-derived hyperpolarizing factor(s): Updating the unknown, *Trends Pharmacol. Sci.* **18**:252–256.

Mombouli, J. V., Illiano, S., Nagao, T., and Vanhoutte, P. M., 1992, The potentiation of bradykinin-induced relaxations by perindoprilat in canine coronary arteries involves both nitric oxide and endothelium-derived hyperpolarizing factor, *Circ. Res.* **71:**137–144.

Mombouli, J.-V., Bissiriou, I., and Vanhoutte, P. M., 1996, Bioassay of endothelium-derived hyperpolarizing factor: Is endothelium-derived depolarizing factor a confounding element?, in: *Endothelium-Derived Hyperpolarizing Factor*, Vol. 1 (P. M. Vanhoutte, ed.), Harwood Academic Publishers, Amsterdam, pp. 51–57.

Moncada, S., and Vane, J. R., 1979, Pharmacology and endogenous roles of prostaglandin endoperoxides, thromboxane A2 and prostacyclin, *Pharmacol. Rev.* **30:**293–331.

Moncada, S., Gryglewski, R. J., Bunting, S., and Vane, J. R., 1976, An enzyme isolated from arteries transforms prostaglandin endoperoxides to an unstable substance that inhibits platelet aggregation, *Nature* **263:**663–665.

Moncada, S., Herman, A. G., Higgs, E. A., and Vane, J. R., 1977, Differential formation of prostacyclin (PGX or PGI$_2$) by layers of the arterial wall. An explanation for the antithrombotic properties of vascular endothelium, *Thromb. Res.* **11:**323–344.

Moncada, S., Palmer, R. J. M., and Higgs, E. A., 1991, Nitric oxide: Physiology, pathophysiology, and pharmacology, *Pharmacol. Rev.* **43:**109–142.

Mügge, A., Lopez, J. A. G., Piegors, D. J., Breese, K. R., and Heistad, D. D., 1991, Acetylcholine-induced vasodilatation in rabbit hindlimb in vivo is not inhibited by analogues of L-arginine, *Am. J. Physiol.* **260:**H242–H247.

Murphy, M. E., and Brayden, J. E., 1995a, Apamin-sensitive K$^+$ channels mediate an endothelium-dependent hyperpolarization in rabbit mesenteric arteries, *J. Physiol.* **489:**723–734.

Murphy, M. E., and Brayden, J. E., 1995b, Nitric oxide hyperpolarisation of rabbit mesenteric arteries via ATP-sensitive potassium channels, *J. Physiol.* **486:**47–58.

Mutafova-Yambolieva, V. N., and Keef, K. D., 1997, Adenosine-induced hyperpolarization in guinea-pig coronary artery involves A$_{2B}$ receptors and K$_{ATP}$ channels, *Am. J. Physiol.* **42:**H2687–H2695.

Nagao, T., and Vanhoutte, P. M., 1992a, Characterization of endothelium-dependent relaxations resistant to nitro-L-arginine in the porcine coronary artery, *Br. J. Pharmacol.* **107:**1102–1107.

Nagao, T., and Vanhoutte, P. M., 1992b, Hyperpolarisation as a mechanism for endothelium-dependent relaxations in the porcine coronary artery, *J. Physiol.* **445:**355–367.

Nagao, T., Illiano, S. C., and Vanhoutte, P. M., 1992a, Heterogeneous distribution of endothelium-dependent relaxations resistant to N^G-nitro-L-arginine in rats, *Am. J. Physiol.* **263:**H1090–H1094.

Nagao, T., Illiano, S. C., and Vanhoutte, P. M., 1992b, Calmodulin antagonists inhibit endothelium-dependent hyperpolarisation in the canine coronary artery, *Br. J. Pharmacol.* **107:**382–386.

Najibi, S., and Cohen, R. A., 1995, Enhanced role of K$^+$ channels in relaxations of hypercholesterolemic rabbit carotid artery to NO, *Am. J. Physiol.* **38:**H805–H811.

Nakashima, M., and Vanhoutte, P. M., 1995, Isoproterenol causes hyperpolarization through opening of ATP-sensitive potassium channels in vascular smooth muscle of the canine saphenous vein, *J. Pharmacol. Exp. Ther.* **272:**379–384.

Nakashima, M., Mombouli, J.-V., Taylor, A. A., and Vanhoutte, P. M., 1993, Endothelium-dependent hyperpolarisation caused by bradykinin in human coronary arteries, *J. Clin. Invest.* **92:**2867–2871.

Nazario, B., Hu, R. M., Pedram, A., Prins, B., and Levin, E. R., 1995, Atrial and brain natriuretic peptides stimulate the production and secretion of C-type natriuretic peptide from bovine aortic endothelial cells, *J. Clin. Invest.* **95:**1151–1157.

Needleman, P., Jakshik, B., and Johnson, E. M., 1973, Sulfhydryl requirement for relaxation of vascular smooth muscle, *J. Pharmacol. Exp. Ther.* **187:**324–331.

Nees, S., Gerbes, A. L., Willershausen-Zonnchen, B., and Gerlach, E., 1980, Purine metabolism in cultured coronary endothelial cells, *Adv. Exp. Med. Biol.* **122:**25–30.

Nelson, M. T., and Quayle, J. M., 1995, Physiological roles and properties of potassium channels in arterial smooth muscle, *Am. J. Physiol.* **268:**C799–C822.

Nelson, M. T., Huang, Y., Brayden, J. E., Hescheler, J., and Standen, N. B., 1990, Arterial dilations in response to calcitonin gene related peptide involve activation of K$^+$ channels, *Nature,* **344:**770–773.

Nishiyama, M., Hashitani, H., Fukuta, H., Yamamoto, Y., and Suzuki, H., 1998, Potassium channels activated in the endothelium-dependent hyperpolarization in guinea-pig coronary artery, *J. Physiol. (London)* **510:**455–465.

Node, K., Kitazake, M., Kosaka, H., Minamino, T., Sato, H., Kuzuya, T., and Hori, M., 1997, Roles of NO and Ca^{2+}-activated K$^+$ channels in coronary vasodilatation induced by 17-beta-estradiol in ischemic heart failure, *FASEB J.* **11**:793–799.

Nossaman, B. D., Kaye, A. D., Feng, C. J., and Kadowitz, P. J., 1997, Effects of charybdotoxin on responses to nitrovasodilators and hypoxia in the rat lung, *Am. J. Physiol.* **16**:L787–L791.

Ohlmann, P., Martinez, M. C., Schneider, F., Stoclet, J. C., and Andriantsitohaina, R., 1997, Characterization of endothelium-derived relaxaing factors released by bradykinin in human resistance arteries, *Br. J. Pharmacol.* **121**:657–664.

Olanrewaju, H. A., Hargittai, P. T., Lieberman, E. A., and Mustafa, S. J., 1995, Role of endothelium in hyperpolarisation of coronary smooth muscle by adenosine and its analogues, *J. Cardiovasc. Pharmacol.* **25**:234–239.

Olanrewaju, H. A, Hargittai, P. T, Lieberman, E. M., and Mustafa, S. J., 1997, Effect of ouabain on adenosine receptor-mediated hyperpolarization in porcine coronary artery smooth muscle, *Eur. J. Pharmacol.* **322**:185–190.

Oltman, C. L., Weintraub, N. L., VanRollins, M., and Dellsperger, K. C., 1998, Epoxyeicosatrienoic acids and dihydroxyeicosatrienoic acids are potent vasodilators in the canine coronary microcirculation, *Circ. Res.* **83**:932–939.

Onoue, H., and Katuzic, Z. S., 1997, Role of potassium channels in relaxations of canine middle cerebral arteries induced by nitric oxide donors, *Stroke* **28**:1264–1270.

Ozaka, T., Doi, Y., Kayashima, K., and Fujimoto, S., 1997, Weibel–Palade bodies as a storage site of calcitonin gene related peptide and endothelin-1 in blood vessels of the rat carotid body, *Anat. Rec.* **247**:388–394.

Pacicca, C., von der Weid, P., and Beny, J. L. 1992. Effect of nitro-L-arginine on endothelium-dependent hyperpolarisations and relaxations of pig coronary arteries, *J. Physiol.* **457**:247–256.

Palmer, R. M. J., Ferridge, A. G., and Moncada, S., 1987, Nitric oxide release accounts for the biological activity of endothelium-derived relaxing factor, *Nature* **327**:524–526.

Palmer, R. M. J., Ashton, D. S., and Moncada, S., 1988, Vascular endothelial cells synthesize nitric oxide from L-arginine, *Nature* **333**:664–666.

Parkington, H. C., Tare, M., Tonta, M. A., and Coleman, H. A., 1993, Stretch revealed three components in the hyperpolarisation of guinea-pig coronary artery in response to acetylcholine, *J. Physiol* **465**:459–476.

Parkington, H. C., Tonta, M., Coleman, H., and Tare, M., 1995, Role of membrane potential in endothelium-dependent relaxation of guinea-pig coronary arterial smooth muscle, *J. Physiol.* **484**:469–480.

Parkington, H. C, Tare, M., and Hammarstrm, A. K. M., 1996, The role of endothelium-derived prostacyclin in regulating tone in vascular smooth muscle, in: *Endothelium-Derived Hyperpolarizing Factor* (P. M. Vanhoutte, ed.), Harwood Academic Publishers, Amsterdam, pp. 57–64.

Pascoal, I. F., and Umans, J. G., 1996, Effect of pregnancy on mechanisms of relaxation in human omental microvessels, *Hypertension* **28**:183–187.

Pataricza, J., Toth, G. K., Penke, B., Hohn, J., and Papp, J. G., 1995, Effect of selective inhibition of potassium channels on vasorelaxing response to cromakalim, nitroglycerin and nitric oxide of canine coronary arteries, *J. Pharm. Pharmacol.* **47**:921–925.

Peng, W., Hoidal, J. R., and Farrukh, I. S., 1996, Regulation of Ca^{2+} activated K$^+$ channels in pulmonary vascular smooth muscle cells—role of nitric oxide, *J. Appl. Physiol.* **81**:1264–1272.

Petersson, J., Zygmunt, P. M., Brandt, L., and Högestätt, E. D., 1995, Substance P-induced relaxation and hyperpolarisation in human cerebral arteries, *Br. J. Pharmacol.* **115**:889–894.

Petersson, J., Zygmunt, P. M., and Högestätt, E. D., 1997, Characterization of the potassium channels involved in EDHF-mediated relaxation in cerebral arteries, *Br. J. Pharmacol.* **120**:1344–1350.

Pfister, S. L., Spitzbarth, N., Nithipatikom, K., Edgemond, W. S., and Campbell W. B., 1999, Endothelium-derived eicosanoids from lipoxygenase relax the rabbit aorta by opening potassium channels, in: *Endothelium-Dependent Hyperpolarizations* Vol. 2 (P. M. Vanhoutte, ed.), Harwood Academic Publishers, Amsterdam, 1999, pp. 17–28.

Plane, F., and Garland, C. J., 1994, Smooth muscle hyperpolarization and relaxation to acetylcholine in the rabbit basilar artery, *J. Autonom. Nerv. Syst.* **49**:S15–S18.

Plane, F., Pearson, T., and Garland, C. J., 1995, Multiple pathways underlying endothelium-dependent relaxation in the rabbit in isolated femoral artery, *Br. J. Pharmacol.* **115**:31–38.

Plane, F. Hurrell, A. Jeremy, J. Y., and Garland, C. J., 1996, Evidence that potassium channels make a major contribution to SIN-1-evoked relaxation of rat isolated mesenteric artery, *Br. J. Pharmacol.* **119**:1557–1562.

Plane, F., Holland, M., Waldron, G. J., Garland, C. J., and Boyle, J. P., 1997, Evidence that anandamide and EDHF act via different mechanisms in the rat isolated mesenteric arteries, *Br. J. Pharmacol.* **121:** 1509–1511.

Plane, F., Wiley, K. E., Jeremy, J. Y., Cohen, R. A., and Garland, C. J., 1998, Evidence that different mechanisms underlie smooth muscle relaxation to nitric oxide and nitric oxide donors in the rabbit isolated carotid artery. *Br. J. Pharmacol.* **123:**1351–1358.

Pomerantz, K., Sinterose, A., and Ramwell, P., 1978, The effect of prostacyclin on the human umbilical artery, *Prostaglandins* **15:**1035–1044.

Popp, R., Bauersachs, J., Sauer, E., Hecker, M., Fleming, I., and Busse, R., 1996, A transferable, β-naphthoflavone-inducible, hyperpolarizing factor is synthesized by native and cultured porcine coronary endothelial cells, *J. Physiol (London)* **497:**699–709.

Prasad, K., and Bharadwaj, L. A., 1996, Hydroxyl radical—a mediator of acetylcholine induced vascular relaxation, *J. Mol. Cell. Cardiol.* **28:**2033–2041.

Pratt, P. F., Edgemont, W. S., Hillard, C. J., and Campbell, W. B., 1998, N-Arachidonylethanolamide relaxation of bovine coronary arteries is not mediated by CB1 cannabinoid receptor, *Am. J. Physiol.* **274:**H375–H381.

Price, J. M., and Hellermann, A., 1997, Inhibition of cyclic-GMP mediated relaxation in small rat coronary arteries by block of Ca^{++}-activated K^+ channels, *Life Sci.* **61:**1185–1192.

Prior, H. M., Webster, N., Quinn, K., Beech, D. J., and Yates, M. S.,1998, K^+-induced dilation of a small renal artery: No role for inward rectifier K^+ channels, *Cardiovasc. Res.* **37:**780–790.

Quignard, J.-F., Chataigneau, T., Corriu, C., Duhault, J., Félétou, M., and Vanhoutte, P. M, 1999a, Effects of SIN-1 on potassium channels of vascular smooth muscle cells of the rabbit aorta and guinea-pig carotid artery, in: *Endothelium-Dependent Hyperpolarizations Factor* (P. M. Vanhoutte, ed.), Harwood Academic Publishers, Amsterdam, pp. 193–200.

Quignard, J.-F., Félétou, M., Thollon, C., Vilaine, J. P., Duhault, J., and Vanhoutte, P. M, 1999b, Potassium ions and endothelium-derived hyperpolarizing factors in guinea-pig carotid and porcine coronary arteries. *Br. J. Pharmacol.* **127:**27–34.

Quignard, J.-F., Chataigneau, T., Corriu, C., Duhault, J., Félétou, M., and Vanhoutte, P. M, 1999c, Potassium channels involved in EDHF-induced hyperpolarization of the smooth muscle cells of the isolated guinea-pig carotid artery. in: *Endothelium-Derived Hyperpolarization Factor* (P. M. Vanhoutte, ed.), Harwood Academic Publishers, Amsterdam, pp. 201–208.

Quilley, J., Fulton, D., and McGiff, J. C., 1997, Hyperpolarizing factors, *Biochem. Pharmacol.* **54:**1059–1070.

Randall, M. D., and Kendall, D. A., 1997, Involvement of a cannabinoid in endothelium-derived hyper-Qpolarizing factor-mediated coronary vasorelaxation, *Eur. J. Pharmacol.* **335:**205–209.

Randall, M. D., and Kendall, D. A., 1998, Anandamide and endothelium-derived hyperpolarizing factor act via a common vasorelaxant mechanism in rat mesentery. *Eur. J. Pharmacol.* **346:**51–53.

Randall, M. D., Alexander, S. P. H., Bennett, T., Boyd, E. A., Fry, J. R., Gardiner, S. M., Kemp, P. A., Mcculloch, A. I., and Kendall, D. A., 1996, An endogenous cannabinoid as an endothelium-derived vasorelaxant, *Biochem. Biophys. Res. Commun.* **229:**114–120.

Randall, M. D., Mcculloch, A. I., and Kendall, D. A., 1997, Comparative pharmacology of endothelium-derived hyperpolarizing factor and anandamide in rat isolated mesentery, *Eur. J. Pharmacol.* **333:**191–197.

Rapacon, M., Mieyal, P., McGiff, J. C., Fulton, D., and Quilley, J., 1996, Contribution of calcium-activated potassium channels to the vasodilator effect of bradykinin in the isolated, perfused kidney of the rat, *Br. J. Pharmacol.* **118:**1504–1508.

Rees, D. D., Palmer, R. M. J., Hodson, H. F., and Moncada, S., 1989, A specific inhibitor of nitric oxide formation from L-arginine attenuates endothelium-dependent relaxation, *Br. J. Pharmacol.* **96:**418–424.

Richard, V., Tanner, F. C., Tschudi, M. R., and Lüscher, T. F., 1990, Different activation of L-arginine pathway by bradykinin, serotonin, and clonidine in coronary arteries, *Am. J. Physiol.* **259:**H1433–H1439.

Robertson, B. E., Schubert, R., Hescheler, J., and Nelson, M. T., 1993, cyclic-GMP-dependent protein kinase activates Ca-activated K channels in cerebral artery smooth muscle cells, *Am. J. Physiol.* **265:**C299–C303.

Rosenblum, W. I., 1987, Hyodoxyl radical mediates the endothelium-dependent relaxation produced by bradykinin in mouse cerebral arterioles, *Circ. Res.* **61:**601–603.

Rosolowski, M., and Campbell, W. B., 1993, Role of PGI_2 and EETs in the relaxation of bovine coronary arteries to arachidonic acid, *Am. J. Physiol.* **264:**H327–H335.

Rosolowski, M., and Campbell, W. B., 1996, Synthesis of hydroxyeicosatetraenoic (HETEs) and epoxyeicosatrienoic acids (EETs) by cultured bovine coronary endothelial cells, *Biochim. Biophys. Acta* **1299:**267–277.

Rubanyi, G.,M., and Vanhoutte, P. M., 1986, Oxygen-derived free radicals, endothelium, and responsiveness of vascular smooth muscle, *Am. J. Physiol.* **250:**H815–H821.

Rubanyi, G. M., and Vanhoutte, P. M., 1987, Nature of endothelium-derived relaxing factor: Are there two relaxing mediators?, *Circ. Res.* **61:**II61–II67.

Ruiz, E., and Tejerina, T., 1998, Relaxant effects of L-citrulline in rabbit vascular smooth muscle, *Br. J. Pharmacol.* **125:**186–192.

Schubert, R., Serebryakov, N. V., Engel, H., and Hopp, H. H., 1996, Iloprost activates KCa channels of vascular smooth muscle cells: Role of cyclic-AMP-dependent proteine kinase, *Am. J. Physiol.* **271:**C1203–C1211.

Schubert, R., Serebryakov, N. V., Mewes, H., and Hopp, H. H., 1997, Iloprost dilates rat small arteries: Role of K$_{ATP}$ and K$_{Ca}$ channel activation by cyclic-AMP-dependent protein kinase, *Am. J. Physiol.* **272:**H1147–H1156.

Sedaa, K. O., Bjur, R. A., Shinozuka, K., and Westfall, D. P., 1990, Nerve and drugs-induced release of adenine nucleosides and nucleotides from rabbit aorta, *J. Pharmacol. Exp. Ther.* **252:**1060–1067.

Sheridan, B. C., McIntyre, R. C., Meldrum, D. R., and Fullerton, D. A., 1997, K$_{ATP}$ channels contribute to beta and adenosine receptor-mediated pulmonary vasorelaxation, *Am. J. Physiol.* **17:**L950–L956.

Shimizu, S., and Paul, R. J., 1998, The endothelium-dependent, substance P relaxation of porcine coronary arteries resistant to nitric oxide synthesis inhibition is partially mediated by 4-aminopyridine-sensitive voltage-dependent K$^+$ channels, *Endothelium* **5:**287–295.

Shimokawa, H., Flavahan, N. A., Lorenz, R. R., and Vanhoutte, P. M., 1988, Prostacyclin releases endothelium-derived relaxing factor and potentiates its action in coronary arteries of the pig, *Br. J. Pharmacol.* **95:**1197–1203.

Shin, J. H., Chung, S., Park, E. J., Uhm, D. Y., and Suh, C. K., 1997, Nitric oxide directly activates calcium-activated potassium channels from rat brain reconstituted into planar lipid bilayer, *FEBS Lett.* **415:**299–302.

Shinozuka, K., Hashimoto, M., Bjur, R. A., Westfall, W. P., and Hattori, K., 1994, In vitro studies of release of adenine nucleotides and adenosine from rat vascular endothelium in response to α_1-adrenoceptor stimulation, *Br. J. Pharmacol.* **113:**1203–1208.

Shryock, J. C., Rubio, R., and Berne, R. M., 1988, Release of adenosine from pig aortic endothelial cells during hypoxia and metabolic inhibition, *Am. J. Physiol.* **254:**H223–H229.

Siegel, G., Stock, G., Schnalke, F., and Litza, B., 1987, Electrical and mechanical effects of prostacyclin in canine carotid artery, in: *Prostacyclin and Its Stable Analogue Iloprost* (R. J. Gryglewski and G. Stock, eds.), Springer-Verlag, Berlin, pp. 143–149.

Siegel, G., Mironneau, J., Schnalke, F., Schroder, G., Schulz, B. G., and Grote, J., 1990, Vasodilatation evoked by K$^+$ channel opening, *Prog. Clin. Biol. Res.* **327:**229–306.

Siegel, G., Emden, J., Wenzel, K., Mironneau, J., and Stock G., 1992, Potassium channel activation and vascular smooth muscle, *Adv. Exp. Med. Biol.* **311:**53–72.

Simonsen, U., Garcia-Sacristan A., and Prieto, D., 1997, Apamin-sensitive K$^+$ channels involved in the inhibition of acetylcholine-induced contractions in lamb coronary small arteries, *Eur. J. Pharmacol.* **329:**153–163.

Smits, P., Williams, S. B., Lipson, D. E., Banitt, P., Ronge, G. A., and Creager, M. A., 1995, Endothelial release of nitric oxide contributes to the vasodilator effect of adenosine in humans, *Circulation* **92:**2135–2141.

Standen, N. B., Quayle, J. M., Davies, N. W., Brayden, J. E., Huang, Y., and Nelson, M. T., 1989, Hyperpolarizing vasodilators activate ATP-sensitive K$^+$ channels in arterial smooth muscle, *Science* **245:**177–180.

Sundquist, T., 1991, Bovine aortic endothelial cells release hydrogen peroxide, *J. Cell. Physiol.* **148:**152–156.

Suzuki, H., Chen, G., Yamamoto, Y., and Miwa, K., 1992, Nitroarginine-sensitive and insensitive components of the endothelium-dependent relaxation in the guinea-pig carotid artery, *Jpn. J. Physiol.* **42:**335–347.

Taguchi, H., Heistad, D. D., Chu, Y., Rios, C. D., Ooboshi, H., and Faraci, F. M., 1996, Vascular expression of inducible nitric oxide synthase isoform associated with activation of Ca^{++}-dependent K$^+$ channels, *J. Pharmacol. Exp. Ther.* **279:**1514–1519.

Taniguchi, J., Furukawa, K. I., and Shigekawa, M., 1993, Maxi K$^+$ channels are stimulated by cyclic guanosine monophosphate-dependent protein kinase in canine coronary artery smooth muscle cells, *Pflügers Arch. Eur. J. Physiol.* **423:**167–172.

Tare, M., Parkington, H. C., Coleman, H. A., Neild, T. O., and Dusting, G. J., 1990, Hyperpolarisation and relaxation of arterial smooth muscle caused by nitric oxide derived from the endothelium, *Nature* **346**:69–71.

Taylor, H. J., Chaytor, A. T., Evans, W. H., and Griffith, T. M., 1998, Inhibition of the gap junctional component of endothelium-dependent relaxations in rabbit iliac artery by 18β-glycyrrhetinic acid, *Br. J. Pharmacol.* **125**:1–3.

Taylor, S. G., Southerton, J. S., Weston, A. H., and Baker, J. R. J., 1988, Endothelium-dependent effects of acetylcholine in rat aorta: A comparison with sodium nitroprusside and cromakalim, *Br. J. Pharmacol.* **94**:853–863.

Urakami-Harasawa, L., Shimokawa, H., Nakashima, M., Egashira, K., and Takeshita, A., 1997, Importance of endothelium-derived hyperpolarizing factor in human arteries, *J. Clin. Invest.* **100**:2793–2799.

Van de Voorde, J., Vanheel, B., and Leusen, I., 1992, Endothelium-dependent relaxation and hyperpolarisation in aorta from control and renal hypertensive rats, *Circ. Res.* **70**:1–8.

Vanhoutte, P. M., 1998, An old-timer makes a come-back, *Nature* **396**:213–216.

Vanhoutte, P. M., and Félétou, M., 1996, Conclusion: Existence of multiple EDHF(s)?, in: *Endothelium-Derived Hyperpolarizing Factor*, Vol. 1 (P. M. Vanhoutte ed.), Harwood Academic Publishers, Amsterdam, pp. 303–307.

von der Weid, P.-Y., 1998, ATP-sensitive K$^+$ channels in smooth muscle cells of guinea-pig lymphatics: Role in nitric oxide and β-adrenoceptor agonist-induced hyperpolarizations, *Br. J. Pharmacol.* **125**:17–22.

Wallerstedt, S. M., and Bodelsson, M., 1997, Endothelium-dependent relaxations by substance P in human isolated omental arteries and veins: Relative contribution of prostanoids, nitric oxide and hyperpolarisation, *Br. J. Pharmacol.* **120**:25–30.

Wang, R., 1998, Resurgence of carbon monoxide: An endogenous gaseous vasorelaxing factor, *Can. J. Physiol. Pharmacol.* **76**:1–15.

Wang, R., and Wu, L. Y., 1997, The chemical modification of KCa channels by carbon monoxide in vascular smooth muscle cells, *J. Biol. Chem.* **272**:8222–8226.

Wang, R., Wang, Z. Z., and Wu, L. Y., 1997a, Carbon-monoxide-induced vasorelaxation and the underlying mechanisms. *Br. J. Pharmacol.* **121**:927–934.

Wang, R., Wu, L. Y., and Wang, Z. Z., 1997b, The direct effect of carbon monoxide on KCa channels in vascular smooth muscle cells, *Pflügers Arch. Eur. J. Physiol.* **434**:285–291.

Wei, C. M., Hu, S., Miller, V. M., and Burnett, J. C., 1994, Vascular actions of C-type natriuretic peptide in isolated porcine coronary arteries and coronary vascular smooth muscle cells, *Biochem. Biophys. Res. Commun.* **205**:765–771.

Weidelt, T., Boldt, W., and Markwardt, F. 1997, Acetylcholine-induced K$^+$ currents in smooth muscle of intact rat small arteries, *J. Physiol. (London)* **500**:617–630.

Weintraub, N. L., Stephenson, A. L., Sprague, R. S., and Lonigro, A. J., 1999, Role of phospholipase A2 in EDHF-mediated relaxation of the porcine coronary artery, in: *Endothelium-Dependent Hyperpolarizations*, Vol. 2 (P. M. Vanhoutte, ed.), Harwood Academic Publishers, Amsterdam, pp. 97–108.

Wellman, G. C., Bonev, A. D., Nelson, M. T., and Brayden, J. E., 1996, Gender differences in coronary artery diameter involve estrogen, nitric oxide and Ca^{2+}-dependent K$^+$ channels, *Circ. Res.* **79**:1024–1030.

Wellman, G. C., Quayle, J. M., and Standen, N. B., 1998, ATP-sensitive K$^+$ channel activation by calcitonin gene related peptide and protein kinase A in pig coronary arterial smooth muscle, *J. Physiol (London)* **507**:117–129.

White, R., and Hiley, C. R., 1997, A comparison of EDHF-mediated responses and anandamide-induced relaxations in the rat isolated mesenteric artery, *Br. J. Pharmacol.* **122**:1573–1584.

White, R., and Hiley, C. R., 1998, The actions of some cannabinoid receptor ligands in the rat isolated mesenteric artery, *Br. J. Pharmacol.* **125**:533–541.

Wise, H., and Jones, R. L., 1996, Focus on prostacyclin and its novel mimetics, *Trends Pharmacol. Sci.* **17**:17–21.

Yamamoto, Y., Fukuta, H., Nakahira, Y., and Suzuki, H., 1998, Blockade by 18β-glycyrrhetinic acid of intercellular electrical coupling in guinea-pig arterioles. *J. Physiol (London)* **511**:501–508.

Yamanaka, A., Ishikawa, K., and Goto, K., 1998, Characterization of endothelium-dependent relaxation independent of NO and prostaglandins in guinea-pig coronary artery, *J. Pharmacol. Exp. Ther.* **285**:480–489.

Zanzinger, J., Czachurski, J., and Seller, H., 1996, Role of calcium-dependent K$^+$ channels in the regulation of arterial and venous tone by nitric oxide in pigs, *Pflügers Arch. Eur. J. Physiol.* **432**:671–677.

Zhang, H., Stockbridge, N., Weir, B., Krueger, C., and Cook, D., 1991, Glibenclamide relaxes vascular smooth muscle constriction produced by prostaglandin F$_{2\alpha}$, Eur. J. Pharmacol. **195**:27–35.

Zhao, Y. J., and Wang, J., 1996, Pulmonary vasoconstriction effects of prostacyclin in rats: Potential role of thromboxane receptors, J. Appl. Physiol. **81**:2595–2603.

Zhao, Y. J., Wang, J., Rubin, L. J., and Yuan, X. J., 1997, Inhibition of Kv and K$_{Ca}$ channels antagonizes NO-induced relaxation in pulmonary artery, Am. J. Physiol. **41**:H904–H912.

Zou, A. P., Fleming, J. T., Falck, J. R., Jacobs, E. R., Gebremedhin, D., Harder, D. R., and Roman, R. J., 1996, 20-Hydroxyeicosatetraenoic acid is an endogenous inhibitor of the large conductance Ca^{++}-activated K$^+$ channel in renal arterioles, Am. J. Physiol. **270**:R228–R237.

Zygmunt, P. M., and Högestätt, E. D., 1996, Endothelium-dependent hyperpolarization and relaxation in the hepatic artery of the rat, in: Endothelium-Derived Hyperpolarizing Factor, Vol. 1 (P. M. Vanhoutte, ed.), Harwood Academic Publishers, Amsterdam, pp. 191–202.

Zygmunt, P. M., Grundemar, L., and Högestätt, E. D., 1994a. Endothelium-dependent relaxation resistant to Nω-nitro-L-arginine in the rat hepatic artery and aorta, Acta Physiol. Scand. **152**:107–114.

Zygmunt, P. M., Högestätt, E. D., and Grundemar, L., 1994b, Light-dependent effects of zinc protoporphyrin IX on endothelium-dependent relaxation resistant to Nω-nitro-L-arginine, Acta Physiol. Scand. **152**:137–143.

Zygmunt, P. M., Edwards, G., Weston, A. H., Davis, S. C., and Högestätt, E. D., 1996, Effects of cytochrome P450 inhibitors on EDHF-mediated relaxation in the rat hepatic artery, Br. J. Pharmacol. **118**:1147–1152.

Zygmunt, P. M., Edwards, G., Weston, A. H., Larsson, B., and Högestätt, E. D., 1997a, Involvement of voltage-dependent potassium channels in the EDHF-mediated relaxation of rat hepatic artery. Br. J. Pharmacol. **121**:141–149.

Zygmunt, P. M., Högestätt, E. D., Waldeck, K., Edwards, G., Kirkup A. J., and Weston, A. H., 1997b, Studies on the effects of anandamide in rat hepatic artery, Br. J. Pharmacol. **122**:1679–1686.

Zygmunt, P. M., Plane F., Paulsson, M., Garland, C. J., and Högestätt, E. D., 1998, Interactions between endothelium-derived relaxing factors in the rat hepatic artery: Focus on regulation of EDHF, Br. J. Pharmacol. **124**:992–1000.

Part VII

Potassium Channels in Cardiac Disease

Chapter 35

Overview: The Role of Potassium Channels in Cardiac Arrhythmias

Albert J. D'Alonzo, Paul C. Levesque, and Michael A. Blanar

1. INTRODUCTION

Potassium channels play an important role in the repolarization process of the cardiac action potential and, hence, determine action potential duration and myocardial refractoriness. Recent molecular, genetic, and electrophysiological studies in humans and in animal models have provided insights into the role of K^+ channels in arrhythmogenesis. Changes in the functional expression or biophysical properties of K^+ channels caused by pathophysiologic conditions, congenital mutations (inherited long QT syndrome), or pharmacologic intervention (acquired long QT syndrome) cause abnormal repolarization of the action potential, increased electrical dispersion, and an increased propensity for arrhythmia.

As shown in Fig. 1, K^+ currents that contribute to normal ventricular repolarization in the human myocardium include I_{to}, I_{Kr}, I_{Ks}, and I_{K1} (Sakmann *et al.*, 1983; Krapivinsky *et al.*, 1995; Wang *et al.*, 1995; Deal *et al.*, 1996; Nabauer and Kaab, 1998). Changes in K^+ currents have been noted under certain experimental or disease conditions including heart failure, ischemia, atrial fibrillation, and myocardial infarction. The majority of studies have reported a reduction in I_{to} and I_{K1} currents (Table I). Other currents either are affected minimally or have not been studied in detail.

The distribution of K^+ channels varies not only within the different chambers of the heart but also across the myocardial walls (Antzelevitch *et al.*, 1991; Li *et al.*, 1998). The heterogeneous distribution of K^+ channels from cell to cell contributes to differences in action potential duration (APD), which give the heart an inherent dispersion of repolarization. Dispersion can be measured using APD, QT interval, and

Albert J. D'Alonzo, Paul C. Levesque, and Michael A. Blanar • Department of Metabolic and Cardiovascular Drug Discovery, Bristol-Myers Squibb Pharmaceutical Research Institute, Princeton, New Jersey 08543-4000.

Potassium Channels in Cardiovascular Biology, edited by Archer and Rusch. Kluwer Academic/Plenum Publishers, New York, 2001.

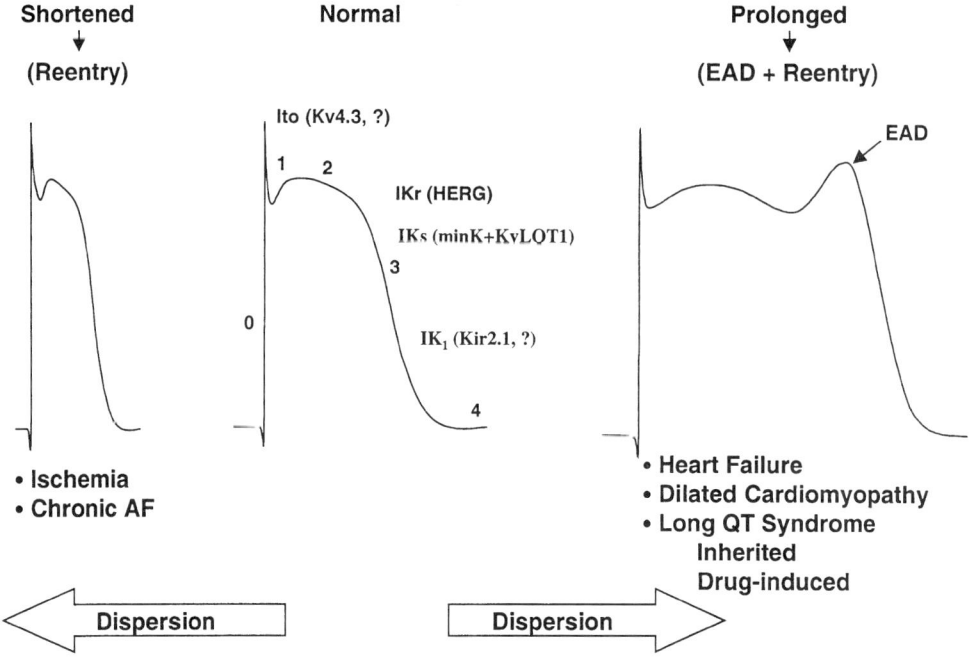

Figure 1. Action potential repolarization results from activation and inactivation of several ion channel conductances, primarily those involving K^+ channels. Immediately following the upstroke of the action potential (phase 0), a transient outward current (I_{to}) is activated, which is responsible for the notch in the action potential (phase 1). During the plateau (phase 2), the delayed rectifier currents (I_{Kr} and I_{Ks}) are activated and contribute to repolarization during phase 3. The inward rectifier current (I_{K1}) also is activated in phase 3, which allows full repolarization of the action potential (phase 4). Under certain pathophysiologic conditions, the action potential duration can either be prolonged (e.g., heart failure, dilated cardiomyopathy, or long QT syndrome) or shortened [e.g., ischemia or chronic atrial fibrillation (AF)]. Abnormal changes in the action potential duration increase the degree of electrical dispersion within the myocardium. Prolongation of the action potential facilitates the occurrence of early afterdepolarizations (EADs), which can initiate torsades de pointes. Shortening of the action potential can facilitate the development of reentrant arrhythmias and sudden cardiac death.

recovery time of the T-wave or refractory period of the myocardium; it also can be expressed as the difference, standard deviation, or variance of these values recorded from multiple sites (Ogawa *et al.*, 1991; Zaputovic *et al.*, 1997; Ducceschi *et al.*, 1998; Goldeli *et al.*, 1998). Increases in dispersion of refractoriness have been associated with arrhythmogenesis in animals (Michelson *et al.*, 1980; Gough *et al.*, 1985; Wetstein *et al.*, 1985; Kowey *et al.*, 1991; Ogawa *et al.*, 1991; D'Alonzo *et al.*, 1995) and humans (Morgan *et al.*, 1992; Haberman *et al.*, 1993; Puljevic *et al.*, 1997; Goldeli *et al.*, 1998).

Modulation of K^+ channels, through either congenital mutations or administration of pharmacologic agents, can potentially be proarrhythmic by either prolonging or shortening the APD (see Fig. 1). Triggered arrhythmias and reentrant arrhythmias are two major categories into which arrhythmias may be classified. Triggered arrhythmias, which may occur when K^+ currents are decreased, can initiate reentrant arrhythmias.

Table I
Alterations of Potassium Currents in Myocardial Pathophysiology

Model[a]	Species	APD[b]	I_{to}	I_{K1}	Other currents	Reference
A. Heart failure/dilated cardiomyopathy						
Hereditary CM	Hamster	↑	↓			Thuringer et al., 1996
Pacing-induced HF	Dog	↑	↓	↓		Kaab et al., 1996
Terminal HF	Human		↓			Nabauer et al., 1993
Terminal HF	Human	↑	±↓[a]	±↓		Wettwer et al., 1993
Terminal HF	Human		↓	±		Kaab et al., 1998
Pacing-induced HF	Rabbit		↓	±		Rozanski et al., 1997
Epinephrine-induced CH	Rat		↓			Mészáros et al., 1996
Hereditary CM	Hamster		↓	↓	↓I_K	Lodge and Normandin, 1997
Terminal HF	Human		↓			Beuckelmann et al., 1993
HF	Human (atria)	↑		↓	↓$I_{K_{Ach}}$	Koumi et al., 1994
Hypertrophy	Rabbit	↑	↓	↓	↓I_{Kur}	McIntosh et al., 1998
Hypertrophy	Rabbit	↑	↓	↓		Gillis et al., 1998
B. Atrial fibrillation						
Rapid pacing	Dog	↓	↓	±	±I_K	Yue et al., 1997
Chronic AF	Human		↓		↓I_{Kur}	Van Wagoner et al., 1997
C. Ischemia						
Computer modeling	—	↓			↑$I_{K_{ATP}}$	Shaw and Rudy, 1997
Decreased pH	Rat		↓		↑$I_{K_{ATP}}$	Xu and Rozanski, 1997
Hypoxia/ischemia	Guinea pig	↓				Faivre and Findlay, 1990
Hypoxia	Guinea pig				↑$I_{K_{ATP}}$	Pan and Zhou, 1997
D. Myocardial infarction						
48 h post infarction	Dog		↓			Jeck et al., 1995
3 wk post infarction	Rat		↓		±I_{Kur}	Gidh-Jain et al., 1996
48 h post infaction	Dog		↓	↓	↑I_K	Pinto and Boyden, 1998
16 wk post infarction	Rat		↓			Rozanski et al., 1998

[a] Abbreviations: AF, Atrial fibrillation; CH, cardiac hypertrophy; CM, cardiomyopathy; HF, heart failure.
[b] APD, Action potential duration.
[c] ±, no change.

Triggered arrhythmias present as monomorphic tachyarrhythmias (single site) or polymorphic tachyarrhythmias (multiple sites), which can precipitate ventricular fibrillation (el-Sherif et al., 1988; el-Sherif, 1991; Pogwizd and Corr, 1992). Reentrant arrhythmias, which may occur under nonpathophysiologic conditions (Frazier et al., 1989), are oscillatory myocardial electrical circuits that may be facilitated by a decrease in the APD (Uchida et al., 1999). A single circuit can be responsible for monomorphic tachyarrhythmias, whereas multiple circuits appear as polymorphic tachyarrhythmias (Pertsov et al., 1993; Schmitt et al., 1998). Multiple reentrant arrhythmias can lead to fibrillation (Abildskov, 1994). The normal myocardium has uniform anisotropic characteristics; (that is, the conduction of electrical activity proceeds faster in the longitudinal direction and slower in the transverse direction, in part owing to differential gap junction density. When the heart moves toward nonuniform anisotropic properties during disease (Papageorgiou et al., 1996; Chen, 1997) or following drug treatment (Coromilas et al., 1995; Lacroix et al., 1998), reentrant arrhythmias can be initiated and maintained more easily (Allessie et al., 1990; Spach and Josephson, 1994; Gray et al., 1996). Thus, reestablishment of uniform repolarization of the myocardium, through modulation of K^+ channel expression, should have antiarrhythmic activity.

In this chapter, we will review some of the more recent experimental approaches (e.g., human genetic linkage analysis, genetically engineered cellular and mouse models, and animal models of arrhythmia) that have increased our understanding of how K^+ channel modulation may contribute to the development of arrhythmias. Further elucidation of the role of specific K^+ ion channels in arrhythmogenesis will be crucial for the development of improved treatment strategies.

2. GENETIC MODELS OF ARRHYTHMIA

2.1. Inheritable Arrhythmias of Humans: The Long QT Syndrome

In an attempt to define the molecular mechanisms of human cardiac arrhythmogenesis, much work has focused on the study of the molecular basis of inheritable human cardiac arrhythmias, such as the long QT syndrome (LQTS). The hallmark of LQTS is a prolonged QT interval of the electrocardiogram (ECG) (>440 ms), which can be manifested as a genetic defect in the function of a cardiac ion channel (inherited long QT) or can be induced with pharmacologic agents (acquired long QT) (Keating and Sanguinetti, 1996; Sanguinetti et al., 1996a,b; Wang et al., 1998). Excessive prolongation of the QT interval can lead to triggered arrhythmias. Multiple triggered arrhythmias can present themselves as a polymorphic ventricular tachycardia called torsades de pointes. Agents that inhibit endogenous cardiac K^+ currents are capable of initiating torsades de pointes-like responses, and the use of these agents in experimental animals has formed the basis for many of the models of acquired LQTS.

Prolongation of the QT interval in the ECG of patients with congenital LQTS reflects delayed repolarization of the action potential, a function performed mainly by delayed rectifier K^+ currents. Thus, genes encoding K^+ channels are believed to be candidate genes for LQTS. Not surprisingly, the genes mutated at four LQTS-linked loci have been shown to encode either K^+ channel proteins or K^+ channel accessory subunits (for recent reviews, see Ackerman et al., 1998; Vincent, 1998). The genes identified at the LQT1 and LQT2 loci encode voltage-regulated K^+ (Kv) channels of the delayed rectifier type, KCNQ1 (KvLQT1) (Barhanin et al., 1996; Sanguinetti et al., 1996b; Wang et al., 1996; Yang et al., 1997) and human ether-a-go-go-related gene (HERG) (Jiang et al., 1994; Curran et al., 1995; Trudeau et al., 1995), respectively. LQT5 encodes KCNE1 (minK, I_{Ks}) (Schulze-Bahr et al., 1997b; Splawski et al., 1997), an accessory subunit that associates with the KCNQ1 channel (Barhanin et al., 1996; Sanguinetti et al., 1996a; Yang et al., 1997). Recently, a HERG accessory subunit, encoded by the KCNE2 gene, has been described, mutations in which are associated with LQTS (Abbott et al., 1999).

Mutations in HERG are associated with chromosome 7-linked LQTS (LQT2) (Curran et al., 1995; Benson et al., 1996; Sanguinetti et al., 1996a; Satler et al., 1996; Li *et al.*, 1997; Tanaka *et al.*, 1997). HERG encodes the rapidly activating component of the cardiac delayed rectifier K$^+$ current, I_{Kr} (Sanguinetti *et al.*, 1995; Trudeau *et al.*, 1995). This inward-rectifying current plays a major role in the repolarization of the human cardiac action potential. Interestingly, voltage protocols used to simulate premature depolarizations [i.e., early afterdepolarizations (EADs)] have demonstrated that fast recovery from inactivation allows I_{Kr} to carry robust outward current at depolarized voltages, suggesting that an inherent function of the HERG channel may be to protect against premature-depolarization-induced arrhythmias (Smith *et al.*, 1996). This characteristic of the HERG channel may provide an explanation for the propensity of some therapeutic drugs to cause torsades de pointes (e.g., terfenadine, cisapride, and sertindole).

All HERG-associated mutations characterized to date reduce I_{Kr}. Reduction of I_{Kr} would be expected to delay cardiomyocyte repolarization and result in a subsequent increase in the cardiac APD, thereby resulting in the observed clinical phenotype. Various LQTS-associated mutations in the HERG channel lead to a reduction in the I_{Kr} current by (1) significant reduction of the membrane-associated channel protein either by truncation or by alteration of HERG cellular trafficking (Furutani *et al.*, 1999); (2) dominant-negative inhibition of the wild-type HERG channel protein (Sanguinetti *et al.*, 1996a); or (3) effects on the kinetic properties of the HERG-encoded current (e.g., increased deactivation) (Chen *et al.*, 1999).

Mutations in KCNQ1 are associated with chromosome 11-linked long QT syndromes (Romano–Ward syndrome; Jervell and Lange-Nielsen syndrome) (Wang *et al.*, 1996). KCNQ1, in combination with KCNE1, encodes the slowly activating component of the delayed rectifier K$^+$ current, I_{Ks} (Barhanin *et al.*, 1996; Sanguinetti *et al.*, 1996b; Yang *et al.*, 1997). Most of the disease-causing mutations in KCNQ1 have been functionally expressed and have been demonstrated to decrease I_{Ks} (Chouabe *et al.*, 1997; Shalaby *et al.*, 1997; Mohammad-Panah *et al.*, 1999). As with I_{Kr}, I_{Ks} also plays a role in the repolarization of the human cardiac action potential. I_{Ks} does not inactivate during maintained depolarizations and deactivates slowly upon repolarization. Owing to these inherent properties, I_{Ks} current would be expected to play a greater role in repolarization at fast heart rates, and thus I_{Ks} current has been considered a potential target for class III antiarrhythmic drug development (Selnick *et al.*, 1997).

It is well known that I_{Ks} current amplitude is increased by sympathetic stimulation (Sanguinetti *et al.*, 1991). The sympathetic-induced increase in repolarizing I_{Ks} current may counterbalance stimulation of depolarizing Ca^{2+} currents, thereby permitting increased Ca^{2+} entry without significant and potentially arrhythmogenic prolongation of the action potential. Interestingly, increasing sympathetic tone in LQT1 patients by administration of isoproterenol (Compton *et al.*, 1998) or in animals by treatment with I_{Ks} blockers (Shimizu and Antzelevitch, 1998) results in excessive prolongation of the cardiac action potential and induction of torsades de pointes.

KCNE1 and KCNE2 colocalize to human chromosome 21q22.1 (Chevillard *et al.*, 1993; Abbott *et al.*, 1999); these genes encode small, membrane-spanning proteins that associate with the pore-forming subunits, KCNQ1 (Barhanin *et al.*, 1996; Sanguinetti *et al.*, 1996b; Yang *et al.*, 1997) and HERG (Abbott *et al.*, 1999), respectively. Mutations in these genes diminish K$^+$ currents carried by the pore-forming subunits, either by slowing activation or by speeding deactivation time constants (Schulze-Bahr *et al.*, 1997a; Splawski *et al.*, 1997; Tyson *et al.*, 1997; Abbott *et al.*, 1999).

2.2. Potassium Channels in Engineered Cellular and Mouse Models

In order to investigate the roles of specific K^+ channels in cardiac electrophysiology and in the pathogenesis of arrhythmias, recent research has focused on the generation of various genetically engineered models in both isolated cells and mice.

To determine if the mouse heart might provide a useful model of human LQTS, Babij *et al.* (1998) created transgenic mice that selectively express a LQTS-associated HERG mutation. Subsequent electrophysiological analyses demonstrated that I_{Kr} was not detectable in these transgenic mice. Although this mutation in humans was associated with a particularly severe form of LQTS, the ECG measurements in mice, including QT intervals, were unchanged. Prolongation of APD was observed only in isolated cardiomyocytes at slow rates and at room temperature, whereas at faster rates and at physiologic temperatures, no differences in APD were observed. Although this study demonstrated that human inherited mutations in HERG are capable of suppressing I_{Kr} current in an *in vivo* mouse model, it also highlights the interspecies differences in physiologic functions of cardiac K^+ channels.

Using a different approach to address the role of HERG/I_{Kr} in cardiac excitability, Nuss *et al.* (1999) overexpressed human HERG via adenoviral gene delivery in cultures of primary rabbit ventricular cardiomyocytes. HERG overexpression resulted in the enhancement of I_{Kr}, without affecting the expression of other K^+ currents. This enhancement of I_{Kr} shortened the APD, decreased the susceptibility to EADs, and increased the effective refractory period. This study demonstrates the importance of HERG/I_{Kr} in cardiac repolarization and may help to explain the propensity for arrhythmias in patients in whom I_{Kr} is suppressed either pharmacologically or congenitally.

To determine the function of the KCNE1-encoded subunit *in vivo*, a targeted disruption of the KCNE1 gene was engineered to generate null mutant mice (Charpentier *et al.*, 1998; Drici *et al.*, 1998). Although the homozygous KCNE1 knockout mice appeared to recapitulate the inner ear defect observed in the Jervell and Lange-Nielsen syndrome, action potential parameters from the ventricles of KCNE1 null mice were indistinguishable from those recorded for wild-type mice. At normal physiologic heart rates, no significant change in the QT interval was observed. The failure to reproduce a long QT syndrome phenotype in these mice may be the result of the continued presence of KCNQ1, which may provide sufficient repolarizing current in the absence of KCNE1, or may indicate that I_{Ks} plays a minimal role in cardiac repolarization in a species having a fast basal heart rate.

Utilizing a novel approach to the generation of a dominant-negative transgenic mouse model of long QT and ventricular arrhythmias, London *et al.* (1998a) disrupted the functional expression of the entire Kv1 family of voltage-gated K^+ channels. Overexpression of a truncated Kv1.1 polypeptide (Folco *et al.*, 1997) substantially reduced outward K^+ currents in ventricular cells derived from transgenic mice, possibly by interfering with normal cellular membrane trafficking of Kv1 channel family members. Action potentials and QT intervals, recorded from isolated transgenic cardiomyocytes and surface electrocardiogram, respectively, were prolonged significantly. In telemetered transgenic mice, an increase in the frequency of premature beats and episodes of ventricular tachycardia was observed. Therefore, despite the differences between murine and human cardiac electrophysiology, this model demonstrates the potential utility of genetically engineered mice in the investigation of the mechanisms by which prolongation of the QT interval contributes to the genesis of ventricular arrhythmias.

Following a similar approach, Barry et al. (1998) probed the molecular and functional characterization of the cardiac transient outward current, I_{to}. Cardiac-specific cell-surface expression of a nonconducting point mutant of Kv4.2 in mice selectively eliminated the transient outward current, thereby demonstrating that Kv4.2 encodes I_{to}. Elimination of I_{to} in these transgenic mice resulted in significant prolongation of the cardiac APD and QT interval. Interestingly, cardiomyocytes derived from these transgenic mice exhibited a sustained outward K$^+$ current, not observed in control littermates, suggesting that alterations in the expression of endogenous ion channels may result in compensatory remodeling. Surprisingly, spontaneous arrhythmias were not observed in these transgenic mice even though they exhibited significantly longer APDs than those recorded from the Kv1 dominant-negative mice generated by London et al. (1998b). These data suggest that other factors besides prolonged ventricular repolarization may contribute to the development of ventricular arrhythmias.

Targeted gene disruption by homologous recombination has also been used to associate a K$^+$ channel with its respective endogenous cardiac current. Specifically, deletion of Kir3.4 (GIRK4) abolished cardiac muscarinic-gated K$^+$ currents ($I_{K(ACh)}$; (Wickman et al., 1998). Although vagal stimulation reduced heart rate less effectively in the knockout mice than in littermate controls, confirming the role of $I_{K(ACh)}$ in the vagal control of heart rate in the mouse, remarkably, the resting heart rate in $I_{K(ACh)}$-deficient mice was not different from that observed in wild-type littermates.

3. ANIMAL MODELS OF ARRYTHMIA

3.1. Acute Ischemia

During acute ischemia, cellular levels of ATP are diminished. This, in turn, relieves the inhibition of ATP-sensitive K$^+$ (K$_{ATP}$) channels (for details, see Chapters 15 and 19). Measurements of K$^+$ channel activity under acute ischemic conditions have shown that activation of K$_{ATP}$ channel current ($I_{K_{ATP}}$) contributes to the majority of electrophysiological effects that are observed (Benndorf et al., 1992). To better understand the role of this channel in cardiac physiology, agents that open or activate K$_{ATP}$ channels have been developed. These agents increase K$^+$ conductance and reduce APD (Gross and Auchampach, 1992b). Activation of $I_{K_{ATP}}$ reduces the propensity for triggered arrhythmias but may increase the likelihood of reentrant arrhythmias (D'Alonzo et al., 1994; Baczkó et al., 1997; Uchida et al., 1999). The proarrhythmic effects of K$^+$ channel openers also may be indirectly linked to increased catecholamine release (D'Alonzo et al., 1998). In contrast, selective blockers of sarcolemmal K$_{ATP}$ channels may be useful in preventing ischemia-induced arrhythmias (Billman et al., 1998; Gögelein et al., 1998).

It is well established that following myocardial infarction there is an increased probability of cardiac arrhythmias and the potential for sudden cardiac death. There are many models of acute ischemia that have been developed in the rat (Mileham et *al.*, 1994; Akao *et al.*, 1997; Garlid *et al.*, 1997; Grover, 1997; Xu and Rozanski, 1997), dog (Pearlman *et al.*, 1980; Gross and Auchampach, 1992a; Billman, 1994; Grover *et al.*, 1995; Sabbah *et al.*, 1995; Vanoli *et al.*, 1995), guinea pig (Faivre and Findlay, 1990; Pan and Zhou, 1997; Novalija and Stowe, 1998; Shigematsu *et al.*, 1998; Southworth *et al.*, 1998), pig (Shen *et al.*, 1996; Rioufol *et al.*, 1997; Rohmann *et al.*, 1997), ferret (Elliott *et al.*, 1989; Galiñanes and Hearse, 1990; Bethell *et al.*, 1998), and cat (Stramba-Badiale *et al.*, 1995). These models encompass the use of regional ischemia, global ischemia, microsphere embolization, low-flow ischemia, changes in pH, ischemic-like solutions, hypoxia, occlusion/reperfusion, and permanent coronary occlusion models.

There are several common animal models of myocardial infarction (Nelson *et al.*, 1990; Amsterdam *et al.*, 1993; Ovize *et al.*, 1995; Chen *et al.*, 1996; Rohmann *et al.*, 1997). These models look at either short-term ischemia (1–3 h) followed by reperfusion and recovery (5 h to days of recovery) or permanent occlusion of a coronary vessel followed by recovery for hours to weeks. Short-term models of ischemia, i.e., recovery within 24 hours, are used to evaluate agents that protect against arrhythmias and/or limit infarct size. These models can be variable due to differences in the degree of ischemic insults, which varies among species (Shen and Vatner, 1996) owing to collateral circulation (Maxwell *et al.*, 1987) as well as differences in the ionic currents that underlie the cardiac action potential (Varró *et al.*, 1993).

Only a few studies have reported changes in K^+ conductance in animal models of acute ischemia. These studies include a 48-h-postinfarction model in the dog (Jeck *et al.*, 1995; Pinto and Boyden, 1998) and a several-week-postinfarction model in the rat (Gidh-Jain *et al.*, 1996; Rozanski *et al.*, 1998). In the dog model, results were obtained using subendocardial Purkinje fibers. In the ischemic region, changes in these fibers included marked action potential prolongation and reduced I_{to} density (Jeck *et al.*, 1995). In addition, it was observed that the inward rectifier current (I_{K1}) was reduced, accounting for an associated reduction in resting membrane potential (Pinto and Boyden, 1998). These Purkinje fibers also displayed an increased delayed rectifier current (I_K). In the rat model, expression of mRNA for the putative I_{to} current components (Kv1.4, Kv2.1, and Kv4.2) was decreased substantially three weeks after myocardial infarction, whereas expression of Kv1.2 and Kv1.5 was unchanged (Gidh-Jain *et al.*, 1996). These results correlate well with electrophysiological results that demonstrate reductions in APD and I_{to} current levels (Gidh-Jain *et al.*, 1996; Rozanski *et al.*, 1998).

3.2. Heart Failure and Hypertrophy

Cardiac hypertrophy associated with chronic heart failure is characterized by decreased left ventricular function, reduced exercise tolerance, impaired quality of life, and increased mortality. Approximately one-half of patients in severe heart failure [class IV of the New York Heart Association (NYHA) classification] die from sudden cardiac death within five years, presumably mediated by ventricular tachyarrhythmias (Birgersdotter-Green *et al.*, 1991). A reduction in repolarizing cardiac K^+ currents and an associated prolongation of the APD are believed to predispose the myocardium to increased electrical instability and an enhanced propensity for arrhythmogenesis.

In heart failure, the current density of I_{to} is markedly reduced (Beuckelmann *et al.*, 1993; Nabauer *et al.*, 1993). This reduction in current density may be explained by the downregulation of Kv4.3, the gene product that underlies I_{to} (Kaab *et al.*, 1998). In addition, abnormal neurohormonal action may modulate expression of K⁺ channels. For example, I_{K1} is inhibited by protein kinase A (PKA)-mediated phosphorylation, and this modulation is significantly reduced in human ventricular myocytes isolated from failing hearts, as compared to nonfailing hearts (Koumi *et al.*, 1995). This downregulation is caused by increased amounts of neurotransmitter as a part of a complex series of changes in the sympathetic nervous system (Bristow *et al.*, 1990). These changes also may affect the regulation of other ion channels.

Rapid ventricular pacing is used to develop heart failure in dogs (Williams *et al.*, 1994; Holzgrefe *et al.*, 1996; Friedman *et al.*, 1998; Moe *et al.*, 1998; Solomon *et al.*, 1998), pigs (Spinale *et al.*, 1989; Chow *et al.*, 1990; Spinale *et al.*, 1990; Tuininga *et al.*, 1996; Cox *et al.*, 1997), sheep (Rademaker *et al.*, 1995, 1996), and rabbits (Ezzaher *et al.*, 1991; Freeman and Colston, 1992; Spinale *et al.*, 1994). In myocytes isolated from failing dog hearts, in which failure is induced by 3–4 weeks of rapid ventricular pacing, the peak of the Ca^{2+}-independent transient outward current (I_{to}) is reduced dramatically compared to that in nonfailing (unpaced) control hearts (Kaab *et al.*, 1996). Whereas no significant differences in I_{to} kinetics or single-channel conductance are observed, the number of functional I_{to} channels is reduced. In addition to a reduction in I_{to}, there also is a marked reduction of I_{K1}. Changes in I_{to} and I_{K1} may explain the prolongation of APD and the reduced resting potential observed under these conditions. Similar reductions in I_{to} are observed in failing rabbit hearts following 2–3 weeks of rapid ventricular pacing (Rozanski *et al.*, 1997), but no change in I_{K1} is observed under these conditions. Reductions in I_{to} during heart failure are observed widely across species (Table II).

In hypertension-induced cardiac hypertrophy, there is a prolongation of the cardiac APD and an increased risk of proarrhythmia (Kulan *et al.*, 1998). Two models of renovascular hypertension have been used to study cardiac hypertrophy: (1) two-kidney, one-clip (2K-1C) rats, which exhibit a high-renin hypertension with a normal plasma volume; and (2) one-kidney, one-clip (1K-1C) rats, which show normal renin serum levels but a raised plasma volume. Expression of Kv4.2 and Kv4.3 mRNAs was diminished by more than 50% in ventricles of 2K-1C and 1K-1C rats, with no changes observed in the expression levels of Kv1.2, Kv1.4, Kv1.5, Kv2.1, or KvLQT1 (KCNQ1) mRNAs (Kulan *et al.*, 1998). These effects are likely to be mediated primarily by an increase in cardiac afterload (Takimoto *et al.*, 1997) and are not unanticipated, because changes in cardiac afterload have been shown to produce significant effects on cardiac repolarization (Dean and Lab, 1989, 1990).

3.3. Atrial Fibrillation

Atrial fibrillation is characterized by disorganized atrial depolarizations and ineffective atrial contraction due to multiple reentrant wavelets. This arrhythmia occurs typically in patients with coronary vascular disease and can present as a paroxysmal (sporadic episodes) and/or persistent form. Although atrial fibrillation usually results

Table II
Drugs Associated with QT Interval Prolongation

Class	Drug	Chemical name	Channel	Reference(s)
Antihistamines	Seldane	Terfenadine	$I_{Kr}, I_{Ks}, I_{to}, I_{Kur}, I_{Ca}, I_{Na}$	Rampe et al., 1993; Crumb et al., 1995; Hanrahan et al., 1995; Ming and Nordin, 1995; Salata et al., 1995; Yang et al., 1995a
	Hismanol	Astemizole	I_{Kr}	Sakemi and VanNatta, 1993; Salata et al., 1995; Vorperian et al., 1996
	Benadryl	Diphenhydramine	I_{Kr}	Zareba et al., 1997; Khalifa et al., 1999
Antibiotics	Iaxin, Zithromax, E-Mycin, EES, Erypeds	Erythromycin	I_{Kr}	Nattel et al., 1990; Brandriss et al., 1994; Wong and Windle, 1995; Antzelevitch et al., 1996
	Bactrim, Septra, Proloprim	Trimethoprim and sulfamethoxazole	???	Lopez et al., 1987
	Pentam, i.v.	Pentamidine	???	Wharton et al., 1987; Eisenhauer et al., 1994; Cardoso et al., 1997; Girgis et al., 1997
		Spiramycin	???	Stramba-Badiale et al., 1997; Verdun et al., 1997
Antiarrhythmics	Quinidine, Quindex	Quinidine	I_{Kr}, I_{to}, I_{Kur}	Fieldman et al., 1977; Roden et al., 1986, 1988; Clark et al., 1995; Wang et al., 1995
	Duraquin, Quiniglute	Procainamide	???	Piergies et al., 1987; Ellenbogen et al., 1993
	Tikosyn	Dofetilide	I_{Kr}	Gwilt et al., 1991; Demolis et al., 1996; Lande et al., 1998

			Channels	References
		Disopyramide		Sadanaga et al., 1993; Furushima et al., 1998
	Pronestyl	Sotalol	I_{Kr}	Uematsu et al., 1994; Fiset et al., 1997
	Betapace	d-Sotalol	I_{Kr}	Campbell, 1987; Funck-Brentano, 1993
	Vascor	Bepridil	I_{K1}, I_K, I_{to}	Hollingshead et al., 1992; Prystowsky, 1992; Coumel et al., 1993; Chouabe et al., 1998
	Stedicor	Azimilide	$I_{Kr}, I_{Ks}, I_{Ca}, I_{Na}$	Fermini et al., 1995; Busch et al., 1997; Groh et al., 1997; Yao and Tseng, 1997; Busch et al., 1998; Gintant, 1998
	Cordarone	Amiodarone	$I_{Kr}, I_{Ks}, I_{Na}, I_{Ca}$	Morgan et al., 1991; Lazzara, 1993; Kodama et al., 1997
		Lidoflazine	???	Hanley and Hampton, 1983; Fazekas and Szekeres, 1990
	Corvert	Ibutilide	I_{Kr}	Yang et al., 1995b; Naccarelli et al., 1996
		Almokalant	I_{Kr}	Carlsson et al., 1993; Carmeliet, 1992a,b; Darpo et al., 1995; Farkas et al., 1996; Darpo and Edvardsson, 1997; Houltz et al., 1998
Gastrointestinal agents	Propulsid	Cisapride	I_{Kr}	Bran et al., 1995; Lupoglazoff et al., 1997; Rampe et al., 1997; Khongphatthanayothin et al., 1998
Antifungal agents	Nizoral	Ketoconazole	I_{Kr}, I_{kur}	Zimmermann et al., 1992; Honig et al., 1993; Smith, 1994; Tran, 1994; Janeira, 1995; Dumaine et al., 1998
	Diflucan	Fluconazole	???	Albengres et al., 1998
	Sporanox	Itraconazole	???	

Table II (*Continued*)

Class	Drug	Chemical name	Channel	Reference(s)
Psychotropic agents	Elavil, Norpramine, Viractil, Compazine, Stelazine, Thorazine, Mellaril, Etrafon	Amitriptyline	I_{to}	Dumovic et al., 1976; Nishimoto et al., 1994; Casis and Sanchez-Chapula, 1998
	Trilafon	Phenothiazine	???	Papaceit et cl., 1990; Napolitano et al., 1994
	Haldol	Haloperidol	I_{Kr}	Kriwisky et al., 1990; Metzger and Friedman, 1993; Suessbrich et al., 1997
		Imipramine	???	Alderton, 1995
	Sinequin	Doxepin	???	Baker et al., 1997
		Sertindole	I_{Kr}	Rampe et al., 1998
	Orap	Pimozide	???	Krähenbuhl et al., 1995
		Thioridazine	I_{Kr}, I_{Ks}	Drolet et al., 1999
Diuretics	Lozol	Indapamide	I_{Ks}, I_{to}, I_{Kur}	Guzzini et al., 1990; Turgeon et al., 1994; Li et al., 1996; Lu et al., 1998
Antiserotinergic agents		Ketanserin	I_{to}, I_{sus}	Arfiero et al., 1990; Zhang et al., 1994; Frishman et al., 1995
		Zimeldine	???	Liljeqvist and Edvardsson, 1989
Antimalarial agents		Halofantrine	???	Baty et al., 994; Toivonen et al., 1994; Ebert et al., 1998
Antispasmodic agents (urinary incontinence)	Mictrol	Terodiline	I_{Kr}	van der Klauw et al., 1992; Jones et al., 1998
Hypocholesterolemic agents	Lorelco	Probucol	???	Browne et al., 1984
Anticancer drugs	Nolvadex	Tamoxifen	I_{Kr}, I_{Ca}	Trump et al., 1992; Liu et al., 1998

from underlying cardiac disease, it also can occur in normal individuals following emotional stress, exercise, surgery, or acute alcohol intoxication. A common feature of atrial fibrillation is a reduction in the atrial APD or refractory period and an increased dispersion of electrical properties. These changes can occur in the presence of high parasympathetic or sympathetic tone (Coumel, 1994, 1996; Olsson, 1996; Liu and Nattel, 1997; Chen *et al.*, 1998; Olgin *et al.*, 1998; Wen *et al.*, 1998).

Investigators have mimicked atrial arrhythmias in animal models using surgical (Frame *et al.*, 1986; Jalil *et al.*, 1997), chemical (Mizuno and Ogawa, 1981; Winslow, 1981; Fukuda *et al.*, 1983; Satoh and Zipes, 1998), nerve stimulation (Schuessler *et al.*, 1992; Geddes *et al.*, 1996), talc-induced pericarditis (Sokoloski *et al.*, 1997), rapid electrical pacing (Peters *et al.*, 1994; Morillo *et al.*, 1995; Elvan *et al.*, 1996; Shinbane *et al.*, 1997; Yue *et al.*, 1997; Wood *et al.*, 1998), and electrical stimulation models (Wijffels *et al.*, 1995). Despite these many animal models of atrial fibrillation, only several studies have looked at changes in K⁺ conductances in relationship to the disease (Van Wagoner *et al.*, 1997; Yue *et al.*, 1997). The changes observed in these studies, however, are not consistent with a decrease in either APD or refractory period, which are the hallmark findings of atrial fibrillation. Thus, additional studies will be needed to determine the contribution of specific K⁺ channels to the pathogenesis of this conduction disorder.

3.4. Drug-Induced QT Prolongation

Numerous models have been developed to chemically induce torsades de pointes. For example, some models use Na⁺ channel activators to mimic the Na⁺ channel defect in SCN5A long QT syndrome or use cesium (a nonselective K⁺ channel blocker) to reproduce polymorphic ventricular tachyarrhythmias with characteristics of torsades de pointes. These models of torsades de pointes have been reviewed recently (Eckardt *et al.*, 1998). Interestingly, many of these pharmacologic models target K⁺ channels that are associated with hereditary LQTS, for example, the HERG channels that underlie I_{Kr}. Drug-induced LQTS results from the interaction of pharmacologic agents with one or more of the major repolarizing currents of the cardiac action potential (Table II). Although many antiarrhythmic agents were designed to block I_K, there are many noncardiac prescription drugs that also interact with this channel. Drugs that block I_{Kr} also are frequently used to develop and characterize models of torsades de pointes (Carlsson *et al.*, 1990; Vos *et al.*, 1995; Zabel *et al.*, 1997; Volders *et al.*, 1998; D'Alonzo *et al.*, 1999). These animal models have proven to be highly predictive of agents later shown to be proarrhythmic in humans.

4. CONCLUSION

Taken together, these studies highlight the importance of K⁺ channels in cardiac repolarization and help explain the propensity for inherited arrhythmias in patients in whom specific K⁺ currents are suppressed. Despite the differences between murine and human cardiac electrophysiology, genetically engineered mouse models are proving useful in investigations of mechanisms underlying the generation of ventricular arrhythmias and confirm the central and important role of K^+ channels in the regulation of normal cardiac electrophysiology and pathophysiology. Continued investigations of inheritable arrhythmias and utilization of animal models will clarify further the role of K^+ channels in cardiac arrhythmogenesis.

REFERENCES

Abbott, G. W., Sesti, F., Splawski, I., Buck, M. E., Lehmann, M. H., Timothy, K. W., Keating, M. T., and Goldstein, S. A., 1999, MiRP1 forms I_{Kr} potassium channels with HERG and is associated with cardiac arrhythmia, *Cell* **97**:175–187.

Abildskov, J. A., 1994, Additions to the wavelet hypothesis of cardiac fibrillation, *J. Cardiovasc. Electrophysiol.* **5**:553–559.

Ackerman, M. J., Schroeder, J. J., Berry, R., Schaid, D. J., Porter, C. J., Michels, V. V., and Thibodeau, S. N., 1998, A novel mutation in KVLQT1 is the molecular basis of inherited long QT syndrome in a near-drowning patient's family, *Pediatr. Res.* **44**:148–153.

Akao, M., Otani, H., Horie, M., Takano, M., Kuniyasu, A., Nakayama, H., Kouchi, I., Sasayama, S., and Murakami T., 1997, Myocardial ischemia induces differential regulation of K_{ATP} channel gene expression in rat hearts, *J. Clin. Invest.* **100**:3053–3059.

Albengres, E., Le Lout, H., and Tillement, J. P., 1998, Systemic antifungal agents: Drug interactions of clinical significance, *Drug Saf.* **18**:83–97.

Alderton, H. R., 1995, Tricyclic medication in children and the QT interval: Case report and discussion, *Can. J. Psychiatry* **40**:325–329.

Allessie, M. A., Schalij, M. J., Kirchhof, C. J., Boersma, L., Huyberts, M., and Hollen, J., 1990, Electrophysiology of spiral waves in two dimensions: The role of anisotropy, *Ann. N.Y. Acad. Sci.* **591**:247–256.

Amsterdam, E. A., Pan, H. L., Rendig, S. V., Symons, J. D., Fletcher, M. P., and Longhurst, J. C., 1993, Limitation of myocardial infarct size in pigs with a dual lipoxygenase–cyclooxygenase blocking agent by inhibition of neutrophil activity without reduction of neutrophil migration, *J. Am. Coll. Cardiol.* **22**:1738–1744.

Antzelevitch, C., Sicouri, S., Litovsky, S. H., Lukas, A., Krishnan, S. C., Di Diego, J. M., Gintant, G. A., and Liu, D. W., 1991, Heterogeneity within the ventricular wall: Electrophysiology and pharmacology of epicardial, endocardial, and M cells, *Circ. Res.* **69**:1427–1449.

Antzelevitch, C., Sun, Z. Q., Zhang, Z. Q., and Yan, G. X., 1996, Cellular and ionic mechanisms underlying erythromycin-induced long QT intervals and torsade de pointes, *J. Am. Coll. Cardiol.* **28**:1836–1848.

Arfiero, S., Ometto, R., and Vincenzi, M., 1990, Prolongation of the QT interval and torsade de pointes caused by ketanserin, *Gazz. Ital. Cardiol.* **20**:869–872.

Babij, P., Askew, G. R., Nieuwenhuijsen, B., Su, C. M., Bridal, T. R., Jow, B., Argentieri, T. M., Kulik, J., DeGennaro, L. J., Spinelli, W., and Colatsky, T. J., 1998, Inhibition of cardiac delayed rectifier K^+ current by overexpression of the long-QT syndrome HERG G628S mutation in transgenic mice, *Circ. Res.* **83**:668–678.

Baczkó, I., Leprán, I., and Papp, J. G., 1997, KATP channel modulators increase survival rate during coronary occlusion–reperfusion in anaesthetized rats, *Eur. J. Pharmacol.* **324**:77–83.

Baker, B., Dorian, P., Sandor, P., Shapiro, C., Schell, C., Mitchell J., and Irvine, M. J., 1997, Electrocardiographic effects of fluoxetine and doxepin in patients with major depressive disorder, *J. Clin. Psychopharmacol.* **17**:15–21.

Barhanin, J., Lesage, F., Guillemare, E., Fink, M., Lazdunski, M., and Romey, G., 1996, K_VLQT1 and IsK (minK) proteins associate to form the I_{Ks} cardiac potassium current, *Nature* **384**:78–80.

Barry, D. M., Xu, H., Schuessler, R. B., and Nerbonne, J. M., 1998, Functional knockout of the transient outward current, long-QT syndrome, and cardiac remodeling in mice expressing a dominant-negative Kv4 alpha subunit, *Circ. Res.* **83**:560–567.

Baty, C. J., Sweet, D. C., and Keene, B. W., 1994, Torsades de pointes-like polymorphic ventricular tachycardia in a dog, *J. Vet. Int. Med.* **8**:439–442.

Benndorf, K., Bollmann, G., Friedrich, M., and Hirche H., 1992, Anoxia induces time-independent K$^+$ current through K$_{ATP}$ channels in isolated heart cells of the guinea-pig, *J. Physiol. (London)* **454**:339–357.

Benson, D. W., MacRae, C. A., Vesely, M. R., Walsh, E. P., Seidman, J. G., Seidman, C. E., and Satler, C. A., 1996, Missense mutation in the pore region of HERG causes familial long QT syndrome, *Circulation* **93**:1791–1795.

Bethell, H. W., Vandenberg, J. I., Smith, G. A., and Grace, A. A., 1998, Changes in ventricular repolarization during acidosis and low-flow ischemia, *Am. J. Physiol.* **275**:H551–H561.

Beuckelmann, D. J., Näbauer M., and Erdmann, E., 1993, Alterations of K$^+$ currents in isolated human ventricular myocytes from patients with terminal heart failure, *Circ. Res.* **73**:379–385.

Billman, G. E., 1994, Role of ATP sensitive potassium channel in extracellular potassium accumulation and cardiac arrhythmias during myocardial ischaemia, *Cardiovasc. Res.* **28**:762–769.

Billman, G. E., Englert, H. C., and Schölkens, B. A., 1998, HMR 1883, a novel cardioselective inhibitor of the ATP-sensitive potassium channel. Part II: Effects on susceptibility to ventricular fibrillation induced by myocardial ischemia in conscious dogs, *J. Pharmacol. Exp. Ther.* **286**:1465–1473.

Birgersdotter-Green, U., Rosenqvist, M., and Ryden, L., 1991, Effect of congestive heart failure treatment on incidence and prognosis of ventricular tachyarrhythmias, *J. Cardiovasc. Pharmacol.* **17**:S53–S58.

Bran, S., Murray, W. A., Hirsch, I. B., and Palmer, J. P., 1995, Long QT syndrome during high-dose cisapride, *Arch. Intern. Med.* **155**:765–768.

Brandriss, M. W., Richardson, W. S., and Barold, S. S., 1994, Erythromycin-induced QT prolongation and polymorphic ventricular tachycardia (torsades de pointes): Case report and review, *Clin. Infect. Dis.* **18**:995–998.

Bristow, M. R., Hershberger, R. E., Port, J. D., Gilbert, E. M., Sandoval, A., Rasmussen, R., Cates, A. E., and Feldman, A. M., 1990, Beta-adrenergic pathways in nonfailing and failing human ventricular myocardium, *Circulation* **82**:I12–25.

Browne, K. F., Prystowsky, E. N., Heger, J. J., Cerimele, B. J., Fineberg, N., and Zipes, D. P., 1984, Prolongation of the QT interval induced by probucol: Demonstration of a method for determining QT interval change induced by a drug, *Am. Heart J.* **107**:680–684.

Busch, A. E., Busch, G. L., Ford, E., Suessbrich, H., Lang, H. J., Greger, R., Kunzelmann, K., Attali, B., and Stuhmer, W., 1997, The role of the IsK protein in the specific pharmacological properties of the I$_{Ks}$ channel complex, *Br. J. Pharmacol.* **122**:187–189.

Busch, A. E., Eigenberger, B., Jurkiewicz, N. K., Salata, J. J., Pica, A., Suessbrich, H., and Lang F., 1998, Blockade of HERG channels by the class III antiarrhythmic azimilide: Mode of action, *Br. J. Pharmacol.* **123**:23–30.

Campbell, T. J., 1987, Cellular electrophysiological effects of D- and DL-sotalol in guinea- pig sinoatrial node, atrium and ventricle and human atrium: Differential tissue sensitivity, *Br. J. Pharmacol.* **90**:593–599.

Cardoso, J. S., Mota-Miranda, A., Conde, C., Moura, B., Rocha-Goncalves, F., and Lecour, H., 1997, Inhalatory pentamidine therapy and the duration of the QT interval in HIV-infected patients, *Int. J. Cardiol.* **59**:285–289.

Carlsson, L., Almgren, O., and Duker G., 1990, QTU-prolongation and torsades de pointes induced by putative class III antiarrhythmic agents in the rabbit: Etiology and interventions, *J. Cardiovasc. Pharmacol.* **16**:276–285.

Carlsson, L., Abrahamsson, C., Andersson, B., Duker, G., and Schiller-Linhardt, G., 1993, Proarrhythmic effects of the class III agent almokalant: Importance of infusion rate, QT dispersion, and early afterdepolarisations, *Cardiovasc. Res.* **27**:2186–2193.

Carmeliet, E., 1993a, Use-dependent block and use-dependent unblock of the delayed rectifier K$^+$ current by almokalant in rabbit ventricular myocytes, *Circ. Res.* **73**:857–868.

Carmeliet, E., 1993b, Use-dependent block of the delayed K$^+$ current in rabbit ventricular myocytes, *Cardiovasc. Drugs Ther.* **7** (Suppl. 3):599–604.

Casis, O., and Sanchez-Chapula, J. A., 1998, Mechanism of block of cardiac transient outward K$^+$ current (I$_{to}$) by antidepressant drugs, *J. Cardiovasc. Pharmacol.* **32**:527–534.

Charpentier, F., Merot, J., Riochet, D., Le Marec, H., and Escande, D., 1998, Adult KCNE1-knockout mice exhibit a mild cardiac cellular phenotype, *Biochem. Biophys. Res. Commun.* **251**:806–810.

Chen, C., Chen, L., Fallon, J. T., Ma, L., Li, L., Bow, L., Knibbs, D., McKay, R., Gillam, L. D., and Waters, D. D., 1996, Functional and structural alterations with 24-hour myocardial hibernation and recovery after reperfusion: A pig model of myocardial hibernation, *Circulation* **94**:507–516.

Chen, J., Zou, A., Splawski, I., Keating, M. T., and Sanguinetti, M. C., 1999, Long QT syndrome-associated mutations in the Per-Arnt-Sim (PAS) domain of HERG potassium channels accelerate channel deactivation, *J. Biol. Chem.* **274**:10113–101138.

Chen, P. S., 1997, Computerized mapping of ventricular fibrillation, *Heart Vessels Suppl.* **12**:224–227.

Chen, Y. J., Chen, S. A., Tai, C. T., Wen, Z. C., Feng, A. N., Ding, Y. A., and Chang, M. S., 1998, Role of atrial electrophysiology and autonomic nervous system in patients with supraventricular tachycardia and paroxysmal atrial fibrillation, *J. Am. Coll. Cardiol.* **32**:732–738.

Chevillard, C., Attali, B., Lesage, F., Fontes, M., Barhanin, J., Lazdunski, M., and Mattei, M. G., 1993, Localization of a potassium channel gene (KCNE1) to 21q22.1-q22.2 by in situ hybridization and somatic cell hybridization, *Genomics* **15**:243–245.

Chouabe, C., Neyroud, N., Guicheney, P., Lazdunski, M., Romey, G., and Barhanin, J., 1997, Properties of KvLQT1 K^+ channel mutations in Romano–Ward and Jervell and Lange-Nielsen inherited cardiac arrhythmias, *EMBO J.* **16**:5472–5479.

Chouabe, C., Drici, M. D., Romey, G., Barhanin, J., and Lazdunski, M., 1998, HERG and KvLQT1/IsK, the cardiac K^+ channels involved in long QT syndromes, are targets for calcium channel blockers, *Mol. Pharmacol.* **54**:695–703.

Chow, E., Woodard, J. C., and Farrar, D. J., 1990, Rapid ventricular pacing in pigs: An experimental model of congestive heart failure, *Am. J. Physiol.* **258**:H1603–H1605.

Clark, R. B., Sanchez-Chapula, J., Salinas-Stefanon, E., Duff, H. J., and Giles, W. R., 1995, Quinidine-induced open channel block of K^+ current in rat ventricle, *Br. J. Pharmacol.* **115**:335–343.

Compton, S. J., Zhang, L., G. M. V., Timothy, K. W., Sanguinetti, M. C., Green, L., Lux, R. L., and Mason, J. W., 1998, Adrenergic Ca^{2+} AND K^+ currents in patients with KvLQT1 mutations, *Circulation (Supple.)* **98**:1776.

Coromilas, J., Saltman, A. E., Waldecker, B., Dillon, S. M., and Wit, A. L., 1995, Electrophysiological effects of flecainide on anisotropic conduction and reentry in infarcted canine hearts, *Circulation* **91**:2245–2263.

Coumel, P., 1994, Paroxysmal atrial fibrillation: Role of autonomic nervous system, *Arch. Mal. Coeur Vaiss.* **87** (Spec. No. 3):55–62.

Coumel, P., 1996, Autonomic influences in atrial tachyarrhythmias, *J. Cardiovasc. Electrophysiol.* **7**:999–1007.

Coumel, P., Leclercq, J. F., Naditch, L., and Pellerin, D., 1993, Evaluation of drug-induced QT interval modifications in dynamic electrocardiography: The case of bepridil, *Fundam. Clin. Pharmacol.* **7**:61–68.

Cox, M. H., O, S. J., de Gasparo, M., Mukherjee, R., Hewett, K. W., and Spinale, F. G., 1997, Myocardial electrophysiological properties in the presence of an AT1 angiotensin II receptor antagonist, *Basic Res. Cardiol.* **92**:129–138.

Crumb, W. J., Jr., Wible, B., Arnold, D. J., Payne, J. P., and Brown, A. M., 1995, Blockade of multiple human cardiac potassium currents by the antihistamine terfenadine: Possible mechanism for terfenadine-associated cardiotoxicity, *Mol. Pharmacol.* **47**:181–190.

Curran, M. E., Splawski, I., Timothy, K. W., Vincent, G. M., Green, E. D., and Keating, M. T., 1995, A molecular basis for cardiac arrhythmia: HERG mutations cause long QT syndrome, *Cell* **80**:795–803.

D'Alonzo, A. J., Darbenzio, R. B., Hess, T. A., Sewter, J. C., Sleph, P. G., and Grover, G. J., 1994, Effect of potassium on the action of the K_{ATP} modulators cromakalim, pinacidil, or glibenclamide on arrhythmias in isolated perfused rat heart subjected to regional ischaemia, *Cardiovasc. Res.* **28**:881–887.

D'Alonzo, A. J., Sewter, J. C., Darbenzio, R. B., and Hess, T. A., 1995, Effects of dofetilide on electrical dispersion and arrhythmias in post-infarcted anesthetized dogs, *Basic Res. Cardiol.* **90**:424–434.

D'Alonzo, A. J., Zhu, J. L., Darbenzio, R. B., Dorso, C. R., and Grover, G. J., 1998, Proarrhythmic effects of pinacidil are partially mediated through enhancement of catecholamine release in isolated perfused guinea-pig hearts, *J. Mol. Cell. Cardiol.* **30**:415–423.

D'Alonzo, A. J., Zhu, J. L., and Darbenzio, R. B., 1999, Effects of class III antiarrhythmic agents in an in vitro rabbit model of spontaneous torsades de pointe, *Eur. J. Pharmacol.* **369**:57–64.

Darpo, B., and Edvardsson N., 1997, Effects of almokalant, a class III antiarrhythmic agent, on supraventricular, reentrant tachycardias. Almokalant Paroxysmal Supraventricular Tachycardia Study Group, *Cardiovasc. Drugs Ther.* **11**:499–508.

Darpo, B., Vallin, H., Almgren, O., Bergstrand, R., Insulander, P., and Edvardsson, N., 1995, Selective Ik blocker almokalant exhibits class III — specific effects on the repolarization and refractoriness of the human heart: A study of healthy volunteers using right ventricular monophasic action potential recordings, *J. Cardiovasc. Pharmacol.* **26**:530–540.

Deal, K. K., England, S. K., and Tamkun, M. M., 1996, Molecular physiology of cardiac potassium channels, *Physiol. Rev.* **76**:49–67.

Dean, J. W., and Lab, M. J., 1989, Arrhythmia in heart failure: Role of mechanically induced changes in electrophysiology, *Lancet* **i**:1309–1312.

Dean, J. W., and Lab, M. J., 1990, Regional changes in ventricular excitability during load manipulation of the in situ pig heart, *J. Physiol. (London)* **429**:387–400.

Demolis, J. L., Funck-Brentano, C., Ropers, J., Ghadanfar, M., Nichols, D. J., and Jaillon, P., 1996, Influence of dofetilide on QT-interval duration and dispersion at various heart rates during exercise in humans, *Circulation* **94**:1592–1599.

Drici, M. D., Arrighi, I., Chouabe, C., Mann, J. R., Lazdunski, M., Romey, G., and Barhanin, J., 1998, Involvement of IsK-associated K$^+$ channel in heart rate control of repolarization in a murine engineered model of Jervell and Lange-Nielsen syndrome, *Circ. Res.* **83**:95–102.

Drolet, B., Vincent, F., Rail, J., Chahine, M., Deschnes, D., Nadeau, S., Khalifa, M., Hamelin, B. A., and Turgeon, J., 1999, Thioridazine lengthens repolarization of cardiac ventricular myocytes by blocking the delayed rectifier potassium current, *J. Pharmacol. Exp. Ther.* **288**:1261–1268.

Ducceschi, V., Sarubbi, B., D'Andrea, A., Liccardo, B., Briglia, N., Carozza, A., Marmo, J., Santangelo, L., Iacono, A., and Cotrufo M., 1998, Increased QT dispersion and other repolarization abnormalities as a possible cause of electrical instability in isolated aortic stenosis, *Int. J. Cardiol.* **64**:57–62.

Dumaine, R., Roy, M. L., and Brown, A. M., 1998, Blockade of HERG and Kv1.5 by ketoconazole, *J. Pharmacol. Exp. Ther.* **286**:727–735.

Dumovic, P., Burrows, G. D., Vohra, J., Davies, B., and Scoggins, B. A., 1976, The effect of tricyclic antidepressant drugs on the heart, *Arch. Toxicol.* **35**:255–262.

Ebert, S. N., Liu, X. K., and Woosley, R. L., 1998, Female gender as a risk factor for drug-induced cardiac arrhythmias: Evaluation of clinical and experimental evidence, *J. Womens Health* **7**:547–557.

Eckardt, L., Haverkamp, W., Borggrefe, M., and Breithardt, G., 1998, Experimental models of torsade de pointes, *Cardiovasc. Res.* **39**:178–193.

Eisenhauer, M. D., Eliasson, A. H., Taylor, A. J., Coyne, P. E., Jr., and Wortham, D. C., 1994, Incidence of cardiac arrhythmias during intravenous pentamidine therapy in HIV-infected patients, *Chest* **105**:389–395.

Ellenbogen, K. A., Wood, M. A., and Stambler, B. S., 1993, Procainamide: A perspective on its value and danger, *Heart Dis. Stroke* **2**:473–476.

Elliott, A. C., Smith, G. L., and Allen, D. G., 1989, Simultaneous measurements of action potential duration and intracellular ATP in isolated ferret hearts exposed to cyanide, *Circ. Res.* **64**:583–591.

el-Sherif, N., 1991, Early afterdepolarizations and arrhythmogenesis: Experimental and clinical aspects, *Arch. Mal. Coeur Vaiss.* **84**:227–234.

el-Sherif, N., Zeiler, R. H., Craelius, W., Gough, W. B., and Henkin, R., 1988, QTU prolongation and polymorphic ventricular tachyarrhythmias due to bradycardia-dependent early afterdepolarizations: Afterdepolarizations and ventricular arrhythmias, *Circ. Res.* **63**:286–305.

Elvan, A., Wylie, K., and Zipes, D. P., 1996, Pacing-induced chronic atrial fibrillation impairs sinus node function in dogs: Electrophysiological remodeling, *Circulation* **94**:2953–2960.

Ezzaher, A., el Houda Bouanani, N., Su, J. B., Hittinger, L., and Crozatier, B., 1991, Increased negative inotropic effect of calcium-channel blockers in hypertrophied and failing rabbit heart, *J. Pharmacol. Exp. Ther.* **257**:466–471.

Faivre, J. F., and Findlay, I., 1990, Action potential duration and activation of ATP-sensitive potassium current in isolated guinea-pig ventricular myocytes, *Biochim. Biophys. Acta* **1029**:167–172.

Farkas, A., Leprán, I., and Papp, J. G., 1996, Effect of almokalant, a specific inhibitor of I$_{Kr}$, on myocardial ischaemia–reperfusion induced arrhythmias in rabbits, *Acta Physiol. Hung.* **84**:281–282.

Fazekas, T., and Szekeres, L., 1990, Analysis of the proarrhythmic action of lidoflazine (Clinium), *Acta Physiol. Hung.* **75**:229–236.

Fermini, B., Jurkiewicz, N. K., Jow, B., Guinosso, P. J., Jr., Baskin, E. P., Lynch, J. J., Jr., and Salata, J. J., 1995, Use-dependent effects of the class III antiarrhythmic agent NE-10064 (azimilide) on cardiac repolarization: Block of delayed rectifier potassium and L-type calcium currents, *J. Cardiovasc. Pharmacol.* **26**:259–271.

Fieldman, A., Beebe, R. D., and Sing Sum Chow, M., 1977, The effect of quinidine sulfate on QRS duration and QT and systolic time intervals in man, *J. Clin. Pharmacol.* **17**:134–139.

Fiset, C., Drolet, B., Hamelin, B. A., and Turgeon, J., 1997, Block of I$_{Ks}$ by the diuretic agent indapamide modulates cardiac electrophysiological effects of the class III antiarrhythmic drug *dl*-sotalol, *J. Pharmacol. Exp. Ther.* **283**:148–156.

Folco, E., Mathur, R., Mori, Y., Buckett, P., and Koren, G., 1997, A cellular model for long QT syndrome. Trapping of heteromultimeric complexes consisting of truncated Kv1.1 potassium channel polypeptides and native Kv1.4 and Kv1.5 channels in the endoplasmic reticulum, *J. Biol. Chem.* **272**:26505–26510.

Frame, L. H., Page, R. L., and Hoffman, B. F., 1986, Atrial reentry around an anatomic barrier with a partially refractory excitable gap: A canine model of atrial flutter, *Circ. Res.* **58**:495–511.

Frazier, D. W., Wolf, P. D., Wharton, J. M., Tang, A. S., Smith, W. M., and Ideker, R. E., 1989, Stimulus-induced critical point: Mechanism for electrical initiation of reentry in normal canine myocardium, *J. Clin. Invest.* **83**:1039–1052.

Freeman, G. L., and Colston, J. T., 1992, Myocardial depression produced by sustained tachycardia in rabbits, *Am. J. Physiol.* **262**:H63–H67.

Friedman, P. A., Foley, D. A., Christian, T. F., and Stanton, M. S., 1998, Stability of the defibrillation probability curve with the development of ventricular dysfunction in the canine rapid paced model, *Pacing Clin. Electrophysiol.* **21**:339–351.

Frishman, W. H., Huberfeld, S., Okin, S., Wang, Y. H., Kumar, A., and Shareef, B., 1995, Serotonin and serotonin antagonism in cardiovascular and non-cardiovascular disease, *J. Clin. Pharmacol.* **35**:541–572.

Fukuda, H., Goto, M., Kondo, M., Uchigasaki, T., and Sekino, Y., 1983, Effects of lorcainide, a new antiarrhythmic agent, on experimental cardiac arrhythmias, *Folia Pharmacol. Jpn.* **81**:115–126.

Funck-Brentano, C., 1993, Pharmacokinetic and pharmacodynamic profiles of *d*-sotalol and *d,l*-sotalol, *Eur. Heart J.* **14** (Suppl. H):30–35.

Furushima, H., Niwano, S., Chinushi, M., Ohhira, K., Abe, A., and Aizawa, Y., 1998, Relation between bradycardia dependent long QT syndrome and QT prolongation by disopyramide in humans, *Heart* **79**:56–58.

Furutani, M., Trudeau, M. C., Hagiwara, N., Seki, A., Gong, Q., Zhou, Z., Imamura, S., Nagashima, H., Kasanuki, H., Takao, A., Momma, K., January, C. T., Robertson, G. A., and Matsuoka, R., 1999, Novel mechanism associated with an inherited cardiac arrhythmia: Defective protein trafficking by the mutant HERG (G601S) potassium channel, *Circulation* **99**:2290–2294.

Galiñanes, M., and Hearse, D. J., 1990, Species differences in susceptibility to ischemic injury and responsiveness to myocardial protection, *Cardioscience* **1**:127–143.

Garlid, K. D., Paucek, P., Yarov-Yarovoy, V., Murray, H. N., Darbenzio, R. B., D'Alonzo, A. J., Lodge, N. J., Smith, M. A., and Grover, G. J., 1997, Cardioprotective effect of diazoxide and its interaction with mitochondrial ATP-sensitive K$^+$ channels: Possible mechanism of cardioprotection, *Circ. Res.* **81**:1072–1082.

Geddes, L. A., Hinds, M., Babbs, C. F., Tacker, W. A., Schoenlein, W. E., Elabbady, T., Saeed, M., Bourland, J. D., and Ayers, G. M., 1996, Maintenance of atrial fibrillation in anesthetized and unanesthetized sheep using cholinergic drive, *Pacing Clin. Electrophysiol.* **19**:165–175.

Gidh-Jain, M., Huang, B., Jain, P., and el-Sherif, N., 1996, Differential expression of voltage-gated K$^+$ channel genes in left ventricular remodeled myocardium after experimental myocardial infarction, *Circ. Res.* **79**:669–675.

Gillis, A. M., Geonzon, R. A., Mathison, H. J., Kulisz, E., Lester, W. M., and Duff, H. J., 1998, The effects of barium, dofetilide and 4-aminopyridine (4-AP) on ventricular repolarization in normal and hyper-trophied rabbit heart, *J. Pharmacol. Exp. Ther.* **285**:262–270.

Gintant, G. A., 1998, Azimilide causes reverse rate-dependent block while reducing both components of delayed-rectifier current in canine ventricular myocytes, *J. Cardiovasc. Pharmacol.* **31**:945–953.

Girgis, I., Gualberti, J., Langan, L., Malek, S., Mustaciuolo, V., Costantino, T., and McGinn, T. G., 1997, A prospective study of the effect of I.V. pentamidine therapy on ventricular arrhythmias and QTc prolongation in HIV-infected patients, *Chest* **112**:646–653.

Gögelein, H., Hartung, J., Englert, H. C., and Schölkens, B. A., 1998, HMR 1883, a novel cardioselective inhibitor of the ATP-sensitive potassium channel. Part I: Effects on cardiomyocytes, coronary flow and pancreatic beta-cells, *J. Pharmacol. Exp. Ther.* **286**:1453–1464.

Goldeli, O., Dursun, E., and Komsuoglu, B., 1998, Dispersion of ventricular repolarization: A new marker of ventricular arrhythmias in patients with rheumatoid arthritis, *J. Rheumatol.* **25**:447–450.

Gough, W. B., Mehra, R., Restivo, M., Zieler, R. H., and El-Sherif, N., 1985, Reentrant ventricular arrhythmias in the late myocardial infarction period in the dog: 13. Correlation of activation and refractory maps, *Circ. Res.* **57**:432–442.

Gray, R. A., Pertsov, A. M., and Jalife, J., 1996, Incomplete reentry and epicardial breakthrough patterns during atrial fibrillation in the sheep heart, *Circulation* **94**:2649–2661.

Groh, W. J., Gibson, K. J., and Maylie, J. G., 1997, Comparison of the rate-dependent properties of the class III antiarrhythmic agents azimilide (NE-10064) and E-4031: Considerations on the mechanism of reverse rate-dependent action potential prolongation, *J. Cardiovasc. Electrophysiol.* **8**:529–536.

Gross, G. J., and Auchampach, J. A., 1992a, Blockade of ATP-sensitive potassium channels prevents myocardial preconditioning in dogs, *Circ. Res.* **70:**223–233.

Gross, G. J., and Auchampach, J. A., 1992b, Role of ATP dependent potassium channels in myocardial ischaemia, *Cardiovasc. Res.* **26:**1011–1016.

Grover, G. J., 1997, Pharmacology of ATP-sensitive potassium channel (K$_{ATP}$) openers in models of myocardial ischemia and reperfusion, *Can. J. Physiol. Pharmacol.* **75:**309–315.

Grover, G. J., D'Alonzo, A. J., Parham, C. S., and Darbenzio, R. B., 1995, Cardioprotection with the K$_{ATP}$ opener cromakalim is not correlated with ischemic myocardial action potential duration, *J. Cardiovasc. Pharmacol.* **26:**145–152.

Guzzini, F., Baroffio, R., Coppetti, D., and Gasparini, P., 1990, Severe ventricular arrhythmia secondary to indapamide-induced hypopotassemia, *Clin. Ter.* **135:**283–287.

Gwilt, M., Arrowsmith, J. E., Blackburn, K. J., Burges, R. A., Cross, P. E., Dalrymple, H. W., and Higgins, A. J., 1991, UK-68,798: A novel, potent and highly selective class III antiarrhythmic agent which blocks potassium channels in cardiac cells, *J. Pharmacol. Exp. Ther.* **256:**318–324.

Haberman, R. J., Rials, S. J., Stohler, J. L., Marinchak, R. A., and Kowey, P. R., 1993, Evidence for a reexcitability gap in man after treatment with type I antiarrhythmic drugs, *Am. Heart J.* **126:**1121–1126.

Hanley, S. P., and Hampton, J. R., 1983, Ventricular arrhythmias associated with lidoflazine: Side-effects observed in a randomized trial, *Eur. Heart J.* **4:**889–893.

Hanrahan, J. P., Choo, P. W., Carlson, W., Greineder, D., Faich, G. A., and Platt, R., 1995, Terfenadine-associated ventricular arrhythmias and QTc interval prolongation. A retrospective cohort comparison with other antihistamines among members of a health maintenance organization, *Ann. Epidemiol.* **5:**201–209.

Hollingshead, L. M., Faulds, D., and Fitton, A., 1992, Bepridil: A review of its pharmacological properties and therapeutic use in stable angina pectoris, *Drugs* **44:**835–857.

Holzgrefe, H., Max, J. M., and D'Alonzo, A. J., 1996, Rate-dependent changes in aortic pressure and flow during tachycardia-induced heart failure: Mechanisms and implications, in: *Pathophysiology of Tachycardia-Induced Heart Failure* (F. G. Spinale, ed.), Futura Publishing Co., Armonk, New York, pp. 153–172.

Honig, P. K., Wortham, D. C., Zamani, K., Conner, D. P., Mullin, J. C., and Cantilena, L. R., 1993, Terfenadine–ketoconazole interaction: Pharmacokinetic and electrocardiographic consequences, *J. Am. Med. Assoc.* **269:**1513–1518.

Houltz, B., Darpo, B., Edvardsson, N., Blomstrom, P., Brachmann, J., Crijns, H. J., Jensen, S. M., Svernhage, E., Vallin, H., and Swedberg, K., 1998, Electrocardiographic and clinical predictors of torsades de pointes induced by almokalant infusion in patients with chronic atrial fibrillation or flutter: A prospective study, *Pacing Clin. Electrophysiol.* **21:**1044–1057.

Jalil, E., Laflamme, M., and Kus, T., 1997, Effects of procainamide on the excitable gap composition in a canine model of atrial flutter, *Can. J. Physiol. Pharmacol.* **75:**1–8.

Janeira, L. F., 1995, Torsades de pointes and long QT syndromes, *Am. Fam. Physician* **52:**1447–1453.

Jeck, C., Pinto, J., and Boyden, P., 1995, Transient outward currents in subendocardial Purkinje myocytes surviving in the infarcted heart, *Circulation* **92:**465–473.

Jiang, C., Atkinson, D., Towbin, J. A., Splawski, I., Lehmann, M. H., Li H., Timothy, K., Taggart, R. T., Schwartz, P. J., and Vincent, G. M., et al., 1994, Two long QT syndrome loci map to chromosomes 3 and 7 with evidence for further heterogeneity, *Nat. Genet.* **8:**141–147.

Jones, S. E., Ogura, T., Shuba, L. M., and McDonald, T. F., 1998, Inhibition of the rapid component of the delayed-rectifier K$^+$ current by therapeutic concentrations of the antispasmodic agent terodiline, *Br. J. Pharmacol.* **125:**1138–1143.

Kaab, S., Nuss, H. B., Chiamvimonvat, N., O'Rourke, B., Pak, P. H., Kass, D. A., Marban, E., and Tomaselli, G. F., 1996, Ionic mechanism of action potential prolongation in ventricular myocytes from dogs with pacing-induced heart failure, *Circ. Res.* **78:**262–273.

Kaab, S., Dixon, J., Duc, J., Ashen, D., Näbauer, M., Beuckelmann, D. J., Steineck, G., McKinnon, D., and Tomaselli, G. F., 1998, Molecular basis of transient outward potassium current downregulation in human heart failure: A decrease in Kv4.3 mRNA correlates with a reduction in current density, *Circulation* **98:**1383–1393.

Keating, M. T., and Sanguinetti, M. C., 1996, Molecular genetic insights into cardiovascular disease, *Science* **272:**681–685.

Khalifa, M., Drolet, B., Daleau, P., Lefez, C., Gilbert, M., Plante, S., O'Hara, G. E., Gleeton, O., Hamelin, B. A., and Turgeon, J., 1999, Block of potassium currents in guinea pig ventricular myocytes and lengthening of cardiac repolarization in man by the histamine H1 receptor antagonist diphenhydramine, *J. Pharmacol. Exp. Ther.* **288**:858–865.

Khongphatthanayothin, A., Lane, J., Thomas, D., Yen, L., Chang, D., and Bubolz, B., 1998, Effects of cisapride on QT interval in children, *J. Pediatr.* **133**:51–56.

Kodama I., Kamiya K., and Toyama J., 1997, Cellular electropharmacology of amiodarone, *Cardiovasc. Res.* **35**:13–29.

Koumi, S., Arentzen, C. E., Backer, C. L., and Wasserstrom, J. A., 1994, Alterations in muscarinic K^+ channel response to acetylcholine and to G protein-mediated activation in atrial myocytes isolated from failing human hearts, *Circulation* **90**:2213–2224.

Koumi, S., Backer, C. L., Arentzen, C. E., and Sato, R., 1995, β-Adrenergic modulation of the inwardly rectifying potassium channel in isolated human ventricular myocytes: Alteration in channel response to β-adrenergic stimulation in failing human hearts, *J. Clin. Invest.* **96**:2870–2881.

Kowey, P. R., Friehling, T. D., Sewter, J., Wu, Y., Sokil, A., Paul, J., and Nocella, J., 1991, Electrophysiological effects of left ventricular hypertrophy: Effect of calcium and potassium channel blockade, *Circulation* **83**:2067–2075.

Krähenbuhl, S., Sauter, B., Kupferschmidt, H., Krause, M., Wyss, P. A., and Meier, P. J., 1995, Case report: Reversible QT prolongation with torsades de pointes in a patient with pimozide intoxication, *Am. J. Med. Sci.* **309**:315–316.

Krapivinsky, G., Gordon, E. A., Wickman, K., Velimirovic, B., Krapivinsky, L., and Clapham, D. E., 1995, The G-protein-gated atrial K^+ channel IKACh is a heteromultimer of two inwardly rectifying K^+-channel proteins, *Nature* **374**:135–141.

Kriwisky, M., Perry, G. Y., Tarchitsky, D., Gutman, Y., and Kishon, Y., 1990, Haloperidol-induced torsades de pointes, *Chest* **98**:482–484.

Kulan, K., Ural, D., Komsuoglu, B., Agacdiken, A., Goldeli, O., and Komsuoglu, S. S., 1998, Significance of QTc prolongation on ventricular arrhythmias in patients with left ventricular hypertrophy secondary to essential hypertension, *Int. J. Cardiol.* **64**:179–184.

Lacroix, D., Delfaut, P., Adamantidis, M., Cardinal, R., Klug, D., Kacet, S., and Dupuis, B., 1998, Differential effects of quinidine, flecainide, and cibenzoline on anisotropic conduction in the isolated porcine heart, *J. Cardiovasc. Electrophysiol.* **9**:55–69.

Lande, G., Maison-Blanche, P., Fayn, J., Ghadanfar, M., Coumel, P., and Funck-Brentano, C., 1998, Dynamic analysis of dofetilide-induced changes in ventricular repolarization, *Clin. Pharmacol. Ther.* **64**:312–321.

Lazzara, R., 1993, Antiarrhythmic drugs and torsade de pointes, *Eur. Heart J.* **14** (Suppl. H):88–92.

Li, G. R., Feng, J., Yue, L., Carrier, M., and Nattel, S., 1996, Evidence for two components of delayed rectifier K^+ current in human ventricular myocytes, *Circ. Res.* **78**:689–696.

Li, G. R., Feng, J., Yue, L., and Carrier, M., 1998, Transmural heterogeneity of action potentials and I_{to1} in myocytes isolated from the human right ventricle, *Am. J. Physiol.* **275**:H369–377.

Li, X., Xu, J., and Li, M., 1997, The human $\Delta1261$ mutation of the HERG potassium channel results in a truncated protein that contains a subunit interaction domain and decreases the channel expression, *J. Biol. Chem.* **272**:705–708.

Liljeqvist, J. A., and Edvardsson, N., 1989, Torsade de pointes tachycardias induced by overdosage of zimeldine, *J. Cardiovasc. Pharmacol.* **14**:666–670.

Liu, L., and Nattel, S., 1997, Differing sympathetic and vagal effects on atrial fibrillation in dogs: Role of refractoriness heterogeneity, *Am. J. Physiol.* **273**:H805–H816.

Liu, X. K., Katchman, A., Ebert, S. N., and Woosley, R. L., 1998, The antiestrogen tamoxifen blocks the delayed rectifier potassium current, I_{Kr}, in rabbit ventricular myocytes, *J. Pharmacol. Exp. Ther.* **287**:877–883.

Lodge, N. J., and Normandin, D. E., 1997, Alterations in I_{to1}, I_{Kr} and I_{k1} density in the BIO TO-2 strain of syrian myopathic hamsters, *J. Mol. Cell. Cardiol.* **29**:3211–3221.

London, B., Jeron, A., Zhou, J., Buckett, P., Han, X., Mitchell, G. F., and Koren, G., 1998a, Long QT and ventricular arrhythmias in transgenic mice expressing the N terminus and first transmembrane segment of a voltage-gated potassium channel, *Proc. Natl. Acad. Sci. U.S.A.* **95**:2926–2931.

London, B., Wang, D. W., Hill, J. A., and Bennett, P. B., 1998b, The transient outward current in mice lacking the potassium channel gene Kv1.4, *J. Physiol. (London)* **509**(Pt. 1):171–182.

Lopez, J. A., Harold, J. G., Rosenthal, M. C., Oseran, D. S., Schapira, J. N., and Peter, T., 1987, QT prolongation and torsades de pointes after administration of trimethoprim-sulfamethoxazole, *Am. J. Cardiol.* **59**:376–377.

Lu, Y., Yue, L., Wang, Z., and Nattel, S., 1998, Effects of the diuretic agent indapamide on Na$^+$, transient outward, and delayed rectifier currents in canine atrial myocytes, *Circ. Res.* **83:**158–166.

Lupoglazoff, J. M., Bedu, A., Faure, C., Denjoy, I., Casasoprana, A., Cezard, J. P., and Aujard, Y., 1997, Long QT syndrome under cisapride in neonates and infants, *Arch. Pediatr.* **4:**509–514.

Maxwell, M. P., Hearse, D. J., and Yellon, D. M., 1987, Species variation in the coronary collateral circulation during regional myocardial ischaemia: A critical determinant of the rate of evolution and extent of myocardial infarction, *Cardiovasc. Res.* **21:**737–746.

McIntosh, M. A., Cobbe, S. M., Kane, K. A., and Rankin, A. C., 1998, Action potential prolongation and potassium currents in left-ventricular myocytes isolated from hypertrophied rabbit hearts, *J. Mol. Cell. Cardiol.* **30:**43–53.

Mészáros, J., Ryder, K. O., and Hart G., 1996, Transient outward current in catecholamine-induced cardiac hypertrophy in the rat, *Am. J. Physiol.* **271:**H2360–H2367.

Metzger, E., and Friedman, R., 1993, Prolongation of the corrected QT and torsades de pointes cardiac arrhythmia associated with intravenous haloperidol in the medically ill, *J. Clin. Psychopharmol.* **13:**128–132.

Michelson, E. L., Spear, J. F., and Moore, E. N., 1980, Electrophysiologic and anatomic correlates of sustained ventricular tachyarrhythmias in a model of chronic myocardial infarction, *Am. J. Cardiol.* **45:**583–590.

Mileham, K. A., Northover, B. J., Winter, A. W., Hundal, S. P., Brammar, W. J., and Conley E. C., 1994, Competitive titration for probing low-abundance ion channel mRNA molecules in normal and regionally-ischaemic heart tissue, *Mol. Cell. Probes* **8:**161–176.

Ming, Z., and Nordin, C., 1995, Terfenadine blocks time-dependent Ca^{2+}, Na$^+$, and K$^+$ channels in guinea pig ventricular myocytes, *J. Cardiovasc. Pharmacol.* **26:**761–769.

Mizuno, K., and Ogawa, K., 1981, Increased concentration of plasma cyclic GMP during aconitine-induced atrial fibrillation in dogs and paroxysmal atrial fibrillation in patients, *J. Cardiovasc. Pharmacol.* **3:**1211–1220.

Moe, G. W., Albernaz, A., Naik, G. O., Kirchengast, M., and Stewart, D. J., 1998, Beneficial effects of long-term selective endothelin type A receptor blockade in canine experimental heart failure, *Cardiovasc. Res.* **39:**571–579.

Mohammad-Panah, R., Demolombe, S., Neyroud, N., Guicheney, P., Kyndt, F., van den Hoff, M., Bar, I., and Escande, D., 1999, Mutations in a dominant-negative isoform correlate with phenotype in inherited cardiac arrhythmias, *Am. J. Hum. Genet.* **64:**1015–1023.

Morgan, J. M., Lopes, A., and Rowland, E., 1991, Sudden cardiac death while taking amiodarone therapy: The role of abnormal repolarization, *Eur. Heart J.* **12:**1144–1147.

Morgan, J. M., Cunningham, D., and Rowland, E., 1992, Dispersion of monophasic action potential duration: Demonstrable in humans after premature ventricular extrastimulation but not in steady state, *J. Am. Coll. Cardiol.* **19:**1244–1253.

Morillo, C. A., Klein, G. J., Jones, D. L., and Guiraudon, C. M., 1995, Chronic rapid atrial pacing: Structural, functional, and electrophysiological characteristics of a new model of sustained atrial fibrillation, *Circulation* **91:**1588–1595.

Nabauer, M., and Kaab, S., 1998, Potassium channel down-regulation in heart failure, *Cardiovasc. Res.* **37:**324–334.

Nabauer, M., Beuckelmann, D. J., and Erdmann, E., 1993, Characteristics of transient outward current in human ventricular myocytes from patients with terminal heart failure, *Circ. Res.* **73:**386–394.

Naccarelli, G. V., Lee, K. S., Gibson, J. K., and VanderLugt, J., 1996, Electrophysiology and pharmacology of ibutilide, *Am. J. Cardiol.* **78:**12–16.

Napolitano, C., Priori, S. G., and Schwartz, P. J., 1994, Torsade de pointes. Mechanisms and management, *Drugs* **47:**51–65.

Nattel, S., Ranger, S., Talajic, M., Lemery, R., and Roy, D., 1990, Erythromycin-induced long QT syndrome: Concordance with quinidine and underlying cellular electrophysiologic mechanism, *Am. J. Med.* **89:**235–238.

Nelson, C. V., Angelakos, E. T., Bonner, R. A., and Hodgkin, B. C., 1990, Electrocardiographic effects of experimental myocardial infarction in pigs, *J. Electrocardiol.* **23:**137–145.

Nishimoto, M., Hashimoto, H., Ozaki, T., Taguchi, T., Ohara, K., and Nakashima, M., 1994, Effects of imipramine and amitriptyline on intraventricular conduction, effective refractory period, incidence of ventricular arrhythmias induced by programmed stimulation, and on electrocardiogram after myocardial infarction in dog, *Arch. Int. Pharmacodyn. Ther.* **328:**39–53.

Novalija, E., and Stowe, D. F., 1998, Prior preconditioning by ischemia or sevoflurane improves cardiac work per oxygen use in isolated guinea pig hearts after global ischemia, *Adv. Exp. Med. Biol.* **454**:533–542.

Nuss, H. B., Marban, E., and Johns, D. C., 1999, Overexpression of a human potassium channel suppresses cardiac hyperexcitability in rabbit ventricular myocytes, *J. Clin. Invest.* **103**:889–896.

Ogawa, S., Furuno, I., Satoh, Y., Yoh, S., Saeki, K., Sadanaga, T., Katoh, H., and Nakamura, Y., 1991, Quantitative indices of dispersion of refractoriness for identification of propensity to re-entrant ventricular tachycardia in a canine model of myocardial infarction, *Cardiovasc. Res.* **25**:378–383.

Olgin, J. E., Sih, H. J., Hanish, S., Jayachandran, J. V., Wu, J., Zheng, Q. H., Winkle, W., Mulholland, G. K., Zipes, D. P., and Hutchins G., 1998, Heterogeneous atrial denervation creates substrate for sustained atrial fibrillation, *Circulation* **98**:2608–2614.

Olsson, S. B., 1996, Atrial fibrillation—new aspects on mechanism and treatment, *J. Intern. Med.* **239**:3–15.

Ovize, M., Aupetit, J. F., Rioufol, G., Loufoua, J., André-Fouët, X., Minaire, Y., and Faucon, G., 1995, Preconditioning reduces infarct size but accelerates time to ventricular fibrillation in ischemic pig heart, *Am. J. Physiol.* **269**:H72–H79.

Pan, S. J., and Zhou, Z. N., 1997, Effect of acute hypoxia on ATP-sensitive potassium current in ventricular myocytes of guinea-pig, *Acta Physiolo. Sin.* **49**:73–78.

Papaceit, J., Moral, V., Recio, J., de Ferrer, J. M., Riva, J., and Bayes de Luna, A., 1990, Severe heart arrhythmia secondary to magnesium depletion: Torsade de pointes, *Rev. Esp. Anestesiol. Reanim.* **37**:28–31.

Papageorgiou, P., Monahan, K., Boyle, N. G., Seifert, M. J., Beswick, P., Zebede, J., Epstein, L. M., and Josephson, M. E., 1996, Site-dependent intra-atrial conduction delay: Relationship to initiation of atrial fibrillation, *Circulation* **94**:384–389.

Pearlman, A. S., Engler, R. L., Goldstein, R. A., Kent, K. M., and Epstein, S. E., 1980, Relative effects of nitroglycerin and nitroprusside during experimental acute myocardial ischemia, *Eur. J. Cardiol.* **11**:295–313.

Pertsov, A. M., Davidenko, J. M., Salomonsz, R., Baxter, W. T., and Jalife, J., 1993, Spiral waves of excitation underlie reentrant activity in isolated cardiac muscle, *Circ. Res.* **72**:631–650.

Peters, R. W., Weiss, D. N., Carliner, N. H., Feliciano, Z., Shorofsky, S. R., and Gold M. R., 1994, Overdrive pacing for atrial flutter, *Am. J. Cardiol.* **74**:1021–1023.

Piergies, A. A., Ruo, T. I., Jansyn, E. M., Belknap, S. M., and Atkinson, A. J., Jr., 1987, Effect kinetics of *N*-acetylprocainamide-induced QT interval prolongation, *Clin. Pharmacol. Ther.* **42**:107–112.

Pinto, J. M., and Boyden, P. A., 1998, Reduced inward rectifying and increased E-4031-sensitive K^+ current density in arrhythmogenic subendocardial Purkinje myocytes from the infarcted heart, *J. Cardiovasc. Electrophysiol.* **9**:299–311.

Pogwizd, S. M., and Corr B., 1992, The contribution of nonreentrant mechanisms to malignant ventricular arrhythmias, *Basic Res. Cardiol.* **87**:115–129.

Prystowsky, E. N., 1992, Effects of bepridil on cardiac electrophysiologic properties, *Am. J. Cardiol.* **69**:63D–67D.

Puljevic, D., Smalcelj, A., Durakovic, Z., and Goldner, V., 1997, QT dispersion, daily variations, QT interval adaptation and late potentials as risk markers for ventricular tachycardia, *Eur. Heart J.* **18**:1343–1349.

Rademaker, M. T., Fitzpatrick, M. A., Charles, C. J., Frampton, C. M., Richards, A. M., Nicholls, M. G., and Espiner, E. A., 1995, Central angiotensin II AT1-receptor antagonism in normal and heart-failed sheep, *Am. J. Physiol.* **269**:H425–H432.

Rademaker, M. T., Charles, C. J., Espiner, E. A., Frampton, C. M., Nicholls, M. G., and Richards, A. M., 1996, Natriuretic peptide responses to acute and chronic ventricular pacing in sheep, *Am. J. Physiol.* **270**:H594–H602.

Rampe, D., Wible, B., Brown, A. M., and Dage R. C., 1993, Effects of terfenadine and its metabolites on a delayed rectifier K^+ channel cloned from human heart, *Mol. Pharmacol.* **44**:1240–1245.

Rampe, D., Roy, M. L., Dennis, A., and Brown, A. M., 1997, A mechanism for the proarrhythmic effects of cisapride (Propulsid): High affinity blockade of the human cardiac potassium channel HERG, *FEBS. Lett.* **417**:28–32.

Rampe, D., Murawsky, M. K., Grau, J., and Lewis, E. W., 1998, The antipsychotic agent sertindole is a high affinity antagonist of the human cardiac potassium channel HERG, *J. Pharmacol. Exp. Ther.* **286**:788–793.

Rioufol, G., Ovize, M., Loufoua, J., Pop, C., André-Fouät, X., and Minaire, Y., 1997, Ventricular fibrillation in preconditioned pig hearts: Role of K_{ATP}^+ channels, *Am. J. Physiol.* **273**:H2804–H2810.

Roden, D. M., Thompson, K. A., Hoffman, B. F., and Woosley, R. L., 1986, Clinical features and basic mechanisms of quinidine-induced arrhythmias, *J. Am. Coll. Cardiol.* **8:**73A–78A.

Roden, D. M., Bennett, P. B., Snyders, D. J., Balser, J. R., and Hondeghem, L. M., 1988, Quinidine delays I_K activation in guinea pig ventricular myocytes, *Circ. Res.* **62:**1055–1058.

Rohmann, S., Fuchs, C., and Schelling, P., 1997, In swine myocardium, the infarct size reduction induced by U-89232 is glibenclamide sensitive: Evidence that U-89232 is a cardioselective opener of ATP-sensitive potassium channels, *J. Cardiovasc. Pharmacol.* **29:**69–74.

Rozanski, G. J., Xu, Z., Whitney, R. T., Murakami, H., and Zucker, I. H., 1997, Electrophysiology of rabbit ventricular myocytes following sustained rapid ventricular pacing, *J. Mol. Cell. Cardiol.* **29:**721–732.

Rozanski, G. J., Xu, Z., Zhang, K., and Patel, K. P., 1998, Altered K$^+$ current of ventricular myocytes in rats with chronic myocardial infarction, *Am. J. Physiol.* **274:**H259–H265.

Sabbah, H. N., Sharov, V. G., Lesch, M., and Goldstein S., 1995, Progression of heart failure: A role for interstitial fibrosis, *Mol. Cell. Biochem.* **147:**29–34.

Sadanaga, T., Ogawa, S., Okada, Y., Tsutsumi, N., Iwanaga, S., Yoshikawa, T., Akaishi, M., and Handa, S., 1993, Clinical evaluation of the use-dependent QRS prolongation and the reverse use-dependent QT prolongation of class I and class III antiarrhythmic agents and their value in predicting efficacy, *Am. Heart J.* **126:**114–121.

Sakemi, H., and VanNatta, B., 1993, Torsade de pointes induced by astemizole in a patient with prolongation of the QT interval, *Am. Heart J.* **125:**1436–1438.

Sakmann, B., Noma, A., and Trautwein, W., 1983, Acetylcholine activation of single muscarinic K$^+$ channels in isolated pacemaker cells of the mammalian heart, *Nature* **303:**250–253.

Salata, J. J., Jurkiewicz, N. K., Wallace, A. A., Stupienski, R. F., Jr., Guinosso, P. J., Jr., and Lynch, J. J., Jr., 1995, Cardiac electrophysiological actions of the histamine H1-receptor antagonists astemizole and terfenadine compared with chlorpheniramine and pyrilamine, *Circ. Res.* **76:**110–119.

Sanguinetti, M. C., Jurkiewicz, N. K., Scott, A., and Siegl, P. K., 1991, Isoproterenol antagonizes prolongation of refractory period by the class III antiarrhythmic agent E-4031 in guinea pig myocytes: Mechanism of action, *Circ. Res.* **68:**77–84.

Sanguinetti, M. C., Jiang, C., Curran, M. E., and Keating, M. T., 1995, A mechanistic link between an inherited and an acquired cardiac arrhythmia: HERG encodes the I_{Kr} potassium channel, *Cell* **81:**299–307.

Sanguinetti, M. C., Curran, M. E., Spector, P. S., and Keating, M. T., 1996a, Spectrum of HERG K$^+$-channel dysfunction in an inherited cardiac arrhythmia, *Proc. Natl. Acad. Sci. U.S.A.* **93:**2208–2212.

Sanguinetti, M. C., Curran, M. E., Zou, A., Shen, J., Spector, P. S., Atkinson, D. L., and Keating, M. T., 1996b, Coassembly of K$_V$LQT1 and minK (IsK) proteins to form cardiac I_{Ks} potassium channel, *Nature* **384:**80–83.

Satler, C. A., Walsh, E. P., Vesely, M. R., Plummer, M. H., Ginsburg, G. S., and Jacob, H. J., 1996, Novel missense mutation in the cyclic nucleotide-binding domain of HERG causes long QT syndrome, *Am. J. Med. Genet.* **65:**27–35.

Satoh T., and Zipes, D. P., 1998, Cesium-induced atrial tachycardia degenerating into atrial fibrillation in dogs: Atrial torsades de pointes?, *J. Cardiovasc. Electrophysiol.* **9:**970–975.

Schmitt, H., Wit, A. L., Coromilas, J., and Waldecker, B., 1998, Mechanisms for spontaneous termination of monomorphic, sustained ventricular tachycardia: Results of activation mapping of reentrant circuits in the epicardial border zone of subacute canine infarcts, *J. Am. Coll. Cardiol.* **31:**460–472.

Schuessler, R. B., Grayson, T. M., Bromberg, B. I., Cox, J. L., and Boineau, J. P., 1992, Cholinergically mediated tachyarrhythmias induced by a single extrastimulus in the isolated canine right atrium, *Circ. Res.* **71:**1254–1267.

Schulze-Bahr, E., Haverkamp, W., Wedekind, H., Rubie, C., Hordt, M., Borggrefe, M., Assmann, G., Breithardt, G., and Funke, H., 1997a, Autosomal recessive long-QT syndrome (Jervell Lange-Nielsen syndrome) is genetically heterogeneous, *Hum. Genet.* **100:**573–576.

Schulze-Bahr, E., Wang, Q., Wedekind, H., Haverkamp, W., Chen, Q., Sun, Y., Rubie, C., Hordt, M., Towbin, J. A., Borggrefe, M., Assmann, G., Qu, X., Somberg, J. C., Breithardt, G., Oberti, C., and Funke, H., 1997b, KCNE1 mutations cause Jervell and Lange-Nielsen syndrome [letter], *Nat. Genet.* **17:**267–268.

Selnick, H. G., Liverton, N. J., Baldwin, J. J., Butcher, J. W., Claremon, D. A., Elliott, J. M., Freidinger, R. M., King, S. A., Libby, B. E., McIntyre, C. J., Pribush, D. A., Remy, D. C., Smith, G. R., Tebben, A. J., Jurkiewicz, N. K., Lynch, J. J., Salata, J. J., Sanguinetti, M. C., Siegl, P. K., Slaughter, D. E., and Vyas, K., 1997, Class III antiarrhythmic activity in vivo by selective blockade of the slowly activating cardiac delayed rectifier potassium current I_{Ks} by (R)-2-(2,4-trifluoromethyl)-N-[2-oxo-5-phenyl-1-(2,2,2-trif-luoroethyl)-2,3-dihydro-1H-benzo[e][1,4]diazepin-3-yl]acetamide, *J. Med. Chem.* **40**:3865–3868.

Shalaby, F. Y., Levesque, P. C., Yang, W. P., Little, W. A., Conder, M. L., Jenkins-West, T., and Blanar, M. A., 1997, Dominant-negative KvLQT1 mutations underlie the LQT1 form of long QT syndrome, *Circulation* **96**:1733–1736.

Shaw, R. M., and Rudy, Y., 1997, Electrophysiologic effects of acute myocardial ischemia: A theoretical study of altered cell excitability and action potential duration, *Cardiovasc. Res.* **35**:256–272.

Shen, Y. T., and Vatner, S. F., 1996, Differences in myocardial stunning following coronary artery occlusion in conscious dogs, pigs, and baboons, *Am. J. Physiol.* **270**:H1312–H1322.

Shen, Y. T., Kudej, R. K., Bishop, S. P., and Vatner, S. F., 1996, Inotropic reserve and histological appearance of hibernating myocardium in conscious pigs with ameroid-induced coronary stenosis, *Basic Res. Cardiol.* **91**:479–485.

Shigematsu, S., Sato, T., and Arita, M., 1998, Class I antiarrhythmic drugs alter the severity of myocardial stunning by modulating ATP-sensitive K^+ channels in guinea pig ventricular muscles, *Naunyn-Schmiedeberg's Arch. Pharmacol.* **357**:283–290.

Shimizu, W., and Antzelevitch, C., 1998, Cellular basis for the ECG features of the LQT1 form of the long-QT syndrome: Effects of beta-adrenergic agonists and antagonists and sodium channel blockers on transmural dispersion of repolarization and torsade de pointes, *Circulation* **98**:2314–2322.

Shinbane, J. S., Wood, M. A., Jensen, D. N., Ellenbogen, K. A., Fitzpatrick, A. P., and Scheinman, M. M., 1997, Tachycardia-induced cardiomyopathy: A review of animal models and clinical studies, *J. Am. Coll. Cardiol.* **29**:709–715.

Smith, P. L., Baukrowitz, T., and Yellen, G., 1996, The inward rectification mechanism of the HERG cardiac potassium channel, *Nature* **379**:833–836.

Smith, S. J., 1994, Cardiovascular toxicity of antihistamines, *Otolaryngol. Head Neck Surg.* **111**:348–354.

Sokoloski, M. C., Ayers, G. M., Kumagai, K., Khrestian, C. M., Niwano, S., and Waldo, A. L., 1997, Safety of transvenous atrial defibrillation: Studies in the canine sterile pericarditis model, *Circulation* **96**:1343–1350.

Solomon, S. B., Nikolic, S. D., Glantz, S. A., and Yellin, E. L., 1998, Left ventricular diastolic function of remodeled myocardium in dogs with pacing-induced heart failure, *Am. J. Physiol.* **274**:H945–H954.

Southworth, R., Shattock, M. J., Hearse, D. J., and Kelly, F. J., 1998, Developmental differences in superoxide production in isolated guinea-pig hearts during reperfusion, *J. Mol. Cell. Cardiol.* **30**:1391–1399.

Spach, M. S., and Josephson, M. E., 1994, Initiating reentry: The role of nonuniform anisotropy in small circuits, *J. Cardiovasc. Electrophysiol.* **5**:182–209.

Spinale, F. G., Schulte, B. A., and Crawford, F. A., 1989, Demonstration of early ischemic injury in porcine right ventricular myocardium, *Am. J. Pathol.* **134**:693–704.

Spinale, F. G., Carabello, B. A., Schulte, B. A., and Crawford, F. A., Jr., 1990, Wavefront myocyte injury and relationship to function in right ventricular ischemia, *Am. J. Physiol.* **258**:H292–H304.

Spinale, F. G., Eble, D. M., Mukherjee, R., Johnson, W. S., and Walker, J. D., 1994, Left ventricular and myocyte structure and function following chronic ventricular tachycardia in rabbits, *Basic Res. Cardiol.* **89**:456–467.

Splawski, I., Tristani-Firouzi, M., Lehmann, M. H., Sanguinetti, M. C., and Keating, M. T., 1997, Mutations in the hminK gene cause long QT syndrome and suppress I_{Ks} function, *Nat. Genet.* **17**:338–340.

Stramba-Badiale, M., Pessano, P., Kirchengast, M., and Schwartz, P. J., 1995, Effects of the potassium channel blocking agent ambasilide on ventricular arrhythmias induced by acute myocardial ischemia and sympathetic activation, *Am. Heart J.* **129**:549–556.

Stramba-Badiale, M., Nador, F., Porta, N., Guffanti, S., Frediani, M., Colnaghi, C., Grancini, F., Motta, G., Carnelli, V., and Schwartz, P. J., 1997, QT interval prolongation and risk of life-threatening arrhythmias during toxoplasmosis prophylaxis with spiramycin in neonates, *Am. Heart J.* **133**:108–111.

Suessbrich, H., Schonherr, R., Heinemann, S. H., Attali, B., Lang, F., and Busch, A. E., 1997, The inhibitory effect of the antipsychotic drug haloperidol on HERG potassium channels expressed in *Xenopus* oocytes, *Br. J. Pharmacol.* **120**:968–974.

Takimoto, K., Li, D., Hershman, K. M., Li, P., Jackson, E. K., and Levitan, E. S., 1997, Decreased expression of Kv4.2 and novel Kv4.3 K^+ channel subunit mRNAs in ventricles of renovascular hypertensive rats, *Circ. Res.* **81**:533–539.

Tanaka, T., Nagai, R., Tomoike, H., Takata, S., Yano, K., Yabuta, K., Haneda, N., Nakano, O., Shibata, A., Sawayama, T., Kasai, H., Yazaki, Y., and Nakamura, Y., 1997, Four novel KVLQT1 and four novel HERG mutations in familial long-QT syndrome, *Circulation* **95**:565–567.

Thuringer, D., Deroubaix, E., Coulombe, A., Coraboeuf, E., and Mercadier, J. J., 1996, Ionic basis of the action potential prolongation in ventricular myocytes from Syrian hamsters with dilated cardiomyopathy, *Cardiovasc. Res.* **31**:747–757.

Toivonen, L., Viitasalo, M., Siikamaki, H., Raatikka, M., and Pohjola-Sintonen, S., 1994, Provocation of ventricular tachycardia by antimalarial drug halofantrine in congenital long QT syndrome, *Clin. Cardiol.* **17**:403–404.

Tran, H. T., 1994, Torsades de pointes induced by nonantiarrhythmic drugs, *Conn. Med.* **58**:291–295.

Trudeau, M. C., Warmke, J. W., Ganetzky, B., and Robertson, G. A., 1995, HERG, a human inward rectifier in the voltage-gated potassium channel family, *Science* **269**:92–95.

Trump, D. L., Smith, D. C., Ellis, P. G., Rogers, M. P., Schold, S. C., Winer, E. P., Panella, T. J., Jordan, V. C., and Fine, R. L., 1992, High-dose oral tamoxifen, a potential multidrug-resistance-reversal agent: Phase I trial in combination with vinblastine, *J. Natl. Cancer Inst.* **84**:1811–1816.

Tuininga, Y. S., De Langen, C. D., Crijns, H. J., Wiesfeld, A. C., Mook, P. H., Bel, K. J., and Lie, K. I., 1996, Electrophysiological, rate dependent, and autonomic effects of the class III antiarrhythmic almokalant after myocardial infarction in the pig, *Pacing Clin. Electrophysiol.* **19**:802–810.

Turgeon, J., Daleau, P., Bennett, P. B., Wiggins, S. S., Selby, L., and Roden, D. M., 1994, Block of I_{Ks}, the slow component of the delayed rectifier K$^+$ current, by the diuretic agent indapamide in guinea pig myocytes, *Circ. Res.* **75**:879–886.

Tyson, J., Tranebjaerg, L., Bellman, S., Wren, C., Taylor, J. F., Bathen, J., Aslaksen, B., Sorland, S. J., Lund, O., Malcolm, S., Pembrey, M., Bhattacharya, S., and Bitner-Glindzicz, M., 1997, IsK and KvLQT1: Mutation in either of the two subunits of the slow component of the delayed rectifier potassium channel can cause Jervell and Lange-Nielsen syndrome, *Hum. Mol. Genet.* **6**:2179–2185.

Uchida, T., Yashima, M., Gotoh, M., Qu, Z., Garfinkel, A., Weiss, J. N., Fishbein, M. C., Mandel, W. J., Chen, P. S., and Karagueuzian, H. S., 1999, Mechanism of acceleration of functional reentry in the ventricle: Effects of ATP-sensitive potassium channel opener, *Circulation* **99**:704–712.

Uematsu, T., Kanamaru, M., and Nakashima, M., 1994, Comparative pharmacokinetic and pharmacodynamic properties of oral and intravenous ($+$)-sotalol in healthy volunteers, *J. Pharm. Pharmacol.* **46**:600–605.

van der Klauw, M. M., van Rey, F. J., and Stricker, B. H., 1992, Polymorph ventricular tachycardia with torsades de pointes caused by administration of terodiline (Mictrol), *Ned. Tijdschr. Geneeskd.* **136**:91–93.

Vanoli, E., Hull, S. S., Jr., Adamson, P. B., Foreman, R. D., and Schwartz, P. J., 1995, K$^+$ channel blockade in the prevention of ventricular fibrillation in dogs with acute ischemia and enhanced sympathetic activity, *J. Cardiovasc. Pharmacol.* **26**:847–854.

Van Wagoner, D. R., Pond, A. L., McCarthy, P. M., Trimmer, J. S., and Nerbonne, J. M., 1997, Outward K$^+$ current densities and Kv1.5 expression are reduced in chronic human atrial fibrillation, *Circ. Res.* **80**:772–781.

Varró, A., Lathrop, D. A., Hester, S. B., Nánási, P. P., and Papp, J. G., 1993, Ionic currents and action potentials in rabbit, rat, and guinea pig ventricular myocytes, *Basic Res. Cardiol.* **88**:93–102.

Verdun, F., Mansourati, J., Jobic, Y., Bouquin, V., Munier, S., Guillo, P., Pages, Y., Boschat, J., and Blanc, J. J., 1997, Torsades de pointes with spiramycine and metiquazine therapy. Apropos of a case, *Arch. Mal. Coeur Vaiss.* **90**:103–106.

Vincent, G. M., 1998, The molecular genetics of the long QT syndrome: Genes causing fainting and sudden death, *Annu. Rev. Med.* **49**:263–274.

Volders, P. G., Sipido, K. R., Vos, M. A., Kulcsár, A., Verduyn, S. C., and Wellens, H. J., 1998, Cellular basis of biventricular hypertrophy and arrhythmogenesis in dogs with chronic complete atrioventricular block and acquired torsade de pointes, *Circulation* **98**:1136–1147.

Vorperian, V. R., Zhou, Z., Mohammad, S., Hoon, T. J., Studenik, C., and January, C. T., 1996, Torsade de pointes with an antihistamine metabolite: Potassium channel blockade with desmethylastemizole, *J. Am. Coll. Cardiol.* **28**:1556–1561.

Vos, M. A., Verduyn, S. C., Gorgels, A. P., Lipcsei, G. C., and Wellens, H. J., 1995, Reproducible induction of early afterdepolarizations and torsade de pointes arrhythmias by *d*-sotalol and pacing in dogs with chronic atrioventricular block, *Circulation* **91**:864–872.

Wang, Q., Curran, M. E., Splawski, I., Burn, T. C., Millholland, J. M., VanRaay, T. J., Shen, J., Timothy, K. W., Vincent, G. M., de Jager, T., Schwartz, P. J., Toubin, J. A., Moss, A. J., Atkinson, D. L., Landes, G. M., Connors, T. D., and Keating, M. T., 1996, Positional cloning of a novel potassium channel gene: KVLQT1 mutations cause cardiac arrhythmias, *Nat. Genet.* **12**:17–23.

Wang, W., Xia, J., and Kass, R. S., 1998, MinK–KvLQT1 fusion proteins, evidence for multiple stoichiometries of the assembled IsK channel, *J. Biol. Chem.* **273:**34069–34074.

Wang, Z., Fermini, B., and Nattel, S., 1995, Effects of flecainide, quinidine, and 4-aminopyridine on transient outward and ultrarapid delayed rectifier currents in human atrial myocytes, *J. Pharmacol. Exp. Ther.* **272:**184–196.

Wen, Z. C., Chen, S. A., Tai, C. T., Huang, J. L., and Chang, M. S., 1998, Role of autonomic tone in facilitating spontaneous onset of typical atrial flutter, *J. Am. Coll. Cardiol.* **31:**602–607.

Wetstein, L., Mark, R., Kelliher, G. J., Friehling, T., O'Connor, K. M., and Kowey, P. R., 1985, Arrhythmia inducibility and ventricular vulnerability in a chronic feline infarction model, *Am. Heart J.* **110:**955–960.

Wettwer, E., Amos, G., Gath, J., Zerkowski, H. R., Reidemeister, J. C., and Ravens, U., 1993, Transient outward current in human and rat ventricular myocytes, *Cardiovasc. Res.* **27:**1662–1669.

Wharton, J. M., Demopulos, P. A., and Goldschlager, N., 1987, Torsade de pointes during administration of pentamidine isethionate, *Am. J. Med.* **83:**571–576.

Wickman, K., Nemec, J., Gendler, S. J., and Clapham, D. E., 1998, Abnormal heart rate regulation in GIRK4 knockout mice, *Neuron* **20:**103–114.

Wijffels, M. C., Kirchhof, C. J., Dorland, R., and Allessie, M. A., 1995, Atrial fibrillation begets atrial fibrillation: A study in awake chronically instrumented goats, *Circulation* **92:**1954–1968.

Williams, R. E., Kass, D. A., Kawagoe, Y., Pak, P., Tunin, R. S., Shah, R., Hwang, A., and Feldman, A. M., 1994, Endomyocardial gene expression during development of pacing tachycardia-induced heart failure in the dog, *Circ. Res.* **75:**615–623.

Winslow, E., 1981, Hemodynamic and arrhythmogenic effects of aconitine applied to the left atria of anesthetized cats: Effects of amiodarone and atropine, *J. Cardiovasc. Pharmacol.* **3:**87–100.

Wong, C. B., and Windle, J., 1995, Erythromycin induced torsades de pointes, *Nebr. Med. J.* **80:**285–286.

Wood, M. A., Caponi, D., Sykes, A. M., and Wenger, E. J., 1998, Atrial electrical remodeling by rapid pacing in the isolated rabbit heart: Effects of Ca^{++} and K^+ channel blockade, *J. Interv. Card. Electrophysiol.* **2:**15–23.

Xu, Z., and Rozanski, G. J., 1997, Proton inhibition of transient outward potassium current in rat ventricular myocytes, *J. Mol. Cell. Cardiol.* **29:**481–490.

Yang, T., Prakash, C., Roden, D. M., and Snyders, D. J., 1995a, Mechanism of block of a human cardiac potassium channel by terfenadine racemate and enantiomers, *Br. J. Pharmacol.* **115:**267–274.

Yang, T., Snyders, D. J., and Roden, D. M., 1995b, Ibutilide, a methanesulfonanilide antiarrhythmic, is a potent blocker of the rapidly activating delayed rectifier K^+ current (I_{Kr}) in AT-1 cells: Concentration-, time-, voltage-, and use-dependent effects, *Circulation* **91:**1799–1806.

Yang, W. P., Levesque, P. C., Little, W. A., Conder, M. L., Shalaby, F. Y., and Blanar, M. A., 1997, KvLQT1, a voltage-gated potassium channel responsible for human cardiac arrhythmias, *Proc. Natl. Acad. Sci. U.S.A.* **94:**4017–4021.

Yao, J. A., and Tseng, G. N., 1997, Azimilide (NE-10064) can prolong or shorten the action potential duration in canine ventricular myocytes: Dependence on blockade of K, Ca, and Na channels, *J. Cardiovasc. Electrophysiol.* **8:**184–198.

Yue, L., Feng, J., Gaspo, R., Li, G. R., Wang, Z., and Nattel, S., 1997, Ionic remodeling underlying action potential changes in a canine model of atrial fibrillation, *Circ. Res.* **81:**512–525.

Zabel, M., Hohnloser, S. H., Behrens, S., Li, Y. G., Woosley, R. L., and Franz, M. R., 1997, Electrophysiologic features of torsades de pointes: Insights from a new isolated rabbit heart model, *J. Cardiovasc. Electrophysiol.* **8:**1148–1158.

Zaputovic, L., Mavric, Z., Zaninovic-Jurjevic, T., Matana, A., and Bradic, N., 1997, Relationship between QT dispersion and the incidence of early ventricular arrhythmias in patients with acute myocardial infarction, *Int. J. Cardiol.* **62:**211–216.

Zareba, W., Moss, A. J., Rosero, S. Z., Hajj-Ali, R., Konecki, J., and Andrews, M., 1997, Electrocardiographic findings in patients with diphenhydramine overdose, *Am. J. Cardiol.* **80:**1168–1173.

Zhang, Z. H., Boutjdir, M., and el-Sherif, N., 1994, Ketanserin inhibits depolarization-activated outward potassium current in rat ventricular myocytes, *Circ. Res.* **75:**711–721.

Zimmermann, M., Duruz, H., Guinand, O., Broccard, O., Levy, P., Lacatis, D., and Bloch, A., 1992, Torsades de pointes after treatment with terfenadine and ketoconazole, *Eur. Heart J.* **13:**1002–1003.

Chapter 36

The Molecular Basis of the Long QT Syndrome

Martin Tristani-Firouzi and Michael C. Sanguinetti

1. INTRODUCTION

Sudden cardiac death is an important cause of cardiovascular mortality in the United States (Kannel *et al.*, 1987; Willich *et al.*, 1987). Cardiac arrhythmias, specifically ventricular tachyarrhythmias, are presumed to be responsible for the majority of these deaths. Recent advances in molecular genetics have allowed the identification of several genes responsible for one form of inherited arrhythmia, the long QT syndrome (LQTS). These findings have furthered our understanding of the molecular basis of the cardiac action potential and offered insights into structure–function relationships of cardiac ion channels.

LQTS is a heterogeneous disorder of myocellular repolarization, characterized by prolongation of the QT interval on the body surface electrocardiogram. The QT interval is defined as the interval between the onset of ventricular depolarization (QRS complex) and termination of repolarization (end of the T-wave). Affected individuals are susceptible to the development of ventricular tachyarrhythmias, which are manifested clinically by recurrent syncope (loss of consciousness), seizures, and sudden death. LQTS is particularly insidious in that it affects primarily young, otherwise healthy individuals, and, not infrequently, sudden cardiac death is the presenting symptom. Without treatment, the risk of sudden cardiac death for LQTS-affected individuals approaches 50% (Ackerman and Clapham, 1997).

Two inherited patterns of LQTS exist: an autosomal dominant form, called Romano–Ward syndrome, and an autosomal recessive form associated with congenital sensorineural deafness, called Jervell and Lange-Nielsen syndrome. The most common form of LQTS is acquired, induced by various drugs, or associated with heart failure (Roden *et al.*, 1996; Schwartz, 1997).

Martin Tristani-Firouzi and Michael C. Sanguinetti ● Department of Pediatrics and Medicine, Eccles Program in Human Molecular Biology and Genetics, University of Utah and Primary Children's Medical Center, Salt Lake City, Utah 84113.

Potassium Channels in Cardiovascular Biology, edited by Archer and Rusch. Kluwer Academic/Plenum Publishers, New York, 2001.

Recently, molecular genetic studies have identified mutations in genes encoding cardiac ion channels as the cause of arrhythmia susceptibility in inherited LQTS. Mutations in at least five genes have been implicated in LQTS, including *KvLQT1* (*KCNQ*1), *minK* (*KCNE*1), and *HERG*, which encode cardiac K$^+$ channel subunits; *SCN5A*, which encodes the Na$^+$ channel α-subunit; and an unidentified gene on chromosome 4. The identification of these genes has greatly advanced our understanding of the structure and function of ion channels responsible for the cardiac action potential and the molecular basis of arrhythmia susceptibility. This chapter describes the search for the molecular basis of LQTS, the cellular consequences of mutations in cardiac ion channel genes, the influence of mutations on clinical phenotype, and the development of genotype-specific therapy.

2. CARDIAC ACTION POTENTIALS AND THE QT INTERVAL

The configuration of the cardiac action potential is determined by time- and voltage-dependent ion channels. (Fig. 1A) The upstroke of the action potential (phase 0) is initiated by activation of voltage-gated Na$^+$ channels. The depolarizing influence of Na$^+$ channels activates voltage-dependent L-type Ca^{2+} currents and several distinct K$^+$ currents that modulate phases 1–3 of the action potential. In humans, the rapid, initial repolarization (phase 1) is due to activation of the transient outward K$^+$ current (I_{to}), together with inactivation of Na$^+$ current. A fine balance between a small inward current through L-type Ca^{2+} channels and a small outward current through the delayed rectifier K$^+$ channels defines the plateau phase (phase 2).

The duration of the plateau phase is a critical determinant of Ca^{2+} influx and myocardial refractoriness. The relatively long duration of cardiac action potentials is principally due to the unique properties of the slowly and rapidly activating cardiac delayed rectifier currents, I_{Ks} and I_{Kr}, respectively. (Fig. 1A). I_{Kr} is characterized by rapid, voltage-dependent inactivation that proceeds at a rate faster than the rate of activation, such that the current rectifies (that is, the magnitude decreases) at more depolarized potentials. The pronounced rectification of I_{Kr}, together with the slow onset of I_{Ks} activation, results in little repolarizing current during the plateau phase.

Time-dependent inactivation of L-type Ca^{2+} channels, in the presence of a sustained outward K$^+$ current (the ultrarapid delayed rectifier current, I_{Kur}) and increasing activation of I_{Ks}, ultimately tips the electrical balance in favor of net outward current. As the myocyte begins to repolarize, I_{Kr} recovers from inactivation and contributes further repolarizing current (phase 3). The rapid rate of recovery from inactivation but slow rate of deactivation of I_{Kr} is critical to the termination of the action potential. In addition, voltage-dependent recovery of block of inward rectifier channels by intracellular polyamines and Mg^{2+} provides further terminal repolarizing current.

Action potential prolongation, with the resultant QT interval prolongation, occurs in the setting of either increased depolarizing or decreased repolarizing current. (Figure1B). Because little net current flows during the plateau phase, small changes in outward or inward conductance can exert dramatic effects on action potential duration. Thus, mutations in genes encoding the cardiac delayed rectifier K$^+$ channels prolong

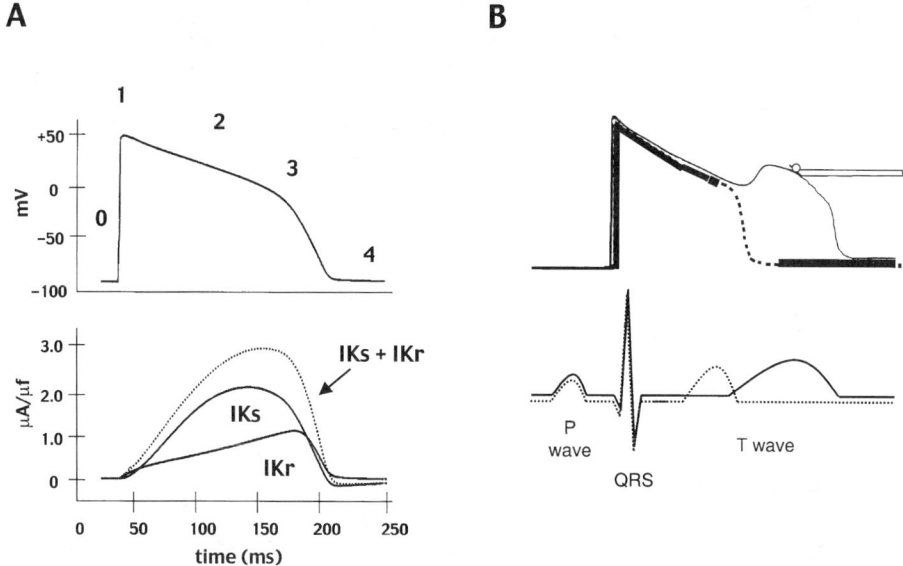

Figure 1. (A) The contribution of delayed rectifier K$^+$ currents to a guinea pig ventricular action potential. *Top*: Membrane potential and the phases (0–4) of the action potential (see text for details). *Bottom*: I_{Ks}, the slowly activating delayed rectifier, provides the majority of repolarizing current during the plateau (phase 2) and early terminal phase (phase 3). I_{Kr}, the rapidly activating delayed rectifier current, is mostly inactivated at depolarized potentials (phases 1 and 2), but recovers from inactivation rapidly during repolarization (phase 3). (Reprinted, with permission, from Zeng *et al.*, 1995.) B, Relationship of the cardiac action potential to the body surface electrocardiogram. The dashed curve represents a normal cardiac action potential, with the corresponding electrocardiogram depicted below. The QT interval is defined as the period between the onset of the QRS complex and the end of the T-wave. Lengthening of the cellular action potential (solid line) causes prolongation of the QT interval. Action potential prolongation can result in a secondary depolarization, called an early afterdepolarization (EAD), which is presumed to initiate ventricular tachycardia.

action potential duration by decreasing repolarizing current during the plateau phase. Conversely, mutations in the gene encoding the cardiac Na$^+$ channel increase depolarizing current during the plateau phase and thereby prolong action potential duration. It is probable that there remain additional undiscovered processes that alter the delicate balance between depolarizing and repolarizing current and thereby contribute to LQTS.

3. MOLECULAR GENETIC APPROACHES TO THE IDENTIFICATION OF LQTS GENES

The identification of genes responsible for arrhythmia susceptibility in LQTS was based initially upon linkage analyses of large kindreds with autosomal dominant forms of LQTS, followed by candidate gene analysis or positional cloning. Using restriction-fragment-length polymorphism and Southern blot analysis in a large LQTS kindred, genetic linkage was established near the Harvey *ras*-1 (H ras-1) locus on the short arm of chromosome 11 (11p15.5) (Keating *et al.*, 1991). Shortly thereafter, the genetic heterogeneity of LQTS became apparent as several groups reported LQTS families whose electrophysiological abnormalities were not linked to chromosome 11 (Benhorin *et al.*, 1993; Curran *et al.*, 1993; Towbin *et al.*, 1994). The identification of loci on chromosome 7 (7q35-36), chromosome 3 (3p21-24), and chromosome 4 (4q25-27) prompted the nomenclature LQT1–4 to describe the LQTS loci in temporal order of identification. To date, five loci and four genes have been associated with LQTS. Additional arrhythmia susceptibility genes await discovery, as several LQTS families are unlinked to the known loci and some sporadic cases lack mutations in the known LQTS genes.

Once LQTS loci were identified, a candidate gene approach was utilized to define the specific genes involved. Candidate genes included both previously cloned genes that mapped to the area of interest and genes hypothesized to play an important physiological role in the disease process. It was postulated that mutations in cardiac ion channel genes or channel regulatory genes were responsible for the repolarization abnormalities of LQTS. In 1994, Warmke and Ganetzky identified *HERG*, the human *ether-a-go-go*-related gene, from a high-stringency screen of human hippocampus cDNA using a mouse homolog of the *Drosophila ether-a-go-go* (*eag*) K^+ channel gene and localized the gene to chromosome 7 (Warmke and Ganetzky, 1994). Investigating *HERG* as a candidate gene, Keating and colleagues localized *HERG* to chromosome 7q35-36 and discovered multiple, distinct mutations in *HERG* in chromosome 7-linked (LQT2) kindreds (Curran *et al.*, 1995). *De novo HERG* mutations from sporadic cases also occur. Heterologous expression studies subsequently revealed that *HERG* encodes the α-subunit of the I_{Kr} potassium channel (Sanguinetti *et al.*, 1995; Trudeau *et al.*, 1995), thus establishing a link between mutations in *HERG*, reduced repolarizing current during the cardiac action potential, and the predisposition to ventricular tachyarrhythmias.

A similar candidate gene approach was used to identify the LQT3 gene. In 1992, the human cardiac Na^+ channel gene, *SCN5A*, was cloned (Gellens *et al.*, 1992), and it was subsequently mapped to chromosome 3p21 (George *et al.*, 1995), making *SCN5A* a candidate gene for LQT3. Mutation analysis of *SCN5A* revealed a 9-base-pair (bp) in-frame deletion in two unrelated LQT3 families (Q. Wang *et al.*, 1995). This results in the deletion of three highly conserved amino acids, lysine 1505–proline 1506–lysine 1507 (ΔKPQ), located in the cytoplasmic linker between domains III and IV, a region known to be important for Na^+ channel inactivation.

Several promising candidate LQT1 genes, including H ras-1, have been eliminated as possibilities by direct DNA sequencing or by linkage analysis (Russell *et al.*, 1995). Therefore, Wang and co-workers embarked upon a positional cloning approach which successfully led to the discovery of *KvLQT1*, a novel cardiac K^+ channel gene (Q. Wang *et al.*, 1996a). DNA polymorphisms at the 11p15.5 locus were used to refine the position of the LQT1 gene to an interval of approximately 700 kilobases. Screening of a yeast artificial chromosome (YAC) containing this interval led to the identification of a region with high sequence homology to members of the *Shaker* K^+ channel gene family. A partial cDNA clone was isolated from a human cDNA library, spanning a portion of the putative S1 domain, S2–S6, a stop codon and a long 3'- flanking region. This gene was originally named *KvLQT1*, because it encodes a putative voltage-gated K^+ (Kv) channel and mutations in this gene cause LQT1 (Q. Wang *et al.*, 1996a).

Heterologous expression studies demonstrated that *KvLQT1* encodes a novel K^+ channel which, when coexpressed with the *minK* gene (minimal K channel gene), induces a current similar to cardiac I_{Ks} (Barhanin *et al.*, 1996; Sanguinetti *et al.*, 1996b).

The discovery of minK as an integral component of the I_{Ks} channel suggested the possibility that *minK* (*KCNE*) is a LQTS gene. Families with LQTS unlinked to the known LQTS loci were screened for mutations in *minK* (Splawski *et al.*, 1997b). Two mutations in *minK* cosegregated with the disease in affected family members but were absent in unaffected family members and healthy controls.

4. MOLECULAR MECHANISMS OF LQTS MUTATIONS

LQTS is a genetically heterogeneous disorder caused by numerous mutations in several cardiac ion channel genes. Individual mutations alter channel function by a variety of mechanisms, including effects due to haploinsufficiency, dominant-negative effects, "gain of function" effects, and altered gating properties. Haploinsufficiency refers to mutations in which the affected individual expresses $\sim 50\%$ of the normal level of functional channel protein. K^+ channel subunits encoded by the mutant allele are incapable of coassembling with normal subunits to form heteromultimeric channels. The net effect is that only subunits encoded by the normal allele form functional channels, resulting in one-half the number of channels normally expressed by unaffected individuals.

In contrast to haploinsufficiency mutations, genes with dominant-negative mutations encode subunits that coassemble with wild-type subunits but reduce the function of the resulting heteromultimeric channel. (Fig. 2) Kv channels are composed of four independent subunits. In the setting of a dominantly inherited K^+ channel mutation, the likelihood of a tetramer composed of four wild-type subunits is $\frac{1}{16}$. Thus, in the case where only one mutant subunit is sufficient to destroy channel function, overall channel function will be reduced by $\frac{15}{16}$ or 94%. Mutations within the pore region tend to cause severe dominant-negative effects (Sanguinetti *et al.*, 1996a). Given the highly organized structure of the K^+ channel ion-conducting pathway (Doyle *et al.*, 1998), minor disruptions caused by a single mutant pore subunit may result in loss of ion permeation. Alternatively, some dominant-negative effects result from protein-processing defects that cause trapping of heteromultimeric complexes within the endoplasmic reticulum and enhance the degradation of the protein (Folco *et al.*, 1997; London *et al.*, 1998; Zhou *et al.*, 1998). Most LQTS-associated K^+ channel mutations cause variable degrees of dominant-negative suppression.

"Gain of function" mutations increase the current conducted by the affected ion channel. Mutations of this type in *SCN5A* destabilize the inactivation gate of the Na^+ channel, resulting in a small fraction of mutant channels remaining open during the plateau phase of the action potential (Bennett *et al.*, 1995). To date, only *SCN5A* mutations are known to exhibit this behavior, although, conceivably, "gain of function" mutations in L-type Ca^{2+} channels could also cause LQTS.

Alteration of the gating properties of ion channels is an additional mechanism whereby LQTS mutations can enhance arrhythmia susceptibility. For example, several missense mutations in *KvLQT1* (Chouabe *et al.*, 1997; Z. Wang *et al.*, 1999) and *minK*

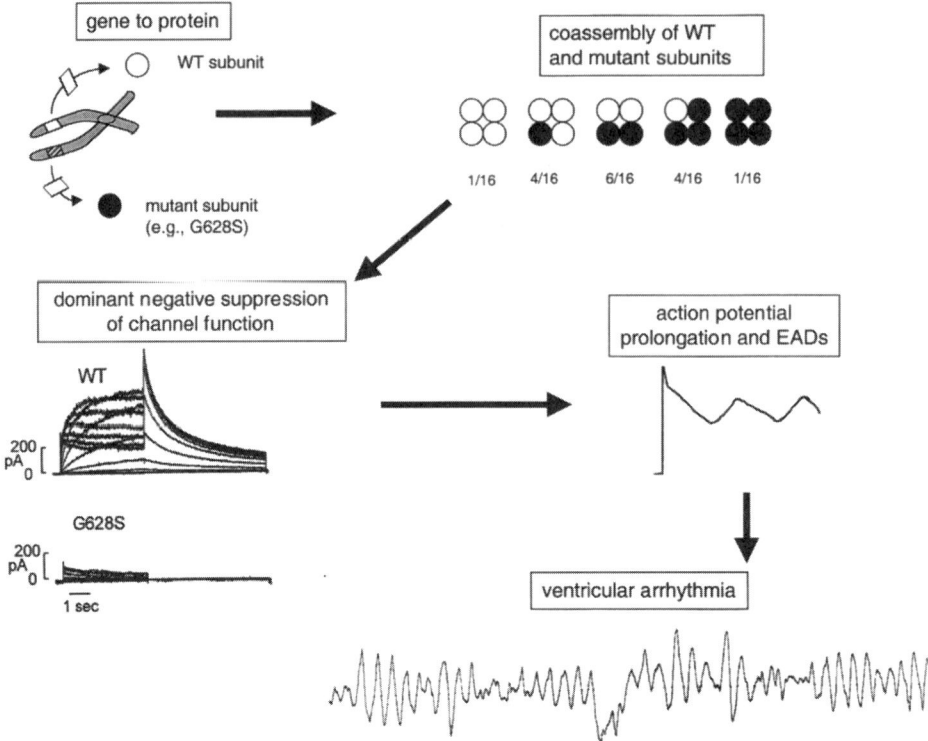

Figure 2. Schematic representation of the mechanism whereby a cardiac K^+ channel gene mutation leads to ventricular tachycardia. *Top left*: Wild-type (WT) and mutant alleles are transcribed and translated into WT (open circle) and mutant (filled circle) K^+ channel subunits, respectively. The mutant subunit in this example is the HERG pore mutant G628S. *Top right*: K^+ channels are formed by independent assembly of four channel subunits. The fraction of channels composed of homo- and heterotetramers assuming equal mixing and matching of WT and mutant subunits is indicated. *Middle left*: HERG pore mutant G628S causes dominant-negative suppression of HERG channel function. WT and G628S HERG currents were elicited by 4-s depolarizing pulses between -70 and $+40$ mV. Tail currents were elicited by repolarization to -70 mV. (Reprinted, with permission, from Zhou, 1998.) *Middle right*: Reduced repolarizing current caused by mutations in cardiac K^+ channel genes leads to action potential prolongation and early afterdepolarizations (EADs). EADs are the presumed trigger for ventricular tachyarrhythmias. *Bottom* : Electrocardiogram of an individual with long QT syndrome and *torsade de pointes* arrhythmia. (Kindly provided by Dr. S. Etheridge, University of Utah.)

(Splawski *et al.*, 1997b) cause a *depolarizing* shift in the voltage dependence of channel activation. Thus, at any given potential during the action potential, a smaller fraction of mutant channels are activated, compared to wild-type channels. Similarly, a *hyperpolarizing* shift in the voltage dependence of inactivation caused by a mutation in *HERG* decreases the number of channels that are available to open (Nakajima *et al.*, 1998). Finally, alterations in the kinetic properties of K^+ channels can influence the amount of repolarizing current during phase 3. Rapid recovery from inactivation and slow deactivation of I_{Kr} channels provide a significant amount of repolarizing current during phase 3. Under conditions of rapid heart rate, "cumulative" activation of I_{Ks}

contributes to shortening of action potential duration because I_{Ks} fails to deactivate completely during shorter diastolic intervals. Thus, mutations that accelerate the deactivation rate of I_{Kr} (Chen *et al.*, 1999) or I_{Ks} (Splawski *et al.*, 1997b) retard the repolarization process and lengthen action potential duration.

5. MUTATIONS IN *KVLQT1*, THE GENE ENCODING THE α-SUBUNIT OF I_{Ks} CHANNELS

Although LQT1 was the first locus linked to LQTS, it was the last gene to be cloned and characterized electrophysiologically. *KvLQT1* encodes a protein of 676 amino acids (Splawski *et al.*, 1998a) and is expressed in the pancreas, heart, lung, kidney, placenta, adrenal gland, thyroid gland, and inner ear (Q. Wang *et al.*, 1996a; Yang *et al.*, 1997). Interestingly, Northern analysis demonstrates the highest expression of mRNA in the pancreas. Despite this finding, exocrine or endocrine pancreatic abnormalities have not been described for individuals with homozygous mutations in *KvLQT1*. Two *KvLQT1* isoforms have been reported, one of which encodes a full-length protein that combines with minK to form functional I_{Ks} channels. A second isoform encodes a truncated KvLQT1 protein that coassembles with wild-type KvLQT1 subunits and causes dominant-negative suppression of I_{Ks} function (Demolombe *et al.*, 1998; Jiang *et al.*, 1997). Although cardiac expression of the truncated clone is greater than that of the full-length isoform, an *in vivo* association of the encoded proteins has yet to be demonstrated. The possible role of the truncated isoform in the pathogenesis of LQTS remains unknown.

As of 1998, 78 mutations within the *KvLQT1* coding sequence had been reported (de Jager *et al.*, 1996; Donger *et al.*, 1997; Kanters, 1998; Li *et al.*, 1998; Neyroud *et al.*, 1997a, 1998; Russell *et al.*, 1996; Splawski et al., 1999; Tanaka *et al.*, 1997), representing 45% of published LQTS mutations. Most mutations tend to occur within the six transmembrane-spanning domains or pore region, although several C-terminal mutations have been described. Strong dominant-negative effects have been reported for missense mutations scattered throughout the pore region and transmembrane segments (Chouabe *et al.*, 1997; Shalaby *et al.*, 1997; Z. Wang *et al.*, 1999; Wollnik *et al.*, 1997). Other missense mutations cause a depolarizing shift in the voltage dependence of I_{Ks} activation. A pronounced effect on activation gating is exerted by the C-terminal mutant R555C, which shifts the isochronal I_{Ks} activation curve by $+30$ mV (Chouabe *et al.*, 1997).

Many *KvLQT1* mutations that reduce I_{Ks} by dominant-negative suppression are inherited in an autosomal dominant fashion (Chouabe *et al.*, 1997; Shalaby *et al.*, 1997; Wollnik *et al.*, 1997). Mutations that cause haploinsufficiency tend to be inherited in a recessive manner (Chouabe *et al.*, 1997; Wollnik *et al.*, 1997). This has led to the hypothesis that haploinsufficiency mutations cause the Jervell and Lange-Nielsen syndrome, whereas dominant-negative mutations cause the Romano–Ward syndrome. Although this is generally true, we have characterized several haploinsufficiency *KvLQT1* mutations associated with a dominant mode of inheritance (Z. Wang *et al.*, 1999).

6. MUTATIONS IN KVLQT1-LIKE CHANNELS ALSO CAUSE DISEASE

Recently, three orthologs of *KvLQT1*—*KCNQ2, KCNQ3*, and *KCNQ4*—have been cloned and characterized. The gene products of *KCNQ2* and *KCNQ3* coassemble to form heteromultimeric channels (Yang *et al.*, 1998) that underlie the M-current in neurons of the central and peripheral nervous system (H. S. Wang *et al.*, 1998). Mutations in these orthologs cause two forms of inherited epilepsy (Biervert *et al.*, 1998; Charlier *et al.*, 1998; Singh *et al.*, 1998). Like mutations in *KvLQT1*, mutations in *KCNQ2* and *KCNQ3* increase cellular excitability by reducing the repolarizing current during the action potential. *KCNQ4* is expressed in sensory outer hair cells of the inner ear. Mutations in this gene cause an autosomal dominant form of deafness (Kubisch *et al.*, 1999).

7. MUTATIONS IN *minK*, THE GENE ENCODING THE β-SUBUNIT OF I_{Ks} CHANNELS

The *minK* gene (*KCNE1, IsK*) was originally cloned from a rat kidney cDNA library using a functional expression assay. This gene encodes a 130-amino-acid protein (129 amino acids in the human protein) which is predicted to form a single transmembrane domain with an extracellular amino and intracellular carboxy terminus (Takumi *et al.*, 1988). When expressed in *Xenopus* oocytes, *minK* induces a slowly activating K^+ current that shares many biophysical and pharmacological properties with cardiac I_{Ks} (Takumi *et al.*, 1988). The unique topology of minK led to the hypothesis that, rather than forming a pore structure by itself, minK is, in fact, a component of a K^+ channel with properties similar to I_{Ks} (for reviews, see (Kaczmarek and Blumenthal, 1997; Sanguinetti and Keating, 1997; Sanguinetti and Zou, 1997). This hypothesis was confirmed by the discovery that minK and KvLQT1 proteins coassemble to form a heteromultimeric I_{Ks} channel (Barhanin *et al.*, 1996; Sanguinetti *et al.*, 1996b). When minK interacts with KvLQT1 proteins, the resultant current (I_{Ks}) is 7-fold larger in amplitude. The activation of the minK–KvLQT1 heteromultimer is slowed by an order of magnitude, and the voltage dependence of activation is shifted by $+20$ mV, compared to current produced by KvLQT1 alone. Thus, minK serves as a Kvβ subunit, modulating the function of the pore-forming KvLQT1 α-subunits.

minK is located on chromosome 21q22.1-22.2 and contains three exons (Splawski *et al.*, 1998). LQTS-associated mutations in *minK* are rare, with only three mutations reported (Splawski *et al.*, 1997b; Tyson *et al.*, 1997b). Two different mutations (S74L, D76N) in a conserved region of the putative intracellular domain reduce I_{Ks}, but do so by distinct mechanisms. As illustrated in Fig. 3, the D76N *minK* mutation causes dominant-negative suppression of I_{Ks}, shifts the voltage dependence of activation by $+16$ mV, and increases the rate of channel deactivation (Splawski *et al.*, 1997b). The S74L mutation similarly alters the voltage dependence and kinetics of I_{Ks} gating but does not cause dominant-negative suppression. Each of these mechanisms would be predicted to reduce repolarizing current during the plateau and terminal phases of repolarization.

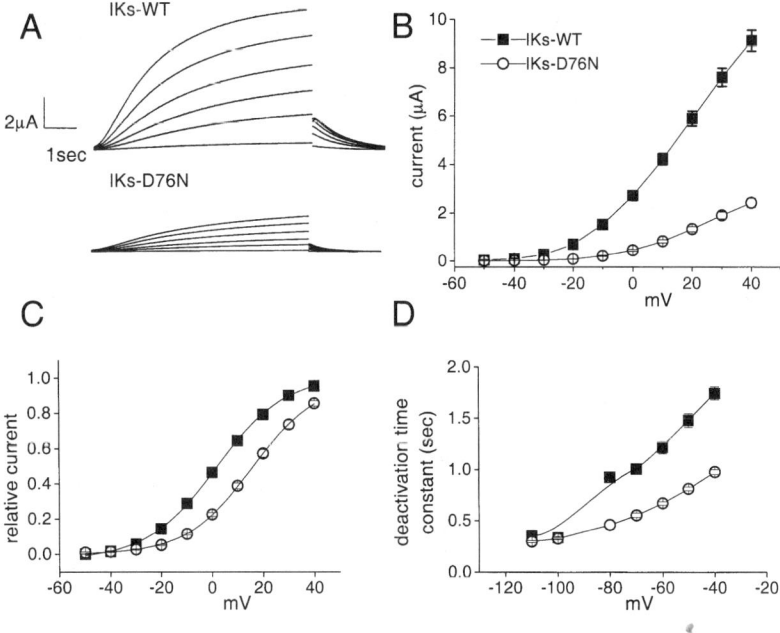

Figure 3. D76N*minK* mutation reduces I_{Ks} by multiple mechanisms. (A) and (B) D76N *minK* causes dominant-negative suppression of I_{Ks}. I_{Ks} was elicited by depolarizations of 7.5-s duration from a holding potential of -80 mV to test potentials of -40 to $+40$ mV. (C) D76N shifts the voltage dependence of activation by $+16$ mV compared to $I_{Ks\text{-}WT}$. The smooth curves are best fits of normalized tail currents to a Boltzmann function. (D) D76N increases the rate of I_{Ks} deactivation. I_{Ks} was activated by a 5-s pulse to $+20$ mV, and tail currents were measured at the indicated potentials. Tail currents were fitted to a single-exponential function. Each of these effects would reduce net repolarizing current during the cardiac action potential. (Reprinted, with permission, from Splawski *et al.*, 1997b.)

8. RECESSIVE MUTATIONS IN *KVLQT1* OR *MINK* CAUSE LQTS AND CONGENITAL SENSORINEURAL HEARING LOSS

Jervell and Lange-Nielsen syndrome represents a severe form of LQTS, characterized by marked QT prolongation in association with sensorineural hearing loss. A critical role for minK in Jervell and Lange-Nielsen syndrome was initially proposed prior to the discovery that minK was a component of the I_{Ks} channel. Null mutant mice with a targeted disruption in *minK* displayed inner ear abnormalities, including deafness and vestibular dysfunction (Vetter *et al.*, 1996). Histologically, the inner ear defects appeared strikingly similar to those reported in Jervell and Lange-Nielsen syndrome. It is now apparent that recessive mutations in either *minK* or *KVLQT1* can result in Jervell and Lange-Nielsen syndrome (Duggal *et al.*, 1998; Neyroud *et al.*, 1997; Schulze-Bahr *et al.*, 1997; Splawski *et al.*, 1997a; Tyson *et al.*, 1997). Both *minK* and *KvLQT1* are expressed in the stria vascularis of the inner ear (Fritzsch and Beisel, 1998), where they play a critical role in K$^+$ homeostasis. Following transduction of the acoustic signal, K$^+$ is transported back into the endolymph through I_{Ks} channels in the stria vascularis. The absence of I_{Ks} in the inner ear causes congenital deafness, and its absence in the heart causes marked QT prolongation. Deafness is recessively inherited in Jervell and Lange-Nielsen syndrome. Predisposition to arrhythmia is dominantly inherited, but expressed with variable penetrance. Thus, parents of a Jervell and Lange-Nielsen child are at risk for sudden death (Splawski *et al.*, 1997a).

9. MUTATIONS IN *HERG*, THE GENE ENCODING THE α-SUBUNIT OF I_{Kr} CHannels

The genomic structure of *HERG* consists of 16 exons spanning 55 kilobases that encodes a protein of 1159 amino acids (Splawski *et al.*? 1998b). *HERG* is predominately expressed in the heart (Curran *et al.*, 1995) but also has been detected in nervous tissue (Arcangeli *et al.*, 1997; Pennefather *et al.*, 1998; Warmke and Ganetzky, 1994). Heterologous expression studies confirmed that *HERG* encodes the α-subunit of the cardiac I_{Kr} channel (Sanguinetti *et al.*, 1995). Several alternatively spliced variants of *HERG* exist (Lees-Miller *et al.*, 1997b; London *et al.*, 1997b), and heteromultimers formed by distinct *HERG* isoforms have biophysical properties that more closely mimic cardiac myocyte I_{Kr} than homomultimeric HERG channels (London *et al.*, 1997b).

To date, 81 mutations in *HERG* have been described, accounting for approximately 45% of the known LQTS mutations (Splawski *et al.*, 1998a, 1999). The majority of *HERG* mutations are missense mutations; however, deletions and frameshift mutations have been reported. The molecular mechanisms whereby *HERG* mutations reduce I_{Kr} are typical of those described for *KvLQT1* mutations, including haploinsufficiency, dominant-negative effects, and altered channel gating and kinetics. For example, an in-frame deletion in the putative S3 domain (ΔI500–F508) encodes a protein that causes haploinsufficiency, presumably because it fails to coassemble with normal HERG subunits (Sanguinetti *et al.*, 1996a). Most missense *HERG* mutations cause a spectrum of dominant-negative effects. The pore mutation G628S causes a severe reduction in I_{Kr} (Sanguinetti *et al.*, 1996a). A mouse transgene overexpressing G628S HERG subunits completely abolishes I_{Kr} (Babij *et al.*, 1998). G628S mutant subunits form mature channels that apparently reach the plasma membrane (Zhou, 1998) but presumably are incapable of ion permeation. Several HERG mutants (Y611H, V822M) are improperly processed and undergo rapid degradation in the endoplasmic reticulum (Zhou, 1998). Examples of mutations that induce gating abnormalities include A614V and V630L. (Fig. 4). These mutants, when expressed with wild-type *HERG*, shift the voltage dependence of inactivation of the resulting heteromultimeric channels by -10 and -20 mV, respectively (Nakajima *et al.*, 1998).

Several *HERG* mutations cluster within the amino terminus, suggesting an important functional role for this region. The HERG amino terminus is believed to interact with the cytoplasmic linker between domains S4 and S5 to slow the rate of channel deactivation (J. Wang *et al.*, 1998). Truncation of the amino terminus, or splice variants that reduce its length, greatly accelerates the rate of channel deactivation (Lees-Miller *et al.*, 1997a; London *et al.*, 1997a; Spector, 1996). The three-dimensional structure of the HERG N-terminus has been resolved by X-ray crystallography and contains a region sharing structural homology with PAS (Per-Arnt-Sim) domains (Morais *et al.*, 1998). PAS domains are basic helix–loop–helix (bHLH) structures that have been shown to be the site of protein–protein interactions for factors that function

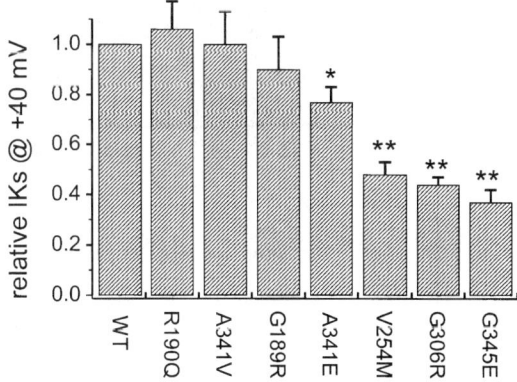

Figure 4. V630L *HERG* mutation reduces I_{Kr} by several mechanisms. (A) and (B) V630L *HERG* mutation causes dominant-negative suppression of channel function. Wild type (WT) and V630L HERG currents were elicited by 4-s depolarizing pulses between −40 and +40 mV. Tail currents were measured at −70 mV. (C) Current–voltage relationships for WT and V630L HERG. (D) V630L HERG shifts the voltage dependence of the HERG channel availability (inactivation) by +16 mV. Voltage dependence of HERG inactivation was assessed using the indicated triple-pulse protocol. The peak of the inactivating current during the third pulse was plotted as a function of the voltage of the second pulse, then fitted with a Boltzmann function. Dominant-negative suppression and a hyperpolarizing shift in the voltage dependence of channel inactivation would reduce the number of available HERG channels during the action potential. (Reprinted, with permission, from Nakajima *et al.*, 1998).

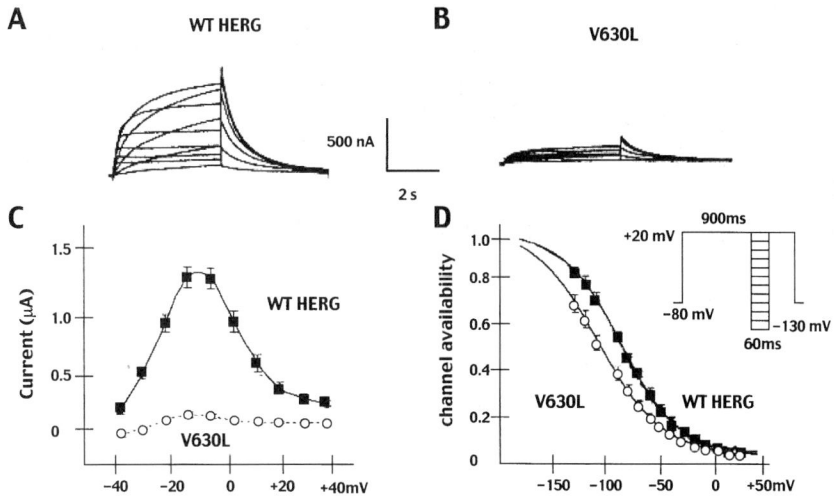

Figure 5. Mutations in the HERG PAS domain increase the rate of channel deactivation. (A) Structure of the PAS domain located in the HERG N-terminus and position of the long QT-associated mutations. (Diagram courtesy of Dr. J. Morais Cabral and R. MacKinnon, Rockefeller University.) (B) Deactivating currents elicited by test pulses between −120 and −50 mV following a 1-s activating pulse to +40 mV. (Reprinted, with permission, from Chen *et al.*, 1999).

in sensing and signal transduction. Several *HERG* mutations are localized to the hydrophobic face of the presumptive PAS domain, in a region that may interact with the S4–S5 cytoplasmic linker. These mutations markedly accelerate the rate of HERG deactivation (Chen *et al.*, 1999), presumably by interfering with binding of the PAS domain to its receptor. (Fig. 5).

The finding that *HERG* encodes the α-subunit of I_{Kr} channels provides a mechanistic link between the inherited and acquired forms of LQTS. A reduction in I_{Kr}, either by *HERG* mutations or by drug-induced blockade (Jurkiewicz and Sanguinetti, 1993), prolongs action potentials and thus predisposes to ventricular arrhythmia. The most common cause of acquired LQTS is blockade of HERG channels. Medications that block HERG channels include the methanesulfonanilide class III antiarrhythmics (Spector *et al.*, 1996), the histamine receptor H1 antagonist terfenadine (Roy *et al.*, 1996, Suessbrich, 1996), and the gastrointestinal prokinetic agent cisapride (Mohammad *et al.*, 1997, Rampe, 1997).

10. MUTATIONS IN *SCN5A*, THE GENE ENCODING THE α-SUBUNIT OF CARDIAC NA$^+$ CHANNELS

Mutations in the gene encoding the cardiac Na$^+$ channel, *SCN5A*, will only be briefly discussed, as the focus of this book is the role of K$^+$ channels in cardiovascular biology. *SCN5A* mutations cause "gain of function" effects that result in an increase in depolarizing current during the action potential. Some, but not all, mutations occur in regions known to be important for Na$^+$ channel inactivation (An, 1998; Kambouris *et al.*, 1998; Makita *et al.*, 1998; Splawski *et al.*, 1999; Q. Wang *et al.*, 1995). For example, the ΔKPQ mutation results from a three amino-acid deletion in a highly conserved region proposed to be involved in channel inactivation. ΔKPQ mutant Na$^+$ channels fail to inactivate completely with depolarization (Bennett *et al.*, 1995). Although the fraction of sustained inward current reflects the contribution of only 3–4% of available channels (Bennett *et al.*, 1995; D. W. Wang *et al.*, 1996), it is sufficient to account for the observed QT prolongation in affected individuals. Other characterized *SCN5A* mutations also alter the inactivation properties of the Na$^+$ channel, but act through distinct molecular mechanisms (An, 1998; Chandra *et al.*, 1998; Dumaine *et al.*, 1996; Makita *et al.*, 1998; D. W. Wang *et al.*, 1996a).

11. CELLULAR MECHANISMS OF TORSADE DE POINTES ARRHYTHMIA

LQTS-disease-causing mutations do not cause arrhythmias *per se* but rather provide the substrate for the induction of arrhythmogenesis. Individuals with LQTS are susceptible to a unique polymorphic ventricular tachycardia, characterized by undulating QRS complexes (torsades de pointes or "twisting around a point"). The paroxysmal nature of torsades de pointes suggests that a trigger is required to induce the arrhythmia in the setting of abnormal repolarization. The cellular basis of this trigger may be early afterdepolarizations (EADs), caused by reopenings of L-type Ca^{2+} channels during the terminal phase of repolarization. In a canine model of drug-induced

LQT3, El-Sherif *et al.*, (1996) demonstrated the initiation of polymorphic ventricular tachycardia from a subendocardial focus, presumably from an EAD-triggered beat within the Purkinje system. Owing to heterogeneity of repolarization across the ventricular wall, functional conduction block develops as the subendocardial triggered activity encounters adjacent, refractory myocardium. The functional block allows circulating, reentrant wave fronts to develop within the ventricular wall that subsequently proceed around the right and left ventricular chambers. Variations in wave front orientation and/or site of triggered activity accounts for the polymorphic electrocardiographic features (El-Sherif *et al.*, 1996). Ultimately, the polymorphic ventricular tachycardia degenerates into ventricular fibrillation, the presumed cause of sudden cardiac death in LQTS.

12. GENOTYPE PREDICTION OF CLINICAL PHENOTYPE

Careful phenotypic analysis of individuals harboring mutations in *KvLQT1*, *HERG*, and *SCN5A* reveals several clinical patterns typical of each genotype. Genotype–phenotype correlation studies have demonstrated that the various forms of LQTS display characteristic electrocardiograms, distinct arrhythmia-inducing phenomena, and varying clinical course. However, tremendous phenotypic variability exists, even within families carrying the identical mutation. In addition, at least 10% of individuals carrying a mutation in *KvLQT1* have what is considered to be a normal QT interval (Vincent *et al.*, 1992a). With these caveats in mind, a discussion of some general phenotypic characteristics of each genotype will follow.

The electrocardiograms of LQT1 and LQT2 individuals tend to display broad-based, prolonged T-waves or low-amplitude T-waves. The electrocardiograms of individuals with *SCN5A* mutations are characterized by a long isoelectric ST segment, followed by peaked, symmetric T-waves in the mid-precordial leads (Moss, 1995). In a pharmacological model of LQT2 and LQT3, the distinctive T-wave morphology was explained by differential transmural heterogeneity in action potential prolongation (Shimizu and Antzelevitch, 1997). The I_{Kr} blocker *d*-sotalol (used to mimic LQT2) caused preferential prolongation of action potentials in the midmyocardial M-cell region. The broad-based T-wave morphology in this LQT2 model was presumed to be due to repolarization occurring in the epi- and endocardium, which was less affected by *d*-sotalol treatment. ATX-II, an agent that slows Na^+ channel inactivation (mimicking LQT3), caused transmural action potential prolongation. The transmural delay in repolarization likely accounts for the isoelectric period immediately following the QRS complex.

Individuals with mutations in *KVLQT1* are particularly prone to arrhythmias in the setting of abrupt, autonomic stimuli. In a study of LQT1 subjects, sudden cardiac death was preceded by exercise, fright, or anxiety in 12 of 13 cases (Vincent *et al.*, 1992). Adrenergic stimulation normally shortens cardiac action potentials, in part by stimulating I_{Ks}. Individuals with a defect in I_{Ks} channels may be unable to appropriately shorten their QT interval in the face of adrenergic stimulation, thereby increasing their risk of arrhythmia in this setting. In contrast to LQT1, the risk of cardiac events under conditions of bradycardia, such as at night or during rest, appears to be greater for

LQT3 individuals (Schwartz *et al.*, 1995). Furthermore, marked shortening of the QT interval in response to exercise has been reported for individuals with *SCN5A* mutations (Schwartz *et al.*, 1995).

The age of onset and frequency of symptoms vary among the three genotypes. Individuals with mutations in *KVLQT1* and *HERG* tend to have more symptoms and present at an earlier age (childhood and adolescence) than those with *SCN5A* mutations (Zareba *et al.*, 1998). Although individuals with *SCN5A* mutations tend to have fewer cardiac events, the percentage of lethal cardiac events is significantly higher among these individuals, compared to those with mutations in either *KVLQT1* or *HERG*. Ultimately, however, there is no difference in cumulative mortality through age 40 among these three genotypes.

Based on our current understanding of genotype–phenotype correlation, knowledge of the genotype can only predict limited aspects of the clinical LQTS phenotype. The extensive phenotypic variation may be partly due to the diverse effects of individual mutations within each genotype. Ideally, one could predict the severity of disease *in vivo* based upon a functional assay of the expressed mutant genes in a heterologous expression system. (Fig 6). Initial attempts to make such a correlation have not been

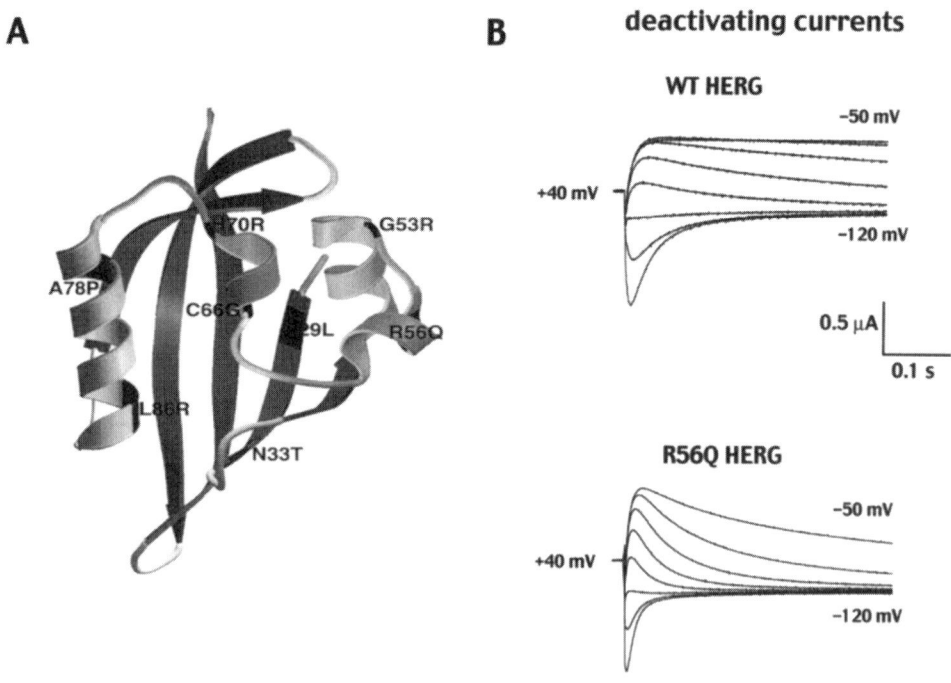

Figure 6. *KVLQT1* mutations cause a spectrum of I_{Ks} dysfunction. Summary of I_{Ks} suppression resulting from coexpression of wild-type (WT) and mutant *KVLQT1*, in the presence of *minK*. Data were normalized relative to average current at +40 mV in cells expressing WT I_{Ks}. WT I_{Ks} (generated from 4 ng of WT *KVLQT1 cRNA*) was compared to I_{ks} (generated from 4 ng of WT plus 4 ng of mutant *KVLQT1 cRNA*). R190Q, A341V, and G189R mutations caused haploinsufficiency. V254M, G306R, and G345E caused variable degrees of dominant-negative suppression of I_{Ks}. (Reprinted, with permission, from Wang *et al.*, 1999.)

successful. For example, the *KvLQT1* mutation V254M has a strong dominant-negative effect on I_{Ks} when expressed in *Xenopus* oocytes, whereas a second mutation, Λ341V, causes haploinsufficiency and, hence, less suppression of I_{Ks} (Z. Wang *et al.*, 1999). One might predict, based on the strong dominant-negative effect, that individuals with the V254M mutation would be more symptomatic. However, there was no statistical difference in corrected QT interval or the presence of symptoms between individuals harboring the V254M versus the A341V mutation. Both mutations are associated with a high incidence of symptoms and sudden death (Z. Wang *et al.*, 1999). Given the heterogeneity within and between families carrying the same mutation, factors other than the single ion channel gene mutation must contribute to the clinical phenotype. Several explanations could account for the lack of correlation between channel function in the heterologous expression system and clinical phenotype. Epigenetic factors, such as heart rate variability, serum electrolyte concentrations, and compensatory changes in other ion channel or regulatory genes likely play an important role in determining the incidence and severity of symptoms in gene carriers. Finally, the physiological consequences of mutations determined in oocytes may not accurately simulate the functional defect in human cardiac myocytes.

13. GENOTYPE-SPECIFIC THERAPY

The identification of LQTS arrhythmia susceptibility genes with distinct physiological consequences offers the possibility of tailoring genotype-specific therapy. One approach to gene-specific therapy takes advantage of the paradoxical effect of extracellular K^+ ions on the magnitude of I_{Kr}. In the case of most K^+ channels, an increase in extracellular K^+ concentration decreases the chemical driving force for outward flux of K^+ and therefore decreases the amplitude of outward current. Paradoxically, an increase in extracellular K^+ concentration actually increases the amplitude of I_{Kr} by shifting the voltage dependence of inactivation toward more depolarized potentials (Sanguinetti and Jurkiewicz, 1992). Based on this knowledge, Compton and colleagues tested the effect of K^+ supplementation on QT interval in patients with *HERG* mutations. An increase in serum K^+ concentration of 1.4 ± 0.7 meq/l decreased the corrected QT interval by 24% and normalized QT dispersion in 7 subjects (Compton *et al.*, 1996).

K^+ channel agonists offer an additional means of genotype-specific therapy for LQTS. A novel benzodiazepine has been reported to activate I_{Ks} and shorten action potentials in guinea pig cardiac myocytes (Salata *et al.*, 1998). The agonist effect of this compound is due to a hyperpolarizing shift in the voltage dependence of I_{Ks} activation. Although this compound has yet to be tested in humans, it offers promise as a therapeutic modality to augment I_{Ks} in LQTS-affected individuals.

Genotype-specific therapy for LQT3 is directed at minimizing the depolarizing influence of mutant Na^+ channels. Schwartz and colleagues tested the hypothesis that treatment with a Na^+ channel blocker, mexiletine, would suppress the inactivation-resistant inward current and shorten the QT interval in patients with *SCN5A* mutations. As predicted, mexiletine shortened the QT interval of a small cohort of LQT3, but not LQT2, individuals (Schwartz *et al.*, 1995).

Preliminary advances in molecular-directed therapy for LQTS have been made, but genotype-specific therapy is still in its infancy. It remains to be proven whether normalization of QT interval with drugs or K^+ supplementation will decrease the incidence of cardiac events and improve survival. The challenge for the future is to identify what factors influence the predisposition to sudden cardiac death in the setting of abnormal repolarization. Understanding these epigenetic factors may lead to novel therapeutic strategies. An additional challenge is to apply the knowledge gained by our understanding of the molecular nature of LQTS to the more common forms of ventricular arrhythmias, such as those occurring in the setting of myocardial infarction, ventricular hypertrophy, or congestive heart failure.

REFERENCES

Ackerman, M. J., and Clapham, D. E., 1997, Ion channels—basic science and clinical disease, *N. Engl. J. Med.* **336:**1575–1586 [published erratum appears in *N. Engl. J. Med.* **337:**579 1997].

An, R., 1998, Novel LQT-3 mutation affects Na^+ channel activity through interactions between α- and $\beta1$-subunits, *Circ. Res.* **83:**141–146.

Arcangeli, A., Rosati, B., Cherubini, A., Crociani, O., Fontana, L., Ziller, C., Wanke, E., and Olivotto, M., 1997, HERG- and IRK-like inward rectifier currents are sequentially expressed during neuronal development of neural crest cells and their derivatives, *Eur. J. Neurosci.* **9:**2596–2604.

Babij, P., Askew, G. R., Nieuwenhuijsen, B., Su, C. M., Bridal, T. R., Jow, B., Argentieri, T. M., Kulik, J., DeGennaro, L. J., Spinelli, W., and Colatsky, T. J., 1998, Inhibition of cardiac delayed rectifier K^+ current by overexpression of the long-QT syndrome HERG G628S mutation in transgenic mice, *Circ. Res.* **83:**668–678.

Barhanin, J., Lesage, F., Guillemare, E., Fink, M., Lazdunski, M., and Romey, G., 1996, KvLQT1 and IsK (minK) proteins associate to form the I_{Ks} cardiac potassium current, *Nature* **384:**78–80.

Benhorin, J., Kalman, Y. M., Medina, A., Towbin, J., Rave-Harel, N., Dyer, T. D., Blangero, J., MacCluer, J. W., and Kerem, B., 1993, Evidence of genetic heterogeneity in the long QT syndrome, *Science* **260:**1960–1961.

Bennett, P. B., Yazawa, K., Makita, N., and George, A. L., 1995, Molecular mechanism for an inherited cardiac arrhythmia, *Nature* **376:**683–685.

Biervert, C., Schroeder, B. C., Kubisch, C., Berkovic, S. F., Propping, P., Jentsch, T. J., and Steinlein, O.K., 1998, A potassium channel mutation in neonatal human epilepsy, *Science* **279:**403–406.

Chandra, R., Starmer, C. F., and Grant, A. O., 1998, Multiple effects of KPQ deletion mutation on gating of human cardiac Na^+ channels expressed in mammalian cells, *Am. J. Physiol.* **274:**H1643–H1654.

Charlier, C., Singh, N. A., Ryan, S. G., Lewis, T. B., Reus, B. E., Leach, R. J., and Leppert, M., 1998, A pore mutation in a novel KQT-like potassium channel gene in an idiopathic epilepsy family. *Nat. Genet.* **18:**53–55.

Chen, J., Zou, A., Splawski, I., Keating, M. T., and Sanguinetti, M. ., 1999, Long QT syndrome-associated mutations in the PAS domain of HERG potassium channels accelerate channel deactivation. *J. Biol. Chem.* **274:**10113–10118.

Chouabe, C., Neyroud, N., Guicheney, P., Lazdunski, M., Romey, G., and Barhanin, J., 1997, Properties of KvLQT1 K^+ channel mutations in Romano-Ward and Jervell and Lange-Nielsen inherited cardiac arrhythmias, *EMBO J.* **16:**5472–5479.

Compton, S. J., Lux, R. L., Ramsey, M. R., Strelich, K. R., Sanguinetti, M. C., Green, L. S., Keating, M. T., and Mason, J. W., 1996, Genetically defined therapy in inherited long QT syndrome: Correction of abnormal repolarization by potassium, *Circ.* **94:**1018–1022.

Curran, M., Atkinson, D., Timothy, K., Vincent, G. H., Moss, A., Leppert, M., and Keating, M., 1993, Locus heterogeneity of autosomal dominant long QT syndrome, *J. Clin. Invest.* **92:**799–803.

Curran, M. E., Splawski, I., Timothy, K. W., Vincent, G. M., Green, E. D., and Keating, M. T., 1995, A molecular basis for cardiac arrhythmia: *HERG* mutations cause long QT syndrome, *Cell* **80:**795–803.

Schulze-Bahr, E., Wang, Q., Wedekind, H., Haverkamp, W., Chen, Q., and Sun, Y., 1997, KCNE1 mutations cause Jervell and Lange-Nielsen syndrome [letter], *Nat. Genet.* **17:**267–268.

Schwartz, P. J. 1997. The long QT syndrome, *Curr. Probl. Cardiol.* **22:**297–351.

Schwartz, P. J., Priori, S. G., Locati, E. H., Napolitano, C., Cantu, F., Towbin, J. A., Keating, M. T., Hammoude, H., Brown, A. M., Chen, L.-S. K., and Colatsky, T. J., 1995, Long QT syndrome patients with mutations of the *SCN5A* and *HERG* genes have differential responses to Na$^+$ channel blockade and to increases in heart rate *Circulation* **92:**3381–3386.

Shalaby, F. Y., Levesque, P. C., Yang, W. P. Little, W. A., Conder, M. L., Jenkins-West, T., and Blanar, M. A., 1997, Dominant-negative KvLQT1 mutations underlie the LQT1 form of long QT syndrome [see comments], *Circulation* **96:**1733–1736.

Shimizu, W., and Antzelevitch, C., 1997, Sodium channel block with mexiletine is effective in reducing dispersion of repolarization and preventing torsade de pointes in LQT2 and LQT3 models of the long-QT syndrome, *Circulation* **96:**2038–2047.

Singh, N. A., Charlier, C., Stauffer, D., DuPont, B. R., Leach, R. J., Melis, R., Ronen, G. M., Bjerre, I., Quattlebaum, T., Murphy, J. V., McHarg, M. L., Gagnon, D., Rosales, T. O., Peiffer, A., Anderson, V. E., and Leppert, M., 1998, A novel potassium channel gene, KCNQ2, is mutated in an inherited epilepsy of newborns, *Nat. Genet.* **18:**25–29.

Spector, P., 1996, Fast inactivation causes rectification of the I_{Kr} channel, *J. Gen. Physiol.* **107:**611–619.

Spector, P. S., Curran, M. E., Keating, M. T., Sanguinetti, M. C., 1996, Class III antiarrhythmic drugs block HERG, a human cardiac delayed rectifier K$^+$ channel: Open-channel block by methanesulfonanilides, *Circ. Res.* **78:**499–503.

Splawski, I., Timothy, K. W., Vincent, G. M., Atkinson, D. L., and Keating, M. T., 1997a. Molecular basis of the long-QT syndrome associated with deafness [see comments], *N. Engl. J. Med.* **336:**1562–1567.

Splawski, I., Tristani-Firouzi, M., Lehmann, M. H., Sanguinetti, M. C., and Keating, M. T., 1997b, Mutations in the hminK gene cause long QT syndrome and suppress I_{Ks} function, *Nat. Genet.* **17:**338–340.

Splawski, I., 1999, Arrhythmia-associated mutations in KVLQT1, Herg, minK and SCN5A. *Hum. Mol. Genet.,* (in preparation).

Splawski, I., Shen, J., Timothy, K. W., Vincent, G. M. Lehmann, M. H., and Keating, M. T., 1998, Genomic structure of three long QT syndrome genes: KVLQT1, HERG, and KCNE1, *Genomics.* **51:**86–97.

Suessbrich, H., 1996, Blockade of HERG channels expressed in *Xenopus* oocytes by the histamine receptor antagonists terfenadine and astemizole, *FEBS Lett.* **385:**77–80.

Takumi, T., Ohkubo, H., and Nakanishi, S., 1988, Cloning of a membrane protein that induces a slow voltage-gated potassium current, *Science* **242:**1042–1045.

Tanaka, T., Nagai, R., Tomoike, H., Takata, S., Yano, K., Yabuta, K., Haneda, N., Nakano, O., Shibata, A., Sawayama, T., Kasai, H., Yazaki, Y., and Nakamura, Y., 1997, Four novel KVLQT1 and four novel HERG mutations in familial long-QT syndrome. *Circulation* **95:**565–567.

Towbin, J. A., Li, H., Taggart, R. T., Lehmann, M. H., Schwartz, P. J., Satler, C. A., Ayyagari, R., Robinson, J. L., Moss, A., and Hejtmancik, J. F., 1994, Evidence of genetic heterogeneity in Romano-Ward long QT syndrome: Analysis of 23 families. *Circulation* **90:**2635–2644.

Trudeau, M. C., Warmke, J. W., Ganetzky, B., and Robertson, G. A., 1995, HERG, a human inward rectifier in the voltage-gated potassium channel family, *Science* **269:**92–95.

Tyson, J., Tranebjaerg, L., Bellman, S., Wren, C., Taylor, J. F., Bathen, J., Aslaksen, B., Sorland, S. J., Lund, O., Malcolm, S., Pembrey, M., Bhattacharya, S., and Bitner-Glindzicz, M., 1997, IsK and KvLQT1: Mutation in either of the two subunits of the slow component of the delayed rectifier potassium channel can cause Jervell and Lange-Nielsen syndrome, *Hum. Mol. Genet.* **6:**2179–2185.

Vetter, D. E., Mann, J. R., Wangemann, P., Liu, J., McLaughlin, K. J., Lesage, F., Marcus, D. C., Lazdunski, M., Heinemann, S. F., and Barhanin, J., 1996, Inner ear defects induced by null mutation of the isk gene, *Neuron* **17:**1251–1264.

Vincent, G. M., Timothy, K., Leppert, M., and Keating, M., 1992a, The spectrum of symptoms and QT intervals in carriers of the gene for the long QT syndrome, *N. Engl. J. Med.* **327:**846–852.

Wang, D. W., Yazawa, K., George, A. L., and Bennett, P. B., 1996. Characterization of human cardiac Na$^+$ channel mutations in the congenital long QT syndrome. *Proc. Natl. Acad. Sci. U.S.A.* **93:**13200–13205.

Wang, H. S., Pan, Z., Shi, W., Brown, B. S., Wymore, R. S., Cohen, I. S., Dixon, J. E., and McKinnon, D., 1998. KCNQ2 and KCNQ3 potassium channel subunits: Molecular correlates of the M-channel [see comments], *Science* **282:**1890–1893.

Wang, J., Trudeau, M. C., Zappia, A. M., and Robertson, G. A., 1998, Regulation of deactivation by an amino terminal domain in human ether-a-go-go-related gene potassium channels, *J. Gen. Physiol.* **112**:637–647.

Wang, Q., Shen, J., Splawski, I., Atkinson, D., Zhizhong, L., Robinson, J., Moss, A., Towbin, J., and Keating, M., 1995, *SCN5A* mutations associated with an inherited cardiac arrhythmia, long QT syndrome, *Cell* **80**:805–811.

Wang, Q., Curran, M. E., Splawski, I., Burn, T. C., Millholland, J. M., VanRaay, T. J. Shen, J., Timothy, K. W., Vincent, G. M., de Jager, T., Schwartz, P. J. Toubin, J. A., Moss, A. J., Atkinson, D. L., Landes, G. M., Connors, T. D., and Keating, M. T., 1996, Positional cloning of a novel potassium channel gene: KVLQT1 mutations cause cardiac arrhythmias. *Nat. Genet.* **12**:17–23.

Wang, Z., Tristani-Firouzi, M., Xu, Q., Liu, M. Keating, M. T., and Sanguinetti, M. C., 1999, Functional effects of mutations in KvLQT1 that cause long QT syndrome, *J. Cardiovasc. Electrophysiol.* **10**:817–826.

Warmke, J., and Ganetzky, B., 1994, A family of potassium channel genes related to *eag* in *Drosophila* and mammals, *Proc. Natl. Acad. Sci. U.S.A.* **91**:3438–3442.

Willich, S., Levy, D., Rocco, M., Tofler, G., Stone, P., and Muller, J., 1987, Circadian variation in the incidence of sudden cardiac death in the Framingham heart study population, *Am. J. Cardiol.* **60**:801–806.

Wollnik, B., Schroeder, B. C. Kubisch, C., Esperer, H. D., Wieacker, P., and Jentsch, T. J., 1997, Pathophysiological mechanisms of dominant and recessive KVLQT1 K$^+$ channel mutations found in inherited cardiac arrhythmias, *Hum. Mol. Genet.* **6**:1943–1949.

Yang, W. P., Levesque, P. C., Little, W. A., Conder, M. L., Shalaby, F. Y., and Blanar, M. A., 1997, KvLQT1, a voltage-gated potassium channel responsible for human cardiac arrhythmias, *Proc. Natl. Acad. Sci. U.S.A.* **94**:4017–21.

Yang, W. P., Levesque, P. C., Little, W. A., Conder, M. L., Ramakrishnan, P., Neubauer, M. G., and Blanar, M. A., 1998, Functional expression of two KvLQT1-related potassium channels responsible for an inherited idiopathic epilepsy, *J. Biol. Chem.* **273**:19419–19423.

Zareba, W., Moss, A. J., Schwartz, P. J., Vincent, G. M., Robinson, J. L., Priori, S. G., Benhorin, J., Locati, E. H., Towbin, J. A., Keating, M. T., Lehmann, M. H., and Hall, W. J., 1998, Influence of genotype on the clinical course of the long-QT syndrome. International Long-QT Syndrome Registry Research Group, *N. Engl. J. Med.* **339**:960–965.

Zeng, J., Laurita, K. R., Rosenbaum, D. S., and Rudy, Y., 1995, Two components of the delayed rectifier K$^+$ current in ventricular myocytes of the guinea pig type: Theoretical formulation and their role in repolarization, *Circ. Res.* **77**:140–152.

Zhou, Z., 1998, Properties of HERG channels stably expressed in HEK 293 cells studied at physiological temperature, *Biophys. J.* **74**:230–241.

Zhou, Z., Gong, Q., Epstein, M. L., and January, C. T., 1998, HERG channel dysfunction in human long QT syndrome: Intracellular transport and functional defects, *J. Biol. Chem.* **273**:21061–21066.

Chapter 37

Altered K^+ Channel Expression in the Hypertrophied and Failing Heart

Koichi Takimoto and Edwin S. Levitan

1. INTRODUCTION

Hypertrophied and failing hearts display abnormal electrophysiological properties. Commonly, myocytes from these diseased hearts exhibit a prolongation of action potential duration. The most consistently detected change in ionic currents that could produce this property is a decrease in the transient outward K^+ current (I_{to}). Recent studies have revealed that two Kv4 subfamily channel subunits, Kv4.2 and Kv4.3, constitute cardiac I_{to} channels. Furthermore, expression of these subunits is decreased in hypertrophied and failing hearts. In this chapter, we will summarize these electrophysiological and molecular biological findings. Possible mechanisms leading to the downregulation of I_{to} channels and consequences of the altered I_{to} channel expression in arrthythmogenesis also will be discussed.

2. ELECTROPHYSIOLOGICAL ALTERATIONS IN MYOCYTES FROM HYPERTROPHIED AND FAILING HEARTS

Sudden cardiac death is the main cause of mortality in patients with cardiac hypertrophy and heart failure. In the case of chronic heart failure patients, the annual mortality rate is 30–50%, and nearly half of the patients die suddenly. The occurrence of sudden death seems unrelated to the severity of the disease; a high incidence of death is seen at an early stage of the disease (Tomaselli *et al.*, 1994). Furthermore, ambulatory electrocardiographic monitoring and other measurements have pointed to

Koichi Takimoto and Edwin S. Levitan ● Department of Pharmacology, University of Pittsburgh, Pittsburgh, Pennsylvania 15261.

Potassium Channels in Cardiovascular Biology, edited by Archer and Rusch. Kluwer Academic/Plenum Publishers, New York, 2001.

the possibility that ventricular tachyarrhythmias are mainly responsible for the high incidence of sudden cardiac death in this group of patients (Pye and Cobbe, 1992; Tomaselli et al., 1994; Nabauer and Kaab, 1998). Left ventricular hypertrophy is another condition that significantly increases the incidence of sudden cardiac death (Kannel, 1991). Cardiac hypertrophy develops as a biological adaptive response to normalize wall stress produced by various overloading conditions, such as hypertension, obesity, diabetes, and elimination of a part of the heart muscle by myocardial infarction. However, left ventricular hypertrophy causes death in nearly 30% of patients within 5 years particularly when it is associated with electrocardiographic changes with a 3- to 5-fold increase in the incidence of sudden death (Kannel, 1991). Hence, an electrophysiological abnormality may significantly contribute to the high incidence of sudden cardiac death in patients with hypertrophy and heart failure.

Because of the significance of electrophysiological abnormalities in sudden cardiac death, a large number of studies have examined electrophysiological alterations in the myocardium from hypertrophied and failing hearts. Studies with experimental models of hypertrophy have shown that a common change found in hypertrophied myocytes is a prolongation of action potential duration (Hart, 1994; Tomaselli et al., 1994; Boyden and Jeck, 1995; Nabauer and Kaab, 1998). This has been observed in hypertrophied myocytes produced by a variety of conditions in many animal species. A prolongation of action potentials may be due to an increase in inward currents or a decrease in outward currents. In agreement with this notion, many studies have shown a decrease in I_{to} (Table I). For instance, a chronic increase in pressure load produced by abdominal aortic ligation significantly reduces macroscopic I_{to} amplitude by a mechanism not involving changes in single-channel conductance or open-state probability (Benitah et al., 1993; Tomita et al., 1994). Likewise, hypertrophied ventricular myocytes secondary to pulmonary artery constriction (Potreau et al., 1995), deoxycorticosterone acetate (DOCA)/salt-induced hypertension (Coulombe et al., 1994), genetically determined hypertension (Cerbai et al., 1994), or increased growth hormone secretion (Xu and Best, 1991) or after myocardial infarction (Qin et al., 1996) also show a significant decrease in I_{to} density. Thus, a reduction in I_{to} density is, in many cases, associated with the observed prolongation of action potential duration in hypertrophied myocytes.

Similar electrophysiological alterations are seen in the myocardium from failing hearts. Despite the fact that the etiology of heart failure is diverse, a prolongation of action potential is consistently observed in myocytes from patients with heart failure. For example, action potential duration is found to be significantly longer in myocytes from patients with dilated cardiomyopathy and ischemic cardiomyopathy (Beuckelmann et al., 1993). A significant increase in action potential duration also is seen in myocytes from several animal models, such as hereditary cardiomyopathy in hamsters (Thuringer et al., 1996) and pacing-induced heart failure in dogs (Kaab et al., 1996) and rabbits (Rozanski et al., 1997). Furthermore, as is the case for hypertrophied myocytes, many studies have shown that a decrease in I_{to} is associated with the detected action potential prolongation (Table I). Thus, a prolongation of action potential and an associated reduction in I_{to} are commonly observed in myocytes from failing hearts. Taken together, it is striking that hypertrophied and failing hearts with diverse conditions show a remarkable similarity in their electrical abnormality.

Table I

Alterations in Ionic Currents in Myocytes from Hypertrophied and Failing Hearts

Condition	Species	Changes in currents[a]	Reference
Terminal heart failure (transplant patients)	Human	$I_{to}\downarrow$, $I_{Ki}\downarrow$, I_{Kir}	Beuckelmann et al., 1993
Terminal heart failure (transplant patients)	Human	$I_{to}\downarrow$	Wettwer et al., 1994
Dilated cardiomyopathy	Human	$I_{Ki}\downarrow$	Koumi et al., 1995
Pacing-induced heart failure	Dog	$I_{to}\downarrow$	Kaab et al., 1996
Pacing-induced heart failure	Rabbit	$I_{to}\downarrow$	Rozanski et al., 1997
Hereditary cardiomyopathy	Hamster	$I_{Ca}\downarrow$, $I_{to}\downarrow$	Thuringer et al., 1996
Hereditary cardiomyopathy	Hamster	$I_{to}\downarrow$, $I_K\downarrow$, $I_{Ki}\downarrow$	Lodge and Normandin, 1997
Left coronary artery ligation	Dog	$I_{to}\downarrow$	Lue and Boyden, 1992
Chronic myocardial infarction	Rat	$I_{to}\downarrow$	Rozanski et al., 1998
Post-myocardial infarction	Rat	$I_{to}\downarrow$	Qin et al., 1996
Renovascular hypertension	Rat	$I_{Ca}\uparrow$	Keung, 1989
Renovascular hypertension	Rat	$I_{to}\uparrow$	Li and Keung, 1994
Perinephritis	Rabbit	$I_{to}\downarrow$, $I_{Ki}\downarrow$	McIntosh et al., 1998
Pulmonary artery constriction	Cat	$I_K\downarrow$, $I_{Ki}\uparrow$	Kleiman and Houser, 1989
Pulmonary artery constriction	Ferret	$I_{to}\downarrow$	Potreau et al., 1995
Growth hormone overproduction	Rat	$I_{to}\downarrow$	Xu and Best, 1991
Abdominal aorta ligation	Rat	$I_{to}\downarrow$	Benitah et al., 1993
Abdominal aorta ligation	Rat	$I_{to}\downarrow$	Tomita et al., 1994
Abdominal aorta constriction	Cat	$I_{to}\downarrow$	Kleiman and Houser, 1988
DOCA salt-induced hypertension	Rat	$I_{to}\downarrow$	Coulombe et al., 1994
Spontaneous hypertension	Rat	$I_{to}\downarrow$	Cerbai et al., 1994
Spontaneous hypertension	Rat	$I_{Ki}\downarrow$	Brooksby et al., 1993
Chronic high-altitude exposure	Rat	$I_{Ca}\downarrow$, $I_{to}\downarrow$	Chouabe et al., 1997

[a] I_{to}, Transient outward K$^+$ current; I_{Ku}, I_{Ca}, voltage-gated Ca^{2+} current; I_K, delayed rectifier K$^+$ current; I_{Kir}, inward rectifier K$^+$ current.

3. DECREASED EXPRESSION OF Kv4 SUBFAMILY CHANNEL SUBUNIT mRNAs IN HYPERTROPHIED AND FAILING HEARTS—A MOLECULAR MECHANISM UNDERLYING THE REDUCTION IN I_{to}

One of the essential steps for elucidating the molecular mechanism underlying the reduction in I_{to} in hypertrophied and failing hearts is to understand the molecular architecture of cardiac I_{to} channels. Studies from many laboratories have provided strong evidence indicating that the two Kv4 subfamily channel genes, Kv4.2 and Kv4.3, encode a large fraction of cardiac I_{to} channels. First, heterologous expression of Kv4.2 or Kv4.3 generates channels with kinetics and pharmacological properties similar to those of the native I_{to} channel. The current inactivates during depolarization and is sensitive to 4-aminopyridine and flecainide, but not to tetraethylammonium (Baldwin et al., 1991; Serodio et al., 1994; Dixon et al., 1996; Tsaur et al., 1997; Yeola and Snyder, 1997). Furthermore, recent studies have provided more direct evidence for the hypothesis that Kv4 subfamily genes encode I_{to} channels in rat cardiomyocytes: application of antisense oligonucleotides specific for Kv4.2 or Kv4.3 mRNA to neonatal or young adult rat ventricular myocytes (Fiset *et al.*, 1997) or viral vector-mediated overexpression of dominant-negative Kv4 subfamily subunits in adult rat ventricular myocytes (Johns *et al.*, 1997) selectively suppresses I_{to}, but not the noninactivating delayed rectifier current. The degree of I_{to} reduction by either method is 60–70%, indicating that a large fraction of the current is encoded by the two Kv4 subfamily genes. Finally, transgenic mice that specifically express a dominant-negative form of Kv4 subfamily subunits in the heart were generated (Barry *et al.*, 1998), and ventricular myocytes from these animals were found to contain significantly smaller I_{to} current. Therefore, Kv4.2 and Kv4.3 subunits appear to constitute I_{to} channels in mouse and rat cardiac myocytes.

Although these two Kv4 subfamily channel subunits are likely to encode a large portion of I_{to} in cardiac myocytes in many animal species, the contribution of each Kv4 subfamily gene product may differ significantly from one species to another. In rat heart, Kv4.2 mRNA and protein are significantly expressed. Furthermore, Kv4.2 mRNA, but not Kv4.3 mRNA, is differentially expressed across the left ventricular wall, and this differential Kv4.2 mRNA expression is well correlated with I_{to} density (Dixon and McKinnon, 1994; Dixon *et al.*, 1996). Thus, Kv4.2 is likely to contribute significantly to the formation of I_{to} channels in the rat myocardium. In contrast, Kv4.2 mRNA is less significant or even undetectable in cardiac myocytes of several other species, including humans, guinea pig, and dog. Therefore, whereas I_{to} channels are formed by Kv4.2 and Kv4.3 subunits in rat cardiac myocytes, Kv4.3 homomeric channels may be the predominant form of I_{to} channels in some other species.

The finding that I_{to} current density is reduced without changes in current kinetics in the myocardium from hypertrophied and failing hearts indicates that the number of functional I_{to} channels is decreased. Perhaps the simplest explanation for the decrease in I_{to} channels is a reduction in expression of the pore-forming α-subunits. In fact, it has been established that expression of K^+ channel α-subunit genes is regulated by hormones, neurotransmitters, neuronal activity, and other conditions in many cell types to produce changes in excitability (Levitan and Takimoto, 1998). In addition to these physiological stimuli, hypertrophy and heart failure might also alter expression of K^+ channel genes. Indeed, the results from several laboratories have indicated that Kv4 subfamily channel mRNAs and proteins are reduced in the myocardium from hypertrophied and failing hearts. First, decreases in Kv2.1 and Kv4.2 mRNAs and proteins were detected in hypertrophied myocardium following experimentally induced myocardial infarction (Gidh-Jain *et al.*, 1996). In this study, the mRNA level of the newly identified Kv4.3 gene was not examined. We have found that renovascular hypertension reduces expression of ventricular Kv4.2 and Kv4.3 mRNAs (Takimoto *et al.*, 1997). The levels of these two Kv4 subfamily channel subunit mRNAs were decreased in two different animal models of renovascular hypertension. The fact that channel downregulation occurred in a model with high blood renin–angiotensin and one with normal levels suggests that an elevation in blood renin–angiotensin is not essential. No changes in other Kv channel α-subunit mRNAs, including Kv2.1, KvLQT1, and Kv1 subfamily channel mRNAs, were detected. Thus, renovascular hypertension appears to specifically downregulate α-subunit mRNAs of the Kv4 subfamily, a finding that fits well with the observed electrophysiological alteration. Finally, recent work has shown a significant reduction in Kv4.3 mRNA level in myocardium from terminal heart failure patients

(Kaab *et al.*, 1998). The expression of other Kv α- and β-subunit mRNAs was unaltered. Hence, a reduction in I_{to} density in hypertrophied and failing hearts is likely to be the result of a decreased expression of Kv4 subfamily channels. These findings again point to a striking similarity, at the molecular level, in the alterations produced by hypertrophy and heart failure.

4. MECHANISMS FOR THE DECREASED EXPRESSION OF Kv4 SUBFAMILY CHANNEL mRNAs I_{to} CHANNELS IN HYPERTROPHIED AND FAILING HEARTS

Although it has become evident that the expression of Kv4 subfamily channel subunit mRNAs is downregulated in hypertrophied and failing hearts, the mechanisms by which the expression of these channel transcripts is decreased remain obscure. Given the diverse pathophysiological and morphological alterations associated with hypertrophy and heart failure, it may be difficult to assume one common mechanism for the reduced channel mRNA expression. Therefore, multiple stimuli and conditions may lead to the decreased channel expression. It is known that many hormones and growth factors, as well as mechanical forces, induce cardiac myocyte hypertrophy. These factors include the α_1-adrenergic agonist phenylephrine, angiotensin II, endothelin-1, and insulin-like growth factor-1. The finding that hypertrophied myocytes produced by various conditions consistently contain a smaller I_{to} component suggests that many of these stimuli may also downregulate expression of Kv4 subfamily channels. Indeed, a recent study reported that Kv4.2 protein level is lowered in neonatal ventricular myocytes cultured for three days in the presence of phenylephrine, endothelin-1, or insulin-like growth factor-1 (Guo *et al.*, 1998). These data support the notion that activation of multiple signaling pathways leads to reductions in Kv4.2 channels. They also imply that the decrease in Kv4.2 channel expression may be secondary to hypertrophy. However, we have detected rapid decreases in Kv4.2 and Kv4.3 mRNAs by angiotensin II in cultured neonatal myocytes within several hours (unpublished observation). Therefore, the levels of channel mRNAs change before myocyte hypertrophy develops. Thus, the detected decrease in Kv4 subfamily channel mRNAs is not simply the result of hypertrophy. Rather, the hormone may simultaneously regulate a set of genes including the Kv4 subfamily genes. It is certainly necessary to consider that these results with neonatal myocyte culture systems may not fully represent hypertrophy-induced alterations in the adult myocardium. However, a large number of papers on hypertrophy-induced gene regulation reveal that alterations in gene expression in these culture systems are very similar to those seen in adult myocardium. Thus, it is possible that the decrease in channel mRNAs occurs *in vivo* in a much shorter time scale before hypertrophy becomes apparent. Taken together with the finding that several hormones and growth factors reduce expression of Kv4.2 proteins, the downregulation of Kv4 subfamily mRNAs may be produced by various conditions as a part of initial hypertrophic responses.

Finally, it is notable that a reduction in I_{to} in hypertrophied myocytes produced by chronic myocardial infarction is reversed by culturing myocytes with an activator of pyruvate dehydrogenase (Rozanski *et al.*, 1998). Furthermore, an inhibitor of this metabolic enzyme was found to decrease I_{to} in myocytes from control animals. Pyruvate dehydrogenase, present as a large complex in mitochondria, is a rate-limiting enzyme of glucose oxidation. Therefore, altering the activity of this enzyme significantly influences the metabolic state of myocytes. Hence, these results indicate that the metabolic state of myocytes may control the level of expression of I_{to} channels. Moreover, alterations in I_{to} channel expression occurred over a period of hours, rather than minutes, suggesting that altered gene expression is involved. This last set of experiments is particularly interesting because the metabolic state may be commonly influenced in hypertrophied and failing hearts. In summary, the complex nature of the cardiac diseases makes it difficult to elucidate the mechanism underlying the decreases in expression of Kv4 subfamily channel genes and I_{to}. Yet, molecular tools identified in the last several years, as well as some new findings, are likely to provide ways to uncover the mechanisms controlling the expression of the Kv4 channel genes and I_{to} channels.

5. REGION-SPECIFIC EXPRESSION OF I_{to} AND ITS ALTERATIONS BY HYPERTROPHY AND HEART FAILURE

Despite the consistent observation that I_{to} channels are decreased in the myocardium from hypertrophied and failing hearts, the issue of whether the decrease in I_{to} creates arrhythmogenic conditions remains unclear. Indeed, theoretical and experimental data indicate that altering I_{to} may have less significant effects on after-repolarization or prolongation of action potential duration than might be expected (Priebe and Beuckelmann, 1998). It can be said that since I_{to} plays a major role in the early phase of repolarization, the time course of action potentials should be altered. In this regard, a transgenic animal study using a dominant-negative Kv4 subfamily α-subunit (Barry et al., 1998) demonstrated that a reduction in I_{to} prolonged action potential duration, at least in mice. The myocardium of these animals also exhibited a longer QT interval in surface echocardiogram. However, these transgenic mice do not develop spontaneous ventricular arrhythmias, supporting the notion that a decrease in I_{to} itself is not arrhythmogenic. In contrast, a decrease in I_{Kur}, the ultrarapidly activating component of cardiac delayed rectifier K^+ current that is associated with the Kv1 gene family has been linked to ventricular arrhythmias. Mice that express dominant-negative Kv1 subfamily subunits have an increased frequency of premature ventricular beats, ventricular arrhythmias, and spontaneous ventricular tachyarrhythmias (London et al., 1998). Therefore, the mice with a reduced I_{to} may be less susceptible to arrhythmias than those with decreased I_{Kur}. This difference in the effects of decreased K^+ currents may be due to the fact that altering I_{to} does not markedly influence later phases of repolarization, which can lead to early afterdepolarizations to produce triggered activity. I_{to} is likely to play a more significant role in myocytes from mice than from humans, because mouse myocytes exhibit triangular, short-duration action potentials. This consideration further curtails the involvement of the decreased I_{to} in the production of arrhythmias in patients with hypertrophy and heart failure.

An important feature of cardiac I_{to} channels is that this current is differentially distributed in the heart. Action potentials recorded from different parts of the heart

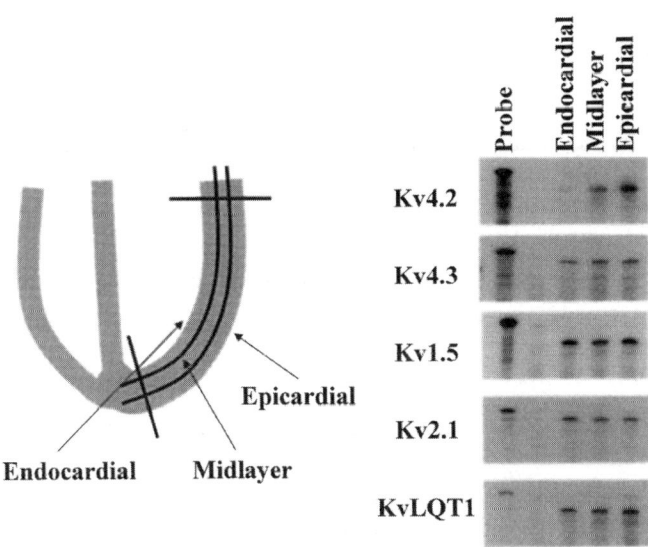

Figure 1. Expression of Kv channel α-subunit mRNAs across the left ventricular wall of the rat heart. The left ventricular free wall was dissected into three layers of equal thickness. Ribonuclease protection assays were used to determine levels of Kv α-subunit mRNAs. Note that Kv4.2 mRNA is differentially distributed in a gradient across the wall.

significantly differ in height, duration, and shape. The differential distribution of I_{to} plays a major role in the production of these differences in action potential waveform (Clark *et al.*, 1993; Liu *et al.*, 1993; Nabauer *et al.*, 1996). In the left ventricular free wall, action potential duration is shorter in the epicardial region or outer side of the wall. This difference in action potential duration is, in large part, accounted for by the differential expression of I_{to} across the left ventricular wall, with myocytes in the epicardial layer expressing the channel at higher levels than endocardial myocytes. The Kv4.2 mRNA also is differentially expressed in a gradient across the left ventricular wall of the rat heart (Fig. 1) (Dixon and McKinnon, 1994). Thus, expression of the Kv4 subfamily channel genes and I_{to} is differentially controlled in distinct locations and is important for setting action potential waveforms. These findings raise the possibility that hypertrophy and heart failure may significantly affect this region-specific expression of the Kv4 subfamily channel genes and I_{to} channels.

It is predicted that variations in action potential duration cause dispersion of repolarization and refractoriness (Tomaselli *et al.*, 1994; Boyden and Jeck, 1995). Thus, a nonuniform reduction in I_{to} may create arrhythmogenic conditions. Despite the importance of the regional differences in action potential waveform, relatively few studies have examined region-selective changes in action potentials or ionic currents in diseased myocardium. Studies in myocytes from heart failure patients have provided controversial results. One study reported that the difference in I_{to} density between endocardial and epicardial myocytes observed in normal hearts is significantly decreased in hearts from patients with heart failure (Nabauer *et al.*, 1996), suggesting that I_{to} is selectively reduced in myocytes from the epicardium. However, another study of myocytes from patients with terminal heart failure showed that I_{to} is decreased only in myocytes from the endocardium (Wettwer *et al.*, 1994). This inconsistency may be due to the variability in pathological conditions of the tissues used. In this sense, animal models may provide better systems to test region-selective effects of hypertrophy and heart failure on the I_{to} channel. In isoproterenol-induced hypertrophied rat myocardium, the regional difference in action potential duration at 25% repolarization across the left ventricular wall is significantly reduced, suggesting epicardial region-selective reduction in I_{to} (Shipsey *et al.*, 1997). This region-selective change in action potential duration contributed to the inversion of the T-wave in the electrocardiogram, indicating that the sequence of repolarization is altered. We also have tested the effect of blood pressure on the differential Kv4.2 mRNA expression across the left ventricular wall (Fig. 2). Our previous study revealed that administration of captopril, an inhibitor of angiotensin-converting enzyme (ACE), to normal rats decreases blood pressure and upregulates ventricular Kv4.2 mRNA (Takimoto *et al.*, 1997). We have administered this drug to rats to examine how the reduction in blood pressure might affect the differential Kv4.2 mRNA expression across the left ventricular wall (Fig. 2). It appears that this drug regimen specifically increases Kv4.2 mRNA levels in the epicardial region, but not the endocardial region. This differential enhancement by the antihypertensive agent increases the gradient of Kv4.2 mRNA levels across the left ventricular wall. The detected epicardial-selective effect could be due to the difference in wall stress. Alternatively, there could be two independent mechanisms of Kv4.2 gene expression for endocardial and epicardial regions, the latter of which is sensitive to changes in blood pressure. In any event, the results suggest that hypertension may specifically decrease epicardial Kv4.2 mRNA, leading to a reduction in the difference in channel mRNA level across the left ventricular wall. Hence, hypertrophy and heart failure may alter the

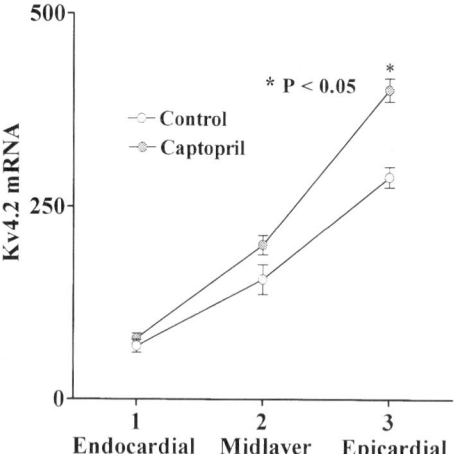

Figure 2. Administration of captopril specifically increases Kv4.2 mRNA in the epicardial region. Rats were given captopril (0.8 mg/ml) or regular water (Control) in their drinking water for 7 weeks. Kv4.2 mRNA levels in three layers of the left ventricular free wall were determined by performing ribonuclease protection assays. Note that the drug administration increased Kv4.2 mRNA in the epicardial region, but not in the endocardial region.

differential expression of Kv4 subfamily channel genes and I_{to}. Altered distribution of I_{to} channels may constitute an important electrophysiological substrate for the production of arrhythmias.

Finally, it also is important to note that heterogeneous alterations in I_{to} and other K$^+$ channel levels within a small region of ventricular tissue may significantly participate in arrhythmogenesis. As the evidence for dramatic regulation of I_{to} and other voltage-gated K$^+$ channels accumulates, it appears quite possible that expression of these K$^+$ channels may be altered due to differences in surrounding conditions. This is particularly true for heart failure patients, the majority of whom have reduced coronary artery perfusion and/or damaged heart muscle. At present, there has been no example of such microanatomical alterations in K$^+$ gene expression; however, this is certainly an area that requires more investigation.

6. CONCLUSIONS AND FUTURE DIRECTIONS

A large number of studies have pointed to a similarity in the electrophysiological alterations in myocardium from hypertrophied and failing hearts. Myocytes from these diseased hearts commonly display a prolongation of action potential duration and an associated reduction in I_{to} density, presumably due to the decreased expression of the Kv4 subfamily mRNAs. However, the mechanism and the consequences of the decreased expression of I_{to} channels are not well understood. Recent insights into the genetic basis of I_{to} and other cardiac K$^+$ channels makes it possible to describe the observed electrophysiological alterations in molecular terms. These investigations may elucidate the mechanisms by which hypertrophy and heart failure decrease Kv4 subfamily channels and by which region-selective channel expression is achieved. The use of transgenic animals has already revealed that a uniform reduction in I_{to} is not arrhythmogenic. Thus, heterogeneous alterations in I_{to} channels may play a role in arrhythmogenesis. Molecular analysis for Kv4 subfamily channel expression also may provide an insight into how disease-associated changes and region-selective regulation interact to control channel expression.

REFERENCES

Baldwin, T. J., Tsaur, M.-L., Lopez, G. A., Jan, Y. N., and Jan, L. Y., 1991, Characteristics of a mammalian cDNA for an inactivating voltage-sensitive K$^+$ channel, *Neuron* 7:471–483

Barry, D. M., Xu, H., Shuessler, R. B., and Nerbonne, J. M., 1998, Functional knockout of the transient outward current, long-QT syndrome, and cardiac remodeling in mice expressing a dominant-negative Kv4 α subunit, *Circ. Res.* 83:560–567.

Benitah, J.-P., Gomez, A. M., Bailly, P., da Ponte, J.-P., Berson, G., Delgado, C., and Lorente, P., 1993, Heterogeneity of the early outward current in ventricular cells isolated from normal and hypertrophied rat hearts, *J. Physiol. (London)* 469:111–138.

Beuckelmann, D. J., Nabauer, M., and Erdmann, E., 1993, Alterations of K$^+$ currents in isolated human ventricular myocytes from patients with terminal heart failure, *Circ. Res.* 73:379–385.

Boyden, P. A., and Jeck, C. D., 1995, Ion channel function in diseases, *Cardiovasc. Res.* 29:312–318.

Brooksby, P., Levi, A. J., and Jones, J. V., 1993, The electrophysiological characteristics of hypertrophied ventricular myocytes from the spontaneously hypertensive rats, *J. Hypertens.* 11:611–622.

Cerbai, E., Barbieri, M., Li, Q., and Mugelli, A., 1994, Ionic basis of action potential prolongation of hypertrophied cardiac myocytes isolated from hypertensive rats of different ages, *Cardiovasc. Res.* **28:**1180–1187.

Chouabe, C., Espanosa, L., Megas, P., Chakir, A., Rougier, O., Freminet, A., and Bonvallet, R., 1997, Reduction of $I_{Ca,L}$ and I_{to1} density in hypertrophied right ventricular cells by simulated high altitude in adult rats, *J. Mol. Cell. Cardiol.* **29:**193–206.

Clark, R. B., Bouchard, R. A., Salinas-Stefanon, E., Sanchez-Chapula, J., and Giles, W. R., 1993, Heterogeneity of action potential waveforms and potassium currents in rat ventricle, *Cardiovasc. Res.* **27:**1795–1799.

Coulombe, A., Momtaz, A., Richer, P., Swynghedauw, B., and Coraboeuf, E., 1994, Reduction of calcium-independent transient outward potassium current density in DOCA salt hypertrophied rat ventricular myocytes, *Pflügers Arch.* **427:**47–55.

Dixon, J. E., and McKinnon, D., 1994, Quantitative analysis of potassium channel mRNA expression in atrial and ventricular muscle of rats, *Circ. Res.* **75:**252–260.

Dixon, J. E., Shi, W., Wang, H.-S., McDonald, C., Yu, H., Wymore, R. S., Cohen, I. S., and McKinnon, D., 1996, Role of the Kv4.3 K^+ channel in ventricular muscle: A molecular correlate for the transient outward current, *Circ. Res.* **79:**659–668.

Fiset, C., Clark, R. B., Shimoni, Y., and Giles, W. R., 1997, *Shal*-type channels contribute to the Ca^{2+}-independent transient outward K^+ current in rat ventricle, *J. Physiol. (London)* **500:**51–64.

Gidh-Jain, M., Huang, B., Jain, P., el-Sherif, N. 1996, Differential expression of voltage-gated K^+ channel genes in left ventricular remodeled myocardium after experimental myocardial infarction *Circ. Res.* **79**(4):669–675.

Guo, W., Kamiya, K., Hojo, M., Kodama, I., and Toyama, J., 1998, Regulation of Kv4.2 and Kv1.4 K^+ channel expression by myocardial hypertrophic factors in cultured newborn rat ventricular cells, *J. Mol. Cell. Cardiol.* **30:**1449–1455.

Hart, G., 1994, Cellular electrophysiology in cardiac hypertrophy and failure, *Cardiovasc. Res.* **28:**933–946.

Johns, D. C., Nuss, H. B., and Marban, E., 1997, Suppression of neuronal and cardiac transient outward currents by viral gene transfer of dominant-negative Kv4.2 constructs, *J. Biol. Chem.* **272:**31598–31603.

Kaab, S., Nuss, H. B., Chiamvimonvat, N., O'Rourke, B., Pak, P. H., Kass, D. A., Marban, E., and Tomaselli, G. F., 1996, Ionic mechanism of action potential prolongation in ventricular myocytes from dogs with pacing-induced heart failure, *Circ. Res.* **78:**262–273.

Kaab, S., Dixon, J., Duc, J., Shen, D., Nabauer, M., Beuckelmann, D. J., Steinbeck, G., McKinnon, D., and Tomaselli, G. F., 1998, Molecular basis of transient outward potassium current downregulation in human heart failure: A decrease in Kv4.3 mRNA correlates with a reduction in current density, *Circulation* **98:**1383–1393.

Kannel, W. B., 1991, Left ventricular hypertrophy as a risk factor: The Framingham experience, *J. Hypertens.* **9:**S3–S9.

Keung, E. C., 1989, Calcium current is increased in isolated adult myocytes from hypertrophied rat myocardium, *Circ. Res.* **64:**753–763.

Kleiman, R. B., and Houser, S. R., 1988, Calcium currents in normal and hypertrophied isolated feline ventricular myocytes, *Am. J. Physiol.* **255:**H1434–H1442.

Kleiman, R. B., and Houser, S. R., 1989, Outward currents in normal and hypertrophied feline ventricular myocytes, *Am. J. Physiol.* **256:**H1450–H1461.

Koumi, S., Backer, C. L., and Arentzen, C. E., 1995, Characterization of inwardly rectifying K^+ channel in human myocytes: Alterations in channel behavior in myocytes isolated from patients with idiopathic dilated cardiomyopathy, *Circulation* **92:**164–174.

Levitan, E. S., and Takimoto, K., 1998, Dynamic regulation of K^+ channel gene expression in differentiated cells, *J. Neurobiol.* **37:**60–68.

Li, Q., and Keung, E. C., 1994, Effects of myocardial hypertrophy on transient outward current, *Am. J. Physiol.* **266:**H1738–H1745.

Liu, D.-W., Gintant, G. A., and Antzelevitch, C., 1993, Ionic bases for electrophysiological distinctions among epicardial, midmyocardial, and endocardial myocytes from the free wall of the canine left ventricle, *Circ. Res.* **72:**671–687.

Lodge, N. J., and Normandin, D. E., 1997, Alterations in I_{to1}, I_{Kr} and I_{K1} density in the BIO TO-2 strain of Syrian myopathic hamsters, *J. Mol. Cell. Cardiol.* **29:**3211–3221.

London, B., Jeron, A., Zhou, J., Buckett, P., Han, X., Mitchell, G. F., and Koren, G., 1998, Long QT and ventricular arrhythmias in transgenic mice expressing the N terminus and first transmembrane segment of a voltage-gated potassium channel, *Proc. Natl. Acad. Sci. U.S.A.* **95**:2926–2931.

Lue, W.-M., and Boyden, P. A., 1992, Abnormal electrical properties of myocytes from chronically infarcted canine heart: Alterations in V_{max} and the transient outward current, *Circulation* **85**:1175–1188.

McIntosh, M. A., Cobbe, S. M., Kane, K. A., and Rankin, A. C., 1998, Action potential prolongation and potassium currents in left-ventricular myocytes isolated from hypertrophied rabbit hearts, *J. Mol. Cell. Cardiol.* **30**:43–53.

Nabauer, M., and Kaab, S., 1998, Potassium channel down-regulation in heart failure, *Cardiovasc. Res.* **37**:324–334.

Nabauer, M., Beuckelmann, D. J., Uberfuhr P., and Steinbeck, G., 1996, Regional differences in current density and rate-dependent properties of the transient outward current in subepicardial and subendocardial myocytes of human left ventricle, *Circulation* **93**:168–177.

Potreau, D., Gomez, J.-P., and Fares, N., 1995, Depressed transient outward current in single hypertrophied cardiomyocytes isolated from the right ventricle of ferret heart, *Cardiovasc. Res.* **30**:440–448.

Pricbc, L., and Beuckelmann, D. J., 1998, Simulation study of cellular electric properties in heart failure, *Circ. Res.* **82**:1206–1223.

Pye, M. P., and Cobbe, S. M., 1992, Mechanisms of ventricular arrthythmias in cardiac failure and hypertrophy, *Cardiovasc. Res.* **26**:740–750.

Qin, D., Zhang, Z., Caref, M. B., Boutjdir, M., Jain, P., and El-Sherif, N., 1996, Cellular and ionic basis of arrhythmias in postinfarction remodeled ventricular myocardium, *Circ. Res.* **79**:461–473.

Rozanski, G. J., Xu, Z., Whitney, R. T., Murakami, H., and Zucher, I. H., 1997, Electrophysiology of rabbit ventricular myocytes following sustained rapid ventricular pacing, *J. Mol. Cell. Cardiol.* **29**:721–732.

Rozanski, G. J., Xu, Z., Zhang, K., and Patel, K. P., 1998, Altered K$^+$ current of ventricular myocytes in rats with chronic myocardial infarction, *Am. J. Physiol.* **274**:H259–H265.

Serodio, P., Kentos, C., and Rudy, B., 1994, Identification of molecular components of A-type channels activating at subthreshold potentials, *J. Neurophysiol.* **72**:1516–1529.

Shipsey, S. J., Bryant, S. M., and Hart, G., 1997, Effects of hypertrophy on regional action potential characteristics in the rat left ventricle: A cellular basis for T-wave inversion? *Circulation* **96**:2061–2068.

Takimoto, K., Li, D., Hershman, K. M., Li, P., Jackson, E. K., and Levitan, E. S., 1997, Decreased expression of Kv4.2 and novel Kv4.3 K$^+$ channel subunit mRNAs in ventricles of renovascular hypertensive rats, *Circ. Res.* **81**:533–539.

Thuringer, D., Coulombe, A., Deroubix, E., Coraboeuf, E., and Mercadier, J. J., 1996, Ionic basis of the action potential prolongation in ventricular myocytes from Syrian hamsters with dilated cardiomyopathy, *J. Mol. Cell. Cardiol.* **28**:387–401.

Tomaselli, G. F., Beuckelmann, D. J., Calkins, H. G., Berger, R. D., Kessler, P. D., Lawrence, J. H., Kass, D., Feldman, A. M., and Marban, E., 1994, Sudden cardiac death in heart failure: The role of abnormal repolarization, *Circulation* **90**:2534–2539.

Tomita, F., Bassett, A. L., Myerburg, R. J., and Kimura, S., 1994, Diminished transient outward currents in rat hypertrophied ventricular myocytes, *Circ. Res.* **75**:296–303.

Tsaur, M.-L., Chou, C.-C., Shih, Y.-H., and Wang, H. L., 1997, Cloning, expression and CNS distribution of Kv4.3, an A-type-K$^+$ channel α subunit, *FEBS Lett.* **400**:215–220.

Wettwer, E., Amos, G. J., Posival, H., and Ravens, U., 1994, Transient outward current in human ventricular myocytes of subepicardial and subendocardial origin, *Circ. Res.* **75**:473–482.

Xu, X., and Best, P. M., 1991, Decreased transient outward K$^+$ current in ventricular myocytes from acromegalic rats, *Am. J. Physiol.* **260**:H935–H942.

Yeola, S. W., and Snyder, D. J., 1997, Electrophysiological and pharmacological correspondence between Kv4.2 current and rat cardiac transient outward current, *Cardiovasc. Res.* **33**:540–547.

Chapter 38

Role of ATP-Sensitive K$^+$ Channels in Cardiac Preconditioning

Garrett J. Gross

1. INTRODUCTION

Ischemic preconditioning (IPC) is a phenomenon by which a brief period of ischemia or hypoxia can result in an adaptive cardioprotective effect when the myocardium is subjected to a more prolonged ischemic insult. IPC occurs in all animals studied as well as in humans and has two phases, an early phase that lasts for 1–2 h following the IPC stimulus and a delayed phase or second window of protection (SWOP) that appears 12–24 h after IPC and lasts for 48–72 h (Przyklenk and Kloner, 1998). A number of mediators (i.e., adenosine, bradykinin, and opioids) and signaling pathways [i.e., tyrosine kinase (TK) and protein kinase C (PKC)] are involved in triggering the transduction of the preconditioning stimulus to the appropriate end effector in the heart. The end effector is most likely an enzyme, a small heat-shock protein (HSP), or an ion channel. The majority of evidence suggests that the ATP-sensitive potassium channel (K$_{ATP}$ channel) is the end effector of both early and delayed IPC and pharmacologically induced preconditioning, and information will be presented to support this hypothesis. Recently, controversy has arisen as to whether the K$_{ATP}$ channel subtype is responsible for the preconditioning effect is the surface or sarcolemmal channel (sarc K$_{ATP}$) or the mitochondrial channel (mito K$_{ATP}$), and evidence will be presented both for and against a role for these channel subtypes in classical or delayed IPC. Theories will also be presented concerning the mechanisms by which opening of the sarc K$_{ATP}$ or mito K$_{ATP}$ channels produces the cardioprotective effect of IPC. Finally, a comment will be made concerning future directions and questions that still need to be answered as to the role and function of the K$_{ATP}$ channel in cardiac preconditioning.

Garrett J. Gross • Department of Pharmacology & Toxicology, The Medical College of Wisconsin, Milwaukee, Wisconsin 53226

Potassium Channels in Cardiovascular Biology, edited by Archer and Rusch. Kluwer Academic/Plenum Publishers, New York, 2001.

1.1. K$_{ATP}$ Channels

ATP-sensitive potassium channels were first discovered by Noma (1983) in isolated membrane patches prepared from guinea pig and rat ventricular myocytes. These channels are inhibited by ATP and to a lesser extent by ADP. K$_{ATP}$ channels have also been identified in a variety of other tissues, including skeletal muscle, smooth muscle, brain, and pancreas, where these channels are involved in insulin secretion (Aguilar-Bryan et al., 1998). In all cell types, these channels appear to couple changes in metabolism to membrane electrical activity. At least two major subunits of these channels have been cloned, and, when reconstituted, they have been found to consist of two kinds of proteins, inwardly rectifying potassium channels (Kir6.x) and sulfonylurea receptors (SUR). These channels form heteromultimers of Kir6.x and SUR subunits which are members of the ATP-binding cassette superfamily. Two Kir6.x and SUR genes have been discovered, and these channel subunits form different combinations depending on the tissue involved (Aguilar-Bryan et al., 1998). SUR1 and Kir6.2 form the pancreatic β-cell channel, SUR2A and Kir6.2 form the cardiac and skeletal muscle channel, and it is thought that SUR2B and Kir6.1 may form the vascular smooth muscle channel, although this is less certain (Okuyama et al., 1998). The mitochondrial K$_{ATP}$ channel has not been cloned although it is thought to consist of SUR and Kir6.x subunits, similar to the sarcolemmal channel (Paucek et al., 1992).

The regulation of the sarcolemmal K$_{ATP}$ channel has been the most intensively investigated, and a number of factors have been identified that are thought to be involved in the complex regulation of this channel (Kersten et al., 1998). Some of these factors are diagrammatically illustrated in Fig. 1 and include ATP, various nucleotide diphosphates, G-proteins, and PKC. In addition, there are a number of drugs, termed potassium channel openers (KCOs), that directly open this channel, such as nicorandil, cromakalim, pinacidil, and bimakalim, and several antagonists that may bind to diverse sites on the channel to inhibit its opening, including glibenclamide, 5-hydroxydecanoic acid, U37883A, and HMR 1883 or HMR 1098. Most of the functional studies in which a role for the sarcolemmal K$_{ATP}$ channel in ischemic preconditioning has been demonstrated are based on the use of these pharmacological probes.

1.2. Ischemic Preconditioning

IPC was first described by Murry et al. (1986) in the canine heart. These investigators found that four 5-min occlusion periods interspersed with 10 min of reflow prior to a prolonged 40-min coronary artery occlusion resulted in a marked reduction in infarct size when assessed 72 h post IPC as compared to infarct size in a nonpreconditioned group of dogs. Subsequently, studies in a number of laboratories with a number of animal models and species have uniformly demonstrated the dramatic cardioprotective effect of IPC (Przyklenk and Kloner, 1998), whereby the severity of myocardial necrosis is reduced. Evidence that IPC also occurs in humans was found when assessing injury on the basis of sequential changes in the ST segment of the electrocardiogram (Deutsch et al., 1990). Because of the reproducibility and efficacy with which IPC reduces ischemic injury, many studies have been carried out to

A. CLOSED STATE

B. OPEN STATE

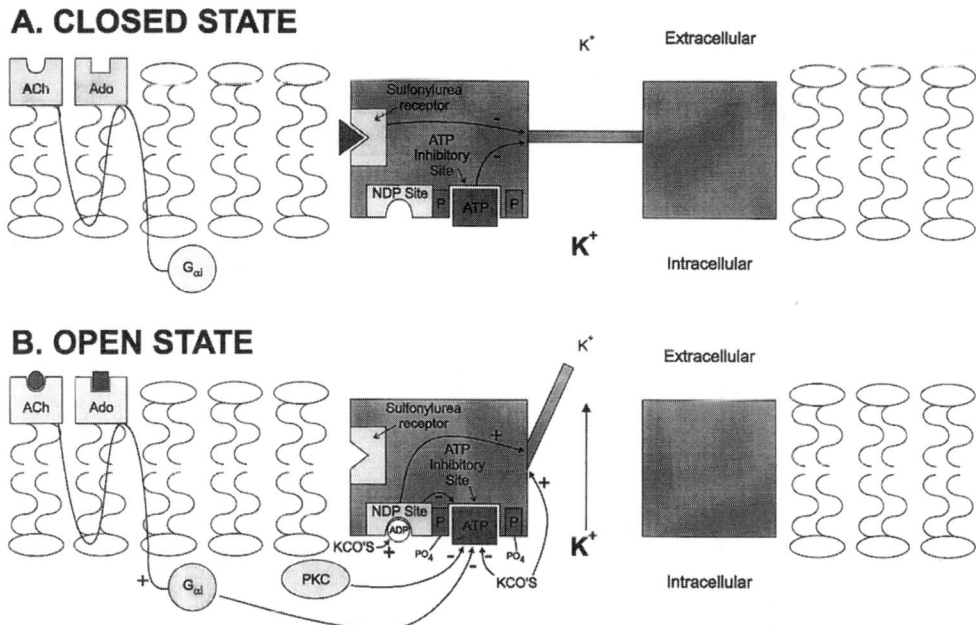

Figure 1. Schematic diagram of the regulation of sarc K$_{ATP}$ channels. The K$_{ATP}$ channel is normally closed (A) in response to increases in intracellular ATP concentrations. Channel opening is inhibited by binding of ATP to the ATP inhibitory site. Drug binding to the sulfonylurea receptor also decreases the open-state probability of the channel. In contrast, nucleotide diphosphates (NDPs), such as ADP, antagonize ATP-induced inhibition of channel opening, an action that requires occupation of a phosphorylation site (P) by inorganic phosphate (PO$_4$) and produces opening of the K$_{ATP}$ channel (B). Acetylcholine (ACH) and adenosine (Ado) enhance channel opening via stimulation of membrane receptors coupled to inhibitory G (G$_i$) proteins. Activated G$_{i\alpha}$ and protein kinase C (PKC) antagonize the inhibitory effect of ATP on the channel. Potassium channel openers (KCOs) enhance opening of the K$_{ATP}$ channel via a direct action on the channel, either by augmenting the stimulatory effects of NDPs on channel opening or by antagonizing the inhibitory effects of ATP on the channel. (Reproduced, with permission, from Kersten et al., 1998.)

determine the cellular mechanisms responsible for the potent cardioprotective effects of IPC with the hope of pharmacologically reproducing this phenomenon for therapeutic benefit. In this regard, a number of endogenous subtances, e.g., adenosine (Liu et al., 1991), opioids (Schultz et al., 1995), and signaling pathways, i.e., tyrosine kinase (Fryer et al., 1998) and PKC (Liu et al., 1994), have been proposed to be involved in triggering the preconditioning signaling cascade; however, the overwhelming majority of evidence suggests that the K$_{ATP}$ channel is a major component of IPC and may serve as the end effector protein in this response. It was originally proposed that the sarc K$_{ATP}$ channel was responsible for the protection observed following IPC; however, more recent evidence suggests that the mito K$_{ATP}$ channel may be the one that mediates IPC-induced cardioprotection (Garlid et al., 1997; Liu et al., 1998). Delayed preconditioning has been less well studied than classical IPC from a mechanistic point of view, although recent studies by Hoag et al. (1997), Pell et al. (1997), and Joyeux et al. (1998) all suggest a role for the K$_{ATP}$ channel in delayed preconditioning following heat shock, however, no studies have been performed to determine the importance of the sarc versus the mito K_{ATP} channel in the second window of protection. In the remainder of this chapter, evidence will be presented concerning the role of the sarc versus the mito K_{ATP} channel in classical IPC, and potential mechanisms by which opening of these channel subtypes may exert a cardioprotective effect will be discussed.

2. K_{ATP} CHANNELS AND CLASSICAL PRECONDITIONING

2.1. Animal Studies

The evidence supporting a role for the K_{ATP} channel in acute or classical IPC in animals is well established and is based primarily on the use of selective pharmacological agonists and antagonists of the channel. Gross and Auchampach (1992) were the first to suggest an important role for K_{ATP} channels in IPC in the canine heart. These investigators showed that glibenclamide antagonized the effect of IPC, whereas the K_{ATP} channel opener aprikalim mimicked the cardioprotective effect of IPC (Fig. 2). Subsequently, Auchampach *et al.* (1992) found that an intracoronary infusion of a different K_{ATP} channel antagonist, 5-hydroxydecanoic acid (5-HD), also blocked the effect of IPC in the dog heart. That the K_{ATP} channel appeared to be a possible downstream event in the signaling cascade involved in IPC was suggested by recent studies of Yao *et al.* (1997), who showed that administration of glibenclamide 60 min after the 5-min period of IPC, when its protective effect is still manifest, blocked IPC. Kitakaze *et al.* (1996) and Grover *et al.* (1992) also showed a role for K_{ATP} channels in IPC in anesthetized dogs. A number of studies in dogs performed with endogenous substances that are thought to be triggers of a preconditioning-like response and several

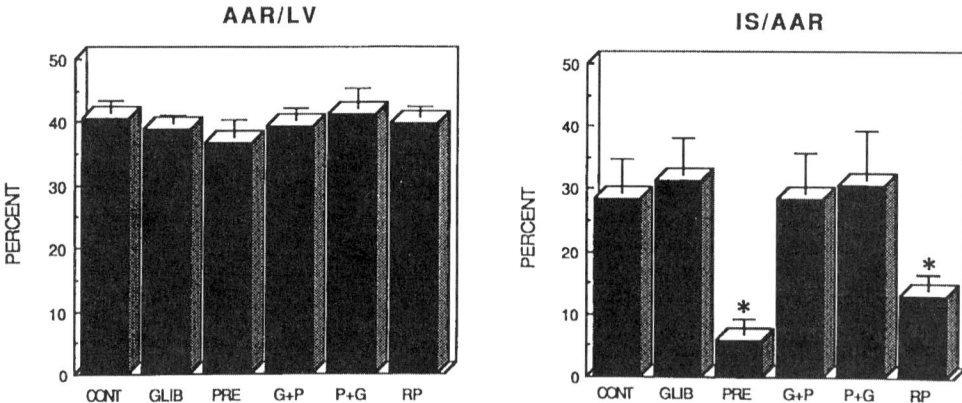

Figure 2. Graphs illustrating the effect of different treatments on the area at risk (AAR), expressed as a percentage of the left ventricular (LV) weight (AAR/LV), and infarct size (IS), expressed as a percentage of the area at risk (IS/AAR). There were no differences between groups in AAR/LV; however, preconditioning (PRE) and aprikalim (RP) produced significant reductions in IS/AAR as compared to the control (CONT) nonpreconditioned group. Pretreatment with glibenclamide 10 min before preconditioning (G + P) or immediately after preconditioning (P + G) completely abolished the protective effect. Glibenclamide had no effect on IS/AAR in the absence of preconditioning (GLIB). (Reproduced, with permission, from Gross and Auchampach, 1992.)

direct openers of the K_{ATP} channel also support a role for K_{ATP} in IPC. Mizumura *et al.* (1995) showed that pretreatment with two K_{ATP} openers, bimakalim and nicorandil, mimicked the cardioprotective effect of IPC in nonhypotensive doses. Interestingly, these compounds also reduced infarct size when administered primarily during reperfusion, which suggests that the K_{ATP} channel may also be important in alleviating reperfusion injury as well as in protecting during the ischemic period. Yao and Gross (1993a, 1994a) also demonstrated that acetylcholine and adenosine acting on their muscarinic M_2 and adenosine A_1 receptors could mimic the cardioprotective effect of IPC, an effect which was totally blocked by 5-HD. Auchampach and Gross (1993) reported that several adenosine receptor antagonists blocked the cardioprotective effect of IPC and that dipyridamole and PD 81723, an adenosine A_1 receptor allosteric enhancing agent (Mizumura *et al.*, 1996), both potentiated the effect of adenosine or IPC to produce a cardioprotective effect. These results obtained in dogs clearly support a role for an adenosine A_1 receptor–K_{ATP} channel link in IPC. That this is a G_i-protein-mediated response was suggested by a study of Miura *et al.* (1997) in which they observed that pertussis toxin blocked the cardioprotective effect of IPC in canine myocardium.

Most studies in other animal species also support a role for the K_{ATP} channel in IPC. In the pig, Schulz *et al.* (1994) showed that glibenclamide blocked the protective effect of IPC at a dose that had no effect on infarct size in nonpreconditioned hearts. Rohmann *et al.* (1994) also showed that glibenclamide blocked IPC in pig hearts and that the K_{ATP} channel opener bimakalim mimicked IPC in this species. It should be noted that unlike the dog, the pig has a poor collateral circulation, and thus these findings suggest that IPC is not dependent on changes in collateral blood flow.

As discussed above, a role for K_{ATP} channels in IPC in large animal hearts is unequivocal. However, there has been some controversy concerning the role of K_{ATP} channels as mediators of IPC in smaller animals such as rabbits or rats, although the majority of studies support an important role in these species as well. In rabbits, Thornton *et al.* (1993) originally found that three doses of glibenclamide did not block IPC; however, Toombs *et al* (1993) showed that glibenclamide was able to block IPC. Because the major difference between these two studies was in the anesthetic used, pentobarbital versus ketamine and xylazine, it was thought that the effect of glibenclamide was anesthetic dependent, and a study of Haessler *et al.* (1994) seemed to support this hypothesis. However, recent studies of Kouchi *et al.* (1998) in pentobarbital-anesthetized rabbits demonstrated that glibenclamide completely abolished the cardioprotective effect of IPC and calcium-induced preconditioning (CPC) when the glibenclamide was administered 45 min prior to the prolonged ischemic period, but not when it was administered 20 min before. Similar results were recently obtained in rats: when glibenclamide was administered 5 min prior to IPC, it was shown to not block IPC, but when it was administered 30 min prior to IPC, it completely blocked the protective effects of IPC in anesthetized rats (Schulz *et al.*, 1997). Thus, in rats and rabbits at least, it appears that the timing of glibenclamide administration is more important in determining whether this compound can block IPC than the anesthetic used.

Studies in isolated cardiac myocytes also support a role for K_{ATP} channels in IPC which is independent of any anesthetic. Liang (1996) used a chick ventricular myocyte model to mimic IPC in which myocytes were preconditioned by a 5-min period of hypoxia prior to a second 90-min period of hypoxia. IPC in this model resulted in 64%

and 66% reductions in the amount of creatine kinase (CK) released and myocytes killed, respectively, as compared to nonpreconditioned myocytes. These protective effects of IPC were blocked by the simultaneous inclusion of two K_{ATP} antagonists, glibenclamide and 5-HD, during the hypoxic preconditioning period. Administration of the K_{ATP} opener pinacidil mimicked the effect of hypoxia-induced preconditioning in this model, an effect blocked by either glibenclamide or 5-HD. An adenosine receptor antagonist, 8-sulfophenyltheophylline (8-SPT), blocked the protective effect of hypoxia-induced preconditioning in the myocytes, and an adenosine A_1 receptor agonist, 2-chloro-N^6-cyclopentyladenosine (CCPA), also mimicked IPC, and this effect was antagonized by glibenclamide, 5-IID, and 8-SPT, which suggests that an A_1 receptor–K_{ATP}-linked mechanism was responsible for the preconditioning-like effect observed in the isolated chick myocyte, similar to previous data obtained in whole animals and isolated organs. Thus, it seems clear that the K_{ATP} channel is an important component of myocardial preconditioning produced by ischemia, hypoxia, or specific endogenous pharmacological agonists.

2.2. Human Studies

Deutsch et al. (1990) were the first to suggest that preconditioning occurs in the human myocardium. These investigators found that the ST segment shifts observed on the electrocardiogram (EKG) of patients undergoing repeated balloon inflations during percutaneous transluminal coronary angioplasty (PTCA) became smaller during sequential occlusions and that less pain was experienced and less lactate released into the coronary sinus during later occlusion periods. Kloner et al. (1995) and Ottani et al. (1995) both showed that patients with more frequent bouts of angina demonstrated a smaller infarct size following an acute myocardial infarction. Taken together, all of these data suggest that IPC exists in human myocardium subjected to repeated brief episodes of ischemia.

The first paper to suggest a role for K_{ATP} channels in preconditioning in humans was presented by Tomai et al. (1994). These investigators showed that IPC produced by repeat occlusions during coronary angioplasty was blocked by glibenclamide. Subsequently, more definitive evidence for K_{ATP} channel involvement in IPC in humans was presented by Speechly-Dick et al. (1995) and Cleveland et al. (1997a,b). In human right atrial trabeculae, Speechly-Dick et al. (1995) produced preconditioning by 3 min of simulated ischemia 7 min prior to 90 min of hypoxic buffer perfusion and 120 min of reperfusion. These investigators found that the protection against contractile dysfunction produced by simulated ischemia was induced by activation of PKC and opening of the K_{ATP} channel and that these protective effects were blocked by the K_{ATP} antagonist glibenclamide. In a similar model, Cleveland et al. (1997a) found that adenosine mimicked the effect of IPC in human tissue and that the effect of adenosine was blocked by glibenclamide, which suggests that adenosine receptor stimulation is linked to the K_{ATP} channel, as was previously shown in various animal studies. Finally, in an important clinical study, Cleveland et al., (1997b) showed that human myocardium obtained from patients on long-term oral hypoglycemic treatment could not be preconditioned by simulated ischemia, whereas patients on insulin therapy could be protected (Fig. 3). These data suggested that IPC in human myocardium is mediated

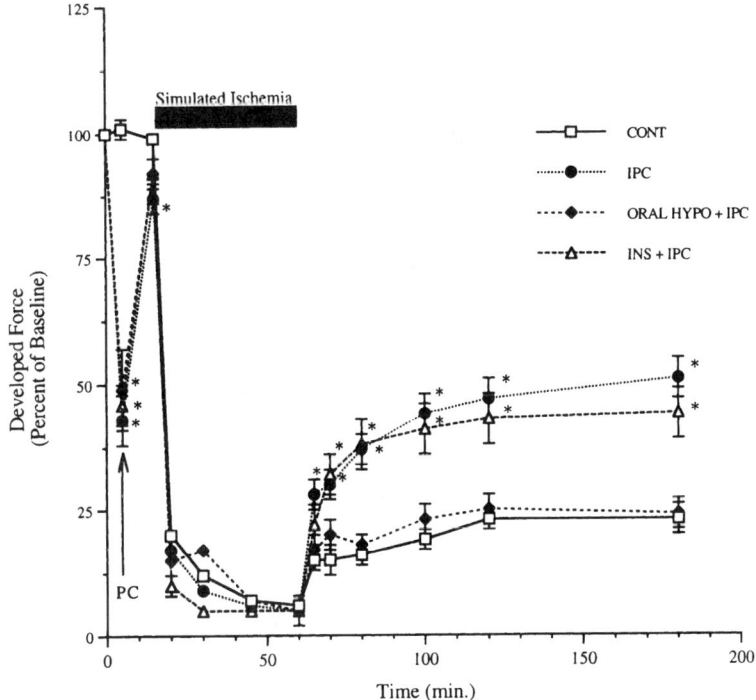

Figure 3. Effects of ischemic preconditioning in trabeculae with and without long-term exposure to oral hypoglycemic agents. Preconditioned trabeculae from patients without oral hypoglycemic exposure (IPC, $n = 5$) show increased recovery of DF relative to trabeculae not given a preconditioning stimulus (CONT, $n = 16$) after simulated ischemic injury. Conversely, trabeculae from patients with long-term exposure to oral hypoglycemic agents (Oral Hypo + IPC, $n = 7$), are not functionally protected by ischemic preconditioning stimulus. Insulin therapy (Ins + IPC, $n = 4$), does not inhibit protection by preconditioning. *$P < 0.05$ vs. control. (Reproduced, with permission, from Cleveland *et al.*, 1997b.)

via K$_{ATP}$ channels and that long-term inhibition of these channels by oral hypoglycemic agents such as glibenclamide may lead to potentially deleterious consequences. Thus, taken together, most animal and human studies clearly suggest an important role for the K$_{ATP}$ channel in mediating the potent cardioprotective effect of IPC observed in all species. In the next section, the evidence for and against a role for the sarc K$_{ATP}$ versus the mito K$_{ATP}$ channel will be presented, along with a discussion of the potential mechanisms by which these two channels might produce cardioprotection.

3. SUBTYPE of K$_{ATP}$ CHANNEL THAT MEDIATES ISCHEMIC PRECONDITIONING

3.1. Sarcolemmal K$_{ATP}$ Channels

Although antagonists of the K$_{ATP}$ channel block the protective effects of IPC and agonists of the channel mimic its effects, the precise mechanism by which opening of

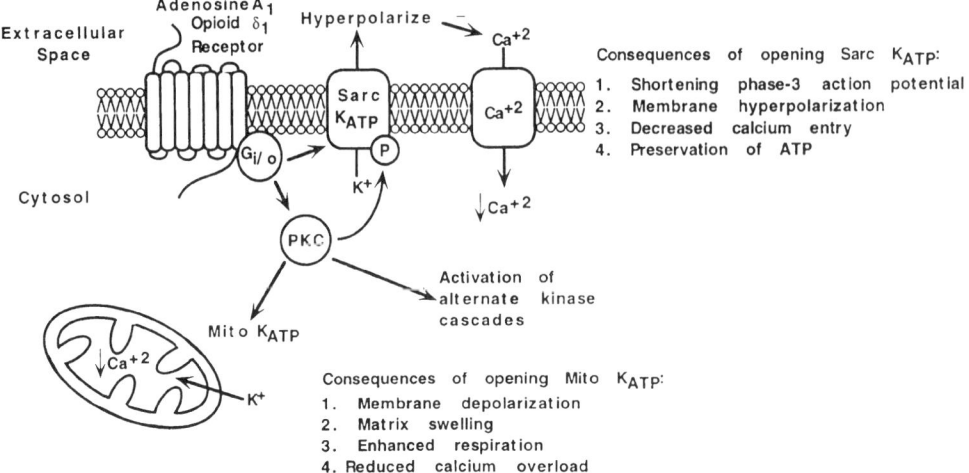

Figure 4. Schematic diagram demonstrating proposed mechanisms by which opening sarc K_{ATP} or mito K_{ATP} channels might produce a cardioprotective effect.

the K_{ATP} channel mediates its cardioprotective effect remains uncertain. Noma (1983) originally hypothesized that opening of the sarc K_{ATP} channel produced by hypoxia, ischemia, or K_{ATP} channel openers would enhance the shortening of the cardiac action potential (AP) by accelerating phase 3 repolarization, which would result in an inhibition of calcium entry into the cell via L-type channels and prevent calcium overload (Fig. 4). In addition, membrane hyperpolarization would also inhibit calcium entry and slow or prevent the reversal of the sodium–calcium exchanger, which normally extrudes calcium in exchange for sodium. The result of these actions would be a reduction in calcium overload during ischemia and early reperfusion and increased cell viability. Indeed, a number of early studies seemed to support this theory. Initial studies by Cole *et al.* (1991), using an isolated guinea pig right ventricular wall preparation, showed that the K_{ATP} channel antagonist glibenclamide inhibited the shortening of the AP during ischemia and resulted in a poor recovery of ventricular function following reperfusion as compared to a control group. Similarly, these investigators also found that the K_{ATP} channel opener pinacidil accelerated the shortening of the AP during ischemia, which resulted in an enhanced recovery of ventricular function during reperfusion. Subsequently, Tan *et al.* (1993) found that preconditioning a guinea pig papillary muscle with a brief period of ischemia or with a K_{ATP} channel opener prolonged the time to electrical uncoupling and that these effects were associated with an enhanced shortening of the AP. Similarly, Yao *et al.* (1993c) and Yao and Gross (1994c) found that there was an association between enhanced recovery of function in stunned myocardium in the presence of a K_{ATP} channel opener, aprikalim, and enhanced shortening of the AP and that another K_{ATP} channel opener, bimakalim, lowered the threshold for IPC and that this effect was also associated with an enhanced rate of AP shortening in dog hearts. Finally, Schulz *et al.* (1994), using anesthetized pigs as a model of IPC, found that IPC resulted in an enhanced shortening of the AP during the prolonged ischemic period in pigs and that this was associated

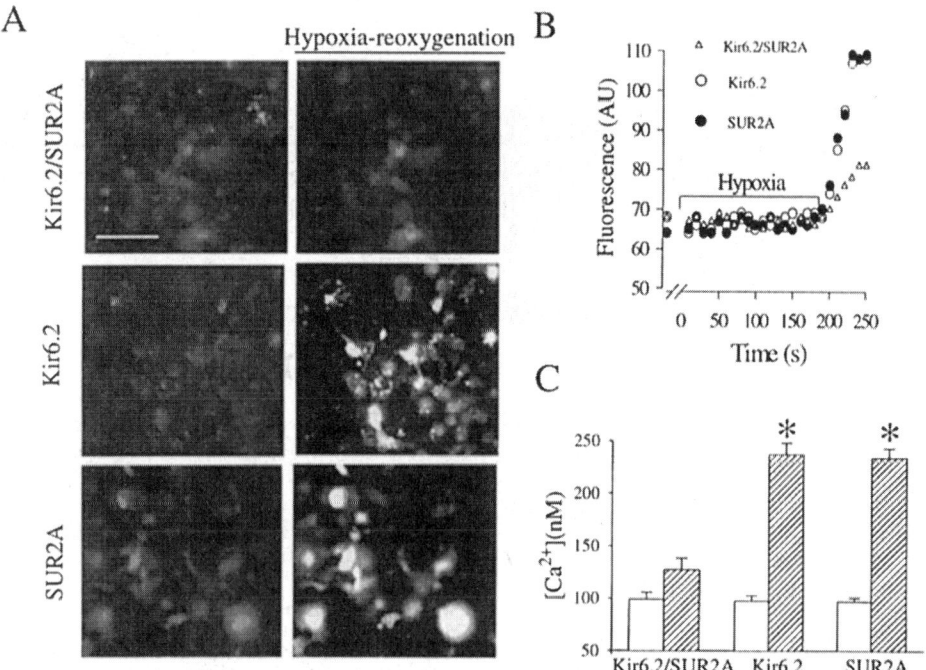

Figure 5. Pinacidil protects COS-7 cells cotransfected with Kir6.2/SUR2A against hypoxia–reoxygenation. (A), Epifluorescent digital images before (*left*) and after (*right*) hypoxia–reoxygenation in fluo-3-loaded cells transfected with channel subunits as indicated and exposed to pinacidil (10 μM). Bar = 90 μm. (B), Superimposed traces showing fluo-3 fluorescence plotted as a function of time in cells shown in (A). (C) Average changes in intracellular Ca^{2+} concentration before (open bars) and after (hatched bars) hypoxia-reoxygenation in cells transfected with channel subunits as indicated and treated with 10 μM pinacidil. Bars represent means $\pm$ SEM ($n = 4$); *$p < 0.01$. (Reproduced, with permission, from Jovanovic *et al.*, 1998.)

with a marked cardioprotective effect as compared to controls. However, the enhanced shortening of the AP was not impressive (approximately 10%) and seemed unlikely to account for the magnitude of cardioprotection seen. In this regard, Jovanovic *et al.* (1998) recently transfected COS-7 cells with the two subunits thought to constitute the sarc K$_{ATP}$ in heart, SUR2A and Kir6.2. In K$_{ATP}$-deficient COS-7 cells, they found that exposure of these cells to 3 min of chemical hypoxia produced by dinitrophenol (DNP) and subsequent reoxygenation produced marked calcium loading. Similar results were obtained in cardiac myocytes expressing the native K$_{ATP}$ channel and exposed to DNP. However, when both subunits of the sarc K$_{ATP}$ channel, SUR2A/Kir6.2, were cotransfected in these COS-7 cells, the addition of the K$_{ATP}$ opener pinacidil attenuated the calcium loading produced by DNP exposure (Fig. 5). Similar results were obtained with pinacidil in cardiac myocytes expressing the native sarc K$_{ATP}$ channel. These results suggested that sarc K$_{ATP}$ channel proteins may have cytoprotective properties when combined to form a functional channel and that this protection can occur independently of APD shortening. The mechanism by which these proteins combine to produce a cardioprotective effect is unknown and requires further study.

The first study to suggest that an enhanced shortening of the AP as a result of sarc K_{ATP} activation was not the mechanism responsible for the cardioprotective effect of IPC or K_{ATP} channel openers was published by Yao and Gross in 1994. These investigators found that a low dose of the K_{ATP} opener bimakalim, which did not enhance AP shortening, produced a cardioprotective effect equal to that of two higher doses of bimakalim that produced an enhanced shortening of the AP (Yao and Gross, 1994b). These authors suggested that the involvement of an intracellular site of action may help to explain the efficacy of bimakalim in reducing infarct size independent of AP shortening. Subsequently, Grover et al. (1995) found that there was no correlation between AP shortening and cardioprotection in dogs following cromakalim administration and that dofetilide, a class III antiarrhythmic drug, which prevented AP shortening in preconditioned hearts, did not antagonize the cardioprotective effect of IPC (Grover et al., 1996). In addition, studies in isolated cardiac myocytes showed a role for the K_{ATP} channel in mediating the protective effects of K_{ATP} channel openers or IPC in the absence of a ventricular action potential (Armstrong et al., 1995). Taken together, these studies suggest that the sarc K_{ATP} channel may not be the site of action responsible for the cardioprotective effect of IPC and K_{ATP} channel openers and indicate the possible involvement of an intracellular site of action.

3.2. Mitochondrial K_{ATP} Channels

Inoue et al. (1991) first identified an ATP-sensitive K^+ channel in the inner mitochondrial membrane (mito K_{ATP}) in rat liver by patch-clamping giant mitoplasts prepared from rat liver mitochondria. These authors found that the mito K_{ATP} channel had several characteristics similar to those of the sarc K_{ATP}, in that it was reversibly inactivated by ATP applied to the matrix side and inhibited by glibenclamide. Subsequently, Paucek et al. (1992) isolated and partially purified a mito K_{ATP} channel from beef heart mitochondria that had several characteristics similar to those of the sarc K_{ATP} channel. However, the function of these channels appears to be intimately involved in matrix volume control as opposed to electrical activity in the case of sarc K_{ATP}. In this regard, opening of mito K_{ATP} leads to membrane depolarization, matrix swelling, accelerated respiration, and slowing of ATP synthesis (Fig. 4). Interestingly, these mito K_{ATP} channels are only sensitive to inhibitors of the channel such as 5-HD or glibenclamide when Mg^{2+}, ATP, and a pharmacological or physiological K_{ATP} opener such as diazoxide or GTP are present (Jaburek et al., 1998).

Evidence that the mito K_{ATP} channel is important in IPC was first presented by Garlid and Grover's groups in 1997. These investigators used the K_{ATP} channel opener diazoxide as a pharmacological tool to demonstrate the importance of mito K_{ATP} as the cardioprotective channel responsible for the beneficial effects of K_{ATP} openers and IPC. They found in reconstituted bovine heart mitochondria that diazoxide opened mitochondrial K_{ATP} with a $K_{1/2}$ of 0.8 μM while opening sarc K_{ATP} at 800 μM (Garlid et al., 1997). Subsequently, they observed that diazoxide, at concentrations (5–20 μM) that would not be expected to open sarc K_{ATP} channels, improved functional recovery in isolated rat hearts subjected to global ischemia and reperfusion and produced an increase in time to contracture equal to that produced by a nonselective K_{ATP} opener, cromakalim, at similar concentrations. The effects of diazoxide and cromakalim were blocked by the K_{ATP} channel antagonists glibenclamide and 5-HD, which confirmed that these agents were acting via K_{ATP} channels. Furthermore, cromakalim and diazoxide were found to be potent activators of K^+ fluxes in reconstituted rat mitochondria, with $K_{1/2}$ values of 1.1 and 0.49 μM, respectively. These results suggested that diazoxide, and perhaps other K_{ATP} channel openers, interacts with the mito K_{ATP} channel to produce cardioprotection.

In agreement with the results of Garlid et al. (1997), recent studies by Liu et al. (1998) have also demonstrated that diazoxide is a selective mito K_{ATP} opener and suggest that 5-HD is a selective mito K_{ATP} channel inhibitor. To determine whether diazoxide is a selective mito K_{ATP} opener, these investigators measured flavoprotein fluorescence (oxidation correlating with mitochondrial depolarization) and sarc K_{ATP} current (I_{KATP}) in isolated rabbit ventricular myocytes. They found that diazoxide produced a reversible oxidation of the flavoproteins with an EC_{50} of 27 μM. This concentration of diazoxide did not affect the sarc K_{ATP} channel, which confirmed the selectivity of this K_{ATP} opener for the mito K_{ATP} channel. Subsequently, they found that diazoxide at 50 μM produced a 50% reduction in the number of rabbit myocytes killed in a model of simulated ischemia. These results are in close agreement with those of Garlid et al. (1997) and further suggest that the mito K_{ATP} channel is the mediator of cardioprotection produced by K_{ATP} openers.

To further substantiate a role for the mito K_{ATP} channel in mediating IPC, Sato et al. (1998) recently attempted to demonstrate that PKC, a known modulator of the K_{ATP} channel in IPC, could have a link to the mito K_{ATP} channel. These investigators showed that diazoxide (100 μM) produced a marked increase in flavoprotein oxidation in isolated rabbit ventricular myocytes, with no effect on I_{KATP}. The phorbol ester PMA had no effect on flavoprotein fluorescence by itself but potentiated and accelerated the effect of diazoxide to activate mito K_{ATP} (Fig. 6). The inactive phorbol ester 4α-phorbol was without effect, and the effect of diazoxide and PMA + diazoxide was blocked by 5-HD. These authors also demonstrated that 5-HD was a selective inhibitor of mito K_{ATP} in their model. That 5-HD is a selective blocker of the mito K_{ATP} channel is subject to some debate, however, because several studies in other models have suggested that 5-HD can block the shortening of the cardiac AP during ischemia (Moritani et al., 1994) and inhibit K^+ efflux from the cell during hypoxia (Sakamoto et al., 1998), an effect which would be expected to result from blocking of the sarc K_{ATP} channel. Diazoxide has also been shown to have effects on mitochondrial metabolism and membrane potential that are independent of its effect on K_{ATP} channels (Grimmsmann and Rustenbeck, 1998) so one needs to be cautious when suggesting that a drug is selective for a given biological event. Nevertheless, in spite of these caveats, the evidence presented by Garlid et al. (1997), Liu et al. (1998), and Sato et al. (1998) is convincing and suggests that the mito K_{ATP} channel may be an important component in mediating the cardioprotective effects of K_{ATP} channel openers and IPC. In this regard, Gogelein et al. (1998) have recently described the pharmacology of a new K_{ATP} channel antagonist, HMR 1883, which appears to be a cardioselective sarc K_{ATP} antagonist. Interestingly, these authors have preliminary data to suggest that this compound does not block IPC in rabbit hearts at doses that block the shortening of the cardiac AP (Linz et al., 1998). In addition, preliminary results from Marban's laboratory (unpublished observations) suggest that this compound has no effect on mito K_{ATP} channels,

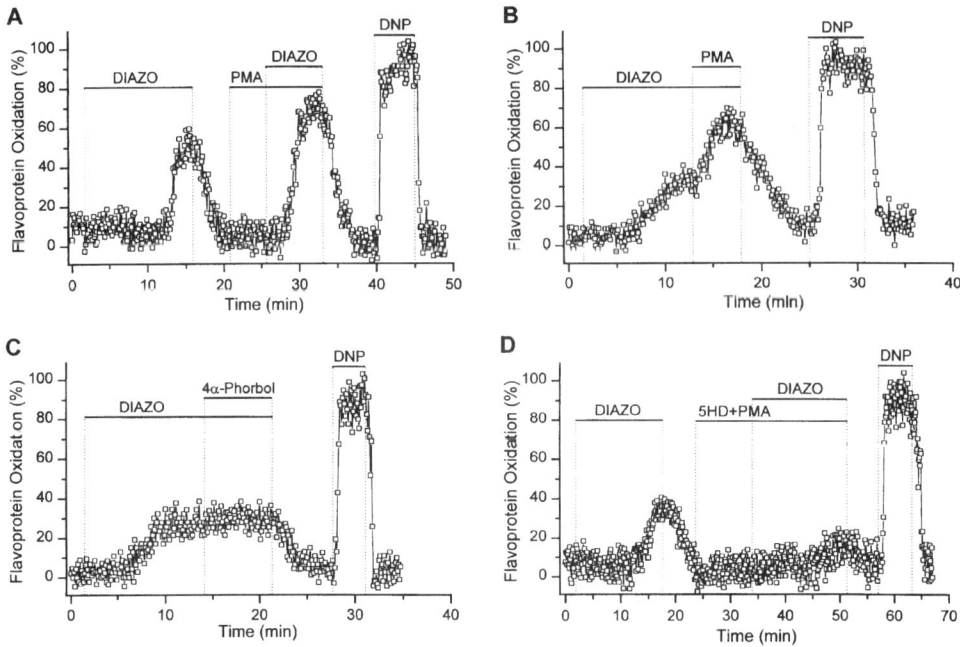

Figure 6. Effect of diazoxide (DIAZO) on flavoprotein fluorescence. (A) and (B), DIAZO (100 μM) induced flavoprotein oxidation and its potentiation by PMA (100 μM). (C) 4α-Phorbol (100 μM) did not affect the oxidative effect of DIAZO. (D) 5-HD (2 mM) completely inhibited the oxidative effect of DIAZO in the presence of PMA. The flavoprotein fluorescence was calibrated by exposing the cells to DNP (100 μM) at the end of the experiments. Bar indicates periods during which the cells were exposed to drug. (Reproduced, with permission, from Sato *et al.*, (1998.)

as assessed by changes in flavoprotein fluorescence. Since Grover *et al.* (1995) have previously described a K_{ATP} channel opener, BMS-180448, which produced a gliben-clamide-reversible cardioprotective effect in isolated guinea pig hearts that was independent of AP shortening, these results suggest that it is possible to synthesize site-specific K_{ATP} modulators that would have a greater therapeutic window than those currently available.

4. K_{ATP} CHANNELS IN DELAYED PRECONDITIONING

Delayed preconditioning against infarction has been reported 12–24 h following a preconditioning stimulus and has been primarily shown to occur in dogs (Kuzuya *et al.*, 1993) and rabbits (Marber *et al.*, 1993). A number of pathophysiological stressors such as ischemia (Marber *et al.*, 1993), heat shock (Marber *et al.*, 1993), an adenosine A_1 receptor agonist (Baxter *et al.*, 1994), free radicals (Zhou *et al.*, 1996), and the nontoxic endotoxin derivative monophosphoryl lipid A (MLA) have been shown to produce delayed preconditioning. Evidence that the K_{ATP} channel may be an end effector in delayed preconditioning was first reported by Mei *et al.* (1996). These authors found that MLA produced a dose-related reduction in myocardial infarct size in dogs that was associated with an enhanced shortening of the monophasic AP, which suggested that MLA may be increasing sarc K_{ATP} activation. However, the protective effect of MLA was completely antagonized by both glibenclamide and 5-HD. Similar results were obtained in rabbits by Elliott *et al.* (1996). That 5-HD blocked the effect of MLA suggests that the mito K_{ATP} channel may also be involved in its cardioprotective effect. Obviously, more studies are necessary to determine the relative importance of the sarc versus the mito K_{ATP} channel in mediating the protective effect of MLA.

Further evidence to support a role for K_{ATP} channels in delayed preconditioning has been suggested by three recent studies using heat stress as the preconditioning stimulus. Hoag *et al.* (1997) and Pell *et al.* (1997) both subjected rabbits to heat stress (42°C for 15 min) and found a reduction in infarct size and an increase in heat shock protein (HSP 72) expression. In both studies, glibenclamide and 5-HD completely abolished the heat shock-induced reduction in infarct size. Similary, Joyeux *et al.* (1998) found that the reduction in infarct size observed in isolated rat hearts induced by heat stress was also blocked by glibenclamide and 5-HD. Although these studies suggest an important role for K_{ATP} channels in several forms of delayed preconditioning, further studies are still needed with ischemia *per se* and an adenosine A_1 receptor agonist to investigate a role for K_{ATP} channels in delayed preconditioning.

5. SUMMARY AND FUTURE DIRECTIONS

Taken together, the results obtained thus far clearly suggest that the K_{ATP} channel is the end effector protein responsible for the cardioprotection observed following either acute or delayed preconditioning. However, more studies are needed to determine the signaling pathways involved in modulating channel activity so that enhanced K_{ATP} occurs as an adaptive response to a variety of preconditioning stimuli: Molecular cloning of the K_{ATP} channel subunits SUR and Kir6.x has shown that there are a number of subtypes which are regulated differently and have their own unique pharmacology. Further studies with the cloned channels are needed to better understand how these channels are regulated under normal and ischemic conditions. Hopefully, the mito K_{ATP} channel will be cloned in the near future, which will allow us to determine more accurately the role for the mito versus the sarc K_{ATP} channel in the cardioprotection produced by K_{ATP} openers and IPC. Another challenge will be to clearly determine the mechanisms by which opening of either the sarc K_{ATP} or the mito K_{ATP} channel, or both, produces beneficial effects in the cardiac myocyte.

REFERENCES

Aguilar-Bryan, L., Clement, J. P., IV, Gonzalez, G., Kunjilwar, K., Babenko, A., and Bryan, J., 1998, Toward understanding the assembly and structure of K_{ATP} channels, *Physiol. Rev.* **78**:227–245.

Armstrong, S. C., Liu, G. S., Downey, J. M., and Ganote, C. E., 1995, Potassium channels and preconditioning of isolated rabbit cardiomyocytes: Effects of glyburide and pinacidil, *J. Mol. Cell. Cardiol.* **27**:1765–1774.

Auchampach, J. A., and Gross, G. J., 1993, Adenosine A_1 receptors, K_{ATP} channels and ischemic preconditioning in dogs, *Am. J. Physiol.* **224**:H1327–H1336.

Auchampach, J. A., Grover, G. J., and Gross, G. J., 1992, Blockade of ischemic preconditioning in dogs by the novel ATP dependent potassium channel antagonist 5-hydroxydecanoate, *Cardiovasc. Res.* **26**:1054–1062.

Baxter, G. F., Marber, M. S., Patel, V. C., and Yellon, D. M., 1994, Adenosine receptor involvement in a delayed phase of protection 24 hours following ischemic preconditioning, *Circulation* **90**:2993–3000.

Cleveland, J. C., Jr., Meldrum, D. A. Rowland, R. T., Banerjee, A., and Harken, A. H., 1997a, Adenosine preconditioning of human myocardium is dependent upon the ATP-sensitive K^+ channel, *J. Mol. Cell. Cardiol.* **29**:175–182.

Cleveland, J. C., Jr., Meldrum, D. R., Cain, B. S., Banerjee, A., and Harken A. H., 1997b,Oral sulfonylurea hypoglycemic agents prevent ischemic preconditioning in human myocardium, *Circulation* **96**:29–32.

Cole, W. G., McPherson, C. D., and Sontag, D., 1991, ATP-regulated K^+ channels protect the myocardium against ischemia-reperfusion damage, *Circ. Res.* **69**:571–581.

Deutsch, E., Berger, M., Kussmaul, W. G., Hirshfeld, J. W., Jr., Herrmann, H. C., and Laskey, W. K., 1990, Adaptation to ischemia during percutaneous transluminal coronary angioplasty: Clinical, hemodynamic, and metabolic features, *Circulation* **82**:2044–2051.

Elliott, G. T., Comerford, M. L., Smith, J. R., and Zhao, L., 1996, Myocardial ischemia/reperfusion protection using monophosphoryl lipid A is abrogated by the ATP-sensitive potassium channel blocker, glibenclamide, *Cardiovasc. Res.* **32**:1071–1080.

Fryer, R. M., Schultz, J. J., Hsu, A. K., and Gross, G. J., 1998, Pretreatment with tyrosine kinase inhibitors partially attenuates ischemic preconditioning in rat hearts, *Am. J. Physiol.* **275**:H2009–H2015.

Garlid, K. D., Paucek, P., Yarov-Yarovy, V., Murray, H. N., Darbenzio, R. B., D'Alonso, A. J., Lodge, N. J., Smith, M. A., and Grover, G. J., 1997, Cardioprotective effect of diazoxide and its interaction with mitochondrial ATP- sensitive K^+ channels: Possible mechanism of cardioprotection, *Circ. Res.* **81**:1072–1082.

Gogelein, H., Hartung, J., Englert, H. C., and Scholkens, B., 1998, HMR 1883, a novel cardioprotective inhibitor of the ATP-sensitive potassium channel. Part I: Effects on cardiomyocytes, coronary flow and pancreatic β-cells, *J. Pharmacol. Exp. Ther.* **286**:1453–1464.

Grimmsmann, T., and Rustenbeck, I., 1998, Direct effects of diazoxide on mitochondria in pancreatic β-cells and on isolated liver mitochondria, *Br. J. Pharmacol.* **123**:781–788.

Gross, G. J., and Auchampach, J. A., 1992, Blockade of ATP-sensitive potassium channels prevents myocardial preconditioning in dogs, *Circ. Res.* **70**:223–233.

Grover, G. J., Sleph, P. G., and Dzwonczyk, S., 1992, Role of myocardial ATP-sensitive potassium channels in mediating preconditioning in the dog heart and their possible interaction with adenosine A_1 receptors, *Circulation* **86**:1310–1316.

Grover, G. J., D'Alonso, A. J., Hess, T., Sleph, P. G., and Darbenzio, R. B., 1995, Glyburide-reversible cardioprotective effect of BMS-180448 is independent of action potential shortening, *Cardiovasc. Res.* **30**:731–738.

Grover, G. J., D'Alonso, A. J., Dzwonczyk, S., Parham, C. S., and Darbenzio, R. B., 1996, Preconditioning is not abolished by the delayed rectifier K^+ blocker dofetilide, *Am. J. Physiol.* **271**:H1207–H1214.

Haessler, R., Kuzume, K., Chien, G. L., Wolff, R. A., Davis, R. F., and Van Winkle, D. M., 1994, Anaesthetics alter the magnitude of infarct limitation by ischaemic preconditioning, *Cardiovasc. Res.* **28**:1574–1580.

Hoag, J. B., Qian, Y. Z., Nayeem, M. A., D'Angelo, M., and Kukreja, R. C., 1997, ATP-sensitive potassium channel mediates delayed ischemic protection by heat stress in rabbit heart, *Am. J. Physiol.* **273**:H2458–H2464.

Inoue, I., Nagase, H., Kishi, K., and Higuti, T., 1991, ATP-dependent K^+ channel in the mitochondrial inner membrane, *Nature* **352**:244–247.

Jaburek, M., Yarov-Yarovoy, V., Paucek, P., and Garlid, K. D., 1998, State-dependent inhibition of the mitochondrial K_{ATP} channel by glyburide and 5-hydroxydecanoate, *J. Biol. Chem.* **273**:13578–13582.

Jovanovic, A., Jovanonvic, S., Lorenz, E., and Terzic, A., 1998, Recombinant cardiac ATP-sensitive K^+ channel subunits confer resistance to chemical hypoxia–reoxygenation injury, *Circulation* **98**:1548–1555.

Joyeux, M., Godin-Ribuot, D., and Ribuot, C., 1998, Resistance to myocardial infarction induced by heat stress and the effect of ATP-sensitive potassium channel blockade in the rat isolated heart, *Br. J. Pharmacol.* **123**:1085–1088.

Kersten, J. R., Gross, G. J., Pagel, P. S., and Warltier, D. C., 1998, Activation of adenosine triphosphate-regulated potassium channels: Mediation of cellular and organ protection, *Anesthesiology* **88**:495–513.

Kitakaze, M. Minamino, T., Node, K., Komamura, K., Shinozaki, Y., Chujo, M., Mori, H., Inoue, M., Hori, M., and Kamada, T., 1996, Role of activation of ectosolic 5'-nucleotidase in the cardioprotection mediated by opening of K$^+$ channels, *Am. J. Physiol.* **270:**H1744–H1756.

Kloner, R. A., Shook, T., Przyklenk, K., Davis, V. G., Iunio, L. Mathews, R. V., Burstein, S., Gibson, C. M., Poole, W. K., Cannon, C. P., McCabe, C. H., and Braunwald, E., 1995, Previous angina alters in-hospital outcome in TIMI 4: A clinical correlate to preconditioning, *Circulation* **91:**37–47.

Kouchi, I., Murakami T., Nawada, R., Akao, M., and Sasayama, S., 1998, K$_{ATP}$ channels are common mediators of ischemic and calcium preconditioning in rabbits, *Am. J. Physiol.* **274:**H1106–H1112.

Kuzuya, T., Hoshida, S., Yamashita, N., Fuji, H., Oe, H., Hori, M., Kamada, T., and Tada, M., 1993, Delayed effects of sublethal ischemia on the acquisition of tolerance to ischemia, *Circ. Res.* **72:**1293–1299.

Liang, B. T., 1996, Direct preconditioning of cardiac ventricular myocytes via adenosine A$_1$ receptor and K$_{ATP}$ channel, *Am. J. Physiol.* **271:**H1769–H1777.

Linz, W., Jung, O., Scholkens, B., and Englert, H., 1998, Different effects of K$_{ATP}$ channel blockers on ischemic preconditioning, *J. Mol. Cell. Cardiol.* **30:**A18 (abstract).

Liu, G. S., Thornton, J. D., Van Winkle, D. M., Stanley, A. W. H., Olsson, R. A., and Downey, J. M., 1991, Protection against infarction afforded by preconditioning is mediated via A$_1$ adenosine receptors in rabbit heart, *Circulation* **84:**350–356.

Liu, Y., Ytrehus, K., and Downey, J. M., 1994, Evidence that translocation of protein kinase C is a key event during ischemic preconditioning of rabbit myocardium, *J. Mol. Cell. Cardiol.* **26:**661–668.

Liu, Y., Sato, T., O'Rourke, B., and Marban, E., 1998, Mitochondrial ATP-dependent potassium channels: Novel effectors of cardioprotection, *Circulation* **97:**2463–2469.

Marber, M. S., Latchman, D. S., Walker, J. M., and Yellon, D. M., 1993, Cardiac stress protein elevation 24 hours after brief ischemia or heat stress is associated with resistance to myocardial infarction, *Circulation* **88:**1264–1272.

Mei, D. A., Elliott, G. T., and Gross, G. J., 1996, K$_{ATP}$ channels mediate late preconditioning against infarction produced by monophosphoryl lipid A, *Am. J. Physiol.* **271:**H2723–H2729.

Miura, K., Kano, S., Nakai, T., Satoh, K., Hoshi, K., and Ichihara, K., 1997, Inhibitory effects of glibenclamide and pertussis toxin on the attenuation of ischemia-induced myocardial acidosis following ischemic preconditioning in dogs, *Jpn. Circ. J.* **61:**709–714.

Mizumura, T., Nithipatikom, K., and Gross, G. J., 1995, Bimakalim, an ATP-sensitive potassium channel opener, mimics the effects of ischemic preconditioning to reduce infarct size, adenosine release and neutrophil function in dogs. *Circulation* **92:**1236–1245.

Mizumura, T., Auchampach, J. A., Linden, J., Bruns, R. F., and Gross, G. J., 1996, PD 81723, allosteric enhancer of the A$_1$ adenosine receptor, lowers the threshold for ischemic preconditioning in dogs, *Circ. Res.* **79:**415–423.

Moritani, K., Miyazaki, T., Miyoshi, S., Asanagi, M., Zhao, L. S., Mitamura, H., and Ogawa, S., 1994, Blockade of ATP-sensitive potassium channels by 5-hydroxydecanoate suppresses monophasic action potential shortening during regional myocardial ischemia, *Cardiovasc. Drugs Ther.* **8:**749–756.

Murry, C. E., Jennings, A. B., and Reimer, K. A., 1986, Preconditioning with ischemia: A delay of lethal cell injury in ischemic myocardium, *Circulation* **74:**1124–1136.

Noma, A., 1983, ATP-regulated K$^+$ channels in cardiac muscle, *Nature* **305:**147–148.

Okuyama, Y., Yamada, M., Kondo, C., Satoh, E., Isomoto, S., Shindo, T., Horio, Y., Kitakaze, M., Hori, M., and Kurachi, Y., 1998, The effects of nucleotides and potassium channel openers on the SUR 2A/Kir6.2 complex K$^+$ channel expressed in a mammalian cell line, HEK 293 T cells, *Pflügers Arch.* **435:**595–603.

Ottani, F., Galvani, M., Ferrini, D., Sorbello, F., Limonetti, P., Pantoli, D., and Rusticali, F., 1995, Prodromal angina limits infarct size: A role for ischemic preconditioning, *Circulation* **91:**291–297.

Paucek, P., Mironova, G., Mahdi, F., Beavis, A. D., Woldegiorgis, G., and Garlid, K. D., 1992, Reconstitution and partial purification of the glibenclamide-sensitive, ATP-dependent K$^+$ channel from rat liver, and beef heart mitochrondria, *J. Biol. Chem.* **267:**26062–26069.

Pell, T. J., Yellon, D. M., Goodwin, R. W., and Baxter, G. F., 1997, Myocardial ischemic tolerance following heat stress is abolished by ATP-sensitive potassium channel blockade, *Cardiovasc. Drugs Ther.* **11:**679–686.

Przyklenk, K., and Kloner, R. A., 1998, Ischemic preconditioning: Exploring the paradox, *Prog. Cardiovasc. Dis.* **40:**517–547.

Rohmann, S., Weygandt, H., Schelling, P., Soei, L. K., Verdouw, P. D., and Lues, I., 1994, Involvement of ATP-sensitive potassium channels in preconditioning protection, *Basic Res. Cardiol.* **89:**563–576.

Sakamoto, K., Yamazaki, J., and Nagao, T., 1998, 5-Hydroxydecanoate selectively reduces the initial increase in extracellular K^+ in ischemic guinea-pig heart, *Eur. J. Pharmacol.* **348**:31–35.

Sato, T., O'Rourke, B., and Marban, E., 1998, Modulation of mitochondrial ATP-dependent K^+ channels by protein kinase C, *Circ. Res.* **83**:110–114.

Schultz, J. J., Rose, E., Yao, Z., and Gross, G. J., 1995, Evidence for involvement of opioid receptors in ischemic preconditioning in rat hearts, *Am. J. Physiol.* **268**:H2157–2161.

Schultz, J. J., Yao, Z., Cavero, I., and Gross, G. J., 1997, Glibenclamide-induced blockade of ischemic preconditioning is time dependent in intact rat heart, *Am. J. Physiol.* **272**:H2607–H2615.

Schulz, R., Rose, J., and Heusch, G., 1994, Involvement of activation of ATP-dependent potassium channels in ischemic preconditioning in swine, *Am. J. Physiol.* **267**:H1341–H1352.

Speechly-Dick, M. E., Grover, G. J., and Yellon, D. M., 1995, Does ischemic preconditioning in the human involve protein kinase C and the ATP-dependent K^+ channel? Studies of contractile function after simulated ischemia in an atrial in vitro model, *Circ. Res.* **77**:1030–1035.

Tan, H. L., Mazon, P., Verberne, H. J., Sleeswijk, M. E., Coronel, R., Opthof, T., and Janse, M., 1993, Ischaemic preconditioning delays ischemia-induced cellular electrical uncoupling in rabbit myocardium by activation of ATP-sensitive potassium channels, *Cardiovasc. Res.* **27**:644–651.

Thornton, J. D., Thornton, C. S., Sterling, D. L., and Downey, J. M., 1993, Blockade of ATP-sensitive potassium channels increases infarct size but does not prevent preconditioning in rabbit hearts, *Circ. Res.* **72**:44–49.

Tomai, F., Crea, F., Gaspardone, A., Versaci, F., De Paulis, R., Penta de Peppo, A., Chiariello, L., and Gioffre, P. A., 1994, Ischemic preconditioning during coronary angioplasty is prevented by glibenclamide, a selective ATP-sensitive K^+ channel blocker, *Circulation* **90**:700–705.

Toombs, C. F., Moore, T. L., and Shebuski, R. J., 1993, Limitation of infarct size in the rabbit by ischaemic preconditioning is reversible with glibenclamide, *Cardiovasc. Res.* **27**:617–622.

Yao, Z., and Gross, G. J., 1994b, Effects of the K_{ATP} channel opener bimakalim on coronary blood flow, monophasic action potential duration and infarct size in dogs., *Circulation* **89**:1769–1775.

Yao, Z., and Gross, G. J., 1993a, Role of nitric oxide, muscarinic receptors and the ATP-sensitive K^+ channel in mediating the effects of acetylcholine to mimic preconditioning in dogs, *Circ. Res.* **73**:1193–1201.

Yao, Z., and Gross, G. J., 1994a, A comparison of adenosine-induced cardio-protection and ischemic preconditioning in dogs, *Circulation* **89**:1229–1236.

Yao, Z., and Gross, G. J., 1994b, Activation of ATP-sensitive potassium channels lowers the threshold for ischemic preconditioning in dogs, *Am. J. Physiol.* **267**:H1888–H1894.

Yao, Z., Auchampach, J. A., Pieper, G. M., and Gross, G. J., 1993b, Cardioprotective effect of monophosphoryl lipid A, a novel endotoxin analogue, in dogs, *Cardiovasc. Res.* **27**:832–838.

Yao, Z., Cavero, I., and Gross, G. J., 1993c, Activation of cardiac K_{ATP} channels: an endogenous protective mechanism during repetitive ischemic, *Am. J. Physiol.* **264**:H495–H504.

Yao, Z., Mizumura, T., Mei, D. A., and Gross, G. J., 1997, K_{ATP} channels and memory of ischemic preconditioning in dogs: Synergism between adenosine and K_{ATP} channels. *Am. J. Physiol.* **272**:H334–H342.

Zhou, X., Zhai, X., and Ashraf, M., 1996, Direct evidence that initial oxidative stress triggered by preconditioning contributes to second window of protection by endogenous antioxidant enzyme in myocytes, *Circulation* **93**:1177–1184.

Chapter 39

Therapeutic Potential of ATP-Sensitive K$^+$ Channel Openers in Cardiac Ischemia

Gary J. Grover

1. INTRODUCTION

ATP-sensitive K$^+$ (K$_{ATP}$) channels were originally described in cardiac myocytes (Noma, 1983), and since their discovery, they have been shown in brain, smooth muscle, kidneys, pancreatic β-cells, and skeletal muscle (Spruce *et al.*, 1985; de Weille *et al.*, 1988; Treherne and Ashford, 1991; Edwards and Weston, 1993). K$_{ATP}$ channels have also been termed metabolically regulated channels because of their linkage to the metabolic state of the cell. These channels are inhibited by physiologic concentrations of ATP. As the ATP concentration falls, the channel open-state probability increases, although it is doubtful whether the degree of ATP reduction necessary for opening this channel would ever be seen under physiologic conditions [see review by Edwards and Weston (1993)]. K$_{ATP}$ channels are also regulated by pH, fatty acids, sulfhydryl-redox state, nitric oxide (NO), various nucleotides, G-proteins, and numerous ligands (Kim and Clapham, 1989; Kirsch *et al.*, 1990; Edwards and Weston, 1993; Coetzee *et al.*, 1995; Ming *et al.*, 1997). They are thought to serve as a link between the metabolic state of the cell and secretory activity, or electromechanical coupling in muscle. K$_{ATP}$ channels are also found in the inner mitochondrial membrane (Inoue *et al.*, 1991), and these will be described in more detail later in this chapter.

Pharmacologic openers and blockers of K$_{ATP}$ channels were developed many years ago, and their physiologic activity was known long before their mechanism of action was discovered. Sulfonylurea antidiabetic agents such as glibenclamide (glyburide) have long been used to treat type II diabetes and are now known to work through blockade of K$_{ATP}$ channels in insulin-secreting cells [see review by Edwards and Weston (1993)]. Blockade of these channels causes insulin secretion secondary to membrane depolarization.

Gary J. Grover • Division of Metabolic Diseases, Bristol-Myers Squibb Pharmaceutical Research Institute, Princeton, New Jersey 08543-4000.

Potassium Channels in Cardiovascular Biology, edited by Archer and Rusch. Kluwer Academic/Plenum Publishers, New York, 2001.

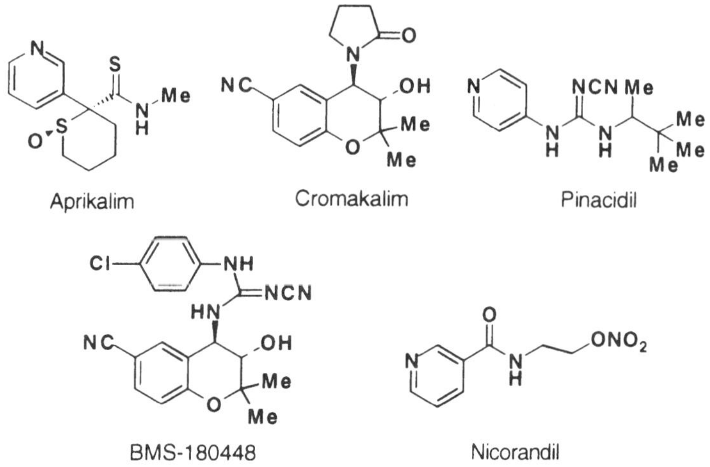

Figure 1. Chemical structures of the major classes of K_{ATP} channel openers.

Several chemical classes of compounds that were known to be potent smooth muscle relaxants, but whose mechanism of action was previously unknown, have been shown to be K_{ATP} channel openers. K_{ATP} openers have been found in diverse chemotypes, as diagrammed in Fig. 1. These agents relax smooth muscle by membrane hyperpolarization (Quast and Cook, 1989), an effect which is abolished by high external K^+ concentrations (distinguishing them from calcium channel blockers). As potent vascular smooth muscle relaxants, there was early excitement over their potential use as antihypertensive agents, but their profile of action did not allow ready distinction from already existing antihypertensive drugs. Other potential uses for K_{ATP} channel openers based on their smooth muscle relaxant properties include the treatment of asthma and urinary incontinence and the promotion of hair growth. These potential indications will not be discussed further in this chapter.

Although K_{ATP} channels were originally described in cardiac ventricular myocytes, their function in normal heart is not well understood. Because of their metabolic regulatory mechanisms, a potential role in the ischemic myocardium was proposed. Some investigators suggested that K_{ATP} channels contributed to the K^+ leaking out of ischemic myocardium (Venkatesh *et al.*, 1991). This "injury current" forms the basis for the ST-segment deviations observed during an ischemic episode. This led several investigators to propose that K_{ATP} channel openers would further destabilize the ischemic myocardium and increase the probability of reentrant arrhythmias. K_{ATP} channel openers such as cromakalim and pinacidil shorten action potential duration (APD), and this activity is enhanced in ischemic tissue (D'Alonzo *et al.*, 1992; Cole *et al.*, 1991). It was also proposed, however, that K_{ATP} channel openers might exert protective effects secondary to their ability to reduce calcium influx through APD shortening. Extensive studies done by numerous investigators have shown that K_{ATP} channel openers exert a direct cardioprotective effect; although the molecular and pharmacologic mechanisms for this protection are just beginning to become apparent, it does not appear to be related to APD shortening (as discussed in detail in the next section of this chapter). Therefore, the therapeutic potential of this class of agents has yet to be fully explored. In this chapter, I will review the pharmacologic profile for the cardioprotective effects of K_{ATP} channel openers, discuss what is known about the mechanism of action of these agents, and provide the latest information regarding the development of K_{ATP} channel openers for potential use in treating acute myocardial ischemia. In this chapter, I will be presenting the data and discussion with a historical perspective, and therefore the order of presentation of findings will, in large part, be on a chronological basis. In this way, the reader can gain a better understanding of how we have arrived at our current understanding of the pharmacology of these cardioprotective drugs.

2. CARDIOPROTECTIVE PROFILE FOR K_{ATP} CHANNEL OPENERS: EARLY STUDIES

Early work by Gross and colleagues showed that nicorandil, a nicotinamide nitrate, protected ischemic myocardium in several animal models (Lamping and Gross, 1985). At that time, this compound was known to be a NO donor, but it was not appreciated that it was also a K_{ATP} channel opener. The results reported by Gross and colleagues were extremely interesting to us once this activity of nicorandil became apparent and suggested the possibility that K_{ATP} channel opening might exert a protective effect. Several years after the original studies by Gross and colleagues on nicorandil, more selective K_{ATP} channel openers were discovered, enabling us to discern more clearly a role for the K_{ATP} channel in cardioprotection. Electrophysiological and physiological studies indicated that these more selective K_{ATP} channel openers activated both smooth muscle and cardiac K_{ATP} channels (Escande et al., 1989; Cook and Quast, 1990).

Early studies showed that structurally distinct K_{ATP} channel openers such as cromakalim, bimakalim, aprikalim, and P-1075 protected isolated rat hearts subjected to ischemia and reperfusion [see review by Grover (1994)]. Ischemic damage was assessed by postischemic recovery of contractile function, necrosis [judged by histology and surrogate markers such as lactate dehydrogenase (LOH) release], and an increased time to the onset of ischemic contracture (Monticello et al., 1996; Grover and Atwal, 1995). Numerous investigators have shown protective effects in isolated, ischemic rat heart preparations as well as in rabbit and guinea pig hearts (Tan et al., 1993; Galinanes et al., 1992; Ford et al., 1998; Cole et al., 1991) and in human atrial tissue ex vivo (Cleveland et al., 1997; Speechly-Dick et al., 1995). The K_{ATP} channel openers appear to work best when given before ischemia in isolated hearts, although minor protective effects have been reported when they are given during reperfusion (Grover et al., 1990; Gomoll et al., 1997; Mizumura et al., 1995). These data combined with the observed increase in the time to the onset of ischemic contracture suggest that these agents are working primarily during ischemia per se, although some direct effect on reperfusion injury cannot be completely ruled out. Initial studies were done in isolated heart preparations because the K_{ATP} channel openers available at the time were potent vasodilators and thus interpretation of direct cardioprotective effects in vivo would be clouded by hypotensive activity.

Although work with isolated hearts allowed investigators to show a direct cardioprotective effect for K_{ATP} channel openers, it was still possible that these drugs provided protection indirectly by increasing coronary blood flow. However, it was subsequently shown that K_{ATP} channel openers were cardioprotective even when given under constant-flow conditions before, during, and after global ischemia in isolated rat hearts, suggesting a lack of importance for coronary vasodilation (Grover et al., 1990). Further evidence showing a direct protective effect came from the work Armstrong et al. (1995) for pinacidil in isolated rabbit ventricular myocytes subjected to hypoxia and reoxygenation. Studies by Liang (1996) also showed protective effects for K_{ATP} channel openers in isolated chick myocytes. The protective concentration range for K_{ATP} channel openers in vitro is between 1 and $10 \mu M$, with the exception of the cyanoguanidine P-1075, which is structurally related to pinacidil and unusually potent in cardiac tissue as well as smooth muscle (Grover, 1994).

An intriguing aspect of the pharmacologic profile for K_{ATP} channel openers was that they exerted cardioprotective effects at doses or concentrations that did not depress contractile function (Grover et al., 1990, 1991). This finding was very interesting, as it distinguished K_{ATP} channel openers from calcium channel blockers, which are generally cardiodepressant at cardioprotective doses. The original hypothesis that K_{ATP} channel openers protect the ischemic myocardium by shortening APD and thereby reducing calcium entry implied that these drugs should exert cardiodepressant effects. In fact, Cole et al. (1991) showed that a high concentration of pinacidil shortened APD and reduced cardiac function in both ischemic and normal myocardium, although ischemic tissue was significantly more sensitive to this activity. These data, of course, did not prove that APD shortening was necessary for cardioprotection. The finding in this study that ischemic conditions increased the sensitivity of myocardial tissue to pinacidil was confirmed by our group using cromakalim (D'Alonzo et al., 1992).

Despite a lack of correlation between cardioprotection and depression of cardiac function, K_{ATP} channel openers consistently increased the time to the onset of ischemic contracture (Ohta et al., 1991; Grover and Atwal, 1995). Cardiac contracture represents rigor bond formation secondary to ATP depletion. Therefore, it was likely that K_{ATP} channel openers were conserving ATP in ischemic tissue. Studies by several laboratories clearly showed that K_{ATP} channel openers significantly conserved ATP during ischemia and enhanced the recovery of ATP upon reperfusion (McPherson et al., 1993; Grover et al., 1991, 1997). ATP conservation was observed at concentrations of the particular K_{ATP} openers that did not cause depressed contractile function, suggesting more efficient use of oxygen. One study showed that a cromakalim analog significantly improved efficiency of oxygen utilization during ischemia and reperfusion (Grover et al., 1997). Interestingly, unlike calcium channel blockers, K_{ATP} channel openers exerted additional beneficial effects beyond that induced by depolarizing cardioplegic solutions (Pignac et al. 1994; Grover and Sleph, 1995), further suggesting that the protective effects of K_{ATP} channel openers are exerted through a mechanism independent of a "cardioplegic" effect.

The studies cited above established a cardioprotective effect for structurally distinct K_{ATP} channel openers. However, although these results suggested that these drugs conferred cardioprotection through a K_{ATP} channel-related mechanism, these studies did not furnish definitive proof. The effect of specific pharmacologic blockers of K_{ATP} channels on the protective activity of K_{ATP} channel openers was tested in numerous models. Glyburide, which blocks K_{ATP} channels in most tissues, completely abolished the cardioprotective effects of structurally distinct K_{ATP} channel openers such as aprikalim, cromakalim, and pinacidil [see review by Grover (1994)]. As would be expected, glyburide also abolished the coronary vasodilator activity of these agents. Another channel blocker, sodium 5-hydroxydecanoate (5-HD), also abolished the cardioprotective effects of K_{ATP} channel openers but did not abolish the vasodilator effects of these agents (McCullough et al., 1991). We originally hypothesized that 5-HD was selective for cardiac ischemic tissue, and, thus, this finding was the first indication that the cardioprotective mechanism for K_{ATP} channel openers may be selective for ischemic or cardiac tissue. This idea will be developed later in this chapter. Notably, K_{ATP} channel blockers alone do not affect the severity of ischemia when given within a reasonable dose range (Grover, 1994). The important point is that pharmacologic blockers of K_{ATP} channels have been universally shown to abolish the protective effects of K_{ATP} channel openers. These blocking drugs appear to selectively abolish the cardioprotective effects of K_{ATP} channel openers, because they are without effect on mechanistically distinct cardioprotective agents such as sodium channel blockers, calcium channel blockers, and sodium/hydrogen exchange inhibitors (Sargent et al., 1993).

Although these in vitro studies suggested an excellent profile of cardioprotection, the effects of vasodilating K_{ATP} channel openers in models of ischemia in vivo were not as clear. We could not find a distinct window of protection for classical vasodilating K_{ATP} channel openers such as cromakalim when administered systemically, although clear cardioprotection was observed when the agents were given directly into the ischemic coronary artery (Grover et al., 1990). Several investigators showed protective effects of cromakalim when administered intravenously, although the window of protection was small (Mizumura et al., 1995; Auchampach et al., 1991, 1992). Cardioprotection in vivo has been observed as reduced infarct size and reduced postischemic stunning in several species (Auchampach et al., 1991, 1992; Toombs et al., 1992). In all of these in vivo studies, K_{ATP} channel blockers abolished these cardioprotective effects. An interesting study from Cavero's group showed that a vasodilating K_{ATP} channel opener resulted in cardiac lesions, and this was due to the vasodilating effect of this agent (Belin et al., 1996).

3. DRUG DEVELOPMENT OF K_{ATP} CHANNEL OPENERS FOR TREATMENT OF ACUTE MYOCARDIAL ISCHEMIA: DEVELOPMENT OF NONVASODILATING K_{ATP} CHANNEL OPENERS

The studies discussed earlier in this chapter suggest a profile of activity for K_{ATP} channel openers which is distinct from that of agents such as calcium antagonists. K_{ATP} channel openers appear to protect the ischemic myocardium without marked effects on cardiac function. Whereas this is a distinct advantage, these drugs also possess liabilities such as profound vasodilator activity as well as APD-shortening activity. For example, the hypotensive effect of K_{ATP} channel openers could further compromise blood-flow-limited tissue as well as cause coronary steal. Furthermore, APD shortening could increase the propensity for reentrant arrhythmias, although no clear consensus has emerged from animal studies (Chi et al., 1990; Vegh et al., 1997; de la Coussaye et al., 1993; D'Alonzo et al., 1998). Although there has been no consensus on such proarrhythmic effects, it is nevertheless prudent to try to minimize such a potentially lethal side effect. These were the issues facing us when we began thinking about developing a novel K_{ATP} channel opener designed specifically for treating acute myocardial ischemia. None of the preexisting K_{ATP} channel openers were designed specifically with myocardial ischemia in mind. Thus, we decided that the first priority was to tackle the issue of separation of vasodilator versus cardioprotective activity.

Drug development for K_{ATP} channel openers at this early phase was hampered by a lack of knowledge about the molecular mechanism of cardioprotection as well as the molecular biology of the K_{ATP} channel. Therefore, it was necessary to design a bioassay that would be predictive of the cardioprotective activity versus vasodilator activity of these agents. Our initial efforts to assess the cardioprotective activity of K_{ATP} channel openers were focused on measuring APD changes in papillary muscle. Unfortunately, we did not find APD shortening in ischemic or nonischemic tissue to be correlated with cardioprotective activity. This finding was disturbing on two counts. First, it suggested that this efficient bioassay method for screening for the potential cardioprotective effect of new K_{ATP} channel openers would not be predictive. Second, it implied that an increase in sarcolemmal K^+ currents also was not predictive of cardioprotection. The first issue was immediately critical to our drug development program, because we needed a reasonably high-throughput test to establish structure–activity relationships (SAR). The second issue is currently being addressed, as will be discussed more thoroughly later in this chapter.

Because K_{ATP} channels had not been cloned at that time, and hence were not available as a bioassay tool to assess the activity of K_{ATP} channel openers, we decided to use an ischemic heart assay. In particular, we observed that K_{ATP} channel openers caused a concentration-dependent increase in the time to onset of ischemic contracture in globally ischemic isolated rat hearts and that this effect was very reproducible. This effect is readily measured as the time necessary to observe the first 5 mm Hg increase in end diastolic pressure during global ischemia. Using this protocol, we were able to screen compounds at a rate of approximately 40/week. The cardioprotective potency was expressed as an EC_{25}, which was defined as the concentration of the test compound causing a 25% increase in the time to contracture. The cardioprotective potency was compared to vasorelaxant potency as measured using preconstricted rat aortas (Atwal et al., 1993).

In our compound deck, we already had numerous compounds that had been tested for K_{ATP} channel-induced vasorelaxant activity. From this series, we determined the cardioprotective potency (EC_{25}) of potent vasodilator compounds as well as structurally related compounds that were not potent vasorelaxants. We rapidly found that there was little correlation between vasorelaxant potency and cardioprotective potency. Interestingly, we found distinct SAR for glyburide-reversible cardioprotective activity (Atwal et al., 1993, 1995, 1996). Because previous studies showed that the ability of glyburide or 5-HD to abolish the cardioprotective effects of K_{ATP} channel openers is specific, we used this as a criterion for compound selection. This turned out to be critical because we found that minor structural modifications of some chemotypes turned these compounds into potent calcium channel blockers, which are not affected by glyburide (Grover et al., 1993).

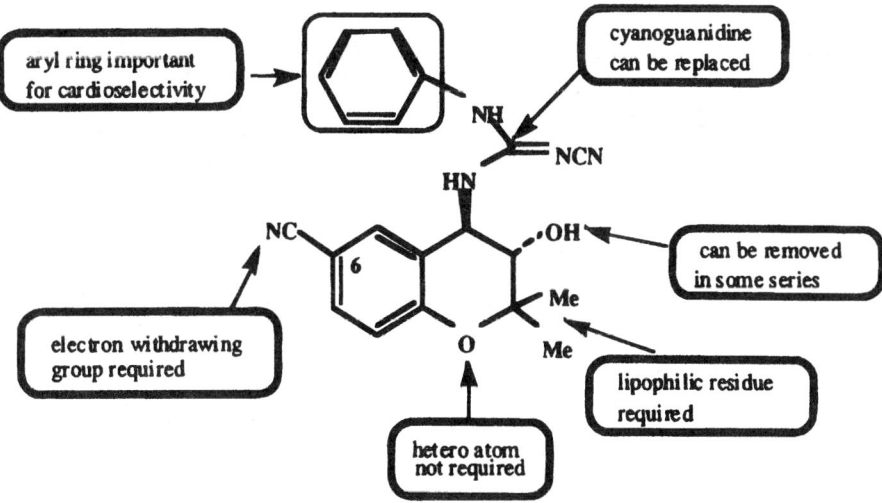

Figure 2. A summary of the structure–activity relationships for the cardioprotective potency of benzo-pyranylcyanoguanidines.

We constructed detailed SAR for a series of benzopyranylcyanoguanidines (Atwal *et al.*, 1993, 1995, 1996). Rather than show detailed chemistry, I have summarized these findings in Fig. 2. For more detailed information on SAR for this series of compounds, the reader is referred to several publications on this subject listed above. Overall, we found that vasodilator potency could be readily reduced while retaining cardioprotective potency. We were never successful in enhancing cardioprotective potency beyond that seen for standard K_{ATP} channel openers such as pinacidil and cromakalim. These studies led to the selection of BMS-180448, which was >100-fold more selective for cardioprotection versus vasodilation relative to cromakalim. The *in vitro* cardioprotective and vasorelaxant potency data for selected reference agents and BMS-180448 are shown in Table I. This compound had >60% oral bioavailability and a 7-h terminal half-life in rats. Such a compound is amenable to oral dosing in a once-per-day regimen, which is ideal for patient compliance and treatment (Grover and Atwal, 1995).

BMS-180448 also protected ischemic canine myocardium *in vivo* and had a significantly greater window of efficacy than vasodilating K_{ATP} channel openers (Grover and Atwal, 1995). The loss of vasodilator activity for BMS-180448 did not impair its cardioprotective activity *in vivo*, and this activity was completely abolished by glyburide and 5-HD. Although the cardioprotective effect of BMS-180448 was apparently related to K_{ATP} channel activation, it had little effect on APD during ischemia and had little effect on sarcolemmal K^+ currents (Grover *et al.*, 1995a). These properties were viewed positively from the clinical perspective, because they suggested that BMS-180448 may exhibit a reduced propensity for reentrant arrhythmias, but they caused us to further question our understanding of the mechanism of action of these agents, as will be discussed later in the chapter. Unfortunately, BMS-180448 could not be advanced as a drug because of liver toxicity noted in phase II trials.

Table I

Cardioprotective and Vasorelaxant Potencies of Various Agents
in Isolated Rat Hearts and Methoxamine-Constricted Rat Aorta

Agent	Cardioprotection $(EC_{25}, \mu M)^a$	Vasorelaxation $(IC_{50}, \mu M)^b$
Bimakalim	5.8	0.09
Aprikalim	2–3	0.04
Levcromakalim	2.3	0.03
Pinacidil	3–4	0.07
BMS-180448	2.5	3.6

$^a EC_{25}$, Concentration for 25% increase in time to contracture in isolated, globally
ischemic rat hearts.
$^b IC_{50}$, Concentration for 50% relaxation of methoxamine-constricted rat aorta.

Ultimately, agents were synthesized that retained glyburide-reversible antiischemic
activity but were totally devoid of vasodilator and APD shortening activity (Rovnyak
et al., 1997). However, although we were able to significantly increase the selectivity for
ischemic myocardium of drugs from our compound deck, we were never able to
increase cardioprotective potency. Clearly, cardioprotective potency at nanomolar
levels would be desirable to reduce the possibility of toxicity. The lack of a specific
molecular-based screening method was probably the limiting factor for our SAR.
However, recent studies are beginning to shed some light on the mechanism of action
of K_{ATP} channel openers, and a more rational approach to drug design is becoming
more evident.

4. RECENT INSIGHTS: CARDIOPROTECTIVE MECHANISM OF ACTION

Our early attempts to develop a K_{ATP} channel opener specifically designed for
treating myocardial ischemia were hampered by a lack of knowledge of the molecular
mechanism of cardioprotection. Little was known about the molecular biology of the
K_{ATP} channels. The lack of a specific molecular target as a focus for our screening efforts
probably impaired our ability to significantly enhance cardioprotective potency, although we were able to achieve great increases in drug selectivity for ischemic
myocardium compared to vascular smooth muscle. While this work was going on, data
were emerging from several fronts suggesting that K_{ATP} channel openers were not
working through a mechanism related to sarcolemmal potassium currents.

As discussed earlier, we found that we could not use myocardial APD shortening
as a surrogate marker for cardioprotection. Gross's group published an interesting
paper showing that a low dose of bimakalim which protected ischemic canine
myocardium had no effect on epicardial APD shortening; higher doses of bimakalim
did shorten APD (Yao and Gross, 1994). We confirmed these results by showing that
a dose of the Kir (delayed rectifier potassium channel) blocker dofetilide that completely abolished the APD-shortening activity of cromakalim did not abolish its cardioprotective activity (Grover *et al.*, 1995b). In this study, glyburide abolished the

cardioprotective effect of cromakalim. Interestingly, many of the "cardioselective" agents coming out of our SAR determinations were devoid of APD-shortening activity, as well as of effects on whole-cell myocyte potassium currents and single-K_{ATP}-channel currents (Grover *et al.*, 1995a). These results raised the possibility that the cardioprotective site for K_{ATP} channel openers was distinct from the classical sarcolemmal K_{ATP} channel. While these findings were interesting, they cast considerable doubt on the hypothesis that K_{ATP} channel openers conferred cardioprotection by acting on sarcolemmal K_{ATP} channels.

Fortunately, several pieces of information came together from molecular biology laboratories and from pharmacology laboratories which shed light on this question. Studies from Garlid's group showed that bovine cardiac or liver mitochondrial K_{ATP} channels were activated by classical K_{ATP} channel openers such as cromakalim and bimakalim (Garlid *et al.*, 1996; Paucek *et al.*, 1992). These experiments were performed using mitochondrial K_{ATP} channels that were reconstituted into lipid bilayers. Interestingly, these investigators were able to compare sarcolemmal and mitochondrial K_{ATP} channel activation using this assay. Agents such as cromakalim activated both sarcolemmal and mitochondrial K_{ATP} channels in the low-micromolar range ($K_{1/2}$ values of 18 and 2 μM, respectively), which is similar to the cardioprotective concentration range. An interesting anomaly was diazoxide, a well-known K_{ATP} channel opener with potent vasodilator activity. This agent also opened bovine cardiac mitochondrial K_{ATP} channels in the low-micromolar range ($K_{1/2} = 0.4\,\mu M$), whereas it only opened reconstituted sarcolemmal K_{ATP} channels close to the millimolar range ($K_{1/2} = 855\,\mu M$). This interesting result led to a collaboration between our group and Garlid's group to determine if diazoxide could protect ischemic rat myocardium. Diazoxide also was found to open rat and bovine cardiac mitochondrial K_{ATP} channels in the low-micromolar range (Garlid *et al.*, 1997) (Fig. 3). Within this concentration range, diazoxide exerted profound cardioprotective activity in isolated rat hearts subjected to 25 min of global ischemia followed by 30 min of reperfusion (Fig. 4). Diazoxide also increased the time to onset of ischemic contracture, and its potency was similar to that seen for cromakalim (Fig. 5); this effect was abolished by glyburide and 5-HD. Also, diazoxide had little effect on cardiac APD or whole-cell myocyte potassium currents (Fig. 6), confirming the earlier results from Garlid's laboratory showing a lack of effect of diazoxide on reconstituted cardiac sarcolemmal K_{ATP} channels. These data strongly suggest that mitochondrial K_{ATP} channels possess the relevant cardioprotective site for K_{ATP} channel openers.

Liu *et al.* (1998) have recently confirmed that diazoxide confers cardioprotection in isolated rabbit myocytes, thereby supporting the hypothesis that mitochondrial K_{ATP} channels are important in mediating cardioprotection. Additional studies in the same model also showed that activation of protein kinase C (PKC) potentiates the action of diazoxide on mitochondrial K_{ATP} channels (Sato *et al.*, 1998). These data provide evidence for linkage between ischemic preconditioning, PKC activity or translocation, and mitochondrial K_{ATP} channels (Downey and Cohen, 1995). (See Chapter 38 for more details on these studies.)

Interestingly, recent findings from molecular biology laboratories provide a basis for beginning to understand the differing pharmacologic potency of diazoxide on K_{ATP} channels expressed in different tissues. Sarcolemmal K_{ATP} channels are a complex of two different proteins, with one subunit being an inwardly rectifying K^+ (Kir) channel

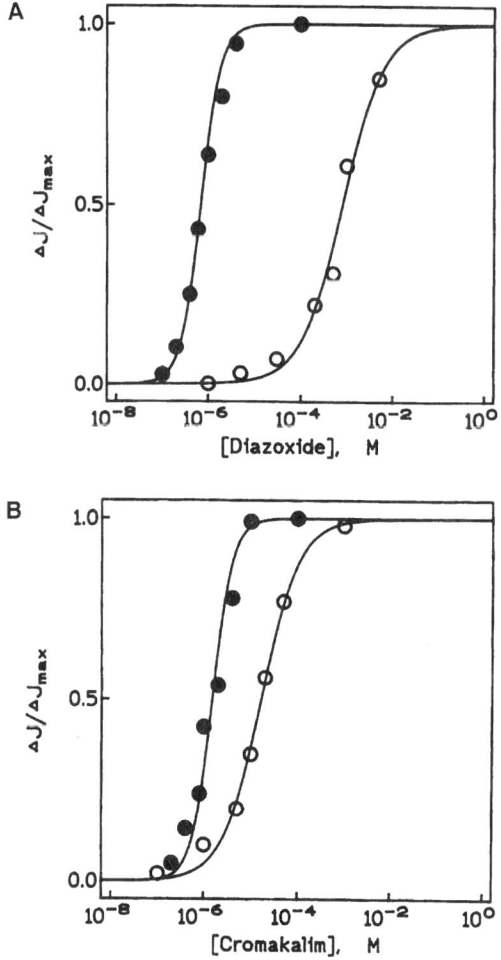

Figure 3. Activation of K^+ flux by diazoxide (A) or cromakalim (B) in K_{ATP} channels from bovine heart mitochondria and sarcolemma. Relative K^+ flux ($\Delta J/\Delta J_{max}$) is plotted versus concentration of drug added to the assay. The figure shows relative fluxes from cardiac mitochondrial K_{ATP} channels (●) and sarcolemmal K_{ATP} channels (○) in response to diazoxide or cromakalim. Observed $K_{1/2}$ values for cromakalim were $1.6 \pm 0.1\,\mu M$ for mitochondrial channels and $18 \pm 2\,\mu M$ for sarcolemmal channels. $K_{1/2}$ values for diazoxide were $0.8 \pm 0.03\,\mu M$ for mitochondrial K_{ATP} channels and $840 \pm 25\,\mu M$ for sarcolemmal K_{ATP} channels. (Reproduced, with permission, from Garlid et al., 1997.)

and the other being the sulfonylurea receptor (SUR) (Ashcroft, 1996; Inagaki et al., 1995, 1996). SUR is the protein conferring a regulatory role to ATP, as well as conferring the sensitivity of the channel to pharmacologic agents. The cardiac sarcolemmal K_{ATP} channel is a complex of Kir6.2 and SUR2, while the pancreatic channel is composed of Kir6.2 and SUR1 (Ashcroft, 1996; Inagaki et al., 1995, 1996). Inagaki et al. (1996) showed that diazoxide opened pancreatic K_{ATP} channels whereas it had no effect on cardiac K_{ATP} channels. These results correlate well with those of the pharmacologic studies discussed above.

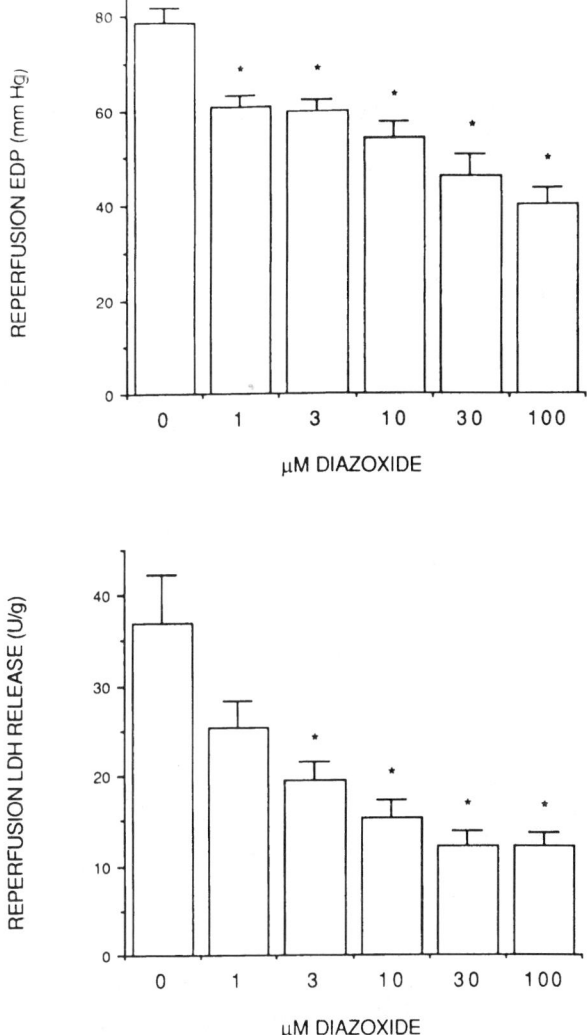

Figure 4. Effect of diazoxide on the release of lactate dehydrogenase (LDH) or reperfusion left ventricular end diastolic pressure (EDP) from ischemic/reperfused rat hearts. The hearts were pretreated for 10 min with diazoxide and then were rendered totally ischemic for 25 min. This was then followed by 30 min of reperfusion without drug. LDH release was reduced in a concentration-dependent manner by diazoxide. Reperfusion EDP (contracture) was also reduced in a concentration-dependent manner by diazoxide. Asterisks denote significance ($p < 0.05$) compared to vehicle. (Reproduced, with permission, from Garlid *et al.*, 1997.)

While the mitochondrial K_{ATP} hypothesis seems to be generating some interest in the scientific community, it is still not clear how opening of this channel can exert cardioprotective effects. Current hypotheses concentrate on a potential inhibition of mitochondrial calcium overload (Liu *et al.*, 1998). Regardless of how mitochondrial K_{ATP} channels confer protection, proof of the importance of these channels for cardioprotection may be helpful for future drug design. Use of the reconstituted

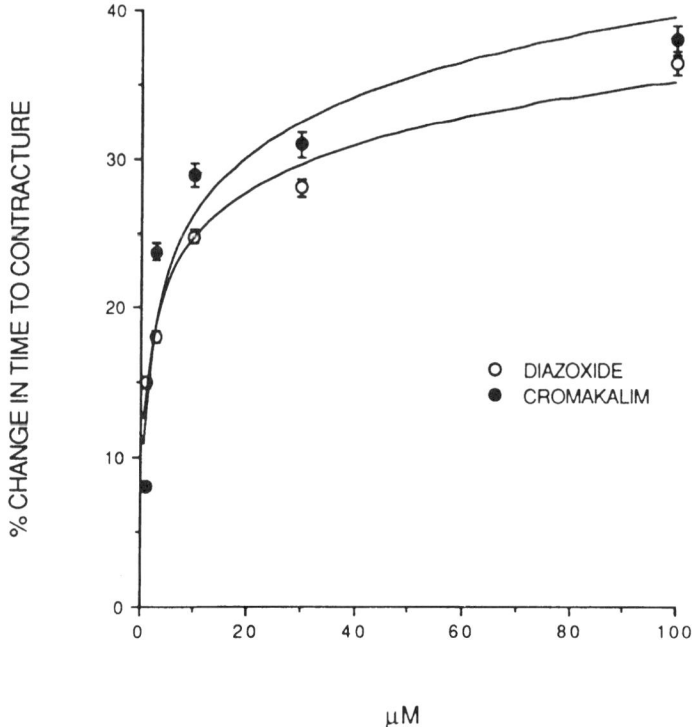

Figure 5. Effect of increasing concentrations of cromakalim (●) or diazoxide (○) on the time to onset of contracture during global ischemia in isolated rat hearts. Time to contracture is defined as the time (min) during ischemia in which a 5 mm Hg increase in end diastolic pressure is observed. (Reproduced, with permission, from Garlid *et al.*, 1997.)

channels for drug development may be difficult, but it may be possible to streamline this assay for drug development (although this is a kinetic assay, which will reduce the throughput of such a screen). It also would be useful to have the mitochondrial K_{ATP} channel cloned, so that a specific and high-throughput screen could be developed, perhaps a receptor binding assay. With such a high-throughput screen in place, it might be possible not only to increase drug selectivity for specific K_{ATP} channel heteromultimers, but also to greatly increase potency. Compared to the isolated rat heart as a bioassay, a more specific assay may be helpful in increasing cardioprotective potency. Significant increases in potency may also reduce the likelihood of "nonmechanism-based" toxicity.

Development of a specific mitochondrial K_{ATP} channel opener may also reduce the potential for proarrhythmic activity, which is observed with agents that reduce APD. This could greatly increase the therapeutic window for this class of agents. BMS-180448, which has little effect on cardiac APD, has no proarrhythmic activity in animal models and may, in fact, exert indirect antiarrhythmic effects secondary to its cardioprotective activity.

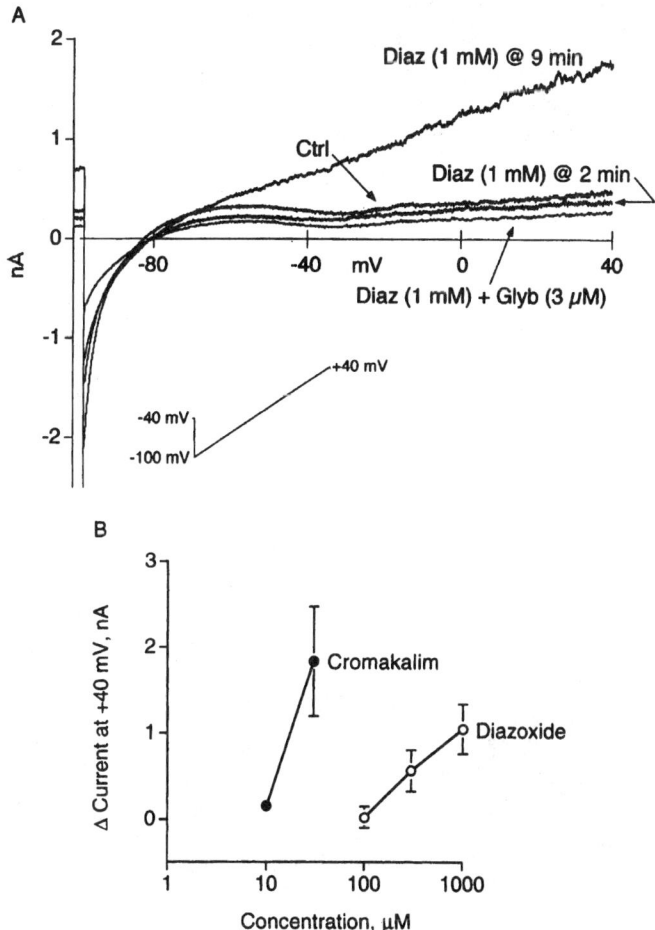

Figure 6. Effects of diazoxide on whole-cell currents recorded from isolated rat ventricular myocytes. (A) Under control (Ctrl) conditions, stimulation of isolated voltage-clamped myocytes with a voltage ramp from -100 to $+40$ mV (46.7 mV/s) evoked an N-shaped current–voltage response. In the example shown, addition of 1 μM diazoxide (Diaz) produced a small decrease in control currents at 2 min, followed by the development of a robust current at voltages positive to ~ -85 mV by 9 min. Subsequent addition of glyburide (Glyb; 3 μM) fully inhibited the activated current. (B) Mean increase in current, measured at $+40$ mV, produced by 100 ($n = 6$), 300 ($n = 18$), and 1000 μM diazoxide ($n = 8$). The effects of 10 ($n = 8$) and 30 μM ($n = 7$) cromakalim are shown for comparison.

5. POTENTIAL CLINICAL APPLICATIONS FOR CARDIOPROTECTIVE K_{ATP} CHANNEL OPENERS

There has been a recent explosive growth of interest in potential clinical applications for K_{ATP} channel openers. This is due, in part, to the findings from several laboratories linking preconditioning to K_{ATP} channels in animal models as well as in humans (Auchampach and Gross, 1993; Tomai *et al.*, 1994; Yellon *et al.*, 1998). Preconditioning is viewed as an endogenous protective mechanism intrinsic not only to heart, but to other tissues, including brain and skeletal muscle. Recent data also suggest that the cardioprotective effect of preconditioning is abolished by K_{ATP} channel blockers but is not correlated with APD shortening, suggesting a potential involvement of mitochondrial K_{ATP} channels (Grover *et al.*, 1996; Hamada *et al.*, 1998).

Much of the work on K_{ATP} channel openers in relation to ischemia has been performed in the heart, and, of course, there is a large unmet need for novel antiischemic or cardioprotective agents. Probably the greatest unmet need is for an agent that will effectively reduce myocardial damage during severe ischemia resulting in infarction. Although much work has been done in experimental models of infarction, few useful therapeutic agents have been developed based on this work. K_{ATP} channel openers may be useful for infarct size limitation, although such compounds should have little or no hypotensive effects and no APD-shortening activity. Unfortunately, infarct size reduction in humans is difficult to measure, and most pharmaceutical companies are loathe to perform such studies in early (phase II) clinical trials. This is because of the lack of surrogate end points such that, at the present time, mortality must be used as the end point, which makes phase II trials prohibitively expensive and time-consuming. The need for additive protection above that seen for fibrinolytic therapy will further "raise the bar" for showing efficacy for such agents.

Another way of attacking this problem is to determine the effect of test compounds in coronary artery bypass patients in whom recovery of contractile function and other readily measured surrogate markers can be used. This trial would be more cost-effective, and some companies may be willing to use positive results from such a trial as a justification for beginning a longer, more expensive infarction trial. There are experimental data to suggest possible utility for K_{ATP} channel openers as adjunctive therapy for bypass surgery (Pignac *et al.*, 1994; Grover and Sleph, 1995). Originally, it was thought that the high K^+ concentrations in cardioplegic solutions would inhibit the protective effects of K_{ATP} channel openers, but the intracellular mechanism of action of these drugs may explain why they retain their efficacy even in high K^+ solutions. Because experimental data show that treatment with K_{ATP} channel openers is more efficacious when begun before cardioplegic arrest, systemic administration may be necessary for effective therapy, and, therefore, compounds with little vasodilator as well as minimal APD-shortening activity would be necessary. However, most large companies would not be interested in pursuing the surgery market alone because it is a relatively small niche market.

The easiest and most cost-effective phase II clinical trials are for chronic stable angina pectoris. This is probably also the largest potential market for anti-ischemic agents. Unfortunately, it is also the most difficult ischemic disease to model in animals. It is presently not clear whether K_{ATP} channel openers would be useful for this disease. Nicorandil is being used in Europe, although this agent also has nitric oxide donor activity, which is well known to exert antianginal effects. Unpublished studies from our laboratory in canine models of effort-induced ischemia suggest that selective K_{ATP} channel openers such as BMS-180448 or cromakalim do not exert significant protective effects in such models of reversible ischemia. Obviously, more work is required to address this issue.

Whereas sarcolemmal K_{ATP} channels are expressed in numerous tissue types, the actual expression level of mitochondrial K_{ATP} channels in various tissues is presently unknown. Mitochondrial proteins tend to be highly conserved, and it is possible that many tissue types contain these channels. K_{ATP} channel openers are known to protect ischemic/reperfused brain as well as skeletal muscle tissue (Pang *et al.*, 1997a,b; Heurteaux *et al.*, 1995). In addition, preconditioning exerts profound protection in these tissues, and data suggest that K_{ATP} channels are important components in mediating this protective activity (Heurteaux *et al.*, 1995; Pang *et al.*, 1997a,b). Also, there are currently no truly efficacious drugs available for treating peripheral vascular disease, and it is certainly possible that K_{ATP} channel openers could be effective. The clinical trials for this indication are relatively straightforward and cost-effective. However, it is not known whether compounds developed with myocardial ischemia in mind would be optimal for skeletal muscle or cerebral ischemia.

6. SUMMARY AND CONCLUSIONS

The recent convergence of the pharmacologic work on K_{ATP} channel openers with findings suggesting a critical role for this channel in preconditioning has greatly increased interest in the role of K_{ATP} channels in acute myocardial ischemia. K_{ATP} channel openers may, in fact, simulate an endogenous protective mechanism not only for the heart, but for other tissues as well. We have come a long way from 10 years ago, when serious arguments raged over the novel and then unexpected finding that K_{ATP} channel openers exerted such profound cardioprotective effects. Hopefully, the general agreement on the cardioprotective effects of K_{ATP} channel openers combined with increasing knowledge of the mechanism for this cardioprotection will further stimulate research, which may result in useful therapeutics for treating acute ischemic conditions.

REFERENCES

Armstrong, S., Liu, G., Downey, J. and Ganote, C., 1995, Potassium channels and preconditioning of isolated rabbit cardiomyocytes: Effects of glyburide and pinacidil, *J. Mol. Cell. Cardiol.* **27**:1765–1774.

Ashcroft, F., 1996, Fresh insights into the interactions of drugs with ATP-sensitive K^+ channels, ID, *Research Alert* **2**:43–37.

Atwal, K., Grover, G., Ahmed, S., Ferrara, F., Harper, T., Kim, K., Sleph, P., Dzwonczyk, S., Russell, A., Moreland, S., McCullough, J. R., and Normandin, D. E., 1993, Cardioselective anti-ischemic ATP-sensitive potassium channel openers, *J. Med. Chem.* **36**:3971–3974.

Atwal, K. S., Grover, G. J., Ferrara, F. N., Ahmed, S. Z., Sleph, P. G., Dzwonczyk, S., and Normandin, D. E., 1995, Cardioselective antiischemic ATP-sensitive potassium channel openers. 2. Structure–activity studies on benzopyranylcyanoguanidines: Modification of the benzopyran ring, *J. Med. Chem.* **38**:1966–1973.

Atwal, K. S., Ferrara, F. N., Ding, C. Z., Grover, G. J., Sleph, P. G., Dzwonczyk, S., Baird, A. J., and Normandin, D. E., 1996, Structure–activity studies on benzopyranylcyanoguanidines: Replacement of the benzopyran portion, *J. Med. Chem.* **39**:306–313.

Auchampach, J., and Gross, G., 1993, Adenosine A_1 receptors, K_{ATP} channels, and ischemic preconditioning in dogs, *Am. J. Physiol.* **264**:H1327–H1336.

Auchampach, J. A., Maruyama, M., Cavero, I., and Gross, G. J., 1991, The new K^+ channel opener aprikalim (RP 52891) reduces experimental infarct size in dogs in the absence of hemodynamic changes, *J. Pharmacol. Exp. Ther.* **259**:961–967.

Auchampach, J., Maruyama, M., Cavero, I., and Gross, G., 1992, Pharmacological evidence for a role of ATP-dependent potassium channels in myocardial stunning, *Circulation* **86**:311–319.

Belin, V., Hodge, T., Picaut, P., Jordan, R., Algate, C., Gosselin, S., Nohynek, G., and Cavero, I., 1996, The myocardial lesions produced by the potassium channel opener aprikalim in monkeys and rats are prevented by blockade of cardiac α-adrenoceptors, *Fundam. Appl. Toxicol.* **31**:259–267.

Chi, L., Uprichard, A. C. G., and Lucchesi, B. R., 1990, Profibrillatory actions of pinacidil in a conscious canine model of sudden coronary death, *J. Cardiovasc. Pharmacol.* **15**:452–464.

Cleveland, J. C., Meldrum, D., Rowland, R., Banerjee, A., and Harken, A., 1997, Adenosine preconditioning of human myocardium is dependent upon the ATP-sensitive K$^+$ channel, *J. Mol. Cell. Cardiol.* **29**:175–182.

Coetzee, W., Nakamura, T., and Faivre, J., 1995, Effects of thiol-modifying agents on K$_{ATP}$ channels in guinea pig ventricular cells, *Am. J. Physiol.* **269**:H1625–1633.

Cole, W., McPherson, C., and Sontag, D., 1991, ATP-regulated K$^+$ channels protect the myocardium against ischemia/reperfusion damage, *Circ. Res.* **69**:571–581.

Cook, N. S., and Quast, U., 1990, Potassium channel pharmacology, in: *Potassium Channels* (N. S. Cook, ed.), Ellis Harwood London, pp. 181–231.

D'Alonzo, A., Darbenzio, R., Parham, C., and Grover, G., 1992, Effects of intracoronary cromakalim on postischaemic contractile function and action potential duration, *Cardiovasc. Res.* **26**:1046–1053.

D'Alonzo, A. J., Darbenzio, R., Hess, T. A., Zhu, J. L., Parham, C. S., and Grover, G. J., 1998, Effects of BMS-191095, a nonvasodilating K$_{ATP}$ opener, on hemodynamics, electrophysiology, cardioprotection, and arrhythmias in naive and myocardial infarcted dogs, *J. Mol. Cell. Cardiol.* **30**:415–423.

de la Coussaye, J. E., Eldedjam,J.-J., and Bruelle, P., 1993, Electrophysiologic and arrhythmogenic effects of the potassium channel agonist BRL 38227 in anesthetized dogs, *J. Cardiovasc. Pharmacol.* **22**:722–730.

de Weille, J., Schmid-Antomarchi, H., Fosset, M., and Lazdunski, M., 1988, ATP-sensitive K$^+$ channels that are blocked by hypoglycemia-inducing sulfonylureas in insulin-secreting cells are activated by galanin, a hyperglycemia-inducing hormone, *Proc. Natl. Acad. Sci. U.S.A.* **85**:1312–1316.

Downey, J. M., and Cohen, M. V., 1995, Signal transduction in ischemic preconditioning, *Z. Kardiol.* **4**:77–86.

Edwards, G., and Weston, A., 1993, The pharmacology of ATP-sensitive potassium channels, *Annu. Rev. Pharmacol. Toxicol.* **33**:597–637.

Escande, D., Thuringer, D., Le Guern, S., Courtcix, J., Laville, M., and Cavero, I., 1989, Potassium channel openers act through an activation of ATP-sensitive K$^+$ channels in guinea-pig cardiac myocytes, *Pflügers Arch.* **414**:669–675.

Ford, W. R., Lopaschuk, G. D, Schulz, R., and Clanahan, A. S., 1998, K$_{ATP}$ channel activation: Effects on myocardial recovery from ischaemia and role in the cardioprotective response to adenosine A$_1$ receptor stimulation, *Br. J. Pharmacol.* **124**:639–646.

Galinanes, M., Shattock, M., and Hearse, D., 1992, Effects of potassium channel modulation during global ischaemia in isolated rat heart with and without cardioplegia, *Cardiovasc. Res.* **26**:1063–1068.

Garlid, K. D., Paucek, P., Yarov-Yarovoy, V., Sun, X., and Schindler, P. A., 1996, The mitochondrial K$_{ATP}$ channel as a receptor for potassium channel openers, *J. Biol. Chem.* **271**:8796–8799.

Garlid, K. D., Paucek, P., Yarov-Yarovoy, B., Murray, H. N. M., Darbenzio, R. B., D'Alonzo, A. J., Lodge, N. J., Smith, M. A., and Grover, G. J., 1997, Cardioprotective effect of diazoxide and its interaction with mitochondrial ATP-sensitive potassium channels: Possible mechanism of cardioprotection, *J. Biol. Chem.* **271**:8796–8799.

Gomoll, A. W., Roth, R. A., Swillo, R. E., Baird, A. J., Sargent, C. S., Behling, R. W., Malone, H. J., and Grover, G. J., 1997, Effect of timing of treatment on the glyburide-reversible cardioprotective effects of BMS-180448, *J. Pharmacol. Exp. Ther.* **281**:24–33.

Grover, G., 1994, Protective effects of ATP-sensitive potassium-channel openers in experimental myocardial ischemia, *J. Cardiovasc. Pharmacol.* **24**:S18–S27.

Grover, G., and Atwal, K. S., 1995, BMS-180448, a glyburide-reversible cardioprotective agent with minimal vasodilator activity, *Cardiovasc. Drug Rev.* **13**:123–136.

Grover, G., and Sleph, P., 1995, Protective effect of K$_{ATP}$ openers in ischemic rat hearts treated with a potassium cardioplegic solution, *J. Cardiovasc. Pharmacol.* **26**:698–706.

Grover, G. J., D'Alonzo, A., Hess, T., Sleph, P., and Darbenzio, R., 1995a, Glyburide-reversible cardioprotective effect of BMS-180448 is independent of action potential shortening, *Cardiovasc. Res.* **30**:731–738.

Grover, G., D'Alonzo, A., Parham, C., and Darbenzio, R., 1995b, Cardioprotection with the K$_{ATP}$ opener cromakalim is not correlated with ischemic myocardial action potential duration, *J. Cardiovasc. Pharmacol.* **26**:145–152.

Grover, G. J., D'Alonzo, A. J., Dzwonczyk, S., Parham, C. S., and Darbenzio, R. B., 1996, Preconditioning is not abolished by the delayed rectifier K$^+$ blocker dofetilide, *Am. J. Physiol.* **271:**H1207–H1214.

Grover, G. J., Dzwonczyk, S., Parham, C., and Sleph, P.,1990, The protective effects of cromakalim and pinacidil on reperfusion function and infarct size in isolated perfused rat hearts and anesthetized dogs, *Cardiovasc. Drugs Ther.* **4:**465–474.

Grover, G., Newburger, J., Sleph, P., Dzwonczyk, S., Taylor, S., Ahmed, S., and Atwal, K., 1991, Cardioprotective effects of the potassium channel opener cromakalim: Stereoselectivity and effects on myocardial adenine nucleotides, *J. Pharmacol. Exp. Ther.* **257:**156–162.

Grover, G. J., Dzwonczyk, S, Sleph, P. G., Malone, H., and Behling, R. W., 1997, Cardioprotective effects of the ATP-sensitive potassium channel opener BMS-180448: Functional and energetic considerations, *J. Cardiovasc. Pharmacol.* **29:**28–38.

Grover, G. J., Sleph, P. G., Dzwonczyk, S., and McCullough, J. R., 1993, Cardioprotective effects of a novel calcium channel antagonist related to the structure of cromakalim, *J. Pharmacol. Exp. Ther.* **267:**102–107.

Hamada, K., Yamazaki, J., and Nagao, T., 1998, Shortening of action potential duration is not a prerequisite for cardiac protection by ischemic preconditioning or a K$_{ATP}$ channel opener, *J. Mol. Cell. Cardiol.* **30:**1369–1379.

Heurteaux, C., Lauritzen, I., Widmann, C., and Lazdunski, M., 1995, Essential role of adenosine, adenosine A1 receptors, and ATP-sensitive K$^+$ channels in cerebral ischemic preconditioning, *Proc. Natl. Acad. Sci. U.S.A.* **92:**4666–4670.

Inagaki, N., Gonoi, T., Clement J. P., Namba, N., Inazawa, J., Gonzalez, G., Aguilar-Bryan, L., Seino, S., and Bryan, J., 1995, Reconstitution of IK$_{ATP}$: An inward rectifier subunit plus the sulfonylurea receptor, *Science* **270:**1166–1170.

Inagaki, N., Gonoi, T., Clement, J., Wang, C., Aguilar-Bryan, L., Bryan, J., and Seino, S., 1996, A family of sulfonylurea receptors determines the pharmacological properties of ATP-sensitive K$^+$ channels, *Neuron* **16:**1011–1017.

Inoue, I., Nagase, H., Kishi, K., and Higuti, T., 1991, ATP-sensitive K$^+$ channel in the mitochondrial inner membrane, *Nature* **352:**244–247.

Kim, D., and Clapham, D., 1989, Potassium channels in cardiac cells activated by arachidonic acid and phospholipids, *Science* **244:**1174–1176.

Kirsch, G., Codina, J., Birnbaumer, L., and Brown, A., 1990, Coupling of ATP-sensitive K$^+$ channels to A$_1$ receptors by G proteins in rat ventricular myocytes, *Am. J. Physiol.* **259:**H820–H826.

Lamping, K., and Gross, G. J.,1985, Improved recovery of myocardial segmental function following a short coronary occlusion in dogs by nicorandil, a potential new antianginal agent, and nifedipine, *J. Cardiovasc. Pharmacol.* **7:**158–166.

Liang, B. T., 1996, Direct preconditioning of cardiac ventricular myocytes via adenosine A$_1$ receptor and K$_{ATP}$ channel, *Am. J. Physiol.* **271:**H1769–H1777.

Liu, Y., Sato, T., O'Rourke, B., and Marban, E., 1998, Mitochondrial ATP-dependent potassium channels: Novel effectors of cardioprotection?, *Circulation* **97:**2463–2469.

McCullough, J., Normandin, D., Conder, M., Sleph, P., Dzwonczyk, S., and Grover, G., 1991, Specific block of the anti-ischemic actions of cromakalim by sodium 5-hydroxydecanoate, *Circ. Res.* **69:**949–958.

McPherson, C., Pierce, G., and Cole, W., 1993, Ischemic cardioprotection by ATP-sensitive K$^+$ channels involves high-energy phosphate preservation, *Am. J. Physiol.* **265:**H1809–H1818.

Ming, Z., Parent, R., and Lavallee, M., 1997, β2-Adrenergic dilation of resistance coronary vessels involves K$_{ATP}$ channels and nitric oxide in conscious dogs, *Circulation* **95:**1568–1576.

Mizumura, T., Nithipatikam, K., and Gross, G. J., 1995, Bimakalim, an ATP-sensitive potassium channel opener, mimics the effects of ischemic preconditioning to reduce infarct size, adenosine release, and neutrophil function in dogs, *Circulation* **92:**1236–1245.

Monticello, T., Sargent, C., McGill, J., Barton, D., and Grover, G., 1996, Amelioration of ischemia/reperfusion injury in isolated rats hearts by the ATP-sensitive potassium channel opener BMS-180448, *Cardiovasc. Res.* **31:**93–101.

Noma, A., 1983, ATP-regulated K$^+$ channels in cardiac muscle, *Nature* **305:**147–148.

Ohta, H., Jinno, Y., Harada, K., Ogawa, N., Fukushima, H., and Nishikori, K., 1991, Cardioprotective effects of KRN2391 and nicorandil on ischemic dysfunction in perfused rat heart, *Eur. J. Pharmacol.* **204:**171–177.

Pang, C. Y., Neligan, P., Zhong, A., Xu, H., and Forrest, C. R., 1997a, Effector mechanism of adenosine in acute ischemic preconditioning of skeletal muscle against infarction, *Am. J. Physiol.* **273**:R887–R895.

Pang, C. Y., Neligan, P., Xu, H., Zhong, A., Hopper, R., and Forrest, C. R., 1997b, Role of ATP-sensitive K^+ channels in ischemic preconditioning of skeletal muscle against infarction, *Am. J. Physiol.* **273**:H44–H51.

Paucek, P., Mironova, G., Mahdi, F., Beavis, A. D., Woldegiorgis, G., and Garlid, K. D., 1992, Reconstitution and partial purification of the glibenclamide-sensitive, ATP-dependent K^+ channel from rat liver and beef heart mitochondria, *J. Biol. Chem.* **267**:26062–26069.

Pignac, J., Bourgouin, J., and Dumont, L., 1994, Cold cardioplegia and the K^+ channel modulator aprikalim (RP 52891): Improved cardioprotection in isolated ischemic rabbit hearts, *Can. J. Physiol. Pharmacol.* **72**:126–132.

Quast, U., and Cook, N. S., 1989, Moving together: K^+ channel openers and ATP-sensitive K^+ channels, *Trends Pharmacol. Sci.* **10**:431–435.

Rovnyak, G. C., Ahmed, S. C., Ding, C. Z., Dzwonczyk, S., Ferrara, F. N., Humphreys, W. G., Grover, G. J., Santafianos,, R.,, Atwal, K. A., Baird, A. J., McLaughlin,L. G., Normandin, D., Sleph, P. G., and Traeger, S. C., 1997, Cardioselective antiischemic ATP-sensitive potassium channel openers. 5. Identification of 4-(N-aryl)-substituted benzopyran derivatives with high selectivity, *J. Med. Chem.* **40**:24–34.

Sargent, C. A., Smith, M. A., Dzwonczyk, S., Sleph, P. G., and Grover, G. J., 1993, Effect of potassium channel blockade on the anti-ischemic actions of mechanistically diverse agents, *J. Pharmacol. Exp. Ther.* **259**:97–103.

Sato, T., O'Rourke, B., and Marban, E., 1998, Modulation of mitochondrial ATP-dependent K^+ channels by protein kinase, *Circ. Res.* **83**:110–114.

Speechly-Dick, M., Grover, G., and Yellon, D., 1995, Does ischemic preconditioning in the human involve protein kinase C and the ATP-dependent K^+ channel? Studies of contractile function after simulated ischemia in an atrial in vitro model, *Circ. Res.* **77**:1030–1035.

Spruce, A., Standen, N., and Stanfield, P., 1985, Voltage-dependent ATP-sensitive potassium channels of skeletal muscle membrane, *Nature* **316**:736–738.

Tan, H., Mazon, P., Verberne, H., Sleeswijk, M., Coronel, R., Opthof, T., and Janse, M., 1993, Ischaemic preconditioning delays ischaemia induced cellular electrical uncoupling in rabbit myocardium by activation of ATP sensitive potassium channels, *Cardiovasc. Res.* **27**:644–651.

Tomai, F., Crea, F., Gaspardone, A., Versaci, F., De Paulis, R., Penta de Peppo, A., Chiariello, L., and Gioffre, P., 1994, Ischemic preconditioning during coronary angioplasty is prevented by glibenclamide, a selective ATP-sensitive K^+ channel blocker, *Circulation* **90**:700–705.

Toombs, C. F., Norman, N. R., Groppi, V. E., Lee, K. S., Gadwood, R. C., and Shebuski, R. J., 1992, Limitation of myocardial injury with the potassium channel opener cromakalim and the nonvasoactive analog U-89,232: Vascular vs. cardiac actions in vitro and in vivo, *J. Pharmacol. Exp. Ther.* **263**:1261–1268.

Treherne, J., and Ashford, M., 1991, The regional distribution of sulphonylurea binding sites in rat brain, *Neuroscience* **40**:523–531.

Vegh, A. Papp, J. G., Gyorgy, K., Kaszala, K., Parratt, J. R., 1997, Does the opening of ATP-sensitive K^+ channels modify ischaemia-induced ventricular arrhythmias in anaesthetized dogs?, *Eur. J. Pharmacol.* **333**:33–38.

Venkatesh, N., Lamp, S. T., and Weiss, J. N., 1991, Sulfonylureas, ATP-sensitive K^+ loss during hypoxia, ischemia, and metabolic inhibition in mammalian ventricle, *Circ. Res.* **74**:623–629.

Yao, Z., and Gross, G., 1994, Effects of the K_{ATP} channel opener bimakalim on coronary blood flow, monophasic action potential duration, and infarct size in dogs, *Circulation* **89**:1769–1775.

Yellon, D. M., Baxter, G. F., Garcia-Dorado, D., Heusch, G., and Sumeray, M. S., 1998, Ischaemic preconditioning: Present position and future directions, *Cardiovasc. Res.* **37**:21–33.

Part VIII

Potassium Channels in Vascular Disease

Chapter 40

Altered Expression and Function of Kv Channels in Primary Pulmonary Hypertension

Jason Xiao-Jian Yuan and Lewis J. Rubin

1. INTRODUCTION

1.1. Pathophysiology and Pathogenesis of Primary Pulmonary Hypertension

Primary pulmonary hypertension (PPH) is a progressive, fatal disease that is characterized by a sustained elevation of pulmonary arterial pressure and pulmonary vascular resistance from an unknown cause. The incidence of PPH is about 1–2 per million in the general population, and the mean life expectancy after diagnosis is 2.5 years. Although PPH can occur in individuals of all ages and both genders, it predominantly affects women (Rubin, 1997).

Vasoconstriction, vascular wall remodeling (media and intimal proliferation), vascular injury, and *in situ* thrombosis all contribute to the increased pulmonary vascular resistance in PPH patients (Palevsky *et al.*, 1989; Pietra, 1997; Tanaka *et al.*, 1996; Voelkel *et al.*, 1997). Medial hypertrophy, suggesting active vasoconstriction and smooth muscle proliferation, is the earliest and most consistent pathological finding (Wood, 1958; Wagenvoort, 1960). Smooth muscle stretch and elevated arterial pressure secondary to vasoconstriction are promoters of smooth muscle cell hypertrophy and hyperplasia (Hishikawa *et al.*, 1994; Kolpakov *et al.*, 1995). These lines of evidence suggest that intrinsic abnormalities of the pulmonary vascular smooth muscle cells may be present and important in the pathogenesis of PPH, although impaired synthesis of nitric oxide (NO) (Dinh-Xuan *et al.*, 1991; Giaid and Saleh, 1995; Steudel *et al.*, 1997), enhanced production of endothelin (Giaid *et al.*, 1993), and imbalanced metabolites of prostacyclin (PGI_2) and thromboxane (Christman *et al.*, 1992) also are associated with PPH.

Jason Xiao-Jian Yuan and Lewis J. Rubin ● Division of Pulmonary and Critical Care Medicine; Department of Medicine, University of California—San Diego, San Diego, California 92103-8382.

Potassium Channels in Cardiovascular Biology, edited by Archer and Rusch. Kluwer Academic/Plenum Publishers, New York, 2001.

1.2. K^+ Channel Regulation of Vasoconstriction and Vascular Remodeling

In pulmonary arterial smooth muscle cells (PASMC), the cytosolic free Ca^{2+} concentration ($[Ca^{2+}]_i$) may increase by Ca^{2+} influx through Ca^{2+} channels in the plasma membrane (Nelson et al., 1990; Fleischmann et al., 1994; Yuan, 1998) or by agonist-mediated release and Ca^{2+}-induced Ca^{2+} release from intracellular Ca^{2+} stores (Berridge, 1993; Blaustein, 1993). A rise in $[Ca^{2+}]_i$ initiates cross-bridge cycling and contraction and may rapidly (50–300 ms) increase nuclear Ca^{2+} to promote cell proliferation by moving quiescent cells into the cell cycle and by propelling the proliferating cells through mitosis (Allbritton et al., 1994). Thus, in addition to triggering vasoconstriction, increased $[Ca^{2+}]_i$ may also play an important role in the hypertrophy of small pulmonary arteries and muscularization of pulmonary arterioles.

Membrane potential (E_m) is an important determinant of $[Ca^{2+}]_i$, because the L-type Ca^{2+} channels (Nelson et al., 1990; Yuan, 1998) that are sensitive to the clinically used Ca^{2+} channel blockers nifedipine and diltiazem, (Rich et al., 1992) are activated by membrane depolarization. In addition to raising $[Ca^{2+}]_i$ by promoting Ca^{2+} influx through voltage-gated Ca^{2+} channels (Nelson et al., 1990; Fleischmann et al., 1994) membrane depolarization also stimulates Ca^{2+} release by facilitating inositol 3,4,5-trisphosphate (IP_3) production in vascular smooth muscle cells (Ganitkevich and Isenberg, 1993; Kukuljan et al., 1994). Furthermore, membrane depolarization is also associated with inhibition of cell apoptosis, which would act to enhance cell survival and thus facilitate cell proliferation (Yu et al., 1997). The effective use of the Ca^{2+} channel blockers in PPH patients (Rich et al., 1992) suggests that an elevated $[Ca^{2+}]_i$ in PASMC due to the opening of voltage-gated Ca^{2+} channels may contribute to the development and maintenance of PPH.

Transmembrane K^+ permeability is a key determinant of E_m when the K^+ gradient across the plasma membrane is constant. At least four classes of K^+ channels may contribute to this permeability in vascular smooth muscle membranes: voltage-gated K^+ (Kv) channels, large-conductance Ca^{2+}-activated K^+ (BK_{Ca}) channels, ATP-sensitive K^+ (K_{ATP}) channels, and inward rectifier K^+ (Kir) channels (Nelson and Quayle, 1995). Under resting conditions, K^+ permeability through Kv channels is thought to primarily determine the level of E_m in PASMC (Archer et al., 1998; Evans et al., 1996; Post et al., 1995; Yuan, 1995; 1998). Thus, E_m is directly related to the level of whole-cell Kv current ($I_{K(V)}$), which is determined by $I_{K(V)} = N \times i_{Kv} \times P_{open}$, where N is the number of membrane Kv channels; i_{Kv} is the single-channel Kv current; and P_{open} is the steady-state open probability of the Kv channel. When Kv channels close (i_{Kv} or P_{open} is decreased) or Kv channel expression declines (N is decreased), E_m becomes less negative and depolarization occurs as a result of decreased $I_{K(V)}$. When Kv channels open (i_{Kv} or P_{open} is increased) or Kv channel expression rises (N is increased), E_m becomes more negative and hyperpolarization occurs as a result of increased $I_{K(V)}$.

1.3. A Proposed Pathogenic Hypothesis of PPH

A common hypothesis for the etiology and pathogenesis of PPH is that vasoconstriction and cell proliferation use overlapping signaling processes that result in parallel intracellular events. In particular, a rise in $[Ca^{2+}]_i$ is thought to activate myosin light-chain kinase to cause contraction and accelerate the cell cycle to promote proliferation. Therefore, vasoconstriction and vascular remodeling involving the proliferation of fibroblasts, smooth muscle cells, and endothelial cells may constitute two linked phenomena in the etiology of PPH in the pulmonary circulation (Voelkel et al., 1997).

It is possible that PPH originates, in part, from a downregulation of Kv channel expression in the vascular smooth muscle membranes of pulmonary arteries or arterioles. The resultant decrease of Kv channel activity and $I_{K(V)}$ availability is thought to promote membrane depolarization and thereby augment Ca^{2+} influx through voltage-gated Ca^{2+} channels to increase $[Ca^{2+}]_i$. Excessive pulmonary vasoconstriction and persistent PASMC proliferation may ensue. In contrast, endothelium-derived relaxing factors (EDRF), including nitric oxide and prostacyclin, may exert their vasodilator effects, in part, by activating K^+ channels to mediate membrane hyperpolarization, thereby reducing voltage-gated Ca^{2+} influx and decreasing the $[Ca^{2+}]_i$ in PASMC.

2. MOLECULAR IDENTITY AND ELECTROPHYSIOLOGICAL CHARACTERIZATION OF Kv CHANNELS IN HUMAN PULMONARY ARTERIAL SMOOTH MUSCLE CELLS

2.1. Molecular Identity

Voltage-gated K^+ (Kv) channels are heteromultimeric tetramers composed of two structurally distinct subunits, the pore-forming α-subunit and the regulatory β-subunit (Isom et al., 1994; Chandy and Gutman, 1995). Although expression of α-subunits alone is enough to generate Kv channels possessing many features of the corresponding channels in situ, studies on native channels have confirmed that the biophysical properties of Kv channels encoded by certain α-subunits are dramatically altered by association with β-subunits. For example, association of Kv channel β-subunits with α-subunits confers the fast (N-type) inactivation to slowly or non-inactivating delayed rectifier Kv channels (Rettig et al., 1994; England et al., 1995). In addition to changing the kinetic properties of Kv channel α-subunits, Kv channel β-subunits also may reduce $I_{K(V)}$ by acting as open-channel blockers (De Biasi et al., 1997).

In PASMC from patients with non-pulmonary-hypertensive diseases (NPH) and pulmonary hypertension secondary to cardiopulmonary diseases (SPH), three Kv channel α-subunit gene transcripts (Kv1.2, Kv1.4, and Kv1.5) have been identified using reverse-transcription polymerase chain reaction (RT-PCR) analysis (Fig. 1A). A Kv channel β-subunit (Kvβ1.1) gene transcript and a *Slowpoke* (hSlo) gene encoding the BK_{Ca} channel (McCobb et al., 1995) also are expressed in human PASMC (data not shown). With the use of specific antibodies, Kv1.2, Kv1.4, and Kv1.5 channel α-subunit proteins have consistently been detected using Western blot analysis (Fig. 1B). These data suggest that multiple Kv channel α- and β-subunits are expressed in human PASMC (Archer et al., 1998; Yuan et al., 1998c).

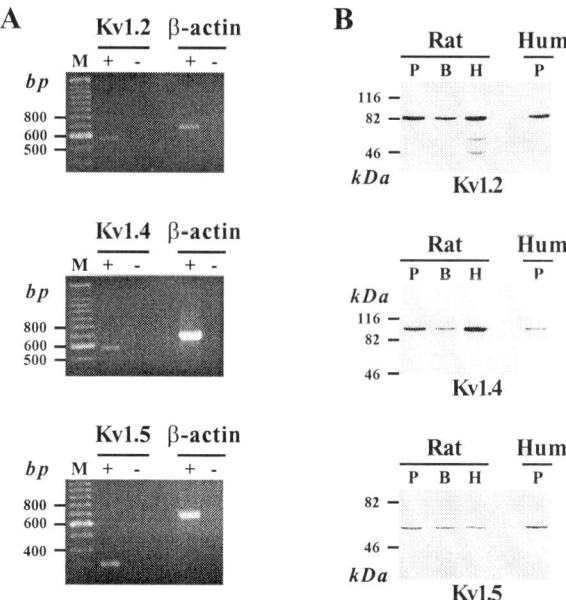

Figure 1. Molecular identity of Kv channels in human PASMC. (A) PCR-amplified products displayed in an agarose gel represent Kv1.2 [565 base pairs (bp)], Kv1.4 (570 bp), and Kv1.5 (300 bp). β-Actin (661 bp) transcripts were used as control. M, Marker. Reverse transcription (RT) was performed in the presence (+) or absence (−) of reverse transcriptase. (B) Western blot analysis of Kv1.2, Kv1.4, and Kv1.5 in rat PASMC (P), brain (B), and heart (H) and in human PASMC (HumP). Immunoblots were incubated with affinity-purified anti-Kv1.2, anti-Kv1.4, and anti-Kv1.5 polyclonal antibodies. Molecular mass markers are indicated on the left in kilodaltons.

2.2. Electrophysiological and Pharmacological Properties

Whole-cell $I_{K(V)}$ can be isolated for patch-clamp measurement in human PASMC by bathing the cells in Ca^{2+}-free solution (Fig. 2A, left) and minimizing BK_{Ca} current ($I_{K(Ca)}$) and K_{ATP} current (I_{KATP}) by including EGTA (10 mM) and ATP (5 mM) in a Ca^{2+}-free pipette solution for cell dialysis. The current–voltage (I–V) relationship reveals that whole-cell $I_{K(V)}$ is activated at about -40 mV (Fig. 2A, right). The current is rapidly activated with a time constant of 4.45 ms at $+80$ mV. Based on the decay kinetics of the current elicited by depolarization to $+80$ mV, the current is composed of a rapidly inactivating component (resembling the A-type $I_{K(V)}$), a slowly inactivating component, and a steady-state component corresponding to the delayed rectifier ($I_{K(V)}$) channels (Fig. 2A, inset). Extracellular application of 4-aminopyridine (4-AP; 5 mM), a blocker of Kv channels, significantly reduces $I_{K(V)}$ (Fig. 2B, inset), and the inhibitory effect of 4-AP is reversible.

Among the identified Kv channel gene transcripts in human PASMC, Kv1.4 is the only channel that shows a *Shaker*-like rapid (N-type) inactivation because it contains

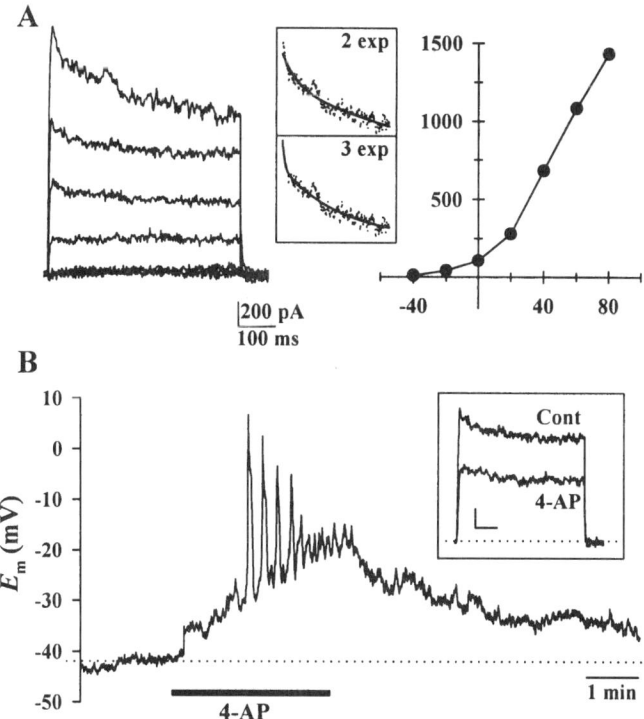

Figure 2. Whole-cell $I_{K(V)}$ and the effects of 4-aminopyridine (4-AP) on $I_{K(V)}$ and E_m in human PASMC. (A) *Left*: a family of currents elicited by depolarizing the cell from a holding potential of -70 mV to a series of command potentials ranging from -40 to $+80$ mV. *Middle*: Two- (2 exp) and three-component (3 exp) exponential fits (smooth lines superimposed on the current records) of inactivation of $I_{K(V)}$ elicited by a 900-ms voltage step to $+80$ mV. *Right*: $I-V$ curve assembled from the data shown in the left panel. (B) E_m measured before, during, and after application of 4-AP (5 mM). *Inset*: Representative current traces elicited by depolarizing the cell to $+60$ mV, recorded before (Cont) and during (4-AP) application of 4-AP. Vertical and horizontal bars denote 200 pA and 100 ms, respectively.

the N-terminal inactivation "ball" (Chandy and Gutman, 1995). The remaining Kv channels, Kv1.2 and Kv1.5, do not exhibit rapid N-type inactivation. However, association of the Kv1.2 and Kv1.5 channels with Kv β-subunits (e.g., Kvβ1.1) confers rapid N-type inactivation to the slowly or non-inactivating Kv1 channels (Rettig *et al.*, 1994; England *et al.*, 1995). Therefore, the rapidly inactivating, A-type $I_{K(V)}$ in human PASMC is probably conferred by the Kv1.4 channel and/or Kv1.2 and Kv1.5 channels associated with β-subunits (e.g., Kvβ1.1). The slowly inactivating and the steady-state $I_{K(V)}$ also may be attributed at least in part to the delayed rectifier Kv channels, Kv1.2 and Kv1.5 (Archer *et al.*, 1998; Yuan *et al.*, 1998c). In addition, $I_{K(Ca)}$ in human PASMC may be endowed by the BK$_{Ca}$ channels, products of the hSlo gene family (McCobb *et al.*, 1995).

2.3. Kv Channel Regulation of E_m and $[Ca^{2+}]_i$

By blocking the Kv channels that are active under resting conditions, 4-AP significantly and reversibly depolarizes PASMC and elicits Ca^{2+}-dependent action potentials (Fig. 2B) that are mediated by the voltage-gated opening of L-type Ca^{2+} channels. Consistent with its inhibitory effect on $I_{K(V)}$ and depolarizing effect on E_m, 4-AP also significantly and reversibly increases $[Ca^{2+}]_i$ in PASMC (Fig. 3). However, inhibition of BK_{Ca} channels by charybdotoxin (ChTX) or inhibition of K_{ATP} channels by glibenclamide (Gli) has a negligible effect on resting E_m and $[Ca^{2+}]_i$ in human PASMC (Fig. 3) (Yuan, 1995). The inability of ChTX and glibenclamide to increase $[Ca^{2+}]_i$ suggests that BK_{Ca} and K_{ATP} channels are inactive in PASMC at resting E_m levels, whereas Kv channels are active and participate in the regulation of resting E_m and $[Ca^{2+}]_i$.

2.4. Single-Channel $I_{K(V)}$ and $I_{K(Ca)}$

In cell-attached membrane patches of human PASMC superfused with Ca^{2+}-free (Fig. 4A) and Ca^{2+}-containing (Fig. 4B) solutions, a small-amplitude K^+ current and a large-amplitude K^+ current are elicited by depolarizing these respective patches to positive potentials. The slope conductance of the large amplitude current, calculated from $I-V$ curves, is between 195 and 250 pS (average slope conductance of 217 ± 8 pS), whereas the conductance of the small amplitude current ranges between 31 and 65 pS (average slope conductance of 50 ± 5 pS). In some cell-attached patches of PASMC superfused with the Ca^{2+}-free solution, K^+ current through several smaller conductance channels (slope conductances between 15–30 pS) also is observed at positive potentials (data not shown). Based on the single channel conductance, the small amplitude K^+ currents represent $I_{K(V)}$ through Kv channels (Fig. 4A), whereas the large amplitude K^+ currents correspond to $I_{K(Ca)}$ through BK_{Ca} channels (Fig. 4B). These channel types appear to be highly expressed in both animal and human PASMC (Albarwani *et al.*, 1994; Archer *et al.*, 1994; Peng *et al.*, 1996; Post *et al.*, 1995; Yuan *et al.*, 1996).

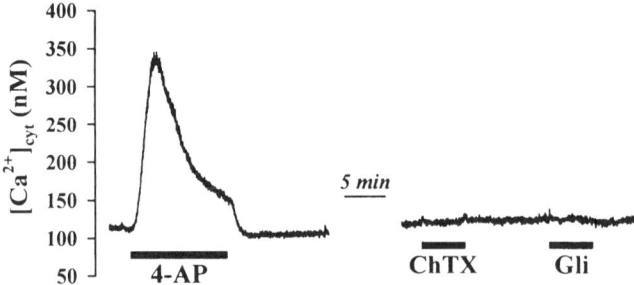

Figure 3. Effects of the K^+ channel blockers 4-aminopyridine (4-AP), charybdotoxin (ChTX), and glibenclamide (Gli) on $[Ca^{2+}]_i$ ($[Ca^{2+}]_{cyt}$) in human PASMC. $[Ca^{2+}]_i$ was measured in the peripheral area of the cells superfused with solutions containing 4-AP (5 mM), ChTX (50 nM), and Gli (10 μM).

A (SPH-PASMC, HP = +80 mV)

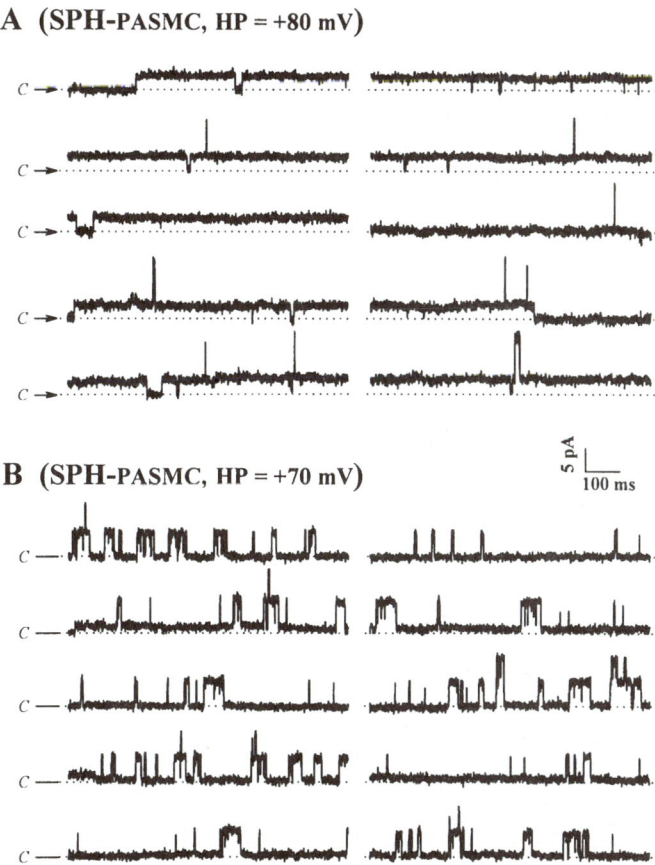

B (SPH-PASMC, HP = +70 mV)

Figure 4. Single-channel $I_{K(V)}$ and $I_{K(Ca)}$ in cell-attached membrane patches of PASMC from a patient with SPH. (A) Representative small-amplitude $I_{K(V)}$ and large-amplitude $I_{K(Ca)}$, elicited at a holding potential (Hp) of $+80$ mV in Ca^{2+}-free bath solution. (B) Representative large-amplitude $I_{K(Ca)}$, elicited at $+70$ mV in Ca^{2+}-containing bath solution. The horizontal arrows and "C" denote the current level when the channels are closed.

3. COMPARISON OF Kv CHANNELS IN PULMONARY ARTERIAL SMOOTH MUSCLE CELLS FROM PATIENTS WITH PRIMARY AND SECONDARY PULMONARY HYPERTENSION

3.1. Reduced Level of α-Subunit mRNA in PASMC from PPH Patients

Findings in PASMC from patch-clamp, fluorescence microscopy, and molecular biology experiments have been compared among patient groups delineated by diagnosis and baseline clinical characteristics: (a) normal subjects without cardiopulmonary diseases (organ donors); (b) patients showing cardiopulmonary diseases (e.g., lung cancer) without pulmonary hypertension (NPH); (c) patients with pulmonary hyper-

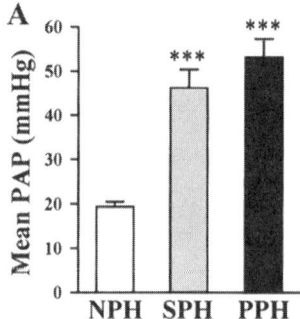 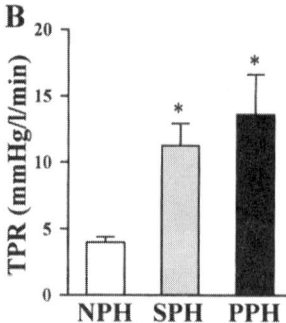

Figure 5. Mean pulmonary arterial pressure (PAP) (A) and total pulmonary resistance (TPR) (B) in patients from whom PASMC were isolated. NPH, Patients with non-pulmonary-hypertension diseases; SPH, patients with secondary pulmonary hypertension; PPH, patients with primary pulmonary hypertension. Data are means $\pm$ SE;*,$P < 0.05$, ***$P < 0.001$ vs NPH.

tension secondary to cardiopulmonary diseases (SPH); and (d) patients diagnosed with primary pulmonary hypertension (PPH) on the basis of the clinical criteria of the National Institutes of Health Register on PPH (Rubin, 1997).

Notably, PPH and SPH patients share similar clinical and hemodynamic characteristics. For example, mean pulmonary arterial pressure (PAP) and total pulmonary resistance (TPR) are comparable in patients with PPH and SPH but are significantly higher than in NPH patients (Fig. 5). Other hemodynamic parameters that are comparable in SPH and PPH patients include cardiac output (4.1 ± 0.3 vs. 4.8 ± 1.0 l/min, $P = 0.45$) and mean right atrial pressure (8.8 ± 1.5 vs. 13.2 ± 1.9 mm Hg, $P = 0.10$). However, although clinical similarities exist, the etiology of PPH and SPH appears to be disparate at least in some respects, including the regulation of gene expression of Kv channels in PASMC from SPH and PPH patients (Yuan et al., 1998a,b).

In this regard, it appears that the mRNA level for the α-subunit of the Kv channel is reduced in PASMC from patients with PPH. Indeed, K$^+$ channel gene transcription and expression are hypothesized to mediate long-term electrophysiological changes in a variety of tissues and cells. Figure 6 shows that defects in gene transcription of Kv1.4 and Kv1.5 channels in human PASMC may contribute to the development of PPH. PASMC of patients with PPH show a significantly lower mRNA level of the Kv1.5 α-subunit compared to PASMC of donors and NPH and SPH patients (Fig. 6A). Furthermore, the mRNA level of the Kv1.4 α-subunit also is significantly lower in PASMC of patients with PPH than in PASMC of donors or patients with NPH and SPH (data not shown). In the same individuals, the mRNA level of the accessory β-subunit, Kvβ1.1, is comparable in PASMC from all patient groups (Fig. 6B). An invariant mRNA of β-actin, used as an internal control, also shows a stable level in PASMC from the different patient groups (Figs. 6A, B). These data indicate that a decreased mRNA expression of Kv1.4 and Kv1.5 channels may be an intrinsic feature of PASMC from PPH patients. As will be discussed, this downregulation of the Kv1.4 and Kv1.5 channel proteins may alter $I_{K(V)}$, E_m and $[Ca^{2+}]_i$ in PASMC from PPH patients to contribute to the pathogenesis of the disease.

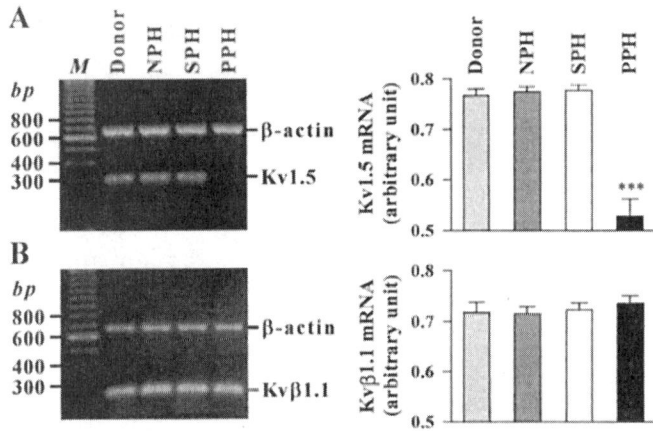

Figure 6. Kvl.5 and Kvβ1.1 mRNA levels in PASMC from normal subjects (Donor) and patients with non-pulmonary-hypertension diseases (NPH), secondary pulmonary hypertension (SPH), and primary pulmonary hypertension (PPH). (A) and (B) *Left*: PCR-amplified products displayed in agarose gels for Kvl.5 [300 base pairs (bp)], Kvβ1.1 (237 bp), and β-actin (661 bp) in PASMC from Donor and patients with NPH, SPH, and PPH. *M*, Molecular markers. *Right*: Summarized data that were normalized to amount of β-actin are expressed as means ± SE.***P < 0.001 vs. SPH (Modified, with permission, from Yuan *et al.*, 1998b.)

3.2. Reduced $I_{K(V)}$ in PASMC from PPH Patients

The whole-cell $I_{K(V)}$ that regulates E_m and $[Ca^{2+}]_i$ is determined, in part, by the number of functional Kv channels expressed in PASMC. Inhibition of Kv1.4 and Kv1.5 channel transcription by antisense oligodeoxynucleotides specifically targeted to the Kv1.4 and Kv1.5 channel genes decreases the level of $I_{K(V)}$ (Feng *et al.*, 1997; Chung *et al.*, 1995). Indeed, in PASMC from PPH patients, the amplitudes of the transient and steady-state $I_{K(V)}$ are both significantly lower than in similar cells from SPH patients (Fig. 7C). Furthermore, $I_{K(V)}$ in PASMC from patients with PPH inactivates more rapidly (Fig. 7A, B). Taken together, these molecular and electrophysiological findings are consistent with the concept that the reduced $I_{K(V)}$ amplitude and accelerated $I_{K(V)}$ inactivation in PASMC from PPH patients may be related to the attenuated expression of Kv1.4 and Kv1.5 α-subunits concurrent with unaltered expression of the regulatory Kvβ1.1-subunit.

Notably, native Kv channels in PASMC may include both α-tetramer channels (composed of the same or different subtypes of Kv1 α-subunits) and α/β heteromultimers. In the latter channel complex, the association of the α-subunits with regulatory β-subunits accelerates Kv channel inactivation (Rettig *et al.*, 1994; England *et al.*, 1995) and promotes open-state block of the Kv channel (De Biasi *et al.*, 1997). Hence, if the number of β-subunits is unchanged in PASMC of PPH patients (Fig. 6B), but the number of α-subunits is decreased (Fig. 6A), whole-cell $I_{K(V)}$ may be reduced because the number of Kv channel pores in the plasma membrane is decreased. Furthermore, an accelerated rate of $I_{(KV)}$ inactivation may result (Fig. 7), because most of the

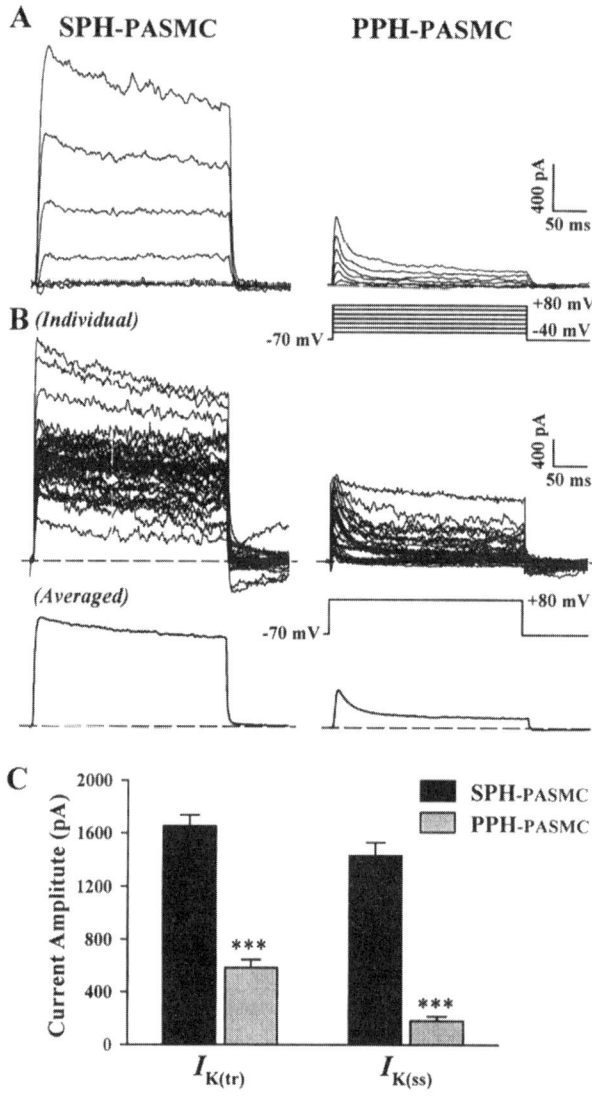

Figure 7. Whole-cell $I_{K(V)}$ in PASMC from SPH and PPH patients. (A) Families of currents in a SPH-PASMC (*left*) and a PPH-PASMC (*right*). (B) Individual (upper panels) and averaged (lower panels) currents from 41 SPH-PASMC (*left*) and 42 PPH-PASMC (*right*). (C) Amplitudes of the currents elicited at a patch potential of $+80$ mV, measured at 10–50 ms for the transient $I_{K(V)}$ ($I_{K(tr)}$) and at 250–290 ms for the steady-state $I_{K(V)}$ ($I_{K(ss)}$). The duration of the test pulse was constant at 300 ms. Data are expressed as means $\pm$ SE. ***$P < 0.001$ vs. SPH-PASMC. (Modified, with permission, from Yuan *et al.*, 1998a.)

remaining functional Kv channels would now represent α/β-heteromultimers that are susceptible to the kinetic influence of the regulatory β-subunit to speed channel inactivation, as diagrammed in Fig. 8.

Interestingly, single-channel findings in isolated membrane patches of PASMC support the concept that Kv channels and the level of $I_{K(V)}$ are reduced in PPH. In the

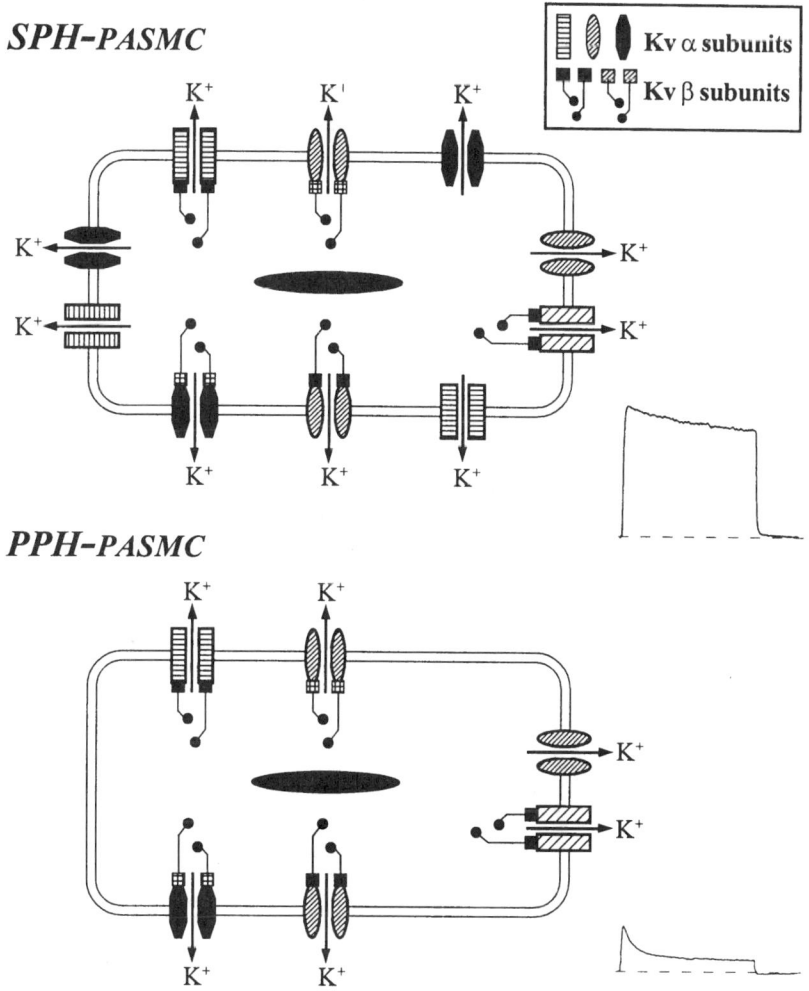

Figure 8. Schematic diagram showing the possible mechanisms involved in the decrease of whole-cell $I_{K(V)}$ in PPH-PASMC. The number of Kv channel α-subunits (e.g., Kv1.4 and Kv1.5) in PPH-PASMC may be lower than in SPH-PASMC, which may account for the reduced amplitude of total $I_{K(V)}$. Because of the reduced number of Kv channel α-subunits and the unchanged number of Kv channel β-subunits, an increased proportion of Kv channel α/β heteromultimers may account for the accelerated inactivation of $I_{K(V)}$.

same cells, the level of $I_{K(Ca)}$ attributed to BK_{Ca} channels is not altered (Fig. 9). Actually, PASMC from patients with PPH showed more high-amplitude events representing BK_{Ca} channels (Fig. 9B). In all PASMC of SPH patients studied at $+90$ mV, $I_{K(V)}$ was observed in 73% of the patches, whereas it was observed in only 12% of patches from PPH patients. In contrast, $I_{K(Ca)}$ was observed in 91% and 100% of patches from SPH and PPH patients, respectively (Fig. 9C). These data suggest that BK_{Ca} channel expression is maintained in PASMC membranes of PPH patients, and hence this channel may represent an available therapeutic target for "potassium channel openers"

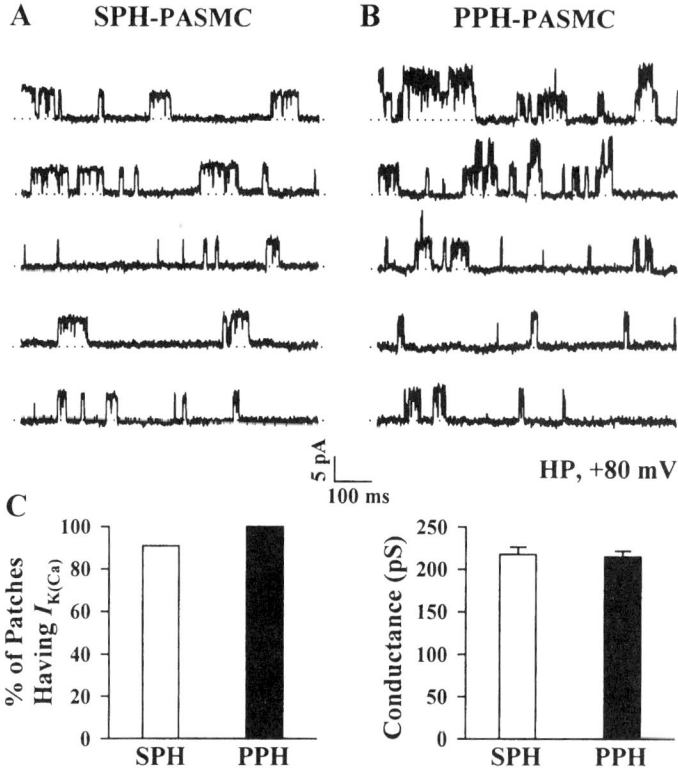

Figure 9. (A) and (B) Comparable large-conductance $I_{K(Ca)}$ in cell-attached membrane patches of PASMC from SPH and PPH patients, respectively. (C) Percentage of patches from SPH- and PPH-PASMC in which $I_{K(Ca)}$ was present (*left*) and averaged slope conductance of $I_{K(Ca)}$ in SPH- and PPH-PASMC (*right*).

designed to interfere with the membrane depolarization in PASMC induced by the downregulation of Kv channels.

3.3. Depolarized E_m and Increased $[Ca^{2+}]_i$ in PASMC from PPH Patients

PASMC from PPH patients show more depolarized E_m (-20 ± 2 mV) compared to E_m levels (-43 ± 1 mV) in similar cells from SPH patients (Yuan *et al.*, 1998a). This finding may be related to the reduced Kv channel expression and $I_{K(V)}$ in the PASMC of PPH patients. There is some evidence that this sustained depolarization may activate L-type Ca^{2+} channels to chronically elevate $[Ca^{2+}]_i$. Indeed, studies on the kinetics of L-type Ca^{2+} current and its relationship to $[Ca^{2+}]_i$ demonstrate that the long-term maintenance of E_m between -35 mV and -20 mV (a voltage range in which Ca^{2+} channel inactivation is incomplete and channel activation threshold is reached) activates L-type Ca^{2+} channels sufficiently to cause a sustained increase in $[Ca^{2+}]_i$ (Fleischmann *et al.*, 1994; Nelson *et al.*, 1990). Indeed, PASMC from PPH patients show significantly higher resting $[Ca^{2+}]_i$ than PASMC from SPH patients (Fig. 10B).

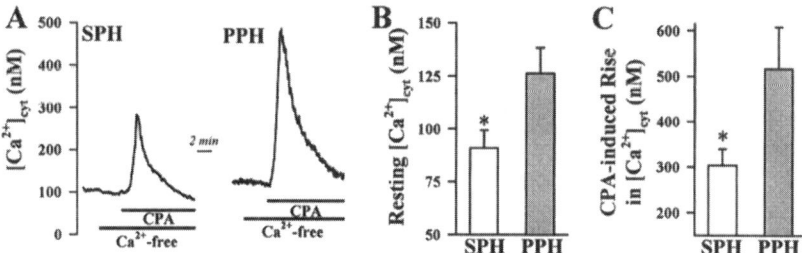

Figure 10. Resting $[Ca^{2+}]_i$ and cyclopiazonic acid (CPA)-induced rise in $[Ca^{2+}]_i$ in PASMC from SPH and PPH patients. (A) Representative record of $[Ca^{2+}]_i$ in a SPH-PASMC (*left*) and a PPH-PASMC (*right*). (B) and (C) Summarized data (means $\pm$ SE) showing resting $[Ca^{2+}]_i$ and the CPA (10 μM)-induced increase in $[Ca^{2+}]_i$ in the absence of extracellular Ca^{2+}, respectively. *$P < 0.05$ vs. PPH.

The ratio of the cytosolic free Ca^{2+} ($[Ca^{2+}]_i$) to stored Ca^{2+} in the sarcoplasmic reticulum ($[Ca^{2+}]_{SR}$) ranges between 1:10,000 and 1:50,000. Thus, very small changes in $[Ca^{2+}]_i$ are associated with relatively large changes in $[Ca^{2+}]_{SR}$ (Blaustein, 1993). Resting $[Ca^{2+}]_i$ in PASMC of patients with PPH is about 25% higher than in PASMC from SPH patients, which may significantly increase $[Ca^{2+}]_{SR}$ owing to Ca^{2+} sequestration mediated by Ca^{2+}-ATPase in the sarcoplasmic reticular membrane (Blaustein, 1993). Indeed, Ca^{2+} mobilization from intracellular stores by cyclopiazonic acid (CPA) is significantly greater in PASMC from patients with PPH than in cells of SPH patients (Fig. 10A, C). Because agonist-induced vasoconstriction and mitogenesis are triggered by an initial release of Ca^{2+} from sarcoplasmic reticulum, the higher $[Ca^{2+}]_{SR}$ in PPH patients may contribute to the augmented agonist-mediated pulmonary vasoconstriction observed in this disease (Brink *et al.*, 1988). Membrane depolarization also facilitates the production of inositol trisphosphate, an important second messenger that links membrane receptors to Ca^{2+}-release channels (IP_3 receptors) in the sarcoplasmic reticulum. Thus, the depolarized E_m in PASMC of patients with PPH also may act to promote agonist- and mitogen-mediated increases in $[Ca^{2+}]_i$. Furthermore, intracellularly stored $[Ca^{2+}]$ is linked to the control of smooth muscle cell growth. The emptying of the IP_3-receptor-sensitive Ca^{2+} pool in the sarcoplasmic reticulum maintains cells in the quiescent state (G_0 phase), whereas refilling of the Ca^{2+} stores resumes normal progression into the cell cycle by stimulating the transitions from the G_0 to the G_1 phase and from the G_1 to the S phase (Magnier-Baubil *et al.*, 1996; Short *et al.*, 1993). Thus, the higher $[Ca^{2+}]_{SR}$ in PASMC of patients with PPH also may contribute to the increased cell proliferation and vascular remodeling observed in this disease.

Thus, in comparison with PASMC from patients with SPH, PASMC from PPH patients show an attenuated mRNA expression of Kv1.4 and Kv1.5 channels, a lower amplitude of $I_{K(V)}$, a more depolarized resting E_m, a higher resting $[Ca^{2+}]_i$, and an increased intracellularly stored $[Ca^{2+}]_{SR}$. These changes attributable to Kv channel alteration apparently are not secondary to chronic pulmonary hypertension, because the SPH and PPH patients from whom PASMC were isolated both showed elevated mean pulmonary arterial pressure and total pulmonary resistance, but Kv channel alterations were observed only in PASMC from patients with PPH. Thus, it appears that the decreased mRNA levels of the Kv1.4 and Kv1.5 α-subunits are a unique feature of PASMC from PPH patients. The inhibition of Kv channel expression and the resultant reduction of Kv channel activity in PASMC from PPH patients are hypothesized to cause membrane depolarization, thereby increasing $[Ca^{2+}]_i$ and triggering pulmonary vasoconstriction, PASMC proliferation, and vascular smooth muscle cell hypertrophy and hyperplasia.

4. CONCLUSION: POSSIBLE ETIOLOGICAL MECHANISMS OF PPH

PPH appears to represent a complex disease characterized by pulmonary hypertension, anomalous vascular constriction and remodeling, and PASMC proliferation. Clearly, multiple parallel mechanisms may contribute to this disease, including impaired endothelium-dependent pulmonary relaxation, an imbalance in the ratio of vasoconstrictor and vasodilator stimuli, and an altered Kv channel subunit expression. This chapter has outlined the evidence that a defect in the mRNA expression of Kv channels in PASMC may result in a reduced number of membrane Kv channels and resultant decrease of whole-cell $I_{K(V)}$. Subsequently, membrane depolarization may activate voltage-gated Ca^{2+} channels in the plasma membrane of PASMC of PPH patients, thereby promoting Ca^{2+} influx to raise $[Ca^{2+}]_i$ and $[Ca^{2+}]_{SR}$. Vasoconstriction and proliferation of PASMC may ensue, leading to medial hypertrophy and a thickening of the pulmonary arteries. The consequent increase in pulmonary vascular resistance is postulated to elevate pulmonary arterial pressure and cause chronic pulmonary hypertension.

ACKNOWLEDGMENTS

The authors would like to express their appreciation to the PPH Cure Foundation, the PPH Research Foundation, and the National Institutes of Health (HL-54043, HL-64945, and HL-02659) for their gracious support of this work. Dr. J. X.-J. Yuan is an Established Investigator of the American Heart Association. We appreciate sincerely J. Wang, M.D., A.M. Aldinger, B.S., M. Juhaszova, Ph.D., J. V. Conte, Jr., M.D., J.E. Seiden, Sc.M., A.E. Bakst, M.D., E. Weiner, M.D., S. P. Gaine, M.D., C. Soriano, M.D., C.L. Bailey, M.D., S.S. McDaniel, B.A., and O. Plaloshyn, M.S., for their great contributions to the study.

REFERENCES

Albarwani, S., Robertson, B. E., Nye, P. C. G., and Kozlowski, R. Z., 1994, Biophysical properties of Ca^{2+}- and Mg^{2+}-ATP-activated K^+ channels in pulmonary arterial smooth muscle cells isolated from the rat, *Pflügers Arch.* **428**:446–454.

Allbritton, N. L., Ozncea, E., Kuhn, M. A., and Meyer, T., 1994, Source of nuclear calcium signals, *Proc. Natl. Acad. Sci. U.S.A.* **91**:12458–12462.

Archer, S. L., Huang, J. M. C., Hampl, V., Nelson, D. P., Schultz, P. J., and Weir, E. K., 1994, Nitric oxide and cGMP cause vasorelaxation by activation of a charybdotoxin-sensitive K channel by cGMP-dependent protein kinase, *Proc. Natl. Acad. Sci. U.S.A.* **91**:7583–7587.

Archer, S. L., Souil, E., Dinh-Xuan, A. T., Schremmer, B., Mercier, J. C., El Yaagoubi, A., Nguyen-Huu, L., Reeve, H. L., and Hampl, V., 1998, Molecular identification of the role of voltage-gated K^+ channels, Kv1.5 and Kv2.1, in hypoxic pulmonary vasoconstriction and control of resting membrane potential in rat pulmonary artery myocytes, *J. Clin. Invest.* **101**:2319–2330.

Berridge, M. J., 1993, Inositol triphosphate and calcium signaling, *Nature* **361**:315–325.

Berridge, M. J., 1995, Calcium signalling and cell proliferation, *BioEssays* **17**(6):491–500.

Blaustein, M. P., 1993, Physiological effects of endogenous ouabain: Control of intracellular Ca^{2+} stores and cell responsiveness, *Am. J. Physiol.* **264**:C1367–C1387.

Brink, C., Cerrina, C., Labat, C., Beley, J., and Benveniste, J., 1988, The effects of contractile agonists on isolated pulmonary arterial and venous muscle preparations derived from patients with primary pulmonary hypertension, *Am. Rev. Respir. Dis.* **137** (Part 2):A106.

Chandy, K. G., and G. A. Gutman., 1995, Voltage-gated K^+ channel genes, in: *Handbook of Receptors and Channels: Ligand- and Voltage-Gated Ion Channels* (R. A. North, ed.), CRC Press, Boca Raton, Florida, pp. 1–71.

Christman, B. W., McPherson, C. D., Newman, J. H., King, G. A., Bernard, G. R., Groves, B. M., and Loyd, J. E., 1992, An imbalance between the excretion of thromboxane and prostacyclin metabolites in pulmonary hypertension, *N. Engl. J. Med.* **327**:70–75.

Chung, S., Saal, D. B., and Kaczmarek, L. K., 1995, Elimination of potassium channel expression by antisense oligonucleotides in a pituitary cell line, *Proc. Natl. Acad. Sci. U.S.A.* **92**:5955–5959.

De Biasi, M., Wang, Z., Accili, E., Wible, B., and Fedida, D., 1997, Open channel block of human heart hKv1.5 by the β-subunit hKvβ1.2, *Am. J. Physiol.* **272**:H2932–H2941.

Dinh-Xuan, A. T., Higenbottam, T. W., Clelland, C. A., Pepke-Zaba, J., Cremona, G., Butt, A. Y., Large, S. R., Wells, F. C., and Wallwork, J., 1991, Impairment of endothelium-dependent pulmonary-artery relaxation in chronic obstructive lung disease, *N. Engl. J. Med.* **324**:1539–1547.

England, S. K., Uebele, V. N., Kodali, J., Bennett, R. P., and Tamkun, M. M., 1995, A novel K^+ channel β-subunit (hKvβ1.3) is produced via alternative mRNA splicing, *J. Biol. Chem.* **270**:28531–28534.

Evans, A. M., Osipenko, O. N., and Gurney, A. M., 1996, Properties of a novel K^+ current that is active at resting potential in rabbit pulmonary artery smooth muscle cells, *J. Physiol.* **496**:407–420.

Feng, J., Wible, B., Li, G.-R., Wang, Z., and Nattel, S., 1997, The antisense oligodeoxynucleotides directed against Kv1.5 mRNA specifically inhibit ultrarapid delayed rectifier K^+ current in cultured adult human atrial myocytes, *Circ. Res.* **80**:572–579.

Fleischmann, B. K., Murray, R. K., and Kotlikoff, M. I., 1994, Voltage window for sustained elevation of cytosolic calcium in smooth muscle cells, *Proc. Natl. Acad. Sci. U.S.A.* **91**:11914–11918.

Ganitkevich, V. Y., and Isenberg, G., 1993, Membrane potential modulates inositol 1,4,5-triphosphate-mediated Ca^{2+} transients in guinea-pig coronary myocytes, *J. Physiol.* **470**:35–44.

Giaid, A., and Saleh, D., 1995, Reduced expression of endothelial nitric oxide synthase in the lungs of patients with pulmonary hypertension, *N. Engl. J. Med.* **333**:214–221.

Giaid, A., Yanagisawa, M., Langleben, D., Michel, R. P., Levy, R., Shennib, H., Kimura, S., Masaki, T., Duguid, W. P., and Stewart D. J., 1993, Expression of endothelium-1 in the lungs of patients with pulmonary hypertension, *N. Engl. J. Med.* **328**:1732–1739.

Hishikawa, K., Nakaki, T., Marumo, T., Hayashi, M., Suzuki, H., Kata, R., and Saruta, T., 1994, Pressure promotes DNA synthesis in rat cultured vascular smooth muscle cells, *J. Clin. Invest.* **93**:1975–1980.

Isom, L. L., DeJongh, K. S., and Catterall, W. A., 1994, Auxiliary subunits of voltage-gated ion channels, *Neuron* **12**:1183–1194.

Kolpakov, V., Rekhtar, M.D., Gordon, D., Wang, W.H., and Kulik, T.J., 1995, Effect of mechanical forces on growth and matrix protein synthesis in the in vitro pulmonary artery: Analysis of the role of individual cell types, *Circ. Res.* **77**:823–831.

Kukuljan, M., Rojas, E., Catt, K. J., and Stojilkovic, S. S., 1994, Membrane potential regulates inositol 1,4,5-triphosphate-controlled cytoplasmic Ca^{2+} oscillations in pituitary gonadotrophs, *J. Biol. Chem.* **269**:4860–4865.

Magnier-Gaubil, C., Herbert, J. M., Quarck, R, Papp, B., Corvazier, E., Wuytack, F., Levy-Toledano, S., and Enouf, J., 1996, Smooth muscle cell cycle and proliferation: Relationship between calcium influx and sarco-endoplasmic reticulum Ca^{2+} ATPase regulation, *J. Biol. Chem.* **271**:27788–27794.

McCobb, D. P., Fowler, N. L., Featherstone, T., Lingle, C. J., Saito, M., Krause, J. E., and Salkoff, L., 1995, A human calcium-activated potassium channel gene expressed in vascular smooth muscle, *Am. J. Physiol.* **269**:H767–H777.

Nelson, M. T., and Quayle, J. M., 1995, Physiological roles and properties of potassium channels in arterial smooth muscle, *Am. J. Physiol.* **268**:C799–C822.

Nelson, M. T., Patlak, J. B., Worley, J. F., and Standen, N. B., 1990, Calcium channels, potassium channels, and voltage-dependence of arterial smooth muscle tone, *Am. J. Physiol.* **259**:C3–C18.

Palevsky, H. I., Schloo, B. L., Pietra, G. G., Weber, K. T., Janicki, J. S., and Fishman, A. P., 1989, Primary pulmonary hypertension, vascular structure, morphometry, and responsiveness to vasodilator agents, *Circulation* **80**:1207–1221.

Peng, W., Karwande, S. V., Hoidal, J. R., and Farrukh, I. S., 1996, Potassium currents in cultered human pulmonary arterial smooth muscle cells, *J. Appl. Physiol.* **80**:1187–1196.

Pietra, G. G., 1997, The pathology of primary pulmonary hypertension, in: *Primary Pulmonary Hypertension* (L. J. Rubin and S. Rich, eds.), Marcel Dekker, New York, pp. 19–61.

Post, J. M., Gelband, C. H., and Hume, J. R., 1995, $[Ca^{2+}]_i$ inhibition of K^+ channels in canine pulmonary artery: Novel mechanisms for hypoxia-induced membrane depolarization, *Circ. Res.* **77**:131–139.

Rettig, J., Heinemann, S. H., Wunder, F., Lorra, C., Parcej, D. N., Dolly, J. O., and Pongs, O., 1994, Inactivation properties of voltage-gated K^+ channels altered by presence of β-subunit, *Nature* **369**:289–294.

Rich, S., Kaufmann, E., and Levy, P. S., 1992, The effect of high doses of calcium-channel blockers on survival in primary pulmonary hypertension, *N. Engl. J. Med.* **327**:76–81.

Rubin, L. J., 1997, Primary pulmonary hypertension, *N. Engl. J. Med.* **336**:111–117.

Short, A. D., Bian, J., Ghosh, T. K., Waldron, R. T., Rybak, S. L., and Gill, D. L., 1993, Intracellular Ca^{2+} pool content is linked to control of cell growth, *Proc. Natl. Acad. Sci. U.S.A.* **90**:4986–4990.

Steudel, W., Ichinose, F., Huang, P. L., Hurford, W. E., Jones, R. C., Bevan, J. A., Fishman, M. C., and Zapol, W. M., 1997, Pulmonary vasoconstriction and hypertension in mice with targeted disruption of the endothelial nitric oxide synthase (NOS 3) gene, *Circ. Res.* **81**:34–41.

Tanaka, Y., Schuster, D. P., Davis, E. C., Patterson, G. A., and Botney, M. D., 1996, The role of vascular injury and hemodynamics in rat pulmonary artery remodeling, *J. Clin. Invest.* **98**:434–442.

Voelkel, N. F., Tuder, R. M., and Weir, E. K., 1997, Pathophysiology of primary pulmonary hypertension: From physiology to molecular mechanisms, in: *Primary Pulmonary Hypertension* (L. J. Rubin and S. Rich, eds.), Marcel Dekker, New York, pp. 83–129.

Wagenvoort, C. A., 1960, Vasoconstriction and media hypertrophy in pulmonary hypertension, *Circulation* **22**:535–546.

Wood, P., 1958, Pulmonary hypertension with special reference to the vasoconstrictive factor, *Br. Heart J.* **20**:557–570.

Yu, S. P., Yeh, C.-H., Sensi, S. L., Gwag, B. J., Canzoniero, L. M. T., Farhangrazi, Z. S., Ying, H. S., Tian, M., Dugan, L. L., and Choi, D. W., 1997, Mediation of neuronal apoptosis by enhancement of outward potassium current, *Science* **278**:114–117.

Yuan, X.-J., 1995, Voltage-gated K^+ currents regulate resting membrane potential and $(Ca^{2+})_i$ in pulmonary arterial myocytes, *Circ. Res.* **77**:370–378.

Yuan, X.-J., 1998, Mechanisms of hypoxic pulmonary vasoconstriction: The role of oxygen-sensitive voltage-gated potassium channels, in: *Oxygen Regulation and Ion Channels and Gene Expression* (J. Lopez-Barneo and E. K. Weir, eds.), Futura Publishing Company, Armonk, New York, pp. 207–233.

Yuan, X.-J., Tod, M. L., Rubin, L. J., and Blaustein, M. P., 1996, NO hyperpolarizes pulmonary artery smooth muscle cells and decreases the intracellular Ca^{2+} concentration by activating voltage-gated K^+ channels, *Proc. Natl. Acad. Sci. U.S.A.* **93**:10489–10494.

Yuan, J. X.-J., Aldinger, A. M., Huhaszova, M., Wang, J., Conte, J. V., Gaine, S. P., Orens, J. B., and Rubin, L. J., 1998a, Dysfunctional voltage-gated K^+ channels in pulmonary artery smooth muscle cells of patients with primary pulmonary hypertension, *Circulation* **98**:1400–1406.

Yuan, X.-J., Wang, J., Juhaszova, M., Gaine, S. A., and Rubin, L. J., 1998b, Attenuated K^+ channel gene transcription in primary pulmonary hypertension, *Lancet* **351**:726–727.

Yuan, X.-J., Wang, J., Juhaszova, M., Golovina, V. A., and Rubin, L. J., 1998c, Molecular basis and function of voltage-gated K^+ channels in pulmonary arterial smooth muscle cells, *Am. J. Physiol.* **274**:L621–L635.

Chapter 41

Anorectic Drugs and the Vasculature

Evangelos D. Michelakis and E. Kenneth Weir

1. INTRODUCTION

The anorectics are drugs used to suppress appetite in patients treated for obesity. More than 3 million prescriptions were filled in 1997 alone for dexfenfluramine (dex), recently the most commonly prescribed anorectic drug (Voelkel *et al.*, 1997). Their use has been associated with life-threatening diseases ranging from primary pulmonary hypertension (PPH) (Abenhaim *et al.*, 1996), strokes (Schwitter *et al.*, 1992), and heart attacks (Bailie, 1991; Evrard *et al.*, 1990) to severe cardiac valvular disease (Connolly *et al.*, 1997; Khan *et al.*, 1988). Although dex is now withdrawn from the market, the interest of the scientific community in this drug remains strong for two main reasons. First, there are now new anorectic drugs, already approved for the treatment of obesity, such as sibutramine, with very similar structures and possibly similar mechanisms of action to dex (Heal *et al.*, 1998; McNeely and Goa, 1998; Van Gaal *et al.*, 1998), and thus physicians and scientists need to be alert to anorectic-related vascular complications. Second, dex might prove to be an excellent tool for establishing pulmonary and even systemic models of experimental hypertension. One of the main reasons that PPH remains a fatal disease with no cure is the lack of animal models. As will be discussed in this chapter, the complications caused by anorectic drugs are a "paradise" for the vascular biologist, because they involve the three most important cells involved in vascular disease: the vascular smooth muscle cell (SMC), the platelet, and the endothelium (Rubin, 1997). This chapter will summarize the evidence that suggests that K^+ channel inhibition plays a critical role in the mechanism of action and the complications related to anorectic drug use.

Evangelos D. Michelakis ● Department of Medicine, University of Alberta, Edmonton, Alberta, Canada T6G 2B7. *E. Kenneth Weir* ● Department of Medicine, VA Medical Center, Minneapolis, Minnesota 55417.

Potassium Channels in Cardiovascular Biology, edited by Archer and Rusch. Kluwer Academic/Plenum Publishers, New York, 2001.

2. ANORECTIC DRUGS AND THE PULMONARY VASCULATURE

Between 1968 and 1973, physicians in central Europe noticed an impressive 10-fold increase in the incidence of PPH. An association between PPH and the anorectic drug aminorex, first released in 1965, was quickly established (Kay *et al.*, 1971). One to two percent of patients exposed to aminorex developed PPH (odds ratio of 52:1) (Mlczoch, 1984). Despite extensive efforts, the mechanism of this drug-induced pulmonary hypertension was never ascertained.

In April 1996, the U.S. Food and Drug Administration (FDA) approved the use of dex for the treatment of obesity. An association between PPH and fenfluramine (dex's parent racemic compound) was suspected since 1981, mostly on the basis of single case reports (Atanassoff *et al.*, 1992; Douglas *et al.*, 1981; Gaul *et al.*, 1982; McMurray *et al.*, 1986; Pouwels *et al.*, 1990). In 1993, Brenot *et al.* provided the first major retrospective analysis which linked PPH with fenfluramine use, but the FDA approved the use of the fenfluramines for the treatment of obesity in 1996. Interestingly, several months after the approval, the results of the International PPH Study (IPPHS) were published (Abenhaim *et al.*, 1996). This was a case-controlled study that showed a statistically significant difference in the incidence of PPH among patients taking anorexigens, the most common of which were fenfluramine and dex. It showed an odds ratio of 23:1 for PPH associated with anorectic drug use for more than 3 months, with strong implications of a dose–response effect for longer periods of use. At the same time, it showed a disappointingly low efficacy in weight loss for short-term (less than 3 months) use (Abenhaim *et al.*, 1996).

The pharmacological mechanisms of action of aminorex and the fenfluramines are thought to be quite different. Aminorex is a catechol derivative, and it exerts its anorectic effect through the release of norepinephrine (Mlczoch, 1984). The fenfluramines work by concurrently increasing the release and inhibiting the reuptake of serotonin in the central nervous system synapses (McTavish and Heel, 1992). It is interesting that two drugs with different mechanisms of action led to the development of indistinguishable forms of PPH. The histologic picture of anorexigen-induced pulmonary hypertension is identical to that of the plexogenic form of PPH (Rubin, 1997; Voelkel *et al.*, 1997). Furthermore, there are long-used drugs with sympathomimetic activity (such as amphetamines) or with somewhat similar effects on serotonin metabolism to those of dex [such as fluoxetine (Stark *et al.*, 1985)] that are not associated with vascular complications, at least to the same degree. Could the two drugs, aminorex and dex, share a common mechanism of action that could be involved in the pathogenesis of PPH? If so, this mechanism(s) should, in theory, account for the three main pathogenic processes of PPH: vasoconstriction, thrombosis, and cell proliferation, all of which occur in the small pulmonary arteries (Rubin, 1997). In this regard, an association between abnormalities in K^+ channels in pulmonary arterial smooth muscle cells and PPH has recently been shown. Yuan *et al.* (1998a) showed that pulmonary arterial smooth muscle cells from patients with PPH have significantly decreased levels of expressed voltage-gated K^+ (Kv) channels, which are associated with a reduced outward K^+ current compared to controls, as measured by the patch-clamp technique. In the same cells, the resting membrane potential (E_m) was depolarized. Below, we review the data linking anorectic drug actions with K^+ channel abnormalities and will discuss how K^+ channel inhibition is involved and might at least partially explain the vasoconstrictive, thrombotic, and proliferative aspects of PPH induced by anorectic therapies.

3. ANORECTIC DRUGS AND THE SYSTEMIC VASCULATURE

Looking back in the literature, a number of case reports initially suggested a link between anorectic drugs and peripheral vascular complications such as stroke (Schwitter *et al.*, 1992), myocardial ischemic syndromes (Bailie, 1991; Evrard *et al.*, 1990), mesenteric ischemia (Schembre and Boynton, 1997), and hypertension (Mabadeje, 1974). Even very recently, a case report in the *New England Journal of Medicine* by Marinella and Berrettoni (1997) described a patient who developed digital ischemia shortly after taking dex. An extensive medical workup suggested vasospasm. The digital ischemia improved when dex was discontinued. However, when the patient started taking the drug again, against medical advice, he quickly developed gangrene, resulting in amputation (Marinella and Berrettoni, 1997). Despite clinical evidence, such as this case, suggesting that the anorectic drugs may be linked to vascular disease, the effects of these drugs in the systemic vasculature have not been comprehensively studied. Because obesity is so prevalent (VanItallie, 1996) and these drugs are so commonly prescribed, even a weak association between anorectic drugs and the development of diseases like hypertension might prove to have a significant public health impact. The fact, however, that obesity itself can cause hypertension makes the study of such an association very difficult. The decrease in blood pressure associated with weight loss might mask any anorectic-induced hypertension. In the study by Abenhaim *et al.* (1996), the patients taking anorectic drugs were somewhat more likely to have systemic hypertension when compared with controls (adjusted odds ratio of 2:1; 95% confidence interval, 0.7–6). This is a provocative finding but did not reach statistical significance. Furthermore, the patients using anorectic drugs had lost approximately 10% of their weight, which should be associated with blood pressure decrease rather than a tendency for hypertension. Below, we present data showing that dex can be a systemic vasoconstrictor in the rat and that these effects may be, at least in part, mediated by the inhibition of Kv channels in the vascular smooth muscle cells of systemic arteries.

4. ANORECTIC DRUGS AND THE HEART VALVES

Connolly *et al.* (1997) first reported that up to 30% of anorectic drug users had echocardiographic evidence of heart valve damage, often requiring surgery. This was quickly confirmed by others (Khan *et al.*, 1998). Immediately following publication of the report by Connolly *et al.*, the fenfluramines were withdrawn from the market. The macroscopic and histologic appearance of these valvular lesions resembled the right-sided valvular lesions seen in patients diagnosed with the serotonin-secreting carcinoid tumor. In this chapter, we will present evidence suggesting that dex causes serotonin release from human platelets, possibly through a mechanism involving Kv channel inhibition.

5. ROLE OF K⁺ CHANNELS IN VASCULAR TONE

In vascular smooth muscle cells, the membrane potential is primarily controlled by K^+ channels (Nelson and Quayle, 1995). Activation of K^+ channels results in membrane hyperpolarization, whereas membrane depolarization occurs when K^+ channel activity is reduced. Depolarization from resting E_m levels (-40 to -60 mV) in vascular smooth muscle cells to more positive potentials activates the voltage-gated Ca^{2+} channels, leading to influx of extracellular Ca^{2+}. Increased intracellular Ca^{2+} $[Ca^{2+}]_i$) is involved in important cellular functions such as smooth muscle cell contraction and proliferation, as well as platelet activation. Thus, E_m is a major determinant of $[Ca^{2+}]_i$, although other factors such as agonist-induced increases in $[Ca^{2+}]_i$ utilizing intracellular Ca^{2+} release sources (e.g., the sarcoplasmic reticulum) also play a significant role in regulating the availability of Ca^{2+} for these processes.

In general, K^+ channel pharmacological blockers are vasoconstrictors, whereas K^+ channel "openers" are vasodilators (Nelson and Quayle, 1995). However, the expression profile of K^+ channel gene families and channel subtypes is extremely diverse in smooth muscle membranes and also, although not extensively studied, in the plasma membranes of platelets (de Silva et al., 1997; Mahaut-Smith, 1995; Mahaut-Smith et al., 1990; Maruyama, 1987). One of the recent advances in vascular biology has been the definition of the expression pattern of K^+ channels not only among different vascular beds, but also along the same arterial tree (Archer et al., 1996; Michelakis et al., 1995). This has important implications for understanding the activity and side effects of drugs that act through modulation of K^+ channel gating or expression levels (Archer et al., 1996; Michelakis et al., 1995).

6. ANORECTIC DRUGS INHIBIT K⁺ CURRENT IN VASCULAR SMOOTH MUSCLE

Aminorex, fenfluramine, and dex inhibit K^+ (Kv) current in vascular smooth muscle cells from rat pulmonary resistance arteries (Weir et al., 1996). Figure 1A shows cumulative data from conventional whole-cell patch-clamp recordings on freshly isolated smooth muscle cells from rat resistance (5th division) pulmonary arteries. At micromolar concentrations, dex reversibly and significantly inhibits K^+ current. Figure 2A shows similar effects of dex on freshly isolated smooth muscle cells from rat renal arteries (3rd to 4th division) (Michelakis et al., 1998b). Dex has also been shown to inhibit whole-cell K^+ current in freshly isolated smooth muscle cells from rabbit and human ductus arteriosus (Reeve et al., 1997).

In the pulmonary and renal artery smooth muscle cells, there is no significant difference in the amount of K^+ current inhibition between conventional and amphotericin-perforated (cytoplasm-intact) patch-clamp recordings. These findings suggest that an intracellular second-messenger system might not be necessary for dex to inhibit K^+ current.

Because the magnitude of dex-induced inhibition of K^+ current is very similar to that of inhibition by 4-aminopyridine (4-AP), which is a specific Kv channel inhibitor at low concentrations ($\leqslant 1$ mM) in pulmonary and renal arterial smooth muscle cells, it is likely that the anorectic agents inhibit one or more Kv channels. Indeed, Patel et

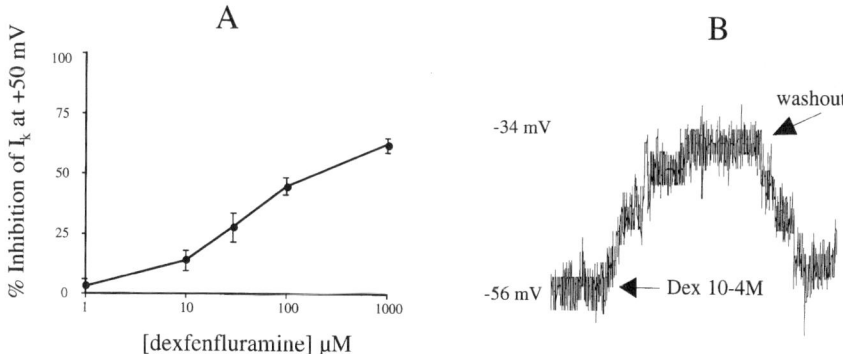

Figure 1. (A) Dex inhibits Kv current and depolarizes the smooth muscle cells of rat resistance pulmonary artery. A logarithmic dose response is shown. Inhibition of K$^+$ current is evident at the relatively low dose of 10 μM. Values are means $\pm$ SEM ($n = 10$). (B) Dex (10^{-4} M) significantly depolarizes the smooth muscle cells of rat pulmonary artery.

al. (1997) recently showed that dex inhibits a particular Kv channel α-subunit (Kv2.1) expressed in *Xenopus* oocytes.

Figures 1B and 2B show whole-cell recordings of E_m in the current-clamp mode. Dex causes significant depolarizations in both pulmonary and renal arterial smooth muscle cells (Michelakis *et al.*, 1998b; Weir *et al.*, 1996). This finding suggests that dex

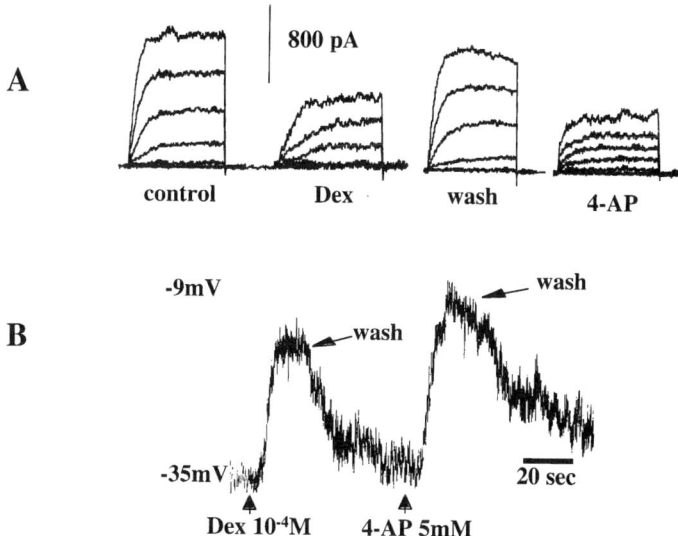

Figure 2. Dex inhibits K$^+$ current and depolarizes isolated rat renal artery smooth muscle cells. (A) Representative tracings are shown from whole-cell patch-clamp recordings. Current was elicited by 20-mV voltage steps from a holding potential of -70 mV to $+70$ mV. Note the similar degrees of K$^+$ current inhibition K$^+$ and depolarization (*top*) caused by dex (0.1 mM) and 4-aminopyridine (4-AP; 5 mM). This suggests that dex and 4-AP might share a common mechanism of action, i.e., inhibition of Kv channels. (B) Dex (10^{-4} M) significantly depolarizes smooth muscle cells of rat renal artery (3rd division renal artery) in a manner very similar to the 4-AP-induced depolarization.

Table 1

Direct Evidence of Kv Channel Inhibition by Dex Fenfluramine from Various Cell Types

Cell type	Technique	Reference
Rat pulmonary artery VSM[a]	Whole-cell patch clamping	Weir et al., 1996
Rat renal, carotid, and basilar artery VSM	Whole-cell patch clamping	Michelakis et al., 1998b
Human ductus arteriosus VSM	Whole-cell patch clamping	Reeve et al., 1997
Rat myocardial cells	Whole-cell patch clamping	Hu et al., 1998
Rat taste cells	Whole-cell patch clamping	Hu et al., 1998
Rat megakaryocytes	Whole-cell patch clamping	Weir et al., 1998
Ooyctes with expressed Kv2.1	Single-cell patch clamping	Patel et al., 1997

[a] VSM, Vascular smooth muscle.

may inhibit one or more Kv channels that contribute importantly in the control of E_m in these cells. Interestingly, Kv2.1 has been identified as a major determinant in the control of E_m in the pulmonary vasculature and in the initiation of hypoxic pulmonary vasoconstriction (HPV) (Archer et al., 1998b; Yuan et al., 1998b). In summary, aminorex, fenfluramine, and dex inhibit a 4-AP-sensitive K^+ current in vascular smooth muscle cells of various arteries and in several species (Table 1).

Dex-induced depolarization would be expected to open voltage-gated Ca^{2+} channels, increase Ca^{2+} influx, and trigger a rise in $[Ca^{2+}]_i$, resulting in vasoconstriction. Indeed, we recently reported preliminary experiments showing a dex-induced rise in $[Ca^{2+}]_i$ in human cultured pulmonary artery smooth muscle cells, as detected by fura-2 fluorometry (Reeve et al., 1998). Although most of the increase in $[Ca^{2+}]_i$ resulted from Ca^{2+} influx through voltage-gated Ca^{2+} channels, a significant amount of Ca^{2+} also was released from intracellular sources (Reeve et al., 1998). The latter event might be due to "Ca^{2+}-induced Ca^{2+} release" following the initial Ca^{2+} influx, or, alternatively, dex may release Ca^{2+} from intracellular stores through a second-messenger system, independent of its effects on the membrane Kv channels.

Aminorex and the fenfluramines are also potent pulmonary vasoconstrictors in the perfused lung (Weir et al., 1996) and in the pulmonary circulation in vivo (Naeije et al., 1996). As shown in Fig. 3, when given as a bolus, these drugs acutely increase the pulmonary vascular resistance in the perfused rat lung model (Weir et al., 1996) and also increase systemic vascular resistance in the anesthetized rat (Michelakis et al., 1998b). The steady-state serum levels of these drugs following bolus administration are similar to the therapeutic levels of fenfluramine and dex in human subjects treated for obesity. Furthermore, Naeije et al. (1996) showed that oral administration of dex for 20 days, which resulted in plasma levels similar to those recommended for patients, increased pulmonary vascular resistance in normal dogs (Fig. 4).

Interestingly, very soon after the fenfluramines were withdrawn from the market, two other anorectic agents, fluoxetine (Prozac) and phentermine, rapidly became very popular for the treatment of obesity ("ProPhen") (Anchors, 1997). Fluoxetine, one of the most commonly prescribed drugs on the market, is thought to exert its mood-stabilizing and anorectic effects through inhibition of serotonin reuptake in the central nervous system (Stark et al., 1985). However, unlike the fenfluramines, it does not concurrently enhance serotonin release. Notably, in neurons, serotonin release is

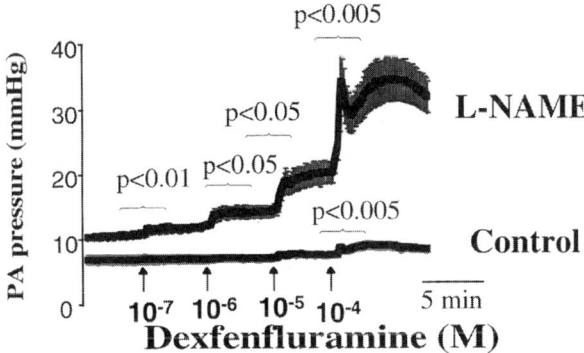

Figure 3. Dex increases pulmonary vascular resistance, especially when nitric oxide synthase is inhibited by N-nitro-L-arginine methyl ester (L-NAME; 5×10^{-5} M). In the isolated, perfused rat lung, dex causes concentration dependent increases in pulmonary artery (PA) pressure, which reflect increases in the pulmonary vascular resistance, because the flow in the perfused lung remains unchanged. This effect is markedly enhanced by L-NAME. Values are mean $\leqslant \pm$SEM ($n = 5$ for control and L-NAME groups). (Reprinted, with permission, from Weir *et al.*, 1996).

dependent on Ca^{2+} influx through voltage-gated Ca^{2+} channels. This release of neuro transmitter is blocked by low concentrations of 4-AP (Cinquanta *et al.*, 1997; Gobbi *et al.*, 1993, 1995; Miyamoto *et al.*, 1990; Pei *et al.*, 1995), suggesting that serotonin release, in contrast to reuptake of this amine, is dependent on the level of E_m in serotonin-secreting neurons. When given intravenously, fluoxetine does not increase pulmonary

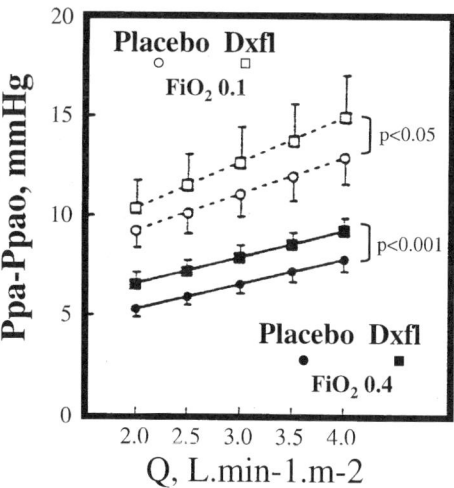

Figure 4. Dex increases pulmonary vascular resistance *in vivo*. Composite plots of mean pulmonary artery pressure (Ppa) minus pulmonary artery occluded pressure (Ppao) at several levels of cardiac output (Q) after 20 days of oral intake of placebo in 12 dogs and of dexfenfluramine (Dxfl; 1.5 mg/kg) in 12 other dogs in hyperoxia (FiO$_2$ 0.4) and in hypoxia (FiO$_2$ 0.1). (Ppa–Ppao)/Q plots shift to higher pressures after dexfenfluramine. (Reprinted, with permission, from Naeije *et al.*, 1996. Copyright American Lung Association.)

vascular resistance in the perfused lung model (Reeve *et al.*, 1999), and although it inhibits K^+ current in pulmonary artery smooth muscle cells, it does not cause depolarization (Reeve *et al.*, 1999). This suggests that fluoxetine inhibits one or more Kv channels, but unlike the channels inhibited by aminorex and the fenfluramines, these channels may not be important in regulating the level of E_m in pulmonary artery smooth muscle cells. For the same reason, fluoxetine may not depolarize neurons and, therefore, not cause serotonin *release* in the central nervous system.

Abnormalities in the levels of smooth muscle cell K^+ channel current or channel protein expression have been reported in both systemic and pulmonary hypertension (Liu *et al.*, 1988; Martens and Gelband, 1998; Rusch and Runnells, 1994; Yuan *et al.*, 1998a). In the pulmonary circulation, Yuan *et al* (1998a) recently showed that pulmonary artery smooth muscle cells from patients with PPH show a decreased 4-AP-sensitive K^+ current and are depolarized when compared to cells from control subjects or patients with secondary hypertension. In addition, the levels of expressed Kv1.5 α-subunit, but not those of a K^+ channel β-subunit, are significantly decreased in these patients. Thus, it is possible that patients predisposed to develop anorectic-induced complications might have dysfunctional or underexpressed K^+ channels in the vascular muscle membranes of the pulmonary circulation.

7. ANORECTIC DRUGS AND THE ENDOTHELIUM

Inhibition of nitric oxide synthase (NOS) significantly potentiates the pressor effects of anorectic drugs. In isolated rat lungs perfused with *N*-nitro-L-arginine methyl ester (L-NAME) to inhibit the synthesis of nitric oxide by endothelial cells, the pressor responses to dex are significantly stronger and evident at lower concentrations of dex ($\leqslant 10^{-6}$ M) that are similar to the plasma levels in patients treated with dex (Weir *et al.*, 1996) (Fig. 3). Oral administration of L-NAME to rats for 3 days also enhances the systemic vascular resistance and systemic blood pressure responses to dex (Michelakis *et al.*, 1998b). Hence, it is possible that endothelial dysfunction may nonspecifically potentiate the pressor responses to a number of unrelated vasoconstrictors. However, in our perfused lung and anesthetized rat model, NOS inhibition significantly potentiates the constrictor effects of dex but not those of angiotensin II.

Although there are no data showing an increase in the plasma levels of serotonin in dex-treated patients, it is possible that serotonin levels are locally and transiently increased as the platelets, which are the major carriers of serotonin in the blood, interact with the endothelium. The vascular effects of serotonin are mediated mainly by the endothelial cell S_1 receptor and the smooth muscle cell S_2 receptor (Frishman *et al.*, 1995). Whereas the S_2 receptor mediates smooth muscle cell contraction and vasoconstriction, the S_1 receptor mediates vasodilation, possibly through the release of nitric oxide and prostaglandin I_2 (Fig. 5) (Frishman *et al.*, 1995). The net effect of serotonin in the vasculature reflects the balance between the two receptors, as shown in Fig. 5. Interestingly, it has been shown that in pigs fed with L-NAME there is a decrease in the S_1/S_2 ratio of expressed receptors that might contribute to the more pronounced vasoconstrictive effects of serotonin in the coronary arteries of these pigs (Kadokami *et al.*, 1996). Therefore, in patients with endothelial

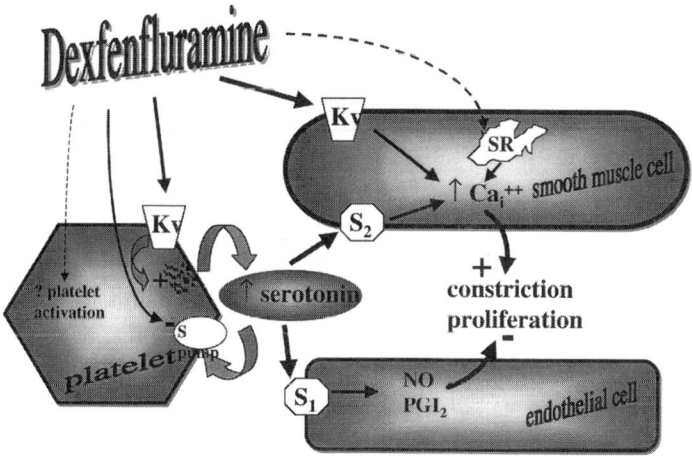

Figure 5. Potential mechanisms of dex-induced vasoconstriction. Kv, Voltage-gated K^+ channels; S, platelet serotonin transporter; S1 and S2, serotonin receptors; SR, sarcoplasmic reticulum); PGI_2, prostaglandin I_2.

dysfunction, the S_2-mediated vasoconstrictive effects of serotonin may predominate.

Patients with PPH show increased levels of NO in their exhaled breath as measured with a chemiluminesence technique (Archer *et al.*, 1998a). This could reflect a compensatory increase in NO production. However, in patients with pulmonary hypertension induced by anorectic drugs, breath NO levels are not increased (Archer *et al.*, 1998a). This might be due to the extensive background endothelial dysfunction that might be necessary for the development of anorectic-induced vascular complications. The abnormal endothelial function present in several conditions associated with obesity, such as hypertension and diabetes (Forte *et al.*, 1997; Graier *et al.*, 1996; Kirchner *et al.*, 1993), may be a further predisposing factor in the development of drug-induced PPH. In addition, there is new evidence suggesting that patients with primary (but not secondary) pulmonary hypertension (Lee *et al.*, 1998), as well as drug-induced PPH (Tuder *et al.*, 1998), might have primary endothelial dysfunction, since endothelial cells within the plexiform lesions from these patients are monoclonal.

8. ANORECTIC DRUGS AND PLATELETS

Platelets are likely to play a central role in the pathogenesis of the vascular complications associated with anorectic drugs. First, thrombosis is an important part of the histologic picture of PPH, and activated platelets are implicated in the vasospasm that is a complication of strokes and myocardial infarcts. Second, platelets are the major carriers of serotonin in the blood, and serotonin has been directly implicated in the pathogenesis of PPH (Herve *et al.*, 1995) and the mechanism of action of fenfluramine and dex in the central nervous system (McTavish and Heel, 1992).

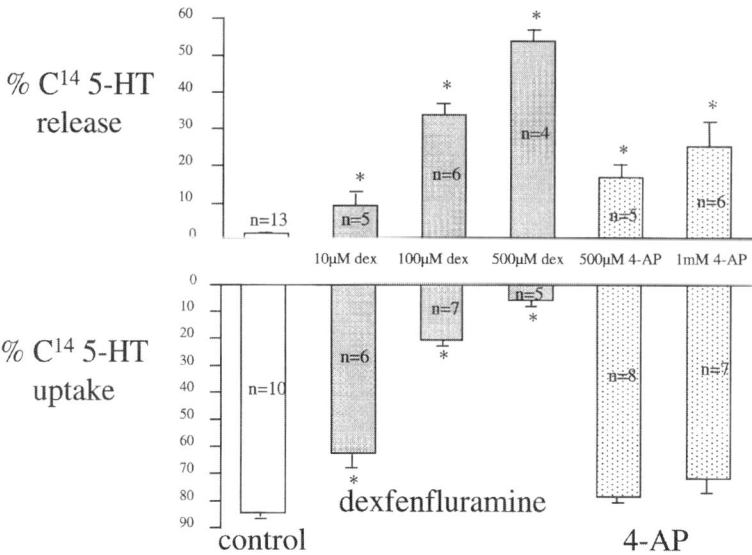

% C^{14} 5-HT release

% C^{14} 5-HT uptake

control   dexfenfluramine   4-AP

Figure 6. Dex increases serotonin release and inhibits serotonin reuptake from human platelets. Serotonin release and reuptake in fresh platelets from human volunteers were measured using radiolabeled serotonin (C^{14} 5-HT). 4-Aminopyridine (4-AP) increases serotonin release but does not appear to be involved in the serotonin reuptake process. Dex may have a dual effect on platelets: to increase serotonin release and decrease serotonin reuptake through inhibition of the serotonin transporter.

Fenfluramine and dex cause release of serotonin from platelets (Michelakis *et al.*, 1998a), but are K$^+$ channels involved in this process? In this regard, Kv and Ca^{2+}-activated (K$_{Ca}$) K$^+$ channels are expressed in platelets (de Silva *et al.*, 1997; Mahaut-Smith, 1995; Mahaut-Smith, *et al.*, 1990; Maruyama, 1987), and 4-AP causes serotonin release from human platelets, similar to the dex-induced release of this amine from platelet preparations (Fig. 6) (Michelakis *et al.*, 1998a). Furthermore, the two different mechanisms of enhancing serotonin release that exist in the brain (i.e., an E_m-dependent release and an E_m-independent reuptake of the amine) also may exist in the platelets. Although the former mechanism has not been explored in platelets, a membrane protein involved in serotonin reuptake by platelets has been identified and cloned (Hranilovic *et al.*, 1996). To investigate the effect of dex on serotonin release and reuptake in platelets, and the potential modulation of this process by Kv channels, we used radiolabeled serotonin ([^{14}C]5-HT) to measure both serotonin release and reuptake in freshly isolated human platelets from healthy volunteers (Michelakis *et al.*, 1998a). Serotonin levels were measured in drug-free medium following incubation with different doses of dex and 4-AP. As shown in Fig. 6, dex increases the release and inhibits the reuptake of serotonin from human platelets at clinically relevant doses (Michelakis *et al.*, 1998a). In contrast, 4-AP, at doses specific for Kv channel inhibition, increases the release of serotonin but does not affect reuptake, which occurs via the serotonin membrane protein transporter (Michelakis *et al.*, 1998a). These data suggest, for the first time, that an E_m-dependent pathway for serotonin release might be present in human platelets. It will be very important to define the role that E_m and Kv channels

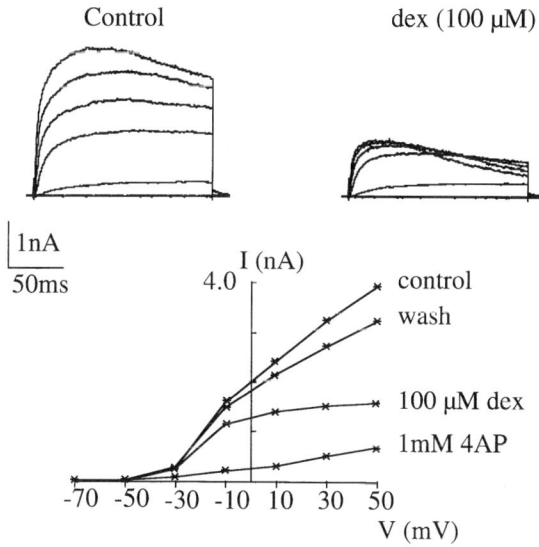

Figure 7. Dex inhibits K⁺ current in rat megakaryocytes. *Top*: Representative traces from whole-cell patch-clamp recordings, showing K⁺ currents elicited by 20-mV voltage steps from a holding potential of −70 mV to +50 mV in control experiments (*left*) and 1.5 min after dex (*right*). *Bottom*: Current–voltage relationship before, during, and after run-on of dex. Also shown is the effect of 1 m*M* 4-AP on the current after the dex was washed out. (Modified, with permission, from Weir *et al.*, 1998.)

might play in platelet activation and release of platelet-derived vasoactive factors, a well-known component of ischemic syndromes.

These data provide indirect evidence that dex and 4-AP may increase serotonin release in human platelets, possibly by a common mechanism of Kv channel inhibition. More direct evidence comes from patch-clamping experiments in megakaryocytes (the precursors of platelets), which show a significant inhibition of K⁺ current by dex (Weir *et al.*, 1998). Figure 7 shows that in freshly isolated rat megakaryocytes, K⁺ current is largely carried by 4-AP-sensitive Kv channels and that this current is reversibly inhibited by dex at clinically relevant doses (Weir *et al.*, 1998). Thus, the possibility exists that dex inhibits one or more platelet Kv channels, leading to depolarization. This might lead to opening of voltage-gated Ca²⁺ channels, increase in [Ca²⁺]ᵢ, and exocytosis of membrane-bound serotonin granules, in a similar manner as in neurons (Fig. 5). Although the presence of L-type Ca²⁺ channels in platelets remains controversial (Doyle and Ruegg, 1985; Pales *et al.*, 1991), the presence of other voltage-gated Ca²⁺ channels, such as O- or P-type, has not been studied. Alternatively, membrane depolarization may activate several enzymes directly involved in second-messenger systems, such as protein kinase C (PKC) (Kong *et al.*, 1991; Wakade *et al.*, 1991), which may release Ca²⁺ from intracellular stores to trigger exocytosis of serotonin as well as possible platelet activation. Independent of the above pathway, dex may increase plasma serotonin by inhibiting the serotonin transporter in platelets (as well as in neurons) and decreasing serotonin reuptake. To our knowledge, there are no systematic studies of plasma (as opposed to serum/blood) levels of scrotonin in patients with and without dex-induced complications.

The platelet storage pool disease (a human disease characterized by abnormal platelet handling of serotonin and high plasma serotonin levels) has been associated with PPH. Ketanserin (an S_2 receptor blocker) may reverse the pulmonary hypertension in patients afflicted with this disorder (Herve *et al.*, 1990). An animal model with similarities to this disease in humans, the fawn-hooded rat, also shows low platelet and elevated plasma serotonin levels, abnormal endothelial function, and severe systemic hypertension (Sato *et al.*, 1992). These animals develop significant pulmonary hypertension as well, if exposed to the mildly decreased ambient-air oxygen levels of Denver's altitude (Sato *et al.*, 1992). Thus a link has been established between abnormal platelet handling of serotonin, vascular endothelial dysfunction, and the development of hypertension and anorectic agents that cause the abnormal release of serotonin from platelets and vasoconstriction may make patients vulnerable to pulmonary and systemic hypertension.

9. ANORECTIC DRUGS AND MYOCARDIAL CELLS

Hu *et al.* (1998) very recently reported that dex inhibits a Kv current in freshly isolated rat myocardial cells, as detected by the whole-cell patch-clamp technique. This inhibition of K^+ current is significant at dex concentrations as low as 30 μM. Although possible effects of dex on the E_m profile of myocardial cells were not studied, these investigators suggested that dex-induced inhibition might contribute to the increase in right ventricular pressure in patients treated with this anorectic drug, through prolongation of the action potential in the ventricular myocytes (Hu *et al.*, 1998). Representative tracings from their work are shown in Fig. 8.

10. QUESTIONS FOR THE FUTURE

Patients predisposed to develop cardiovascular complications from anorectic drug use might have (a) abnormal metabolism and pharmacokinetics of the drugs, (b) abnormal K^+ channels in the pulmonary and systemic smooth muscle cells and in platelets. (c) abnormal serotonin handling by platelets and endothelial cells, or (d) abnormal endothelial function, primary or secondary (smoking, hypertension, diabetes, or viral infections, such as HIV). In the vascular smooth muscle cells of these patients, anorectic drugs may cause additional K^+ channel inhibition, thereby enhancing depolarization and further increasing $[Ca^{2+}]_i$. In the arterial smooth muscle cells, these changes may induce vasoconstriction and smooth muscle cell proliferation (Fig. 5), whereas in platelets, release of serotonin and possibly platelet activation may ensue (Fig. 5). Notably, serotonin may potentiate the direct effects of dex on smooth muscle cells (Fig. 5), possibly by inhibiting K^+ channels (Hevers and Hardie, 1995; Stefani *et al.*, 1990; Wang *et al.*, 1992), and vasoconstriction may be more pronounced secondary to dysfunctional endothelium and/or a decrease in the S_1/S_2 receptor ratio. By the same mechanism, serotonin secreted from platelets may contribute to the proliferative fibroblast response and thickening of the mitral and aortic valves seen in patients taking anorectic drugs.

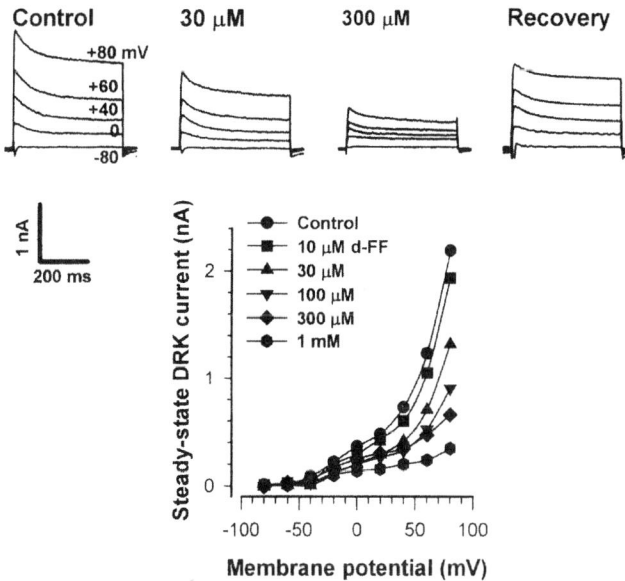

Figure 8. Dex inhibits whole-cell K$^+$ current in freshly dispersed rat myocardial cells. *Top:* Whole-cell K$^+$ currents elicited by 20 mV depolarizing steps from a holding potential of −80 m*M* were progressively blocked by 30 m*M* and 300 m*M* dexfenfluramine (*d*-FF). After washout of *d*-FF, K$^+$ current completely recovered. *Bottom:* Steady-state current voltage relationships for delayed rectifier K$^+$ (DRK) current show concentration block by *d*-FF of K$^+$ current at all membrane potential values. This K$^+$ current was inhibited by 4-AP, providing further evidence that dex inhibits one or more Kv channels. (Reprinted, with permission, from Hu *et al.*, 1998.)

In summary, the predisposing factors as well as the possible mechanisms of action of anorectic drugs are multifactorial. The complications are rare but severe. In this regard, the scientific community needs to be alert to potential anorectic-induced vascular complications because the problem of obesity is widespread and more anorectic agents will be developed to treat it.

REFERENCES

Abenhaim, L., Moride, Y., Brenot, F., Rich, S., Benichou, J., Kurz, X., Higenbottam, T., Oakley, C., Wouters, E., Aubier, M., Simoneau, G., and Begaud, B., 1996, Appetite-suppressant drugs and the risk of primary pulmonary hypertension, *N. Engl. J. Med.* **335**:609–616.

Anchors, M., 1997, *Safer than Fen-Phen*, Prima Publishing, Rocklin, California.

Archer, S. L., Huang, J. M. C., Reeve, H. L., Hampl, V., Tolarova, S., Michelakis, E. D., and Weir, E. K., 1996, Differential distribution of electrophysiologically distinct myocytes in conduit and resistance arteries determines their response to nitric oxide and hypoxia, *Circ. Res.* **78**:431–442.

Archer, S., Djaballah, K., Humbert, M., Weir, E., Fartoukh, M., Simonneau, G., and Dinh-Xuan, A., 1998a, Nitric oxide deficiency in pulmonary hypertension associated with use of the anorectic agents fenfluramine and dexfenfluramine, *Am. J. Resp. Crit. Care Med.* **158**:1061–1067.

Archer, S., Souil, E., Dinh-Xuan, A., Schremmer, B., Mercier, J., Al Yaagoubi, A., Nguyen-Huu, L., Reeve, H., and Hampl, V., 1998b, Molecular identification of the role of voltage-gated K channels, Kv1.5 and Kv2.1, in hypoxic pulmonary vasoconstriction and control of resting membrane potential in rat pulmonary artery myocytes, *J. Clin. Invest.* **101**:2319–2330.

Atanassoff, P., Weiss, E., Schnid, E., and Tornic, M., 1992, Pulmonary hypertension and dexfenfluramine, *Lancet* **339**:436.

Bailie, T., 1991, Fenfluramine-induced unstable angina pectoris, *Can. J. Pharmacol.* **44**:211–215.

Brenot, F., Herve, P., Petitpretz, P., Parent, F., Duroux, P., and Simonneau, G., 1993, Primary pulmonary hypertension and fenfluramine use, *Br. Heart J.* **70**:537–541.

Cinquanta, M., Frittoli, E., and Gobbi, M., 1997, Further evidence of Ca^{++} dependent, exocytotic-like serotonin release induced by *d*-fenfluramine. *Pharmacol. Res.* **35**:439–442.

Connolly, H. M., Crary, J. L., McGoon, M. D., Hensrud, D. D., Edwards, B. S., Edwards, W. D., and Schaff, H. V., 1997, Valvular heart disease associated with fenfluramine phentermine, *N. Engl. J. Med.* **337**:581–588.

de Silva, H., Carver, J., and Aronson, J., 1997, Pharmacological evidence of calcium-activated and voltage-gated potassium channels in human platelets, *Clin. Sci.* **93**:249–255.

Douglas, J., Munro, A., Kitchin, H., Muir, A., and Proudfoot, A., 1981, Pulmonary hypertension and fenfluramine, *Br. Med. J.* **283**:881–883.

Doyle, V., and Ruegg, U., 1985, Lack of evidence for voltage dependent calcium channels on platelets, *Biochim. Biophys. Acta* **127**:161–167.

Evrard, P., Allaz, A. F., and Urban, P., 1990, Myocardial infarction associated with the use of dexfenfluramine, *Br. Med. J.* **301**:345.

Forte, P., Copland, M., Smith, L. M., Milne, E., Sutherland, J., and Benjamin, N., 1997, Basal nitric oxide synthesis in essential hypertension, *Lancet* **349**:837–842.

Frishman, W. H., Huberfeld, S., Okin, S., Wang, Y., Kumar, A., and Shareef, B., 1995, Serotonin and serotonin antagonism in cardiovascular and non-cardiovascular disease, *J. Clin. Pharmacol.* **35**:541–572.

Gaul, G., Blazek, E., Deutsch, E., and Heeger, H., 1982, Ein Fall von chronischer pulmonaler hypertonie nach Fenfluramineinnahme, *Wien. Klin. Wochenschr.* **22**:618–621.

Gobbi, M., Frittoli, E., Mennini, T., and Garattini, S., 1993, Evidence of an exocytotic like release of [^{3}H]5-HT induced by *d*-fenfluramine in rat hippocampal synaptosomes, *Eur. J. Pharmacol.* **238**:9–17.

Gobbi, M., Crespi, D., and Mennini, T., 1995, Effects of fluoxetine on basal and $K^{(+)}$-induced tritium release from synaptosomes preloaded with [^{3}H]serotonin, *Life Sci.* **56**:785–791.

Graier, W. F., Simecek, S., Kukovetz, W. R., and Kostner, G. M., 1996, High D-glucose-induced changes in endothelial Ca^{2+}/EDRF signaling are due to generation of superoxide anions, *Diabetes* **45**:1386–1395.

Heal, D., Cheetham, S., Prow, M., Martin, K., and Buckett, W., 1998, A comparison of the effects on central 5-HT function of sibutramine hydrochloride and other weight-modifying agents, *Br. J. Pharmacol.* **125**:301–308.

Herve, P., Drouet, L., Dosquet, C., Launay, J., Rain, B., Simonneau, G., Caen, J., and Duroux, P., 1990, Primary pulmonary hypertension in a patient with a familial platelet storage pool disease: Role of serotonin, *Am. J. Med.* **89**:117–120.

Herve, P., Launay, J., Scrobonaci, M., Brenot, F., Simonneau, G., Petitpretz, P., Poubeau, P., Cerrina, J., Duroux, P., and Drouet, L., 1995, Increased plasma serotonin in primary pulmonary hypertension, *Am. J. Med.* **99**:249–254.

Hevers, W., and Hardie, R. C., 1995, Serotonin modulates the voltage dependence of delayed rectifier and Shaker potassium channels in *Drosophila* photoreceptors, *Neuron* **14**:845–856.

Hranilovic, D., Lesch, K., Ugarkovic, D., Cicin-Sain, L., and Jernej, B., 1996, Identification of serotonin transporter mRNA in rat platelets, *J. Neural Transm.* **103**:957–963.

Hu, S., Wang, S., Gibson, J., and Gilbertson, T., 1998, Inhibition of delayed rectifier K^+ channels by dexfenfluramine (Redux), *J. Pharmacol. Exp. Ther.* **287**:480–486.

Kadokami, T., Egashira, K., Kuwata, K., Fukumoto, Y., Kozai, T., Yasutake, M., Kuga, T., Shimokawa, H., Sueishi, K., and Takeshita, A., 1996, Altered serotonin receptor subtypes mediate coronary microvascular hyperreactivity in pigs with chronic inhibition of NO synthesis, *Circulation* **94**:182–189.

Kay, J., Smith, P., and Heath, D., 1971, Am.inorex and the pulmonary circulation, *Thorax* **26**:262–270.

Khan, M., Herzog, C., St Peter, J., Hartley, G., Madlon-Kay, R., Dick, C., Asinger, R., and Vessey, J., 1998, The prevalence of cardiac valvular insufficiency assessed by transthoracic echocardiography in obese patients treated with appetite-suppressant drugs, *N. Engl. J. Med.* **339**:713–718.

Kirchner, K. A., Scanlon, P. H., Dzielak, D. J., and Hester, R. L., 1993, Endothelium-derived relaxing factor responses in DOCA-salt hypertensive rats, *Am. J. Physiol.* **265**:R568–R572.

Kong, S., Choy, Y., Fung, K., and Lee, C., 1991, Membrane depolarization induces protein kinase C translocation and voltage operated calcium channel opening in PU5-1.8 cells: Protein kinase C as a negative feedback modulator for calcium signalling, *Second Messengers Phosphoproteins* **13**:117–130.

Lee, S., Shroyer, K., Markham, N., Cool, C., Voelkel, N., and Tuder, R., 1998, Monoclonal endothelial cell proliferation is present in primary but not secondary pulmonary hypertension, *J. Clin. Invest.* **101**:927–934.

Liu, Y., Hudetz, A., Knaus, H., and Rusch, N., 1998, Increased expression of Ca^{++} sensitive K^+ channels in the cerebral microcirculation of genetically hypertensive rats: Evidence for their protection against cerebral vasospasm, *Circ. Res.* **82**:729–737.

Mabadeje, A. F., 1974, Fenfluramine-induced hypertension, *West Afr. J. Pharmacol. Drug Res.* **2**:54P.

Mahaut-Smith, M., 1995, Calcium-activated potassium channels in human platelets, *J. Physiol (London)* **484**:15–24.

Mahaut-Smith, M., Rink, T., Collins, S., and Sage, S., 1990, Voltage-gated potassium channels and the control of membrane potential in human platelets, *J. Physiol. (London)* **428**:723–735.

Marinella, M. and Berrettoni, B., 1997, Digital necrosis associated with dexfenfluramine, *N. Engl. J. Med.* **337**:1776–1777.

Martens, J., and Gelband, C., 1998, Ion channels in vascular smooth muscle: Alterations in essential hypertension, *Proc. Soc. Exp. Biol. Med.* **218**:192–203.

Maruyama, Y., 1987, A patch-clamp study of mammalian platelets and their voltage-gated potassium current, *J. Physiol. (London)* **391**:467–485.

McMurray, J., Bloomfield, P., and Miller, H., 1986, Irreversible pulmonary hypertension after treatment with fenfluramine, *Br. Med. J.* **292**:239–240.

McNeely, W., and Goa, K., 1998, Sibutramine: A review of its contribution to the management of obesity, *Drugs* **56**:1093–1124.

McTavish, D., and Heel, R., 1992, Dexfenfluramine: A review of its pharmacologic properties and therapeutic potential in obesity, *Drugs* **43**:713–733.

Michelakis, E., Reeve, H., Huang, J., Tolarova, S., Nelson, D., Weir, E., and Archer, S., 1995, Potassium channel diversity in vascular smooth muscle cells, *Can. J. Physiol. Pharmacol.* **75**:889–897.

Michelakis, E., Johnson, G., Leis, L., Archer, S., and Weir, E., 1998a, Dexfenfluramine and 4-aminopyridine (an inhibitor of voltage-gated potassium channels) increase serotonin release from human platelets, *Am. J. Respir. Crit. Care. Med.* **157**:A588.

Michelakis, E., Weir, E., Nelson, D., and Archer, S., 1998b, Dexfenfluramine causes systemic hypertension through inhibition of voltage-sensitive potassium channels, *J. Mol. Cell Cardiol.* **30**:A756.

Miyamoto, J. K., Uezu, E., Yusa, T., and Terashima, S., 1990, Efflux of 5-HIAA from 5-HT neurons: A membrane potential dependent process, *Physiol. Behav.* **47**:767–772.

Mlczoch, J., 1984, *Drug and Dietary Induced Pulmonary Hypertension*, in: *Pulmonary Hypertension* E. K. Weir and J. T. Reeves (eds.), Futura Publishing Company, Mount Kisco, NY, New York, pp. 341–360

Naeije, R., Maggiorini, M., Delcroix, M., Leeman, M., and Melot, C., 1996, Effects of chronic dexfenfluramine treatment on pulmonary hemodynamics in dogs, *Am. J. Respir. Crit. Care Med.* **154**:1347–1350.

Nelson, M., and Quayle, J., 1995, Physiological roles and properties of potassium channels in arterial smooth muscle cells, *Am. J. Physiol.* **37**:C799–C822.

Pales, J., Palacios-Araus, L., Lopez, L., and Gual, L., 1991, Effects of dihydropyridines and inorganic calcium blockers on aggregation and on intracellular free calcium in platelets, *Biochim. Biophys. Acta* **1064**:169–174.

Patel, A., Lazdunski, M., and Honore, E., 1997, Kv2.1/Kv9.3, a novel ATP-dependent delayed-rectifier K channel in oxygen-sensitive pulmonary artery myocytes, *EMBO J.* **16**:6615–6625.

Pei, Q., Leslie, R. A., Grahame-Smith, D. G., and Zetterstrom, T. S., 1995, 5-HT efflux from rat hippocampus in vivo produced by 4-aminopyridine is increased by chronic lithium administration, *Neuroreport* **6**:716–720.

Pouwels, H., Smeets, J., Cheriex, E., and Wouters, E., 1990, Pulmonary hypertension and fenfluramine, *Eur. Respir. J.* **3**:606–607.

Reeve, H., Tolarova, S., Michelakis, E., Archer, S., and Weir, E., 1997, Effects of the anorectic agent dexfenfluramine and 4-aminopyridine on the ductus arteriosus during development, *Circulation* **96**:I–245.

Reeve, H., Archer, S., and Weir, E., 1998, Dexfenfluramine releases intracellular calcium in rat pulmonary artery smooth muscle cells, *Circulation* **98**:I–341.

Reeve, H., Nelson, D., Archer, S., and Weir, E., 1999, Effects of fluoxetine, phentermine and venlafaxine on pulmonary arterial pressure and electrophysiology, *Am. J. Physiol.* **276**:L213–L219.

Rubin, L., 1997, Primary pulmonary hypertension, *N. Engl. J. Med.* **336**:111–117.

Rusch, N., and Runnells, A., 1994, Remission of high blood pressure reverses arterial potassium channel alterations, *Hypertension* **23**:941–945.

Sato, K., Webb, S., Tucker, A., Rabinovitch, M., O'Brien, R., McMurtry, I., and Stelzner, T., 1992, Factors influencing the idiopathic development of pulmonary hypertension in the fawn-hooded rat, *Am. Rev. Respir. Dis.* **145**:793–797.

Schembre, D., and Boynton, K., 1997, Letter to the editor, *N. Engl. J. Med.* **336**:510–511.

Schwitter, J., Agosti, R., Ott, P., Kalman, A., and Waespe, W., 1992, Small infarctions of cochlear, retinal and encephalic tissue in young women, *Stroke* **23**:903–907.

Stark, P., Fuller, R., and Wong, D., 1985, The pharmacologic profile of fluoxetine, *J. Clin. Psychiatry* **46**(3 Pt. 2):7–13 .

Stefani, A., Surmeier, D. J., and Kitai, S. T., 1990, Serotonin enhances excitability in neostriatal neurons by reducing voltage-sensitive potassium currents, *Brain Res.* **529**:354–357.

Tuder, R., Radisavljevic, Z., Shroyer, K., Polak, J., and Voelkel, N., 1998, Monoclonal endothelial cells in appetite suppressant-associated pulmonary hypertension, *Am. J. Respir. Crit. Care Med.* **158**:1999–2001.

Van Gaal, L., Wauters, M., and De Leeuw, I., 1998, Anti-obesity drugs: what does sibutramine offer? An analysis of its potential contribution to obesity treatment, *Exp. Clin. Endocrinol Diabetes* **106** (Suppl. 2):35–40.

VanItallie, T. B., 1996, Prevalence of obesity, *Endocrinol. Met. Clin. North Am.* **25**:887–905.

Voelkel, N. F., Clarke, W. R., and Higenbottam, T., 1997, Obesity, dexfenfluramine, and pulmonary hypertension: A lesson not learned?, *Am. Respir. Crit. Care Med.* **155**:786–788.

Wakade, T., Bhave, S., Bhave, A., Malhotra, R., and Wakade, A., 1991, Depolarizing stimuli and neurotransmitters utilize separate pathways to activate protein kinase C in sympathetic neurons., *J. Biol. Chem.* **266**:6424–6428.

Wang, Y., Strahlendorf, J. C., and Strahlendorf, H. K., 1992, Serotonin reduces a voltage dependent transient outward potassium current and enhances excitability of cerebellar Purkinje cells, *Brain Res.* **571**:345–349.

Weir, E. K., Reeve, H. L., Huang, J., Michelakis, E., Nelson, D. P., Hampl, V., and Archer, S. L., 1996, Anorexic agents aminorex, fenfluramine and dexfenfluramine inhibit potassium current in rat pulmonary vascular smooth muscle and cause pulmonary vasoconstriction, *Circulation* **94**:2216–2220.

Weir, E., Reeve, H., Johnson, G., Michelakis, E., Nelson, D., and Archer, S., 1998, A role for potassium channels in smooth muscle cells and platelets in the etiology of primary pulmonary hypertension, *Chest* **114** (3 Suppl.):200S–204S.

Yuan, J., Aldinger, A., Juhaszova, M., Wang, J., Conte, J., Gaine, S., Orens, J., and Rubin, L., 1998a, Dysfunctional voltage gated potassium channels in the pulmonary artery smooth muscle cells of patients with primary pulmonary hypertension, *Circulation* **98**:1400–1406.

Yuan, X., Wang, J., Golovina, V., and Rubin, L., 1998b, Molecular basis and function of voltage-gated K^+ channels in pulmonary arterial smooth muscle cells, *Am. J. Physiol.* **274**:L621-635.

Chapter 42

Induction of Ca^{2+}-Activated K^+ Channel Expression during Systemic Hypertension: Protection against Pathological Vasoconstriction

Marcie G. Berger and Nancy J. Rusch

1. INTRODUCTION

During the past 15 years, patch-clamp studies have characterized a diverse population of K^+ channel types in vascular smooth muscle cells. Prominent among these is the high-conductance, Ca^{2+}-activated K^+ channel (BK_{Ca} channel), which appears to be ubiquitously expressed in arterial smooth muscle membranes. The BK_{Ca} channel in some vascular beds may act together with other K^+ channel types to set the level of resting membrane potential in the arterial smooth muscle cells. During vascular excitation, the further activation of BK_{Ca} channels may provide a powerful pathway to hyperpolarize the arterial smooth muscle cells, thereby limiting voltage-gated Ca^{2+} influx and buffering vasoconstriction in small arteries and resistance vessels. Thus, under physiological conditions, the BK_{Ca} channel acts as a homeostatic mechanism to counteract arterial constriction and maintain blood flow to critical organs and tissues.

With this in mind, this chapter will focus on the homeostatic role of the vascular BK_{Ca} channel in the pathogenesis of systemic hypertension, a pathological state characterized by an abnormally high level of arterial tone (Folkow, 1982). Recent evidence suggests that BK_{Ca} channel expression in vascular smooth muscle membranes increases during chronic hypertension, thereby limiting the development of pathological vascular tone and preventing vasospasm in critical regulatory beds such as the cerebral circulation (Rusch et al., 1992; England et al., 1993; Liu et al., 1994; Rusch and Runnells, 1994; Liu et al., 1995; Martens and Gelband, 1996; Liu et al., 1997, 1998). The enhanced

Marcie G. Berger ● Department of Cardiology, Sinai Samaritan Medical Center, Milwaukee, WI 53233   *Nancy J. Rusch* ● Departments of Pharmacology, Cardiovascular Research Center, The Medical College of Wisconsin, Milwaukee, Wisconsin 53226. This effort was supported by National Heart, Lung and Blood Institute grant R01 HL-59238 from the National Institutes of Health.

Potassium Channels in Cardiovascular Biology, edited by Archer and Rusch. Kluwer Academic/Plenum Publishers, New York, 2001.

expression of BK_{Ca} channels also may provide an amplified target for vasodilator drugs designed to open the BK_{Ca} channel in arterial smooth muscle membranes to elicit vascular relaxation (Berger and Rusch, 1999). In view of these observations, this chapter will (a) briefly cite the current knowledge regarding the molecular and physiological profile of the BK_{Ca} channel in the vasculature, (b) review early evidence implicating altered BK_{Ca} channel current in the pathogenesis of systemic hypertension, and (c) discuss new studies which argue for a pressure-induced increase in the expression of BK_{Ca} channel α-subunits in arterial smooth muscle membranes as a homeostatic mechanism to regulate vascular tone in hypertension. The initial chapters in this book have provided comprehensive reviews of the molecular, pharmacological, and functional properties of the BK_{Ca} channel and have discussed methods for its detection in the cardiovascular system. The reader is referred to Chapters 4 and 13 for background information.

2. EXPRESSION AND FUNCTION OF BK_{Ca} CHANNELS IN THE VASCULATURE

The high-conductance, Ca^{2+}-activated K^+ channel, called the "BK_{Ca} channel" to acknowledge the high-amplitude unitary K^+ currents associated with its open state, is densely expressed in vascular smooth muscle membranes. As such, it is uniquely positioned to influence the membrane potential of the vascular smooth muscle cell, and hence, regulate the level of vascular tone. Whereas many other ion channel types originate from multiple gene families, the pore-forming α-subunit of the BK_{Ca} channel appears to arise from a single gene family (McCobb et al., 1995). However, a high level of alternative splicing of the common primary transcript along with the subsequent tetrameric formation of the α-subunit complex may generate significant phenotypic diversity (Shen et al., 1994; Lagrutta et al., 1994; Knaus et al., 1995; McCobb et al., 1995). As reviewed in Chapter 4 of this book, the N-terminal region of the BK_{Ca} channel α-subunit shares partial homology with the Shaker voltage-gated channels, showing six putative transmembrane domains (S1–S6) and a highly conserved pore region between S5 and S6 (McCobb et al., 1995). Domain S4 appears to possess an intrinsic voltage sensor, which is associated with gating charge movement and subsequent channel opening independent of the concentration of intracellular calcium (Stefani et al., 1997). The Ca^{2+} sensitivity of the BK_{Ca} channel appears to be conferred by the extra four transmembrane domains at the C-terminal region of the α-subunit and by close association of the α-subunit with a regulatory β-subunit, which increases the Ca^{2+} sensitivity of the BK_{Ca} channel by as much as 10-fold (Toro et al., 1991; Knaus et al., 1994; Cui et al., 1997; McManus et al., 1995; Meera et al., 1996; Schreiber and Salkoff, 1997). Thus, in any given circulatory bed, the levels of expression and association of the α- and β-subunits represent fundamental events that will determine the open-state probability of the BK_{Ca} channel, and hence its functional contribution to the regulation of vascular tone.

Phenotypic profiles of the BK_{Ca} channel in aorta and in cerebral, coronary, mesenteric, and skeletal muscle arteries describe a noninactivating or slowly inactivating channel with a single-channel conductance between 200 and 300 pS (Toro et al., 1991; England et al., 1993; Wang et al., 1993; Sansom and Stockand, 1994; Jackson et *al.*, 1996; Tanaka *et al.*, 1997; Jackson and Blair, 1998; Liu *et al.*, 1998). Values calculated for K, the parameter derived from Boltzmann analyses that describes the slope of the voltage–activation relationship, have been similar for BK$_{Ca}$ channels regardless of the type of vascular smooth muscle studied. Thus, BK$_{Ca}$ channels expressed in different vascular beds appear to show a common level of voltage sensitivity for channel activation (Toro *et al.*, 1991; Scornick and Toro, 1992; Dopica *et al.*, 1994; Carl *et al.*, 1996; Liu *et al.*, 1998). However, a wide range of half-maximal activation values ($V_{1/2}$) derived from Boltzmann analyses have been reported for these same channels, indicating that BK$_{Ca}$ channels located in different vascular beds show highly variable Ca^{2+} sensitivities or Ca^{2+} set points (Toro *et al.*, 1991; Sansom and Stockand, 1994; Jackson *et al.*, 1996; Jackson and Blair, 1998). Although the precise molecular events underlying the phenotypic heterogeneity of native BK$_{Ca}$ channels are unknown, this functional difference between BK$_{Ca}$ channels from different circulatory regions may be related to the expression of tissue-specific α-subunit isoforms or to variable levels of α- and β-subunit association in different vascular tissues (Rusch *et al.*, 1996; Rusch and Liu, 1997; Tanaka *et al.*, 1997).

Because the open-state probability of BK$_{Ca}$ channels is regulated primarily by changes in membrane potential and internal Ca^{2+} concentration, membrane depolarization and rises in cytosolic free Ca^{2+} act synergestically to enhance the open-state probability of the BK$_{Ca}$ channel during vascular excitation (Nelson *et al.*, 1990; Carl *et al.*, 1996; Tanaka *et al.*, 1997). The small size and high input resistance of vascular smooth muscle cells, coupled with the large unitary amplitude of BK$_{Ca}$ channel current, may enable BK$_{Ca}$ channels to efficiently buffer depolarization and limit the final level of arterial constriction (Nelson *et al.*, 1990). In this regard, a number of studies indicate that BK$_{Ca}$ channels can effectively regulate the level of resting membrane potential and tone in small cannulated arteries, perfused at physiological distending pressures to mimic *in situ* blood pressure levels. Under these conditions, pharmacological block of BK$_{Ca}$ channels by iberiotoxin or charybdotoxin depolarizes small arteries from the cerebral, coronary, and skeletal muscle circulations, triggering vasoconstriction (Berczi *et al.*, 1992; Brayden and Nelson, 1992). This phenomenon is illustrated in Fig. 1, which shows images of a small rat cerebral artery (top panel) and human coronary artery (lower panel) obtained before and after vessel exposure to the specific BK$_{Ca}$ channel blocker iberiotoxin. In both types of arteries, pharmacological block of vascular BK$_{Ca}$ channels by iberiotoxin triggers a pronounced vasoconstriction. Studies such as these imply that K$^+$ efflux mediated by the tonic opening of BK$_{Ca}$ channels buffers native vasoconstrictor pathways, thereby permitting small arteries to maintain an optimal diameter for the perfusion of distal tissues.

3. EARLY EVIDENCE FOR AUGMENTED BK$_{Ca}$ CHANNEL CURRENT IN HYPERTENSION

Arterial muscle membranes exposed to high blood pressure *in situ* show a generalized increase in membrane ionic permeability, including an enhanced permeability to the potassium ion. In pioneering experiments almost 25 years ago, radioisotopic and ion-exchange methods first detected an augmented transmembrane influx of

Rat Cerebral Artery

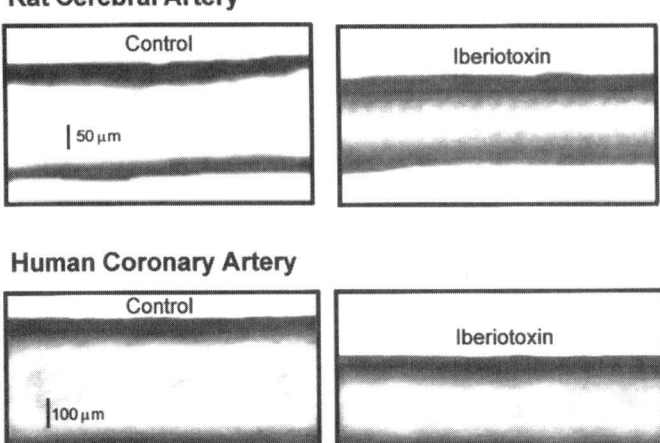

Human Coronary Artery

Figure 1. Photographic illustrations of a small, cannulated rat cerebral artery (*top*) and human coronary artery (*bottom*). Arteries were pressurized at 80 mm Hg and superfused with drug-free physiological salt solution (Control) or solution containing 100 nM iberiotoxin. Pharmacological block of BK_{Ca} channels by iberiotoxin induced a pronounced constriction of both types of small arteries. (K. Gauthier, M. G. Berger, and N. J. Rusch, unpublished data.)

Na^+ and Ca^{2+} in vascular smooth muscle cells from hypertensive rats. This influx of excitatory cations was electrically buffered by an augmented efflux of K^+ ions from the same smooth muscle cells (Friedman and Friedman, 1976; Jones, 1973, 1974; Jones and Hart, 1975). Several laboratories observed this pattern of enhanced transmembrane Na^+, Ca^{2+}, and K^+ flux in aorta and renal and caudal arteries from rats with genetic, renal, and salt-sensitive forms of hypertension (Friedman and Friedman, 1976; Garwitz and Jones, 1982a,b; Jones, 1973, 1974; Jones and Hart, 1975; Jones, 1983). These findings of a generalized increase in membrane ionic permeability in arteries from hypertensive rats with divergent genetic and endocrine profiles led to the theory that these ionic abnormalities constitute a "universal defect" of vascular smooth muscle membranes exposed to hypertension (Kwan, 1985). According to this theory, the increased Ca^{2+} influx may produce the abnormally high Ca^{2+}-dependent tone observed in arteries of hypertensive animals, as well as the increased peripheral vascular resistance in animals and humans with essential hypertension (Rusch and Hermsmeyer, 1993). In contrast, the concurrent augmentation of K^+ efflux observed in vascular smooth muscle cells from hypertensive animals may provide a protective mechanism to hyperpolarize the vascular smooth muscle cell membrane, thereby reducing voltage-gated Ca^{2+} influx and promoting vascular relaxation (Jones, 1983). The persistence of an enhanced K^+ efflux at temperatures as low as 2°C pointed to an increased "passive" flux of K^+ through membrane ion channels, rather than a modification in active membrane ionic transport mechanisms, as the underlying mechanism for augmented K^+ current (Friedman and Friedman, 1976; Jones, 1973, 1974, 1983). Later experiments indicated that this increased K$^+$ efflux in arteries of hypertensive rats was normalized by removal of extracellular Ca^{2+} and also by pharmacological block of voltage-gated Ca^{2+} channels (Smith and Jones, 1991). Taken together, these findings implicated a Ca^{2+}-dependent K$^+$ channel as the homeostatic pathway for enhanced K$^+$ efflux in vascular smooth muscle cells in hypertension. Almost a decade later, the advent of patch-clamp and molecular biology techniques would confirm the identity of the BK$_{Ca}$ channel as the mediating protein of this ionic alteration.

Since these discoveries, microelectrode studies comparing the electrical properties of isolated arteries from normotensive and hypertensive rats have supported the theory that an enhanced K$^+$ efflux may normalize the level of resting membrane potential in arteries exposed to hypertension. In particular, although Na$^+$ and Ca^{2+} influx into the vascular smooth muscle cell is increased in hypertension, which should depolarize the cell membrane, the level of resting membrane potential in arteries from hypertensive rats appears to be normal (Hermsmeyer, 1976; Abel *et al.*, 1981; Hermsmeyer, 1982; Stekiel *et al.*, 1986; Longhurst *et al.*, 1988; Lamb and Webb, 1989). For example, normal values for resting membrane potential between -52 and -56 mV have been reported in caudal arterial segments from several rat models of genetic and salt-induced hypertension, including the spontaneously hypertensive rat (SHR) (Hermsmeyer, 1976), the stroke-prone SHR (Lamb and Webb, 1989), the salt-sensitive Dahl rat (Abel *et al.*, 1981), and the salt-sensitive deoxycorticosterone (DOCA) rat (Hermsmeyer, 1982; Longhurst *et al.*, 1988). Similarly, the resting membrane potential in vascular smooth muscle cells from the mesenteric artery of normotensive Wistar Kyoto (WKY) rats and SHR is reported to show comparable values of -56 and -57 mV, respectively (Stekiel *et al.*, 1986). Likewise, the reported membrane potential of -52 mV in mesenteric arteries from salt-sensitive DOCA hypertensive rats is the same as values obtained in arteries of age-matched, normotensive rats (Longhurst *et al.*, 1988). Thus, it appears that vascular smooth muscle cells can activate homeostatic, hyperpolarizing mechanisms to maintain normal membrane potential values despite an augmented inward cation current driving depolarization. New evidences suggests that the upregulation of BK$_{Ca}$ channels in the vascular smooth muscle membrane may provide one such pathway for maintaining resting membrane potential and, hence, buffering vascular excitation during hypertension.

Clearly, tension-recording studies have demonstrated an amplified role for BK$_{Ca}$ channels in buffering Ca^{2+}-dependent tone in arteries of hypertensive animals. For example, pharmacological block of BK$_{Ca}$ channels by tetraethylammonium (TEA) elicits a strong Ca^{2+}-dependent constriction in vascular segments of aorta from SHR with genetic hypertension, whereas aortic segments from normotensive WKY rats do not constrict (Rusch *et al.*, 1992; Rusch and Runnells, 1994). Likewise, aortic segments from rats exposed to a surgically induced increase in blood pressure for 2 weeks constrict strongly in response to TEA, whereas aortic segments exposed to a normal blood pressure level do not (Rusch *et al.*, 1992). Similarly, carotid, femoral, renal, and mesenteric arteries from SHR also show Ca^{2+}-dependent constrictions in response to TEA, charybdotoxin, or iberiotoxin, drugs which reduce the open-state probability of BK$_{Ca}$ channels (Asano *et al.*, 1993a,b; Kolias *et al.*, 1993; Rusch *et al.*, 1996; Thompson *et al.*, 1987). Under the same conditions, similar vessels from normotensive animals do not constrict. Hence, high blood pressure appears to augment K$^+$ efflux through BK$_{Ca}$

channels in arterial smooth muscle cells, and these cells appear to rely extensively on an increased resting K^+ current to prevent pathological depolarization and constriction. Apparently, arteries from normotensive animals do not require this compensatory mechanism to maintain relaxation.

4. MOLECULAR MECHANISMS OF BK_{Ca} CHANNEL UPREGULATION IN AORTIC SMOOTH MUSCLE CELLS

Patch-clamp techniques have further identified the BK_{Ca} channel as the common, single-channel pathway underlying the enhanced K^+ efflux in vascular smooth muscle cells from hypertensive rats (England et al., 1993; Liu et al., 1994; Rusch and Runnells, 1994; Liu et al., 1995; Martens and Gelband, 1996; Liu et al., 1997, 1998). As shown by the sample traces in Fig. 2A and 2B, reports from several laboratories indicate that the level of whole-cell K^+ current through iberiotoxin-sensitive BK_{Ca} channels is higher in aortic smooth muscle cells from SHR and stroke-prone SHR compared to WKY rats (Rusch et al., 1992; England et al., 1993; Liu et al., 1994, 1997; Rusch and Runnells, 1994). This finding has been confirmed in aortic smooth muscle cells from rats with renal and salt-induced forms of hypertension (Rusch et al., 1992; Liu et al., 1995). Because the cell membrane area is not significantly increased in single aortic smooth muscle cells from SHR compared to WKY animals, a pressure-induced hypertrophy of the vascular smooth muscle cell cannot explain the increased K^+ current through BK_{Ca} channels in hypertension (Liu et al., 1995, 1997, 1998). Rather, the level of whole-cell K^+ current through BK_{Ca} channels closely corresponds to the blood pressure level of the host animal, as evidenced by the rapid normalization of elevated BK_{Ca} current in aortic smooth muscle cells of hypertensive rats after blood pressure is lowered by antihypertensive agents (Rusch and Runnells, 1994). Thus, the BK_{Ca} channel likely represents a pressure-sensitive membrane protein in arterial smooth muscle cells, which can be dynamically regulated by the in situ blood pressure level of the host animal. Detailed single-channel comparisons between BK_{Ca} channels expressed in WKY and SHR aortic smooth muscle membranes have demonstrated that the single-channel conductance of the BK_{Ca} channel is not altered in hypertension, but that more open-state events are observed in single cell-free membrane patches of SHR (England et al., 1993). The latter findings imply that the amplified BK_{Ca} channel current in aortic smooth muscle membranes from SHR may relate to either (1) an increased expression of the BK_{Ca} channel, which would provide more unitary pathways for K^+ efflux in hypertension, or (2) a change in the regulation of the BK_{Ca} channel, perhaps related to alterations in the pore-forming α-subunit or its membrane-associated regulatory proteins.

Exploring these possibilities, recent studies using a sequence-specific antibody directed against the S9/S10 linker of the α-subunit of the BK_{Ca} channel have confirmed an increased expression of this pore-forming subunit in aortic smooth muscle cells from hypertensive rats (Liu et al., 1997). The Western blot in Fig. 2C shows a sample comparison of the density of 125-kDa immunoreactive bands obtained from aortic smooth muscle membranes isolated from a single SHR and a single WKY rat. These distinct bands correspond to the known molecular mass of the BK_{Ca} channel α-subunit.

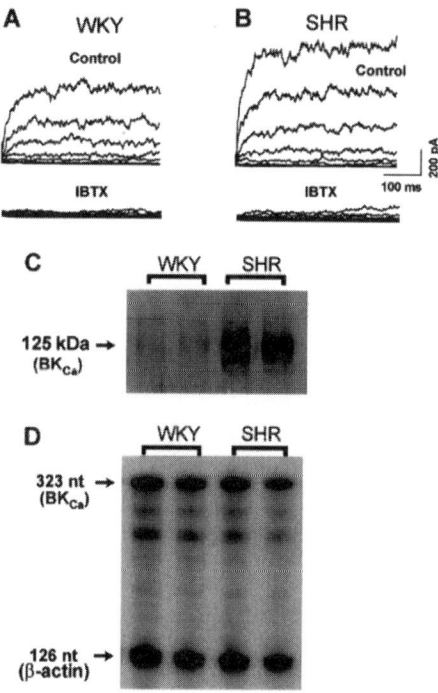

Figure 2. (A) and (B) Original traces of whole-cell K$^+$ currents in WKY and SHR aortic smooth muscle cells. Currents were elicited by 10-mV depolarizing steps from -60 mV to 0 mV in control solution (top traces) and after block of BK$_{Ca}$ channels by 100 nM iberiotoxin (IBTX). SHR cells had a larger component of IBTX-sensitive, BK$_{Ca}$ current than WKY cells. (C) Western blot comparing the expression of the BK$_{Ca}$ channel α-subunit between aortic smooth muscle membranes from WKY rats (lanes 1 and 2) and SHR (lanes 3 and 4). The density of the 125-kDa band corresponding to the α-subunit was greater in SHR. (D) Ribonuclease protection assay of aortic tissue from WKY rats (lanes 1 and 2) and SHR (lanes 3 and 4). No difference in the expression of the 323-nt protected fragment, corresponding to the expected size of the protected mRNA for the α-subunit of the BK$_{Ca}$ channel, was detected between SHR and WKY rat aorta. The 126-nt band represents mRNA expression for the internal standard, β-actin. (Adopted and modified from Liu *et al.*, 1997, with permission.)

As detailed by Liu *et al.* (1997), the expression level of the α-subunit of the BK$_{Ca}$ channel appears to be 2- to 3-fold higher in aortic smooth muscle membranes from hypertensive rats compared to normotensive control animals. In the same preparations, the molecular mechanisms by which the vascular smooth muscle cells in the SHR detect the signal of elevated blood pressure and transduce this signal into an overexpression of BK$_{Ca}$ channel α-subunits remain unclear. However, initial studies indicate that transcript levels for the BK$_{Ca}$ channel α-subunit do not appear to be increased in aortic smooth muscle cells of the SHR, suggesting that posttranscriptional events may be involved (Liu *et al.*, 1997). As shown in Fig. 2D, comparison of mRNA levels by ribonuclease protection assays indicates that the transcript level for the BK$_{Ca}$ channel α-subunit protein (associated with the 323-nt band) is unchanged in aortic smooth muscle cells from SHR compared to WKY rats. In the same preparation, the transcript

Formation of BK$_{Ca}$ Channels

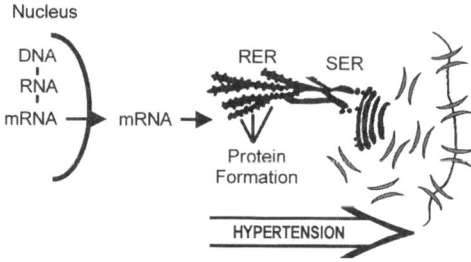

Figure 3. Fundamental steps in the formation of BK$_{Ca}$ channels. Initial evidence indicates that hypertension affects posttranscriptional events, resulting in an upregulation of BK$_{Ca}$ channel expression.

level for the internal protein standard β-actin (associated with the 126-nt band) also is similar between rat strains. As diagrammed in Fig. 3, it is inferred from these findings that the increased expression of the BK$_{Ca}$ channel α-subunit during chronic hypertension may not involve gene regulation, but may be mediated by altered posttranscriptional events, including enhanced translational or translocational processes in the smooth muscle cell or reduced protein turnover in the cytoplasmic membrane.

Although many of the fundamental events that govern the expression of membrane proteins are still unknown, recent studies have clearly indicated that elevations in the level of *in situ* blood pressure can alter protein expression in the vascular smooth muscle membrane. For example, the density of many types of membrane proteins involved in vascular excitation, including membrane-delineated receptors, GTP-binding proteins, and ionic exchangers, are reported to be altered in the cytoplasmic membrane of arterial smooth muscle cells during hypertension (Papp *et al.*, 1993; Li *et al.*, 1994; Sickowski *et al.*, 1994; Kanagy and Webb, 1996; Wang *et al.*, 1996). Thus, elevated blood pressure levels may remodel the molecular composition of the membrane protein population, with a corresponding shift in the mechanisms that regulate cell excitation. Although altered gene regulation likely mediates some of these changes in the expression of ion transport proteins, changes in posttranscriptional regulation also appear to represent an important physiological adaptive mechanism. Indeed, changes in the events regulating the trafficking and membrane turnover of proteins may profoundly influence the expression levels of many important membrane-delineated ion transport proteins, including glucose and H^+-ATPase transporters, renal water channels, and cAMP-sensitive Cl^- channels (Bradbury and Bridges, 1994; Smith, 1995). Similarly, early evidence indicates that changes in posttranscriptional events may mediate the overexpression of the BK$_{Ca}$ channel α-subunit during chronic hypertension, although further studies using a variety of experimental approaches will be required to pinpoint the molecular events that transduce the physical stimulus of high blood pressure into changes in protein expression (Liu *et al.*, 1997, 1998; Berger and Rusch, 1999).

5. FUNCTIONAL IMPLICATIONS OF BK$_{Ca}$ CHANNEL UPREGULATION IN THE MICROCIRCULATION

Most studies characterizing vascular BK$_{Ca}$ channels in hypertension have used the rat aorta as a model, because its large size provides adequate tissue to comprehensively study BK$_{Ca}$ channel expression by different methods (Rusch et al., 1992; England et al., 1993; Liu et al., 1994; Rusch and Runnells, 1994; Liu et al., 1995, 1997). Importantly, however, studies focusing on the expression and function of BK$_{Ca}$ channels during hypertension also have been extended recently to resistance arteries (Martens and Gelband, 1996; Liu et al., 1998; Berger and Rusch, 1999). In particular, the observation that the expression of the BK$_{Ca}$ channel α-subunit is upregulated in aortic smooth muscle membranes from SHR has been confirmed in multiple other circulatory beds from this hypertensive rat model (Liu et al., 1998; Berger and Rusch, 1999). For example, the Western blot in Fig. 4 shows that the density of the 125-kDa band corresponding to the α-subunit of the BK$_{Ca}$ channel is higher in cerebral, mesenteric, and coronary resistance arteries from SHR compared to WKY rats. These data suggest that the upregulation of BK$_{Ca}$ channel α-subunits in vascular smooth muscle membranes during chronic hypertension may represent a universal adaptive response of in situ vascular beds, including large- and small-caliber blood vessels. Patch-clamped vascular smooth muscle membranes from the cerebral and renal microcirculations of SHR also show augmented levels of K$^+$ current through BK$_{Ca}$ channels, corresponding to the enhanced expression of pore-forming α-subunits during hypertension (Liu et al., 1998). Furthermore, in the cerebral circulation, the single-channel properties of the BK$_{Ca}$ channel, including unitary conductance and voltage and Ca^{2+} sensitivity, are similar between SHR and WKY cells (Liu et al., 1998). This latter finding implies that the higher level of BK$_{Ca}$ current observed in vascular smooth muscle cells of SHR is not mediated by an altered phenotypic profile of single BK$_{Ca}$ channels, but instead by an increased expression of the BK$_{Ca}$ channel α-subunit.

Importantly, recent imaging studies of in situ cerebral arterioles suggest that the upregulation of the BK$_{Ca}$ channel α-subunit may limit the development of pathological levels of vascular tone in this circulatory bed (Paterno et al., 1997; Liu et al., 1998). In this regard, the photographs in Fig. 5 demonstrate that pharmacological block of BK$_{Ca}$

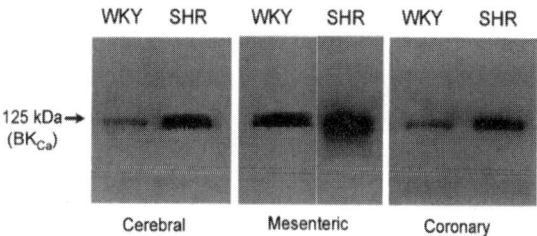

Figure 4. Western blot showing BK$_{Ca}$ channel α-subunit expression in smooth muscle membranes from mesenteric, coronary, and cerebral resistance arteries from WKY rats and SHR. The density of the 125-kDa immunoreactive band, corresponding to the α-subunit was higher in SHR membranes. (Reproduced from Berger and Rusch, 1999, with kind permission from Kluwer Academic Publishers.)

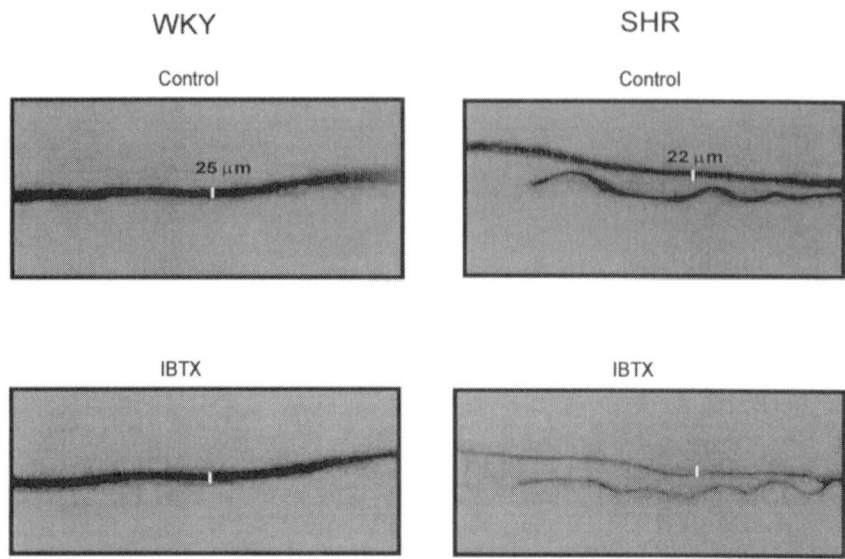

Figure 5. Effect of topical application of 10 nM iberiotoxin (IBTX) on the diameters of WKY rat (*left frames*) and SHR (*right frames*) pial arterioles. Arterioles on the surface of the cerebral cortex were observed through *in situ* cranial windows in anesthetized animals. Block of BK_{Ca} channels by IBTX mildly constricted the WKY arterioles but triggered severe constriction of the SHR arterioles. (Reproduced from Liu *et al.*, 1998, with permission.)

channels by iberiotoxin mildly constricts cerebral arterioles of WKY rats (left frames) but triggers severe constriction in arterioles of SHR (right frames). These findings provide initial compelling evidence for K^+ efflux through BK_{Ca} channels as a powerful homeostatic mechanism that is activated during hypertension to limit smooth muscle cell excitability in the cerebral microcirculation. Thus, SHR cerebral microvessels, which regulate blood flow to the cerebral cortex, appear to rely extensively on an increased resting K^+ current through BK_{Ca} channels to prevent pathological depolarization and vasoconstriction. The vascular adaptive mechanism may prevent closure of small vessels and avert episodes of cerebral ischemia (Liu *et al.*, 1998).

Interestingly, the pressure-induced upregulation of BK_{Ca} channel α-subunit expression in vascular smooth muscle cells may be unique to the BK_{Ca} channel protein, because other K^+ channels that contribute to the control of vessel diameter do not appear to be upregulated in hypertension (Kitazono *et al.*, 1993; Martens and Gelband, 1996). For example, the amplitude of whole-cell K^+ current mediated by voltage-gated K^+ channels in renal vascular smooth muscle cells of the SHR is reduced, rather than increased, during chronic hypertension (Martens and Gelband, 1996). Similarly, vasodilator responses mediated by ATP-sensitive K^+ channels appear to be attenuated in cerebral arteries of stroke-prone SHR compared to WKY rats (Kitazono *et al.*, 1993). Notably, BK_{Ca} channel expression and function is enhanced in vascular smooth muscle cells of these same blood vessels (Liu *et al.*, 1998; Berger and Rusch, 1999). Hence, the stimulus of high blood pressure appears to be linked preferentially to an increased expression of the BK_{Ca} channel α-subunit.

6. THERAPEUTIC TARGETING OF VASCULAR BK$_{Ca}$ CHANNELS IN HYPERTENSION

An augmented vascular resistance is a universal finding in animal and human forms of essential hypertension (Folkow, 1982). Hence, new vasodilator drugs which relieve this pathological vascular tone may have widespread therapeutic applicability as antihypertensive agents. In this regard, several properties of the BK$_{Ca}$ channel indicate that it may represent a promising molecular target for vasodilator drug therapies designed to alleviate hypertension. First, several new K$^+$ channel openers that preferentially activate BK$_{Ca}$ channels have been discovered recently, suggesting the feasibility of specifically targeting amino acid sequences unique to the α- or β-subunits of the BK$_{Ca}$ channel with vasodilator drugs (McManus et al., 1993; Gribkoff et al., 1997). Second, phenotypic differences have been reported between BK$_{Ca}$ channels expressed in nonvascular and vascular tissue, and between BK$_{Ca}$ channels expressed in different vascular beds. In particular, highly variable levels of channel sensitivity to activation by intracellular Ca^{2+} have been described for cloned and native BK$_{Ca}$ channels (Atkinson et al., 1991; Adelman et al., 1992; Butler et al., 1993; Jackson et al., 1996; Wang et al., 1996; Jackson and Blair, 1998; Liu et al., 1998). This functional diversity is thought to relate, at least in part, to alternative splicing of the primary transcript of the BK$_{Ca}$ channel α-subunit, with the subsequent generation of multiple splice variants that assemble into different subtypes of tetrameric α-subunits (Lagrutta et al., 1994; Knaus et al., 1995; McCobb et al., 1995; Shen et al., 1994). Notably, the localization of these subtypes of BK$_{Ca}$ channels to different vascular beds may permit the selective targeting of vasodilator drugs to specific circulations. Moreover, the recent description of a BK$_{Ca}$ channel subtype in the microcirculation of skeletal muscle that shows unique Ca^{2+}-sensitive properties may be significant, because these microcirculatory vessels contribute heavily to the pathological vasoconstriction that underlies hypertension (Jackson et al., 1996; Jackson and Blair, 1998). Hence, drugs that preferentially activate BK$_{Ca}$ channels in smooth muscle cell membranes may effectively reduce blood pressure levels. Finally, the pressure-induced upregulation of BK$_{Ca}$ channel α-subunits in arterial smooth muscle membranes may amplify the number of binding sites available to new antihypertensive drugs designed to activate BK$_{Ca}$ channels, thereby augmenting their therapeutic potency. Furthermore, because α-subunit expression is proportional to the blood pressure signal (Rusch and Runnells, 1994), the lowering of blood pressure by effective drug treatment also would act to reduce the number of BK$_{Ca}$ channels available for subsequent drug binding. Thus, adverse side effects of antihypertensive therapies, such as orthostatic hypotension, may be minimized by taking advantage of the proportional relationship between the level of in situ blood pressure and BK$_{Ca}$ channel expression.

7. SUMMARY

In the past few years, there has been an increasing awareness that ion channel populations in the vascular smooth muscle membrane demonstrate remodeling of their native channel subunits during cardiovascular disease states. As the molecular composition of these ion channel populations shifts, the type of ion channel families that regulate the excitability of the vascular smooth muscle cells also may be profoundly altered. In systemic hypertension, as summarized in this chapter, an increased expression of the BK_{Ca} channel α-subunit in vascular smooth muscle membranes appears to be a universal event mediated by posttranscriptional regulation. Functionally, this upregulation of the pore-forming α-subunit may provide a homeostatic mechanism by which BK_{Ca} channels can counteract the pathological vascular tone that form the basis of hypertensive disease. As such, BK_{Ca} channels may assist other physiological feedback mechanisms of the body that act to buffer further rises in blood pressure levels, which, in the absence of compensatory mechanisms, could lead to malignant hypertension. Finally, in critical vascular beds such as the cerebral microcirculation, the upregulation of the BK_{Ca} channel α-subunit appears to empower this channel with the ability to buffer vasoconstriction and prevent vasospasm. Hence, the development of vasodilator drugs targeted to this highly expressed and functionally relevant K^+ channel type may provide new therapies for normalizing regional tissue perfusion and lowering blood pressure levels in hypertensive pathologies.

REFERENCES

Abel, P. W., Trapani, A., Matsuki, N., Ingram, M. J., Ingram, F. D., and Hermsmeyer, K., 1981, Unaltered membrane properties of arterial muscle in Dahl strain genetic hypertension, *Am. J. Physiol.* **241:**H224–H227.

Adelman, J. P., Shen, K.-Z., Kavanaugh, M. P., Warren, R. A., Wu, Y., Lagrutta, A., Bond, C. T., and North, R. A., 1992, Calcium-activated potassium channels expressed from cloned complementary DNAs, *Neuron* **9:**209–216.

Asano, M., Masuzawa-Ito, K., Matsuda, T., Imaizumi, Y., Watanabe, M., and Ito, K., 1993a, Functional role of Ca^{2+}-activated K^+ channels in resting state of carotid arteries from SHR, *Am. J. Physiol.* **265:**H843–H851.

Asano, M., Masuzawa-Ito, K., and Matsuda, T., 1993b, Charybdotoxin-sensitive K^+ channels regulate the myogenic tone in the resting state of arteries from spontaneously hypertensive rats, *Br. J. Pharmacol.* **108:**214–223.

Atkinson, N. S., Robertson, G. A., and Ganetzky, B., 1991, A component of calcium-activated potassium channels encoded by the *Drosophila slo* locus, *Science* **253:**551–555.

Berczi, V., Stekiel, W. J., Contney, S. J., and Rusch, N. J., 1992, Pressure-induced activation of membrane K^+ current in rat saphenous artery, *Hypertension* **19:**725–729.

Berger, M. G., and Rusch, N. J., 1999, Voltage and calcium-gated potassium channels: Functional expression and therapeutic potential in the vasculature, in: *Perspectives in Drug Discovery and Design* (J. M. Sabatier and H. Darbon, eds.), Kluwer Academic, Dordrecht, The Netherlands, Vol. 15/16, pp. 313–332.

Bradbury, N. A., and Bridges, R. J., 1994, Role of membrane trafficking in plasma membrane solute transport, *Am. J. Physiol.* **267:**C1–C24.

Brayden, J. E., and Nelson, M. T., 1992, Regulation of arterial tone by activation of calcium-dependent potassium channels, *Science* **356:**532–535.

Butler, A., Tsunoda, S., McCobb, D. P., Wei, A., and Salkoff, L., 1993, *mSlo*, a complex mouse gene encoding "maxi" calcium-activated potassium channels, *Science* **261:**221–224.

Carl, A., Lee, H. K., and Sanders, K. M., 1996, Regulation of ion channels in smooth muscles by calcium, *Am. J. Physiol.* **271:**C9–C34.

Cui, J., Cox, D. H., and Aldrich, R. W., 1997, Intrinsic voltage dependence and Ca^{2+} regulation of *mslo* large conductance Ca-activated K^+ channels, *J. Gen. Physiol.* **109:**647–673.

Dopica, A. M., Kirber, M. T., Singer, J. J., and Walsh, J. V., 1994, Membrane stretch directly activates large conductance Ca^{2+}-activated K$^+$ channels in mesenteric artery smooth muscle cells, *Am. J. Hypertens.* **7:**82–89.

England, S. K., Wooldridge, T. A., Stekiel, W. J., and Rusch, N. J., 1993, Enhanced single-channel K$^+$ current in arterial membranes from genetically hypertensive rats, *Am J. Physiol.* **264:**H1337–H1345.

Folkow, B., 1982, Physiological aspects of primary hypertension, *Physiol. Rev.* **62:**347–504.

Friedman, S. M., and Friedman, C. L., 1976, Cell permeability, sodium transport, and the hypertensive process in the rat, *Circ. Res.* **39:**433–441.

Garwitz, E. T., and Jones, A. W., 1982a, Aldosterone infusion into the rat and dose dependent changes in blood pressure and arterial ionic transport, *Hypertension* **4:**374–381.

Garwitz, E. T., and Jones, A. W., 1982b, Altered arterial ion transport and its reversal in the aldosterone hypertensive rat, *Am. J. Physiol.* **243:**H929–H933.

Gribkoff, V. K., Starrett, J. E., and Dworetzky, S. I., 1997, The pharmacology and molecular biology of large-conductance calcium-activated (BK) potassium channels, *Adv. Pharmacol.* **37:**319–348.

Hermsmeyer, K., 1976, Electrogenesis of increased norepinephrine sensitivity of arterial vascular muscle in hypertension, *Circ. Res.* **38:**362–367.

Hermsmeyer, K., 1982, Electrogenic ion pumps and other determinants of membrane potential in vascular muscle (the 1982 Bowditch Lecture), *Physiologist* **25:**454–465.

Jackson, W. F., and Blair, K. L., 1998, Characterization and function of Ca^{2+}-activated K$^+$ channels in arteriolar muscle cells, *Am. J. Physiol.* **274:**H27–H34.

Jackson, W. F., Blair, K. L., and Rusch, N. J., 1996, Large conductance Ca^{2+}-activated K$^+$ channels in arteriolar muscle cells have low Ca^{2+}-sensitivity. *Microcirculation* **3:**92 (Abstract).

Jones, A. W., 1973, Altered ion transport in vascular smooth muscle from spontaneously hypertensive rats, *Circ. Res.* **33:**563–572.

Jones, A. W., 1974, Reactivity of ion fluxes in rat aorta during hypertension and circulatory control, *Fed. Proc.* **33:**133–137.

Jones, A. W., 1983, Arterial tissue cations, in: *Hypertension, Physiology and Treatment* (J. Genest, O. Kuchel, P. Hamet, and M. Cantin, eds.), McGraw-Hill, New York, pp. 488–497.

Jones, A. W., and Hart, R. G., 1975, Altered ion transport in aortic smooth muscle during deoxycorticosterone acetate hypertension in the rat, *Circ. Res.* **37:**333–341.

Kanagy, N. L., and Webb, R. C., 1996, Increased responsiveness and decreased expression of G proteins in deoxycorticosterone hypertension, *Hypertension* **27:**740–745.

Kitazono, T., Heistad, D. D., and Faraci, F. M., 1993, ATP-sensitive potassium channels in the basilar artery during chronic hypertension, *Hypertension* **22:**677–681.

Kolias, T. J., Chai, S., and Webb, R. C., 1993, Potassium channel antagonists and vascular reactivity in stroke-prone spontaneously hypertensive rats, *Am. J. Hypertens.* **23:**1077–1082.

Knaus, H.-G., Folander, K., Garcia-Calvo, M., Garcia, M. L., Kaczorowski, G. J., Smith, M., and Swanson, R., 1994, Primary sequence and immunological characterization of β-subunit of high conductance Ca^{2+}-activated K$^+$ channel from smooth muscle, *J. Biol. Chem.* **269:**17274–17278.

Knaus, H.-G., Eberhart, A., Koch, R. O. A., Munujos, P., Schmalhofer, W. A., Warmke, J. W., Kaczorowski, G. J., and Garcia, M. L., 1995, Characterization of tissue-expressed α subunits of the high conductance Ca^{2+}-activated K$^+$ channel, *J. Biol. Chem.* **270:**22434–22439.

Kwan, C. Y., 1985, Dysfunction of calcium handling by smooth muscle in hypertension, *Can. J. Physiol. Pharmacol.* **63:**366–374.

Lagrutta, A., Shen, K., North, R. A., and Adelman, J. P., Functional differences among alternatively spliced variants of *Slowpoke*, a *Drosophila* calcium-activated potassium channel, *J. Biol. Chem.* **269:**20347–20351.

Lamb, F. S., and Webb, R. C., 1989, Regenerative electrical activity and arterial contraction in hypertensive rats, *Hypertension* **13:**70–76.

Li, P., Zou, A. P., Al-Kayed, N. J., Rusch, N. J., and Harder, D. R., 1994, Guanine nucleotide-binding proteins in aortic smooth muscle from hypertensive rats, *Hypertension* **23:**914–918.

Liu, Y., Jones, A. W., and Sturek, M., 1994, Increased barium influx and potassium current in stroke-prone spontaneously hypertensive rats, *Hypertension* **23:**1091–1095.

Liu, Y., Jones, A. W., and Sturek, M., 1995, Ca^{2+}-dependent K$^+$ current in arterial smooth muscle cells from aldosterone-salt hypertensive rats, *Am. J. Physiol.* **269:**H1246–H1257.

Liu, Y., Pleyte, K. A., Knaus, H.-G., and Rusch, N. J., 1997, Increased expression of Ca^{2+}-sensitive K$^+$ channels in aorta of hypertensive rats, *Hypertension* **30:**1403–1409.

Liu, Y., Hudetz, A., Knaus, H.-G., and Rusch, N. J., 1998, Increased expression of Ca^{2+}-sensitive K^+ channels in the cerebral microcirculation of genetically hypertensive rats. Evidence for their protection against cerebral vasospasm, *Circ. Res.* **82:**729–737.

Longhurst, P. A., Rice, P. J., Taylor, D. A., and Fleming, W. W., 1988, Sensitivity of caudal arteries and the mesenteric vascular bed to norepinephrine in DOCA-salt hypertension, *Hypertension* **12:**133–142.

Martens, J. R., and Gelband, C. H., 1996, Alterations in rat interlobar artery membrane potential and K^+ channels in genetic and nongenetic hypertension, *Circ. Res.* **79:**295–301.

McCobb, D. P., Natalie, L. F., Featherstone, T., Lingle, C. J., Saito, M., Krause, J. E., and Salkoff, L., 1995, A human calcium-activated potassium channel gene expressed in vascular smooth muscle, *Am. J. Physiol.* **269:**H767–H777.

McManus, O. B., Harris, G. H., Giangiacom, K. M., Feigenbau, P., Reube, J. P., Addy, M. E., Burka, J. F., Kaczorowski, G. J., and Garcia, M. L., 1993, An activator of calcium-dependent potassium channels isolated from a medicinal herb, *Biochemistry* **32:**6128–6133.

McManus, O. B., Helms, L. M. H., Pallanck, L., Ganetsky, B., Swanson, R., and Leonard, R. J., 1995, Functional role of the β subunit of high conductance calcium-activated potassium channels, *Neuron* **14:**645–650.

Meera, P., Wallner, M., Jiang, Z., and Toro, L., 1996, A calcium switch for the functional coupling between α (*hslo*) and β subunits ($K_{V,Ca}\beta$) of maxi K channels, *FEBS Lett.* **382:**84–88.

Nelson, M. T., Patlak, J. B., Worley, J. F., and Standen, N. B., 1990, Calcium channels, potassium channels, and voltage-dependence of arterial smooth muscle tone, *Am. J. Physiol.* **259:**C3–C18.

Papp, B., Corvazier, E., Magnier, C., Kovacs, T., Bourdeau, N., Levy-Toledano, S., Bredoux, R., Levy, B., Poitevin, P., Lompre, A. M., Wuytack, F., and Enouf, J., 1993, Spontaneously hypertensive rats and platelet Ca^{2+}-ATPases: Specific upregulation of the 97-kDa isoform, *Biochem. J.* **295:**685–690.

Paterno, R., Heistad, D. D., and Faraci, F. M., 1997, Functional activity of Ca^{2+}-dependent K^+ channels is increased in basilar artery during chronic hypertension, *Am. J. Physiol.* **272:**H1287–H1291.

Rusch, N. J., and Hermsmeyer, K., 1993, Vascular muscle calcium channels in hypertension, in: *Ionic Transport in Hypertension: New Perspectives* (A. Coca, ed.), CRC Press, Boca Raton, Florida, pp. 197–227.

Rusch, N. J., and Liu, Y., 1997, Potassium channels in hypertension: Homeostatic pathways to buffer arterial contraction, *J. Lab. Clin. Med.* **130:**245–251.

Rusch, N. J., and Runnells, A. M., 1994, Remission of high blood pressure reverses arterial potassium channel alterations, *Hypertension* **23:**941–945.

Rusch, N. J., De Lucena, R. G., Wooldridge, T. A., England, S. K., and Cowley, A. W., 1992, A Ca^{2+}-dependent K^+ current is enhanced in arterial membranes of hypertensive rats, *Hypertension* **19:**301–307.

Rusch, N. J., Liu, Y., and Pleyte, K. A., 1996, Mechanisms for regulation of arterial tone by Ca^{2+}-dependent K^+ channels in hypertension, *Clin. Exp. Pharmacol. Physiol.* **23:**1077–1082.

Sansom, S. C., and Stockand, J. D., 1994, Differential Ca^{2+}-sensitivities of BK(Ca) isochannels in bovine mesenteric vascular smooth muscle, *Am. J. Physiol.* **266:**C1182–C1189.

Schreiber, M., and Salkoff, L., 1997, A novel calcium-sensing domain in the BK channel, *Biophys. J.* **73:**1355–1363.

Scornik, F. S., and Toro, L., 1992, U46619, a thromboxane A_2 agonist, inhibits K_{Ca} channel activity for pig coronary artery, *Am. J. Physiol.* **262:**C708–C713.

Shen, K., Lagrutta, A., Davies, N. W., Standen, N. B., Adelman, J. P., and North, R. A., 1994, Tetraethylammonium block of *Slowpoke* calcium-activated potassium channels expressed in *Xenopus* oocytes: Evidence for tetrameric channel formation, *Pflügers Arch.* **426:**440–445.

Sickowski, M., Davies, D. E., and Ng, L. L., 1994, Sodium–hydrogen antiporter protein in normotensive Wistar-Kyoto rats and spontaneously hypertensive rats, *J. Hypertens.* **12:**775–781.

Smith, A. E., 1995, Treatment of cystic fibrosis based on understanding CFTR, *J. Inher. Metab. Dis.* **18:**508–516.

Smith, J. M., and Jones, A. W., 1991, Calcium antagonists inhibit elevated potassium efflux from aorta of aldosterone-salt hypertensive rats, *Hypertension* **15:**78–83.

Stefani, E., Ottolia, M., Noceti, F., Olcese, R., Wallner, M., Latorre, R., and Toro, L., 1997, Voltage-controlled gating in a large conductance Ca^{2+}-sensitive K^+ channel (*hslo*), *Proc. Natl. Acad. Sci. U.S.A.* **94:**5427–5431.

Stekiel, W. J., Contney, S. J., and Lombard, J. H., 1986, Small vessel membrane potential, sympathetic input, and electrogenic pump rate in SHR, *Am. J. Physiol.* **250:**C547–C556.

Tanaka, Y., Meera, P., Song, M., Knaus, H.-G., and Toro, L., 1997, Molecular constituents of maxi K$_{Ca}$ channels in human coronary smooth muscle: Predominant $\alpha + \beta$ subunit complexes, *J. Physiol.* **502:**545–557.

Thompson, L. P., Bruner, C. A., Lamb, F. S., King, C. M., and Webb, R. C., 1987, Calcium influx and vascular reactivity in systemic hypertension, *Am. J. Cardiol.* **59:**29A–34A.

Toro, L., Vaca, L., and Stefani, E., 1991, Calcium-activated potassium channels from coronary smooth muscle reconstituted in lipid bilayers, *Am. J. Physiol.* **260:**H1779–H1789.

Wang, Y., and Mathers, D. A., 1993, Ca^{2+}-dependent K^{+} channels of high conductance in smooth muscle cells isolated from rat cerebral arteries, *J. Physiol.* **462:**529–545.

Wang, D. H., Du, Y., and Yao, A., 1996, Regulation of the gene encoding angiotensin II receptor in vascular tissue, *Microcirculation* **3:**237–239.

Wei, A., Solaro, C., Lingle, C., and Salkoff, L., 1994, Calcium sensitivity of BK-type K$_{Ca}$ channels determined by a separable domain, *Neuron* **13:**671–681.

Chapter 43

Antisense Approaches and the Modulation of Potassium Channel Function in the Cardiovascular System

Craig H. Gelband

1. INTRODUCTION

Potassium channels are key regulators of resting membrane potential and tightly control the patterns of electrical excitability in cardiac myocytes and vascular smooth muscle cells. Under typical conditions of physiological variability *in vivo*, the level of K^+ channel expression in cardiovascular membranes appears to be dynamically regulated to permit normal tissue-specific electrophysiological function. However, in some cardiovascular pathologies, as reviewed in other chapters in this book, alterations in K^+ channels appear to contribute to, rather than buffer, abnormalities in cell excitation patterns. For example, mutations in myocardial K^+ channels are linked to cardiac arrhythmias and sudden death (see Chapters 17 and 36), and in the pulmonary vasculature, the downregulation of K^+ channels may contribute to the etiology of primary pulmonary hypertension (see Chapters 27 and 40). Thus, it is not surprising that tremendous efforts are being directed toward understanding the molecular mechanisms that govern the gene and protein expression levels of K^+ channel subunits, as well as the phenotypic profile of K^+ channels in cardiac myocytes and vascular smooth muscle cells. Similarly, new therapeutic approaches are being considered to treat ion channel abnormalities in cardiovascular disease states.

In this regard, traditional agents which modulate ion channel function are used to treat cardiac arrhythmias, hypertension, diabetes, and vasospastic diseases. Most of these agents are relatively reliable and affordable, and their short duration of action has permitted their use in short-term therapy. However, there are major limitations and disadvantages of traditional drug therapies. The effects of these drugs are short-lived,

Craig H. Gelband • Department of Physiology, College of Medicine, University of Florida, Gainesville, Florida 32610.

Potassium Channels in Cardiovascular Biology, edited by Archer and Rusch. Kluwer Academic/Plenum Publishers, New York, 2001.

and they have to be administered on a regular basis. In addition, traditional therapeutic agents produce significant side effects, which create major compliance issues with patients. Furthermore, traditional agents do not always reverse or prevent morphological and pathophysiological complications associated with the primary disease state and hold little promise for an actual cure.

In light of this, many investigators have begun to explore gene therapy as an alternative approach. On a conceptual level, this approach may offer major advantages over traditional drug therapy. It may eliminate the issue of compliance, because long-term and permanent therapeutic effects are possible. Genetic control also may be influential in providing cures rather than continued therapies, if appropriate target genes are identified and their expression is regulated. Finally, gene therapy may be the key to minimizing or eliminating side effects inherent with traditional therapy. The objectives of this chapter are to (1) introduce the principles of gene therapy as they relate to antisense approaches, (2) review recent studies in which antisense approaches have been used to elucidate K^+ channel function, and (3) discuss the potential usefulness of gene therapy in preventing or reversing ion channel alterations associated with systemic hypertension.

2. THE ANTISENSE APPROACH

Inhibition of protein expression by molecular suppression of messenger RNA (mRNA) at a genetic level has been accomplished by the use of antisense technology (Fig. 1). In a normal cell, mRNA is synthesized in the nucleus and is transported across the nuclear membrane into the cytoplasm, where it undergoes translation involving ribosomes. Protein produced as a result of translation is appropriately processed and compartmentalized to exert its physiological action. Antisense technology takes advantage of the fact that the antisense orientation of RNA for a given protein hybridizes with the active mRNA in the cytoplasm. As a result, normal translation of the mRNA is blocked, mRNA is degraded, and no formation of the relevant protein occurs. Thus, antisense treatment inhibits translation and decreases the levels of active protein. Antisense approaches have been used successfully to alter the expression of K^+ channels in *Xenopus* oocytes, neuronal cell lines, and cultured cardiac atrial and ventricular myocytes. *In vivo*, full-length antisense cDNA for the angiotensin type 1 receptor (AT_1 receptor) has been demonstrated to produce long-term antihypertensive effects in a genetic rat model of hypertension and thereby prevent pathological ion channel alterations associated with chronic blood pressure elevation (Iyer *et al.*, 1996; Martens *et al.*, 1998; Gelband *et al.*, 1999). Thus, K^+ channel populations in cardiac and vascular smooth muscle membranes may be direct targets of antisense therapy, or they may be modified by gene therapies directed against diseases that entail their involvement.

In either situation, the key to successful antisense gene therapy is to identify an ideal vector, which can deliver a given gene into the relevant target tissues. The objective is to express the transgene in an appropriate quantity and for an adequate duration to exert its beneficial effects. The time, quantity, and tissue specificity will, of course, vary depending on the gene of interest and the pathophysiological condition it

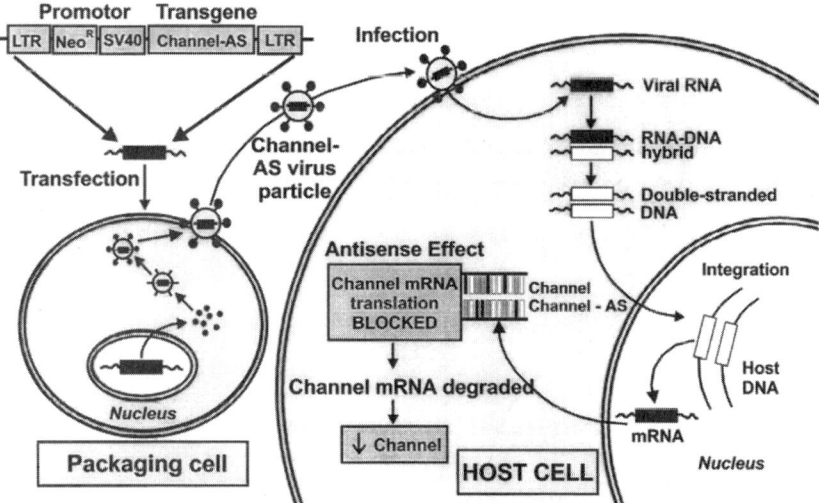

Figure 1. Mechanism of viral-mediated antisense gene therapy. The transgene, in this case an ion channel antisense (AS), is inserted into the virus or retrovirus genome. This retrovirus has long-terminal repeats (LTR), a neomycin-resistance gene (Neo), and an SV40 promoter. The virus enters the packaging cell, which packages and releases the viral particles. Viral particles infect the host cell and produce channel antisense, which hybridizes with the target mRNA to block translation.

is designed to correct. Thus, the definition of an ideal vector is not a general one, but is specific to a particular physiological or pathophysiological situation. For example, an ideal vector for long-term or permanent control of K$^+$ channel expression should have the following characteristics: (i) it should be able to integrate into a specific location of the chromosome for the long-term expression of a given transgene; (ii) the vector genome should be easily manipulated in order to introduce large segments of DNA; (iii) it should be conveniently produced in large quantities containing high concentrations of viral particles [$>1 \times 10^8$ colony-forming units (cfu)/ml]; (iv) it should be constructed so that tissue-specific expression is appropriately regulated; and (v) it should elicit no or a very minimal immune response. In contrast, if one is interested in altering K$^+$ channel expression transiently, for example, to downregulate K$^+$ channels in proliferating smooth muscle cells implicated in restenosis after angioplasty, an ideal vector would not be one that produces a long-term effect. Instead, the ideal vector should cause an immediate and robust transduction of a relevant gene in a transient fashion and be specific for the restenotic site.

Viral vectors have become tremendously popular vehicles to deliver genes, because they are efficient in delivering a transgene into the host tissue. Viral genomes that are humanized, highly efficient, and simple for nonvirologists to use are now being developed. In a humanized viral vector, all the genes and promoters that have the slightest chance of expressing in humans are deleted. As a result, humanized viral particles are practically incapable of producing and exerting adverse effects in humans. In addition, they have the inherent advantage of recognizing specific receptors on host tissues, thus providing an efficient mechanism for their delivery across the plasma membrane. Both RNA and DNA viral vectors are available, with each having its own advantages. A retroviral vector that has a high infection efficiency of transduction and integrates into DNA for long-term expression of the transgene, is sometimes preferred. Retroviral vectors are excellent in their ability to infect dividing cells but have limited ability to infect nondividing cells. The use of a retroviral vector has permitted the successful transduction of membrane receptor proteins in neurons and astroglial and vascular smooth muscle cells *in vitro* (Lu *et al.*, 1995, 1997), and the antisense may be expressed on a long-term basis (Lu and Raizada, 1995; Lu *et al.*, 1995; Iyer *et al.*, 1996; Martens *et al.*, 1998; Lu *et al.*, 1998; Gelband *et al.*, 1999).

3. USE OF ANTISENSE OLIGONUCLEOTIDES TO MODIFY K⁺ CHANNEL EXPRESSION *In Vitro*

In vitro, antisense technologies have been used to elucidate the function of K^+ channels in cardiovascular tissues. In the heart, where a number of different K^+ channel genes are expressed, K^+ channels play an exquisite role in the regulation of the action potential configuration. Antisense oligodeoxynucleotides (ASODN) to various K^+ channels have been used to investigate if specific channels underlie the various components of the repolarizing phase of the cardiac action potential. The specific K^+ current subtypes that are postulated to play a role in the cardiac action potential include the inward rectifier K^+ current (I_{Kir}), the ultrarapid (I_{Kur}), rapid (I_{Kr}), and slowly activating (I_{Ks}) voltage-gated K^+ (Kv) currents, the ATP-sensitive K^+ current (I_{KATP}), and the transient outward K^+ current (I_{to}). Antisense technologies directed against many of the K^+ channel subtypes have been useful in elucidating their contributions to the repolarizing K^+ current of cardiac muscle.

For example, Feng *et al.* (1997) used antisense targeted to Kv1.5 to investigate the putative role of this channel in contributing to the ultrarapid K^+ current, I_{Kur}, in the atrial myocardium. Indeed, antisense ASODNs against Kv1.5 caused a significant reduction in I_{Kur} amplitude in human atrial cells when compared to similar cells treated with missense to Kv1.5. When ASODNs against Kir2, a member of the inward rectifier family of K^+ channels, was applied to rat ventricular myocytes, not only was the amplitude of I_{Kir} reduced but the 21-pS channel that underlies I_{Kir} was suppressed (Nakamura *et al.*, 1998). ASODNs against the sulfonylurea receptors SUR1 and SUR2 have been noted to dramatically reduce the K^+ current component associated with K_{ATP} channels in rat ventricular myocytes (Yokoshiki *et al.*, 1999). A similar approach has provided insight into the distinct K^+ channel subtypes that underlie I_{to}, the current that is responsible for the "notch" or fast repolarization phase of atrial and ventricular action potentials. In these studies, Wang *et al.* (1999) used antisense to Kv1.4, Kv4.2, and Kv4.3 to elucidate the functional contributions of these rapidly inactivating K^+ channels to I_{to}. It was found that all three channels underlie I_{to} in rabbit atrial myocytes, whereas only ASODN targeted to Kv4.3 affected the level of I_{to} in human atrial cells. Finally, ASODN against minK, a K^+ channel that combines with HERG or KvLQT1 to form I_{Kr} and I_{Ks}, specifically inhibited both types of current (Yang *et al.*, 1995; McDonald *et al.*, 1997). Therefore, antisense approaches are helping to elucidate the functional roles of K^+ channel subtypes in particular cell systems.

However, antisense methodologies have yet to be used as a tool to identify the K$^+$ channel subtypes that underlie the distinct components of K$^+$ efflux in vascular smooth muscle cells.

4. K$^+$ CHANNEL ALTERATIONS IN CARDIOVASCULAR PATHOLOGIES: POTENTIAL TARGETS OF ANTISENSE THERAPIES

As summarized above, the antisense approach has been used primarily to alter K$^+$ channel gene expression *in vitro* and thereby identify specific K$^+$ channel phenotypes that contribute to the excitability patterns of cultured myocardial cells. However, to be therapeutically useful, antisense therapies targeting K$^+$ channels ultimately will need to be directed toward the treatment of cardiovascular pathologies in which the normal patterns of K$^+$ channel expression are disrupted *in vivo*. In this regard, several disease-specific profiles of K$^+$ channel expression have been described in the atrial and ventricular myocardium and in the vasculature. Gene therapy may represent one therapeutic avenue for normalizing these aberrant channel profiles. Indeed, as discussed in a later section of this chapter, it appears that anomalous patterns of ion channel expression in pathogenic states such as systemic hypertension may be restored to near-normal profiles by gene therapy, including antisense treatment that secondarily restores typical patterns of K$^+$ channel expression by eliminating primary pathogenic stimuli.

In the heart, one might imagine that antisense therapies may be useful for minimizing the expression of K$^+$ channel genes encoding mutant channel subunits that promote cardiac arrhythmias and sudden death. Similarly, antisense therapy may provide a useful tool for restoring normal patterns of excitation to myocardial cells possessing native K$^+$ channels that represent normal proteins but are expressed in abnormal densities or patterns during disease states. For example, as reviewed elsewhere in this book (Chapters 17 and 36), the molecular basis of the long QT syndrome includes mutations in several different genes encoding cardiac K$^+$ channels that result in myocardial repolarization disorders and susceptibility to ventricular tachyarrhythmias and sudden death. Antisense suppression of these gene products may normalize repolarization abnormalities by reducing mutant K$^+$ channel α-subunits that otherwise may incorporate into defective heteromultimeric channels to reduce the amplitude of repolarizing K$^+$ currents or alter their gating properties (Sanguinetti *et al.*, 1996; Folco *et al.*, 1997; London *et al.*, 1998). In contrast, an upregulation of voltage-gated K$^+$ channels is suggested to contribute to the shortened atrial refractory period and abbreviated action potential duration that exists in atrial myocytes during chronic atrial fibrillation, although patch-clamp studies have not confirmed this hypothesis (for details, see Chapter 17). Thus, identification of the K$^+$ channel subtypes involved in mediating this dysrhythmia will be required before specific antisense therapies can be developed to target this disorder.

In the vasculature, the recent link between K$^+$ channel activation and the proliferation of vascular smooth muscle cells may provide an important future target of antisense therapy. In nonvascular and vascular cell types, the activation of K$^+$ channels is associated with an increased Ca^{2+} influx resulting from the enhanced electrical driving force established by membrane hyperpolarization. This influx of Ca^{2+}, in turn, is thought to act as a messenger to initiate the mitogenic signal cascade that triggers cell proliferation (Tsien et al., 1982; DeCoursey et al., 1984; Neylon et al., 1994; Lepple-Wienhues et al., 1996; for a review, see Wonderlin and Strobl, 1996). Similarly, serum growth factors appear to activate K^+ channels to induce factor-stimulated mitogenesis (Wang et al., 1997). Hence, cultured rat aortic smooth muscle cells that rapidly proliferate express a high level of Ca^{2+}-dependent K^+ channel activity and an enhanced level of hyperpolarization that is blocked by charybdotoxin (Neylon et al., 1994). Thus, an early event in the initiation of smooth muscle cell proliferation may be an upregulation of K^+ channels in the cell plasma membrane, and antisense therapy that lowers the expression level of these regulatory channels may impede the anomalous proliferation of vascular muscle cells in conditions including atherosclerosis and post-balloon angioplasty.

Additionally, an abnormal profile of K^+ channel expression appears to be involved in the pathogenesis of several forms of vasoconstrictor disease (for a review, see Berger and Rusch, 1999). Indeed, other chapters in this book (Chapters 27 and 40) describe in detail the disease-specific shifts in K^+ channel expression that may occur in the vasculature of animals with hypoxia-induced pulmonary hypertension, in human subjects with primary pulmonary hypertension, and in genetic and renal models of chronic systemic hypertension in rats. In these hypertensive diseases, an increased vascular tone appears to be linked to a downregulation of the pore-forming α-subunits of Kv channels in the vascular smooth muscle membranes, perhaps concurrent with an increased activity of voltage-gated Ca^{2+} channels. Although a compensatory increase in the membrane expression of Ca^{2+}-activated K^+ channels may buffer the development of further vascular tone, the final excitability level of the vascular smooth muscle cell apparently is enhanced, resulting in vasoconstriction. Clearly, antisense therapies that could act to normalize these abnormal ion channel profiles may restore typical patterns of electrical excitability to the vasculature, and thereby reduce the enhanced vasoconstrictor tone that is the hallmark finding in hypertensive disease.

5. VOLTAGE-GATED K^+ (Kv) CHANNELS IN VASCULAR SMOOTH MUSCLE MEMBRANES

As reviewed elsewhere in this book (Chapters 18 and 26), K^+ channels in vascular smooth muscle membranes represent the products of multiple gene families. Of these, the Kv gene family comprises the largest and most diverse class of known ion channels, and Kv channels appear to be ubiquitously expressed in vascular smooth muscle. They have been identified and their currents characterized in the portal, coronary, cerebral, renal, pulmonary, and mesenteric vasculatures (Beech and Bolton, 1989; Benham and Bolton, 1986; Bonnet et al., 1991; Gelband and Hume, 1992; Hume and Leblanc, 1989; Jones, 1973; Martens and Gelband, 1996; Okabe et al., 1987; Smirnov and Aaronson, 1992b; Toro and Stefani, 1987; Wilde and Lee, 1989).

Importantly, alterations in the transcriptional regulation of the Kv gene family may be involved in the primary etiology of several cardiovascular pathologies, or these alterations may occur as a secondary contributor to disease. However, because a single smooth muscle cell expresses multiple Kv channel subtypes and, furthermore, K⁺ channel distributions differ between vascular beds and even along the length of a given arterial tree (Albarwani *et al.*, 1995; Evans *et al.*, 1996), the issue of identifying distinct Kv channel gene subtypes to target with antisense approaches is extremely complex. Furthermore, within each gene family, alternative splicing may give rise to many channel subtypes, and Kv channel subunits from the same gene family may form a heterogeneous population of multimeric functional structures (Jan and Jan, 1992). This high level of Kv channel diversity complicates the identification of functionally relevant Kv channel components that may act to regulate the excitability level in vascular smooth muscle cells.

Despite this intricate expression system, pharmacological studies indicate that Kv channels, sensitive to block by the drug 4-aminopyridine, represent an important mechanism for membrane hyperpolarization in vascular smooth muscle cells. Owing to the high input resistance of vascular smooth muscle cells, the opening of only a small number of Kv channels may limit membrane depolarization and facilitate vasorelaxation (Nelson and Quayle, 1995). Kv channels also help to establish the negative resting membrane potential of arterial muscle cells. These channels show a steep relationship between membrane potential and maintained open-state probability. As determined from the overlap of mean activation and inactivation curves, Kv channels provide a significant, sustainable outward K⁺ current at the resting membrane potential of vascular smooth muscle cells (Nelson and Quayle, 1995). Thus, pharmacological block of Kv channels leads to membrane depolarization in current-clamped vascular smooth muscle cells as well as constriction of intact vessels (Gelband and Hume, 1995; Knot and Nelson, 1995; Post *et al.*, 1995; Smirnov and Aaronson, 1992b; Yuan, 1995). For example, the bath application of 4-aminopyridine to provide pharmacological block of Kv channels causes membrane depolarization (~ 20 to 25 mV) of single smooth muscle cells isolated from rat renal resistance arteries, and the same drug contracts isolated segments of rat renal arteries (Martens and Gelband, 1996). Vasoconstrictor substances such as angiotensin II and histamine and hypoxia also may contract vascular smooth muscle by closing Kv channels (Ishikawa *et al.*, 1993; Post *et al.*, 1995; Gelband and Hume, 1995; Martens and Gelband, 1998). Thus, at least in some vascular beds, Kv channels appear to significantly contribute to the resting membrane potential of the vascular smooth muscle cells, and the downregulation of these channels would be expected to result in the development of abnormal vascular tone.

It is important to realize, however, that the mechanism by which Kv channels buffer vasoconstriction is by mediating membrane hyperpolarization, which, in turn, reduces voltage-gated Ca^{2+} influx through L-type Ca^{2+} channels in the plasma membrane. The open-state probability of L-type Ca^{2+} channels is highly dependent on membrane potential, and the overlapping nature of the L-type Ca^{2+} channel activation and inactivation curves predicts a small sustained Ca^{2+} influx at the levels of resting membrane potentials observed in arteries and arterioles *in situ*. Therefore, the L-type Ca^{2+} channel is thought to account for the steady-state levels of cytosolic Ca^{2+}, $[Ca^{2+}]_i$, necessary to maintain vascular tone, and Kv channels, membrane potential, and Ca^{2+} influx are intricately linked to the regulation of vessel diameter and resting vascular tone (Martens and Gelband, 1998; Fleischmann *et al.*, 1994; Smirnov and Aaronson, 1992).

6. VASCULAR Kv AND Ca^{2+} CHANNEL ALTERATIONS ARE COEXPRESSED IN HYPERTENSION

Essential hypertension in both humans and animals involves a gradual and sustained increase in total peripheral resistance. This fundamental reduction in vessel diameter apparently reflects an increased contractile state of the arterial smooth muscle cells, because resting vascular tone is elevated and the contractile responses to a number of physiological stimuli are enhanced in hypertension. These changes may contribute to the elevation of blood pressure. In the literature, a multitude of evidence suggests that these contractile disturbances during hypertension are associated with altered ionic permeabilities of the vascular smooth muscle cells (Jones, 1973; Martens and Gelband, 1998). For example, patch-clamped vascular smooth muscle cells isolated from several circulatory beds of the spontaneously hypertensive rat (SHR) show increased L-type Ca^{2+} current densities compared to similar cells from normotensive Wistar Kyoto (WKY) rats (Rusch and Hermsmeyer, 1988; Ohya et al., 1993; Wilde et al., 1994; Cox and Lozinskaya, 1995; Lozinskaya and Cox, 1997). Thus, an increased voltage-gated Ca^{2+} current appears to be a common finding in arterial muscle membranes from the SHR, which is sometimes regarded as an animal model of human essential hypertension. Concurrently, at least in some vascular smooth muscle cells of the SHR, an increased membrane density of outward Ca^{2+}-dependent K$^+$ current exists. Single-channel studies have attributed this finding to a higher expression level of large-conductance, Ca^{2+}-sensitive K$^+$ channels (BK$_{Ca}$ channels). The elevated K$^+$ current through BK$_{Ca}$ channels is thought to be due to alterations in posttranscriptional channel processing that enhance protein expression and may represent a secondary protective response to promote cell hyperpolarization. Thus, the upregulation of BK$_{Ca}$ channels may act to buffer further increases in voltage-gated Ca^{2+} influx and vascular tone during hypertensive disease (England et al., 1993; Liu et al., 1997).

However, to date, most fluorescent Ca^{2+} imaging and electrophysiological studies of vascular ionic changes in essential hypertension have been performed on vascular segments or isolated vascular muscle cells of larger arteries. Conflicting results still exist regarding changes in vascular contractility, resting membrane potential, [Ca^{2+}]$_i$, and the onset of ion channel alterations (Bohr, 1974; Hermsmeyer, 1976; Lamb and Webb, 1989; Stekiel, 1989; Rusch et al., 1992; Martens and Gelband, 1996; Ohya et al., 1996; Martens and Gelband, 1998). These differences may be due, at least in part, to the differences in the vascular beds studied. Mulvany (1990) and Mulvany and Aalkjaer (1990) also suggest that the increased contractility observed in vascular beds from hypertensive models must reside in the resistance vessels, whereas larger conduit vessels play a less significant role in setting peripheral vascular resistance. Thus, studies of clinically relevant resistance vessels may provide special insight into the role of ion channel alterations in the development and maintenance of essential hypertension.

In this regard, vascular smooth muscle cells enzymatically isolated from renal resistance arteries of WKY rats and SHR appear to provide a useful model for studying ion channel changes during hypertension. As shown in Fig. 2, these single cells from WKY rats and from SHR maintain an elongated, spindle shape. Furthermore, they reversibly contract to vasoactive contractile agonists, although SHR cells contract with a higher sensitivity. For example, Table I shows that the EC$_{50}$ values (effective drug concentrations causing half-maximal contraction) for the α_1-adrenergic receptor agon-

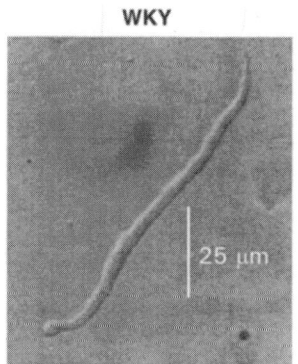
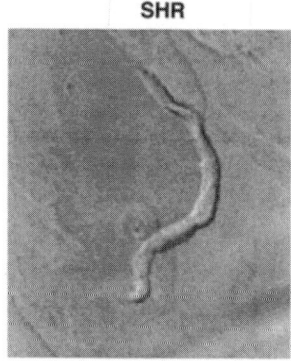

WKY

SHR

25 μm

Figure 2. Representative micrographs of vascular smooth muscle cells enzymatically isolated from renal resistance arteries of WKY rats and SHR.

ist phenylephrine are 380 nM and 202 nM in renal smooth muscle cells of WKY rats and SHR, respectively. Similarly, SHR cells show a greater contractile sensitivity to the depolarizing agent KCl than WKY arteries, as reflected by the significantly lower EC_{50} value for KCl in the SHR preparation. This finding is consistent with the elevated basal and agonist-stimulated $[Ca^{2+}]_i$ that exists in the renovascular smooth muscle cells of the hypertensive animal and also may reflect a loss of the vasodilating influence of the endothelial cells (Martens and Gelband, 1996).

At least two ion channel alterations in the renovascular smooth muscle cells of SHR are thought to contribute potentially to this increased contractile sensitivity. First, the density of K^+ current through Kv channels is suppressed in SHR compared to WKY rat renal vascular smooth muscle cells, as observed in Fig. 3. In these studies, charybdotoxin (ChTx) and niflumic acid were used to inhibit BK_{Ca} and Cl^- currents, respectively, and thereby permit the preferential measurement of Kv current elicited by

Table I
AT$_1$ R-AS Gene Delivery Prevents Alterations in Vascular Reactivity of Developing Renal Resistance Arteries

Species and treatment[a]	Blood pressure (mm Hg)	EC$_{50}$[b]	
		Phenylephrine (nM)	KCl (mM)
WKY	112	380	34.8
SHR	192*	202*	24.0*
SHR + AT$_1$R-AS	124**	476**	34.0**
SHR + LNSV	208*	132*	26.5*

[a] WKY, Wistar Kyoto rats; SHR, spontaneously hypertensive rat pups; SHR + AT$_1$R-AS, SHR pups treated with LNS virus containing AT$_1$ receptor antisense; SHR + LNSV, SHR pups treated with LNS virus alone.
[b] $n = 6$ vessels from 3 animals.
*$P < 0.01$ vs. WKY.
**$P < 0.01$ vs. SHR.

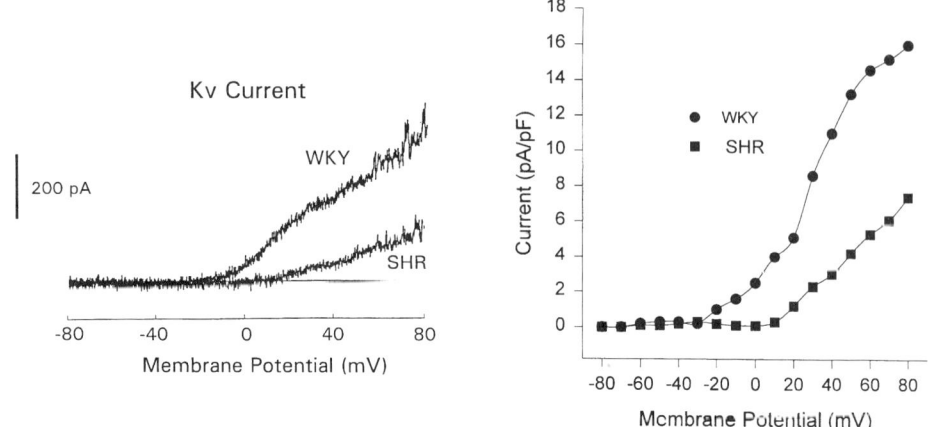

Figure 3. Voltage-gated K$^+$ (Kv) currents recorded from patch-clamped renal vascular smooth muscle cells of WKY rats and SHR. (A) Representative outward K$^+$ current recorded from WKY and SHR cells during a ramp depolarization from a holding potential of -80 mV to $+80$ mV. (B) Current–voltage relationships for Kv current recorded from WKY and SHR cells normalized to cell capacitance. Maximal Kv current amplitude elicited during stepwise depolarizations was plotted as a function of membrane potential ($n = 8$).

a depolarizing voltage ramp from a holding potential of -80 mV to a maximal potential of $+80$ mV (Fig. 3, left). The resulting current–voltage plot of the Kv current density in WKY rat and SHR renal arterial muscle cells reveals about a 2- to 3-fold higher Kv density in the SHR cells (Fig. 3, right). Additionally, the resting membrane potential of current-clamped cells from SHR was approximately 20 mV more depolarized than that of cells from WKY rats, consistent with a loss of hyperpolarizing K$^+$ current. The second ionic alteration observed in the renovascular muscle cells of SHR is an augmented L-type Ca^{2+} current density. In this regard, Fig. 4 shows that patch-clamped renovascular smooth muscle cells from SHR showed an increased voltage-gated Ca^{2+} current compared to that recorded from cells from WKY rats. Thus, it appears that a reduced level of Kv current, coupled to an enhanced level of voltage-gated L-type Ca^{2+} current, may act to enhance Ca^{2+}-dependent tone in small renal arteries of SHR (Martens and Gelband, 1998). Similar to findings with other SHR arteries (Berger and Rusch, 1999), BK$_{Ca}$ channels appear to upregulate in SHR renal arteries, which would act to reduce cell excitability. However, the final remodeled ion channel profile in the SHR arterial smooth muscle membrane appears to favor vascular activation, which may influence the regulation of renal vascular resistance in the hypertensive host.

7. PREVENTION OF HYPERTENSION AND ION CHANNEL ALTERATIONS BY RETROVIRAL-MEDIATED DELIVERY OF AT$_1$ RECEPTOR ANTISENSE

The diagnostic measure of hypertension is an elevation in blood pressure, which represents the product of cardiac output and total peripheral vascular resistance. Of

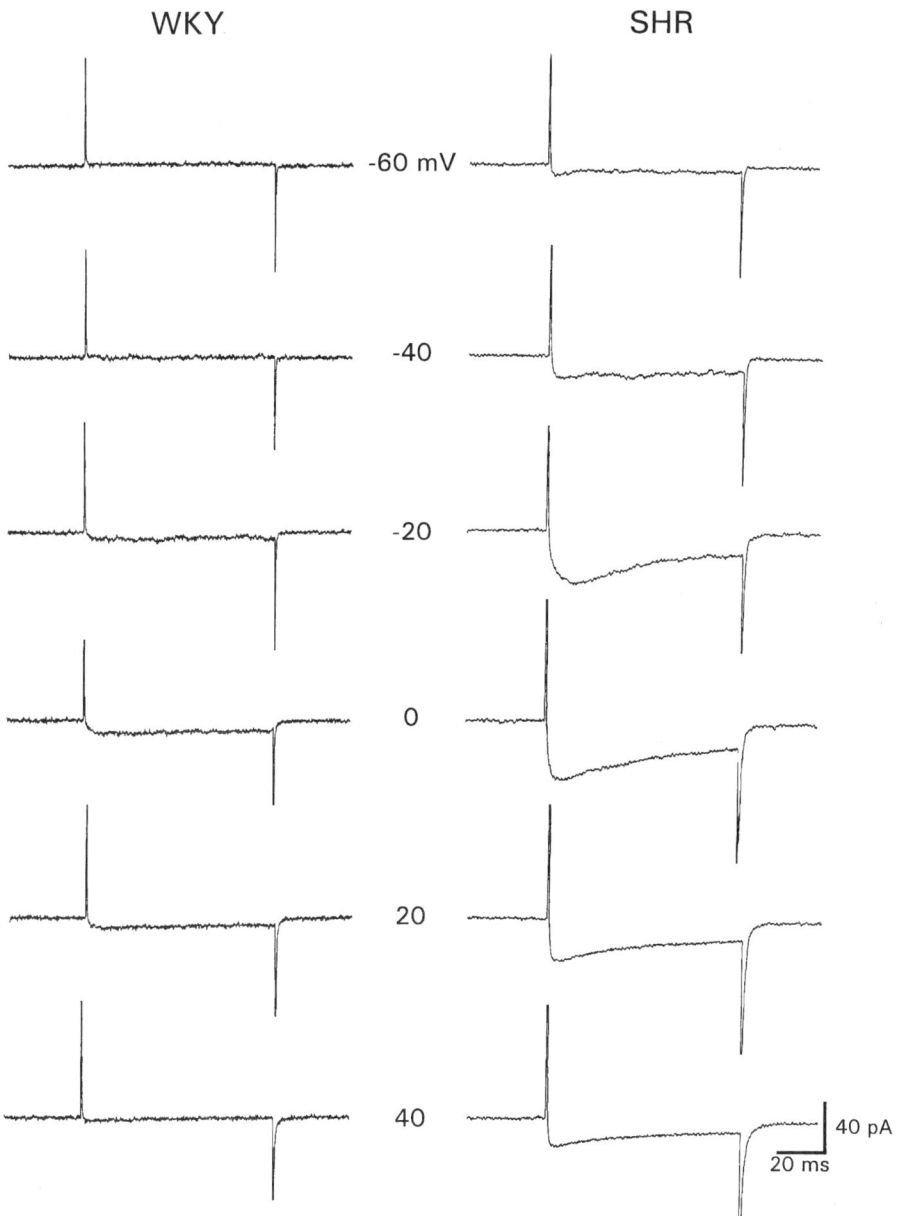

Figure 4. Ca^{2+} currents in vascular smooth muscle cells from WKY rats and SHR. Inward Ca^{2+} currents were recorded using the patch-clamp technique in vascular smooth muscle cells dissociated from renal resistance arterioles and bathed in a solution containing 2.5 mM calcium.

these two factors, essential hypertension is characterized by a near normal cardiac output, but the presence of an elevated resistance. Total peripheral resistance, in turn, is inversely proportional to vessel diameter, particularly the diameter of the small arteries and arterioles. Thus, ultimately the dynamic regulation of vessel diameter, total peripheral resistance, and blood pressure will be governed by the contractile state of the vascular smooth muscle cells within the vessel walls.

For this reason, some genes relevant to either vasodilation or vasoconstriction have been explored for their usefulness in the control of hypertension. Studies by Chao and her associates (Chao et al., 1998; Lin et al., 1999) have targeted genes relevant to vasodilator pathways and have shown that the delivery of genes for kallikrein or atrial natriuretic peptide to Dahl salt-sensitive rats transiently lowers systemic blood pressure and reverses salt-induced renal injury. In contrast, our laboratory has targeted the renin–angiotensin system (RAS) at the genetic level for the control of hypertension and the reversal of ion channel alterations (Martens et al., 1998). Notably, traditional agents which either inhibit the formation of angiotensin II (Ang II) [(e.g., renin and angiotensin converting enzyme inhibitors (ACE inhibitors)] or the actions of Ang II (e.g., AT_1 receptor antagonists) already represent successful antihypertensive drugs in wide clinical use. These agents are highly reliable and easily affordable, and their short duration of action makes them an excellent choice for reversible short-term therapy for high blood pressure, although the antihypertensive effect of these drugs is short-lived and they produce significant side effects. These traditional agents also lower blood pressure in the SHR model of genetic hypertension. For example, the maintenance of SHR pups on an ACE inhibitor for an extended time period prevents the development of high blood pressure (Regan et al., 1997).

Based on the successful antihypertensive action of traditional drug therapy directed at RAS, we have delivered AT_1 receptor ASODN to SHR pups to determine if (i) its long-term expression prevents the development of high blood pressure and (ii) the pathological and ionic abnormalities associated with this disease also are minimized (Gelband et al., 1999; Iyer et al., 1996; Martens et al., 1998). Indeed, Table I shows that a single intracardiac injection of viral particles containing AT_1 receptor ASODN effectively reduces high blood pressure in SHR by a maximum of 80 mm Hg. Approximately a 30 mm Hg reduction in blood pressure persisted in the antisense ASODN-treated SHR (SHR + AT_1R-AS) compared to the untreated hypertensive animals (SHR), or compared to control SHR treated with the LNS virus alone (SHR + LNSV). Furthermore, the selectivity of this antihypertensive effect is comparable to that reported for treatment with ACE inhibitors or AT_1 receptor antagonists, in that SHR show a significant decrease in blood pressure whereas little antihypertensive effect is observed in the normotensive WKY rats (Lu et al., 1997).

Treatment of neonatal SHR with the AT_1 receptor antisense vector also completely prevents, on a long-term basis, the contractile abnormalities observed in renal smooth muscle cells of SHR (Gelband et al., 1999; Martens and Gelband, 1998; Martens et al., 1998). Table I shows that the lower EC_{50} values for phenylephrine and KCl observed in renal arteries of untreated SHR are normalized or near-normalized by antisense treatment, as evidenced by the higher EC_{50} values for these vasoconstrictor agonists in the vascular cells of the antisense-treated animals. Importantly, this prevention of contractile abnormalities is associated with the normal phenotypic expression of Kv and L-type Ca^{2+} channel currents. In fact, AT_1 receptor antisense treatment appears

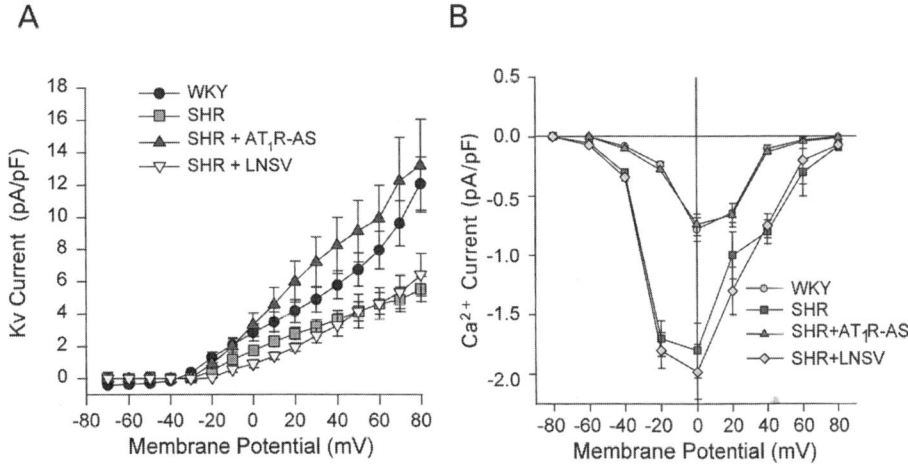

Figure 5. Current–voltage relationships for Kv (A) and Ca^{2+} (B) current density in renal smooth muscle cells of normotensive WKY rats and of untreated SHR, SHR treated with antisense (AT$_1$R-AS), and SHR treated with the LNS virus alone (SHR + LNSV). These experiments demonstate that AT$_1$ receptor antisense therapy prevents the alterations in Kv and Ca^{2+} current density in renal smooth muscle cells of SHR.

to effectively prevent the reduction in Kv current density and upregulation of L-type Ca^{2+} current that is typical of renal arterial muscle membranes of untreated SHR (Gelband *et al.*, 1999; Martens and Gelband, 1998). In this regard, the current–voltage relationships in Figs. 5A, B illustrate the effects of AT$_1$ receptor antisense gene therapy on renal arterial Kv and Ca^{2+} current densities, respectively. Figure 5A shows that the renovascular smooth muscle cells of SHR treated as neonates with AT$_1$ receptor ASODN (SHR + AT$_1$R-AS) show Kv current densities comparable to those observed in cells of normotensive WKY. In contrast, Kv current densities in the vascular muscle cells of untreated SHR and of control SHR treated with virus alone (SHR + LNSV) are clearly suppressed. Similarly, Fig. 5B shows that an early intervention of antisense therapy prevents the elevated L-type Ca^{2+} current densities observed in vascular muscle cells of untreated SHR or of control SHR treated with virus alone. Thus, a single administration of AT$_1$ receptor ASODN by a retroviral-mediated delivery system to SHR pups attenuates hypertension on a long-term basis in adult SHR. Furthermore, it prevents ion channel alterations associated with increased contractile responses, which may be involved in the pathogenesis of increased vascular tone. This antisense treatment may have advantages over traditional drug therapy, because only a single injection is needed for a long-term antihypertensive effect, and it fully prevents the renal pathophysiological changes associated with hypertensive disease. Furthermore, no visible side effects are observed in the treated animals based on weight, morphological, and behavioral parameters (Gelband *et al.*, 1999; Martens and Gelband 1998). Thus, it is possible that antisense gene therapy may represent a sound approach for preventing the development of high blood pressure on a long-term basis in some forms of hypertension. Ion channel abnormalities also appear to be prevented, although this may represent a secondary effect of blood pressure lowering. Hence, studies using antisense therapy targeted directly to ion channel abnormalities will be required to evaluate the effectiveness of this strategy for correcting primary ion channel defects in cardiovascular cell membranes.

8. CAN ANTISENSE GENE THERAPY REVERSE HYPERTENSION?

In order for a gene therapy approach to be applicable to the treatment of human hypertension, it should have the capability to reverse hypertension once it is established. Notably, this requirement may not be necessary if a reliable genetic marker to predict hypertension is identified in the future. In this regard, initial findings in Table II show that the intracardiac administration of 1×10^9 cfu viral particles in adult SHR results in a rapid and persistent reduction in blood pressure which lasts as long as 36 days in the treated animals (SHR + AT_1R-AS). This antihypertensive effect is not observed in the WKY rat or in control SHR treated with virus alone (SHR + LNSV). Furthermore, the EC_{50} values in Table II indicate that the enhanced contractile sensitivity to phenylephrine and KCl observed in the renal smooth muscle cells of untreated SHR are normalized in cells of adult SHR treated with AT_1 receptor antisense (Katovich et al., 1999). These observations indicate, for the first time, that the reversal of high blood pressure by antisense gene therapy is associated with a parallel reversal of contractile abnormalities in the vasculature, providing support for the concept of using antisense gene therapy in adult animals. Similarly, these studies raise the possibility that the future use of channel-directed antisense therapy may reverse the abnormalities in ion channel remodeling that appear to contribute to conduction defects in the myocardium and to the development of anomalous vascular tone in hypertensive disease.

Table II
AT$_1$ R-AS Gene Delivery Reverses Alterations in Vascular Reactivity of
Renal Resistance Arteries

		EC_{50}[b]	
Species and treatment[a]	Blood pressure (mm Hg)	Phenylephrine (nM)	KCl (mM)
WKY	110 ± 11	358 ± 15	37 ± 2
SHR	195 ± 13*	110 ± 13*	20 ± 3*
SHR + AT$_1$R-AS	122 ± 22**	366 ± 18**	30 ± 2**
SHR + LNSV	216 ± 18*	92 ± 11*	20 ± 2*

[a] WKY, Wistar Kyoto rats; SHR, spontaneously hypertensive rats; SHR + AT$_1$R-AS, adult SHR treated with LNS virus containing AT$_1$ receptor antisense; SHR + LNSV, adult SHR treated with LNS virus alone.
[b] $n = 24$ rings from 6 animals for each group.
*$P < 0.01$ vs. WKY.
**$P < 0.01$ vs. SHR.

9. SUMMARY

Antisense gene therapy is an exciting therapeutic approach for the future control of excitability in cardiac and vascular smooth muscle cells. Experiments *in vitro* and *in vivo* indicate that antisense therapies effectively alter the level of expression of K$^+$ channels in cultured myocardial cells and change the density and properties of K$^+$ current and the action potential configuration in atrial and ventricular myocytes. In vascular smooth muscle cells, the antisense approach has not been used to clarify the contribution of distinct K$^+$ channel subtypes to macroscopic K$^+$ current, but this approach clearly will prove valuable in the future. Indeed, as outlined in this chapter, antisense therapy involving transgenes for disease-causing systems can reverse hypertension and can also reverse K$^+$ channel abnormalities involved in the final full expression of this disease. Particularly, reversal of chronic hypertension by the suppression of AT$_1$ receptors by a retroviral vector containing AT$_1$ receptor ASODN results in a long-term attenuation of high blood pressure and also restores normal levels of Kv current to renovascular smooth muscle cells in which these channels are downregulated as part of the disease process. Thus, the gene therapy approach is conceptually sound and experimentally feasible and represents a potential new therapy for the control of ion channel disorders. However, new viral vectors with high transduction efficiency for the transgene, whose expression can be controlled on demand by exogenous chemicals, must be tested for their usefulness in the delivery of antisense to target tissues expressing K$^+$ channel disorders. Furthermore, physiological experiments must be carried out to establish if antisense therapy directed against primary K$^+$ channel disorders results in the desired therapeutic effect *in vivo* and, concurrently, if the gene therapy can be administered with only minimal adverse effects.

ACKNOWLEDGMENTS

This work was supported by a grant from the National Institutes of Health (HL-52189) and the Council for Tobacco Research. I would also like to acknowledge Dr. Nancy Rusch for her editorial insight.

REFERENCES

Albarwani, S., Heinert, G., Turner, I. L., and Kozlowski, R. Z., 1995, Differential K$^+$ channel distribution in smooth muscle cells isolated from the pulmonary arterial tree of the rat, *Biochem. Biophys. Res. Commun.* **208:**183–189.

Beech, D. J., and Bolton, T. B., 1989, A voltage-dependent outward current with fast kinetics in single smooth muscle cells isolated from rabbit portal vein, *J. Physiol. (London)* **412:**397–414.

Benham, C. D., and Bolton, T. B., 1986, Spontaneous transient outward currents in single visceral and vascular smooth muscle cells of the rabbit, *J. Physiol. (London)* **381:**385–406.

Berger, M. B., and Rusch, N. J., 1999, Voltage and calcium-gated potassium channels: Functional expression and therapeutic potential in the vasculature, in: *Perspectives in Drug Discovery and Design* (J. M. Sabatier and H. Dorbon, eds.), Kluwer Academic Publishers, Dordrecht, The Netherlands, Vol. 15/16, pp. 313–332.

Bohr, D. F., 1974, Reactivity of vascular smooth muscle from normal and hypertensive rats: Effect of several cations, *Fed. Proc.* **33:**127–132.

Bonnet, P., Rusch, N. J., and Harder, D. R., 1991, Characterization of an outward K$^+$ current in freshly dispersed cerebral arterial muscle cells, *Pflügers. Arch.* **418:**292–296.

Chao, J., Zhang, J. J., Lin, K. F., and Chao, L., 1998, Adenovirus-mediated kallikrein gene delivery reverses salt-induced renal injury in Dahl salt-sensitive rats, *Kidney Int.* **54:**1250–1260.

Cox, R. H., and Lozinskaya, I. M., 1995, Augmented calcium currents in mesenteric artery branches of the spontaneously hypertensive rat, *Hypertension* **26:**1060–1064.

DeCoursey, T. E., Chandy, K. G., Gupta, S., and Cahalan, M. D., 1984, Voltage-gated K$^+$ channels in human T lymphocytes: A role in mitogenesis?, *Nature* **307:**465–468.

England, S. K., Wooldridge, T. A., Stekiel, W. J., and Rusch, N. J., 1993, Enhanced single-channel K$^+$ current in arterial membranes from genetically hypertensive rats, *Am. J. Physiol.* **264:**H1337–H1345.

Evans, A. M., Osipenko, O. N., and Gurney, A. M., 1996, Properties of a novel K$^+$ current that is active at resting potential in rabbit pulmonary artery smooth muscle cells, *J. Physiol. (London)* **496:**407–420.

Feng, J., Wible, B., Li, G. R., and Nattel, S., 1997, Antisense oligodeoxynucleotides directed against Kv1.5 mRNA specifically inhibit ultrarapid relayed rectifier K$^+$ current in cultured adult human atrial myocytes, *Circ. Res.* **80:**572–579

Fleischmann, B. K., Murray, R. K., and Kotlikoff, M. I., 1994, Voltage window for sustained elevation of cytosolic calcium in smooth muscle cells, *Proc. Natl. Acad. Sci. U.S.A.* **91:**11914–1198.

Folco, E., Mathur, R., Mori, Y., Buckett, P., and Koren, G., 1997, A cellular model for long QT syndrome. Trapping of heteromultimeric complexes consisting of truncated Kv1.1 potassium channel polypeptides and native Kv1.4 and Kv1.5 channels in the endoplasmic reticulum, *J. Biol. Chem.* **272:**26505–26510.

Gelband, C. H., and Hume, J. R, 1992, Ionic currents in single smooth muscle cells of the canine renal artery, *Circ. Res.* **71:**745–758.

Gelband, C. H., and Hume, J. R., 1995, [Ca^{2+}]$_i$ inhibition of K$^+$ channels in canine renal artery. Novel mechanism for agonist-induced membrane depolarization, *Circ. Res.* **77:**121–130.

Gelband, C. H., Reaves, P. Y., Evans, J., Wang, H., Katovich, M. J., and Raizada, M. K., 1999, Angiotensin II type 1 receptor antisense gene therapy prevents altered renal vascular calcium homeostasis in hypertension, *Hypertension* **33:**360–365.

Hall, J. E, Brands, M. W., and Henegar, J. R., 1999, Angiotensin II and long-term arterial pressure regulation: The overriding dominance of the kidney, *J. Am. Soc. Nephrol.* **10**(Suppl. 12):S258–S265.

Hermsmeyer, K., 1976, Electrogenesis of increased norepinephrine sensitivity of arterial vascular muscle in hypertension, *Circ. Res.* **38:**362–267.

Hume, J. R., and Leblanc, N., 1989, Macroscopic K$^+$ currents in single smooth muscle cells of the rabbit portal vein, *J. Physiol. (London)* **413:**49–73.

Ishikawa, T., Hume, J. R., and Keef, K. D., 1993, Modulation of K$^+$ and Ca^{2+} channels by histamine H1-receptor stimulation in rabbit coronary artery cells, *J. Physiol. (London)* **468:**379–400.

Iyer, S. N., Lu, D., Katovich, M. J., and Raizada, M. K., 1996, Chronic control of high blood pressure in the spontaneously hypertensive rat by delivery of angiotensin type 1 receptor antisense, *Proc. Natl. Acad. Sci. U.S.A.* **93:**9960–9965.

Jan, L. Y., and Jan, Y. N., 1992, Structural elements involved in specific K$^+$ channel functions, *Annu. Rev. Physiol.* **54:**537–555.

Jones, A. W., 1973, Altered ion transport in vascular smooth muscle from spontaneously hypertensive rats: Influences of aldosterone, norepinephrine, and angiotensin, *Circ. Res.* **33:**563–572.

Katovich, M. J., Gelband, C. H., Reaves, P. Y., Wang, H., Dang, H., and Raizada, M. K., 1999, Reversal of hypertension by retroviral-mediated (LNSV) delivery of angiotensin II type 1 receptor antisense (AT$_1$ R-AS) in the adult spontaneously hypertensive rat (SHR). *Am. J. Physiol.* **277:**H1260–H1264.

Knot, H. J., and Nelson, M. T., 1995, Regulation of membrane potential and diameter by voltage-dependent K$^+$ channels in rabbit myogenic cerebral arteries, *Am. J. Physiol.* **269:**H348–H355.

Krieger, J. E., and Dzau, V. J., 1991, Molecular biology of hypertension, *Hypertension* **18:**S13–S17.

Lamb, F. S., and Webb, R. C., 1989, Regenerative electrical activity and arterial contraction in hypertensive rats, *Hypertension* **13:**70–76.

Lepple-Wienhues, A., Berweck, S., Bohmig, M., Leo, C. P., Meyling, B., Garbe, C., and Wiederholt, M., 1996, K$^+$ channels and the intracellular calcium signal in human melanoma cell proliferation, *J. Membr. Biol.* **151:**149–157.

Lin, K. F., Chao, J., and Chao, L., 1999, Atrial natriuretic peptide gene delivery reduces stroke-induced mortality rate in Dahl salt-sensitive rats, *Hypertension* **33**:219–224.

Liu, Y., Plcyte, K., Knaus, H.-G., and Rusch, N. J., 1997, Increased expression of Ca^2-sensitive K$^+$ channels in aorta of hypertensive rats, *Hypertension* **30**:1403–1409.

London, B., Jeron, A., Zhou, J. Buckett, P., Han, X., Mitchell, G. F., and Koren, G., 1998, Long QT and ventricular arrhythmias in transgenic mice expressing the N terminus and first transmembrane segment of a voltage-gated potassium channel, *Proc. Natl. Acad. Sci. U.S.A.* **95**:2926–2931.

Lozinskaya, I. M., and Cox, R. H., 1997, Effects of age on Ca^{2+} currents in small mesenteric artery myocytes from Wistar-Kyoto and spontaneously hypertensive rats, *Hypertension* **29**:1329–1336.

Lu, D., and Raizada, M. K., 1995, Delivery of angiotensin II type 1 receptor antisense inhibits angiotensin action in neurons from hypertensive rat brain, *Proc. Natl. Acad. Sci. U.S.A.* **92**:2914–2918.

Lu, D., Yu, K., and Raizada, M. K., 1995, Retrovirus-mediated transfer of an angiotensin type I receptor (AT$_1$-R) antisense sequence decreases AT-Rs and angiotensin II action in astroglial and neuronal cells in primary cultures from the brain, *Proc. Natl. Acad. Sci. U.S.A.* **92**:1162–1166.

Lu, D., Raizada, M. K., Iyer, S., Reaves, P., Yang, H., and Katovich, M. J., 1997, Losartan versus gene therapy: Chronic control of high blood pressure in spontaneously hypertensive rats, *Hypertension* **30**:363–370.

Lu, D., Yang, H., and Raizada, M. K., 1998, Attenuation of ANG II actions by adenovirus delivery of AT receptor antisense in neurons and SMC, *Am. J. Physiol.* **274**:H719–H727.

Martens, J. R., and Gelband, C. H., 1996, Alterations in rat interlobar artery membrane potential and K$^+$ channels in genetic and nongenetic hypertension, *Circ. Res.* **79**:295–301.

Martens, J. R., and Gelband, C. H., 1998, Ion channels in vascular smooth muscle: Alterations in essential hypertension, *Proc. Soc. Exp. Biol. Med.* **218**:192–203.

Martens, J. R., Reaves, P. Y., Lu, D., Katovich, M. J., Berecek, K. H., Bishop, S. P., Raizada, M. K., and Gelband, C. H., 1998, Prevention of renovascular and cardiac pathophysiological changes in hypertension by angiotensin II type 1 receptor antisense gene therapy, *Proc. Natl. Acad. Sci. U.S.A.* **95**:2664–2669.

McDonald, T. V., Yu, Z., Ming, Z., Paima, E., Meyers, M. B., Wang, K. W. Goldstein, S. A., and Fishman, G. I., 1997, A minK–HERG complex regulates the cardiac potassium current I(Kr), *Nature* **388**:289–292.

Mulvany, M. J., 1990, Structure and function of small arteries in hypertension, *J. Hypertens. Suppl.* **8**:S225–S232.

Mulvany, M. J., and Aalkjaer, C., 1990, Structure and function of small arteries, *Physiol. Rev.* **70**:921–961.

Nakamura, Y. T., Artman, M., Rudy, B., and Coetzee, W. A., 1998, Inhibition of rat ventricular IK1 with antisense oligonucleotides targeted to Kir2.1 mRNA, *Am. J. Physiol.* **274**:H892–H900.

Nelson, M. T., and Quayle, J. M., 1995, Physiological roles and properties of potassium channels in arterial smooth muscle, *Am. J. Physiol.* **268**:C799–C822.

Neylon, C. B., Avdonin, P. V., Larsen, M. A., and Bobik, A., 1994, Rat aortic smooth muscle cells expressing charybdotoxin-sensitive potassium channels exhibit enhanced proliferative responses, *Clin. Exp. Pharmacol. Physiol.* **21**:117–120.

Ohya, Y., Abe, I., Fujii, K., Takata, Y., and Fujishima, M., 1993, Voltage-dependent Ca^{2+} channels in resistance arteries from spontaneously hypertensive rats, *Circ. Res.* **73**:1090–1099.

Ohya, Y., Abe, I., Fujii, K., and Fujishima, M., 1996, Membrane channels in smooth muscle cells of arteries from hypertensive rats, in: *Smooth Muscle Excitation* (T. B. Bolton and T. Tomita, eds.), Academic Press, San Diego, pp. 459–465.

Okabe, K., Kitamura, K., and Kuriyama, H., 1987, Features of 4-aminopyridine sensitive outward current observed in single smooth muscle cells from the rabbit pulmonary artery, *Pflügers Arch.* **409**:561–568.

Post, J. M., Gelband, C. H., and Hume, J. R., 1995, [Ca^{2+}]$_i$ inhibition of K$^+$ channels in canine pulmonary artery: Novel mechanism for hypoxia-induced membrane depolarization, *Circ. Res.* **77**:131–139.

Raizada, M. K., Katovich, M. J., Wang, H., and Gelband, C. H., 1999, Is antisense gene therapy a step in the right direction in the control of hypertension? *Am. J. Physiol.* **277**:H423–H432.

Regan, C. P., Bishop, S. P., and Berecek, K. H., 1997, Early, short-term treatment with captopril permanently attenuates cardiovascular changes in spontaneously hypertensive rats, *Clin. Exp. Hypertens.* **19**:1161–1177.

Rusch, N. J., and Hermsmeyer, K., 1988, Calcium currents are altered in the vascular muscle cell membrane of spontaneously hypertensive rats, *Circ. Res.* **63**:997–1009.

Rusch, N. J., De Lucena, R. G., Wooldridge, T. A., England, S. K., and Cowley, A. W., Jr., 1992, A Ca^{2+}-dependent K$^+$ current is enhanced in arterial membranes of hypertensive rats, *Hypertension* **19**:301–307.

Sanguinetti, M. C., Curran M. E., Zou, A., Shen, J., Spector, P. S., Atkinson, D. L., and Keating, M. T., 1996, Spectrum of HERG K$^+$ channel dysfunction in an inherited cardiac arrhythmia, *Proc. Natl. Acad. Sci., U.S.A.* **93**:2208-2212.

Smirnov, S. V., and Aaronson, P. I., 1992a, Ca^{2+} currents in single myocytes from human mesenteric arteries: Evidence for a physiological role of L-type channels, *J. Physiol. (London)* **457**:455–475.

Smirnov, S. V., and Aaronson, P. I., 1992b, Ca^{2+}-activated and voltage-gated K$^+$ currents in smooth muscle cells isolated from human mesenteric arteries, *J. Physiol. (London)* **457**:431–454.

Stekiel, W. J., 1989, Electrophysiological mechanisms of force development by vascular smooth muscle membrane in hypertension, in: *Blood Vessel Changes in Hypertension* (K. M. L. W. Lee, ed.), Boca Raton, CRC Press, Boca Raton, Florida, pp. 127–170.

Toro, L., and Stefani, E., 1987, Ca^{2+} and K$^+$ current in cultured vascular smooth muscle cells from rat aorta, *Pflügers Arch* **408**:417–419.

Tsien, R. Y., Pozzan, T., and Rink, T. J., 1982, T-cell mitogens cause early changes in cytoplasmic free Ca^{2+} and membrane potential in lymphocytes, *Nature* **295**:68–71.

Wang, L., Xu, B., White, R. E., and Lu, L., 1997, Growth factor-mediated K$^+$ channel activity associated with human myeloblastic ML-1 cell proliferation, *Am. J. Physiol.* **273**:C1657–C1665.

Wang, Z., Feng, J., Shi, II., Pond, A., Nerbonne, J. M., and Nattel, S., 1999, Potential molecular basis of different physiological properties of transient outward K$^+$ current in rabbit and human atrial myocytes, *Circ. Res.* **84**:551–561.

Wilde, D. W., and Lee, K. S., 1989, Outward potassium currents in freshly isolated smooth muscle cell of dog coronary arteries, *Circ. Res.* **65**:1718–1734.

Wilde, D. W., Furspan, P. B., and Szocik, J. F., 1994, Calcium current in smooth muscle cells from normotensive and genetically hypertensive rats, *Hypertension* **24**:739–746.

Wonderlin, W. F., and Strobl, J. S., 1996, Potassium channels, proliferation and G1 progression, *J. Membr. Biol.* **154**:91–107.

Yamazaki, T., and Yazaki, Y., 1999, Role of tissue angiotensin II in myocardial remodeling induced by mechanical stress, *J. Hum. Hypertens.* **12**:S43–S47.

Yang, T., Kuppershmidt, S., and Roden, D. M., 1995, Anti-minK antisense decreases the amplitude of the rapidly activating cardiac delayed rectifier K$^+$ current. *Circ. Res.* **77**:1246–1253.

Yokoshiki, H., Sunagawa, M., Seki, T. M., and Sperelakis, N., 1999, Antisense oligodeoxynucleotides of sulfonylurea receptors inhibit ATP-sensitive K$^+$ channels in cultured neonatal rat ventricular cells, *Pflügers Archiv.* **437**:400–408.

Yuan, X. J., 1995, Voltage-gated K$^+$ currents regulate resting membrane potential and [Ca^{2+}]$_i$ in pulmonary arterial myocytes, *Circ. Res.* **77**:370–378.

Index